Honeybees of Asia

H.R. Hepburn • S.E. Radloff
Editors

Honeybees of Asia

Springer

Editors
Professor Dr. H.R. Hepburn
Department of Zoology and Entomology
Rhodes University
Grahamstown 6140, South Africa
r.hepburn@ru.ac.za

Professor Dr. S.E. Radloff
Department of Statistics
Rhodes University
Grahamstown 6140, South Africa
s.radloff@ru.ac.za

ISBN 978-3-642-16421-7 e-ISBN 978-3-642-16422-4
DOI 10.1007/978-3-642-16422-4
Springer Heidelberg Dordrecht London New York

Cover illustration: Pollen forager of *Apis cerana* on an ornamental flower (*Portulaca oleracea*) in the centre of Hangzhou (Zhejiang, China). Photo: Nikolaus Koeniger

Cover design: WMXDesign GmbH, Heidelberg, Germany

Printed on acid-free paper

Springer is part of Springer Science+Business Media (www.springer.com)

In Memoriam

Eva Widdowson Crane

1912–2007

Preface

Studies on the biology of honeybees stem from ancient times, in both Asia and Europe. However, published scientific works on the honeybees of both regions gained unanticipated momentum on the heels of World War II and were boosted exponentially by Sputnik a decade later. Since that time, 95% of all publications on Asian and 99% on European honeybees were published. We believe that the publication of the Ruttner's monographs (1988, 1992) was further major stimuli for research on Asian honeybees. Having just brought extraordinary clarity to the "real" honeybees (*Apis cerana*, *Apis dorsata*, *Apis florea*, and *Apis mellifera*), soon after *Apis koschevnikovi*, *Apis andreniformis*, *Apis laboriosa*, and *Apis nigrocincta* reappear in the literature. Some 50% of all literature on Asian honeybees follows publication of Ruttner's classic work. Another major impetus for increased research on honeybees in Asia undoubtedly stems from the rather thorough cover given to this literature by Eva Crane and colleagues through some 50 odd years of Apicultural Abstracts.

Interestingly, the lion's share of work on Asian honeybees is also historically postcolonial in origin. It has also very largely resulted from the joint efforts of Asian and Western scientists working in tandem. On the Asian side, this year, 2010, also sees the 10th international conference of the Asian Apicultural Association, a body that has both stimulated Asian colleagues and made Western ones warmly received. Perusal of recent apicultural literature shows that East–West scientific alliances are increasing rapidly and bearing substantial fruit.

This volume is presented as a monograph. Monographs are usually understood to be complete and detailed expositions of a subject at an advanced level. While we believe that we have achieved this end through the inclusion of chapters by specialists in the field, it must be pointed out that while each chapter shows a reasonable depth of understanding, nonetheless they clearly indicate chasms in our knowledge of the honeybees of Asia. Much presented here is completely new and has, as yet, not been published in journals. Compared with the literature on western honeybees, that for Asia reveals very thin coverage for honeybee physiology, biochemistry, genetics, and pathology. This volume is a status quo report of what is known, and we fervently hope that this collation will provide stimuli to broaden the base of the biology of the Asian honeybees.

Grahamstown, South Africa
January 2011

H.R. Hepburn
S.E. Radloff

Contents

Contributors

D.P. Abrol Division of Entomology, Sher-e-Kashmir University of Agricultural Sciences and Technology, Chatha, Jammu (J&K) 180 009, India, dharam_abrol@rediffmail.com

Richard T. Corlett Department of Biological Sciences, National University of Singapore, Singapore, Singapore 117543, corlett@nus.edu.sg

R.M. Crewe Social Insect Research Group, Department of Zoology and Entomology, University of Pretoria, Pretoria 0002, South Africa, robin.crewe@up.ac.za

Orawan Duangphakdee Ratchaburi Campus, King Mongkut's University of Technology, Thonburi, Bangkok 10140, Thailand, Orawan.dua@kmutt.ac.th

Michael S. Engel Division of Entomology, Natural History Museum, University of Kansas, 501 Crestline Drive-Suite #140, Lawrence, KS 66049-2811, USA, msengel@ku.edu

Ingemar Fries Department of Ecology, Swedish University of Agricultural Sciences, Uppsala 750 07, Sweden, ingemar.fries@ekol.slu.se

Stefan Fuchs Institut für Bienenkunde (Polytechnische Gesellschaft), FB Biowissenschaften, Goethe-Universität Frankfurt am Main, Karl von Frisch Weg 2 61440, Germany, s.fuchs@bio.uni-frankfurt.de

H.R. Hepburn Department of Zoology and Entomology, Rhodes University, Grahamstown 6140, South Africa, r.hepburn@ru.ac.za

Thomas Hoetzl Institute of Zoology, University of Graz, Graz, Austria, frank.weihmann@uni-graz.at

Gerald Kastberger Institute of Zoology, University of Graz, Graz, Austria, gerald.kastberger@uni-graz.at

Nikolaus Koeniger Institut für Bienenkunde, (Polytechnische Gesellschaft) Fachbereich Biowissenschaften Goethe Universität, Frankfurt a.M. Karl-von-Frisch-Weg 2, 61440 Oberursel, Germany, nikolaus.koeniger@bio.uni-frankfurt.de

Gudrun Koeniger Institut für Bienenkunde (Polytechnische Gesellschaft), Fachbereich Biowissenschaften, Goethe Universität, Frankfurt a.M, Karl-von-Frisch-Weg 2, 61440, Oberursel, Germany, gudrun.koeniger@bio.uni-frankfurt.de

Chariya Lekprayoon Center of Excellence in Biodiversity, Department of Biology, Faculty of Sciences, Chulalongkorn University, Bangkok 10330, Thailand, Chariya.l@chula.ac.th

Uma Partap Honeybees Project, Sustainable Livelihoods and Poverty Reduction Programme, International Centre for Integrated Mountain Development (ICIMOD), PO Box 3226, Kathmandu, Nepal, upartap@icimod.org

Mananya Phiancharoen King Mongkut's University of Technology Thonburi, Ratchaburi Campus, Bangkok 10140, Thailand, manaya.phi@kmutt.ac.th

Christian W.W. Pirk Social Insect Research Group, Department of Zoology and Entomology, University of Pretoria, Pretoria 0002, South Africa, cwwpirk@zoology.up.ac.za

Sarah E. Radloff Department of Statistics, Rhodes University, Grahamstown 6140, South Africa, s.radloff@ru.ac.za

Deborah Smith Department of Ecology and Evolutionary Biology/Entomology, University of Kansas, Haworth Hall, 1200 Sunnyside Ave, Lawrence, KS 66045, USA, debsmith@ku.edu

Catherine L. Sole Social Insect Research Group, Department of Zoology and Entomology, University of Pretoria, Pretoria 0002, South Africa, clsole@zoology.up.ac.za

Ken Tan Eastern Bee Research Institute of Yunnan, Agricultural University, Heilongtan, Kunming, Yunnan Province, People's Republic of China, eastbee@public.km.yn.cn

Jürgen Tautz BEE Group, Biozentrum, Universität Am Hubland, 97074 Würzburg, Germany, Tautz@biozentrum.uni-wuerzburg.de

Natapot Warrit Center of Excellence in Entomology, Department of Biology, Faculty of Sciences, Chulalongkorn University, Bangkok 10330, Thailand; Department of Entomology, National Museum of Natural History, Smithsonian Institution, Washington, DC 20013, USA, Natapot.w@chula.ac.th

Frank Weihmann Institute of Zoology, University of Graz, Graz, Austria, frank.weihmann@uni-graz.at

Ming-Xian Yang Department of Zoology and Entomology, Rhodes University, Grahamstown 6140, South Africa; Department of Zoology, Sichuan Agricultural University, Yaan, People's Republic of China, damoguyan2006@yahoo.com.cn

Chapter 1
The Asian Species of *Apis*

Sarah E. Radloff, H.R. Hepburn, and Michael S. Engel

1.1 Introduction

The number of species of honeybees recognised over the last two and a half centuries has varied quite considerably, following the original descriptions of *Apis mellifera* (1758) by Linnaeus and *Apis florea* (1787), *Apis cerana* (1793) and *Apis dorsata* (1793) by Fabricius. In the nineteenth century, Frederick Smith (1854–1871) described some 20 additional species, often based on single specimens; only his taxa *Apis andreniformis* (1858) and *Apis nigrocincta* (1861), however, survived in honeybee systematics. Contemporaneously, Gerstäcker (1863) published the first comprehensive phylogenetic and taxonomic treatise on *Apis,* and reduced all previously described forms (except *A. andreniformis* and *A. nigrocincta*, which he either missed or ignored) to only the original four Linnean and Fabrician species. Although Smith (1865) subsequently presented his case for seven species, the views of Gerstäcker (1863) prevailed into the twentieth century (Koschevnikov 1900–1905; Enderlein 1906; von Buttel-Reepen 1906).

Matters then rested for another half century, until Maa (1953) published an abstruse monograph in which he introduced some 24 species of honeybees within four genera. These taxa have subsequently been almost totally ignored in the apicultural literature, and the historically older views of Gerstäcker (1863) have endured until relatively recently. During the years leading up to the publication of Ruttner's (1988) monograph, a search for East Asian honeybees (probably stimulated by Maa's

S.E. Radloff
Department of Statistics, Rhodes University, Grahamstown 6140, South Africa
e-mail: s.radloff@ru.ac.za

H.R. Hepburn
Department of Zoology and Entomology, Rhodes University, Grahamstown 6140, South Africa
e-mail: r.hepburn@ru.ac.za

M.S. Engel
Division of Entomology, Natural History Museum, University of Kansas, 501 Crestline Drive-Suite #140, Lawrence, KS 66049-2811, USA
e-mail: msengel@ku.edu

H.R. Hepburn and S.E. Radloff (eds.), *Honeybees of Asia*,
DOI 10.1007/978-3-642-16422-4_1, © Springer-Verlag Berlin Heidelberg 2011

original paper) ensued, with *Apis laboriosa* re-announced (Sakagami et al. 1980), *A. andreniformis* re-established (Wu and Kuang 1986, 1987; Kuang 1983), *Apis koschevnikovi* rediscovered (Mathew and Mathew 1988; Rinderer 1988) and *A. nigrocincta* re-entering the scene (Hadisoesilo and Otis 1996). Finally, *Apis nuluensis* was described as a new species (Tingek et al. 1996). When Ruttner (1992) subsequently published his natural history of honeybees, he included *A. laboriosa*, *A. andreniformis* and *A. koschevnikovi* alongside the "traditional" four species. In the most recent taxonomy of honeybees, Engel (1999) applied a phylogenetic species concept and accordingly regarded *A. laboriosa* and *A. nuluensis* as synonyms of *A. dorsata* and *A. cerana*, respectively – a view that has not been widely accepted by apiculturists, who have tended to employ alternate species concepts (that is, either the biological species or the evolutionary species concepts). Even now, the number of recognised species of honeybees remains in a state of flux.

Conceptualisation of species recognition also changed through the centuries, from the Platonic concept, exemplified by Linnaeus, to the slow introduction of the idea of a biological species, developed by Poulton (1908), Rensch (1929) and Dobzhansky (1937) and subsequently widely promulgated by Huxley (1940) and Mayr (1942). Indeed, today there are as many concepts for species recognition as there are putative honeybee species, and the very system by which we recognise biological units in nature is fiercely debated (e.g., Wheeler and Meier 2000). Moreover, honeybee researchers have focussed almost exclusively on the oldest of the currently used species concepts, the biological species concept.

Nonetheless, whether a species is diagnosed by population phenomena (the biological species concept), evolutionary lineages (the evolutionary species concept) or genealogical descent (the phylogenetic species concept), classification still requires that species-specific characteristics be brought to bear in the circumscription of species. Likewise, there have been several phylogenetic analyses conducted (Deodikar 1960; Sakai et al. 1986; Sheppard and Berlocher 1989; Alexander 1991; Garnery et al. 1991; Smith 1991; Petrov 1992; Willis et al. 1992; Engel and Schultz 1997; Engel 1999; Raffiudin and Crozier 2007; cf. Chap. 2), all based implicitly on the correctness of the named species.

Following the non-Linnean views of DuPraw (1964), however, coupled with the idea that sub-specific categories are untenable in a contiguous population (Wilson and Brown 1951), Hepburn and Radloff attempted to bypass the problem of classification by designating statistically defined populations of honeybees under the new coinage of "morphoclusters" (Hepburn et al. 2001a, b, 2005; Radloff et al. 2005a, b, c, 2010). They have since accepted the arguments of Engel (personal communication) that "morphoclusters" are really statistically defined "subspecies" to which they had been inconsistently applying trinomial names. Here, we report the results of a full multivariate morphometric analysis of the Asian species of *Apis* and correct the classification of *Apis* in accordance with the rules of the International Code of Zoological Nomenclature.

The systematics of honeybees has also undergone a paradigm shift as earlier evolutionary taxonomic methods and systems of organisation have become passé, having been replaced by the contemporary emphasis on populations, the statistical distribution of morphological characters and the reconstruction of evolutionary lineages. Moreover, there has been no diagnostic account of the Asian species of *Apis* since Maa (1953). Here, we present the analyses of the currently recognised species of *Apis*: *A. andreniformis*, *A. cerana*, *A. dorsata*, *A. florea*, *A. koschevnikovi*, *A. laboriosa*, *A. mellifera*, *A. nigrocincta* and *A. nuluensis* (noting that *laboriosa* and *nuluensis* are valid only under the antiquated biological species concept). We combine metrical and descriptive morphological characters, DNA characteristics (cf. Chap. 4), behaviour and nesting (cf. Chap. 6) so as to holistically define honeybee species and more easily identify them, either in an equipped laboratory or under field conditions.

1.2 The Dwarf Honeybees

1.2.1 Identification of Apis andreniformis *and* Apis florea

The distinctness of both *A. florea* and *A. andreniformis* as unequivocal, valid biological species is now well established and rests on the cumulative knowledge of the morphology of drone genitalia (Lavrekhin 1935; Ruttner 1975, 1988; Kuang and Li 1985; Wu and Kuang 1986, 1987; Wongsiri et al. 1990; Chen 1993; Patinawin and Wongsiri 1993), differences in nest structure (Thakar and Tonapi 1962; Dung et al. 1996; Rinderer et al. 1996; cf. Chap. 6), chemical profiles of beeswax (Aichholz and Lorbeer 1999, 2000; cf. Chap. 6), morphometrics (Jayavasti and Wongsiri 1992; Rinderer et al. 1995), allozyme polymorphism (Nunamaker et al. 1984; Li et al. 1986; Gan et al. 1991), mtDNA sequence divergences (Smith 1991; Willis et al. 1992; Nanork et al. 2001; cf. Chap. 4), flight (Radloff et al. 2001; cf. Chap. 13), timing of mating flights (Rinderer et al. 1993; Otis et al. 2001; cf. Chap. 8), sexual selection (Baer 2005) and niche differences (Oldroyd et al. 1992; Booncham et al. 1995; Rinderer et al. 2002; cf. Chap. 6). Several of these differences contribute to the complete reproductive isolation between the two species (Koeniger and Koeniger 1991, 2000, 2001; Otis 1991; Dung et al. 1996; cf. Chap. 8).

Unfortunately, accurate identifications of the dwarf honeybees in the older literature are often difficult to assess because the worker bees are morphologically similar and the species are sympatric over a wide area that extends from northeastern India to Indochina (Otis 1996; cf. Chap. 3). Some of the historical confusion between *A. florea* and *A. andreniformis* stems from the fact that their classification is based on workers, which do not show great morphological differentiation. Moreover, the descriptions and taxonomic keys of Maa (1953) were based on very limited numbers of specimens, and some of the purported differences between the two species become blurred if many workers of a colony are analysed.

The most reliable characteristics to rapidly distinguish *A. florea* and *A. andreniformis* are as follows: in drones, the "thumb" of the bifurcated basitarsus of the hind leg, which in *A. florea* is much longer than that of *A. andreniformis* (Ruttner 1988); the structure of the endophallus (Lavrekhin 1935; Wongsiri et al. 1990; Koeniger 1991; cf. Chap. 8); the cubital index in worker bees, which, at about 3 in *A. florea*, is significantly less than that in *A. andreniformis*, which is at about 6; the jugal-vannal ratio of the hindwing, which, at about 75 in *A. florea* is greater than that of *A. andreniformis*, at about 65; the abdominal tergite 2, which in *A. andreniformis* is deeply punctate, unlike that in *A. florea*; and the marginal setae on the hind tibiae, which in *A. florea* are usually entirely white, while those in *A. andreniformis* are dark-brown to blackish, in sclerotised, non-callow individuals.

Several subspecies, varieties, and nationes of *A. florea,* first described by Fabricius (1787), have been described over the last two centuries (Engel 1999). *A. andreniformis* was described by Smith (1858) as a species distinct from *A. florea* (Fabricius 1787) but was usually included among the varieties or subspecies of the latter for nearly a century, until its re-establishment as a species by Maa (1953). Although *A. andreniformis* was often considered a subspecies of *A. florea*, no sub-specific taxa have ever been proposed for *A. andreniformis*. Unfortunately, an unspecifiable number of specimens of *A. andreniformis* may have been misidentified as *A. florea* during this period. All named forms were eventually resolved into colour variants from widely separated localities (Dover 1929). Subsequently, Maa (1953) synonymised all previous such taxa of earlier workers (Gerstäcker 1863; Enderlein 1906; von Buttel-Reepen 1906; Cockerell 1911; Dover 1929), and no sub-specific categories of *A. florea* have been proposed since then (Hepburn et al. 2005).

The mistaken notion that abdominal tergites 1 and 2 of *A. florea* are reddish and other segments at least partially reddish, while those of *A. andreniformis* are uniformly black, still permeates the literature. However, an inspection of several hundred workers from several different colonies of each species quickly demonstrates the extreme variation in pigmentation. This precludes these characters as a useful distinguishing trait – a point actually recognised rather long ago (Drory 1888; Dover 1929). Finally, the combs of the two species are very different (Rinderer et al. 1996; cf. Chap. 6). Full bibliographies of the literature on *A. florea* and *A. andreniformis* are given in Hepburn and Hepburn (2005, 2009), respectively; cf. Chap. 20).

1.2.2 **Apis andreniformis** ***F. Smith (1858)***

A. andreniformis, the smallest of the honeybees, has been studied far less than *A. florea*. To date, there has been a single univariate morphometric comparison of *A. andreniformis* from southeastern Thailand and Palawan Island in the Philippines (Rinderer et al. 1995). These two widely separated populations (~3,000 km) differed only in a few characters that related to wing and metatarsal lengths, which indicates that it is likely a very homogeneous species. Likewise, estimates of the mtDNA haplotype divergence within the species was about 2% for *A. florea* and 0.5% for *A. andreniformis*, indicating rather homogeneous populations in both cases (Smith 1991; cf. Chap. 4).

The only published multivariate morphometric analysis of this species is the recent study of Rattanawannee et al. (2008), who collected 67 colonies throughout Thailand – 30 of which were for morphometric analysis and the remaining 37 for DNA polymorphism. Twenty characters were used to assess morphometric variation. Principal component analysis yielded four factor scores, which, when plotted, formed a single group, supported by a dendrogram generated from the cluster analysis. Using linear regression analysis, Rattanawannee et al. (2008) demonstrated the clinal pattern of morphometric characters, wherein body size decreases from west to east, associated with decreasing altitude, while it increases from south to north, associated with increasing altitude. Genetic variation, however, based on the sequence analysis of the cytochrome oxidase subunit b, yielded two groups – a result taken as tentative, pending more extensive analyses across the whole area of distribution of *A. andreniformis* (cf. Chap. 3).

1.2.3 Apis florea *Fabricius (1787)*

Several univariate morphometric studies on regional or country bases have appeared through the years, but they have not affected the taxonomy of the species. In the first multivariate morphometric analysis of *A. florea*, Ruttner (1988) had only limited material, from geographically non-contiguous regions. Although the data were insufficient for a comprehensive analysis, Ruttner (1988) demonstrated geographic variability and obtained three morphoclusters for *A. florea*. Recently, Tahmasebi et al. (2002) analysed *A. florea* and defined two morphoclusters from a geographical continuum in Iran. Combining their data with that of Ruttner (1988) and Mogga and Ruttner (1988), they also reported three morphoclusters for all *A. florea*; but again, a lack of geographical contiguity applies to these data as well. A multivariate study of the *A. florea* of Thailand has also been conducted (Chaiyawong et al. 2004). The raw data of Ruttner (1988), Tahmasebi et al. (2002), Mogga and Ruttner (1988) and Chaiyawong et al. (2004) were included in a subsequent study in which previous gaps in the distribution had been filled, finally allowing a comprehensive morphometric database for *A. florea* over its entire distribution to be compiled (Hepburn et al. 2005).

Principal component, discriminant and cluster analyses using the single linkage (nearest neighbour) procedure were carried out and produced a dendrogram of three main clusters (Fig. 1.1). Phenetically, cluster 1 initially linked colonies from Myanmar and Thailand, followed by Cambodia and finally Northern Vietnam; cluster 2 initially linked colonies from Oman, North India and Nepal, followed by those from South India; cluster 3 linked colonies from Iran and Pakistan; while clusters 2 and 3 linked colonies from Southern Vietnam (Fig. 1.1).

Radloff and Hepburn (1998, 2000) and Hepburn et al. (2001b) established empirically that the greater the sampling distances between localities, the greater

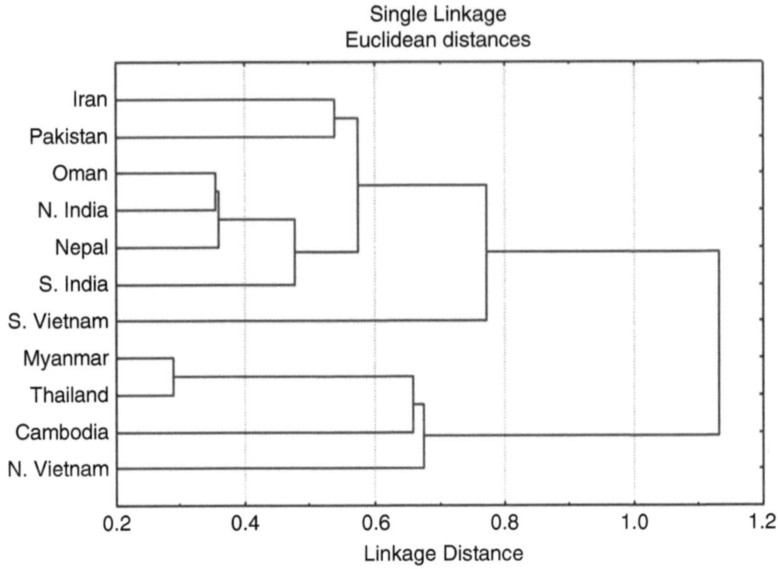

Fig. 1.1 Hierarchical clustering dendrogram for *Apis florea*, derived from single linkage clustering on morphometric characters: length of femur (5); length of tibia (6); length of metatarsus (7); tergite 3; longitudinal (9); tergite 4; longitudinal (10); length of forewing (17); wing angle G18 (25), averaged for countries. The original coded numbers assigned to these characters by Ruttner (1988)

the likelihood that artefactual morphoclusters would emerge in multivariate analyses. Conversely, where between-group variation is larger than within-group variation, biometric subgroups falling within smaller geographic domains may be swamped and obscured. Radloff et al. (2003b) also established the statistical significance of both colony sample size and individual bee sample size to studies of honeybee populations. These principles are particularly useful in the analyses of previous studies of *A. florea* and explain why Tahmasebi et al. (2002) defined two morphoclusters when they analysed the *A. florea* of Iran. Combining their data with that of Ruttner (1988) and Mogga and Ruttner (1988), Radloff et al. (2003b) reported three morphoclusters. In both studies, however, there was still a lack of geographical contiguity in the samples and each of the three groups was separated by intervals of about 3,000 km. When Hepburn et al. (2005) analysed the bees from the whole spectrum of localities sampled, the clinal nature of the morphometric measurements of the species became readily apparent. Precisely this same pattern was obtained in studies of *A. cerana* (Radloff et al. 2010).

On a mesoscale level, there have been several regional studies of morphometric variation in *A. florea* in India and Iran, representing sampling intervals of about 3,000 km. In northwestern India and eastern Pakistan, extending along a north–south transect between 25° and 32°N latitude, a transition in the populations occurs. There are significant interlocality differences in both the mean values of morphometric characters and their coefficients of variation, for most characters measured (Narayanan et al. 1960; Bhandari 1983; Sharma 1983) – implying heterogeneity in the population. Likewise, at hotter, drier and lower latitudes, *A. florea* are smaller than those at cooler and higher latitudes, leading to the proposition of possibly different ecotypes associated with climate at particular latitudes (Narayanan et al. 1960; Bhandari 1983). There are, however, alternative views on this point (Sharma 1983). Within a sample from India, Hepburn et al. (2005) obtained a strong, significant positive correlation between altitude and the principal component variables that reflect size. This pattern might benefit from additional attention.

Tahmasebi et al. (2002) reported an analysis of *A. florea* from 26 localities in Iran and obtained two morphoclusters: a western group of larger bees at higher latitudes (29–34°) and a lower latitude group of smaller bees to the east (<29° latitude). In the study of Hepburn et al. (2005), one morphocluster with two indistinct clusters of smaller eastern and larger western bees were noted. Here, the distributional variation in morphometric characters is clinal: northwestern bees are larger than southeastern ones (Özkani et al. 2009). In the final analysis, *A. florea* is a single species comprised of three discernible morphoclusters. The northwestern-most bees comprise a morphocluster that is statistically quite distinct from that to the southeast; but they are not isolated. Rather, they are joined by large areas of intermediate forms, resulting in a continuous cline in morphometric traits within this panmictic species.

1.3 The Medium-Sized Bees

1.3.1 Identification of Apis cerana, Apis koschevnikovi, Apis nigrocincta *and* Apis nuluensis

The sympatric occurrence of *A. cerana* with other medium-sized bees, *A. koschevnikovi*, *A. nigrocincta* and *A. nuluensis*, in southeastern Asia, unfortunately means that an indeterminable amount of previous "*A. cerana*" literature may inadvertently include data derived from other species (Hepburn et al. 2001a). To assist in overcoming this problem, we list metric characters that, in combination, separate these four species as follows: firstly, the cubital indexes of the forewings, which are 3.9 for *A. cerana*, 7.2 for *A. koschevnikovi*, 3.7 for *A. nigrocincta* and 2.4 for *A. nuluensis* – quickly separating paired comparisons for all, with the exception of an *A. cerana* and *A. nigrocincta* option. To separate this combination (*A. cerana* from *A. nigrocincta*), three measurements may be used: the length of the basal portion of the radial cell of the forewing, which is 1.2 mm in *A. cerana* and 1.8 mm in *A. nigrocincta*; the length of the apical portion of the radial cell, which is 1.8 mm in *A. cerana* and 1.1 mm in *A. nigrocincta*; and the length of the labial palp, which is 1.8 mm in *A. cerana* and 3.7 mm in *A. nigrocincta*.

1.3.2 Apis cerana *Fabricius (1793)*

Over the last two decades, great strides have been made following Ruttner's (1988) first multivariate analysis of this species. Subsequent authors used Ruttner's interpretations of *A. cerana* as a new baseline and concentrated on morphoclusters derived from multivariate analyses on a microscale level (Muzaffar and Ahmad 1989; Pesenko et al. 1989; Rinderer et al. 1989; Otis and Hadisoesilo 1990; Singh et al. 1990; Sulistianto 1990; Szabo 1990; Ono 1992; Verma 1992; Verma et al. 1994; Hadisoesilo and Otis 1996; Fuchs et al. 1996; Damus and Otis 1997; Sylvester et al. 1998) as well as on a more regional, mesoscale level (Yang 1986, 2001; Peng et al. 1989; Diniz-Filho et al. 1993; Damus and Otis 1997; Tilde et al. 2000; Hepburn et al. 2001a, b; Kuang 2002; Radloff and Hepburn 2002; Smith 2002; Tan et al. 2003; Radloff et al. 2003a, 2005a, b, c).

Historically, unravelling the structural complexity of *A. cerana* (Fabricius 1793) has been a continuous process, the details of which were recently given by Radloff et al. (2010). They reported the first multivariate morphometric analysis of *A. cerana* across its full geographical range and identify the statistically definable morphoclusters and subcluster populations within them. Principal component (PC) plots, using both the first and second PC scores and the first and third PC scores, did not reveal distinct morphoclusters. However, a substructuring of the PC plots was obtained by introducing local labelling and running a hierarchical cluster analysis, using the mean scores for PC 1 to 3 to identify homogeneous morphoclusters. This approach revealed six main morphoclusters, which were defined (Radloff et al. 2010) as follows (cf. Fig. 3.3):

1. Morphocluster I, "Northern *cerana*", which extends from northern Afghanistan and Pakistan through northwest India, across southern Tibet, northern Myanmar, China and northeasterly into Korea, far eastern Russia and Japan. Six subclusters or populations are morphometrically discernible within this morphocluster (a) an "Indus" group in Afghanistan, Pakistan and Kashmir; (b) a "Himachali" group in Himachal Pradesh, India; (c) an "Aba" group in Ganshu and Sichuan provinces in China, northern China and Russia; (d) a subcluster in central and eastern China; (e) a "southern *cerana*" subcluster in southern Yunnan, Guangdong, Guangxi and Hainan in China and (f) a "*japonica*" group in Japan and Korea.
2. Morphocluster II, "Himalayan *cerana*", which includes the bees of northern India and some of southern Tibet and Nepal. Two subclusters are discernible within this morphocluster: the bees of the northwest, which are termed the "Hills" group, and those of the northeast, termed the "Ganges" group (cf. Figs. 3.1 and 3.3).
3. Morphocluster III, "Indian plains *cerana*", which occurs across the plains of central and southern India and Sri Lanka as a fairly uniform population, long known as "plains *cerana*" in this subcontinent (cf. Figs. 3.1 and 3.3).
4. Morphocluster IV, "Indo-Chinese *cerana*", which forms a compact group in Myanmar, northern Thailand, Laos, Cambodia and southern Vietnam (cf. Figs. 3.1 and 3.3).

5. Morphocluster V, "Philippine *cerana*", which is restricted to the Philippines, but with the exclusion of most of Palawan Island, which instead groups with morphocluster VI. Within these islands, there are subclusters, and these bees are termed after the major island groups located there: "Luzon" bees, "Mindanao" bees and "Visayas" bees. The latter two subclusters show closer morphometric similarity than the former (cf. Figs. 3.1 and 3.3).
6. Morphocluster VI, "Indo-Malayan *cerana*", which extends from southern Thailand, through Malaysia and Indonesia. This large area consists of a rather morphometrically uniform bee, below the South China Sea. Three subclusters are discernible within this morphocluster: (a) Palawan (Philippines) and Borneo bees; (b) Malay Peninsula, Sumatera and some Sulawesi bees; and (c) Indonesia (Java, Bali, Irian Jaya, some Sulawesi and Sumatera) bees (cf. Figs. 3.1 and 3.3).

We must now consider how these results relate to earlier geographically large-scale analyses. When all of the mesoscale morphoclusters of Radloff et al. (2010) are compared with the new macroscale results, the only discrepancies are that, in the former, (1) the bees of the Philippines were included with those of Indonesia and Borneo; and (2) the bees of Japan are now placed in the Northern Asia morphocluster of the latter. However, there are differences between the mapped morphocluster results of Ruttner (1988) and Damus and Otis (1997) and those of Radloff et al. (2010). These discrepancies are best explained by the sampling differences in each study, which affected the degree of morphometric discrimination of the honeybees of Japan.

Ruttner (1988) had access to only a very small sample of large *A. cerana* from China and none from Russia. The only morphocluster I bees available to him were from the far northwest of the *A. cerana* range (Afghanistan and Pakistan) and some 6,000 km distant from Japan – the bees of which form a subcluster in a continuum of *A. cerana* morphocluster I. Gaps in the sampling inevitably resulted in the differences between Afghani and Japanese *A. cerana* being artefactually magnified. The dataset of Damus and Otis (1997) was based on the much smaller bees of the more southerly oceanic islands (Philippines, Indonesia, Borneo, etc.) with the same effect.

Returning to the matter of sampling, many thorough multivariate studies of *A. cerana*, sampled at a microscale basis, had been published; but, with the advantage of hindsight, the effects of limited sampling are evident. An important series of papers was published on sub-Himalayan *A. cerana*; however, the areas sampled were widely separated, and the net result was discrimination of seven distinct morphoclusters (Singh et al. 1990; Verma 1992; Verma et al. 1994). When the original data from all these papers were subsequently combined into a much larger dataset in collaboration with those authors, and for which the previous geographical gaps were filled, the newer multivariate analysis (now on a geographical continuum in the sub-Himalayan region) yielded only four morphoclusters for the same region – two of which contained biometric subclusters (Hepburn et al. 2001b).

Analysis of the *A. cerana* of the western sub-Himalayas yielded an additional Hindhu Kush morphocluster, bringing the Himalayan string of morphoclusters to five (Radloff et al. 2005a). The analysis found that high variance domains occur at the edges of the morphoclusters and biometric subclusters. The bees decrease in size from west to east, but increase in size with increasing altitude. When analyses were subsequently extended from Afghanistan to Vietnam, covering all of southern-mainland Asia, scores from the principal components analysis yielded five statistically identifiable morphoclusters (Radloff et al. 2005b). At this continental resolution, the five morphoclusters previously obtained in the regional analyses of the Himalayan string (Hepburn et al. 2001b; Radloff et al. 2005a) were reduced to three, which were also coherently distributed with the different climatic zones of the region (Radloff et al. 2005b).

In a parallel series of studies on the *A. cerana* of China, Tan et al. (2002, 2003) showed that bees from the northern high-altitude areas of Yunnan Province were clearly larger and darker and showed similarities to samples from Beijing, Nepal and northern India, whereas bees from southern Yunnan clustered with the bees of Thailand and Vietnam. These results were completely consistent with those of Radloff et al. (2005b) for the bees of southern Yunnan. Morphometric analyses of *A. cerana* from oceanic Asia yielded two distinct morphoclusters, bringing the then total number of morphoclusters to seven (Radloff et al. 2005c). On completion of the above series of regional mesoscale studies, the newly formed comprehensive dataset for all *A. cerana* was subjected to multivariate morphometric analysis. The final result was that six distinct morphoclusters of *A. cerana* were obtained, as discussed above (Radloff et al. 2010; cf. Fig. 3.3).

1.3.3 Apis koschevnikovi *Enderlein (1906)*

A. koschevnikovi was originally described by Enderlein (1906) as "*Apis indica* variety *koschevnikovi*" and by von Buttel-Reepen (1906) as "*Apis mellifica indica* variety *koschevnikovi*". Authorship for this species has however been formally assigned to Enderlein (Engel 1999) as *A. koschevnikovi* Enderlein (1906), in accordance with nomenclatural practice. With few exceptions (Maa 1953; Goetze 1964), there were no accounts of *A. koschevnikovi* until its rediscovery eight decades later in Borneo (Mathew and Mathew 1988; Rinderer 1988; Tingek et al. 1988). However, *A. koschevnikovi* had indeed been widely collected in the Sundaland region of Southeast Asia during the interim, as evidenced by collections in various museums (Otis 1996). In a recent flurry of publications (Hepburn and Hepburn 2008), it has been established that *A. koschevnikovi* is a morphometrically distinct species (Tingek et al. 1988; Rinderer et al. 1989; Ruttner et al. 1989; Sulistianto 1990; Hadisoesilo et al. 1999), reproductively isolated (Koeniger et al. 1996c) and differing in both nuclear and mitochondrial DNA regions (Arias et al. 1996; Takahashi et al. 2002; Raffiudin and Crozier 2007) from other species of *Apis*, with which it has a sympatric distribution.

Although most characters of length are some 10–15% greater in worker honeybees of *A. koschevnikovi* than in *A. cerana* (Rinderer et al. 1989; Sulistianto 1990), these species may be confused in alcohol-preserved specimens that do not show the natural reddish-yellow brightness of the former. Multivariate analyses of *A. koschevnikovi* samples from Malaysia, Borneo and Indonesia clearly established that this species is comprised of a single morphocluster (Hadisoesilo et al. 2008). Moreover, the morphocluster can be delimited with as few as 12 morphological characters. It would also appear to be a very homogeneous species, in comparison with *A. cerana*, over the same area of distribution, because the average coefficient of variation in *A. koschevnikovi* is 1.8%, while in *A. cerana*, it is 4.3% for the same characters (Hadisoesilo et al. 2008).

1.3.4 **Apis nigrocincta** ***F. Smith (1861)***

The life history of *A. nigrocincta* F. Smith (1861) is curiously similar to that of *A. koschevnikovi*. Described as a new species by F. Smith (1861), it remained virtually unreported, with a few exceptions, for more than a century, until it was re-examined in the 1990s. In the first instance, Hadisoesilo et al. (1995) detected two distinct groups of honeybees in Sulawesi, Indonesia. A discriminant analysis of these bees showed one group to be *A. cerana* and the other as neither *A. cerana* nor *A. koschevnikovi*. Moreover, these then unidentified bees appeared similar to *A. nigrocincta* when compared to the holotype. In rapid succession, the Guelph group confirmed that the unknown bees were indeed *A. nigrocincta* and that they occur in the Philippines as well (Damus and Otis 1997). Further multivariate analyses confirmed that *A. nigrocincta* occurred in western Sulawesi, Mindanao Island in the Philippine chain and on Sangihe Island, situated between the two (Damus and Otis 1997).

Studies of drone flight times further supported the status of *A. nigrocincta* as a species distinct from *A. cerana* (Hadisoesilo and Otis 1996; Otis et al. 2001). Interestingly, they found no differences in the drone genitalia of *A. nigrocincta* and *A. cerana*. The reality of *A. nigrocincta* as a valid species continued to grow when it was shown that the cappings of drone cells in *A. nigrocincta* lacked the well-known pore that is present in *A. cerana* (Hadisoesilo and Otis 1998). Jayavasti and Wongsiri (1992) were able to differentiate *A. nigrocincta* and *A. cerana* on the basis of sting morphology, while Keeling et al. (2001) established species-specific differences in the mandibular gland pheromones of queens. The species was also recognised in taxonomic studies of *Apis* by Engel (1999).

Shortly afterwards, the separation of these species through mtDNA analyses (Smith et al. 2000), receptor gene sequences (Raffiudin and Crozier 2007) as well as new haplotypes for the non-coding region of mtDNA (Takahashi et al. 2002) confirmed the *A. nigrocincta* species. More recent analyses of nuclear and mitochondrial DNA sequences further support the validity of *A. nigrocincta* (Arias and Sheppard 2005). Finally, Raffiudin and Crozier (2007) supported *A. nigrocincta* as a valid species on the basis of general biology, DNA, acoustics, waggle dance and combs.

Only in the last decade have we acquired sufficient evidence to consider *A. nigrocincta* as a reasonably well-defined valid species. Hadisoesilo and Otis (1996) and Otis et al. (2001) demonstrated that, although sympatric with *A. cerana*, *A. nigrocincta* is reproductively isolated from other Asian *Apis* species in the timing of its mating flights, is distinguishable from other *Apis* species in morphometric analyses (Hadisoesilo et al. 1995; Hadisoesilo and Otis 1996) and differs in mtDNA haplotypes (Smith et al. 2000, 2003). However, until very recently, its known distribution was limited to Indonesia and the Philippines (Otis 1996). Interestingly, Otis (1996) suggested that *A. nigrocincta* might have been derived from China, because it shares closer similarities with *A. cerana* from the mainland than from the southwest.

1.3.5 Apis nuluensis *Tingek et al. (1996)*

Just over a decade ago, Tingek et al. (1996) collected bees at flowers on Gunung Emas at an altitude of about 2,000 m, which appeared distinctly different from *A. cerana* and *A. koschevnikovi*. They conducted morphometric measurements on these blackish bees, using most of Ruttner's (1988) characters, and showed that they differed significantly from *A. cerana* and *A. koschevnikovi* workers and drones (with which they are sympatric), and accordingly described these bees as a new species, *A. nuluensis*. More extensive measurements were reported by Fuchs et al. (1996), who found that, in a principal component analysis plotting the first three of the axes derived from principal components, *A. nuluensis* was clearly separated from the other sympatric Asian *Apis* species. Moreover, a hierarchic cluster analysis of group centroids in canonical function space clearly showed that *A. nuluensis* is quite distinctly separated from the other species.

While the above remarks are restricted to inferences based entirely on morphometrics, other biological observations were soon brought to bear on the legitimacy of *A. nuluensis* as a distinct species under the biological species concept. Koeniger et al. (1996a, b) observed that the drone mating flight period was temporally completely isolated from those of *A. cerana* and *A. koschevnikovi*. Although there is a very small window of temporal overlap between *A. nuluensis* and *A. cerana,* the physical differences between the two would be adequate to obviate any hetero-specific mating. This separation in time is a pre-mating barrier that provides complete reproductive isolation among the honeybees with which it is sympatric (Koeniger et al. 1996a, b; cf. Chap. 8).

Although *A. nuluensis* was initially proposed on the basis of morphological and behavioural characters, Arias et al. (1996) analysed variable sites for the ND2 mitochondrial gene as well as for the intron of EF-1α – the results of which indicate that *A. nuluensis* and *A. cerana* are closely related or even that the former was derived from the latter, which challenges the validity of the species under more modern species concepts, such as the phylogenetic species concept. They concluded that *A. nuluensis* diverged from *A. cerana* more recently than did

A. koschevnikovi. Using a slightly different approach, Takahashi et al. (2002) and Tanaka et al. (2001) investigated the haplotypes for the non-coding region of mitochondrial DNA and reached essentially the same conclusion as Arias et al. (1996). Similarly, on the basis of morphometrics, Fuchs et al. (1996) concluded that *A. nuluensis* shares a greater similarity with *A. cerana* than with *A. koschevnikovi*. *A. nuluensis* is thus far known only from montane forests on the Gunung Emas in Sabah State, Malaysian Borneo. This area is at the northeastern tip of mountain ranges that extend continuously for about 1,000 km to the southwest, along a spine of mountains that extends two-thirds the length of Borneo. The region is remote, sparsely inhabited and not readily accessible. It seems highly likely that *A. nuluensis* occurs along this spine.

1.4 The Giant Honeybees

1.4.1 Apis dorsata *Fabricius (1793)*

The classification of the giant honeybees, *A. dorsata* and *A. laboriosa,* has long been problematical. The former was described by Fabricius in 1793 and various forms were introduced between then and the time of Maa (1953). Maa recorded the various synonymies that had previously arisen and then reshaped and split the species into *A. breviligula* (one specimen from the Philippines), *A. binghami* (Sulawesi, formerly the Celebes) and *A. dorsata* (the wider distribution as known today). Over the next three decades, however, none of the names proposed by Maa (1953) appeared in the apicultural literature in any form other than "*A. dorsata*".

The next important discussion of these bees was that of Ruttner (1988), who noted that the standard deviations of several morphometric characters, representing widely separated localities were very small indeed so that *A. dorsata* appeared very homogeneous. He further argued that differences regarded by some as species-specific in the *A. dorsata* group are of the same order of magnitude as those used to discriminate subspecies of *A. mellifera*. Acknowledging some unusual aspects of the biology of *A. laboriosa*, Ruttner (1988) nonetheless was not prepared to recognise this bee as a clear-cut species, especially in the light of the report that no differences could be found in the male genitalia of *A. dorsata* and what purported to be "*A. laboriosa*" (McEvoy and Underwood 1988). He did, however, support the subspecies of *A. d. binghami*, *A. d. breviligula* and *A. d. dorsata*. *A. d. breviligula* is a conspicuously short-tongued bee of the Philippines, whose behaviour differs in important respects from *A. d. dorsata*. Congregations of several nests, common in areas of the latter, do not occur in those of the former; likewise, seasonal migration, also common in the former, is absent from the latter (Morse and Laigo 1968). *A. d. binghami* is a long-tongued, long-winged form, also isolated at the periphery of *A. dorsata* distribution in Sulawesi. Whether peripheral isolates should be considered as taxonomically distinct is a matter that is open to debate (Lo et al. 2010).

The history of works on *A. dorsata* once again illuminates the problem of sample size. When the spectrum of sampling has been wide, even though it contains many geographical gaps, it may be that a species seems rather homogeneous; and so it appeared to Ruttner. Prior to Ruttner (1988), however, many smallish and preliminary investigations had been reported. Most such studies emanated from India (Ratnam 1939; Deodikar 1959a, b, Deodikar et al. 1977; Trehan and Singh 1961; Jain 1967; Kshirsagar 1969; Sharma 1983; Bhandari 1983; Mujumdar and Kshirsagar 1986; Singh et al. 1990) and revolved around populations of northwest India, where great variations in altitude occur. All of these studies on the morphometrics and population structure of *A. dorsata* demonstrated that the populations sampled showed significant interlocality variation, which attests to the heterogeneity of these bees. Similar results were reported elsewhere (Kuang 1986). Unfortunately, there has not been any comprehensive multivariate morphometric analysis over the entire range of *A. dorsata* to date. However, it may well eventuate that the inferences from nuclear and mitochondrial DNA sequence data will prove more informative than those derived from morphometrics (Arias and Sheppard 2005; cf. Chap. 4).

In any event, over the last century, there have been only three "pre-biological species" taxonomic systematists (Enderlein 1906; von Buttel-Reepen 1906; Maa 1953) and two post-Huxley systematists (Daly 1985; Engel 1999) within honeybee systematics. Engel (1999) is the only contemporary systematist working on honeybees who presents both usage views: one in which there exists only *A. dorsata* and the other in which there exist *A. d. binghami*, *A. d. breviligula* and *A. d. dorsata*. The practice among honeybee biologists has, however, been to use the trinomial epithet as a tool on which to simply apply their names, based on inferences about the magnitude of differences they encounter. In these circumstances, the post-Ruttner apicultural literature abounds with the names *A. d. binghami*, *A. d. breviligula* and *A. d. dorsata*, as well as *A. laboriosa*. More recently, the names *A. binghami*, *A. breviligula* and *A. dorsata*, as well as *A. laboriosa*, are beginning to appear in common usage within the literature, which may reflect a growing consensus on the matter under certain species concepts (Lo et al. 2010). It would appear that total evidence and quantitative approaches, uniting multiple, independent lines of evidence, will be needed in place of morphometrics in the circumscription of species for this particular group of bees.

1.4.2 **Apis laboriosa** *F. Smith (1871)*

Like other lesser-known species of honeybees, the Himalayan *A. laboriosa* remained virtually unreported for a century after its original description by F. Smith (1871). While von Buttel-Reepen (1906) listed it as a subspecies of *A. dorsata*, Maa (1953) effectively resurrected its species status. More recently, Engel (1999) referred to *A. laboriosa* somewhat equivocally as *A. dorsata laboriosa* but did not accord it species status when applying a phylogenetic species concept.

Under both the biological and evolutionary species concepts, this form is considered a valid species and a recognised taxonomical entity. Real interest in *A. laboriosa* gained momentum following a major morphometric and biogeographical analysis by Sakagami et al. (1980). They established unequivocally that it was different from *A. dorsata* in 96 of 103 different morphometric measurements but, surprisingly, remained somewhat equivocal as to its taxonomical status. Li (1984), Chen (1993) and Trung et al. (1996) also distinguished the two species morphologically. McEvoy and Underwood (1988) argued, somewhat tenuously, that *A. laboriosa* and *A. dorsata* are sound species, based on the fact that no morphologically intermediate forms were known. These two species are very rarely sympatric, with *A. dorsata* usually occurring below altitudes of 1,500 m and *A. laboriosa* between altitudes of 2,500 and 4,000 m (Roubik et al. 1985; Allen 1995; Otis 1996; Thapa et al. 2001).

Nonetheless, a general consensus that *A. laboriosa* is a well-defined species under the biological species concept, developed only after (1) Li et al. (1986) and Kuang and Li (1988) clearly separated *A. laboriosa* from *A. dorsata* and other *Apis* species by their esterase isozyme profiles; (2) Underwood (1990) showed that *A. laboriosa* and *A. dorsata* are reproductively separated by drone mating flight times; (3) Blum et al. (2000) reported that no common chemical constituents were found in analyses of the cephalic and abdominal secretions of *A. laboriosa* and *A. dorsata*; (4) Aichholz and Lorbeer (1999, 2000) showed that the chemical profile of *A. laboriosa* beeswax differs unequivocally from that of all other *Apis*; (5) Kirchner et al. (1996) showed that, unlike *A. dorsata*, there is no acoustic component of the waggle dance in *A. laboriosa* and (6) Woyke et al. (2008) identified their differences in defensive behaviour. Sequence divergence between *A. laboriosa* and *A. dorsata* was consistent with behavioural data and supports the species status of *A. laboriosa* under the biological species concept (cf. Chap. 4).

1.5 Conclusion

Phylogenetic analyses strongly supported the basic topology that is recoverable from morphometric analysis, which groups the honeybees into three major clusters: giant bees (*A. dorsata*, *A. binghami* and *A. laboriosa*), dwarf bees (*A. andreniformis* and *A. florea*) and cavity-nesting bees (*A. mellifera*, *A. cerana*, *A. koschevnikovi*, *A. nuluensis* and *A. nigrocincta*). The clade of Asian cavity-nesting bees, however, included paraphyletic taxa. Exemplars of *A. cerana* collected from divergent portions of its range were less related to each other than were the sympatric taxa, *A. cerana*, *A. nuluensis* and *A. nigrocincta*. Nucleotide sequence divergence between allopatrically distributed western (*A. mellifera*) and eastern (*A. cerana*, *A. koschevnikovi*, *A. nigrocincta* and *A. nuluensis*) cavity-nesting species (being around 18% for the mitochondrial gene and 10–15% for the nuclear intron) suggested an earlier divergence for these groups than previously estimated from both morphometric and behavioural studies.

This latter finding necessitates a re-evaluation of the hypothesised origin of extant European, African and West Asian *A. mellifera*. In addition, the growing evidence of honeybee diversity in the geological past is not only expanding the total number of species but also forcing a reconsideration of global *Apis* biogeography. By example, the recent discovery of fossil honeybees in North America expands the lineage natively into the New World (Engel et al. 2009). The discovery of giant honeybees in Japan during the Miocene, demonstrates how, under changing climates, lineages considerably expanded their historical ranges (Engel 2006). Perhaps most interestingly, the diversity of basal fossil species currently suggests a more western origin for the honeybees, with a subsequent invasion and rapid radiation across Asia, which resulted in the remarkable array of species and challenging forms we see today.

References

Aichholz R, Lorbeer E (1999) Investigation of combwax of honeybees with high-temperature gas chromatography and high-temperature gas chromatography-chemical ionization mass spectrometry. I. High-temperature gas chromatography. J Chromatogr A 855:601–615

Aichholz R, Lorbeer E (2000) Investigation of combwax of honeybees with high-temperature gas chromatography and high-temperature gas chromatography-chemical ionization mass spectrometry. II. High-temperature gas chromatography-chemical ionization mass spectrometry. J Chromatogr A 883:75–88

Alexander BA (1991) Phylogenetic analysis of the genus *Apis* (Hymenoptera: Apidae). Ann Entomol Soc Am 84:137–149

Allen MF (1995) Bees and beekeeping in Nepal. Bee World 76:185–194

Arias MC, Sheppard WS (2005) Phylogenetic relationships of honey bees (Hymenoptera: Apinae; Apini) inferred from nuclear and mitochondrial DNA sequence data. Mol Phylogenet Evol 37:25–35

Arias MC, Tingek S, Kelitu A, Sheppard WS (1996) *Apis nuluensis* Tingek, Koeniger and Koeniger, 1996 and its genetic relationship with sympatric species inferred from DNA sequences. Apidologie 27:415–422

Baer B (2005) Sexual selection in *Apis* bees. Apidologie 36:187–200

Bhandari VC (1983) Biometrical studies on *Apis florea* F and *Apis dorsata* F of northwestern India. Thesis, Himachal Pradesh University, Simla

Blum MS, Fales HM, Morse RA, Underwood BA (2000) Chemical characters of two related species of giant honeybees (*Apis dorsata* and *A. laboriosa*): possible ecological significance. J Chem Ecol 26:801–807

Booncham U, Wongsiri S, Thirakupt K (1995) Niche differences between *A. florea* and *A. andreniformis* in dry-evergreen forest, Thailand. In: International Conference on Tropical Bees Environment, Pedu Lake, Kedah, p 33

Chaiyawong T, Deowanish S, Wongsiri S, Sylvester HA, Rinderer TE, de Guzman L (2004) Multivariate morphometric study of *Apis florea* in Thailand. J Apic Res 43:123–127

Chen YC (1993) Apiculture in China. Agricultural Scitech Publishing House of China, Beijing

Cockerell TDA (1911) Descriptions and records of bees XXXV. Ann Mag Nat Hist 8:310–319

Daly HV (1985) Insect morphometrics. Annu Rev Entomol 30:415–438

Damus MS, Otis GW (1997) A morphometric analysis of *Apis cerana* F. and *Apis nigrocincta* Smith populations from Southeast Asia. Apidologie 28:309–323

Deodikar GB (1959a) Geographical distribution based on morphological characters (structure of genitalia and crossability among different bee species). Indian Bee J 23:60–61

Deodikar GB (1959b) Some taxonomic problems in honeybees. 1. The concept of supra-generic grouping. Bee World 40:121–124

Deodikar GB (1960) Some indications on the phylogenetic interrelations among bees. Bee Genet Inf Bull 1:15–16

Deodikar GB, Ghatge AL, Phadke RP, Mahindre DB, Kshirsagar K, Muvel KS, Thakar CV (1977) Nesting behaviour of Indian honeybees. III. Nesting behaviour of *Apis dorsata* Fab. Indian Bee J 39:1–12

Diniz-Filho JAF, Malapsina O, Pignata MIB (1993) Geographic variation in *Apis cerana indica* F.: a spatial autocorrelation analysis of morphometric patterns. J Apic Res 32:65–72

Dobzhansky TG (1937) Genetics and the origin of species. Columbia University Press, New York

Dover C (1929) Wasps and bees in the Raffles Museum, Singapore. Bull Raffles' Museum, Singapore 2:43–70

Drory E (1888) From my notebook: notes on bees from a trip around the world. Bienenvater 20:20–30 [in German]

Dung PX, Thai PH, Trung LQ (1996) Some comparative characteristics between *Apis florea* and *Apis andreniformis* in Minh Hai Province. In: Proceedings of the 3rd Asian apiculture association, Hanoi, p 62

DuPraw EJ (1964) Non-Linnean taxonomy. Nature 202:849–852

Enderlein G (1906) New honeybees and contribution to the distribution of the genus *Apis*. Stett Entomol Ztg 67:331–334 [in German]

Engel MS (1999) The taxonomy of recent and fossil honey bees (Hymenoptera: Apidae; *Apis*). J Hymenopt Res 8:165–196

Engel MS (2006) A giant honey bee from the Middle Miocene of Japan (Hymenoptera: Apidae). Am Mus Novit 3504:1–12

Engel MS, Schultz TR (1997) Phylogeny and behavior in honey bees (Hymenoptera: Apidae). Ann Entomol Soc Am 90:43–53

Engel MS, Hinojosa-Diaz IA, Rasnitsyn AP (2009) A honey bee from the Miocene of Nevada and the biogeography of *Apis* (Hymenoptera: Apidae; Apini). Proc Calif Acad Sci 60: 23–38

Fabricius JC (1787) Mantissa insectorum. Hafniae, Proft [in Latin, cited from Maa 1953]

Fabricius JC (1793) Entomologia systematica emendata et aucta. Secundum classes, ordines, genera, species adjectis synonimis, locis, observationibus, descriptionibus. Hafniae, Proft [in Latin, cited from Maa 1953]

Fuchs S, Koeniger N, Tingek S (1996) The morphometric position of *Apis nuluensis* Tingek, Koeniger and Koeniger, 1996 within cavity-nesting honeybees. Apidologie 27:397–405

Gan YY, Otis GW, Mardan MB, Tan SG (1991) Allozyme diversity in Asian *Apis*. In: Smith DR (ed) Diversity in the genus *Apis*. Westview, Boulder, pp 117–130

Garnery L, Vautrin D, Cornuet JM, Solignac M (1991) Phylogenetic relationships in the genus *Apis* inferred from mitochondrial DNA sequence data. Apidologie 22:87–92

Gerstäcker CEA (1863) On the geographical distribution and varieties of the honeybee, with remarks upon the exotic honeybees of the old world. Ann Mag Nat Hist 11(270–283): 333–347

Goetze GKL (1964) The honeybee under natural and beekeeping conditions. Part 1. Systematics, reproduction and heritability. Monogr Angew Entomol 19:1–20 [in German]

Hadisoesilo S, Otis GW (1996) Drone flight times confirm the species status of *Apis nigrocincta* Smith, 1861 to be a species distinct from *Apis cerana* F., 1793, in Sulawesi, Indonesia. Apidologie 27:361–369

Hadisoesilo S, Otis GW (1998) Differences in drone cappings of *Apis cerana* and *Apis nigrocincta*. J Apic Res 37:11–15

Hadisoesilo S, Otis GW, Meixner M (1995) Two distinct populations of cavity-nesting honeybees (Hymenoptera: Apidae) in South Sulawesi, Indonesia. J Kans Entomol Soc 68:399–407

Hadisoesilo S, Meixner M, Ruttner F (1999) Geographic variation within *Apis koschevnikovi* Buttel-Reepen, 1906, in Borneo. Treubia 31:299–306

Hadisoesilo S, Raffiudin R, Susanti W, Atmowidi T, Hepburn C, Radloff SE, Fuchs S, Hepburn HR (2008) Morphometric analysis and biogeography of *Apis koschevnikovi* Enderlein (1906). Apidologie 39:495–503

Hepburn HR, Smith DR, Radloff SE, Otis GW (2001) Infraspecific categories of Apis cerana: morphometric, allozymal and mtDNA diversity. Apidologie 32:3–23.

Hepburn HR, Hepburn C (2005) Bibliography of *Apis florea*. Apidologie 36:377–378

Hepburn HR, Hepburn C (2008) Bibliography of *Apis koschevnikovi*. Apidologie 39:507

Hepburn HR, Hepburn C (2009) Bibliography of *Apis andreniformis*. Apidologie 40:3

Hepburn HR, Smith DR, Radloff SE, Otis GW (2001a) Infraspecific categories of *Apis cerana*: morphometric, allozymal and mtDNA diversity. Apidologie 32:3–23

Hepburn HR, Radloff SE, Verma S, Verma LR (2001b) Morphometric analysis of *Apis cerana* populations in the southern Himalayan region. Apidologie 32:435–447

Hepburn HR, Radloff SE, Otis GW, Fuchs S, Verma LR, Ken T, Chaiyawong T, Tahmasebi G, Ebadi R, Wongsiri S (2005) *Apis florea*: morphometrics, classification and biogeography. Apidologie 36:359–376

Huxley J (1940) The new systematics. Clarendon, Oxford

Jain PC (1967) Morphometrics of bees *Apis dorsata* F., *Apis indica* F., *Apis florea* F., *Megachile lanata* F., *Trigona smithii* B. (Hymenoptera: Apidae) and their foraging habits at Udaipur. Thesis, University of Udaipur

Jayavasti S, Wongsiri S (1992) Scanning electron microscopy analysis of honeybees (*Apis florea, Apis dorsata, Apis cerana, Apis mellifera, Apis andreniformis* and *Apis koschevnikovi*) stings. Recent Adv Toxicol Res 2:193–204

Keeling CI, Otis GW, Hadisoesilo S, Slessor KN (2001) Mandibular gland component analysis in the head extracts of *Apis cerana* and *Apis nigrocincta*. Apidologie 32:243–252

Kirchner WH, Dreller C, Grasser A, Baidya D (1996) The silent dances of the Himalayan honeybee, *Apis laboriosa*. Apidologie 27:331–339

Koeniger G (1991) Diversity in *Apis* mating systems. In: Smith DR (ed) Diversity in the genus *Apis*. Westview, Boulder, pp 199–212

Koeniger N, Koeniger G (1991) An evolutionary approach to mating behaviour and drone copulatory organs in *Apis*. Apidologie 22:581–590

Koeniger N, Koeniger G (2000) Reproductive isolation among species of the genus *Apis*. Apidologie 31:313–339

Koeniger N, Koeniger G (2001) Behavioural mating barriers among sympatric species of the genus *Apis*. In: Proceedings of the 7th international conference on apiculture tropical bees, Chiang Mai, pp 79–85

Koeniger N, Koeniger G, Gries M, Tingek S, Kelitu A (1996a) Observations on colony defense of *Apis nuluensis* Tingek, Koeniger and Koeniger, 1996 and predatory behaviour of the hornet, *Vespa multimaculata* Pérez, 1910. Apidologie 27:341–352

Koeniger N, Koeniger G, Gries M, Tingek S, Kelitu A (1996b) Reproductive isolation of *Apis nuluensis* Tingek, Koeniger and Koeniger, 1996 by species-specific mating time. Apidologie 27:353–360

Koeniger N, Koeniger G, Tingek S, Kelitu A (1996c) Interspecific rearing and acceptance of queens between *Apis cerana* Fabricius, 1793 and *Apis koschevnikovi* Buttel-Reepen, 1906. Apidologie 27:371–380

Koschevnikov GA (1900–1905) Materials on the natural history of the honeybee. Soc Amateurs Sci Nat 1–2 [cited from Alapatov 1929]

Kshirsagar KK (1969) Preliminary observations on the biometry of *Apis dorsata* Fab. queen. Indian Bee J 31:12–16

Kuang BY (1983) Studies on the small black honeybee. Chin Apic 6:10 [in Chinese]

Kuang BY (1986) The species of *Apis* in China. Apic China 5:7–9 [in Chinese]

Kuang BY (2002) Evolution and classification of Apini. J Bee 2:1–2 [in Chinese]

Kuang BY, Li YC (1985) The genus *Apis* in China. Chinese Beekeep 76:7–9 [in Chinese]

Kuang BY, Li YO (1988) Studies on Apini. Sci Agric Sinica 21:85–90 [in Chinese]

Lavrekhin FA (1935) Contribution to the study of variation of the honeybee. 2. Comparative biometric characteristics of the sexual appendages of the drones belonging to different forms of the genus *Apis*. Zool Zh 14:655–663 [in Russian]
Li S, Meng YP, Chang JT, Li JH, He SY, Kuang BY (1986) A comparative study of esterase isozymes in 6 species of *Apis* and 9 genera of Apoidea. J Apic Res 25:129–133
Li YQ (1984) Discussions on the taxonomy of six species of *Apis*. J Bee 4:19–21 [in Chinese]
Linnaeus C (1758) Systema naturae per regna tria naturae, secundum classes, ordines, genera, species, cum characteribus, differentiis, synonymis, locis. Salviae Holmiae Stockholm [in Latin, cited from Maa 1953]
Lo NH, Gloag RS, Snderson DL, Oldroyd BP (2010) A molecular phylogeny of the genus *Apis* suggests that the Giant Honey Bee of the Philippines, *A. breviligula* Maa, and the Plains Honey Bee of southern India, *A. indica* Fabricius, are valid species. Syst Entomol 35:226–233
Maa T (1953) An inquiry into the systematics of the tribus *Apidini* or honeybees (Hym.). Treubia 21:525–640
Mathew S, Mathew K (1988) The 'red' bees of Sabah. Newsl Beekeep Trop Subtrop Countries 12:10
Mayr E (1942) Systematics and the origin of species. Columbia University Press, New York
McEvoy MV, Underwood BA (1988) The drone and species status of the Himalayan honeybee, *Apis laboriosa* (Hymenoptera: Apidae). J Kans Entomol Soc 61:246–249
Mogga J, Ruttner F (1988) *Apis florea* in Africa: source of the founder population. Bee World 69:100–103
Morse RA, Laigo FM (1968) Honeybees in the Philippines. Philipp Biota 3:3–4
Mujumdar S, Kshirsagar KK (1986) Morphometric characterization of *Apis dorsata* Fabr. workers. Indian Bee J 48:25–29
Muzaffar N, Ahmad R (1989) Distribution and competition of *Apis* spp. in Pakistan. In: Proceedings of the 4th international conference on apiculture in tropical climates, Cairo, pp 449–452
Nanork P, Deowanish S, Wongsiri S (2001) Mitochondrial DNA variability of dwarf honeybee (*Apis florea* Fabricius (1787)) in Thailand by using PCR-RFLP technique. In: Proceedings of the 7th international conference on tropical bees, Chiang Mai, pp 341–345
Narayanan ES, Sharma PL, Phadke KG (1960) Studies on biometry of the Indian bees II. Tongue length, wing hooks, worker-brood-cell size and thickness of comb of *Apis florea* F. at Pusa (Bihar). Indian Bee J 22:81–88
Nunamaker RA, Wilson WT, Ahmad R (1984) Malate dehydrogenase and non-specific esterase isoenzymes of *Apis florea*, *A. dorsata*, and *A. cerana* detected by isoelectric focusing. J Kans Entomol Soc 57:591–595
Oldroyd BP, Rinderer TE, Wongsiri S (1992) Pollen resource partitioning by *Apis dorsata, Apis cerana, Apis andreniformis* and *Apis florea* in Thailand. J Apic Res 31:3–7
Ono M (1992) The Asian honeybees (*Apis* spp.). Honeybee Sci 13:19–22
Otis GW (1991) A review of the diversity of species within *Apis*. In: Smith DR (ed) Diversity in the genus *Apis*. Westview, Boulder, pp 29–49
Otis GW (1996) Distributions of recently recognized species of honeybees (Hymenoptera: Apidae; *Apis*) in Asia. J Kans Entomol Soc 69:311–333
Otis GW, Hadisoesilo S (1990) Honeybee survey of South Sulawesi. J Penelitian Kehutanan 4:1–3
Otis GW, Hadisoesilo S, Sugiyani, Cooper AM (2001) Using social bees as indicators of habitat quality in Sulawesi, Indonesia. In: Proceedings of the 7th international conference on tropical bees, Chiang Mai, pp 313–317
Özkani A, Gharlek MM, Özden B, Kandemir I (2009) Multivariate morphometric study on *Apis florea* distributed in Iran. Turk J Zool 33:93–102
Patinawin S, Wongsiri S (1993) Male genitalia of honey bees. In: Connor LJ, Rinderer TE, Sylvester HA, Wongsiri S (eds) Asian apiculture. Wicwas, Cheshire, Connecticut, pp 110–116
Peng YS, Nasr ME, Locke SJ (1989) Geographical races of *Apis cerana* Fabricius in China and their distribution. Review of recent Chinese publications and a preliminary statistical analysis. Apidologie 20:9–20

Pesenko YA, Lelei AS, Radchenko VG, Filatkin GN (1989) Chinese wax-bee, *Apis cerana cerana* F. (Hymenoptera: Apidae) in the Far East of the USSR. Entomol Obozr 68:527–548 [in Russian]

Petrov P (1992) Distribution and phylogeny of the honeybee genus *Apis* (Hymenoptera: Apidae). Uspekhi Sovremennoi Biologii 112:359–372 [in Russian]

Poulton EB (1908) Essays on evolution. Clarendon, Oxford

Radloff SE, Hepburn HR (1998) The matter of sampling distance and confidence levels in the subspecific classification of honeybees, *Apis mellifera* L. Apidologie 29:491–501

Radloff SE, Hepburn HR (2000) Population structure and morphometric variance in the *Apis mellifera scutellata* group of honeybees in Africa. Genet Mol Biol 23:305–316

Radloff SE, Hepburn HR (2002) Multivariate morphometric analysis of the *Apis cerana* populations of southern Asia. In: Proceedings of the 14th international congress, IUSSI, Sapporo, p 209

Radloff SE, Hepburn HR, Otis GW (2001) Flight machinery of honey bees (Hymenoptera: Apidae; *Apis*). In: Proceedings of the 7th international conference on tropical bees, Chiang Mai, pp 163–168

Radloff SE, Hepburn HR, Koeniger G (2003a) Comparison of flight design of Asian honeybee drones. Apidologie 34:353–358

Radloff SE, Hepburn HR, Bangay L (2003b) Quantitative analysis of intercolonial and intracolonial morphometric variance in honeybees, *Apis mellifera* and *Apis cerana*. Apidologie 34:339–352

Radloff SE, Hepburn HR, Fuchs S (2005a) The morphometric affinities of *Apis cerana* of the Hindu Kush and Himalayan regions of western Asia. Apidologie 36:25–30

Radloff SE, Hepburn HR, Hepburn C, Fuchs S, Otis GW, Sein MM, Aung HL, Pham HT, Tam DQ, Nuru A, Tan K (2005b) Multivariate morphometric analysis of the *Apis cerana* populations of mainland southern Asia. Apidologie 36:127–139

Radloff SE, Hepburn HR, Fuchs S, Otis GW, Hadisoesilo S, Hepburn C, Ken T (2005c) Multivariate morphometric analysis of the *Apis cerana* populations of oceanic Asia. Apidologie 36:475–492

Radloff SE, Hepburn C, Hepburn HR, Fuchs S, Hadisoesilo S, Tan K, Engel MS, Kuznetsov V (2010) Population structure and classification of *Apis cerana*. Apidologie 41:589–601

Raffiudin R, Crozier RH (2007) Phylogenetic analysis of honeybee behavioral evolution. Mol Phylogenet Evol 43:543–552

Ratnam R (1939) A preliminary study in the biometric variations in the Indian honeybees. Madras Agric J 27:432–439

Rattanawannee A, Channchao C, Wongsiri S (2008) Morphometric and genetic variation of small dwarf honeybees *Apis andreniformis* Smith, 1858 in Thailand. Insect Sci 14:451–460

Rensch B (1929) The concept of geographical races and the problem of speciation. Gebrüder Borntraeger, Berlin [in German]

Rinderer TE (1988) The rediscovery of *Apis koschevnikovi*. Am Bee J 128:807

Rinderer TE, Koeniger N, Tingek S, Mardan M, Koeniger G (1989) A morphological comparison of the cavity dwelling honeybees of Borneo *Apis koschevnikovi* (Buttel-Reepen, 1906) and *Apis cerana* (Fabricius, 1793). Apidologie 20:405–411

Rinderer TE, Oldroyd BP, Wongsiri S, Sylvester HA, de Guzman LI, Potichot S, Sheppard WS, Buchmann SL (1993) Time of drone flight in four honeybee species in south-eastern Thailand. J Apic Res 32:27–33

Rinderer TE, Oldroyd BP, Wongsiri S, Sylvester HA, de Guzman LI, Stelzer JA, Riggio RM (1995) A morphological comparison of the dwarf honeybees of southeastern Thailand and Palawan, Philippines. Apidologie 26:387–394

Rinderer TE, Wongsiri S, Kuang BY, Liu JS, Oldroyd BP, Sylvester HA, de Guzman LI (1996) Comparative nest architecture of the dwarf honeybees. J Apic Res 35:19–26

Rinderer TE, Oldroyd BP, de Guzman LI, Wattanachaiyingcharoen W, Wongsiri S (2002) Spatial distribution of the dwarf honeybees in an agroecosystem in southeastern Thailand. Apidologie 33:539–543

Roubik DW, Sakagami SF, Kudo I (1985) A note on distribution and nesting of the Himalayan honeybee *Apis laboriosa* Smith (Hymenoptera: Apidae). J Kans Entomol Soc 58:746–749
Ruttner F (1975) A metatarsal clasping structure in the *Apis* drone. Entomol Germ 2:22–29
Ruttner F (1988) Biogeography and taxonomy of honeybees. Springer, Berlin
Ruttner F (1992) Naturgeschichte der Honigbienen. Ehrenwirth, Munchen
Ruttner F, Kauhausen D, Koeniger N (1989) Position of the red honey bee, *Apis koschevnikovi* (Buttel-Reepen 1906), within the genus *Apis*. Apidologie 20:395–404
Sakagami SF, Matsumura T, Ito K (1980) *Apis laboriosa* in the Himalaya, the little known world largest honeybee (Hymenoptera: Apidae). Insecta Matsumurana 19:47–77
Sakai S, Hoshiba H, Hoshiba E (1986) Integrated taxonomic information on honeybees. In: Proceedings of the 30th International Congress Apicultural, Nagoya, pp 134–139
Sharma PC (1983) Morphometric studies on *Apis florea* F. and *Apis dorsata* F. of Himachal Pradesh and Punjab. Thesis, Himachal Pradesh University
Sheppard WS, Berlocher SH (1989) Allozyme variation and differentiation among four *Apis* species. Apidologie 20:419–431
Singh MP, Verma LR, Daly HV (1990) Morphometric analysis of the Indian honeybee in the northeast Himalayan region. J Apic Res 29:3–14
Smith DR (1991) Mitochondrial DNA and honeybee biogeography. In: Smith DR (ed) Diversity in the genus *Apis*. Westview, Boulder, pp 131–176
Smith DR (2002) Biogeography of *Apis cerana*: Southeast Asia and the Indo-Pakistan subcontinent. In: Proceedings of the 16th international congress, IUSSI, Sapporo, p 209
Smith DR, Villafuerte L, Otis G, Palmer MR (2000) Biogeography of *Apis cerana* F. and *A. nigrocincta* Smith: Insights from mtDNA studies. Apidologie 31:265–280
Smith DR, Palmer MR, Otis GW, Damus M (2003) Mitochondrial DNA and AFLP markers support species status of *Apis nigrocincta*. Insectes Soc 50:185–190
Smith F (1858) Catalogue of the hymenopterous insects collected at Sarawak, Borneo; Mount Ophir, Malacca; and at Singapore, by A.R. Wallace. Proc Linn Soc Lond 2:42–130
Smith F (1861) Descriptions of new species of hymenopterous insects collected by A.R. Wallace at Celebes. Proc Linn Soc Lond 5:57–93
Smith F (1865) On the species and varieties of the honey-bees belonging to the genus *Apis*. Ann Mag Nat Hist 15:372–380
Smith F (1871) A catalogue of the aculeate Hvmenoptcra and Ichneumonidae of India and the eastern Archipelago. J Linn Soc 11:285–415
Sulistianto A (1990) Morphometric analysis of Indonesian honeybees. Thesis, University of Wales College, Cardiff
Sylvester HA, Limbipichai K, Wongsiri S, Rinderer TE, Mardan MB (1998) Morphometric studies of *Apis cerana* in Thailand and the Malaysian peninsula. J Apic Res 37:137–145
Szabo TI (1990) Morphometric characteristics of *Apis cerana* from Sri Lanka. Apidologie 21:505–509
Tahmasebi G, Ebadi R, Tajabadi N, Akhoundi M, Faraji S (2002) The effects of geographical and climatic conditions on the morphological variation and separation of Iranian small honeybee (*Apis florea* F.) populations. J Sci Tech Agric Nat Res 6:169–176
Takahashi JI, Nakamura J, Sasaki M, Tingek S, Akimoto SI (2002) New haplotypes for the non-coding region of mitochondrial DNA in cavity-nesting honey bees *Apis koschevnikovi* and *Apis nuluensis*. Apidologie 33:25–31
Tan K, Fuchs S, Ruiguang Z (2002) Morphometrical characterization of *Apis cerana* in Yunnan Province of China. In: Proceedings of the 14th international congress, IUSSI, Sapporo, p 208
Tan K, Fuchs S, Koeniger N, Ruiguang Z (2003) Morphological characterization of *Apis cerana* in the Yunnan province of China. Apidologie 34:553–561
Tanaka H, Roubik DW, Kato M, Liew F, Gunsalam G (2001) Phylogenetic position of *Apis nuluensis* of northern Borneo and phylogeography of *A. cerana* as inferred from mitochondrial DNA sequences. Insectes Soc 48:44–51

Thakar CV, Tonapi KV (1962) Nesting behaviour of Indian honeybees. II. Nesting habits and comb cell differentiation in *Apis florea* Fab. Indian Bee J 24:27–31

Thapa R, Shrestha R, Manandhar DN, Bista S, Kafle B (2001) Beekeeping in Nepal. In: Proceedings of the 7th international conference on tropical bees, Chiang Mai, pp 409–413

Tilde AC, Fuchs S, Koeniger N, Cervancia CR (2000) Morphometric diversity of *Apis cerana* Fabr. within the Philippines. Apidologie 31:249–264

Tingek S, Mardan M, Rinderer TE, Koeniger N, Koeniger G (1988) Rediscovery of *Apis vechti* (Maa, 1953): the Saban honey bee. Apidologie 19:97–102

Tingek S, Koeniger G, Koeniger N (1996) Description of a new cavity-nesting species of *Apis* (*Apis nuluensis* n. sp.) from Sabah, Borneo, with notes on its occurrence and reproductive biology. Senckenbergiana Biol 76:115–119

Trehan KN, Singh H (1961) Measurements of the tongue and of the worker and drone cells of *Apis dorsata* Fab. in the Punjab. Indian Bee J 23:31–36

Trung LQ, Dung PX, Ngan TX (1996) A scientific note on first report of *Apis laboriosa* F Smith, 1871 in Vietnam. Apidologie 27:487–488

Underwood BA (1990) The behavior and energetics of high-altitude survival by the Himalayan honeybee, *Apis laboriosa*. Thesis, Cornell University, Ithaca

Verma LR (1992) Species and genetic diversity in Himalayan honeybees. In: Verma LR (ed) Honeybees in mountain agriculture. Oxford and IBH, New Delhi, pp 39–49

Verma LR, Mattu VK, Daly HV (1994) Multivariate morphometrics of the Indian honeybee in the northwest Himalayan region. Apidologie 25:203–223

von Buttel-Reepen H (1906) Contributions to the systematics, biology as well as the historical and geographical distribution of honeybees (*Apis mellifica* L.) their variability and other *Apis* species. Mitt Zool Mus Berlin 3:117–201 [in German]

Wheeler QD, Meier R (eds) (2000) Species concepts and phylogenetic theory. Columbia University Press, New York

Willis LG, Winston ML, Honda BM (1992) Phylogenetic relationships in the honeybee (genus *Apis*) as determined by the sequence of the cytochrome oxidase II region of mitochondrial DNA. Mol Phylogenet Evol 1:169–178

Wilson EO, Brown WL (1951) The subspecies concept and its taxonomic implication. Syst Zool 2:97–111

Wongsiri S, Limbipichai K, Tangkanasing P, Mardan M, Rinderer TE, Sylvester HA, Koeniger G, Otis GW (1990) Evidence of reproductive isolation confirms that *Apis andreniformis* (Smith, 1858) is a separate species from sympatric *Apis florea* (Fabricius, 1787). Apidologie 21:47–52

Woyke J, Wilde J, Wilde M, Sivaram V, Cervancia C, Nagaraja N, Reddy M (2008) Comparison of defense body movements of *Apis laboriosa*, *Apis dorsata dorsata* and *Apis dorsata breviligula* honey bees. J Insect Behav 21:481–494

Wu YR, Kuang B (1986) A study of the genus *Micrapis* (Apidae). Zool Res 7:99–102 [in Chinese]

Wu YR, Kuang B (1987) Two species of small honeybee: a study of the genus *Micrapis*. Bee World 68:153–155

Yang GH (1986) The distribution of the eastern honeybee (*Apis cerana* Fab.) in China and an analysis of its subspecies. J Yunnan Agric Univ 12:202–206 [in Chinese]

Yang GH (2001) The biology of *Apis cerana*. GH Yang, Beijing [in Chinese]

Chapter 2
Phylogeny of the Genus *Apis*

Nikolaus Koeniger, Gudrun Koeniger, and Deborah Smith

2.1 The Early Historical Background

In the middle of eighteenth century, under the influence of Linnaeus's then new approach to classification (1758), a stream of exotic specimens of an unexpected diversity began to reach western museums and collections (Fabricius 1787). This flow of new material continued in the nineteenth century, the time of the great explorers and collectors like Alfred Russell Wallace (Smith 1858, 1865). Among these mainly tropical animals, honeybees received special attention, and based on colouration, size and other morphological characters, the specimens were classified and named, resulting in more than 100 different names, many of which referred to single worker bees deposited in the major collections of natural history museums.

Coincidently during that period, drastic changes in beekeeping took place. The introduction of the movable frame hive around 1850 started a new epoch in the relationship between beekeepers and their bees. A new guild of scientific bee-keepers and beekeeping scientists began to spearhead research. Personalities like Johann Dzierzon (1849), Baron August von Berlepsch (1873) and others, in close cooperation with professional biologists like Prof. B.T.E. von Siebold (1856), described the biological ground plan of *A. mellifera* and extended knowledge of colony structure and honeybee biology. The "acclimatisation" movement, whose goal it was to augment and "improve" the fauna and flora of a region by introducing

N. Koeniger
Institut für Bienenkunde, (Polytechnische Gesellschaft) Fachbereich Biowissenschaften Goethe Universität, Frankfurt a.M. Karl-von-Frisch-Weg 2, D61440 Oberursel, Germany
e-mail: nikolaus.koeniger@bio.uni-frankfurt.de

G. Koeniger
Institut für Bienenkunde, Karl-von-Frisch-Weg 2, 61440 Oberursel, Germany
e-mail: gudrun.koeniger@bio.uni-frankfurt.de

D. Smith
Department of Ecology and Evolutionary Biology/Entomology, University of Kansas, Haworth Hall, 1200 Sunnyside Ave, Lawrence, KS 66045, USA
e-mail: debsmith@ku.edu

H.R. Hepburn and S.E. Radloff (eds.), *Honeybees of Asia*,
DOI 10.1007/978-3-642-16422-4_2,

and gradually acclimatising non-native species, also contributed to our understanding of honeybee biology.

Honeybee colonies from several southern regions, not only Mediterranean countries like Italy, Greece and Cyprus but also Egypt and other African regions, were imported to Europe (and honeybees from Europe were exported to Africa). The results of these importations were similar in all cases, where indigenous *A. mellifera* populations were naturally present. The imported exotic honeybees interbred with the local populations producing fully fertile hybrids. These practical experiences demonstrated beyond any doubt that African and European honeybee populations belonged to one species, *A. mellifera*, and that the impressive variability in colouration, morphology and behaviour among African, Middle East and European honeybees were differences within one species.

2.2 von Buttel-Reepen to Maa: 1900–1953

2.2.1 H. von Buttel-Reepen (1906)

The conflict between the numerous species names found in the collections of several natural history museums and the perfect hybridisation of African and European honeybees was solved by von Buttel-Reepen (1906) in his groundbreaking treatise on the taxonomy and phylogeny of honeybees. Based on his solid practical beekeeping experience, he apparently applied a rigid "biological species definition" much earlier than published by Mayr (1942). von Buttel-Reepen reduced the number of *Apis* species to three. The Asian species *Apis dorsata* and *Apis florea* were recognised as species and the numerous later named specimens were categorised as subspecies or even variations of subspecies. It might be of interest to mention that *Apis laboriosa* (Smith 1865, 1871) was listed as a subspecies of *A. dorsata*, while *Apis andreniformis* (Smith 1858) was presented as a variation of the subspecies *A. florea florealis* (Horne 1870).

The cavity-dwelling honeybees were then summarised under the species name *Apis mellifica* L., which is an invalid synonym for *A. mellifera*. Accommodating the diversity within this species, von Buttel-Reepen divided *A. mellifera* into three subspecies:

1. *Apis m. indica*
   Mainly Asian honeybees were listed in this subspecies, which were named as variations. Among them were *Apis koschevnikovi* (von Buttel-Reepen 1906) and *Apis nigrocincta* (Smith 1861), which regained their species status much later.
2. *A. m. unicolor*
   This subspecies contained mainly African honeybees as variations; but, *Apis cerana* (Fabricius 1793) is also found in this group (mislabelled specimens). According to F. Ruttner (personal communication), several specimens originating

in the Cameron highlands of Malaysia had been relabelled with "Cameroon" (Africa).

3. *A. m. mellifera*
   This subspecies contained the European populations as variations (*A. m. m. ligustica* Spinola 1806, *A. m. m. carnica* Pollmann 1879 etc.).

It must be noted that Gerstäcker (1862) had presented a systematic review, which was reprinted as an introduction to the monograph of von Buttel-Reepen. Gerstäcker was already aware of the natural distribution of the honeybee species. He correctly listed all western bees (Africa, Europe and Middle East) as subgroups of *A. mellifera* and he recognised three additional Asian species: *A. dorsata*, *A. florea* and *Apis indica*. Ironically, von Buttel-Reepen, who had a better and broader access to international collections, "corrected" Gerstäcker and reported details of (mislabelled) specimens of *A. cerana* and *A. koschevnikovi* from Africa.

Besides his classifications, von Buttel-Reepen presented his phylogenetic considerations in the form of a tree diagram (Fig. 2.1), which was mainly based on his perception of the evolution of social behaviour starting from solitary bees up to the highly social honeybees.

von Buttel-Reepen regarded Meliponini as the sister group of the Apini. The Apini then were split into two groups. The cavity-dwelling honeybee species (remaining in cavity) with an uncertain position for *A. cerana* (dotted line) represented a first branch. Without presenting any arguments, von Buttel-Reepen regarded the European *A. mellifera* as the most advanced form of honeybees.

As a second branch, the common ancestor for the open-nesting species, *A. dorsata* and *A. florea*, were separated from the cavity-dwelling bees, and after a short common evolution, diverged into the dwarf bees and the giant bees. He classified *A. dorsata* as the most primitive form of honeybees, because *A. dorsata*, like the meliponines, rears workers and drones in brood cells of similar size (an early form of "out-group" comparison). In his tree diagram (Fig. 2.1), von Buttel-Reepen indicated the "primitive" status of *A. dorsata* by the length of the lines (distance to top); so the most advanced stage (top) is reached by the cavity-dwelling species. *A. florea* was given a medium position while *A. dorsata* has reached "only" the level of Meliponini and Bombini. (To facilitate further reading, we have always used the nomenclature of honeybee species as is valid today. *A. indica*, which was used by many older authors, is replaced by *A. cerana*, etc.).

2.2.2 A.S. Skorikov (1929)

Skorikov (1929) presented a revision of the genus *Apis*, referring to Ashmead (1904), who had subdivided *Apis* into two genera: *Megapis* and *Apis*. Skorikov recognised the subgenus *Micrapis* within the genus *Apis* and suggested maintaining *Apis* in its broad sense, as a genus encompassing all living honeybees, and dividing it into three subgenera:

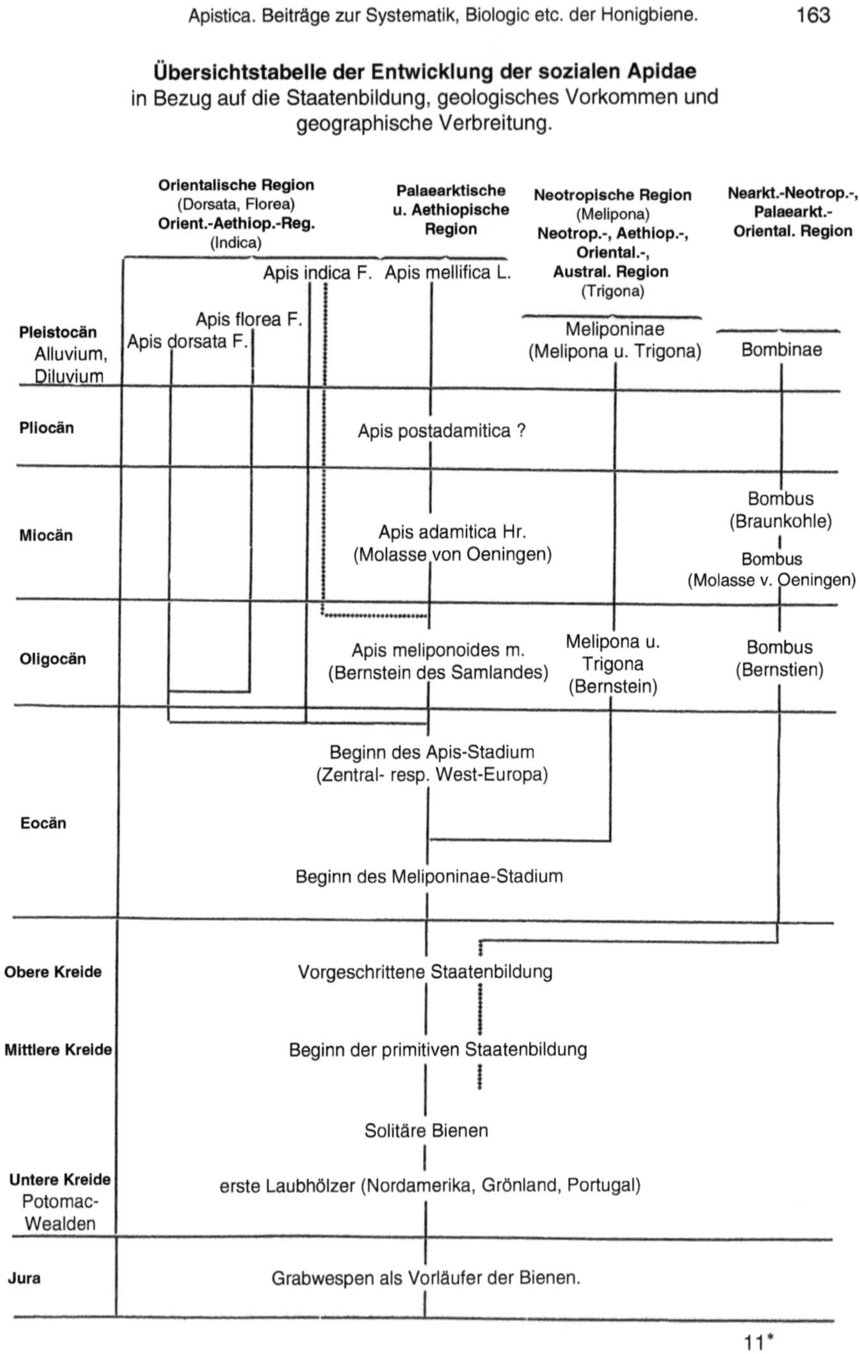

Fig. 2.1 The tree diagram of von Buttel-Reepen (1906) showing the development of sociality in Bombinae, Meliponinae and Apinae and the relationships among the honeybee species

Genus *Apis*

1. Subgenus *Megapis* (Ashmead 1904)

   Diagnosis:
   - Drones are equal in size to workers.
   - Special-sized drone cells do not occur.
   - Special queen cells do not exist (shown to be incorrect by subsequent observations).
   - Combs are attached below a thick tree branch, or any protruding structure of a building, about 60 cm long and 35 cm broad.
   - The distance between the peripheral ocelli is larger than the distance of these ocelli to the rim of the compound eye.
   - The bees are very large.
   - This is the most primitive form.
   - Only one species known: *A.* (*M.*) *dorsata* F. 1793.

2. Subgenus *Micrapis* (Ashmead 1904)

   Diagnosis:
   - Drones are larger than workers.
   - There are special queen cells.
   - The distance between the peripheral ocelli is much larger than the distance of these ocelli to the rim of the compound eye.
   - Drone has an original appendix at the tarsus of the hind leg.
   - The bees are very small.
   - Only one species known: *A.* (*M.*) *florea* F. 1787.

3. Subgenus *Apis* (Linnaeus 1758)

   Diagnosis:
   - Multiple parallel combs in tree cavities.
   - Three kinds of comb cells like *A. florea*.
   - The distance between the peripheral ocelli is usually not larger than the distance of these ocelli to the rim of the compound eye.

   Section 1: Asian cavity-nesting honeybees.
   Section 2: African subspecies of *A. mellifera*.
   Section 3: Middle East and European subspecies of *A. mellifera*.

2.2.3 T.C. Maa 1953

Maa's classification of honeybees (1953) was based mainly on the morphological analysis of single bees. He divided the genus *Apis* into three genera with two subgenera and 24 species:

1. Genus *Megapis* (Ashmead 1904) comprised four species
2. Genus *Apis* (L. 1758) with two subgenera (Subgenus *Sigmatapis* (Maa 1953) with nine species; Subgenus *Apis* s. str. with seven species)
3. *Micrapis* (Ashmead 1904) with two species

Maa's system of 24 "morpho-species" collided with experiences in the West where the impressive intraspecific variation of *A. mellifera* and the successful hybridisation among subspecies were common knowledge. As a result, his system was not accepted and his valuable contributions to the classification of the honeybees in Asia, and in particular, Southeast Asia were ignored. Only much later, with the rediscovery of *A. koschevnikovi* (Tingek et al. 1988) and *A. andreniformis* (Wu and Kuang 1987), were Maa's merits generally acknowledged. Maa's phylogenetic ideas are only precise in regard to his subgenus *Megapis* (*A. dorsata* group): "The genus *Megapis* beyond any doubt includes the most primitive forms" (p 633). However, the positions of the other honeybee groups remained undecided. Further, Maa presented a table of several morphological and other characters, which were classified as a "generalised" stage or "specialised" stage. There are, however, no arguments or any reasoning about those classifications.

2.3 Lindauer to Engel and Schultz: 1956–1997

2.3.1 M. Lindauer (1956)

The next great contribution to the phylogeny of honeybees was presented by Martin Lindauer (1956), who went to India and Sri Lanka to study the dance language of Asian honeybee species. Lindauer was familiar with the questionable systematic status of *A. cerana,* which was listed as a subspecies of *A. mellifera* (Schmiedeknecht 1907; Friese 1923). His observations of *A. cerana* in Sri Lanka convinced him that *A. cerana* was not a subspecies of *A. mellifera* but was a "good" species.

Sri Lankan *A. cerana* were significantly smaller in comparison to *A. mellifera.* Lindauer further argued that absconding behaviour and the perforated capping of the drone brood cell were characters that are not found in *A. mellifera*, and he clearly supported the species status of *A. cerana.* In his observations on the dance language, Lindauer found no differences between *A. cerana* and *A. mellifera* in the communication of the direction to a food source. "A dancing *A. mellifera* bee should have no difficulties to communicate the right direction to the food source to an *A. cerana* worker and vice versa" (verbal translation of the German text; 1956, pp 526–527). And, indeed, this proposition has recently been confirmed experimentally (cf. Chap. 19).

The dance frequency of *A. cerana* in relation to foraging distance, however, was much lower compared to *A. mellifera.* He concluded that food communication by

A. cerana was more precise at closer distances, and in his training experiments, Lindauer was not able to train *A. cerana* beyond 750 m while European *A. mellifera* have a normal flight radius of 3 km. Next, Lindauer studied the dance behaviour of *A. florea*. He noted: "As already mentioned, the elongated honey cells are built on a horizontal platform (above the nesting branch), which is further perfected by the backs of bees which rest on top and that is the exclusive dancing area" (verbal translation of the German text; 1956, p 532).

Lindauer argued that the communication of direction is significantly simpler on a horizontal dance platform than on a vertical comb. On a horizontal platform, the bee can point directly to the goal during the waggle run and the complicated transfer of the angle to the sun direction into the field of gravity – as found in *A. mellifera* and *A. cerana*– is not required. So Lindauer argued that the "simpler" dance of *A. florea* was phylogenetically older than the dances on vertical combs. In a series of experiments, he changed the natural position of the *A. florea* comb and covered the dance platform to induce foragers to dance on a vertical surface. The foragers of *A. florea,* however, did not dance on a vertical surface. He further blocked the bees' view of the sun and the sky, which blocked dancing behaviour or resulted in waggle runs with changing directions. In the end, Lindauer concluded that his experiments had proven that *A. florea* needs a horizontal platform and a direct view of the sky for dance communication, and he suggested that the *A. florea* dance is an ancestral precursor of dance communication in *A. mellifera* (cf. Chap. 14).

Lindauer next turned to *A. dorsata* and reported: "Concerning the communication of direction to a food source, at a first glance seems to be identical with communication of *A. cerana* and *A. mellifera*: the marked foragers (*A. dorsata*) danced on the vertical comb and transposed the direction to the goal in the field of gravity in the same way as *A. mellifera* does" (verbal translation of the German text; 1956, p 344). He argued that the dances of *A. dorsata* are restricted to a dance area from which the sun or the sky is visible. He concluded that the dance of *A. dorsata* holds an intermediate position between the *A. florea* dance on a horizontal platform and the *A. mellifera* dance on a vertical comb in the darkness, without a view of the sky. At the end, Lindauer tried to block an *A. dorsata* colony's view of the sky by putting up a cloth. He saw that dancing stopped at once, but an immediate colony defence event forced him to make a fast retreat. He argued about *A. dorsata*: "The thought must not be rejected, that a dance on a vertical surface under the open sky in comparison to a dance in the dark hive means a simplification" (verbal translation of the German text; 1956, pp 545–546). (The careful and balanced wording in the original German must not be overlooked). Lindauer was fully aware of the experimental limitations and his intention was to propose an idea or an hypothesis.

Lindauer had presented a simple, plausible and suggestive scenario of the evolution of dance communication within the genus *Apis*: The most primitive stage of dance communication is found in *A. florea*, where the bees dance on a horizontal surface in daylight pointing directly to the food source. The next evolutionary step is present in *A. dorsata*, where bees already transpose the (horizontal) direction to the food source in the field of gravity on the vertical comb. But

A. dorsata still "needs" daylight for its dances. The most derived condition is found in *A. cerana* and *A. mellifera*, where dance communication became independent of day light and functions in the darkness of the nest cavity.

Together with the brilliant work of Karl von Frisch (1965), Lindauer's hypothesis was widely published and reported in many major Zoology textbooks. It became an example for evolution of behaviour *par excellence*. Without any doubt, honeybee evolution gained general and scientific popularity, and Lindauer's hypothesis had a profound effect on the further ideas of *Apis* phylogeny. We will return to Lindauer's work later.

2.3.2 G.B. Deodikar, C.V. Thakar (1966) and H. Fahrenhorst (1977)

Deodikar and co-workers presented cytogenetic studies of Indian honeybees and reported that *A. florea* and *A. dorsata* had $n = 8$ chromosomes, while the chromosome number of *A. cerana* was $n = 16$ (Deodikar and Thakar 1966; Thakar and Deodikar 1966). They suggested that *A. cerana* evolved by polyploidy from ancestral forms like *A. florea* with $n = 8$ to its current status of $n = 16$. Since reduction in chromosome numbers in the course of evolution is unlikely (except by Robertsonian fusions), this was very strong evidence supporting Lindauer's scenario. About 10 years later, however, Fahrenhorst (1977) presented indisputable histological evidence that the four *Apis* species, *A. florea*, *A. dorsata*, *A. cerana* and *A. mellifera* uniformly had $n = 16$ chromosomes. Thus chromosome number ultimately provided no information on the phylogenetic relationships among these *Apis* species.

2.3.3 R. Jander, U. Jander (1970) and E. Horn (1975)

Support for Lindauer's hypothesis came from Jander and Jander (1970). They presented a detailed study of gravity orientation (geotaxis) in bees. They identified two types of directional orientation to gravity: "Walking bees with progeotaxis reduce their angle of orientation with increasing slope of the substrate, whereas bees with metageotaxis keep it constant. In classifying bees according to their type of gravity orientation, the genus *Apis* can be split into two groups according to type of orientation to gravity. *A. dorsata*, *A. cerana* and *A. mellifera* are characterised by metageotaxis, *A. florea* and the remaining bees by progeotaxis" (1970, p 355).

In 1975, E. Horn reported that bees generally possess two systems of gravity reception, the pedal proprioreceptors and the petiolus organ. In *Bombus terrestis*, *A. mellifera* and *A. cerana* both systems are involved in orientation. He reported: "The position of *A. florea* is remarkable. This bee does not transpose the direction of the food source into the field of gravity (Horn 1975). It dances on a horizontally inclined dance base and indicates the direction of the food source relative to the sun. This bee has lost the ability for gravity reception by pedal proprioreceptors" (1975, p 687). In regard to the work of the Janders (1970), these results did not favour their interpretation that the progeotaxis of *A. florea* was a more primitive and basal form of orientation compared to the metageotaxis of *A. cerana* and *A. mellifera*. Instead, Horn's results seemed to indicate that the simpler form of gravity orientation of *A. florea* might represent a derived condition within the honeybees.

2.3.4 N. Koeniger 1976

In 1976, N. Koeniger pointed to some uncertainty about the polarity of Lindauer's generally accepted hypothesis on the evolution of dance communication in honeybees. He argued that in contrast to Lindauer's ideas, cavity-dwelling might be an ancestral condition in the genus *Apis*. Bombini and Meliponini generally build their nests in cavities, and the close relationship between these groups and the Apini is well established. He argued that the ancestral honeybees might have nested in cavities. This would contradict Lindauer's assumption of the ancestral position of the open-nesting *A. florea*. Koeniger further claimed that, referring to the paper of Horn (1975), the dance of *A. florea* could also be a derived simplification of the dance pattern of the three other *Apis* species. In the end, Koeniger presented some arguments about the polarity of honeybee evolution: originating from a cavity-dwelling ancestor there might have been two different directions of selection. First, cavity dwelling honeybees like *A. cerana* and *A. mellifera* could extend their distribution to subtropical and temperate conditions, where colony reproduction would require large swarms that are forced to find a nest cavity within a short time. Under tropical conditions, however, small swarms resulting from high colony reproduction may survive for a longer period outside and such honeybees could exploit seasonal differences in nectar availability. At the end of this development, the transition to open-nesting forms might have taken place to avoid the limitations, which result from the scarcity of nest cavities. Today, effects of both selective directions can be observed by comparing the tropical African races of *A. mellifera* with races of *A. mellifera* from temperate Europe. Koeniger concluded that the phylogeny of the honeybees was not yet settled and that more studies were required. These studies should be focussed on characters of honeybees, which could be compared to homologous characters within the bumblebees and stingless bees.

2.3.5 F. Ruttner (1988)

In his classic treatise, Ruttner (1988) presented a detailed and critical review of the literature on honeybee evolution and summarised: "No clear conclusions can be derived from these observations about the evolution of the *Apis* species. Therefore, the question of the evolution of honeybees will be investigated from all available data on palaeontology, historic and morphometric taxonomy and ecology, which were considered essential in the study of general evolution of Apidae" (1988, p 26). In his approach to *Apis* phylogeny, Ruttner analysed the wing venation of fossil and extant honeybee species, together with two species of *Bombus*, and demonstrated a close relationship between *A. dorsata* and *Apis armbrusteri* when he compared 16 wing venation angles in a PCA. By introducing size-related characters, however, *A. dorsata* was clearly separated from the smaller *A. armbrusteri*.

Since Ruttner did not consider the importance of size-related morphometric characters in honeybee evolution and refer to size independent similarity with *A. dorsata* in wing venation angles, he postulated: "unchanged morphology indicates unchanged biology. Therefore, it is very likely that *dorsata*-like *A. armbrusteri* and *Synapis* were tropical open-air-nesting honeybees" (1988, p 30). Further, the extinction of *Apis* in Europe with the beginning of the ice ages is, according to Ruttner, another indication that these fossil honeybees were open-nesting species: "Confined to the ecological conditions of the tropics, this 'conservative' *Apis* type became extinct in Europe as a consequence of climatic deterioration at the end of the Tertiary" (1988, p 36). In a second step, according to Ruttner's hypothesis, the first cavity-dwelling honeybees appeared in the late Pliocene or early Pleistocene, and the separation of *A. cerana* and *A. mellifera* happened rather recently.

2.3.6 B.A. Alexander (1991)

In his cladistic analysis of the honeybees, Alexander (1991) included several morphological characters and character states of worker bees and drones, including sclerites of the male genitalia. His analysis confirmed again the same phylogenetic tree already suggested by many previous authors, including Lindauer and Ruttner. The "newly" rediscovered species, *A. koschevnikovi* and *A. andreniformis* clustered according to their expected positions: *A. koschevnikovi* within the cavity-dwelling species and *A. andreniformis* near to *A. florea*. In his discussion, Alexander carefully examined the question whether or not the dance of *A. florea* and *A. andreniformis*, performed on a horizontal surface, was an ancestral character for the genus, and he concluded that his analysis was ambiguous in this regard:

> If dancing on a horizontal surface is plesiomorphic for the genus *Apis*, then one need only postulate a single transition to vertical dancing, in the common ancestor of the *dorsata* group and the cavity-nesting species. If vertical dancing is the plesiomorphic condition, the one transition to horizontal dancing took place in the common ancestor of *A. florea* and *A. andreniformis* (1991, p 146).

Alexander's cladogram did not support the hypothesis that the extant cavity-nesting species are the basal lineages within *Apis*. However, this cladogram did not unequivocally indicate whether the ancestral *Apis* were cavity-nesting (Alexander

1991). Alexander's analysis also showed the male morphology to be highly derived within *Apis*: "The drones of *A. florea* and *A. andreniformis* have more unquestionably derived features of external morphology (antennae, hind basitarsus) than any other species in the genus. That is, in overall appearance, the drones of the ancestral *Apis* may have looked more like *mellifera* than *florea*. Determining the polarity of external morphological characters of drones is relatively straightforward, because an out-group comparison can clearly establish the plesiomorphic condition" (1991, p 146).

To summarise we quote Alexander:

> However, it is best to keep in mind that other possible explanations could be equally consistent with our current hypothesis of phylogenetic relationship among the species of *Apis*. In particular, it is worth considering that the architectural and behavioural components of the ancestral dance language may not have been retained in an unmodified form in any of the extant species of honeybees (1991, p 147).

2.4 Molecular Approaches to Honeybee Phylogeny

The application of molecular genetic data to the study of honeybee variation and phylogeny began in the late 1980s and early 1990s with mitochondrial DNA restriction site polymorphisms and mitochondrial DNA sequence data (Moritz et al. 1986; Smith and Brown 1988, 1990; Cornuet and Garnery 1991; Crozier et al. 1991; Garnery et al. 1991, 1992; Hall and Smith 1991; Sheppard et al. 1991a, b; Smith 1991a, b; Smith et al. 1991; Willis et al. 1992). The application of mitochondrial DNA data to the study of honeybees soon received a tremendous boost from the work of R. and C. Crozier (Crozier et al. 1989; Crozier and Crozier 1992, 1993), who sequenced the complete mitochondrial genome of *A. mellifera ligustica*. Besides providing insight into the evolution of animal mitochondrial genomes, these sequences enabled other researchers to design primers for the amplification of mitochondrial genes by means of the polymerase chain reaction (PCR) and carry out comparative studies of other honeybee populations and species. Mitochondrial DNA sequences (cf. Chap. 4) have remained the primary source of molecular data for studies of honeybee phylogeny, though single copy nuclear gene sequences are now making contributions to the storey (e.g. Arias and Sheppard 2005). Allozyme, microsatellite and DNA-fingerprint polymorphisms have also been applied to aspects of *Apis* phylogeny but have had much smaller impact.

2.4.1 Major Lineages Within Apis

The three main lineages within *Apis* – dwarf, giant and cavity-nesting honeybees – are supported by molecular studies, just as they are by morphology and behaviour

(Cornuet and Garnery 1991; Garnery et al. 1991; Engel and Schultz 1997; Arias and Sheppard 2005; Raffiudin and Crozier 2007; cf. Chap. 1). Studies from 1990s onward have steadily added additional species and genes to the analyses and have taken advantage of more sophisticated techniques for analysis. Cornuet and Garnery (1991) used partial sequences of the large subunit of mitochondrial RNA gene (lsrRNA or 16S rRNA), and the protein-coding mitochondrial gene Cytochrome Oxidase I (COI). Their data set included two subspecies of *A. mellifera*, *A. cerana* (from India or Sri Lanka), *A. dorsata* and *A. florea*, with *Bombus lucorum* to root tree, and their analyses employed maximum parsimony and neighbour joining methods. They found *A. florea* to be basal within this group of *Apis* species, sister to a clade including giant and cavity-nesting honeybees, and *A. dorsata* sister to a clade containing both cavity-nesting species. Their phylogeny can be summarised as: *Bombus* (*A. florea* (*A. dorsata* (*A. cerana* + *A. mellifera*))).

Shortly thereafter, Garnery et al. (1991) produced an additional analysis of these species (with an additional out-group, *Xylocopa violacea*), using a partial sequence of another mitochondrial protein-coding gene, Cytochrome Oxidase II (COII). Neighbour-joining and parsimony analyses produced different topologies, and their consensus tree gave very little resolution of species relationships, except to group the two cavity-nesting species together. Willis et al. (1992) used the complete sequence of COII to analyse a larger set of species: *A. florea* (from Thailand); *A. andreniformis, A. cerana* and *A. dorsata* (from peninsular Malaysia) and *A. mellifera* (Crozier et al. 1989) and *A. koschevnikovi* (from Sabah, Borneo, Malaysia). An ichneumonid wasp, *Exeristes roborator* (Liu and Beckenbach 1992, misspelled as *Excristes*) was used as an out-group. This analysis did not recover the three main lineages. *A. cerana* was basal within the genus *Apis*, sister to a clade containing the other *Apis* species. The two dwarf bee species grouped together, but *A. koschevnikovi* appeared as a sister to the dwarf bees, rather than grouping with the other cavity nesters. Their phylogeny can be summarised as *Exeristes* (*cerana* (*mellifera* (*dorsata* (*koschevnikovi* (*florea* + *andreniformis*))))).

Cameron et al. (1992) provided an early review of the use of molecular data in hymenopteran systematics, and included as an example an analysis of relationships among *Apis* species (*A. florea, A. dorsata, A. mellifera, A. cerana, A. koschevnikovi*) with examples of Meliponini, Bombini and Xylocopini as out-groups. Their maximum parsimony analysis of 500–600 base pairs of the small subunit of mitochondrial ribosomal RNA (16S rDNA) produced two equally parsimonious trees, one of which produced the expected topology of (out-groups) (*A. florea* (*A. dorsata* (*A. mellifera* (*A. cerana* + *A. koschevnikovi*)))), and one which paired *A. mellifera* and *A. dorsata* as sister taxa, and *A. cerana* and *A. koschevnikovi* as sister taxa, with *A. florea* basal in the genus. Engel and Schultz (1997) carried out a combined or total evidence analysis of *Apis* species relationships using all of the suitable data sets then available: COII (Garnery et al. 1991; Willis et al. 1992; Crozier and Crozier 1993), 16S rDNA (Cameron et al. 1992), and adult and larval morphological characters (Torchio and Torchio 1975; Alexander 1991; Koeniger et al. 1991). Among their goals was to distinguish among data sets in terms of which were most informative and which were too "noisy" to be useful in this context. They determined that among the molecular data sets, 16S is more informative at the level of *Apis* species, while COII is evolving too rapidly to provide high quality information about relationships among *Apis* species. Maximum parsimony analyses using total evidence, 16S sequence + morphology, 16S sequence alone, or morphology alone all produced the same topology, placing dwarf bees basal to the *A. dorsata* and the cavity-nesting lineages: (*A. andreniformis* + *A. florea*) (*A. dorsata* (*A. mellifera* (*A. cerana* + *A. koschevnikovi*))).

Leelamanit et al. (2004) examined *A. florea*, *A. dorsata*, *A. cerana* (all from India), *A. laboriosa* (from Nepal) and *A. mellifera* (a Japanese cultivar and a sequence from Genbank), using the mitochondrial protein-coding gene NADH dehydrogenase subunit 4 (ND4) as a source of data, and a *Drosphila* as an out-group. Their maximum likelihood analysis consistently grouped *A. mellifera* and *A. cerana* together, and analysis of nucleotide (but not amino acid) sequences placed *A. florea* in a basal position, but the positions of the two giant bee species remained unresolved.

Arias and Sheppard (2005) examined the mitochondrial protein-coding gene, NADH dehydrogenase subunit 2 (ND2), and an intron in the nuclear protein-coding gene, elongation factor 1-alpha (EF1-alpha), from *A. florea*, *A. andreniformis*, *A. dorsata*, *A. d. binghami*, *A. d. laboriosa*, *A. mellifera*, *A. koschevnikovi*, *A. nuluensis* and *A. nigrocincta*. The stingless bee, *Melipona bicolor*, was used to root their trees. Their analyses of both ND2 and EFI-alpha intron (employing parsimony, maximum likelihood and neighbour-joining methods) all produced the same basic topologies. The three major branches, dwarf, giant and cavity-nesting bees, were recovered with dwarf bees appearing basally and the giant and cavity-nesting bees appearing as sister lineages.

Relations among the cavity-nesting species were not completely resolved, as discussed below. Their topologies can be summarised as: *Melipona* ((A. *andreniformis* + *A. florea*) ((*A. dorsata*, *A. d. laboriosa*, *A. d. binghami*) (*A. mellifera* (*A. koschevnikovi* (*A. cerana*, *A. nigrocincta*, *A. nuluensis*))))). One issue that might complicate the interpretation of this study is the finding by Danforth and Ji (1998) that there are two nuclear copies of EF 1-alpha (F1 and F2) in honeybees, and probably in all major bee families. Arias and Sheppard (2005) used primers that would preferentially amplify a portion of the F1 copy (Walldorf and Hovemann 1990) but do not indicate whether they were aware of the existence of two gene copies, or indicate how they ensured that they compared homologous copies in each species.

Raffiudin and Crozier (2007) employed sequences of the mitochondrial large subunit ribosomal RNA, COII, ND2 and one nuclear protein-coding gene, inositol 1, 4, 5 triphosphate receptor (itpr) gathered from *A. florea*, *A. andreniformis*, *A. dorsata*, *A. laboriosa*, *A. d. binghami*, *A. mellifera*, *A. koschevnikovi*, *A. nigrocincta*, *A. cerana*, *A. nuluensis*. *A. laboriosa* was collected in Nepal, while all of the other Asian species were collected in Thailand, Malaysia or Indonesia. They employed three out-group species, *B. terrestris*, *M. bicolor* and *Trigona fimbriata*. Sequences were analysed using maximum parsimony and Bayesian methodologies (Huelsenbeck and Ronquist 2001). Their analyses recovered the three main lineages, with dwarf bees basal, and the giant bee clade (*A. dorsata*, *A. laboriosa* and *A. d. binghami*) sister to a clade containing all the cavity-nesting honeybees. The Bayesian analysis of the complete data set can be summarised as: (out-groups) ((*A. florea* + *A. andreniformis*) ((*A. laboriosa* (*A. d. binghami* + *A. dorsata*)) (*A. mellifera* (*A. koschevnikovi* (*A. nuluensis* (*A. cerana* + *A. nigrocincta*)))))).

2.5 Consensus Tree Structure

During the last few decades, the number of recognised honeybee species increased considerably. Surprisingly, however, the increase of honeybee species from three species in the time of von Buttel-Reepen (1906) "by 3 times" to the nine generally recognised honeybee species of today, did not result in an increase of biological variation in a similar range. Instead the overall variation among the extant species did not change much, because the newly recognised species distinctly "clustered" with the three species recognised by von Buttel-Reepen (1906) with the effect that the "new" species clearly fell in three taxonomic groups. We discuss the tree topology of the three subgenera according to Skorikov (1929):

1. Subgenus *Megapis* (Ashmead 1904) that comprises all extant giant honeybee species.
2. Subgenus *Micrapis* (Ashmead 1904) with the extant dwarf honeybee species.
3. Subgenus *Apis* (*sensu stricto*) (L. 1758) with all extant species, which have medium-sized worker bees.

The consensus tree topology resulted from very different approaches to honeybee evolution. First, Lindauer (1956), studying dance communication in Sri Lanka, suggested a basal position for *Micrapis* and a monophyletic origin of *Megapis* and *Apis* (vertical dances). Koeniger (1976) accepted the tree topology of Lindauer (1956) (but not the polarity!). Ruttner (1988) and Alexander (1991), based on a broad variety of morphological, behavioural and paleontological data, also supported the same tree topology. The following authors (Engel and Schultz 1997; Arias and Sheppard 2005; Oldroyd and Wongsiri 2006; Raffiudin and Crozier 2007) included a fast growing amount of DNA sequence data in their reconstruction of *Apis* phylogeny and confirmed again the consensus tree topology (Fig. 2.2). Summarising the overwhelming evidence and the fact that this tree topology has remained unchanged for more than 50 years of intensive research, this consensus tree of *Apis* is one of the best supported results of phylogenetic research.

2.6 Polarity of *Apis* Evolution

The structure or topology of the phylogenetic tree was never really challenged (as far as we know). The polarity or the evolutionary scenario, however, remained a matter of dispute and controversy. Lindauer (1956) suggested that the horizontal

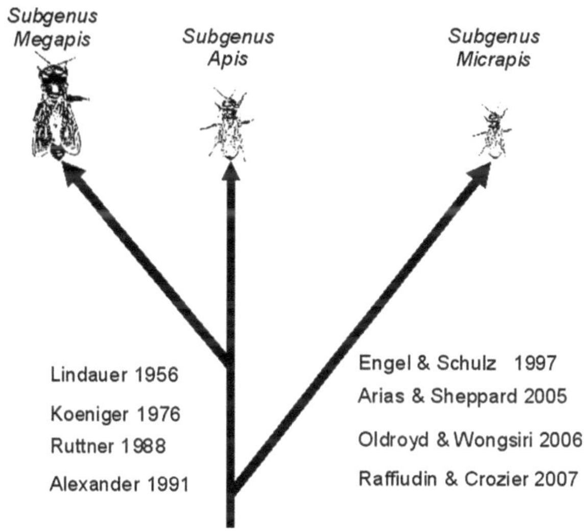

Fig. 2.2 The consensus tree topology

dance of *A. florea* represents the basal condition for the living *Apis* species. The next evolutionary stage (according to Lindauer) is the vertical dance: in the open as in *A. dorsata*. The last, most derived condition is the vertical dance in the darkness of the hive found in cavity-nesting bees (Fig. 2.3).

To facilitate the further discussion, we included the origin of the genus *Apis* (Fig. 2.3) in our considerations. According to Michener (2000), the tribe of Apini belongs to the monophyletic group of corbiculate bees, together with the Euglossini, the Bombini and the Meliponini, and obviously the basal condition for all corbiculate bees is cavity nesting.

The vast majority of publications on honeybee evolution present the conclusion that the ancestor of modern honeybees abandoned cavity nesting (inherited from their corbiculate ancestors) and switched to the open-nesting condition and horizontal dances directed to the food source. The only recent contradiction was presented by Raffiudin and Crozier (2007), who included several behavioural characters along with sequence data in their maximum parsimony and Bayesian analyses (see Sect. 2.4). Regarding the orientation of the dance, they concluded that the ancestor of modern *Apis* species probably had "a silent and vertical directional waggle dance and a single comb to its nest" (2007, p 550). This conclusion contradicted Lindauer's generally accepted hypothesis, based on the basal position of the horizontal waggle dance of *A. florea* pointing directly to the food source (Fig. 2.3). There is, however, no comment by the authors concerning their controversial conclusion.

After transition to open nesting (and to a horizontally directed waggle dance) the ancestral *Micrapis* branched away leaving the other bees (*Megapis* + *Apis*) as a next monophyletic group. At this stage, the ancestral honeybees developed the

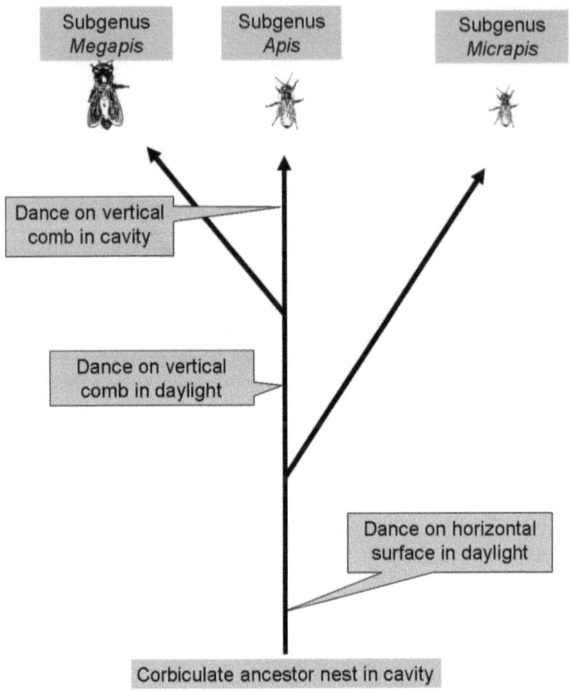

Fig. 2.3 The generally accepted polarity for the evolution of honeybee dances (Lindauer 1956). The labels indicate evolutionary changes in dance communication

vertical dance still nesting in the open. A bit later the ancestral *Megapis* separated from still open-nesting forms. After this separation, the ancestor of the subgenus *Apis* regained access to a nest cavity and gave origin to the extant cavity-nesting honeybee species. Summarising, we have three changes of dance communication and nesting mode in this scenario (Fig. 2.3).

The alternative scenario of honeybee evolution (Koeniger 1976) postulated that the vertically directed waggle dance in a dark nest cavity as a basal condition for the genus *Apis* maintaining cavity-nesters from the corbiculate origin straight to the extant cavity-dwelling honeybee species (Fig. 2.4). There are two more changes in this scenario: Firstly, the ancestral *Micrapis* (after leaving the nest cavity) built its single comb around a small twig constructing a platform on top of the nest. Under these conditions, *Micrapis* simplified the dance communication and developed a horizontally directed waggle pointing directly to the food source.

Secondly, (independent of *Micrapis*) the ancestor of *Megapis* turned to open nesting and fixed its single comb under a large branch or cliff. These bees then performed their waggle dances in daylight and in the field of gravity on the vertical surface of their clustering nestmates. All together, there are three changes in dance behaviour and nesting mode in the above scenario.

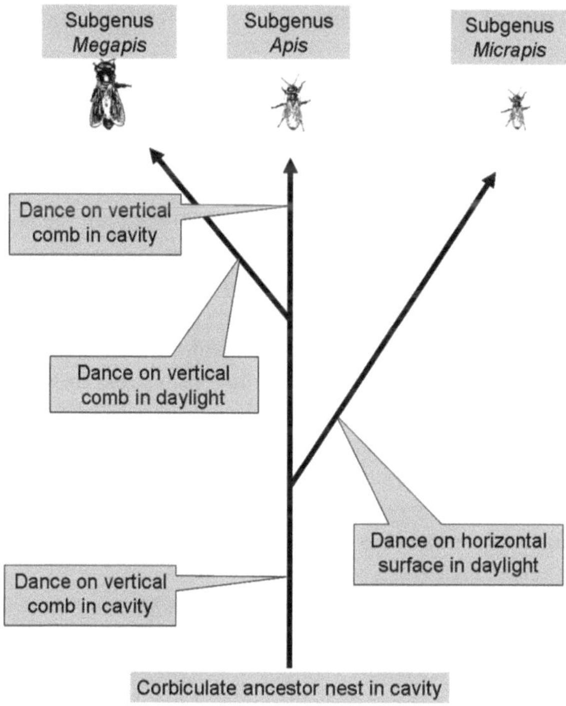

Fig. 2.4 The alternative polarity of honeybee evolution. The labels indicate evolutionary changes in dance communication and mode of nesting

2.6.1 Comparison of the General and the Alternative Scenario

In the literature review, several arguments were already highlighted, which were brought forward in favour of the general scenario and in opposition to the alternative concept. We do not want to repeat that discussion here again. Comparing both scenarios, we will focus instead on a more general discussion of honeybee evolution and introduce some new arguments.

At the beginning, the immense molecular evidence for the structure of the consensus tree must be underlined. Therefore, no alternative tree topology will be considered in our comparison. Further, by parsimonious analysis, both scenarios are equally probable (Alexander 1991). The undisputable evidence for a more derived drone morphology of *A.* (*Micrapis*) *florea* and a more basal position of the *A. mellifera* drone can be understood as an argument in favour of the alternative scenario. The scenario of honeybee evolution, however, is based on characters of workers. Further, it is evident that natural selection for workers is different to the selection of sexuals in honeybees (Koeniger and Koeniger 1991).

2.7 Evolution of Worker Size in *Apis*, the Ultimate Scenario

In contrast to solitary insects, social insects practise a more or less pronounced division of labour, which is based in many species on a sterile worker caste and reproductives (Wilson 1971). In the genus *Apis* the body sizes of sexuals (queen and drones) remain relatively constant whereas the size of worker bees (independent of reproductive requirements) shows a higher variation among the species (Fig. 2.5) (Koeniger et al. 1993).

According to Seeley et al. (1982), the size of the worker is of high selective significance. Surprisingly, up to now, worker size has not played a key role in recent discussions of *Apis* phylogeny; although it was in the focus of early honeybee biologists (Ashmead 1904; Skorikov 1929) and the traditional systematic classification of the genus *Apis* is based on worker size: *Micrapis* (comprises species with small workers), *Apis* (medium-sized workers) and *Megapis* ("giant" workers). According to the consensus tree structure, all three groups are monophyletic units (and since evolution does not jump) and each stage must be connected by arguable transitions to the next stage. Therefore, species (*Micrapis*) with small workers must evolve via ancestors (*Apis*) having medium-sized workers to become *Megapis*, a giant worker bee. It is not parsimonious to put a species with dwarf bees near to the root and next "jump" to giant bees and at the end reach medium-sized worker bees. A more logical and more likely approach would be to keep medium-sized worker bees throughout the tree, then *Micrapis* and later *Megapis* is derived from ancestral *Apis* medium-sized bees. Transitions to small and to large workers must have happened via medium-sized workers. That is indispensable. At this point, the question arises whether or not worker size in *Apis*

Fig. 2.5 Size differences are greater in workers (*left*) than drones (*right*) *A. dorsata* (*above*) *A. andreniformis* (*middle*) *A. cerana* (*below*). Specimens from Borneo

is an "independent" character or belongs to a syndrome of several other features. In other words: How tightly is worker size integrated in a functional framework or the entity of the subgenus?

2.7.1 Worker Size and Mode of Nesting

Honeybees, as all organisms, are functional entities and this limits the freedom of combinations, which are possible without interference with survival (adaptive value). It is obvious, for example that the mode of nesting has far reaching importance on the biology of a honeybee species, which exceeds that of dance communication. Many behavioural characters of honeybees are closely linked to the nest site. Colony defence, migration, seasonal dispersal, swarming, foraging range and many more characters are confounded and must not be treated like different incidents, which have occurred independently from each other during evolution.

As mentioned before, there are five recognised species of cavity-nesting bees with medium-sized workers (*Apis*), two species of dwarf bees (*Micrapis*) and at least two or even four species of giant bees (*Megapis*) (Fig. 2.6).

Open-nesting species have either large or small workers; and cavity-nesting species have medium-sized workers. But how has natural evolution linked "worker size" to "mode of nesting"? Moving out of a cavity, a colony has to face an increased predatory pressure because in an open-nesting situation the protective cavity walls are lost. Looking at the extant species there are two alternative routes open to cope with higher predatory and parasitic pressure.

Fig. 2.6 Worker bees of the nine recognised honeybee species. Open-nesting species have either large workers (*A. laboriosa*, *A. dorsata*) or small workers (*A. andreniformis*, *A. florea*). Cavity dwelling species have medium-sized workers (*A. mellifera*, *A. koschevnikovi*, *A. nuluensis*, *A. nigrocincta* and *A. cerana*)

Fig. 2.7 Comb of *A. andreniformis* is presented for comparison of size by co-author (G.K.)

Fig. 2.8 *A. andreniformis* colony hidden in the canopy of a mango tree

2.7.2 The Micrapis Syndrome

Firstly, the bee colonies can become small and hide (Fig. 2.7). Many colonies of *A. florea* and *A. andreniformis* are about the size of a leaf and are hidden (Fig. 2.8). The small workers fly nearly soundlessly and cease foraging whenever a larger animal approaches (SenSarma et al. 2002).

When detected by a predator the bees first try to defend the nest. However, very often their stings are too short (Fig. 2.9) and the defence force too weak to repel, for example a tree shrew or a large hunting *Vespa*. Then the colony takes to the wing and absconds (Koeniger et al. 2010). Because of their mode of nesting, these bees quickly find a hidden twig around which they build a new comb. The survival strategy of a dwarf bee colony depends on hiding and, in case of attack by a dominant predator, retreat to one of uncountable small-nesting branches. Further, the small worker size and the limited flight range do not allow long distance migrations. However, dispersal within a limited range enables these bees to select seasonally adapted nesting sites and helps these small colonies to outlast periods of adverse conditions and food scarcity (Tirgari 1971). To summarise, the small size of its worker seems to be a fundamental feature of the *Micrapis* syndrome like the horizontally directed waggle dance, the unique comb construction, several characters mentioned above and numerous more.

Fig. 2.9 Comb of *A. dorsata* is presented for comparison of size by co-author (G.K.)

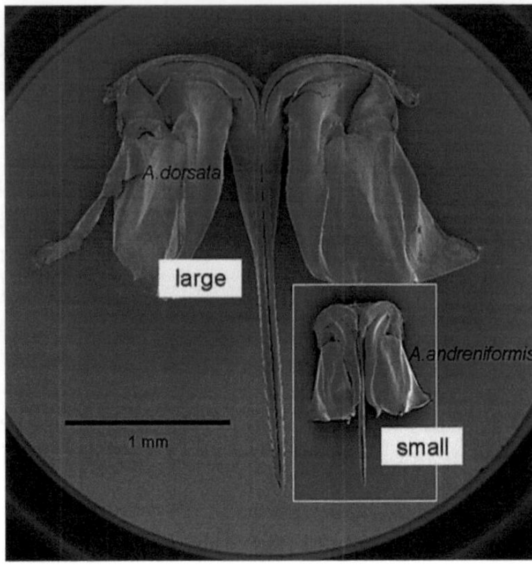

Fig. 2.10 The long sting of *A. dorsata* (*left*) penetrates almost any natural protection. The small sting of *A. andreniformis* (*right*) is less effective defending the colony against larger predators

2.7.3 The Megapis Syndrome

In *Megapis,* an alternative strategy to counterbalance the loss of "protective cavity walls" is found. The colonies and worker bees are large (Fig. 2.10). The long stings of giant bees penetrate almost any natural protection (Fig. 2.9). In many parts of Asia, these colonies aggregate in bee trees or under cliffs. The huge combs are exposed and, in contrast to even larger species of the Southeast Asian rainforest, the nests of these honeybees are not hidden (Seeley 1985). They rely on their formidable defence and their effective alarm communication (Koeniger et al. 1979). An *A. dorsata* colony defends itself at all costs and very soon neighbouring colonies join in the defence, which is nearly invincible. Further, the large worker size allows long range flights and migrations (Koeniger and Koeniger 1980). Like migratory bird species, these bees can leave the area whenever seasonal conditions deteriorate and return when the seasonal nectar flow starts again. Large workers as well as the huge open combs hanging under a support and the vertical waggle dance in day light are fundamental parts of the *Megapis* syndrome.

2.7.4 The Apis Syndrome (Cavity-Nesting Subgenus)

Cavity-nesters must communicate via their dance language in the darkness of the cavity. They possess vertically oriented, multiple parallel combs, pendant from the ceiling of the cavity and their dances are performed in the bee ways/space (that is the space between the surfaces of two neighbouring combs). Protected by the walls

Fig. 2.11 A small group of workers guard the entrance of a nest cavity occupied by a colony of *A. nuluensis*

of the cavities, colony defence is focused at the nest entrance and is usually performed by a small group of guard bees (Fig. 2.11).

The large majority of the workers are engaged in other activities and generally 50% or more of the workers collect pollen and nectar (Seeley et al. 1982). So foraging efficiency seems to be a major driving force in worker size of cavity-nesters and has resulted in medium-sized, high-energy worker bees (Dyer 1991).

2.7.5 The Ultimate Scenario of Honeybee Evolution

Worker size in the genus *Apis* varies within a wide range. However, the size variation among the extant species within each subgenus is rather small, and within these small limits, worker size in honeybees seems to be irrevocably linked to an entire syndrome. There is no question that the honeybee colony's existence depends on the functional relations within the syndrome, which is a complex framework of many known, and still very many unknown, biological characters. Therefore, each evolutionary change of a single character was either detrimental or integrated in a gradual reconstruction of the whole functional system resulting in the three different syndromes which are found today.

Furthermore, both ways to deal with the unprotected situation of the open-nesting colony seems to have evolved in a kind of disruptive selection, which resulted in the *Micrapis* syndrome at one end and the alternative *Megapis* syndrome at the other.

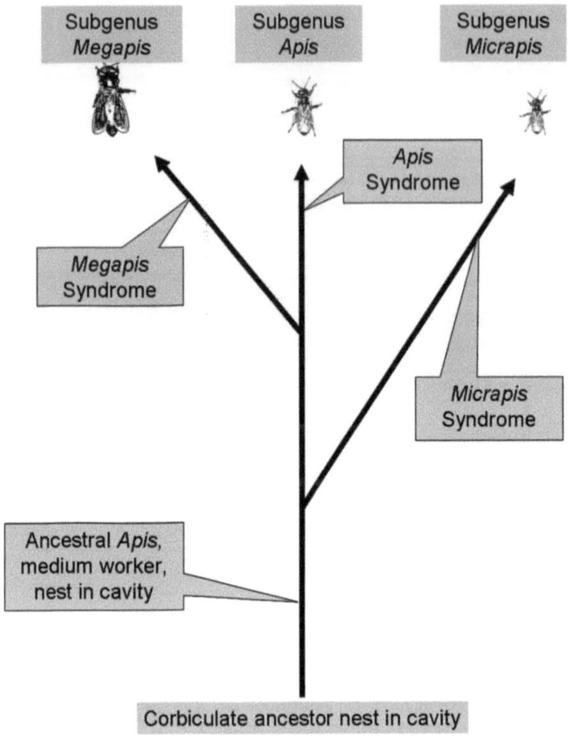

Fig. 2.12 The "ultimate" scenario of honeybee evolution. The labels indicate evolutionary transitions

Without any doubt, the horizontal dance on top of the comb represents an essential part of the *Micrapis* syndrome. However, the generally accepted scenario of *Apis* phylogeny (Fig. 2.3) postulates an evolutionary transition of an ancestral *Micrapis* syndrome (horizontal dance in the open) to the *Megapis* syndrome (indicated by a vertical dance in the open), which seems to be rather unlikely.

Turning to the alternative scenario (Fig. 2.4), there is the cavity-nesting syndrome with medium-sized worker bees (ancestral *Apis*) in a basal position. From that condition, gradual and arguable functional transitions to open-nesting and smaller worker bees do not pose any difficulties, and finally lead to the *Micrapis* syndrome of the extant dwarf bees. Further up in the consensus tree the ancestor of *Megapis* separated from ancestral cavity-nesters with medium-sized workers (Fig. 2.12).

2.8 Worker Size and the Fossil Record

The functional relationship between worker size and its fundamental role in the syndromes of the subgenera opens new perspectives in understanding the fossil record of honeybee evolution. Based on the size parameters of the extant species

Table 2.1 Fossil honeybees of medium size (ancestral *Apis*)

Taxon	Epoch (Location)	Author
(a) The *henshawi* group (*Synapis* Cockerell)		
henshawi[a]	Oligocene (Europe)	Cockerell
vetusta[b]	Oligocene (Europe)	Engel
cuenoti[a]	Oligocene (Europe)	Theobald
petrefacta[a]	Miocene (China)	Riha
longitibia[a]	Miocene (China)	Zhang
miocenica[a]	Miocene (China)	Hong
(b) The *armbrusteri* group (*Cascapis* Engel)		
armbrusteri[a]	Miocene (Europe)	Zeuner
nearctica[c]	Miocene (North America)	Engel et al.

[a]Engel (1998)
[b]Engel (1999)
[c]Engel et al. (2009)

(Ruttner 1988; Oldroyd and Wongsiri 2006), the fossil honeybees (workers) can be assigned to three ancestral precursors of the extant subgenera:

Megapis: Forewing length larger than 11 mm; body length larger than 16 mm.

Micrapis: Forewing length less 7 mm; body length less than 10 mm.

Apis: Forewing length from 7.5 to 10 mm; body length from 10 to 15 mm.

This assignment to an ancestral syndrome has far-reaching biological implications. According to the previously presented arguments, worker size, mode of nesting and food communication are closely associated. Therefore, in contrast to the previous dogma of *Apis* evolution, we conclude that the medium-sized fossil honeybees (Table 2.1) were cavity-dwelling species with a vertical buzzing dance in the dark nest cavity.

Not long ago, the first fossil giant honeybee, *Apis lithohermaea* was described by Engel (2006) from the middle Miocene of Japan. The size of this bee (body length 18 mm; forewing 11.3 mm) associated this fossil with the ancestor of the subgenus *Megapis* (Engel 2006).

The presented discussion and conclusions clearly indicate that the biology of this fossil bee was most probably very similar to extant *A. dorsata*. Thus, *A. lithohermaea* was most probably a fossil representative of ancestral *Megapis*. The dating of this fossil bee to middle Miocene fits nicely to the overall scenario presented here. The first middle-sized fossil honeybees (Table 2.1) are placed in Oligocene, and are considerably older, indicating that cavity-dwelling and the ancestral *Apis* syndrome may hold a more basal position compared to the ancestral *Megapis* syndrome.

References

Alexander BA (1991) Phylogenetic analysis of the genus *Apis (*Hymenoptera: Apidae). Ann Entomol Soc Am 84:137–149

Arias MC, Sheppard WS (2005) Phylogenetic relationships of honey bees (Hymenoptera: Apinae; Apini) inferred from nuclear and mitochondrial DNA sequence data. Mol Phylogenet Evol 37:25–35

Ashmead WH (1904) Remarks on honeybees. Proc Entomol Soc Wash 6:120–123

Cameron SA, Derr JN, Austin AD, Woollwy JB, Wharton RA (1992) The application of nucleotide sequence data to phylogeny of Hymenoptera: a review. J Hymenopt Res 1:63–79

Cornuet JM, Garnery L (1991) Mitochondrial DNA variability in honeybees and its phylogeographic implications. Apidologie 22:627–642

Crozier RH, Crozier YC (1992) The Cytochrome-b and ATPase genes of honeybee mitochondrial DNA. Mol Biol Evol 9:474–482

Crozier RH, Crozier YC (1993) The mitochondrial genome of the honeybee *Apis mellifera:* complete sequence and genome organization. Genetics 133:97–117

Crozier RH, Crozier YC, Mackinlay AG (1989) The CO-I and CO-II region of honeybee mitochondrial DNA: evidence for variation in insect mitochondrial evolutionary rates. Mol Biol Evol 6:399–411

Crozier YC, Koulianos S, Crozier RH (1991) An improved test for Africanized honeybee mitochondrial DNA. Experientia 47:968–969

Danforth BN, Ji SQ (1998) Elongation factor-1 alpha occurs as two copies in bees: implications for phylogenetic analysis of EF-1 alpha sequences in insects. Mol Biol Evol 15:225–235

Deodikar GB, Thakar CV (1966) Cyto-genetics of Indian honeybees and bearing on taxonomic and breeding problems. Indian J Genet 36A:386–393

Dyer FC (1991) Comparative studies of dance communication: analysis of phylogeny and function. In: Smith DR (ed) Diversity in the genus *Apis*. Westview, Boulder, pp 177–198

Dzierzon J (1849) Neue verbesserte Bienenzucht des Pfarrers Dzierzon. Publ. by author [in German]

Engel MS (1998) Fossil honey bees and evolution in the genus *Apis* (Hymenoptera: Apidae). Apidologie 29:265–281

Engel MS (1999) The taxonomy of recent and fossil honey bees (Hymenoptera: Apidae; *Apis*). J Hymenopt Res 8:165–196

Engel MS (2006) A giant honey bee from the Middle Miocene of Japan (Hymenoptera: Apidae). Am Mus Novitat 3504:1–12

Engel MS, Schultz TTR (1997) Phylogeny and behaviour in honey bees (Hymenoptera: Apidae). Ann Entomol Soc Am 90:43–53

Engel MS, Hinojosa-Diaz IA, Rasnitsyn AP (2009) A honey bee from the Miocene of Nevada and the biogeography of *Apis* (Hymenoptera: Apidae). Proc Calif Acad Sci 60:23–38

Fabricius JC (1787) Supplemental list of insects setting out recently discovered species together with their generic characteristics, specific distinguishing features, emendations and remarks. Hafniae, Proft, Copenhagen [in Latin]

Fabricius JC (1793) Systematic description of nature in accordance with the three natural kingdoms, according to classes, orders, genera and species with their distinguishing characteristics, alternative names and habitats. Hafniae, Proft Copenhagen [in Latin]

Fahrenhorst H (1977) Nachweis übereinstimmender Chromosomen-Zahlen (n=16) bei allen 4 *Apis*-Arten. Apidologie 8:89–100 [in German]

Friese H (1923) Die europäischen Bienen (Apidae). Walter de Gruyter, Berlin

Garnery L, Vautrin D, Cornuet JM, Solignac M (1991) Phylogenetic relationships in the genus *Apis* inferred from mitochondrial DNA sequence data. Apidologie 22:87–92

Garnery L, Cornuet JM, Solignac M (1992) Evolutionary history of the honey bee *Apis mellifera* inferred from mitochondrial DNA analysis. Mol Ecol 1:145–154

Gerstäcker CEA (1862) Über die die geographische Verbreitung und die Abänderungen der Honigbiene nebst Bemerkungen über die ausländischen Honigbienen der alten Welt. Festschrift XI Wanderversammlungdeutsch. Bienenwirthe. Potsdam, Kraemer [in German]

Hall HG, Smith DR (1991) Distinguishing African and European honeybee matrilines using amplified mitochondrial DNA. Proc Natl Acad Sci 88:4548–4552

Horn EP (1975) Mechanisms of gravity processing by leg and abdominal gravity receptors in bees. J Insect Physiol 21:673–679

Horne C (1870) Notes on the habits of some hymenopterous insects of the north-west provinces of India. Trans Zool Soc Lond 7:161–196

Huelsenbeck JP, Ronquist F (2001) MRBAYES: Bayesian inference of phylogenetic trees. Bioinformatics 17:754–755

Jander R, Jander U (1970) Über die Phylogenie der Geotaxis innerhalb der Bienen (Apoidea). Z Vgl Physiol 66:355–368 [in German]

Koeniger N (1976) Neue Aspekte der Phylogenie innerhalb der Gattung *Apis*. Apidologie 7:357–366 [in German]

Koeniger N, Koeniger G (1980) Observations and experiments on migration and dance communication of *Apis dorsata* in Sri Lanka. J Apic Res 19:21–34

Koeniger N, Koeniger G (1991) An evolutionary approach to mating behaviour and drone copulatory organs in *Apis*. Apidologie 22:581–590

Koeniger N, Weiss J, Maschwitz U (1979) Alarm pheromones of the sting in the genus *Apis*. J Insect Physiol 25:467–476

Koeniger G, Koeniger N, Mardan M, Otis GW, Wongsiri S (1991) Comparative anatomy of male genital organs in the genus *Apis*. Apidologie 22:539–552

Koeniger G, Koeniger N, Mardan M, Wongsiri S (1993) Variance in weight of sexuals and workers within and between four *Apis* species (*Apis florea*, *Apis dorsata*, *Apis cerana* and *Apis mellifera*). In: Proceedings of the 1st international conference on Asian honeybees bee mites, Bangkok, Wicwas, Cheshire, pp 104–109

Koeniger N, Koeniger G, Tingek S (2010) Honey bees of Borneo – exploring the center of *Apis* diversity. Nat. History Publ. (Borneo) Kota Kinabalu pp 262

Leelamanit W, Neelasaewee S, Boonyom R, Panyim S, Hayashi T, Yasue H, Amano K (2004) The NADH dehydrogenase genes of *Apis mellifera*, *A. cerana*, *A. dorsata*, *A. laboriosa* and *A. florea*: sequence comparison and genetic diversity. J Anim Genet 31:3–12

Lindauer M (1956) Über die Verständigung bei indischen Bienen. Z Vgl Physiol 38:521–557 [in German]

Linnaeus C (1758) Systema naturae per regna tria naturae: secundum classes, ordines, genera, species, cum characteribus, differentiis, synonymis, locis. 10th edn. Holmiae (Laurentii Salvii) [in Latin]

Liu H, Beckenbach A (1992) Evolution of the mitochondrial cytochrome oxidase II gene among 10 orders of insects. Mol Phylogenet Evol 1:41–52

Maa T (1953) An inquiry into the systematics of the tribus *Apidini* or honeybees (Hym.). Treubia 21:525–640

Mayr E (1942) Systematics and the origin of species. Columbia, New York

Michener CD (2000) The bees of the world. John Hopkins University Press, Baltimore, London

Moritz RFA, Hawkins CF, Crozier RH, Mackinley AG (1986) A mitochondrial DNA polymorphism in honeybees (*Apis mellifera* L). Experientia 42:322–324

Oldroyd BP, Wongsiri S (2006) Asian honey bees. Harvard University Press, Cambridge

Pollmann A (1879) Wert der verschiedenen Bienenrassen und deren Varietäten, 2nd edn. Voigt, Berlin

Raffiudin R, Crozier RH (2007) Phylogenetic analysis of honey bee behavioral evolution. Mol Phylogenet Evol 43:543–552

Ruttner F (1988) Biogeography and taxonomy of honey bees. Springer, Berlin

Schmiedeknecht O (1907) Die Hymenopteren Mitteleuropas. Fischer, Jena [in German]

Seeley TD (1985) Honeybee ecology. Princeton University Press, Princeton, NJ

Seeley TD, Seeley RH, Akratanakul P (1982) Colony defence strategies of the honeybee in Thailand. Ecol Monogr 52:43–63

SenSarma M, Fuchs S, Werber C, Tautz J (2002) Worker piping triggers hissing for coordinated colony defence in the dwarf honeybee *Apis florea*. Zoology 105:215–223

Sheppard WS, Rinderer TE, Mazzoli JA, Stelzer JA, Shimanuki H (1991a) Gene flow between African-derived and European derived honey bee populations in Argentina. Nature 349:782–784

Sheppard WS, Soares AAE, Dejong D, Shimanuki H (1991b) Hybrid status of honey bee populations near the historic origin of Africanization in Brazil. Apidologie 22:643–652

Skorikov AS (1929) Eine neue Basis für eine Revision der Gattung *Apis* L. Rep Appl Entomol Leningrad 4:249–264 [in German]

Smith F (1858) Catalogue of the hymenopterous insects collected at Sarawak, Borneo; Mount Ophir, Malacca; and at Singapore, by A.R. Wallace. Proc Linn Soc Lond 2:42–130

Smith F (1861) Descriptions of new species of hymenopterous insects collected by A.R. Wallace at Celebes. Proc Linn Soc Lond 5:57–93

Smith F (1865) On the species and varieties of the honey-bees belonging to the genus *Apis*. Ann Mag Nat Hist 15:372–380

Smith F (1871) A catalogue of the aculeate Hvmenoptcra and Ichneumonidae of India and the eastern Archipelago. J Linn Soc 11:285–415

Smith DR (1991a) African bees in the Americas - insights from biogeography and genetics. Trends Ecol Evol 6:17–21

Smith DR (1991b) Mitochondrial DNA and honey bee biogeography. In: Smith DR (ed) Diversity in the genus *Apis*. Westview, Boulder, pp 131–176

Smith DR, Brown WM (1988) Polymorphisms in mitochondrial DNA of European and Africanized honeybees (*Apis mellifera*). Experientia 44:257–260

Smith DR, Brown WM (1990) Restriction endonuclease cleavage site and length polymorphisms in mitochondrial DNA of *Apis mellifera mellifera* and *A. m. carnica* (Hymenoptera: Apidae). Ann Entomol Soc Am 83:81–88

Smith DR, Palopoli MF, Taylor BR, Garnery L, Cornuet JM, Solignac M, Brown WM (1991) Geographical overlap of 2 mitochondrial genomes in Spanish honeybees (*Apis mellifera iberica*). J Hered 82:96–100

Spinola M (1806) Insectorum Liguriae species novae aut rariores, quas in agro Ligustico nuper detexit descripsit, et iconibus illustravit adjecto catalogo specierum auctoribus jam enumeratarum, quae in cadem regione passim occurrunt. Yves Gravier, Genuae [in Latin]

Thakar CV, Deodikar GB (1966) Chromosome number in *Apis florea* Fab. Curr Sci 7:186

Tingek S, Mardan M, Rinderer TE, Koeniger N, Koeniger G (1988) Rediscovery of *Apis vechti* (Maa 1953): the Saban honey bee. Apidologie 19:97–102

Tirgari S (1971) Biology and behavioural characteristics of the Iranian dwarf honeybee. Proc Int Beekeep Congr 23:344–345

Torchio PF, Torchio DM (1975) Larvae of the Apidae (Hymenoptera: Apoidea), part 1. Apini, *Apis*. Agric Exp Stn. Utah State Univ Res Rep 10:1–36

von Berlepsch A (1873) Die Biene und ihre Zucht mit beweglichen Wabenin Gegenden ohne Spätsommertracht. J. Schneider, Mannheim [in German]

von Buttel-Reepen H (1906) Contributions to the systematics biology as well as the historical and geographical distribution of honeybees (*Apis mellifica* L) their variability and other *Apis* species. Mitt Zool Mus Berl 3:117–201 [in German]

von Frisch K (1965) Tanzsprache und Orientierung der Bienen. Springer, Berlin [in German]

von Siebold BTE (1856) Wahre Parthenogenese bei Schmetterlingen und Bienen. Engelmann, Leipzig [in German]

Walldorf U, Hovemann BT (1990) *Apis mellifera* cytoplasmic elongation factor 1-alpha (EF-1-alpha) is closely related to *Drosophila melanogaster* EF-1-alpha. FEBS Lett 267:245–249

Willis LG, Winston ML, Honda BM (1992) Phylogenetic relationships in the honeybee (genus *Apis*) as determined by the sequence of the cytochrome oxidase II region of mitochondrial DNA. Mol Phylogenet Evol 1:169–178

Wilson EO (1971) The insect societies. Harvard University Press, Cambridge

Wu Y, Kuang B (1987) Two species of small honeybee: a study of the genus *Micrapis*. Bee World 68:153–155

Chapter 3
Biogeography

H.R. Hepburn and Sarah E. Radloff

3.1 Introduction

Asia is defined here as that region to the east of a north–south line through the Ural Mountains, down to the Arabian Sea (Peel et al. 2007; Fig. 3.1). The climatic zones of this large area are those defined by the Köppen–Geiger climate classification system (Köppen 1936). This system is an out-growth of plant geographical studies, beginning in the nineteenth century and is predicated on the idea that major differences in naturally occurring vegetation accurately reflect the effects of climate. It is the most widely used system today and has been periodically edited and revised, most recently by Peel et al. (2007) (cf. Peel et al. 2007, Supplementary Material Section).

While it may seem odd that a climate system should ultimately derive from plant geography, atmospheric scientists have compared contemporary global climate model data with Köppen's maps and found a discrepancy of only about 0.5% (Kalvova et al. 2003). By the same token, Qian et al. (2003) have clearly shown that phytogeographical patterns in eastern Asia are closely related to latitude, which co-varies with various climatic variables. Qian et al. (2003) also concluded that: "Evolutionary processes such as the adaptation of plants to cold climates and current and past land connections are likely responsible for the observed latitudinal patterns." The combined results of these biogeographical and atmospheric studies provide a ready justification for using the Köppen–Geiger system in biogeographical studies of honeybees (Fig. 3.1).

Map illustrations in most publications on the distributions of honeybee species are routinely delineated by their outer boundary limits. This practice is somewhat unfortunate, because it predisposes the reader to think that the species is somehow

H.R. Hepburn
Department of Zoology and Entomology, Rhodes University, Grahamstown 6140, South Africa
e-mail: r.hepburn@ru.ac.za

S.E. Radloff
Department of Statistics, Rhodes University, Grahamstown 6140, South Africa
e-mail: s.radloff@ru.ac.za

H.R. Hepburn and S.E. Radloff (eds.), *Honeybees of Asia*,
DOI 10.1007/978-3-642-16422-4_3, © Springer-Verlag Berlin Heidelberg 2011

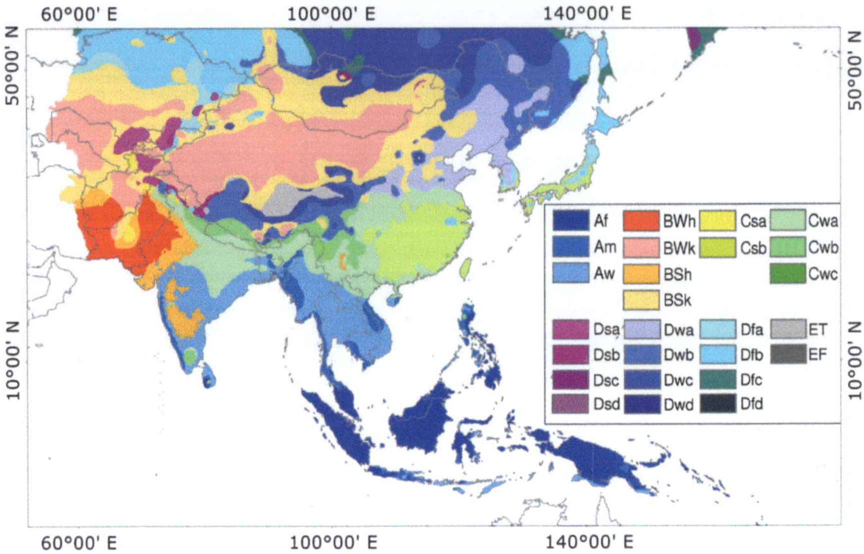

Fig. 3.1 The Köppen–Geiger climate classification system. A – *Tropical*: f: rainforest, m: monsoon, w: savannah; B – *Arid*: W: desert, S: steppe; h: hot, k: cold; C – *Temperate*: s: dry summer, w: dry winter, f: without dry season; a: hot summer, b: warm summer, c: cold summer; D – *Cold*: s: dry summer, w: dry winter, f: without dry season; a: hot summer, b: warm summer, c: cold summer, d: very cold winter; E – *Polar*: T: tundra, F: frost

contiguous over the entire enclosed area. In actual fact, of the 40 Mkm^2 of Asian land mass, calculations of the relative percentages of land mass that are occupied by the main climate types yield the following climatic regions: cold (43.8%), arid (23.8%), tropical (16.3%), temperate (12.3%) and polar (3.8%) (Peel et al. 2007). The arid, cold and polar climatic regions are essentially devoid of honeybees, thus, excluding water masses, leaving roughly only an area of 28.6% (about 11 Mkm^2) suitable for sustaining bees.

Biogeographical inferences about honeybees have been closely tied to prevailing concepts of species in historical terms as well as with the actual amount of data available at any given time. Indeed, the literature shows an early phase of entirely typological thinking (Maa 1953), more recently replaced by the biological species concept (DuPraw 1964; Ruttner 1988). Another consideration is that precursors of contemporary *Apis* species have been described from the Eocene epoch of the lower Quaternary period, while recognisably modern honeybees first appear in the Pleistocene (Zeuner and Manning 1976; Engel 1998; Grimaldi and Engel 2005; Engel et al. 2009), so that speciation and radiation of modern *Apis* species must be considered in a Pleistocene context.

Moreover, given the palaeoclimatic history for this period (Zubakov and Borzenkova 1990; Goudie 1992; Sloan et al. 1996; Voris 2000), it is now possible to parsimoniously infer how the distributions of honeybee species may have arisen. Here we present new and updated distribution maps for the Asian honeybees and consider how current distributions might reflect the speciation history of these bees in light of palaeoclimatic change. In the distribution maps given here, all known localities from which each species has been collected and reported in the literature (Hepburn and Hepburn 2005, 2006, 2007, 2008, 2009) have been plotted in maps as discrete points. With few exceptions (very probably Afghanistan and possibly Myanmar), the resulting geographical voids simply indicate that there are few to no bees in those areas. Quaternary climatic data relevant to southern Asia were derived from Zubakov and Borzenkova (1990), Goudie (1992), Sloan et al. (1996) and Voris (2000).

3.2 The Dwarf Honeybees: *Apis andreniformis* and *Apis florea*

3.2.1 Apis andreniformis

The distributions of the dwarf honeybees, *A. andreniformis* and *A. florea*, are shown in Fig. 3.2. *A. andreniformis* extends from the foothills of the Himalayas in Sikkim and Bhutan, south-eastward through Myanmar, Yunnan Province in southern China, Thailand and Indochina, then southerly through Malaysia and most of the Sundaland islands of Indonesia as well as on Palawan in the Philippines. It is primarily a lowland species and the frequency distributions of altitude values for 115 cited localities in the literature show that 83% were found below 1,000 m and the balance between 1,000 and 2,500 m.

In the north of its range, the climate is wet and dry tropical, with alternating wet and dry seasons which give way southward to a tropical wet climate with heavy year-round precipitation (Fig. 3.1). A strip of tropical wet climate also edges the western border of Myanmar and eastern border of Vietnam. Most of the area is of tropical rainforest climate; however, tropical dry forest covers central Myanmar and most of central Thailand. *A. andreniformis* is essentially a tropical species of Sundaland, west of the Wallace and Merrill Lines. Of equal interest is where *A. andreniformis* does not occur: Sulawesi and those islands of the Philippines beyond the Sulu and Celebes seas; the former is separated from Borneo by the Makassar Straight and the latter from Palawan by the Sulu Sea. Both of these seas are over a kilometre deep and no land connections existed between them during the Tertiary period (Moss and Wilson 1998).

3.2.2 Apis florea

A. florea is extremely widespread, extending some 7,000 km from its eastern-most extreme in Vietnam and south-eastern China, across mainland Asia, along and below the southern flanks of the Himalayas, westwards to the Plateau of Iran and south-westerly into Oman (Hepburn et al. 2005; Fig. 3.2). This constitutes some 70° of longitude East (40°–110°) and nearly 30° of latitude North (6°–34°). Variations in altitude range from sea-level to about 2,000 m. *A. florea* has also been introduced in

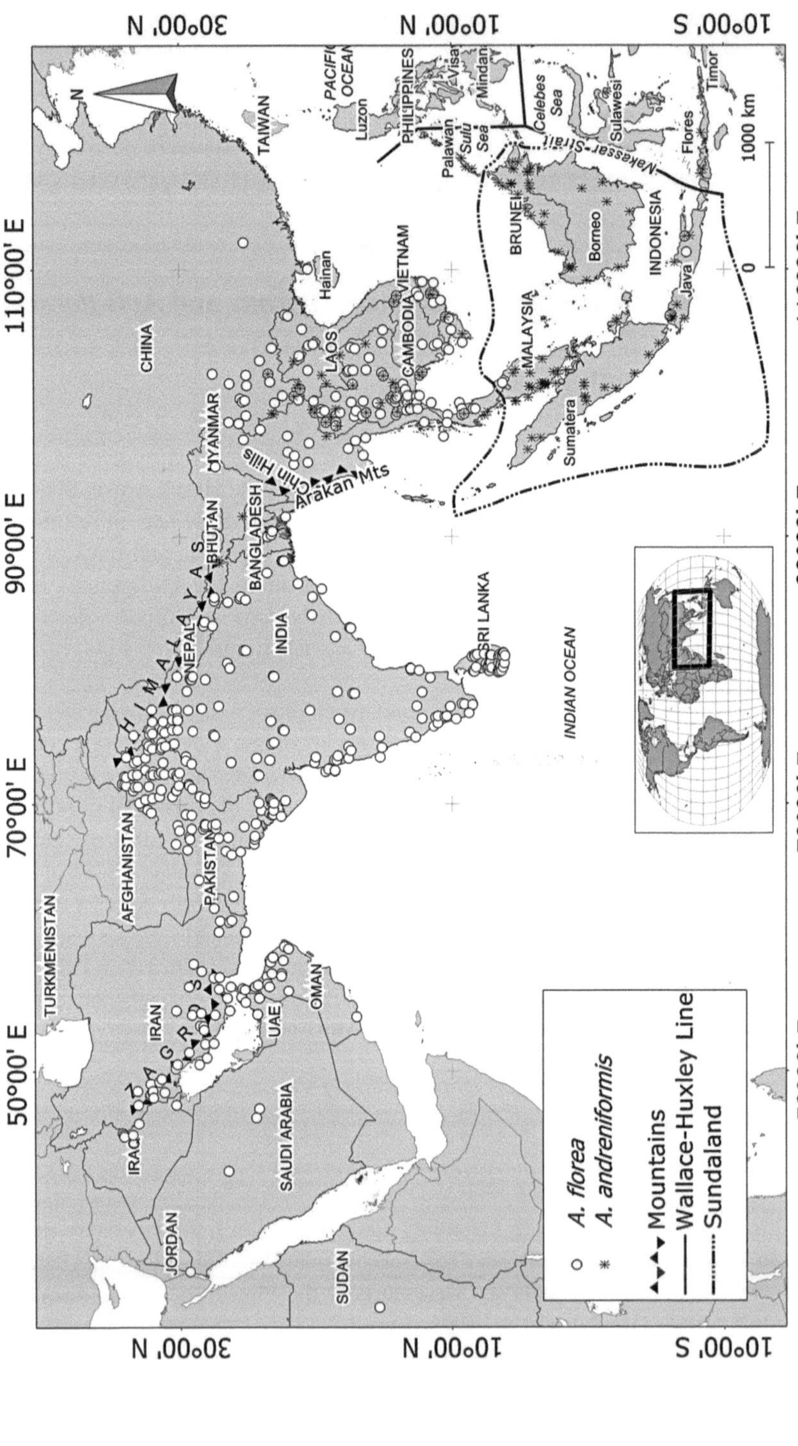

Fig. 3.2 Distribution of the dwarf honeybees, *Apis andreniformis* and *Apis florea*

historical times in Saudi Arabia, Sudan, and Java, Indonesia (Maa 1953; Hepburn et al. 2005). Its climatic range is much broader than that of *A. andreniformis* and includes both desert and semi-arid zones in the west, which gradually change to subtropical, tropical dry, tropical wet, and dry and tropical wet climate in the southeast (Fig. 3.2).

The climatic zones of *A. florea* change from tropical rainforest in the east to savannah at a line more or less following the natural border between Pakistan and India in the west. Further westwards, there is a transition from tropical to subtropical steppe and semi-desert. There is a seasonal rainfall gradient across the range of *A. florea* from the eastern tropical to the western subtropical and semi-desert climates of the west. There is also a gradient of mean daily temperature extremes that increase from the eastern tropical zone to the western subtropical zone. In terms of honeybee movement and dispersal, mean daily temperatures exceed 15°C throughout the year in all but China, Nepal, areas of northern India, Pakistan and Iran, where temperatures are too low for flight for between 3 and 6 months of the year. The combined rainfall and temperature changes from east to west are reflected in the swarming and migration biology of *A. florea* (Hepburn et al. 2005; Hepburn 2006; cf. Chap. 7).

Over this large expanse, reproductive swarming overlaps and occurs at the end of the dry season (March to May). In many areas, there is also a second swarming phase, before the onset of the drier season of the year (August to October), associated with migration. Therefore, despite climatic and phytogeographical variation, there are no seasonal barriers to gene-flow within what is actually a very large panmictic population. Nevertheless, *A. florea* and *A. andreniformis* do not occur on most offshore island groups, which may reflect on their dispersal abilities (Otis 1996). This, coupled with synchronous reproductive seasons across its distribution, must mitigate against regional differentiation of these species into recognisable subspecies or races. Like *A. andreniformis*, *A. florea* is also a lowland species, and the frequency distributions of 99 localities cited in the literature show that 85% occur below 1,000 m, with the balance between 1,000 and 2,500 m (Hepburn et al. 2005). Although there is an impression in the literature that *A. andreniformis* is, on average, distributed at somewhat higher elevations than *A. florea* (Otis 1996; Wongsiri et al. 1996), there is no statistically significant difference in their frequency distributions with respect to altitude (Hepburn et al. 2005).

In a recent discussion of the origin of these two species, Oldroyd and Wongsiri (2006) suggested that speciation between *A. andreniformis* and *A. florea* must have pre-dated island formation, because both occur on mainland Asia and Java. This suggestion is misinformed: first, because the oldest *Apis* fossils only date from the mid-Tertiary period, while the islands in question were extant, at least as far back to the Triassic period (about 200 Ma); second, *A. florea* has only been collected around the port area of Djakarta and on ships, which points to an accidental introduction, such as the ones in Sudan and Jordan. If we posit a common and widespread pre-*florea-to-andreniformis* split before the Pleistocene, there is a convenient barrier caused by the first Pleistocene, Pre-Pastonian glacial period, from 800 to 1,300 ka (McMillan 2005; Gibbard et al. 2007). The divergence of *A. andreniformis* and *A. florea* into distinct species can parsimoniously be attached to this time. As the Himalayas extend eastwards, their southern flank is the Arakan

Yoma, which defines the western border of Myanmar. This arc extends southward for nearly 1,000 km, from near Manipur in Bengal to southern Myanmar, and would have been ice-bound during glacial periods – any one of which could have provided a barrier for the evolving proto-populations of *A. andreniformis* and *A. florea*.

In recent years, the dwarf honeybee, *A. florea*, has been steadily expanding westwards, both naturally and accidentally, via global transportation (Mogga and Ruttner 1988; Hepburn et al. 2005). *A. florea* is now well established in Iraq, Oman and Yemen and has recently been detected in central Saudi Arabia, Sudan and Jordan (Mogga and Ruttner 1988; Glaiim 1992; Hepburn et al. 2005; Haddad et al. 2008). Throughout this expansion along the Arabian Peninsula and into Africa, it has proven to be a highly successful coloniser, well adapted to hot, arid conditions of both urban and rural landscapes and seemingly unaffected by competition from any local *A. mellifera* (El-Shafie et al. 2002).

Most recently, *A. florea* has become established in the area around Aqaba, Jordan (Haddad et al. 2008). Subsequently, Haddad et al. (2009) carried out a principal component analysis on the Jordanian and Sudanese samples, in combination with others from the western half of the *A. florea* distribution, to identify the probable source of origin of the founder population in Jordan. A cluster analysis showed a dendrogram of two main clusters: one linking samples from Pakistan and Sudan, then Oman and Jordan and then Saudi Arabia and finally Iran; a second cluster linking colonies from southern India and Sri Lanka (cf. Chap. 1).

It is almost certain that the *A. florea* in Jordan arrived through human transportation, because the region is surrounded by vast expanses of barren wasteland that precludes gradual expansion of the natural geographic range from the northeast. It is very probable that the route of entry was by sea into the port of Aqaba in the north-eastern hook of the Red Sea. And, indeed, there is a very large shipping trade along the Red Sea and Arabian Sea between Oman and Jordan (Haddad personal communication). The *A. florea* in Sudan were first discovered in Khartoum but have since spread along the Nile River (Mogga personal communication). We believe that it is highly probable that the Jordanian and Sudanese populations will eventually converge in the Sinai Peninsula.

3.3 The Medium-Sized Honeybees: *Apis cerana*, *Apis koschevnikovi*, *Apis nigrocincta* and *Apis nuluensis*

3.3.1 Apis cerana

A. cerana naturally occurs on the Asian landscape, encompassing a series of climatic zones from tropical moist rainforest, wet–dry tropical savannah, mid-latitude steppe, dry mid-latitude grasslands, moist continental deciduous forest and taiga (Fig. 3.1; cf. Muller 1982; Peel et al. 2007). It is of interest to observe how the distributions of the different morphoclusters, and possibly of subclusters within them, may be related to these various physiographic and climatic zones (Fig. 3.3; cf. Chap. 1).

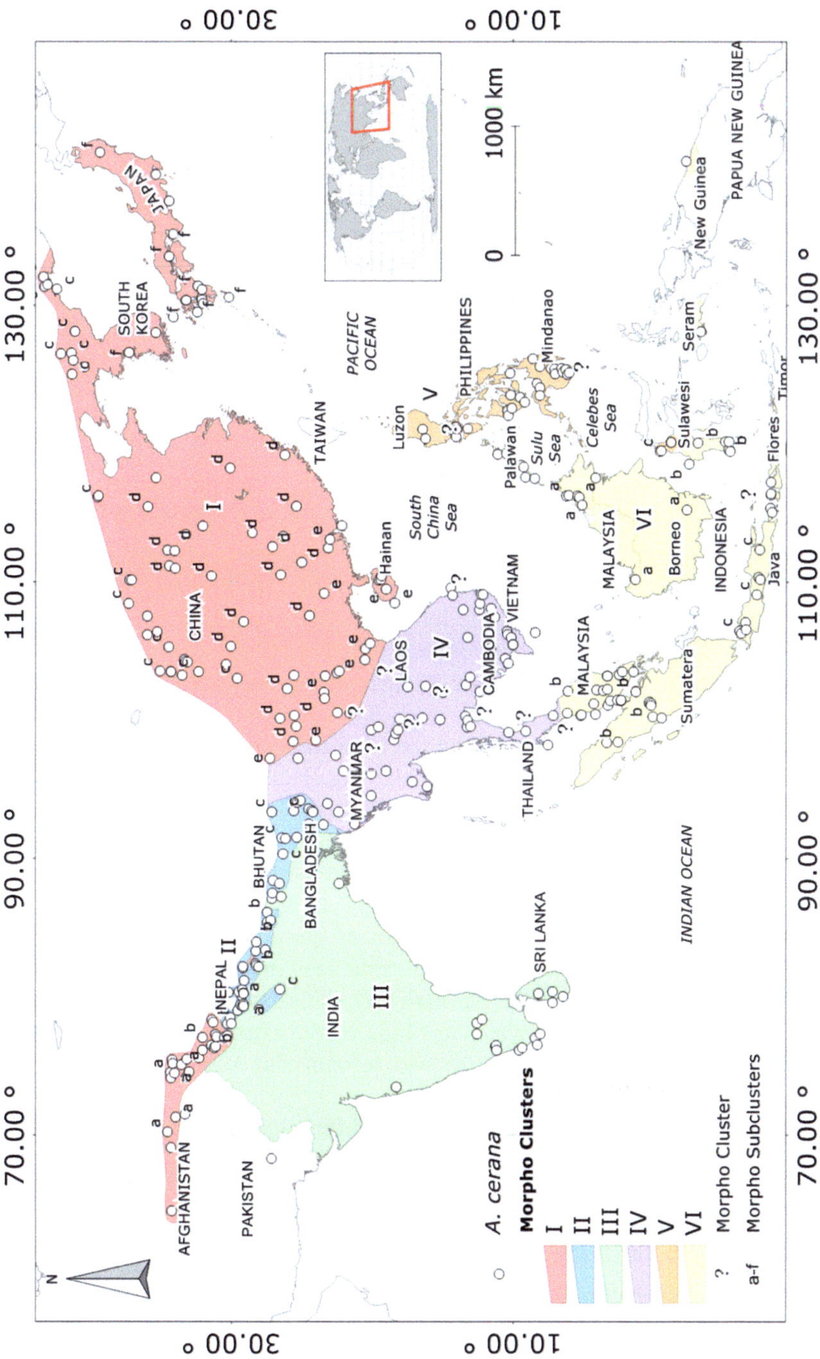

Fig. 3.3 Distribution of *Apis cerana*

The Philippine V and Indo-Malayan VI morphoclusters both occur below the South China Sea in the tropical wet climatic zone, yet are morphometrically differentiated; however, reproductive isolation among them can be interpreted in a geological context (Voris 2000).

During the last 250 KBP of the Pleistocene, the Philippines were physically separated from the Indo-Malayan region for about 75% of the time (Voris 2000) – clearly allowing differentiation through genetic drift. Differentiation within the Philippine group, as biometric subclusters within the major island systems, similarly reflects that they too were physically separated from each other for 70% of the last 250 KBP. In stark contrast, the bees of the Indo-Malayan morphocluster occur in the Indonesian chain of islands that have been connected for 50% of the last 250 KBP and nearly 70% of the last 17 KBP (Voris 2000), which clearly forms a genetically more panmictic population. These geological data are completely consistent with the Euclidean linkage distances between these two oceanic morphoclusters (Radloff et al. 2010).

Proceeding to the continental mainland, land-links between the Indo-Malayan VI and southeast Asian IV morphoclusters, resulting from lowered sea levels, have shown nearly 50% of the last 250 KBP and nearly 70% of the last 17 KBP, while there have never been land connections between them and the Indian subcontinent. Here the Euclidean linkage distances provide a trend in the expected direction. Nonetheless, there is a sharp transition between the Indo-Malayan VI and the Southeast Asian morphocluster IV, from a tropical, evergreen rainforest to mixed moist, deciduous forest with a wet and dry tropical climate, to the northeast of the Isthmus of Kra in the northern Thai–Malay Peninsula (Hughes et al. 2003). The rainforests actually extend up the western side of the Isthmus, along the Chin Hills and Arakan Yoma forests, running between Myanmar and Bangladesh, the Naga Hills of India, the highlands of northern Laos and Chuor Phnum of Cambodia. These ecological discontinuities could well provide physiographic and ecological barriers between the Southeast Asian IV and the adjoining eastern edge of the sub-Himalayan morphocluster II.

The sub-Himalayan II and southeast Asian IV morphoclusters are adjacent in terms of their Euclidean linkage distances, subtend northern morphocluster I along the Himalayas, and then give way to a complex range of mountains (Hengduanshan, Wuliangshan, Taniantawengshan, Nushan and Qingshuilangshan), which turn southwards along the present Myanmar–China border (Radloff et al. 2010). The sub-Himalayan morphocluster contains two recognisable subclusters: the larger Indus and smaller Ganges populations. The morphometric distinctness of these two populations arises from the fact that they differ significantly in the seasonal timing of reproductive swarming (Hepburn et al. 2001).

Northern morphocluster I extends over the largest land areas where *A. cerana* occurs and consists of five biometrically definable subclusters (Fig. 3.3). The Indus population I occurs in a tropical, wet and dry climate, while its north-western neighbour, the Himachali population, occurs in the subtropical steppe. These climatic differences provide a degree of reproductive isolation between the populations because of seasonal differences in swarming, so reducing the possibility of gene flow between them (Sharma 1960; Saraf 1985; Muzaffar and Ahmed 1989).

The Himachali population is also largely isolated from the eastern (north and south) Chinese subclusters by the intervening subarctic climate of the Tibetan highlands. Moving eastwards, the next populations of morphocluster I are the northerly Aba, southeastern and Japonica biometric subclusters. The Aba area is in a subtropical mountain system, while that of the southeastern and Japonica subclusters are situated in subtropical humid forest. The Aba and southeastern subclusters have only a partial overlap in seasonal swarming (Gong 1983; Liu and Lai 1990; Zhu 1996).

Furthermore, the Chinese literature (Yang 1986, 2001; Zhen-Ming et al. 1992; Chen 1993) refers to several populations of bees in this morphocluster based on a combination of some morphometric data, physiology, pigmentation, cell size, low temperature tolerance and foraging (Yang 2001) – not identifiable, however, in the recent work of Radloff et al. (2010). The population groups include: highland and lowland populations on Hainan Island; an eastern Chinese honeybee with Guangxi, Guangdong, Hunan, Yungui Plateau, north China and Changbai mountains ecotypes; a south Yunnan group; and a south Tibetan bee. These proposed groups of bees could, however, not be resolved in the analyses of Radloff et al. (2010) nor are they supported statistically in the original Chinese literature; however, they remain in current use in China. Recent attempts to resolve these uncertainties are beginning to appear (Ling et al. 2008; Ling-Juan and Ken 2008).

As a final observation, it can be noted that *A. cerana* has been introduced to areas where it does not naturally occur. In the nineteenth century, it was introduced, from maritime Russia, into the north-eastern provinces, Heilongjiang and Jilin, China (Chen 1993). Likewise, it has been introduced into Ambon Island, to the east of Sulawesi (Maa 1953), as well as into Iran (Gassparian 1977). That *A. cerana* is an invasive species is evident from the fact that it was introduced from Java into Papua New Guinea in the late 1970s and now occurs throughout New Guinea (Anderson 1989). From there it has reached islands in the Torres Strait (Dunn 1992) and, even more recently, New Britain and the Solomon Islands (Anderson 2005 personal communication). *A. cerana* from Papua New Guinea are morphometrically indistinguishable from those of Java. It has also been intercepted several times at Darwin, Brisbane and near Perth, Australia. A GARP (Generic Attribute Registration Protocol) analysis (Diniz-Filho personal communication) indicates that all of the area on or east of the Great Dividing Range in Australia, as well as the southwest, would be suitable for *A. cerana* when it finally penetrates that continent.

3.3.2 Apis koschevnikovi

The known geographical distribution of *A. koschevnikovi* is limited to the Malay Peninsula and Borneo (Malaysia and Indonesia) (Fig. 3.4). The numerical weight of altitudinal distributions for *A. koschevnikovi* recovered from the literature provides values mostly below 1,200 m, which raises the question of whether this is due to it being more difficult for collectors to reach higher elevations, or if the bees are actually

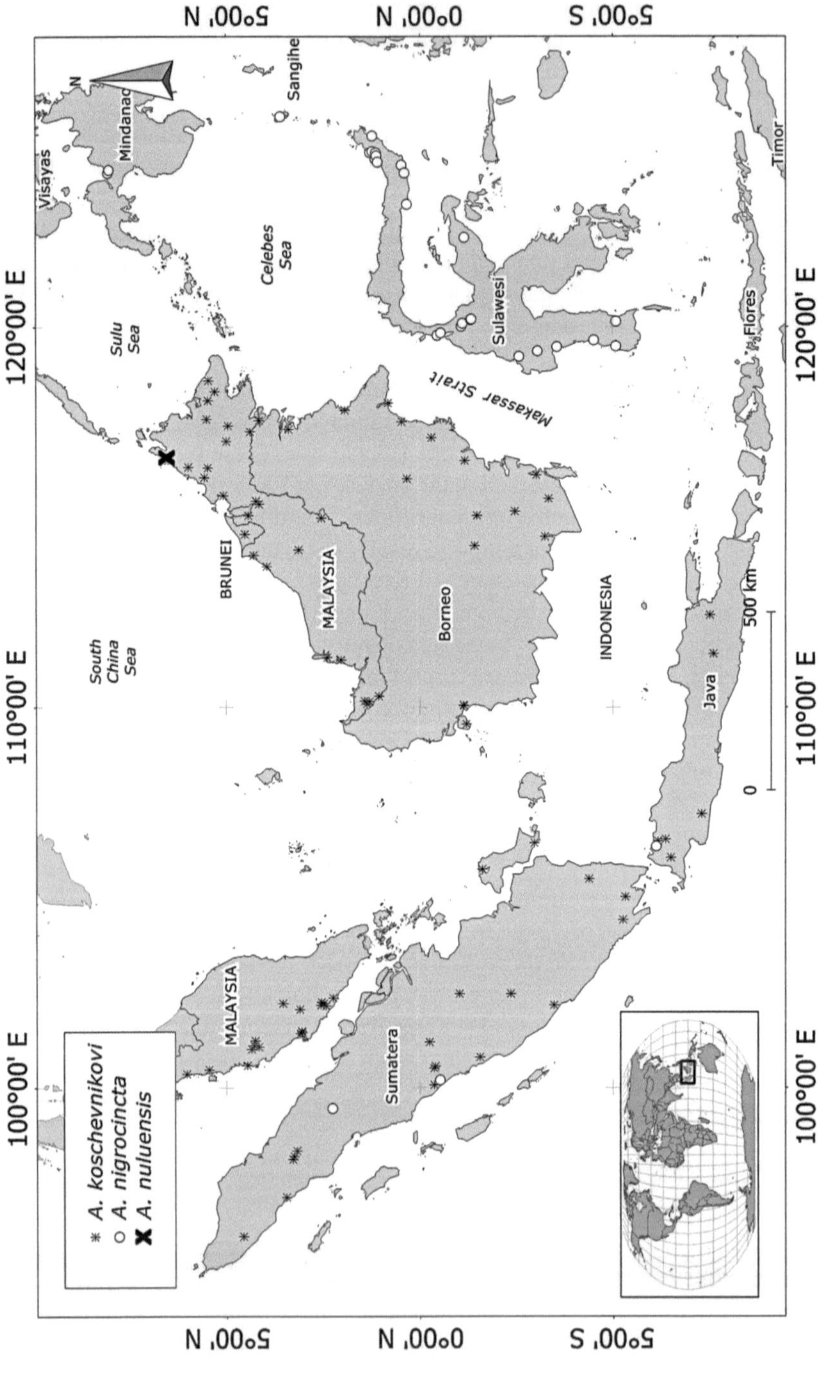

Fig. 3.4 Distribution of *Apis koschevnikovi*, *Apis nigrocincta* and *Apis nuluensis*

scarcer at these higher elevations. The higher elevations in Borneo are particularly remote and inaccessible. Bearing in mind that the islands of the south China Sea were often connected to the mainland during much of the Pleistocene (Voris 2000), and that *A. koschevnikovi* occurs in tropical evergreen rainforests, which extend into Thailand and Myanmar to the west, and Cambodia and Vietnam to the east (Fig. 3.1), its absence from the latter appears puzzling at first sight (Hadisoesilo et al. 2008).

However, Hughes et al. (2003) found a significant transition between northern Indochinese and southern Sundaic (Indo-Malay) avifauna assemblages, just north of the Isthmus of Kra in the Thai–Malay Peninsula (over which *A. koschevnikovi* must have passed). This area is associated with a change from wet seasonal evergreen dipterocarp rainforest to mixed, moist deciduous forest, north of the Isthmus of Kra. Hughes et al. (2003) also reviewed the climatological and ecological factors associated with the forest types and hypothesised that the avian transition tracks the northern phytogeographical boundary and that hypothetical seaways at the end of the Oligocene could account for the development of both phytogeographical and avifaunal transitions as well. Utilising intraspecific molecular variation in fresh water crustaceans, de Bruyn et al. (2005) provided unambiguous evidence in support of the hypothesis of Hughes et al. (2003). To the extent that these interpretations are sound, they could very conveniently explain why *A. koschevnikovi* does not occur in tropical evergreen rainforests, which extend into Thailand, Myanmar, Cambodia and Vietnam (Hadisoesilo et al. 2008).

It is apparent that the range of *A. koschevnikovi* is diminishing, because it is now either poorly represented in, or absent from, several areas from which it has been recorded decades ago (Otis 1996). This has been attributed to habitat changes, resulting from deforestation, and the establishment of tea, oil palm, rubber and coconut plantations (Otis 1996; Eltz 2004). A final point of interest is that throughout its range, *A. koschevnikovi* is sympatric with *A. cerana*; however, the former is a denizen of the primaeval evergreen rainforests of Sundaland, while the latter is abundant in secondary forests, agricultural and even urban settings.

3.3.3 Apis nigrocincta *and* Apis nuluensis

Only very little can be said about the biogeography of either of these species. *A. nigrocincta* only occurs in Sulawesi and neighbouring small islands and on Mindanao in the Philippine group (Fig. 3.4). Otis (1996) suggested that *A. nigrocincta* might have derived from China, because it shares closer morphological similarities with *A. cerana* from the mainland than with those of the southwest. Nonetheless, *proto-nigrocincta* could well have come from Borneo during the Pleistocene glaciations and *A. cerana* may have returned there during an interglacial period.

A. nuluensis has thus far only been encountered in the mountainous areas of Sabah State, Borneo, Malaysia; given what is known, however, it is probably widespread along the Crocker Range, deep into Borneo (Fig. 3.4). Tanaka et al.

(2001) were able to support the suggestion that *A. nuluensis* is closely related to *A. cerana* using mtDNA data. They also suggested that this species may have derived from an isolated population of honeybees that colonised Borneo during Plio-Pleistocene glaciations. They also inferred that, based on the divergence rate of mtDNA and the fact that geographic radiation of *A. cerana* occurred in unison with divergence, this could well have led to the speciation of *A. nuluensis*.

3.4 The Giant Honeybees *Apis dorsata* and *Apis laboriosa*

3.4.1 Apis dorsata

A. dorsata, sensu lato, occurs from western India, eastwards below the Himalayas, throughout continental Asia and oceanic Asia, including the Philippines and Sulawesi, Indonesia (Fig. 3.5). The climatic zones cover the arid west of India, through tropical wet and dry, and tropical rainforest. In terms of altitudinal distribution, for 220 published localities, 85% were below 1,000 m and the balance between 1,000 and 3,000 m. *A. laboriosa* mainly occurs between 2,500 and 4,000 m. There is a significant difference in the altitudinal distributions of *A. dorsata* and *A. laboriosa*, and, while they are partially sympatric, they appear to be seasonally isolated (Roubik et al. 1985; Otis 1996; Thapa et al. 2001).

Although *A. dorsata* and *A. laboriosa* are closely related, they have enormously different dispersal characteristics. *A. dorsata* occurs on all the Philippine islands, and Palawan and the Calamian island groups were often connected during glacial periods; however, neither was ever connected to Mindoro, Panay, Negros or Cebu. During glacial, however, Palawan was only about 70 km from Mindoro (Diamond and Gilpin 1983). Similarly, *A. dorsata* also occurs in Sulawesi, which is about 100 km from Borneo, and where early *proto-dorsata* could possibly have rafted. Mardan (1989) reported that *A. dorsata* crosses the Straight of Malacca between Sumatera and the Malay Peninsula, which is a distance of about 50 km. The northeast monsoon could assist the passage of *A. dorsata*, either in flight or rafting to lands never connected, but these remain unverified conjectures. The seasonal migrations of *A. laboriosa* have now been documented (Underwood 1990; Thapa et al. 2001; Woyke et al. 2001), and analyses of their population dynamics begun in Nepal (Woyke et al. 2003; Joshi et al. 2004). The seasonal migrations are altitudinal in nature and reflect the biogeography of these montane bees.

3.4.2 Apis laboriosa

Although it has been stated that the natural range of distribution of *A. laboriosa* barely overlaps that of *A. dorsata* and then only in different seasons (Otis 1996;

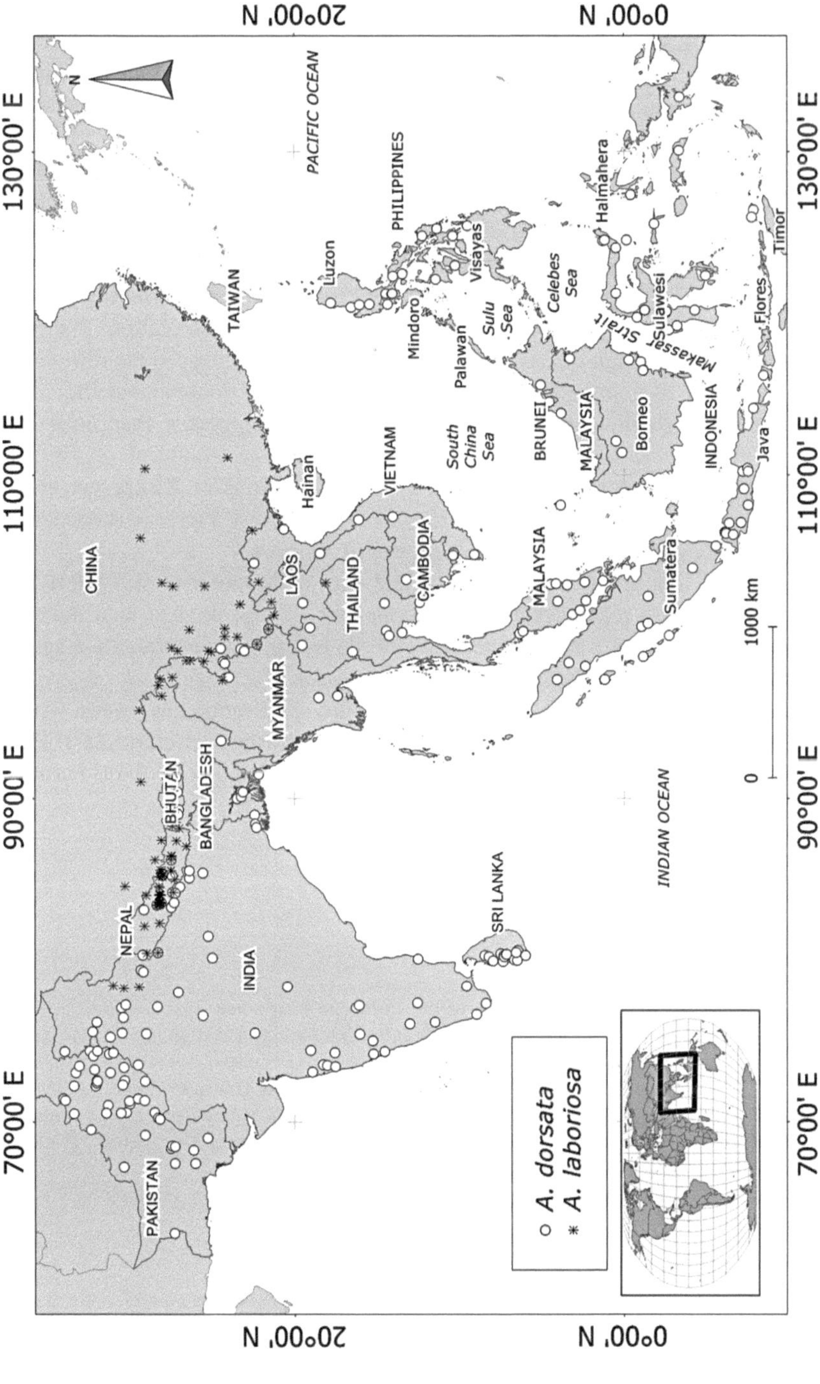

Fig. 3.5 Distribution of the giant honeybees, *Apis dorsata* and *Apis laboriosa*

Underwood 1990), in a study of insect diversity along an altitudinal gradient, Joshi et al. (2008) collected both species at the same sites in Nepal. *A. laboriosa* is continuous from Uttar Pradesh to northern Laos (Fig. 3.5). *A. laboriosa* is distributed in the lower reaches of the Himalayas and extends from north-western Nepal along the mountains, through Bhutan, Sikkim, north-eastern India, Myanmar and southern China, northern Laos and Vietnam (Sakagami et al. 1980; Roubik et al. 1985; Otis 1996; Takahashi and Nakamura 2003; Fig. 3.5). The climatic zones include subtropical to the east and highland to the west. The vegetation is tropical, moist, deciduous forest.

To date, there have been no suggestions as to what barriers could have separated *proto-dorsata* populations so as to foster any divergence into the distinct species of *A. dorsata* and *A. laboriosa*. There are, nonetheless, several considerations that bear on this problem. Although differences caused by aggregation exceed the differentiation attributable to geographic region, Paar et al. (2004) documented the characteristics of migration and colony aggregations and suggested that there are significant differences in population structure between geographical regions. A second point of considerable interest were the observations of Kastberger et al. (2001) that migratory colonies of *A. dorsata* do not fly across a river in Assam, and further postulated that rivers can represent genetic barriers.

If we couple this biology to the geological data of Mugnier et al. (2003) and Sakai et al. (2006), a real physical barrier emerges. The geological data indicate that, about 3 million years ago, convergence of the Indian plate eventually resulted in the raising of the lower Himalayas and the formation of a gigantic lake system in Kathmandu valley between the Mahabharat and Shivapuri mountain ranges. Given the migratory behaviour of contemporary *A. dorsata* (Kastberger et al. 2001; Paar et al. 2004), there could well have been a barrier associated with these massive geological changes.

References

Anderson DL (1989) Mites and pathogens of the European honey bee *Apis mellifera* and the eastern hive bee *Apis cerana* in Papua New Guinea. Entomology Division, DSIR, Auckland, New Zealand

Chen Y (1993) Apiculture in China. Agricultural Scitech Publishing House of China, Beijing

de Bruyn M, Nugroho E, Hossain MM, Wilson JC, Mather PB (2005) Phylogeographic evidence for the existence of an ancient biogeographic barrier: the Isthmus of Kra seaway. Heredity 94:370–378

Diamond JM, Gilpin ME (1983) Biogeographic umbilici and the origin of the Philippine avifauna. Oikos 41:307–321

Dunn KJ (1992) Exotic Asian bee detected in Torres Strait. Bee Briefs 9:18–19

DuPraw EJ (1964) Non-Linnean taxonomy. Nature 202:849–852

El-Shafie HAF, Mogga JB, Basedow T (2002) Studies on the possible competition for pollen between the honey bee, *Apis mellifera sudanensis*, and the imported dwarf honey bee *Apis florea* (Hymenoptera: Apidae) in North-Khartoum (Sudan). J Appl Entomol 126:557–562

Eltz T (2004) Spatio-temporal variation of apine bee attraction to honeybaits in Bornean forests. J Trop Ecol 20:317–324
Engel MS (1998) Fossil honey bees and evolution in the genus *Apis* (Hymenoptera: Apidae). Apidologie 29:265–281
Engel MS, Hinojosa-Diaz IA, Rasnitsyn AP (2009) A honey bee from the Miocene of Nevada and the biogeography of *Apis* (Hymenoptera: Apidae; Apini). Proc Calif Acad Sci 60:23–38
Gassparian S (1977) Studies on *Apis indica cerana* in the eastern part of Iran. In: Proceedings of the 26th International Apicultural Congress, Apimondia, Adelaide, pp 293–296
Gibbard PL, Boreham S, Cohen KM, Moscariello A (2007) Global chronostratigraphical correlation table for the last 2.7 million years, v. 2007b. Subcommission on quaternary stratigraphy. Department of Geography, University of Cambridge, Cambridge, England
Glaiim MK (1992) First definitive record of *Apis florea* in Iraq. Beekeep Dev 24:3
Gong YF (1983) The natural beekeeping conditions and honeybee races in China. J Fujian Agric Coll 12:241–249
Goudie AS (1992) Environmental change, 3rd edn. Oxford University Press, Oxford
Grimaldi D, Engel MS (2005) Evolution of the insects. Cambridge University Press, Cambridge
Haddad N, de Miranda JR, Bataehna A (2008) Discovery of *Apis florea* in Aqaba Jordan. J Apic Res 47:173–174
Haddad N, Fuchs S, Hepburn HR, Radloff SE (2009) *Apis florea* in Jordan: source of the founder population. Apidologie 40:508–512
Hadisoesilo S, Raffiudin R, Susanti W, Atmowidi T, Hepburn C, Radloff SE, Fuchs S, Hepburn R (2008) Morphometric analysis and biogeography of *Apis koschevnikovi* Enderlein (1906). Apidologie 39:495–503
Hepburn HR (2006) Absconding, migration and swarming in honeybees: An ecological and evolutionary perspective. In: Kipyatkov VE (ed) Life cycles in social insects – behaviour, ecology and evolution. St. Petersburg University Press, St. Petersburg, pp 121–136
Hepburn HR, Hepburn C (2005) Bibliography of *Apis florea*. Apidologie 36:377–378
Hepburn HR, Hepburn C (2006) Bibliography of *Apis cerana*. Apidologie 37:651–652
Hepburn HR, Hepburn C (2007) Bibliography of the giant honeybees, *Apis dorsata* Fabricius (1793) and *Apis laboriosa* F. Smith (1871). Apidologie 38:219–220
Hepburn HR, Hepburn C (2008) Bibliography of *Apis koschevnikovi*. Apidologie 39:507
Hepburn HR, Hepburn C (2009) Bibliography of *Apis andreniformis*. Apidologie 40:3
Hepburn HR, Radloff SE, Verma S, Verma LR (2001) Morphometric analysis of *Apis cerana* populations in the southern Himalayan region. Apidologie 32:435–448
Hepburn HR, Radloff SE, Otis GW, Fuchs S, Verma LR, Tan K, Chaiyawong T, Tahmasebi G, Wongsiri S (2005) *Apis florea*: morphometrics, classification and biogeography. Apidologie 36:359–376
Hughes JB, Round PD, Woodruff DS (2003) The Indochinese-Sundaic faunal transition at the Isthmus of Kra: an analysis of resident forest bird species distributions. J Biogeogr 30: 569–580
Joshi SR, Ahmad F, Gurung MB (2004) Status of *Apis laboriosa* populations in Kaski district, western Nepal. J Apic Res 43:176–180
Joshi SR, Kumar K, Arya M (2008) Assessment of insect diversity along an altitudinal gradient in Pinderi forests of Western Himalaya, India. J Asia Pac Entomol 11:5–11
Kalvova J, Halenka T, Bezpalcova K, Nemesova I (2003) Köppen climate types in observed and simulated climates. Stud Geophys Geod 47:185–202
Kastberger G, Sharma DK, Hüttinger E, Kranner G (2001) Morphometric studies of *Apis dorsata* in Assam, India. In: Proceedings of the 7th international conference on tropical bees, Chiang Mai, p 129
Köppen W (1936) Das geographisce System der Klimate. In: Köppen W, Geiger G (eds) Handbuch der Klimatologie 1. C. Gebr, Borntraeger, Germany, pp 1–44 [in German]
Ling Y, Ting J, Min L, Wen-Bin B, Guo-Hong C (2008) Genetic diversity and relationship between genetic distance and geographical distance of 6 *Apis cerana cerana* populations in China. Res J Anim Sci 2:183–187

Ling-Juan L, Ken T (2008) Morphology and taxonomy of *Apis cerana* in Deqin. J Yunnan Agric Univ 23:230–232 [in Chinese]
Liu CS, Lai YS (1990) Swarming in *Apis cerana cerana*. Apic Sci Tech 3:2–4 [in Chinese]
Maa TC (1953) An inquiry into the systematics of the tribus Apidini or honey bees (Hym.). Treubia 21:525–640
Mardan MB (1989) Thermoregulation in the Asiatic giant honeybee *Apis dorsata* (Hymenoptera: Apidae). Thesis, University of Guelph
McMillan AA (2005) A provisional quaternary and neogene lithostratigraphic framework Great Britain. Netherland J Geosci 84:87–107
Mogga JB, Ruttner F (1988) *Apis florea* in Africa: source of the founder population. Bee World 69:100–103
Moss SJ, Wilson EJ (1998) Biogeographic implications of the tertiary palaeogeographic evolution of Sulawesi and Borneo. In: Hall R, Holloway JD (eds) Biogeography and geological evolution of SE Asia. Backhuys, Leiden, pp 133–163
Mugnier JL, Huyge P, Leturmy P, Jouanne F (2003) Episodicity and rates of thrust-sheet motion in the Himalayas, Western Nepal. Am Assoc Petrol Geol Mem 82:1–24
Muller MJ (1982) Selected climatic data for a global set of standard stations for vegetation science. Junk, The Hague
Muzaffar N, Ahmed R (1989) Distribution and competition of *Apis* spp. in Pakistan. In: Proceedings of the 4th International Conference on Apicultural in Tropical Climates, Cairo, pp 449–452
Oldroyd BP, Wongsiri S (2006) Asian honey bees. Harvard University Press, Cambridge
Otis GW (1996) Distributions of recently recognized species of honey bees (Hymenoptera: Apidae; *Apis*) in Asia. J Kans Entomol Soc 69:311–333
Paar J, Oldroyd BP, Hüttinger E, Kastberger G (2004) Genetic structure of an *Apis dorsata* population: The significance of migration and colony aggregation. J Hered 95:119–126
Peel MC, Finlayson BL, McMahon TA (2007) Updated world map of the Köppen-Geiger climate classification. Hydrol Earth Syst Sci 11:1633–1644
Qian H, Song JS, Krestov P, Guo Q, Wu Z, Shen X, Guo X (2003) Large-scale phytogeographical patterns in East Asia in relation to latitudinal and climatic gradients. J Biogeogr 30: 129–141
Radloff SE, Hepburn C, Hepburn HR, Fuchs S, Hadisoesilo S, Tan K, Engel MS, Kuznetsov V (2010) Population structure and classification of *Apis cerana*. Apidologie 41:589–601
Roubik DW, Sakagami SF, Kudo I (1985) A note on distribution and nesting of the Himalayan honeybee *Apis laboriosa* Smith (Hymenoptera: Apidae). J Kans Entomol Soc 58:746–749
Ruttner F (1988) Biogeography and taxonomy of honeybees. Springer, Berlin
Sakagami SF, Matsumura T, Ito K (1980) *Apis laboriosa* in Himalaya, the little known world largest honeybee (Hymenoptera: Apidae). Insecta Matsumurana 19:47–77
Sakai H, Sakai H, Yahagi W, Fujii R, Hayashi T, Upreti BN (2006) Pleistocene rapid uplift of the Himalayan frontal ranges recorded in the Kathmandu and Siwalik basins. Palaeogeogr Palaeoclimatol Palaeoecol 241:16–27
Saraf SK (1985) Beekeeping in Gurez, Jammu and Kashmir, India. Indian Bee J 47:36–37
Sharma PL (1960) Observations on the swarming and mating habits of the Indian honeybee. Bee World 41:121–125
Sloan LC, Crowley TJ, Pollard D (1996) Modeling of middle Pliocene climate with the NCAR GENESIS general circulation model. Mar Micropaleontol 27:51–61
Takahashi JI, Nakamura J (2003) A scientific note on levels of polyandry of 2 queens of the Himalayan giant honeybee, *Apis laboriosa*. Apidologie 34:191–192
Tanaka H, Roubik DW, Kato M, Liew F, Gunsalam G (2001) Phylogenetic position of *Apis nuluensis* of northern Borneo and phylogeography of *A. cerana* as inferred from mitochondrial DNA sequences. Insectes Soc 48:44–51
Thapa R, Shrestha R, Manandhar DN, Bista S, Kafle B (2001) Beekeeping in Nepal. In: Proceedings of the 7th international conference on tropical bees, Chiang Mai, pp 409–413

Underwood BA (1990) Seasonal nesting cycle and migration patterns of the Himalayan honeybee *Apis laboriosa*. Natl Geogr Res 6:276–290

Voris HK (2000) Maps of pleistocene sea levels in Southeast Asia: shorelines, river systems and time durations. J Biogeogr 27:1153–1167

Wongsiri S, Lekprayoon C, Thapa R, Thirakupt K, Rinderer TE, Sylvester HA, Oldroyd BP, Booncham U (1996) Comparative biology of *Apis andreniformis* and *Apis florea* in Thailand. Bee World 77:23–35

Woyke J, Wilde J, Wilde M (2001) A scientific note on *Apis laboriosa* winter nesting and brood rearing in the warm zone of the Himalayas. Apidologie 32:601–602

Woyke J, Wilde J, Wilde M (2003) Periodic mass flights of *Apis laboriosa* in Nepal. Apidologie 34:121–128

Yang G (1986) The distribution of the eastern honeybee (*Apis cerana* Fab.) in China and an analysis of its subspecies. J Yunnan Agric Univ 12:202–206 [in Chinese]

Yang G (2001) The biology of *Apis cerana*. GH Yang, Beijing [in Chinese]

Zeuner FE, Manning FG (1976) A monograph on fossil bees (Hymenoptera: Apoidea). Bull Brit Mus (Nat Hist) Geol 27:151–268

Zhen-Ming J, Guanhang Y, Shuangxiu H, Shikui L, Zaijin R (1992) The advancement of apicultural science and technology in China. In: Verma LR (ed) Honeybees in mountain agriculture. Oxford & IBH, New Delhi, pp 133–147

Zhu YY (1996) The causes of swarming in *Apis cerana cerana*. J Bee (China) 10:23 [in Chinese]

Zubakov VA, Borzenkova II (1990) Global palaeoclimate of the late Cenozoic. Elsevier, New York

Chapter 4
Asian Honeybees and Mitochondrial DNA

Deborah R. Smith

4.1 Introduction

Animal cells, like those of nearly all eukaryotes, contain mitochondria; these organelles, sometimes called "the cell's powerhouse", are responsible for aerobic respiration and possess their own small, circular DNA chromosome – the mtDNA. Mitochondrial DNA (mtDNA) is now one of the primary sources of data used in the study of animal population biology, biogeography and phylogeny (cf. reviews by Avise et al. 1987; Moritz et al. 1987, 2000), and it has played a major role in the study of intra- and inter-specific variation in honeybees. Animal mtDNA is particularly popular because it is ubiquitous, easy to obtain, small, contains a consistent set of genes and has portions evolving at different rates. The use of animal mtDNA in population biology and systematics got off to an early start in part because it was relatively easy to extract and purify in large quantities – an important feature prior to the advent of the polymerase chain reaction, PCR. The presence of many mitochondria per cell, particularly in insect flight muscle, gives high copy number for mitochondrial genes; the circular configuration of the mitochondrial chromosome makes it possible to separate it from linear nuclear chromosomes (Wright et al. 1983) and the small size of these chromosomes – approximately 16,000 bp in honeybees (Crozier and Crozier 1993) – reduces the chances of breakage during purification.

All animal cells contain mitochondria, and (with minor exceptions) the same genes are present in the mtDNA of all animals. There are 13 genes coding for proteins associated with aerobic metabolism: the cytochrome oxidases (abbreviated COI, COII and COIII, or cox1, cox2, cox3), cytochrome b or cytb, ATPases 6 and 8 and the NADH dehydrogenases (abbreviated ND1, ND2, ND3, ND4L, ND4, ND5 and ND 6). There are also genes associated with protein production: components of ribosomes used within the organelles (the small and large subunits, abbreviated

D.R. Smith
Department of Ecology and Evolutionary Biology/Entomology, University of Kansas, Lawrence, KS 66045, USA
e-mail: debsmith@ku.edu

H.R. Hepburn and S.E. Radloff (eds.), *Honeybees of Asia*,
DOI 10.1007/978-3-642-16422-4_4, © Springer-Verlag Berlin Heidelberg 2011

ssrRNA or 12S rRNA and lsrRNA or 16S rRNA, respectively) and 22 (give or take one or two) transfer RNA or tRNA genes that are used within the mitochondrion (Simon et al. 1994). This means that one can confidently compare the same genes across many different animal taxa, though one must be aware of nuclear copies of mitochondrial genes, or NUMTs, which are fairly common in honeybees (Behura 2007; Pamilo et al. 2007).

Judicious selection of the mtDNA gene or regions to be compared can provide relatively fast or slowly evolving sequences to suit a particular problem. Another major advantage of animal mtDNA is the compilation of a large comparative data base, built up as researchers have sequenced mitochondrial genes in a variety of organisms. Finally, among animals, the paternal mtDNA is rarely passed on to offspring; this maternal inheritance is useful in distinguishing maternal and paternal parents in cases of hybridisation.

Research on honeybee mtDNA began relatively early, with studies of *Apis mellifera* (Moritz and Hawkins 1985; Moritz et al. 1986; Smith and Brown 1988); most of the earliest studies were geared towards the population biology of Africanised bees in the Americas (Smith and Brown 1988; Hall and Muralidharan 1989; Smith et al. 1989; Arias and Nobrega 1991; Crozier et al. 1991; Hall and Smith 1991; Rinderer et al. 1991; Sheppard et al. 1991a, b; Smith 1991a, b) and, subsequently, differentiation among the European and African populations and subspecies of *A. mellifera* (Smith 1988; Smith and Brown 1990; Cornuet and Garnery 1991; Garnery et al. 1991, 1992; Smith et al. 1991, 1997; Arias and Sheppard 1996, 2005; Sheppard et al. 1996; de la Rua et al. 1998, 2001a, b, 2004, 2005, 2006; Meixner et al. 2000; Canovas et al. 2008; Munoz et al. 2009; Shaibi et al. 2009). These studies continue to the present day (e.g. Sheppard and Smith 2000; Pinto et al. 2003, 2004, 2005; Schneider et al. 2004).

It was not long after mtDNA studies of *A. mellifera* were initiated that researchers turned their eyes to the mother-lode of *Apis* species found in Asia (Cornuet and Garnery 1991; Garnery et al. 1991; Smith 1991a, b; Willis et al. 1992). The first study of geographic variation in the mtDNA of Asian honeybees employed restriction enzyme digests of intact mitochondrial chromosomes (Smith and Hagen 1996) or genomic DNA (Deowanish et al. 1996) extracted from bee tissues kept fresh or frozen at −80°C. As DNA extraction methods improved, researchers were able to make use of bee samples stored in ethanol or other preservatives, greatly simplifying collection and transport of specimens, and making some older, archived specimens available to genetic analysis. When PCR and DNA sequencing methods became easier and more widely available, it became standard practice to amplify portions of the mitochondrial genome, and either sequence them or digest them with restriction enzymes.

These studies surveyed restriction site polymorphisms and later the DNA sequences of particular genes or regions of the mitochondrial genome. Sequencing studies in particular received a tremendous boost from a few key research projects: the publication of the complete sequence of the honeybee mitochondrial genome (Crozier et al. 1989; Crozier and Crozier 1992, 1993), an analysis of a non-coding region in the honeybee mitochondrial genome (Cornuet et al. 1991) and the publication of a compilation of primers useful in PCR amplification of insect mitochondrial genes (Simon et al. 1994). The most "popular" mitochondrial genes for the study of *Apis* variation and phylogeography have been COI, COII and a non-coding intergenic region (Garnery et al. 1991; Willis et al. 1992; Smith and Hagen 1996, 1999; de la Rua et al. 2000; Smith et al. 2000, 2003, 2004, 2005; Tanaka et al. 2001, 2003; Takahashi et al. 2002, 2007; Oldroyd et al. 2006; Tan et al. 2006, 2007; Warrit et al. 2006). In honeybees, the region from the 3′ end of COI to the 5′ end of COII includes a non-coding sequence (Cornuet et al. 1991), which is particularly useful for detecting intra-specific variation as it is assumed that mutations can accumulate in the non-coding sequence without impairing the fitness of the bee. Other studies have also made use of ND2 (Arias et al. 1996; Tan et al. 2006; Raffiudin and Crozier 2007), ND4 (Leelamanit et al. 2004), 16S rRNA (Engel and Schultz 1997; Sittipraneed et al. 2001; Tanaka et al. 2001; Raffiudin and Crozier 2007), cytb (Insuan et al. 2007), and ATPase 6 and 8 (Songram et al. 2006).

The basic question – Is there an intra-specific genetic structure in the Asian honeybees species as there is in *A. mellifera*? – was quickly followed by a series of research programmes: Does geographic variation in mtDNA match subspecies or species designations of earlier authors? Does mtDNA variation provide insights into the historic and prehistoric distribution of *Apis* species? Do the dwarf, giant and cavity-nesting honeybee lineages show similar biogeographical patterns, or does each lineage show an idiosyncratic pattern of differentiation? Can mtDNA be used to infer phylogenetic relationships among *Apis* species? The phylogeny of *Apis* species is addressed in Chap. 2; here we focus on questions related to the intra-specific variation and phylogeography of the Asian honeybees.

4.2 MtDNA and the Cavity-Nesting Bees

The cavity-nesting honeybees include the western species *A. mellifera* and the Asian species *Apis cerana*, *Apis koschevnikovi*, *Apis nigrocincta* and *Apis nuluensis*. Among the Asian species, *A. cerana* has the most extensive geographic range, from Japan and the Primorski region of Siberia in the northeast, south to the Philippines and Indonesia, and west to India, Pakistan and Afghanistan (Ruttner 1988; Hepburn et al. 2001; Radloff et al. 2010; cf. Chap. 3). The overwhelming majority of research on mtDNA of Asian honeybees has focussed on this species, primarily with the goal of characterising local populations (Sihanuntavong et al. 1999; de la Rua et al. 2000; Smith et al. 2000, 2003, 2004, 2005; Sittipraneed et al. 2001; Oldroyd et al. 2006; Songram et al. 2006; Tan et al. 2006, 2007; Warrit et al. 2006; Takahashi et al. 2007) and understanding the large-scale biogeography of this widespread species (Smith 1991a, b; Deowanish et al. 1996; Smith and Hagen 1996, 1999; Smith et al. 2000; Hepburn et al. 2001).

The other species have more restricted ranges: *A. nuluensis* is confined to mountainous regions of the island of Borneo (Tingek et al. 1996), *A. nigrocincta* to the islands of Sulawesi, Sangihe (Indonesia) and Mindanao (Philippines) (Damus and Otis 1997; Otis and Hadisoesilo 1999) and *A. koschevnikovi* to the tropical evergreen forest regions of Borneo, Sumatra and portions of the Malay Peninsula (Mathew and Mathew 1988; Tingek et al. 1988; Otis 1996; Hadisoesilo et al. 2008). Most studies of mtDNA in these species have emphasised genetic differentiation between these geographically restricted species and the widespread *A. cerana*, and phylogenetic relationships among the cavity-nesting species (Arias et al. 1996; Smith et al. 2000, 2003; Tanaka et al. 2001; Takahashi et al. 2002; Raffiudin and Crozier 2007).

4.2.1 MtDNA of Apis cerana

It is now possible to see the large framework of *A. cerana* phylogeography as indicated by mtDNA, although many questions remain to be addressed. Studies comparing samples collected over large portions of *A. cerana*'s distribution (Smith 1991a, b; Deowanish et al. 1996; Smith and Hagen 1996, 1999; Smith et al. 2000) showed intra-specific mtDNA variation with strong geographic structure. Since the mid-1990s, most studies of *A. cerana* mtDNA diversity have included the non-coding intergenic region; approximately 60 different non-coding sequences have been reported. Each *A. cerana* non-coding sequence is named for the place it was first observed, though they may occur in other areas. In several populations of *A. cerana* and *A. nigrocincta*, some or all individuals examined lack most of the non-coding region. These will be discussed later in Sect. 4.6. It should also be emphasised that sequence variation occurs in coding, as well as non-coding regions, though at a lower frequency.

Four main "lineages" or groups of closely related mtDNA *A. cerana* haplotypes have been observed: (1) the "Mainland Asian" group, (2) the "Sundaland" group, (3) the "Oceanic Philippines" group and (4) a "Yellow Indian" group. Continuing studies have sampled additional geographic locations, and examined variation within the four main lineages. Smaller units within these major groups are sometimes recognised as separate units, such as the bees from Palawan and Japan.

4.2.1.1 The Mainland Asian Group

The mainland Asian group of mtDNA haplotypes has been found in bees from Japan, Korea, the Primorski region of Siberia, China, Taiwan, Vietnam, Cambodia, Laos, northern Thailand (north of ~11°N latitude), Burma (Myanmar), Nepal, India and Pakistan. Approximately 30 different haplotypes (based on sequences of the non-coding region) belonging to the Mainland Asian group have been described in the literature (Smith et al. 2004, 2005; Oldroyd et al. 2006; Tan et al. 2006, 2007; Warrit et al. 2006; Takahashi et al. 2007). All are very similar, most differing by only a single base substitution or insertion/deletion; however, they differ tremendously in their frequency and distribution (Smith et al. 2005; Tan et al. 2007).

A single haplotype called "Japan1" (Smith and Hagen 1996) is the most common and most widespread. It occurs from Japan, where it is virtually the only haplotype observed (Smith and Hagen 1996; Takahashi et al. 2007), and China, where it is the most frequently encountered haplotype, to the west of Burma, and south of Vietnam, Cambodia and Laos, where it is merely one of a diverse assemblage of related haplotypes (Hepburn et al. 2001; Smith et al. 2004, 2005; Tan et al. 2006, 2007). The higher mitochondrial diversity in Indochina suggests that this region may have served as a refuge for *A. cerana* during Pleistocene glaciations if northern Asia was dominated by habitat inhospitable to honeybees. For example, today few or no Chinese *A. cerana* occur in cold meadow or steppe habitat (Hou 1983; Yang 1984a, b; Peng et al. 1989; Tan et al. 2006), which may have been more widespread in the past. When conditions became more hospitable to honeybees, rapid population expansion out of Southeast Asia could have carried the Japan1 haplotype across northern Asia.

The Japan1 haplotype is also conspicuously absent from some Mainland Asian populations. The *A. cerana* of Thailand have been studied extensively (Deowanish et al. 1996; Sihanuntavong et al. 1999; Sittipraneed et al. 2001; Imjongjirak et al. 2004; Warrit et al. 2006) but the Japan1 haplotype has not been observed; instead, the bees of northern Thailand carry the haplotype ThaiN1 (named after the place where it was first observed), which differs from Japan1 by a single insertion/deletion, to the near exclusion of other haplotypes. The reasons for this are not clear but may be solved with further investigation of past climates. Thus far, samples collected from southern India have revealed a set of haplotypes not known outside of India, but the Japan1 haplotype has not been found. However, the vast Indo-Pakistan subcontinent is severely under-sampled, so a description of the honeybee mtDNA of this region would be premature.

4.2.1.2 The Sundaland

The Sundaland mitochondrial group of mtDNA haplotypes has been found on the islands of Palawan, Borneo, Indonesia (Sumatra, Java, Bali, Lombok, Timor, Flores and Sulawesi), as well as southern Thailand (south of ~11°N latitude), peninsular Malaysia and in a small area of Burma. The Indonesian Archipelago and the islands of Borneo and Sulawesi lie on (Sumatra, Java, Bali, Borneo, Palawan) or adjacent to (Flores, Timor, Sulawesi) the shallow Sunda continental shelf. During periods of glaciations, global sea levels dropped below the present level (BPL) as water was sequestered in polar ice caps and glaciers, and during such times, portions of the Sunda shelf were exposed as dry land. For example, 21,000 years ago during the last glacial maximum, sea levels were approximately 120 m BPL, and the islands of Sumatra, Java, Bali and possibly Palawan were united in a large dry land extension of the Southeast Asian mainland. Lombok, Timor and Flores were probably isolated by Borneo and narrow channels, while Sulawesi remained isolated by a deep channel, the Makassar strait (Sathiamurthy and Voris 2006). Over the past 250,000 years, the Earth has experienced multiple cycles of glaciation and deglaciation during which the islands of the Sunda shelf have been alternately united and isolated. It is worth noting that sea levels attained maximum low levels for relatively short periods of time but have been 30–40 m BPL for much longer periods (Voris 2000). At 40 m BPL, most of the Sunda shelf was submerged, but the islands of Sumatra, Java and Borneo remained connected to one another and to the Malay Peninsula by land bridges, and at 30 m BLP, Sumatra and Java remained connected to the Malay Peninsula, but Borneo was isolated (Voris 2000).

Much of what we know about the bees of the Sunda region stems from research by Otis, his colleagues and students (Otis 1991, 1996; Damus and Otis 1997; Otis and Hadisoesilo 1999; Smith et al. 2000, 2003; Hadisoesilo et al. 2008) and from collections that they freely made available to other researchers (e.g. Smith and Hagen 1996). However, considering the complex geological history and biogeography of the Sunda region this area is still, with the exception of Borneo (Tanaka et al. 2003), rather poorly sampled in terms of mtDNA diversity. However, enough is known to make broad comparisons with the Mainland Asian group.

Unlike the situation in the Mainland Asian group of haplotypes, there is no single widespread haplotype. Instead, groups of populations share the same or similar haplotypes. In populations of the Malay Peninsula (Thailand south of 11°N latitude and peninsular Malaysia), the Malay1 haplotype is most common, with an additional four haplotypes (ThaiS1, Malay2, 3 and 4) present at lower frequency. The haplotype Samui1 has been found only on the Thai island of Samui (Smith and Hagen 1996, 1999; Warrit et al. 2006). The Malay1 haplotype has also been found in one specimen from Sumatra (Smith unpublished). Populations on the islands of Java, Bali, Lombok, Timor and Flores share a set of related haplotypes: Java1 in Java, Bali, Timor and Flores, Bali1 in Bali and Lombok1 in Lombok. A different set of haplotypes is found in Borneo and yet another set in Palawan. This pattern seems consistent with a large population, which was slowly restricted to isolated islands as ocean levels rose, leaving them to diverge in isolation.

Tanaka et al. (2003) carried out a more detailed study of variation in COI of the cavity-nesting bees of Borneo, including *A. cerana*. They found two groups of *A. cerana* haplotypes in Borneo, one with broad geographic distribution in Borneo, and the other with a more limited range in South Kalimantan, in southeastern Borneo. The authors suggested that the former group of *A. cerana* has maintained large, continuous populations on Borneo. The authors also reported results of an unpublished analysis that included more Sundaland *A. cerana*. This study indicated the haplotype group found in South Kalimantan is phylogenetically closer to *A. cerana* from Java and Bali than to the other, more widespread haplotype group on Borneo. Indeed, one of their South Kalimantan sequences is reported to be identical to one from central Java. This indicates that there was secondary colonisation of Borneo by *A. cerana* from the Indonesian islands sometime after Bornean and Java–Bali populations differentiated. Although the bees may have colonised under their own power, it is also possible that they were transported – deliberately or inadvertently – by humans.

A subset of Sundaland haplotypes (the "Palawan" group) has a somewhat more complicated geographic distribution. Two haplotypes, Palawan1 and Palawan2, have only been found on the island of Palawan, but two related haplotypes, Cebu1 and MindanaoP, have been found on the Oceanic Philippine islands of Cebu and Mindanao, respectively (de la Rua et al. 2000; Smith et al. 2000). One possible scenario is that these bees reached the Oceanic Philippine islands via "stepping stones" formed by island archipelagos, as described in Sect. 4.2.1.4 on the Oceanic Philippine lineage.

The cavity-nesting bees of Sulawesi, *A. cerana* and *A. nigrocincta*, also present a complex picture. Geologists have long known that Sulawesi was assembled from several smaller fragments of land (Moss and Wilson 1998; Hall 2001), and many organisms show genetic differentiation among populations corresponding to the original "puzzle pieces" that make up the modern island of Sulawesi (deBoer and Duffels 1996; Walton et al. 1997; Wilson and Moss 1999; Hall 2001; Evans et al. 2003; Merker et al. 2009). The distribution of Sulawesi's cavity-nesting species and mitochondrial lineages is not known in enough detail to say whether honeybees show the same patterns. However, the Java1 haplotype of *A. cerana* is present, as well as three haplotypes of *A. nigrocincta*.

4.2.1.3 The Isthmus of Kra and the Boundary Between Mainland and Sundaland Populations

The Mainland Asian and Sundaland mitochondrial lineages meet abruptly in the Malay Peninsula at the Isthmus of Kra (Smith 1991a, b; Smith and Hagen 1996; Sihanuntavong et al. 1999; Sittipraneed et al. 2001; Songram et al. 2006; Warrit et al. 2006). North of 11°N latitude, *A. cerana* carries Mainland mtDNA, while south of this line, *A. cerana* carries Sundaland mtDNA (Warrit et al. 2006). Although there are no obvious physical barriers in the Isthmus of Kra, at least two biogeographic transitions zones are found there: the Kangar-Pattani line, between 6° and 8°N latitude, and the Kra ecotone, between 11° and 13°N latitude (Whitmore 1984; Hughes et al. 2003; Woodruff 2003; Woodruff and Turner 2009). The Kangar-Pattani line marks a transition between Sundaic and Indochinese flora, while the Kra ecotone is associated with the change from wet seasonal evergreen dipterocarp forest in the south to mixed moist deciduous forest in the north (Hughes et al. 2003). The Kra ecotone has been documented as a biogeographic transition zone in *A. cerana* mtDNA lineages and 152 species of birds (Hughes et al. 2003), though in other groups, such as mammals, no single "line of transition" can be documented (Woodruff and Turner 2009). Woodruff (2003) suggests that the distribution of animal groups across the Kra ecotone is associated with modern distribution of vegetation types in the Isthmus of Kra and that the distribution of vegetation types may be the result of frequent marine transgressions during the Miocene (24–13 mya) and late Pliocene (5.5–4.5 mya).

4.2.1.4 The Oceanic Philippines Group

The Oceanic Philippines group of mitochondrial lineages includes mtDNA haplotypes from the Philippine islands excluding Palawan. While Palawan was intermittently connected to Borneo and through Borneo, to the Asian mainland, the so-called oceanic Philippine islands have apparently never had a dry land connection to Borneo or to the Asian mainland. During the mid and late Pleistocene periods of low sea level, many of the oceanic islands of the Philippines were joined into larger units, or "mega-islands": Greater Luzon, Greater Mindanao (which includes the modern islands of Mindanao, Samar, Leyte and Bohol) and Greater Negros–Panay (which includes modern Negros, Panay, Cebu and Masbate). Two island chains may form "stepping stones" between Borneo and the oceanic islands of the Philippines: the Palawan chain between Borneo and Mindoro, and the Sulu Archipelago between Borneo and Mindanao (Heaney 1985, 1986; Heaney et al. 2005).

As was the case with the Sundaland samples, the distribution of the mtDNA haplotypes in Philippine *A. cerana* seems to be strongly influenced by Pleistocene geological history. The mtDNA haplotypes found here are highly divergent from those of other *A. cerana*, as might be expected if these bees have been isolated from other *A. cerana* populations since they colonised these islands. There is also unusually high divergence among haplotypes within the group (de la Rua et al. 2000; Smith et al. 2000), which may stem from the isolation of populations on the several Pleistocene mega-islands. A "Luzon" collection of haplotypes includes two haplotypes from Luzon and one similar sequence found on Mindanao, and a "Mindanao" collection of haplotypes included two haplotypes from Mindanao and a related haplotype from the island of Negros. As noted above, two "Palawan" group sequences also occurred on the islands of Cebu and Mindanao. A comprehensive survey of mtDNA variation in *A. cerana* of the Philippine islands probably still has much to reveal about the biogeography of Philippine honeybees.

4.2.1.5 The Yellow Indian Group

The Yellow Indian group of mitochondrial haplotypes is found in India, Sri Lanka and possibly the Andaman Islands. Indian apiculturalists have long recognised two forms of *A. cerana* in India, variously called the Black, or Hill bees and the Yellow, or Plains bees. Smith (1991a, b) used restriction enzyme polymorphisms to show that the mitochondrial genomes of the two types differed, and later, sequencing of the non-coding region (Smith and Hagen 1996) confirmed that the two strains are different. The mtDNA of the Black bees sampled by Smith and Hagen (1996) belongs to the Mainland Asian group, although none of the haplotypes found in India have been observed outside the Indo-Pakistan subcontinent. Oldroyd et al. (2006) also demonstrated the presence of different mitochondrial genomes in Black and Yellow Indian bees, and additionally demonstrated differences in the time of drone mating flights. Lo et al. (2010) presented additional evidence from both mitochondrial and nuclear genes that the Yellow Indian bees represent a valid species. The Yellow Indian haplotype has also been found on Sri Lanka and the Andaman Islands. Verma (personal communication) noted that bees have been transported from the Indian mainland to the Andamans; the source of imported bees and the identity of native Andaman bees should be ascertained. At the moment, the precise ranges of the Black and Yellow Indian *A. cerana* are not known; most sampling has taken place in Karnataka State in the south of India.

Why are there two distinct forms of *A. cerana* (possibly two species) in India? One possibility is that the Himalayan Mountains and smaller mountain ranges to the east have partially isolated the bees of the Indian subcontinent from the surrounding areas. *A. cerana* does not occur in the Himalayas, and during Pleistocene glaciations, the bees of the Indian subcontinent may have been cut off from populations to the east (through what are now Assam, Bangladesh and Myanmar) and west (through what is now Pakistan) by cold dry grassland habitats. The ancestors of the Yellow bees could have diverged in isolation. When inter-glacial conditions returned, the Indian continent could have been colonised by a second wave of Black *A. cerana* from mainland Asia. At present, this is purely speculative, and other scenarios, such as colonisation of India by a population of Sundaland bees moving along the Asian coastline into India and Sri Lanka, are also possible.

This question might be solved by a robust phylogenetic analysis of the relationships of the Yellow Indian bees to other *A. cerana* lineages, though the non-coding sequence is of little help in this context. *Apis* mtDNA is AT rich (84.9% A+T in *A. mellifera*; Crozier and Crozier 1993), and the non-coding even more so (91% A+T in Sundaland *A. cerana*); the non-coding sequence of the Yellow Indian bees takes this trend to 99% A+T; there is only a single "C" and no "G" (Smith and Hagen 1996). It is not possible to align reliably the non-coding sequences of the Yellow Indian bees with those of other *A. cerana* populations (with only two bases and the possibility of insertions and deletions, alignments become arbitrary). Thus, the Yellow Indian bees were not included in earlier analyses of the relationships among *A. cerana* mtDNA haplotypes.

For a number of reasons, a comprehensive analysis of phylogenetic relationships among *A. cerana* mitochondrial lineages has not been completed. First, many geographic regions have not been sampled or need more intensive sampling (e.g. the Indo-Pakistan region). In some cases, data sets used by different authors are not congruent (e.g. some only using COII, some primarily using COI). In addition, the widespread use of the non-coding intergenic region to detect variation and recognise mitochondrial lineages has a distinct drawback for phylogenetic analysis: because the non-coding region evolves extremely rapidly, it is generally not possible to align non-coding sequences of *A. cerana* with those of other cavity-nesting species (with the exception of *A. koschevnikovi*), much less the dwarf and giant honeybees. Indeed, the non-coding sequences of the Yellow Indian *A. cerana* can only be partially aligned with those of other *A. cerana* as the non-coding sequences of the Yellow Indian bees consist almost exclusively of "A" and "T" (there is one C, and no Gs). As a result, rooting a tree of the *A. cerana* mtDNA lineages cannot be done using the non-coding sequences alone.

Studies in progress (Smith unpublished) using the sequence of the COI and COII genes from *A. cerana*, other cavity nesting-bees, and the giant and dwarf bees to create a rooted tree of mtDNA haplotypes have produced two placements of the Yellow Indian bees. Both show unequivocally that the Black Indian *A. cerana* are members of the Mainland Asian group. One analysis indicates the Yellow Indian mtDNA lineage is a sister group to the remaining Mainland, Sundaland and Oceanic Philippine lineages, which would favour the first scenario. The other analysis supports three main mtDNA lineages – a Yellow Indian group, a Luzon–Mindanao or Oceanic Philippine group, and a Mainland Asian + Sundaland group – without resolving branching order among the three. The Mainland Asian + Sundaland group contains two well supported branches, the Mainland Asian haplotypes and the Sundaland haplotypes.

4.3 MtDNA Variation in Other Asian Cavity-Nesting Species

Much less research has been devoted to the mitochondrial genomes of the other cavity-nesting bees of Asia, *A. koschevnikovi*, *A. nuluensis* and *A. nigrocincta*; most studies were undertaken to determine the species status of these bees (e.g. Smith et al. 2003) or phylogenetic relationships among the cavity-nesting bees (Arias et al. 1996; Smith et al. 2000; Tanaka et al. 2001; Arias and Sheppard 2005; Raffiudin and Crozier 2007), though at least one study (Tanaka et al. 2003) focusses on genetic and geographic variation in the cavity-nesting bees of Borneo.

Phylogenetic studies of cavity-nesting bees have included different subsets of the *Apis* species, with that of Raffiudin and Crozier (2007) having the most complete coverage: ten *Apis* taxa (both species and subspecies), including all the recognised cavity-nesting species, *A. mellifera*, *A. cerana*, *A. koschevnikovi*, *A. nuluensis* and *A. nigrocincta*. These studies also made use of different mitochondrial and nuclear genes. The mitochondrial genes used by various authors include COI (Tanaka et al. 2001), COII (Tanaka et al. 2001; Raffiudin and Crozier 2007), 16S rRNA (Tanaka et al. 2001; Raffiudin and Crozier 2007) and ND2 (Arias et al. 1996; Tanaka et al. 2001; Arias and Sheppard 2005; Raffiudin and Crozier 2007). The general consensus of these phylogenetic analyses of the cavity-nesting bees is that *A. koschevnikovi* is sister to a clade containing *A. cerana*, *A. nigrocincta* and *A. nuluensis* and that there is a close relationship between these three species; some studies have found that *A. cerana* samples constitute a monophyletic species, while others have found either *A. nuluensis*, *A. nigrocincta* or both nested within *A. cerana*, rendering *A. cerana* paraphyletic.

Arias et al. (1996) examined the phylogenetic position of *A. nuluensis* within *Apis*, using an intron in a nuclear EF-1 alpha gene and mitochondrial ND2 gene as sources of sequence data. They employed samples of *A. cerana* from Borneo and Sulawesi (Sundaland mitochondrial lineage) and Sri Lanka (Yellow Indian mitochondrial lineage), as well as *A. nuluensis*, *A. koschevnikovi* and *A. mellifera*. Analysis of the ND2 sequences placed *A. koschevnikovi* as sister to the remaining

Asian cavity-nesting species and nested *A. nuluensis* within *A. cerana* thus: *A. mellifera* (*A. koschevnikovi* (Sri Lanka *A. cerana* (Borneo *A. cerana* (Sulawesi *A. cerana* + *A. nuluensis*)))). The sequences generated from the EF-1alpha intron retained *A. cerana* as a monophyletic group of populations: *A. mellifera* (*A. koschevnikovi* (All *A. cerana* + *A. nuluensis*)). In a later study (Arias and Sheppard 2005) in which they employed the same genes and taxa with the addition of *A. nigrocincta*, most of their analyses placed *A. mellifera* as a sister to the Asian cavity-nesting species and placed *A. koschevnikovi* as sister to a clade containing the other Asian species: *A. mellifera* (*A. koschevnikovi* (*A. nuluensis*, *A. cerana*, *A. nigrocincta*)). The relationships among *A. nuluensis*, *A. cerana* and *A. nigrocincta* were not well resolved. Most of their analyses showed that some *A. cerana* were more closely related to either *A. nuluensis* or *A. nigrocincta* than to other *A. cerana*, rendering *A. cerana* paraphyletic, although the relationships differed among analyses. However, the Sundaland Sulawesi and Borneo *A. cerana* consistently grouped together, to the exclusion of the Yellow Indian *A. cerana* from Sri Lanka.

Tanaka et al. (2001) also investigated the position of *A. nuluensis*, using mitochondrial 16S rRNA, CO1 and COII genes and including samples of *A. mellifera*, *A. koschevnikovi* and *A. nuluensis* from Borneo, and samples of *A. cerana* from Borneo (Sundaland lineage), Taiwan, northern Thailand, the Primorski region of Siberia, Korea and Japan (all Mainland Asian lineage). They concluded that COII did not provide adequate resolution for this level of resolution (i.e. that it was evolving too rapidly). Analyses of their COI sequences either placed *A. nuluensis* as sister to a monophyletic *A. cerana* and recovered the two mitochondrial lineages represented in their *A. cerana* samples (*A. nuluensis* (Mainland *A. cerana* + Sundaland *A. cerana*)); or left them as an unresolved trichotomy: (*A. nuluensis* + Mainland *A. cerana* + Sundaland *A. cerana*). Using only their COI data set, the placements of *A. mellifera* and *A. koschevnikovi*, as well as *Apis dorsata* and *A. florea*, were poorly resolved. Using their 16S rRNA data set, they found *A. koschevnikovi* sister to (*A. nuluensis* + *A. cerana*) and *A. nuluensis* sister to monophyletic *A. cerana*: *A. koschevnikovi* (All *A. cerana* + *A. nuluensis*). The placement of *A. mellifera*, *A. dorsata* and *A. florea* were still poorly resolved. Smith et al. (2000, 2003) considered the relationship of *A. nigrocincta* to samples representing the major mitochondrial lineages within *A. cerana*. They found that in unrooted trees based on the non-coding intergenic sequences, *A. nigrocincta* shared a branch with Oceanic Philippine *A. cerana*, particularly those from the Mindanao group of haplotypes.

Raffiudin and Crozier (2007) carried out the most complete analysis to date of species relationships within *Apis*, including all widely recognised species (*A. florea*, *A. andreniformis*, *A. dorsata*, *A. mellifera*, *A. koschevnikovi*, *A. cerana*, *A. nuluensis* and *A. nigrocincta*) as well as some taxa whose species status is not universally accepted (*A. laboriosa* and *A. d. binghami*). However, geographic coverage was somewhat limited; *A. cerana* was represented only by samples from Borneo (Sundaland mtDNA lineage). They employed three mitochondrial genes, 16S rRNA, COII and ND2, as well as a nuclear gene for inositol 1,4,5 triphosphate receptor (*itpr*). Their analysis recovered cavity-nesters as a single clade with the following relationships: *A. mellifera* (*A. koschevnikovi* (*A. nuluensis* (*A. cerana* + *A. nigrocincta*))). However, since only a single population of *A. cerana* was included, it is not possible to detect whether *A. nuluensis* or *A. nigrocincta* branch within *A. cerana* to give the paraphyletic *A. cerana* found by other studies.

Tanaka et al. (2001, 2003) also investigated the genetic and geographic variation of *A. koschevnikovi* populations on Borneo. Their work revealed three groups of haplotypes consisting, roughly, of northern, eastern and northeastern + western populations. The authors note that because the distribution of *A. koschevnikovi* in Asia indicates a strong preference for rainforest habitats, the genetic differentiation observed on Borneo suggests that Borneo's rainforests were fragmented into at least three major blocks at some time in the past, leading to the observed differentiation among *A. koschevnikovi* populations on the island today.

4.4 MtDNA and the Giant Honeybees

The giant honeybees include one widespread species, *A. dorsata*, and three populations whose species status has been a matter of debate (Ruttner 1988): *A. d. laboriosa*, *A. d. binghami* and *A. d. breviligula*. The Himalayan bee was described under the name *Apis laboriosa* by F. Smith (1871); later, Maa (1953) erected the genus *Megapis* for the giant honeybees and recognised four species: *A. laboriosa*, *A. binghami*, *A. breviligula* and *A. dorsata*. The Himalayan bee was recognised as a distinct species by Sakagami et al. (1980), Roubik et al. (1985) and McEvoy and Underwood (1988) among others, though more recently Engel (1999) has argued against raising any of these populations to species status. Here, I will follow Engel (1999) in using *A. d. dorsata*, *A. d. laboriosa*, *A. d. binghami* and *A. d. breviligula* to refer to these bees, although some or all of these may come to be recognised as valid species.

Apis dorsata dorsata occurs from Pakistan, India, Sri Lanka through Southeast Asian and southern China, Indonesia and the Philippines. According to the review by Otis (1996), *A. d. laboriosa* occurs in Nepal, Bhutan, the mountains of north western India, northern Burma (Myanmar), southwestern China and northern Laos; as he predicted, it also occurs in the mountain regions of northern Vietnam (Trung et al. 1996). The other two distinctive giant bee populations are *A. d. binghami*, the giant bee of Sulawesi (Indonesia), and *A. d. breviligula*, the giant honeybee found on the oceanic islands of the Philippines.

As of this writing, there is no comprehensive study of the phylogeography of giant bees using mtDNA data, although several studies have each examined subsets of the giant bee lineage. Smith (1991a, b) surveyed the entire mitochondrial genome of giant bees collected from Pakistan, India (Bangalore, Karnataka State and the Andaman Islands), northern Thailand, Malaysia (peninsular Malaysia and Sabah, Borneo) and Indonesia (Sulawesi) using six 6-base restriction enzymes. A distance phenogram produced from estimates of sequence divergence among the samples showed three main branches: Sulawesi (corresponding to *A. d. binghami*); a group including Pakistan, the Andaman Islands, northern Thailand, peninsular Malaysia and Borneo; and, somewhat unexpectedly, a third group containing the samples from southern India. No bees from the Philippines were included.

Since that initial study, sequences of mitochondrial genes have been generated for regional studies of giant honeybee mtDNA diversity and for larger phylogenetic studies (i.e. phylogeny of *Apis* species). Tanaka et al. (2003) carried out a regional study of the *A. d. dorsata* of Borneo; they collected samples from multiple locations in Sabah, Sarawak and Kalimantan and screened them for variation in the mitochondrial COI gene. Although they detected sequence variation, there was no geographic structuring of the *A. d. dorsata* populations sampled on Borneo. Similarly, Insuan et al. (2007) carried out a regional study of Thai *A. d. dorsata* using samples collected from north, northeast, central, peninsular Thailand and Samui island. They carried out restriction enzyme digests of PCR amplified fragments of the COI–COII region, ATPase 6&8, Cytb and 16S rRNA; this revealed six composite haplotypes, with the common haplotype present at frequencies of 92–100% in each region and no geographic structure in distribution of rarer haplotypes.

Other investigations of giant bee mtDNA have been in the context of larger phylogenetic studies of *Apis* species. Most include just a single representative (individual, location or population) of the giant bees (Cameron 1991, 1993; Cornuet and Garnery 1991; Garnery et al. 1991; Cameron et al. 1992; Willis et al. 1992; Whitfield and Cameron 1998; Cameron and Mardulyn 2001; Leelamanit et al. 2004), though a few include multiple examples of giant bees. Arias and Sheppard (2005) sequenced mitochondrial ND2 and nuclear EF1-ά genes of *A. d. dorsata* from Thailand and Sri Lanka, *A. d. binghami* from Sulawesi, and *A. d. laboriosa* from India and Nepal, as well as dwarf and cavity-nesting bees. Raffiudin and Crozier (2007) sequenced the mitochondrial ND2 and COII genes, and the nuclear inositol 1,4,5 triphosphate receptor gene in all the dwarf and cavity-nesting species, well as *A. d. dorsata* from Borneo, *A. d. laboriosa* from Nepal and *A. d. binghami* from Sulawesi. All trees from both studies placed *A. d. laboriosa* as sister to a clade containing *A. d. dorsata* and *A. d. binghami*. In Arias and Sheppard's (2005) study, which included *A. d. dorsata* from two locations, *A. d. binghami* was nested within *A. dorsata* as follows: *A. d. laboriosa* (Sri Lankan *A. dorsata* (Thai *A. dorsata* + *A. d. binghami*)).

Smith and L. S. Villafuerte (unpublished) sequenced a portion of COI, $tRNA^{LEU}$, the short non-coding sequence and the 5′ end of COII of giant bees from Pakistan, India (Karnataka, Assam, Andaman Islands), northern Thailand, peninsular Malaysia, Sulawesi, Borneo and the Philippines (Luzon). Preliminary analyses indicate four groups of giant bees: *A. d. dorsata* from Pakistan, Andaman Islands, Thailand, peninsular Malaysia and Borneo; *A. d. binghami* from Sulawesi; *A. d. breviligula* from Luzon; and as in the 1991 survey, *A. d. dorsata* from India as a distinct group. Thus far, relationships among these four are not resolved, and *A. d. laboriosa* has not been included. Ruttner has noted that by morphometric measures, *A. d. dorsata* is "surprisingly uniform" (1988, p 117), and even Maa (1953), "always ready to create new taxa" (Ruttner 1988, p 117), did not suggest that the southern India bees merited unusual recognition. To my knowledge, the only suggestion that Indian *A. d. dorsata* might differ from other *A. d. dorsata* is from Sharma and Thakur (1999), who proposed the name *A. d. indica* (replaced with *A. d. dorsata*, Engel 2003) for giant bees from the Doon Valley in north central India.

Most recently, Lo et al. (2010) carried out a phylogenetic analysis of *Apis*, which included examples of *A. d. dorsata* from Malaysian Borneo and the Philippine island of Palawan, *A. d. laboriosa* from Nepal, *A. d. binghami* from Sulawesi and *A. d. breviligula* from several islands in the Philippines. This study examined both mitochondrial (COII, 12S rRNA) and nuclear (*itpr*, or inositol 1,4,5 triphosphate receptor) genes. In their tree, all the *A. d. breviligula* samples grouped together and, interestingly, formed a sister clade to the remaining giant bees sampled. This configuration implies that if the giant bees reached the oceanic Philippines via a dispersal event, this must have occurred early in the history of giant bees, and that the Philippine radiation of giant bees is as old as the radiation of giant bees in the rest of Asia.

It is notoriously difficult to achieve consensus on the species status of isolated island populations, and the giant honeybees provide a very good case study of this problem (cf. Chap. 1). The populations described by various authors as separate species – *A. d. laboriosa*, *A. d. binghami*, *A. d. breviligula* as well as the giant bees of India – show evidence of genetic differentiation in mitochondrial and nuclear genes. The giant bees *A. d. binghami* on Sulawesi and *A. d. breviligula* in the oceanic Philippine islands differ in appearance from the other giant bees, both *A. d. binghami* and *A. d. breviligula* having generally dark black integument with anterior bands of white pubescence on abdominal tergite (Maa 1953) in contrast to the more brown and yellow or orange colouration of mainland Asian *A. d. dorsata*. The Philippine giant bees differ behaviourally from the other giant bees, in that they do not form the large nest aggregations on trees or buildings seen in *A. d. dorsata* and the Indian giant bees and they have been shown to comprise a monophyletic branch within *Apis* and within the giant bee lineage. Both *A. d. binghami* and *A. d. dorsata* are clearly genetically isolated from other giant bee populations by virtue of their geographic isolation, and for some species concepts, this is enough to confer species status.

4.5 MtDNA and the Dwarf Honeybees

The dwarf honeybees comprise two species, *A. florea* and *Apis andreniformis*, with broadly overlapping ranges. The range of *A. andreniformis* includes Nepal, south-western China, Burma, Laos, Cambodia, Vietnam and Thailand and continues into peninsular Malaysia, Sumatra, Java, and the smaller islands of Indonesia, Borneo and Palawan; the range of *A. florea* extends from Saudi Arabia and eastern Oman, southern Iran, Pakistan, Sri Lanka and India, and overlaps the range of *A. andreniformis* in Nepal, Bangladesh, Burma (Myanmar), southwestern China, Laos, Cambodia and Vietnam, and southernmost Thailand (Ruttner 1988; Otis 1991, 1996; Hepburn et al. 2005; Hepburn and Radloff 2010; cf. Chap. 3). Dwarf bees are absent from the island of Sulawesi and the Oceanic islands of the Philippines. In the last century, *A. florea* has expanded its range (undoubtedly with human assistance) westward to Sudan (Lord and Nagri 1987; Mogga and Ruttner 1988), and more recently, to Jordan and Israel (Haddad et al. 2008, 2009; Moritz et al. 2009).

No comprehensive survey of mtDNA variation has been published for the dwarf bees. Smith (1991a, b) surveyed the entire mitochondrial genome of *A. florea* collected from India (Bangalore, Karnataka State) and northern Thailand, and *A. andreniformis* collected from peninsular Malaysia and Sabah, Borneo, using six 6-base restriction enzymes. This study only confirmed that the mitochondrial genomes of the two dwarf bee species differed and that sequence divergence between the mitochondrial genomes of the two dwarf bee species (5–6%) was similar to that observed among *A. cerana* populations (3–6%) and less that that observed among lineages of giant bees (6–11%), using the same restriction site measures for all samples.

As is the case for giant honeybees, sequences of mitochondrial genes have been generated for regional studies of dwarf honeybee mtDNA diversity and for larger phylogenetic studies of *Apis* species. Rattanawannee et al. (2007) screened samples of *A. andreniformis* from 34 localities in Thailand and four samples from Tenom, Borneo and Malaysia for sequence variation in a 520 bp fragment of the *cytb* gene. Although they detected sequence variation, they found no geographic structure within Thailand, though in a neighbour-joining tree generated from the sequence data all samples from Borneo were clustered together, and nested within the Thai sequences. Numerous other authors have used exemplars of dwarf bees in larger phylogenetic studies (e.g. Cameron 1991, 1993; Garnery et al. 1991; Cameron et al. 1992; Willis et al. 1992; Whitfield and Cameron 1998; Tanaka et al. 2003; Arias and Sheppard 2005; Raffiudin and Crozier 2007). Arias and Sheppard (2005) included samples of *A. florea* from Thailand and Sri Lanka, and *A. andreniformis* from Borneo and Thailand. Not unexpectedly, their phylogenetic trees recovered each of the two dwarf bees as a separate lineage, but unlike earlier studies using restriction enzymes, this data set did not show unusually low divergence between the dwarf bee species relative to *A. cerana* and the giant bee populations. Higgs et al. (2010) pointed out that polymorphisms in the mitochondrial 16S rRNA gene could be used to discriminate *A. florea* and *A. andreniformis* in Thailand.

Smith (unpublished) and J. Palikij (unpublished) sequenced a portion of COI, $tRNA^{LEU}$, the short non-coding sequence and the 5′ end of COII of *A. florea* collected from Sudan, Saudi Arabia, Israel, Jordan, Iran, Pakistan, southern India, Sri Lanka, China, Thailand and Cambodia, and *A. andreniformis* from China, Thailand, peninsular Malaysia, Borneo and the Philippines (Palawan). Preliminary analyses and comparison with published sequences (e.g. Garnery et al. 1991) showed an eastern and western lineage within *A. florea*, with the boundary between Bangladesh and Burma. The two *A. florea* populations may have been separated by episodes of Pleistocene glaciation and diverged in isolation. This analysis also showed two branches within *A. andreniformis*, one on mainland Asia and the other on Sundaic islands. These two *A. andreniformis* lineages were probably separated by rising sea levels during warmer Pleistocene inter-glacials. The divergence between the *A. florea* lineages is greater than divergence between *A. andreniformis* lineages, suggesting that the split between the *A. florea* lineages predates that between the two *A. andreniformis* lineages.

These data also contribute to investigations into the origins of *A. florea* in Israel, Jordan and Sudan. The COI, leucine tRNA, non-coding and COII sequences are identical in samples from Israel and Jordan; for Sudan samples, only sequences from the leucine-tRNA, non-coding region and COII have been generated, but those are identical to the Israeli and Jordan samples. All definitely belong to the western mtDNA branch of *A. florea*, and all differ from the *A. florea* found in Saudi Arabia. More specifically, the bees in Sudan, Israel and Jordan group with sequences from Iran and Pakistan and match exactly sequences from samples collected in eastern Iran. Given the sparse sampling in Iran and Pakistan, it is safest to suggest that the populations of *A. florea* in Sudan, Israel and Jordan originated in Iran, Pakistan or both. Since the Sudan population was established before the Jordanian–Israeli population, it is also conceivable that the Sudan population gave rise to the Jordan–Israeli population.

4.6 Structure of the Non-coding Region

The non-coding sequence lying between the COI and $_{UUR}$tRNA^{LEU} genes on the 5′ side and COII on the 3′ side is a major source of data for mitochondrial variation in cavity-nesting *Apis*. This element was first noted as a series of tandem repeats in restriction enzyme studies of *A. mellifera* mtDNA. Its location in the mitochondrial genome, DNA sequence and structure were first described by Cornuet et al. (1991), who present evidence that it arose from partial duplications of the adjacent $_{UUR}$tRNA^{LEU}, and in the case of *A. mellifera*, additional duplication of a portion of the 3′ end of COI (Cornuet et al. 1991). This non-coding sequence is present in some form in all extant honeybee species.

Honeybee mtDNA is known for its base composition bias towards A and T, but the *Apis* non-coding region takes base composition bias to a new level, being up to 100% A+T. In a single-stranded molecule (generated, for instance, during DNA replication), the 5′ and 3′ ends of the sequence can base pair with each other to form a double-stranded "stem" similar to that seen in the classic "clover-leaf" 3-D structures of tRNAs (Fig. 4.1). The stem includes highly conserved sequences found in nearly all *Apis* mtDNAs examined (Figs. 4.1 and 4.2). The non-coding sequences of the dwarf and giant honeybee coding regions are quite short: 28–39 bp in *A. florea*, 32 bp in *A. andreniformis* and 28–29 bp in *A. d. dorsata*. In these bees, the non-coding region can be folded into a simple stem and loop structure (Fig. 4.2). In the Asian cavity-nesting species such as *A. cerana*, the non-coding sequence is approximately 95–100 bases long, and in at least some cases, it can fold and self pair to form a clover-leaf structure similar to that of a tRNA (Cornuet et al. 1991;

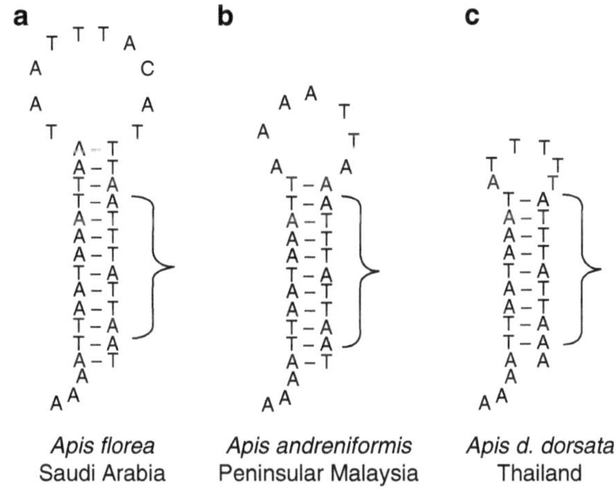

Fig. 4.1 Non-coding intergenic sequences from the mitochondrial genomes of dwarf and giant honeybees, showing the stem and loop structures that can form in single-stranded DNA. (**a**) Western lineage *Apis florea* from Saudi Arabia; (**b**) mainland lineage of *Apis andreniformis* from peninsular Malaysia; (**c**) *Apis dorsata* from Thailand. *Brackets* enclose conserved stem sequences

Smith et al. 2000; Fig. 4.2). The *A. mellifera* non-coding region is much larger (196 to at least 642 bp) and more complex in structure, including repeating elements with sequence similarity to the 3′ end of COI and the adjacent leucine tRNA (Cornuet et al. 1991).

In some cavity-nesting honeybees, most of the non-coding region is absent, leaving behind the stem sequences and a tiny "loop" similar to the non-coding region found in dwarf and giant honeybee mtDNAs (Fig. 4.2). Short non-coding sequences have been found in *A. cerana* from Taiwan, the Philippines (Smith and Hagen 1996; Smith et al. 2000) and China (Shi Wei personal communication), *A. nigrocincta* from Sulawesi (Smith and Hagen 1996) and *A. nuluensis* from Borneo (Takahashi et al. 2002). Although in all cases the clover-leaf portion has been lost and the stem sequences retained, the exact break points and the bases remaining differ. The scattered geographic occurrence of these deletions, the fact that they occur in different species and in different mitochondrial lineages and the sequence variation among them indicate that these are independent losses.

4.7 Some Broader Biogeographic Implications

Mitochondrial DNA (mtDNA) surveys have revealed the outlines of honeybee historical biogeography and indicate areas that will reward additional investigation. Mitochondrial DNA variation in *A. florea*, *A. d. dorsata* and *A. cerana* indicate

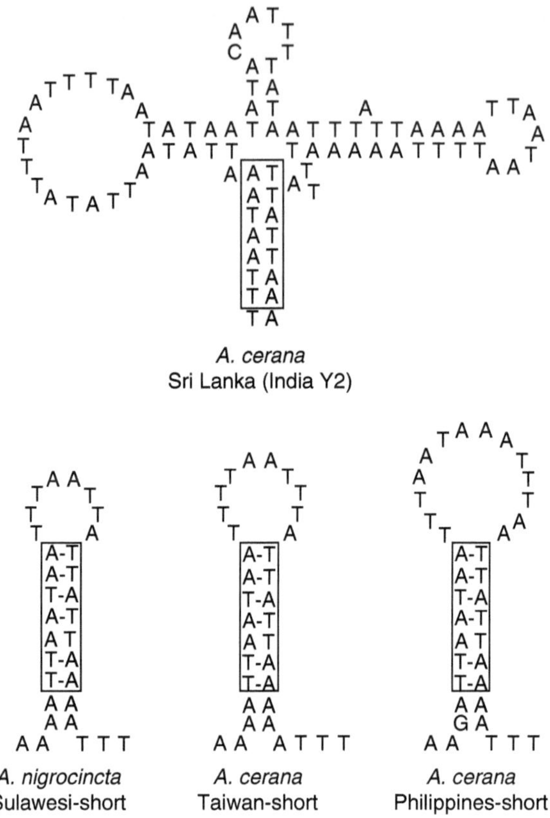

Fig. 4.2 Non-coding intergenic sequence from the mitochondrial genome of Yellow Indian *Apis cerana* (*above*) and three examples of *A. cerana* and *A. nigrocincta* in which most of the non-coding sequence has been lost (*below*). *Boxes* surround stem sequences, which are a subset of the conserved stem sequences in the dwarf and giant honeybee non-coding sequences (reproduced from Smith et al. 2000)

a biogeographic break between India and eastern Asia. Hepburn and Radloff (2011) inferred from the modern distributions of *A. florea* and *A. andreniformis* that during the early Pleistocene (0.8–1.3 mya) glaciation the eastern Himalayas and lower ranges to the east and south (the Chin Hills, the Arakan Yoma in western Burma, and Shan in Thailand and Laos) could have served as a physical and climatic barrier separating the incipient dwarf bee species. The observations on mitochondrial variation in *A. florea*, *A. d. dorsata* and *A. cerana* suggest that during subsequent episodes of glaciation these ranges may have isolated honeybee populations in India from the rest of Asia, leading to their subsequent genetic divergence.

In the case of *A. cerana*, the early population isolated in India may have given rise to the Yellow or Plains bees. During warmer inter-glacials, *A. cerana* populations from eastern Asia could have "reinvaded" India and given rise to the Hill or Black *A. cerana* of India, which belong to the Mainland Asian haplotype group.

These hypotheses can be addressed by a survey of the genetic variation and diversity of honeybees in India, and by more sampling in the region from northeast India, through Bangladesh and Myanmar, into south western China. The Mainland Asian lineage of *A. cerana* was itself probably strongly affected by Pleistocene glaciation. The low haplotype diversity across seen in most populations of the Mainland Asian mitochondrial lineage, the ubiquity of a single mtDNA haplotype (the so-called Japan 1 haplotype) and the high haplotype diversity observed in Indochina populations (Smith et al. 2005) all suggest a relatively recent and rapid population expansion from a refuge population in southeast Asia across north Asia.

There is also a lack of study of the genetic variation among the honeybees found on the islands of the Sunda shelf, particularly Sumatra, Java, Bali and the smaller islands of the Indonesian Archipelago. This is surprising considering that the modern study of diversity in Asian honeybees began with work in this area (e.g. Damus and Otis 1997; Otis and Hadisoesilo 1999; Smith et al. 2000, 2003; Hadisoesilo et al. 2008). This region was strongly affected by changes in sea-level mediated by cycles of Pleistocene glaciation; studies of the effects of repeated joining and isolation of the islands of the Sunda shelf on the genetic diversity of *Apis* species are still in their early stages.

The Philippine Islands are a hotspot for biological diversity, due largely to their long isolation and the periodic formation and fragmentation of "mega-islands" caused by changes in sea level. This is reflected in the extreme divergence of the Philippine giant bees, *A. d. breviligula*, and the most recent suggestion that they be raised to species status (Lo et al. 2010), and by the high mtDNA sequence divergence between the Philippine and other *A. cerana* populations. More detailed study of inter-island variation in Philippine honeybees as well as their historical relationship to other populations is merited.

4.8 MtDNA, Morphometrics and *Apis* Phylogeny

Mitochondrial DNA, nuclear genes, and morphometrics do not always paint the same picture of *Apis* diversity and biogeography, although they are in broad agreement. *Apis* mtDNA, like that of most animals, is typically maternally inherited without recombination; gene flow mediated by movement of males (drones) is therefore largely invisible when viewed through the lens of mtDNA. Thus, what appear to be sharp boundaries between populations belonging to different mtDNA lineages may be much "fuzzier" when the nuclear genome, inherited from both paternal and maternal sources, is examined. Second, mtDNA may be a good record of past population genetic events, including the movement of populations, but it reveals little or nothing about adaptation of populations to their local environments. Thus, while the mtDNA of bees from northern Asia may seem monotonously similar, their morphology and behaviours can vary strikingly among different habitats (Tan et al. 2006, 2007; cf. Chap. 1).

Although the phylogeny of the honeybee species is understood reasonably well, intra-specific phylogeography, particularly for *A. cerana* and *A. dorsata*, is still an area of active research. Some apparent cases of paraphyly among species of cavity-nesting bees and the history of their diversification may become clearer if the mitochondrial lineages within *A. cerana* – whether one considers them species or not – are taken into consideration.

References

Arias MC, Nobrega FG (1991) Location of genes in *Apis mellifera scutellata* derived mitochondrial DNA of Africanized honeybees. Apidologie 22:611–619

Arias MC, Sheppard WS (1996) Molecular phylogenetics of honeybee subspecies (*Apis mellifera* L) inferred from mitochondrial DNA sequence. Mol Phylogenet Evol 5:557–566

Arias MC, Sheppard WS (2005) Phylogenetic relationships of honeybees (Hymenoptera: Apinae; Apini) inferred from nuclear and mitochondrial DNA sequence data. Mol Phylogenet Evol 37:25–35

Arias MC, Tingek S, Kelitu A, Sheppard WS (1996) *Apis nulensis* Tingek, Koeniger and Koeniger, 1996 and its genetic relationship with sympatric species inferred from DNA sequences. Apidologie 27:415–422

Avise JC, Arnold J, Ball RM, Bermingham E, Lamb T, Neigel JE, Reeb CA, Saunders NC (1987) Intraspecific phylogeography – the mitochondrial DNA bridge between populations genetics and systematics. Annu Rev Ecol Syst 18:489–522

Behura SK (2007) Analysis of nuclear copies of mitochondrial sequences in honeybee (*Apis mellifera*) genome. Mol Biol Evol 24:1492–1505

Cameron SA (1991) A new tribal phylogeny of the Apidae inferred from mitochondrial DNA sequences. In: Smith DR (ed) Diversity in the genus *Apis*. Westview, Boulder, pp 71–88

Cameron SA (1993) Multiple origins of advanced eusociality in bees inferred from mitochondrial DNA sequences. Proc Natl Acad Sci 90:8687–8691

Cameron SA, Mardulyn P (2001) Multiple molecular data sets suggest independent origins of highly eusocial behavior in bees (Hymenoptera: Apinae). Syst Biol 50:194–214

Cameron SA, Derr JN, Austin AD, Woollwy JB, Wharton RA (1992) The application of nucleotide sequence data to phylogeny of Hymenoptera: a review. J Hymenopt Res 1:63–79

Canovas F, de la Rua P, Serrano J, Galian J (2008) Geographical patterns of mitochondrial DNA variation in *Apis mellifera iberiensis* (Hymenoptera: Apidae). J Zool Syst Evol Res 46:24–30

Cornuet JM, Garnery L (1991) Mitochondrial DNA variability in honeybees and its phylogeographic implications. Apidologie 22:627–642

Cornuet JM, Garnery L, Solignac M (1991) Putative origin and function of the intergenic region between COI and COII of *Apis mellifera* L mitochondrial DNA. Genetics 128:393–403

Crozier RH, Crozier YC (1992) The Cytochrome-b and ATPase genes of honeybee mitochondrial DNA. Mol Biol Evol 9:474–482

Crozier RH, Crozier YC (1993) The mitochondrial genome of the honeybee *Apis mellifera* – complete sequence and genome organization. Genetics 133(1):97–117

Crozier RH, Crozier YC, Mackinlay AG (1989) The CO-I and CO-II region of honeybee mitochondrial DNA: evidence for variation in insect mitochondrial evolutionary rates. Mol Biol Evol 6:399–411

Crozier YC, Koulianos S, Crozier RH (1991) An improved test for Africanized honeybee mitochondrial DNA. Experientia 47:968–969

Damus MS, Otis GW (1997) A morphometric analysis of *Apis cerana* F and *Apis nigrocincta* Smith populations from southeast Asia. Apidologie 28:309–323 de la Rua P, Serrano J, Galian J (1998) Mitochondrial DNA variability in the Canary Islands honeybees (*Apis mellifera* L.). Mol Ecol 7:1543–1547

de la Rua P, Simon UE, Tilde AC, Moritz RFA, Fuchs S (2000) MtDNA variation in *Apis cerana* populations from the Philippines. Heredity 84:124–130

de la Rua P, Galian J, Serrano J, Moritz RFA (2001a) Genetic structure and distinctness of *Apis mellifera* L. populations from the Canary Islands. Mol Ecol 10:1733–1742

de la Rua P, Galian J, Serrano J, Moritz RFA (2001b) Molecular characterization and population structure of the honeybees from the Balearic islands (Spain). Apidologie 32:417–427

de la Rua P, Jimenez Y, Galian J, Serrano J (2004) Evaluation of the biodiversity of honeybee (*Apis mellifera*) populations from eastern Spain. J Apis Res 43:162–166

de la Rua P, Hernandez-Garcia R, Jimenez Y, Galian J, Serrano J (2005) Biodiversity of *Apis mellifera iberica* (Hymenoptera: Apidae) from northeastern Spain assessed by mitochondrial analysis. Insect Syst Evol 36:21–28

de la Rua P, Galian J, Pedersen BV, Serrano J (2006) Molecular characterization and population structure of *Apis mellifera* from Madeira and the Azores. Apidologie 37:699–708

deBoer AJ, Duffels JP (1996) Historical biogeography of the cicadas of Wallacea, New Guinea and the West Pacific: a geotectonic explanation. Palaeogeogr Palaeoclimatol Palaeoecol 124:153–177

Deowanish S, Nakamura J, Matsuka M, Kimura K (1996) MtDNA variation among subspecies of *Apis cerana* using restriction fragment length polymorphism. Apidologie 27:407–413

Engel MS (1999) The taxonomy of recent and fossil honey bees (Hymenoptera: Apidae; *Apis*). J Hymenopt Res 8:165–196

Engel MS (2003) A replacement name in *Apis* (Hymenoptera: Apidae). J Kans Entomol Soc 76:71

Engel MS, Schultz TR (1997) Phylogeny and behavior in honeybees (Hymenoptera: Apidae). Ann Entomol Soc Am 90:43–53

Evans BJ, Supriatna J, Andayani N, Setiadi MI, Cannatella DC, Melnick DJ (2003) Monkeys and toads define areas of endemism on Sulawesi. Evolution 57:1436–1443

Garnery L, Vautrin D, Cornuet JM, Solignac M (1991) Phylogenetic relationships in the genus *Apis* inferred from mitochondrial DNA sequence data. Apidologie 22:87–92

Garnery L, Cornuet JM, Solignac M (1992) Evolutionary history of the honeybee *Apis mellifera* inferred from mitochondrial DNA analysis. Mol Ecol 1:145–154

Haddad N, de Miranda JR, Bataena A (2008) Discovery of *Apis florea* in Aqaba, Jordan. J Apis Res 47:172–173

Haddad N, Fuchs S, Hepburn HR, Radloff SE (2009) *Apis florea* in Jordan: source of the founder population. Apidologie 40:508–512

Hadisoesilo S, Raffiudin R, Susanti W, Atmowidi T, Hepburn C, Radloff SE, Fuchs S, Hepburn HR (2008) Morphometric analysis and biogeography of *Apis koschevnikovi* Enderlein (1906). Apidologie 39:495–503

Hall R (2001) Cenozoic reconstructions of SE Asia and the SW Pacific: changing patterns of land and sea. In: Metcalfe I, Smith JMB, Morwood M, Davidson ID (eds) Faunal and floral migrations and evolution in SE Asia-Australasia. A.A. Balkema (Swets & Zeitlinger Publishers), Lisse, pp 35–56

Hall HG, Muralidharan K (1989) Evidence from mitochondrial DNA that African honeybees spread as continuous maternal lineages. Nature 339:211–213

Hall HG, Smith DR (1991) Distinguishing African and European honeybee matrilines using amplified mitochondrial DNA. Proc Natl Acad Sci 88:4548–4552

Heaney LR (1985) Zoogeographic evidence for middle and late Pleistocene land bridges to the Philippine Islands. Mod Quat Res SE Asia 9:127–143

Heaney LR (1986) Biogeography of mammals in SE Asia: estimates of rates of colonization, extinction and speciation. Biol J Linn Soc 28:127–165

Heaney LR, Walsh JS, Peterson AT (2005) The roles of geological history and colonization abilities in genetic differentiation between mammalian populations in the Philippine archipelago. J Biogeogr 32:229–247

Hepburn HR, Radloff SE (2011) Biogeography of the dwarf honeybees, *Apis andreniformis* and *A. florea*. Apidologie (in press)
Hepburn HR, Smith DR, Radloff SE, Otis GW (2001) Infraspecific categories of *Apis cerana*: morphometric, allozymal and mtDNA diversity. Apidologie 32:3–23
Hepburn HR, Radloff SE, Otis GW, Fuchs S, Verma LR, Ken T, Chaiyawong T, Tahmasebi G, Ebadi R, Wongsiri S (2005) *Apis florea*: morphometrics, classification and biogeography. Apidologie 36:359–376
Higgs JS, McHale M, Oldroyd BP (2010) A scientific note on a rapid method for the molecular discrimination of *Apis andreniformis* and *A. florea*. Apidologie 41:96–98
Hou HY (1983) Vegetation of China with reference to its geographical distribution. Ann Mo Bot Gard 70:509–548
Hughes JB, Round PD, Woodruff DS (2003) The Indochinese-Sundaic faunal transition at the Isthmus of Kra: an analysis of resident forest bird species distributions. J Biogeogr 30:569–580
Imjongjirak C, Ngewsra L, Pramual C, Insuan S, Pala-Or K, Klinbunga S, Sittipraneed S (2004) Development of a simple and rapid method for typing population origins of the honeybee (*Apis cerana*) in Thailand. J Apis Res 43:17–20
Insuan S, Deowanish S, Klinbunga S, Sittipraneed S, Sylvester HA, Wongsiri S (2007) Genetic differentiation of the giant honeybee (*Apis dorsata*) in Thailand analyzed by mitochondrial genes and microsatellites. Biochem Genet 45:345–361
Leelamanit W, Neelasaewee S, Boonyom R, Panyim S, Hayashi T, Yasue H, Amano K (2004) The NADH dehydrogenase genes of *Apis mellifera*, *A. cerana*, *A. dorsata*, *A. laboriosa* and *A. florea*: sequence comparison and genetic diversity. J Anim Genet 31:3–12
Lo N, Gloag RS, Anderson DL, Oldroyd BP (2010) A molecular phylogeny of the genus *Apis* suggests that the giant honeybee of the Philippines, *A. breviligula* Maa, and the plains honeybee of southern India, *A. indica* Fabricius, are valid species. Syst Entomol 35:226–233
Lord WG, Nagri SK (1987) *Apis florea* discovered in Africa. Bee World 68:39–40
Maa TC (1953) An inquiry into the systematics of the tribus *Apidini* or honeybees (Hym.). Treubia 21:525–640
Mathew S, Mathew K (1988) The "red" bees of Sabah. Newsl Beek Trop Subtrop 12:10
McEvoy MV, Underwood BA (1988) The drone and species status of the Himalayan honeybee, *Apis laboriosa* (Hymenoptera: Apidae). J Kans Entomol Soc 61:246–249
Meixner MD, Arias MC, Sheppard WS (2000) Mitochondrial DNA polymorphisms in honeybee subspecies from Kenya. Apidologie 31:181–190
Merker S, Driller C, Perwitasari-Farajallah D, Pamungkas J, Zischler H (2009) Elucidating geological and biological processes underlying the diversification of Sulawesi tarsiers. Proc Natl Acad Sci 106:8459–8464
Mogga J, Ruttner F (1988) *Apis florea* in Africa: source of the founder population. Bee World 69:100–103
Moritz RFA, Hawkins CF (1985) Isolation of mitochondrial DNA of the honeybee (*Apis mellifera*). Apidologie 16:223–225
Moritz RFA, Hawkins CF, Crozier RH, Mackinley AG (1986) A mitochondrial DNA polymorphism in honeybees (*Apis mellifera* L). Experientia 42:322–324
Moritz C, Dowling TE, Brown WM (1987) Evolution of animal mitochondrial DNA – relevance for population biology and systematics. Ann Rev Ecol Syst 18:269–292
Moritz C, Patton JL, Schneider CJ, Smith TB (2000) Diversification of rainforest faunas: an integrated molecular approach. Annu Rev Ecol Syst 31:533–563
Moritz RFA, Haddad N, Bataieneh A, Shalmon B, Hefetz A (2009) Invasion of the dwarf honeybee *Apis florea* into the near East. Biol Invasions 12:1093–1099
Moss SJ, Wilson MEJ (1998) Biogeographic implications from the Tertiary palaeogeographic evolution of Sulawesi and Borneo. In: Hall R, Holloway JD (eds) Biogeography and geological evolution of SE Asia. Backhuys, Leiden, pp 165–196
Munoz I, Dall'Olio R, Lodesani M, de la Rua P (2009) Population genetic structure of coastal Croatian honeybees (*Apis mellifera carnica*). Apidologie 40:617–626

Oldroyd BP, Reddy MS, Chapman NC, Thompson GL, Beekman M (2006) Evidence for reproductive isolation between two colour morphs of cavity nesting honeybees (*Apis*) in south India. Insectes Soc 53:428–434
Otis G (1991) A review of the diversity of species within *Apis*. In: Smith DR (ed) Diversity in the genus *Apis*. Westview, Boulder, pp 29–49
Otis GW (1996) Distributions of recently recognized species of honeybees (Hymenoptera: Apidae; Apis) in Asia. J Kans Entomol Soc 69(Suppl):311–333
Otis GW, Hadisoesilo S (1999) Insights into honeybee biology from *Apis nigrocincta* of Indonesia. In: Hoopingarner R, Connor LJ (eds) Apiculture for the 21st Century. Wicwas, Cheshire, pp 69–79
Pamilo P, Viljakainen L, Vihavainen A (2007) Exceptionally high density of NUMTs in the honeybee genome. Mol Biol Evol 24:1340–1346
Peng YS, Nasr ME, Locke SJ (1989) Geographical races of *Apis cerana* Fabricius in China and their distribution – review of recent Chinese publications and a preliminary statistical analysis. Apidologie 20:9–20
Pinto MA, Johnston JS, Rubin WL, Coulson RN, Patton JC, Sheppard WS (2003) Identification of Africanized honeybee (Hymenoptera: Apidae) mitochondrial DNA: validation of a rapid polymerase chain reaction-based assay. Ann Entomol Soc Am 96:679–684
Pinto MA, Rubink WL, Coulson RN, Patton JC, Johnston JS (2004) Temporal pattern of Africanization in a feral honeybee population from Texas inferred from mitochondrial DNA. Evolution 58:1047–1055
Pinto MA, Rubink WL, Patton JC, Coulson RN, Johnston JS (2005) Africanization in the United States: replacement of feral European honeybees (*Apis mellifera* L.) by an African hybrid swarm. Genetics 170:1653–1665
Radloff SE, Hepburn C, Hepburn HR, Fuchs S, Haisoesilo S, Tan K, Engel MS, Kuznetsov V (2010) Population structure and classification of *Apis cerana*. Apidologie. doi:10.1051/apido/2010008
Raffiudin R, Crozier RH (2007) Phylogenetic analysis of honeybee behavioral evolution. Mol Phylogenet Evol 43:543–552
Rattanawannee A, Chanchao C, Wongsiri S (2007) Morphometric and genetic variation of small dwarf honeybees *Apis andreniformis* Smith, 1858 in Thailand. Insect Sci 14:451–460
Rinderer TE, Stelzer JA, Oldroyd BP, Buco SM, Rubink WL (1991) Hybridization between European and Africanized honeybees in the neotropical Yucatan peninsula. Science 253:309–311
Roubik DW, Sakagami SF, Kudo I (1985) A note on distribution and nesting of the Himalayan honeybee *Apis laboriosa* Smith (Hymenoptera: Apidae). J Kans Entomol Soc 58:746–749
Ruttner F (1988) Biogeography and taxonomy of honeybees. Springer, Berlin
Sakagami SF, Matsumura T, Ito K (1980) *Apis laboriosa* in Himalaya, the little known world's largest honeybee (Hymenoptera: Apidae). Insecta Matsumurana 19:44–77
Sathiamurthy E, Voris HK (2006) Maps of Holocene sea level transgression and submerged lakes on the Sunda shelf. Nat Hist J Chulalongkorn Univ Suppl 2:1–43
Schneider SS, Hoffman GD, Smith DR (2004) The African honeybee: factors contributing to a successful biological invasion. Annu Rev Entomol 49:351–376
Shaibi T, Munoz I, Dall'Olio R, Lodesani M, de la Rua P, Moritz RFA (2009) *Apis mellifera* evolutionary lineages in northern Africa: Libya, where orient meets occident. Insectes Soc 56:293–300
Sharma V, Thakur ML (1999) Morphometric characterization of giant honeybee *Apis dorsata* Fab. (Hymenoptera: Apidae) from Doon Valley. Ann For 7:125–135
Sheppard WS, Smith DR (2000) Identification of African-derived bees in the Americas: a survey of methods. Ann Entomol Soc Am 93:159–176
Sheppard WS, Rinderer TE, Mazzoli JA, Stelzer A, Shimanuki H (1991a) Gene flow between African-derived and European derived honeybee populations in Argentina. Nature 349:782–784
Sheppard WS, Soares AEE, Dejong D, Shimanuki H (1991b) Hybrid status of honeybee populations near the historic origin of Africanization in Brazil. Apidologie 22:643–652

Sheppard WS, Rinderer TE, Meixner MD, Yoo HR, Stelzer JA, Schiff NM, Kamel SM, Krell R (1996) Hinfl variation in mitochondrial DNA of old world honey bee subspecies. J Hered 87:35–40

Sihanuntavong D, Sittipraneed S, Klinbunga S (1999) Mitochondrial DNA diversity and population structure of the honeybee, *Apis cerana*, in Thailand. J Apic Res 38:211–219

Simon C, Frati F, Beckenbach A, Crespi B, Liu H, Flook P (1994) Evolution, weighting, and phylogenetic utility of mitochondrial gene seqeunces and a compilation of conserved polymerase chain reaction primers. Ann Entomol Soc Am 87:651–701

Sittipraneed S, Sihanuntavong D, Klinbunga S (2001) Genetic differentiation of the honeybee (*Apis cerana*) in Thailand revealed by polymorphism of a large subunit of mitochondrial ribosomal DNA. Insectes Soc 48(3):266–272

Smith F (1871) Descriptions of some new insects collected by Dr Anderson during the expedition to Yunnan. Proc Zool Soc Lond 1871:244–249

Smith DR (1988) Mitochondrial DNA polymorphisms in five Old World subspecies of the honeybee and in New World hybrid populations. In: Needham G, Page RE (eds) In: Proceedings of the international conference Africanized honeybees and bee mites. Ellis Horwood, Chichester, pp 303–312

Smith DR (1991a) African bees in the Americas – insights from biogeography and genetics. Trends Ecol Evol 6:17–21

Smith DR (1991b) Mitochondrial DNA and honeybee biogeography. In: Smith DR (ed) Diversity in the genus *Apis*. Westview, Boulder, pp 131–176

Smith DR, Brown WM (1988) Polymorphisms in mitochondrial DNA of European and Africanized honeybees (*Apis mellifera*). Experientia 44(3):257–260

Smith DR, Brown WM (1990) Restriction endonuclease cleavage site and length polymorphisms in mitochondrial DNA of *Apis mellifera mellifera* and *A. m. carnica* (Hymenoptera: Apidae). Ann Entomol Soc Am 83:81–88

Smith DR, Hagen RH (1996) The biogeography of *Apis cerana* as revealed by mitochondrial DNA sequence data. J Kans Entomol Soc 69:294–310

Smith DR, Hagen RH (1999) Phylogeny and biogeography of *Apis cerana* subspecies: testing alternative hypotheses. In: Hoopingarner R, Connor L (eds) Apiculture for the 21st Century. Wicwas, Cheshire, pp 60–68

Smith DR, Taylor OR, Brown WM (1989) Neotropical Africanized honeybees have African mitochondrial DNA. Nature 339:213–215

Smith DR, Palopoli MF, Taylor BR, Garnery L, Cornuet JM, Solignac M, Brown M (1991) Geographical overlap of 2 mitochondrial genomes in Spanish honeybees (*Apis mellifera iberica*). J Hered 82:96–100

Smith DR, Slaymaker A, Palmer M, Kaftanoglu O (1997) Turkish honeybees belong to the east Mediterranean mitochondrial lineage. Apidologie 28:269–274

Smith DR, Villafuerte L, Otis G, Palmer MR (2000) Biogeography of *Apis cerana* F. and *A. nigrocincta* Smith: insights from mtDNA studies. Apidologie 31:65–279

Smith DR, Palmer MR, Otis G, Damus M (2003) Mitochondrial DNA and AFLP markers support species status of *Apis nigrocincta*. Insectes Soc 50:185–190

Smith DR, Warrit N, Hepburn HR (2004) *Apis cerana* from Myanmar (Burma): unusual distribution of mitochondrial lineages. Apidologie 35:637–644

Smith DR, Warrit N, Otis G, Thai PH, Tam DQ (2005) A scientific note on high variation in the non-coding mitochondrial sequences of *Apis cerana* from South East Asia. J Apic Res 44:197–198

Songram O, Sittipraneed S, Klinbunga S (2006) Mitochondrial DNA diversity and genetic differentiation of the honeybee (*Apis cerana*) in Thailand. Biochem Genet 44:256–269

Takahashi J, Nakamura J, Sasaki M, Tingek S, Akimoto S (2002) New haplotypes for the non-coding region of mitochondrial DNA in cavity-nesting honeybees *Apis koschevnikovi* and *Apis nuluensis*. Apidologie 33:25–31

Takahashi J, Yoshida T, Takagi T, Akimoto S, Woo K, Deowanish S, Hepburn R, Nakamura J, Matsuka J (2007) Geographic variation in the Japanese islands of *Apis cerana japonica* and in *A. cerana* populations bordering its geographic range. Apidologie 38:335–340

Tan K, Meixner MD, Fuchs S, Zhang X, He SY, Kandemir I, Sheppard WS, Koeniger N (2006) Geographic distribution of the eastern honeybee, *Apis cerana* (Hymenoptera: Apidae) across ecological zones in China: morphological and molecular analyses. Syst Biodivers 4:473–482

Tan K, Warrit N, Smith DR (2007) Mitochondrial DNA diversity of Chinese *Apis cerana*. Apidologie 38:238–246

Tanaka H, Roubik DW, Kato M, Liew F, Gunsalam G (2001) Phylogenetic position of *Apis nuluensis* of northern Borneo and phylogeography of *A. cerana* as inferred from mitochondrial DNA sequences. Insectes Soc 48:44–51

Tanaka H, Suka T, Kahono S, Samejima H, Mohamed M, Roubik DW (2003) Mitochondrial variation and genetic differentaition in honeybees (*Apis cerana*, *A. koschevnikovi* and *A. dorsata*) of Borneo. Tropics 13:107–117

Tingek S, Mardan M, Rinderer TE, Koeniger N, Koeniger G (1988) Rediscovery of *Apis vechti* (Maa, 1953) – the Saban honeybee. Apidologie 19:97–102

Tingek S, Koeniger N, Koeniger G (1996) Description of a new cavity dwelling species of *Apis* (*Apis nuluensis*) from Sabah, Borneo with notes on its occurrence and reproductive biology (Hymenoptera: Apoidea; Apini). Sencken Biol 76:115–119

Trung LQ, Dung PX, Ngan TX (1996) A scientific note on first report of *Apis laboriosa* F Smith, 1871 in Vietnam. Apidologie 27:487–488

Voris HK (2000) Maps of Pleistocene sea levels in Southeast Asia: shorelines, river systems and time durations. J Biogeogr 27:1153–1167

Walton C, Butlin RK, Monk KA (1997) A phylogeny for grasshoppers of the genus *Chitaura* (Orthoptera: Acrididae) from Sulawesi, Indonesia, based on mitochondrial DNA sequence data. Biol J Linn Soc 62:365–382

Warrit N, Smith DR, Lekprayoon C (2006) Genetic subpopulations of *Varroa* mites and their *Apis cerana* hosts in Thailand. Apidologie 37:19–30

Whitfield JB, Cameron SA (1998) Hierarchical analysis of variation in the mitochondrial 16S rRNA gene among Hymenoptera. Mol Biol Evol 15:1728–1743

Whitmore TC (1984) Tropical rain forests of the Far East, 2nd edn. Oxford Scientific Publications, Clarendon, Oxford

Willis LG, Winston ML, Honda BM (1992) Phylogenetic relationships in the honeybee (genus *Apis*) as determined by the sequence of the cytochrome oxidase II region of mitochondrial DNA. Mol Phylogenet Evol 1:169–178

Wilson MEJ, Moss SJ (1999) Cenozoic palaeogeographic evolution of Sulawesi and Borneo. Palaeogeogr Palaeoclimatol Palaeoecol 145:303–337

Woodruff DS (2003) Neogene marine transgressions, palaeogeography and biogeographic transitions on the Thai-Malay Peninsula. J Biogeogr 30:551–567

Woodruff DS, Turner LM (2009) The Indochinese-Sundaic zoogeographic transition: a description and analysis of terrestrial mammal species distributions. J Biogeogr 36:803–821

Wright JW, Spolsky C, Brown WM (1983) The origin of the parthenogenetic lizard Cnemidophorus laredoensis inferred from mitochondrial DNA analysis. Herpetologica 39:410–416

Yang GH (1984a) The survey of the resource of the Chinese honeybee. Zhongguo Yangfeng 3:4–7 [in Chinese]

Yang GH (1984b) The survey of the resource of the Chinese honeybee II. Zhongguo Yangfeng 6:16–19 [in Chinese]

Chapter 5
Genetic Considerations

Catherine L. Sole and Christian W.W. Pirk

5.1 General Introduction

We define genetic considerations in the context of this book in terms of the role and effects of genetics, especially with regard to mating frequency, behaviour (in the form of the waggle dance, aggregation and migration) and the coefficient of genetic variation with respect to morphological characters. Thus, we do not address the role of genetics on topics which are addressed in other chapters.

5.2 Mating Frequency

The success of eusocial insects is thought to be attributed to high intracolonial relatedness. However, some of the most highly eusocial insects – for example, the genus *Apis* – are characterised by low genetic relatedness, which in their case is the result of a high mating frequency, in contrast to high numbers of queens like in the ant *Formica exsecta* (Katzerke et al. 2006). Species with moderate levels of polyandry (6–10 matings) have been shown to have decreased brood variability (Page 1980; Page and Metcalf 1982) and lower worker–queen conflict over sex ratios within a colony (Moritz 1985; Queller 1993), while an increased number of matings (more than 6–10) have been shown to have a minimal effect on these two factors (Fuchs and Moritz 1998).

Reports differ as to the number of matings of virgin honeybee queens, with differences occurring at both intra- and interspecific levels within the genus *Apis*. *Apis dorsata* queens were shown by Wattanachaiyingcharoen et al. (2003) to mate with up to 100 drones on up to six successive mating flights (Tan et al. 1999), with an effective mating frequency of between 26.9 and 88.5, which falls within an

C.L. Sole and C.W.W. Pirk
Social Insect Research Group, Department of Zoology and Entomology, University of Pretoria, Pretoria 0002, South Africa
e-mail: clsole@zoology.up.ac.za; cwwpirk@zoology.up.ac.za

H.R. Hepburn and S.E. Radloff (eds.), *Honeybees of Asia*,
DOI 10.1007/978-3-642-16422-4_5, © Springer-Verlag Berlin Heidelberg 2011

Table 5.1 Observed number of matings and sperm count in the different *Apis* species

Species	Observed mating frequency Mean ± SE	Sperm count
A. mellifera	17.32 ± 1.10[a]	11.95×10^6 [b]
		11.40×10^6 [c]
		10.62×10^6 [d]
		7.54×10^6 [e]
		8.54×10^6 [f]
		8.60×10^6 [g]
		7.08×10^6 [h]
A. florea	8.00 ± 1.64[i]	0.44×10^6 [j]
	16.00 ± 2.50[k]	
A. dorsata	54.90 ± 31.46[l]	2.46×10^6 [m]
A. andreniformis	13.50 ± 4.51[l]	0.13×10^6 [m]
A. cerana	18.80 ± 5.56[l]	4.66×10^6 [n]
A. koschevnikovi	16.20 ± 10.53[l]	2.13×10^6 [o]
A. laboriosa	22.00 ± 1.41[l]	No data available
	20.19 ± 6.07[p]	
A. nigrocincta	54.00 ± 11.52[l]	No data available
A. nuluensis	No data available	1.30×10^6 [q]

[a]Kraus et al. (2005)
[b]Schlüns et al. (2003)
[c]Rinderer et al. (1985)
[d]Elbassiouny (1992)
[e]Duay et al. (2002)
[f]Moritz (1981)
[g]Rinderer et al. (1999)
[h]Berg and Koeniger (1990)
[i]Oldroyd et al. (1995)
[j]Koeniger et al. (1989)
[k]Palmer and Oldroyd (2001)
[l]as stated in Tarpy et al. (2004)
[m]Koeniger et al. (1990)
[n]Woyke (1975)
[o]Koeniger and Koeniger (1991, 2000); Tarpy et al. (2004)
[p]Paar et al. (2004b)
[q]Koeniger et al. (1996); Koeniger and Koeniger (2000)

earlier range as suggested by Moritz (1995). At the opposite end of the scale, we have *Apis florea* that has been shown to have a mating frequency of between 5 and 14 (Oldroyd et al. 1995), which was later raised to 16 by Palmer and Oldroyd (2001). *Apis andreniformis* and *A. florea*, therefore, fall at the lower end, with *Apis cerana*, *Apis koschevnikovi* and *Apis laboriosa* in the middle and *A. dorsata* and *Apis nigrocincta* at the upper end of mating frequencies. To date no data exists for the mating frequency of *Apis nuluensis* (Table 5.1).

Mating frequency is defined as the number of times a virgin female mates with a drone, i.e. the number of sexual partners a queen honeybee may have, after which the sperm is stored in the spermatheca for the fertilisation of eggs. The number of matings a female insect undertakes can have an effect on the fecundity of the female (Ridley 1988; Fig. 5.1). Interestingly, the nuptial flights of drones and queens differ

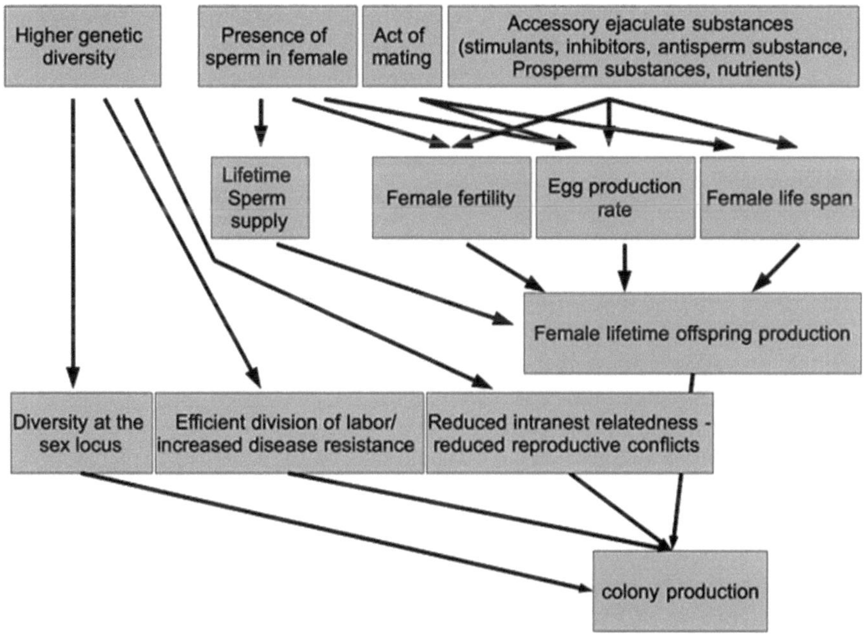

Fig. 5.1 Effects of multiple mating which could act directly on the queen as well as on colony productivity

between honeybee species and may be a mechanism of sympatric and temporal spatial separation, especially common in Asian honeybees (Koeniger and Koeniger 2000), resulting in speciation.

5.2.1 *Why Polyandry?*

Eusociality is generally characterised by co-operative brood care, overlapping generations and reproductive division of labour (Wilson 1971). The question therefore remains: why do the species of the genus *Apis* have such a range of mating frequencies? The amount of matings comprise a variety of factors, for example, number and duration of mating flights, the act of insemination, volume of ejaculate, quantity and variability of sperm and seminal proteins (Richard et al. 2007). With only a single laying queen in a honeybee colony, we know that worker reproduction is uncommon and that workers mainly gain inclusive fitness through the reproduction of their queen mother, drone father, half-sisters, super-sisters and brothers. Once the queen starts ovipositing she never mates again, so the amount of sperm stored in her spermathecae can be a limitation to her fitness. Is this therefore the reason a honeybee queen mates multiply? However, we do know that a certain number of matings is needed to fill her spermatheca and, once this is completed, egg-laying commences (Tarpy and Page 2000; Schlüns et al. 2005a). Is spermatheca size therefore the reason for multiple matings and is spermatheca size related to the size of the queen? *Apis dorsata* are the largest Asian honeybees in size and have the most matings per queen, while *A. florea*, nearly the smallest, with the fewest matings. Yet the number of sperm received by a queen honeybee per mating is also limited by the quantity of sperm a drone can provide. The next question that comes to mind is: are the numbers of matings controlled by the sperm provider, i.e. the drone? We know that *A. dorsata* drones produce 2.4 million sperm/ejaculate, while *A. florea* and *A. adreniformis* produce at least 5 and 18 times less sperm than that, respectively (Table 5.1). However, this does not indicate how many matings are needed to fill the spermatheca and whether this may be the defining explanation for polyandry, added to which 97% of the semen is expelled after mating (Koeniger and Koeniger 2000). Another point to be raised is do all drones transfer sperm or is there a certain amount of unsuccessful matings, i.e. mating but no sperm transfer? Would this be a plausible explanation for multiple matings as "a safe guard" against poor fertility? If a queen exhibits low/poor fertility she will start laying drones, resulting in colony death (Schlüns et al. 2005a).

It has also been hypothesised that the number of partners a queen has makes her more attractive to the worker force behind her (Richard et al. 2007), i.e. her physiology, pheromone profile and behaviour differ after multiple matings. This in turn could have long-term effects for overall colony fitness in that she has workers that are responsive to both her needs and those of the colony. Furthermore, studies have shown that a genetically diverse colony has added benefits in terms of increased disease resistance (Hamilton 1987; Sherman et al. 1988; Schmid-Hempel 1998; Baer and Schmid-Hempel 1999; Tarpy 2003), homozygosity at the sex determining locus (Page 1980; Ratnieks 1990; Crozier and Pamilo 1996; Tarpy and Page 2002) and effective regulation of adaptation to environmental variables and therefore overall increased colony survival (Page et al. 1995; Jones et al. 2004). Disease is a problem for virtually all organisms, consequently one of the results of natural selection would be adaptations to minimise the negative effects of disease. Genetically distinct worker groups, or subfamilies, through multiple matings, may result in different alleles with resistance to different diseases (Tarpy and Seeley 2006) on which genetic drift can then act. With genetic differences occurring within a single colony one could assume that increased survival of the colony as a whole would be a result, as the entire colony would not die from a single disease but only the individuals showing susceptibility.

The mechanism defining the sex determination system of honeybees involves a single gene with many alleles (Beye et al. 2003). The haplodiploid genetic structure of honeybees dictates that female honeybees have two copies of this gene, while male honeybees or drones have a single copy (and are generally laid from unfertilised eggs). Fertilised eggs that are heterozygous at the locus for the sex-determining gene develop into females; however, if a fertilised egg that is homozygous for the sex determining locus is laid, it will develop into a sterile male that will not reach maturity (Woyke 1963, 1965). The existence of a high proportion of these "diploid

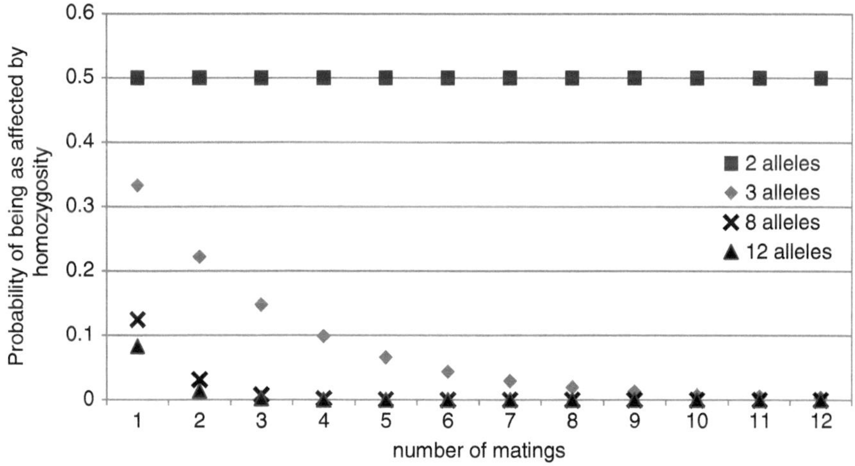

Fig. 5.2 On the *y*-axis, the probability of being affected by homozygosity at the sex locus as in the case with only two alleles (*indicated by squares*) is shown. It shows that the probability is reduced with increasing matings and increased number of alleles. Moreover, increasing the number of matings to three with three different alleles (*indicated by diamonds*) results in a similar reduction, as does increasing the number of alleles to eight (*indicated by crosses*), and remain singly mated. Clearly, from an evolutionary point of view, it seems simpler to adjust the number of matings as opposed to increasing the number of alleles, especially when sufficient potential mates are around (for example in drone congregation areas)

drones" results in low brood viability, which in turn will have a negative impact on the colony as a whole (Page 1980; Tarpy and Page 2002). Therefore, increased numbers of matings insure that a single queen (colony) will be less affected by potential homozygosity at the sex locus (Fig. 5.2 gives an example of the effect of the number of matings as compared to the number of alleles).

Regulation with respect to environmental variation can be explained in that there are genotypically determined differences in behaviour (Page et al. 1995). A genetically diverse colony could therefore be considered to be able to operate more efficiently under changing environmental conditions. That a colony with diverse genotypes may be able to respond better to environmental changes than one with fewer genotypes, consequently increasing the overall fitness advantage, lends support to the prediction. Workers may, therefore, be allocated to different tasks under different or changing environmental conditions, for example, when a colony may be under stress (Kolmes et al. 1989; Robinson and Page 1989a). Tasks, which may benefit a colony, are the collection of diverse pollen, increased foraging range and rates, stabilising of colony behaviour, overwintering survival, food storage, colony growth (increased brood rearing and therefore an increased worker force) and defence behaviour (Robinson and Page 1989b; Page et al. 1995; Mattila and Seeley 2007; Seeley and Tarpy 2007). All of this demonstrates that colony fitness and productivity are enhanced through genetic diversity. However, the question remains: how does the colony create the balance between enough genetic diversity and too much genetic diversity which could negatively impact on a colony?

We have given a number of reasons for multiple matings or polyandry as no definite, single answer seems to be available or can be agreed upon. The adaptive significance of polyandry still seems to be under debate. From the above discussions, we can clearly see that even though a distinction exists between proximate and the evolutionary mechanisms of multiple matings, or extreme polyandry, they are linked. To fully understand polyandry, one needs to interpret one with the other in mind. Our thoughts therefore relating to this are: do we still need to get to the bottom of this? = yes; are these systems so complex that we will never truly understand them? = probably; or are they more simple than we expect? = even we are not sure. Which ever answer one chooses or whatever thoughts one may have on this subject, the more we learn, the more questions we ask and the closer we hopefully get to the truth or what we perceive as the truth. What we can say about polyandry in honeybees is that it evolved and is maintained due to its many selective advantages, which result in an overall increased fitness and this outweighs the costs involved.

5.3 Coefficient of Genetic Variation and Morphological Characters

The coefficient of variation is defined as the degree of variation from one data series to another, i.e. standard deviation divided by the mean. Phenotypic variation can be measured by means of the coefficient of genetic variation, which provides a quantitative measure of the genetic stability of the morphological characters measured. The idea behind this section of the chapter is whether we can correlate phenotypic variation as defined by morphometric measurements – e.g. the length of the hind tibia – to genetic stability (for examples of morphometric characters measured, see Table 5.2).

Akahira and Sakagami (1959) measured morphometric characters in two species of honeybees, *A. cerana* and *A. mellifera.* For both species, drones (males), virgin queens and workers from the same mother queen were analysed as well as drones derived from the laying workers (the laying workers were the offspring of the same mother queen as the three previously mentioned groups). All individuals were raised in the same colony so as to remove the effect of environmental variation. They hypothesised that by calculating the coefficients of variation for the morphometric measurements any differences occurring among the groups could therefore be correlated to the genetic composition within the group. Their hypothesis could not been tested at that time due to the lack of the necessary statistical methods. Later, Pirk et al. (2005) re-analysed their data using more updated statistical methods and found that *A. mellifera* queen mother-produced workers and drones, virgin queen-produced drones and worker-produced drones all differed significantly with respect to their coefficients of variation, which is unusual. *Apis cerana*, on the other hand, for workers and drones derived from the same queen mother as well as drones

Table 5.2 Morphometric measurements for the following characters where measured by Akahira and Sakagami (1959)

Characters measured
Distance between upper margin of vertex and labrum
Maximum distance between outer margins of the eyes
Length of flagellum
Distance between anterior and posterior margins of mesonotum
Distance between anterior and posterior margins of scutellum
Length hind tibia
Length hind basitarsus
Width of hind basitarsus
Distance between outer margin of third axillary sclerite of forewing and wing tip
Distance between cross-point of subdiscoidal and discoidal veins of forewing and wing tip
Maximum width forewing
Maximum length radial cell
Distance between cross-point of medial and submarginal veins of hindwing to wing tip
Distance from base cubital cell to wing tip
Maximum width hind wing
Radial, medial and vannal vein length in hind wing
Length of right and left costal veins with hamuli
Number hamuli on right and left hind wings
Number hamuli on cubital index

derived from laying workers of the same queen mother, showed no significant variation for the coefficients of variation.

Why is this so? For honeybees and other social insects, it is well documented that not all mates contribute equally to the worker offspring, added to which, if a queen mates with related males, this would further reduce the number of effective matings (Pamilo 1993). Reduced effective matings would result in reduced worker variation; furthermore, several traits and characters are inherited by dominant-recessive inheritance (Rothenbuhler et al. 1968; Woyke 1978) (see Fig. 5.3 for a detailed explanation of dominant-recessive inheritance reproduced from Pirk et al. 2005). In trying to explain the significant differences in coefficients of variation from the different laying individuals with regard to the *A. mellifera* drones, one needs to think in the terms of Mendelian inheritance. Take the simple example of a queen with the genotype (= phenotype) W(hite)B(lack) for a character that mates with a drone with the dominant allele G(rey), this would result in uniform workers with WG, BG as the genotype but Y as the phenotype. However, if both types of workers arrhenotokously produce males, the males will carry W, B or G alleles with the frequency 25% W, 25% B and 50% G. Therefore, by inheriting dominant alleles of a gene from their father worker genetic variation would be reduced but that of their haploid progeny not. The flip side to this is that if the dominant alleles were inherited from the female side, the queen would therefore have to be homozygous; if she was heterozygous more variability would be expected in her worker produced population. This in turn would mean no significant difference between queen-laid drones and those of laying workers; however, this was not the case in *A. mellifera* (Pirk et al. 2005). Therefore, the only way that significant differences in the

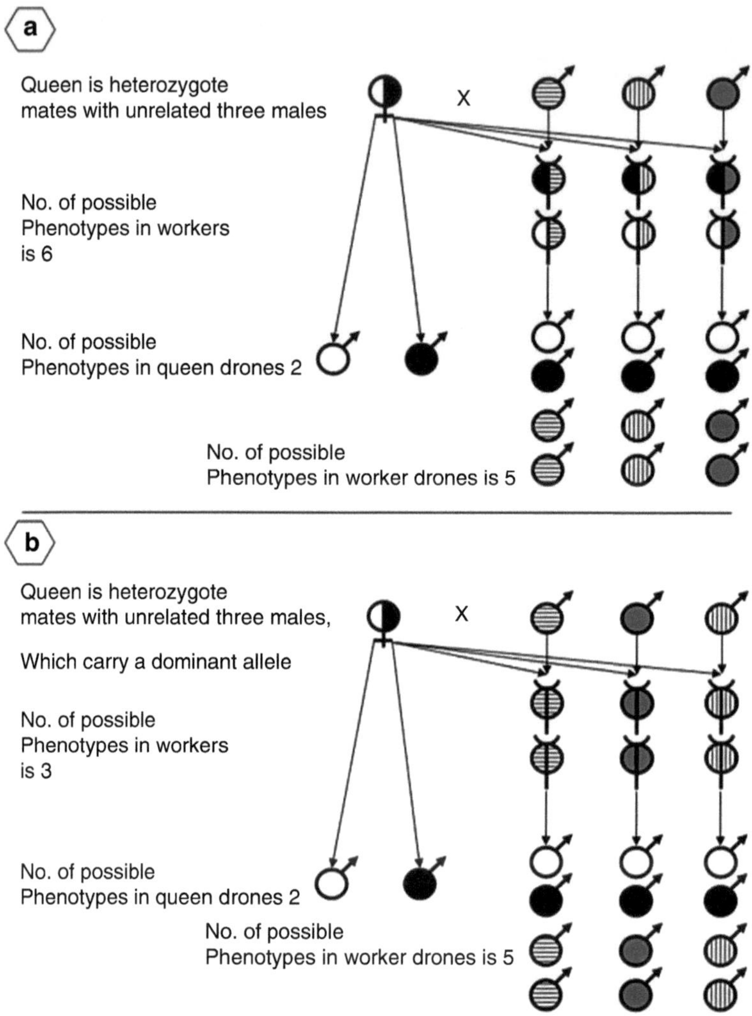

Fig. 5.3 The effect of dominant-recessive inheritance on variation. The numbers of potential permutations within the three groups (workers, drones and LW drones) are shown. (**a**) Queen mates with unrelated drones, (**b**) queen mates with unrelated drones, which carry a dominant allele

coefficients of variation could be obtained would be if workers inherited dominant alleles from the male side or if they were inbred (Maynard Smith 1989). Leading on from this we can therefore conclude that calculating the coefficient of variability from morphometric data does not serve as an index of genetic stability. Interestingly, this conclusion may have consequences for the idea of using morphometric measurements to distinguish closely related species or sub-species (cf. Chap. 1).

With regard to *A. cerana* drones, despite the lack of significant differences in the coefficients of variation between the groups, when ranking the coefficients of variation and comparing the different groups, the situation in *A. cerana* becomes comparable to the situation in *A. mellifera*. Consider two characters, TB and WW, which are highly heritable (Oldroyd and Moran 1983). Variation will also therefore be higher in worker offspring than in the workers themselves, supporting the idea of the inheritance of dominant genes in *A. cerana* (Pirk et al. 2005). Interestingly, when Pirk et al. (2005) compared coefficients of variation between *A. mellifera* and *A. cerana* for worker laid drones alone, the coefficients of variation were significantly different. This can be explained in terms of mating frequency between the species and that *A. mellifera* have on average a higher effective paternity frequency than *A. cerana* (Palmer et al. 2001).

5.4 Genetic Basis of Behaviour

5.4.1 Waggle Dance

With the evolution of group living or sociality comes the need for communication amongst group members. Communication between colony members comes in different forms, be they chemical cues – such as pheromones – or the physical guiding of nest mates to a site, as seen in stingless bees. With this in mind, honeybee species of the genus *Apis* communicate through body movement, by means of a dance, such things as where a good food source or nest site can be found (Dyer 2002; cf. Chaps. 6 and 14). Through a quick and efficient communication system a resource too big for a single individual can be exploited thereby out competing one's competition, be it at an inter- or intraspecific level. Workers from different patrilines vary in the extent to which they express behaviours that may contribute to the discovery and usage of a food resource and thereby enhance the efficiency and productivity of the whole colony (Mattila and Seeley 2007; Mattila et al. 2008). The presence therefore of multiple patrilines will increase the effectiveness of a colony's ability to communicate with one another. Indeed, Mattila et al. (2008) found that genetic variability in *A. mellifera* correlated with the precision of foraging-related communication signals, such as the waggle dance. Better communication means increased foraging rates and exploitation of patchy and ephemeral food resources. We also know that different workers within a colony respond differently to different stimuli, it therefore follows that a colony with many patrilines should have a strong stimulus to response link (Mattila et al. 2008), i.e. increased diversity leads to an increase in waggle dance behaviour as well as better overall response to the dance and interpretation thereof. Dwarf honeybees (*A. florea* and *A. andreniformis*) have been shown to be ancestral or "basal" in honeybee phylogenetics (Alexander 1991; Garnery et al. 1991; Cameron et al. 1992; Willis et al. 1992; Engel and Schultz 1997; Tanaka et al. 2001; Arias and Sheppard 2005; Raffiudin and Crozier 2007; cf. Chap. 2). It would therefore appear that ancestral communication developed in a form of honeybee such as that seen in *A. florea* (Lindauer 1956; Dyer 2002; Raffiudin and Crozier 2007), suggesting that the indicated benefits of a genetically diverse worker force found in *A. mellifera* are also present in all Asian honeybees.

5.4.2 Aggregations and Migration

We can distinguish two distinct groups of honeybees with regards to their nesting behaviour: on the one hand, the cavity-nesting species as seen in *A. cerana* and on the other hand, the open nesting species such as *A. dorsata*, *A. florea*, *A. andreniformis* and *A. laboriosa* (cf. Chap. 6). The open nesting species are often found to occur in nest aggregations, several nests within close proximity, e.g. located on the same tree. We know from previous studies (Koeniger and Koeniger 1980; Crane et al. 1993) that when absconding, *A. dorsata* can migrate up to 200 km from any single nesting site. Not only do they move up to 200 km but also manage to find their way back to the same nesting site during a return migration which occurs later in the year (Neumann et al. 2000; Paar et al. 2000). This suggests that aggregations are not random and do not represent a random sample of the population. Indeed, in the case of *A. dorsata*, there is evidence that these aggregations show less genetic variation than that between different aggregations (Paar et al. 2004a). It has been suggested that these aggregations function as distinct breeding units (Paar et al. 2004a). However, Kraus et al. (2005) showed that drone congregation areas (DCAs) are used by several subpopulations ensuring genetic flow between them and reducing the negative effect of inbreeding by increasing the effective population size. The observed closer genetic relatedness within an aggregation may be a result of limited nesting places and their home-site fidelity.

In case of the dwarf honeybee, *A. florea*, results indicate that aggregations are unrelated (Wattanachaiyingcharoen et al. 2008); however, this species absconds and migrates over much smaller distances (cf. Chap. 7). The high degree of variation in the relatedness among aggregations of honeybee colonies of the different species could be due to non-adaptive causes like a patchy distribution of nest sites or formation resulting from interactive factors not related to the environment (Czárán and Bartha 1992; Ball 1999) or to their behavioural and migratory similarities/differences with respect to their degree of home site fidelity.

5.5 Concluding Comments

Most, if not all, of the conclusions we have drawn regarding genetic characterisations have very little or no phylogenetic basis and are based on comparisons of Asian honeybees to what we see in *A. mellifera*. All this indicates that there is still a lot to be completed on Asian honeybees and that we do not have a clear understanding of what we are working on. Could this be the base for much of the variation seen? Or, could it be environmentally driven? Should we therefore not take a step back and define the "what" before the "how"?

References

Akahira Y, Sakagami SF (1959) Notes on the differences in some external characteristics between Japanese and European honeybees (Studies on the Japanese Honeybee *Apis cerana cerana* Fabr. III). Annot Zool Jpn 32:35–42

Alexander BA (1991) Phylogenetic analysis of the genus *Apis* (Hymenoptera: Apidae). Ann Entomol Soc Am 84:137–149

Arias MC, Sheppard WS (2005) Phylogenetic relationships of honey bees (Hymenoptera: Apinae; Apini) inferred from nuclear and mitochondrial DNA sequence data. Mol Phylogenet Evol 37:25–35

Baer B, Schmid-Hempel P (1999) Experimental variation in polyandry affects parasite loads and fitness in a bumble-bee. Nature 397:151–154

Ball P (1999) The self-made tapestry: pattern formation in nature. Oxford University Press, Oxford

Berg S, Koeniger N (1990) Larger drones (*Apis mellifera*) have more offspring. In: Proceedings of German Zoological Society 83rd Meeting. Frankfurt am Main, Gustav Fischer, p 614

Beye M, Hasselmann M, Fondrk R, Page R, Omholt S (2003) The gene csd is the primary signal for sexual development in the honeybee and encodes an SR-Type protein cell. Cell 114:419–429

Cameron SA, Derr JN, Austin AD, Woolley JB, Wharton RA (1992) The application of nucleotide sequence data to phylogeny of the Hymenoptera: a review. J Hymenopt Res 1:63–79

Crane E, van Luyen V, Mulder V, Ta TC (1993) Traditional management system for *Apis dorsata* in submerged forests in southern Vietnam and central Kalimantan. Bee World 74:27–40

Crozier RH, Pamilo P (1996) Evolution of social insect colonies: sex allocation and kin selection. Oxford University Press, New York

Czárán T, Bartha S (1992) Spatiotemporal dynamic models of plant populations and communities. Trends Ecol Evol 7:38–42

Duay P, De Jong D, Engels W (2002) Decreased flight performance and sperm production in drones of the honey bee (*Apis mellifera* L.) slightly infested by *Varroa destructor* mites during pupal development. Genet Mol Res 1:227–232

Dyer FC (2002) The biology of the dance language. Annu Rev Entomol 47:917–949

Elbassiouny AMM (1992) Comparative studies on the carniolan honey bee race and its crosses. Thesis, Ain Shams University, Cairo

Engel MS, Schultz TR (1997) Phylogeny and behavior in honey bees (Hymenoptera: Apidae). Ann Entomol Soc Am 90:43–53

Fuchs S, Moritz RFA (1998) Evolution of extreme polyandry in the honeybee. Behav Ecol Sociobiol 9:269–275

Garnery L, Vautrin D, Cornuet JM, Solignac M (1991) Phylogenetic relationships in the genus *Apis* inferred from mitochondrial DNA sequence data. Apidologie 22:87–92

Hamilton WD (1987) Kinship, recognition, disease, and intelligence: constraints of social evolution. In: Ito Y, Brown JL, Kikkawa J (eds) Animal societies: theory and facts. Japanese Scientific Society, Tokyo, pp 81–102

Jones JC, Myerscough MR, Graham S, Oldroyd BP (2004) Honeybee nest thermoregulation: diversity promotes stability. Science 305:402–404

Katzerke A, Neumann P, Pirk CWW, Bliss P, Moritz RFA (2006) Seasonal nestmate recognition in the ant *Formica exsecta*. Behav Ecol Sociobiol 61:143–150

Koeniger N, Koeniger G (1980) Observations and experiments on migration and dance communication of *Apis dorsata* in Sri Lanka. J Apic Res 19:21–34.

Koeniger N, Koeniger G (1991) An evolutionary approach to mating behavior and drone copulatory organs in *Apis*. Apidologie 22:581–590

Koeniger N, Koeniger G (2000) Reproductive isolation among species of the genus *Apis*. Apidologie 31:313–339

Koeniger N, Koeniger G, Wongsiri S (1989) Mating and sperm transfer in *Apis florea*. Apidologie 20:413–418

Koeniger G, Koeniger N, Mardan M, Punchihewa RWK, Otis GW (1990) Numbers of spermatozoa in queens and drones indicate multiple mating of queens in *Apis andreniformis* and *Apis dorsata*. Apidologie 21:281–286

Koeniger G, Koeniger N, Tingek S (1996) A scientific note on weight and sperm numbers of drones of *Apis nuluensis* Tingek, Koeniger and Koeniger, 1996. Apidologie 27:427–428

Kolmes SA, Winston ML, Fergusson LA (1989) The division of labour among worker honeybees (Hymenoptera: Apidae): the effects of multiple patrilines. J Kans Entomol Soc 62:80–95

Kraus F, Koeniger N, Tingek S, Moritz R (2005) Temporal genetic structure of a drone congregation area of the giant Asian honeybee (*Apis dorsata*). Naturwissenschaften 92:578–581

Lindauer M (1956) Über die Verständigung bei indischen Bienen. Z Vgl Physiol 38:521–557 [in German]

Mattila HR, Seeley TD (2007) Genetic diversity in honey bee colonies enhances productivity and fitness. Science 317:362–364

Mattila HR, Burke KM, Seeley TD (2008) Genetic diversity within honeybee colonies increases signal production by waggle-dance foragers. Proc R Soc B 275:809–816

Moritz RFA (1981) Der Einfluss der Inzucht auf die Fitness der Drohnen von *Apis mellifera carnica*. Apidologie 12:41–55

Moritz RFA (1985) The effects of multiple mating on the worker-queen conflict in *Apis mellifera* L. Behav Ecol Sociobiol 16:375–377

Moritz RFA (1995) High degree of polyandry in *Apis dorsata* detected by DNA microsatellite variability. Behav Ecol Sociobiol 37:357–363

Neumann P, Koeniger N, Koeniger G, Tingek S, Kryger P, Moritz RFA (2000) Home-site fidelity in migratory honeybees. Nature 406:474–475

Oldroyd BP, Moran C (1983) Heritability of worker characters in the honey bee. Aust J Biol Sci 36:323–332

Oldroyd BP, Smolenski AJ, Cornuet JM, Wongsiri S, Estoup A, Rinderer TE, Crozier RH (1995) Levels of polyandry and intracolonial genetic relationships in *Apis florea*. Behav Ecol Sociobiol 37:329–335

Paar J, Oldroyd BP, Kastberger G (2000) Giant honeybees return to their nest sites. Nature 406:475

Paar J, Oldroyd BP, Huettinger E, Kastberger G (2004a) Genetic structure of an *Apis dorsata* population: the significance of migration and colony aggregation. J Hered 95:119–126

Paar J, Oldroyd BP, Huettinger E, Kastberger G (2004b) Levels of polyandry in *Apis laboriosa* from Nepal. Insectes Soc 51:212–214

Page RE (1980) The evolution of multiple mating behavior by honey bee queens (*Apis mellifera*). Genetics 96:263–273

Page RE, Metcalf RA (1982) Multiple mating, sperm utilization, and social evolution. Am Nat 119:263–281

Page RE, Robinson GE, Fondrk MK, Nasr ME (1995) Effects of worker genotype diversity on honey bee colony development and behaviour (*Apis mellifera* L.). Behav Ecol Sociobiol 36:387–396

Palmer K, Oldroyd BP (2001) Mating frequency in *Apis florea* revisited (Hymenoptera: Apidae). Insectes Soc 48:40–43

Palmer KA, Oldroyd BP, Franck P, Hadisoesilo S (2001) Very high paternity frequency in *Apis nigrocincta*. Insectes Soc 48:327–332

Pamilo P (1993) Polyandry and allele frequency differences between the sexes in the ant, *Formica aquilonia*. Heredity 70:472–480

Pirk CWW, Hepburn HR, Radloff SE (2005) Coefficients of genetic variation in honeybees: Akahira and Sakagami (1959) revisited. Honeybee Sci 26:153–158 [in Japanese]

Queller DC (1993) Genetic relatedness and its components in polygynous colonies of social insects. In: Keller L (ed) Queen number and sociality in insects. Oxford University Press, Oxford, pp 132–152

Raffiudin R, Crozier RH (2007) Phylogenetic analysis of honey bee behavioral evolution. Mol Phylogenet Evol 43:543–552

Ratnieks FLW (1990) The evolution of polyandry by queens in the social Hymenoptera: the significance of the timing of removal of diploid males. Behav Ecol Sociobiol 26:343–348

Richard F-J, Tarpy DR, Grozinger CM (2007) Effects of insemination quantity on honey bee queen physiology. PLoS ONE 2(10):e980. doi:10.1371/journal.pone.0000980

Ridley M (1988) Mating frequency and fecundity of insects. Biol Rev 63:509–549

Rinderer TE, Collins AM, Pesante D (1985) A comparison of Africanized and European drones: weights, mucus gland and seminal vesicle weights, and counts of spermatozoa. Apidologie 16:407–412

Rinderer TE, de Guzman LI, Lancaster VA, Delatte GT, Stelzer JA (1999) Varroa in the mating yard: the effect of *Varroa jacobsoni* and Apistan on drone honey bees. Am Bee J 139:134–139

Robinson GE, Page RE (1989a) Genetic basis for division of labour in an insect society. In: Breed MD, Page RE (eds) The genetics of social evolution. Westview, Boulder, pp 61–80

Robinson GE, Page RE (1989b) Genetic determination of nectar foraging, pollen foraging, and nest-site scouting in honey bee colonies. Behav Ecol Sociobiol 24:317–323

Rothenbuhler WC, Kulincevic JM, Kerr WE (1968) Bee genetics. Annu Rev Entomol 2:413–438

Schlüns H, Schlüns EA, van Praagh J, Moritz RFA (2003) Sperm numbers in drone honeybees (*Apis mellifera*) depend on body size. Apidologie 34:577–584

Schlüns H, Moritz RFA, Neumann P, Kryger P, Koeniger G (2005a) Multiple nuptial flights, sperm transfer and the evolution of extreme polyandry in honeybee queens. Anim Behav 70:125–131

Schmid-Hempel P (1998) Parasites in social insects. Princeton University Press, Princeton, NJ

Seeley TD, Tarpy DR (2007) Queen promiscuity lowers disease within honeybee colonies. Proc R Soc B 274:67–72

Sherman PW, Seely TD, Reeve HK (1988) Parasites, pathogens, and polyandry in social Hymenoptera. Am Nat 131:602–610

Smith JM (1989) Evolutionary genetics. Oxford University Press, Oxford

Tan NQ, Mardan M, Thai PH, Chinh PH (1999) Observations on multiple mating flights of *Apis dorsata* queens. Apidologie 30:339–346

Tanaka H, Roubik DW, Kato M, Liew F, Gunsalem G (2001) Phylogenetic position of *Apis nuluensis* of northern Borneo and phylogeography of *A. cerana* as inferred from mitochondrial DNA sequences. Insectes Soc 48:44–51

Tarpy DR (2003) Genetic diversity within honeybee colonies prevents severe infections and promotes colony growth. Proc R Soc B 270(1510):99–104

Tarpy DR, Page RE (2000) No behavioral control over mating frequency in queen honey bees (*Apis mellifera* L.): implications for the evolution of extreme polyandry. Am Nat 155:820–827

Tarpy DR, Page RE (2002) Sex determination and the evolution of polyandry in honey bees (*Apis mellifera*). Behav Ecol Sociobiol 52:143–150

Tarpy DR, Seeley TD (2006) Lower disease infections in honeybee (*Apis mellifera*) colonies headed by polyandrous vs monandrous queens. Naturwissenschaften 93:195–199

Tarpy DR, Nielsen R, Nielsen DI (2004) A scientific note on the revised estimates of effective paternity frequency in *Apis*. Insectes Soc 51:203–204

Wattanachaiyingcharoen W, Oldroyd BP, Wongsiri S, Palmer K, Paar J (2003) A scientific note on the mating frequency of *Apis dorsata* Fabricius. Apidologie 34:85–86

Wattanachaiyingcharoen W, Wongsiri S, Oldroyd BP (2008) Aggregations of unrelated *Apis florea* colonies. Apidologie 39:531–536

Willis LG, Winston ML, Honda BM (1992) Phylogenetic relationships in the honeybee (genus *Apis*) as determined by the sequence of the cytochrome oxidase II region of mitochondrial DNA. Mol Phylogenet Evol 1:169–178

Wilson EO (1971) The insect societies. Harvard University Press, Cambridge

Woyke J (1963) What happens to diploid drone larvae in a honeybee colony. J Apic Res 2:73–75

Woyke J (1965) Do honeybees eat diploid drone larvae because they are in worker cells? J Apic Res 15:105–123

Woyke J (1975) Natural and instrumental insemination of *Apis cerana indica* in India. J Apic Res 14:153–159

Woyke J (1978) Biology of reproduction and genetics of the honeybee. Final Technical Report for USDA, Bee Culture Division, Warsaw Agricultural University, Warsaw, p 381

Chapter 6
Biology of Nesting

Mananya Phiancharoen, Orawan Duangphakdee, and H.R. Hepburn

6.1 Introduction

Nesting is critical to the survival, homeostasis and stability of honeybee colonies and provides an arena in which colony growth unfolds in the annual colony cycle of swarming, reproduction and migration (cf. Chaps. 7 and 8). The Asian honeybees comprise three groups with respect to nesting: the dwarf honeybees, *Apis andreniformis* and *Apis florea*, are open-nesting species; the medium-sized species, *A. cerana*, *A. koschevnikovi*, *A. nigrocincta* and *A. nuluensis*, are cavity-nesting bees; and the giant honeybees, *Apis dorsata* and *Apis laboriosa*, are also open-nesting species. The ways in which the nests of these species are attached to a substrate further differentiate them (1) there are no horizontal surfaces in the nests of the medium-sized and the giant honeybees, so communication using the dance language occurs in the vertical mode for them; (2) while in the dwarf species, dance language communication occurs in the horizontal mode (cf. Chap. 14). Both open-air nesting and cavity-dwelling nesting sites further constrain the honeybees in terms of colony defence (cf. Chaps. 17 and 18) and micro-environmental homeostasis (cf. Chap. 19). Of equal importance is the nature of the nesting sites, mode of comb construction and the physico-chemical properties of the actual construction materials.

M. Phiancharoen and O. Duangphakdee
Ratchaburi Campus, King Mongkut's University of Technology, Thonburi, Bangkok 10140, Thailand
e-mail: manaya.phi@kmutt.ac.th; orawan.dua@kmutt.ac.th

H.R. Hepburn
Department of Zoology and Entomology, Rhodes University, Grahamstown 6140, South Africa
e-mail: r.hepburn@ru.ac.za

H.R. Hepburn and S.E. Radloff (eds.), *Honeybees of Asia*,
DOI 10.1007/978-3-642-16422-4_6,

6.2 Nesting Sites

6.2.1 The Dwarf Honeybees

The nests of *A. florea* are often found in disturbed areas, compared to those of *A. andreniformis*, and occur in urban areas and those with intensive agricultural activity as well as in savanna (Free 1981; Booncham et al. 1995). The nests of *A. florea* are usually more exposed to sunlight and often have one surface of the comb exposed to direct sunlight for several hours a day. *A. florea* are more likely to nest in diverse places such as high up in tall trees and on the walls and eaves of buildings in southeast Asia, while in arid western Asia, they commonly nest in caves and buildings as well (Whitcombe 1984; Mossadegh 1990). *A. andreniformis* nest throughout southeast Asia and are most commonly observed in and near undisturbed, mixed deciduous and evergreen forests. Their nesting habitat is usually a dark and shady place (20–35% sun), well hidden and widely spaced.

A. florea and *A. andreniformis* usually build a single-comb nest in shrubs, bushes and small trees, but double-comb nests have occasionally been reported for *A. florea* (Douglas 1886; Akratanakul 1977; Free 1981; Whitcombe 1984). The comb is almost always parallel with the main branch to which it is attached. The brood area is situated below the branch while honey is stored in cells around and above the branch. Sometimes there may be two or three twigs passing through the comb of an *A. florea* nest, particularly where the twigs are rather thin. However, this has not been reported for *A. andreniformis*. The nests of *A. florea* are typically about 0.5–10 m above ground, but in towns and cities they are commonly found up to 15 m (Wongsiri et al. 1996). *A. florea* are very adaptable in nesting and seem to find suitable sites under extremely varied conditions.

A. florea have a strong tendency to orient the principal axis of the comb facing north in Thailand (Akratanakul 1977), but to the southwest in Oman (Whitcombe 1984). Studies in Kanchanaburi, northwestern Thailand, have shown that *A. florea* have a wide range of nesting habitats and food sources, estimated from the number of tree species used for nesting. The diameters of nesting branches are also significantly greater in *A. florea*: 1.23 $\pm$ 0.55 cm diameter than in *A. andreniformis*: 0.89 $\pm$ 0.31 cm (Wongvilas et al. 2009), and this is supported by observations that the mature nests of the former are significantly greater than those of the latter (Rinderer et al. 1996).

Neither *A. florea* nor *A. andreniformis* form nest aggregations in the same tree or shrub (Akratanakul 1977). However, they can achieve a relatively high concentration of nests/ha (Wattanachaiyingcharoen et al. 2008; Hepburn et al. 2010). Nesting density of *A. andreniformis* remains unreported; however, casual observations in Sabah State, Borneo (Duangphakdee personal observation) and Sumatera, Indonesia (Hepburn personal observation), indicate that they are far more diffusely distributed. However, Rinderer et al. (2002) reported that colonies of *A. andreniformis* and *A. florea* have a tendency to be located near nests of their own species in southeastern Thailand. *A. andreniformis* and *A. florea* select similar nest sites, but the spatial correlations of these sites were significantly negative, indicating that colonies may avoid areas containing nests of the other species. It is rare to find the nests of *A. florea* in the same tree with other honeybee species; but, curiously, a nest of *A. florea* was once seen in the same tree with one of *A. dorsata*, the former was about 6 m from the ground and the latter 13 m (Phiancharoen personal observation). However, *A. florea* build nests aggregated near each other, forming spatial clumps (Rinderer et al. 2002; Wattanachaiyingcharoen et al. 2008).

6.2.2 The Cavity-Nesting Honeybees

The cavity-nesting bees show preferences for nesting sites, and these vary within races and among *Apis* species. Among the sympatric cavity-nesting species of Indonesia, different species nest in distinctly different habitats. *A. cerana* mainly nest in agricultural or disturbed areas, while *A. nigrocincta* nest more deeply in forests (Hadisoesilo 1997). Similarly, *A. koschevnikovi* typically occur in primaeval forests, while *A. cerana* occur mostly in secondary forests, agricultural and urban areas in Peninsular Malaysia (Otis and Hadisoesilo 1996). It is not evident that these species specifically avoid aggregated nest sites, but some reports are suggestive to the contrary (Hadisoesilo 1997; Bakker 1999). There is no published information on nesting by *A. nigrocincta*, *A. koschevnikovi* and *A. nuluensis*.

The nest cavity volume of *A. cerana* is usually about 10–15 l, but ranges from 4.5 to 97 l (Inoue et al. 1990; Oldroyd and Wongsiri 2006). The nest entrances are usually about 1–2 m above ground, but they seem to have no real preference for height because nests can also be many metres above ground or in cavities within the ground. Entrance sizes range between 2 and 100 cm^2 (Seeley et al. 1982; Inoue et al. 1990; Oldroyd and Wongsiri 2006). Such studies are extremely few, but Bakker (1999) reported that *A. nigrocincta* may be less specific in its choice of nest sites.

Nest density is probably related to topographical variations and the availability of profitable forage. The nest density for *A. mellifera* ranges from 0.5 to 7.8 nests per km^2, whereas nest density in tropical bees is greater. For example, Inoue et al. (1990) measured nest density of *A. cerana* in Padang, Sumatra and found 22 nests/km^2 with a mean distance of about 100 m between nests. Aggregations of nests are not well known in cavity-nesting bees; nevertheless, Rinderer et al. (2002) suggested that there is a tendency to form aggregated nests in *A. cerana*.

6.2.3 The Giant Honeybees

Unlike other *Apis*, the giant bees, *A. dorsata* and *A. laboriosa*, build exposed and easily visible nests. *A. dorsata* build nests in inaccessible places and stand out in their surrounds like vertical rock faces (hence "rock bee" in India) in cuts along hill profiles, tall man-made structures such as water towers and buildings and the higher branches of remarkably tall trees. Unusual nests only 0.1 m above ground have also been observed (Phiancharoen personal observation). Because *A. dorsata* build their nests on remarkably tall trees, *Kompassia* being one such tree that regularly serves as a nesting site for *A. dorsata*, the trees have acquired a local Thai name "ton phung" or "yuan phung" which means "bee tree".

A. dorsata build their combs in a north–south direction, minimising exposure to strong wind and sunlight (Deodikar et al. 1977; Reddy and Reddy 1993). *A. laboriosa* have never been reported to nest on the branches of trees (Roubik et al. 1985; Underwood 1986), but there is simply no information as to whether they are able to do so. Colonies of *A. dorsata* re-use preferred trees after an absence of several months (Paar et al. 2000; Neumann et al. 2000). Although odour and tactile or chemical cues associated with the material of old combs seem more likely to guide the final choice, previous work has suggested that visual information is used by migratory colonies to relocate places where nesting has proven successful (Paar et al. 2000; Neumann et al. 2000).

Giant honeybees vary quite considerably in their nesting habits and relative nest densities. *A. dorsata* and *A. laboriosa* are very gregarious species and 20–30 nests in a single tree are fairly common for the former as are cliff overhangs for the latter. Reports include a range of 67–256 colonies per tree for *A. dorsata* (Butani 1950; Lindauer 1957; Singh 1962; Deodikar et al. 1977). Oddly, Morse and Laigo (1969) found almost no aggregations in the Philippines. In an extensive survey of *A. laboriosa* at 54 cliff sites in western Nepal, Joshi et al. (2004) reported an average aggregation of six nests per cliff with a range of 1–37.

A. dorsata colonies nest gregariously; however, placing empty combs in previously occupied trees, or on nearby trees of the same species, did not attract more swarms. The same number of colonies that left trees returned to previously occupied trees (Liu et al. 2007). Although it is believed that few individuals probably live long enough to make a return journey to their original site, some colonies nonetheless return exactly to their former trees (Paar et al. 2000; Neumann et al. 2000; Liu et al. 2007). Because the longevity of workers has not been determined under field conditions, it may prove that there really is nothing "magical" about returning to their original nests following migration in *A. dorsata*.

A. laboriosa are the largest species of *Apis* and are distributed along the Himalayas from Nepal to Vietnam (cf. Chap. 3). It builds exposed nests under rock ledges in deep, vertical river valleys, most commonly at 1,200–3,500 m (Roubik et al. 1985; Underwood 1986) and seems confined to areas higher than 2,500 m in central and western Himalayas. Those nests at 1,200–2,000 m could possibly be occupied throughout the year, but nest sites above 2,800 m are occupied only 4 months in summer (Underwood 1990). By late November, dropping temperatures make even the lower altitude cliff sites unsuitable for colony survival, and the colonies must migrate to the forest near the ground where they remain as combless winter clusters until late January (Underwood 1990). Those that nest below 1,200 m are reported not to migrate (Woyke et al. 2001).

6.3 Nest Construction

6.3.1 Comb Structure

6.3.1.1 The Dwarf Honeybees

Both *A. florea* and *A. andreniformis* usually have a single exposed comb, typically situated on a single branch usually in a shady location (Figs. 6.1 and 6.2). Akratanakul (1977) stated that when the bees start to build, they encircle twigs with hexagons; but it is geometrically impossible to encircle a regular cylinder with hexagons, and close examination of such combs reveals a combination of various other polygons (Hepburn personal observation). The size of the mature nest from crown to lower edge of the comb is about 16.2 cm in length in *A. florea* and 13.7 cm in *A. andreniformis*, and the width is about 16.9 cm in the former and 12.2 cm in the latter. The mature nests of *A. florea* vary in size and several colonies 35–40 cm in length and 25–35 cm in width have been reported (Akratanakul 1977). Mature combs of *A. florea* are always larger than those of *A. andreniformis* (Rinderer et al. 1996), which build a single, comparatively small comb from 0.7 to 0.9 m in surface area on branches or twigs of shrubs about 3–4 m above ground (Seeley 1985; Wongsiri et al. 1996; Koeniger and Koeniger 2000).

A honey storage area extends above and around the supporting branch with the brood area below. In the nests of both *A. florea* and *A. andreniformis*, pollen is stored below the honey, just above the brood. Drone cells are found at the lower margin of the nest below worker brood cells, and swarm queen cells are found protruding vertically from the bottom of the brood nest. The combs of both species have brood and pollen storage areas, with worker cells from one side meeting worker cells from the other side and drone cells meeting drone cells (Wongsiri et al. 1996). The recurved honey storage cells on the crown above the twig or branch are

Fig. 6.1 Mature comb of *Apis andreniformis* with drone and queen cells Photo: S. Wongvilas

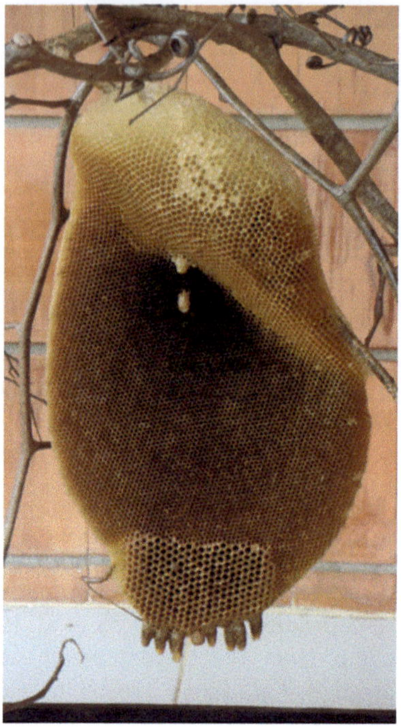

Fig. 6.2 Single-comb nest of *Apis florea* Photo: M. Phiancharoen

much longer than the brood cells. The comb structures of the honey storage areas of *A. andreniformis* and *A. florea* are fundamentally different in that the comb of *A. andreniformis* has a midrib both above and below the twig; *A. florea* have a midrib only in the brood area below the twig (Rinderer et al. 1996).

Drone cells of *A. florea* are notably larger than those of workers and may extend to 1/3 of the total brood area. Drone cells, which have been constructed under queenless conditions, are noticeably irregular. *A. florea* emergency queen cells are appressed to the brood area of the comb, while swarm queen cells skirt the bottom of the comb as in *A. mellifera* (Sakagami and Yoshikawa 1973). The queen cells of *A. andreniformis* are 1.24 ± 0.26 cm deep and 0.54 ± 0.08 cm in diameter, while worker cells are 0.76 ± 0.14 cm deep and 0.27 ± 0.11 cm in diameter. The queen cells of *A. florea* are 1.41 ± 0.15 cm deep and 0.47 ± 0.09 cm in diameter and worker cells 0.93 ± 0.07 cm deep with 0.29 ± 0.15 cm diameter. The depth and width of worker cells and the width of drone cells are all significantly larger in nests of *A. florea* than those of *A. andreniformis*. The depth of drone cells is similar for both *A. florea* and *A. andreniformis* (Rinderer et al. 1996; Wongsiri et al. 1996).

Rinderer et al. (1996) compared the proportional relationships between the nests of *A. andreniformis* and *A. florea* and showed that they are similar, but that *A. andreniformis* is about 13% smaller. Branch thickness is estimated at about 1 cm greater in diameter for *A. florea* than in *A. andreniformis*. Wongsiri et al. (1997) reported that the height and width of the *A. andreniformis* brood area are about 25% and 16% smaller, respectively, than that of the *A. florea* brood area. The honey storage area of an *A. andreniformis* nest is narrower and not as high or wide as in an *A. florea* nest. The depth and width of worker and drone cells are all significantly larger in *A. florea* than *A. andreniformis*. *A. andreniformis* combs with drone cells will soon produce queen cells, and this is taken as the criterion of a "mature nest" (Rinderer et al. 1996).

Other distinctive features of lower combs are geometric dislocations causing irregularities at the transition from worker brood cells where they are enlarged to accommodate drones cells. Both *A. florea* and *A. andreniformis* use a sticky resin ("sticky band") smeared around the supporting branch on both sides of the comb, which is a very effective ant repellent (Duangphakdee et al. 2005a). The sticky bands of *A. florea* appear thicker than those of *A. andreniformis* (Duangphakdee personal observation). The worker cells of *A. florea* also vary geographically with latitude: S India averages 1,560 dm^2, United Provinces 1,240 dm^2 and Punjab 1,190 dm^2 (Mehta 1948; Muttoo 1956; Thakar and Tonapi 1962).

6.3.1.2 The Cavity-Nesting Honeybees

The cavity honeybees, *A. cerana*, *A. koschevnikovi*, *A. nuluensis* and *A. nigrocincta*, build multiple comb nests usually in dark cavities such as the hollows of the trunks of live or dead trees, underneath roofs, water jars, and caves (Fig. 6.3). Normally, the nesting sites of cavity-dwelling honeybees are close (4–5 m) to the ground. This habit facilitates keeping them in specially constructed containers such as wooden boxes, clay pots, logs, and wall openings. The several combs are built in parallel and a uniform distance, the bee space, is maintained between them. The brood combs consist of cells of two sizes: smaller ones for the worker brood and larger ones for drones. The queen cells are built on the lower edges of the combs (Akratanakul 1987). Honey is stored in the upper part of the combs, but also in the outer combs, adjacent to the cavity walls. The worker cells are larger than the dwarf honeybees and smaller than giant honeybees. The size of worker cells is 4.3–4.8 mm in *A. cerana* and 4.5 mm in *A. koschevnikovi* and *A. nigrocincta* (Tingek et al. 1996), but there is no information for *A. nuluensis*. The depths of

Fig. 6.3 A colony of *Apis cerana* in a clay pot
Photo: M. Phiancharoen

the cells are 1.01 cm in *A. cerana* (Inoue et al. 1990) and 1.02 cm in *A. koschevnikovi* (Oldroyd and Wongsiri 2006). The drone cells of *A. cerana* are 4.7–5.3 mm in diameter, which are larger than worker cells as in most other *Apis* species. The drone cells of the cavity-nesting bees protrude above the surface of worker cells. The brood cells of drones have a central pore in *A. cerana* and *A. koschevnikovi* which is absent from *A. nigrocincta* (Hadisoesilo and Otis 1998), but there is no information for *A. nuluensis*. Worker cell size in *A. cerana* varies geographically and is larger in colder regions: Japan: 4.7–4.8 mm, Kashmir: 4.9 mm, High Himalaya: 4.9 mm, sub-Himalaya: 4.7 mm, central India: 4.5 mm, southern India: 4.3 mm and Philippines: 3.6–4.0 mm (Crane 1993).

6.3.1.3 The Giant Honeybees

The giant honeybees build a single open-nesting comb, hanging underneath huge tree branches of tall trees, high cliffs, buildings or water tanks (Fig. 6.4). Ramachandran (1939) stated that all cells in the combs of *A. dorsata* are of uniform size, without any distinction between worker and drone cells, and this has subsequently been confirmed many times (Singh 1947; Schwarz 1948; Butler 1954; Muttoo 1956; Lindauer 1957). The combs of *A. dorsata* are more or less semicircular. Nectar and honey are usually stored in the highest part of the comb. Pollen is stored next to the honey, and the rest of the comb contains brood. The average width of an *A. dorsata* comb is 92.4 ± 26.2 cm (n = 158), ranging from 43 to 162 cm and height 47.8 ± 11.4 cm (n = 146), ranging from 23 to 90 cm. Honey storage is not distributed equally in the comb. About 3–4 weeks after inception of the nest, a colony might store 4.09 ± 2.56 kg (n = 152) but can reach a peak of 15.7 kg. The thickness of the comb containing honey is highly variable, ranging from the normal thickness of worker brood comb 3 cm to 19 cm in areas containing large honey stores (Tan 2007). Worker cells of *A. dorsata* are 4.50–4.75 cells per 2.5 cm and 10 cells measured in a line across their parallel sides cover 5.64–5.35 cm (Thakar and Tonapi 1961).

The mean depth of the worker cells on either side of the median septum varies from 16 to 19 mm, that of drone cells 19.0 ± 0.4 mm (n = 90). The average volume of a worker brood cell is about 4 cc (Thakar and Tonapi 1961; Tan 2007).

Fig. 6.4 A mature comb of *Apis dorsata* Photo: M. Phiancharoen

Fig. 6.5 Dots of wax on branch above comb of *Apis dorsata* form "micro-awnings" that deflect rain drops Photo: M. Phiancharoen

The drone cells have raised cappings, while the worker brood cells are fairly uniformly aligned with undifferentiated cappings. But drone cells in *A. dorsata* do not have dome-like capping as in *A. mellifera* or *A. florea*. Drone cells in *A. cerana*, *A. mellifera* and *A. florea* are differentiated in both height and cross-section, whereas the drone cells of *A. dorsata* are differentiated by height alone.

Viswanathan (1950) first observed differentiated queen cells in *A. dorsata*. Thakar and Tonapi (1961) noted that queen cells appear as pendant projections, two-thirds from the base within the comb body with only one third projecting outside the comb edge. Their average depth is 2.5 cm, the diameter of the mouth is 0.6 cm and the volume is approximately 10 cc (Thakar and Tonapi 1961). Tan (2007) also reported that all natural queen cells were at the lower edge of the comb. There is a strong positive relationship between the size of the colony and the number of the queen cells reared. A fully developed comb of an *A. dorsata* colony is 120.9 $\pm$ 17.6 cm ($n = 32$) in width, 56.6 $\pm$ 5.3 cm ($n = 32$) in height and may have 9.7 $\pm$ 2.4 ($n = 32$) queen cells. There are two types of queen cells: long, narrow ones and short, broad queen cells. For the long and narrow queen cells, the maximum inner depth and width are 28.1 $\pm$ 0.7 mm ($n = 16$) and 8.5 $\pm$ 0.4 mm ($n = 14$), respectively, resulting in a height-to-width ratio of up to 3.3. For the short and broad queen cells, the maximum inner depth and width were 22.6 $\pm$ 1.6 mm ($n = 13$) and 12.4 $\pm$ 1.4 mm ($n = 13$) (Tan 2007). The unique nests of *A. dorsata* have small drops of wax on the attachment branch just above the comb which function as "micro-awnings" to deflect raindrops (Koeniger and Phiancharoen personal observation; cf. Fig. 6.5).

6.3.1.4 Comb Construction in Mixed-Species Colonies

Yang et al. (2010) studied comb construction in mixed-species colonies of *A. cerana* and *A. mellifera* headed by heterospecific queens (cf. Chap. 19). Two types of cell size foundation, *A. cerana* and *A. mellifera*, were put in the mixed-species colonies and in pure *A. cerana* and *A. mellifera* colonies. Comb building on the two types of foundation differed between the groups. The mixed-species colonies did not discriminate between the wax types, but the *A. cerana* cell size foundation was modified during comb building. In pure *A. mellifera* colonies, the *A. cerana* cell size was modified, and the queens were reluctant to lay eggs in such cells. In pure *A. cerana* colonies, the *A. mellifera* cell size was extended without any modification, but these cells were used either for rearing drone brood or for food storage. The principal elements of comb building behaviour common to both species are considered as conserved after speciation. The use of mixed colonies is now established as an important probe to explain the social mechanisms driving comb construction and to illuminate other behavioural traits that evolved prior to speciation (cf. Chap. 19).

6.3.2 Unusual Nests and Man-Made Nesting Sites

Deforestation for agriculture and the infrastructure of towns has led to serious declines in honeybees in several regions, a trend unlikely to decline in future. One interesting consideration is the extent to which bees can utilise man-made structures. *A. cerana* build nests on exposed sites with reduced light intensity such as inside houses, in cupboards and cabinets and roof corners. Unusual nests include those (1) built without a dancing platform; (2) some attached vertically instead of horizontally; (3) double combs in single-comb bees (*A. florea*: Douglas 1886; Akratanakul 1977; Free 1981; *A. dorsata*: Morse and Laigo 1969); (4) nests of queenless colonies and (5) in recessed window frames like *A. dorsata* often do.

Rafter beekeeping is a practical way of sustainable beekeeping with *A. dorsata*. The beekeeper uses artificial supports as a nest site for *A. dorsata*, called a "rafter" because it slopes like the rafters of a house. Rafter beekeeping is well known in Vietnam, Indonesia, Malaysia and little Andaman in India. The earliest report of rafter beekeeping comes from the area of the Kapuas River, Kalimantan, Indonesia (van Lijnden and Groll 1851). A rafter made from odourless hard wood about 1.6–2.3 m length is supported on poles at each end and set at a 25–30° angle similar to that of natural colonies. The rafters are left in position permanently, and incoming swarms are likely to settle and build combs there. This is done in November–December and, 2 or 3 months later, beekeepers harvest the honey and wax (Mol 1933).

6.4 Building Materials

6.4.1 Beeswax: Chemical and Physical Properties

The chemical composition and physical properties of beeswaxes have been pursued for centuries, and the earlier works have been documented (Grün and Halden 1929). The first studies of Asian beeswaxes appeared just a century ago (Hooper 1904;

Bellier 1906; Büchner 1906; Ueno 1915; Roberts and Islip 1922; Ikuta 1931, 1934) and variously record the physical constants (specific gravity, melting point, acid and saponification values, etc.) for *A. cerana*, *A. dorsata*, *A. florea* and *A. mellifera*. As knowledge of the hydrocarbon, alcohol and acid fractions of beeswaxes developed, two points important to honeybee biology emerged. First, Phadke et al. (1969) re-examined the physical constants for the beeswaxes of *A. cerana*, *A. dorsata*, *A. florea* and *A. mellifera* and showed each to be extremely homogenous as demonstrated by very small standard deviations in the values of the samples measured. Shortly after, Narayana (1970) and Phadke et al. (1971) both saw that carbon chain length was on average shorter in the three Asian beeswaxes than in *A. mellifera* wax and that this would account for the lower melting points of the Asian beeswaxes.

Progress in wax chemistry advanced greatly with the then newly available techniques of both thin-layer and gas–liquid methods of chromatography. Tulloch (1980) summarised earlier works on *A. cerana*, *A. dorsata*, *A. florea* and *A. mellifera* and concluded that the three Asian beeswaxes more closely resemble each other than any of the *A. mellifera* wax, a point subsequently confirmed by Brand-Garnys and Sprenger (1988). With even more sophisticated gas-chromatographic methods, Aichholz and Lorbeer (1999, 2000) and Aichholz et al. (2000) examined the waxes of *A. andreniformis*, *A. cerana*, *A. dorsata*, *A. florea*, *A. laboriosa* and *A. mellifera* and showed that they are complex mixtures of homologous neutral lipids containing a range of 20–64 carbon length molecules.

Aichholz et al. (2000) investigated the waxes with high temperature gas chromatography and obtained a characteristic elution pattern for the waxes of each honeybee species, confirming the earlier analyses of Tulloch (1980) and Brand-Garnys and Sprenger (1988). Following Tulloch (1980), Aichholz et al. (2000) defined the major compound families as those exceeding 5% of the total, so that alkanes, alkenes, free fatty acids, monoesters, diesters and hydroxy-monoesters are the major compound families and fatty alcohols and hydroxydiesters minor constituents (Table 6.1). There are notable species-specific differences in the waxes

Table 6.1 The major compound families of the comb waxes of *A. andreniformis*, *A. florea*, *A. cerana*, *A. mellifera*, *A. dorsata* and *A. laboriosa* (Aichholz and Lorbeer 1999)

Compound family	*A. andreniformis*	*A. florea*	*A. cerana*	*A. mellifera*	*A. dorsata*	*A. laboriosa*
Alkanes total	18.5	12.5	11.4	12.8	10.8	10.8
Alkenes total	5.9	7.5	7.4	2.9	0.6	5.3
Diene total	3.4	–	–	–	–	–
Hydrocarbons total	27.8	20	18.8	15.7	11.4	16.1
Fatty acids total	2.6	0.8	3.6	18	4.9	4.3
Fatty alcohols total	–	0.4	1.8	0.6	–	–
Monoesters total	27.5	41.1	33.4	40.8	36.9	37.5
Hydroxymonoesters total	13.6	9.1	18.1	9.2	23.3	23.6
Diesters total	12.9	15.7	12.2	7.4	11.9	8.8
Hydroxydiesters total	3.9	2.3	3	–	1.4	1.1
Esters total	57.9	68.2	66.7	57.4	73.5	71
Total	88.3	89.4	90.9	91.7	89.8	91.4

Table 6.2 Comparison of the compound composition of derivatised combwax of *A. mellifera*, *A. cerana*, *A. florea*, *A. andreniformis*, *A. dorsata*, and *A. laboriosa* by GC–FID analysis (Aichholz and Lorbeer 1999)

Structure	Peak	*A. mellifera*	*A. cerana*	*A. florea*	*A. andreniformis*	*A. dorsata*	*A. laboriosa*
Alkane C23	1	0.4	0	0	1.1	0.4	0.3
Alkane C25	3	1.5	0.9	1.5	7	4.3	3.8
Alkane C27	10	6.2	8.2	6.3	4.9	3.6	3.6
Alkane C29	17	2.6	2.3	3	2.8	1.2	1.7
Alkane C31	22	1.5	0	1.2	1.8	0.9	1
Alkane C33	26	0.3	0	0.5	0.5	0.4	0.4
Alkane C35	30	0.3	0	0	0.4	0	0
Alkene C27	8	0	0	0.6	0.5	0	0
Alkene C29	16	0	0.6	1	1	0	0
Alkene C31	21	0.8	0	2.3	0	0	0.3
Alkene C33	25	2.1	0.4	3	0	0.6	1.9
Alkene C35	29	0	5.4	0.6	1	0	1.7
Alkene C37	34	0	1	0	1.4	0	0.8
Alkene C39	38	0	0	0	1.3	0	0.6
Alkene C41	41	0	0	0	0.7	0	0
Fatty acid C20	13	1.1	0	0	0.8	0.8	0
Fatty acid C22	19	0.7	0	0	0	0.3	0.4
Fatty acid C24	24	6	0	0	0	1.4	0.7
Fatty acid C26	27	2.1	0.5	0	0	0	0
Fatty acid C28	31	2.6	1.2	0.4	0.5	0	0
Fatty acid C30	35	2.1	1.9	0.4	0.4	0	0
Fatty acid C32	39	1.6	0	0	0.2	0.3	0.6
Fatty acid C34	43	1.5	0	0	0.3	1.4	1.8
Fatty acid C36	46	0.3	0	0	0.4	0.7	0.8
Fatty alcohol C33	32	0.3	1.8	0.4	0	0	0
Fatty alcohol C35	36	0.3	0	0	0	0	0
Diene C35	28	0	0	0	0.4	0	0
Diene C37	33	0	0	0	0.9	0	0
Diene C39	37	0	0	0	1.1	0	0
Diene C41	40	0	0	0	1	0	0
Diester C54	67	0	0	0	0	1	0.6
Diester C54	68	1.2	0	0.7	0.7	5.6	4.1
Diester C56	69	0	0	0	0	1	0.9
Diester C56	70	1.2	0.6	1	1	2.4	2
Diester C58	72	0	0	0.8	0.6	0.5	0.3
Diester C58	73	1.4	2.3	5.2	4.2	1	0.9
Diester C60	75	0	1.1	1.1	0.9	0	0
Diester C60	76	2	5.3	4.2	3.4	0.4	0
Diester C62	78	0	0.7	0.7	0	0	0
Diester C62	79	1.2	1.6	1.7	1.6	0	0
Diester C64	81	0.4	0.6	0.3	0.5	0	0
Hydroxydiester C50	71	0	0.7	0	0.4	1	0.7
Hydroxydiester C52	74	0	0	0	0.6	0.4	0.4
Hydroxydiester C54	77	0	1	1.1	1.6	0	0
Hydroxydiester C56	80	0	1	0.6	0.9	0	0
Hydroxydiester C58	82	0	0.3	0.6	0.4	0	0
Hydroxymonoester C40	48	0	0	0	0.4	3.3	2.3

(*continued*)

Table 6.2 (continued)

Structure	Peak	*A. mellifera*	*A. cerana*	*A. florea*	*A. andreniformis*	*A. dorsata*	*A. laboriosa*
Hydroxymonoester C40	49	0.9	0	0	0.4	9.6	8.4
Hydroxymonoester C42	51	0	0	0	0	4	4.5
Hydroxymonoester C42	52	0.8	0.4	0.4	0.8	2.5	2.6
Hydroxymonoester C44	54	0	2.8	0	0	1.3	1.3
Hydroxymonoester C44	55	1.8	0	3.3	4.3	0.5	0.6
Hydroxymonoester C46	57	0.9	9.2	0	0	0.4	0.4
Hydroxymonoester C46	58	2.3	0	2.9	4.7	0.3	0.4
Hydroxymonoester C48	61	0.6	4.4	0	0	0.3	0.5
Hydroxymonoester C48	62	1.6	0	1.5	1.9	0.5	0.9
Hydroxymonoester C50	64	0	0.5	0	0.8	0.3	0.7
Hydroxymonoester C50	65	0.3	0.8	0.7	0	0.3	0.5
Hydroxymonoester C52	66	0	0	0.3	0.3	0	0.5
Monoester C38	42	0	0	0	0	0.5	0.7
Monoester C40	44	6.6	0.7	1.5	1.3	26.8	24.9
Monoester C42	47	4.6	0.9	3.4	1.5	4.7	4.5
Monoester C44	50	5.7	4.8	9.7	7.7	0.7	1
Monoester C46	53	11.9	23.7	17	10.7	0.9	1.6
Monoester C48	56	9	2.2	7.3	4.7	1.7	2.7
Monoester C50	60	2.6	0.6	1.8	1.3	1.2	1.6
Monoester C54	63	0.4	0.5	0.4	0.3	0.4	0.5

among honeybee species (Table 6.2), but all share a complex mixture of homologous neutral lipids in common: C_{25}–C_{29} alkanes, C_{40}–C_{54} monoesters, C_{42}–C_{52} hydroxymonoesters and C_{56}–C_{58} diesters (Aichholz and Lorbeer 1999; Aichholz et al. 2000). At this point in time, our knowledge of the composition of the waxes of all honeybee species is nearly on a par; however, pathways of synthesis remain available only for *A. mellifera* (Hepburn 1986; Hepburn et al. 1991). Given what is known of species-specific composition (Table 6.2), there is considerable opportunity for biochemical studies of beeswaxes in future.

Differences in the relative amounts of the major families of compounds in the waxes could be expected to be reflected in the physical, mechanical properties of the waxes. Buchwald et al. (2006) recently reported the results of a comparative study of the mechanical properties of several different beeswaxes (*A. andreniformis*, *A. cerana*, *A. dorsata* and *A. mellifera*) and measured, among other things, the relative stiffness and resilience of the waxes. Because the mechanical properties of any structure result from both the intrinsic chemical nature of a material as well as its structural form, it is obviously desirable, but experimentally extremely difficult to work with whole comb specimens. So, Buchwald et al. (2006) compromised by eliminating structure and simply measured the behaviour of wax cylinders under compression.

Although compression testing is not biologically appropriate for extrapolation to whole combs which are actually tension members with a relatively complex structure, the results of such measurements have heuristic value in trying to relate mechanical behaviour to differences in the major compound families of comb waxes given in Table 6.3. Resilience represents the amount of energy required to deform the test material until it begins to fail irrecoverably. Stiffness is simply the rate of change of stress per unit strain. Figure 6.6 shows that the wax of *A. dorsata* is significantly stiffer than the others. The waxes of *A. cerana* and *A. dorsata* do not significantly differ but are significantly more resilient than that of the intermediate *A. mellifera*, which in turn is more resilient than that of *A. andreniformis*.

Hepburn et al. (2009) calculated the weighted frequency distributions of the compounds of Table 6.2 to determine the average chain length of each type of wax as shown in Table 6.3. Although there is a trend suggesting that the waxes of the dwarf bees have the largest chain length, there are no significant differences among the waxes. This is surprising because, as a general rule, stiffness, strength, yield stress and other properties increase with increasing carbon chain length in polymers (Salamone 1996) but that relationship does not hold for beeswaxes.

Based on their results, Buchwald et al. (2006) showed that the giant honeybees, which build single but very large combs, are indeed the stiffest and most resilient of all honeybee species. They must also sustain the weight of the brood nest. And because of their setting high in trees, they are well exposed to possible wind damage. The multiple combs of the medium-sized bees (*A. cerana*, *A. koschevnikovi*, *A. nigrocincta* and *A. nuluensis*) are usually constructed in cavities with multiple attachment sites so that the load of nest contents is distributed over several points of attachment. The dwarf bees make very small single combs, and they are seldom exposed. At the end of the day, it must be remembered that mature combs are not beeswax. They are fibre-re-enforced with increasingly more silk deposition by succeeding generations of brood (Hepburn and Kurstjens 1988). And, while our knowledge of the beeswaxes advances that of the silk fraction is thus far restricted to *A. mellifera* (Hepburn et al. 1979).

In a different study on wax discrimination by honeybees, Hepburn et al. (2009) performed a cluster analysis of beeswax composition based on the data of Aichholz and Lorbeer (1999) (Table 6.2) to assess their relative affinities as measured by the Euclidean distances using the unweighted pair-group centroid amalgamation rule. A parsimonious unweighted pair-group analysis based on the distributions of the chemical constituents for 82 elution peaks of the derivatised comb waxes of *A. mellifera*, *A. cerana*, *A. florea*, *A. andreniformis*, *A. dorsata* and *A. laboriosa* is shown in Fig. 6.7. The giant honeybee group (*A. dorsata* and *A. laboriosa*) is clearly segregated from the other species as are the dwarf species (*A. andreniformis* and *A. florea*), while *A. mellifera* is placed close to its sister-group, *A. cerana*.

Table 6.3 Weighted frequency distributions for carbon chain length variation in the waxes of *A. mellifera*, *A. cerana*, *A. florea*, *A. andreniformis*, *A. dorsata* and *A. laboriosa* calculated from the data of Table 6.2

Structure	*A. mellifera*		*A. cerana*		*A. florea*		*A. andreniformis*		*A. dorsata*		*A. laboriosa*	
	Comp. (%)	Wt. freq.	Comp. (%)	Wt. freq.	Comp. (%)	Wt. freq.	Comp. (%)	Wt. freq.	Comp. (%)	Wt. freq.	Comp. (%)	Wt. freq.
C20	1.1	7.9	0.0	0.0	0.0	0.0	0.8	6.0	0.8	5.9	0.0	0.0
C22	0.7	5.5	0.0	0.0	0.0	0.0	0.0	0.0	0.3	2.4	0.4	3.2
C23	0.4	3.3	0.0	0.0	0.0	0.0	1.1	9.5	0.4	3.4	0.3	2.5
C24	6.0	51.8	0.0	0.0	0.0	0.0	0.0	0.0	1.4	12.3	0.7	6.1
C25	1.5	13.5	0.9	8.2	1.5	13.8	7.0	65.4	4.3	39.5	3.8	34.3
C26	2.1	19.6	0.5	4.7	0.0	0.0	0.0	0.0	0.0	0.0	0.0	0.0
C27	6.2	60.2	8.2	80.4	6.9	68.8	5.4	54.5	3.6	35.7	3.6	35.1
C28	2.6	26.2	1.2	12.2	0.4	4.1	0.5	5.2	0.0	0.0	0.0	0.0
C29	2.6	27.1	2.9	30.5	4.0	42.8	3.8	41.2	1.2	12.8	1.7	17.8
C30	2.1	22.7	1.9	20.7	0.4	4.4	0.4	4.5	0.0	0.0	0.0	0.0
C31	2.3	25.7	0.0	0.0	3.5	40.1	1.8	20.9	0.9	10.3	1.3	14.6
C32	1.6	18.4	0.0	0.0	0.0	0.0	0.2	2.4	0.3	3.5	0.6	6.9
C33	2.7	32.1	2.2	26.4	3.9	47.5	0.5	6.2	1.0	12.1	2.3	27.4
C34	1.5	18.4	0.0	0.0	0.0	0.0	0.3	3.8	1.4	17.5	1.8	22.1
C35	0.6	7.6	5.4	68.6	0.6	7.8	1.8	23.5	0.0	0.0	1.7	21.5
C36	0.3	3.9	0.0	0.0	0.0	0.0	0.4	5.4	0.7	9.3	0.8	10.4
C37	0.0	0.0	1.0	13.4	0.0	0.0	2.3	31.8	0.0	0.0	0.8	10.7
C38	0.0	0.0	0.0	0.0	0.0	0.0	0.0	0.0	0.5	7.0	0.7	9.6
C39	0.0	0.0	0.0	0.0	0.0	0.0	2.4	35.0	0.0	0.0	0.6	8.4
C40	7.5	108.0	0.7	10.2	1.5	22.1	2.1	31.4	39.7	583.6	35.6	514.1
C41	0.0	0.0	0.0	0.0	0.0	0.0	1.7	26.0	0.0	0.0	0.0	0.0
C42	5.4	81.6	1.3	19.8	3.8	58.9	2.3	36.1	11.2	172.9	11.6	175.9
C44	7.5	118.8	7.6	121.4	13.0	211.1	12.0	197.3	2.5	40.4	2.9	46.1
C46	15.1	250.0	32.9	549.4	19.9	337.9	15.4	264.7	1.6	27.0	2.4	39.9
C48	11.2	193.5	6.6	115.0	8.8	155.9	6.6	118.4	2.5	44.1	4.1	71.1
C50	2.9	52.2	2.6	47.2	2.5	46.1	2.5	46.7	2.8	51.4	3.5	63.2
C52	0.0	0.0	0.0	0.0	0.3	5.8	0.9	17.5	0.4	7.6	0.9	16.9
C54	1.6	31.1	1.5	29.4	2.2	43.9	2.6	52.5	7.0	138.9	5.2	101.4

(*continued*)

Table 6.3 (continued)

Structure	*A. mellifera*		*A. cerana*		*A. florea*		*A. andreniformis*		*A. dorsata*		*A. laboriosa*	
	Comp. (%)	Wt. freq.	Comp. (%)	Wt. freq.	Comp. (%)	Wt. freq.	Comp. (%)	Wt. freq.	Comp. (%)	Wt. freq.	Comp. (%)	Wt. freq.
C56	1.2	24.2	1.6	32.5	1.6	33.1	1.9	39.8	3.4	70.0	2.9	58.6
C58	1.4	29.2	2.6	54.7	6.6	141.3	5.2	112.7	1.5	32.0	1.2	25.1
C60	2.0	43.2	6.4	139.4	5.3	117.4	4.3	96.4	0.4	8.8	0.0	0.0
C62	1.2	26.8	2.3	51.8	2.4	54.9	1.6	37.1	0.0	0.0	0.0	0.0
C64	0.4	9.2	0.6	13.9	0.3	7.1	0.5	12.0	0.0	0.0	0.0	0.0
Total (%)	91.7		90.9		89.4		88.3		89.8		91.4	
Mean		39.7		43.9		44.4		42.5		40.9		40.7
SD		55.7		98.4		73.9		58.4		104.8		92.4

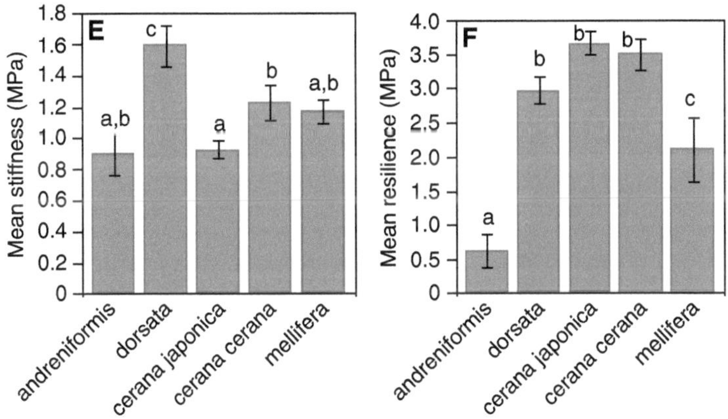

Fig. 6.6 Comparisons of stiffness (E) and resilience (F) (mean ± SE) among the waxes of *A. andreniformis*, *A. cerana*, *A. dorsata* and *A. mellifera*. *Differing letters* indicate groups that are statistically different (from Buchwald et al. 2006)

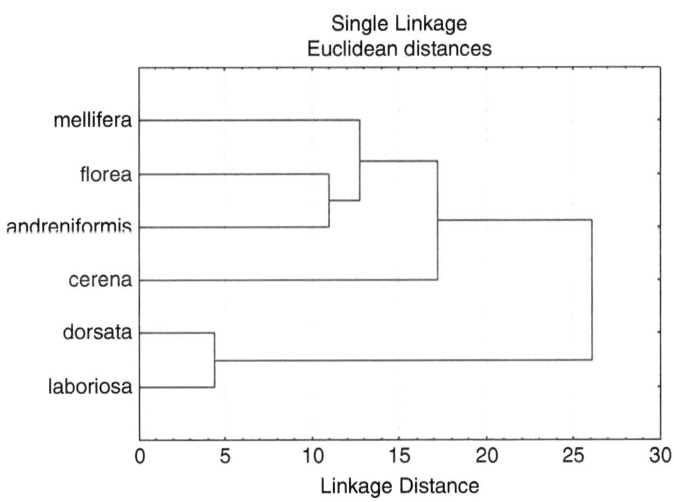

Fig. 6.7 Hierarchical clustering dendrogram derived from single linkage clustering on compound composition of derivatised comb wax

The Euclidean distances for the beeswaxes present a very similar picture which is consistent with recent analyses of *Apis* species in which three distinct clusters of sister groups result from morphometric (Alexander 1991) behavioural (Raffiudin and Crozier 2007) as well as DNA sequence analyses (Arias and Sheppard 2005): (1) dwarf bees (*A. florea* and *A. andreniformis*), (2) giant honeybees (*A. dorsata* and *A. laboriosa*) and (3) a cluster consisting of the medium-sized bees (*A. cerana*, *A. koschevnikovi*, *A. mellifera*, *A. nigrocincta* and *A. nuluensis*). In any event, the close proximity of the beeswax unweighted pair-groups to those based on DNA and morphometrics suggests that the wax glands and their products of secretion were highly conserved features during honeybee evolution (Fig. 6.7).

6.4.2 Resins

The dwarf honeybees, *A. florea* and *A. andreniformis*, utilise plant resins as nest materials. They apply a band of sticky resin on the twigs supporting the comb (Fig. 6.8). These bands are about 2.8 ± 2.1 cm wide with a range of 0.5–10.05 cm and trap any small animals attempting to cross into a colony (Seeley et al. 1982). A band is built on both sides of the twigs, but there is a strong tendency for the bands to be thicker at the side proximal to the tree trunk than the distal side of the twig tip (Duangphakdee personal observation).

The sticky bands of *A. florea* and *A. andreniformis* have clear-cut repellent properties against the weaver ants, *Oecophylla smaragdina* (Duangphakdee et al. 2005b). Repair and re-enforcement of the sticky bands by *A. florea* is strongly correlated with intrusions of weaver ants into the colony (Duangphakdee et al. 2005b). Duangphakdee (2006) performed some preliminary GC-MS analyses that give a rough idea of the chemical constituents of sticky bands. The resins consist of more than 50 compounds (Fig. 6.9) and the most abundant are steriods (30.32%) and a triterpene (amyrin – 45.72%).

6.5 Wax Salvage

Absconding or migrating colonies commonly occur in tropical honeybees in Africa (Hepburn and Radloff 1998) and especially in Asia (Oldroyd and Wongsiri 2006).

Fig. 6.8 Sticky band, an ant-repellent resin on branch extending to an *A. florea* nest Photo: Duangphakdee

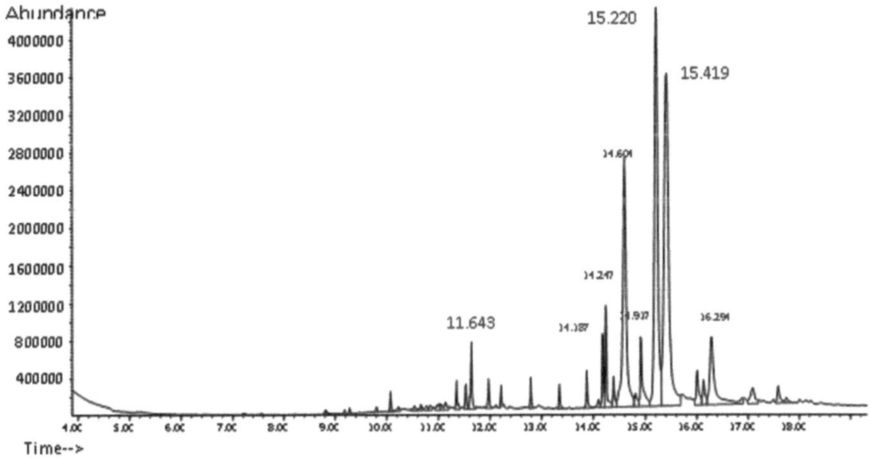

Fig 6.9 Chromatogram of the sticky band material collected at *A. florea* nests in Thailand. Preliminary identifications based on the WILEY 7N library database. To consider only the main constituents: $R_t = 11.643$: no satisfying library match; $R_t = 14.187$: phenol derivative; $R_t = 14.247$: phenol derivative; $R_t = 14.604$: triterpene (amyrin); $R_t = 14.917$: heptacosane; $R_t = 15.220$: steroid (cyclolanostentol); $R_t = 15.419$: triterpene (amyrin); $R_t = 16.294$: steroid (cyclolanostentol); R_t = retention time (Duangphakdee 2006)

To cannibalise wax from a deserted nest and reuse it in the construction of a new nest can be cost effective (Pirk et al. 2010) but is only known for *A. florea* among the Asian species (Akratanakul 1977; Dutton and Free 1979; Wongsiri et al. 1996; cf. Chap. 13). There are no reports of heterospecific salvage. Because honeybees can distinguish olfactory differences between combs of different colonies of the same species (Breed et al. 1988), the question arises to what extent is there behavioural plasticity for wax choice among honeybee species. Do they discriminate among waxes that they might naturally encounter (as in the southeast Asian species) compared with waxes foreign to them? Finally, what are the cladistic relationships of beeswaxes of different sister groups and whether these similarities and differences can be related to wax choice in different species of honeybees?

Hepburn et al. (2010) investigated whether *A. florea* would preferentially choose to salvage wax from their own, original natal combs over that of other conspecific combs and whether the bees would salvage wax from facsimiles of *A. florea* combs fashioned from the waxes of *A. cerana*, *A. dorsata* and *A. mellifera*. The results demonstrated that *A. florea* wax-salvaging foragers show a significantly greater preference for natal comb than non-natal combs and did not collect any wax from heterospecific combs. This means that *A. florea* can clearly discriminate combs from different colonies intraspecifically and interspecifically.

The hydrocarbons of honeybee comb wax vary significantly between colonies, and this variation is explained in part by genetic differences among colonies (Breed et al. 1988, 1995; Sasaki et al. 2000; Wilde et al. 2001). Wax salvage behaviour probably reflects genetic differences for this trait in *A. florea* not shown in *A. andreniformis*. In comparisons between similar samples of heterospecific wax utilisation within combs (Hepburn et al. 2009), small pieces of wax foundation of *A. florea*, *A. cerana*, *A. mellifera* and *A. dorsata* were inserted in the middle of *A. florea* combs. All the *A. florea* colonies unequivocally accepted *A. cerana*, *A. dorsata* and *A. florea* wax inserts and built on these foundations, but rejected the *A. mellifera* wax inserts. This contrast obviously indicates that sensory discrimination of waxes by *A. florea* is exercised in the field but not in the nest, a context in which it ought not to be necessary in the absence of heterospecific nest parasitism by other honeybee species (Hepburn et al. 2010).

References

Aichholz R, Lorbeer E (1999) Investigation of combwax of honeybees with high-temperature gas chromatography and high-temperature gas chromatography-chemical ionization mass spectrometry. I. High-temperature gas chromatography. J Chromatogr A 855:601–615

Aichholz R, Lorbeer E (2000) Investigation of combwax of honeybees with high-temperature gas chromatography and high-temperature gas chromatography-chemical ionization mass spectrometry. II. High-temperature gas chromatography-chemical ionization mass spectrometry. J Chromatogr A 883:75–88

Aichholz R, Lorbeer E, Pechhacker H, Hüttinger E (2000) The chemical compounds of beeswax from *Apis* species. In: Proceedings of 4th Asian apiculture association conference, Kathmandu, pp 152–154

Akratanakul P (1977) The natural history of the dwarf honey bee, Apis florea F. in Thailand. Thesis, Cornell University, Ithaca

Akratanakul P (1987) Honeybee diseases and enemies in Asia: a practical guide. FAO Serv Bull 68(5):1–51

Alexander BA (1991) A cladistic analysis of the genus *Apis*. In: Smith DR (ed) Diversity in the genus *Apis*. Westview, Boulder, pp 1–28

Arias MC, Sheppard WS (2005) Phylogenetic relationships of honey bees (Hymenoptera: Apinae; Apini) inferred from nuclear and mitochondrial DNA sequence data. Mol Phylogenet Evol 37:25–35

Bakker DR (1999) Foraging and habitat selection by two species of honey bee near Lore Lindu national park in Sulawesi. Thesis, University of Guelph, Indonesia

Bellier J (1906) Beeswax from Annam. Ann Chem Anal 11:366–368

Booncham U, Wongsiri S, Thirakupt K (1995) Niche differences between *A. florea* and *A. andreniformis* in dry-evergreen forest, Thailand. In: International conference on tropical bees environment, Pedu Lake, Kedah, p 33

Brand-Garnys EE, Sprenger J (1988) Beeswax new aspects of a classic raw material for cosmetics. Seifen, Öele, Fette, Wachse 114:547–552 (in German)

Breed MD, Williams KR, Fewell JH (1988) Comb wax mediates the acquisition of nest-mate recognition cues in honey bees. Proc Natl Acad Sci 85:8766–8769

Breed MD, Garry MF, Pearce AN, Hibbard BE (1995) The role of wax comb in honey bee nestmate recognition. Anim Behav 50:489–498

Büchner G (1906) On Indian Ghedda-wax. Chem-Ztg 43:528–529 (in German)

Buchwald R, Breed MD, Greenberg AR, Otis G (2006) Interspecific variation in beeswax as a biological construction material. J Exp Biol 209:3984–3989

Butani DK (1950) An *Apis dorsata* colony in New Delhi. Indian Bee J 12:115

Butler CG (1954) The world of the honeybees. Collins, London

Crane E (1993) Current status of research on Asian honey bees. In: Connor LJ, Rinderer TE, Sylvester HA, Wongsiri S (eds) Asian apiculture. Cheshire, Connecticut, pp 19–41

Deodikar GB, Ghatge AL, Phadke RP, Mahindre DB, Kshirsagar KK, Muvel S, Thakar SS (1977) Nesting behaviour of Indian honeybees. III. Nesting behaviour of Apis dorsata Fab. Indian Bee J 39:1–12
Douglas JC (1886) The hive-bees indigenous to India and the introduction of the Italian bee. J Asiatic Soc Bengal II Nat Sci 55:83–96
Duangphakdee O (2006) Biologically active compounds worker bees use to repel ants. Thesis, Chulalongkorn University, Bangkok
Duangphakdee O, Koeniger N, Deowanish S, Hepburn HR, Wongsiri S (2005a) Ant repellent resins of honeybees and stingless bees. Insect Soc 56:333–339
Duangphakdee O, Koeniger N, Koeniger G, Wongsiri S, Deowanish S (2005b) Reinforcing a barrier – a specific social defense of the dwarf honeybee (*Apis florea*) released by the weaver ant (*Oecophylla smaragdina*). Apidologie 36:505–512
Dutton RW, Free JB (1979) The present status of beekeeping in Oman. Bee World 60:176–185
Free JB (1981) Biology and behaviour of the honeybee *Apis florea*, and possibilities for beekeeping. Bee World 62:46–59
Grün A, Halden W (1929) Analysis of fats and waxes. von Julius Springer, Berlin (in German)
Hadisoesilo S (1997) A comparative study of two species of cavity-nesting honey bees of Sulawesi. Thesis, University of Guelph, Indonesia
Hadisoesilo S, Otis GW (1998) Differences in drone cappings of *Apis cerana* and *Apis nigrocincta*. J Apic Res 37:11–15
Hepburn HR (1986) Honeybees and wax. Springer, Berlin
Hepburn HR, Kurstjens SP (1988) The combs of the honeybee as composite materials. Apidologie 19:25–36
Hepburn HR, Radloff SE (1998) Honeybees of Africa. Springer, Berlin
Hepburn HR, Chandler HD, Davidoff MR (1979) Extensometric properties of insect fibroins: the green lacewing cross-beta, honeybee alpha-helical and greater waxmoth parallel-beta conformations. Insect Biochem 9:69–77
Hepburn HR, Bernard RTF, Davidson BC, Muller WJ, Lloyd P, Kurstjens SP, Vincent SL (1991) Synthesis and secretion of beeswax in honeybees. Apidologie 22:21–36
Hepburn HR, Radloff SE, Duangphakdee O, Phiancharoen M (2009) Interspecific utilization of wax in comb building by honeybees. Naturwissenschaften 96:719–723
Hepburn R, Duangphakdee O, Phiancharoen M, Radloff S (2010) Comb wax salvage by the red dwarf honeybee, *Apis florea* F. J Insect Behav 23:159–164
Hooper D (1904) Indian bees'-wax. Agric Ledger 7:201–238
Ikuta H (1931) The investigation of Japanese beeswax. Anal Chem 3:430–437
Ikuta H (1934) The investigation of Japanese beeswax (III): composition of hydroxy fatty acid. Analyst 59:161–162
Inoue T, Adri S, Salmah S (1990) Nest site selection and reproductive ecology of the Asian honey bee, *Apis cerana indica*, in Central Sumatra. In: Sakagami SF, Ohgushi R, Roubik DW (eds) Natural history of social wasps and bees in equatorial Sumatra. Hokkaido University Press, Sapporo, pp 219–232
Joshi SR, Ahmad F, Gurung MB (2004) Status of *Apis laboriosa* populations in Kaski district, western Nepal. J Apic Res 43:176–180
Koeniger N, Koeniger G (2000) Reproductive isolation among species of the genus *Apis*. Apidologie 31:313–339
Lindauer M (1957) Communication among the honeybees and stingless bees of India. Bee World 38(3–14):34–39
Liu F, Roubik DW, He D, Li J (2007) Old comb for nesting site recognition by *Apis dorsata*? Field experiments in China. Insect Soc 54:424–426
Mehta VK (1948) Apis florea. Indian Bee J 10(79):81
Mol GA (1933) Collecting wax and honey in the lake region of Western Borneo. Landbouw Buitenzorg 2:80–86 (in Dutch)
Morse R, Laigo FM (1969) *Apis dorsata* in the Philippines. Philippine Association of Entomologists, Laguna

Mossadegh MS (1990) Nesting behaviour of *Apis florea* F. (Hymenoptera: Apidae) in Khuzestan, Iran. In: Proceedings of the 11th international congress, IUSSI, Bangalore, pp 669–670
Muttoo RN (1956) Facts about beekeeping in India. Bee World 37:125–133
Narayana N (1970) Studies in Indian honeys and bees waxes. Central Bee Research Institute, Poona
Neumann P, Koeniger N, Koeniger G, Tingek S, Kryger P, Moritz RFA (2000) Home-site fidelity in migratory honeybees. Nature 406:474–475
Oldroyd BP, Wongsiri S (2006) Asian honey bees. Harvard University Press, Cambridge
Otis GW, Hadisoesilo S (1996) *Apis nigrocincta*, a previously unrecognized honey bee species from Indonesia: lessons in speciation. In: Proceedings of 3rd Asian apiculture association conference, Hanoi, pp 2–6
Paar J, Oldroyd B, Kastberger G (2000) Giant honeybees return to their nest sites. Nature 406:475
Phadke RP, Nair KS, Nandekar KU (1969) Indian beeswaxes. I. Their physico-chemical constants. Indian Bee J 31:52–55
Phadke RP, Nair KS, Nandekar KU (1971) Indian beeswaxes. II. The nature of their chemical constituents. Indian Bee J 33:3–5
Pirk CWW, Crous KL, Duangphakdee O, Radloff SE, Hepburn HR (2010) Economics of comb wax salvage by the red dwarf honeybee, *Apis florea*. J Comp Physiol (in press)
Raffiudin R, Crozier RH (2007) Phylogenetic analysis of honey bee behavioral evolution. Mol Phylogenet Evol 43:543–552
Ramachandran S (1939) Beekeeping in South India. Superintendent Government Press, Madras, Bull 37:1–12
Reddy SM, Reddy CC (1993) Studies on the distribution of nests of giant honey bee (*Apis dorsata* F.). Indian Bee J 55:36–39
Rinderer TE, Wongsiri S, Kuang BY, Liu JS, Oldroyd BP, Sylvester HA, de Guzman LI (1996) Comparative nest architecture of the dwarf honey bees. J Apic Res 35:19–26
Rinderer TE, Oldroyd BP, de Guzman LI, Wattanachaiyingcharoen W, Wongsiri S (2002) Spatial distribution of the dwarf honey bees in an agroecosystem in southeastern Thailand. Apidologie 33:539–543
Roberts OD, Islip HT (1922) The constants of Indian beeswax. Analyst 47:246–251
Roubik DW, Sakagami SF, Kudo I (1985) A note on distribution and nesting of the Himalayan honey bee *Apis laboriosa* Smith (Hymenoptera: Apidae). J Kans Entomol Soc 58:746–749
Sakagami SF, Yoshikawa K (1973) Additional observations on the nest of the dwarf honeybee, *Apis florea* (Hymenoptera: Apidae). Kontyû 41:217–219
Salamone JC (1996) Polymeric materials encyclopedia. CRC, Boca Raton
Sasaki M, Fujiwara M, Numaguchi Y (2000) Proboscis extension reflex (PER) for bee odour and non-volatile surface wax components: analysis of nestmate discrimination. In: Proceedings of the 4th Asian apiculture association conference, Kathmandu, p 37
Schwarz HF (1948) Stingless bees (Meliponidae) of the western hemisphere. Bull Am Mus Nat Hist 90:1–536
Seeley TD (1985) Honeybee ecology: a study of adaptation in social life. Princeton University Press, Princeton
Seeley TD, Seeley RH, Akratanakul P (1982) Colony defense strategies of honeybees in Thailand. Ecol Monogr 52:43–63
Singh S (1947) The honeybees of India: the abc and xyz of bee culture. AI Root Company, Medina, Ohio, pp 560–562
Singh S (1962) Beekeeping in India. Indian Council of Agricultural Research, New Delhi
Tan NQ (2007) Biology of *Apis dorsata* in Vietnam. Apidologie 38:221–229
Thakar SS, Tonapi KV (1961) Nesting behaviour of Indian honeybees. I. Differentiation of worker, queen and drone cells on the combs of *Apis dorsata* Fab. Bee World 42:61–62, 72
Thakar SS, Tonapi KV (1962) Nesting behaviour of Indian honeybees. II. Nesting habits and comb cell differentiation in Apis florea Fab. Indian Bee J 24:27–31
Tingek S, Koeniger G, Koeniger N (1996) Description of a new cavity nesting species of *Apis* (*Apis nuluensis* n. sp.) from Sabah, Borneo, with notes on its occurrence and reproductive biology (Insecta: Hymenoptera: Apoidea; Apini). Senckenbergiana Biol 76:115–119

Tulloch AP (1980) Beeswax – composition and analysis. Bee World 61:47–62
Ueno S (1915) Corean beeswax. Analyst 40:343–344
Underwood BA (1986) The natural history of *Apis laboriosa* Smith in Nepal. Thesis, Cornell University, Ithaca
Underwood BA (1990) The seasonal cycle of nesting and migration by the Himalayan honey bee, *Apis laboriosa*. In: Proceedings of the 11th international congress, IUSSI, p 117
van Lijnden OWC, Groll J (1851) Observations on the Kapuas River basin area. Naturk Tijdschr Nederlandsch-Indie 2:537–636 [in Dutch]
Viswanathan H (1950) Note on *Apis dorsata* queen cells. Indian Bee J 12:55
Wattanachaiyingcharoen W, Wongsiri S, Oldroyd BP (2008) Aggregations of unrelated *Apis florea* colonies. Apidologie 39:531–536
Whitcombe RP (1984) The biology of *Apis* spp. in Oman with special reference to *Apis florea* Fab. Thesis, University of Durham, Durham
Wilde J, Woyke J, Wilde M (2001) Rearing honey bee queens on cups of *Apis mellifera* and *Apis dorsata* wax. In: Proceedings of the 7th international conference on tropical bees, Chiang Mai, pp 207–220
Wongsiri S, Thapa R, Oldroyd BP, Burgett MD (1996) A magic bee tree: Home of *Apis dorsata*. Am Bee J 136:196–199
Wongsiri S, Lekprayoon C, Thapa R, Thirakhupt K, Rinderer TE, Sylvester HA, Oldroyd BP, Booncham U (1997) Comparative biology of *Apis andreniformis* and *Apis florea* in Thailand. Bee World 78:23–35
Wongvilas S, Deowanish S, Wongsiri S (2009) Habitat of red dwarf honeybee (*Apis florea*) and the black dwarf honeybee (*Apis andreniformis*) in Kanchanaburi Province. The King Mongkut's Institute of Technology, Ladkrabang Academic Conference and Exhibition, Thailand, pp 28–35
Woyke J, Wilde J, Wilde M (2001) A scientific note on *Apis laboriosa* winter nesting and brood rearing in the warm zone of the Himalayas. Apidologie 32:601–602
Yang MX, Tan K, Radloff SE, Phiancharoen M, Hepburn HR (2010) Comb construction in mixed species colonies of honeybees *Apis cerana* and *Apis mellifera*. J Exp Biol 213:1659–1664

Chapter 7
Absconding, Migration and Swarming

H.R. Hepburn

7.1 Introduction

Striking differences among honeybees include the extent of resource investment in reproductive swarming in comparison with absconding and migration. These differences are ascribable to evolution in regions of very different climates and the seasonality of flowering in tropical and temperate regions. These are the essential perquisites that both allow and drive honeybees to abandon their maternal nests and to establish new ones. The phenomenon of nest abandonment by absconding/migration is ubiquitous in Asian and African honeybees.

In the temperate zone, honeybees have only brief periods of weather favourable for swarming and establishing new colonies and nests and to conserve large stores to survive winter; few abscond or migrate (Martin 1963). In sharp contrast, tropical climates with perennial flowering facilitate swarming, absconding and migration. However, predation pressure is severe in many tropical areas (Toumanoff 1939; Morse and Nowogrodzki 1990; Oldroyd and Wongsiri 2006) and must also select for greater colony mobility through absconding and migration. Honeybees that construct a single comb in the open are subject to constraints different in kind and quality from those that affect the multiple-comb, cavity-dwelling species. Thus, thermal homeostasis is also a continuous selective pressure affecting the distribution and seasonal cycles of such colonies.

Absconding, migration and swarming in honeybees are all strategies correlated with climate and the seasonal phenology of flowering across the range of temperate and tropical regions. Reproductive swarming almost always occurs in the ascendant phase of colony growth, driven by pollen/nectar flows (Seeley 1985; Winston 1991; Crane 1993; Hepburn and Radloff 1998). Migration is a resource-related, seasonal movement of tropical honeybee colonies, usually without the production of queens and drones. It maximises the colonisation of new areas and provides a spatial refuelling cycle clearly driven by r-selection. It is a direct evolutionary alternative to

H.R.Hepburn
Department of Zoology and Entomology, Rhodes University, Grahamstown 6140, South Africa
e-mail: r.hepburn@ru.ac.za

H.R. Hepburn and S.E. Radloff (eds.), *Honeybees of Asia*,
DOI 10.1007/978-3-642-16422-4_7,

the K-selected (Verhulst 1838; Pianka 1970) hoarding characteristic of temperate bees (Hepburn 2006). The primary stimuli for prepared absconding or migration are chronic resource depletion, deterioration of the microclimate of the nest or continuous disturbance by pests or predators.

To migrate/abscond, bees must have sufficient flight fuel and energy reserves to construct new combs at a new site. Absconding colonies typically engorge honey, accelerate wax production, reduce oviposition and consume eggs and young larvae so conserving protein. Absconding/migration may be beneficial to the survival, dispersal and propagation of honeybees, but imposes serious difficulties for beekeeping in the tropics. This has stimulated observations and experiments over the last century or so that are beginning to provide a biological basis for understanding absconding, migration and reproductive swarming in the broader context of insect sociality (Wilson 1971, 2000). This review synthesises the relevant literature and seeks underlying principles for these behaviours as life history phenomena for Asian honeybees.

7.2 Terminology

The terminology applied to explain the stimuli and mechanisms that result in honeybees abandoning their nests (swarming, absconding and migration) has developed very slowly over many decades. Historically, most interpretations derive from inferences based on natural history observations, very few comparative in nature and little from direct experimental intervention. Thus, much of the literature on absconding, migration and swarming is anecdotal and the terminology conflated (Lipinski 2001). Lack of clarity in the past is attributable to the fact that nest abandonment by honeybee colonies covers a spectrum of overlapping phenomena without definite boundaries (Nakamura 1993, 1994, 1995; Hepburn 2006; Hepburn and Radloff 1998).

The terminology used here is as follows. Reproductive swarming is defined as the movement of at least one queen and part of a honeybee colony from the maternal nest to an entirely new site for colony reproduction and may also involve migration. This includes the prime swarm and all afterswarms or casts (sensu Janscha 1771), and there has always been a wide consensus on this usage even before the "queen" was actually recognised as such (Ménendez de Torres 1586). Migrations of honeybees are easily distinguished from swarming colonies because in the former there are no queen cells or drones on the combs (Woyke 1976). In its simplest, broadest sense, migration is the seasonally predictable movement of a whole colony from one region to another; a return journey is not prerequisite but may be the case (Neumann et al. 2000; Paar et al. 2000).

Given suitable flight temperatures and perennial flowering in tropical climates, migration is an evolutionary alternative to the temperate strategy of massive hoarding (Sakagami 1960; Woyke 1976; Nakamura 1993; Zhang 1996; Hepburn 2006). Distinctions between migration and absconding made in the past were based on the tacit and completely arbitrary assumption that migration involved very many colonies and absconding but few. Absconding is also the movement of a whole colony away from its maternal nest site to another place. But there are two different kinds of absconding: prepared absconding and simple absconding. The inferred causes of prepared absconding (functionally synonymous with migration) are often intertwined, but seem to resolve into severe, chronic disturbance and/or predation, declining quality of the nest site and resource depletion both in the nest and abroad in the field. Restricted size of the nest cavity, coupled to overheating and unavailability of surface water, may also induce absconding (Hepburn 2006).

The elements of predictability mostly relate to diminishing colony stores because of seasonally declining forage, unfavourable microclimatic conditions and the sustained, chronic disturbances of pests and predators, any or all of which are reflected in the behaviour of a colony preparing to abandon its combs. Simple absconding is also termed "acute disturbance" and is a well-known reaction to events of a disastrous nature or to the severe onslaughts for which abandoning the nest is an immediate response. "Chronic disturbance absconding" is an example of prepared absconding of longer duration and results in changes in brood conditions in a colony for several days to weeks before absconding (Nakamura 1993, 1995). For some weeks prior to absconding, such colonies have little honey or brood, foraging is reduced and their combs are chewed. In the narrow sense, simple absconding is used as the desertion of the maternal nest by the whole colony in response to acute disturbances such as fire, floods, heavy predation and the interventions of beekeepers or honey hunters. Typically, small, newly settled colonies abscond at the least disturbance, larger ones require major interference. Both are less prone to abscond in the ascendant phase of a nectar-pollen flow which sustains colony growth than in the descendant phase (Pokhrel et al. 2006). It remains to be seen how colony size and ascendant colony growth in relation to disturbance can be satisfactorily quantified.

Migration or prepared absconding is reflected in the behaviour of a colony preparing to abandon its combs. In this context, the terms "nomadic absconding" (Crane 1990), "prepared absconding" (Woyke 1978; Nakamura 1993, 1995), "undisturbed absconding" (Mutsaers 1991) and "chronic disturbance absconding" (Nakamura 1993, 1995) are synonymous with "migration" because all involve preparations for the departure of colonies. Simple absconding, also "acute disturbance", is the reaction to disaster. "Chronic disturbance absconding" is an example of prepared absconding of longer duration and for some weeks prior to absconding such colonies have little honey or brood and foraging is reduced (Nakamura 1993, 1995). The element of distance moved by a migrating/absconding colony is relative and subjective in nature: a relocation of only 20 m by *Apis florea*, 20 km by *Apis cerana* or 200 km by *Apis dorsata* all equate to migration.

7.3 Seasonal Absconding/Migration

Seasonal migration is characteristic of *A. cerana*, *A. florea*, *A. andreniformis*, *A. dorsata*, *A. laboriosa* and African (not Eurasian) *A. mellifera* (information for other species is lacking); but this behaviour is dampened at colder latitudes (Lavrekhin

Fig. 7.1 Relationship between the preparation period for prepared absconding or migration and distance eventually moved for four types of colony movement (after Nakamura 1993)

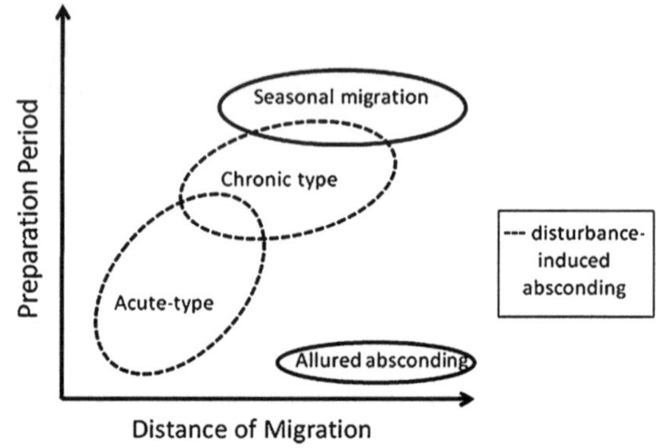

and Pankova 1975; Yang and Xu 1982; Kuznetsov 2002). Migrations are seasonally predictable consequences of resource depletion or dearth, a widespread decline in available pollen and nectar usually associated with high temperatures, extreme aridity or, conversely, extended rains or cold. Changes in microclimatic conditions of the nest and the increased advent of pests/predators are very often coincident with and even amplify the effects of resource depletion. Resource depletion as a stimulus for prepared absconding and migration is a conspicuous and recurrent theme in the literature for all species of honeybees (except Eurasian races of *A. mellifera*). The essential differences in the relationships between the distance of colony movement (migration) and the length of preparation time before movement and nest abandonment can be visualised diagrammatically (Fig. 7.1).

The climatic conditions that lead to seasonal dearth, and then to migration, obviously vary. *A. cerana* migrate during periods of high temperatures and again during the dearth of the dry season (Koeniger and Koeniger 1980; Ahmed and Muzaffar 1987; Dulta et al. 1988; Punchihewa et al. 1990; Pokhrel et al. 2006) or following the abatement of the heavy rains (Olsson 1989). A complex pattern occurs in the mountains of Sichuan Province, China, where most colonies migrate/abscond every year. The region is a highly convoluted landscape where subtropical and temperate climatic regimes both occur within short distances, depending on the altitude. So, some species of flowers are in bloom for a long time but in different places, and *A. cerana* migrate continually (Chen 1995).

Amongst mountain honeybees, the case of *A. laboriosa* is a particularly interesting one because after its original description from China (Smith 1871), it was really explored for the first time a century later in Nepal (Sakagami et al. 1980). These authors noted the complex topography and climate of the region which has intrigued many because it is physically, but not seasonally, sympatric with *A. dorsata* (Underwood 1991; Allen 1995; Roubik et al. 1985; Otis 1996; Thapa et al. 2000). In any event, Underwood (1987) (et seq.) established that this species swarms reproductively in April–May in the low hills and then migrates vertically to the sub-alpine cliffs at 2,500–3,500 m where it settles between May and October and migrates vertically down to the low hills before the colder months of November to March.

This has been independently confirmed by Thapa (1998) who further noted that there is a low hills deme of *A. laboriosa*, which does not migrate at all. This suggests the beginnings of population differentiation in this species. Underwood (1990) further observed that colonies preparing to migrate discontinued brood rearing and greatly reduced and then also discontinued pollen forage, which is characteristic of preparation for migration (prepared absconding, cf. Punchihewa et al. 1990). Finally, during the final few days before migration, all flight activity was reduced. Similarly, at this time foragers could not be recruited to feeding dishes although during summer this was routine practise. Finally, the bees began their migration dances indicating very long distances followed by workers exhibiting "schwirrlaeuferinnen" (Lindauer 1956 – a reasonably understandable English term could be "buzz-runner", N. Koeniger personal communication). The key factor for survival of *A. laboriosa* in the harsh climate of the Himalayas is seasonal migration.

Cycles of absconding/migration and swarming by *A. florea* are also linked to forage availability and high temperatures, seasonal changes in flowering, dearth and the depletion of resources (Sandhu and Singh 1960; Thakar and Tonapi 1962; Aggarwal and Kapil 1988; Sihag 2000). In a general summary of the biology of these little bees in Thailand, Deowanish et al. (2001) reported that both *A. florea* and *A. andreniformis* abscond in response to increased exposure to heat/sunlight and during dearth. Tirgari et al. (1969) and Tirgari (1971) noted that in Iran, *A. florea* migrated to sites with maximum exposure to sunshine in late autumn and in spring returned to dense foliage where the combs were not directly exposed to sun. *A. florea* abscond 2–3 times a year usually due to a combination of resource depletion and adverse microclimatic conditions. Similar observations were reported by Mossadegh and Birjandi (1986) and Mossadegh (1990).

Dutton and Free (1979) and Free (1981) regarded dearth as the stimulus for prepared absconding by *A. florea* in Oman, possibly exacerbated by unusually high temperatures and abetted by pests and predators. They reported that *A. florea* abscond/migrate during unfavourable conditions to cultivated lands near villages and then return to the mountains at the end of summer. Some colonies remain permanently at the village sites. Essentially, the same observations were made for *A. florea* in Bangladesh (Nash and Murrell 1981) and Thailand (Akratanakul 1977). In Oman, Whitcombe (1984) plotted the seasonal cycles of *A. florea* in relation to temperature regimes and floral cycles. Absconding or migration occur in summer (May to July) and wanes in September; but not all colonies abscond/migrate (Table 7.1).

On the southernmost of the Andaman Islands off the coast of India, Ahmed (1989) observed that *A. dorsata* colonies were abundant around Port Blair, but with the onset of the torrential monsoon rains and high winds between May and November, they migrate into deep forests. Similarly, Reddy (1983) showed that the temporal pattern of migration by *A. dorsata* is related to the monsoon which determines rainy and dry seasons. Mardan (1989) agreed and also suggested that *A. dorsata* crossed the 40 km wide Strait of Malacca from Sumatra to southwestern Malaysia.

Table 7.1 Phenology (standardised monthly frequencies) of *A. florea* colonies in relation to flowering in Oman (Whitcombe 1984)

	J	F	M	A	M	J	J	A	S	O	N	D	*n*	Mean±SD
90% Bee cover on comb	31	11	133	56	65	98	−64	−206	−159	−43	−23	98	99	74.42 ± 14.76
Worker brood	−174	−34	78	78	78	16	33	−17	78	78	22	−236	96	86.08 ± 17.82
Adult drones	−126	−82	−63	159	189	110	−33	−3	−126	−3	24	−48	96	33.67 ± 26.58
Drone brood	−102	43	174	139	161	−41	−62	−49	−102	−1	−58	−102	96	23.25 ± 22.81
Q cells and recent swarms	−73	−73	37	42	293	−10	−31	−73	−73	53	−21	−73	101	13.92 ± 19.12
Absconding	−125	61	−70	−118	54	116	150	54	−125	−125	47	81	191	35.17 ± 14.54
Flowering (Bee) Mtns, Wadis, Plains	102	80	196	131	−92	−114	−56	−71	−100	−42	−71	37	238	19.83 ± 13.89
Flowering (Bee) coastal, irrigated	13	33	223	133	−58	63	13	−78	−168	−58	−58	−58	321	26.75 ± 9.98

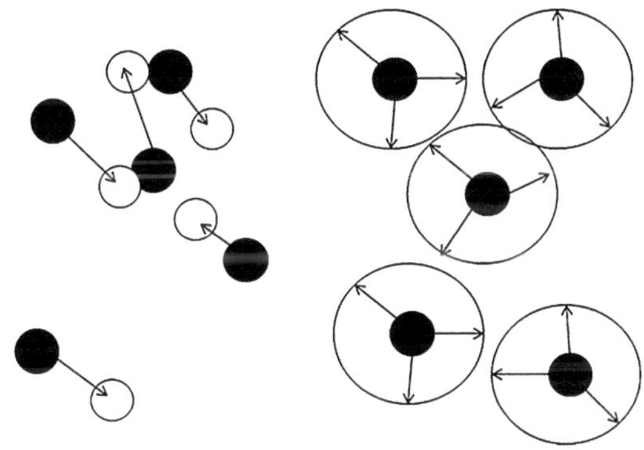

Fig. 7.2 Foraging strategies of absconding (*left*) and stationary (*right*) honeybee colonies. *Black circles* colony nest site, *white circles* forage sites (after Nakamura 1993)

A. dorsata colonies migrate annually in response to resource depletion, allowing them to track seasonally varying sources of forage in different areas (Dyer and Seeley 1994). Colonies arrived at their study site at the end of the dry season, swarmed but stayed in the area until the onset of the rains. Absconding/migrations by thousands of colonies of *A. dorsata* coincide with the monsoon, which determines the dry and rainy seasons, hence the onset of resource depletion (Deodikar et al. 1977; Koeniger and Koeniger 1980; Thapa 1998). Ahmed (1989) showed that migration in *A. dorsata* is facultative and he delayed absconding/migration by re-queening colonies and then providing supplemental feeding. In an analysis of hundreds of colonies of *A. dorsata*, Deodikar et al. (1977) reported that they migrate to the hills in summer and to the plains during the monsoon in response to declining forage or resource depletion, interpretations endorsed by Venkatesh and Reddy (1989) and Ahmed (1989).

In the case of *A. laboriosa*, Underwood (1990) concluded that the effects of altitude are rather marked because these bees are apparently unable to survive the cold of winter on cliffs above 1,675 m but return to higher altitudes with warmer weather. This species is unique in that these bees do not build comb nor rear brood during winter. In spring, the bees migrate back to their traditional cliff sites and begin the annual cycle again.

Absconding and migration under tropical conditions would be advantageous to small, weak colonies, which may not be able to effectively expand their forage areas or to scout for new but patchily distributed resources. Resource-related absconding, often termed "seasonal migration", is a tropical strategy to survive relatively long periods of dearth or the distribution of forage resources at very wide distances apart. When resources deteriorate, the weaker absconding colonies abandon depleted areas and move on. The stronger stationary colonies simply expand the area in which they forage, small incremental increases in radius resulting in very large area gains for little additional foraging distance incurred (Fig. 7.2; Hepburn 2006).

While an absconding or migrating colony could strike out in a random direction, essentially leaving their future resource fate to chance, directions of real colony movements to survive dearth periods will likely vary under different types of vegetative

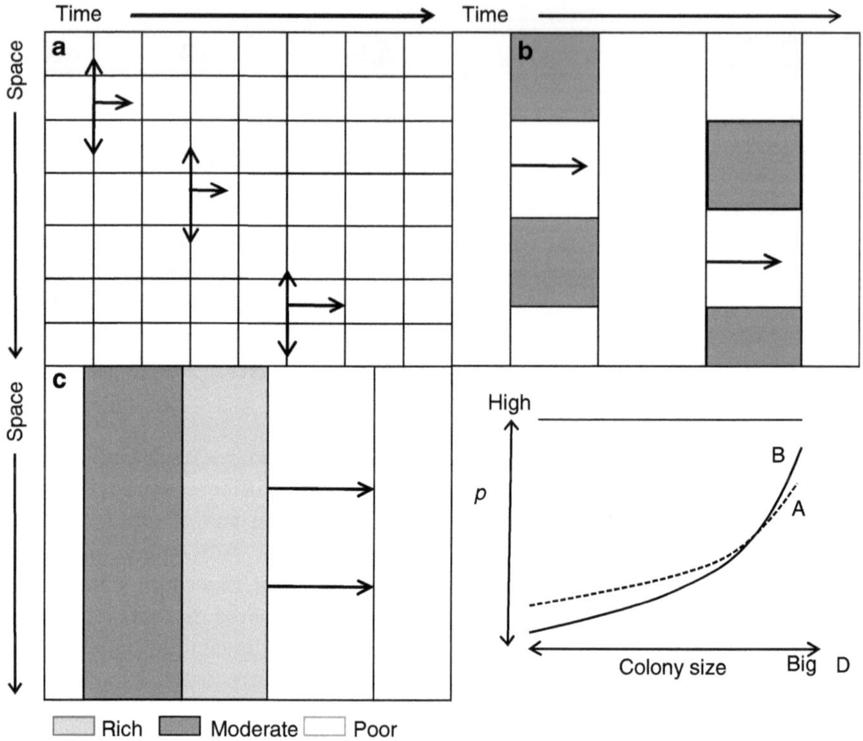

Fig. 7.3 Direction of colony movements during dearth under different resource distribution patterns. (**a**) Patchy resource distribution under aseasonal tropical conditions; (**b**) Large patch distribution of resources under seasonal conditions; (**c**) temperate, seasonal resource distribution; (**d**) Probability of colony movement in relation to colony size, where low *p* is migratory, high *p* is stable and stationary with **a**, **b** and **c** as above (after Nakamura 1993)

resource distributions under differing climatic conditions. In tropical areas where resources are patchily distributed but seasonality is low, a draughts-board solution of colony movement enhances the odds of finding suitable new forage areas. Under tropical and subtropical conditions where vegetative resources occur in larger patches and where seasonal conditions alternate between wet and dry phases, a direction strategy similar to the tropics is preferred, but distance will be greater. In the final scenario of temperate conditions with seasonality and a cold winter, direction of movements is more limited and distances all the greater (Fig. 7.3; Hepburn 2006).

These three strategies can be combined into a probability plot (Fig. 7.4). The relationship between disturbance and the potential for absconding is illustrated in (Fig. 7.5). The literature for *A. florea* also reflects considerable argument as to primary or secondary causes of nest abandonment. It is abundantly clear that there are several sources of stimuli that drive absconding and migration and that no specific one need be unique nor need the several stimuli be mutually exclusive.

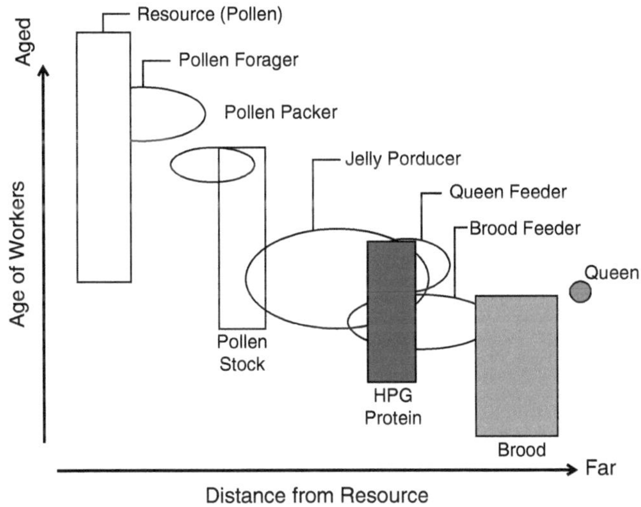

Fig. 7.4 Distance from resources in relation to worker polyethism. *Solid bars* indicate functional buffers within the colony (after Nakamura 1993)

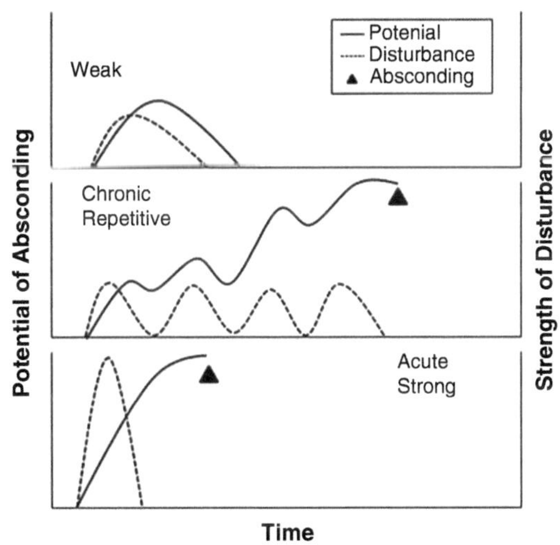

Fig. 7.5 Relationship between intensity of disturbance and absconding potential (after Nakamura 1993)

It appears quite improbable as well as unnecessary that a single factor could explain the observed movements of honeybees when there is preparation before such a move. The workers eat older larvae to produce bee milk to feed the queen and other larvae. Other attempts to prevent absconding/migration by supplementary feeding and manipulation of the combs were also successful (Lin 1977; Olsson 1989; Wei 1994).

7.3.1 Resource Depletion: Pollen

The role of pollen in prepared absconding or migration by *A. cerana* was quantitatively investigated by Punchihewa et al. (1990) in field studies in Sri Lanka. They noted, like Woyke (1976), that one of the distinct characteristics of a colony preparing to abscond is a decline in the numbers of returning foragers carrying pollen. They observed that preparations are taken by a colony before absconding and that it is not instantaneous as in disturbance-induced absconding. The preparative stages are phasic and gradually build up and can be observed externally by measuring a progressive reduction in pollen foragers over time.

Punchihewa et al. (1990) developed what they termed a "colony performance index" (CPI) defined as the total number of returning foragers bearing pollen loads divided by the periods of observations (in seconds) times the total number of returning foragers which is then multiplied by 100 to obtain an index as a percentage. In the course of preparative absconding, the CPI index gradually falls to zero on the actual day of absconding. A simple experiment further illustrates the significance of pollen for prepared absconding. When colonies were transported to pollen-poor sites, workers commenced eating older larvae and stopped feeding younger ones; however, the queens continued to lay eggs (Woyke 1976). Thus, during periods of pollen dearth, the workers greatly reduce brood rearing.

These authors also noted that once the absconding impulse is activated, notable changes in colony behaviour occur. The bees no longer defend the nest against intruding predators/parasites such as wax moths and ants and likewise cease to rear brood although the queen continues to lay eggs. Thus wax moth infestations, broodlessness, lack of stores and reduced pollen forage can be observed in such colonies preparing to abscond; however, Punchihewa et al. (1990) regarded these developments as secondary events associated with absconding and not as primary stimuli. Although no hard numerical data, such as the CPI which measures changes in numbers of foragers returning with pollen that Punchihewa et al. (1990) developed for *A. cerana*, have been published for *A. florea*, declining pollen forage has been noted here also (Phianchareon and Dunagphakdee personal communication).

7.3.2 Resource Depletion: Brood

The obvious effects of resource depletion can be directly measured as changes in honeybee colonies. Dulta et al. (1988) fortnightly recorded the demographic changes in 10 strong and 10 weak colonies of *A. cerana*. In all of the weaker colonies, which duly absconded/migrated, honey and pollen stores, eggs, open and sealed brood decreased by nearly 90%. In contrast, the 10 stronger colonies, which did not abscond had increased honey stores and had only small decreases in pollen stores and brood area. The sequence of demographic changes recorded in *A. cerana* from tropical Sri Lanka (Punchihewa et al. 1990) mirror those from India (Dulta et al. 1988; Pokhrel et al. 2006).

Turning to *A. florea*, Akratanakul (1977) reported from Thailand that about a fortnight before two of five absconding colonies also stopped brood rearing and on desertion of the nests, neither eggs nor brood was left behind. Likewise, Akratanakul (1977) noted that before departure the colony waited for capped brood to emerge. This colony was presumed to have absconded owing to resource depletion. These observations accord with those of Woyke (1976) in India. Another consequential aspect of a brood stimulus is that the presence of brood apparently inhibits/delays the actual onset of absconding by *A. florea* in Oman (Dutton and Free 1979).

The interpretations as to cause are supported by experiments to avert absconding/migration. Woyke (1976) gave two combs of brood and pollen to a colony of *A. cerana* and over the next 2 days the bees ate all of the brood but did not forage for pollen. The colony was then given two additional combs of brood of all ages following which they began collecting pollen and rearing brood and absconding was averted. The principal conclusions reached were that during a poor pollen flow, the queens continue to lay more eggs resulting in more larvae than nurse bees can feed and that workers eat older larvae and produce bee milk to feed the queen and other larvae.

Other ad hoc attempts to prevent absconding/migration (effectively to nullify the dearth) include supplementary feeding with sugar and reducing the numbers of combs before the bees stopped rearing brood but before the pollen flow ended (Olsson 1989). This same approach has become routine practise with *A. cerana* in China (Wei 1994; Lin 1995; Chen 1995, 1996a, b). Alternatively, as in Sri Lanka, colonies can be "artificially migrated" by beekeepers to new areas with ample sources of forage thus obviating the stimulus of localised resource depletion (Koeniger 1976).

Brood plays an important role both as a stimulus and inhibiting factor in relation to absconding/migration. Resource-related absconding or seasonal migration is a strategy to overcome long periods of dearth. However, brood cannibalism, which is common in *A. cerana*, only overcomes short periods of dearth or small fluctuations in forage resources (Nakamura 1993, 1994, 1995). The significance of brood to ameliorate protein shortages is largely related to colony size. There are clear links between brood rearing and "prepared" absconding and it is evident that broodless colonies were the most likely to abscond (Woyke 1976, 1978). Singh and Sharma (1943) considered that broodlessness in *A. cerana* forebodes absconding.

In another experiment, Raj and Channabasavanna (1983) dequeened six broodless and eight broodright colonies and simply moved their queens about 2 m away from their hives. All broodless bees rejoined their queens but only one broodright colony did so. At the end of summer, four broodless colonies given brood did not abscond, while three of four other broodless colonies not given brood did abscond. In the case of *A. dorsata*, Koeniger and Koeniger (1980) concluded that the migrations are essentially driven by climatically induced resource depletion. The experimental evidence for this is that colonies kept in flight rooms in Oberursel exhibited the usual preparations for migration: declining brood rearing a few times a year. They concluded that seasonal migration in *A. dorsata* seems to be genetically fixed.

Resource-related absconding (which is usually seen as seasonal migration) is considered a strategy to overcome long periods of dearth, while cannibalism is adapted to overcome shorter periods of dearth or small fluctuations in forage resources (Nakamura 1993, 1994, 1995). The rationale for this interpretation is that a honeybee colony has several metabolic buffers to reduce the effects of fluctuation in the normal field resources. This includes all pollen stocks, the secretions of the hypopharyngeal glands of the workers, as well as brood which can be recycled to usable form through cannibalism. The spatial relationships of protein sources as buffer zones within the colony are shown diagrammatically in Fig. 7.4.

7.4 Predation Pressure

Pests and predators as well as robbing amongst colonies of *A. cerana* are generally regarded as major contributory factors for prepared absconding throughout Asia (Sakagami 1960; Leong 1987; Verma 1992; Wei 1994; Chen 1995, 1996a, b). Olsson (1989) demonstrated the impact of pests and predators by narrowing hive entrances to restrict the entry of wasps and beetles. Reduction in the numbers of combs and replacement of old with new combs also reduces the chance of wax moths gaining a deleterious foothold in the nest. Likewise, uniting all weak colonies greatly reduced absconding, in part, by providing a stronger defence force against natural enemies.

However, Punchihewa et al. (1990) observed that once the "absconding impulse" sets in, a colony no longer defends nests against pests and does not rear brood even though queen continues to lay eggs. So the broodlessness and wax moths are considered secondary. The full extent of predatory forces acting against Asian honeybees may not be quantifiable but its impact is striking (cf. Toumanoff 1939). Often interactive effects leading to migratory absconding include gradually deteriorating field resources which lead to reduced brood production, hence colony size, providing sufficient imbalance to allow the takeover and total nest destruction by wax moth.

Akratanakul (1977) concluded that absconding/migration in *A. florea* is driven by two main stimuli: resource depletion and serious disturbances by various pests and predators (man, ants and wax moths). When he continuously provided supplementary feeding and pollen was still abundant, a combination of weaver ant and wax moth infestations apparently precipitated their absconding. Moreover, Seeley (1983) thought that nest site selection was largely driven by predator avoidance criteria. While the pests and predators of *A. dorsata* are numerous, there are actually fewer reported links between them and absconding/migration in these bees than might be expected. On the one hand, both Ahmed (1989) and Viraktamath (1989) reported that infestations of wax moths lead to absconding in *A. dorsata*; on the other, Mahindre (1997) thought that a wax moth infestation is not a sufficient stimulus to induce prepared absconding in *A. dorsata* colonies but honey hunters are. Nonetheless, Viraktamath (1989) reported that some 34/56 combs were infested with wax moths before absconding/migrating, and after abandonment, all 56 nests were infested with wax moths, which he regarded as a contributory cause of absconding.

A reasonable compromise interpretation would be that colonies of honeybees preparing to abscond generally reduce their defensiveness against intruders at which time any wax moths present in the nests might begin to expand their activities and numbers and, by the time the colony actually abandons the nest, the wax moth population flourishes and is regarded as a cause of absconding on a "post hoc ergo propter hoc" basis. Kastberger et al. (1999) have elegantly documented the behaviour of honey buzzard attacks on *A. dorsata* colonies, and while these birds certainly evoke a massive defence attack, the bees are not induced to abscond. In the case of *A. laboriosa*, the broodless winter cluster remains close to the ground and the bees are often attacked by drongos and thrushes, which induce some absconding; otherwise migration is temperature and resource related (Underwood 1990).

7.5 Microenvironmental Effects

The interrelationships between variables that may lead to prepared absconding/migration are often intertwined, especially in the case of climatic factors which drive dearth as well as more directly affect the microclimatic conditions of the open-air or cavity-nesting species. Absconding intensity in *A. cerana* is closely related to ecological and climatic conditions: it is most intense in dry, lowland areas and becomes less acute in wetter areas and with increasing altitude in Sri Lanka (Punchihewa et al. 1990). Environmental effects are particularly evident where different climatic systems converge. So, in Sichuan, China, where very weak colonies cannot effectively thermoregulate, they normally abscond and often amalgamate with others (Chen 1995).

More direct effects of microclimate on the absconding impulse are seen in colonies of *A. cerana* kept in hives or natural nest cavities. All of the hived colonies attempted to abscond while none in natural nest cavities did so (Punchihewa et al. 1990). This difference was attributed to microclimatic conditions in the hives. The evidence for this conclusion was that a reduced CPI in hived bees could be experimentally increased by control of temperature and humidity regulation in hives. This has the effect of averting a low CPI index, which activates the absconding impulse. Moreover, Punchihewa et al. (1990) were able to artificially reduce the CPI of a colony by constant disturbance, smoke and high temperatures.

Depreciation of nest quality may result from inundation by rain, overheating, too small a cavity size and fire. Presumably, the efficacy of absconding has had a marked effect on the probability of colony survival in the drier areas, thus selecting against absconding. Nest site selection is of fundamental importance from a microclimatic and thermoregulatory stand point, especially those stimuli which perturb the microclimatic environment of the nest. For example, *A. florea* move to sites providing maximum sunshine in autumn and in spring to sites with reduced insolation (Tirgari 1971; Mossadegh 1990). *A. florea* that dwell in caves move more deeply into them at the beginning of the hot season and, when cooler conditions return, move closer to the opening again (Free 1981). Similar movements have been noted for nest positions in bushes. Nonetheless, Seeley (1983) thought that nest site selection was largely driven by predator avoidance criteria.

Subsequent to absconding, colonies of *A. florea* may cluster for several days at an interim site near the old nest and over the next few days scout bees search for new nest sites (Akratanakul 1977). There are few data for the distances travelled by such colonies, but Soman and Chawda (1996) reported that in 1 year many colonies were found about 20–25 km from where the bees normally occur. Tirgari (1971) observed that colonies of *A. florea* moved only 5–400 m and Whitcombe (1984) observed that absconding colonies only moved short distances of 4–65 m while reproductive swarms moved 25–800 m. *A. dorsata* very evidently move considerable distances in the course of migration; but, as noted by Koeniger and Koeniger (1980), it is extremely difficult to follow migrating swarms in dense jungle or to estimate distances travelled. Nevertheless, they observed that swarms may take as much as a month to reach their final destinations and that they stop frequently along the way, perhaps then moving as far as 100–200 km, a view shared with Lindauer (1956).

Desertion of the nest and successful re-establishment elsewhere requires that the bees have sufficient flight fuel and reserves of energy to construct new combs when they have settled at a new nest site. Engorgement is a normal step in preparing to abscond/migrate in *A. m. scutellata* (Otis et al. 1981) as well as for *A. cerana* in China (Chen 1995). Reduction in egg-laying commonly precedes migration and so conserves protein. Likewise the worker bees commonly eat even sealed brood in the colony before leaving the nest. Thus full preparation for migration includes elevated wax production, increased protein uptake and massive fuel intake.

Because colonies kept in flight rooms at Oberursel, Germany exhibited the usual preparations for migration, declining brood rearing a few times a year, Koeniger and Koeniger (1980) concluded that seasonal migration in *A. dorsata* might be genetically fixed. This is reminiscent of sheltered migratory birds exhibiting zugenruhe. It has also been mooted that there is a genetic basis for absconding/migration in *A. florea* (Koeniger et al. 1982; Whitcombe 1984), but there is not much experimental evidence to assess this hypothesis beyond intuition. By the same token, Muttoo (1957) reported that several colonies of *A. cerana* neither absconded nor migrated in an 11-year period. In a breeding programme using recurrent selection, the tendency for both swarming and absconding could be reduced in *A. cerana* (Deodikar and Thakar 1966) and in *A. mellifera* (Foti 1980). (Nonetheless, it is extremely noteworthy that *A. mellifera* honeybees of temperate European origin have not acquired the absconding/migratory traits even four centuries after their introduction to the subtropical New World.)

The genetic variability for the "prepared absconding" trait is evidenced by honeybee reaction to the seasonal nature of pollen income cycles, which might start and stop during any particular season. Parallel observations are that some colonies preparing to abscond did not collect pollen or rear brood even if pollen flow conditions improved in the area. Yet, other colonies in the same apiary quickly responded to available pollen and reached peak development. So, after such a pollen flow diminished, some colonies immediately ceased brood rearing and prepared to abscond, while others did not. These constitute significant and probably genetically based differences among *A. cerana* colonies exposed to the same environmental conditions in the same apiary (Woyke 1976; Nakamura 1993, 1994, 1995).

7.6 Swarming and Migration

The significance of resource depletion as the general driving force for large-scale migratory movements is also reflected in *A. cerana* (Ahmed 1989; Nakamura 1993). The situation in *A. florea* is somewhat different: in some regions, migration is apparently driven by resource depletion (Thakar and Tonapi 1962) and in others as a response to changes in ambient temperature in the microenvironment of the nest (Tirgari 1971). It would appear that migration is a reasonably clearly defined phenomenon among tropical honeybees where resource depletion is usually the immediate stimulus for movement; it is not directly tied to swarming as such even though the movements deliver colonies to more favourable localities where reproductive swarming may eventually occur.

In China, Fang (1981) employed spring management techniques to prevent swarming in *A. cerana*. This included dividing the bees by groups of "laying colonies", "rearing colonies" and "caretaker colonies". By suitably manipulating and interchanging combs of eggs and brood among these colonies, none attempted to swarm; only those that were actual queen-rearing colonies attempted swarming. So factors affecting absconding are decidedly complex and not necessarily attributable to a single cause. Recent studies continue along this line in China (Zuyin 1991; Shi 1992; Chen 1995, 1996a, b).

Seasonal absconding/migration by *A. florea* contrasts sharply with reproductive swarming in this species. Dutton and Free (1979) and Free (1981) observed in Oman that brood rearing begins in February and peaks in April, with drones being produced towards the end of February, a second peak of brood production occurs in autumn. Little brood is produced in the hot summer months. *A. florea* swarm during the ascendant phase of colony growth during March and April and sometimes have a second peak in September–October. The old queen leaves with the first swarm and, sometimes, as many as eight swarms a year emanate from the same colony (Tirgari 1971). A similar cycle occurs in Iran where brood rearing begins in April and peaks in May when queens are produced and swarming occurs. Likewise, there may be a second phase of swarming in autumn.

Brood rearing starts again at the end of August and continues until the end of November. Drones were reared in mid-March and queen cells produced. No brood is produced during December–January. Analysis of the weather patterns for both

Oman and Iran suggests that the two critical periods for swarming in spring and then sometimes in autumn could be associated with changes in environmental temperatures and or changes in day length. Swarming occurs from mid-April to May. Interestingly, some 24% of swarms were attracted to deserted combs for which old comb wax could well be an attractant.

As to distance swarmed, Tirgari et al. (1969) observed that three colonies of *A. florea* moved between 5 and 400 m between nest sites in Iran. Whitcombe (1984) observed that seven absconding colonies only moved short distances of 4–65 m; reproductive swarming colonies further: 25–800 m. By contrast *A. dorsata* very evidently move considerable distances in the course of migration; but as noted by Koeniger and Koeniger (1980) who tried tracking their movements in Sri Lanka, it is extremely difficult to follow migrating swarms in dense rain forest and jungle or to estimate distance travelled. Nevertheless, they observed that swarms may take as much as a month to reach their final destinations and that along the way they frequently stop. They speculated that the bees move as far as 100–200 km. This estimate of 100 km is supported by Hamlyn-Harris (1902) and Lindauer (1956) and is also implicit in other studies where the presence or absence of bees is noted over the changing seasons. Further observations on flight capacity are reported in Chap. 13.

7.7 Dance Language in Absconding/Migration

The classical work of von Frisch (1967) defined the primary elements of the dance language of honeybees; work greatly extended for Asian honeybees by Lindauer (1956). That a modified dance language is associated with absconding/migrating was demonstrated by Koeniger and Koeniger (1980). They recorded the migration dances of *A. dorsata*, which entail only waggle dances of very long durations. The bees do not return to their starting point at the end of a run but begin anew elsewhere. They do not complete the total cycle of the usual waggle dance, thus separating a migration dance from the waggle dance associated with foraging. Essentially, the same observations were made for absconding colonies of *A. florea* (Free 1981; Dyer and Seeley 1994) and *A. cerana* (Sasaki 1991).

Sasaki (1991) precisely defined the absconding dance in *A. cerana* based on video recordings. Initially, the dance seemed to be performed by few, specific workers within the colony but the majority of bees eventually also danced without first having scouted. The dance consisted of arrhythmic, slow and long waggle runs but lacking the return runs of foraging dances. The duration of the dances and intervals between dances were irregular and indicated twice the estimated distance for foraging in *A. cerana*. When this colony absconded, it settled in a nearby tree. Sasaki (1991) concluded that the function of this dance is indeed to mobilise the whole colony to abscond. Because of the short distance from the maternal nest to an interim site at which the bees first settled and the distance discrepancy indicated by the waggle dance, it is probable that only direction to a new nesting site is encoded in the absconding dance; the slow tempo does not necessarily mean long distance migration.

Because of slight errors in reading the relative position of the sun near its zenith results in very large orientation errors in the waggle dances of other honeybees in the tropics, Duangphakdee et al. (2009) very recently demonstrated that the frequency distribution of absconding by *A. florea* with respect to time was found to be bimodal with a pronounced lull at noonday, which in turn is related to the altitude angle of the sun. Such colonies largely avoid flying off between 12:00 and 13:00 on the one hand and their preferred departure angle of the sun is between 55° and 65°, on the other. Given the difficulties of taking an accurate reading of the sun at angles ±6° of the sun's zenith (resulting in a 1 h loss around noon) and the 2 h required to reach consensus over the final direction to be flown, the bees are simply left with two time windows, morning and afternoon, in which to abscond, and indeed some 90% of the red dwarf honeybee colonies do so. By the same token, the frequency of waggle dancers is also bimodal (Duangphakdee et al. 2009; cf. Chap. 14).

7.8 Post-Absconding Behaviour

In Thailand, subsequent to absconding, colonies of *A. florea* may cluster for several days at an interim site a few metres from the parental nest. After settling in an interim bush, over the next few days scout bees search for new nest sites as indicated by waggle dances performed on the horizontal surface of the cluster (Akratanakul 1977; Duangphakdee and Phiancharoen pers comm.). Although, Whitcombe (1984) never observed the use of interim sites during absconding by *A. florea* in Oman, Tirgari (1971), Dutton and Free (1979) and Seeley (1983) all reported that after absconding, if the new nest site is close to the old one, bees recover wax and honey from old nests in Oman and Thailand, respectively.

A pair of interesting papers on migration in *A. dorsata* from different authors appeared in the same issue of a journal. In both cases, Neumann et al. (2000) working in insular Malaysia (Sabah) and Paar et al. (2000) in Assam used DNA microsatellite analysis to establish that migrating colonies of *A. dorsata* actually returned to the precise nest site from which they had previously migrated. This experimentally confirms the long-standing claims that this is so by local honey-hunters in both countries. Because of the time intervals between the migratory events, it is assumed that none of the workers would have lived long enough to work on memory for the return flight. In both cases, amazement was expressed on the grounds that only the queen would have known both old and new nest sites because all of the workers making the first journey would have died by old age attrition. This is an entirely gratuitous assumption because the average life span of *A. dorsata* workers is simply unknown. Against this, in an extensive study of *A. dorsata* in the Philippines, Morse and Laigo (1969) described swarming and reported that they could stimulate absconding with heavy smoking, but do not even mention migration.

Omani beekeepers believe that *A. florea* that migrate may also return to the very same nest (Whitcombe 1984). Whether there is memory involved or simply coincidence has not been established; however, the odour of comb wax could well serve as a pheromonal attractant for such bees. Koeniger and Koeniger (1980) observed that migrating *A. dorsata* colonies also rested at interim sites for a couple of days. Two or three hours before the swarm was air-borne, flight activity dropped from 10–40 to 1–2 per minute following which another type of dance, the migration dance was observed. Shortly before the swarm moved off, there were hissing sounds and within half a minute the bees moved off.

While the compass directional component of a migratory colony can be measured from the migration dance, it was recently shown that unexpected barriers exist for migratory routes. Kastberger showed that the Brahmaputra River is a major divide and that few colonies traverse this river in Assam. Other circumstantial evidence comes from Mardan (1989) who reports the appearance of large numbers of *A. dorsata* colonies in areas along the southwestern coast of Malaysia previously with only relatively few colonies. He thought that these honeybees derived from the northeastern coast of Sumatera, which is only 40 km away across the narrow Straits of Malacca.

Amalgamation of colonies occurs in *A. cerana* (Nakamura 1993, 1995). Small colonies often amalgamate with other colonies on the move ("allured absconding" of Nakamura 1993, 1994, 1995) or enter the nests of other colonies and suffer a degree of slaughter that usually accompanies such a course (Singh and Sharma 1943). Amalgamation of small swarms (migratory colonies and/or reproductive swarms) into conglomerates provides an immediate solution to building up colony population and reducing pro rata costs of survival, inclusive fitness not withstanding. Significant differences in colony behaviour are associated with the degree of genotypic diversity: the greater the diversity, the greater the averaging effect on phenotypic behaviour (Page et al. 1995). Amalgamation increases the patrilines of the new conglomerates.

Evolutionarily, it also has the effect of directing the energies of non-viable units towards survival of a large social unit whatever the implications for inclusive fitness. It has been established experimentally that genetic variability among workers influences the division of labour (Calderone and Page 1988, 1992). Page et al. (1995) further showed that significant differences in colony behaviour are associated with the degree of genotypic diversity: the greater the diversity, the greater the averaging effect on phenotypic behaviour.

7.9 Reproductive Swarming

Surprisingly little of substance has been published on reproductive swarming by Asian honeybees and most of the literature simply records anecdotal seasonal information. The combined records on swarming for *A. cerana* suggest a wide variation from one population to another: monophasic swarming at higher temperate latitudes and altitudes as in *A. laboriosa* (Underwood 1992), then clearly biphasic swarming in *A. florea* nearer and below the Tropic of Cancer (Dutton and Free 1979; Free 1981; Aggarwal and Kapil 1988; Mossadegh 1990) and appear nearly aseasonal in equatorial regions (Lindauer 1956; Inoue et al. 1990).

The extent of variability in swarming by *A. cerana* is readily illustrated by reports that in southeast China they are prone to swarming irrespective of colony size, congestion or food supply (Kellogg 1941; Gong 1983), a view supported by observations that even small colonies of *A. cerana* swarm on the plains and hills of India (Crane 1993). Indeed, a major conclusion of a detailed analysis of *A. cerana* in Sumatra was that small colonies swarm and neither broodnest congestion nor colony size can be operative stimuli for swarming (Hadisoesilo 1990).

The extent to which swarming is genetically determined is not known; however, Deodikar and Thakar (1966) demonstrated that they could select against swarming by *A. cerana* in India. Others have claimed that beekeeping management techniques can be used to reduce swarming (Fang 1981; Saraf 1985). Against this, in studies of *A. cerana* in Vietnam, Chinh et al. (1996) observed positive correlations between available forage, colony growth and production of reproductives. They concluded that the synchronised production of drones and queens is defined by nutrient flow into the colony. In a high flow, the colony accelerates growth and when large enough, drones and queens are produced, and eventually the colony swarms.

Reproductive swarming in temperate zone honeybees varies with climate and season and more particularly with weather and the availability of forage. When adverse circumstances arise, swarming may be either postponed or aborted entirely. Reproductive swarming in the honeybees of tropical and subtropical Asia and Africa is just as seasonal as in temperate zones, but "season" acquires a different meaning. The sharply defined categories "spring–summer–autumn–winter" of the temperate regions usually resolve into alternating wet and dry seasons in tropical climates; but mountainous areas are often more temperate. It is evident that climatic regime greatly influences swarming behaviour irrespective of the species. The propensity for and frequency of swarming varies among colonies on a regional and often very local level, almost always associated with the sequence: rainfall > flowering > swarming. If the usual season for swarming is preceded by drought and dearth, swarming may be reduced or even totally inhibited.

At the opposite end of the spectrum, unseasonably heavy rains may provide the insertion of an atypical "extra" swarming period. In the more tropical areas of its distribution, *A. cerana* may swarm virtually continuously throughout the year as in southern China (Yang and Xu 1982) or may be biphasic without any obvious seasonal rhythms as occurs in Sumatera (Inoue et al. 1990) and southern India (Venkatesh and Reddy 1989). By contrast, in more temperate regions, swarming is very clearly related to a single well-defined season: spring in Japan (Yoshida 1988, 1997) or summer in northern India (Goyal 1978). In an extensive study of nearly 350 colonies of *A. dorsata* at five sites in India, Venkatesh and Reddy (1989) showed that reproductive swarming is biphasic. The subtropical *A. laboriosa* of Nepal has a single swarming period during April and May (Underwood 1990). Swarming in *A. florea* is often biphasic (Free 1981) but may also occur in a single spring season (Soman and Chawda 1996; Akratanakul 1977). Yet, in very tropical Sri Lanka, swarming is continuous throughout the year (Lindauer 1956).

In southern Japan, *A. cerana* typically swarm just once a year from April to August (Okada 1993; Sasaki et al. 1995; Saji 1996; Sugahara 2000). But biphasic swarming has also been recorded for *A. cerana* in Japan (Sugahara 2000). On the great plains of India, swarming appears confined to April–May (Sharma 1960; Goyal 1978). As one proceeds northwards to the cold hills of the Kashmir Valley, swarming is shifted later in the year to June and July (Saraf 1985). Biphasic swarming by *A. cerana* in China has also been noted (Zhang 1996; Zou 1996).

The literature on *A. florea* suggests that biphasic seasonal swarming is a common characteristic throughout the range of the red dwarf bee. In Oman, *A. florea* swarming occurs during the major phase of colony growth, March–April, but there may also be a second phase of swarming in September–October (Dutton and Free 1979; Free 1981). To the northeast of Iran, Mossadegh (1990) studied *A. florea* extensively for over a decade and recorded two peaks, in May during winter and July during spring. Similarly, Aggarwal and Kapil (1988) noted that swarming occurs during major colony growth (March–April) but may also occur again in September–October. Other brief reports on *A. florea* include those of Franssen (1932), Millen (1944), Ghatge (1949), Akratanakul (1977), Soman and Chawda (1996) and Sihag (2000).

In a very recent study of swarming dynamics, Huang et al. (2009) employed microsatellite markers to determine whether workers can consciously decide to leave or to stay in the colony during swarming, and if they can, what may be an underlying cause. Using three colonies of *A. cerana* under natural swarming conditions, they determined the genetic relatedness and subfamily composition of workers that either did or did not swarm. Their results showed that subfamily distribution of workers in swarms and those remaining in the colony were significantly different from random decision making. Huang et al. (2009) concluded that workers "prefer" to stay with their super sister virgin queens. These data are the first to indicate that genetic relatedness may affect workers making non-random decisions during swarming.

The two giant honeybees, *A. dorsata* and *A. laboriosa*, contrast sharply in swarming behaviour. Venkatesh and Reddy (1989) monitored some 346 colonies of *A. dorsata* in southern India over 2 years and observed that swarming is biphasic with peaks over October–December and March–April. They also concluded that swarming also led to migration because about 300 of their colonies swarmed and then migrated over this period. On the other hand, *A. laboriosa* nest on cliff edges in aggregations at high elevations during the short Himalayan summer and swarm in April–May. They abandon the cliffs and migrate to lower elevations in autumn (November–December) where they cluster without building combs and return to the cliffs in late January–February (Underwood 1992).

The apicultural literature is heavily dominated by studies of temperate *A. mellifera* and among the fundamental k-selection tenets of this literature is that crowding of the nest is a major stimulus for reproductive swarming. While this may well correlate with the issue of the so-called prime swarm, it cannot apply to subsequent swarms from the same colony when population pressure has become considerably reduced. This idea does not hold up to myriad observations on Asian (Takuno 1994; Sharma 1960) and African (Hepburn and Radloff 1998) species and races of honeybees. Kellogg (1941) noted in Fukien Province in the southeast of China that *A. cerana* are prone to swarming, irrespective of colony size, congestion or food supply. A major conclusion is that small *A. cerana* colonies swarm, and broodnest congestion and colony size do not provide convincing explanations as to the cause (Hadisoesilo 1990; Inoue et al. 1990).

7.10 Discussion

The biology of Asian and African species of honeybees repeatedly demonstrates that migration and absconding cannot be rigorously defined, nor the skein of interacting variables quantified. There are no definite boundaries to discriminate these absconding movements; they all overlap in their appearances and processes. Nonetheless, Nakamura (1993) has captured the differences in the relationships between distance of colony movement (migration) and the length of preparation time before movement and nest abandonment can be visualised diagrammatically (Fig. 7.1). Although this depiction was prepared for *A. cerana* by Nakamura (1993), it is generally quite applicable to all species of *Apis*. This diagram accounts for the lack of precisely definable boundaries for the movements of honeybees. The literature overwhelmingly supports three primary stimuli for prepared absconding or migration (when these two terms are synonymised) (1) resource depletion – dearth from either extended dry or wet (monsoon) conditions resulting in depletion of sufficient forage resources; (2) deterioration of the physical microclimate of the nest that bears on nest abandonment; and (3) the persistent infestation and/or disturbance by pests and predators. The intensity and frequency of occurrence of any of these three major stimuli are also likely to be seasonal (and therefore predictable) in nature. Disturbance absconding which is a virtually immediate reaction to severe disturbance and for which there is no nest or colony preparation include natural catastrophes such as fires, floods, hail-storms, or any drastic disturbance such as may be associated with beekeeping (Whitcombe 1984; Wei 1994; Chen 1995).

References

Aggarwal K, Kapil RP (1988) Observations on the effect of queen cell construction on *Euvarroa sinhai* infestation in drone brood of *Apis florea*. In: Needham GR, Page RE, Delfinado-Baker M, Bowman CE (eds) Africanized honey bees and bee mites. Ellis Horwood, Chichester, pp 404–408

Ahmed R (1989) A note on the migration of *Apis dorsata* in the Andaman and Nicobar Islands. Bee World 70:62–65

Ahmed R, Muzaffar N (1987) Some biological observations on the wild colonies of the Oriental bee *Apis cerana* F. in Pakistan. In: Proceedings of 31st international apiculture congress Apimondia, Warsaw, pp 182–183

Akratanakul P (1977) The natural history of the dwarf honey bee, *Apis florea* F. in Thailand. Thesis, Cornell University, Ithaca

Allen MF (1995) Bees and beekeeping in Nepal. Bee World 76:85–194

Calderone NW, Page RE (1988) Genotypic variability in age polyethism and task specialization in the honey bee, Apis mellifera (Hymenoptera: Apidae). Behav Ecol Sociobiol 22:17–25

Calderone NW, Page RE (1992) Effects of interactions among genotypically diverse nestmates on task specialization by foraging honey bees (Apis mellifera). Behav Ecol Sociobiol 30:219–226

Chen SN (1995) Absconding in *Apis cerana cerana*. Apic Sci Tech 2:16, in Chinese

Chen YH (1996a) Reasons for and the prevention of swarming in *Apis cerana cerana*. J Bee 7:18, in Chinese

Chen XG (1996b) The best method to prevent swarming and absconding of *Apis cerana cerana*. Apic Sci Tech 5:17, in Chinese

Chinh PH, Lap PV, Thang DV, Hanh PD, Ngoc PV, Chinh TX, Toan TV (1996) Evaluation of a closed population breeding programme with *Apis cerana cerana* after six years of selection. In: Proceedings of 3rd Asian apiculture association conference, Hanoi, p 47

Crane E (1990) Bees and beekeeping: science, practice and world resources. Heinemann Newnes, Oxford

Crane E (1993) Current status of research on Asian honey bees. In: Connor LJ, Rinderer TE, Sylvester HA, Wongsiri S (eds) Asian apiculture. Wicwas, Cheshire, pp 19–41

Deodikar GB, Thakar CV (1966) Utilization and improvement of the local floras as bee pasturage. Indian Bee J 28:11–16

Deodikar GB, Ghatge AL, Phadke RP, Mahindre DB, Kshirsagar KK, Muvel S, Thakar CV (1977) Nesting behaviour of Indian honeybees. III. Nesting behaviour of Apis dorsata Fab. Indian Bee J 39:1–12

Deowanish S, Wattanachaiyingcharoen W, Wongsiri S, Oldroyd BP, Leepitakrat S, Rinderer TE, Sylvester HA (2001) Biodiversity of dwarf honey bees in Thailand. In: Proceedings of 7th international conference on tropical climates, Chiang Mai, pp 97–103

Duangphakdee O, Radloff SE, Pirk CWW, Hepburn R (2009) Sun angle time windows for absconding by the dwarf honeybee, *Apis florea*. J Insect Physiol 55:1009–1012

Dulta PC, Rana BS, Verma LR, Mattu VK (1988) Absconding behaviour of the Indian honey bee. Indian Bee J 50:67

Dutton RW, Free JB (1979) The present status of beekeeping in Oman. Bee World 60:176–185

Dyer FC, Seeley TD (1994) Colony migration in the tropical honey bee *Apis dorsata* F. (Hymenoptera: Apidae). Insectes Soc 41:129–140

Fang WA (1981) Spring management for controlling swarming by Chinese honeybees (*Apis cerana*). Zhongguo Yangfeng 2:5, in Chinese

Foti N (1980) Data on heritability of the quiet queen supersedure in the Carpathian bee (*Apis mellifera carpatica*). In: Proceedings of 27th international congress on apiculture, Bucharest, pp 247–252

Franssen CJH (1932) Notes on *Micrapis florea* F. Natuur Tijd Nederlandsch-Indie 92:55–63, in Dutch

Free JB (1981) Biology and behaviour of the honeybee, *Apis florea*, and possibilities for beekeeping. Bee World 62:46–59

Ghatge A (1949) Some research work on *Apis florea*. Indian Bee J 11(5–6):9

Gong YF (1983) The natural beekeeping conditions and honeybee races in China. J Fujian Agric Coll 12:241–249, in Chinese

Goyal NP (1978) Performance of *Apis mellifera* and *A. cerana indica* as observed in the Punjab plains. Indian Bee J 40:3–5

Hadisoesilo S (1990) A preliminary study on the swarm of *Apis cerana* at Kuok, Riau province, Indonesia. Bull Penelitian Kehutanan 5:357–363

Hamlyn-Harris R (1902) *Apis dorsata* F. considered in the light of domestication. Entomol Rec J Var 14:321

Hepburn HR (2006) Absconding, migration and swarming in honeybees: an ecological and evolutionary perspective. In: Kipyatkov VE (ed) Life cycles in social insects – behaviour, ecology and evolution. St. Petersburg University Press, St. Petersburg, pp 121–136

Hepburn HR, Radloff SE (1998) Honeybees of Africa. Springer, Berlin

Huang Q, Zeng ZJ, Yan WY, Huang K (2009) Decision making plasticity of swarming honeybees (*Apis cerana cerana*). Res J Biotechnol 4

Inoue T, Adri S, Salma S (1990) Nest site selection and reproductive ecology of the Asian honey bee, *Apis cerana indica*, in Central Sumatra. In: Sakagami SF, Ohgushi RI, Roubik DW (eds) Natural history of social wasps and bees in equatorial Sumatra. Hokkaido University Press, Sapporo, pp 219–232

Janscha A (1771) Abhandlung von Schwärmen der Bienen. Kurzböck, Vienna, in German

Kastberger G, Attenborough R, Reddish P (1999) The magic trees of Assam. Nat Geogr, ZDF, ORF & epo-film, Vienna

Kellogg CR (1941) The Chinese bee. Bee World 11:82

Koeniger N (1976) The Asiatic honeybee *Apis cerana*. In: Proceedings of the 1st international conference on apiculture in tropical climates, London, pp 47–49

Koeniger N, Koeniger G (1980) Observations and experiments on dance communication in *Apis florea* in Sri Lanka. J Apic Res 19:21–34

Koeniger N, Koeniger G, Punchihewa RWK, Fabritius M (1982) Observations and experiments on dance communication in *Apis florea* in Sri Lanka. J Apic Res 21:45–52

Kuznetsov VN (2002) The Chinese wax bee *Apis cerana cerana* F. (Hymenoptera, Apidae) in the Primorski Territory. Institute for Biology and Soil Sciences, Far East Branch, Russian Academy of Sciences, Vladivostok, in Russian

Lavrekhin FA, Pankova SV (1975) Biology of the honeybee colony (*Apis* spp.) in Russia. Kolos, Moscow, in Russian

Leong CT (1987) Apiculture – annual report of the research branch, department of agriculture for the year 1986. Ministry of Agriculture and Community Development, Sarawak, Malaysia

Lin ML (1977) The prevention of swarming and migration of *Apis cerana*. Apic China 1:16, in Chinese

Lin ML (1995) The prevention of swarming and migration of *Apis cerana cerana*. Apic China 1:16, in Chinese

Lindauer M (1956) Communication among the honeybees and stingless bees of India. Zeit Vergl Physiol 38:521–557, in German

Lipinski Z (2001) Essence and mechanism of nest abandonment by honeybee swarms. Z. Lipinski, Olsztyn

Mahindre DB (1997) *Apis dorsata* – are they migratory? Indian Bee J 3:184–188

Mardan MB (1989) Thermoregulation in the Asian giant honeybee, *Apis dorsata* (Hymenoptera: Apidae). Thesis, University of Guelph, Guelph

Martin O (1963) Communication in honeybee swarms. Insectes Soc 10:13–42, in German

Ménendez de Torres L (1586) Brief tract on the cultivation and rearing of honeybee colonies. (reprinted 1983) Torrejón del Rey, AMUVARI, Spain (cited from Crane 1990, in Spanish)

Millen TW (1944) Apis florea. Indian Bee J 6:106–108

Morse RA, Laigo FM (1969) *Apis dorsata* in the Philippines. Monogr Philipp Assoc Entomol 1:1–96

Morse RA, Nowogrodzki R (eds) (1990) Honeybee pests, predators and diseases. Cornell University Press, Ithaca

Mossadegh MS (1990) Nesting behaviour of *Apis florea* F. (Hymenoptera: Apidae) in Khuzestan, Iran. In: Proceedings of 11th international congress, IUSSI, Bangalore, pp 669–670

Mossadegh MS, Birjandi AK (1986) *Euvarroa sinhai* Delfinado and Baker (Acarina: Mesostigmata): a parasite mite on *Apis florea* F. in Iran. Am Bee J 126:684–685

Mutsaers M (1991) Bees in their natural environment in southwestern Nigeria. Nigerian Field 56:3–18

Muttoo RN (1957) Some so called 'peculiarities' of behaviour of Indian honeybees as compared to the European bees. Indian Bee J 19:62–64

Nakamura J (1993) Regulatory system in a honeybee colony toward resource deterioration. Thesis, Tamagawa University, Tokyo, in Japanese

Nakamura J (1994) Absconding as an adaptive behaviour of the Asian honey bee, *Apis cerana*. Bull Fac Agric Tamagawa Univ 34:81–102, in Japanese

Nakamura J (1995) Brood cannibalism in colonies of the Asian honeybee, *Apis cerana*, as response to resource fluctuation. Tamagawa Univ Res Rev 1:33–48 [in Japanese]

Nash WT, Murrell DC (1981) Beekeeping in Bangladesh. Am Bee J 121:352–356

Neumann P, Koeniger N, Koeniger G, Tingek S, Kryger P, Moritz RFA (2000) Home-site fidelity in migratory bees. Nature 406:474–475

Okada I (1993) Swarm of the Japanese honeybee, *Apis cerana japonica* Radoszkowski. Honeybee Sci 14:61–72, in Japanese

Oldroyd BP, Wongsiri S (2006) Asian honey bees. Harvard University Press, Cambridge

Olsson J (1989) *Apis cerana*: management methods for prevention of absconding. In: Proceedings of 4th international conference on apiculture in tropical climates, Cairo, pp 453–457

Otis GW (1996) Distributions of recently recognized species of honey bees (Hymenoptera: Apidae; *Apis*) in Asia. J Kans Entomol Soc 69:311–333

Otis GW, Winston ML, Taylor OR (1981) Engorgement and dispersal of Africanized honeybee swarms. J Apic Res 20:3–12

Paar J, Oldroyd BP, Kastberger G (2000) Giant honeybees return to their nest sites. Nature 406:475

Page RE, Robinson GE, Fondrk MK, Nasr ME (1995) Effects of genotypic diversity on honeybee colony development and behaviour (*Apis mellifera* L). Behav Ecol Sociobiol 36:387–396

Pianka ER (1970) On r- and K-selection. Am Natural 104:592–597

Pokhrel S, Thapa RB, Neupane FP, Shrestha SM (2006) Absconding behaviour and management of *Apis cerana* honeybee in Chitwan, Nepal. J Inst Agric Anim Sci 27:77–86

Punchihewa RWK, Koeniger N, Howpage D (1990) Absconding behaviour of *Apis cerana* in Sri Lanka. In: Proceedings of 11th international congress, IUSSI, Bangalore, pp 106–107

Raj J, Channabasavanna GP (1983) Queen recognition, rearing and colony building by the Indian honeybee (*Apis cerana indica*). In: Proceedings of 2nd international conference on apiculture in tropical climates, New Delhi, pp 378–384

Reddy CC (1983) Studies on the nesting behaviour of *Apis dorsata* F. In: Proceedings of 2nd international conference on apiculture in tropical climates, New Delhi, pp 391–397

Roubik DW, Sakagami SF, Kudo I (1985) A note on distribution and nesting of the Himalayan honey bee *Apis laboriosa* Smith (Hymenoptera: Apidae). J Kans Entomol Soc 58:746–749

Saji O (1996) Ethnoecology of bee and man. II. Traditional beekeeping in Ishizuchi rim region on Shikoku Island. Honeybee Sci 17:49–60, in Japanese

Sakagami SF (1960) Preliminary report on the specific difference of behaviour and other ecological characters between European and Japanese honeybees. Acta Hymenopt 1:171–198

Sakagami SF, Matsumura T, Ito K (1980) *Apis laboriosa* in the Himalaya, the little known world largest honeybee (Hymenoptera: Apidae). Insecta Matsumurana 19:47–77

Sandhu AS, Singh S (1960) The biology and brood rearing activities of the little honeybee (*Apis florea* Fabricius). Indian Bee J 22:27–35

Saraf SK (1985) Beekeeping in Gurez, Jammu and Kashmir, India. Indian Bee J 47:36–37

Sasaki M (1991) Absconding dance in *Apis cerana japonica*: a slow and long tail-waggling motivates the whole colony to translocate. In: Proceedings of 11th international congress, IUSSI, Bangalore, pp 125–126

Sasaki M, Ono M, Yoshida T (1995) Some biological aspects of the north-adapted eastern honeybee, *Apis cerana japonica*. In: Kevan PG (ed) The Asiatic hive bee: Apiculture, biology and role in sustainable development in tropical and subtropical Asia. Enviroquest, Cambridge, Ontario, pp 59–78

Seeley TD (1983) The ecology of temperate and tropical honeybee societies. Am Sci 71:264–272
Seeley TD (1985) Honeybee ecology. Princeton University Press, Princeton
Sharma PL (1960) Observations on the swarming and mating habits of the Indian honeybee. Bee World 41:121–125
Shi D (1992) Swarm prevention in *Apis cerana cerana*. J Bee 4:12, in Chinese
Sihag RC (2000) Ecobiology of the little honeybee (*Apis florea)* in semi and subtropical climates of India. In: Proceedings of 4th Asian apiculture association conference, Kathmandu, pp 46–49
Singh S, Sharma PL (1943) Absconding among Indian bees, *Apis indica*. Indian Bee J 5:98–101
Smith F (1871) Descriptions of some new insects collected by Dr Anderson during the Expedition to Yunnan. In: Proceedings of zoological society of London, pp 244–249
Soman AG, Chawda SSA (1996) Contribution to the biology and behaviour of the dwarf bee, *Apis florea* F. and its economic importance in Kutch, Gujarat, India. Indian Bee J 58:81–88
Sugahara M (2000) Feral colonies of Japanese honey bees, *Apis cerana japonica* and their life history. 2. Natural nests and swarming. Honeybee Sci 21:35–39
Takuno Y (1994) Traditional beekeeping on Tsushima Island. Honeybee Sci 15:59–68 [in Japanese]
Thakar CV, Tonapi KV (1962) Nesting behaviour of Indian honeybees II. Nesting habits and comb cell differentiation in Apis florea Fab. Indian Bee J 24:27–31
Thapa RB (1998) Colony migration of the giant honeybee, *Apis dorsata* Fabr. Thesis, Chulalongkorn University, Bangkok
Thapa R, Wongsiri S, Oldroyd BP, Prawan S (2000) Migration of *Apis dorsata* in northern Thailand. In: Proceedings of 4th Asian apiculture association conference, Kathmandu, pp 39–43
Tirgari S (1971) On the biology and manipulation of *Apis* (*Micrapis*) *florea* F. in Iran. In: Proceedings of 23rd international beekeepers Congress, pp 330–332
Tirgari S, Polous H, Foroughi H, Fakhari F, Nabavit MA (1969) On the occurrence and biology of little bee (*Micrapis florea* F.) and the first account of its migration (Hymenoptera: Apidae). In: Proceedings of national congress on entomology and pest control (Iran), p 4, in Persian
Toumanoff C (1939) The enemies of honeybees. Imprimerie d'Extreme-Orient, Hanoi, in French
Underwood BA (1987) In search of the world's largest honeybee. Glean Bee Cult 115 (234–236):245
Underwood BA (1990) The behavior and energetics of high-altitude survival by the Himalayan honey bee, *Apis laboriosa*. Thesis, Cornell University, Ithaca
Underwood BA (1991) Thermoregulation and energetic decision-making by the honeybees *Apis cerana, Apis dorsata* and *Apis laboriosa*. J Exp Biol 157:19–34
Underwood BA (1992) Seasonal cycle of the Himalayan honey bee, *Apis laboriosa*. Honeybee Sci 13:1–8, in Japanese
Venkatesh G, Reddy CC (1989) Rates of swarming and absconding in the giant honey bee, *Apis dorsata* Buttel-Reepen (1906). Proc Indian Acad Sci 6:425–430
Verhulst PF (1838) Notice sur la loi que la population pursuit dans son accroissement. Corres Math Phys 10:113–121 [in French]
Verma LR (1992) Honeybees in mountain agriculture. Oxford and IBH, New Delhi
Viraktamath S (1989) Nesting behaviour of rockbees, *Apis dorsata* Fabr. in Raichur, Karnataka. Indian Bee J 51:133–134
von Frisch K (1967) The dance language and orientation of bees. Belknap Press of Harvard University, Cambridge
Wei CZ (1994) Reasons for and prevention of absconding in *Apis cerana cerana*. Apic China 2:23, in Chinese
Whitcombe RP (1984) The biology of *Apis* spp. in Oman with special reference to *Apis florea* Fab. Thesis, University of Durham, Durham
Wilson EO (1971) The insect societies. Harvard University Press, Cambridge
Wilson EO (2000) Sociobiology: the new synthesis twenty-fifth anniversary edition. Harvard University Press, Cambridge

Winston ML (1991) The biology of the honey bee. Harvard University Press, Cambridge
Woyke J (1976) Brood-rearing efficiency and absconding in Indian honeybees. J Apic Res 15:133–143
Woyke J (1978) Biology of reproduction and genetics of the honeybee. Bee Culture Division, Warsaw Agricultural University, Warsaw
Yang GH, Xu SY (1982) The distribution and variations of the Chinese honey bee. Zhongguo Yangfeng 3:17–19, in Chinese
Yoshida T (1988) Japanese honeybee, ecology and its rearing method. Honeybee Sci 19:17–26, in Japanese
Yoshida T (1997) Japanese honey bee, ecology and rearing methods IV. Honeybee Sci 18:165–174, in Japanese
Zhang JG (1996) The effects of swarming leading to the old queen leaving the hive in *Apis cerana cerana*. Apic Sci Tech 4:17, in Chinese
Zou XH (1996) Swarming in *Apis cerana cerana*. J Bee 1:18, in Chinese
Zuyin X (1991) An experiment to control the swarming impulse in *Apis cerana* and to increase honey yields during the spring nectar flow. Apic China 50:3–5, in Chinese

Chapter 8
Comparative Reproductive Biology of Honeybees

Gudrun Koeniger, Nikolaus Koeniger, and Mananya Phiancharoen

8.1 Introduction

As a hymenopteran heritage, honeybees have haplo-diploid sex determination. This system of sex determination depends on a single sex locus and inbreeding has a severe negative impact on the viability of the offspring (cf. Chap. 5). Thus, natural selection favoured several behavioural mechanisms to prevent mating among related drones and queens. Altogether, the "general aim" of honeybee mating can be best understood as an optimised mode "of avoiding relatives".

Furthermore, in higher solitary animals (as well as in solitary Apoidea) sexual selection and ecological, physiological adaptations are often counteractive and must be balanced by natural selection resulting in an evolutionarily stable equilibrium, which is a trade off among all selective pressures. This trade off puts strict limits on diversity and specialisation of individuals. Social insects, and in particular honeybees, the evolution of a complex division of labour, have overcome these narrow constraints and resulted in an amazing individual diversity. At the cost of forfeiting its reproductive abilities, the honeybee worker is highly adapted and shaped for its functions within the colony and for foraging. On the other hand, neither the queen nor the drone is able to forage or to take part in any social activity. Instead, queens and drones are restricted to and specialised for their reproductive

G. Koeniger
Institut für Bienenkunde (Polytechnische Gesellschaft), Fachbereich Biowissenschaften, Goethe Universität, Frankfurt a.M, Karl-von-Frisch-Weg 2, 61440 Oberursel, Germany
e-mail: gudrun.koeniger@bio.uni-frankfurt.de

N. Koeniger
Institut für Bienenkunde (Polytechnische Gesellschaft), Fachbereich Biowissenschaften, Goethe Universität, Frankfurt a.M, Karl-von-Frisch-Weg 2, 61440 Oberursel, Germany
e-mail: nikolaus.koeniger@bio.uni-frankfurt.de

M. Phiancharoen
King Mongkut's University of Technology Thonburi, Ratchaburi Campus, Bangkok 10140, Thailand
e-mail: mananya_p@yahoo.com

H.R. Hepburn and S.E. Radloff (eds.), *Honeybees of Asia*,
DOI 10.1007/978-3-642-16422-4_8, © Springer-Verlag Berlin Heidelberg 2011

functions. Thus, it is no exaggeration: the highly developed honeybee society (together with their method of sex determination) has shaped the behaviour and physiology of queens and drones and resulted in several unique, dramatic and even bizarre events which will be addressed in this chapter.

The focus of this book is Asian honeybees comprising eight species. The ninth species, the western *Apis mellifera* of Africa, Europe and the near East, however, is one of the best studied insects, and scientific data on *A. mellifera* biology surpass all Asian honeybees by far. If a complete and general account on honeybee mating and reproduction is intended, the state of the art requires the mandatory inclusion of *A. mellifera* in our descriptions and discussions of reproduction. Furthermore, this decision is justified by the similarities and close relatedness within the cavity-dwelling honeybees, the subgenus *Apis*, which unites *A. mellifera* with four other Asian honeybee species.

8.2 Mating System

Since the early days of science (von Berlepsch 1873), it has been well known that neither a worker bee nor a drone or a queen can exist without the social context of the colony. This is a main argument in favour of the concept of "super-organism" (Gerstung 1910; Moritz and Southwick 1992; Moritz and Fuchs 1998; Hölldobler and Wilson 2009) comparable to body parts of a higher organism – each individual member of a highly developed insect society has lost the ability to survive without its colony. Nevertheless, the individual honeybee and especially queens and drones are still subject to natural selection. Individual selection, however, is tightly interwoven with a complex selection at colony level. Therefore, a description and discussion of honeybee reproductive biology is inevitably caught in a dilemma for which clear evolutionary definitions are not easily available. So our arguments and examples are mainly based on literature, our research experiences and knowledge of Asian honeybees. With this chapter's first section on the "mating system", we present a kind of preface which will enable the readers to get a general idea of *Apis* reproductive biology and to locate sections of their special interests.

8.2.1 Social Structure and Male-Biased Sex Ratio

Generally, honeybee colonies are monogynous, having only one queen which lives together with thousands of workers or, to put it the other way around, each honeybee queen regularly "needs" a colony of several hundreds or thousands of workers. In terms of reproduction, a successful establishment of a daughter queen is inevitably linked to investment in large numbers of worker bees and only colonies, which are big enough for swarming (colony fission), can invest in daughter queens. In comparison, drone production does not involve anything more than the rearing and maintenance of individual drones. Therefore, the "costs" per drone are less than 1% of the investment for a single queen. In conclusion, the mongynous social structure of *Apis*, in combination with colony multiplication by swarming, has resulted in an extremely drone-biased sex ratio (cf. Chap. 5 and Sect. 8.3.4).

8.2.2 Panmictic Drone Congregation Areas and Species-Specific Daily Mating Flight Periods

Apis drones meet at drone congregation areas (DCA) in large numbers and these drone congregations start independently of the queens (Ruttner and Ruttner 1965; Koeniger and Koeniger 2000). Drones fly their waiting loops at the DCA long before the first queen starts out for a mating flight. The landmark which guides the drones, and later the queen, to the DCA seems to be prominent tree canopies in Asian honeybees. The landmarks for *A. mellifera* DCAs are still unknown. The number of drones commuting daily between a DCA ranges from several hundreds to about 16,000. Further, a large number of colonies in the surrounding area also contribute to the drone population of a DCA. In conclusion, the *Apis* queen meets a large number of drones from many colonies at a DCA. Thus the negative consequences of inbreeding in honeybees have resulted in a most panmictic mating behaviour.

Mating season and reproductive swarming in honeybees depend on the availability of ample pollen and nectar. Honeybee swarms are very vulnerable to food scarcity. Therefore, the mating season of honeybees generally depends on flowering cycles (cf. Chap. 7) and under sympatric conditions, honeybee species produce drones and queens simultaneously. The encounter of alien sex partners (drones and queens of different *Apis* species) is avoided by well separated species-specific, daily mating periods (cf. Sects. 8.3 and 8.9).

8.2.3 Queen Polyandry

In colonies of *Apis* species, the number of effective paternities ranges from about 12 to 18. There are, however, exceptions in each subgenus. In the subgenus *Apis*, *Apis nigrocincta* queens have a significantly higher mating frequency. In *Megapis*, *Apis dorsata* colonies also have a higher paternity number, and in *Micrapis*, *Apis florea* colonies have a lower paternity number. A change of paternity numbers has evolved three times independently, and the balance between the costs of additional mating and the benefits resulting from a high number of paternities on the colony level seems to be subject to a high selective pressure (cf. Sect. 8.9). In conclusion, the asymmetry of mating behaviour between the sexes – the ultimate monogamous drone and the extremely polyandrous queen – has reached an extremely high level in the genus *Apis* (cf. Sect. 8.8).

8.2.4 Drone–Drone Competition

The extremely male-biased sex ratio has resulted in a high rate of competition among drones. The drone's sensory input is optimised towards fast queen recognition. The huge compound eyes cover the whole upper part of the head. The number of ommatidia in drones is much higher than in workers or queens and the spatial resolution is well-developed. These functional adaptations are more or less uniform throughout the genus *Apis*. Adaptations of queens to flight speed and detection of pheromones are, in comparison to their male counterparts, less expressed. Thus, selective pressure caused by the male-biased sex ratio has been focussed on the drones, while queens remain more or less unchanged (besides the unquestionable adaptations to high egg production).

Video observations of competing drones of Asian species and *A. mellifera* did not reveal evidence for direct combat among drones chasing after a queen. Observations of colliding drones were rare and regularly resulted in the exclusion of both drones from the comet of pursuers. Competition of *Apis* drones seems to be focussed on fast flight: to be the first when a queen is ready to mate, or to manoeuvre to the best position behind the queen and maintain it whenever she is still engaged in copulation (cf. Sect. 8.5).

8.2.5 Drone Monogamy

The extreme drone-biased sex ratio in the genus *Apis* significantly reduces a drone's chance to successfully mate. The probability to mate even a second time is very close to zero. So natural selection has favoured drones which performed a "total investment" whenever they got a rare chance to meet a young queen. Common among all *Apis* species is the general anatomy of the male genitalia, which consist mainly of a densely folded membrane packed inside the drone's abdomen. Therefore, the organ – specific to the genus *Apis* – has got its scientific name: endophallus (internal phallus). In the course of copulation, the organ is inflated by the drone's haemolymph, and the eversion of this huge endophallus inevitably causes the drone's loss of motility and finally the drone's death. In this regard, the *Apis* drone is a "one-way shuttle designed for sperm transfer" (cf. Sect. 8.7).

8.2.6 Transfer of Spermatozoa

Though the genital tract of the *Apis* queen is well equipped to initiate the ejaculation and the drone's separation, it generally corresponds closely to the female anatomy of Apoidea and other Hymenoptera (Snodgrass 1956). Apparently anatomical adaptations to typical copulation in *Apis* (during flight) are more conspicuous in drones than in queens. Drones of the subgenera *Apis* and *Megapis* deposit their spermatozoa in the oviducts of the queen. In the following hours, the spermatozoa are moved backward from the oviducts to the vagina from where more than 90% of the received sperm masses are expelled. During this process, spermatozoa of many drones pass the orifice of the spermaduct (ductus spermaticus) and eventually enter the spermatheca. Thus the spermatheca is loaded with spermatozoa of many drones, which results in an increase of genetic variability among the workers of the colony. In *Micrapis*, the tip of the drone's endophallus is placed directly at the orifice of the sperm duct. The spermatozoa can move directly into the spermatheca. Drones have fewer sperm than a spermatheca can hold and each drone can deposit about half of its sperm load. Thus also in *Micrapis*, the queen's spermatheca contains spermatozoa of several drones (cf. Sect. 8.7).

8.2.7 Mating Sign: Post-mortem Co-operation of Apis *Drones?*

Video recordings demonstrated that drones (*A. mellifera*) remove the mating sign by pulling it out with a special hairy field of the endophallus (Koeniger 1986). This is present in drones of all *Apis* species. The debate, however, about the purpose (adaptive advantage) of the mating sign – this "costly" drone product – is still in progress. The idea was put forward that the mating sign might facilitate queen recognition by successive drones. Thus, after being successful in the competition with fellow drones, a drone seems to cooperate "post-mortem" with his successors by marking the queen with a conspicuous mating sign in order to reduce the mating-flight time (cf. Sects. 8.5 and 8.6).

8.3 Rearing Drones and Queens and Their Sex Ratio

8.3.1 Drone Rearing

In *Apis*, drones are generally reared from unfertilised eggs. Because drone rearing is costly and drones do not contribute to the economy of the honeybee colony, the regulation of drone production is a major focal point of natural selection and the main prerequisites for this regulation are different sizes of the comb cell. In dwarf honeybees (subgenus *Micrapis*) and in the cavity-dwelling honeybees (subgenus *Apis*), there are two distinct cell sizes, worker bees are reared in smaller comb cells and drones in the larger comb cells (cf. Chap. 6).

Combs in *Apis* are a static and a more or less permanent element. Once the comb is built, the relation of worker cells and drone cells is fixed and cannot be changed unless new combs are built. In the subgenera *Apis* and in *Micrapis*, drone and worker brood cells were counted. In *A. mellifera*, comparison of drone and worker cells in the reproductive season revealed that on average, 5,100 are drone brood and 31,000 worker brood cells, which correspond to 14% of drone brood of the total brood area (Weiss 1962) while Allen (1965) measured about 16%. Considering the total area of combs including cells with honey, drone cell area comprises about 10% in average. In *Apis cerana* and *Apis koschevnikovi*, it is about 11% (Punchihewa 1994; Koeniger unpublished data).

Besides brood rearing, comb cells serve many purposes in the subgenera *Apis* and at times when no drones are reared, drone cells can be filled with honey and serve for food storage. In this way, worker bees have the possibility to effectively use the comb and either filling them with honey or emptying them for drone brood production; the drone cells are adjusted to the seasonal requirements of the colony.

In *Micrapis*, the situation is different. Generally, at the end of the swarming season, colonies of *Apis andreniformis* and *A. florea* abscond and leave the old comb. The swarms start building a new comb, and during the initial period of colony growth, the bees will build worker cells only and continually extend the comb down and sideward. At the beginning of the next flowering season, drone cells are added at the lower edge of the comb and the downward extension of the comb consists mainly of drone cells (cf. Chap. 6). The number is adjusted to the strength of the colony. The queen will lay eggs and a batch of drones is reared and, with a delay of 10–20 days, queen cells are also built and with the sealing of the first queen cells the first swarm with the mother queen leaves the comb. From this moment no further eggs are available and after the last brood has emerged, the last daughter queen with all remaining workers and drones leaves the old comb. The number of worker cells of deserted combs in *A. andreniformis* colonies ($n = 10$) averaged 4,900 and 5,500 in *A. florea* ($n = 6$) and that of drone cells 390 and 570, respectively. Accordingly, in *A. andreniformis* 7.4% of the brood is drone brood, and in *A. florea* it is 9.4% (personal observations). In *Micrapis*, honey is stored exclusively in the top portion of the comb and in contrast to cavity-dwelling species which also use drone cells for storage.

A further regulation of drone production is practised by the queen. Only during a limited period will queens oviposit in the larger drone cells. To perceive the size of the cell, the queen carries out an inspection of the comb cell before oviposition. The head together with the front legs are introduced into the comb cell, and during this cell inspection, the queen seems to measure the size of the cell (Koeniger 1970). In cases of large drone cells, the queen blocks Bresslau's sperm pump, so that no sperm is added to the egg during its passage through the oviduct and this unfertilised egg will eventually develop into a drone. In *A. cerana* and in *A. koschevnikovi*, we observed the queen's cell inspection regularly and the behavioural pattern is nearly identical to the cell inspection of an *A. mellifera* queen. Therefore, we conclude that the recognition of drone cells might be similar in *A. koschevnikovi* and *A. cerana*. Observations on queens of *A. nuluensis* and *A. nigrocinta* are not yet available. The same holds true for *A. andreniformis* and *A. florea*.

The conditions of drone production are different in *Megapis*. In *A. dorsata* and *A. laboriosa*, all brood cells are the same size. Accordingly, a size-related designation of cells for drone brood does not exist though Tan (2007) could show that queens

Fig. 8.1 Difference in body size drones (*below*) and workers (*above*) of Asian honeybees. *From left to right*: *A. koschevnikovi*, *A. nuluensis*, *A. cerana*, *A. andreniformis* and *A. dorsata*

prefer to lay drone eggs into brood cells slightly larger than the average. Only after sealing can drone brood be recognised because of a higher cell capping. In further contrast to *Micrapis* and the cavity-nesting *Apis*, in *A. dorsata*, drone brood cells are scattered irregularly among the worker brood throughout the comb (Tan 2007). Thus the deposition of unfertilised eggs seems to be entirely under control of the queen. There are no data on the number of drone brood. Therefore, we calculate the number of drone cells by assuming that, like in the other species, about 10% from the total brood cells are drone brood cells. The brood area of colonies averaged at $93 \times 50\ \text{cm}^2 = 4{,}650\ \text{cm}^2$ (personal observation; Tan 2007) which equals about 30,400 cells. Accordingly, about 3,000 drones are produced.

The relationship between drone body size and worker body size differs among Asian honeybee species. In the subgenus *Micrapis*, drones are much larger than workers (in *A. andreniformis* about 5 to 1), in *Megapis* drones are of similar size (about 1.2 to 1) and the subgenus *Apis* holds a medium position (Fig. 8.1, Table 8.1).

A last but very important point of regulation of drone production is cannibalism of eggs and young larvae by worker bees. There is also sound experimental evidence available for *A. cerana* (Nakamura 1995), but there is no doubt that cannibalism of drone brood plays an important role in all species of honeybees.

Drones are produced and maintained seasonally, only during the time when colonies can support them and when virgin queens are potentially available. The season matches the time of reproductive swarms (cf. Chap. 10) and is connected in all species to the blooming season. The peak of drone rearing precedes the rearing of queens. In *A. cerana*, appearance of drone brood is described as a "timely indication for the start of the reproductive season" (Punchihewa 1994).

Table 8.1 Absolute weight (mg) of colony members

Species	Worker (*n*)	Drone (*n*)	Young queen (*n*)	Old queen (*n*)
A. florea	25.5 ± 2.6 (25)	77.6 ± 2.6 (25)	86 (5)	95 (2)
A. dorsata	127.8 ± 7.8 (21)	155.7 ± 8.5 (34)	272.0 ± 40.1 (9)[a]	290 (2)
A. cerana	60.4 ± 9.0 (25)	83.4 ± 5.3 (37)	122 ± 13 (25)[b]	
A. melliferra ligustica	85.8 ± 11.8 (25)	211.1 ± 17.6 (25)	202 (3)	240 (2)

[a]Tan (2007); [b]Lap et al. (1992); Koeniger et al. (1993)

Table 8.2 Development time (days) of queens and drones from oviposition to hatching

Species	Queen	Drone	Worker	Reference
A. florea		22.5	20.6	Sandhu and Singh (1960)
A. dorsata	16.5	23.7	19.7	Tan (2007)
A. cerana	14–15	23–24	20	Punchihewa (1994)
A. koschevnikovi	15	23	20	Unpublished data
A. mellifera	16	24	21	Winston (1987)

Drones need from 22.5 to 24 days to develop, 2 or 3 days longer than worker bees and about 8 days longer than queens (Table 8.2). This development seems to be independent of the imago's weight (Table 8.1).

8.3.2 Queen Rearing

Queens in honeybees are not genetically determined. Young female larvae in an early stage are bipotent and can develop into either worker bees or queen bees. The route of development depends on cell type and nutrition (Winston 1987). In the case of queen loss, colonies feed young female larvae in worker cells with royal jelly. They reconstruct and change a worker cell into an emergency queen cell (Fig. 8.2; cf. Chap. 6) from which fully functional young queens will emerge. Emergency queen cells are known from *A. andreniformis*, *A. florea*, *A. cerana*, *A. koschevnikovi* and *A. mellifera*. The rearing of emergency queens seems to be a mandatory response to queen loss in the subgenus *Apis*. In *Micrapis*, however, queen loss often results in the dispersal of the colony, most often in *A. andreniformis* (Koeniger et al. 2000; Oldroyd and Wongsiri 2006).

Royal jelly as the agent of queen determination does not seem to be identical but similar across species borders of closely related species. Young larvae of *A. cerana* grafted in colonies of *A. koschevnikovi* were successfully raised to imagines and a low percentage, also vice versa (Koeniger et al. 1996c). Although early experiments to rear *A. cerana* queens in *A. mellifera* colonies failed (Ruttner 1988), the practical difficulties to do this have now been overcome and these two species can produce queens heterospecifically (cf. Chap. 19).

For queen rearing, workers build special cone-shaped queen cells at the lower edge of the comb so that queen cells are vertically orientated with the opening below. Thus they contrast with the other brood cells of the comb, with a more or less

Fig. 8.2 Sealed emergency queen cells on an *A. florea* comb. After queen removal *A. florea* colonies often feed young female larvae in worker cells with royal jelly and change worker cells into emergency queen cells

horizontal longitudinal axis. The queen cells are rather uniform within honeybees (Fig. 8.3). Only the size of queen cells differs according to the size of the queens (Table 8.1): dwarf bees (*Micrapis*) have smaller queen cells than the cavity-nesting species. Queen cells of *A. dorsata* and *A. laboriosa* have the largest diameter. The number of queen cells per colony depends apparently on colony strength and season but is similar among the honeybee species. They range from 6 to 10 in the dwarf and giant honeybees and from 7 to 15 in the cavity-nesting species (only few data are known for the Asian cavity-dwelling species). Generally, several queen cells are constructed almost simultaneously and upon completion the queen lays eggs in all of them. With the rapid growth of the queen larva, the cell walls are extended downwards. The queen cells are sealed between 7 and 9 days after the larvae hatch from eggs. That seems to be the signal for the old queen to leave the nest together with a fair amount of workers.

The developmental time of the three castes is rather similar among the species studied so far, in spite of the large size differences among the species. Drones need the longest period and queens develop fastest; workers are in a medium position of roughly 20 days. This seems to be a principal character for the genus *Apis* (Table 8.2).

After the prime swarm with the mother queen has left, the development of the daughter queens ends within a short period. In the cavity-dwelling species

Fig. 8.3 Queen cells are generally cone-shaped and at the lower edge of the comb. The *A. andreniformis* queen cell (*right*) still has the capping, which was opened by the emerging queen. The neighbouring queen cell (*2nd from right*) shows a hole at the side, which indicates that this cell was attacked and the queen was killed by her sister before emerging

(*A. cerana*, *A. mellifera* and *A. koschevnikovi*), the first emerged queen starts "piping" (Otis et al. 1995). The thorax is pressed on the surface of the comb and vibrates causing a piping sound. The already fully developed sisters answer to this signal also by vibrating the thorax which, because of the structure of the queen cell, results in a tooting sound. As long as the piping and tooting continues, all other young queens remain in their cells and are fed by workers through small slits at the lower end of the queen cell. As soon as the next swarm with the first young queen has left the nest, the next queen or queens will emerge. Now piping and tooting can start again until a further swarm with a second young queen has left. At the end when not enough worker bees are left to form a swarm, a young queen will eliminate all "surplus" young queens either by killing her sisters in the cell or, if some managed to emerge, in a fight until only one young queen is left. The signal (pheromone) which triggers the fight between the young queens is not yet indentified. Experiments, however, demonstrated that queen recognition and fighting behaviour seem to be similar in all species tested so far. Young queens of *A. florea* were attacked and fought with queens of *A. mellifera* or *A. cerana* (Pflugfelder et al. 2004). In case several *A. florea* queens emerge at the same time, often inferior queens were observed to leave the colony and wait near the nest. Worker bees feed them from time to time. If the queen on the comb does leave with a swarm, one of these inferior queens will enter the comb and be accepted by the remaining bees (Koeniger unpublished observation; Phiancharoen unpublished observation).

8.3.3 Sexual Maturation

After emergence queens are not sexually mature. The chemical composition of the spermathecal fluid still undergoes changes and takes more than 3 days in *A. mellifera* to fully develop (Klenk et al. 2004). *A. mellifera* queens begin mating flights only at the age of 5–6 days. Sexual maturation in the dwarf honeybees is faster as queens begin mating flights already at 3 days old (*A. andreniformis*, Koeniger et al. 2000; *A. florea*, Phiancharoen unpublished observation). Queens may leave with an after-swarm before sexual maturation is completed.

Drones also undergo a period of sexual maturation. Young drones are found in the brood rearing area where they are fed by the worker bees. In contrast to the larval developmental time, sexual maturation seems quite different among the *Apis* species. Sexual maturation in drones can be best measured by the migration of spermatozoa from the testes to an enlarged part of the vas deferens, the vesiculae seminales, where they are stored for the lifetime of a drone. As in *A. mellifera*, drones of the Asian species do not produce additional sperm after emerging from the brood cell and sperm number is therefore fixed. After emergence, the vesiculae seminales are empty (unpublished observation). Within a few days spermatozoa are transferred continually from the testes to the vesiculae siminales. In *A. dorsata*, the process needs about 12 days, the main migration occurs from the third to the eighth day (Tan et al. 1996). Data on the duration of sperm transfer to the vesiculae seminales are still missing for other species but may be faster. For example, the mean age for the first mating flights was 5.6 ± 1.3 days in *A. florea* drones, while it was 11.1 ± 4.8 days in *A. mellifera* (USA) (Rueppell et al. 2005) and 12 days in Europe (Drescher 1969).

In addition, during sexual maturation, several accessory glands produce the substances which later form a mating sign. The cornua of endophalli of freshly emerged drones are whitish in *A. dorsata*, *A. andreniformis*, *A. florea*, *A. cerana*, *A. koschevnikovi* and *A. mellifera* and become orange only after the cornual gland starts secretion within the next few days (personal observation). Also the bulbus glands are active, the secretion forms a ring-like structure in mature *A. mellifera*, *A. cerana*, *A. koschevnikovi* and *A. dorsata* drones. *A. mellifera* drones have, in addition, chitin plates, while in *A. florea* the secretory vesicles remain scattered within the cuticle (Moors 2010).

8.3.4 The Bias in Sex Ratio

Complete data on the ratio of adult queens, drones and worker bees in the Asian *Apis* species are still lacking. So, we counted drone and queen cells of abandoned combs of the dwarf and giant honeybee species. *A. andreniformis* colonies ($n = 10$) constructed an average of 390 drone and six queen cells *A. florea* ($n = 6$) drone cells averaged at 570 and ten queen cells. Assuming that only one set of drones and queens are produced per season, the sex ratio of queens and drones is only about 1–60 (Table 8.3).

In *A. andreniformis*, only two to five queen cells had emerged (recognised by a typical cell opening which was cut by an emerging queen), and in *A. florea*, three to

Table 8.3 Ratio of drone and queen cells

Species (*n* colonies)		Drone cells (calculated)	Queen cells	Emerged queens	Ratio cells
A. mellifera (Europe)	(4)	5,100	10	1–4	1/510
	(1)	5,600			
A. cerana	(2)	900	15	1–5	1/60
A. koschevnikovi	(1)	700	7	1–2	1/100
A. dorsata (Malaysia)	(6)	(3,300)	6	?	(1/550)
A. dorsata (Vietnam)	(9)	(3,340)	9.7	?	(1/300)
A. andreniformis	(10)	390	6	2–5	1/65
A. florea	(6)	570	10	3–4	1/57

four cells had emerged while the other cells showed signs of destruction. Observations of four colonies revealed that from each colony only two virgin queens were able to start a new colony. These results would change the bias towards the drones by nearly five times. In three cases, the combs were abandoned within 2 weeks, one queen produced one set of brood before leaving (Phiancharoen unpublished observation). It seems that more queens are always reared than in the end are provided with a swarm.

At the edge of six empty *A. dorsata* combs, we found an average of 6 queen cups in Malaysia, Borneo (Fig. 8.3; unpublished observation) and 9.7 in Vietnam (Tan 2007). As mentioned above, we calculated about 3,000 drone cells. Thus the sex ratio would be roughly 1/500 in Borneo and 1/300 in Vietnam, which is in a similar range as *A. mellifera*.

For the hive-nesting Asian species, there are only data for one *A. koschevnikovi* and two *A. cerana* colonies (unpublished observation). In these cases, the sex ratio is about or even below 1/100.

These data demonstrate that the sex ratio differs significantly from 1:1 and that the high sex bias of queen and drone brood cells towards males is a general characteristic in the genus *Apis*. This bias in sex ratio can be partly explained by the different costs between queen and drone production. The costs of rearing queens and drones and maintaining them until mating are only slightly different. After mating, the drones die but each queen founding a new colony requires thousands of worker bees.

This leads to the question of whether it is more profitable for spreading the colonies genes by investing in cheap drones than in expensive queens. Evolution surely has resulted in a balance between these two targets but data are still not available. The ratio differs among the *Apis* species from about 1/60 to 1/500. The significance of the "conflict over sex ratio" (Trivers and Hare 1976; Crozier and Pamilo 1996) in honeybees is difficult to assess. Because of the high polyandry of queens and the different proportions of patrilines present in the colony, different interests in the sex ratio by worker bees and queens seem to be of minor importance since natural selection in the genus *Apis* seems to operate mainly on colony level (Hölldobler and Wilson 2009).

8.4 Mating Flight Periods

8.4.1 Season

Mating season in honeybees and reproductive swarming depend on the availability of ample pollen and nectar. At the beginning, large numbers of workers must be reared. Colonies must reach a high population of workers before swarms are issued. Swarms need immediate access to nectar for comb building and pollen for brood rearing, because they have no other reserves than the content of their honey crops. Whenever bad environmental conditions over an extended period limit foraging success, the survival of a honeybee swarm is at risk. Honeybee swarms are very vulnerable to food scarcity. Therefore, the mating season of honeybees generally depends on seasonal blooming cycles (cf. Chap. 7). Accordingly, in Sri Lanka (Koeniger and Wijayagunesekera 1976), in Thailand (Rinderer et al. 1993) and in Borneo (Koeniger et al. 1996b), swarming seasons of Asian honeybees overlap and sympatric *Apis* species produce drones and queens simultaneously. Because of the uniform mode of colony multiplication by swarming within the genus, there seems to be no "evolutionary flexibility" and the reproductive season among sympatric honeybee species is tightly synchronised.

Another cause for similar reproductive seasons among honeybees seems to be the environmental requirements for mating flights. Especially in temperate conditions, mating flights are weather dependent. Strong winds, low temperatures and clouds are known to prevent mating flights of *A. mellifera* queens (Ruttner 1956a, b). So mating and swarming are restricted to a season when favourable weather conditions are most likely to occur.

8.4.2 Time of Mating Flights

8.4.2.1 The "Allopatric Situation" of *A. cerana*

Among the Asian honeybee species, *A. cerana* has the most extensive natural distribution. In consequence, it overlaps with many of the other Asian *Apis* species (Ruttner 1988; cf. Chap. 3). However, regionally, there are large areas where *A. cerana* is the only honeybee. Within the Asian continent, these areas are mainly in the northern part of their range, in mountain ranges and in the Japanese islands (with the exception of Hokkaido). A daily *A. cerana* mating period from 12:30 to 16:00 has been reported from Bihar in North India (Sharma 1960). Drones of *A. cerana indica* (originating from the mountains of the North West Frontier Province of Pakistan) flew between 12:00 and 15:30 in Germany (Ruttner 1973). Verma (1991) observed mating flights of *A. c. indica* queens in the Shimla Hills (north India) between 12:30 and 15:30. In Japan, drones of *A. c. japonica* flew from 13:15 to 17:00, and successful mating flights of *A. c. japonica* queens occurred between 14:35 and 16:35 (Yoshida et al. 1994; Yoshida 1995).

In all of the various places where *A. cerana* occurs as the only *Apis* species, drone flight periods last for more than 3 h and start shortly after noon. The time and overall duration of drone and queen flights and the timing during the early afternoon seem to be typical for honeybees in allopatric conditions. Under these conditions, the data are nearly identical in *A. mellifera* and in *A. cerana.*

8.4.2.2 The "Sympatric Situation"

The daily flight period differs under sympatric conditions with two to four indigenous *Apis* species. Then mating flights occur in a limited time slot, which is different across the species but may differ with different regions (Table 8.4a). *A. andreniformis* drones have a uniform flight period beginning after 12:00 and ending shortly before 14:00 in Borneo, peninsular Malaysia and Thailand as well (Rinderer et al. 1993; Koeniger et al. 2000).

A. florea has a much wider distribution and the daily flight periods of drones show remarkable differences (Table 8.4a). In Sri Lanka, drones flew about 2 h

Table 8.4a Drone flight periods of sympatric Asian honeybee species.

Species	Sri Lanka	Thailand	Sabah, Borneo	Sulawesi
A. andreniformis	–	12:15–13:45	12:00–13:45	
A. florea	12:00–14:30	14:00–16:45[a]	–	
		13:00–15:30[b]		
		13:45–15:30[c]		
A. cerana	16:15–17:15	15:15–17:30	14:00–16:15	12:45–15:00
A. nigrocincta				15:00–17:30
A. koschevnikovi	–	–	16:45–18:30	
A. dorsata	18:00–18:45	18:15–18:45	18:15–19:05	

See also review Otis et al. (2001)
[a]Rinderer et al. (1993)
[b]Buangwangpang (2009)
[c]Koeniger et al. (1989)

Table 8.4b Queen flight periods of sympatric Asian honeybee species.

Species	Sri Lanka	Thailand	Sabah, Borneo	Vietnam
A. andreniformis			12:33–13:10[c]	
A. florea		14:04–16:45[b]		
A. cerana	16:15–16:55[a]			
A. nigrocincta				
A. koschevnikovi			16:45–18:30[d]	
A. dorsata				18.14–14.51[e]

[a]Punchihewa (1992)
[b]Koeniger et al. (1989)
[c]Koeniger et al. (2000)
[d]Koeniger et al. (1994a)
[e]Tan et al. (1999)

earlier than in Thailand. According to Koeniger et al. (1989), the drone flight window of *A. florea* in Bangkok was between 13:45 and 15:30. In Chiang Mai, more than 97% of the drones started after 13:00 and 93% returned before 15:30 (Buawangpang et al. 2009). Southeast of Bangkok, in the overlap zone with *A. andreniformis*, the drone flight period occurred 30–45 min later (Table 8.4a) and ended only at 16:45 (Rinderer et al. 1993). This resulted in an overlap with the *A. cerana* drone flight period for 90 min – the only case so far observed in a sympatric situation. The differences in the mating period of *A. florea* in Thailand may be explained with different times of the year or different weather conditions during data collection or with the presence of *A. andreniformis*.

A. cerana exhibits an even higher variability in drone flight periods. While in Sabah (east Borneo) it starts at 14:00 (Koeniger et al. 1994b), in Thailand (southeast of Bangkok) it starts shortly after 15:00 (Rinderer et al. 1993) and in Sri Lanka only after 16:00 (Koeniger and Wijayagunesekera 1976). In Sulawesi, the start of drone flight is the earliest with 12:45 (Hadisoesilo and Otis 1996). These differences cannot be explained by collecting the data at different times of a year.

The drone flight period of *A. dorsata* seems uniform across Asia; drones start their flights shortly after sunset (Koeniger and Wijayagunesekera 1976; Koeniger et al. 1994c). Near the equator, the shift of sunset during the year is marginal. Further north (Nepal), the change of sunset time during the year exceeds an hour and the drone flight period is shifted accordingly (Woyke et al. 2001).

Queens of all species only start for mating flights at the peak of the drone flight period thus maximising the chance to meet many drones and thus minimising the duration of their individual flights (Table 8.4b). The period of queen flights covers 30–45 min while it lasts generally more than 90 min for drones (except in *A. dorsata*).

8.4.2.3 Situation in Allopatric but Overlapping Habitats

Some *Apis* species have special habitats on mountains where an overlap with the lowland species occurs only in a thin borderline: examples are *A. nuluensis* and *A. cerana* in Sabah, Malaysia and *A. laboriosa* and *A. dorsata* in Nepal. The open-nesting giant honeybee species *A. laboriosa*, adapted to the high altitudes of the Himalayas, has a drone flight period between 12:20 and 14:20 (Underwood 1990). Likewise, *A. nuluensis* is the only honeybee species in the mountains of Borneo above 1,700 m, and its drone flight period is between 10:45 and 13:15 (Koeniger et al. 1996b). Drone flights were recorded only from two *A. nuluensis* and one *A. laboriosa* colony. There is need of additional confirmation.

8.4.3 Time Signals for Mating Flights

The isolated species-specific daily mating periods of sympatric Asian honeybees must be triggered by well-defined time signals. In *A. mellifera*, it is well documented that the internal clock of workers is synchronised on a colony level (Frisch and Koeniger 1994). Whether or not the signal for mating flight activity is based on

Table 8.5 Individual drone flight duration (min)

Species	Duration (min)	Reference
A. florea	31 (±9, SD)	Buawangpang et al. (2009)
A. cerana	20	Punchihewa (1992)
	15	Burgett et al. (2007)
A. koschevnikovi	20	Unpublished data
A. dorsata	20	Tan et al. (1999)
A. mellifera	16–31	Drescher (1969)

internal signals of individual drones or queens or, alternatively, whether a general signal of the colony initiates the mating period was addressed by cross-fostering experiments. Sealed drone brood was partially exchanged between colonies of *A. cerana* and colonies of *A. koschevnikovi*, resulting in colonies with a mixed drone population: *A. cerana* and *A. koschevnikovi* drones were accepted in alien as well in conspecific colonies. Observations of drone flights showed a clear result. The drone flight periods of *A. cerana* drones and *A. koschevnikovi* from the same colony were significantly different. Drones of both species flew according to their species-specific periods whether or not they were kept in alien or conspecific colonies. Thus the period of mating flight might not be determined by the colony. The internal clock of an individual drone seems to trigger the mating flight period (Koeniger et al. 1994b).

8.4.4 Duration of Individual Mating Flights

Across all species drones leave the colony for 20–30 min (Table 8.5) (*A. florea*, Buawangpang et al. 2009; *A. cerana*, Woyke 1975; Punchihewa 1992; *A. dorsata*, Tan et al. 1999; *A. koschevnikovi*, Koeniger personal observation). Individual flight duration seems independent of the size of queens or drones (Koeniger et al. 1993), of the flight speed (Koeniger et al. 2005) and even of the aerodynamic characteristics of the drones (Radloff et al. 2003). So we assume that the provisioning of the drones for the mating flight is regulated by the amount of food a drone can take before starting.

The duration of individual queen flights is more variable and may last from 15 up to 30 min (*A. koschevnikovi*, Koeniger et al. 1994a; *A. dorsata*, Tan et al. 1999; *A. cerana*, Woyke 1975; Punchihewa et al. 1990). In *A. mellifera*, it depends mainly on the rate of successful copulations which queens could achieve (Schlüns et al. 2005; Koeniger and Koeniger 2007). We assume that the same mechanism holds true for the other species.

8.5 Drone Congregation Areas and Flight Behaviour of Drones

8.5.1 Drone Congregation Areas

It has been known for about 100 years that *A. mellifera* drones have a rendezvous place where they circle at a height of 10–30 m above distinct areas, the so-called drone congregation areas (DCAs) (review Koeniger and Koeniger 2000). DCAs were also detected for four Asian *Apis* species as well.

While *A. mellifera* drones assemble over open areas, drones of Asian species use prominent trees as landmarks for their DCA (Fig. 8.4). The characteristics of the actual location at the trees vary among the species and, as known for *A. cerana*, even within its subspecies.

In Sri Lanka, *A. cerana indica* drones gather in close proximity to the trees (Fig. 8.5). These drones restrict their flight to an open space near the canopies of trees and do not follow a (caged) queen far into the open space either above or at the side of the canopy (Punchihewa et al. 1990). In Borneo, drones assemble (like in Sri Lanka) near canopies of trees (Koeniger et al. 1998). In Japan, however, drones of *A. c. japonica* congregate in the open air high above prominent trees (Yoshida and Yamazaki 1993; Fujiwara et al. 1994). In Germany, *A. cerana* drones originating from northern Pakistan visited a DCA in an open valley (Ruttner et al. 1972; Ruttner 1973).

Taken together, DCAs of *A. cerana* show a high degree of variability. Thus the specific features of these DCAs may be mainly the result of local adaptations to environmental factors. For example, avoiding predators, such as birds (*Merops* sp., etc.), by flying near tree canopies might have a higher selective advantage under tropical conditions than in the mountains of northern Pakistan. Considering the limited data available (in comparison to *A. mellifera*), we expect that an even wider range of differences among the DCAs of *A. cerana* may become apparent with ongoing research on Asian honeybees.

Fig. 8.4 Prominent trees are landmarks for drones and queens of *A. dorsata* (A.d.), *A. koschevnikovi* (A.k.) and *A. cerana* (A.c.) for mating places, but each species has a different meeting location at the trees

Fig. 8.5 A dummy queen tied to a long stick is moved within the DCA. Many *A. cerana* drones (*small black objects*) are chasing the queen dummy

Fig. 8.6 *A. koschevnikovi* drones compete for dummy queens under the canopies of lower trees at a height from 3 to 10 m. The drones are flying slowly (at less than 2.6 m/s) around the dummy queen. A drone trying to mount the dummy has a green mark on his thorax. He came from a colony which was about 1,500 m distant from the DCA

Drone congregations of *A. koschevnikovi* were regularly observed to occur under a thick cover of vegetation (Fig. 8.6). The height above the ground of different DCAs varied between 1.5 and 12 m in the only region that has been studied, in the Tenom district (Sabah) (Koeniger and Koeniger 2000). At present, there is no information available on the DCAs of the other two cavity-dwelling honeybees, *A. nuluensis* and *A. nigrocincta*.

In Borneo, drones of *A. dorsata* congregate under the canopy of tall emergent trees (Fig. 8.7). The imminent, tall tree tops seem to serve as visual landmarks, and applying this criterion, several "new" *A. dorsata* DCAs have been located (Koeniger et al. 1994c). The drones of *A. dorsata* assemble under the umbrella of the canopy and do not follow the dummy queen moved into the open air. Further, drone attraction showed a maximum of 3–5 m below the canopy. The height above the ground ranged between 10 and 35 m depending on the size of the tree. Recently, several *A. dorsata* DCAs have been detected under tall trees in Sri Lanka

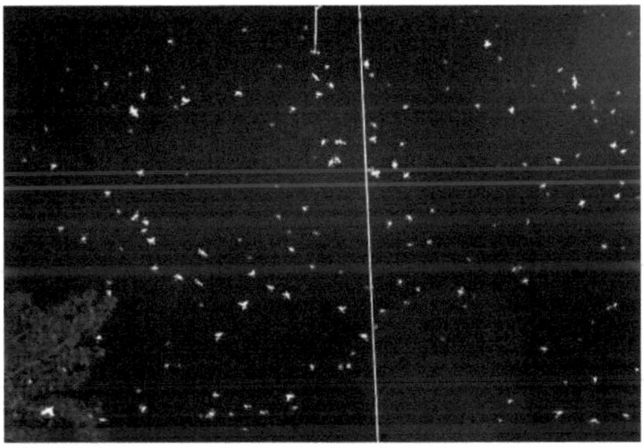

Fig. 8.7 *A. dorsata* drones attracted to a dummy queen (*centre upper edge*). At dusk large numbers of *A. dorsata* drones (more than 5,000) assemble at a height of 20 m below the branches of tall trees where they circle and wait for young queens. These "drone congregation areas" (DCA) are visited by drones from many colonies from the surrounding area. DNA-analysis show that *A. dorsata* drones caught at this DCA originated from more than 50 different mother colonies at distances of several kilometres

(Punchihewa unpublished data). The drone congregation area of *A. laboriosa* remains undiscovered.

Recently, the first description of a DCA of *A. florea* in India was described (Nagaraja and Brockmann 2009). The drones of *A. florea* assemble at DCAs close to trees at a height of 2–4 m. The DCAs of *A. andreniformis* have not yet been found.

It is not surprising that among allopatric species, DCAs show similarities. Some convincing evidence for these similarities came from Ruttner (1973). In Germany, drones from imported *A. c. indica* colonies (which originated from Pakistan) were caught together with *A. m. carnica* at the same DCA and at the same time. On the other hand, in Japan where differences between the DCAs of *A. mellifera* and *A. c. japonica* could be shown, drones of both species where present at either DCA during the overlapping time of drone flight (Yoshida and Yamazaki 1993).

8.5.2 Drone Competition at the Drone Congregation Area

Drones of most species (except *A. dorsata*) leave the colony twice or even three times per day and the mean of total flights are 13.6 days in *A. mellifera* (Rueppell et al. 2005) and 9.1 days in *A. florea* (Buawangpang et al. 2009). For a cohort of 20 *A. florea* males, all flights could be observed and the mean life flight time was summed up to 6h and 38min on 14.8 flights. There are no such data for the other species. But considering these data and the description of multiple flights per drone from other species (own observations) as a basis we hypothesise that the drones' mean total life flight time of a species exceeds more than five times that of the queens' in the respective species. This hypothesis leads to the conclusion that the queen/drone ratio in a DCA is five times more biased towards drones than calculated for the ratio of brood cells. Calculating the reduction in virgin queens because of queen competition, the bias is even more enhanced.

The extremely male-biased sex ratio at a DCA has resulted in a high competition among drones. The flight apparatus is well developed (Radloff et al. 2003). The sensory input is optimised towards fast queen recognition. The drone antenna has a large number of olfactory perception pore plates geared to detect queen pheromones. Drones of *A. mellifera* have about seven times more pore plates than worker bees (Esslen and Kaissling 1976). The huge compound eyes cover the whole upper part of the head. The number of ommatidia is much higher than in workers or queens and the spatial resolution is well developed (Seidl 1980; Vallet and Cole 1993). These functional adaptations are more or less uniform throughout the genus *Apis* (Radloff et al. 2003). The adaptation of queens, however, to flight speed and detection of pheromones are, in comparison to their male counterparts, less expressed. Thus selective pressure caused by the male-biased sex ratio was focussed on the drones and queens remained more or less "unchanged".

When a single queen arrives at a DCA, it is crucial for a drone to recognise the queen quickly, faster than competitors. In many insect species, male insects are attracted by sex pheromones released by receptive females. The mode of orientation via pheromones is most effective when females are stationary and the male can follow the gradient of increasing odour. However, the situation at the DCA is different. During flight, the virgin queen changes her flight path constantly to stay within the DCA. Furthermore, drones fly near the queen wafting the pheromones irregularly. Thus the sex pheromone will indicate the presence of a queen, but is not sufficient to detect the queen's position. For *A. mellifera*, it is known that visual cues play an important role in queen recognition (van Praagh et al. 1980; Vallet and Cole 1993; Gries and Koeniger 1996). *A. mellifera* drones follow butterflies or stones (Ruttner and Ruttner 1965) and similarly drones of *A. cerana* and *A. koschevnikovi* were observed to follow a hornet, *Vespa affinis* (Koeniger et al. 1994d). We cannot exclude that the hornet used additional cues such as odours to attract *A. cerana* drones because it was exploiting the drones' behaviour by catching a pursuing drone.

The visual location of the queen may include various cues. From a more distant position, drones may recognise the dense group of drones following a queen and thus could be attracted from a large distance. Video analysis of drone behaviour pursuing a tethered queen of three Asian species (Koeniger et al. 2005) revealed that the comet-shaped formation of drones in direct pursuit changed within seconds. Further, often all the group suddenly left the queen as if following a fellow drone. For orientation nearer to the queen colour plays a role. In *A. mellifera*, objects with different colours to those of a drone were more attractive. Further, queens marked with the bright white and orange mating sign (Koeniger 1990) attracted more drones than unmarked ones. Since drones of the Asian cavity-nesting species also

Table 8.6 Median numbers of drones in a drone comet (quartiles)

A. cerana $n = 8$	*A. koschevnikovi* $n = 11$	*A. dorsata* $n = 11$	*A. mellifera* $n = 16$
28	9	27	31
(21> < 37)	(6> <14)	(9> < 37)	(20> <41)

produce mating signs (Koeniger et al. 2001), we assume a similar mechanism in these species.

Drones of the species *A. cerana*, *A. koschevnikovi*, *A. dorsata* and *A. mellifera*, which pursued a dummy queen moving in a circular course, flew in a comet-shaped formation below and behind the queen. The median numbers of drones in such a comet ranged from nine (*A. koschevnikovi*) to 31 drones (*A. mellifera*) (Table 8.6). Video observations of competing drones of Asian species and *A. mellifera* did not reveal evidence for direct combat among drones chasing after a queen. Collisions of drones were rare and regularly resulted in the exclusion of both drones from the comet of pursuers. Thus drone density behind the queen and distance to the queen has to be adjusted to avoid collisions between drones. The density of drones depends on their flight speed which differs between the species and seems to be a limiting factor which restricts the number of pursuing drones compared to the several hundreds or even thousands of drones present at the DCA. The median of flight speed near the queen ranged from 2.6 m/s (*A. koschevnikovi*) to 4.1 m/s (*A. dorsata*).

Competition of *Apis* drones seems to be focussed to fly fast in order to be the first when a queen is ready to mate and equally important to manoeuvre to the best position behind the queen and maintain it whenever she is still engaged in copulation. This seems to be difficult: the median duration of a drone's presence in the mating comet did not exceed 2 s. Either by overtaking or leaving/entering the comet, drones seem to compete for more promising positions in the pursuing comet. Good visual perception and the ability of high acceleration are necessary to reach and maintain such a position. Drones of all species can accelerate from 10 to 20 m/s^2 (Table 8.7). Drones keeping a position within a space of not more than 2,000 cm^3 behind the queen were able to grasp the dummy and these successful drones often managed to stay longer near the queen than their fellow drones (Koeniger et al. 2005).

8.5.3 Panmictic DCAs

Apis drones meet at a DCA in large numbers and fly their waiting loops before queens start for the mating flights. The number of drones commuting daily between the colony and a DCA ranges from about 16,000 (*A. mellifera*) to several hundreds in Asian species. The flight range per drone varies from 1 to 2 km for *A. cerana* and *A. koschevnikovi* (Tingek, personal communication), up to 5 km and more in

Table 8.7 Flight characteristics of drones in a mating comet

	A. cerana	*A. koschevnikovi*	*A. dorsata*	*A. mellifera*
Fresh weight (mg)	80	103	155	224
Quartiles	74 > <82	98 > < 107	151 > < 159	98 > < 107
n	31	30	26	35
Speed (m/s)	3.7	2.6	4.1	3.2
Quartiles	3.5 > < 4	2.6 > < 2.9	3.8 > < 4.7	2.8 > < 3.7
n	75	45	95	98
Stay of successful drones in comet (s)	1.0	3.5	2.9	1.0
Quartiles	0.9 > < 1.4	1.3 > < 7.3	1.9 > < 3.7	0.9 > < 1.7
n	11	15	11	24
Acceleration (m/s^2)	10	10	20	10
Quartiles	4 > < 20	5 > < 20	10 > < 35	5> < 15
n	113	323	171	455

A. mellifera (Ruttner and Ruttner 1966). Further, a large number of colonies in the surrounding area contribute to the drone population of a DCA. On one DCA of *A. dorsata*, drones of 53 colonies were estimated by microsatellite analysis (Kraus et al. 2005a, b), and in *A. mellifera* more than 200 colonies had sent their drones to one DCA (Baudry et al. 1998). In conclusion, the *Apis* queen meets a large number of drones from very many colonies at a DCA. Among the terrestrial animals, the *Apis* DCA is one of the best panmictic systems.

8.6 Mating on the Wing: A Technical Problem

Mating on the wing can be technically compared to material transfer from a shuttle to a space station. The technical procedure requires a step-by-step approach. First the shuttle must manoeuvre to the correct position near the lock of the space station and then it is connected to the outside of the space station. After further adjustment of its position, the shuttle docks securely to the station so its lock is placed on the still closed lock of the station. Then the touching surfaces of both objects must be sealed before the lock of the station and the gate of the shuttle open. Now the shuttle must be internally connected to the station and the material transferred from the shuttle into the space station. At the end, the internal connection must be removed and the locks closed before the shuttle separates from the space station. The docking mechanism and lock at the space station are then free for the next shuttle. This technical description may help to understand the different phases of honeybee mating. One must, however, not forget that natural copulation high in the air by a fast flying honeybee queen and a pursuing drone is more complicated than the above space operation.

8.6.1 Initial Docking

The behaviour of free flying drones behind a rotating mounted queen or a dummy queen was recorded and analysed (Gries and Koeniger 1996; Gries 1997; Koeniger et al. 2005). The drones of the cavity-dwelling species (*A. koschevnikovi*, *A. cerana* and *A. mellifera*) pursue the dummy queen generally from behind and below. From there the successful drone flies upward and accelerates until the drone's thorax touches the dorsal surface of the dummy queen's abdomen. Legs 1 (front) and legs 2 hold the abdomen laterally. Legs 3 then grasp the abdomen from below. The inner surfaces of the tibiae and basitarsi of legs 3 press the queen's abdomen under the drone's thorax, holding the queen actively in a dorso-ventral brace (Fig. 8.8). Thus the initial docking is accomplished and the still flying drone is securely attached to the surface of the queen's abdomen. Now the drone can adjust his position precisely until he is able, by bending his abdomen, to place the tip with the genital orifice inside the open sting chamber of the queen (Fig. 8.8).

Drones of *A. dorsata* grasp the abdomen in a similar way as described above but the connection is further supported by "an adhesive organ" at the inner surface of basitarsus 3. The organ has a 0.5 mm thick brush consisting of robust, palm like hairs (Ruttner 1988; Fig. 8.9).

The structure of the basitarsus 3 of the *A. florea* drone, however, is most remarkable. It is bifurcated and a "thumb" together with the main trunk of the basitarsus forms a clasp organ with stiff bristles on the inner surface of the basitarsal thumb and smooth plumose hairs opposite (Fig. 8.9). It fits exactly to the queen's hind tibia (Ruttner 1988). The clasp of *A. andreniformis* drones is similar, only the "thumb" is shorter compared to *A. florea* drones (Wu and Kuang 1986).

In conclusion, the initial docking of the *Apis* drone to the flying queen is achieved by the drone's legs, which grasp the queen's abdomen. In the cavity-dwelling species no "special" organs support a strong connection, while *A. dorsata* drones have an adhesive organ at the inner surface of hind leg's basitarsus. In *A. florea* and *A. andreniformis*, the basitarsus of the drone's hind leg is bifurcated forming a clasp organ to hold the tibia of queen's hindlegs (Fig. 8.9).

Fig. 8.8 An *A. mellifera* drone successfully holds a tethered queen in a dorso-ventral clasp. Because of the fast wing beat the wings appear blurred

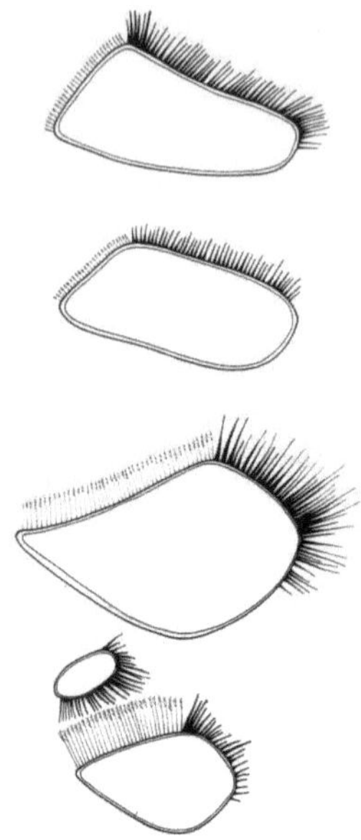

Fig. 8.9 Cross section through the basitarsus of the drone's hind leg. The hair on the inner surface of *A. dorsata* consist of a 0.5 mm thick brush of robust, palm like hairs, *A. florea* drones have plumose hairs and opposite, a thumb with stiff bristles (A. m.: *A. mellifera*, A. c.: *A. cerana*, A. d.: *A. dorsata*, A. f.: *A. florea*) after Ruttner (1988)

8.6.2 Internal Connection of Drone and Queen

As a result of the initial docking, the flying drone places his genital opening into the open sting chamber of the flying queen. The soft tissue of the wall of the queen's sting chamber smoothly clings to the surface around the tip of the drone's abdomen. Then with a sudden contraction of the abdominal muscles the drone presses its haemolymph in the endophallus. The endophallus consists only of a thin chitinous membrane which is densely folded in the abdomen prior to copulation (Koeniger et al. 1991; Patinawin and Wongsiri 1993). The pressure from the haemolymph inflates the endophallus turning it inside out (eversion). Using only its own force, the drone can evert only half of the endophallus, then he becomes paralysed and the motionless body loses its grip and is forced about 90° backward by the form of the inflated endophallus (Fig. 8.10) (Koeniger 1986).

In this phase, the half everted endophallus terminally forms a thin tube (cervix) which contains the sperm, while the basal part is extended by large masses of mucus and firmly fills the queen's sting chamber and vagina like a cork in the neck of a bottle (Fig. 8.11).

The anchorage of the paralysed drone is probably further supported by the queen's complex system of muscles consisting of several ring (sphincter) and longitudinal groups of muscles at the vagina and median oviduct and a conspicuous muscular valve fold in the vagina (Camargo and Mello 1970; Figs. 8.12 and 8.13). The short, thick cornua of the endophallus with its "orange coloured" sticky and greasy secretion may further contribute to strengthen the attachment. All these

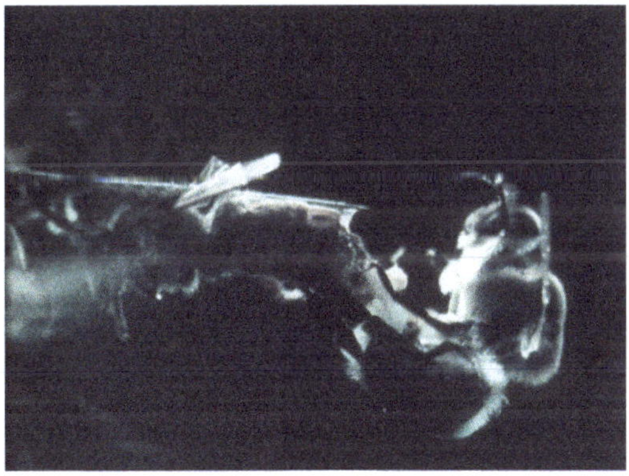

Fig. 8.10 The paralysed drone lost its grip on the queen and the connection between the queen and drone is now secured internally by the half everted endophallus

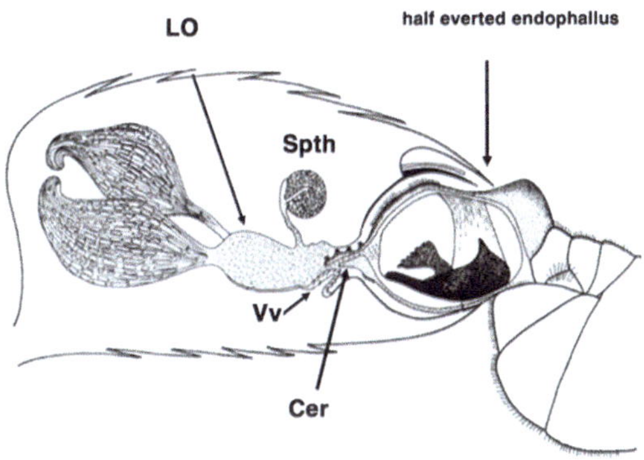

Fig. 8.11 Anatomy of a copulating pair. The endophallus is half everted, the cervix (*Cer*) is still a thin tube and enters the median oviduct for sperm transfer behind the valvula vaginalis (*Vv*). The basal part is filled by a large mass of mucus strengthening the internal connection between queen and drone. *LO*: lateral oviducts filled with sperm; *Spth*: spermatheca connected to the median oviduct

characters guarantee a sufficiently strong connection between the flying queen and the paralysed drone.

This mechanism of relying solely on the internal connection is proven for *A. mellifera*. Because of the similar endophallus morphology in the cavity-nesting species (Figs. 8.14 and 8.15), we suggest that it is similar in all species of the subgenus *Apis*.

In *A. florea* and *A. andreniformis*, drones also lose mobility after eversion of their endophalli, but these species developed an elegant technique to keep an external anchor point to the queen. The forceps-like clasps at the metatarsus of the drone's hind leg hook to the hind legs of queen (Fig. 8.16) (Ruttner 1988) and keep the pair locked together supported by the sticky secretion of the cornua (Koeniger and Koeniger 1991). Thus these drones do not need to produce mucus to enlarge and strengthen the endophallus and their male accessory glands are reduced (Koeniger et al. 1991). The queen's genital muscular system is similar to that of *A. mellifera* (Camargo 1972) and will support the connection.

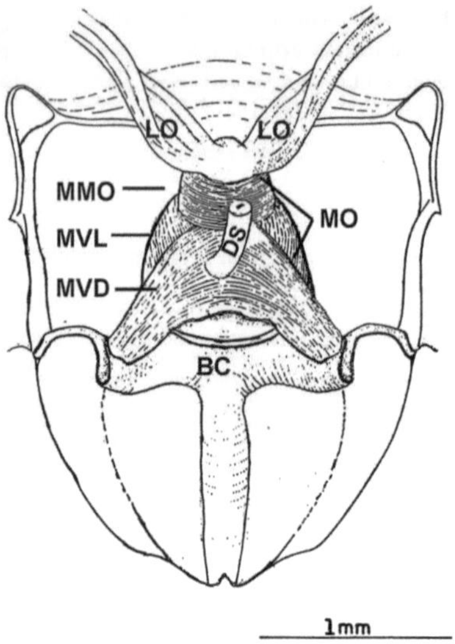

Fig. 8.12 Dorsal view of the muscle system of the genital duct of *A. florea* (after Camargo 1972). *BC*: bursa copulatrix, *DS*: ductus spermaticus, *LO*: lateral oviduct, *MO*: median oviduct, *MMO*: muscle of median oviduct, *MVD*: dorsal muscle of vagina, *MVL*: lateral muscle of vagina, *V*: vagina

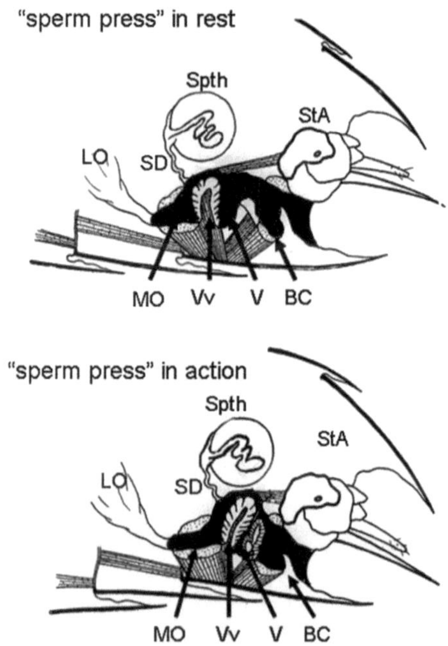

Fig. 8.13 Muscle action of bursa copulatrix, median oviduct and valve fold of *A. mellifera* during the filling of the spermatheca (after Ruttner 1956a, b). Abbreviations: *BC* bursa copulatrix, *DS* ductus spermaticus, *LO* lateral oviduct, *MO* median oviduct, *Spth* spermatheca, *StA* sting apparatus, *V* vagina, *Vv* vaginal valve fold

A. dorsata drones also have a special organ to keep an external bond: the adhesive organ at the hind leg functions as a hook-and-loop fastener beyond the drone's active grasping of the queen (Ruttner 1988). Queens of *A. dorsata*,

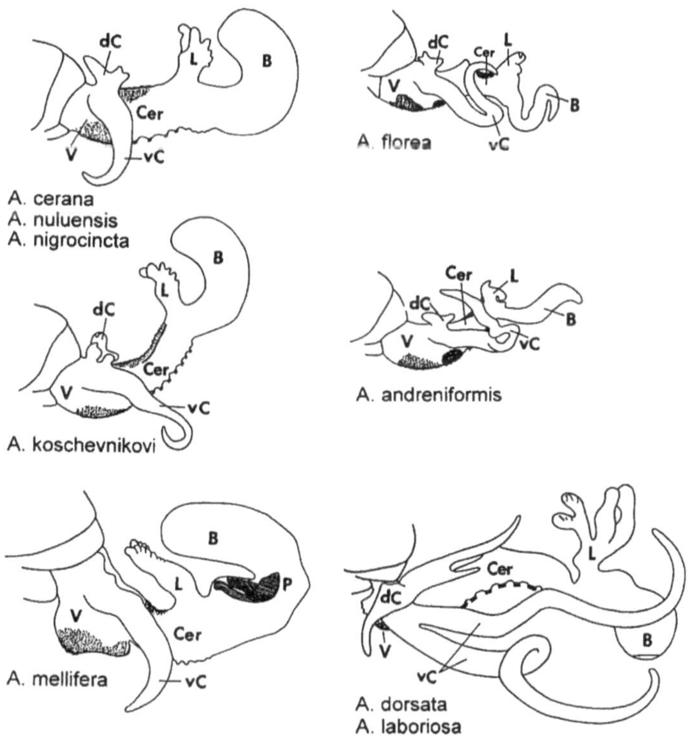

Fig. 8.14 Everted endophalli of *Apis* spp., *lateral view*. *B* bulbus, *Cer* cervix, *P* chitinous plate of the bulbus, *dC* dorsal cornua, *L* lobe, *V* vestibulum, *vC* ventral cornua, *dotted area*: hairy fields

however, lack the muscle system at the median oviduct and also at the valve fold (Camargo 1972). Connection of the *A. dorsata* pair during the copulation may be achieved by the basal part of the endophallus filled with mucus, the extreme length of the thin tube-like cervix equipped with many bristles and the long curled sticky cornua (Figs. 8.14 and 8.15) which we observed in each of 25 copulations of *A. dorsata* drones inside the queen. These observations do not support a suggestion of Woyke (2001) that the cornua of the *A. dorsata* endophallus are attached to the surface of the queen's abdomen.

To sum up, we want to state that the connection of the immobile drone is mainly achieved internally by the endophallus in the subgenus *Apis*. The *Megapis* and as well as the species of *Micrapis* have special organs to keep an additional external anchor point but their mechanisms are different. To come back to our comparison with the space shuttle, there the strong connection at the outside of the space station is kept until all material transfer is achieved and undocking can start. In the honeybees, transfer of spermatozoa and mating sign is more complicated because of the double function of the endophallus – as a coupling element and as well as a transfer system of sperm and mating sign.

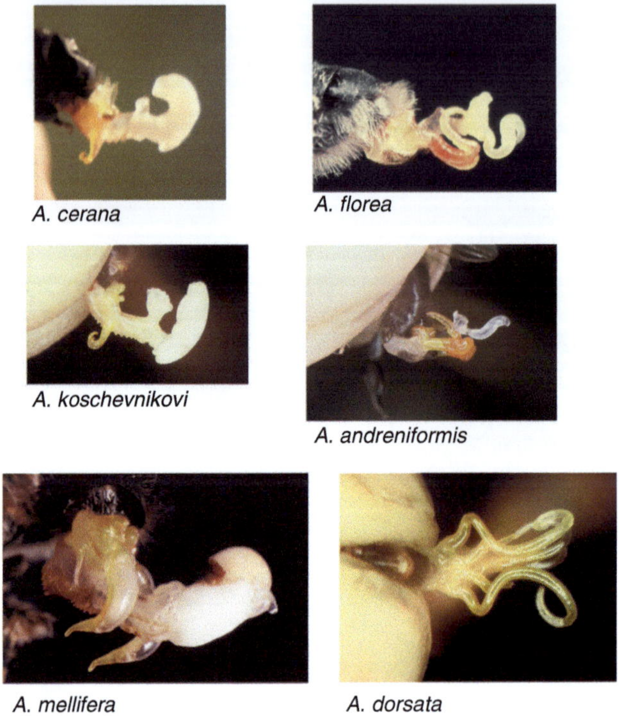

Fig. 8.15 Everted endophalli induced by pressing hemolymph from the abdomen by hand. Mucus and sperm remain inside the endophallus at its distal part (*whitish*), while a sticky orange secretion covers the outside of the horn like appendages at its anterior part (photos: *A. f.*, *A. k.*, *A. a.*, and *A. d.* Gard Otis)

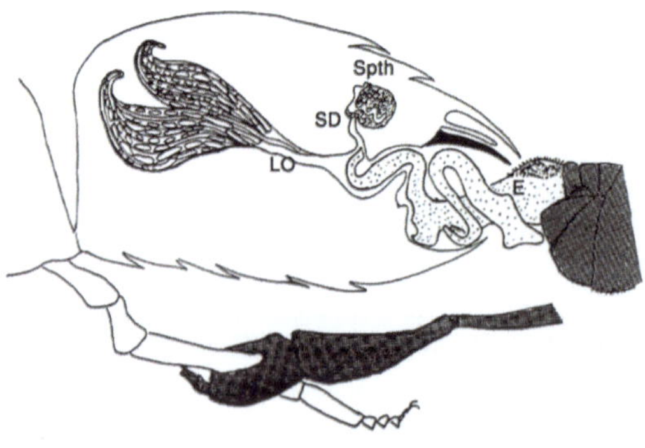

Fig. 8.16 Hypothetical reconstruction of the anatomy of *A. florea* in copula. The "thumb" of the drone's hind leg strengthens the connection between queen and drone. The tip of the endophallus (*E*) is placed by the drone directly at the orifice of the ductus spermaticus (*SD*) to the spermatheca (*Spth*). *LO* lateral oviducts

8.6.3 Deposition of Sperm and Decoupling

The drone is immobile after insertion of the half everted endophallus and is no longer in command of its muscles. Thus the sperm transfer and separation of the pair is initiated by the queen and it happens within a split second. The queen contracts her genital muscles and those of the sting chamber. First, the sperm transfer from the thin tubular cervix into the oviducts is completed and the oviduct is probably closed by the vaginal valve fold.

Almost at the same time the full eversion of endophallus and cornua follows with dramatic effects. In the cavity-nesting species, the mucus is still inside the endophallus and firmly fills the sting chamber. The thin membrane of the endophallus is pealed off during full eversion. First, the thin tubular cervix opens its folds by unzipping the initial hairy breech (Woyke 2008) and thus can glide over the mucus, followed by the last part of the endophallus, the bulbus. At the same time, the cornua are fully everted and push the endophallus (with the immobile drone!) out of the queen (Fig. 8.17), leaving their sticky orange secretion together with the mucus in the sting chamber, the so-called mating sign. In conclusion, the above procedure does not involve a rupture of the drone's endophallus.

In *Micrapis*, decoupling of the pair seems easier. The queen turns the hind leg in a way that the thumb-like clasp is detached. Simultaneously, she loosens the grip of the endophallus with her genital muscles and the drone will fall down. Only the cornual secretion is left in the sting chamber (Koeniger et al. 2000).

A different mechanism must be used by *A. dorsata*. Uncoupling the hind legs of the drone from the queen's abdomen may be achieved by pushing them off with the queen's hind legs. In the process of full eversion, the long curled cornua may push the drone back (Fig. 8.18). It cannot yet be decided if the mucus is left in the sting chamber similar as in *A. mellifera* or if it will be removed together with the endophallus. In five queens returning from 14 mating flights, no mating signs

Fig. 8.17 The thin membrane of the endophallus (*A. mellifera*) successively peels off the mucus and bulbus secretions which are left behind together with the orange cornual secretion as mating sign in the queen' sting chamber

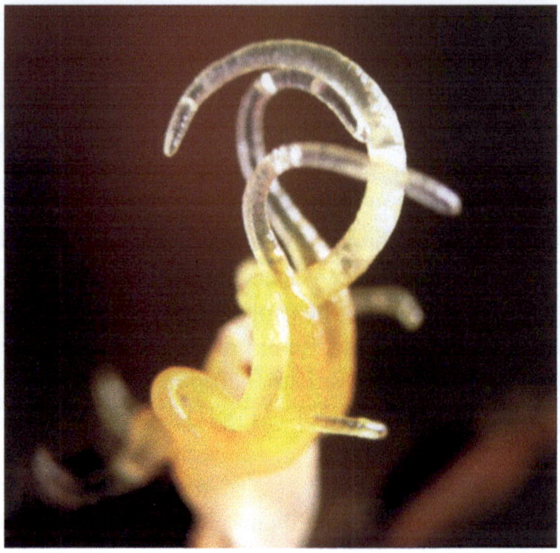

Fig. 8.18 In *A. dorsata*, the lower part of the cornua is covered with the sticky secretion of the cornual gland, but the tips are empty. This may be interpreted that when the cornua are fully everted, they may assist in pushing back the drone after sperm transfer

were seen (Tan et al. 1999). During the experimentally induced eversion of the endophallus, no mucus was expelled (Woyke 2001).

In *Apis*, endophalli do not rupture during the uncoupling of the pair. Thus the detachment process must not be understood as an autotomy (like the sting). Sperm, and later male gland secretions, are transferred to the queen. However, the drone will die after falling to the ground because its haemolymph remains irreversibly in the everted endophallus.

Returning to the initial comparison, undocking of a space shuttle is principally different from the separation of the drone from the queen. Before a shuttle separates from the station the internal connection is hauled inboard, the locks are closed and only then are the external connections detached. Many space shuttles can be reused several times; the drone, however, is a kind of "one way shuttle", once his sperm is transferred, he is bound to die. As discussed earlier, this might be due the male-biased sex ratio in honeybees.

8.7 Mating Sign

During copulation, drones transfer in addition to sperm some structured secretions, the mating sign (Fig. 8.19), which is deposited in the queen's sting chamber. Queens of cavity-nesting species return from mating flights with a mating sign that protrudes from the sting chamber (Fig. 8.20). It is generally removed only after entering the nest by rubbing the abdomen on the comb or with help by workers and is carried out of the hive within a couple of minutes after return. These mating signs were described for *A. mellifera* (Woyke and Ruttner 1958), *A. cerana* (Woyke 1975) and *A. koschevnikovi* (Koeniger et al. 1994a). In the free nesting dwarf honeybee *A. andreniformis*, the structure of a mating sign does not contain white

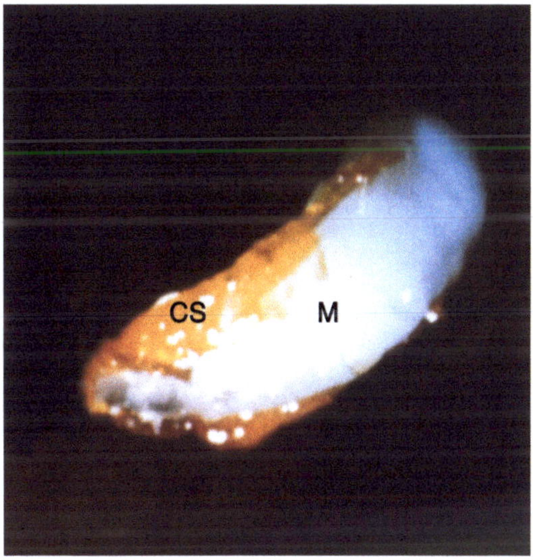

Fig. 8.19 Mating sign of *A cerana*. *M* mucus, *CS* corneal secretion

Fig. 8.20 An *A. koschevnikovi* queen returns from a successful mating flight to the nest entrance. The mating sign protrudes from her sting chamber

mucus but only two stripes of an orange coloured secretion (Fig. 8.21; Koeniger et al. 2001). In *A. florea* (Koeniger et al. 1989) and *A. dorsata*, queens as well no male secretions were detected in the queen (Tan et al. 1996) though *A. dorsata* drones have well developed mucus glands (MG) and in both species the endophalli produce the orange coloured corneal secretion.

To explore the composition of mating signs of *Apis* species, it is helpful to look at the anatomy and structure of endophalli (Fig. 8.22). The mucus is produced by the male accessory glands (mucus glands MG), the connective substances connecting the chitin plates in *A. mellifera* originate from the bulbus gland and the orange coloured secretions from the cornual glands (CG) (Koeniger et al. 1989; Koeniger et al. 1996a). MG are well developed in all drones of the hive-nesting species and in

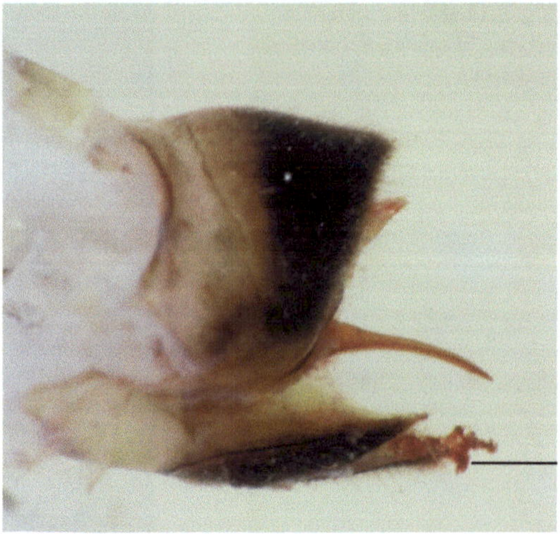

Fig. 8.21 The mating sign of *A. andreniformis* consists (arrow) only of the cornual secretion

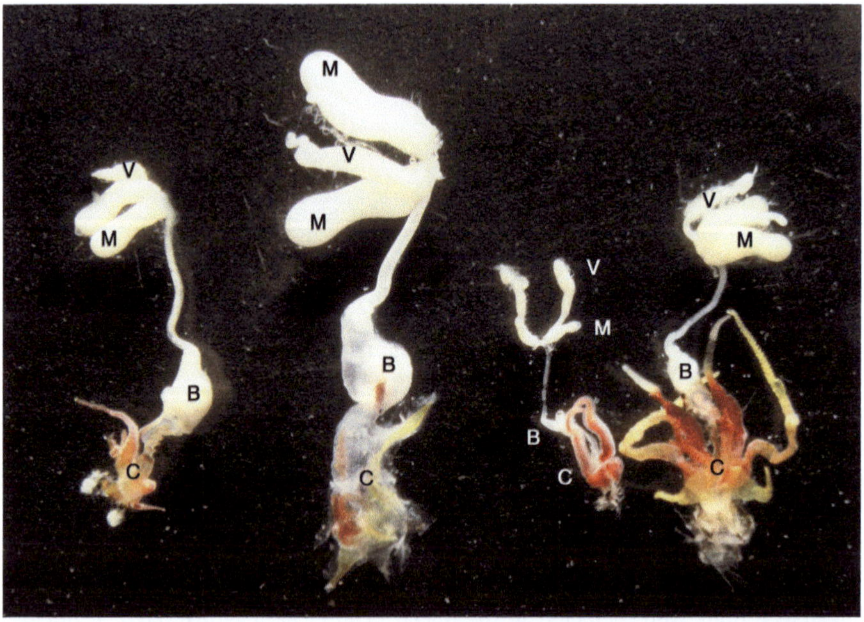

Fig. 8.22 Male reproductive organs in situ. *From left to right*: *A. dorsata*, *A. florae*, *A. mellifera*, *A. cerana*. Abbreviations: *B* bulbus, *C* Cornua with its orange coloured gland secretion, *M* Mucus gland, *V* vesiculae seminales. Note the tiny mucus glnads of *A. florae*

A. dorsata (Koeniger et al. 1991): For the male reproductive organs of *A. laboriosa*, no differences from *A. dorsata* were found (McEvoy and Underwood 1988). By contrast, in *A. florea* and *A. andreniformis*, MG are tiny (Fig. 8.22). Likewise, in these species, the bulbus is not as thick as in the others.

According to these data, a mating sign composed of mucus and secretions of the bulbus and cornual gland are formed by *Apis* and maybe also by *Megapis* species. But the bulbus gland is not yet studied in detail and the function of its secretion is not yet understood. We suggest that it may form a tube like structure enclosing the soft mucus.

8.7.1 Mating Sign: Post-mortem Cooperation also in Asian Species?

Until 1950, the mating sign was interpreted as a plug to prevent further mating. The function as a mating plug can be set aside. First, it does not prevent multiple mating during one flight. For *A. mellifera*, several matings during one flight were proven by Woyke (1960), for *A. cerana* by Woyke (1975) and for *A. koschevnikovi* by Koeniger et al. (1994a). At the end of mating, every drone leaves its own sign (Woyke and Ruttner 1958; Koeniger 1986). In *A. mellifera*, drones remove the mating sign by pulling it out with a special hairy field of the endophallus. This hairy field is also present in the species of the subgenus *Apis*. Queens of *A. cerana* and *A. koschevnikovi* also return from one mating flight with a mating sign and their oviducts are filled with the sperm of about five to ten drones (Koeniger et al. 1994a). Obviously, drones of these species too are able to remove the mating sign of their predecessors.

Second, the mating sign does not prevent sperm loss. Queens of *A. mellifera*, *A. cerana* and *A. florea* have a muscular vaginal valve (valvula vaginalis, Camargo and Mello 1970; Camargo 1972; Ruttner et al. 1973) which actively closes (or opens) the oviduct (Ruttner 1956a, b) and thus regulates the rejection of sperm. Further, mating signs in *A. mellifera* are removed within seconds during mating flights by drones and only few minutes after the last mating in the colony while the filling of the spermatheca takes several hours (Woyke 1988). A similar mechanism may function in the other species of the subgenus *Apis*. In *Micrapis*, the sperm is deposited at the orifice of the spermaduct to the spermatheca by the pointed endophallus. Here the valvula vaginales in addition may regulate the positioning of the male copulation organ. In *Megapis*, such a valvula vaginalis is not present.

As mentioned above, the function of mating signs was long debated. Much later, video recordings demonstrated that drones (*A. mellifera*) remove the mating sign by pulling it out with a special hairy field of the endophallus (Koeniger 1986). However, the debate about the purpose (adaptive advantage) of the mating sign – this "costly" drone product – is still in progress and some arguments shall be presented here. A major problem of drones in the middle of a large group of fast moving competitors is to recognise a fast flying young queen.

Attempts of drones to copulate with fellow drones occur regularly (Fig. 8.23). Based on these observations, the idea was put forward that the mating sign might mark the young queen and facilitate queen recognition for successive drones. A set of experiments supported the queen recognition hypothesis. Dummy queens with

Fig. 8.23 Confusion. The drone (*far right*) attempts copulation by bending its hind legs to grasp the drone ahead

an attached mating sign were significantly more attractive than the control dummies (Koeniger 1990). But why should a successful dead drone support his successor? First of all, after mating the queen carries the sperms of the previous drone. In this situation, selection will eliminate drones whenever the queen with "his" sperm does not reach the safety of the colony. Drones, however, would benefit whenever they succeed to reduce the duration of queen's risky mating flight. An obvious way to "end" the queen's flight would be to seal her vagina. Apparently, the mating sign does not prevent further mating (see above). Why? The answer is the high adaptive value of the genetic diversity within the *Apis* colony and evolution has favoured extreme polyandry of *Apis* queens. Drones benefit by supporting the next male to mate the queen in fast succession. Thus, after being successful in the competition with fellow drones, a drone seems to cooperate "post-mortem" with his successor by marking the queen with a conspicuous mating sign in order to reduce the mating-flight time. In *A. cerana* and *A. koschevnikovi* and even in *A. andreniformis*, parts of the mating signs protrude out of the sting chamber (Koeniger et al. 1994a, 2001).

8.8 Sperm Transfer

In the genus *Apis*, total numbers of spermatozoa transferred to the queen from an average of 1.7 million in *A. andreniformis* to an average of 107 million spermatozoa in *A. mellifera*. The numbers of spermatozoa in the queen's spermatheca vary from an average of 0.78 mio in *A. andreniformis* to 4.73 mio in *A. mellifera*. So the ratio (of total spermatozoa deposited in the queen) reaching the spermatheca is about 50% in *A. andreniformis* and less than 5% in *A. mellifera*. Thus within the genus *Apis*, spermatozoa deposition by drones and the transfer of spermatozoa within the genital system of the queen vary significantly among the species, which is the topic of this section. A summary of sperm numbers and relations is presented in Table 8.8.

Annotation: the numbers differ from those of Koeniger and Koeniger (2000) and Oldroyd and Wongsiri (2006) because new data on sperm numbers and number of patrilines were included.

Table 8.8 Number of spermatozoa per drone, per spermatheca and per drone retained in the spermatheca

Species	Spermat. per drone	Spermat. in spermatheca[a]	Effective paternity[b]	Spermat. per drone in spermatheca	% Spermat. per drone in spermatheca
A. andreniformis	0.13	0.78	13.3 ± 1.7	0.05	45.1
A. florea	0.43	1.05	7.9 ± 3.33	0.13	30.9
A. cerana	1.1	1.35	14.1 ± 3.93	0.096	8.7
A. koschevnikovi	1.7	2.13	13.3 ± 10.38	0.16	9.4
A. mellifera	9.2	4.73	11.6 ± 7.90	0.40	4.4
A. dorsata	1.6	3.94	44.2 ± 27.15	0.089	5.5

[a]Koeniger and Koeniger (2000); [b]See Table 8.9

8.8.1 Sperm Deposition by the Drone

The endophalli of *Apis* drones can be classified into three types. In the subgenus *Apis* (cavity-dwelling species), sperm is ejaculated when the short cervix is still a thin tubule and spermatozoa are brought into the median and lateral oviducts behind the vaginal valve (Fig. 8.11). The queen probably actively bends the valve fold and also must relax the sphincter-like muscles of the median oviduct. Mucus only appears later during eversion of the bulbus.

In the *Megapis* drones, the cervix of the endophallus is extremely long (Fig. 8.14) and spermatozoa are ejected through the cervix before the eversion of the bulbus starts (Woyke 2001). The long cervix of the *Megapis* endophallus probably reaches the lateral oviducts during ejaculation of spermatozoa. Also, in *Megapis*, the excretion of mucus is linked with the eversion of the bulbus which occurs at the last stage of the eversion of the endophallus. Thus in both subgenera (*Apis* and *Megapis*) during natural mating excretion of mucus is temporally separated and occurs after sperm deposition.

In *Micrapis*, the mode of sperm deposition is strikingly different. The spermatozoa are ejaculated only after the complete eversion of the endophallus at the pointed tip of the tubular bulbus. Further, a more active reaction of the queen's genital muscle system may play an important role to allow placing the tip of the endophallus exactly at the orifice of the spermathecal duct (Fig. 8.16).

8.8.2 Distribution of Individual Sperm Portions

In the subgenus *Apis*, sperm of the following drones will push the sperm of their predecessors further into the lateral oviducts. But the sperm of different drones in the oviducts does not remain in different well-defined layers as described by Moritz (1993). The walls of oviducts are not solid (like a glass syringe for instrumental insemination). Instead, the walls of oviducts are thin, flexible and folded. Thus the oviducts are actually extended by the sperm. In addition, the two lateral oviducts may be filled by different drones.

Fig. 8.24 Distribution of different dyes in the balloon. Order of colours filled into the balloon: *red*, *white*, *brown*, *yellow*, *blue*, *green*. There is no regular layer of the portions. The red first portion was displaced outward, the green last portion is mainly in the centre

A simple model was used to simulate the physical conditions for sperm transfer. A small rubber balloon was consecutively filled with differently coloured portions of a semifluid (with a consistency similar to honeybee sperm). As a result, portions were neither layered nor evenly mixed, but each coloured portion was distributed unevenly, sometimes as a compact part in the centre, sometimes as a stretched mass along the walls (Fig. 8.24). The red first portion was displaced to the margin, while the last and green portion is mainly in the centre. According to the above model, the sperm bundles of the drones are also unevenly distributed in the subgenus *Apis* and in *Megapis* as well.

In conclusion, because of the dilatability of the lateral oviducts, the position of the sperm bundles per drone is incidental and neither a regular first nor a last male have a preferential place for the filling process of the spermatheca. This is in line with the results of Franck et al. (2002) who could not find a last male advantage in naturally mated queens of *A. mellifera*.

8.8.3 Transfer to the Spermatheca

The anatomical structure of the queen's genital duct plays an essential role during sperm transfer. *A. mellifera*, *A. cerana* and *A. florea* queens possess a bursa copulatrix and a complicated muscle system stretching from the vagina to lateral oviducts (Fig. 8.12). It consists of (1) a muscular valvula vaginales, (2) several muscles inserting at the vagina and (3) a muscle surrounding the median oviduct cranial of the orifice of the spermathecal duct. The genital duct of giant honeybee queens differs from the other species. Queens of *A. dorsata* lack a bursa copulatrix as well as the muscular system of the vagina and median oviduct, and the valve fold has no muscles (Camargo and Mello 1970; Camargo 1972). We suggest that the queens which have been studied represent the typical character for each of the three subgenera of honeybees.

The transfer of spermatozoa from the oviducts to the spermatheca has only been studied in detail in *A. mellifera* (Ruttner and Koeniger 1971; Gessner and Ruttner 1977). Obviously, this process is a combination of active movements of the queen's muscles and motility of the spermatozoa. Spermatozoa of *A. mellifera* as well as those of *A. cerana* and *A. dorsata* have the tendency to accumulate with their heads against walls (personal observation). During the transport of the sperm from the oviducts in the direction of the vagina by the muscles of the lateral oviducts, some spermatozoa reach the orifice of the spermathecal duct. Then the muscular system of the spermathecal duct (Bresslau's sperm pump) and the valvula vaginalis play a crucial role. The former sucks the spermatozoa into the spermatheca and the latter regulates the ejection of superfluous spermatozoa.

In cavity-nesting honeybee species, all spermatozoa which accumulate at the dorsal side of the oviducts will have a chance to enter the spermathecal duct during their transportation back to the vagina. Spermatozoa positioned in the centre of oviduct or even further away from the orifice of the spermathecal duct are ejected through the vagina. This mechanism may explain how small bundles of spermatozoa of many mates are present in the spermatheca. The differences per drone may result from the specific position of the respective sperm bundle. In *A. cerana* and *A. koschevnikovi*, about 9% of the spermatozoa reach the spermatheca. In *A. dorsata*, the percentage is only 5.5% though drones have about the same number of spermatozoa. This low percentage may be explained by the missing muscular system of the genital ducts and by the large spermatheca which can hold up to four million spermatozoa compared to 1.4–2.1 in *A. cerana* and *A. koschevnikovi*.

The filling process is different in the dwarf honeybees. The number of spermatozoa per drone is low and obviously the deposition at the orifice of the spermathecal duct allows an instant migration into spermatheca before the next mating (Koeniger et al. 2000). Accordingly, a much higher percentage of spermatozoa per drone (*A. florea* 31% and *A. andreniformis* 45%) reach the spermatheca. To which extent the sperm is displaced by sperm of the next drone at the orifice before it can enter the spermatheca is not known. Maybe because of the higher number of spermatozoa in *A. florea* only a lower percentage of spermatozoa can enter the spermaduct than in *A. andreniformis* before it is displaced by the endophallus and sperm from the next drone. In any case, queens of the dwarf honeybee have many spermatozoa in their spermatheca when they return from mating flights (Koeniger et al. 1989, 2000).

8.8.4 Sperm Storage

In the genus *Apis*, queens store living spermatozoa for several years. Queens provide specialised morphological structures for sperm storage. The spermathecae are equipped with a huge gland and surrounded by a dense tracheal net. Amazingly, little is known about how queens are able to keep sperm cells viable over such a long period of time. Research has started only for *A. mellifera*. Glandular secretions, produced by the spermathecal glands, are found in the spermatheca (Klenk et al. 2004). These secretions contain proteins, metabolites and other chemicals. In *A. mellifera*, several proteins have been proposed to be responsible for this storage (Collins et al. 2006). In addition, high K+ concentrations and the high pH of the spermathecal fluid have been proposed to lower the metabolic rate of sperm during storage (Verma 1973). A systematic analysis of these female-derived proteins has been conducted recently combined by a proteomic identification of these proteins and their metabolic network (Baer et al. 2009).

The technique of instrumental insemination has permitted the study of heterospecific sperm transfer and storage. The longevity was reduced within days when the cross-inseminated pair was composed of different subgenera (cf. Sect. 8.9).

8.9 Polyandry of Honeybee Queens

Polyandry of queens in the genus *Apis* was discovered applying several different classical methods:

1. Offspring analysis of genetically marked queens and drones (Roberts 1944; Taber 1954 and others).
2. Comparison of sperm volumes of drones with the sperm volumes in the oviducts of queens returning from mating flights (Triasko 1956a, b; Woyke 1960).
3. Determination and frequency of diploid drones in a population (Adams et al. 1977).

The estimations with these methods ranged from 7 to 17 successful matings per queen. The use of modern, molecular methods started only about 20 years ago and resulted in a similar range of mating frequencies.

For the SE Asian species two methods have mainly been used – comparisons of sperm numbers in freshly mated queens with sperm numbers of drones or using DNA microsatellites. Sperm counting allowed an estimation of the minimal number of mating of queens per mating flight. The total number of patrilines per queen can be determined only by analysing the paternity frequency of a colony using DNA microsatellites. A review on eight species of *Apis* was published by Tarpy et al. (2004), Paar et al. (2004), Oldroyd and Wongsiri (2006) and Takahashi et al. (2008). In Table 8.9, we combine the currently known estimates of paternity within honeybee colonies.

Differences in mating frequencies were widely accepted as typical for each species in spite of small colony sample sizes and restricted sampling sites. Thus far (2010) only four colonies of *A. cerana* (Oldroyd et al. 1998), *A. koschevnikovi* (Rinderer et al. 1998) and *A. nigrocincta* (Otis and Hadisoesilo 1996) from a single site were analysed. Recently, more data were published for *A. andreniformis* and *A. dorsata* (Table 8.10). Only from the European species *A. mellifera*, more than 100 colonies were studied, but mainly from honeybees living under temperate conditions.

Table 8.9 Current estimates of paternity within honeybee colonies. Included for each species is sample size of colonies, the average worker sample size (n), observed insemination number (k), and mean effective paternity frequency (*me*) corrected for non-detection and non-sampling errors

Species	Colony sample size	Workers (n) (Mean ± SD)	k (Mean ± SD)	*me* (Mean ± SD)
A. andreniformis	14	60 ± 15.1	13.5 ± 4.51	13.3 ± 5.7
A. cerana	4	74 ± 19.4	18.8 ± 5.56	14.1 ± 3.93
A. dorsata	23	140 ± 94.6	54.9 ± 31.46	44.2 ± 27.15
A. florea	11	81 ± 56.5	11.6 ± 5.02	7.9 ± 3.33
A. koschevnikovi	4	74 ± 51.1	16.2 ± 10.53	13.3 ± 10.38
A. laboriosa	7	111.4 ± 66.7	34.4 ± 14.8	17.4 ± 6.6
A. mellifera	123	61 ± 72.6	12.0 ± 6.32	11.6 ± 7.90
A. nigrocincta	4	159 ± 17.6	54.0 ± 11.52	40.3 ± 23.39

Table 8.10 The numbers of colonies and workers (n), number of observed fathers (k) and effective paternity frequency (*me*) in *A. andreniformis* colonies from Malaysia and Thailand

Colony ID	Colony (n)	Workers (n)	k (Mean ± SD)	*me* (Mean ± SD)
A. andreniformis Kuala Lumpur	6	200	13.5 ± 1.9	11.7 ± 1.5
A. andreniformis Borneo	4	200	20.5 ± 2.4	17.6 ± 1.7
A. andreniformis Thailand	4	60 ± 15.1	13.5 ± 4.5	10.5 ± 1.9

The published sample size of *A. mellifera* is about 123 with an overall mean effective paternity frequency of 12. Splitting the original data of different samples reveals a large variance in queen mating within the same locality among colonies. It is discussed that honeybee queens modulate their number of nuptial flights according to the environmental conditions which is supported by data of Neumann et al. (1999): in *A. m. carnica*, the effective paternity frequency averaged to 20.4 ± 1.7 (n = 16) on the mainland but was only 11.8 ± 1.2 (n = 13) on an island. The large genetic variance for polyandry in honeybees and its high heritability further indicates that polyandry may be favoured by balanced selection between individual queen and colony level (Kraus et al. 2005a, b). Also significant differences in their effective paternity frequencies were demonstrated on a subspecies level (Franck et al. 2000). They range from 9.3 ± 3 in *A. m. sicula* to 34 ± 14.2 in *A. m. capensis* and according to Kraus et al. (2004) 21.1 ± 8.2 in *A. m. carnica* and 40.9 ± 14.1 *A. m. capensis.*

Significant differences also occurred in studies of *A. dorsata* in different regions. While in north and central Thailand, the effective mating frequency was 63.0 ± 5.7 (n = 13) (Wattanachaiyingcharoen et al. 2003), it was 25.6 ± 11.6 (n = 6) in Borneo (Malaysia) (Moritz et al. 1995) and 20.0 ± 6.6 in south eastern part of Thailand (n = 4) (Oldroyd et al. 1996). A similar situation was found in recent studies with *A. andreniformis*: the effective mating frequency of queens in the population from the Malay Peninsula (Takahashi et al. 2008) and the Thai population of this species (Oldroyd et al. 1997) were 11.7 and 10.5, respectively, differing significantly from queens in the Borneo population with 17.6 effective mating frequencies ($P < 0.01$, t-test; Table 8.10; Takahashi et al. 2008).

Including recent results (until 2009) in most *Apis* species, the number effective paternity ranges from about 12 to 18, with one exception each of the main groups. *A. florea* have lower (7.9 ± 3.33) and *A. dorsata* have higher (44.2 ± 27.15) frequencies and the cavity-dwelling species *A. nigrocincta* (40.3 ± 23.4). These data suggest that the high level of patrilines is a common trait for the genus *Apis* and is modified only by one species per subgenus. Thus, we suggest that hypotheses for the evolution and maintenance of extreme polyandry in *A. mellifera* also hold good for the Asian species. Several hypotheses are published, such as sperm need, task specialisation, sperm selection, reproductive conflict, resistance to parasites and diseases (Sherman et al. 1988; Palmer and Oldroyd 2000; Tarpy and Page 2001; Oldroyd and Wongsiri 2006).

Schlüns et al. (2005) re-analysed published data by applying a sample size calibration method to survey the differences and similarities in paternity skew among seven species in the genus *Apis*. The frequent patrilines appeared to be similarly distributed in all tested species. The rare patrilines are those which contribute most to the deviation from homogeneity. Omitting the rare patrilines in the correspondence analysis, the differences of frequency distribution of the nine most frequent patrilines were no longer significant. Thus there is an astonishing similarity in the most frequent patrilines, in spite of the significantly different mating frequencies and the different modes of semen transfer in the dwarf honeybees and different numbers of spermatozoa in drones from 0.14 to more than 10 million (Koeniger and Koeniger 2000; Palmer and Oldroyd 2000).

The paternity skew among honeybee species still differs significantly, particularly due to the rare patrilines. Fuchs and Moritz (1998) suggest that some genes for any worker specialisation confer an advantage on colony fitness only when they are rare. This would require a stable mix of sperm from a few drones which contribute that trait and many which do not. To ensure both specific, low within-colony proportions of "rare specialist" genes and to reduce random variation of these proportions would require mating with high numbers of drones. The quantitative implementation shows that moderate to very high numbers of matings are required to exploit colony advantages from genotypic allocation of workers to rare tasks. Extreme polyandry could thus result from colony selection dependent on the intracolonial frequency of rare genetic specialists.

In any case, the high polyandry of queens is a remarkable common character of the genus *Apis* which results in a very low genetic relatedness of 0.27–0.29 among workers within one colony.

8.10 Reproductive Isolation

Achieving reproductive isolation is an evolutionary "key event", which marks a point of independent genetic development and divergence. The speciation within the genus *Apis* probably took its course when *Apis* populations became geographically isolated (Ruttner 1988; Oldroyd and Wongsiri 2006; Koeniger et al. 2010; cf.

Chaps. 2 and 3). With enough time, genetic drift and natural selection toward locally adapted types led to the genetic divergence of both populations and with it several incompatibilities between the separated populations occurred. Traditionally, isolation mechanisms are categorised according to their temporal relation to fertilisation as pre-zygotic or post-zygotic barriers (Mayr 1963). Examples of post-zygotic barriers are non-reproductive hybrids or non-viable hybrids or incompatibility between egg and spermatozoa. Pre-zygotic barriers may function by non-fitting sex organs or by behavioural differences which prevent the mating between alien sex partners.

The reunion of formerly isolated *Apis* species has probably resulted in a high selective pressure on the concerned populations. As a model for the first stage of reunion, the interactions between imported *A. cerana* and native *A. mellifera* in Europe will be discussed. Drones and queens of *A. mellifera* and imported *A. cerana* left the colony at the same time and *A. cerana* drones were captured at the *A. mellifera* DCA (Ruttner 1973). Obviously also *A. cerana* queens visited the DCA, and *A. mellifera* drones were attracted by their sex pheromones (Ruttner and Kaissling 1968). Mating of hetereo-specific pairs ended with the loss of queens. Thus, under early sympatric conditions, selection strongly favours mechanisms of isolation which avoid such disastrous encounters. Finally, hazardous heterospecific contacts between sexuals of both species must be totally blocked at a behavioural level. Though there are differences in the bouquet of sex pheromones (Plettner et al. 1997), drones of the three tested Asian species (as well as *Apis mellifera*) copulated with the same type of dummy marked with 9-oxodec-trans-2-enoic acid (9-ODA) (Shearer et al. 1970; Koeniger et al. 2010) which was proven to be the main component of the queen's sex pheromone (Gary 1962). Accordingly, sex pheromones do not function as a reliable mating barrier.

8.10.1 Different Daily Mating Periods Are Complete Mating Barriers

As described in Sect. 8.2, mating among different sympatric honeybee species does not occur. The separation of species-specific daily mating periods blocks the meeting of alien sex partners. But what were the environmental requirements for a selective shift of the daily mating period? Generally, the daily time of mating flights is subject to natural selection, too. Leaving the safety of the nest for mating poses a risk for the queen and for the colony as well. Selection should favour queens which fly at the most convenient (warmest?) and stable weather conditions. At least in the northern regions of Asia and on higher mountains obviously noon is the time with the best mating flight conditions. In these areas, *Apis* mating always occurs in the early afternoon. The same is true for all races of *A. mellifera* in temperate zones. In conclusion, under temperate climatic conditions a shift of the mating period towards daytime with suboptimal weather conditions may have a highly negative effect because of high queen losses. In consequence under those climatic conditions, the survival of honeybees may depend on an optimal daily mating period and time-sharing mating periods may not be possible. Thus the occurrence of sympatric *Apis* species seems to require warm weather conditions and a relatively constant day length. This is true for southern Asia with its tropical and subtropical climatic conditions, and accordingly sympatric *Apis* species are restricted to those regions. Tropical Africa, however, has been reached only by *Apis mellifera* more recently (Ruttner 1988).

8.10.2 Heterospecific Transfer and Storage of Spermatozoa

The occurrence of further reproductive isolation mechanism reflects the independent development of two species. The technique of instrumental insemination permitted the study of heterospecific sperm transfer and storage. *A. mellifera* queens were inseminated each with about eight million spermatozoa from *A. mellifera*, *A. cerana*, *A. dorsata* or *A. florea* drones. The number of spermatozoa reaching the spermatheca was the same. These data suggest that either physiology of the genital duct and its fluid is similar across all tested species or the drones' seminal fluid contains enough substances which enable the spermatozoa to reach the spermatheca. Within the next 4 weeks, motility of spermatozoa of *A. mellifera* and *A. cerana* did not change and remained at nearly 100%. The motility of *A. florea* spermatozoa decreased to 33.9% and motility of *A. dorsata* spermatozoa decreased to 26% after 4 weeks (Phiancharoen et al. 2004).

In the case of closely related (sister species), after cross insemination of *A. cerana* queens with sperm of *A. koschevnikovi* and vice versa (Koeniger and Koeniger 2000; Phiancharoen unpublished data), spermatozoa entered the spermatheca at about the same percentage as after natural mating (Table 8.11; Koeniger and Koeniger 2000; Palmer and Oldroyd 2000). In all cross inseminations between *A. cerana* and *A. koschevnikovi*, spermatozoa in the spermatheca were viable when queens were dissected 40 days after insemination. The physiological conditions in spermathecae of cavity-nesting species enabled only a short survival spermatozoa of the dwarf and giant honeybees.

Table 8.11 Percentage of sperm reaching the spermatheca after cross insemination

Queen species (*n*)	Drone species	% Sperm in spth
A. cerana (15)	*A. cerana*	9.2
A. cerana (10)	*A. koschevnikovi*	8.1
A. koschevnikovi (4)	*A. koschevnikovi*	9.3
A. koschevnikovi (4)	*A. cerana*	9.1
A. koschevnikovi (2)	*A. dorsata*	2.2 and 5.5

8.10.3 Fertilisation of Eggs and Hybrids (post-zygotic)

In all tests of cross insemination across *Apis* species, the embryo died or no reproductive hybrids were obtained. Ruttner and Maul (1983) cross-inseminated queens of both *A. cerana* and *A. mellifera* successfully, but no larvae developed from the eggs. Histological studies revealed that the eggs were fertilised. During the first 3 days of development, the cells disintegrated and the embryo died during the blastoderm stage. No offspring emerged after insemination of *A. mellifera* queens with sperm of a representative species of each of the three subgenera. Calculations based on non-hatching eggs showed that at the beginning of oviposition the same percentage of *A. mellifera* eggs were fertilised by spermatozoa from *A. cerana* and *A. florea* while it was much less by *A. dorsata* spermatozoa (Phiancharoen et al. 2004).

This was different after the interspecific insemination of seven *A. cerana* queens with *A. koschevnikovi* sperm. In all cases, larvae developed. We could study the emerged offspring from only two queens. In both cases, we obtained hybrids with mixed drone- and worker-like characters (Koeniger and Koeniger 2000).

In conclusion, sperm longevity in alien spermathecae, successful fertilisation of eggs and finally the grade of development of hybrids reflect the genetic distance and follow the phylogenetic tree. Complete reproductive isolation under sympatric conditions is achieved in a very early stage and resulted in avoiding heterospecific contacts, and this is based on separated mating periods. Importations of allopatric honeybee species regularly result in heterospecific copulations with deadly consequences.

References

Adams J, Rothman ED, Kerr WE, Paulino ZL (1977) Estimation of sex alleles and queen matings from diploid male frequencies in a population of *A. mellifera*. Genetics 86:583–596

Allen MD (1965) The effect of a plentiful supply of drone comb on colonies of honeybees. J Apic Res 4:109–119

Baer B, Heazlewood JL, Taylor NL, Eubel H, Millar AH (2009) The seminal fluid proteome of the honeybee *Apis mellifera*. Proteomics 9:2085–2097

Baudry E, Solignac M, Garnery L, Gries M, Cornuet JM, Koeniger N (1998) Relatedness among honeybees (*Apis mellifera*) of a drone congregation. Proc R Soc Lond B 265:2009–2014

Buawangpang N, Sukumalanand P, Burgett M (2009) *Apis florea* drone flight: longevity and flight performance. Apidologie 40:20–25

Burgett DM, Titayavav M, Sukumalonand P (2007) The drone mating flight of the eastern honeybee (*A. cerana* F.): duration, temporal period and inter-flight period. Nat Hist Bull Siam Soc 55:99–104

Camargo JMF (1972) Preliminary notes on the reproductive system of *Apis dorsata* and *A. florea* queens. In: Anonymous (ed) Homeuagem à Warwick E Kerr. Faculdade de Filosofia. Ciências e Letras, Rio Claro, pp 47–56 [in Portuguese]

Camargo JMF, Mello MLS (1970) Anatomy and histology of the genital tract, spermatheca, spermathecal duct and glands of *Apis mellifica* queens (Hymenoptera: Apidae). Apidologie 1:351–374

Collins AM, Caperna TJ, Williams V, Garett WM, Evans JD (2006) Proteomic analyses of male contributions to honeybee sperm storage and mating. Insect Mol Biol 15:541–549

Crozier RH, Pamilo P (1996) Evolution of social insect colonies: sex allocation and kin selection. Oxford University Press, Oxford
Drescher W (1969) Die Flugaktivität von Drohnen der Rasse *Apis mellifera carnica* und *Apis mellifera ligustica* in Abhängigkeit von Lebensalter und Witterung. Z Bienenforsch 9:390–409 [in German]
Esslen J, Kaissling KE (1976) Zahl und Verteilung antennaler Sensillen bei der Honigbiene (*Apis mellifera* L.). Zoomorphologie 83:227–251 [in German]
Franck P, Koeniger N, Lahner G, Crewe R, Solignac M (2000) Evolution of extreme polyandry: an estimate of mating frequency in two African honeybee subspecies, A. mellifera monticola and A. mellifera scutellata. Insectes Soc 47:364–370
Franck P, Solignac M, Vautrin D, Cornuet JM, Koeniger G, Koeniger N (2002) Sperm competition and last-male precedence in the honeybee. Anim Behav 64:503–509
Frisch B, Koeniger N (1994) Social synchronisation of the activity rhythms of honeybees within the colony. Behav Ecol Sociobiol 35:91–98
Fuchs S, Moritz RFA (1998) Evolution of extreme polyandry in the honeybee *Apis mellifera* L. Behav Ecol Sociobiol 9:269–275
Fujiwara S, Miura H, Kumagai T, Sawaguchi T, Naya S, Goto KT, Suzuki K (1994) Drone congregation of *Apis cerana japonica* in an open area over larger trees (*Zelkovia serrata*). Apidologie 25:331–337
Gary NE (1962) Chemical mating attractants in the queen honeybee. Science 136:773–774
Gerstung F (1910) Der Bien und sein Zucht. Fritz Pfenningstorff, Berlin [in German]
Gessner B, Ruttner F (1977) Transfer der Spermien in die Spermatheka der Bienenkönigin. Apidologie 8:1–18 [in German]
Gries M (1997) Vergleichende Untersuchungen zum Flugverhalten von Drohnen der Gattung *Apis* bei der Königinnenverfolgung. Thesis, Universität Frankfurt am Main [in German]
Gries M, Koeniger N (1996) Straight forward to the queen: pursuing honeybee drones (*Apis mellifera* L.) adjust their body axis to the direction of the queen. J Comp Physiol A 179:539–544
Hadisoesilo S, Otis GW (1996) Drone flight times confirm the species status of *Apis nigrocincta* Smith, 1861 to be a species distinct from *Apis cerana* F. 1793, in Sulawesi, Indonesia. Apidologie 27:361–369
Hölldobler B, Wilson EO (2009) The superorganism, the beauty, elegance and strangeness of insect societies. W.W. Norton, New York
Klenk M, Koeniger G, Koeniger N, Fasold F (2004) Proteins in spermathecal gland secretion and spermathecal fluid and the properties of a 29 kDa protein in queens of *A. mellifera*. Apidologie 35:371–381
Koeniger N (1970) Über die Fähigkeit der Bienenkönigin (*A. mellifera* L.) zwischen Arbeiterinnen- und Drohnenzellen zu unterscheiden. Apidologie 1:115–142 [in German]
Koeniger G (1986) Mating sign and multiple mating in the honeybee. Bee World 67:141–150
Koeniger G (1990) The role of mating sign in honey bees, *Apis mellifera* L.: does it hinder or promote multiple mating? Anim Behav 39:444–449
Koeniger N, Koeniger G (1991) An evolutionary approach to mating behaviour and drone copulatory organs in *Apis*. Apidologie 22:581–590
Koeniger N, Koeniger G (2000) Reproductive isolation among species of the genus *Apis*. Apidologie 31:313–339
Koeniger N, Koeniger G (2007) Mating flight duration of *Apis mellifera* queens: As short as possible, as long as necessary. Apidologie 38:606–611
Koeniger N, Wijayagunesekera HNP (1976) Time of drone flight in the three Asian honeybee species (*Apis cerana*, *Apis florea*, *Apis dorsata*). J Apic Res 15:67–71
Koeniger N, Koeniger G, Wongsiri S (1989) Mating and sperm transfer in *Apis florea*. Apidologie 20:413–418
Koeniger G, Koeniger N, Mardan M, Otis G, Wongsiri S (1991) Comparative anatomy of male genital organs in the genus *Apis*. Apidologie 22:539–552

Koeniger G, Koeniger N, Mardan M, Wongsiri S (1993) Variance in weight of sexuals and workers within and between 4 *Apis* species (*Apis florea*, *Apis dorsata*, *Apis cerana* and *Apis mellifera*). In: Proceedings of 1st international conference on Asian honey bees bee mites, Bangkok, Wicwas Press, Cheshire, pp 104–109

Koeniger G, Koeniger N, Tingek S (1994a) Mating flights, number of spermatozoa, sperm transfer and degree of polyandry in *Apis koschevnikovi* (Buttel-Reepen, 1906). Apidologie 25:224–238

Koeniger G, Koeniger N, Tingek S (1994b) Crossfostered drones of *Apis cerana* Fabricius, 1793 and *Apis koschevnikovi* v. Buttel-Reepen, 1906 fly at their species specific mating times. Insectes Soc 41:73–78

Koeniger N, Koeniger G, Tingek S, Kelitu A, Mardan M (1994c) Drones of *Apis dorsata* Fabricius, 1793 congregate under the canopy of tall emergent trees in Borneo. Apidologie 25:249–264

Koeniger N, Koeniger G, Mardan M (1994d) Mimicking a honeybee queen? *Vespa affinis indosinensis* Pérez 1919 hunts drones of *Apis cerana* F. 1793. Ethology 98:149–153

Koeniger G, Hänel H, Wissel M, Herth W (1996a) Cornual gland of the honey bee drone (*Apis mellifera* L.): structure and secretion. Apidologie 27:145–156

Koeniger N, Koeniger G, Gries M, Tingek S, Kelitu A (1996b) Reproductive isolation of *Apis nuluensis* (Tingek, Koeniger and Koeniger 1996), by species specific mating time. Apidologie 27:353–360

Koeniger N, Koeniger G, Tingek S, Kelitu A (1996c) Interspecific rearing and acceptance of queens between *Apis cerana* Fabr. 1793 and *Apis koschevnikovi* Buttel-Reepen 1906. Apidologie 27:371–380

Koeniger N, Tingek S, Koeniger G, Gries M, Kelitu A (1998) Exploring the biodiversity of honeybees. Borneo 4:18–33

Koeniger G, Koeniger N, Tingek S, Kelitu A (2000) Mating flights and sperm transfer in the dwarf honey bee *Apis andreniformis* (Smith, 1858). Apidologie 31:301–311

Koeniger G, Koeniger N, Tingek S (2001) Diversity of the mating sign and its function in the genus *Apis*. In: Proceedings of the 7th international conference on tropical bees, Chiang Mai, pp 87–90

Koeniger N, Koeniger G, Gries M, Tingek S (2005) Drone competition at drone congregation areas in four *Apis* species. Apidologie 36:211–221

Koeniger N, Koeniger G, Tingek S (2010) Honey bee of Borneo, exploring the center of *Apis* diversity. Nat Hist Publ (Borneo), Kota Kinabalu, pp 53–60

Kraus FB, Neumann P, van Praagh J, Moritz RFA (2004) Sperm limitation and the evolution of polyandry in the honeybee (*Apis mellifera* L.). Behav Ecol Sociobiol 55:494–501

Kraus FB, Neumann P, Moritz RFA, Koeniger N, Tingek S (2005a) Using drones for estimating colony number by microsatellite DNA analyses of haploid males in *Apis*. Apidologie 36:223–229

Kraus FB, Neumann P, Moritz RFA (2005b) Genetic variance of mating frequency in the honeybee (*Apis mellifera* L.). Insectes Soc 52:1–5

Lap PV, Chinh PH, Chinh TX (1992) Some biological characteristics of *A. cerana* queen bees in Vietnam. In: Connor LJ, Rinderer TE, Sylvester HA, Wongsiri S (eds) Asian apiculture. Wicwac, Cheshire, pp 117-123

Mayr E (1963) Animal species and evolution. Harvard University Press, Cambridge

McEvoy M, Underwood BA (1988) The drone and species status of the Himalayan honey bee, *Apis laboriosa* (Hymenoptera: Apidae). J Kans Entomol Soc 61:246–249

Moors DL (2010) Functional-morphological study of the male reproductive system of honeybees. Dissertation. Katholieke Universiteit Leuven, pp 138–146

Moritz RFA (1993) Intracolonial relationship in the honey bee colony (*Apis mellifera*): molecular evidence and behavioural consequences. Verh Dtsch Zool Gesells 86:151–158

Moritz RFA, Fuchs S (1998) Organisation of honeybee colonies: characteristics and consequences of a superorganism concept. Apidologie 29:7–22

Moritz RFA, Southwick EE (1992) Bees as superorganisms – an evolutionary reality. Springer, Heidelberg

Moritz RFA, Kryger P, Koeniger G, Koeniger N, Estoup A, Tingek S (1995) High degree of polyandry in *Apis dorsata* queens detected by DNA microsatellite variability. Behav Ecol Sociobiol 37:357–363

Nagaraja N, Brockmann A (2009) Drone congregation areas of red dwarf honeybee, *Apis florea.* Nat Prec.: dl:10101/npre.2009.3955.1

Nakamura J (1995) Brood cannibalism in colonies of Asian honeybee, *A. cerana*, as response to resource fluctuation. Tamagawa Univ Res Rev 1:33–48

Neumann P, Moritz RFA, van Praagh J (1999) Queen mating frequency in different types of honeybee mating apiaries. J Apic Res 38:11–18

Oldroyd BP, Wongsiri S (2006) Asian honey bees. Harvard University Press, Cambridge

Oldroyd BP, Smolenski AJ, Cornuet JM, Wongsiri S, Estoup A, Rinderer TE, Crozier RH (1996) Levels of polyandry and intracolonial genetic relationships in *Apis dorsata* (Hymenoptera: Apidae). Ann Entomol Soc Am 89:276–283

Oldroyd BP, Clifton MJ, Wongsiri S, Rinderer TE, Crozier RH (1997) Polyandry in the genus *Apis*, particularly *Apis andreniformis*. Behav Ecol Sociobiol 40:17–26

Oldroyd BP, Clifton MJ, Parker K, Wongsiri S, Rinderer TE, Crozier RH (1998) Evolution of mating behaviour in the genus *Apis* and an estimate of mating frequency in *A. cerana* (Hymenoptera: Apidae). Ann Entomol Soc Am 91:700–709

Otis GW, Hadisoesilo S (1996) *Apis nigrocincta*, a previously unrecognized honey bee species from Indonesia: Lessons in speciation. In: Proceedings of 3rd Asian apiculture association conference, Hanoi, pp 2–6

Otis GW, Patton K, Tingek S (1995) Piping by queens of Apis cerana Fabricius 1793 and *Apis koschevnikovi* von Buttel-Reepen 1906. Apidologie 26:61–65

Otis GW, Koeniger N, Rinderer TE, Hadisoesilo S, Yoshida T, Tingek S, Wongsiri S, Mardan MB (2001) Comparative mating flight times of Asian honey bees. In: Proceedings of 7th international conference on tropical bees, Chiang Mai, pp 137–141

Paar J, Oldroyd BP, Huettinger E, Kastberger G (2004) Levels of polyandry in *Apis laboriosa* Smith from Nepal. Insectes Soc 51:212–214

Palmer KA, Oldroyd BP (2000) Evolution of multiple mating in the genus *Apis*. Apidologie 33:553–561

Patinawin S, Wongsiri S (1993) Male genitalia of honey bees. In: Connor LJ, Rinderer TE, Sylvester HA, Wongsiri S (eds) Asian apiculture. Wicwas, Cheshire, pp 110–116

Pflugfelder J, Koeniger N, Svatos A, Crewe R (2004) Interspecific aggression among virgin queens (*Apis mellifera, Apis cerana, Apis koschevnikovi* and *Apis florea*) and evidence for a chemical releaser of queen fighting. In: Proceedings of 8th Asian apiculture association conference, Los Banos, p 362

Phiancharoen M, Wongsiri S, Koeniger N, Koeniger G (2004) Instrumental insemination of *Apis mellifera* queens with hetero- and conspecific spermatozoa results in different sperm survival. Apidologie 35:503–511

Plettner E, Otis GW, Wimalaratne PDC, Winston ML, Slessor KN, Pankiw T, Punchihewa RWK (1997) Species- and caste-determined mandibular gland signals in honeybees (*Apis*). J Chem Ecol 23:363–377

Punchihewa RWK (1992) Beobachtungen und Experimente zur Paarungsbiologie von *Apis cerana indica* in Sri Lanka. Thesis, Universität Frankfurt am Main [in German]

Punchihewa RWK (1994) Beekeeping for honey production in Sri Lanka. Sri Lanka Department of Agriculture

Punchihewa RWK, Koeniger N, Koeniger G (1990) Congregation of *Apis cerana indica* drones in the canopy of trees in Sri Lanka. Apidologie 21:201–208

Radloff SE, Hepburn HR, Koeniger G (2003) Comparison of flight design of Asian honeybee drones. Apidologie 34:1–6

Rinderer TE, Oldroyd BP, Wongsiri S, Sylvester HA, De Guzman LI, Potichot S, Sheppard WS, Buchman SL (1993) Time of drone flight in four honeybee species in southeastern Thailand. J Apic Res 32:27–33

Rinderer TE, Stelzer A, Oldroyd BP, Tingek S (1998) Levels of polyandry and intracolonial relationships in *Apis koschevnikovi*. J Apic Res 37:281–287

Roberts WC (1944) Multiple mating of queen bees proved by progeny and flight tests. Glean Bee Cult 72:255–259, 303

Rueppell O, Fondrk MK, Page RE (2005) Biodemographic analysis of male honey bee mortality. Aging Cell 4:13–19

Ruttner F (1956a) Zur Frage der Spermaübetragung bei der Bienenkönigin. Insectes Soc 3:351–359 [in German]

Ruttner F (1956b) The mating of the honeybee. Bee World 37:2–15

Ruttner F (1973) Drohnen von *Apis cerana* Fabr auf einem Drohnensammelplatz. Apidologie 4:41–44 [in German]

Ruttner F (1988) Biogeography and taxonomy of honeybees. Springer, Berlin

Ruttner F, Kaissling KE (1968) Über die interspezifische Wirkung des Sexuallockstoffes von *Apis mellifica* und *Apis cerana*. Zeit Vergl Physiol 59:362–370

Ruttner F, Koeniger G (1971) Die Füllung der Spermatheka der Bienenkönigin – aktive Einwanderung oder passiver Transport der Spermatozoen. Zeit Vergl Physiol 72:411–422 [in German]

Ruttner F, Maul V (1983) Experimental analysis of reproductive interspecies isolation of *Apis mellifera* L. and *Apis cerana* F. Apidologie 14:309–327

Ruttner F, Ruttner H (1965) Untersuchungen über die Flugaktivität und das Paarungsverhalten der Drohnen. II. Beobachtungen an Drohnensammelplätzen. Z Bienenforsch 8:1–9 [in German]

Ruttner F, Ruttner H (1966) Untersuchungen über die Flugaktivität und das Paarungsverhalten der Drohnen. III. Flugweite und Flugrichtung der Drohnen. Z Bienenforsch 8:332–254 [in German]

Ruttner F, Woyke J, Koeniger N (1972) Reproduction in *Apis cerana*. 1. Mating behaviour. J Apic Res 11:141–146

Ruttner F, Woyke J, Koeniger N (1973) Reproduction in *Apis cerana*. 2. Reproductive organs and natural insemination. J Apic Res 12:21–34

Sandhu AS, Singh S (1960) The biology and brood rearing activities of the little honeybee – *Apis florea* Fabricius. Indian Bee J 22:27–35

Schlüns H, Moritz RFA, Neumann P, Kryger P, Koeniger G (2005) Multiple nuptial flights, sperm transfer and the evolution of extreme polyandry in honeybee queens. Anim Behav 70:125–131

Seidl R (1980) Die Sehfelder und Ommatidien-Divergenzwinkel der drei kasten der Honigbien (*A. mellifera)*. Verh Dtsch Zool Ges 367 [in German]

Sharma PJ (1960) Observations on the swarming and mating habits of the Indian honeybee. Bee World 41:121–125

Shearer DA, Boch R, Morse RA, Laigo FM (1970) Occurrence of 9-oxodec-*trans*-2-enoic acid in queens of *Apis dorsata*, *Apis cerana* and *Apis mellifera*. J Insect Physiol 16:1437–1441

Sherman PW, Seeley TD, Reeve HK (1988) Parasites, pathogens, and polyandry in social Hymenoptera. Am Nat 131:602–610

Snodgrass RE (1956) The anatomy of the honey bee. Cornell University Press, London

Taber S (1954) The frequency of multiple mating of queen honey bees. J Econ Entomol 47:995–998

Takahashi J, Hashim SB, Tingek S, Shimizua I, YoshidaT (2008) A scientific note on the levels of polyandry in the black dwarf honeybee *Apis andreniformis* from Malaysia. Apidologie 39:233–234

Tan NQ (2007) Biology of *Apis dorsata* in Vietnam. Apidologie 38:221–229

Tan NQ, Mardan MB, Thai PH, Chinh PH (1996) Some reproductive biology of *Apis dorsata*. In: Proceedings of 3rd Asian apiculture association conference, Hanoi, pp 15–22

Tan NQ, Mardan MB, Thai PH, Chinh PH (1999) Observations on multiple mating flights of *Apis dorsata* queens. Apidologie 30:339–346

Tarpy DR, Page RE (2001) The curious promiscuity of queen honey bees (*Apis mellifera*): evolutionary and behavioural mechanisms. Ann Zool Fennici 38:255–265

Tarpy DR, Nielsen R, Nielsen DI (2004) A scientific note on the revised estimates of effective paternity frequency in *Apis*. Insectes Soc 51:203–204

Triasko VV (1956a) Multiple matings of queen bees. Pchelovodstvo 34:29–31

Triasko VV (1956b) Polyandry in honey bees. XVI International beekeepers congress, Preliminary Science Meeting 233/56

Trivers RL, Hare H (1976) Haplodiploidy and the evolution of the social insects. Science 191:249–263

Underwood BA (1990) Time of drone flight of *Apis laboriosa* Smith, 1871 in Nepal. Apidologie 21:501–504

Vallet AM, Cole JA (1993) The perception of small objects by the drone honeybee. J Comp Physiol A 172:183–188

van Praagh JP, Ribi W, Wehrhan C, Wittmann D (1980) Drone bees fixate the queen with the dorso-frontal part of their compound eyes. J Comp Physiol 136:263–266

Verma LR (1973) An ionic basis for a possible mechanism of sperm survival in the spermatheca of the queen honey bee *Apis mellifera*. Comp Biochem Physiol 44:1325–1331

Verma LR (1991) Beekeeping in integrated mountain development. Oxford and IBH, New Delhi

von Berlepsch A (1873) Die Biene und ihre Zucht auf beweglichen Waben in Gegenden ohne Spätsommertracht, 3rd edn. J. Schneider, Mannheim [in German]

Wattanachaiyingcharoen W, Oldroyd BP, Wongsiri S, Palmer K, Paar S (2003) A scientific note on the mating frequency of *Apis dorsata* Fabricius. Apidologie 34:85–86

Weiss K (1962) Untersuchungen über die Drohnenbruterzeugung im bienenvolk. Arch Bienenkunde 39:1–7 [in German]

Winston ML (1987) The biology of the honey bee. Harvard University Press, Cambridge

Woyke J (1960) Natural and artificial insemination of queen honeybees. Pszcz Zesz Nauk 4:183–275

Woyke J (1975) Natural and artificial insemination of *Apis cerana* in India. J Apic Res 14:153–159

Woyke J (1988) A mathematical model for the dynamics of spermatozoa entry into the spermathecae of instrumentally inseminated queen honeybees. J Apic Res 27:122–125

Woyke J (2001) Eversion of drone endophallus and the probable mating process in *Apis dorsata*. In: Proceedings of the 7th Asian apiculture association conference on tropical bees, Chiang Mai, pp 188–194

Woyke J (2008) Why the eversion of the endophallus of honey bee drone stops at the partly everted stage and significance of this. Apidologie 39:627–636

Woyke J, Ruttner F (1958) An anatomical study of the mating process in the honeybee. Bee World 39:3–18

Woyke J, Wilde J, Wilde M (2001) *Apis dorsata* drone flights, collection of semen from everted endophalli and instrumental insemination of queens. Apidologie 32:407–416

Wu Y, Kuang B (1986) A study of genus *Micrapis* (Apidae). Zool Res 7:99–102

Yoshida T (1995) Differences in mating flight times and drone congregation areas of native *Apis cerana japonica* and introduced *Apis mellifera*. Honeybee Sci 16:57–66

Yoshida T, Yamazaki M (1993) Difference in drone congregation areas of *Apis mellifera* and *Apis cerana japonica* as a reproductive isolation mechanism. In: Connor LK, Rinderer T, Sylvester A, Wongsiri S (eds) Asian apiculture. Wicwas, Cheshire, pp 99–103

Yoshida T, Saito J, Kajigaya N (1994) The mating flight times of native *Apis cerana japonica* Radoszkowski and introduced *Apis mellifera* L. in sympatric conditions. Apidologie 25:353–360

Chapter 9
Pheromones

Christian W.W. Pirk, Catherine L. Sole, and R.M. Crewe

9.1 Introduction

In the words of Blum and Brand (1972), "social insects are a rich source of volatile compounds" (noted originally by René de Réaumur (1926) in *The Natural History of Ants*). Among these volatiles are pheromones, chemicals emitted by the individuals of a species to communicate messages to other individuals of the same species. The behaviour and success of a honeybee colony is often determined by the intensity and frequency to which individuals respond to these stimuli and thereafter perform tasks (Pankiw et al. 1994). Honeybees, being highly social insects, are divided into two female castes within a colony: the queen and several thousand workers. The queen is the main reproductive and head of the colony – which she controls by means of primer/releaser pheromones – and mediates aspects of worker behaviour as well as partially inhibiting worker reproduction.

The workers, on the other hand, perform non-reproductive tasks such as brood rearing, nest guarding and foraging. This caste differentiation ultimately causes behavioural and physiological differences, which in turn result in differences in the composition of pheromones released. These differences can be seen in the composition of the mandibular and Dufour's glands as well as in alarm and aggregation pheromones. Honeybee pheromones can be divided into two broad categories, namely primer and releaser pheromones. Primer pheromones are those that trigger or suppress developmental events while releaser pheromones are known to cause a change in the behaviour of the recipient. This chapter addresses the caste-specific differences in pheromone blends and how this correlates to functionality within the colony.

C.W.W. Pirk and R.M. Crewe
Department of Zoology and Entomology, University of Pretoria, Pretoria 0002, South Africa
e-mail: cwwpirk@zoology.up.ac.za; robin.crewe@up.ac.za

C.L. Sole
Scarab Research Group, Department of Zoology and Entomology, University of Pretoria, Pretoria 0002, South Africa
e-mail: clsole@zoology.up.ac.za

H.R. Hepburn and S.E. Radloff (eds.), *Honeybees of Asia*,
DOI 10.1007/978-3-642-16422-4_9, © Springer-Verlag Berlin Heidelberg 2011

9.2 Mandibular and Dufour's Glands of Asian Honeybees (Primer Pheromones)

9.2.1 The Mandibular Glands of Honeybees

In most cases, both female castes are known to produce a pheromonal blend of several compounds. Mated *Apis mellifera* queens produce two aliphatic compounds, 9-hydroxy-2-(*E*)-decenoic acid (9-HDA) and 9-keto-2-(*E*)-decenoic acid (9-ODA) and two aromatic ones, methyl hydroxybenzoate (HOB) and 4-hydroxy-3-methoxyphenylethanol (HVA) (Winston and Slessor 1992). Workers have 10- hydroxy-2-(*E*)-decenoic acid (10-HDA), 10-hydroxydecanoic acid (10-HDAA) and their respective diacids, 2-(*E*)-decenedioic ($C_{10:1}$ DA) and decanedioic ($C_{10:0}$ DA). Having said this, both castes are able to produce aliphatic compounds of the other castes in small quantities. Queens have some 10-HDA and 10-HDAA (Crewe 1982) while workers have trace amounts of (9-HDA) (Plettner et al. 1995). Queen mandibular pheromone (QMP) is known to elicit both short- and long-term behaviours. Short-term behaviours include retinue formation, swarm stabilisation (Morse 1963; Winston and Slessor 1992) and drone attraction (Butler and Fairey 1964), while in the long term the inhibition of queen rearing is controlled through QMP (Butler 1954a,b; Pain 1954; Winston et al. 1990, 1991).

Compounds found in the mandibular glands of *A. mellifera* workers are known to be involved in food preservation and larval nutrition. As this book is on Asian honeybees, we have limited our discussion on *A. mellifera* and use it to create a starting idea of pheromones in honeybees. Plettner et al. (1997) went on to compare the mandibular compounds of *A. mellifera* with those of *A. dorsata*, *A. florea*, *A. andreniformis* and *A. cerana*. Queens of these four Asian honeybee species contain 9-ODA and 9-HDA while only *A. cerana* contains detectable amounts of HOB. The worker acid, 10-HDA, was found in all four Asian honeybee species, while 10-HDAA was detected only in *A. cerana* and *A. florea*. Workers of all species had 10-HDA, 10-HDAA, $C_{10:0}$ DA and 9-HDA in their mandibular glands, while *A. cerana* was the only species with no detectable $C_{10:1}$ DA and 9-ODA. With respect to quantitative differences in the total production of mandibular gland secretion, *A. cerana* and *A. dorsata* produce substantially more than European *A. mellifera* – 200–300 μg compared to 150–200 μg, respectively (Billen and Morgan 1998).

Plettner et al. (1997) showed quite clearly that differences in mandibular gland secretions occur between open- and cavity-nesting species. Queens of open-nesting species do not contain any aromatic compounds or 10-HDAA, while the workers of cavity-nesting species contain 9-ODA, a typical queen mandibular gland compound. Within the open-nesting group of honeybees, the dwarf bees, *A. florea* and *A. andreniformis*, although closely related phylogenetically (Alexander 1991), show differences within mandibular gland secretions. For example, workers show differences in their 10-HDAA/ $C_{10:0}$ DA profiles while both castes have different profiles for 9-HDA/9-ODA. Interestingly, the queens of the Asian cavity-nesting species,

A. cerana, do not produce HVA, which is necessary for *A. mellifera* to produce full retinue behaviour. As *A. cerana* queens elicit a full retinue response without the HVA compound, one might assume that a primer response is gained either through other chemical cues or additional non-chemical cues, as yet to be identified.

In summary, it appears that all species of *Apis* analysed to date have distinct blends of compounds in the mandibular glands of both castes. Furthermore, within each species, queen and worker variation is evident. Phylogenetic studies (Alexander 1991; Dyer 1991; Raffiudin and Crozier 2007) indicate that the more ancestral, open-nesting bees (cf. Chap. 2) have a simpler queen mandibular blend containing only aliphatic components as well as smaller differences between queen and worker secretions (Plettner et al. 1997). This may be an indication that the components and functionality of the mandibular gland have played a significant role in the evolutionary history of the genus *Apis* (Plettner et al. 1997).

9.2.2 The Dufour's Gland Pheromones

The Dufour's gland was first described by Dufour in 1841 (Trojan 1930) and is attached to the sting apparatus in all female hymenopterans, an area generally used for defensive behaviour by workers and reproduction by queens (Katzav-Gozansky et al. 2002; Martin and Jones 2004). In line with its anatomical placement the Dufour's gland has been hypothesised to have many functions including the toxic enhancement of venom (Carlet 1890), a lubricant for the moving parts of the sting, neutralisation of the excess acid secretion in the sting (Trojan 1930; Kerr and de Lello 1962), a protective coating for the eggs (Billen 1987) and an egg-marking pheromone to aid worker policing (Ratnieks 1995).

Martin and Jones (2004) studied chemical changes in Dufour's gland secretions associated with ovary development in several species of honeybees – *A. mellifera*, *A. cerana*, *A. andreniformis* and *A. florea*. They found that C_{21}–C_{33} hydrocarbons were present in all individuals while C_{28}–C_{38} compounds were only associated with queens, and eicosenol, a C_{20} unsaturated alcohol, is only associated with non-laying workers. At a finer scale, C_{28}–C_{38} esters are associated only with cavity-nesting honeybee queens, while eicosenol, a worker alarm pheromone, is associated only with their non-laying workers. Three main chemical groups secreted by the Dufour's gland are hydrocarbons, esters and eicosenol.

Dufour's gland has been shown to have caste biosynthetic plasticity in *Apis* species (Katzav-Gozansky et al. 1997a; Sole et al. 2002) and is metabolically active throughout an individual's life (Katzav-Gozansky et al. 2000). This chemical plasticity, therefore, raises doubts about there being caste-specific biosynthetic pathways. Katzav-Gozansky et al. (2002) found that the glandular expression of the Dufour's gland chemicals is regulated by two factors, firstly, a social factor, that is the presence or absence of the queen and/or the interactions between nest members; and secondly, worker physiology, that is caste, task and reproductive state of the individual bees. The Dufour's gland exhibits both caste specificity

(Katzav-Gozansky et al. 1997b) and biosynthetic plasticity (Sole et al. 2002) which raises questions as to its function and activity regulation that remain to be answered. Worker-laid eggs covered with queen-like esters were policed as quickly as worker-laid eggs with no queen-like substance on them, which rules out Dufour's gland being used solely in egg-marking (Katzav-Gozansky et al. 2001, 2002).

Based on what is known, it is assumed that, in the presence of a queen, the synthesis of esters in the worker caste is inhibited and, on removal of the queen, workers will produce queen-like esters. As only females with activated ovaries possess esters in their Dufour's gland secretion, a link may exist between ovary development and Dufour's gland secretion (Katzav-Gozansky et al. 2002). However, very little is known and much is in a state of speculation. Katzav-Gozansky et al. (2002) raised the following questions that need addressing: Is neuro-endocrine control involved? What is the nature of the queen signal? Does a single compound in the queens' secretions inhibit production in the workers? Is there a case for gland cross-reactivity (i.e. the mandibular gland secretion of the queen inhibits Dufour's ester production in workers)?

9.3 Alarm, Aggregation and Other Pheromones of Asian Honeybees (Releaser Pheromones)

All known species of honeybees show a well developed and coordinated defensive behaviour (Butler 1954c) involving several behavioural mechanisms to fend off potential threats (e.g. shimmering and hissing – Fuchs et al. 2001, "testudo" behaviour – Pirk et al. 2002, both in *A. florea*) and if this were not sufficient, it escalates to multiple stinging events. This defensiveness was originally described by Huber (1814) who stated that the attacks were enhanced, and probably triggered, by the odour of the venom. However, it has been subsequently shown that neither the venom, the venom glands nor the sac plays a role in producing this odour (cf. Schmidt et al. 1997). Isopentyl acetate (IPA) was originally shown to be an important component of the alarm pheromone (Boch et al. 1962; Free and Simpson 1968), however as chemical methods of analysis improved over the years, more substances (approximately 30) were identified (for details of selected ones see Table 9.1). The amounts of IPA produced vary between species, but do not seem to be correlated with, honeybee size pis *dorsata* produce the most (21.8 μg), followed by *A. mellifera* (1.9 μg) and thereafter by *A. cerana* and *A. florea* with only 0.2 μg (Koeniger et al. 1979). Koeniger et al. (1979) showed that all four species react to the extract of the sting apparatus from each of the species but to a lesser degree and that the reaction could not be correlated with the amount of IPA produced/used.

Schmidt et al. (1997) dissected the sting apparatus of *A. mellifera*, *A. florea*, *A. dorsata* and *A. cerana* and analysed their venoms using GC-MS and showed that they contain large, oily droplets, lacking in the other three species. Furthermore, *A. cerana* seems to possess 50–100 times more eicosenol than *A. mellifera*,

Table 9.1 Alarm pheromone components

Substance	Species	Reference
Iso-pentyl acetate	*A.m., A.c., A.d., A.f.*	Morse et al. (1967), Koeniger et al. (1979)
1- hexanol	*A.m.*	Collins and Blum (1983)
Octyl acetate	*A.m., A.c.*	Schmidt et al. (1997), Wager and Breed (2000)
Butyl acetate	*A.m.*	Hepburn et al. (1994)
1-butanol	*A.m.*	Collins and Blum (1983)
1-octanol	*A.m.*	Collins and Blum (1983)
2-methyl-1-butanol	*A.m.*	Wager and Breed (2000)
3-methyl-1-butanol	*A.m.*	Wager and Breed (2000)
Hexyl acetate	*A.m.*	Bloch et al. (1962)
2-decen-1-yl acetate	*A.d., A.f.*	Veith et al. (1978), Koeniger et al. (1979)
2-nonanol	*A.m., A.c.*	Collins and Blum (1983), Schmidt et al. (1997)
(Z)-11-eicosen-1-ol	*A.m., A.c.*	Pickett et al. (1982), Schmidt et al. (1995)
Eicosenol	*A.m., A.c.*	Schmidt et al. (1997)
Benzyl acetate	*A.m.*	Hepburn et al. (1994)

plus long chain alcohols and hydrocarbons, which were found in oily droplets of the venom. Schmidt et al. (1997) suggested that the oily droplet plays a major role as an alarm pheromone in *A. cerana*. It might also play a role as a carrier for other active components which in turn trigger defence behaviour (Veith et al. 1978; Pickett et al. 1982). The suggestion that it might serve as an alarm component on its own (Pickett et al. 1982) was not supported by results ranking it as inactive in eight of nine assay categories including releasing stinging (Free et al. 1989). It only inhibited the release of the Nasonov pheromones by scenting guards. Additional differences between the two sister species is not only the amount of eicosenol but also the anatomical source as well. In *A. mellifera*, most of it is found in the setose area, whereas in *A. cerana* large quantities are found in the venom itself (Schmidt et al. 1997). Three hypotheses have been put forward to explain its function: it plays a role in marking food sources or marking enemies, or as a carrier facilitating the spreading of aqueous venom toxins. Martin and Jones (2004) proposed that the biosynthesis of esters and eicosenol in Dufour's gland is caste-regulated and, following the argument by Pickett et al. (1982), that it is a worker alarm pheromone.

Another component found in the sting extract, which does not occur in all species, is 2-decen-1-ly acetate (2da). It was only found in the extracts of *A. florea* and *A. dorsata* and has the specific effect of extending the duration of the alarm reaction when presented in combination with IPA compared to pure IPA (Koeniger et al. 1979). The unusual venom of *A. cerana* contains large oily droplets within an otherwise aqueous secretion. Chemical analysis (GC-MS) revealed that the venom oil consists of (Z)-11-eicosen-1-ol (81.2%), other linear alcohols (7.7%) and linear hydrocarbons (11.1%). The eicosenol is present in extremely large quantities, averaging over 250 μg per insect, and is absent, or present in small quantities, in other parts of the sting apparatus. An investigation of the site of eicosenol storage in *A. mellifera* showed it to be absent from the venom and to be associated with the setose area where the more volatile components of the alarm pheromone are stored, as previously shown by others. A third honeybee species, *A. dorsata,* interestingly enough does not contain the alcohol. The function of eicosenol in *A. cerana* is not clear, but may be used in the marking of already stung intruders or to attract foragers to marked floral resources.

9.3.1 Aggregation Pheromones

The components released by the Nasanov gland are partially responsible for individual honeybees staying together in a swarm and forming a cohesive unit. A complex of seven different components (Pickett et al. 1980) is released from the dorsal surface of the abdomen mostly accompanied by fanning behaviour to attract fellow workers (Free 1987) either in swarm preparation or in organising the defence of the colony. Three of the components are in particular responsible for the lemon-like smell – geraniol, geranial and neral. In *A. florea*, exposure of the Nasonov gland appears not to be accompanied by fanning (Free and Williams 1979), added to which no evidence could be found that *A. florea* uses this gland to scent mark foraging sources as Butler (1954c) reported when he observed that workers exposed their Nasonov gland when visiting flowers.

9.4 Conclusion

The success of honeybees is mainly based on their ability to communicate efficiently, which enables them to explore highly temporal food sources and organise a rapid response to disturbances, and which are crucial for the self-organisation of a colony. Chemical communication, particularly in honeybees, is based on the fact that pheromones are released under context-dependent circumstances. For example, who releases the pheromone and who is on the receiving end and which class of pheromones is used will depend on the circumstances (Le Conte and Hefetz 2008). These multi-functional properties of pheromones make them ideal tools to investigate underlying evolutionary mechanisms of speciation.

In the *Apis* complex, *A. mellifera*, *A. cerana*, *A. florea*, *A. dorsata* and *A. nuluensis* give us an insight as to how a neutral mutation might have resulted in different functional adaptations, as all species show similarities but with added significant differences, for example the case of the (*Z*)-11-eicosen-1-ol in *A. cerana*. Moreover, interspecific comparisons of the abundance of different chemical compounds, over 30 of which are involved in an alarm response, could provide us with more information and understanding of the evolution of pheromonal communication and its proximate mechanisms. As Table 9.1 indicates, most of the honeybee pheromonal work has been done with *A. mellifera*. Comprehensive and comparative approaches in investigating the similarities and differences within and between the balance of the Asian honeybee species is lacking and needs further investigation.

References

Alexander B (1991) A cladistic analysis of the genus *Apis*. In: Smith DR (ed) Diversity in the genus *Apis*. Westview, Boulder, pp 1–28

Billen J (1987) New structural aspects of the Dufour's and venom glands in social insects. Naturwissenschaften 74:340–341

Billen JPJ, Morgan ED (1998) Pheromone communication. In: Vander Meer RK, Breed MD, Espelie KE, Winston ML (eds) Pheromone communication in social insects ants, wasps, bees, and termites. Westview, Boulder, pp 3–33

Blum MS, Brand JM (1972) Social insect pheromones: their chemistry and function. Am Zool 12:553–576

Boch R, Shearer DA, Stone BC (1962) Identification of iso-amyl acetate as an active compound in the sting pheromone of the honeybee. Nature 195:1018–1020

Butler CG (1954a) The world of the honeybee. Collins, London

Butler CG (1954b) The importance of 'queen substance' in the life of a honeybee colony. Bee World 35:169–176

Butler CG (1954c) The method an importance of the recognition by a colony of honeybees (*A. mellifera*) of the presence of its queen. Trans R Entomol Soc Lond 105:11–29

Butler CG, Fairey EM (1964) Pheromones of the honeybee: Biological studies of the mandibular gland secretion of the queen. J Apic Res 3:65–76

Carlet G (1890) Mémoire sur le venin et l'aiguillon de l'abeille. Ann Sci Nat Zool 9:1–17 [in French]

Collins AM, Blum MS (1983) Alarm responses caused by newly identified compounds derived from the honeybee sting. J Chem Ecol 9:57–65

Crewe RM (1982) Compositional variability: the key to social signals produced by honeybee mandibular glands. In: Breed MD, Michener CD, Evans HE (eds) The biology of social insects. Westview, Boulder, pp 318–322

Dyer FC (1991) Phylogeny and function of dance communication. In: Smith DR (ed) Diversity in the genus *Apis*. Westview, Boulder, pp 177–198

Fuchs S, Sen Sarma M, Werber C, Tautz J (2001) The cry of the honeybee *Apis florea*. Mtg Euro Sect (IUSSI), Berlin, p 164

Free JB (1987) Pheromones of social bees. Cornell University Press, Ithaca

Free JB, Simpson J (1968) The alerting pheromones of the honeybee. Zeit Vergl Physiol 61:361–365

Free JB, Williams IH (1979) Communication by pheromones and other means in *Apis florea* colonies. J Apic Res 18:16–25

Free JB, Ferguson AW, Simpkins JR (1989) Honeybee responses to chemical components from the worker sting apparatus and mandibular glands in field tests. J Apic Res 28:7–21

Hepburn HR, Jones GE, Kirby R (1994) Introgression between *Apis mellifera capensis* Eschscholtz and *Apis mellifera scutellata* Lepeletier. Apidologie 25:557–565

Huber F (1814) New observations on honeybees. Dadant translation (1926). Dadant, Hamilton

Katzav-Gozansky T, Soroker V, Hefetz A, Cojocaru M, Erdmann DH, Francke W (1997a) Plasticity of caste-specific Dufour's gland secretion in the honeybee (*Apis mellifera* L.). Naturwissenschaften 84:238–241

Katzav-Gozansky T, Soroker V, Hefetz A (1997b) The biosynthesis of Dufour's gland constituents in queens of the honeybee (*Apis mellifera*). Invert Neurosci 3:239–243

Katzav-Gozansky T, Soroker V, Hefetz A (2000) Plasticity in caste-related exocrine secretion biosynthesis in the honey bee (*Apis mellifera*). J Insect Physiol 46:993–998

Katzav-Gozansky T, Soroker V, Ibarra F, Francke W, Hefetz A (2001) Dufour's gland secretion of the queen honeybee (*Apis mellifera*): an egg discriminator pheromone or a queen signal? Behav Ecol Sociobiol 51:76–86

Katzav-Gozansky T, Soroker V, Hefetz A (2002) Honeybees Dufour's gland – idiosyncrasy of a new queen signal. Apidologie 33:525–537

Kerr WE, de Lello E (1962) Sting glands in stingless bees have a vestigial character. J NY Entomol Soc 70:190–214

Koeniger N, Weiss J, Maschwitz U (1979) Alarm pheromones of the sting in the genus *Apis*. J Insect Physiol 25:467–476

Le Conte Y, Hefetz A (2008) Primer pheromones in social hymenoptera. Annu Rev Entomol 53:523–542

Martin SJ, Jones GR (2004) Conservation of biosynthetic pheromone pathways in honeybees *Apis*. Naturwissenschaften 91:232–236

Morse RA (1963) Swarm orientation in honeybees. Science 141:357–358

Morse RA, Shearer DA, Boch R, Benton AW (1967) Observations on alarm-substances in the genus *Apis*. J Apic Res 6:113–118

Pain J (1954) Sur l'ecohormone des reines d'abeilles. Comptes R Acad Sci 239:1869–1870 [in French]

Pankiw T, Winston ML, Slessor KN (1994) Variation in response to honey bee (*Apis mellifera* L.) queen mandibular pheromone (Hymenoptera: Apidae). J Insect Behav 7:1–15

Pickett JA, Williams IH, Smith MC, Martin AP (1980) Nasonov pheromone of honey bee, *Apis mellifica* L. (Hymenoptera: Apidae). Part I. Chemical characterization. J Chem Ecol 6:425–434

Pickett JA, Williams IH, Martin AP (1982) (*Z*)-11-eicosen-1-ol, an important new pheromonal component from the sting of the honey bee, *Apis mellifera* L. J Chem Ecol 8:163–175

Pirk CWW, Hepburn HR, Radloff SE, Erlandsson J (2002) Defense posture in the dwarf honeybee, *Apis florea*. Apidologie 33:289–294

Plettner E, Sutherland GRJ, Slessor KN, Winston ML (1995) Why not be a queen? Regioselectivity in mandibular secretions of honeybee castes. J Chem Ecol 21:1017–1029

Plettner E, Otis GW, Wimalaratne PDC, Winston ML, Slessor KN, Pankiw T, Punchihewa PWK (1997) Species- and caste-determined mandibular gland signals in honeybees (*Apis*). J Chem Ecol 23:363–377

Raffiudin R, Crozier RH (2007) Phylogenetic analysis of honey bee behavioral evolution. Mol Phylogenet Evol 43:543–552

Ratnieks FLW (1995) Evidence for a queen-produced egg-marking pheromone and its use in worker policing in the honeybee. J Apic Res 34:31–37

Schmidt JO, Morgan ED, Oldham NJ, Do Nascimento RR, Dani FR (1997) (*Z*)-11-Eicosen-1-ol, a major component of *Apis cerana* venom. J Chem Ecol 23:1929–1939

Sole CL, Kryger P, Hefetz A, Katzav-Gozansky T, Crewe RM (2002) Mimicry of queen Dufour's gland secretions by workers of *Apis mellifera scutellata* and *A. m. capensis*. Naturwissenschaften 89:561–564

Trojan E (1930) Die Dufoursche Drüse bei *Apis mellifica*. Z Morphol Ökol Tiere 19:678–685 [in German]

Veith J, Weiss J, Koeniger N (1978) A new alarm pheromone (2-decen-1-yl acetate) isolated from the stings of *Apis dorsata* and *Apis florea* (Hymenoptera: Apidae). Experientia 34:423

Wager BR, Breed MD (2000) Does honey bee sting alarm pheromone give orientation information to defensive bees? Ann Entomol Soc Am 93:1329–1332

Winston ML, Slessor KN (1992) The essence of royalty: honey bee queen pheromone. Am Sci 80:374–385

Winston ML, Higo HA, Slessor KN (1990) Effects of various dosages of queen mandibular gland pheromone on the inhibition of queen rearing in the honey bee (Hymenoptera: Apidae). Ann Entomol Soc Am 83:234–238

Winston ML, Higo HA, Colley SJ, Pankiw T, Slessor KN (1991) The role of queen mandibular pheromone and colony congestion in honey bee (*Apis mellifera*) reproductive swarming. J Insect Behav 4:649–660

Chapter 10
Honeybees in Natural Ecosystems

Richard T. Corlett

10.1 Introduction

The importance of honeybees in agriculture (cf. Chaps. 11 and 12) has completely overshadowed their role in natural ecosystems, about which relatively little is known. Even without active human assistance, the extensive high-density monocultures of agriculture play to the honeybee strength of rapid and overwhelming recruitment, but things are likely to be very different in species-rich natural vegetation where a great diversity of other bees are present. Although both the number and quality of non-crop pollination studies in Asia are increasing, it is still necessary to fill large gaps in our current knowledge and understanding of Asian honeybees by extrapolation from studies elsewhere on the only non-Asian honeybee, *Apis mellifera*. This situation is far from ideal, and it is important that the reader is not misled into thinking that we know more about the botany of wild Asian honeybees than we really do.

10.2 The Ecology of Honeybees

The ecology of wild *Apis* species has received surprisingly little attention anywhere in their natural range. The most detailed honeybee research has been done on domesticated colonies of temperate *A. mellifera* visiting crop monocultures, but it is not clear how much the findings from these studies can be extrapolated to other plants, localities and honeybee species. Key points from studies of domesticated honeybees that are of potential relevance to wild bees visiting wild flowers include the following: honeybees forage preferentially near the nest but can fly longer distances (>1 km) rapidly when necessary; individual bees tend to forage for either nectar or pollen from a single plant

R.T. Corlett
Department of Biological Sciences, National University of Singapore, Singapore 117543, Republic of Singapore
e-mail: corlett@nus.edu.sg

H.R. Hepburn and S.E. Radloff (eds.), *Honeybees of Asia*,
DOI 10.1007/978-3-642-16422-4_10, © Springer-Verlag Berlin Heidelberg 2011

species on a single trip, but constancy declines with successive trips; pollen foragers are often the better pollinators because they tend to have more pollen on the body and cannot avoid contact with the anthers; honeybees are unable to "work" some common pollination mechanisms; most pollen from a plant is deposited on the next few flowers, but some can persist for 20 flowers or more; bees leaving the hive carry viable second-hand pollen, which could potentially increase pollination distances and outcrossing; most individual bees are recruited by dance to a particular location, but a variable proportion of bees scouts for new resources on their own; honeybee colonies collect food over a huge area, changing their focus continuously in response to changes in resource availability (Free 1993; Delaplane and Mayer 2000; Beekman et al. 2007). The unresolved question of how much pollen transfer occurs between bees in the nest is key for estimating potential maximum outcrossing distances, since even a scout bee is unlikely to cover more than a few hundred metres, while the colony as a whole may forage over tens of square kilometres.

The little data for African *A. mellifera* suggests that tropical colonies may forage over smaller areas and focus more on pollen than temperate colonies (Schneider 1989), but it is not clear how general these differences are. How are Asian honeybees different from European and African *A. mellifera*? For a start, they are much more diverse. Sympatric small (workers 7–10 mm long), medium (10–14 mm) and large (17–20 mm) species coexist over much of tropical Asia, with a maximum of five species in one area (cf. Chap. 3). The three size groups are monophyletic, with the small bees basal (Raffiudin and Crozier 2007; Lo et al. 2010; cf. Chap. 2), although the oldest fossil honeybee in the region is a large, *dorsata*-like species from Middle Miocene Japan (Engel 2006). The medium-size species are all somewhat smaller than European *A. mellifera* and form smaller colonies (<2,000 workers in *Apis koschevnikovi*; Roubik 2005), but all honeybee species, including the small ones, appear capable of foraging over long distances (>1 km) if necessary, at flight speeds of 5–8 m/s (Oldroyd and Wongsiri 2006). Only *Apis cerana* exhibits the tolerance of seasonal low temperatures shown by northern strains of *A. mellifera* and as a result, it is the first active honeybee on cool mornings in the northern tropics and the only species, apart from the Himalayan *A. laboriosa*, to extend significantly north of the tropics.

In contrast to the other Asian species, the large species, *Apis dorsata* and *Apis laboriosa*, are considerably larger than *A. mellifera* and form bigger colonies (<100,000 workers). Moreover, multiple colonies (<100) are often concentrated at a single site: a single, huge tree in *A. dorsata* or a rock face in *A. laboriosa*. Most colonies of *A. dorsata* appear to be migratory, tracking the availability of floral resources over tens or hundreds of kilometres, either seasonally or on irregular supra-annual cycles. Some colonies appear to cross tens of kilometres of open sea (cf. Chap. 7). Unlike any other honeybee, *A. dorsata* can forage in low light at dawn and dusk, and all night when the moon is at least half full (Dyer 1985; Somanathan et al. 2009). A recent molecular study suggests that the large honeybee of the Philippines (excluding Palawan), which does not form colony aggregations, is a distinct species, *A. breviligula*, and that the form from Sulawesi may also be distinct (Lo et al. 2010; cf. Chap. 1).

10.3 The Botany of Asian Honeybees

10.3.1 Asian Honeybees as Flower Visitors and Pollinators

Research on Asian honeybees has focussed on interactions with crop plants, and most information is more or less crop-specific (cf. Chaps. 11 and 12). The information on wild bees in native forests comes largely from a small number of study sites: most importantly, Lambir, Sarawak (4°N) (Momose et al. 1998; Itioka et al. 2002); Pasoh, Peninsular Malaysia (3°N) (Appanah 1985); Kakachi in the Western Ghats, India (8°N) (Devy and Davidar 2003, 2006); Coromandel Coast, India (12°N) (Selwyn and Parthasarathy 2006); Vientiane, Laos (18°N) (Kato et al. 2008); and Yakushima (30°N) (Yumoto 1987, 1988) and Ashiu (35°N) in Japan (Nagamitsu and Inoue 1999).

The bee-focussed literature suggests that at least the medium and large *Apis* species concentrate on the most rewarding resources, i.e., those with large amounts of easily accessible nectar and/or pollen (Roubik 1989). In Hong Kong, *A. cerana* visitation rates per flower were strongly positively related to the mean number of flowers on a plant across 11 species of *Ilex* (Aquifoliaceae) (Tsang and Corlett 2005). The bees recruit rapidly and in large numbers to mass-flowering trees or large stands of smaller plants (often crops or weedy exotics). They appear to be particularly important in tropical forests as visitors to trees with supra-annual flowering schedules (Momose et al. 1998; Devy and Davidar 2003; Somanathan et al. 2009) and also to the somewhat analogous situation of the mass-flowering, monocarpic Acanthaceae, which flower and die at multiyear intervals (Devy and Davidar 2006; Anitha and Prasad 2007; Sharma et al. 2008). In contrast, plant-focussed studies report medium and large *Apis* species among many visitors to a wide range of other plant species (e.g. Momose et al. 1998; Corlett 2001). Very little is known about the interaction of the small honeybee species with wild plants, but it is apparent that they are rarely the dominant visitors. It is possible, however, that even when not numerically dominant, honeybees of all sizes provide a unique pollination service by dispersing pollen over longer distances than most other bees.

Honeybees do not recruit to single flowers and will ignore profitable sources if a more profitable one is available nearby. As a result, the likelihood of honeybees visiting a particular flower species is determined not just by the characteristics of individual flowers but also by the number and density of flowers of that species available at the same time, plus the availability of other co-flowering species in the vicinity. Honeybees may therefore visit flower types in crop monocultures, where the high density and lack of alternatives makes them attractive, which they would avoid in natural vegetation. For example, cardamom (*Elettaria cardamomum*, Zingiberaceae) is dominated by *Apis* bees when cultivated at high densities but not when growing wild in the forest (Kuriakose et al. 2009). Indeed, there seem to be few flower types that one or more *Apis* species cannot exploit profitably when they are grown in high density monocultures, so crop studies are a poor guide to honeybee flower preferences.

All non-crop studies report that Asian honeybees visit mostly pale-coloured (white, cream, yellow or green), rotate (wheel-shaped), cup or brush-shaped flowers with easily accessible nectar and/or pollen (e.g. Kiew 1993; Momose et al. 1998; Raju et al. 2003; Raju and Rao 2004; Selwyn and Parthasarathy 2006). Spectral reflectance curves are not available for any Asian honeybee flowers, but honeybees are known to use only brightness cues at long range, where an object subtends a visual angle of $<15°$, while depending on their trichromatic colour vision (sensitive from UV to green) at close range (de Ibarra and Vorobyev 2009).

At most Asian sites, classic bilabiate (two-lipped) "bee flowers", including forest understorey orchids, gingers (Zingiberaceae), Marantaceae and Gesneriaceae, are visited largely by non-social bees. In some cases, honeybees are excluded from such flowers by tightly closed petals (Inoue et al. 1990), but in others, they are probably deterred by the small numbers flowering at any one time, which precludes mass recruitment. Honeybees are also excluded from nectarless flowers with apically porose anthers (e.g., *Melastoma*, *Dillenia*), except as pollen scavengers. This is probably because they are unable to "buzz"-extract the pollen. They also cannot reach the nectar in most tubular flowers, although they may collect pollen from them. A variety of other morphological devices are used by some plant species to exclude honeybees, for example, unpalatable nectars (Shuttleworth and Johnson 2009). Nocturnal anthesis, in contrast, excludes most or all bees except *A. dorsata* (Momose et al. 1998; Rao et al. 2001; Devy and Davidar 2003; Somanathan et al. 2009).

Field observations also commonly reveal honeybees visiting flowers that, judging by their size and the spatial separation of anthers and stigma, are targeted at much larger pollinators. The honeybees may often contribute nothing to pollination (see below), but this is not necessarily always the case. In Hong Kong, *A. cerana* visited all six *Rhododendron* species and was the most common visitor to three (Ng and Corlett 2000). Despite the size mismatch and the difficulty these bees had in harvesting the sticky pollen from apically porose anthers, enough honeybees contacted both anthers and stigma to suggest that they may have been more important pollinators than the individually more efficient, but less abundant, larger bees (*Bombus* and *Xylocopa*). *A priori*, the honeybee foraging strategy should also make them poor pollinators of dioecious species, for which constancy to one floral morph would discourage movements between sexes. This may often be the case, particularly where male floral displays are larger or pollen is the major attraction (e.g., *Rhus trichocarpa*; Matsuyama et al. 2009), but all Asian members of the largest woody dioecious clade, *Ilex* (Aquifoliaceae), for which there are records, appear to be pollinated largely by *Apis* bees (Tsang and Corlett 2005; personal observation).

The floral characteristics used in traditional plant taxonomy and identification are summarised in the floral formula, which represents the numbers, relative positions and fusion of floral parts in a standard way. Despite the changes in classification brought about by the availability of molecular sequence data, these floral characteristics still tend to be more or less constant at the family or subfamily level. For flower visitors, however, the most important floral characteristics – the conspicuousness of the flowers and the accessibility of the resources – are only weakly related to the floral formula, since they depend on features such as the size, shape and colour of floral parts, which are not included and can vary widely even within a single genus. Therefore, family membership, with a few exceptions, is not a good predictor of pollination biology.

Asian honeybees have been reported to visit wild species in a wide range of plant families. A few families are clearly overrepresented in these reports, most strikingly the Euphorbiaceae, which they may often fail to pollinate because male and female flowers are separate and only pollen is collected (see below). Other major families where many species are visited by honeybees include the Anacardiaceae, Apocynaceae, Aquifoliaceae, Arecaceae (palms), Asteraceae, Burseraceae, Dipterocarpaceae, Elaeocarpaceae, Fabaceae (legumes), Lamiaceae, Malvaceae (in the broad sense, including Tiliaceae and Sterculiaceae), Myrtaceae, Pentaphylacaceae (including *Eurya*), Rosaceae, Rubiaceae, Rutaceae, Sapindaceae, Sapotaceae and Verbenaceae. Most of these families also have many species that exclude honeybees morphologically or by other means. This is particularly striking in the huge families Fabaceae and Rubiaceae, where many species are visited by honeybees, but many more are not. Many other families include one or a few genera that are visited by honeybees. Common families that honeybees usually avoid include Annonaceae (mostly pollinated by beetles), Araceae (beetles, flies), Bignoniaceae (bats), Clusiaceae (in the narrow sense, excluding Calophyllaceae and Hypericaceae, which honeybees do visit) (various), Loranthaceae (birds, various), Marantaceae (other bees), Melastomataceae (except *Memecylon*, which honeybees visit) (other bees), Myristicaceae (beetles), Phyllanthaceae (various), Polygalaceae (other bees) and Zingiberaceae (other bees).

10.3.2 Asian Honeybees as Thieves and Commensals

Pollen can make plant babies or bee babies but not both. This mismatch between bee and flower interests is highlighted by the significant proportion of resources that honeybees obtain from plants in circumstances that make pollination improbable or impossible. Bees may fail to pollinate any flowers that they are poorly adapted to, but it is usually nectar-gatherers that fail to pollinate since they need not contact anthers. Unlike carpenter bees, Asian honeybees are not often recorded as nectar-robbers from insect-pollinated flowers, but they are frequently observed as "morning after" scavengers on both nectar and pollen from bat-pollinated species with nocturnal anthesis, such as *Durio* and *Musa*. *A. dorsata* takes nectar from the large, red, bird-pollinated flowers of *Bruguiera gymnorhiza* without triggering explosive pollen release from the anthers (Nagarajan et al. 2010). Honeybees are also known to collect hemipteran honeydew from a variety of sources (e.g. Joshi et al. 2000) and to visit extrafloral nectaries. Indeed, rubber (*Hevea brasiliensis*) is an important honey plant in the region because bees forage at nectaries on the leaf stalks (Kevan 1995). *A. dorsata* and *A. florea* have been observed harvesting juice from damaged grapes (cf. Chap. 12).

Pollen theft is more overt, with pollen from wind-pollinated taxa (e.g., grasses, including *Zea mays*, sedges, *Casuarina equisetifolia*, *Quercus* spp.) and the male flowers of monoecious or dioecious species (palms, Euphorbiaceae) making up a significant proportion of all pollen collected, at least in disturbed landscapes (e.g., Suryanarayana et al. 1992; Kiew 1993; Kevan 1995; Nagamitsu and Inoue 1999; Tilde et al. 2003; Roubik 2005; personal observation). It is, however, possible that foraging bees may sometimes assist wind-pollination by facilitating pollen release (Pierre et al. 2010). Scavenging resources from dying flowers will not have a negative impact on the plant and the harvesting of extrafloral nectar and honeydew will only do so if it reduces the resources available for defensive mutualisms with ants. The impacts of non-mutualistic pollen collection will depend on whether or not a reduction in the pollen available for pollination reduces the plant's male success, which is very difficult to assess, but the huge quantities harvested suggest that some degree of negative impact is inevitable. The active removal of pollen deposited by other bees from the stigmas of flowers with inaccessible porous anthers has been reported for introduced *A. mellifera* in Australia (Gross and Mackay 1998).

10.3.3 Deception of Honeybees

Most honeybee workers are directed to their foraging site by dancing scouts, so the opportunities for deception by plants that offer no bee rewards are limited in comparison with solitary bees. However, there are clearly enough honeybee scouts on exploratory trips (or, perhaps, misdirected dance followers) for deception to be a viable strategy, and a scattering of cases of orchids (e.g., Sasaki and Ono 1993) and asclepiads (Wang et al. 2009) routinely pollinated by unrewarded honeybees have been reported from the region. All of them involve pollen assembled into pollinia, which are transported but not consumed by the bees (Roubik 1989).

10.3.4 Resin Collection by Asian Honeybees

In addition to food, honeybees also utilise plants as a source of the resins that are used, in mixtures with beeswax and other secretions called propolis, as sealants and barriers (Duangphakdee et al. 2009). Indeed, resin barriers appear to be essential to the survival of the exposed arboreal nests of dwarf honeybees in the face of the threat from weaver ants (*Oecophylla smaragdina*) and, presumably, other ant species. *A. mellifera* is known to collect resin from the bark and buds of a wide variety of plants in other parts of the world (e.g., Silici et al. 2007), but there have been few such studies in Asia. *A. cerana* in Okinawa collected resin from the sticky surfaces of *Macaranga tanarius* (Euphorbiaceae) fruits (Kumazawa et al. 2008), while *A. mellifera* in southern Japan collected resin from the trunk of *Rhus javanica* (Murase et al. 2008).

10.4 The Importance of Honeybees in Asia

Honeybees appear to be relatively unimportant as pollinators in most forests in Asia (Kato et al. 2008). The major exception is the tropical rainforest canopy, where honeybees, especially *A. dorsata*, are major pollinators of mass-following species, particularly those that do not flower every year. Dominance by itself, however, does not indicate that they have an irreplaceable role, as shown by the ability of moths to functionally replace *A. dorsata* in the pollination of *Dipterocarps tempehes* when the giant honeybee was absent during an episode of weak general flowering at Lambir, Sarawak (Kenta et al. 2004). Honeybees are also important in the dry evergreen forests of eastern India (Selwyn and Parthasarathy 2006). Outside the forest, honeybees are more prominent, particularly in areas that would naturally be forested, and are probably more important, since the diversity of other social bees is much lower in such habitats. This seems to reflect the ability of several *Apis* species, particularly *A. cerana*, to adapt to nesting and foraging in human-dominated landscapes. Three species, *A. cerana*, *A. andreniformis* and *A. dorsata*, are apparently thriving in Singapore, which is largely urbanised. In Hong Kong, *A. cerana*, the only *Apis* present, visited 71 of the 83 woody species (86%) studied and was the numerically dominant visitor on 46 (55%) of these (Corlett 2001). The abundance of *A. cerana* in deforested Asian landscapes may compensate for – indeed, may overcompensate for – the decline in bee diversity following deforestation, as introduced *A. mellifera* appears to do in the Neotropics (Dick et al. 2003). Honeybees are also important throughout the region as pollinators of crop monocultures (cf. Chap. 11).

In an evolutionary context, it has been suggested that the diversity of honeybees in tropical east Asia has contributed to the lower diversity of other bees compared with the Neotropics (Roubik 1989). However, there is no evidence for a similar increase in overall bee diversity in New Guinea, where *Apis* was also absent until recently. Another factor in this difference between Asian and Neotropical bee communities may be the phenomenon of community-wide mass flowering at multiyear intervals in Southeast Asian rain forests, which favours large perennial colonies that are able to forage over very large areas and then either migrate or store enough resources to get them through the lean periods (Corlett 2004).

10.5 Conservation Concerns

Honeybees are vulnerable to both habitat loss and to direct exploitation for honey and wax. The equatorial lowland, forest-dependent, *A. koschevnikovi* appears to be under threat in some areas from the extremely high rates of deforestation within its native range, largely for oil palm (Hadisoesilo et al. 2008), but *A. cerana* replaces it in deforested areas, so it is unlikely that there is any major loss in pollination services. There is also no evidence for a decline in the density of small honeybees following deforestation, and *A. florea* will readily nest in disturbed areas (cf. Chap. 6).

In contrast, *A. dorsata* may be both irreplaceable, particularly for night-flowering canopy trees, and more sensitive to human impacts, as the favourite target of honey-hunters (Crane 1999; Oldroyd and Nanork 2009). Moreover, there may be a "half-empty forest" effect [in the sense of Redford and Feinsinger (2001)] if reduced *A. dorsata* densities reduce local nest-to-nest and bee-to-bee competition, thus reducing bee movements and, as a consequence, outcrossing as well as mean pollen dispersal distances. With *dorsata*-sized *Apis* bees present in the region from at least the Middle Miocene (Engel 2006), there has been ample time for dependence on this species to evolve. *Apis dorsata*, and probably also the understudied *A. laboriosa*, deserves the sort of conservation attention that is normally reserved for charismatic vertebrates.

The only honeybee invasion reported in the region is the establishment of feral colonies of *A. mellifera* on the remote oceanic Bonin (Ogasawara) Islands, where they may be disrupting native pollinator networks (Abe 2006, but see also Abe et al. 2008). Asian honeybees, however, are invasive outside the region, with *A. florea* spreading west, into the Arabian Peninsula and Africa, and *A. cerana* spreading east into New Guinea and the Solomon Islands (cf. Chap. 3). It is likely that both these species have also expanded their ranges within Asia with deforestation. Successful invasions imply the availability of unused resources in human-modified landscapes and/or a competitive advantage over native flower visitors. Unfortunately, there is no information on how these invasive Asian honeybees interact with the local flora and bee fauna. This contrasts with the attention given to invasive *A. mellifera* in the Neotropics (e.g. Roubik and Villanueva-Gutiérrez 2009). The invasion of New Guinea by *A. cerana* is of particular interest and concern as, unlike the Neotropics, this island has a largely Asian lowland flora at the generic and family level, and a relatively impoverished stingless bee fauna (Rasmussen and Cameron 2010).

10.6 Conclusions and Research Needs

The overall impression from the currently available literature is that Asian honeybees are considerably less important as pollinators in undisturbed natural ecosystems than they are in deforested areas and in agricultural systems. This impression is based on very few natural sites, however, and even at these sites, honeybees dominate on some resources. Moreover, even at sites where they are not dominant, honeybees may be important if they disperse pollen further than other bees or are more willing to cross open areas between habitat fragments. Simply replicating the existing types of data across more sites would be useful, but the key research questions will be best answered by a more structured approach. Flower-focussed studies should ideally take place along multi-kilometre transects from deep inside intact forest (or natural non-forest vegetation) into anthropogenic open habitats. Bee-focussed studies will be most useful if they can assess the use of resources by individual colonies over the entire annual (or, in equatorial Southeast Asia, supra-annual) flowering cycle across a range of habitat types. *A. dorsata* and *A. laboriosa*, with their huge, migratory colonies, will be particularly hard to study in this way, but they need to be properly understood within a systemic framework.

Acknowledgements I am grateful to Randall Hepburn, Ng Sai-chit, Anita Tsang and Laura Wong for their advice and assistance at various stages in this project.

References

Abe T (2006) Threatened pollination systems in native flora of the Ogasawara (Bonin) Islands. Ann Bot 98:317–334

Abe T, Makino S, Okochi I (2008) Why have endemic pollinators declined on the Ogasawara Islands? Biodivers Conserv 17:1465–1473

Anitha K, Prasad SN (2007) Mass flowering and pollinators of *Strobilanthes consanguinea* in the Western Ghats, South India. Curr Sci 92:1680–1681

Appanah S (1985) General flowering in the climax rain forests of South-east Asia. J Trop Ecol 1:225–240

Beekman M, Gilchrist AL, Duncan M, Sumpter DJT (2007) What makes a honeybee scout? Behav Ecol Sociobiol 61:985–995

Corlett RT (2001) Pollination in a degraded tropical landscape: a Hong Kong case study. J Trop Ecol 17:155–161

Corlett RT (2004) Flower visitors and pollination in the Oriental (Indomalayan) region. Biol Rev 79:497–532

Crane E (1999) The world history of beekeeping and honey hunting. Duckworth, London

de Ibarra NH, Vorobyev M (2009) Flower patterns are adapted for detection by bees. J Comp Physiol B 195:319–323

Delaplane KS, Mayer DF (2000) Crop pollination by bees. CABI, Oxford

Devy MS, Davidar P (2003) Pollination systems of trees in Kakachi, a mid-elevation wet evergreen forest in Western Ghats, India. Am J Bot 90:650–657

Devy MS, Davidar P (2006) Breeding systems and pollination modes of understorey shrubs in a medium elevation wet evergreen forest, southern Western Ghats, India. Curr Sci 90:838–842

Dick CW, Etchelecu G, Austerlitz F (2003) Pollen dispersal of tropical trees (*Dinizia excelsa*: Fabaceae) by native insects and African honeybees in pristine and fragmented Amazonian rainforest. Mol Ecol 12:753–764

Duangphakdee O, Koeniger N, Deowanish S, Hepburn HR, Wongsiri S (2009) Ant repellent resins of honeybees and stingless bees. Insectes Soc 56:333–339

Dyer FC (1985) Nocturnal orientation by the Asian honey bee, *Apis dorsata*. Anim Behav 33:769–774

Engel MS (2006) A giant honey bee from the Middle Miocene of Japan (Hymenoptera: Apidae). Am Mus Novit No 3504:1–12

Free JB (1993) Insect pollination of crops. Academic, London

Gross CL, Mackay D (1998) Honeybees reduce fitness in the pioneer shrub *Melastoma affine* (Melastomataceae). Biol Conserv 86:169–178

Hadisoesilo S, Raffiudin R, Susanti W, Atmowidi T, Hepburn C, Radloff SE, Fuchs S, Hepburn HR (2008) Morphometric analysis and biogeography of *Apis koschevnikovi* Enderlein (1906). Apidologie 39:495–503

Inoue T, Salmah S, Sakagami SF, Yamane S, Kato M (1990) An analysis of anthophilous insects in central Sumatra. In: Ohgushi R, Sakagami SF, Roubik DW (eds) Natural history of social wasps and bees in equatorial Sumatra. Hokkaido University Press, Sapporo, pp 201–218

Itioka T, Inoue T, Kato M, Nagamitsu T, Momose K, Sakai S, Yumoto T, Mohamad SU, Hamid AA, Yamane SK (2002) Six-year population fluctuation of the giant honeybee *Apis dorsata* F

(Hymenoptera: Apidae) in a tropical lowland dipterocarp forest in Sarawak. Ann Entomol Soc Am 94:545–549

Joshi SR, Pechhacker H, Willam W, von der Ohe W (2000) Physico-chemical characteristics of *Apis dorsata*, *A. cerana* and *A. mellifera* honey from Chitwan district, central Nepal. Apidologie 21:367–375

Kato M, Kosaka Y, Kawakita A, Okuyama Y, Kobayashi C, Phimminith T, Thongphan D (2008) Plant pollinator interactions in tropical monsoon forests in Southeast Asia. Am J Bot 95:1375–1394

Kenta T, Isagi Y, Nakagawa M, Yamashita M, Nakashizuka T (2004) Variation in pollen dispersal between years with different pollination conditions in a tropical emergent tree. Mol Ecol 13:3575–3584

Kevan PG (1995) Bee botany: pollination, foraging and floral calendars. In: Kevan PG (ed) The Asiatic hive bee. Enviroquest, Cambridge, pp 113–116

Kiew R (1993) Characteristics of Malaysian flowers visited by *Apis cerana*. In: Conner LJ, Rinderer TE, Sylvester A, Wongsiri S (eds) Asian apiculture. Wicwas, Cheshire, pp 361–368

Kumazawa S, Nakamura J, Murase M, Miyagawa M, Ahn MR, Fukumoto S (2008) Plant origin of Okinawan propolis: honeybee behavior observation and phytochemical analysis. Naturwissenschaften 95:781–786

Kuriakose G, Sinu PA, Shivanna KR (2009) Domestication of cardamom (*Elettaria cardamomum*) in Western Ghats, India: divergence in productive traits and a shift in major pollinators. Ann Bot 103:727–733

Lo N, Gloag RS, Anderson DL, Oldroyd BP (2010) A molecular phylogeny of the genus *Apis* suggests that the giant honey bee of the Philippines, *A. breviligula* Maa, and the plain honey bee of southern India, *A. indica* Farbricius, are valid species. Syst Entomol 35:226–233

Matsuyama S, Osawa N, Sakimoto M (2009) Generalist pollinators in the dioecious shrub *Rhus trichocarpa* Miq. (Anacardiaceae) and their role in reproductive success. Plant Species Biol 24:215–224

Momose K, Yumoto T, Nagamitsu T, Kato M, Nagamasu H, Sakai S, Harrison R, Itioka T, Hamid AA (1998) Pollination biology in a lowland dipterocarp forest in Sarawak. Malaysia. I. Characteristics of the plant-pollinator community in a lowland dipterocarp forest. Am J Bot 85:1477–1501

Murase M, Kato M, Sun A, Ono T, Nakamura J, Sato T, Kumazawa S (2008) *Rhus javanica* var. *chinensis* as a new plant origin of propolis from Okayama, Japan. Biosci Biotechnol Biochem 72:2782–2784

Nagamitsu T, Inoue T (1999) Differences in pollen sources of *Apis cerana* and *Apis mellifera* at a primary beech forest in central Japan. J Apic Res 38:71–78

Nagarajan B, Krishnamoorthy M, Padmini S, Daniel A (2010) Nectar robbery in bird pollinated mangroves! Curr Sci 98:603–604

Ng SC, Corlett RT (2000) Comparative reproductive biology of the six species of *Rhododendron* (Ericaceae) in Hong Kong, South China. Can J Bot 78:221–229

Oldroyd BP, Nanork P (2009) Conservation of Asian honey bees. Apidologie 40:296–312

Oldroyd BP, Wongsiri S (2006) Asian honey bees. Harvard University Press, Cambridge

Pierre J, Vaissiere B, Vallee P, Renard M (2010) Efficiency of airborne pollen released by honeybee foraging on pollination in oilseed rape: a wind insect-assisted pollination. Apidologie 41:109–115

Raffiudin R, Crozier RH (2007) Phylogenetic analysis of honey bee behavioral evolution. Mol Phylogenet Evol 43:543–552

Raju AJS, Rao SP (2004) The rock honeybee, *Apis dorsata* and its pollination potential in some plants. J Nat Conserv 16:59–65

Raju AJS, Rao SP, Kumari PK (2003) The role of *Apis florea* in the pollination of some plant species in Andhra Pradesh, India. Korean J Apic 18:37–42

Rao SP, Atluri JB, Reddi CS (2001) Intermittent mass blooming, midnight anthesis and rockbee pollination in *Pterocarpus santalinus* (Fabaceae). Nordic J Bot 21:271–276

Rasmussen C, Cameron SA (2010) Global stingless bee phylogeny supports ancient divergence, vicariance, and long distance dispersal. Biol J Linn Soc 99:206–232

Redford KH, Feinsinger P (2001) The half-empty forest: Sustainable use and the ecology of interactions. In: Reynolds J, Mace GM, Redford KH, Robinson JG (eds) Conservation of exploited species. Cambridge University Press, Cambridge, pp 370–399

Roubik DW (1989) Ecology and natural history of tropical bees. Cambridge University Press, Cambridge

Roubik DW (2005) Honeybees in Borneo. In: Roubik DW, Sakai S, Hamid Karim A (eds) Pollination ecology and the rain forest. Springer, New York, pp 89–103

Roubik DW, Villanueva-Gutiérrez R (2009) Invasive Africanized honey bee impact on native solitary bees: a pollen resource and trap nest analysis. Biol J Linn Soc 98:152–160

Sasaki M, Ono M (1993) Pollination ecology of the Japanese honey bee and the oriental orchid. In: Proceedings of 1st international conference on Asian honeybees and bee mites. Wicwas, Cheshire, pp 391–397

Schneider SS (1989) Spatial foraging patterns of the African honey bee *Apis mellifera scutellata*. J Insect Behav 2:505–522

Selwyn MA, Parthasarathy N (2006) Reproductive traits and phenology of plants in tropical dry evergreen forest on the Coromandel coast of India. Biodivers Conserv 15:3207–3234

Sharma MV, Kuriakose G, Shivanna KR (2008) Reproductive strategies of *Strobilanthes kunthianus*, an endemic, semelparous species in southern Western Ghats, India. Bot J Linn Soc 157:155–163

Shuttleworth A, Johnson SD (2009) The importance of scent and nectar filters in a specialized wasp-pollination system. Funct Ecol 23:931–940

Silici S, Unlu M, Vardar-Unlu G (2007) Antibacterial activity and phytochemical evidence for the plant origin of Turkish propolis from different regions. World J Microbiol Biotechnol 23:1797–1803

Somanathan H, Warrant EJ, Borges RM, Wallen R, Kelber A (2009) Resolution and sensitivity of the eyes of the Asian honeybees *Apis florea*, *Apis cerana* and *Apis dorsata*. J Exp Biol 212:2448–2453

Suryanarayana MC, Rao GM, Singh T (1992) Studies on pollen sources for *Apis cerana* Fabr. and *Apis mellifera* L. bees at Muzaffarpur, Bihar, India. Apidologie 23:33–46

Tilde AC, Payawal PC, Cervancia CR (2003) Pollen spectra among natural populations of *Apis cerana* Fabricius in the Philippines. Philipp Agric Sci 86:27–37

Tsang ACW, Corlett RT (2005) Reproductive biology of the *Ilex* species (Aquifoliaceae) in Hong Kong, China. Can J Bot 83:1645–1654

Wang D-K, Sun G-F, Zhai S-H, Guo L-H, Xiao F-H (2009) The floral syndrome and its pollination in *Cynanchum otophyllum* (Asclepiadaceae). Acta Bot Yunnanica 31:8–14

Yumoto T (1987) Pollination systems in a warm temperate evergreen broad-leaved forest on Yaku Island. Ecol Res 2:133–146

Yumoto T (1988) Pollination systems in the cool temperate mixed coniferous and broad-leaved forest zone of Yakushima Island. Ecol Res 3:117–130

Chapter 11
The Pollination Role of Honeybees

Uma Partap

11.1 Introduction

Pollination is an essential ecosystem service that results in gene flow in many crops and natural, wild flora. It is an essential prerequisite for fertilisation and fruit/seed set. Pollination processes are based on the ecological principle of species' inter-relationships or interactions, known as "proto-cooperation", between the plants and their pollinating agents called "pollinators". Pollinators visit the flowers of plants to obtain their food (i.e. nectar and pollen) and, in return, pollinate them. In many cases, it is the result of an intricate relationship between plants and their pollinators, and the reduction or loss of either affects the survival of both.

Pollination benefits society by increasing food security and improving livelihoods by the role it plays in conserving biological diversity in agricultural and natural ecosystems. Reduced agricultural yields and deformed fruit often result from insufficient pollination rather than a deficiency of agricultural inputs such as agrochemicals, agronomic practises and irrigation, etc. In natural ecosystems, visual clues of insufficient pollination are more subtle than in agriculture, but the consequences can be just as severe as the local extinction of a plant species, a noticeable decline in fruit- and seed-eating animals, the loss of vegetation cover and ultimately, if keystone species are involved, the demise of healthy ecosystems and their services. In recent years, the Convention of Biological Diversity (CBD) has recognised pollination as a key driver in the maintenance of biodiversity and ecosystem function.

U. Partap
Honeybees Project, Sustainable Livelihoods and Poverty Reduction Programme, International Centre for Integrated Mountain Development (ICIMOD), PO Box 3226, Kathmandu, Nepal
e-mail: upartap@icimod.org

H.R. Hepburn and S.E. Radloff (eds.), *Honeybees of Asia*,
DOI 10.1007/978-3-642-16422-4_11, © Springer-Verlag Berlin Heidelberg 2011

11.2 Pollination: Prerequisite for Completing the Life Cycles of Plants

11.2.1 Flower: The Reproductive Structure in Plants

Flowering is a vital process in the reproduction of most plants. A flower consists of the organs essential for reproduction. A typical hermaphroditic (bisexual) flower has four parts: calyx, corolla, androecium and gynoecium. The calyx (sepals) is usually green and provides protection to other floral parts in the bud stage. The corolla (petals) is variously coloured. It helps in the attraction of insect pollinators and provides a platform for their landing. The androecium is comprised of stamens – the male sex organs. A stamen consists of a filament and an anther. Anthers produce male spores, called pollen grains, which are released by rupturing open the anther wall after they mature. The female part of the flower is called the gynoecium; it is made up of pistils, also called carpels. A pistil consists of an ovary, having one to many ovules, a style and a stigma. When the pistil is ready to perform reproduction, the stigma becomes receptive by developing a sticky surface to which pollen grains become attached when they come in contact. Many plants, for example, apple, citrus, peach, pear, plum, sunflower, cabbage, cauliflower and mustard, produce hermaphroditic (bisexual) flowers that have both male and female sex organs. However, there are others, such as various cucurbits, that produce monosexual (either male or female) flowers on the same or different branches of the same plant. There are still others, such as kiwifruit, that produce male and female flowers on different plants.

11.2.2 Pollination Processes

Pollination is the transfer of pollen grains from the anther to the stigma of the same flower or another flower of the same plant or another plant of the same species and leads to fertilisation, i.e. the union of male and female nuclei. After pollen grains land on the receptive surface of the stigma, they send pollen tubes down through the style to the ovary. The male nucleus of each pollen tube then unites with the ovule effecting fertilisation. After fertilisation, the ovule and associated tissues develop into seeds and fruit. Pollination is an essential prerequisite for fertilisation and the development of seeds and fruit and is vital for completing the life cycle of plants, thus helps in improving crop productivity and maintaining biodiversity (Partap and Partap 1997).

Plants generally exhibit two pollination modes: self-pollination and cross-pollination. "Self-pollination" is the transfer of pollen from the anther of a flower to the stigma of the same flower, i.e. pollination within a flower, and plants thus pollinated are considered to be self-pollinated. Self-pollination takes place automatically in nature when anthers and stigma are of the same height, both mature at the same time, and fresh pollen comes into contact with a receptive stigma. Self-pollinated crops usually produce plenty of dry pollen. In some cases, self-pollination can lead to a condition called inbreeding depression in which breeding between close relatives may result in a reduction in genetic diversity and the possible expression of negative traits in the population and a loss of ability to evolve in response to a changing environment (McGregor 1976; Free 1993). A significant number of plants, particularly grasses, conifers, and oaks, rely upon the wind to spread pollen. Some of the world's most important food crops, such as rice, wheat and maize, are wind pollinated. A few even use water, such as horned pondweed, hornwort and some species of water starwort. However, the great majority of plants, more than 70% of species, depend on biotic pollinating agents such as insects, birds, bats and other animals to transport the pollen for them.

"Cross-pollination" is the transfer of pollen from the anthers of one flower to the stigma of another flower of the same or a different plant of the same species. When this takes place, the plant is called cross-pollinated. Cross-pollination always requires a pollinator – an agent that carries the pollen grains from the anthers of one flower to the stigma of another flower. When pollen is transferred to a flower of the same plant, cross-pollination is equivalent to self-pollination. It tends to decrease the likelihood of genetic variation in the crop. The other type of cross-pollination takes place when pollen from the flower of one plant is transferred to a flower of another plant of either the same or a different variety. This type of cross-pollination increases the likelihood of genetic variation. Cross-pollination occurs if flowers are unisexual and are borne on either the same or different plants, for example, species belonging to the families Actinidiaceae, Anacardiaceae, Cucurbitaceae, etc.; anthers and stigma are physically excluded, i.e. they are present at different heights. For example, anthers and stigma mature at different times in sunflower and safflower. Others such as onions and tree fruits like jujube, peaches, plums and almonds are self-incompatible (flowers are not fertilised by the pollen of the same variety), for example, many varieties of apples, almonds, pears, etc.

11.2.3 Pollination Requirements of Plants

Depending upon their pollination requirements, plant species are either self-fertile (self-compatible) or self-infertile (self-sterile or self-incompatible). A plant that is fertilised by its own pollen and can produce seed and fruit is called self-fertile or self-compatible, whereas a plant that cannot be fertilised by its own pollen but needs pollen from another plant of the same species is called self-sterile or self-incompatible. Commercial varieties of many fruit crops are self-sterile or self-incompatible, e.g. almonds, apples, plums and cherries. Various vegetable crops, for example, commercial varieties of cabbage, cauliflower, broccoli and radish also exhibit some degree of self-incompatibility. They essentially require cross-pollination to produce seeds and fruit (McGregor 1976; Free 1993) and would not do so unless cross-pollination takes place. Moreover, it is not only self-sterile varieties that benefit from cross-pollination, but self-fertile varieties also produce more and better quality seeds/fruit if they are cross-pollinated (Free 1993).

11.3 Importance of Pollination to Agriculture and Biodiversity

As explained, pollination is vital in completing the life cycles of plants. It ensures better crop yields whether in grain crops, seeds or fruit crops and is also necessary for the maintenance of biodiversity. It helps in soil conservation and soil fertility improvement through enhancing replenishment of soil nutrients, thus helping in conservation of environment and biodiversity. Figure 11.1 explains the importance of pollination in improving livelihoods through enhancing agricultural production and conserving environment and biodiversity.

11.4 Pollinators: Agents of Cross-Pollination

Self-pollinated plants are automatically pollinated when receptive stigmas come into contact with freshly released pollen from the same flower. However, many plants are not pollinated automatically. Moreover, in self-sterile crops, such as

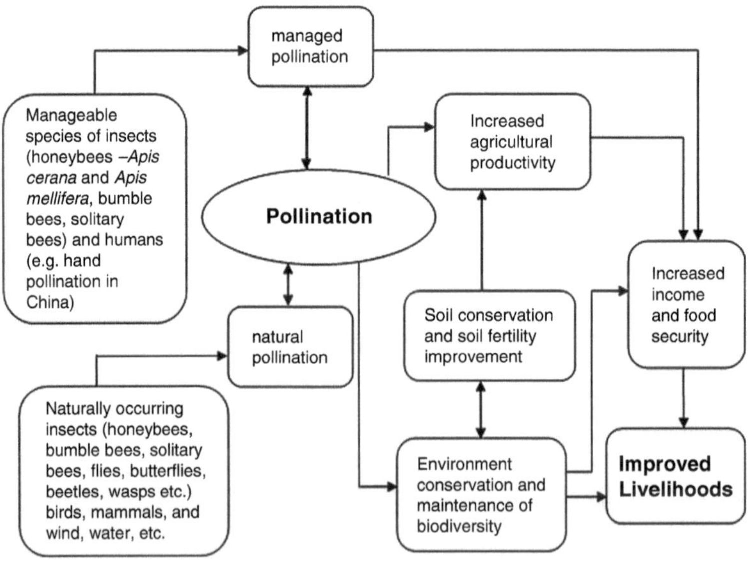

Fig. 11.1 Contribution of pollination to enhancing agricultural productivity and improving rural livelihoods (from Partap 2003a, b)

almonds and apples, flowers cannot be fertilised by pollen of the same flower or even the same plant. Pollen is needed from some other compatible source. Such crops need external agents (pollinators) to help them transfer pollen. Two types of pollinating agents occur in nature: abiotic and biotic.

11.4.1 Abiotic Pollinating Agents

Good examples of abiotic pollinating agents are wind, water and gravity. Many crop plants are successfully pollinated by wind, especially those that produce dry pollen, for example, grain crops such as rice, wheat, maize and millet, and nuts like chestnut, pecan nut and walnut and gymnosperms, etc. Wind pollination, also called anemophily, occurs in plants having specific characteristics that include reduced leaf-surface area, exposed flowers and reduced perianth (sepals and petals); long stamens and sometimes explosive anther dehiscence (pollen release); production of large amounts of pollen; smooth, dry, winged or balloon-shaped pollen grains; lack of nectar and nectaries; and flowers without colour or scent. "Water pollination", also called hydrophily, is only found in some water plants, for example, trapa. Pollination by gravity, also called geophily, is found in self-pollinated crop plants where pollen falls on to receptive stigma of other flowers because of gravity.

11.4.2 Biotic Pollinating Agents

Biotic pollination agents (biotic agents) include insects, birds and various mammals. Biotic pollination – also called zoophily – occurs when animals visit flowers for nectar and/or pollen and incidentally transfer pollen grains from one flower to another flower of the same or another plant. There is some kind of relationship between the pollen vector (pollinators) and the flowers of plants that are pollinated by biotic pollinating agents. They produce flowers of attractive colours and odours producing relatively small amounts of pollen, pollen grains vary in size and external sculpture, and are sticky in nature and flowers having nectaries that produce nectar. It has been estimated that over 75% of the world's crops and over 80% of all flowering plants depend on biotic pollinators. Globally, the annual contribution of pollinators to agricultural crops has been estimated at about US $54 billion (Kenmore and Krell 1998). In the United States, native pollinators are estimated to provide annual pollination services worth US $4.1 billion to agriculture (Prescott-Allen and Prescott-Allen 1990).

Many species of insects, birds, bats, and some non-flying mammals, etc. play an important role in pollination of various plants including cultivated crops. Birds and flying foxes are important pollinators of some plants (Proctor et al. 1996; Buchmann and Nabhan 1996). Buchmann and Nabhan (1996) reported 92 genera of plants belonging to 50 families that depend upon flying foxes for pollination. These authors reported that in Australia, 110 species of birds pollinate about 250 species of plants. Five hundred species of flower thrips, 10,000–15,000 species of wasps, 16 families of butterflies, 45 families of flies, 30 families of beetles, 1,500 species of birds, over 86 species of bats, one species of lemurs, 56–59 species of flying foxes and one species of non-flying mammals such as lemurs have been reported as important pollinators of various plants (Buchmann and Nabhan 1996).

Insect pollination is called entomophily and plants that require insect pollination to set fruit and seed are called entomophilous plants. Different kinds of insects, including bees, flies, beetles, butterflies, midges, moths, wasps and weevils, are important pollinators of many plants. Agricultural and horticultural crops, forage crops, ornamental plants and other wild plants are all effectively pollinated by insects that visit flowers for nectar and/or pollen. Among all the insects, bees are the most effective pollinators of crops and natural flora and are reported to pollinate over 70% of the world's cultivated crops; about 15% of the world's 100 principal crops are pollinated by manageable species of honeybees, bumblebees and solitary bees, while at least 80% are pollinated by other naturally occurring pollinators (Nabhan and Buchmann 1997).

11.5 Declining Pollinator Diversity and Its Impacts on Agricultural Productivity and Biodiversity

In recent years, pollinator populations and diversity have been declining worldwide (Partap and Partap 1997, 2002; Partap et al. 2001; Ahmad et al. 2003; Eardley et al. 2006). Evidence of this decline is available in some intensively cultivated areas of the Himalayan region. Studies carried out in some apple farming areas of the Himalayan region, e.g. Maoxian County of China, Himachal Pradesh Province in India, Balochistan Province in Pakistan, Thimphu and Paro valleys in Bhutan and Jumla district in Nepal, revealed that inadequate pollination has severely affected apple production. The yield has gone down and the quality of fruit was inferior due to inadequate pollination caused by the lack of natural pollinators of apple, which forced farmers to manage pollination of their cash-crops (Partap and Partap 2002). Recently, Ahmad et al. (2003) recorded other evidence of pollinator decline in Kaski district in Nepal. They reported a decline in the number of *Apis laboriosa* nests at eight sites from 182 nests in 1986 to 48 in 2002. They found that while the number of nests declined substantially at three sites, four sites were completely deserted by the bees.

Researchers throughout the world are convinced that the factors causing this decline could be the loss of habitat, with the accompanying decrease in food supplies (nectar and pollen) as a result of decline in pristine areas, land use changes, increase in monoculture-dominated agriculture and the negative impacts of modern agricultural interventions, e.g. use of chemical fertilisers and pesticides (Verma and Partap 1993; Aizen and Fiensinger 1994; Partap and Partap 1997, 2002;

Allen-Wardell et al. 1998; Ahmad et al. 2003). Clearing of forests and grasslands for farming has resulted in the loss of nesting sites and food sources of the pollinators. This has been clearly revealed by our studies carried out in mountain areas of China, India and Pakistan where apple orchards are being extended into the forestland and the grassland (Partap and Partap 2002). Changes in climate might also be affecting insect numbers (Partap and Partap 2002). Other factors, such as insufficient focus and capacities of national institutions and changing economic and social landscapes, have further aggravated this decline in the populations of indigenous honeybees throughout large parts of the Himalayas.

Increases in monocropping have also reduced the diversity of plants providing food for natural pollinators. Earlier, farmers used to grow a variety of crops, which bloomed during different months of the year and provided food for a number of natural insect pollinators, but the transformation of agriculture from traditional mixed-crop farming to high value cash-crop farming in recent years has led to an increase in monocrop-dominated agriculture. Examples of this are available in several pocket areas of the Himalayan region where farmers have switched over to the cultivation of apples and other fruit crops and off-season vegetable farming (Partap and Partap 2002). Further, modern industrial farming techniques such as use of pesticides and other agrochemicals may also be contributing to a decline in natural pollinators, especially wild bees, which could be playing a much greater role in pollinating crops. Studies carried out by ICIMOD (Partap and Partap 2001, 2002) revealed a serious lack of pollinators in apple farming areas of the Himalayan region due to heavy and indiscriminate use of pesticides on apple and other cash-crops. Similar examples are available from many areas around the world. Claire Kremen and her colleagues (quoted in Washington Post, December 16, 2002, p. 10) reported that farms using large amounts of pesticides were also frequented less often by the bees in Sacramento Valley in California (www.uoguelph.ca/~iucn/ accessed on April 16, 2003).

Increases in honey hunting practises and ruthless hunting of the nests of wild honeybees are another reason for the decline in population of indigenous honeybees. For example, in Nepal in the past, honey-hunting formed a part of tradition and a source of income for the communities; at present it is being commercialised and exploited by big contractors and companies (Ahmad et al. 2003). Development of infrastructure and tourism are other contributing factors which may have made adverse impact on populations of wild bees and other pollinators through destruction of the nesting sites and food sources. Introduction of invasive exotic species to an area is yet another factor affecting populations of native species. This may lead to the competition for food, for nesting and transfer of pests and diseases from one species to another.

The introduction of *Apis mellifera* to increase honey production has led to a decline in beekeeping with indigenous *Apis cerana* in several mountain areas. The reasons were complex (1) large scale efforts made by all government, non-government and private entrepreneurs to promote this European bee to all areas without realising that this bee may not be suitable to small-scale beekeeping and to remote and high altitude areas; (2) the introduction of *A. mellifera* without providing proper training on management of this species; (3) transfer of pests and diseases between *A. mellifera* and the indigenous species and vice versa; (4) competition for floral resources; and (5) human factors, e.g. temptation of keeping exotic species to get more honey, etc. (Partap and Partap 1997). The occurrence of forest fires also affects pollinator populations by not only destroying the nesting places and food sources of the pollinators but also killing them. Lastly, changes in climate might also affect insect numbers (Partap and Partap 2001, 2002).

This decline in pollinator populations and diversity presents a serious threat to agricultural production and conservation and maintenance of biodiversity in many parts of the temperate, subtropical and tropical world. Its impacts include reduced agricultural productivity and biodiversity. One indicator of the decline in natural insect pollinators is decreasing crop yields and quality, despite necessary agronomic inputs. Examples can be found in Himachal Pradesh in northwest India, northern Pakistan and parts of China where, despite all agronomic inputs, production and quality of fruit crops such as apples, almonds, cherries and pears is declining (Partap and Partap 2001, 2002; Partap et al. 2001).

Studies further revealed that because of the pollinator decline in parts of China (e.g. Maoxian county), farmers are forced to pollinate their crops, e.g. apple and pears by hand using human pollinators, which is an expensive and time consuming method of pollinating agricultural crops owing to the increased scarcity of labour and costs. Similar stories are also available from other countries, e.g. Pakistan where, disappointed with the very low yield and quality of apples as a result of poor pollination, several farmers in Azad Jammu and Kashmir have chopped out their apple trees and are using this land for other purposes. Thus, there is a pressing need to protect, promote and diversify pollinator resources in several countries of the developing world. This calls for initiating research and extension activities in this direction and developing strategies to conserve pollinator populations. This will require much more understanding of multiple services provided by pollinator diversity and the factors that influence them, including farmers, in order to secure sustained pollinator services in agricultural ecosystems.

11.6 Honeybees as Pollinators of Cultivated Crops and Natural Flora

There are about 25,000 species of bees found in the world. They include honeybees, bumble bees, stingless bees and solitary bees. Bees are the most efficient and most important pollinators of many cultivated crops and wild flora. A large number of bees on a crop ensure good pollination that results in higher yields and better quality produce. Bees visiting the flowers of a crop become conditioned to that particular crop and visit a number of its flowers during single foraging trips. Further, while collecting nectar and pollen, the bee brushes against the anthers of a flower and some pollen grains are picked up by the hairs on its body and head. When the bee visits another flower, some of the pollen grains are captured by the sticky surface of a receptive stigma, thus effecting cross-pollination. Bees have the following characteristics that make them the most important pollinators:

1. Bees are social insects; while other insects collect nectar mostly to satisfy their individual needs, bees collect nectar and pollen to feed their young.
2. Bees have body hairs. When a bee visits a flower some pollen becomes attached to its body and is transferred to the stigma of another flower that the bee visits next, thus accomplishing pollination.
3. Bees show flower constancy: e.g. a foraging bee usually moves from one flower to another of the same species for as long as nectar and pollen are available. Other insects haphazardly visit flowers of different species. This constancy in foraging is important for effective cross-pollination of a particular plant species.
4. Many species of bees, e.g. honeybees and stingless bees, are kept in man-made nests and mass-reared for honey production.
5. Many species of bees can also be managed for crop pollination. The most important are honeybees, bumble bees and solitary bees, e.g. alkali bees, horn-faced bees and leaf-cutter bees.

Honeybees are the most efficient of all the bees as pollinators of crops and natural flora. This is because their body parts are especially modified to pick up pollen grains; they have the potential to work for long hours, show flower constancy and are adapted to different climates (McGregor 1976; Free 1993). Most importantly, some honeybee species can be managed and transported to fields to pollinate crops. Moreover, technology for managing them for production of honey and pollination is available. In addition to this, Free (1993) lists the following characteristics – social and behavioural – that make honeybees the most effective and reliable crop pollinators.

1. Honeybees live in colonies where the young are nursed and fed by adults with a mixture of honey and pollen throughout the year. They work together to supply each other's needs and co-operate to raise offspring. Honeybee colonies are large compared to other bees consisting of 5,000–80,000 individuals depending upon the species.
2. They have the potential for long working hours. They start their foraging early in the morning and cease late in the evening, working many hours a day.
3. Honeybees have evolved a special communication system by which thousands of foragers can be deployed when a good food source is present.
4. They are micro-manipulators of flowers and visit many flowers per unit time compared to other bee species.
5. Some species can be managed in large numbers and moved to crops where and when necessary.
6. Most importantly, honeybees provide honey, beeswax and other bee products.
7. Finally, they are found in different geographical areas and are adapted to different altitudes and climates.

11.7 Principal Honeybee Species Found in Asia

Nine species of honeybees, including the giant honeybees or rock bees (*Apis dorsata* and *Apis laboriosa*), the little honeybee (*Apis florea*), the small dark honeybees (*Apis andreniformis*), the Asian honeybee (*Apis cerana*) and the European honeybee (*Apis mellifera*), are found in Asia. Among these, *A. dorsata*, *A. laboriosa*, *A. florea* and *A. andreniformis* cannot be kept in hives. Honey from these bees is harvested by traditional honey-hunting methods. Other bees, *A. cerana* and *A. mellifera*, are being managed in hives for honey production and pollination.

A. dorsata is found throughout the Asian region up to 2,000 m. It builds single comb nests in the open, on branches of tall trees and tall buildings and chimneys, in shady places during summer and sunny places during winter. As many as 70 or more colonies can be found on a single tree (cf. Chap. 6). This species is migratory in nature; a colony never stays in the same place for more than 6 months. *A. dorsata* produces harvestable amounts of honey and is an important pollinator of many crops and other plants. It nests in low hill areas during winter and migrates to the high hills in summer (cf. Chap. 7).

A. laboriosa is found from 1,200 to 3,500 m in remote mountainous areas of Bhutan, China, India and Nepal (Sakagami et al. 1980; Underwood 1986; Batra 1995). It nests beneath rock overhangs on vertical cliff faces. Colonies are found at a height of at least 10 m above the ground and occur in groups. Like *A. dorsata*, 70 or more colonies can be found at a single cliff site. It is also an important crop pollinator. It is also migratory in nature and a colony does not remain in one place the year round.

A. florea is one of the smallest honeybee species and is called the dwarf bee. It also builds single comb nests on branches of bushes, hedges, small trees, and chimneys, etc. This species is found in the plains and in hilly areas up to 500 m. It is also migratory in nature and a colony seldom stays in one place for more than 6 months. *A. florea* is another pollinator of agricultural crops.

A. andreniformis is found in the tropical region in the south of Hengduan mountain range of China, Thailand and Malaysia where the elevation is lower than 1,000 m. It builds its nests in small trees or bushes. Concrete information on this bee species is very scarce (cf. Chap. 20). Formerly in taxonomy, it was considered as a subspecies of *A. florea*, until 1986 when Wu and Kuang (1987) convincingly identified it as a separate species. This bee was first named "*Micrapis andreniformis*" and migrates twice a year, in February to March and August to September. Generally, *A. andreniformis* is regarded as a species on a relatively low phylogenetic level within the genus (Wu and Kuang 1987; but, cf. Chap. 2).

The Asian honeybee, *A. cerana*, and the European honeybee, *A. mellifera*, can be kept in hives and managed for honey production and crop pollination. They are cavity-nesting, hive bees. *A. cerana*, the Asian hive bee or Himalayan hive bee, is widespread up to 3,000 m throughout Asia. It has a gentle temperament, an industrious nature and good hygiene qualities (Verma 1990). Unlike *A. dorsata*, *A. laboriosa*, *A. andreniformis* and *A. florea* that build single comb nests in the open, *A. cerana* makes multiple parallel combs inside a cavity. Beekeeping with this bee is a common tradition among several countries of Asia. Farmers keep it in traditional fixed-comb hives such as log, wall and earthen-pitcher hives and in movable-frame wooden hives. A colony of *A. cerana* produces 5–20 kg of honey per year and is an excellent crop pollinator. This species has not become popular among commercial beekeepers because of its relatively low honey production and undesirable behavioural traits such as frequent swarming, absconding and robbing habits.

A. mellifera has been imported to Asia for commercial honey production. This species is kept in hives and makes parallel combs. It has gained high popularity among commercial beekeepers because it produces more honey than *A. cerana*, maintains a prolific queen, has low swarming and absconding tendencies and has good honey-gathering qualities. However, beekeeping with this species requires expensive technology and a high degree of chemical treatment to control diseases and parasites to which it is more susceptible (Verma 1990; Ahmad et al. 2002).

Recently, more species of *Apis* have been identified in Asia. These include *Apis koschevnikovi* (also named for a short period as *Apis vechti*) reported from Sabah, Malaysia, *Apis nigrocincta* reported from Philippines and *Apis binghami* reported from Sulawesi islands of Indonesia (cf. Chap. 1).

11.8 Asian Honeybees and Crop Pollination

11.8.1 Crops Pollinated by Honeybees

It has been estimated that the Himalayan region has over six million colonies and nests of indigenous and exotic honeybees that produce over 76,000 metric tonnes of honey every year. The figures on the number of honeybee colonies and honey production for Asia would obviously be very high. The honeybees play an important role in the pollination of various crops and natural flora and enhance vibrancy of floral diversity and agricultural productivity, thus contributing to the income and livelihood of poor mountain communities. A number of crops such as vegetable, fruit and nuts, oilseed, spices and forage and fibre crops either depend on or are benefited by honeybee pollination (Table 11.1).

11.8.2 Advantages of the Asian Honeybee, A. cerana, over the European Honeybee, A. mellifera

In the context of development, beekeeping has always been promoted exclusively as an enterprise for honey production; its role in enhancing agricultural productivity and maintaining biodiversity has always been overlooked. Today, most government and non-government organisations and the private sector are promoting beekeeping to increase honey production. The move towards the introduction of *A. mellifera* is an example of this. However, the work carried out by ICIMOD has proved that

Table 11.1 Crops that are dependent on, or benefit from, honeybee pollination

Crops dependent on bee pollination	Crops benefiting from bee pollination
Fruit and nut crops	
Almond	Apricot (few varieties)
Avocado	Blackberry
Apple (all commercial varieties)	Citrus
Apricot (some varieties)	Peach
Cherry (many varieties)	Persimmon
Kiwifruit	Strawberry
Litchi	
Mango	
Plum (many varieties)	
Pear (many varieties)	
Vegetable seed crops	
Cabbage	Beans
Carrot	Capsicum
Cauliflower	Eggplant
Cucumber	Okra
Onion	Tomato
Pumpkin	
Radish	
Squash	
Turnip	
Vegetable crops	
Bitter gourd	
Bottle gourd	
Muskmelon	
Pumpkins	
Sponge gourd	
Squash	
Watermelon	
Oilseed crops	
Sunflower	Mustard
Niger	Rape
	Cotton
Spice crops	
	Greater cardamom
	Chillies
	Coriander

Source: Ahmad (1987), Partap (1999a, b)

some subspecies of *A. cerana* found in the hills and mountain areas are potentially as productive as *A. mellifera*, but adequate research and development efforts are required to realise this potential. In this context, it is important to mention that a number of resources and time have been invested, and great deal of research efforts made, to increase the productivity of *A. mellifera*. The available evidence indicates that about 100–150 years ago the productivity of *A. mellifera* was as low as that of *A. cerana* today, i.e. almost 2–5 kg/year (Pechhacker et al. 2001). It is due to the large investment in research and development efforts to see a colony of *A. mellifera* producing 30–50 kg/year, today.

Even though *A. mellifera* produces large quantities of honey per colony, the indigenous *A. cerana* has distinct advantages over *A. mellifera*. To produce large quantities of honey, *A. mellifera* requires intensive management, standardised equipment and larger foraging areas. Furthermore, the race of *A. mellifera* introduced to this region is not very suitable for cold, mountainous areas. The bee is also not suitable for the small-scale stationary beekeeping usually practised in mountain areas. The colonies may not be kept at one place throughout the year for profitable beekeeping. The colonies need a lot of sugar syrup to survive during lean periods. In order to obtain profitable harvests of honey from *A. mellifera*, the colonies need to be migrated continuously from plain to hilly and mountainous areas and vice versa with the flowering seasons of honey plants. However, the mountain areas are relatively inaccessible, often lack transport facilities and communication infrastructure. Under these circumstances, migratory beekeeping becomes an expensive, vulnerable and high risk business. *A. mellifera* is prone to diseases, parasitic mites and wasps and needs antibiotics and other chemicals to treat their colonies, which will also contaminate the honey stored by these bees.

On the other hand, *A. cerana* is highly suitable for small-scale stationary beekeeping. This bee is very well adapted to cold mountain conditions and colonies can be kept at the same place throughout the year without much input. Beekeeping with *A. cerana* rarely has any capital requirement and needs simple management techniques. The beekeepers usually catch swarms during the active swarming season, put them in log or box hives and place them near their houses. *A. cerana* has adapted to local diseases, parasites and enemies and does not need any medicines or chemicals to treat them. As a result, honey stored by these bees is pure and free of any chemicals. Moreover, *A. cerana* bees are better pollinators of early blooming crops and flora.

11.9 Overview of Research on Honeybee Pollination

11.9.1 Pollination Efficiency of Honeybees

The pollination efficiency of different insects has been evaluated on the basis of a number of characteristics. It has been recognised that there are inherent differences in the ability of various species to effect pollination in spite of their abundance. Normally, the efficiency of an insect species as a pollinator has been attributed to its foraging behaviour and the amount of loose pollen grains adhering to its body (Bohart and Nye 1960; Free 1993). Pollination indices are calculated on the basis of relative abundance and foraging behaviour (Sharma 1990; Partap and Verma 1992, 1994; Verma and Partap 1993, 1994). Foraging efficiency of honeybees depends upon their foraging behaviour, for example, the duration of foraging period, time spent on the flower and number of flowers visited per minute (i.e. foraging speed and foraging rate), whether foraging as a top-worker or side-worker and preferences for pollen or nectar and the amount of loose pollen carried on the body. Foraging speed (time spent per flower) and foraging rate (number of flowers visited per minute) depend upon the foraging behaviour of insects and floral structure of the crop concerned, particularly the corolla depth (Gilbert 1980). Free and Williams (1973) reported that honeybees spend 131 s per kale flower when collecting pollen loads and 94 s when collecting nectar only.

The number of loose pollen grains on the body of foragers depends upon the size of the bee and the way it worked on the flowers. Normally, the "top-workers" carry more pollen grains on their bodies than the "side-worker" bees. Bohart and Nye (1960) reported that pollen collecting honeybees on carrot bloom literally waded across the head, swished their abdomens back and forth and scrapped the pollen from stamens with their forelegs. The nectar collectors stood higher on the flowers, moved about less and lapped droplets from the exposed nectaries. Nectar robbing in cauliflower bloom by many honeybee foragers has been found in northwest India by Kumar et al. (1994). Side-foraging of cauliflower nectar by *A. cerana* and *A. mellifera* and of okra nectar by *A. cerana* has been reported by Kapoor and Dhaliwal (1989) and Mishra et al. (1987), respectively. Sihag and Rathi (1994) have also suggested a formula for calculating performance scores of insect visitors on different pollination attributes. In addition to pollinators, pollination of a particular crop or plant species also depends upon its attractiveness and rewards offered to the pollinators.

Asian hive bees, *A. cerana*, have been reported to be more efficient pollinators of various fruit and vegetable crops than European *A. mellifera*. In their field experiments on the foraging behaviour of *A. cerana* on cabbage, cauliflowers, broad leaf mustard and lettuce conducted in the Kathmandu valley of Nepal, Partap and Verma (1992, 1994) and Verma and Partap (1993, 1994) reported that the foragers started working on cauliflower and cabbage bloom early in the morning and ceased late in the evening, thus pollinating these crops for comparatively longer hours than *A. mellifera*. Similar results were obtained by Verma and Rana (1994) on apple crop in the Shimla hills of the north-western Indian Himalayas. Further, in *A. cerana*, pollen collectors outnumbered nectar collectors, while in *A. mellifera*, nectar collectors outnumbered pollen collectors, though the ratio between pollen and nectar collectors decreased during the day for both species on broad leaf mustard, cabbage, cauliflower and lettuce in Kathmandu valley of Nepal (Verma and Partap 1993).

Regarding the foraging behaviour of *A. cerana* on lettuce and radish, the bees started their foraging activities early in the morning and ceased late in the evening; foraging for a period of 11.50 + 0.04 h/day on radish and 3.00 + 0.05 h/day on lettuce. The peak foraging hours occurred from 11:00 to 13:00 for radish and 9:00 to 11:00 for lettuce crop. The average duration of a foraging trip was 22.1 + 0.03 min for radish and 15.66 + 0.04 min for lettuce. Furthermore, the bees visited an average of eight, nine and five flowers per minute of radish during different hours of the day and ten flowers per minute of lettuce during morning hours (Verma and Partap 1993; Partap and Verma 1994).

A. cerana worked on the flowers mainly from the top position as "top workers", while *A. mellifera* mostly from the side position "side-workers" in cabbage, cauliflowers and radish crops in Kathmandu valley of Nepal (Partap and Verma 1992, 1994; Verma and Partap 1993, 1994). According to Erickson and Peterson (1979a, b), 14,666 and 9,582 carrot pollen grains were found on the body of pollen gatherers and nectar gatherers, respectively. Priti and Sihag (1997) found that on cauliflower bloom, *A. dorsata* carried the maximum and *Musca domestica* the minimum number of loose pollen grains. Pollen grain carrying capacity varied in different foraging groups of honeybees and also on different crops. Different workers have reported different results, for example, Kumar et al. (1993) found that nectar gatherers carried pollen grains either equal to or greater than pollen gatherers on toria and cauliflower. Other workers have found more pollen grains over the body of pollen gatherers than nectar gatherers on cauliflower (Dhaliwal 1980; Kapoor 1983) and onion (Kumar et al. 1985).

11.9.2 Honeybees as Pollinators of Various Crops

A great deal of experimental research on insect pollination of various crops has been carried out in different parts of the world, especially on type (species) and number of insect visitors, their foraging behaviour on the flowers and their impact on the yield and quality of various agricultural and horticultural crops. The pollination literature is full of such references (McGregor 1976; Free 1993; Kevan 1995). However, most of the research has been done in developed countries of the world where honeybees are being used for the pollination of various crops. The two notable reference books "Insect pollination of cultivated crop plants" by McGregor (1976) and "Insect pollination of crops" by Free (1993) present excellent details of research undertaken worldwide on insect pollination of crops. Most references speak about the role and importance of different species of honeybees in pollinating various crops.

The role of honeybees in pollination of several other crops has also been studied by various scientists in different parts of the world. For example, in Egypt, honeybees are reported as the most important pollinators of *Brassica oleracea* comprising 98% of visitors (Hussein and Abdel-Aal 1982). Similarly, in New Zealand, honeybees comprised 99% of all flower visitors of cauliflower (Forster et al. 1973). In the Indian sub-continent, *A. cerana*, *A. dorsata* and *A. florea* are all important in pollinating cauliflower, although to various extents in different localities. Thirty-four species of insects, including honeybees, visited the cauliflower bloom at Solan, Himachal Pradesh in north-western India (Sharma et al. 1974). *A. cerana* has been reported as one of the most frequent visitors of nine cultivars of okra in north-western Indian Himalayas (Mishra et al. 1987) and onion bloom in the United States (Treherne 1923; Trofimec 1940; Agati 1952; Bohart et al. 1970; Caron et al. 1975). In India, *A. cerana*, *A. dorsata*, *A. florea* and *Trigona iridipennis* were reported as primary insect pollinators of onion (Singh and Dharmwal 1970; Jadhav 1981; Rao and Suryanarayana 1989). Kumar et al. (1985) found that three onion species, *Allium cepa*, *A. fistulosum* and *A. cepa fistulosum*, greatly benefited by insect pollination, especially by *A. cerana* and flies. Carrot flowers are visited by different types of insects (Bohart and Nye 1960); however, pollen gathering honeybees were more valuable than nectar gatherers.

11.9.3 *Impact of Honeybee Pollination on Crop Yield and Quality*

Several studies were specifically undertaken to show the impact of honeybees, particularly *A. cerana* in enhancing the productivity levels of various crops such as fruit and nuts, vegetables, pulses, oilseeds, spices and fibre and forage crops (Deodikar and Suryanarayana 1977; Dulta and Verma 1987; Gupta et al. 1993, 2000; Partap 2000a, b; Partap and Verma 1992, 1994; Partap et al. 2000a, b; Singh et al. 2000; Verma and Partap 1993, 1994). These studies have shown that pollination by *A. cerana* increases fruit and seed set, enhances quality of fruit (shape, size, weight, colour and taste) and seeds and reduces premature fruit drop. The quality of pollination is determined by the number of colonies per unit area, the strength of bee colonies, placement of colonies in the field, the time of placement of bee colonies and the weather conditions. Experiences from pilot experiments have shown that the best results are achieved by placing strong bee colonies free of diseases with large amounts of unsealed brood when the crop is at 5–10% flowering (Free 1993; Verma and Partap 1993).

Research conducted in some countries of the Asian region has proved that honeybee pollination increases the yield and quality of apple in Shimla hills of Himachal Pradesh in North-western India (Dulta and Verma 1987; Gupta et al. 1993), peach, plum, citrus and strawberry in Kathmandu valley of Nepal (Partap 2000a, b; Partap et al. 2000a, b) and kiwi in Shimla hills of Himachal Pradesh, India (Gupta et al. 2000) (Table 11.2). The results of this research also showed that honeybee pollination not only increased the fruit set but also reduced fruit drop in apple, peach, plum and citrus. Results further showed an increase in the fruit juice and sugar content in citrus and reduction in the percentage of misshapen fruits in strawberry (Table 11.2). Looking at Table 11.2, we find an increase of 10% in the

Table 11.2 Impact of honeybee (*Apis cerana*) pollination on fruit productivity

Crop	Experimental site	Increase in fruit set (%)	Increase in fruit weight (%)	Increase in fruit size (length, diameter) (%)	References
Apple	Shimla Hills, HP, India	10	33	15, 10	Dulta and Verma (1987)
Peach	Kathmandu valley, Nepal	22	44	29, 23	Partap et al. (2000a, b)
Plum	Kathmandu valley, Nepal	13	39	11, 14	Partap et al. (2000a, b)
Citrus	Kathmandu valley, Nepal	24	35	9, 35 Premature fruit drop decreased by 46%, fruit juice increased by 68% and sugar contents in juice by 39%	Partap (2000a)
Strawberry	Kathmandu valley, Nepal	112	48	Misshapen fruits decreased by 50%	Partap (2000b)

Table 11.3 Impact of honeybee (*Apis cerana*) pollination on vegetable seed production in Kathmandu valley, Nepal

Crop	Increase in pod setting (%)	Increase in seed setting (%)	Increase in seed weight (%)
Cabbage	28	35	40
Cauliflower	24	34	37
Radish	23	24	34
Broad leaf mustard	11	14	17
Lettuce	12	21	9

Source: Partap and Verma (1992, 1994); Verma and Partap (1993, 1994)

apple fruit set while in the case of strawberries, it is 112% with a fruit weight increase of 33–48%, respectively, showing benefits of perfected pollination by honeybees under experimental conditions at the micro-level.

Studies conducted in the Kathmandu valley of Nepal have also shown that honeybee pollination enhanced seed production and quality of seed in various vegetable crops such as cabbage, cauliflower, radish, broad leaf mustard and lettuce (Partap and Verma 1992, 1994; Verma and Partap 1993, 1994). These crops, when pollinated by *A. cerana*, produced more and heavier seeds that showed a higher percentage of germination than the control plants. These results confirm the usefulness of bee pollination and its role in increasing crop productivity and improving the quality of fruit and seeds (Table 11.3). Scientific evidence from other parts of India confirms that bee pollination improves the yield and quality of other vegetable crops such as asparagus, carrots, onion, turnips and several other crops (Deodikar and Suryanarayana 1977). Honeybee pollination has been reported to increase the productivity of oil rape seeds by 18.7–37.8% and oil extraction could be increased up to10.4% in Manipur, India (Singh et al. 2000). Production was increased by 34.5–36.4% in sunflower and 38.5–43.6% in case of radish seed production (Verma and Partap 1993; Partap and Verma 1994; Singh et al. 2000). In other crops, such as buckwheat, soybean and cotton, the production increased from 23.8 to 77.7% (Kozin 1976).

Researchers in different parts of the world reported different percentages of increase in seed set due to pollinators. Increases of 22–100%, 100–300%, 100–125%, 91–135.4% and 353.5–987.8% in the seed yield have been reported due to assured pollination by bees in radish, cabbage, turnip, carrot and onion crops, respectively (Singh 1997). The results of such studies on important vegetable crops are given below.

An increase of 300% in seed crops of cabbage was reported due to insect pollination by Radchenko (1966). Even greater effects of insect pollination have been reported from India, Pakistan and Bangladesh. Rauala (1972) reported 68 and 9% set in cages with and without honeybees, respectively. Increased seed set (129%) was found, due to presence of honeybees, by Muhammad et al. (1973). Sihag (1986) found that plants caged to exclude insects and plants not caged produced 13 and 978 pods per plant, respectively. Similarly, 9.0 and 11.5 seeds per pod in caged (without insects) and open plots were found by Alam et al. (1987). Tewari and Singh (1983) demonstrated a decrease in seed set and yield with increasing distance from three *A. cerana* colonies. Kumar et al. (1989) found significantly higher seed set and seed weight in open pollinated than in bagged flowers in five cultivars of cauliflower.

Insect pollination increased the number of seeds in a pod (Free 1976) and there was a positive correlation between the number of seeds in a pod and its weight. Tanda (1984, 1985) reported a 19% increase in okra yield in Punjab, India, due to intensive bee pollination. Mishra et al. (1987) found that the weight and length of capsules, and seed number per capsule, were significantly higher in open pollinated than in bagged flowers of okra. In Poland, Woyke (1981) found that onion plots caged without bees, caged with bees and not caged produced 2, 210 and 669 seeds per head, respectively. Also, in India, Kumar et al. (1989) found greater set and yield and better germination of onion seeds from plots caged with bees than from plots caged without bees and open plots (61, 90, 72%) and estimated seed yields of 73, 275, 97 kg/ha, respectively, on plots caged without bees, caged with bees and not caged. Rao and Suryanarayana (1989) also reported higher seed yield in onion due to placement of bee colonies.

11.10 Managing Honeybees for Crop Pollination

The efficiency and effectiveness of a honeybee colony in pollination of various crops depend upon a number of factors (Verma 1990; Free 1993). The quality of pollination is, therefore, determined by the number of colonies per unit area, strength of bee colonies, amount of brood in the colony, number of foraging bees, condition/health of the colony, placement of colonies in the field, time of placement of bee colonies and the weather conditions. Scientists have developed farmer-friendly techniques for using honeybees for pollination (Kozin 1976; Verma 1990; Free 1993; Gupta et al. 1993; Partap 1999a, b). A simple method of crop pollination is to place honeybee colonies in a field when the crop starts flowering. Experience from pilot experiments has shown that the best results are achieved by placing strong bee colonies (having large amounts of unsealed brood and free of diseases) at the time of 5–10% flowering in the crop. However for effective pollination, according to Verma (1990), Free (1993) and Partap (1999a, b), the following things should be kept in mind while using bee colonies for crop pollination.

11.10.1 Strength and Health of the Colony

Large, strong colonies are better pollinators than small colonies because larger colonies have more forager bees. Also, good honey-yielding colonies are more efficient pollinators. Research has shown that a bee colony with 60,000 worker bees produces one-and-a-half times more honey than four colonies with 15,000 workers each (Verma 1990; Free 1993; Verma and Partap 1993). The same is true for pollination. Thus, as far as possible, strong colonies should be used for pollination. The colonies should be well settled, disease-free and have young, laying queens.

The strength of a honeybee colony depends above all on the availability of bee forage and season. Generally, colony strength is poor during winter because of low temperatures and a dearth of bee flora. Thus, when required in early spring for pollination of temperate fruit crops, these colonies are not strong enough to act effectively. To help colonies maintain their strength, bee colonies should be moved to low hill/plain areas during winter because it is warmer and floral sources are available and returned to the hills in spring when temperate fruit and vegetable crops are blooming. Such migration of bee colonies is the practise in Himachal Pradesh in India, North–West Frontier Province (NWFP) and Punjab in Pakistan and northern parts of China (Partap and Partap 2002).

11.10.2 The Number of Colonies Required for Pollination Varies for Different Crops

The number of colonies required for pollination depends on the total number of plants, total number of flowers per plant, attractiveness of the flowers to bees, duration of flowering, amount of nectar and pollen, bee species used, strength of colonies, number of pollen foragers, and amount of unsealed brood in the colonies (Kozin 1976; Verma 1990; Free 1993). In general, three strong colonies of *A. mellifera* per hectare of crop are recommended for adequate pollination (Kozin 1976; Verma 1990; Free 1993). Colonies of *A. cerana* are smaller: a 10-frame full-strength colony of *A. mellifera* is three times larger than a 10-frame full-strength colony of *A. cerana*. However, the foraging rate of *A. cerana* is 1.5 times greater than *A. mellifera*. Therefore, four to five strong colonies of *A. cerana* per hectare of crop are required. The number of colonies required also varies from crop to crop and from season to season for the same crop. The summary of pollination management of different crops is provided in Table 11.4.

11.10.3 Time of Placement of Bee Colonies in the Field

Time of placement of bee colonies is important for high yield and good quality produce. It is important to synchronise flower opening and foraging activity of bees. Freshly migrated colonies are more likely to visit a crop than those in place for a long time. Colonies should be brought to the field when 5–10% of the crop is in bloom. If colonies are placed early, bees will forage on flowers of wild plants nearby, becoming conditioned to these and ignoring the target crop when it blooms. If bees are moved late, they will pollinate only late on less vigorous flowers, resulting in poor yields and low-quality produce. For effective pollination of crops that flower for a short period, such as plum, bees should be moved when plants just start blooming because 50% flowering is achieved within 3–4 days.

Table 11.4 Summary of pollination management of different crops

Crop	Blooming period	Number of colonies per hectare		
		A. mellifera	*A. cerana*	Time of colony placement
Fruit crops				
Almond	Mid-February to mid-March	5–8	10–12	5–10% bloom
Apple	April (7–10 days)	5–8	10–12	5% bloom
Apricot	Mid-February (2–3 weeks)	2–3	4–6	5–10% bloom
Avocado	April–May	5–8	10–12	10–15% bloom
Cherry	February (7–10 days)	2–3	4–6	5% bloom
Citrus	March–April	2–3	4–5	5–10% bloom
Kiwifruit	March–April	8–9	16–20	5–10% bloom
Litchi	March–April	2–3	4–6	5–10% bloom
Mango	February	2–3	4–6	5–10% bloom
Peach	February–March (3–4 weeks)	1–2	2–3	5–10% bloom
Pear	February–March (7–10 days)	5	8–9	5% bloom
Perimmon	March–April (2 weeks)	2–3	4–6	5–10% bloom
Plum	February (1–2 weeks)	2–3	4–6	5% bloom
Strawberry	February–April (2 months)	As many as 15	25	5–10% bloom
Vegetable crops				
Cabbage	February–March	5	8–10	10–15% bloom
Carrot	March–April	5–8	10–12	10–15% bloom
Cauliflower	March–April	5	8–10	10–15% bloom
Cucumber – monoecious	June–September	1	2–3	10–15% bloom
Cucumber – gynoecios	June–September	8	12–16	10–15% bloom
Cucurbits (pumpkin, squash, gourd)	June–September	5–8	10–12	10–15% bloom
Okra	June–September	1–2	2–3	10–15% bloom
Onion	April	5–8	10–12	5–10% bloom
Radish	March–April	2–3	4–6	10–15% bloom
Turnip	February–March	2–3	4–6	5–10% bloom
Oilseed crops				
Mustard	December–January	3–5	5–8	10–15% bloom
Rape	February–March	3–5	5–8	10–15% bloom
Niger	August–September	3–5	6–8	5–10% bloom
Safflower	March–April	5	4–6	5–10% bloom
Sunflower	June	5	8–10	5–10% bloom
Spice crops				
Cardamon	March–April	2–3[a]	4–6	10–15% bloom
Chilli	July–September	2–3[a]	4–6	10–15% bloom
Coriander	February–April	2–3[a]	4–6	10–15% bloom

[a]No specific recommendation; *Source*: Partap (1999a, b)

11.10.4 Distribution of Bee Colonies in the Field

Honeybees prefer to visit sources of nectar and pollen that are near to their colonies (200–300 m). At over 500 m, pollination activity diminishes greatly. For effective pollination, colonies should be placed singly instead of in groups and distributed evenly in the field. Field experience has also shown that the distribution of colonies in an orchard depends upon the pollinators in proportion to the orchard. Gupta et al. (1993) have suggested the placement of varying number of *A. mellifera* colonies on fruit set in apple orchards having different proportions of pollinators. Based on the findings of their experimental research, these authors suggested that in apple orchards with 20% polliniser proportion, bee colonies should be placed in groups of four, and each group separated by 50 m gives maximum fruit set. But with those with a higher polliniser proportion, four to five bee colonies should be distributed in the orchard to give a good fruit set.

11.10.5 Condition of the Colonies to be Used for Pollination

Bees that forage for pollen (pollen collectors) are known to be better pollinators than nectar collectors. Colonies being used for pollination should have large amounts of unsealed brood. This will increase the pollen requirement and the colony will recruit more pollen foragers. The amount of unsealed brood in a colony can be increased by adding frames of unsealed brood from another colony that is not being used for pollination. Pollen collection can also be increased by taking out frames in which the bees have stored pollen.

11.10.6 Attracting Foraging Bees to Crops

For crops that are poor nectar producers and relatively unattractive to honeybees, e.g. kiwifruit, bees should be fed sugar syrup in which a few fresh flowers and pollen of the crop have been soaked for some hours. Feeds should be given at night or early morning before bees go foraging. This increases pollen collection the following day and thus pollination.

11.10.7 Avoiding Competitive Sources of Forage and Crops

Sometimes bees ignore the crop to be pollinated and forage on other more attractive plants, e.g. weeds nearby. In such cases, removal of weeds to avoid competition in attracting bees is necessary, and the use of good agronomic practises for a healthy crop such as proper manuring, irrigation and pesticide application is advisable. Usually, however, weeds are useful as they provide additional nectar and pollen sources. Some crops are less attractive to bees or not at all. In such crops, flowers can be sprayed with a honey or sugar solution to attract more foragers.

11.10.8 Weather Conditions

Weather plays an important role in determining the success of pollination. It affects both the foraging activities of bees and fruit/seed setting by plant. Cool weather and wind affect foraging activities of bees. Therefore, colonies should be placed in sunny, sheltered locations giving protection from wind to encourage maximum flights in spring. Where no natural windbreak is available, provide a temporary wind shelter.

11.11 Payments for the Pollination Services of Honeybees: Examples from Himalayas

As pointed out earlier in this chapter, a wide diversity among species, including agricultural crops, depends on, or is benefited by, honeybee pollination. Honeybees are therefore essential for "food diversity", biodiversity and the maintenance of natural resources. It has been argued that the main significance of honeybees and beekeeping is considered in pollination of crops and natural flora whereas hive products are of secondary value. It has been estimated that the benefit of using honeybees for enhancing crop yields through cross-pollination is much higher than their role as producers of honey and beeswax. Morse and Calderone (2001) showed that the value of honeybee pollination to crop production in the United States is US $14.6 billion annually. Similar estimates have been made for other countries. For example, the value of honeybee pollination in Canadian dollars is 1.2 billion per year in Canadian agriculture (Winston and Scott 1984), about US $150 million in UK (Carreck and Williams 1998) and US $2.3 billion per year in New Zealand (Matheson and Schrader 1987). Cadoret (1992) estimated that the direct contribution of honeybee pollination to increased farm production in 20 Mediterranean countries was US $5.2 billion per year –US $3.2 billion in developing countries and 2 billion in other countries.

Keeping in view the contribution of honeybees to agriculture, honeybees are being used for perfecting pollination, especially in developed countries like the United States, Canada, Europe and Japan that have long used honeybees for pollination of crops such as apples, almonds, pears, plums, cucumbers, melons, watermelons and berries. But the developing countries lag far behind in making use of honeybees for crop pollination. Even though plenty of scientific evidence is now available to prove that honeybee pollination increases the productivity of various cash-crops, the practise of using honeybees for crop pollination does not yet exist in these countries.

However, recently, realising that the fruit industry has been suffering heavily because the service of bees was no longer adequate, local governments, particularly in Himachal Pradesh and Maoxian County (southwestern Sichuan province, China), have been undertaking special efforts to promote managed pollination of apples since late 1990s. While in Himachal the government promoted honeybees for pollination of apples, the government in Maoxian initially promoted "hand pollination". In 1990s, apple orchards in Maoxian county of China were pollinated by hand using "human pollinators". These people were paid 15–20 Yuan per day in addition to providing them food for their pollination services (Partap and Partap 2000, 2002; Partap et al. 2000a, b). Now, even in China the local institutions are also promoting beekeeping for pollination of apples and other crops.

A survey done by the author in apple farming areas of Bhutan, China, India, Nepal and Pakistan revealed that farmers in Himachal Pradesh in north-western Indian Himalayas are using honeybees for apple pollination (Partap 1998). Himachal is the only place where honeybees are being promoted for apple pollination. It is interesting to note that while in the United States, the first colonies of honeybees, *A. mellifera*, were rented for pollination of pears in Virginia in 1895 (Waite 1895) and for apple pollination in 1909 in New Jersey (Morse and Calderone 2001), but in Himachal honeybees were first rented for apple pollination only recently in 1996. Here, farmers are using both *A. cerana* as well as *A. mellifera*, and this has led to the growth of a new vocation. A number of pollination entrepreneurs (beekeepers who rent honeybee colonies for crop pollination) have now started up in the state to complement the official services. Apple farmers have created a heavy demand for supply of honeybees for pollination. Use of beekeeping for pollination of cash-crops is of great benefit not only to the farmers but also to the beekeepers. Beekeepers receive money for the pollination services of honeybees, and farmers' income is increased through boosting crop productivity as a result of pollination services of bees.

Every year tens of beekeepers with hundreds of bee colonies move to various hills and valleys of Himachal Pradesh to provide pollination services to apple farmers and get paid in return for their services. Farmers pay beekeepers IRs 800 (US $18) as a pollination fee for one colony of honeybees during each flowering season. An organised system of payment for pollination services of honeybees exists in Himachal Pradesh. Here the government is promoting private entrepreneurship for bee pollination. It arranges training and demonstrations to apple farmers on the use of honeybees for pollination. Other countries of Asia where the pollination related productivity problems exist can learn from the Himachal experience.

11.12 Promoting Honeybees for Pollination

While much research remains to be done to bring pollination into rural development and land management, the author is of the opinion that there is need for integrating pollination as an input in agriculture and horticulture development packages of practises. This will help promote beekeeping for pollination. However, the challenge is great: in fact, the importance of pollination to agricultural husbandry and biodiversity conservation have not been recognised by policy makers, planners, development workers and farmers in most countries of Asia. There is no conceptual clarity or recognition of the value of pollination. The main constraints to promoting honeybees for managed pollination are lack of awareness and understanding among farmers, extension workers, planners and policy makers about the importance of pollinators and pollination, lack of integrating pollination in agricultural development packages, scarcity of managed colonies of honeybees, and lack of knowledge about conservation, rearing and use of pollinators and their pollination behaviour. Partap (2000a, b, c) and Partap and Partap (2002) discuss the strategies to promote honeybees for pollination. According to them, this would require the following.

11.12.1 Raising Awareness

Lack of awareness at all levels – be it farmers, extension workers and professionals at policy and planning level – is one of the main problems in promoting managed pollination. With a few exceptions, farmers in those areas where there is a pollination problem are not aware of the value of honeybees (including other pollinators) for agricultural production. This is both because beekeeping has always been promoted exclusively as an enterprise for honey production and because cash-crop farming is a new activity in many developing countries, and there is no indigenous knowledge of the need for managed crop pollination for enhancing cash-crop production. Raising awareness at all levels about the importance of managed crop pollination through beekeeping and other pollinators is the first step as part of development efforts.

11.12.2 Including Pollination as a Technological Input to Agricultural Development Packages

Pollination has been overlooked in agricultural development strategies and is not included as a technological input in agricultural development packages. High value agriculture is being promoted in several areas, and extension institutions offer packages of practises for each type of crop, but the importance of managing pollination to achieve higher yields has been overlooked. Thus, farmers have no way of knowing how essential it can be. This weakness in the agricultural extension system needs to be addressed. Since pollination is essential for the production of fruit and seeds, it should be included in agricultural development packages by promoting beekeeping for crop pollination as a "double benefit approach". Thus, the most important step in promoting the wider use of honeybees for crop pollination is to include beekeeping as part of agricultural development efforts. Including managed pollination in agricultural development packages will also help develop strategies to conserve, promote and use other pollinators.

11.12.3 Influencing Thinking About Bees and Beekeeping

Traditional thinking is that beekeeping is for honey production; its role in crop pollination is rarely considered. Today, most government agencies are only engaged in promoting beekeeping for honey production. The move towards introduction of *A. mellifera* to increase honey production is an example of this. Thus, there is a need to change the general mindset about honeybees and beekeeping and to raise awareness about the importance of managed crop pollination.

11.12.4 Strengthening Research and Development Institutions

Managed crop pollination is a relatively new area. There are few institutions with explicit mandates or expertise for research and extension in this area. Most institutions are working only with beekeeping and promoting it as a cottage industry to increase family income through the sale of honey. Promoting the value of honeybees as reliable pollinators of agricultural crops will require special efforts to strengthen research and extension systems. This is necessary in order to underline applied research in key areas of managed crop pollination. Issues such as decline of pollinator populations and the need to conserve them need to be addressed by the institutions.

11.12.5 Human Resources Development and Capacity Building

Lack of knowledge among farmers about the pollination behaviour of honeybees is another constraint hindering the use of honeybees for crop pollination. Even those farmers, who do know that they can use honeybees to increase apple pollination and yield, do not always know how to use bees. Though linked with the institutional strengthening, it requires more focus to build the capacities of individual farmers, development workers and farmer-led organisations that are the agents of change. There is need to train farmers and beekeepers in managing honeybees for crop pollination. There is also a need to develop human resources and build their capacities to initiate activities in the areas of conserving, rearing and using pollinators to improve pollination and thus agricultural productivity.

References

Agati G (1952) Indagini ed osservazioni sulla biologia fiorale della cipolla. Riv Ortoflorofruttic Ital 36:67–77 [in Italian]

Ahmad R (1987) Honeybee pollination of important entomophilous crops. Pakistan Agricultural Research Council, Islamabad

Ahmad F, Partap U, Joshi SR, Gurung MB (2002) Why the HKH region needs the Himalayan honeybee *Apis cerana*. ICIMOD, Kathmandu

Ahmad F, Joshi SR, Gurung MB (2003) The Himalayan cliff bee *Apis laboriosa* and the honey hunters of the Himalayas. ICIMOD, Kathmandu

Aizen MA, Fiensinger P (1994) Habitat fragmentation, native insect pollinators, and feral honeybees in Argentine 'chaco serrano'. Ecol Appl 4:378–392

Alam MZ, Quadir MA, Ali MS (1987) Pollinating behaviour of honeybee. *Apis indica* F. and its influence on seed production of cauliflower. Bangladesh Hortic 15:25–30

Allen-Wardell GP, Bernhardt R, Bitner A (1998) The potential consequences of pollinator declines on conservation of biodiversity and stability of food crop yields. Conserv Biol 12:8–17

Batra SWT (1995) Biology of *Apis laboriosa* Smith, a pollinator of apples at the high altitude in the Greater Himalaya of Garhwal, India. USDA Agricultural Research Series, Tektran

Bohart GE, Nye WP (1960) Insect pollinators of carrots in Utah. Bull Utah Agric Exp Stn 419

Bohart GE, Nye WP, Hawthorn LR (1970) Onion pollination as affected by different levels of pollinator activity. Bull Utah Agric Exp Stn 482

Buchmann S, Nabhan GP (1996) The forgotten pollinators. Island, Washington DC

Cadoret JP (1992) Doorway doses help defeat honeybees' vampire enemy. Ceres 24:8–9

Caron DM, Lederhourse RC, Morse RA (1975) Insect pollinators of onion in New York State. HortScience 10:273–274

Carreck N, Williams IH (1998) The economic value of bees in the UK. Bee World 79:115–123

Deodikar GB, Suryanarayana MC (1977) Pollination in the services of increasing farm production in India. In: Nair PKK (ed) Advances in pollen spore research. Today and Tomorrow, New Delhi, pp 60–82

Dhaliwal HS (1980) Ecology and relative efficiency of principal insect pollinators of *Brassica* seed crop. Thesis, HPKV, COA, Solan, Himachal Pradesh

Dulta PC, Verma LR (1987) Role of insect pollinators on yield and quality of apple fruit. Indian J Hortic 44:274–279

Eardley C, Roth D, Clarke C, Buchmann S, Gemmill B (2006) Pollinators and pollination: a resource book for policy and practice. African Pollinator Initiative, Pretoria

Erickson EH, Peterson CE (1979a) Asynchrony of floral events and other differences in pollinator foraging stimuli between male fertile and male sterile carrot inbreds. J Am Soc Hortic Sci 104:639–643

Erickson EH, Peterson CE (1979b) Problems encountered in the pollination of cytoplasmically male sterile hybrid carrot seed parents. In: Proceedings of 4th international symposium on pollination, Maryland, pp 59–63

Forster IW, Clinch PG, Palmer-Jones T (1973) Observations on the pollination of chou moellier (*Brassica oleracea* L.). NZ J Exp Agric 1:46–48

Free JB (1976) Insect pollination of tropical crops. Report of the Central Association of Beekeepers, UK

Free JB (1993) Insect pollination of crops, 2nd edn. Academic, London

Free JB, Williams IH (1973) The foraging behaviour of honeybees (*Apis mellifera* L.) on Brussels sprout (*Brassica oleracea* L.). J Appl Ecol 10:489–499

Gilbert FS (1980) Flower visiting by hover flies: morphology of mouth parts in relation to feeding on nectar and pollen in some urban species. Ecol Entomol 6:245–265

Gupta JK, Goyal NP, Sharma JP, Gautam DR (1993) The effect of placement of varying numbers of *Apis mellifera* colonies on the fruit set in apple orchards having different proportions of

pollinisers. In: Proceedings of international symposium on pollination in the tropics, Bangalore, pp 197–201
Gupta JK, Rana BS, Sharma HK (2000) Pollination of kiwifruit in Himachal Pradesh. In: Matsuka M, Verma LR, Wongsiri S, Shrestha KK, Partap U (eds) Asian bees and beekeeping: progress of research and development. Oxford and IBH, New Delhi, pp 169–170
Hussein MH, Abdel-Aal SA (1982) Wild and honey bees as pollinators of 10 plant species in Assiut area, Egypt. Z Angew Entomol 93:342–346
Jadhav LD (1981) Role of insects in the pollination of onion (*Allium cepa* L.) in the Maharashtra State, India. Indian Bee J 43:61–63
Kapoor KS (1983) Foraging strategies and pollinating activities of *Apis cerana indica* F. and *Apis mellifera* L. on *Brassica campestris* L. var. *botrytis*. Thesis, HPKV, COA, Solan, Himachal Pradesh
Kapoor KS, Dhaliwal HS (1989) Comparative energetics of *Apis cerana indica* F and *A. mellifera* L. on *Brassica oleracea* L. var *botrytis*. Indian Bee J 51:101–103
Kenmore P, Krell R (1998) Global perspectives on pollination in agriculture and agro ecosystem management. In: Proceedings of the International Workshop on the Conservation and Sustainable use of Pollinators in Agriculture, 7–9 Oct 1998, Sao Paulo
Kevan PG (1995) Insect pollination of economically important plants of tropical and subtropical Asia. In: Kevan PG (ed) The Asiatic honeybee: apiculture, biology and role in sustainable development in tropical and subtropical Asia. Enviroquest, Cambridge, Ontario, pp 129–134
Kozin RB (1976) Pollination of entomophilous agricultural crops by bees. Amerind, New Delhi
Kumar J, Mishra RC, Gupta JK (1985) The effect of mode of pollination on *Allium* species with observations on insects as pollinators. J Apic Res 24:62–66
Kumar J, Mishra RC, Gupta JK (1989) Effect of honeybee pollination on onion (*Allium cepa* L.) seed production. Indian Bee J 51:3–5
Kumar J, Rao KVK, Gupta JK (1993) Amount of loose pollen on the body of bees visiting spring blooming crops. J Insect Sci 6:259–263
Kumar J, Rao KVK, Gupta JK (1994) Pollination efficiency of bees visiting blossom of *Brassica campestris* L. var. *toria* in mid hills of Himachal Pradesh, India. Indian Bee J 56:202–206
Matheson AG, Schrader M (1987) The value of honeybees to New Zealand's primary production. Ministry of Agriculture and Fisheries, Nelson, New Zealand
McGregor SE (1976) Insect pollination of cultivated crop plants. USDA, Washington
Mishra RC, Kumar J, Gupta JK (1987) Effect of mode of pollination on fruit characteristics of okra, *Abelmoschus esculentus* (L.) Moench. Proc Indian Natl Sci Acad 53:157–160
Morse RA, Calderone NW (2001) The value of honeybees as pollinators of US crops in 2000. Bee Cult 128:1–15
Muhammad S, Gondal A, Manzoor UH (1973) Studies on the role of *Apis indica* F. in the pollination of cauliflower (*Brassica oleracea* var. *botrytis* Linn) and radish (*Raphanus sativus* Linn). SU Res J (Sci Ser) 7:87–93
Nabhan GP, Buchmann S (1997) The fraying web of life. In: World resources 2000–2001. UNDP, UNEP, WB, WRI, pp 136–138
Partap U (1998) Successful pollination of apples in Himachal Pradesh. Beekeep Dev 48:6–7
Partap U (1999a) Conservation of endangered Himalayan honeybee, *Apis cerana* for crop pollination. Asian Bee J 1:44–49
Partap U (1999b) Pollination management of mountain crops through beekeeping: trainers' resource book. ICIMOD, Kathmandu
Partap U (2000a) Foraging behaviour of *Apis cerana* on sweet orange (*Citrus sinensis* var. Red Junar) and its impact on fruit production. In: Matsuka M, Verma LR, Wongsiri S, Shrestha KK, Partap U (eds) Asian bees and beekeeping: progress of research and development. Proceedings of 4th Asian Apicultural Association International Conference, Kathmandu, pp 174–177
Partap U (2000b) Pollination of strawberry by the Asian hive bee, *Apis cerana* F. In: Matsuka M, Verma LR, Wongsiri S, Shrestha KK, Partap U (eds) Asian bees and beekeeping: progress of research and development. Oxford and IBH, New Delhi, pp 178–182

Partap U (2000c) Foraging competition between *Apis cerana* and *Apis mellifera* and its impact on crop pollination. In: Matsuka M, Verma LR, Wongsiri S, Shrestha KK, Partap U (eds) Asian bees and beekeeping: progress of research and development. Oxford and IBH, New Delhi, pp 206–208
Partap U (2003a) Improving agricultural productivity and livelihoods through pollination: some issues and challenges. In: Waliyar F, Collette L, Kenmore PE (eds) Beyond the gene horizon. FAO, ICRISAT, Rome and India, pp 118–135
Partap U (2003b) Cash-crops farming in the Himalayas: the importance of pollinator management and managed pollination. In: Biodiversity and the ecosystem approach in agriculture, forestry and fisheries. FAO, Rome, pp 224–247
Partap U, Partap T (1997) Managed crop pollination: the missing dimension of mountain agricultural productivity. Mountain Farming Systems' Discussion Paper, MFS 97. ICIMOD, Kathmandu
Partap U, Partap T (2000) Pollination of apples in China. Beekeep Dev 54:6–7
Partap U, Partap T (2001) Declining apple production and worried Himalayan farmers: promotion of honeybees for pollination. Issues in Mountain Development (IMD). ICIMOD, Kathmandu
Partap U, Partap T (2002) Warning signals from apple valleys of the HKH region: pollination problems and farmers' management efforts. ICIMOD, Kathmandu
Partap U, Verma LR (1992) Floral biology and foraging behaviour of *Apis cerana* on lettuce crop and its impact on seed production. Prog Hortic 24:42–47
Partap U, Verma LR (1994) Pollination of radish by *Apis cerana*. J Apic Res 33:237–241
Partap U, Shukla AN, Verma LR (2000a) Pollination of peach and plum by *Apis cerana*. In: Matsuka M, Verma LR, Wongsiri S, Shrestha KK, Partap U (eds) Asian bees and beekeeping: progress of research and development. Oxford and IBH, New Delhi, pp 171–173
Partap U, Shukla AN, Verma LR (2000b) Comparative attractiveness of broad-leaf mustard, cauliflower and radish to *Apis cerana* in Kathmandu Valley of Nepal. In: Proceedings of the 4th Asian apicultural association international conference, Kathmandu, pp 203–205
Partap U, Partap T, Yoghua H (2001) Pollination failure in apple crop and farmers' management strategies in Hengduan, China. Acta Hortic 561:225–229
Pechhacker H, Joshi SR, Chatt A (2001) Traditional beekeeping in rural areas: a comparison between *Apis cerana* beekeeping in Nepal and *Apis mellifera* beekeeping in Morocco. Bee World 82:99–105
Prescott-Allen R, Prescott-Allen C (1990) How many crops plants feed the world? Conserv Biol 4:365–374
Priti, Sihag RC (1997) Diversity, visitation frequency, foraging behaviour and pollinating efficiency of insect pollinators visiting cauliflower (*Brassica oleracea* L. var. *botrytis* cv. Hazipur Local) blossoms. Indian Bee J 59:230–237
Proctor M, Yeo P, Lack A (1996) The natural history of pollination. Harper Collins, London
Radchenko TG (1966) Role of honeybees as pollinators in increasing the seed crop from cabbage and radish. Bdzhilnitstvo 72–75
Rao GM, Suryanarayana MC (1989) Effect of honeybee pollination on seed yield in onion (*Allium cepa* L.). Indian Bee J 50:9–11
Rauala TS (1972) Pollination studies in cauliflower (*Brassica oleracea* var. *botrytis* L.). J Res Punjab Agric Univ 9:580–585
Sakagami SF, Matsumura T, Ito K (1980) *Apis laboriosa* in the Himalaya, the little known world's largest honeybee (Hymenoptera: Apidae). Insecta Matsumurana NS 19:47–77
Sharma HK (1990) Floral resources and foraging characteristics of hive bees. Thesis, Dr YS Parmar University of Horticulture and Forestry, Solan, Himachal Pradesh
Sharma PL, Dhaliwal HS, Kakar KL (1974) Insect visitors and pollinators of cauliflower (*Brassica oleracea* var. *botrytis*) seed crop bloom. Himachal J Agric Res 2:74–78
Sihag RC (1986) Insect pollination increases seed production in cruciferous and umbelliferous crops. J Apic Res 25:21–126
Sihag RC, Rathi A (1994) Diversity, abundance, foraging behaviour and pollinating efficiency of different bees visiting pigeon pea (*Cajanus cajan* (L.) Mill sp.) blossoms. Indian Bee J 56:187–201

Singh Y (1997) Role of honeybees in farm production, agricultural growth and rural reconstruction in India. Indian Bee J 59:24–30

Singh JP, Dharmwal SS (1970) The role of honey bees in seed setting on onion at Pant Nagar, District Nainital, India. Indian Bee J 32:23–27

Singh MP, Singh KI, Devi CS (2000) Role of *Apis cerana* pollination on yield and quality of rapeseed and sunflower crops. In: Matsuka M, Verma LR, Wongsiri S, Shrestha KK, Partap U (eds) Asian bees and beekeeping: progress of research and development. Oxford and IBH, New Delhi, pp 186–189

Tanda AS (1984) Insect pollination in okra (*Abelmoschus esculentus*). In: Proceedings of the 3rd orient entomology symposium, Kariavattom, India, p 25

Tanda AS (1985) Floral biology, pollen disposal and foraging behaviour of honeybees in okra (*Abelmoschus esculentus*). J Apic Res 24:225–227

Tewari GN, Singh K (1983) Studies on insect pollinators in relation to seed production in cauliflower (*Brassica oleracea* var. *botrytis* L.). Indian Bee J 45:54–55

Treherne RC (1923) The relation of insects to vegetable seed production. Rep Queb Soc Protect Plants Ann Rep 15:47–49

Trofimec NK (1940) The biology and fertilization in *Allium*. Vest Sots Rasteniev Rasten 5:76–86

Underwood BA (1986) The natural history of *Apis laboriosa* Smith in Nepal. Thesis, Cornell University, Ithaca

Verma LR (1990) Beekeeping in integrated mountain development: economic and scientific perspectives. Oxford and IBH, New Delhi

Verma LR, Partap U (1993) The Asian hive bee, *Apis cerana*, as a pollinator in vegetable seed production. ICIMOD, Kathmandu

Verma LR, Partap U (1994) Foraging behaviour of *Apis cerana* on cabbage and cauliflower and its impact on seed production. J Apic Res 33:231–236

Verma LR, Rana RS (1994) Further studies on the behaviour of *Apis cerana* and *Apis mellifera* foraging on apple flowers. J Apic Res 33:175–179

Waite MB (1895) The pollination of pear flowers. Veg Path Bull 5, USDA

Winston ML, Scott CD (1984) The value of bee pollination to Canadian agriculture. Can Beekeep 11:134

Woyke HW (1981) Some aspects of the role of the honeybee in onion seed production in Poland. Acta Hortic 111:91–98

Wu YR, Kuang BY (1987) Two small species of honeybees: a study of genus *Micrapis*. Bee World 68:152–155

Chapter 12
Foraging

D.P. Abrol

12.1 Introduction

The habitats of most animals provide resources of different types that are essential for the species' survival, but these may not necessarily be close together. Accordingly, animals have to bridge some temporal or spatial distances to locate essential resources (Schoener 1971). Foraging can be defined as the search for and acquisition of food. It is one of the most consistent and demanding tasks for any living organism related to its survival. The data on foraging ability and foraging behaviour are therefore necessary to the understanding of population dynamics and the community structure of honeybees, as well as to develop conservation strategies. The foraging behaviour of social insects is especially interesting because individuals do not forage to meet their own nutritional needs; rather, they forage to meet the needs of the colony. Foraging labour is also divided such that some individuals forage for nectar, some for pollen, some carrying both and fewer with water. Foraging requires information of locations, distances between them and the availability of food. The honeybees that consistently return to the same location are called "central-place" foragers in which every foraging trip consists of distinct "outward" and "homeward" paths.

Flowering plants and honeybees have a special relationship in which both benefit from one another; honeybees obtain food and, in turn, facilitate the pollination process. Honeybees visit flowers to collect pollen and nectar, the former being the principal source of protein, fats, vitamins and minerals, which are essential for honeybee growth and development, repairing of worn out tissue and stimulating the development of hypopharyngeal glands. Nectar consists of the carbohydrates, mainly sucrose, fructose and glucose, and is the raw material of honey.

D.P. Abrol
Division of Entomology, Sher-e-Kashmir University of Agricultural Sciences and Technology, Chatha, Jammu (J&K) 180 009, India
e-mail: dharam_abrol@rediffmail.com

H.R. Hepburn and S.E. Radloff (eds.), *Honeybees of Asia*,
DOI 10.1007/978-3-642-16422-4_12, © Springer-Verlag Berlin Heidelberg 2011

The foraging behaviour of bees is a complex phenomenon which depends upon several factors. Besides the physical features of flowers such as colour, shape and odour, the diversity of environmental factors such as temperature, humidity, light, solar radiation, time of the day and nectar flow decisively shape the behaviour of pollinating insects (Visscher and Seeley 1982; Corbet et al. 1993). The activity spectra of both flowers and pollinators coincide in daily and yearly rhythms (Faegri and van der Pijil 1978). The bees and flowers make a unique, mutually compatible system, the functions of which are affected by a set of common environmental factors in a manner that the functional activities of bees and flowers are synchronised.

To achieve this, there are obvious interactions between biotic and abiotic factors such as temperature, light intensity, solar radiation, wind velocity, rains, dew drops, time of day, distance to food sources, the availability of resources and the abundance and concentration of sugar in nectar. Diel patterns of bee visits are related to the changes in the quality and quantity of floral rewards as well as the direct influence of environmental factors. Hence, the complex interactions of biophysical parameters and food rewards determine foraging success (Heinrich 1975). Bertsch (1983) stated that a close dependence between pollinating insects and their flowers can be better understood if the functions of flowers, their adaptations and the needs of flower visitors are known. This would consequently lead to the management of a particular group of pollinators under a particular set of ecological conditions.

Floral patterns and other specialised mechanisms that attract pollinators seem to have evolved in relation to the sensory development of the bees themselves which, in the course of development, gradually perfected their own sensory faculties and physical efficiency to identify colour, odour, iconic numerals and the symmetry of flowers. Differentiation of nectaries accompanying floral evolution became responsible for trophyletic activities of progressively evolving contemporary pollinators which directed specific changes in their award-getting capabilities (Leppik 1977). Thus, co-evolutionary adaptations by the vast variety of angiosperms passing through several niches of time have chiefly been attributed to the mutually interrelated modifications amongst flowers and their pollinators.

Obviously reward systems offered by the flowers play a determinant role besides the interplay of other factors for which visitors' relationships became heavily set between the visitors' energy needs and the quantity of food they can harvest from flowers (Hocking 1953; Heinrich and Raven 1972), which influence the frequency of visits to flowers (Heinrich 1973, 1975). Because the social set up of honeybees requires such a large food supply for their brood, many more times their own energy requirements are secured by repeated visits to flowers (Reddy and Reddy 1984). This can be sustained only if a flower provides sufficient rewards to attract the foragers on the one hand and limit the reward on the other, compelling the visitors to frequent other flowers of the same species (Heinrich 1975). This enables the development of an optimum strategy both by flowers and pollinators for maximum cross pollination (Emlen 1966; Schoener 1971; Cody 1974). Aspects of dance communication and pathogens in relation to foraging are discussed in Chaps. 14 and 15, respectively.

12.2 Foraging Preferences

12.2.1 Foraging in Relation to Reward from Flowers

Colours, shapes and odours of flowers, energy requirements and caloric rewards to visitors offered by flowers also determine whether or not an animal can be a dependable flower visitor (Heinrich and Raven 1972; Heinrich 1975; Abrol 1985, 1986a, b, 1991a, b, c, 1992a, b, 1993, 1995a, b, 1998a, 2000b, 2005a, b, 2007a, b). Reward systems developed by flowers enable bees to make distinctions between them and closely related species or ecotypes. This has resulted in a mutualistic development of a co-evolutionary process and has evolved partnerships between the two (Leppik 1977). The interdependence of pollinating bees and flowers depends much upon their energy requirements and the balance sheet they share with flowers (Heinrich and Raven 1972). Pollinators with high energy requirements may not forage at flowers that provide low caloric rewards (Heinrich 1983).

The energetics of pollinator–plant interactions have been investigated in only few instances, but some interesting information on bumblebees and honeybees has also been generated (Heinrich 1975; Schaffer et al. 1979). Bumblebees and honeybees have been widely used in studying foraging energetics because of their high energy requirements. Their foraging patterns are largely determined by the relationship of energy requirements of flight to nectar rewards in flowers (Heinrich 1975). The high energy requirements of bumblebees and honeybees result, in part, from their social organisation and ability to thermoregulate (Heinrich 1983).

Most bee/plant interactions depend upon the energy needs of the pollinators and the energy available from plants (Heinrich and Raven 1972; Heinrich 1975, 1983; Abrol 1992a). The foraging strategies of bees and allocation of time to various daily activities are probably related to their level of sociality. Honeybees, which are active throughout the year, need a continuous influx of energy for thermoregulation and brood development. In contrast, solitary bees have fewer energy demands due to their individual nesting habits, as they need not incubate brood nor support nestmates. Solitary bees can be oligolectic/polylectic, whereas longer-lived colonies of eusocial bees are largely restricted to polylecty, which can affect their foraging efficiencies at certain floral hosts (Wcislo and Cane 1996). Evidently, the foraging strategies of honeybees could be entirely different from wild solitary bees.

Abrol (2007b) studied the energy used by *Apis cerana* and *Apis mellifera* in relation to various rapeseed, mustard (*Brassica juncea*, *Brassica carinata*, *Brassica campestris* var. *toria*) and umbelliferous crops (*Foeniculum vulgare* and *Coriandrum sativum*). The energy balance was found to vary from one day to another and from crop to crop. In general, the foraging profitability per bee was in the order: *B. carinata* > *B. juncea* > *B. campestris* var. *toria* > *F. vulgare* > *C. sativum* for both species. *B. carinata* appears to be competitively superior as a forage crop over *B. juncea* > *B. campestris* var. *toria* > *F. vulgare* > *C. sativum*. It was further found that the flight activities of *A. cerana* commenced much earlier and ceased later than

A. mellifera. *A. cerana* worked longer in the field than *A. mellifera*. On average *A. cerana* spent 300.0 min foraging, 336.78 min in nest activities and 803.21 min resting, compared to *A. mellifera* which spent less time foraging (284.28 min), nest activities (222.50 min) and more time resting (933.21 min). On average, *A. cerana* made 10–15 foraging trips/day for nectar and the average trip time was between 23.0 and 25.0 min, while *A. mellifera* made 10–12 trips/day and on average spent 26.5 min/trip.

Foraging behaviour by *A. florea* and *A. dorsata*, which have been recorded as versatile pollinators of wide variety of field and fruit crops, varied in time and space. Abrol (1992a) reported that bees with higher energy requirements do not forage on flowers providing low caloric rewards. The flowers of *F. vulgare*, *C. sativum*, *Daucus carota*, *Allium cepa*, *Trigonella foenumgraecum* var. *Kasuri* and *Mangifera indica* were not attractive to *A. dorsata* because the bees could make no profit from these crops; however, *A. florea* was observed in large numbers. This bee, with its smaller size and body weight, is morphologically better adapted to extract maximum caloric reward from the flowers. Because of relatively low energy requirements, *A. florea* was still able to maintain an energy balance and visited the crop in large numbers. Sihag and Kapil (1983) reported that *A. florea* visited *Brassica* crops in larger numbers than *A. dorsata* and attributed this to the different energy demands of the bees. In the case of sunflower, a wide variety of pollinators were equally attracted, though caloric rewards per floret were low, yet each bee species was able to maintain an energy balance. Because the flower head bears a platform, there is no barrier for landing by foragers, and the energetic costs are reduced due to a temporary suspension of hovering flight, and large number of flowers can be visited in rapid succession.

Similarly, *Luffa cylindrica* flowers were visited by almost all kinds of bee pollinators because the flowers are large, with open nectaries and easily accessible to each type of pollinator, irrespective of their body size. An interesting situation was observed in the case of *Medicago sativa* and *Trifolium alexandrium* which bloom simultaneously and compete for pollinators. *M. sativa* has a comparatively higher caloric reward than *T. alexandrium*. Honeybees were attracted to *M. sativa* during the early hours of the day at peak periods of nectar production, but from 11:00 onward, foraging shifted to *T. alexandrium* due to a reduction in quantity of available rewards from *M. sativa*. Evidently, pollinators will accept low-rewarding flowers, when high floral rewards are not available or nearby resources are depleted.

A. florea visited flowers with low caloric rewards whereas *A. dorsata* preferred those with high rewards. This behaviour was largely guided by their energy demands. Generally, the smaller flowers with little nectar are unattractive to large hovering animals such as humming birds and sphinx moths, which are probably not sufficient for their energetic demands. However, insects such as bumble bees which land on inflorescences of *Spiraea latifolia* L. (Rosaceae) and *Solidago canadensis* (Asteraceae) are still able to maintain an energy balance despite minute amounts of nectar in flowers (Heinrich 1972). This is because the energy expended in walking from one flower to another is a hundred times less than an equivalent period of flight and the flowers can be visited in rapid succession (Henrich and Raven 1972).

Abrol (1998a) studied the foraging behaviour of *A. florea* and *A. dorsata* in relation to the energy production pattern of four plant species. The energy produced ranged from 0.35 J/flower/day in *Lantana indica* to 1,031.0 J in *Adhatoda vasica*. However, the structurally more complex flowers of *A. vasica* were less preferred than those of the other species. The bees preferred to forage on *B. juncea*, followed by *B. campestris* var. *toria*, from which they could harvest more energy per unit time. In a similar study, Abrol (2006a) observed the foraging behaviour of *A. mellifera*, *A. cerana*, *A. dorsata* and *A. florea* in relation to energy production rates of *Prunus persica* and a simultaneously blooming weed, *Lepidagathus incurva*. Energy produced ranged from a minimum of 0.64 ± 0.01 J (*L. incurva*)/ flower/ day to a maximum of 1.49 ± 0.14 J/day (*P. persica*). The weed, having a higher nectar sugar concentration and higher flowering density, attracted more bees than the peach blossoms. The foraging rates of bees were much higher on the weed and they could harvest more energy per unit time from the weed flowers. The results indicate that food acquisition efficiency and quality of food determine the foraging decisions of bees. Bees are generally attracted to flowers that are available in large numbers and are more rewarding in terms of energy (Abrol 1993).

Sihag and Rathi (1992) reported that *A. florea* was more attracted to *B. compestris* flowers, providing less caloric rewards, than to *Cajanas cajan* having high caloric rewards. The differential foraging resulted from the fact that bees could easily harvest more energy per unit time from the simple flowers of *B. compestris* than *C. cajan*, the flowers of which are more complex, tunnelled and papilionaceous, making the visitor's approach much more complex. Earlier, Sihag (1990) observed similar behaviour in *A. florea* and *A. dorsata* on *B. chinensis* and *Eruca sativa* wherein the flowers of *B. chinensis* were preferred. Nagamitsu and Inoue (1999) found that pollen collection from seven plant taxa differed between *A. cerana* and *A. mellifera*; *A. cerana* preferred tall trees, while *A. mellifera* favoured short herbs. Similarly, the influence of mustard, radish and clover in detracting honeybee pollination from apples caused a negative impact on apple pollination by *A. florea*. Verma (1995) observed in his study that, when alternate sources were available, *A. mellifera* showed no preference for foraging on apple flowers.

These investigations revealed that bees prefer to forage on more attractive plants and neglect others when they flower at the same time and location, thus affecting the pollination requirements of the less attractive plant species to a greater extent (cf. Chap. 11). The pollinators with high energy requirements may not forage at the flowers which provide insufficient caloric rewards (Heinrich 1983). In general, the size of the flower and caloric rewards in relation to size of visitor and energy requirements seems to be determinants for resource partitioning among the various bees and thus permitting co-existence under similar ecological conditions. The populations of certain species of pollinating bees were found to be a function of their body size as well as the size of the flowers, because the feeding pattern of many animals suffers as a function of their triplicate structure (Heinrich 1979a, b).

12.2.2 Foraging Preferences of Bees in Relation to Nectar–Sugar Components

Sihag and Kapil (1983) found that *A. dorsata*, with high energy requirements, visited sucrose-dominated flowers more frequently than did *A. florea*, which relied mostly on glucose-dominated ones. While honeybees may prefer equiproportioned sugars, Waller (1972) and Abrol (1985) stated that bees prefer nectar with one dominant sugar than the equiproportioned ones. Abrol and Kapil (1991) studied the foraging strategies of honeybees and solitary bees in relation to nectar sugar components of 13 crops plants and found that *A. dorsata* and solitary bees *Megachile lanata*, *M. cephalotes* and *Xylocopa fenstrata* visited sucrose-dominated nectars more frequently than did *A. florea* and *Pithitis smargdula*, which heavily relied on glucose-dominant nectars. Preferences for sucrose- or glucose-dominant nectars indicate that total caloric reward is an important factor determining the attractiveness of foraging on a particular crop (Abrol 1989a, b; Abrol et al. 1989; Abrol and Kapil 1991).

Abrol (1991c) reported that *A. dorsata* was generally most active between 11:00 and 14:00 when temperatures ranged between 24.5 and 34.5°C, relative humidity (RH) between 22.5 and 43.0%, light intensity between 5,100 and 5,500 lx and solar radiation between 60 and 85 mW/cm^2, soil temperature between 22.5 and 32.0°C, nectar–sugar concentration between 32.0 and 45.0% and wind velocity between 2.9 and 4.2 km/h. Increases in temperature above this range resulted in a decline of bee activity. Of the seven factors studied, the direct effects of light intensity, solar radiation and soil temperature were positive and the other factors negative.

12.2.3 Nectar Productivity and Bee Activity

One source of variation in foraging cues between genotypes is that the nectar may provide a variety of stimuli. It has been found to be a very significant parameter that decisively shapes the behaviour and physiology of pollinators in relation to their energy demands (Heinrich 1975; Abrol 1986a, 1990, 1995a, 2007a, b, 2009, 2010). Abrol (1990) studied the energetics of nectar production in 54 cultivars of apple in relation to the foraging behaviour of *A. cerana* and *A. mellifera* and found that cultivars varied greatly in their patterns of nectar production characteristics and pollinator attractiveness. Cultivars providing higher amounts of sugar, a higher nectar concentration and energy were highly attractive to bees, and these were reflected in their population dynamics. In a similar study, Abrol (1989b) and Abrol and Kumar (2009) recorded the foraging preferences of *A. cerana* and *A. mellifera* in relation to nectar production in 13 cultivars of strawberry (*Fragaria* sp.). The results suggest that cultivars with higher caloric rewards had a competitive edge over others in attracting foraging populations of both species. Similar results were obtained in relation to the foraging preferences of *A. mellifera* and *A. cerana* on almond cultivars (Abrol 1995a). Foraging of both *A. cerana* and *A. mellifera* correlated with energy yields.

12.2.4 Floral Constancy

Floral constancy is one of the striking features of honeybees wherein a single trip a honeybee almost invariably collects a reward from just one type of flower (Free 1970). Flower constancy is important for bees because they learn to extract nectar and pollen while visiting the same flower type (Grant 1950). The constancy of bees can be estimated by examination of pollen types on the body after foraging. For example, 99% of the pollen grains in pollen pellets carried by *A. mellifera* have been shown to be from the same species of plant (Stimec et al. 1997). Raj et al. (1993) found that *A. mellifera* and *A. cerana* foraging on rapeseed, *B. campestris* exhibited pollen consistency: 66.78% of the pollen loads collected by *A. mellifera* and 65.45% by *A. cerana* were found to be from a mustard crop. Priti and Sihag (1997) also observed a similar phenomenon in the case of *A. mellifera* and *A. cerana* visiting cauliflower (*B. olearacea* L. var. *botrytis* cv. Hazipur Local) blossoms at Hisar, India.

Choudary (1978) reported that *A. cerana* showed higher floral fidelity on alfalfa (*M. sativa*) compared to *A. florea*, *A. dorsata* and *A. mellifera.* Later, Dhaliwal and Atwal (1986b) and Naim and Bisht (1989) also confirmed strong floral fidelity in *A. cerana* where foragers collected one kind of pollen in a single trip and continued to collect it throughout the day. Honeybees' floral constancy makes them well-suited for pollination purposes, but at any one time, a colony of bees may be collecting pollen from dozens of different plant species. Nataraj et al. (2000) also reported flower constancy in *A. cerana* under Tamilnadu conditions, whereas Thiyagesan et al. (2001a, b) reported on the risk sensitive and central foraging in honeybee *A. cerana.* Devkota and Thapa (2005) found that the pollen loads of *A. cerana* bees in March and June were mostly from coconut (90%), grasses (6%) and jackfruit (4%), while in *A. mellifera* (67%) pollen loads were from coconut (25%) grasses and (10%) from Rubiaceae, thereby indicating floral constancy in the collection of pollen.

12.2.5 Species-Specific Variations in Foraging Activity

Abrol et al. (2005) studied foraging by *A. cerana* and *A. dorsata* on peach flowers in relation to weather conditions and found that the initiation of flights was a dual threshold of temperature and light intensity. Cessation of foraging was controlled by reduction in light intensity for both species, though the temperature was still over the levels required for initiation of foraging activity. The foraging population correlated significantly and positively with ambient temperature, light, solar radiation, nectar sugar concentration and negatively with relative humidity. Only one factor, namely light intensity, directly influenced the flight activity of *A. cerana* while the flight of *A. dorsata* was influenced both by temperature and light intensity. This demonstrates that different, yet closely related, bee species differ in their responses to environmental conditions. The differences in the responses of the bees are species-specific and indicative of their physiological adaptations. For instance, Sihag and Abrol (1986) found that relative humidity and solar radiation were the important factors influencing flight activity for *A. florea*.

Abrol (1992a) studied the foraging strategies of six bee species in relation to 25 crops. *A. dorsata* with its larger size and higher energetic demands preferred *C. cajan*, *P. aculeata*, *P. glabra* and *L. cylindrica* than flowers of *F. vulgare*, *C. sativum*, *D. carota*, *A. cepa*, *T. foengraceum* and *M. indica*. The latter group of flowers was visited throughout the day. With its small size and body weight, *A. florea* is better adapted to extract maximum rewards from these flowers. In the case of *Brassica* crops, *A. dorsata* visited flowers during early hours of the day and at peak periods of nectar production. From 11:00 onward, however, foraging shifted to *T. alexandrium* due to reduction in quantity of reward available from *M. sativa* and *Brassica* crops. *A. florea* commenced activity between 10:00 and 11:00 and was predominant throughout the day. In the case of sunflowers all types of pollinators continued foraging throughout the day. Though the caloric reward per flower is low, each bee species was able to maintain its energy balance.

12.3 Floral Rewards

12.3.1 Nectar

Nectar is the primary floral reward and food source offered by plants to attract pollinators. Nectar variation can be ecologically important because pollinators often exhibit a preference for certain types of nectars over others, which can affect the plants that are pollinated. Variation in nectar traits can also be important evolutionarily. Phenotypic variation in nectar traits can have both a genetic and environmental component. If variation of a nectar trait has a genetic component, the trait can be subjected to selective pressure if it is also heritable and affects plant fitness. If selective pressure is placed on a nectar trait through pollinator preferences for certain nectar types over others, it could result in ethological isolation, which has the potential to promote species divergence or maintenance during secondary contact (Grant 1994).

12.3.2 Nectar Foraging

The strength of a honeybee colony is important for gathering surplus honey. Sharma and Sharma (1950) found that *A. cerana* colonies of 18,000 bees were unable to give any surplus, while colonies of 12,000 bees failed even to supply their own needs; colonies of 20,000 or more gave a surplus. Nectar gathering capacity per 100 bees increased steadily as the number of bees in the colonies increased. The amount of syrup collected by the foragers of this species was measured for sugar concentration of 20, 30, 40, 50 and 60% of their body weight as compared to 82 (Singh 1971). Average nectar loads of *A. cerana* and *A. mellifera* foraging on *Plectranthus rugosus* were 18 and 27 μl, respectively (Gupta et al. 1984). Diwan and Rao (1969) observed that *A. cerana* foragers collected water prior to visiting the white sugar-rings of dehydrated nectar secreted by the exposed nectaries of *Synalenium grantii* Hook.

Mattu and Verma (1985) reported that *A. cerana* in the Shimla hills of the northwestern Himalayas showed a daily peak in nectar collection after that of pollen. Over the whole year there were more nectar collectors than pollen collectors, pollen + nectar collectors or water collectors. But there were considerable variations in the percentages of nectar collectors compared to the pollen, pollen + nectar and water collectors during different seasons. Dhaliwal and Atwal (1986a) observed the effect of age of crop, soil moisture and phosphate fertilisers on nectar production in *Brassica* crops and their effects on the foraging activity of bees. Kapil and Kumar (1975) reported nectar robbing by *A. dorsata* on *B. juncea*. Sihag (1983) observed *A. dorsata* and *A. florea* foraging on the juice of ruptured/damaged grapes (*Vitis vinifera* L.) during a dearth. The grape juice contained 14–20% dissolved sugars and was comparable with nectars of cruciferous crops in having glucose-dominated sugars. Around 10,000–45,000 bees/ha of *A. dorsata* and 20,000–60,000 bees/ha of *A. florea* were attracted to damaged grapes.

12.3.3 Pollen Foraging

Foraging patterns sometimes show an inverse relationship between pollen gathering activity and nectar gathering and non–foraging activity (Reddy 1980a). Bisht and Pant (1968) reported that *A. cerana* collected pollen throughout the year with a maximum from January to March at Delhi. But in the Shimla hills of the northern Himalayas, this bee collected pollen and nectar throughout the year (Mattu and Verma 1985). Foraging activity was greater in summer and autumn than in winter and in monsoon periods. Times of greater flight activity varied from season to season. The correlations between pollen foraging activity and temperature and relative humidity were also given.

Similar observations were reported by Singh (1971) in Saharanpur who also found that foraging hours varied with season of the year. Verma (1983) found that peak pollen collected by *A. cerana* in Jeolikote was between 08:00 and 11:00 in February–March. Proportions of bees in a colony that foraged for pollen, the pollen stores in hive and amount of brood reared followed a similar pattern (Reddy 1980b). Thakur et al. (1982) reported that *A. cerana* foraging on mustard had considerable activity in the morning and *A. mellifera* picked up by 10:00 to 10:30 and remained active till late in the evening. At Pune, *A. cerana* and *A. florea* commenced foraging on onion between 09:00 and 09:30 and 08:15 and 08:40, respectively, and continued till 18:30. Peak foraging by *A. cerana* occurred in the afternoon between 15:00 and 16:00. *A. cerana* foragers started foraging on mustard under mid-hill conditions between 09:00 and 10:00 and stopped foraging beyond 15:00 (Bhalla et al. 1983a; Mattu et al. 1994), but its activity on stone fruit continued up to 16:00 (Bhalla et al. 1983b).

Choudary (1978) analysed 5,200 pollen loads but only 56 (1%) contained pollen from more than one plant species, which showed a high floral fidelity by *A. cerana*. Similar observations were reported by Chaturvedi (1973, 1977), Sharma (1970a, b) and Jhajj and Goyal (1979b). Dhaliwal and Atwal (1986b) reported that *A. florea*, *A. dorsata* and *A. mellifera* foraging on alfalfa (*M. sativa* L.) carried 60.0, 60.0 and 66.6% pure pollen loads. Jhajj and Goyal (1979c, 1986) also reported that in *A. cerana* and *A. mellifera*, flower constancy decreased with time; after 5 days only 20% of the original *A. cerana* and 28% of *A. mellifera* remained flower constant. A change from pollen to nectar foraging was more common than vice-versa.

In caged pollination experiments, Mohan (1973) found that bees did not pollinate sunhemp flowers and collected pollen without touching the stigma. Dhaliwal and Atwal (1986b) also observed this type of behaviour in *A. mellifera*, *A. dorsata* and *A. florea* foraging on alfalfa (*M. sativa*). *A. cerana* collected an average pollen load of 0.019 g from *B. juncea*, *Pyrus malus* and *Zea mays* (Punjabi et al. 1969), but little pollen from maize. Average pollen loads from mustard were 8 mg (Naim and Bisht 1979). Dhaliwal (1970) compared pollen collection by *A. mellifera* reared in *A. cerana* combs or in *A. mellifera* combs. The comb cells size was found to considerably affect the pollen carrying capacities of the foragers. Jhajj and Goyal (1979a) have also reported reversal of pollen foraging to hive-guarding by carbon dioxide anaesthesia.

12.4 Foraging Flight

12.4.1 Speed of Flight

In honeybees, the speed of flight has been found to vary from 20.9 to 25.7 km/h (average 24 km/h) for loaded bees and from 10.9 to 29.0 km/h (average 20 km/h) for empty bees. This is perhaps due to the fact that a bee, on an outward flight, does not make a beeline in all the cases for the source of supply. Bees do not readily forage in wind blowing over 24 km/h. The flight speeds of several honeybees are given in Table 12.1.

12.4.2 Thermoregulation During Foraging

The ability of honeybees to thrive in a vast range of environments, from temperate regions to humid tropical and hot desert habitats, reflects powerful temperature

Table 12.1 Flight speeds and wing-loading of honeybees

Species	Flight speed (ms^{-1})	Wing-loading (Nm^{-1})	Reference
A. florea	4–81	7–28	Dyer and Seeley (1987)
A. cerana	7–17	7–84	Dyer and Seeley (1987)
A. dorsata	8–12	13–02	Dyer and Seeley (1987)
A. mellifera	12–61	14–21	Seeley (1986)

control, not only by whole colonies inside their nests, but also by individual foragers out amidst the flowers. These bees must maintain their body temperature below the lethal upper limit of about 45°C, and while flying, must keep their thoracic temperature above 27°C, the lower limit for steady flight. Foraging bees manage to fulfil these requirements across a range of ambient temperatures from about 5–45°C thus achieving considerable freedom from temperature limitations when collecting resources for the colony. Stop-and-go foraging is expensive in time and energy spent on warm-up, but can still be profitable when working on rich flowers. Once a bee reaches the flowers her behaviour will likely consist of intermittent flights between flowers, thus automatically affording her opportunity for warm-up when perched on two flowers. Honeybees possess an effective cooling system for flight at high temperatures. The primary mechanism for cooling the thorax appears to be to keep the head cool by regurgitating fluid from the honey stomach and holding the extruded nectar triplet between the tongue and mandibles.

Dyer and Seeley (1987) found that thoracic flight temperature in foragers of three Asian honeybee species which, together with the European species *A. mellifera*, span a fivefold range in body mass from the smallest species to the largest. Over a 15°C range in ambient temperature, they found that thoracic flight temperature in each species is strongly dependent upon ambient temperature. However, the temperature gradients (thoracic flight temperature/ambient temperature) at a given temperature do not appear to increase with body size in the four species. The smallest species, *A. florea*, shows the smallest thoracic flight temperature/ambient temperature, but the intermediate-sized *A. cerana* and *A. mellifera* both show a consistently higher thoracic flight temperature/ambient temperature than the largest species, *A. dorsata*. Furthermore, on a mass-specific basis, *A. dorsata* and *A. florea* are more similar to each other than either of to the other two species. This physiological dichotomy among the four species parallels a dichotomy in nesting behaviour and colony demography (Singh et al. 2007).

12.4.3 Foraging Speed and Foraging Rate

Atwal (1970) and Atwal et al. (1970) reported that *A. cerana*, *A. dorsata* and *A. florea* visited 12.7, 16.6 and 17.2 flowers of sarson per min. But *A. cerana* was a faster worker than the others on *Eranthenum*, radish, *Eucalyptus* and berseem. In general, *A. florea* was the slowest forager on *B. juncea* flowers, but were visited by

A. dorsata at 12.3 per min. Under mid-hill conditions and visiting mustard blooms, *A. cerana* foraged 10 flowers per min (Bhalla et al. 1983a; Raj and Rana 1993, 1994) but on stone fruit flowers, the rates were 6.9 per min on peach and 4.9 per min on almond (Bhalla et al. 1983b). On alfalfa, foraging behaviour of honeybees was related to their body weights in order: *A. mellifera* < *A. florea* < *A. dorsata*. On onion blooms, *A. cerana* was the faster flier than *A. florea*, visiting 1.93 and 1.34 umbels per min, respectively. But on an umbel, *A. florea* showed a better foraging rate than *A. cerana* visiting 6.53 and 5.93 flowers per min, respectively. Dhaliwal and Atwal (1986b) compared the tripping efficiency of different bees visiting alfalfa, where the order of efficiency was *A. dorsata* > *A. mellifera* > *A. florea*.

Kapil and Kumar (1975) reported that the temporal change in foraging rate of *A. dorsata* was 9.23 flowers per min in the morning, but 16.0 flowers per min in the evening, and *A. cerana* was a slower forager than *A. mellifera* on *Plectranthus rugosus*. These species foraged 17.5 and 25.8 flowers per min in the morning and 25.0 and 33.6 flowers per min in the evening, respectively (Gupta et al. 1984; Shah and Shah 1989). The concentration and amounts of nectar were responsible for morning and evening differences in the rate of foraging.

Collection of full pollen loads from sarson flowers took longer than for a load of nectar. Diwan and Salvi (1965) and Sihag (1984) have confirmed nocturnal foraging by *A. dorsata* and night flowering plants were generally visited. Jadhav (1982) found that *A. florea* and *A. dorsata* stopped foraging when solar eclipse was at its peak and did not resume even after the eclipse was over, whereas Diwan (1980) found that returning foragers of *Trigona iridipennnis* had difficulty in locating the hive.

Chaudhary and Sihag (2004) reported that *A. dorsata* was the most abundant on onion blossoms followed by *A. florea* and then the dipterous insects. However, the number of *A. florea* was significantly higher on the blossoms of carrot and fennel than *A. dorsata*. Chandel et al. (2004) found that in the case of onion (*A. cepa* L.) flowers, *A. dorsata* proved to be the dominant visitor (7.4 bees/m^2/2 min) and most efficient pollinator visiting on average seven flowers/umbel/visit during peak hours of foraging (12:00–14:00) compared with *A. cerana* (5.4 flower/umbel/visit). *A. florea* and *A. mellifera* visited onion flowers least and ranked fourth and fifth overall in visiting frequency. *A. dorsata* had the maximum foraging period (06:30–18:55) followed by *A. cerana* (06:45–18:30) and *A. mellifera* had the least foraging period (07:25–18:20) on onion seed crop.

Foraging speed (time spent per flower) and foraging rate (number of flowers visited per minute) depend upon the foraging behaviour of the insects and floral structure of the crop concerned, particularly the corolla depth. Free and Williams (1973) reported that honeybees spend 131 s per kale flower when collecting pollen loads and 94 s when collecting nectar only. Higher numbers of flowers of cauliflower were visited per min by *A. dorsata* followed by *A. mellifera*, *A. cerana*, *Eristalis* spp. *Ceratina* spp., *Halictus* sp. and *Lasioglossum* sp. (Kakar 1981, 1983; Selvakumar et al. 2006). Kapil and Kumar (1975) found that, on average, *A. dorsata* visited 12.30 and 10.79 *B. juncea* flowers per minute. In addition, they found that *A. dorsata* on average visited fewer flowers of *B. juncea* during the morning between 09:30 and 11:00 when the temperature was low, but increased with the advance of day to a maximum between 14:30 and 15:45 when the temperature was relatively higher.

Abrol (2000a) studied the foraging rate of *A. mellifera* and *A. cerana* on kiwifruit (*Actinidia deliciosa*) flowers and found that the time spent on a flower averaged 33.0 s for *A. mellifera* and 38.0 s for *A. cerana*, and the numbers of flowers visited/min were 1.8 and 1.6, respectively. Abrol (2006b) found that *A. florea* visited fewer flowers/umbel during the earlier and later part of the day, with a maximum of flowers/umbel/min visited between 12:00 and 14:00. On average, flowers visited/min ranged between 17.13 and 27.31 during different hours of the day with an overall average of 22.78 flowers/min.

Similarly, the number of umbels visited/min ranged between 0.98 and 1.29 with an average of 1.14 umbels/min. Devkota and Thapa (2005) studied the foraging behaviour of *A. cerana* and *A. mellifera* in broccoli blooms under caged and open conditions in Chitwan, Nepal and found that both preferred open plots for foraging. *A. cerana* foraged in significantly higher numbers on broccoli flowers (an average of 11.39 and 12.11 flowers/min) compared to *A. mellifera* (an average of 9.03 and 10.89 flowers/min) under caged and open conditions, respectively. Chandel et al. (2004) reported that *A. dorsata* and *A. cerana* visited 7.5 and 5.4 flower/umbel/visit in case of onion flowers on average.

12.4.4 Flight Patterns

The flight of *A. cerana* resembles that of a fly – rapid, hasty, unpredictably zigzagging – compared to the steady, clumsy flight of European *A. mellifera* (Kellog 1941; Lindauer 1956; Sakagami 1959). This behaviour helps in escaping from flying predators, like hornets and bee-eating birds. It remains to be investigated, however, whether flight patterns are species-specific and whether some tropical *A. mellifera* show the same behaviour. When bees were dislocated 50 m, the speed of homing was found to be quicker for *A. cerana* (192 s) than for *A. m. ligustica* (295 s; Atwal and Dhaliwal 1969).

12.4.5 General Flight Activity and Foraging Range

Bees are the most important pollinators and understanding the scale at which they forage has important ecological implications and conservation applications (Axelrod 1960; Bawa 1990; cf. Chap. 11). Bees are the primary pollinators for most ecological regions of the world. Flight activity of honeybees is an index of their foraging potential for pollination and honey production. Bees generally forage near their hives and longer flights are generally avoided, because the energy required would exceed the value of nectar gathered. Their foraging distances strongly influence the sexual reproduction of most flowering plants and can determine the genetic structure of plant populations (Campbell 1985; Waser et al. 1996). Bee foraging distances also affect agricultural production. Many bees that pollinate crops nest in natural habitats and forage on crops within their daily travel distance (Ricketts 2004). Foraging distances therefore determine the spatial scale at which bees can provide pollination services to crops (Kremen 2005; cf. Chap. 11).

The foraging ranges of pollinators provide information regarding their intrinsic capabilities in gathering nectar and pollen resources that largely depend on their body size and energy needs (Abrol 1986a; Abrol and Kapil 1987), yet are also determined by the quality of reserve fuel carried in flight. The smaller bees with low energy requirements and little reserve fuel may avoid foraging large distances (Hocking 1953). Abrol (1988a) reported that *A. florea* workers forage up to 150 m from their nesting sites. Bees generally visit flowers within a radius of 1–3 km. *A. cerana* normally forages 1–1.5 km, while *A. mellifera* may forage up to 3 km, but if the food sources are limited and if the competition is more, the foragers may go further. The foragers of *A. cerana* have been reported to fly 600–1,040 m against 3–14 km in *A. mellifera*. Evidently, the realistic flight ranges of bees may be less than those values reported in literature.

Lindauer (1956) was not able to train *A. cerana* workers in Sri Lanka farther than 750 m from the hive. During these experiments, the flight distance of bees collecting on natural sources did not exceed 300 m. The average foraging distance on cauliflower was about 400 m and on barberry 600–700 m in the Punjab, India, but uphill foraging distances were much shorter (300 m) than in flat country (Dhaliwal and Sharma 1974). In Kashmir, *A. cerana* was observed to collect pollen from *Crocus sativus* at a distance of 3.75 km (Shah and Shah 1980). *A. florea* are the most important pollinators of field crops in the plains of India and Pakistan. The dwarf bee, *A. florea*, has a short flight range, often less than 100 m from the nest. The maximum distance they can fly from the nest for foraging is often less than 750 m. The dwarf bee is able to survive in very hot and dry climates with ambient temperatures reaching 50°C or more. The flight range of the giant bee, *A. dorsata*, is more than 5 km (Koeniger and Vorwohl 1979). Under normal forage conditions, they have been observed to visit sources 2–3 km away from the nest.

12.4.6 Flight Intensity of Foraging Bees

Wingbeat frequency is an indication of flight intensity and is an important parameter, not only from the aerodynamic and bioenergetic points of view, but also helps in assessing the pollinating efficiency of bees (Heinrich 1975; cf. Chap. 13). The wingbeat frequency is often species-specific. In general, smaller insects have a higher wingbeat frequency than larger ones (Sotavalta 1952) and those with smaller wings, a higher wingbeat frequency than those with long or heavy wings. The smaller the surface area of the wings, the more rapidly the wings must beat to keep the animal in flight (Kammer and Heinrich 1974). Wingbeat frequency is a function of body parameters (cf. Chap. 13).

Abrol (1985) found that the wingbeat frequency of *A. florea* of 12–19 mg ranged between 92 and 106 cps with an average of 101 cps, that of *A. dorsata* 106–129 mg, 111–116 cps with an average of 113.95 cps. Interestingly, smaller bees have higher wingbeat frequency than the larger ones on the basis of g/body weight. *A. florea*, per gram body weight, exhibited 6,688.79 cps compared to *A. dorsata* with 1,162.16 cps/g/body weight – a fivefold difference. *A. cerana* from the plains of northern India (Punjab) showed higher frequencies than yellow *A. mellifera* from Europe: 306 wingbeats per second vs. 235 for workers and 283 vs. 225 for drones (Goyal and Atwal 1977). These frequency differences could well be a mere consequence of wing size differences.

12.5 Weather Factors

The physical environment influences flower visiting, foraging behaviour and the effectiveness of pollination in complex ways. Foraging activities of pollinating insects are under the control of different environmental variables.

12.5.1 Commencement and Cessation of Foraging Activity

Each honeybee species has its specific ecological threshold below which foraging activity does not occur (Reddy 1979). The time of commencement of bee activity varies from day to day and depends upon the attainment of minimum threshold conditions for foraging. Abrol (2006b) studied foraging behaviour by *A. dorsata*, *A. mellifera*, *A. cerana* and *A. florea* on litchi flowers (*Litchi chinensis* Sonn.) in relation to weather factors and found that 16°C, 74% RH, 600 lx light intensity and 10 mW/cm^2 solar radiation for *A. dorsata*; 16°C, 75% RH, 800 1x and 10 mW/cm^2 solar radiation for *A. mellifera*; 15.5°C, 76% RH, 600 lx, 9 mW/cm^2 solar radiation for *A. cerana*; and 18.5°C, 64% RH, 1,700 lx and 20 mW/cm^2 solar radiation for *A. florea* appeared to be minimum thresholds for initiation of flight activity. Cessation of activities in all the honeybee species was governed mainly by declines in values of light intensity and solar radiation, irrespective of other factors. Earlier, Kapil and Kumar (1975) also reported 15–18°C as the minimum threshold temperature for commencement of field activities in honeybees.

Abrol (2006b) studied foraging behaviour by *A. florea* on carrot (*D. carota*) flowers and found that their flight activities commenced between 07:26 and 07:44 in the morning when the temperature ranged from 19.4 to 25°C, RH between 51.0 and 75.0%, 1,300 and 4,700 lx and solar radiation between 38.0 and 68.0 mW/cm^2. Cessation of activities was governed mainly by declines in values of light intensity (500–1,300 lx) and solar radiation (10–43 mW/cm^2) which were appreciably lower at cessation than at the commencement of foraging. He further found that temperature, light intensity and solar radiation were the three most important factors which exerted pronounced and positive direct effects on flight activities of *A. florea*, while relative humidity had a negative direct effect. Direct effects of nectar sugar concentration were positive and negligible.

Sihag (1984) studied the limiting effects of light intensity, ambient temperature and humidity on the commencement and cessation of flight activity of *A. dorsata* and *A. florea*. *A. dorsata* maintained foraging activity when illumination was as low as 0.5–1.0 lx at night, but foraging activity was limited if below 16.5°C or above 38°C and if humidity was below 32% or there were rains. On the other hand, in *A. florea*, illumination below 200 lx, below 18.5°C or above 40°C and humidity below 25% or if there was rain, caused the cessation of flight activity. The highest limits of light intensity for *A. florea* were 6,500 and 7,000 lx, respectively. But before approaching these limits, temperature had also crossed the upper limiting values. However, once begun, the flight activity in *A. florea* was maintained by the direct effects of solar radiation and relative humidity. Temperature, relative humidity and nectar sugar concentration had no direct effect on flight activity (Sihag and Abrol 1986). Dhaliwal and Atwal (1986a) reported 33°C as the optimum field temperature for peak activity of *A. mellifera* on alfalfa.

Abrol (1985) found that *A. dorsata* was physiologically more tolerant of large fluctuations in ambient temperature than *A. florea*, which restricts its activity during unfavourable weather conditions. Cold weather conditions and wind constrain the foraging activity of *A. florea*. It has been observed after a small rain shower in winter, cooling the atmosphere, that the foraging activity of *A. florea* did not resume for many days, even when the temperature was above the threshold level of 15°C. This may be due to poor thermoregulatory capabilities of this species. Lindauer (1957) reported that the habits of *A. florea* are limited by climatic conditions, especially by cold. Worker activity is greater at higher temperatures at which the bees have to expend less energy to raise thoracic temperature to that required for flight. Abrol and Kapil (1987) found that *A. florea* can forage at the extremely high temperature of 43°C and utilise highly concentrated nectar of 68.0%.

Abrol (1985) found that *A. florea* was more sensitive to variations in ambient temperature than *A. dorsata*. This implies that *A. florea* is adapted to work at relatively higher temperature (43°C) and is limited by fluctuations in lower ambient temperature. It is probably advantageous for the small bees to work at higher temperatures because they can efficiently radiate enough heat to the environment by their relatively larger surface area/volume ratio. Similar observations were made by Willmer (1983) who reported that small, lightly coloured species visited isolated flowers and were highly correlated with solar radiations. Kapil and Kumar (1975) found that foraging by *A. dorsata* on *B. juncea* peaked between 21 and 23°C. Abrol (2000a) studied foraging by *A. mellifera* and *A. cerana* on kiwifruit (*Actinidia deliciosa*) and found that peak foraging occurred between 11:00 and 14:00 for *A. mellifera* (24.0–28.5°C, RH 63–73%) and between 10:00 and noon for *A. cerana* (21.5–26°C, RH 72–77%).

A. laboriosa lives in the Himalayas under harsh conditions at altitudes of 1,200 to 3,600 m, where a temperature range of 4.5–23.4°C is encountered. The lower range of temperatures recorded in *A. laboriosa* territory does not occur in *A. dorsata* territory (6°C and 27°C), and the upper temperature range recorded in *A. dorsata* territory does not occur in *A. laboriosa* territory. However, a large temperature range is the same for the active life of both species. According to Dyer and Seeley (1987), *A. dorsata* workers do not fly below 17°C. In a similar study, Woyke et al. (2003a) found that the closely related *A. dorsata* workers started foraging at 18°C and *A. laboriosa* at 10°C. The increase in temperature to 12°C resulted in a 10-fold increase in the number of foragers leaving the nests of both species, and their flight activity reactions to temperature changes were similar. A high correlation between temperatures and the number of foragers was found in both species, below 16°C. The environmental conditions in which the bees have lived for long periods have influenced their behaviour more than their phylogenetic relationship.

According to Lundie (1925), the lowest temperature at which the flights of *A. mellifera ligustica* began was 10°C, but flight normally commenced between 12°C and 14°C in April. Thus, it can be concluded that in a similar temperature range and weather conditions, the starting temperature for foraging and response of flight activity to temperature changes are similar among *A. laboriosa*, *A. m. mellifera* and *A. m. ligustica*. The lower number of foragers flying before and after noon at the same temperature may be due in great part to the direction of meteorological changes. Woyke et al. (2000) found that during an eclipse of the sun, a lower number of *A. mellifera* workers left the hive before the maximal eclipse when the light intensity decreased than after the eclipse when light intensity increased.

Thus, although a strong correlation between temperature and flight activity was found, some other factors, like the direction of changes in light intensity and others, influence flight activity. However, both *A. laboriosa* and *A. mellifera* reacted similarly. Szabo (1980) found a significant correlation between temperature (range 14–24°C) and the flight activity of *A. mellifera*. However, at higher temperatures, no correlation between those two factors was found (Domagała Lipińska 1962; Lee et al. 1987), and above 30°C, flight activity decreased with temperature increases (Gary 1967; Woyke 1992). These differences are explained by Woyke et al. (2003b) who suggested that the correlation between air temperature and flight activity is stronger at lower temperatures, but is weaker, or does not exist, at higher temperatures. Other bees, like *A. cerana* and *A. mellifera*, are already foraging at those temperatures 15°C. This indicates that territories in which the morning temperatures are lower than 18°C are the expansion ones in which *A. dorsata* is not able to take advantage of all food available.

When early morning temperatures are low, *A. dorsata* workers start foraging at higher temperatures. This suggests that more heat is required to warm the outer worker layer of the nest curtain. *A. laboriosa* workers start foraging at 10°C, which is 6–8°C lower than that of *A. dorsata*. Thus, the Himalayan bee can collect food at lower temperatures. The start of foraging, in relation to ambient temperature, and the reaction of flight activity to temperature and other meteorological changes during the day were similar in *A. laboriosa* and *A. mellifera*. A high correlation between the temperature and the number of foragers was found in both species, for morning hours at temperatures below 16°C. Hence, the start of foraging of *A. dorsata* and *A. laboriosa* living in territories characteristic of different temperature ranges occurs (within the same temperature range) at different temperatures. However, both the start of foraging and the response of flight activity to temperature changes are similar in two phylogenetically distant honeybee species, *A. laboriosa* and *A. mellifera*, living in territories of similar temperature range.

12.5.2 Diurnal Trends in Foraging Activity in Relation to Environmental Factors

Foraging activity occurs only when suitable ecological conditions are attained. Kapil and Brar (1971) recorded peak activity of *A. florea* on toria between 21 and 25°C and 50 and 57% RH during November. Benedek and Prener (1972) found that flower visiting rate increased with increasing air temperature. Morning activity of *A. mellifera* was related to nectar flow, but with photoperiod during the afternoon. In general, optimum conditions for bee activity vary from season to season, depending upon the geographical regions, time of the year, melliferous crops or species of bees. Abrol (2006a, b, 2007a, b, c) found that in the case of *A. dorsata*, the important factors are temperature, light intensity and solar radiation; in the case of *A. mellifera*, these three plus RH; in *A. cerana*, light intensity; and solar radiation and RH in *A. florea*.

Using path analysis of ecological factors did not support the widely held contention that changes in nectar sugar concentration are reflected by the spectrum of flower visitors (Corbet 1978a; Abrol 2006a, b, 2007a, b). This may be true under experimentally controlled conditions where all other factors are kept constant and bees have to respond to variations in one factor (nectar sugar concentration) only. But, under field conditions multifarious factors operate simultaneously so that nectar sugar concentration, a biotic factor, like bee activity, is influenced by temperature and radiation and there is a significant correlation between them.

Evidently, the resulting significant relationship between bee visits with nectar sugar is incidental under field conditions. Although nectar sugar concentration is an important factor (Corbet 1978a), under field conditions this is also influenced by other factors simultaneously operating on bee activity. In between the commencement and cessation of foraging, *A. florea* activity was highest on flowers when temperatures ranged between 25 and 38°C and declined at higher temperatures (Abrol 2006a). Similar observations were made by Free (1981) who found that metabolic activity of honeybees increases with increasing temperature and they visit many flowers at that time.

Abrol (2006a) studied the foraging behaviour of *A. dorsata*, *A. mellifera*, *A. cerana* and *A. florea* visiting litchi flowers (*L. chinensis* Sonn.) and found that maximum foraging occurred between 11:00 and 13:00 between 23 and 34°C,

RH between 65 and 87%, light intensity between 2,700 and 6,700 lx, solar radiation between 24 and 35 mW/cm^2 and nectar–sugar concentration between 40 and 68%. However, on cloudy/overcast days, the pattern was altogether different. There was no well defined peak. Foraging was generally low and occurred only when conditions within which foraging occurs were attained.

Neupane et al. (2006) found that the activity of *A. dorsata* foragers was more pronounced on bottlebrush, litchi, citrus and summer squash during the morning when the temperature was lower (19.3°C at 07:30) than in the afternoon (33.1°C at 17:00). The pronounced activity of the bees during the morning was due to the early release of fresh pollen and nectar and a suitable foraging temperature. *Apis dorsata* foragers forage for long distances, showing maximum foraging activity between 25 and 35°C, the distance and range being higher than that of *A. mellifera*.

Sihag (2000a) found that *A. florea* started foraging when ambient temperature exceeded 18°C and continued foraging until temperatures approached 43°C. Maximum foraging activity was between 30 and 40°C. These ranges are higher than those of *A. dorsata*, *A. mellifera* and *A. cerana*. In the case of *A. dorsata*, foraging commenced when temperature surpassed 16°C and continued to about 40°C (Sihag 2000b). Maximum foraging was between 25 and 35°C. Abrol (1992c, d) found that during winter temperature acted as a stimulus for the initiation of flight activity in *A. florea*, *A. dorsata*, *A. cerana* and *A. mellifera*, whereas light intensity provided the minimum threshold during summer. Woyke et al. (2003a) found that workers of *A. laboriosa* neither foraged below 10°C nor on overcast days.

A. cerana can forage at lower temperatures than *A. mellifera* (Verma 1988). Foraging is also greatly affected by changing temperature, e.g. no foraging occurs below 8°C, with some activity between 8 and 16°C, optimum activity between 16 and 32°C and reduced foraging above 32°C (Roberts 1979) but under some circumstances foraging may continue up to temperatures of 42–48°C. Verma and Dulta (1986) found that *A. cerana* started foraging earlier in the morning than *A. mellifera* and ceased later in the evening. Peak activities occurred between 09:00 and 11:30 and 11:00 and 13:29 for *A. cerana* and *A. mellifera*, respectively. An *A. florea* colony begins foraging between 06:00 and 07:30 and continues until dusk, although foraging often diminishes during the hottest part of the day (Akratanakul 1977). Foragers tend to keep close to their colony with a maximum range of 500 m. *A. florea* prefers flowers whose outlines appear subdivided (Dixit 1956), and they collect more pollen from composite species than *A. cerana* and *A. dorsata* do (Phadke 1968).

The individual temperature tolerance of *A. florea* seems to be higher than *A. mellifera*. In the morning, *A. florea* field bees start about 2 h later than *A. mellifera* (18°C vs. 10°C), but continue at temperatures higher than 40°C, when *A. mellifera* stops all flight activities. Therefore, a different daily flight pattern is the result. *Apis dorsata* was also observed to forage and dance during bright nights (Diwan and Salvi 1965). The extrapolated position of the sun (Dyer 1985) is used for compass orientation and not the moon. A proof of nocturnal activity is also the repeatedly observed attraction of *A. dorsata* to street lamps.

A. cerana is reported to be active at lower temperatures than *A. mellifera*. Field activities have been observed in Ussuria at 6–8°C. The same was noticed in

Kashmir (Shah 1980) and China (Oschmann 1961; Gong 1983). In Himachal Pradesh, *A. mellifera* is the more active species during the warm season and *A. cerana* during the cold season (Adlakha and Sharma 1975). *A. cerana* collects nectar form *Plectranthus* blossoms on cool autumn days, while *A. mellifera* does not (Sharma et al. 1980). During cool morning hours, while *A. mellifera* remains still inactive, *A. cerana* robs their nests (Goyal 1974). Again, these behavioural differences might not be species-linked, but could be specific adaptations of certain ecotypes. Altitudinal gradients affect pollination and pollinator's foraging, e.g. humming birds are more effective pollinators at high elevations while bees at middle (Cruden 1972) and moths at middle and low altitudes (Cruden et al. 1976). The initiation of foraging of *A. mellifera* and *A. cerana* was reported to be delayed and ceased earlier with increasing altitude (Verma and Dulta 1986). Wind affects the foraging activity of insects. Wind velocity of 24–34 km/h adversely affects the foraging behaviour of bees (Lundie 1925; Verma 1990).

12.5.3 Size-Related Foraging and Weather Conditions

The large size of *A. laboriosa* workers is almost certainly one of the major adaptations that have enabled this species to survive in temperate climates while the other open-nesting honeybees are confined to the tropics and subtropics. *A. laboriosa* workers are able to forage at least 5–6°C lower than the minimum ambient at which *A. dorsata* workers fly (Dyer and Seeley 1987, 1991; Underwood 1991). This has apparently been accomplished largely through an increase in body size, without resort to creating a higher-powered bee and without a disproportionate increase in thoracic mass, such as seems to have been the case with *A. cerana* in Nepal (cf. Chap. 13).

12.5.4 Nocturnal Foraging

Although the moon's illumination is essential for nocturnal flight, the moon itself is ignored for orienting the dances in *A. dorsata* (Dyer 1985). Rather, bees probably use the sun's position as a reference point for their dances, even if it is below the horizon. This ability may involve an extension of the mechanism that honeybees employ to find the sun on overcast days. Rao et al. (2008) found that *Pterocarpus santalinus* blooms massively in the dry season. The flowers open at the midnight and are bright yellow in colour. They are homogamous and are visited by the rockbee *A. dorsata* on moonlit nights or otherwise from dawn up to 07:30. The natural habitat of *P. santalinus* has a hot dry climate, and the nocturnal flowering and foraging of the rockbee at that time during moonlit nights appear to be an adaptation to avoid the adverse effects of high temperatures during daytime.

12.6 Thresholds and Timing

12.6.1 Factor Compensating Mechanisms

Abrol (1985) found that foraging activity in bees may be a consequence of several interacting factors in which the value of one factor with high values may compensate for lower values of the other factors. He found that *A. florea* commenced foraging where a combination of 19.5°C, 87% RH, 1,700 lx light intensity and 64 mW/cm^2 were attained. However, on another day, activity at higher temperatures operated when other factors were comparatively of less intensity. This implies that higher temperatures can compensate for lower values of other factors and trigger foraging while, in the former case, higher values of light intensity and solar radiation could compensate for a lower temperature and result in the initiation of foraging. Foraging in honeybees is under short duration turn-key control of the physical environment. Each honeybee species is guided by the specific ecological threshold for normal foraging activity whose maintenance differs inter- and intra-specifically depending on their adaptability (Abrol 1985, 1987a, b, c).

12.6.2 Ecological Threshold of Nectar Secretion and Foraging Activity of Honeybees

Co-evolutionary adaptations have resulted in a close correlation between plants and their pollen vectors such that the functional activities of both are more often synchronised. Each bee has its ecological threshold for the initiation of field activity (Abrol and Kapil 1987). In plants too, a characteristic rhythm of nectar secretion occurs with peak periods of nectar production usually synchronised when conditions are optimum for bee activity (Corbet 1978b). Abrol (1985) reported that, in *A. dorsata* and *A. florea*, the onset of flight occurred at 15°C under subtropical conditions of Hisar during winter. Abrol (1986b) found that in flowers of *B. juncea*, nectar secretion occurred between 12 and 15°C thereby indicating that 12–15°C is a minimum threshold temperature for activation of the enzymatic machinery to secrete nectar.

This overlap of activity thresholds of two different groups makes them interdependent, each relying on the other for energy rewards and reproduction, respectively. Abrol et al. (1988) reported that under the temperate conditions of Kashmir, India, where no nectar secretion was observed in flowers of *Eriobotrya japonica* below 8°C, secretion began at 10°C coinciding with initiation flight of *A. cerana*. Similarly, threshold temperatures below which no nectar is produced have been reported for many plants. Basswood (*Tilia americana*) begins to secrete at 18°C (Desmuth 1933), bird cherry (*Prunus laurocerasus*) at 18–20°C, cucumber (*Cucumis sativa*) at 17–21°C (Collision 1973) and *B. juncea* at 12–15°C (Abrol 1986b). Evidently, the physiology of plants has synchronised their functional activities for the benefit of both plant and pollinator.

12.6.3 Sequence and Timing of Honeybee Visits

The visits of bees to flowers depend largely upon the energy requirement reward system and occur in a very predictable and ordered manner (Schlising 1970; Heinrich 1975; Murrell and Nash 1981). Abrol (1987a, b) found that larger bees, like *A. dorsata*, *A. mellifera* and *A. cerana*, with higher energetic demands and superior thermoregulatory capabilities visited flowers early in the morning when the caloric reward is greatest and competition with other pollinators is minimal. Whereas, the smaller ones, such as *A. florea*, with relatively low energy requirements and poor thermoregulatory capabilities, visit flowers in the late morning hours when energy expenditure due to prevailing higher temperature is minimal.

12.6.4 Foraging as a Function of Species-Specific Adaptations

Several species of highly social *Trigona* (Johnson and Hubbell 1974, 1975) and *A. mellifera* (Sakagami 1959; Schaffer et al. 1979) specialise in large productive flower clusters. The large size of their colonies results in large pollen and nectar requirements for colony maintenance and growth, which favours the efficient exploitation of large resource patches (Michener 1974). Abrol (1998b, c) found that under a similar set of climatic conditions, the responses of *A. cerana* and *A. mellifera* were very different. *A. cerana* started flying earlier in the morning and also at a lower temperature, light intensity and solar radiation and also worked for longer periods than *A. mellifera*. Solar radiation directly affected the activity of *A. cerana*, and solar radiation and light intensity directly affected that of *A. mellifera*. Forbes and Cervancia (1994) also reported similar findings for *A. cerana* and *A. mellifera* in the Philippines.

Abrol (2006a) studied foraging of *A. dorsata*, *A. mellifera*, *A. cerana* and *A. florea* visiting litchi flowers and found that all the species differed in their responses to environmental factors prevailing under a similar set of conditions. Of all the factors studied, temperature, light intensity and solar radiation were the three important factors whose influence on foraging was most pronounced.

12.7 Resource Portioning

12.7.1 Competition for Floral Resources

Researchers have recently obtained evidence of competition between bee species in natural situations (cf. Chap. 19). *A. cerana* is sympatric with *A. dorsata* and *A. florea* without any apparent adverse ecological consequences. Chahal et al. (1986) found that *A. florea* showed aggressiveness towards *A. mellifera* both at hives and on flowers. Partap (2000) studied foraging competition between *A. cerana* and *A. mellifera* on four crops blooming simultaneously and found that *A. cerana* foragers were less abundant when *A. mellifera* foragers were in the plots in all the crops. After removal of *A. mellifera*, the number of *A. cerana* foragers increased significantly. Evidently, the presence of *A. mellifera* reduces the number of *A. cerana* foragers in the field.

The two species try to displace each other rather than visiting/pollinating flowers. It is therefore inferred that the presence of both species adversely affects the pollinating efficiency of both bees. Removal of *A. mellifera* resulted in significant increases in time spent per flower, number of flowers visited per minute, weight of pollen loads and number of pollen collectors. Numerous interactions among the species were noted and small individuals generally attacking larger ones. *A. dorsata*, however, was attacked only by *A. cerana* and never by the other two species. In the case of *A. florea*, their small size did not deter them from showing aggression towards other species at a food source. When foraging at dishes containing sugar syrup, *A. florea* was as successful (or more so) than *A. cerana* and *A. dorsata* in maintaining its station and inducing the departure of bees of other species. However, *A. florea* and *A. cerana* workers sometimes forage together at artificial food sources without showing aggression.

12.7.2 Resource Partitioning

In a mutualistic relationship with flowering plants, honeybees gather nectar and pollen from blossoms, yet not all species exploit the same floral resources. This is partly reflected in geographical variation, because honeybee species often live in allopatric ranges which contain different types of vegetation (Schneider and McNally 1992, 1993). In cases of sympatry, resources may be partitioned in other ways, for example, on size. Sharma et al. (2000) studied resource partitioning in *A. mellifera* and *A. cerana* and found that of the 23 plant species studied, *A. mellifera* and *A. cerana* avoided competition by visiting the plants in differing intensities. During spring, *A. mellifera* preferred *Rubus ellipticus* (7.85 bees) and *Prunus armeniaca* (7.6 bees), whereas *A. cerana* was more active on *Malus domestica* (9.91 bees). Partap et al. (2000a) found that *A. cerana* workers started foraging earlier in the morning than *A. mellifera* and, in the evening, *A. mellifera* stopped earlier. Gupta and Reddy (1992) found that foraging activity of *A. mellifera* and *A. cerana* on wild cherry peaked between 11:00 and 14:00, respectively. Body size and tongue length also play a predominant role in determining the resource partitioning of the bees (Table 12.2). For example, Oldroyd et al. (1992a, b) found evidence suggesting that pollen resources were partitioned among *A. dorsata*, *A. cerana*, *A. florea* and *A. andreniformis* based on bee size, with the larger species using the richest pollen resources.

El Shafie et al. (2002) studied 21 pollen sources used by *A. m. sudanensis* and *A. florea* between early December and late March and found some hints of competition

Table 12.2 Tongue length and body weight of different honeybee species

Species	Tongue length (mm)	Reference	Body weight (mg)	Reference
A. florea	3.31	Dyer and Seeley (1987)	22.6 23.4	Underwood (1991), Abrol (1985)
A. dorsata	6.45 6.73	Dyer and Seeley (1987), Sakagami et al. (1980)	118.0 123.0	Underwood (1991), Abrol (1985)
A. cerana	5.40 4.39–5.53	Dyer and Seeley (1987), Mishra and Kumar (1997)	55.0 77.0	Underwood (1991), Abrol (2007a, b, c)
A. mellifera	6.11 5.7–7.2	Dyer and Seeley (1987), Mishra and Kumar (1997)	85.0 113.0	Underwood (1991), Abrol (2007a, b, c)
A. laboriosa	7.05	Sakagami et al. (1980)	165.4	Underwood (1991)

between the two species; however, further studies showed that *A. m. sudanensis* and *A. florea* do coexist. This coexistence is based on different daily rhythms of pollen collection. *A. mellifera* collected pollen of *Acacia seyal*, date palm and onions mainly in the early morning, while *A. florea* started pollen collection later in the morning and ended in the afternoon. But in contrast to *A. mellifera*, *A. florea* collected pollen all day without interruption, even at very high air temperatures. Niche overlap (concerning the times of visits to flowers) between the two species was very low on date palms and of medium importance in *A. seyal*. It is remarkable that in total, *A. florea* was always present in higher numbers than *A. m. sudanensis* on flowers.

In foraging studies in India, Gupta et al. (1984) found that *A. cerana* started foraging on *Plectranthus* in the morning at 06:00, whereas *A. mellifera* was seen only after 07:00. Similarly, *A. mellifera* stopped working after 17:00 and *A. cerana* continued until 18:00. During these observations both species collected only nectar from the flowers of *P. rugosus*. Even though other good sources of pollen were available to the bees, no other nectar sources were observed in the area. Neupane et al. (2006) found that *A. dorsata* preferentially concentrated their visits on litchi flowers in the early morning and then declined towards the later part of the day. The bees preferentially collected more nectar from bottlebrush, litchi flowers and pollen from bottlebrush, citrus and summer squash flowers showing distinct resource partitioning.

However, the flowers of radish were preferentially foraged for pollen by *A. cerana* and *A. mellifera* showing a distinct foraging preference by different honeybee species. Several researchers have reported similar findings. Rao and Lazar (1980) established that *A. dorsata* never visited the flowers of onion during the entire period of its blooming, while *A. cerana* and *A. florea* were observed collecting nectar. Verma and Partap (1994), Partap and Verma (1994) and Partap et al. (2000b) reported that cauliflowers were more attractive to *A. cerana* than radish. Other instances of resource partitioning are reviewed in Chap. 11.

In addition to different times of foraging, food resources may be partitioned spatially. For example, species may forage on different strata. Rinderer et al. (1996)

found that *A. cerana* and *A. koschevnikovi* foraged more frequently at the tops of the yellow flame tree (*Peltophorum pterocarpum*) than at the middle or bottom levels in Sabah, Borneo. *A. dorsata* foraged equally at the top and middle of trees, while *A. andreniformis* foraged equally at all strata. These results suggest that species demonstrate stratum fidelity. Alternatively, nectar and pollen production may be more pronounced at the top and middle of a tree, and certain bees (e.g. *A. cerana* and *A. koschevnikovi*) may competitively exploit the best resources. Instead of solitary bees, *A. cerana*, *A. dorsata* and a few *Trigona* species seemed to be the main pollinators in the upper canopy in Malayan forests (Appanah and Kevan 1995).

Verma and Chauhan (1985) observed insect pollinators on apple trees and found that 40.6% of them foraged between 2 and 3 m from the ground. The rest foraged in roughly equal numbers above 3 m and below 2 m. The higher numbers on the middle branches may indicate stratified foraging, but alternatively may merely reflect a greater number of flowers on the middle branches. Abrol (1988b) reported that *A. cerana* foraged in large numbers on the middle branches of loquat, *E. japonica*, approximately 3 m above ground compared to lower or upper branches.

Differences in proboscis length may be one way in which resources are partitioned between *A. cerana* and *A. nigrocincta*. In morphometric analyses of *A. cerana* and *A. nigrocincta*, samples of each collected from locations only 12 km apart had significantly different proboscis lengths (Hadisoesilo et al. 1995). These two species may handle some of the same flowers with differing amounts of ease because of their different proboscis lengths. More information on foraging preferences of honeybees is urgently needed, particularly for *A. florea* and *A. mellifera*. A number of observations are available regarding different flower preferences between the two species. *A. cerana* is reported to visit a greater variety of plants, including wild species, while *A. mellifera* foraged mainly on *Trifolium* and *Brassica* in Japan (Miyamoto 1958). In Himachal Pradesh, the main nectar source of *A. cerana*, *Plectranthus* sp., is neglected by *A. mellifera* foragers (Goyal 1974). Honey collected from *A. cerana* bees (originating from China) in Germany showed differences in the pollen spectrum compared to *A. mellifera* honey collected simultaneously at the same location (Vorwohl 1968).

The similar size and resource needs of *A. cerana* and *A. nigrocincta* suggest that these species could share similar ecological niches, yet in zones of sympatry they seem to occupy different habitats. Bakker (1999) found that *A. cerana* forages in disturbed habitats and *A. nigrocincta* forages more extensively in forested areas. In *A. cerana*, dancing scouts indicated distances from 10 to 1,420 m, and the average dances for final destinations ranged from 99 to 780 m. Dances of *A. nigrocincta* scouts ranged from 75 to 2,340 m, and final destinations ranged from 140 to 1,920 m indicating that *A. nigrocincta* travelled farther than *A. cerana* when scouting for new nest sites.

Bakker (1999) reported that foragers of *A. cerana* and *A. nigrocincta* concentrated on different pollen resources, suggesting that resource partitioning may occur on a different level but the reasons for preferences for different kinds of pollen are as yet unknown. A comparison of *A. cerana* and *A. nigrocincta* foraging activity throughout the day revealed large variation from day to day both among and between colonies. Resources do not seem to be partitioned by differences in time of foraging; however, slight differences in time of foraging may result from each colony exploiting different resources. He further found that *A. cerana* and *A. nigrocincta* have dance curves with equal slopes, but *A. nigrocincta* started waggle dances at a farther distance from the hive (25 m, compared to 10 m for *A. cerana*). *A. nigrocincta* also had a wider flight range than *A. cerana*, both in distance travelled while foraging and distances of nest sites investigated by swarm scouts; but both species had relatively short foraging ranges. A wider flight range could give *A. nigrocincta* an advantage over *A. cerana*, enabling the former to scout farther for food and for nesting sites. However, he did not find any significant difference in foraging locations of two species.

12.7.3 Influence of Honeybees on Other Bee Species

Nectar robbing during a cauliflower bloom by many honeybee foragers has been found in Solan India by Kumar et al. (1994) (for further discussion, cf. Chap. 19). Side foraging of cauliflower nectar by *A. cerana* and *A. mellifera* and of okra nectar by *A. cerana* has been reported by Kapoor and Dhaliwal (1989) and Mishra et al. (1987), respectively. Kapil and Kumar (1975) reported that few foragers of *A. dorsata* collected nectar from *B. juncea* flowers without actually entering the flowers. Sharma et al. (2001) found that *A. florea*, *A. dorsata*, *A. cerana* and *A. mellifera* all foraged as top workers on flowers of *B. campestris* var. *sarson*, *A. cepa*, *D. carota*, *T. alexandrinum* and *Helianthus annuus*, except that *A. florea* foraged as side workers on *Brassica* flowers.

12.8 Organoleptic Aspects

12.8.1 Foraging in Relation to Flavour, Taste and Colour of Flowers

The actual recognition of a source of food or nectar is influenced by various factors such as colour, flavour and taste. For example, bees can distinguish green, blue, white, yellow and violet flowers, but are insensitive to red ones. The bees can also discriminate between different concentrations of sweetness and varying salt content in flowers. If flowers with different concentrations of sugar are blooming simultaneously, the bees will prefer those which have a higher content. Once the bees have started to work on a certain source for nectar or pollen, they continue to visit it and seldom visit more than one or two species on a single trip. Bees also drink water and prefer stagnant water to fresh. If a pinch of salt is added to the fresh water it also becomes attractive. Foraging behaviour of *A. florea* was studied on *Portulaca grandiflora* flowers having different patches of straw yellow, dark yellow, purple and red flowers (Abrol 2003). Maximum foraging was observed on straw yellow flowers followed by dark yellow, purple and red flowers, respectively.

12.8.2 Foraging and Biochemical Characters of Host Plants

Insects require appreciable amounts of protein, carbohydrate, lipid and nitrogen for growth and development (Bertsch 1983). A direct correlation is expected to exist between the honeybee visits and the biochemical nature of flower parts.

Table 12.3 *Apis florea* visits to *Ipomea staphylina* and *Eucalyptus tereticornis* during different times of day

Time (IST)	Visits/30 min/inflorescence	
	I. staphylina	*E. tereticornis*
06:30–07:00	5	8
07:00–07:30	5	8.4
07:30–08:00	6	8.9
08:30–09:00	6.2	9
09:00–09:30	6	7.5
12:00–12:30	–	1.3
13:00–13:30	–	1
16:00–16:30	1.4	3
16:30–17:00	1.0	2

Source: Nathan et al. (1999)
IST Indian standard time

Table 12.4 Nutritional components of *Ipomea staphylina* and *Eucalyptus tereticornis* flowers. Mean ± SE, $n = 5$.

Biochemical parameters	Anthers	Pollen	Stigma	Petals
Protein (mg/g)				
I. staphylina	92.34 ± 6.85	121.08 ± 8.96	112.40 ± 7.31	89.93 ± 6.92
E. tereticornis	95.58 ± 7.31	120.60 ± 8.96	140.21 ± 10.38	93.68 ± 8.94
Carbohydrate (mg/g)				
I. staphylina	98.42 ± 8.53	111.02 ± 9.18	87.01 ± 7.48	132.48 ± 11.56
E. tereticornis	101.31 ± 8.87	115.65 ± 8.34	91.38 ± 7.34	143.56 ± 9.26
Lipid (mg/g)				
I. staphylina	25.05 ± 2.50	36.81 ± 2.89	33.18 ± 2.75	21.08 ± 2.83
E. tereticornis	31.69 ± 2.47	43.86 ± 3.63	39.15 ± 2.98	27.01 ± 2.10
Nitrogen (%)				
I. staphylina	1.29 ± 0.078	1.08 ± 0.013	1.34 ± 0.021	2.18 ± 0.065
E. tereticornis	1.38 ± 0.07	1.15 ± 0.07	1.49 ± 0.06	2.58 ± 0.08
Water content (%)				
I. staphylina	42.19 ± 2.81	38.03 ± 2.73	55.05 ± 3.92	75.02 ± 4.68
E. tereticornis	45.36 ± 2.94	36.91 ± 2.11	57.90 ± 3.95	69.01 ± 4.80

Source: Nathan et al. (1999)

Nathan et al. (1999) monitored the foraging activity of *A. florea* on *Ipomea staphylina* and *Eucalyptus tereticornis* flowers and found that nectar secretion is minimal in both, thus forcing the bees to make repeated visits to a large number of flowers to satisfy their nutritional demands (Table 12.3). Interestingly, they found that *A. florea* stayed longer on flowers of *I. staphylina* than *E. tereticornis*. The increased rate of visits by *A. florea* to the flowers of *E. tereticornis* compared to *I. staphylina* was attributed to a higher amount of protein, carbohydrate and optimum levels of lipid and nitrogen in *E. tereticornis* (Table 12.4). Furthermore, *A. florea* preferred to visit *E. tereticornis* than *I. staphylina* since *E. tereticornis* had a higher flower density.

12.9 Conclusions

Bees are excellent test animals for optimal foraging theories because energy gained from nectar can be accurately measured. Most energetics research is on bumblebees and honeybees, primarily because of their high energy requirements and because they are easily followed while foraging. More work is needed on the energetics of foraging in other taxa of bees and on the importance of factors other than energy, such as nutritive values of different pollens, in determining bee foraging patterns. No bee species is an absolute generalist forager. All bees specialise, to some extent, through innate or learned preferences for particular plant taxa, foraging times, spatial distributions of flowers, floral structures or floral products. Most studies of competition for floral resources among bee species, especially between honeybees and wild bees, have been inconclusive because competition is difficult to prove in the field. Future studies should include manipulation of forager and resource levels and behavioural observations of the interactions between bees on flowers.

References

Abrol DP (1985) Analysis of biophysical interactions in causing foraging behaviour of some bees – a study in bioenergetics. Thesis, CCS Haryana Agricultural University, Hisar, India

Abrol DP (1986a) Ecophysiological adaptations between pollinating bees and their flowers. Environ Ecol 4:161

Abrol DP (1986b) Differential dilution of nectar by sub-tropical bees. Environ Ecol 4:188–189

Abrol DP (1987a) Influence of thermal and energetic constraints on the pollination activity of *Xylocopa pubescens* Spinola (Hymenoptera: Anthophoridae). Environ Ecol 5:90–93

Abrol DP (1987b) Activity time budgets and pollination potential of *A. cerana indica* workers. Environ Ecol 5:707–709

Abrol DP (1987c) Interrelationships and path coefficient analysis of environmental factors influencing pollination activity of *A. dorsata* F. Trop Ecol 28:147–154

Abrol DP (1988a) Foraging range of subtropical bees, *Megachile flavipes*, *Megachile nana* (Hymenoptera: Megachilidae) and *A. florea* (Hymenoptera: Apidae). J Indian Inst Sci (Biol Sci) 68:43–47

Abrol DP (1988b) Ecology and behaviour of three bee species pollinating loquat (*Eriobotrya japonica* Lindley). Proc Indian Nat Sci Acad B 54:161–163
Abrol DP (1989a) Studies on abundance, diversity, behaviour and importance of native pollinators for crop production. Korean J Apic 4:25–40
Abrol DP (1989b) Studies on ecology and behaviour of insect pollinators frequenting strawberry blossoms and their impact on yield and fruit quality. Trop Ecol 30:96–100
Abrol DP (1990) Energetics of nectar production in some apple cultivars as a predictor of floral choice by honeybees. Trop Ecol 31:116–122
Abrol DP (1991a) Insect pollination and crop production in Jammu and Kashmir. Curr Sci 65:265–269
Abrol DP (1991b) Foraging strategies of honeybees in pollinating apple flowers. J Anim Morphol Physiol 38:109–113
Abrol DP (1991c) Path analysis of environmental factors influencing daily flight activity of *A. dorsata* F. Acta Ecol Gen 12:619–626
Abrol DP (1992a) Bioenergetics in bee-flower interrelationships – an analysis of foraging behaviour. Korean J Apic 7:39–66
Abrol DP (1992b) Energetics of nectar production in some strawberry cultivars as a predictor of floral choice by honeybees. J Biosci 17:41–44
Abrol DP (1992c) Foraging in honeybees *A. cerana indica* F. and *A. dorsata* F. (Hymenoptera: Apidae) – activity and weather conditions. J Indian Inst Sci (Biol Sci) 72:395–401
Abrol DP (1992d) Diel pattern of *A. cerana indica* in *Asparagus officinalis* L. Sci Cult 58:25
Abrol DP (1993) Intervarietal differences in nectar production affecting pollinator attractiveness and fruit set in almond. In: Proceedings of International Symposium on Pollination in Tropics, Bangalore, 8–13 Aug 1993, pp 120–122
Abrol DP (1995a) Energetics of nectar production in some almond cultivars as a predictor of floral choice by honeybees, *A. cerana indica* F. and *A. mellifera* L. (Hymenoptera: Apidae). Proc Indian Nat Sci Acad B 57:127–132
Abrol DP (1995b) Studies on abundance, diversity and behaviour of insect visitors of okra (*Hibiscus esculentus* L.) (Moench) blossoms and their importance as pollinators. Korean J Apic 10:1–4
Abrol DP (1998a) Food acquisition efficiency as a determinant of honeybee foraging decisions. Indian Bee J 60:133–136
Abrol DP (1998b) Foraging in honeybees *A. cerana indica* F. and *A. dorsata* F. (Hymenoptera: Apidae) activity and weather conditions. J Indian Inst Sci 72:395–401
Abrol DP (1998c) Environmental factors influencing flight activity in honeybees *A. cerana* Fab. and *A. mellifera* L. (Hymenoptera: Apidae). Indian Bee J 60:71–75
Abrol DP (2000a) Insect pollinators of kiwifruit (*Actinidia deliciosa*). Indian Bee J 62:55–57
Abrol DP (2000b) Energy budget versus plant pollinator interactions. J Palynol 36:17–36
Abrol DP (2003) Foraging behaviour of honeybee *A. florea* F. as influenced by flower colour on *Portulaca grandiflora*. Sci Cult 69:289–290
Abrol DP (2005a) Pollination energetics. J Asia Pacif Entomol 8:3–14
Abrol DP (2005b) Foraging ecology of insect pollinators visiting cherry flowers and their impact on fruit production. Korean J Apic 20:159–168
Abrol DP (2006a) Diversity of pollinating insects visiting litchi flowers (*Litchi chinensis Sonn.*) and path analysis of environmental factors influencing foraging behaviour of four honeybee species. J Apic Res 45:180–187
Abrol DP (2006b) Factors influencing flight activity of *A. florea* F, an important pollinator of *Daucus carota* L. J Apic Res 45:6
Abrol DP (2007a) Foraging behaviour of *A. mellifera* and *A. cerana* as determined by the energetics of nectar production in different cultivars of *Brassica campestris* var. *toria*. J Apic Sci 51:5–10
Abrol DP (2007b) Differences in sugar metabolism among honeybees, *A. mellifera* and *A. cerana* potentially impact floral attractiveness. J Apic Res 46:181–190

Abrol DP (2007c) Honeybee and rapeseed – a pollinator plant interaction. Adv Bot Res 45: 339–367

Abrol DP (2009) Bees and beekeeping in India, 2nd edn. Kalyani, Ludhiana

Abrol DP (2010) Foraging behavior of *Apis florea* F., an important pollinator of *Allium cepa* L. J Apic Res and Bee World 49(4):318–325

Abrol DP, Kapil RP (1987) Nectar dilution pattern of bees in semi-arid environments. Curr Sci 56:681

Abrol DP, Kapil RP (1991) Foraging strategies of honeybees and solitary bees as determined by nectar sugar components. Proc Indian Nat Sci Acad B 57:127–132

Abrol DP, Kumar A (2009) Foraging activity of *Apis* species on strawberry blossoms as influenced by pesticides. Pak Entomol 31:57–65

Abrol DP, Kapil RP, Sihag RC (1988) Physiological adaptations of subtropical bees in relation to energy intakes. Res Dev Rep 5:68–79

Abrol DP, Bhat AA, Khan AR (1989) Influence of behaviour of *A. cerana indica* Fabr. on cross pollination in apple. Indian Bee J 51:103–104

Abrol DP, Sharma D, Monobrullah M (2005) Abundance and diversity of pollinating insects visiting peach and plum flowers and their impact on fruit production. SKUAST J Res 4:38–45

Adlakha RL, Sharma OP (1975) *A. mellifera* vs *A. indica*. Glean Bee Cult 103:160

Akratanakul P (1977) The natural history of the dwarf honey bee, *Apis florea* F. in Thailand. Thesis, Cornell University, Ithaca

Appanah S, Kevan PG (1995) Bees and the natural ecosystem. In: Kevan PG (ed) The Asiatic hive bee. Enviroquest, Cambridge, Ontario

Atwal AS (1970) Insect pollinators of crops. Punjab Agricultural University Press, Ludhiana

Atwal AS, Dhaliwal GS (1969) Some behavioural characteristics of *A. indica* F and *A. mellifera* L. Indian Bee J 31:1–8

Atwal AS, Bains SS, Singh B (1970) Bee flora for four species of *Apis* at Ludhiana. Indian J Entomol 32:330–334

Axelrod DI (1960) The evolution of flowering plants. In: Tafl S (ed) Evolution after Darwin: the evolution of life, vol 1. University of Chicago Press, Chicago

Bakker DR (1999) Foraging and habitat selection by two species of honeybee near Lore Lindu National Park in Sulawesi Indonesia. Thesis, University of Guelph, Guelph

Bawa KS (1990) Plant-pollinator interactions in tropical rain forests. Annu Rev Ecol Syst 21:399–422

Benedek P, Prener J (1972) Effect of temperature on behaviour and pollinating efficiency of honeybees on winter rape flowers. Angew Entomol 71:120–124

Bertsch A (1983) Nectar production of *Epilobium angustifolium* L at different air humidities; nectar sugar in individual flowers and the optimal foraging theory. Oecologia 59:40–48

Bhalla OP, Verma AK, Dhaliwal HS (1983a) Foraging activity of insect pollinators visiting stonefruits. J Entomol Res 7:91–94

Bhalla OP, Verma AK, Dhaliwal HS (1983b) Insect visitors of mustard bloom (*Brassica campestris* var. *sarson*) their number and foraging behaviour under midhill conditions. J Entomol Res 7:15–17

Bisht DS, Pant NC (1968) Studies on the pollen gathering activity of the Indian honeybee *A. cerana indica* F under Delhi conditions. Indian J Entomol 44:368–372

Campbell DR (1985) Pollen and gene dispersal – the influences of competition for pollination. Evolution 39:418–431

Chahal BS, Brar HS, Gatoria GS, Jhajj HS (1986) Aggressive behaviour of *A. florea* towards *A. mellifera* in hive robbing and in foraging. J Apic Res 25:134–138

Chandel RS, Thakur RK, Bhardwaj NR, Pathania N (2004) Onion seed crop pollination: a missing dimension in mountain horticulture. Acta Hortic (ISHS) 631:79–86

Chaturvedi M (1973) An analysis of honeybee pollen loads from Banthra Lucknow, India. Grana 13:139–144

Chaturvedi M (1977) Further investigation on the pollen analysis of bee loads from Banthra, India. New Bot 4:41–47

Chaudhary N, Sihag RC (2004) Relative abundance diversity and dominance concentration of two honeybee species on three concurrently flowering crops onion (*Allium cepa* L), carrot (*Daucus carota* L) and fennel (*Foeniculum vulgare* L). In: Proceedings of 1st European conference on apidology, Udine, pp 19–23

Choudary RK (1978) Floral fidelity in the Indian honeybee (*A cerana indica*). Indian Bee J 40:33–35

Cody M (1974) Optimization in ecology. Science 183:1156–1164

Collision CH (1973) Nectar secretion and how it affects the bee activity of honeybees in the pollination of hybrid pickling cucumbers *Cucumis melo* L. Thesis, Michigan State University

Corbet SA (1978a) A bee's view of nectar. Bee World 59:25–32

Corbet SA (1978b) Bees and nectar of *Echium vulgare*. In: Richards AJ (ed) The pollination of flowers by insects. Academic, London, pp 21–30

Corbet SA, Fussel M, Ake R, Fraser A, Gunsen E, Salvage A, Smith K (1993) Temperature and the pollinating activity of social bees. Ecol Entomol 18:17–30

Cruden RW (1972) Pollinators in high-elevation ecosystem: relative effectiveness of birds and bees. Science 176:1439–1440

Cruden RW, Kinsman S, Stockhouse RE, Linhart YB (1976) Pollination fecundity and the distribution of multi-flowered plants. Biotropica 8:204–210

Desmuth G (1933) Temperature and nectar secretion. Glean Bee Cult 51:582–584

Devkota FR, Thapa RB (2005) Foraging preference of *A. cerana* F and *A. mellifera* L to broccoli under caged and open conditions in Chitwan. J Inst Agric Anim Sci 26:167–168

Dhaliwal GS (1970) Observations on pollen carrying capacity of *A. indica* F and *A. mellifera* L. Indian J Entomol 32:173–174

Dhaliwal GS, Atwal AS (1986a) Effect of age of crop plant spacing soil moisture and phosphate fertilizers on bee activity on *Brassica* crop. In: Kapil RP (ed) Pollination biology: an analysis. Inter India, New Delhi, pp 91–102

Dhaliwal GS, Sharma PL (1974) Foraging range of the Indian honeybee. J Apic Res 13:137–141

Dhaliwal JS, Atwal AS (1986b) Factors influencing tripping efficiency of bees on alfalfa (*Medicago sativa* L). In: Kapil RP (ed) Pollination biology: an analysis. Inter India, New Dehli, pp 79–90

Diwan VV (1980) Honeybee behaviour during total solar eclipse. Indian Bee J 42:97–99

Diwan VV, Rao SK (1969) Foraging behaviour of *A. indica* F on the flowers of *Synadenium grantii* Hook. Indian Bee J 31:22

Diwan VV, Salvi SR (1965) Some interesting behavioural features of *A. dorsata* Fab. Indian Bee J 27:52

Dixit KK (1956) Differential response to form and pattern in two species of Indian honeybees. J Bombay Nat Hist Soc 54:107–117

Domagała Lipińska A (1962) Daily dynamics of flight of *Apidae* and temperature. Ecol Polska Ser B 8:55–57 [in Polish]

Dyer FC (1985) Nocturnal orientation by the Asian honeybee *A. dorsata*. Anim Behav 33:769–774

Dyer F, Seeley TD (1987) Interspecific comparisons of endothermy in honey-bees (*Apis*): deviations from the expected size-related patterns. J Exp Biol 127:1–26

Dyer FC, Seeley TD (1991) Dance dialects and foraging range in three Asian honeybee species. Behav Ecol Sociobiol 28:227–233

El Shafie HAF, Mogga JBB, Basedow T (2002) Studies on the possible competition for pollen between the honeybee, *A. mellifera sudanensis* and the imported dwarf honeybee *A. florea* (Hym. Apidae) in North-Khartoum (Sudan). J Appl Entomol 126:557–562

Emlen JM (1966) The role of time and energy in food preferences. Am Nat 100:611–617

Faegri K, van der Pijil L (1978) The principles of pollination ecology, 3rd edn. Oxford University Press, New York

Forbes MF, Cervancia CR (1994) Foraging behaviour of *A. cerana* F. and *A. mellifera* L. (Apidae: Hymenoptera) in Majayjay. Philipp J Sci 123:21–27

Free JB (1970) Insect pollination of crops. Academic, London

Free JB (1981) Ecology and behaviour of the honeybee *A. florea* and possibilities for beekeeping. Bee World 62:49–59

Free JB, Williams IH (1973) The foraging behaviour of honeybees (*Apis mellifera* L.) on Brussels Sprout (*Brassica oleracea* L.). J Appl Ecol 10:489–499

Gary NE (1967) Diurnal variations in the intensity of flight activity from honeybee colonies. J Apic Res 6:65–68

Gong YF (1983) The natural beekeeping conditions and honeybee races in China. J Fujian Agric Coll 12:241–249

Goyal NP (1974) *A. indica* indica F and *A. mellifera* L as complimentary to each other for the development of apiculture. Bee World 55:98–101

Goyal NP, Atwal AS (1977) Wingbeat frequency of *A. indica indica* F and *A. mellifera* L. J Apic Res 16:47–48

Grant V (1950) The flower constancy of bees. Bot Rev 16:379–398

Grant V (1994) Modes and origins of mechanical and ethological isolation in angiosperms. Proc Natl Acad Sci 91:3–10

Gupta JK, Reddy MCM (1992) Foraging intensity of insects and nectar of wild cherry *Prunus puddum* Roxb. Entomon 17:91–94

Gupta JK, Mishra RC, Kumar J (1984) *Plectranthus* as forage for *A. cerana indica* F. and *A. mellifera* L. Apidologie 15:75–82

Hadisoesilo S, Otis GW, Meixner M (1995) Two distinct populations of cavity nesting honeybees (Hymenoptera: Apidae) in South Sulawesi Indonesia. J Kans Entomol Soc 68:399–407

Heinrich B (1972) Energetics of temperature regulation and foraging in a bumble bee *Bombus terricola* Kirby. J Comp Physiol 77:49–65

Heinrich B (1973) The energetics of bumblebees. Sci Am 288:96–102

Heinrich B (1975) Energetics of pollination. Annu Rev Ecol Syst 6:139–170

Heinrich B (1979a) Keeping a cool head: honeybee (*A. mellifera*) thermal regulation. Science 205: 1269–1271

Heinrich B (1979b) Bumblebee economics. Harvard University Press, Cambridge

Heinrich B (1983) Insect foraging energetic. In: Jones CE, Little RJ (eds) Handbook of experimental pollination biology. Academic Press, New York, pp 187–215

Heinrich B, Raven PH (1972) Energetics and pollination ecology. Science 176:597–602

Hocking B (1953) The intrinsic range and speed of insect flight. Trans R Entomol Soc Lond 104:223–345

Jadhav LD (1982) Effect of solar eclipse on the foraging activity of honeybees. Biovigyanam 8:157–161

Jhajj HS, Goyal NP (1979a) Comparative behaviour of pollen foragers of *A. cerana indica* and *A. mellifera*. J Apic Res 18:279–284

Jhajj HS, Goyal NP (1979b) Comparative studies on the flower constancy of *A. cerana indica* F. and *A. mellifera* L. In: Proceedings of 4th international symposium on pollination, Maryland, pp 333–339

Jhajj HS, Goyal NP (1979c) Reversal in the normal behavioural sequence of honeybee foragers with carbon dioxide anaesthesia. Indian Bee J 41:17–18

Jhajj HS, Goyal NP (1986) Comparative flower constancy of *A. cerana indica* F. and *A. mellifera* L. on removing their colonies to a new site. In: Kapil RP (ed) Pollination biology: an analysis. Inter India, New Delhi, pp 73–77

Johnson LK, Hubbell SP (1974) Aggression and competition among stingless bees: field studies. Ecology 55:120–127

Johnson LK, Hubbell SP (1975) Contrasting foraging strategies and coexistence of two bee species on a single resource. Ecology 56:1398–1406

Kakar KL (1981) Foraging behaviour of insect pollinators of cauliflower bloom. Indian J Ecol 8:126–130

Kakar KL (1983) Prospects of bees as pollinating agents of cauliflowers. In: Proceedings of 2nd international conference on apiculture in tropical climates, New Delhi, pp 545–556

Kammer AE, Heinrich B (1974) Metabolic rates related to muscle activity in bumble bees. J Exp Biol 61:219–227
Kapil RP, Brar HS (1971) Foraging behaviour of *A. florea* F on *Brassica campestris* var. *toria*. In: Proceedings of 12th international congress on apiculture, Moscow, pp 335–339
Kapil RP, Kumar S (1975) Foraging activity of *A. dorsata* F on *Brassica juncea* Czern and Coss. J Bombay Nat Hist Soc 71:327–331
Kapoor KS, Dhaliwal HS (1989) Comparative energetics of *A. cerana indica* Fabr. and *A. mellifera* L. on *Brassica oleracea* L. var. *botrytis*. Indian Bee J 51:99–101
Kellog CR (1941) Some characteristics of the oriental honeybee *A. cerana* F. in China. J Econ Entomol 34:717–719
Koeniger N, Vorwohl G (1979) Competition for food among four sympatric species of Apini in Sri Lanka (*A. dorsata*, *A. cerana*, *A. florea* and *Trigona iridipennis*). J Apic Res 18:95–109
Kremen C (2005) The ecology and management of crop pollination services. In: Groom MJ, Meffe GK, Carroll CR (eds) Principles of conservation biology, 3rd edn. Sinauer and Associates, Sunderland, p 54
Kumar J, Rao KVK, Gupta JK (1994) Pollination efficiency of bees visiting blossom of *Brassica campestris* L. var. *toria* in mid-hills of Himachal Pradesh India. Indian Bee J 56:202–206
Lee ML, Choi SY, Cho YH (1987) Diurnal activity of the honey bees (*A. mellifera*) at the hive entrance. Korean J Apic 2:117–121
Leppik EE (1977) Floral evolution in relation pollination ecology. Today and Tomorrow, Prizters, New Delhi, pp 104–108
Lindauer M (1956) Uber die Verstandigung bei indischen Bienen. Zeit Vergl Physiol 38:521–557
Lindauer M (1957) Communication among the honeybees and the stingless bees of India. Bee World 38(3–14):34–39
Lundie AE (1925) The flight activities of the honeybee. USDA Bull No 1328:1–38
Mattu VK, Verma LR (1985) Studies on the annual foraging cycle of *A. cerana indica* F in Simla hills of northwest Himalayas. Apidologie 16:1–18
Mattu VK, Chaudhary DK, Kumar L (1994) Foraging ecology of *A. cerana* F. and *A. mellifera* L. in pollinating stone fruit crops. Pest Manag Econ Zool 2:35–39
Michener CD (1974) The social behavior of the bees. Harvard University Press, Cambridge
Mishra RC, Kumar Y (1997) *A. mellifera* L. for Indian apiculture. In: Mishra RC, Garg R (eds) Perspectives in Indian apiculture. Agro-Botanica, Bikaner, pp 119–149
Mishra RC, Kumar J, Gupta JK (1987) Effect of mode of pollination on fruit characteristics of okra, *Abelmoschus esculentus* (L) Moench. Proc Indian Natl Sci Acad B 53:157–160
Miyamoto S (1958) Biological studies on Japanese bees. X Differences in flower relationships between Japanese and European honeybees. Sci Rpts Hyogo Univ Agric Ser Agric 3:99–107
Mohan JKV (1973) Some observations on the bee pollination in sunhemp (*Crotalaria juncea*) and their importance in breeding work. Jute Bull 25:214–216
Murrell DC, Nash WT (1981) Nectar secretion by toria (*Brassica campestris* L. v. *toria*) and foraging behaviour of three *Apis* species on toria in Bangladesh. J Apic Res 20:34–38
Nagamitsu T, Inoue T (1999) Differences in pollen sources of *A. cerana* and *A. mellifera* at primary beach forest in central Japan. J Apic Res 38:71–78
Naim M, Bisht DS (1979) Pollen gathering capacity of the Indian honeybee *A. cerana indica* Fabr. Indian J Entomol 41:295–296
Naim M, Bisht DS (1989) Fidelity and time spent on flowers by *A. cerana* F. foragers. Indian J Entomol 51:222–223
Nataraj SM, Varadharajan K, Thiyagesan K, Sankari A (2000) Flower constancy in the Indian honeybee *A. cerana indica*. In: Proceedings of 8th national Tamil science congress [in Tamil]
Nathan SS, Murugan K, Kumar NS, Jeyabalan D, Muthuraman S (1999) Pollination potential of honeybee *A. florea* Linn. in relation to biochemical properties of host plants. Zoos Print 14:3–4
Neupane KR, Dhakal DD, Thapa RB, Gautam DM (2006) Foraging preference of giant honeybee *A. dorsata* F to selected horticultural crops. Inst Agric Anim Sci 27:87–92

Oldroyd BP, Rinderer TE, Wongsiri S (1992a) Pollen resource partitioning by *A. dorsata*, *A. cerana*, *A. andreniformis*, and *A. florea* in Thailand. J Apic Res 31:3–7

Oldroyd BP, Rinderer TE, Buco SM (1992b) Intra-colonial foraging specialism by honeybees (*A. mellifera*) (Hymenoptera: Apidae). Behav Ecol Sociobiol 30:291–295

Oschmann H (1961) Eine bienenkundliche Reise in die Volksrepublik China. Arch Geflugetz Kleintierkd 10:235–259

Partap U (2000) Foraging competition between *A. cerana* and *A. mellifera* and its impact on crop pollination. In: Proceedings of 4th Asian apicultural association conference, Kathmandu, pp 206–208

Partap U, Verma LR (1994) Pollination of radish by *A. cerana*. J Apic Res 33:237–241

Partap U, Shukla AN, Verma LR (2000a) Comparative foraging behaviour of *A. cerana* and *A. mellifera* in pollinating peach and plum flowers in Kathmandu Valley Nepal. In: Proceedings of 4th Asian apicultural association conference, Kathmandu, pp 193–198

Partap U, Shukla AN, Verma LR (2000b) Comparative attractiveness of broad leaf mustard, cauliflower and radish to *A. cerana* in Kathmandu Valley of Nepal. In: Matsuka M, Verma LR, Wongsiri S, Shrestha KK, Partap U (eds) Asian bees and beekeeping: progress of research and development. Oxford and IBH, New Delhi, pp 203–205

Phadke RP (1968) Studies on Indian honeys. 3. Proximate composition and physico-chemical characterization of honeys from the wild honeybees *A. dorsata*, *A. florea* and *Trigona*. Indian Bee J 30:3–8

Priti S, Sihag RC (1997) Diversity visitation frequency foraging behaviour and pollinating efficiency of insect pollinators visiting cauliflower (*Brassica olearacea* L. var. *botrytis* cv Hazipur Local) blossoms. Indian Bee J 59:230–237

Punjabi AA, Saraf SK, Singh G (1969) Studies on the pollen carrying capacity of the hill strain of the Indian honeybee *A. indica* F. in Kashmir in relation to its body weight. Indian Bee J 31:9–11

Raj D, Rana VK (1993) Foraging behaviour of *A. mellifera* L. and *A. cerana indica* F. at the hive entrance during rapeseed bloom. J Entomol Res 17:105–110

Raj D, Rana VK (1994) Time spent by *A. mellifera* L. and *A. cerana indica* F. foragers on rapeseed bloom. J Entomol Res 18:335–339

Raj D, Rana VK, Devi N (1993) Comparative studies on pollen loads and pollen constancy of *A. mellifera* L. and *A. cerana indica* F. on rapeseed. J Entomol Res17:43–48

Rao GM, Lazar M (1980) Studies on the behaviour and pollination of onion (*Allium cepa* L.). In: Proceedings of 2nd conference on apiculture in tropical climate, New Delhi, pp 580–589

Rao SP, Atluri JB, Reddi CS (2008) Intermittent mass blooming midnight anthesis and rockbee pollination in *Pterocarpus santalinus* (Fabaceae). Nordic J Bot 21:271–276

Reddy CC (1979) The influence of various meteorological factors on the foraging behavior of *A. cerana indica*. In: Proceedings of 4th international symposium on pollination, Maryland, pp 341–345

Reddy CC (1980a) Observations on the annual cycle of foraging and brood rearing by *A. cerana indica* colonies. J Apic Res 19:17–20

Reddy CC (1980b) Studies on the nesting behaviour of *A. dorsata* F. In: Proceedings of 2nd international conference on apiculture in tropical climates, New Delhi, pp 391–397

Reddy CS, Reddy EUS (1984) Bee flower interaction and pollination potential. Proc Indian Acad Sci 93:373–390

Ricketts TH (2004) Tropical forest fragments enhance pollinator activity in nearby coffee crops. Conserv Biol 18:1262–1271

Rinderer TE, Marx BD, Grîes M, Tingek S (1996) A scientific note on stratified foraging by Sabahan bees on the yellow flame tree (*Peltophomm pterocarpum*). Apidologie 27:423–425

Roberts RB (1979) Energetics of cranberry pollination. In: Proceedings of 4th International Symposium on Pollination. Special Miscellaneous Publication No 1, pp 431–440

Sakagami SF (1959) Some interspecific relations between Japanese and European honeybees. J Anim Ecol 28:51–68

Sakagami SF, Matsumura T, Ito K (1980) *A. laboriosa* in Himalaya the little known world largest honeybee (Hymenoptera: Apidae). Insecta Matsumurana NS 19:47–77
Schaffer WM, Jensen DB, Hobbs DE, Gurevitch I, Todd JR, Schaffer MV (1979) Cornpetition foraging energetics and the cost of sociality in three species of bees. Ecology 60:976–987
Schlising RA (1970) Sequence and timing of bee foraging on flowers of *Ipomoea* and *Aniseia* (Convolvalacae). Ecology 51:1061–1067
Schneider SS, McNally LC (1992) Seasonal patterns of foraging activity in colonies of the African honeybee *A. mellifera scutellata* in Africa. Insectes Soc 39:181–193
Schneider SS, McNally LC (1993) Spatial foraging patterns and colony energy status in the African honeybee *A. mellifera scutellata*. J Insect Behav 6:195–210
Schoener TW (1971) Theory of feeding strategies. Annu Rev Ecol Syst 2:369–404
Seeley TD (1986) Social foraging by honeybees: how colonies allocate foragers among patches of flowers. Behav Ecol Sociobiol 19:343–354
Selvakumar P, Sinha SN, Pandita VK (2006) Abundance and diurnal rhythm of honeybees visiting hybrid seed production plots of cauliflower (*Brassica oleracea* var. *botrytis* L.). J Apic Res 45:7–15
Shah AM (1980) Beekeeping in Kashmir. In: Proceedings of 2nd international conference on apiculture in tropical climates, New Delhi, pp 197–204
Shah FA, Shah TA (1980) Flight range of *A. cerana* from Kashmir. Indian Bee J 42:48
Shah FA, Shah TA (1989) Comparative foraging behaviour of *A. cerana* Fabr and *A. mellifera* L during *Isodon rugosus* (Wall) Codd bloom in Kashmir Valley. Indian Bee J 51:138–139
Sharma HK, Gupta JK, Rana BS (2000) Diurnal activity of *A. cerana* F. and *A. mellifera* L. on different flora during spring and honey flow period. Pest Manag Econ Zool 8:151–154
Sharma M (1970a) An analysis of pollen loads of honeybees from Kangra India. Grana 10:35–42
Sharma M (1970b) Studies on the pollen loads of honeybees from Kangra India. J Palynol 6:104–110
Sharma OP, Mishra RC, Dogra GS (1980) Management of *A. mellifera* L. in Himachal Pradesh. In: Proceedings of the 2nd conference on apiculture in tropical climates, New Delhi, pp 205–210
Sharma PL, Sharma AC (1950) Influence of numbers in a colony on the honey gathering capacity of bees (*A. indica* hill variety). Indian Bee J 12:106–107
Sharma SK, Singh JR, Mahla JC (2001) Foraging behaviour of *Apis* spp in semi-arid sub-tropical climate. 3. On flowers of mustard, onion, carrot, berseem and sunflower. Crop Res (Hisar) 21:332–334
Sihag RC (1983) Honeybee forage for grape (*Vitis vinifera* L.) juices during dearth period. Indian Bee J 45:111–112
Sihag RC (1984) Influence of environmental factors on pollination activity of bees. Environ Ecol 2:149–152
Sihag RC (1990) Differential attractiveness of two cruciferous species to honeybees. In: Proceedings of 11th international congress, IUSSI, pp 434–436
Sihag RC (2000a) Ecobiology of the little honeybee (*A. florea*) in semi and subtropical climates of India. In: Proceedings of 4th Asian apiculture association conference, Kathmandu, pp 46–48
Sihag RC (2000b) Ecobiology of the giant honeybee (*A. dorsata*) in semi-arid subtropical climates of India. In: Proceedings of 4th Asian apiculture association conference, Kathmandu, pp 49–52
Sihag RC, Abrol DP (1986) Correlation and path-coefficient analysis of environmental factors influencing flight activity of *A. florea* F. J Apic Res 25:202–208
Sihag RC, Kapil RP (1983) Foraging strategies of honeybees as determined by quality and quantity of nectar. In: Proceedings of 5th international symposium on pollination, Versailles, pp 51–59
Sihag RC, Rathi A (1992) Foraging modes and foraging rates of different bee pollination of pigeon pea (*Cajanus cajan* (L.) Millsp.). In: Proceedings of international symposium on pollination in tropics, pp 93–95
Singh RP, Singh AK, Singh RP (2007) The effect of the availability of bee forage plants and environmental conditions on the nesting of *A. dorsata* Fabr. J Apic Res 46:276–281

Singh Y (1971) Studies on pollen gathering activity of Indian honeybee *A. cerana indica* Fabr. under Saharanpur conditions. Prog Hortic 12:31–38

Sotavalta O (1952) On the difference and variation of the wing stroke frequency of insects in wing mutation and loading experiments and in experiments at sub-atmospheric pressure. Ann Zool Soc Vanamo 15:1–67

Stimec J, Scott-Dupree CD, McAndrews JH (1997) Honeybee *A. mellifera* pollen foraging in southern Ontario. Can Field Nat 11(3):454–456

Szabo TI (1980) Effect of weather factors on honey flight activity and colony weight gain. J Apic Res 19:164–171

Thakur AK, Sharma OP, Garg R, Dogra GS (1982) Comparative studies on foraging behaviour of *A. mellifera* and *A. cerana indica* on mustard. Indian Bee J 44:91–92

Thiyagesan KM, Varadharajan M, Abirami S (2001a) Studies on risk-sensitive foraging in the Indian honeybee (*A. cerana indica*). J Apic Res 40:6–20

Thiyagesan KM, Varadharajan M, Abirami S (2001b) Studies on central place foraging in the Indian honeybee *A. cerana indica*. J Apic Res 40:21–26

Underwood BA (1991) Thermoregulation and energetic decision-making by the honeybees *A. cerana*, *A. dorsata* and *A. laboriosa*. J Exp Biol 157:19–34

Verma LR (1988) Biological and economic characters of *A. cerana indica* F. Honeybee Sci 7:151–157

Verma LR (1990) Beekeeping in integrated mountain development. Oxford and IBH, New Delhi

Verma LR (1995) Pollination ecology of apple orchards by hymenoptrous insects in Matiana temperate zone. In: Kevan PG (ed) The Asiatic hive bee. Enviroquest, Ontario, pp 143–146

Verma LR, Chauhan P (1985) Distribution abundance and diversity of insect pollination in apple orchards of Shimla hills. Indian J Ecol 12:286–292

Verma LR, Dulta PC (1986) Foraging behaviour of *A. cerana indica* and *A. mellifera* in pollinating apple flowers. J Apic Res 25:197–202

Verma LR, Partap U (1994) Foraging behaviour of *A. cerana* on cauliflower and cabbage and its impact on seed production. J Apic Res 33:231–236

Verma SK (1983) Studies on the foraging behaviour of *A. cerana indica* Fabr. in Jeolikot (Nainital, India). Indian Bee J 45:5–7

Visscher PK, Seeley TD (1982) Foraging strategy of honeybee colonies in a temperate deciduous forest. Ecology 63:1790–1801

Vorwohl G (1968) Naturliche Diastaseschwache der Honige von *A. cerana* F. Z Bienenforsch 9:232–236

Waller GD (1972) Evaluating responses of honeybees to sugar solutions using an artificial flower feeder. Ann Entomol Soc Am 65:857–862

Waser NM, Chittka L, Price MV, Williams NM, Ollerton J (1996) Generalization in pollination systems and why it matters. Ecology 77:1043–1060

Wcislo WT, Cane JH (1996) Resource utilization by solitary bees (Hymenoptera: Apoidea) and exploitation by their natural enemies. Annu Rev Entomol 41:257–286

Willmer PG (1983) Thermal constraints on activity pattern of nectar feeding insects. Ecol Entomol 8:455–469

Woyke J (1992) Diurnal flight activity of African bees *A. mellifera adansonii* in different seasons and zones of Ghana. Apidologie 23:107–117

Woyke J, Jasiński Z, Fliszkiewicz C, Woyke H (2000) Flight activity of *A. mellifera* foragers at hive entrance during 86% eclipse of sun. Pszczel Zesz nauk 44:239–252

Woyke J, Wilde J, Wilde M (2003a) Flight activity reaction to temperature changes in *A. dorsata*, *A. laboriosa* and *A. mellifera*. J Apic Sci 47:73–79

Woyke J, Wilde J, Wilde M (2003b) Periodic mass flights of *A. laboriosa* in Nepal. Apidologie 34:121–127

Chapter 13
Energetic Aspects of Flight

H.R. Hepburn, Christian W.W. Pirk, and Sarah E. Radloff

13.1 Introduction

Ultimately, all of the intricate biological processes involved in swarming, migration and absconding, reproduction, seasonal and life cycles of the colonies, foraging and flight, communication by pheromones or the dance language, pollination, disease resistance and defence against predators are entirely predicated on the energetic efficiency of each particular behaviour or system. Thus, the metabolic costs subtending these activities are of paramount importance for the immediate subsistence of honeybee colonies and, ultimately, their evolutionary survival. Yet, studies on the energetics of biological activities and processes in honeybees are, at this point in time, in their nascency.

We know a great deal about changes associated with ovarian activation or inhibition under queenright and queenless conditions (cf. Chaps. 8 and 19), but absolutely nothing of the attendant metabolic costs. While we are beginning to grasp the essential biochemical pathways in the synthesis of beeswax, precise measurements of the metabolic costs of wax synthesis and secretion, much less comb-building, remain opaque and elusive (cf. Chap. 6). Similarly, massive defensive operations elicited by serious predators carry incalculable costs (cf. Chaps. 17 and 18). In physico-chemical terms, biological processes are extremely inefficient and much energy transfer is dissipated as heat loss. Any hopes of measuring the costs of honeybee colonial homeostasis are simply beyond reach.

H.R. Hepburn
Department of Zoology and Entomology, Rhodes University, Grahamstown 6140, South Africa
e-mail: r.hepburn@ru.ac.za

C.W.W. Pirk
Department of Zoology and Entomology, University of Pretoria, Pretoria 0002, South Africa
e-mail: cwwpirk@zoology.up.ac.za

S.E. Radloff
Department of Statistics, Rhodes University, Grahamstown 6140, South Africa
e-mail: s.radloff@ru.ac.za

H.R. Hepburn and S.E. Radloff (eds.), *Honeybees of Asia*,
DOI 10.1007/978-3-642-16422-4_13,

Nonetheless, over the last two or three decades, the previously unbreachable wall of energetics has shown us some crannies and chinks. Physiological measurements, particularly with respect to flight in both foraging and swarming contexts, have been clearly spurred on by the gradual assimilation of the ideas that both individual honeybees as well colonies actually make informed decisions (Seeley 1995). The results of the cumulative studies on the energetics of flight presented in this review are most parsimoniously explained by honeybee decisions based on analysis of relative costs.

While the honeybee thorax powers flight, substantial differences in the ratio of flight engine mass (thorax) to that of fuselage (whole body) have large aerodynamic effects on flight. As a consequence, and from dimensional considerations alone, ultimate flight capacity depends on the maximum excess power available to a honeybee over that required to maintain equilibrium in steady level flight (Hepburn et al. 1998a, b). As it transpires, from an aeronautical engineering perspective (von Mises 1959), worker honeybees of the Asian species of *Apis* comprise three design classes based on wing-loading, engine size and excess power index (EPI) (Radloff et al. 2001). Worker and drone honeybees are both strong flyers, but all they share in common on the wing are those flights associated with cleansing, swarming and absconding (Koeniger 1990).

By contrast, the flights of worker bees are directed towards short distance foraging and carrying relatively heavy loads (nectar and pollen) back to the nest, while the real prowess of drone flight is expressed in covering longer distances to drone congregation areas (cf. Chap. 8). Given the generally high levels of polyandry in honeybees and that a drone can mate but once, drones must be under severe selective pressure for aerodynamic prowess in mating comets, which have been observed in *Apis cerana*, *Apis koschevnikovi* and *Apis dorsata*. Thus, the drone flight mechanism is of interest to considerations of conspecific sexual dimorphism (in swarming or absconding contexts), but particularly in the context of mating.

Comparing the different species of honeybees, we would want to know, among other things: (1) What are the interspecific variations in the flight parameters and characteristics among workers bees, drones and queens? (2) Are there any caste-specific differences within a species? (3) Are there any adaptations in the flight systems in relation to altitude? (4) Are there special requirements for absconding and migration? and (5) Does flight design affect colonial life, and if so, in what ways?

Although the musculature that drives flight in honeybees resides in the thorax (Josephson 2006), different species of honeybee vary quite considerably in overall size and in the relative proportions of their body parts (Dyer and Seeley 1987). So, to make comparisons between bees of different sizes, let alone species, it is convenient to view them as miniature aircraft and examine relationships between the mass of the flight engine with the fuselage and wings. Besides purely physical considerations, the flight muscles of different species vary in the kinds and frequencies of the allozyme, malate dehydrogenase – an enzyme crucial to energy production in the citric acid cycle of mitochondria (Harrison et al. 1996). The importance of this is that different allozymes (phenotypes) of malate dehydrogenase are correlated with differences in the mass-specific metabolic flight rates of honeybees (Coelho and Mitton 1988; Harrison et al. 1996). In our present, somewhat simplified, analysis, certain features that modify honeybee flight, such as changeable wing-beat frequency and amplitude (Nachtigall and Hanauer-Thieser 1994), are held as constants so that honeybees could be considered from a general theory of flight (von Mises 1959).

13.2 Dimensional Considerations of Flight

In the following discussion, our analyses are based solely on dimensional considerations, so that the full flight capacity of a honeybee depends on the EPI of the system, which is the maximum power available over that required to maintain steady, level flight (Radloff et al. 2001). There are two points which must be reiterated: (1) thoracic engine capacity (expressed as horsepower/watts) was estimated by multiplying the measured metabolic capacity by average thoracic mass for each species (based on the data of Dyer and Seeley 1987, 1991a, b; Harrison and Hall 1993) and (2) the EPI is defined as (r^2/W), where W is wing-loading and r the ratio of thoracic mass to total mass (Hepburn et al. 1998a, b, 1999).

The results of such analyses of workers of Asian honeybee species included the means for total body mass, thoracic mass, thorax to whole body mass ratio, wing surface area, body mass ratio, wing-loading and EPI (Table 13.1). There is nearly a sevenfold difference in both thoracic and body mass between the dwarf honeybees, *Apis andreniformis* and *Apis florea*, on the one hand and the giant, *A. dorsata*, on the other. The Asian species can be divided into three significantly different groups: the dwarf bees, *A. andreniformis* and *A. florea*, the medium-sized bees, *A. cerana*, *A. koschevnikovi*, *Apis nigrocincta* and *Apis nuluensis* and lastly, the giant bees, *A. dorsata*. European and African *Apis mellifera* form two more significantly different groups which also differ from the Asian species (Table 13.1). Nonetheless, there is a ratio of unity in the scaling of total body mass to thoracic mass across this large size spectrum in the Asian species.

13.2.1 Scaling Effects in Foragers

The effect of scaling is reduced when mass values are expressed as thorax to whole body mass ratios. The effect of this revaluation is that the original five groups obtained reduce to only two: European *A. mellifera* alone and all the other species together (Table 13.1). The former has the aerodynamically least favourable "engine-to-aircraft" mass ratio of 0.45, while all of the Asian species have a mass ratio close to 0.51 which confers to them a mass-dimensional advantage of about 13%.

In terms of wing surface area, the *Apis* species are divisible into three significantly different groups (Table 13.1). However, there remains a fivefold range in

Table 13.1 Means ± SD of dimensional characteristics of honeybee flight

Species and subspecies	Colony (n)	Whole body mass (mg)	Thoracic mass (mg)	Mass ratio	Wing surface area (mm^2)	Wing-loading (mg/mm^2)	EPI[a]
A. d. binghami	2	42.04 ± 0.05	20.71 ± 0.20	0.49 ± 0.01	116.63 ± 3.12	0.36 ± 0.01	0.82 ± 0.02
A. dorsata	6	33.67 ± 6.53	17.92 ± 4.39	0.53 ± 0.03	104.08 ± 17.99	0.32 ± 0.01	0.93 ± 0.03
A. koschevnikovi	3	13.91 ± 2.55	6.83 ± 0.92	0.51 ± 0.02	45.04 ± 1.06	0.31 ± 0.05	0.93 ± 0.08
A. nigrocincta	4	13.18 ± 1.67	7.13 ± 1.34	0.52 ± 0.02	37.66 ± 1.33	0.35 ± 0.04	0.88 ± 0.05
A. cerana	10	12.21 ± 1.15	6.52 ± 0.66	0.53 ± 0.03	38.53 ± 1.19	0.32 ± 0.03	0.95 ± 0.07
A. florea	17	6.08 ± 1.33	3.16 ± 0.69	0.52 ± 0.02	26.32 ± 2.13	0.23 ± 0.04	1.09 ± 0.10
A. andreniformis	7	4.99 ± 0.79	2.47 ± 0.42	0.49 ± 0.03	24.43 ± 1.06	0.20 ± 0.03	1.10 ± 0.08
A. mellifera:							
Africa	94	18.18 ± 2.67	9.62 ± 0.85	0.54 ± 0.02	49.50 ± 2.12	0.37 ± 0.05	0.90 ± 0.06
Europe	27	26.74 ± 3.24	12.31 ± 1.45	0.45 ± 0.02	57.83 ± 2.55	0.46 ± 0.06	0.64 ± 0.07

Data from Hepburn et al. (1999), Radloff et al. (2001)
[a]Excess power index

wing surface areas (which exhibit a scaling relationship differing from mass) which has important consequences for wing-loading. Wing-loading is the measure of unit mass of the total aircraft to unit area of wing surface. This value is a major element that ultimately determines the fuel requirements of staying aloft on the wing. So, the smaller the wing-loading value, the more aerodynamically favourable a position a worker bee of one species will have against others. For wing-loading, four significantly different classes emerge and the dwarf bees, *A. andreniformis* and *A. florea*, enjoy a 35% lift surface advantage over the other species (Table 13.1).

The EPI is a complex number that integrates all other aerodynamic values and ratios. It must be stressed that, in the discussion that follows, only dimensional aspects are considered, while in reality ultimate flight capacity must include engine combustion efficiency and horsepower (discussed below). It is the EPI that removes the sevenfold variation in mass values and fivefold variation in wing surface area and allows absolute comparisons of these different honeybee species.

On the basis of the EPI, all of the species can be classified into three significantly different groups (Table 13.1). The dwarf bees, *A. andreniformis* and *A. florea*, form a group with a very high EPI of about 1.1. The second and largest group includes the medium-sized and giant species as well as African *A. mellifera* with an average EPI of about 0.9, which is about 18% less than that of the dwarf species. Finally, European *A. mellifera* weigh in with an average EPI of 0.64, which is about 40% worse than the dwarf honeybees and 30% less than the medium-sized Asian group.

These sorts of dimensional comparisons would completely account for the flying capacity of honeybees with two provisos: (1) that all honeybee flight muscles were identical in metabolic activity per unit mass of muscle, and (2) if temperature and altitude, which affect air density, were constants (which, of course, they are not). However, compensatory adjustments can be made for metabolic capacity. The relationship between the EPI and mass-specific metabolic capacity of thoracic flight muscle, adjusted for muscle mass per bee and redefined as "engine capacity" (horsepower in watts), yields a rather striking linearity (Fig. 13.1).

13.2.2 Scaling Effects in Drones

Dimensional analyses of the flight capacity of drones were performed, and the means for total body mass, thorax mass, thorax to whole body mass ratio, wing surface area and wing-loading were measured and the EPI calculated (Radloff et al. 2003). Tests for the comparisons of means showed significant differences among the species. There is a nearly fourfold difference in total body mass between the smallest drones of the dwarf honeybees, *A. andreniformis* and *A. florea*, and the giant bee *A. dorsata* – differences far less than those seen in worker bees. Nonetheless, the drones can be separated into four reasonably distinct total body mass groups. While thoracic mass varied just over threefold (as compared to sevenfold in workers), the drones again fall into four groups.

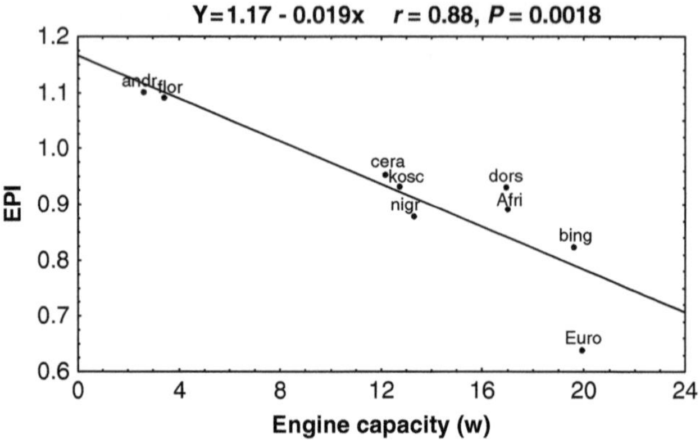

Fig. 13.1 The relationship between the excess power index (EPI) and engine capacity (Watts) for *A. andreniformis*, *A. florea*, *A. cerana*, *A. koschevnikovi*, *A. nigrocincta*, *A. dorsata*, African *A. mellifera*, *A. dorsata binghami* and European *A. mellifera*

The effects of these whole body and thoracic mass differences are quite evident in the "engine-to-aircraft" mass ratios. The dwarf drones of *A. andreniformis* have the greatest engine to body mass, followed by *A. dorsata*, then *A. nigrocincta* and *A. koschevnikovi* and finally *A. cerana* and *A. florea*. However, the thorax to whole body mass ratios vary by about 15% among the species, and the deviation for a mean ratio for all drones is only about 2%, except for *A. andreniformis* and *A. koschevnikovi* (8%).

By contrast, wing surface area varies by over 50% and for this variable the drones form three significantly different classes: *A. dorsata* alone, *A. koschevnikovi* alone and followed by the other species in a single group. Thus, changes in wing surface area among drones of different sizes scale vary differently from mass, and the scaling differences in wing-loading values are evident in wing-loading values. Because wing-loading is a ratio, the magnitude of the differences is reduced such that only two groups emerge: *A. andreniformis* and *A. florea* have significantly lower wing-loading values than the other species. In terms of generating aerodynamic lift, this confers a major advantage of power availability to the dwarf drones. In the case of drones, only two groups emerge: the dwarf honeybees as one group and the other species as a second group. On dimensional grounds alone the dwarf drones enjoy a 20% advantage over the other drones in terms of available excess power.

Paired comparisons of the flight dimensions between conspecific workers and drones show a highly significant sexual dimorphism (Radloff et al. 2001, 2003). Analysing the species one by one, with one single exception, all conspecific pair comparisons of workers and drones significantly differ with respect to the EPI. Mean comparisons of the other flight dimensional characteristics for worker and drones are given in Table 13.2. On average, the drones have an energetically more favourable wing-loading ratio than their corresponding worker bees. Coupled to the mass ratios, this also means that drones have significantly elevated values for the

Table 13.2 Mean comparisons of flight dimensional characteristics between conspecific worker and drone honeybees

Species and subspecies	Whole body mass	Thoracic mass	Mass ratio	Wing surface area	Wing-loading	EPI[a]
A. d. binghami	ns	ns	*	*	*	**
A. dorsata	ns	ns	ns	*	*	**
A. koschevnikovi	**	**	ns	**	ns	ns
A. nigrocincta	**	**	**	**	**	**
A. cerana	**	**	**	**	**	**
A. florea	**	**	**	**	*	**
A. andreniformis	**	**	**	**	*	**

Data from Radloff et al. (2001, 2003)
[a]Excess power index
**$P < 0.01$
*$P < 0.05$
ns not significant

EPI as compared to workers. So although drones of all the Asian species are bigger and heavier than their conspecific workers, they also have a correspondingly and relatively larger wing surface area, hence lower wing-loading and higher excess power indices than worker bees.

13.2.3 Aerodynamic Efficiency

In addition to variable body proportions, intra- and interspecific differences in mass-specific flight metabolism may also affect aerodynamic efficiency (Coelho and Mitton 1988; Harrison et al. 1996). For example, ecotypical adaptations include increasing body size with increasing altitude (Hepburn et al. 1998a, b, 2000), which is well documented for *A. florea* (Tahmasebi et al. 2002; Hepburn et al. 2005) as well as *A. cerana* (Kapil 1956; Diniz-Filho et al. 1993; Hepburn et al. 2000; Radloff et al. 2010), and structural variations in the ratio of indirect and direct flight muscles (Dulta and Verma 1987). The EPI tacitly assumes that all flight muscles are more or less metabolically the same in the production of power per unit mass of muscle and that all bees fly at the same air densities with a constant ambient temperature at the same fixed altitude and barometric pressure. However, this is simply not the case. In African *A. mellifera*, wing surface area increases at a greater rate than whole body mass with increasing altitude (Hepburn et al. 1998a, b), which reduces flight costs. That honeybees can also alter wing-beat frequency and angle of wing attack also affects both flight efficiency and costs (Rothe and Nachtigall 1989).

Detailed information on how flight characteristics can change in these ways is only available for *A. mellifera*, therefore for purposes of discussion these variables are simply treated as constants. This is a reasonable approximation on a priori grounds, because flight efficiency is relatively insensitive to alterations in the aerodynamic parameters at the low Reynolds numbers characteristic of honeybee flight (Hepburn et al. 1998a, b). Despite some flexibility in wing-beat frequency and angle of attack, honeybees are constrained in the ways in which aerodynamic efficiency can be altered. The EPI, for example, can be varied either by changing the thoracic muscle to whole body mass ratio or by increasing wing surface area relative to total mass.

For Asian honeybees, the wing-loading values resolve, more or less, into two groups: the two dwarf honeybees as one group and the other species as the other group. This trend is exactly the same with the EPI values, which indicates greatest aerodynamic efficiency in the dwarf bees, *A. andreniformis* and *A. florea*, followed by the other Asian species and lastly European *A. mellifera*. A remaining relationship of considerable importance is that between the EPI and engine capacity (horsepower), as shown in Fig. 13.1. Although the dwarf honeybees have relatively small engines, they have an extremely high relative wing-loading index. Moreover, as engine capacity increases, it does so inversely to wing-loading so that the least economically successful honeybee aircraft is European *A. mellifera*.

The apparent variation in the aerodynamic relationships described above naturally evokes questions as to how they may have arisen. Two possibilities, not mutually exclusive, come to mind: absconding or migration and thermoregulation. All the tropical honeybee species of Asia and Africa are migratory and notorious for absconding (Woyke 1976, 1977; Otis et al. 1981; Crane 1990; Hepburn and Radloff 1998; cf. Chap. 7). A fundamental characteristic of absconding or migrating honeybees is a high fuel-loading before departing from the home nest. Otis et al. (1981) showed that workers in prime swarms of *A. m. scutellata* in the neotropics carry proportionately larger nectar and honey loads than do European *A. mellifera*. Moreover, absconding *A. m. scutellata* swarms may carry even larger loads exceeding their nett empty weights (Otis et al. 1981). By contrast, migration simply does not occur in European *A. mellifera* (Ribbands 1953) and absconding is rare (Simpson 1958, 1959; Martin 1963). Clearly, highly mobile honeybees require an aerodynamic capacity to facilitate routine absconding and migration, and their EPI values more than compensate for this.

The warm to hot climates that make absconding and migration viable strategies for tropical Asian and African honeybees go hand in hand with reduced energy inputs for colony thermoregulation. The situation is quite different in temperate Europe where the climate is characterised by long periods of cold, unfavourable for either foraging or flight. This climate effectively precludes absconding and migration and imposes heavy metabolic demands for honeybees to achieve colonial homeostasis (Seeley 1985; Moritz and Southwick 1992). This may be achieved through increased body mass without a concomitant increase in wing surface area, but effectively increasing wing-loading. A larger body would obviously reduce the area-to-volume ratio, thus reducing the rate of heat loss. In a physioecological, evolutionary sense, this would amount to a trade-off between aerodynamic ability and colonial homeothermy. Traits that could improve the aerodynamic prospects of temperate *A. mellifera*, particularly improving wing-loading, would work against them in terms of heat loss. These interpretations are consistent with similar conclusions reached by Dyer and Seeley (1987) and Underwood (1991) that there are functional links between energetic constraints on individuals that differ from those affecting whole colonies.

13.2.4 The Problem of Mountains

Although honeybees live in mountains, permanent colonies above 3,000 m are rare (Ruttner 1992). At this elevation, air density, oxygen tension and wind all conspire against the high metabolic, temperature-controlled flight engine of honeybees (Withers 1982; Rothe and Nachtigall 1989; Joos et al. 1997). Because honeybees are miniature aircraft operating at low Reynolds numbers, the aerodynamic power requirement increases with altitude because of reduced air density. This problem can in principle be solved in one of two ways: either by increasing power output or reducing the power requirement (von Mises 1959). The first option is precluded by the physiological inability of honeybees to deliver sufficient oxygen to the flight engine at higher altitudes (Withers 1982; Rothe and Nachtigall 1989; Joos et al. 1997). The only alternative, viable hypothesis to explain how honeybees can fly at higher altitudes is a reduction in the aerodynamic power requirement. Hepburn et al. (1998a) made such an analysis of *A. mellifera* from sea level to about 2,300 m. The results of their morphometric measurements showed a group of low and high altitude phenotypes. Multiple comparison tests showed that the two groups differed significantly with respect to wing surface area, thoracic and whole body mass, wing-loading thorax to whole body mass ratio, as well as their EPI values.

For a given altitude, the aerodynamic power requirement can be modified by varying aerodynamic efficiency or their EPI values. The former would entail changes in wing-beat frequency and/or wing to body geometry, but there is little scope here because of the relative insensitivity of the aerodynamic parameters to these effects at the low Reynolds numbers characteristic of honeybee flight. However, the EPI can be varied over a wide range by changing the proportion of thoracic to total body mass or by increasing wing area relative to total mass. In the study of Hepburn et al. (1998b), it was found that engine mass increased only 6% from low to high altitude, while wing surface increased by 11% in absolute terms. The larger wing area lowers the aerodynamic power requirement relative to power actually available to the bees and this provides the solution for high altitude flight in honeybees.

13.2.5 Fuelling Flight

While measurements of the blood sugars that actually fuel honeybee flight in *A. mellifera* were reported decades ago (Beutler 1936a, b, 1937) and increasingly sophisticated studies have appeared more recently (Harrison and Fewell 2002; Roberts and Elekonich 2005), there have been extremely few such studies on

Asian honeybee species. Although Ray et al. (1968) established the presence of a key enzyme for muscle metabolism in *A. cerana*, the first attempt to relate glycogen content in workers of *A. cerana* was that of Bisht (1983) who showed that glycogen content is proportional to body weight, and that for every mg per body weight there is 32.6 μg glycogen, that during flight, bees consume 0.26 mg per min and for every 1 mg lost there is a reduction of 34.6 μg glycogen.

Authors of the same period were trying to approximate the energetics of foraging, as exemplified by Dulta and Verma (1987) who noted that *A. mellifera* can forage about twice the distance of *A. cerana* – differences which also correspond to body size. This raised the possibility of a relationship between flight distance and size of the thoracic musculature. Dulta and Verma (1987) then gave values for four species as being 8,500 m for *A. dorsata*, 5,100 m for *A. mellifera*, 3,200 m for *A. cerana* (Shah and Shah 1980 give 3,750 m) and 1,200 m for *A. florea*. Dulta and Verma (1987) then measured the dimensions of the principal indirect flight muscles in histological preparations of *A. cerana*, *A. dorsata*, *A. florea* and *A. mellifera* and found that they differed significantly in decreasing size as follows: *A. dorsata* > *A. mellifera* > *A. cerana* > *A. florea*. On the strength of these results, they concluded that flight range was greatest in *A. dorsata* than *A. mellifera*, followed by *A. cerana* and *A. florea*.

Subsequently, Dulta and Verma (1987) provided a quantitative estimation of fuel sources and relevant enzyme activities for the same four species of honeybees over the active summer and inactive winter seasons. The major fuel source is glycogen. And the relative amounts of glycogen, lipids, glycogen phosphorylase, succinic dehydrogenase and α-glycerophosphate dehydrogenase per mg of thoracic flight muscle differed significantly in summer among the series in descending order as follows: *A. dorsata* > *A. mellifera* > *A. cerana* > *A. florea*, while in winter, values were significantly lower than in summer but in the same descending order. This result could be expected because Bisht (1983) had already shown that glycogen content is proportional to body weight in *A. cerana*.

13.2.6 Thermal Considerations

Dyer and Seeley (1987) were the first to investigate thoracic flight temperature in foragers of *A. cerana*, *A. dorsata* and *A. florea* in comparison to *A. mellifera*. These four species span a fivefold range in body mass from the smallest, *A. florea*, to the largest, *A. dorsata*. They found that the thoracic temperatures related to a wide range of ambient temperatures. However, the temperature gradients between thoracic and the ambient did not regularly increase with increasing size of the species as might have been expected from other studies of insect flight. *A. florea*, the smallest of these bees, showed the smallest difference between thoracic and ambient temperature. The two medium-sized species, *A. cerana* and *A. mellifera*, both showed a greater difference between thoracic and ambient temperature than the even larger *A. dorsata*. Moreover, the rate of passive convective heat loss from the thorax scaled linearly and inversely with body size in all four species, but there was no obvious morphological basis for these results.

However, wing-loading and flight speed were disproportionately high in *A. cerana* and *A. mellifera* relative to *A. dorsata* and *A. florea*, and these correlates of heat production suggest that an increased mass-specific metabolic rate in flight might explain these high differences between thoracic and ambient temperature. These results of Dyer and Seeley (1987) raise some fundamental questions about the physiological mechanisms in the energetics of honeybee flight. The authors conclude with an insightful speculation that there may be functional links between the energetic constraints on individuals and on colonies.

This work was extended by Underwood (1991) who showed that honeybees can regulate thoracic heat production during flight, and it appears that they can vary flight metabolism with respect to expected gains. The temperature gradients of *A. laboriosa* workers in flight were about the same as those of *A. mellifera* workers with a body mass only half that of *A. laboriosa*. Underwood (1991) further confirmed the observations of Dyer and Seeley (1987) that the circulatory system of *A. laboriosa* workers shows no striking differences from *A. mellifera* and *A. dorsata*. Underwood (1991) suggested that the relatively low difference between thoracic and ambient temperature in relation to body size in *A. laboriosa* might indicate a relatively low metabolic rate and consequentially low heat production. He supported this inference by an analysis of mass-specific wing-loading. Finally, he concluded that *A. laboriosa* can be grouped with *A. dorsata* and *A. florea* as a relatively low-powered, open-nesting honeybee, in contrast to the more high-powered cavity-nesting *A. cerana* and *A. mellifera*. Dyer and Seeley (1987) measured the flight speed on round trips of individual bees trained to a feeder and found that the averages were 7.28 m/s for *A. cerana*, 7.17 m/s for *A. dorsata*, 4.81 m/s *A. florea* and 7.84 m/s for *A. mellifera*.

13.3 Cost Analysis of Wax Salvage

13.3.1 Introduction

The secretion of wax and construction of combs represent a large metabolic investment by honeybees, so that desertion or loss of the nest, for whatever reason, constitutes an energetically, hefty expenditure (Hepburn et al. 1984). Nonetheless, nest desertion by absconding or migrating colonies is a common feature of tropical honeybees (Hepburn and Radloff 1998; Oldroyd and Wongsiri 2006). Despite the possible cost effectiveness of cannibalising wax from a deserted nest and reusing it in the construction of a new one, this behaviour has been reported thus far only for the red dwarf honeybee, *A. florea* (Akratanakul 1977; Dutton and Free 1979;

Booncham 1995; Wongsiri et al. 1997; Hepburn et al. 2009); but we suspect that it may eventually be observed in the little studied *A. andreniformis* as well.

However, there is a peculiar element to this wax salvage behaviour. Colonies which only abscond a relatively short distance, perhaps less than 200 m, usually return to salvage old nest wax (Booncham 1995), but those which go considerably further, as in seasonal migrations, do not (Seeley et al. 1982). Thus, just as the timing of comb construction in temperate honeybees requires the right balance between the energy costs of construction and the opportunity costs of missed nectar flows (Pratt 2004), so the dwarf honeybees are faced with decisions on the expenditure of energy. Wax salvage would clearly be counter-productive unless the energy input to yield threshold was a profitable one. Pirk et al. (2011) measured the energetic efficacy of wax recovery and provided a first approximation for calculating the energetics of wax salvage behaviour.

13.3.2 Field Observations

Pirk et al. (2011) continuously observed six colonies of *A. florea* until they absconded, following which the distances flown to the new nest sites were measured. When a colony absconded, the comb and supporting twig were weighed and re-hung (Fig. 13.2). When returning, wax-scavenging bees arrived at the comb crown, they were marked with coloured craft paints so that individual bees who were returning

Fig. 13.2 The single comb nest of *Apis florea* consists of a bulbous waxen "crown" replete with honey surrounding a twig, below which is the fan-shaped brood comb. The very white wax surrounding the fan is newly secreted wax; the greater, *yellowish-grey area* consists of capped developing brood, while the empty *brownish* group of cells at centre is brood cells from which an earlier generation has already emerged but in which the queen has not yet laid a new series of eggs. To induce absconding, the comb below the crown is cut away. The *three odd rectangles* at centre are comb inserts from another experiment

Fig. 13.3 The isolated "crown" comb above the twig as it appears immediately after a colony of *Apis florea* has absconded

Fig. 13.4 The isolated "crown" comb above the twig as it appears 1 week after a colony of *Apis florea* has absconded and during which wax-salvaging foragers have cannibalised the wax

more than once for wax collection could be identified. The comb crown was weighed after 24 h (Fig. 13.3) and then 1 week later to determine the amount of wax collected following absconding (Fig. 13.4). Finally, individual foragers laden with salvaged wax were caught and the total weight of wax per bee was measured. After 1 week, the comb crown wax remaining on the twig was removed and the twig weighed to determine the original comb crown weight. The average crop loads and wax loads of workers were measured to obtain relevant data for the energetic calculations, as well as the required handling time to salvage a load of wax.

13.3.3 Energy Considerations

Energy values of samples of beeswax from *A. florea* were subjected to calorimetric analysis and expressed in mega Joules per kg (=J/mg). The energy value for honey is on average 1,272 kJ per 100 g (National Honey Board USA 2010) and bees need about 6.25 kg of honey to produce 1 kg of wax (Weiss 1965). The amount of floral nectar required to produce 6.25 units of honey was calculated based on an average sugar content of nectar (25%) in relation to honey (80%).

The following data were taken as constants for *A. florea*: flight speed (4.81 m/s) and mass-specific metabolic rate (400 W/kg thorax) (Dyer and Seeley 1987). The cost of a single salvaging trip to obtain wax (mJ) and the energetic costs involved in securing that amount of floral nectar necessary to produce the same amount of wax were calculated. Furthermore, the distance where the cost of salvaging wax is equal to the cost of substituting floral nectar to obtain the energetic equivalent of salvaged wax was also calculated. The results on wax salvage by six colonies of *A. florea* were rather variable and colony size varied from about 3,715 bees to 25,200 bees; there was no correlation between colony size and weight of the original comb. The observed absconding distances of the colonies were 51.50 ± 27.86 m to a new nesting site with a range of 15–89 m. Turning to wax salvage of a pre-absconding average net comb crown weight of 79.06 ± 96.11 mg, slightly over half of this wax, 42.18 ± 66.54 mg, was salvaged within 24 h and, by the end of 1st week, only an additional 4.11 mg had been removed. Thus, nearly 60% of the wax available was quickly removed, staving off competition from any other scavenging colonies.

There was no correlation between colony size and the amount of wax taken. In fact, the smallest colony of 3,715 bees salvaged 76.6 mg wax, which is significantly greater than the 7.04 mg salvaged by the largest one of 25,200 bees. The salvaged loads of wax in both corbiculae of five individual foragers ranged from 0.7 to 1.2 mg and averaged 1.12 mg. There was a highly significant correlation between the amount of wax present in the crown at the time of absconding and the amount subsequently salvaged: the greater the amount of crown wax available, the more taken. Furthermore, the workers stayed on average 6.3 ± 2.1 min to salvage one load of wax. The crop content of nectar foragers ranged from 7 to 13 μl with an average of 8.18 ± 2.0 μl.

13.3.3.1 Energetics of Salvaging Wax

The costs involved in salvaging wax by a worker and substituting the salvaged wax with nectar were determined using the following parameter values:

Mass-specific metabolic rate of flying bee (rf) = 0.4 mW/mg (Dyer and Seeley 1987); mass-specific metabolic rate of non-flying bee (rs) = 0.05 mW/mg (Schmid-Hempel et al. 1985); and flight speed (v) = 4.81 m/s (Dyer and Seeley 1987). Parameters measured and calculated by Pirk et al. (2011): mass of bee (mb) = 26.165 mg; mass of wax load (mw) = 1.12 mg; mass of nectar (25%) load (mn) = 9.91 mg; absconding distance (d) = 51.5 m; handling time for wax salvage (tw) = 378 s (6.3 min); handling time for nectar collection (tn) = 159 s; and number of trips to collect nectar for one load of wax (T) = 2.26.

Cost for Salvaging Wax

The energy involved in salvaging wax on a single return trip is given by:

$$E_{\text{wax}} = (E_{\text{wax\_OUT}} + E_{\text{wax\_HANDLE}} + E_{\text{wax\_IN}}), \quad (13.1)$$

where $E_{\text{wax\_OUT}}$ denotes the energy spent on the outbound flight, $E_{\text{wax\_HANDLE}}$ the energy used during handling the wax (getting it into the corbiculae) and $E_{\text{wax\_IN}}$ the energy spent on the return flight with the wax load.

The energy spent by a worker on the outbound flight is given by:

$$E_{\text{wax\_OUT}} = \text{rf} \times \text{mb} \times \frac{d}{v}, \quad (13.1a)$$

giving $E_{\text{wax\_OUT}} = 112.1$ mJ.

The energy used by the worker handling the wax is given by:

$$E_{\text{wax\_HANDLE}} = \text{rs} \times \text{mb} \times \text{tw}, \quad (13.1b)$$

giving $E_{\text{wax\_HANDLE}} = 494.5$ mJ.

The energy spent on the return flight with the wax load is given by:

$$E_{\text{wax\_IN}} = \text{rf} \times \text{mb} \times (1 + 0.01\ \text{mw}) \times \frac{d}{v} \quad (13.1c)$$

where the mass metabolic rate is adjusted by 1% for each mg of additional weight (Schmid-Hempel et al. 1985), giving $E_{\text{wax\_IN}} = 113.3$ mJ.

Therefore, $E_{\text{wax}} = 112.1\ \text{mJ} + 494.5\ \text{mJ} + 113.3\ \text{mJ} = 719.9$ mJ is the cost involved in salvaging wax on a single return trip.

Cost for Collecting the Nectar to Replace the Wax

Because about 6.25 kg are needed to produce 1 kg of wax (Weiss 1965), to replace 1.12 mg of wax, 7 mg of honey must be invested. Setting the sugar concentration of nectar at 25%, and because honey has a sugar concentration of about 80%, 22.4 mg (=7 mg × 80/25) of nectar must be collected to produce 1.12 mg of wax. (For the sake of simplicity, we do not take into account the energy required to ripen nectar into honey.) With an average crop load of 8.81 μl, and the specific weight of a 25% nectar solution being 1.1125, workers carry 9.91 mg of nectar. From this it follows that the workers need to make 2.26 foraging trips to garner 22.4 mg of floral nectar. The collection of the nectar is not instantaneous and the availability of floral nectar clearly affects the time required to collect 9.91 mg. Based on the nectar production measured by Farkas (2008) with an average production of 0.375 mg/h, the required handling time is 159 s to fill the crop.

The energy involved in collecting the nectar on a single return trip is given by:

$$E_{\text{nectar}} = T(E_{\text{nec\_OUT}} + E_{\text{nec\_HANDLE}} + E_{\text{nec\_IN}}), \tag{13.2}$$

where $E_{\text{nec\_OUT}}$ is the energy used on the outbound flight, $E_{\text{nec\_HANDLE}}$ the energy used during handling the nectar and $E_{\text{nec\_IN}}$ the energy invested in the return flight with the nectar load.

The energy spent by a worker on the outbound flight is given by:

$$E_{\text{nec\_OUT}} = \text{rf} \times \text{mb} \times \frac{D}{v}, \tag{13.2a}$$

where D is the average foraging distance. Assuming the same distance as in (13.1a), $D = 51.5$ m, giving $E_{\text{nec\_OUT}} = 112.1$ mJ.

The energy used by the worker during handling the nectar is given by:

$$E_{\text{nec\_HANDLE}} = \text{rs} \times \text{mb} \times \text{tn}, \tag{13.2b}$$

giving $E_{\text{nec\_HANDLE}} = 208.0$ mJ.

The energy spent on the return flight with the nectar load is given by:

$$E_{\text{nec\_IN}} = \text{rf} \times \text{mb} \times (1 + 0.01\,\text{mn}) \times \frac{D}{v}, \tag{13.2c}$$

where the mass metabolic rate is adjusted for each mg of additional weight of nectar by 1% (Schmid-Hempel et al. 1985), giving $E_{\text{nec\_IN}} = 123.2$ mJ.

Therefore, $E_{\text{nectar}} = 2.26$ (112.1 mJ + 208.0 mJ + 123.2 mJ) = 1001.85 mJ is the cost involved in collecting the nectar in one return trip.

Hence, the *net energy gain* = $E_{\text{nectar}} - E_{\text{wax}} = 1001.85$ mJ − 719.89 mJ = 281.96 mJ.

The energy gain for salvaging the wax instead of replacing it with nectar would be 282 mJ, under the condition that average absconding distance and average foraging distance are both 51.5 m.

Setting (13.1) equal to (13.2) and solving for d, the absconding distance, we obtain the maximum distance at which energetically salvaging and collecting nectar are of equivalent cost:

$$d = \frac{(E_{\text{nec}} - E_{\text{wax\_HANDLE}})}{\frac{\text{rf} \times \text{mb}}{v} \times (2 + 0.01 \times \text{mw})} = 115.9\,\text{m}.$$

At this absconding distance, the energy spent in salvaging wax is equal to the energy spent substituting the wax with nectar under the assumption that the average foraging distance is 51.5 m. Using the median foraging distance estimated by Dyer and Seeley (1987) of 268 m, salvaging wax would be energetically beneficial up to a distance of 607 m (Fig. 13.5); however, data from Punchihewa et al. (1985) are

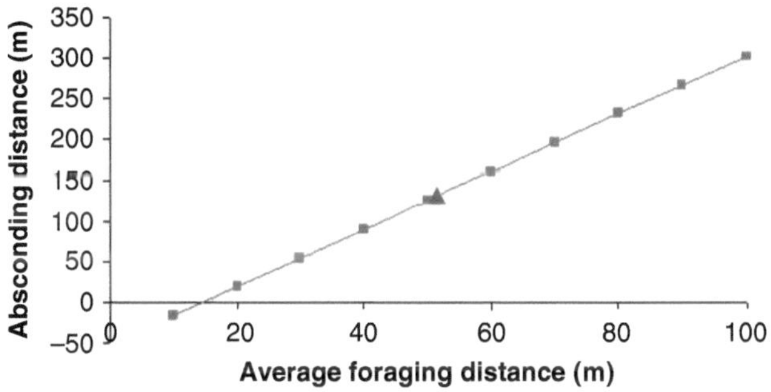

Fig. 13.5 Relationship between average foraging distance and absconding distance – the *triangle* indicates the break-even point. Below the line it pays off to salvage wax, above it pays off to substitute the wax left behind with freshly collected nectar

similar to our observations that the average foraging distance is below 100 m (see Fig. 3 of Punchihewa et al. 1985).

13.3.3.2 Energy Balance

The calculations clearly show that there is a finely tuned energetic trade-off between wax salvaging as compared to fresh wax secretion fuelled with nectar. The observation that *A. florea* colonies of this particular population are difficult to train to a feeder further than 250 m away (personal observation) suggests that the actual average foraging distance might be below 100 m, similar to the ones reported by Punchihewa et al. (1985). Taking this into account, the energetic trade-off explains why *A. florea* colonies do not return to salvage wax if they abscond further than 100 m (Seeley et al. 1982) or 200 m (Booncham 1995). The average foraging distance reported by Dyer and Seeley (1987) at which absconding would be feasible, perhaps up to 600 m, might not contradict the energetic trade-off, but rather reflect the incapacity of researchers to follow an absconding swarm for more than 200 m in the dense forests of South-East Asia (personal observation).

Despite the lack of robust data on average foraging distances, the energetic trade-off between salvaging wax and that energy which must be invested to produce the wax from nectar nicely explains why *A. florea* seems to be the only honeybee species that shows wax-salvaging behaviour. This is because it is only energetically beneficial up to 115 m, assuming an average foraging distance of 51.5 m. The six absconding colonies flew an average distance of 51.50 ± 27.86 m to a new nesting site and, within this limit, there were no correlations between distance moved and the amount of wax salvaged. The distance between the new and old nest site is an important parameter because colonies which abscond further than 100 m away from the natal nest do not return to salvage wax (Seeley et al. 1982). Indeed, the energetic trade-off scenario described here explains that observation, because it is in that case not energetically feasible to do so (Pirk et al. 2011). The proposed trade-off between energy recovery and distance gives a parsimonious explanation why such behaviour is only observed in *A. florea*, despite the fact that worker bees of all honeybee species can be observed salvaging odd bits of abandoned propolis, wax and honey stores.

Indeed, Pirk et al. (2011) also did the calculations for *A. mellifera*, based on data available for this species, with conservative estimates (e.g. for the wax load) and an average foraging distance of 1,000 m (Seeley 1985). The salvaging of wax, according to their calculations, would only be energetically beneficial up to an absconding distance of around 1,100 m, which is well below the observed minimum absconding range of 6,000 m (Schneider 1990). Published data on foraging distances show that *A. florea* workers can travel about 800 m to forage at artificial feeders, and this suggests that absconding is less related to tapping new resources than to microclimatic parameters related to the nest itself. The energetic trade-off explains the underlying reasons why *A. florea* only salvages wax from the old nest if the new nesting site is less than 100 m away: energetically, it pays to recycle.

References

Akratanakul P (1977) The natural history of the dwarf honey bee, *Apis florea* F. in Thailand. Thesis, Cornell University, Ithaca

Beutler R (1936a) Uber den Blutzucker der Bienen. Verh Dtsch Zool Ges Leipzig 38:140–146

Beutler R (1936b) Uber den Blutzucker der Biene (*Apis mellifica*). Naturwissenschaften 31:486–491

Beutler R (1937) Uber den Blutzucker der Bienen. Zeit Vergl Physiol 24:71–115

Bisht DS (1983) Flight energetics of Indian honey bee, *Apis cerana indica*. J Entomol 45:1–5

Booncham U (1995) Niche differences among four sympatric species of honey bees in dry-evergreen forest. Thesis, University of Chulalongkorn, Bangkok [in Thai]

Coelho JR, Mitton JB (1988) Oxygen consumption during hovering is associated with genetic variation of enzymes in honey-bees. Funct Ecol 2:141–146

Crane E (1990) Bees and beekeeping: science, practice and world resources. Heinemann Newnes, Oxford

Diniz-Filho JAF, Malapsina O, Pignata MIB (1993) Geographic variation in *Apis cerana indica* F: a spatial autocorrelation analysis of morphometric patterns. J Apic Res 32:65–72

Dulta PC, Verma LR (1987) Comparative biometric studies on flight muscles of honeybees in the genus *Apis*. J Apic Res 26:205–209

Dutton RW, Free JB (1979) The present status of beekeeping in Oman. Bee World 60:176–185

Dyer FC, Seeley TD (1987) Interspecific comparisons of endothermy in honeybees (*Apis*): deviations from the expected size-related patterns. J Exp Biol 127:1–26

Dyer FC, Seeley TD (1991a) Dance dialects and foraging range in three Asian honey bee species. Behav Ecol Sociobiol 28:227–233

Dyer FC, Seeley TD (1991b) Nesting behavior and the evolution of worker tempo in four honey bee species. Ecology 72:156–170

Farkas A (2008) Nectar production and nectar sugar composition of three oilseed rape (*Brassica napus*) cultivars in Hungary. Acta Hortic 767:275–284

Harrison JF, Fewell JH (2002) Environmental and genetic influences on flight metabolic rate in the honey bee, *Apis mellifera*. Comp Biochem Physiol A 133:323–333

Harrison JF, Hall HG (1993) African-European honeybee hybrids have low nonintermediate metabolic capacities. Nature 363:258–260
Harrison JF, Nielsen DI, Page RE (1996) Malate dehydrogenase phenotype, temperature and colony effects on flight metabolic rate in the honey-bee, *Apis mellifera*. Func Ecol 10:81–88
Hepburn HR, Radloff SE (1998) Honeybees of Africa. Springer, Berlin
Hepburn HR, Hugo JJ, Mitchell D, Nijland MJM, Scrimgeour AG (1984) On the energetic costs of wax production by the African honeybee, *Apis mellifera adansonii*. S Afr J Sci 80:363–368
Hepburn HR, Radloff SE, Steele GR, Brown R (1998a) Dimensional aspects of flight efficiency in honeybee populations of Africa. J Apic Res 37:147–154
Hepburn HR, Youthed C, Illgner P, Radloff SE, Brown RE (1998b) Production of aerodynamic power in mountain honeybees (*Apis mellifera*). Naturwissenschaften 85:389–390
Hepburn HR, Radloff SE, Fuchs S (1999) Flight machinery dimensions of honeybees, *Apis mellifera*. J Comp Physiol B 169:107–112
Hepburn HR, Radloff SE, Oghiakhe S (2000) Mountain honeybees of Africa. Apidologie 31:205–222
Hepburn HR, Radloff SE, Otis GW, Fuchs S, Verma LR, Ken T, Chaiyawong T, Tahmasebi T, Ebadi R, Wongsiri S (2005) *Apis florea*: morphometrics, classification and biogeography. Apidologie 36:359–376
Hepburn R, Duangphakdee O, Phiancharoen M, Radloff SE (2009) Comb wax salvage by the red dwarf honeybee, *Apis florea* F. J Insect Behav 23:159–164
Joos B, Lighton JRB, Harrison JF, Suarez RK, Roberts SP (1997) Effects of ambient oxygen tension on flight performance, metabolism, and water loss of the honeybee. Physiol Zool 70:167–174
Josephson RK (2006) Comparative physiology of insect flight muscle. In: Josephson RK (ed) Nature's versatile engine: insect flight muscle inside and out. Springer, Berlin
Kapil RP (1956) Variations in the biometrical characters of the Indian honeybee (*Apis indica* F). Indian J Entomol 18:440–457
Koeniger G (1990) Multiple mating, mating sign and mate selectivity in honey bees. Adv Invert Reprod 5:483–487
Martin P (1963) The mechanism of colony division in honeybee swarming and a study of the problem of absconding. Insectes Soc 10:13–42
Moritz RFA, Southwick EE (1992) Bees as superorganisms: an evolutionary reality. Springer, Berlin
Nachtigall W, Hanauer-Thieser MM (1994) Flight of the honey bee VII: metabolic power versus flight speed relation. J Comp Physiol B 165:484–489
National Honey Board (2010) USA. http://www.honey.com/honeyindustry/, Cited 22 June 2010
Oldroyd BP, Wongsiri S (2006) Asian honey bees. Harvard University Press, Cambridge
Otis GW, Winston ML, Taylor OR (1981) Engorgement and dispersal of Africanized honeybee swarms. J Apic Res 20:3–12
Pirk CWW, Crous KL, Duangphakdee O, Hepburn HR, Radloff SE (2011) Economics of comb wax salvage by the red dwarf honeybee, *Apis florea*. J Comp Physiol (in press)
Pratt SC (2004) Collective control of the timing and type of comb construction by honey bees (*Apis mellifera*). Apidologie 35:193–205
Punchihewa RWK, Koeniger N, Kevan PG, Gadawski RM (1985) Observations on the dance communication and natural foraging ranges of *Apis cerana*, *Apis dorsata* and *Apis florea* in Sri Lanka. J Apic Res 24:168–175
Radloff SE, Hepburn HR, Otis GW (2001) Flight machinery of honey bees (Hymenoptera: Apidae, *Apis*). In: Proceedings of the 7th international conference on tropical bees, Chiang Mai, pp 163–168
Radloff SE, Hepburn HR, Koeniger G (2003) Comparison of flight design of Asian honeybee drones. Apidologie 34:353–358
Radloff SE, Hepburn C, Hepburn HR, Fuchs S, Hadisoesilo S, Tan K, Engel MS, Kuznetsov V (2010) Population structure and classification of *Apis cerana* Fabricius, 1793. Apidologie 41:589–601

Ray S, Pant NC, Mehrotra KN (1968) Phosphohexose isomerase (D-glucose-6-phosphate ketol-isomerase) from the Indian honeybee, *Apis indica*. Indian J Entomol 30:37–44

Ribbands CR (1953) The behaviour and social life of honeybees. Bee Research Association, London

Roberts SP, Elekonich MM (2005) Muscle biochemistry and the ontogeny of flight capacity during behavioral development in the honey bee, *Apis mellifera*. J Exp Biol 208:4193–4198

Rothe U, Nachtigall W (1989) Flight of the honey bee. J Comp Physiol 158:719–727

Ruttner F (1992) Natural history of honey bees. Ehrenwirth Verlag, Munich, in German

Schmid-Hempel P, Kacelnik A, Houston AI (1985) Honeybees maximize efficiency by not filling their crop. Behav Ecol Sociobiol 17:61–66

Schneider SS (1990) Nest characteristics and recruitment behaviour of absconding colonies of the African honey bee, *Apis mellifera scutellata* (Hymenoptera: Apidae). J Kans Entomol Soc 3:225–240

Seeley TD (1985) Honeybee ecology: a study of adaptation in social life. Princeton University Press, Princeton

Seeley TD (1995) The wisdom of the hive: the social physiology of honey bee colonies. Harvard University Press, Cambridge

Seeley TD, Seeley RH, Akratanakul P (1982) Colony defense strategies of honeybees in Thailand. Ecol Monogr 52:43–63

Shah FA, Shah TA (1980) Flight range of *Apis cerana* from Kashmir. Indian Bee J 42:48

Simpson J (1958) The factors which cause colonies of *Apis mellifera* to swarm. Insectes Soc 5:77–95

Simpson J (1959) Variation in the incidence of swarming among colonies of *Apis mellifera* throughout the summer. Insectes Soc 6:85–99

Tahmasebi G, Ebadi R, Tajbadi N, Akhondi M, Faraji S (2002) The effects of geographical and climatic conditions on the morphological variation and separation of the Iranian small honey-bee (*Apis florea* F) populations. J Sci Tech Agric Nat Res 6:169–176, in Persian

Underwood BA (1991) Thermoregulation and energetic decision-making by the honeybees *Apis cerana*, *Apis dorsata* and *Apis laboriosa*. J Exp Biol 157:19–34

von Mises R (1959) Theory of flight. Dover, New York

Weiss K (1965) Über den Zuckerverbrauch und die Beanspruchung der Bienen bei der Wachserzeugung. Zeits Bienenforschung 8:106–124

Withers PC (1982) The effects of ambient air pressure on oxygen consumption of resting and hovering honeybees. J Comp Physiol B 141:433–437

Wongsiri S, Lekprayoon C, Thapa R, Thirakhupt K, Rinderer TE, Sylvester HA, Oldroyd BP, Booncham U (1997) Comparative biology of *Apis andreniformis* and *Apis florea* in Thailand. Bee World 78:23–35

Woyke J (1976) Brood-rearing efficiency and absconding in Indian honeybees. J Apic Res 15:133–143

Woyke J (1977) Brood rearing and absconding of tropical honeybees. In: Fletcher DJC (ed) African bees: taxonomy, biology and economic use. Apimondia, Pretoria, pp 96–102

Chapter 14
The Dance Language

Orawan Duangphakdee, H.R. Hepburn, and Jürgen Tautz

14.1 Introduction

Karl von Frisch presented the first results of his investigations into the honeybee dance language to the scientific public in 1923 and, nearly half a century later, summarised his investigations in the classic "The orientation and the dance language of honeybees" published in 1967. This work gives a clear picture of the honeybee dances as elements of a symbolic language for Western honeybees, *Apis mellifera*. The fact that an animal as simple as a honeybee can possess a symbolic language and transmit information concerning distance and direction aroused the interests of scientists to study the honeybee dance language. Such studies resulted in the honeybee exhibiting the most sophisticated form of animal communication excepting primates.

The dances of honeybees have been studied intensively since the works of von Frisch, but most studies were focused almost exclusively on *A. mellifera*. Even though more than 1,500 papers have been published on this topic since its discovery, many questions remain unanswered and enigmatic as well. In parallel with these works, we must ask: What has been learned about the characteristics of the dance languages of the Asian species of honeybees? The first significant study of the dance language of Asian honeybees was that of von Frisch (1923) and Lindauer (1956a, b), the latter a student of von Frisch.

O. Duangphakdee
Ratchaburi Campus, King Mongkut's University of Technology Thonburi, Bangkok 10140, Thailand
e-mail: Orawan.dua@kmutt.ac.th

H.R. Hepburn
Department of Zoology and Entomology, Rhodes University, Grahamstown 6140, South Africa
e-mail: r.hepburn@ru.ac.za

J. Tautz
BEE Group, Biozentrum, Universität Am Hubland, Würzburg 97074, Germany
e-mail: Tautz@biozentrum.uni-wuerzburg.de

H.R. Hepburn and S.E. Radloff (eds.), *Honeybees of Asia*,
DOI 10.1007/978-3-642-16422-4_14,

In evolutionary terms, the dance language has been an important adaptation to reduce group costs in searching for essential resources. This is particularly so for Asian honeybees because of the high diversity of suitable nest sites and patchily distributed food sources. The accuracy of the dance language allows the colony to track profitable resources. However, not all dance followers exposed to the information contained in a dance necessarily navigate to the dance-indicated target location. And, for that matter, not all dance followers use the transmitted information that they have apparently received. The dance process permits nestmates to decide whether the information presented to them represents a source that is a truly profitable trade-off of time and energy for floral rewards.

An a priori problem arises about this transmitted information: its accuracy. It is widely believed that the honeybee dance language communication system does not specify a food location with pinpoint accuracy. Indeed, Riley et al. (2005) have performed a very telling experiment to reveal that some recruits leave the hive with prior knowledge of the direction and distance of a food source that they have never previously visited. Even if the information encoded in the dance is highly accurate, most recruits would not reach the intended food sources without the use of odour and visual cues in the final stages of their flights to such floral rewards. The dance language of honeybees, by which foragers inform nestmates of the location of food, incorporates a striking array of individual and colonial adaptations enabling the rapid discovery and exploitation of rich floral resources in a vast area around the nest (von Frisch 1967; Seeley 1985; Seeley and Levien 1987; Dyer and Seeley 1991).

We know that communication by dances has evolved in the genus *Apis*. The stingless bees, nearest to the genus *Apis*, are able to communicate about new food sources and use scent to lead recruits to them – behaviour not resembling *Apis* dances. Nevertheless, this communication is more primitive because there is no distance and direction information (Lindauer 1956a, b). Somewhat later, stingless bees were reported to successfully communicate direction and distance (Jarau et al. 2000), but their information transformation is less accurate than that of *Apis* dances.

Because the dance language has evolved within *Apis*, it seems highly probable that many features common to all the Asian species will have been conserved following speciation. Although basically similar in their social organisation and in the resources they exploit (Koeniger and Vorwohl 1979), the Asian species differ markedly in body size, colony size and nesting behaviour (Dyer and Seeley 1991), and many of their food resources are patchily distributed and widely spread over the tropical forests.

14.2 Variation in the Dances

14.2.1 Dances for Food Sources

When an experienced forager returns to the colony with a nectar or pollen load that is sufficiently attractive to warrant a return to the source, she lands on the comb and may perform two alternative behaviours. If she is already committed to the site, she will dance to inform other dance followers where the food source is. If not yet so committed, she will walk around the colony finding receiver bees to receive her load. Then, after unloading, she will perform dances if her load has been shown sufficient interest by the receiver bees. In which case, such recruits then leave the colony and find the target resource based on received information (Seeley et al. 2006). This is called "language" in honeybees. Languages differ intra- and inter-specifically throughout *Apis*. Moreover, genetic diversity and complexity of nesting sites have resulted in more diversity in the dances of the different species.

14.2.1.1 Round and Waggle Dances

In the long course of his research, von Frisch considered the dances of honeybees as of two types. First, the "round dance" which is performed if the target source is near to the nest, and if so, then the dance contains no information on direction. In this dance, the dancer starts with a run in a straight line and suddenly reverses direction and starts a new cycle. There are often one or two complete circles, but frequently only three-quarters or halves of circles (von Frisch 1967).

Second, the "waggle dance" is performed if the target source is further away from the nest. In these dances, bees will encode information both for distance as well as direction to the target patch. Typically in *A. mellifera*, in a waggle dance, the dancer runs in a straight line and returns in a semicircle to the starting point, then again runs through the same straight line portion and returns in a semicircle in the opposite direction (von Frisch 1967). On the straight part of the run, the bee buzzes its wings and vibrates its abdomen – the so-called "waggle". The angle of the axis of this waggle phase relative to the vertical represents the angle of the goal to the sun's azimuth. If the food source can be found by flying directly towards the sun, the middle, straight line portion, of the dance is perfectly vertical. Any angle to the right or left of the sun is communicated by running at the same angle to the vertical. Distance, meanwhile, is positively correlated with the duration of the waggle. The longer the bee waggles, the further away the food. If the target resource is only a short distance away from the nest, the waggling phase becomes quicker (or "more lively" as von Frisch originally described it).

Since the 1920s, the view that honeybees have two discrete recruitment dances has been widely accepted. Later, there were claims that round dances also encoded distance and direction (Jensen et al. 1997; Oldroyd and Wongsiri 2006). The mechanism of how the dances are interpreted remains frustratingly enigmatic. Gardner et al. (2008) reported that both the round and waggle dances convey information about distance and direction. They concluded that honeybees have just one "adjustable" dance and because the waggle run becomes quicker, the closer the food source, the bee suddenly changes direction so that her dance will become sickle-shaped. As the food source becomes closer to the nest, the dancer is unable to make complete turns, so that what might have been a waggle run becomes disfigured, and the dance will eventually lose the figure of eight and become a round dance. The transition point of change from round dance to waggle dance is highly variable within the same race or species and is highly influenced by genetic and environmental factors (cf. Oldroyd and Wongsiri 2006). Generally, the shape of the course and the tail waggling in *Apis cerana* seem to be identical to those described for *A. mellifera*. In mixed colonies of *A. cerana* and *A. mellifera*, the foragers of both species successfully recruited the other species (Tan et al. 2008).

In the waggle run of *Apis florea*, the bee does not move her body parallel to the comb as consistently as occurs in the waggle runs of *A. mellifera* and *A. cerana*. During a straight run, a bee elevates her abdomen or alternates horizontally and vertically. This vertical jerking of the abdomen also occurs at intervals in round dances (Lindauer 1956a, b). The waggling of *Apis andreniformis* is similar to that of *A. florea* (Duangphakdee personal observation).

In *A. mellifera*, the transition point is genetically and ecologically variable. The change from the round to the waggle dance is 5 m in *A. cerana*, 2.5 m in *Apis dorsata* and less than 1 m in *A. florea* (Towne 1985b). Although the dance is partly sickle-shaped, studies of *A. florea* in Thailand found that the direction component in the dance is still distinctly detectable even at less than 0.5 m (Duangphakdee personal observation). We can evidently assume that maybe a real "round dance" exists in *A. florea*, but that it is somewhat different from that of *A. mellifera*. A curious aspect is that a dance with close similarity to the round dance also occurs in the absconding dance in this bee. Does it serve the same function as in foraging dances of other species? This hypothesis has been suggested but no experiments have shown it to be so.

14.2.1.2 Dance Area

Due to differences in the nest structures of different honeybee species, where dances take place and how they are influenced by the nature of dance floors available to them are highly important (Tautz 1996). This means that the direction component of the dances is constrained by the structure of the nest, specifically the dance floor, so that the bees must dance either vertically or horizontally. There are three major groups of nest construction architecture utilised by cavity-nesting bees, dwarf bees and giant bees (cf. Chap. 6).

Of the cavity-nesting Asian honeybees, *A. cerana* have been most intensively studied. *A. cerana* dance in an enclosed nest cavity near the nest entrance on the vertical plane of the comb. Like *A. mellifera*, they indicate the location of the target resource relative to the sun during the straight portion of the waggle dance. The angle of the waggle phase at an angle relative to the vertical represents the angle of the target food source relative to the sun. Other cavity-nesting species, *Apis koschevnikovi*, *Apis nigrocincta* and *Apis nuluensis*, seem to have similar dance floors and, in passing observations, we observed that *A. koschevnikovi* appeared similar to *A. cerana*. But these species have not been analysed systematically.

The dwarf bees dance on the relatively flat surface of the crown of the comb in a horizontal plane. So the dancer can point directly to the food source, with no need to refer to gravity as in the cavity-nesting bees. Because the crown is a fairly restricted

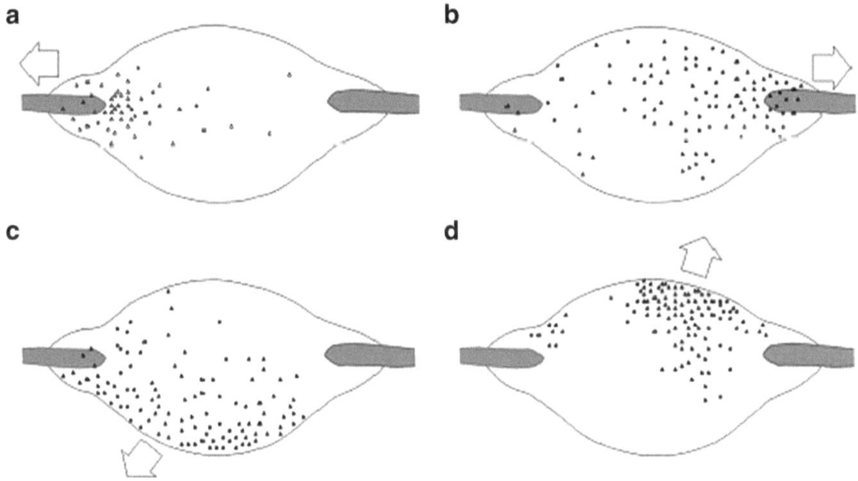

Fig. 14.1 Dance map (*top view*) for *A. florea* where *triangles* represent the centres of the waggle phases for (**a**) dances for forage for 50 waggle phases, (**b–d**) dances for nest sites for 100 waggle phases before colony liftoff. The *arrows* represent the target direction

area of about 40–60 mm^2 of comb surface and often vaulted, some dancers perform the dance beyond the crown, sloping downward on to the vertical surface of the comb overhang. The dance will take place on the portion of the crowded dancing floor closest to the target (Koeniger et al. 1982; Fig. 14.1).

Lindauer (1956a, b) showed that *A. florea* normally dance on a horizontal plain. The experiment of rotating the comb so that the crown is upside down and therefore no longer horizontal caused severely disturbed dancing. However, the bees will automatically find a horizontal floor for dancing. For example, if the crown is upside down, then the bees simply dance on the lower edge of the comb, which was recently turned upward. If the crown is covered by other objects, the dance either becomes disorientated or bees perform dances on the horizontal plane on top of objects covering it. Nevertheless, if they cannot find a horizontal platform, bees will dance such that the straight run points to the goal, as, for example, when bees are forced to dance on a small vertical platform without enough space to make a return run in waggle dancing, they will dance directly out into the air pointing to the target. Like in a swarm cluster, the dwarf bees successfully dance in a vertical plane with a similar rule: the dancers on the top of the cluster point directly to the goal, whereas the dancers on the curtain are on the vertical plane and, instead of referring to gravity like other cavity-nesting and giant bees do, the dance is orientated directly towards the goal (Dyer and Seeley 1985a).

The giant bees, *A. dorsata*, nest in the open on a single large comb (ca. 1 m diameter) attached to a tree branch or other overhanging structure. Their nests are extended vertically beneath supporting branches, so that they have no flat area anywhere in the nest. Because of no horizontal surface, bees perform dances on the vertical plane, on the surface of the curtain of the bees at the mouth area of the comb

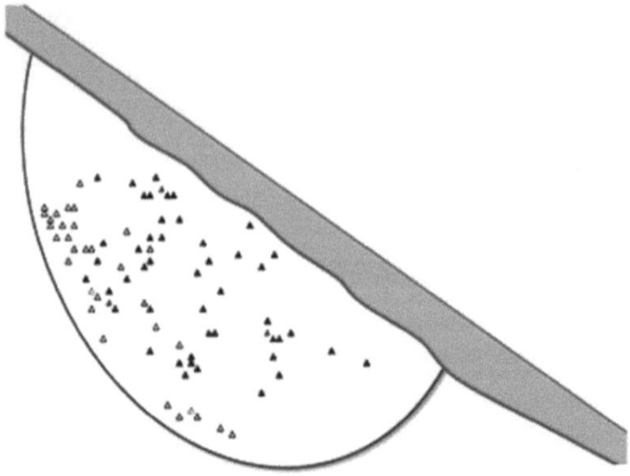

Fig. 14.2 Dance map of *A. dorsata* where *triangles* represent the centres of the waggle phases (*open triangles* represent foraging dances and *closed triangles* represent migration dances)

(Morse and Laigo 1969; Dyer and Seeley 1991; Fig. 14.2). Giant bees encode information about direction similar to cavity-nesting bees, not pointing to the goal directly but dancing relative to the position of the sun or the sun's azimuth (Lindauer 1956a, b).

14.2.1.3 Foraging Range vs. the Dance Tempo

An enduring mystery concerning the dance languages is the existence of geographical and species-specific "dialects" by which bees communicate flight distance (von Frisch 1967; Dyer and Seeley 1991). This is represented by the slope of the distance curve vs. dance tempo which is obviously relative to the ecotype of each study or race of honeybees in which the experiment was conducted. It has been hypothesised that dance dialects in honeybees are adaptively tuned to enhance efficiency of communication over the distances that its foragers typically fly, which is termed the "adaptive tuning hypothesis" (Dyer and Seeley 1991; Oldroyd and Wongsiri 2006). The accepted functional explanation is that the dances of each species are adapted to optimise foraging as related to the typical foraging range of the different populations (Lindauer 1956a, b; Boch 1957; von Frisch 1967; Punchihewa et al. 1985; Towne and Gould 1988; Dyer and Seeley 1991). According to this "adaptive tuning hypothesis" (Dyer and Seeley 1991), geographical races and species exhibit coding correlated with flight distances. According to Lindauer (1956a, b), the flight range of *A. cerana* is about 800 m, *A. florea* 350 m and *A. dorsata* seems to be greater but has not yet been determined.

Esch et al. (2001) showed that honeybees can estimate distance by measuring the optical flow perceived in flight. The environment has a considerable bearing on visual stimulation and perception as the bee flies along a complex and changing environment. The distance so measured will also be encoded in the waggle dance, thus affecting the shape of the dance curve so that comparisons of the same species from landscapes of differing complexity must be considered very carefully. As a consequence, it is not possible to use any standardised calibration curve for the interpretation of dances even for the same colony. Even so, the actual distance signal in the dances should fairly reflect similarities and differences among dialects. Races of *A. mellifera* have different curves of circuit duration vs. flight distance, thus there are specific codes to obtain distance information in racially mixed colonies, even if the dancer's code is specific for different distances (Boch 1956, 1957; Dyer and Seeley 1991).

In comparisons of Asian *Apis* species in Sri Lanka, Lindauer (1956a, b) found that the steepness of the dialect curve was inversely correlated with body size, which in turn is related to potential flight range. In small bees like *A. florea* (short flight range), the slope is very steep, while in large ones like *A. dorsata* (long flight range), the slope of the dialect curve is much less steep. Dialects are often similar among colonies. For example, in Sri Lanka, Lindauer (1956a, b) found that the slopes he measured were similar to those studied by Punchihewa et al. (1985), but often slopes differ among geographical races (Boch 1957).

Dialects are clearly represented among *A. cerana*, *A. dorsata* and *A. florea* and show differences in the slopes of the curves between distance and dance tempo (Punchihewa et al. 1985). However, in a similar study of *A. florea*, *A. cerana* and *A. dorsata* in Thailand, Dyer and Seeley (1991) found no striking dialects. The dialect curve of *A. cerana* and *A. nigrocincta* is very similar, where they are sympatric in Central Sulawesi, but the dance tempo of that in *A. cerana* was more rapid than that reported by Dyer and Seeley (1991) (Bakker 1999). Comparisons of the relative proportions of waggle and round dances as a function of distance did not differ in *A. florea* and *A. mellifera* (SenSarma et al. 2004).

This raises the possibility that the link between ecology and the distance code is more subtle than previously supposed, if indeed such a link exists (Dyer and Seeley 1991), and again the optic flow hypothesis might be considered. Differences in foraging range also show that changing dance tempo reflects changing food distances. For *A. cerana* and *A. florea* foraging at artificial feeders, the dance tempo was observed to decline more rapidly than has been reported for *A. mellifera* as distance to the food source increased. In *A. dorsata*, the change in dance tempo with increasing distance was much slower and similar to that reported for *A. mellifera* (Punchihewa et al. 1985).

14.2.2 Dances for Nest Sites

Finding a new nest site is one of the most important decisions a honeybee colony must make to ensure its reproductive success. For the cavity-nesting bees, the cavity must be of sufficient size to allow full development of the colony while the opening should be sufficiently small that it affords protection from predators and the elements (Franks et al. 2002; Seeley et al. 2006; Oldroyd et al. 2008). The characteristics of the cavity, size and height of the nest entrances must also be suitable (Inoue et al. 1990). Honeybees will search for new nest sites during absconding, migration and reproductive swarming. Reproductive swarming is defined as the movement of at least one queen and part of a honeybee colony from the maternal nest to an entirely new site for colony reproduction. Migration is the seasonally predictable movement of many whole colonies of the same population from one region to another. Absconding is usually restricted to the movement of a whole colony away from its maternal nest site to another place, usually caused by local environmental perturbations (Hepburn 2006; cf. Chap. 7).

The process of nest site selection in *A. mellifera* is well understood (Lindauer 1956a, b; Seeley et al. 2006). In addition to fundamental differences in behaviour and habitat variation, the Asian honeybees have striking differences in dance communication compared to *A. mellifera*. However, all *Apis* species use the same basic information coding system for the dance language. Commonly, nest site selection is presaged by the promotion of an initial stimulation factor, e.g. diminishing resources, habitat changes and disturbances by natural enemies, or sufficient time for a mature colony to produce a daughter colony, following which the scouts start searching for an apposite nest.

By performing waggle dances that indicate the direction of intended movement (Koeniger and Koeniger 1980; Dyer and Seeley 1991, 1994; Oldroyd and Wongsiri 2006), the colony will become informed through the decision making process of scouts (Seeley et al. 2006); and later, the whole colony will become airborne, taking a journey to the most profitable floral source or new nest site.

14.2.2.1 Initiation of Dances and Dance Areas

The dance area is tuned to fundamental differences between migration and foraging. In *A. cerana*, the absconding dance shifts around on various parts of the upper comb, but never in the usual foraging dance area located close to the entrance (Sasaki 1991). The dance area in an *A. dorsata* colony is distinctly different from the dance area used to communicate information about food. Dances for food occur only on a circumscribed area on the surface of the cluster called the “mouth”. Frequently, this area has the greatest light intensity, whereas migration dances can occur anywhere on the cluster surface (Koeniger and Koeniger 1980; Fig. 14.2). However, in the dwarf honeybees, the foraging dance is spread over the top of the crown, whereas dances for nest sites are not as distinct as those for recruiting foragers. Interestingly, the waggle dances for nest sites tend to be performed on the distinct and dense side of the crown toward the direction of the indicated target (Duangphakdee unpublished observation; Fig 14.1).

Waggle dances for nest sites are normally initiated long before colony movement. During migration of *A. dorsata*, the first dance was observed 1–3 days before colony departure. Those colonies observed appeared to leave the maternal nest only 1–2 days after the first migration dance had occurred (Koeniger and Koeniger 1980). The waggle dances for reproductive swarms in *A. florea* start 3–7 days before departing from the maternal nest. Later, after the swarms have left the mother colony and formed temporary clusters, the scouts of the daughter swarms will start to perform dances for a future nest site, a process in which the final decision is achieved within 1–7 days (Phiancharoen personal observation).

Studies of migration in *A. dorsata* (Koeniger and Koeniger 1980) and absconding in *A. florea* (Duangphakdee et al. 2010, unpublished observations) demonstrated that foraging recruitment for a food source is likely to completely stop about an hour before the bees fly off to allow the colony to proceed with nest site selection and reach a consensus decision. In an established swarm of *A. florea*, which had been caged for 3–4 days, the first migration dance could only be observed about half an hour after the colony was released and was able to reach a decision to find a new resting site within 4–24 h (Oldroyd et al. 2008). The first absconding dances in *A. florea* and *A. andreniformis* normally occur 1–2 days before a colony moves. During the stage of polling for a nest site, foraging dances frequently occur. The activity of foraging dances will later become reduced while dancing for a nest site becomes dominant. Until the colony reaches a decision to leave the nest, the last brood has often already joined the cluster. Nevertheless, sometimes a colony leaves a few young bees behind (Duangphakdee et al. 2010, unpublished observations).

14.2.2.2 Information Advertised by the Dances

Typically, dances for nest sites have been reported as waggle dances (Lindauer 1955). However, it has been frequently observed that waggle dances for nest sites are sometimes not complete cycles in several species such as *A. dorsata* (Koeniger and Koeniger 1980), *A. cerana* (Sasaki 1991), *A. florea* and *A. andreniformis* (Duangphakdee personal observation). Sometimes the bees run while wagging their abdomens but without returning in a semicircle to the starting point to complete the cycle. That the dancers only perform the waggle run with no return phase seems to be uniquely used for nest site selection dances.

As the waggle phase contains information about distance, the longer the waggle phase, the longer the distance, while shorter ones refer to nearby targets. Duration of the waggle phase in *A. dorsata* is very distinct from other species in this genus: it can be prolonged, lasting up to 8 s, and this waggle phase could become even more extended towards the last phase of the nest site selection process. This very possibly implies a long distance flight with no specific distance being given (Koeniger and Koeniger 1980). The absconding dances in *A. cerana* are also composed of long waggle runs from 8 to more than 60 s performed in the nest before the whole colony departs from the mother colony. The functional aspect of this may not be the same in *A. dorsata*, as the swarm reported on here moved to a resting place only 40 m away. Sasaki (1991) suggested that the function of a prolonged phase in the dance of *A. cerana* is to stimulate the whole colony to leave the mother nest.

The absconding dances in *A. dorsata* do not appear to attract recruits (Koeniger and Koeniger 1980). At the beginning, there may be only 2–3 dancers which increase to 5–15 dancers after 24 h until the colony moves (W.S. Robinson personal communication). However, most of the dances observed in a reproductive swarm (Oldroyd et al. 2008) and in absconding colonies of the dwarf bees,

Fig. 14.3 The "psuedo-round" dance in swarms of the dwarf honeybees (dance circuits frequently overlap)

A. florea and *A. andreniformis*, clearly attracted recruits similar to those observed for migration dances as in the cavity-nesting bees, *A. mellifera* and *A. cerana*. This suggests that the migration dance of *A. dorsata* is a derived characteristic for this species.

There are also interesting dances which were reported independently by Oldroyd et al. (2008) and Duangphakdee et al. 2010 (unpublished observations) to the effect that a significant number of dances in the swarms of *A. florea* and *A. andreniformis* have either limited or no distance information (which were regarded as "round dances" by Oldroyd et al. 2008). In this dance, bees do not form a fully formed figure of eight. Rather, they dance in random circles, sometimes waggling the abdomen with no specific direction (Fig. 14.3). This kind of dance occurs with sufficient regularity that it ought not be dismissed as just "errors". In established reproductive swarms, the number of these "psuedo-round dances" was about half to four times more frequent than waggle dances and persisted until the last half hour before liftoff (Oldroyd et al. 2008). Similar results were also observed in natural swarms (Duangphakdee unpublished observation).

14.2.2.3 The Targets of the Dances

Like *A. mellifera*, the Asian cavity-nesting bees are likely to have a limited number of potential nest sites. The cavity-nesting bees show a preference for volume and cavity height, size, position and direction of the entrance and evidence of previous occupancy (Seeley and Morse 1978; cf. Chap. 6). The absconding dance in *A. cerana* is performed within the nest. Before leaving the mother colony, the direction of the waggle runs tends to unity in all the dancing bees; nevertheless, it seems that a new nesting site has not yet been specified at this phase. The bees move apart from the mother nest and settle under the branch of a nearby tree and the final nesting site will be decided later in the swarm cluster. Sasaki (1991) suggested that this unity among individuals and the compensation in the angle of the waggling run over time appears to be only a remaining fragment of the originally evolved system and that it has no functional meaning.

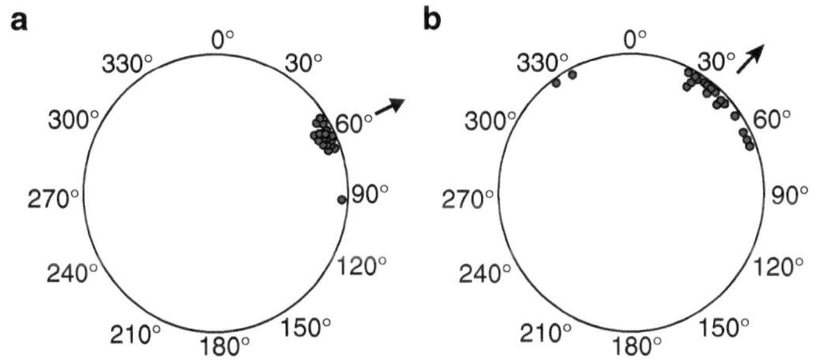

Fig. 14.4 Schematic drawing of the last 20 waggle dances advertising a nest site before liftoff of (**a**) *A. florea*: new nest site located at 70 m from the original nest. The bees in this example dance with a divergence angle of 37° and (**b**) *A. andreniformis*: new nest site located at 68 m from the original nest. The bees in this example dance with a divergence angle of 45°. Each waggle run direction is denoted by a *circle*. The *small arrows* denote the mean direction of waggle runs (outlier is not included). The experiments were conducted in different locations: *A. florea* at Chom Bueng, Thailand (13.61N, 99.60E) and *A. andreniformis* at Tenom, Malaysia (05.11N, 115.95E)

In observations of a swarm of *A. dorsata*, before the swarm flew off, the bees danced facing in the same direction, demonstrating that the colony had reached a consensus decision before becoming airborne (Koeniger and Koeniger 1980) as occurs in *A. mellifera* (Seeley et al. 2006); nevertheless, details of scouting behaviour and reaching a quorum are not yet clear. Similar to *A. cerana*, the real nesting site will be decided later in the swarm cluster. In the case of *A. dorsata* that migrate long distances, they might need to make decisions several times to reach a final nesting site.

In contrast to cavity-nesting species, for an open-nesting species like *A. florea*, it seems that there are more potential sites (many patches/twigs) that could well be equally suitable for building a nest. The final dances cover a broad area, especially with respect to distance, but often direction as well. This suggests that the bees do not advertise a particular twig that they may have visited, but a general direction, because bees still performed dances indicating nest sites in several different directions. This study contends that there is no consensus in the nest site selection process in *A. florea* (Oldroyd et al. 2008). However, in absconding by *A. florea*, the scouts dance with a relatively high angular deviation up to 45°with a target distance less than 100 m and colonies distinctly reach consensus in dancing before they are airborne (Duangphakdee et al. 2010, unpublished observations; Fig. 14.4a). Similarly, colonies of *A. andreniformis* also reach consensus (Duangphakdee personal observation; Fig. 14.4b).

These contradictory results might be explained by the work of Seeley and Visscher (2003) who concluded that colony decision making might be sensing a quorum rather than sensing a consensus. Since the colony decides where to go, the bees will start to produce an audio signal, or "worker piping", to stimulate other bees to warm-up for flight (Seeley and Tautz 2001; Seeley et al. 2003). This warm-up often takes time and is usually long enough for scouts to achieve consensus for a chosen nest site before the entire swarm takes flight. If the bees have actually made a decision, but there is

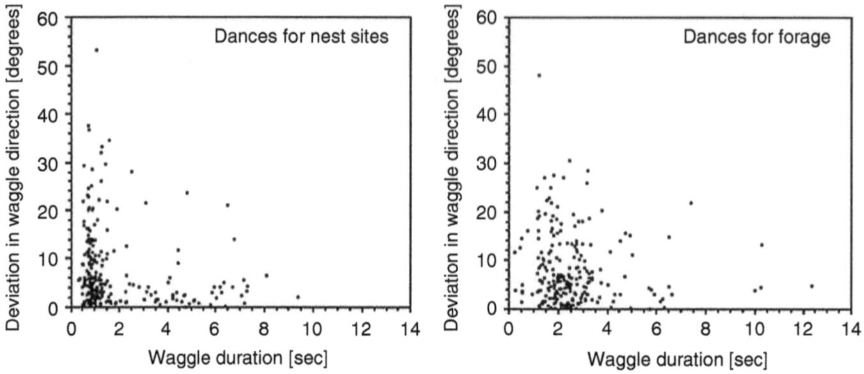

Fig. 14.5 Deviation of waggle dance angles for nest sites (*left*) and foraging (*right*) by dwarf bees, *A. florea* (Beekman et al. 2008)

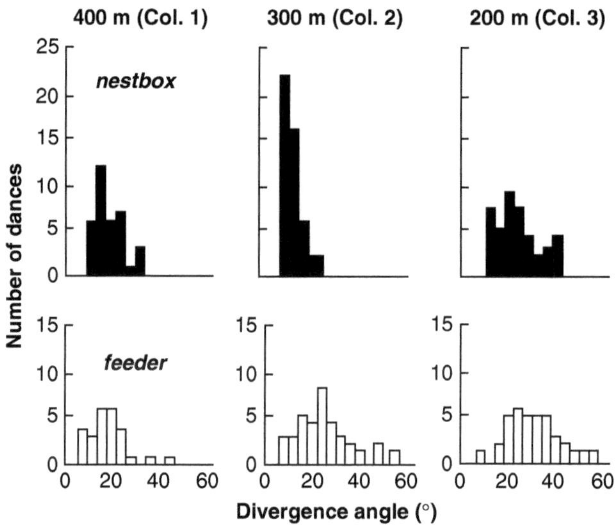

Fig. 14.6 Distribution of divergence of angles in the dances of *A. mellifera* indicating either a nest-box (*filled bar*) or a feeder (*open bar*) at different distances (Weidenmüller and Seeley 1999)

not sufficient time for the scouts to achieve consensus, the colony will probably take off even if dancing scouts are still advertising different nest sites.

According to the tuned-error hypothesis, *A. florea* nest site selection is wide, similar to food sources, so that dance precision for nest sites and food sources is essentially the same irrespective of context (Beekman et al. 2008; Fig. 14.5). In contrast, nest site selection by *A. mellifera* is for a cavity, so that dance precision is necessary. *A. mellifera* dances for nest site cavities are more precise than dances for food source-patches (Weidenmüller and Seeley 1999; Fig. 14.6). However, the tuned-error hypothesis is not favoured in the studies of Towne (1985b) and Tanner and Visscher (2006) who found that there was no difference in dance precision between nest sites and food sources. On the other hand, they also found that bees dance with a greater divergence of angles when dancing on different substrates. These authors explored an alternative idea that changing precision of the dances is not an adaptive modification driven by natural selection but is more likely to be a physical adjustment when they dance on different surfaces.

Comparative studies of absconding by *A. florea* in Ratchaburi, Thailand and *A. andreniformis* in Sabah, Malaysia show slight differences in the time necessary to reach a quorum. *A. andreniformis* seem to reach a decision and then take off faster than *A. florea*. Normally, foraging activity in dwarf bees drops at least 1 h before colony liftoff to permit the dancers to intensively advertise those nest sites to which they are committed. This period is likely to take longer in *A. florea* than *A. andreniformis* until all bees take off to the chosen nest site (Duangphakdee et al. 2010, unpublished observations).

14.3 Compass and Navigation for the Dances

A. mellifera bees use many objects for the orientation of their dances. The sun's position is encoded within the dances of the scouts as a mobile reference point by which to navigate to the goal (von Frisch 1967). When the sun is covered by cloud, bees will estimate the position of the sun by perceiving the plane of vibration of the polarised sky (von Frisch 1967), or they often use visually conspicuous objects for the orientation of dances or flight to reach a goal (von Frisch 1967). The earth's magnetic field influences dances but it is not known whether it serves as a reference for honeybee dances.

Even though Asian honeybees use many objects for navigation in flight and for dance orientation, the position of the sun seems to be the most distinct object which has been reported as encoded symbolically in the dance language in cavity-nesting bees and giant bees. It may be possible that other objects could also be encoded but there are no reports along these lines. In *A. florea*, sun and blue sky provide the actual reference for the dances but they are reported to be able to include landmarks in the symbolic information of their dances (Dyer 1985a). The cavity-nesting and giant honeybees, which dance in the vertical plane, need the sun's azimuth as a reference point to navigate. The dances of *A. cerana* are performed in a dark cavity on the vertical comb, and it is clear that these bees can recognise the position of the sun by visually observing only a small area of blue sky through the small opening of the nest entrance (von Frisch 1949).

The sun is an important clue for the orientation of the dances of honeybees. To translate directional information, those bees being recruited need to read the angle of the waggle run relative to gravity and then translate this into the direction of a food source which requires the sun as a compass. Therefore, the changing position of the sun relative to date and time of day greatly affects the dance in honeybees (Duangphakdee et al. 2009). *A. mellifera*, *A. cerana* and *A. dorsata* compensate relative to the changing of the solar azimuth in their foraging dances. Interestingly, *A. florea* do not shift their dances with the changing of azimuth, but always orient their dances toward the food source. This species seems to have a specialised system of information processing which differs from other species of *Apis* (Dyer 1985a).

Due to differences of the dances' floors, considerable recent interest has been focused on the dwarf bees which dance in a horizontal plane. *A. florea* do not need to adopt an angle relative to gravity, but they can point to a goal directly. There are related reports (Koeniger et al. 1982) demonstrating that the sun is important to the orientation of *A. florea* dances because their dances become disoriented if they are placed in a box or building where they cannot see the sun. In a further experiment, Koeniger et al. (1982) showed that these bees can orientate dances without a visible sky and are able to use a surrogate sun with the aid of a mirror, and by changing the angle of the mirror, so the angles danced by the bees change (Koeniger et al. 1982, Oldroyd and Wongsiri 2006). A similar experiment was done with *A. dorsata*, but changing the apparent position of the sun, or blocking sight of the blue sky did not affect migration dancing (Koeniger and Koeniger 1980). Although the migration dances of *A. dorsata* are not affected by the sun nor is direction encoding affected, without sight of the sun, the intensity of dancing decreases over time. *A. cerana* also have the ability to change direction in absconding dances to compensate for movements of the sun (Sasaki 1991).

It is clear that the sun is particularly important for dance orientation in *A. florea*. This was recently documented by Duangphakdee et al. (2009) who observed the dances of *A. florea* at different times of day during dance recruitment for both food sources and new nest sites. Their results showed that the number of dancing bees was bimodal with respect to time, with a pronounced trough during 12:00–13:00 compared to 10:00–11:00 and 13:00–14:00 (Fig. 14.7). Moreover, the angular accuracy of the waggle dance over time changed. During recruitment for nest sites and food sources in the morning (10:00–11:00) and afternoon (13:00–14:00), only very few dancers performed dances without a direction component, while those around midday (12:00–13:00) showed a significantly greater number of dances lacking a direction component. Even those dances with a direction component contained larger errors in angular accuracy over the noon hour.

We must consider why bees largely avoid dancing at noon. It was first established that the waggle dances of *A. mellifera* become disoriented when the sun is within ±3° of the zenith (Lindauer 1956a, b). Then, New and New (1962) also noted similar difficulties for waggle dancers between ±3° and ±4° at tropical latitudes in that the bees clearly became "confused" in observing the angle of the sun. The closer the sun passes to the zenith, the more rapidly the azimuth changes around noon (Fig. 14.8). This means the time element becomes critical, because the closer the sun passes to the zenith, the more rapidly the azimuth changes at noon, particularly in the tropics. Observations on *A. florea* demonstrated that they had difficulties in taking an accurate reading of the sun at angles ±6° of the sun's zenith and their preferred departure angle of the sun is between 55° and 65° (Duangphakdee et al. 2009).

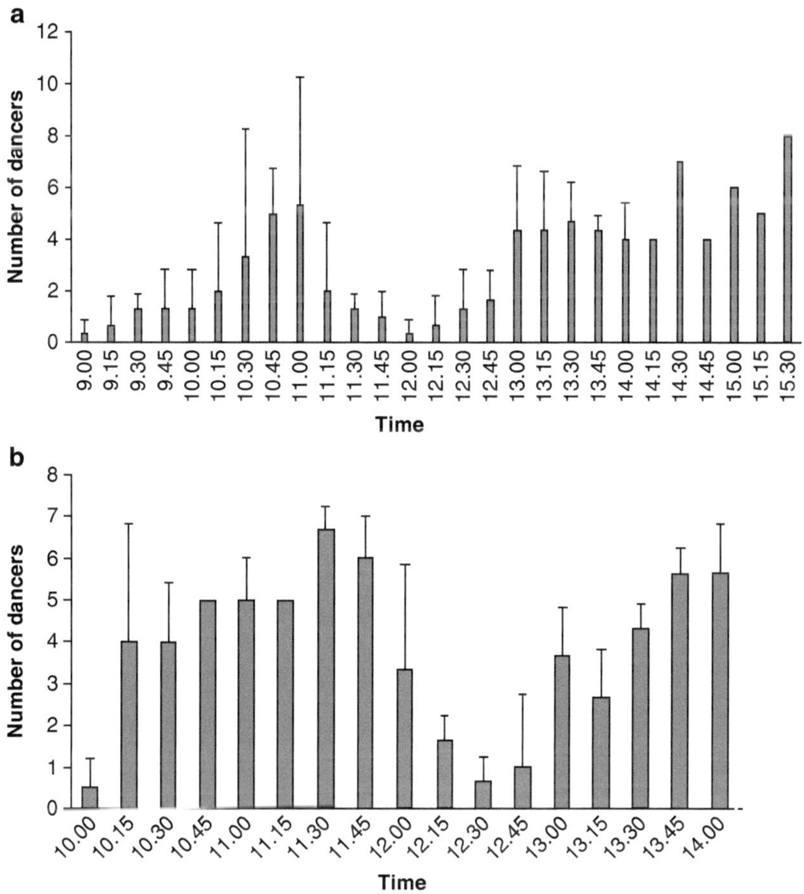

Fig. 14.7 Frequency distribution of *A. florea* dancers for (**a**) a new nest site and (**b**) a food source as a function of time of day (Duangphakdee et al. 2009)

This explains the difficulty of accurately reading the sun angle at the noon hour, and why the waggle dances became disoriented between 12:00 and 13:00. This is related to the fact that the bees largely avoid dancing at noontime, both for finding nest sites after absconding as well as foraging for food sources (Fig. 14.7). The disorientation of the dances becomes especially crucial during the process of finding a new nest site. During this process, the bees need time to poll before reaching a consensus on where to ultimately go (Seeley and Visscher 2003). To achieve a consensus decision takes at least 2 h (Duangphakdee personal observation). As they have difficulty in accurately determining the sun's azimuth, resulting in defective navigational skills around noon, the bees simply avoid the noon hour, but are left with two time windows, morning and afternoon. This is also likely to occur in other species in *Apis* (Koeniger and Koeniger 1980; Sasaki 1991; Oldroyd et al. 2008).

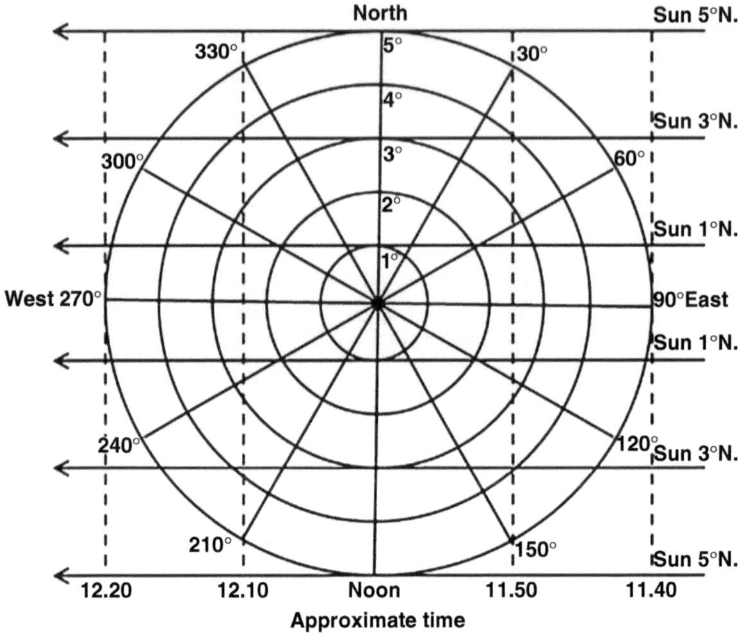

Fig. 14.8 The sun's path near the zenith. *Circles* indicate zenith distances; *radii* indicate azimuths. The closer the sun passes to the zenith, the more rapidly the azimuth changes around noon. The times given are accurate for the equator but vary by a minute or two at other places in the tropics (New and New 1962)

14.3.1 Dancing at Night or in the Dark

Some species of *Apis* have evolved the ability to enhance communication in darkness or at night. In this context of "dance", visual transformation might not be sufficient enough to communicate in the dark. Crepuscular honeybees like *A. mellifera* produce sounds during the dance, which are potentially important in the transmission of information for orientation and the dance circuit of recruited bees (Towne 1985a). Dancers produce airborne sound signals both in round dances (Kirchner et al. 1988) and waggle dances (Michelsen et al. 1986), which enhance information transmission in the darkness of the nest cavity.

This sound signal occurs in the Asian cavity-nesting bees, *A. cerana* (Towne 1985a; Kirchner and Dreller 1993), while in open-nesting bees like *A. dorsata*, the dances were first reported to be silent (Towne 1985a). However, this was subsequently shown to be incorrect (Kirchner and Dreller 1993) because later, *A. dorsata* have been observed to forage and perform dances (Dyer 1985b) at low light intensity such as occurs at half or full moon. The acoustic signal in the dance of *A. dorsata*, therefore, was reconsidered. A decade later, Kirchner and Dreller (1993) discovered that *A. dorsata* use sound in dance communication both during the day and on moon-lit nights. The duration of the dance sound was positively correlated with distance. Not all dances produce sound, and the probability of sound emission during a dance is positively correlated with profitable food sources. The probability of dancers producing sound at night is three times higher than during the day. Consequently, *A. dorsata* are able to emit sounds in dances to communicate distance and direction, as in *A. mellifera*.

Other supporting information comes from observations on the dances of *Apis laboriosa*. Because of their distribution (cf. Chap. 3), the ambient temperature at night is too cold for bees to fly. This species has never been reported to have nocturnal activity. The dances of *A. laboriosa* are entirely silent (Kirchner et al. 1996). *A. florea* dances are also silent (Towne 1985a). This supports the idea that acoustic dance communication may have evolved through selection on dancing at low light intensities, whereas the species which dance with sufficient light intensity do not need sound (Towne 1985a; Kirchner et al. 1996). Moreover, it must be noted that the motion of wing-buzzing, which produces sound, is also more easily recognised at low light intensities.

Although the moon's illumination is essential for nocturnal flight by *A. dorsata*, the moon itself is ignored in dance orientation. Rather, bees probably use the (invisible) sun's position as a reference point for their dances, by estimating the position of the sun which is below the horizon. This ability may involve an extension of the mechanism that honeybees employ to find the sun on overcast days (Dyer 1985a).

14.4 Others Dances

von Frisch (1967) had noted that *A. mellifera* performed dorsoventral abdominal vibrations (DVAV) or jerking dances, but evidence for the function of this dance is somewhat obscure and its real function remains an open question (Fletcher 1975). This kind of dance also occurs in Asian species and they are similar of those *A. mellifera*. The returning foragers of many species – based on our observations of *A. dorsata*, *A. cerana*, *A. koschevnikovi*, *A. florea* and *A. andreniformis* (Duangphakdee personal observation) – frequently performed DVAV dances. Foragers often start waggle dancing for food sources after a DVAV (Duangphakdee personal observation). Newly emerged queens of *A. florea* perform DVAV dances to attract the attention of workers. The DVAV dances of queens are also reported to occur before swarming or mating flights (von Frisch 1967; Winston 1987). The jostling dance, spasmodic dance, buzzing run, shaking dance and trembling dance (von Frisch 1967; Winston 1987) also occur in Asian cavity-nesting and dwarf bees just as in *A. mellifera* (Duangphakdee personal observation) but have not been studied at all.

Another curious observation is that of dance communication by drones of *A. andreniformis*. About 1 h before they fly to a drone congregation area (DCA), mature drones walk upward to the crown area. Before flying, some drones start performing runs in circling loops with their wings extended to the side. These runs appear identical to the round dances of *A. mellifera*. This is the only report of drone dances. This dance is hypothesised to stimulate other drones to synchronise group flight to the DCA (Wongsiri et al. 1996; Oldroyd and Wongsiri 2006). The advantage of taking off as a group perhaps enhances mate location, mating attempts at the DCA and/or protection from predators (Rinderer et al. 1992). The drone dance appears to be similar in form and tempo with the "psuedo-round dance" in absconding *A. florea* and *A. andreniformis* (Duangpkakdee et al. 2010, unpublished observations). If so, the functional meaning of this "pseudo-round dance" might be explained as similar to the dances of drones that are thought to motivate nestmates to take flight.

References

Bakker DR (1999) Foraging and habitat selection by two species of honeybee near Lore Lindu national park in Sulawesi, Indonesia. Thesis, University of Guelph

Beekman M, Gloag RS, Even N, Wattanachaiyingcharoen W, Oldroyd BP (2008) Dance precision of *Apis florea* – clues to the evolution of the honeybee dance language? Behav Ecol Sociobiol 62:1259–1265

Boch R (1956) Die Tänze der Bienen bei nahen und fernen Trachtquellen. Zeit Vergl Physiol 38:136–167, in German

Boch R (1957) Rassenmässige Unterschiede bei den Tänzen der Honigbiene (*Apis mellifera* L.). Zeit Vergl Physiol 40:289–320, in German

Duangphakdee O, Radloff SE, Pirk CWW, Hepburn R (2009) Sun angle time windows for absconding by the dwarf honeybee, *Apis florea*. J Insect Physiol 55:1009–1012

Dyer FC (1985a) Mechanisms of dance orientation in the Asian honeybee *Apis florea* L. J Comp Physiol A 157:183–198

Dyer FC (1985b) Nocturnal orientation by the Asian honeybee, *Apis dorsata*. Anim Behav 33:769–774

Dyer FC, Seeley TD (1991) Nesting behavior and the evolution of worker tempo in four honeybee species. Ecology 72:156–170

Dyer FC, Seeley TD (1994) Colony migration in the tropical honeybees *Apis dorsata* F. (Hymenoptera: Apidae). Insectes Soc 41:129–140

Esch HE, Zhang S, Srinivasen MV, Tautz J (2001) Honeybee dances communicate distances by optic flow. Nature 411:581–583

Fletcher DJC (1975) Significance of dorsoventral abdominal vibration among honey-bees (*Apis mellifera* L.). Nature 256:721–723

Franks NR, Pratt SC, Mallon EB, Britton NF, Sumpter DJT (2002) Information flow, opinion polling and collective intelligence in house-hunting social insects. Philos Trans R Soc Lond B357:1567–1583

Gardner KE, Seeley TD, Calderone NW (2008) Do honeybees have two discrete dances to advertise food sources? Anim Behav 75:1291–1300

Hepburn HR (2006) Absconding, migration and swarming in honeybees: an ecological and evolutionary perspective. In: Kipyatkov VE (ed) Life cycles in social insects – behaviour, ecology and evolution. St. Petersburg University Press, St. Petersburg, pp 121–136

Inoue T, Adri S, Salmah S (1990) Nest site selection and reproductive ecology of the Asian bees, *Apis cerana indica* in central Sumatra. In: Sakagami SF, Ohgushi R, Roubik DW (eds) Natural history of social wasps and bees in equatorial Sumatra. Hokkaido University Press, Sapporo, pp 219–232

Jarau S, Hrncir M, Zucchi R, Barth FG (2000) Recruitment behaviour in stingless bees. Melipona scutellaris and M. quadrifasciata. I. Foraging at food sources differing in direction and distance. Apidologie 31:81–91

Jensen IL, Michelsen A, Lindauer M (1997) On the directional indications of the round dances of honeybees. Naturwissenschaften 84:452–454

Kirchner WH, Dreller C (1993) Acoustical signals in the dance language of the giant honeybees, *Apis dorsata*. Behav Ecol Sociobiol 33:67–72

Kirchner WH, Lindauer M, Michelsen A (1988) Honeybee dance communication. Acoustical indication of direction in round dances. Naturwissenschaften 75:629–630

Kirchner WH, Dreller C, Grasser A, Baidya D (1996) The silent dances of the Himalayan honeybee, *Apis laboriosa*. Apidologie 27:331–339

Koeniger N, Koeniger G (1980) Observations and experiments on migration and dance communication of *Apis dorsata* in Sri Lanka. J Apic Res 19:21–34

Koeniger N, Vorwohl G (1979) Competition for food among four sympatric species of apini in Sri Lanka (*Apis dorsata*, *Apis cerana*, *Apis florea* and *Trigona iridipennis*). J Apic Res 18:95–109

Koeniger N, Koeniger G, Punchihewa RKW, Fabritius M (1982) Observations and experiments on dance communication in *Apis florea* in Sri Lanka. J Apic Res 21:45–52

Lindauer M (1955) Schwarmbienen auf Wohnungssuche. Zeit Vergl Physiol 37:263–324, in German

Lindauer M (1956a) Communication among the honeybees and stingless bees of India. Bee World 38(3–14):34–39

Lindauer M (1956b) On understanding Asian honeybees. Bee World 38:521–557, in German

Michelsen A, Kirchner WH, Lindauer M (1986) Sound and vibrational signals in the dance language of the honeybee, *Apis mellifera*. Behav Ecol Sociobiol 18:207–212

Morse RA, Laigo FM (1969) *Apis dorsata* in the Philippines. Monogr Philipp Assoc Entomol 1:1–96

New DAT, New JK (1962) The dances of honeybees at small zenith distances to the sun. J Exp Biol 39:271–291

Oldroyd BP, Wongsiri S (2006) Asian honeybees. Harvard University Press, Cambridge

Oldroyd BP, Gloag RS, Even N, Wattanachaiyingcharoen W, Beekman M (2008) Nest site selection in the open-nesting honeybee *Apis florea*. Behav Ecol Sociobiol 62:1643–1653

Punchihewa RWK, Koeniger N, Kevan PG, Gadawski RM (1985) Observations on the dance communication and natural foraging ranges of *Apis cerana, Apis dorsata* and *Apis florea* in Sri Lanka. J Apic Res 24:168–175

Riley JR, Gregger U, Smith AD, Reynolds DR, Menzel R (2005) The flight paths of honeybees recruited by the waggle dance. Nature 435:205–207

Rinderer TE, Oldroyd BP, Sylvester AH, Wongsiri S, de Guzman LI (1992) Evolution of bee dances. Nature 360:305

Sasaki M (1991) Absconding dance in *Apis cerana japonica*: a slow and long tail-waggling motivates the whole colony to translocate. In: Veeresh GK, Mallik B, Viraktamath CA (eds) Social insects and the environment. Oxford & IBH, New Delhi, pp 125–126

Seeley TD (1985) Honeybee ecology: a study of adaptation in social life. Princeton University Press, New Jersey

Seeley TD, Levien RA (1987) Social foraging by honeybees: how a colony tracks rich sources of nectar. In: Menzel R, Mercer A (eds) Neurobiology and behavior of the honeybee. Springer, Heidelburg, pp 38–53

Seeley TD, Morse RA (1978) Nest site selection by the honeybee. Insectes Soc 25:323–337

Seeley TD, Tautz J (2001) Worker piping in honeybee swarms and its role in preparing for liftoff. J Comp Physiol 187:667–676

Seeley TD, Visscher PK (2003) Choosing a home: how the scouts in a honeybee swarm perceive the completion of their group decision making. Behav Ecol Sociobiol 54:511–520

Seeley TD, Kleinhenz M, Bujok B, Tautz J (2003) Thorough warm-up before take-off in honeybee swarms. Naturwissenschaften 90:256–260

Seeley TD, Visscher PK, Passino KM (2006) Group decision making in honeybee swarms. Am Sci 94:220–229

SenSarma M, Esch H, Tautz J (2004) A comparison of the dance language in *Apis mellifera carnica* and *Apis florea* reveals striking similarities. J Comp Physiol A 190:49–53

Tan K, Yang MX, Radloff SE, Hepburn HR, Zhang ZY, Luo LJ, Li H (2008) Dancing to different tunes: heterospecific deciphering of the honeybee waggle dance. Naturwissenchaften 95: 1165–1168

Tanner DA, Visscher PK (2006) Do honeybees tune error in their dances in nectar-foraging and house-hunting? Behav Ecol Sociobiol 59:571–576

Tautz J (1996) Honeybee waggle dance: recruitment success depends on the dance floor. J Exp Biol 199:1375–1381

Towne WF (1985a) Acoustic and visual cues in the dances of four honeybee species. Behav Ecol Sociobiol 16:185–187

Towne WF (1985b) The spatial precision and mechanisms of the dance communication of honeybees: experimental and comparative studies. Thesis, Princeton University

Towne WF, Gould JL (1988) The spatial precision of the honeybees' dance communication. J Insect Behav 1:129–155

von Frisch K (1923) Über die 'Sprache' der Bienen. Zool Jahrb Physiol 40:1–186, In German

von Frisch K (1949) Die Polarisation des Himmelslichtes als orientierender Faktor bei den Tanzen der Bienen. Experientia 5:142–148, In German

von Frisch K (1967) The dance language and orientation of bees. Belknap Press of Harvard University, Cambridge

Weidenmüller A, Seeley TD (1999) Imprecision in waggle dances of the honeybee (*Apis mellifera*) for nearby food sources: error or adaptation. Behav Ecol Sociobiol 46:190–199

Winston ML (1987) The biology of the honeybee. Harvard University Press, Cambridge

Wongsiri S, Lekprayoon C, Thapa R, Thirakupt K, Rinderer TE, Sylvester HA, Oldroyd BP, Booncham B (1996) Comparative biology of *Apis andreniformis* and *Apis florea* in Thailand. Bee World 77:23–35

Chapter 15
Diseases of Asian Honeybees

Ingemar Fries

15.1 Introduction

Diseases of Asian honeybee species are largely an unexplored field. Compared to the wealth of information available on honeybee pathology from the European honeybee, *Apis mellifera*, our knowledge of the Asian honeybee species is truly rudimentary. Nevertheless, honeybees of all species share some fundamental properties that are important for host–parasite co-evolution. They all live in colonies with thousands of individuals, often packed closely together, and they all have colony level reproduction through colony fission when they swarm. Thus, they share some common features that make some understanding of honeybee pathology from the European honeybee applicable to other honeybee species.

Understanding more of the pathology of Asian honeybee species is important for several reasons. With an increased awareness of the importance of pollination and biodiversity, conservation of Asian honeybee species becomes essential (cf. Chap. 11). Understanding how pathogens and parasites interact within and between species may be instrumental in preventing unexpected colony losses. This has become increasingly important as intended, as well as unintended, transport of live honeybees over long distances, even between continents, frequently occurs. As we have already seen with the *Varroa destructor* mite, and possibly with the *Nosema ceranae* fungal infections, the impact of one pathogen in one honeybee species may be very different in another closely related host system. Thus, more research on pathogens from Asian honeybee species is warranted to prevent losses from disease in these bees and to identify possible threats from interspecific transmission of pathogens.

I. Fries
Department of Ecology, Swedish University of Agricultural Sciences, Uppsala 750 07, Sweden
e-mail: ingemar.fries@ekol.slu.se

H.R. Hepburn and S.E. Radloff (eds.), *Honeybees of Asia*,
DOI 10.1007/978-3-642-16422-4_15, © Springer-Verlag Berlin Heidelberg 2011

15.2 Honeybee Epidemiology

Honeybees reproduce as individuals when the queen produces offspring. However, to survive, honeybees must also reproduce at the colony level. Colony level reproduction in honeybees occurs when the colonies swarm, increasing the number of colonies by colony fission. Without colony level reproduction, the species would perish because colonies succumb for various reasons. Honeybee colonies are superorganisms (cf. Moritz and Southwick 1992) that consist of individual units (bees) that have no function or survival capacity taken out of their colony context. Conceptually, this can be compared to taking simple neurons in the brain out of their brain context. Individual (unintelligent) bees become a functioning (intelligent) colony when integrated. The simple (unintelligent) neurons become a thinking (at best) brain when integrated into a whole.

Colony level reproduction offers an ideal route for pathogen transmission between colonies. If any of the swarming bees carry pathogens, these are readily transmitted to a new host colony. In fact, all honeybee pathogens known from the European honeybee, *A. mellifera*, can be transferred by adult bees, as well as pathogens that only infect the brood stages (Fries and Camazine 2001). Pathogen transmission as a result of colony level reproduction can be regarded as a vertical transference of pathogens where the parental colony infects the daughter colony. Thus, there is an important route for vertical transmission for all types of pathogens in the honeybee system. This should have important consequences for the development of pathogen virulence in honeybees whether under natural or managed conditions (Fries and Camazine 2001).

Although several factors influence the evolution of virulence in pathogens, the mode of transmission between hosts is crucial (Lipsitch et al. 1996). The different modes of pathogen transmission can be divided into two categories: horizontal and vertical. Horizontal transmission refers to parasite transmission between individuals of the same generation, while vertical transmission refers to that from parent to offspring (Canning 1982). When considering colonies of social insects, we can further subdivide horizontal and vertical transmission into both intra-colonial (between individuals within a colony) and inter-colonial (between individuals from different colonies) components (Table 15.1).

When horizontal and vertical pathogen transmissions are compared, vertical transmission is expected to select for decreased virulence. This is because the pathogen is

Table 15.1 Modes of pathogen transmission within and between honeybee colonies (From Fries and Camazine 2001)

	Horizontal	Vertical
Intra-colony	Worker to brood, worker or drone	Queen to daughter (worker)
		Queen to daughter (queen)
	Drone to worker or drone	Queen to son (drone)
Inter-colony	Worker to worker or drone	Swarming
	Drone to worker or drone (drifting, robbing)	

dependent on host reproduction for its own transmission between hosts. Thus, in this case, the goals of the host and pathogen are aligned and a benign host–parasite relationship is likely to evolve. This trade-off between virulence and mode of transmission has been tested and confirmed experimentally in several different systems (Bull et al. 1991; Herre 1993; Turner et al. 1998). Although the main mode of pathogen transmission appears to be important for moulding host–pathogen relations, the trade-off between virulence and mode of transmission is only one, although important, aspect. Where both horizontal and vertical transmission occur, as would be the case in honeybees, the expected virulence is not simply a function of which mode of transmission predominates (Lipsitch et al. 1995a). Theoretically, it is also clear that if parasites lower the fitness of the host, they will not survive unless there is also some degree of horizontal transmission (Lipsitch et al. 1995b). Nevertheless, pathogens that depend on host reproduction for their transmission are unlikely to become virulent and cause considerable host mortality.

Honeybee pathology is almost entirely focused on managed populations of the European honeybee, *A. mellifera*. Resistance to disease may very well vary between *Apis* species, but the main difference between the pathology of European bees and the Asian honeybee species is most likely that populations of the former remain mainly domesticated in large parts of the world, whereas the latter largely exist as non-managed wild populations. Where wild populations of the European bee do exist, their health status is not well documented, but they appear to be healthier than managed populations of honeybees (Goodwin et al. 1994; Hornitzky et al. 1996; Manning et al. 2007). Because of their importance, honeybee pathogens have been intensively studied in European honeybees and much is known about the causes of disease, the symptoms pathogens produce and how they can be controlled (e.g. cf. Bailey and Ball 1991; Morse and Flottum 1997).

However, the natural history and biological details of specific pathogens are not sufficiently well understood to define host–parasite relations in epidemiological terms (epidemiology, rather than epizootiology, is chosen here because the theories developed for "demos" – people – also encompass animals – "zoo"). In studying the prevalence and transmission of disease agents in social insects, it is obvious that one cannot limit such studies to individual bees nor to individual colonies. It is also necessary to consider the inter-colonial transmission of pathogens. In terms of fitness, the successful transfer of a pathogen's offspring to a new colony is a critical step in its life history. If a parasite or pathogen fails to achieve a foothold in another host colony, the parasite will not increase its reproductive fitness, regardless of how prolific it has been within the individual bee or within the original host colony. However, only a few studies have been concerned with the actual epidemiology of honeybee diseases even in European bees.

It would not be surprising if non-managed populations of honeybees have a better health status as compared to honeybees kept for apicultural purposes. Apiculture will inevitably increase the local colony density with increased opportunities for horizontal transmission between colonies of pathogens through robbing and drifting behaviour. Furthermore, the shifting of material and boxes between colonies during management may also lead to increased horizontal pathogen transmission. Interestingly, local densities of *Apis dorsata* and of *Apis laboriosa* may be as high as in managed apiaries of the European honeybee, because of their tendencies to aggregate colonies in specific trees (Deodikar et al. 1977). It would be interesting to compare if the inter-colonial rates of vertical vs. horizontal pathogen transmission are different in colonies of honeybees that aggregate as compared to colonies that do not, such as *Apis cerana* (Table 15.1) or *Apis florea*.

15.3 Bacterial Infections

15.3.1 American Foulbrood

American foulbrood (AFB) is caused by the spore-forming Gram-positive bacterium *Paenibacillus larvae* (Genersch et al. 2006a). AFB is spread worldwide in European honeybee populations and is generally regarded as the most serious of all brood diseases of this particular honeybee (Hansen and Brødsgaard 1999). AFB is also infective for *A. cerana* (Singh 1961; Chen et al. 2000); however, the validity of the only report from the field of AFB infections in *A. cerana* (Singh 1961) has been questioned (Oldroyd and Wongsiri 2006). Although both *A. mellifera* and *A. cerana* larvae are susceptible to the disease, the Asian honeybee rarely, if ever, shows clinical infections – probably because they remove infected larvae before the larvae are capped (Chen et al. 2000).

It is well documented that European honeybees, with the ability to detect and remove diseased brood, can be completely AFB resistant (Spivak and Gilliam 1998). There are no reports of AFB from honeybee species other than *A. mellifera* and *A. cerana*, although this may reflect a lack of investigation rather than complete host resistance or pathogen specificity. Nevertheless, AFB appears to be a problem for managed populations of the European honeybee only. This could reflect a more developed hygienic behaviour in other species as shown for *A. cerana* (Chen et al. 2000) and/or less susceptible larvae requiring higher spore doses to become infected, as also occurs in *A. cerana* (Chen et al. 2000). Data from AFB infections in European honeybees suggest that the virulence of AFB, being lethal at a colony level in contrast to other honeybee diseases that are shaped by evolution, could be dependent on apicultural practises, and that the pathogen could probably be maintained without causing frequent colony mortality in a natural system (Fries et al. 2006).

Considering that most Asian honeybees exist as wild populations, infections of AFB with a lack of clinical symptoms may be maintained in such populations and mainly spread between colonies through vertical transmission. This could parallel the finding of AFB spores in honey from European honeybees in parts of Africa where clinical symptoms of AFB have never been detected (Hansen et al. 2003). It would be of considerable interest if AFB spores could be retrieved from Asian honeybees and tested for both their larval and colony level virulence, in comparison with isolates from managed European honeybees. It has been demonstrated that different isolates of *P. larvae* differ considerably in their larval virulence, with the most virulent isolates at the larval level and actually least virulent at the colony level (Genersch et al. 2005). Hypothetically, such strains with high virulence at the larval level and low colony level virulence could be maintained without symptoms in honeybee populations, mainly through colony level vertical transmission.

15.3.2 European Foulbrood

European foulbrood (EFB) is caused by the bacterium *Melissococcus plutonius*, Gram-positive, lanceolate cocci (Bailey 1956). EFB is widely distributed in managed populations of the European honeybee and is regionally causing more colony losses than AFB (Wilkins et al. 2007; Roetschi et al. 2008). Bacterial cells of *M. plutonius* are digested with contaminated food and multiply within the midguts of honeybee larvae where it probably competes for nutrients with the host, who eventually dies from starvation (Bailey 1983). A high density of colonies and apiaries has been shown to promote pathogen transmission. Thus, the spatial structure of the host is important for EFB transmission (Belloy et al. 2007). Not surprisingly, problems with EFB in European honeybees are reported primarily from areas with a high colony density, such as Switzerland (Roetschi et al. 2008) and the United Kingdom (Wilkins et al. 2007).

Although reports are few, EFB has been reported to cause considerable damage, both in managed colonies of *A. cerana* (Chinh 1998) and in one colony of *A. laboriosa* (Allen et al. 1990). The first report of EFB in Asian honeybees is from 1971 when the source of infection was assumed to be European honeybee colonies, introduced from the United States of America (Diwan et al. 1971). However, when compared to isolates from European honeybees, this isolate of *M. plutonius* from *A. cerana* could be clearly distinguished using serological typing (Bailey 1974). Thus, *M. plutonius* is probably an endemic infection in Asian honeybees.

There are no reports of EFB from honeybee species other than *A. mellifera*, *A. cerana* and *A. laboriosa*. As for the case with AFB, this probably reflects a lack of investigation rather than host resistance or pathogen specificity. The finding of EFB in managed colonies of *A. cerana*, where colony density is extreme in apiaries as compared to natural conditions, and in *A. laboriosa*, where local colony density in wild populations may also be high, could reflect the pattern observed for European honeybees: symptoms of EFB are associated with high colony density.

15.4 Fungal Infections

15.4.1 Nosema *spp.*

Based on molecular evidence, Microsporidia are now included in the cluster Fungi (Sina et al. 2005). Thus, from a taxonomic point of view, Microsporidia are highly specialised parasitic fungi. Only two microsporidians that have been characterised are known to be infective to honeybees, *Nosema apis* (Zander 1909) and *Nosema ceranae* (Fries et al. 1996). Microsporidians are intracellular parasites with a unique mode of cell invasion (Figs. 15.1 and 15.2). They disperse between hosts as spores which eject a polar filament protruding from the germinating spore when in the right environment, such as the midgut of honeybees in the cases of *N. apis* and *N. ceranae*. The filament must penetrate a host cell membrane into the host cell in order to inject the infective sporoplasm into the host cell cytoplasm where parasite replication, and later spore production, is initiated (Larsson 1986).

N. ceranae from *A. cerana* was first described in samples from the Bee Institute of the Chinese Academy of Agricultural Sciences outside Beijing, China (Fries et al. 1996). There are earlier observations of microsporidian infections in *A. cerana*, referred to as infections by *N. apis* (Singh 1975; Lian 1980; Yakobson et al. 1992); but although there are size differences between the two parasites, they are difficult to differentiate using light microscopy. Thus, these observations may in fact be observations of *N. ceranae*. Cross-infection experiments using *N. ceranae* isolates from China and *N. apis* isolates from Sweden in both *A. cerana* and *A. mellifera* in China, respectively, have demonstrated that

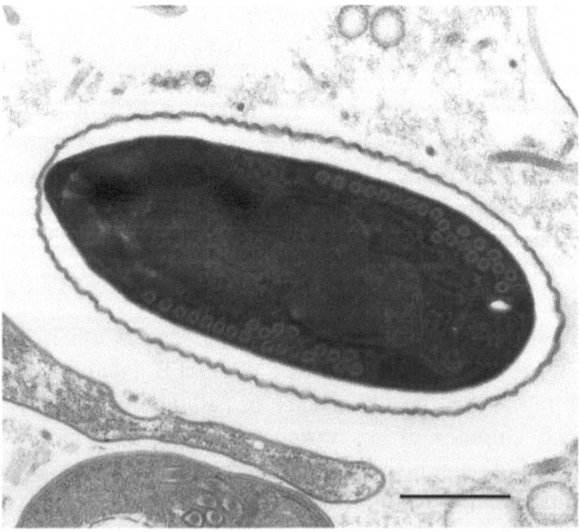

Fig. 15.1 Longitudinal section of a spore of *Nosema ceranae* isolated from *A. cerana*. The *bar* is 1 μm. Photo: I. Fries

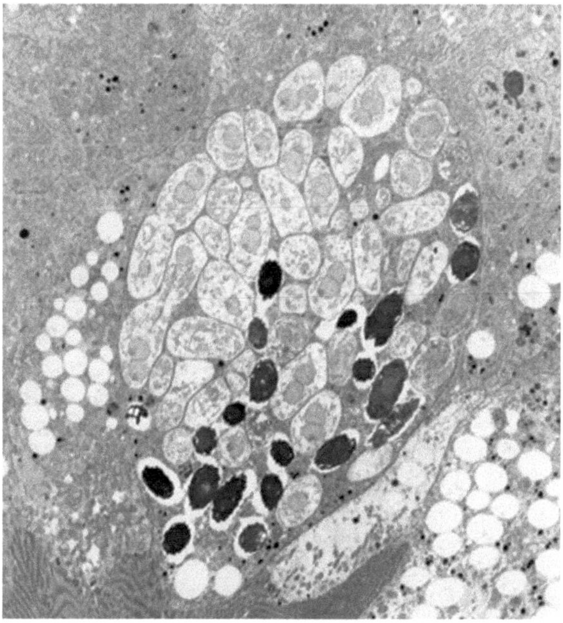

Fig. 15.2 A ventricular cell of *A. cerana* infected by *N. ceranae*, surrounded by healthy tissue. Photo: I. Fries

both parasites are cross-infective across hosts. However, *N. ceranae* develops better in *A. mellifera* as compared to *N. apis* in *A. cerana* (Fries and Feng 1995; Fries 1997). Thus, *N. apis* infections are probably less of a threat to *A. cerana* as compared to *N. ceranae* infections in *A. mellifera*. Indeed, natural infections in the field of *N. apis* in *A. cerana* do exist, but with a much lower prevalence as compared to *N. ceranae* (Chen et al. 2009), whereas *N. ceranae* infections in *A. mellifera* have become widespread and dominant throughout the world in recent years (Higes et al. 2006; Klee et al. 2007). For a review of *N. ceranae* infections in European honeybees, see Fries (2010).

The impact of *N. ceranae* infections on *A. cerana* colonies has not been well studied (Fig. 15.3). The similar lifecycle compared to *N. apis*, with infections of the ventricular cells, suggests that infections are detrimental and may be a threat to profitable beekeeping with this particular honeybee (Fries et al. 1996). If earlier reports of damaging effects and even colony losses from microsporidian infections in *A. cerana* (Singh 1975; Lian 1980) are indeed reports of *N. ceranae* infections, this parasite may need to be controlled in *A. cerana* under certain conditions.

In Asian honeybees, *N. ceranae* has also been detected in *Apis koschevnikovi* using molecular tools in samples from Indonesia (GenBank accession # FJ789802). Thus, it appears likely that this parasite is also infective for other Asian honeybee species. The impact on colony fitness from these infections in Asian honeybees remains unknown. An infection by an unknown microsporidium has been reported

Fig. 15.3 The queen bee of *Apis cerana* is often more dark and uniformly coloured compared to the worker bees. Photo: I. Fries

from *A. florea* and reported to be infective for *A. mellifera* (Böttcher et al. 1973, 1974, 1975). Nothing further is known about this infection.

15.4.2 Chalkbrood

Chalkbrood in honeybees is caused by the heterothallic fungus, *Ascosphaera apis*. The spores are ingested by honeybee larvae with food and the fungal hyphae eventually mummify the infected pupa. For an extensive review on *A. apis* infections in European honeybees, see Aronstein and Murray (2010). Very little is known about chalkbrood in Asian honeybees. The pathogen has been found in *A. cerana* colonies in South Korea (Camazine 1989; Gilliam et al. 1993) and is occasionally found in the same host species in China, mainly in drone brood (Shi Wei personal communication). It appears likely that all *Apis* species can become infected with this fungus because it is less host-specific than most honeybee pathogens. Besides *Apis* spp., *A. apis* has also been isolated from the solitary bees, *Nomandia melanderi* (Rose et al. 1984) and *Xylocopa californica* (Gilliam et al. 1994). However, there are no records of this fungal infection from any other Asian honeybee species, other than from *A. cerana*, and the impact of the infection in that species remains unknown.

15.5 Virus Infections

Five different virus infections have so far been identified from Asian honeybees using serological tools: *Apis* iridescent virus (AIV), black queen cell virus (BQCV), deformed wing virus (DWV), Kashmir bee virus (KBV) and Thai sacbrood virus (TSBV) (Allen and Ball 1996). Undoubtedly, more virus infections will be detected as molecular tools using specific primers become increasingly available.

15.5.1 Apis *Iridescent Virus*

AIV was first detected in dead adult worker bees of *A. cerana* in Kashmir (Bailey et al. 1976). AIV is infective for *A. mellifera* in the laboratory, yet the infection may be limited in nature to *A. cerana* (Allen and Ball 1996); other *Apis* species that occur in Asia however seem likely as alternative hosts (Bailey and Ball 1978). Infections with AIV may cause severe symptoms in colonies of *A. cerana*, causing infected bees to become inactive and lose the ability to fly. This results in inactive clusters of bees in front of infected hives, rendering the name "clustering disease" to the phenomenon – first believed to be caused by tracheal mites (Bailey and Ball 1978). Large colonies of *A. cerana* have been observed to perish within 2 months of becoming visibly infected (Bailey and Ball 1978). Fat body tissue is most frequently infected, but a range of other tissue types become infected as well, including ovaries. The variety of tissues that become infected suggest that the infection could be transmitted via eggs, faeces or gland secretions (Bailey and Ball 1978).

15.5.2 Black Queen Cell Virus

BQCV is associated with *N. apis* infections in *A. mellifera* (Bailey et al. 1983). Thus, it appears likely that BQCV is associated with *N. ceranae* infections in Asian honeybees, although this remains to be verified. To date there is only one record of BQCV infections from Asian honeybees. Using serology, BQCV was detected in diseased drone brood of *A. florea* in Iran (Allen and Ball 1996).

15.5.3 Deformed Wing Virus

In Asian honeybees, DWV has so far been isolated only from *A. cerana* (Allen and Ball 1996). It appears likely that this virus is also infective for all other *Apis* species because it is infective for both *Bombus pascuorum* and *Bombus terrestris* and probably for other bumblebee species as well (Genersch et al. 2006b). The recent attention to DWV infections in European honeybees is due to the devastating effects of the infections when associated with *Varroa destructor* (Bowen-Walker et al. 1999) or *Tropilaelaps mercedesae* (Forsgren et al. 2009) infestations, where the mites serve as mechanical vectors between adult honeybees and brood. The virus also replicates in the mites (Ongus et al. 2004; Dainat et al. 2009), probably increasing vector efficacy. In Asian honeybees, where these mites are endemic, the infestations are kept at low levels by the honeybees. Thus, DWV probably occurs mostly as covert infections, as was the case in European bees before the Asian mites were introduced (Allen and Ball 1996).

15.5.4 Kashmir Bee Virus

The first detection of KBV was from *A. cerana* colonies in Kashmir (Bailey and Woods 1977), and serologically related strains were later found to also be infective for *A. mellifera*. As is the case with DWV, and probably all virus infections that replicate upon injection into honeybee haemolymph, KBV is also vectored by mites that feed on honeybee brood. Indeed, KBV has also been detected in *Varroa jacobsoni* from *A. cerana* (Anderson 1989). KBV infections have not been detected in any other Asian honeybee spp. other than *A. cerana*, but probably exist as covert infections.

15.5.5 Thai Sacbrood Virus

TSBV is serologically related to sacbrood virus in *A. mellifera*, but has distinct physico-chemical properties (Bailey et al. 1982). This virus was first isolated from *A. cerana* colonies in Thailand (Bailey et al. 1982) and appears to be the most widespread and devastating virus infection known from Asian honeybees. Symptoms of sacbrood have been observed also in *A. florea* and in *A. dorsata*, but these observations have not been verified with serological or molecular methods. TSBV infections have been reported to cause considerable losses of *A. cerana* colonies (Verma et al. 1990) and are found in most areas where beekeeping with this honeybee occurs (Allen and Ball 1996). TSBV infections appear to be more detrimental to *A. cerana* colonies as compared to sacbrood infections in *A. mellifera* (Anderson 1995). TSBV and the so-called Chinese sacbrood virus (CSBV), found in *A. cerana* colonies in China, are serologically identical (Allen and Ball 1996; Yan et al. 2009). Molecular data from TSBV from India (GenBank Accession # EU156753) and from CSBV from China (GenBank Accession # AF469603) suggest that they are closely related and can be regarded as different varieties of the same virus. CSBV has been reported to be a serious threat to profitable apiculture with *A. cerana* in China (Dong et al. 1984).

References

Allen M, Ball B (1996) The incidence and world distribution of honeybee viruses. Bee World 77:141–162

Allen MF, Ball BV, Underwood BA (1990) An Isolate of *Melissococcus pluton* from *Apis laboriosa*. J Invertebr Pathol 55:439–440

Anderson DL (1995) Viruses of *Apis cerana* and *Apis mellifera*. In: Kevan PG (ed) The Asiatic hive bee: apiculture, biology, and role in sustainable development in tropical and subtropical Asia. Enviroquest, Cambridge, Ontario, pp 161–170

Anderson DL (1989) Mites and pathogens of the European honeybee *Apis mellifera* and the Eastern hive bee *Apis cerana* in Papua New Guinea. Entomology Division, DSIR, Auckland, p 76

Aronstein KA, Murray KD (2010) Chalkbrood disease in honeybees. J Invertebr Pathol 103:S20–S29

Bailey L (1956) Aetiology of European foulbrood; a disease of the larval honeybee. Nature 178:1130
Bailey L (1974) Unusual type of *Streptococcus pluton* from eastern hive bee. J Invertebr Pathol 23:246–247
Bailey L (1983) *Melissococcus pluton*, the cause of European foulbrood of honeybees (*Apis* spp.). J Appl Bacteriol 55:65–69
Bailey L, Ball BV (1978) *Apis* iridescent virus and clustering disease of *Apis cerana*. J Invertebr Pathol 31:368–371
Bailey L, Ball BV (1991) Honeybee pathology, 2nd edn. Academic, London
Bailey L, Woods RD (1977) Two more small Rna viruses from honeybees and further observations on sacbrood and acute bee paralysis viruses. J Gen Virol 37:175–182
Bailey L, Ball BV, Woods RD (1976) An iridovirus from bees. J Gen Virol 31:459–461
Bailey L, Carpenter JM, Woods RD (1982) A strain of sacbrood virus from *Apis cerana*. J Invertebr Pathol 39:264–265
Bailey L, Ball BV, Perry JN (1983) Association of viruses with two protozoal pathogens of the honeybee. Ann Appl Biol 103:13–20
Belloy L, Imdorf A, Fries I, Forsgren E, Berthoud H, Kuhn R, Charrière J-D (2007) Spatial distribution of *Melissococcus plutonius* in adult honeybees collected from apiaries and colonies with and without symptoms of European foulbrood. Apidologie 38:136–140
Böttcher FK, Hirschfelder H, Mautz D, Weiss K (1973) Die Tätigkeit der Bayerischen Landesanstalt für Bienenzucht in Erlangen im Jahre 1972. Imkerfreund 28:74–84, in German
Böttcher FK, Mautz D, Weiss K (1974) Die Tätigkeit der Bayerischen Landesanstalt für Bienenzucht in Erlangen im Jahre 1973. Imkerfreund 29:70–80, in German
Böttcher FK, Mautz D, Weiss K (1975) Die Tätigkeit der Bayerischen Landesanstalt für Bienenzucht in Erlangen im Jahre 1974. Imkerfreund 30:75–85, in German
Bowen-Walker PL, Martin SJ, Gunn A (1999) The transmission of deformed wing virus between honeybees (*Apis mellifera* L.) by the ectoparasitic mite *Varroa jacobsoni* Oud. J Invertebr Pathol 73:101–106
Bull JJ, Molineux IJ, Rice WR (1991) Selection of benevolence in a host-parasite system. Evolution 45:875–882
Camazine S (1989) Beekeeping in South Korea. Bee World 70:66–74
Canning EU (1982) An evaluation of protozoal characteristics in relation to biological control of pests. Parasitology 84:119–149
Chen YP, Evans JD, Zhou L, Boncristiani H, Kimura K, Xiao TG, Litkowski AM, Pettis JS (2009) Asymmetrical coexistence of *Nosema ceranae* and *Nosema apis* in honeybees. J Invertebr Pathol 101:204–209
Chen Y-W, Wang C-H, An J, Ho K-K (2000) Susceptibility of the Asian honeybee, *Apis cerana*, to American foulbrood, *Paenibacillus larvae larvae*. J Apic Res 39:169–175
Chinh PH (1998) Tai sac brood virus disease control in Vietnam. In: Proceedings of 4th Asian apicultural association international conference, Katmandu, pp 57–59
Dainat B, Ken T, Berthoud H, Neumann P (2009) The ectoparasitic mite *Tropilaelaps mercedesae* (Acari, Laelapidae) as a vector of honeybee viruses. Insectes Soc 56:40–43
Deodikar GB, Ghatge AL, Phadke RP, Mahindre DB, Kshirsagar KK, Muvel S, Thakar SS (1977) Nesting behaviour of Indian honeybees. III. Nesting behaviour of *Apis dorsata*. Indian Bee J 39:1–12
Diwan VV, Kshirsag KK, Rao AVR, Raghunat D, Bhambure CS, Godbole SH (1971) Occurrence of a new bacterial disease of Indian honeybees *Apis indica* F. Curr Sci 40:196–197
Dong BY, Fang YZ, Zhang GC, Guo ZQ, Liu LS (1984) Study of the sacbrood virus. Bee Cult China 3:6–8, in Chinese
Forsgren E, de Miranda JR, Isaksson M, Shi W, Fries I (2009) Deformed wing virus associated with *Tropilaelaps mercedesae* infesting European honeybees (*Apis mellifera*). Exp Appl Acarol 47:87–97

Fries I (1997) Protozoa. In: Morse RA, Flottum K (eds) Honeybee pests, predators and diseases. A.I. Root, Medina, Ohio, pp 59–76

Fries I (2010) *Nosema ceranae* in European honeybees (*Apis mellifera*). J Invertebr Pathol 103: S73–S79

Fries I, Camazine S (2001) Implications of horizontal and vertical pathogen transmission for honeybee epidemiology. Apidologie 32:199–214

Fries I, Feng F (1995) Cross-infectivity of *Nosema apis* in *Apis mellifera* and *Apis cerana*. In: Proceedings of 34th Apimondia international apicultural congress, Lausanne, pp 151–155

Fries I, Feng F, da Silva A, Slemenda SB, Pieniazek NJ (1996) *Nosema ceranae* n. sp. (Microspora, Nosematidae), morphological and molecular characterization of a microsporidian parasite of the Asian honeybee *Apis cerana* (Hymenoptera, Apidae). Eur J Protistol 32: 356–365

Fries I, Lindström A, Korpela S (2006) Vertical transmission of American foulbrood (*Paenibacillus larvae*) in honeybees (*Apis mellifera*). Vet Microbiol 114:269–274

Genersch E, Ashiralieva A, Fries I (2005) Strain- and genotype-specific differences in virulence of *Paenibacillus larvae* subsp *larvae*, a bacterial pathogen causing American foulbrood disease in honeybees. Appl Environ Microbiol 71:7551–7555

Genersch E, Forsgren E, Pentikainen J, Ashiralieva A, Rauch S, Kilwinski J, Fries I (2006a) Reclassification of *Paenibacillus larvae* subsp *pulvifaciens* and *Paenibacillus larvae* subsp *larvae* as *Paenibacillus larvae* without subspecies differentiation. Int J Syst Evol Microbiol 56:501–511

Genersch E, Yue C, Fries I, de Miranda JR (2006b) Detection of deformed wing virus, a honeybee viral pathogen, in bumble bees (*Bombus terrestris* and *Bombus pascuorum*) with wing deformities. J Invertebr Pathol 91:61–63

Gilliam M, Lorenz BJ, Prest DB, Camazine S (1993) *Ascosphaera apis* from *Apis cerana* from South Korea. J Invertebr Pathol 61:111–112

Gilliam M, Lorenz BJ, Buchmann SL (1994) *Ascosphaera apis*, the chalkboard pathogen of the honeybee, *Apis mellifera*, from larvae of a carpenter bee, *Xylocopa californica arizonensis*. J Invertebr Pathol 63:307–309

Goodwin RM, Houten TA, Perry JH (1994) Incidence of American foulbrood infections in feral honeybee colonies in New Zealand. NZ J Zool 21:285–287

Hansen H, Brødsgaard C (1999) American foulbrood: a review of its biology, diagnosis and control. Bee World 80:5–23

Hansen H, Brødsgaard CJ, Kryger P, Nicolaisen M (2003) A scientific note on the presence of *Paenibacillus larvae larvae* spores in sub-Saharan African honey. Apidologie 34:471–472

Herre EA (1993) Population structure and the evolution of virulence in nematode parasites of fig wasps. Science 259:1442–1445

Hornitzky M, Oldroyd BP, Somerville D (1996) *Bacillus larvae* carrier status of swarms and feral colonies of honeybees (*Apis mellifera*) in Australia. Aust Vet J 73:116–117

Higes M, Martín R, Meana A (2006) *Nosema ceranae*, a new microsporidian parasite in honeybees in Europe. J Invertebr Pathol 92:93–95

Klee J, Besana AM, Genersch E, Gisder S, Nanetti A, Dinh Quyet T, Tong Xuan C, Puerta F, Ruz JM, Kryger P, Message D, Hatjina F, Korpela S, Fries I, Paxton RJ (2007) Widespread dispersal of the microsporidian *Nosema ceranae*, an emergent pathogen of the western honeybee, *Apis mellifera*. J Invertebr Pathol 96:1–10

Larsson R (1986) Ultrastructure, function, and classification of Microsporidia. Progr Protistol 1:325–390

Lian CZ (1980) *Nosema* disease in honeybee (*Apis cerana cerana*). Apic China 4:15–16, in Chinese

Lipsitch M, Herre EA, Nowak MA (1995a) Host population structure and the evolution of virulence: a "law of diminishing returns". Evolution 49:743–748

Lipsitch M, Nowak MA, Ebert D, May RM (1995b) The population dynamics of vertically and horizontally transmitted parasites. Proc R Soc Lond B 260:321–327

Lipsitch M, Siller S, Nowak MA (1996) The evolution of virulence in pathogens with vertical and horizontal transmission. Evolution 50:1729–1741
Manning R, Lancaster K, Rutkay A, Eaton L (2007) Survey of feral honeybee (*Apis mellifera*) colonies for *Nosema apis* in Western Australia. Aust J Exp Agric 47:883–886
Moritz RFA, Southwick EE (1992) Bees as superorganisms. Springer, Berlin
Morse RA, Flottum K (eds) (1997) Honeybee pests, predators and diseases. A.I Root, Medina, Ohio
Oldroyd BP, Wongsiri S (2006) Asian honeybees. Harvard University Press, London
Ongus JR, Peters D, Bonmatin JM, Bengsch E, Vlak JM, van Oers MM (2004) Complete sequence of a picorna-like virus of the genus *Iflavirus* replicating in the mite *Varroa destructor*. J Gen Virol 85:3747–3755
Roetschi A, Berthoud H, Kuhn R, Imdorf A (2008) Infection rate based on quantitative real-time PCR of *Melissococcus plutonius*, the causal agent of European foulbrood, in honeybee colonies before and after apiary sanitation. Apidologie 39:362–371
Rose JB, Christensen M, Wilson WT (1984) *Ascosphaera* species inciting chalkbrood in North America and taxonomic key. Mycotaxon 19:41–55
Sina M, Alastair G, Farmer M, Andersen R, Anderson O, Barta J, Bowser S, Brugerolle G, Fensome R, Fredericq S, James T, Karpov S, Kugrens P, Krug J, Lane C, Lewis L, Lodge J, Lynn D, Mann D, Maccourt R, Mendoza L, Moestrup O, Mozley S, Nerad T, Shearer C, Smirnov A, Spiegel F, Taylor M (2005) The new higher level classification of eukaryotes with emphasis on the taxonomy of protists. J Eukaryot Microbiol 52:399–451
Singh S (1961) Appearance of American foulbrood disease in Indian honeybee (*Apis cerana indica* Fabr.). Indian Bee J 27:46–50
Singh Y (1975) *Nosema* in Indian honeybee (*Apis cerana indica*). Am Bee J 115:59
Spivak M, Gilliam M (1998) Hygienic behaviour of honeybees and its application for control of brood diseases and varroa. Part I. Hygienic behaviour and resistance to American foulbrood. Bee World 79:124–134
Turner PE, Cooper VS, Lenski RE (1998) Trade-off between horizontal and vertical modes of transmission in bacterial plasmids. Evolution 52:315–329
Verma LR, Rana BS, Verma S (1990) Observations on *Apis cerana* colonies surviving from Thai sacbrood infestations. Apidologie 21:169–174
Wilkins S, Brown M, Andrew A, Cuthbertson GS (2007) The incidence of honeybee pests and diseases in England and Wales. Pest Manag Sci 63:1062–1068
Yakobson B, Pothichot S, Wongsiri S (1992) Possible transfer of *Nosema apis* from *Apis mellifera* to *Apis cerana*. In: Proceedings of international conference on asian honeybees bee mites, Bangkok, p 97
Yan X, Chen JH, Han RC (2009) Detection of Chinese Sacbrood Virus (CSBV) in *Apis cerana* by RT-PCR Method. Sociobiology 53:687–694
Zander E (1909) Tierische Parasiten als Krankenheitserreger bei der Biene. Münch Bienenztg 31:196–204, in German

Chapter 16
Asian Honeybee Mites

Natapot Warrit and Chariya Lekprayoon

16.1 Introduction

In honeybees, brood parasitism by parasitic mites (Acari) has evolved in every lineage except *A. mellifera* (Eickwort 1994). Despite the fact that all parasitic mites of honeybees are native to Asia, except for one tracheal mite species, we rarely encounter cases where these mites cause serious damage to their native hosts (Boot et al. 1997; Oldroyd 1999). Natural selection for benign host–parasite interactions might play a role in these relationships to allow co-existence between the bees and the mites (Fries and Camazine 2001; cf. Chap. 15). However, the more domesticated *A. mellifera* is not as fortunate as its relatives. As in many cases with horizontal transmission of a parasite to a new host, the host is ill-equipped behaviourally and physiologically to defend itself and may succumb to the novel invader (Lipsitch et al. 1995a, b). Thus, *A. mellifera* falls victim to these mites of Asian origin, which frequently results in the disintegration of colonies.

In Asia, two families of parasitic mesostigmatid mites are represented: Varroidae and Laelapidae. The tracheal mite, *Acarapis woodi*, in another family of parasitic mites from the order Prostigmata, is native to Europe and is considered to be an introduced parasitic species of Asian honeybees. Since the end of World War II, westerners introduced large numbers of *A. mellifera* colonies into Asia for commercial purposes (Crane 1988). These events provided the opportunities for the mites to find new hosts. Most of the time, cross-infections of the mites from their

N. Warrit
Center of Excellence in Entomology, Department of Biology, Faculty of Sciences, Chulalongkorn University, Bangkok, Thailand 10330,
and
Department of Entomology, National Museum of Natural History, Smithsonian Institution, Washington, DC 20013, USA
e-mail: Natapot.w@chula.ac.th

C. Lekprayoon
Center of Excellence in Biodiversity, Department of Biology, Faculty of Sciences, Chulalongkorn University, Bangkok 10330, Thailand
e-mail: Chariya.l@chula.ac.th

H.R. Hepburn and S.E. Radloff (eds.), *Honeybees of Asia*,
DOI 10.1007/978-3-642-16422-4_16, © Springer-Verlag Berlin Heidelberg 2011

native hosts to *A. mellifera* occur as the result of robbing or drifting of foraging bees carrying mites from nearby colonies (Cook 1987; Rath et al. 1991; Fries and Camazine 2001). The moving of mite-infested colonies back to Europe from Asia introduced the parasites that eventually established themselves in apiaries in new locations throughout the world.

The most well-known case of an Asian mite introduction into Europe and the Americas is the notorious *Varroa* mite that switched hosts from its native Asian cavity-nesting species, *Apis cerana* to *Apis mellifera*, which later spread to almost every part of the world (Griffith and Bowman 1981; De Jong et al. 1982; de Guzman et al. 1997, 1998; Sammataro et al. 2000; Zhang 2000). Vast resources and attention have been invested in studies of chemical and biological treatments of *Varroa* on *A. mellifera*, despite numerous reports on the problems of differential resistance of the mites to these treatments (cf. Gerson et al. 1991; Lodesani et al. 1995; Colin et al. 1997; Eischen 1998a, b; Elzen et al. 1998). However, almost a century after the discovery of *Varroa*, we learned that the variations in resistance to *Varroa* stem from the genetic variations in different populations of these superficially similar mites (Anderson and Trueman 2000).

For the past decade, molecular genetics combined with classical taxonomic studies have shed light and revealed a lack of homogeneity in the populations of the parasitic mites of the honeybees in Asia. We are only beginning to learn about these differences and the potential they hold for *A. mellifera*. To understand the biology of Asian bee mites, and hence their control, can best be understood by reference to their native hosts. This chapter attempts to discuss the diversity, not only at the species level, but also the variations in different mite populations on Asian honeybees in the hopes of emphasising the importance of population genetic studies in research on honeybee mite epidemiology. The focus of this chapter is on the parasitic mesostigmatid mites of honeybees, Varroidae and Laelapidae, because they are the major groups of parasitic mites in Asia that also have an impact on beekeeping worldwide.

A list of currently described parasitic mesostigmatid mites and their native hosts in Asia is given in Table 16.1, as well as an identification key to the honeybee mites in Appendix 16.1. To avoid as much duplication as possible of other texts regarding Asian bee mites, we provide only short descriptions of the mites' life history traits for *Varroa* and *Tropilaelaps*. For more details, readers should consult other resources on topics such as life cycles, pathological conditions and treatments (cf. Morse and Nowogrodzki 1990; Sammataro et al. 2000; Webster and Delaplane 2001). In addition, the current situation on the tracheal mites in *A. cerana* in Asia is briefly discussed in Sect. 16.5 along with a section on honeybee non-parasitic mites in Sect. 16.6.

16.2 *Varroa* Mites

One of the prominent parasitic mites that have a long history with beekeepers around the world is the genus *Varroa* (Family Varroidae). The type species of the genus was

Table 16.1 Mesostigmatid species of parasitic honeybee mites, their native hosts and geographic distributions in Asia. Data presented here were compiled from references cited in the text

Mite species	Native hosts	Geographical distributions
Family Varroidae		
Genus Varroa Oudemans (1904)		
V. destructor Anderson and Trueman (2000)	Mainland *A. cerana*[a]	India, Pakistan, Nepal, China, Japan, Korea, Taiwan, northern Thailand, Vietnam
V. jacobsoni Oudemans (1904)	Sundaland *A. cerana*[a]	Central and southern Thailand, Malaysia, Indonesia, Palawan Island
	A. nigrocincta	Sulawesi Island
V. underwoodi Delfinado-Baker and Aggarwal (1987a)	*A. cerana*	Presumably in *A. cerana* range except for China, Vietnam, Philippine Islands
	A. nigrocincta[b]	Sulawesi Island
	A. nuluensis[b]	Mt. Kinabalu, Borneo
V. rindereri de Guzman and Delfinado-Baker (1996)	*A. koschevnikovi*	Malaysia, Indonesia except islands beyond east of the Wallace line
Genus *Euvarroa* Delfinado and Baker (1974)		
E. sinhai Delfinado and Baker (1974)	*A. florea*	The Middle East, Pakistan, India, Nepal, Burma, Indochina, Malay Peninsula
E. wonsirii Lekprayoon and Tangkanasing (1991)	*A. andreniformis*	Burma, Indochina, Malaysia, Indonesia except islands beyond east of the Wallace line, Palawan Island
Family Laelapidae		
Genus *Tropilaelaps* Delfinado and Baker (1961)		
T. mercedesae Anderson and Morgan (2007)	*A. dorsata*	India, Nepal, Burma, south China, Indochina, Malaysia, Indonesia except Sulawesi, Palawan Island
T. clareae Delfinado and Baker (1961)	*A. breviligula*	Philippines except Palawan Island
	A. dorsata binghami	Sulawesi Island
	A. laboriosa	Nepal
T. koenigerum Delfinado-Baker and Baker (1982)	*A. dorsata*	India, Thailand, Borneo
	A. laboriosa	Nepal
T. thaii Anderson and Morgan (2007)	*A. laboriosa*	North Vietnam

[a]*A. cerana* populations in Asia can be categorised as belonging to four major mitochondrial lineages (Smith and Hagen 1996; Smith et al. 2000). Two of the four of these lineages have wide distributions. The so-called "Mainland" *A. cerana* is distributed throughout the India–Pakistan subcontinent, China, Japan, Laos, Cambodia, Vietnam, north and central Thailand. The "Sundaland" population occupies southern Thailand, Malaysia and Indonesia. The other two lineages occur in the Philippines

[b]The records of *V. underwoodi* on *A. nigrocincta* and *A. nuluensis* have never been confirmed. However, the probability of finding *V. underwoodi* on these bees is high (Oldroyd and Wongsiri 2006)

described early in the twentieth century and named *V. jacobsoni* Oudemans 1904, from a colony of *A. cerana* in Java, Indonesia. *Varroa* mites later spread into *A. mellifera* colonies that were introduced into Asia and have since infected many commercial bee colonies around the world due to transcontinental shipments of bee colonies (de Guzman et al. 1997, 1998). *Varroa* has now spread into most parts of the world except for Australia and the Antarctica (Sammataro et al. 2000). A decade ago, New Zealand and Hawaii were considered to be *Varroa*-free zones; however, recent reports have shown that the mites have already become established in those areas (Zhang 2000; HDOA 2010).

Varroa mites are generally elliptical and flattened dorso-ventrally, which provides the mites with the ability to attach themselves inconspicuously to sclerites of bees or to hide between areas where it is difficult for the bees to groom such as the propodeum (Yoder et al. 1999). The colour of *Varroa* ranges from reddish-brown to bright red. In an *A. cerana* colony, *Varroa* can reproduce only in the drone brood cells (Peng et al. 1987; Büchler et al. 1992; Fries et al. 1996; Boot et al. 1997; Rath 1999) because the bees can detect, groom and remove the mites from worker bees and worker brood cells. This bee behaviour reflects a co-adaptation between the host and parasite during their evolutionary history (Oldroyd 1999; Rath 1999; Sasagawa et al. 1999).

All species of the genus *Varroa* are native to Asia. Current knowledge on *Varroa* taxonomy recognises four valid species: *V. jacobsoni* Oudemans, *V. destructor* Anderson and Trueman, *V. underwoodi* (Delfinado-Baker and Aggarwal 1987a) and *V. rindereri* (de Guzman and Delfinado-Baker 1996). However, at least one population in the Philippine Islands has a distinct genetic composition different from others in mainland Asia and Indonesia that may also be valid as a separate species (Anderson and Trueman 2000; Anderson 2004). Only *V. destructor* is reported to colonise and wreak havoc on *A. mellifera* (Anderson and Trueman 2000; Warrit et al. 2006; Navajas et al. 2010). *A. cerana* is the primary host to at least three species of *Varroa* in Asia, except for *V. rindereri*. As to current knowledge, we know that *A. koschevnikovi* is parasitised only with *V. rindereri* (de Guzman and Delfinado-Baker 1996). A congeneric of *A. cerana* in Sulawesi, *A. nigrocincta*, is reported to harbour *V. jacobsoni* (Anderson and Trueman 2000), while there is speculation that *V. underwoodi* may infest this bee, and also *A. nuluensis* on Mt. Kinabalu, Malaysia (Otis and Kralj 2001).

V. jacobsoni was once believed to have as wide a distribution as *A. cerana* ranging throughout Asia; however, before 2000, there were numerous reports on pathological and genetic variations of *V. jacobsoni* in both *A. cerana* and *A. mellifera* worldwide (Moritz and Haenel 1984; Camazine 1986; Delfinado-Baker 1988; Delfinado-Baker and Houck 1989; Ritter et al. 1990; Moretto et al. 1991; Anderson 1994; Eguaras et al. 1995; Kraus and Hunt 1995; Anderson and Sukarsih 1996; De Jong and Soares 1997; de Guzman et al. 1997, 1998; Anderson and Fuchs 1998). Subsequently, Anderson and Trueman (2000) published a breakthrough paper that revealed that *V. jacobsoni* is in fact a complex of at least two species – *V. jacobsoni* sensu stricto (s. str.) and *V. destructor*. This finding was based on genetic variations of the cytochrome *c* oxidase I (COI) gene and the results of reproductive isolation experiments of both mites on *A. cerana* and *A. mellifera*. The study revealed that *V. destructor* is a native parasite of *A. cerana* in mainland Asia.

The natural distribution of *V. destructor* ends in northern Thailand (Anderson and Trueman 2000; Warrit et al. 2006), while further down the Malay Peninsula and the Indonesia archipelago, the mite that reproduces in native *A. cerana* is *V. jacobsoni*. This finding also suggested that the mite species in Asia might have co-evolved with certain *A. cerana* mitochondrial DNA lineages (Anderson and Trueman 2000; Warrit et al. 2006). Only populations of *V. destructor* from Korea and Japan (designated by Anderson and Trueman (2000) as "K" and "J" haplotypes) are known to reproduce successfully in *A. mellifera*. The "K" haplotype was reported to be far more virulent than the "J" counterpart (de Guzman et al. 1997; de Guzman and Rinderer 1999; Anderson and Trueman 2000; Garrido et al. 2003). Navajas et al. (2010) suggested that the unique genetic make-up of Japanese and Korean *V. destructor* may be a factor facilitating their colonising of *A. mellifera* and might have stemmed from a genetic "bottleneck" event in the *Varroa* population in mainland Asia. They also added that the "K" and "J" haplotypes have variants that are well established in *A. mellifera* colonies in China, Taiwan, Vietnam and Thailand that have never been reported in *A. mellifera* in Europe and the Americas and might pose a new threat to apiculture worldwide.

Little is known about the basic biology of the more obscure *V. underwoodi* and *V. rindereri*. It is widely understood that *V. underwoodi* can be found with *A. cerana* throughout its range and sometimes living sympatrically with *V. destructor* or *V. jacobsoni* depending upon the locality of *A. cerana* (Oldroyd and Wongsiri 2006). With the distribution of *A. koschevnikovi* in the Malay Peninsula and the Indonesian archipelago, one can assume that *V. rindereri* is spread along with this bee in the region, except for the islands east of the Wallace line (Otis 1996).

16.2.1 Life Cycle of Varroa

The life cycle of *V. destructor* on *A. mellifera* is well studied and can be related to an understanding of this mite on *A. cerana* (cf. De Jong 1997; Donzé and Guerin 1994, 1997; Sammataro et al. 2000; Oldroyd and Wongsiri 2006). *Varroa*'s life cycle consists of two phases: phoretic and parasitic (Oldroyd and Wongsiri 2006). In the parasitic phase, the gravid adult female mite enters the prepupal brood cell 1 or 2 days before cell-capping by the nurse bees. The mite conceals itself, submerged in the liquid brood food until the cell is capped. The peritreme structure of the mite is used like a snorkel to help the mite breathe under the liquid food (Donzé and Guerin 1997). After the cell is capped and the prepupa is formed, the mite starts feeding on the bee haemolymph. About 60 h after the cell is capped, the mother mite produces her first egg, which develops into a male. All of the subsequent eggs (usually 3 or 4), which are laid at 30 h intervals, are destined to become females.

Immature mites (nymphs) feed on the haemolymph of the prepupa at the site where the mother mite usually feeds (Donzé and Guerin 1994). Male mites require slightly less time than females (5–6 and 6–7 days, respectively) to develop into adults. Mating occurs between siblings of the same brood. The life cycle of the male mite is short compared to the females. The male mite dies in the brood cell after the last female sister is fertilised and before the cell is uncapped. After the brood cell is uncapped, the female mite begins its phoretic phase. However, if the bee colony is not in its reproductive cycle, the mite will seek for a newly emerged adult bee as its short-term host. The mite stays on the bee's metasoma, or the area behind the head, and feeds continuously most of the time. The phoretic phase of the mite is shortened if the infested colony is full of new eggs and larvae that have recently been laid by the queen bee. Le Conte et al. (1989) suggested that, because of the higher levels of fatty acid esters produced by drone larvae, the female mites prefer invading drones more than worker cells.

16.2.2 Impact of Varroa *on* A. cerana

Tewarson et al. (1992) provided the first insights into the life cycle of *V. destructor* on an *A. cerana* colony in India. The mite population growth rate in an *A. cerana* colony is significantly slower than in *A. mellifera* (Boot et al. 1997). Most of the time, the mites infest the drone brood, and it is rare to observe the mites invading worker cells. Thus, the damage to the colony of the mite's native host is not as severe as we observe in *A. mellifera*, where mites are found reproducing successfully in worker cells. The defensive behaviours of *A. cerana* against the mites can explain the low growth rate of the mite population. The worker bees can groom each other to remove mites, a behaviour that is lacking in *A. mellifera* (Peng et al. 1987; Büchler et al. 1992; Fries et al. 1996; Rath 1999), and can detect capped cells that are infested with mites, which are later uncapped and the mites removed or both the bee prepupae and the mites are buried together (Büchler et al. 1992; Fries et al. 1996; Boot et al. 1997; Boecking and Spivak 1999).

16.3 *Euvarroa* Mites

A sister genus to *Varroa*, *Euvarroa* is a genus of parasitic mites that has been reported to colonise the Asian dwarf honeybees. *Euvarroa sinhai* Delfinado and Baker 1974 (the type species of the genus *Euvarroa*; Varroidae) was originally described from its host, *A. florea*, in India (Delfinado and Baker 1974). It can be found throughout the distribution area of *A. florea* in Asia from the Middle East to the Malay Peninsula (Otis 1996). A second species, *E. wongsirii* Lekprayoon and Tangkanasing 1991, was later discovered from *A. andreniformis* in Thailand (Lekprayoon and Tangkanasing 1991) and has a distribution ranging from south China to Palawan Island of the Philippines (Otis 1996). The two mite species share an overlap of distribution in South East Asia. Both *Euvarroa* mites superficially resemble one another but can be easily distinguished from *Varroa* by their more or less similar body lengths and widths. Thus, *Euvarroa* is more triangularly shaped than *Varroa* which is more oval. *E. wongsirii* can be distinguished morphologically from *E. sinhai* by the broader triangular body and anal plate bearing 47–54 long lanceolate setae on a wider posterior end (where *E. sinhai* has 39–40 setae) (Lekprayoon and Tangkanasing 1991, 1993; Morin and Otis 1993).

The biology of *Euvarroa* is similar to that of *Varroa*. The mites are only capable of reproducing in the drone brood cells of their hosts and disperse to other colonies via both drones and worker bees (Akratanakul and Burgett 1976; Mossadegh and Birjandi 1986; Aggarwal 1988; Kapil and Aggarwal 1989; Morin and Otis 1993). As with *Varroa*, *Euvarroa* causes little damage to their endemic host colonies. This may be because of the intensive grooming behaviour of the worker bees and the seasonal presence of drone brood cells that reduces the population of the mite to a minimal. However, evidence suggests that *E. sinhai* can reproduce in worker brood cells of *A. mellifera* (Mossadegh 1990). Also, Koeniger et al. (1993) reported that *E. sinhai* can survive in adult workers of *A. mellifera* and *A. cerana* in Thailand.

However, the impact of the *Euvarroa* infestation in apiary colonies of *A. mellifera* or *A. cerana* has not been determined. There are suggestions about why *Euvarroa* does not colonise *A. mellifera* colonies successfully. Sihag (1988) reported that competition among *Varroa*, *Euvarroa* and *Tropilaelaps* in the same *A. mellifera* colonies in Haryana, India, has an effect that reduces the population of *Euvarroa* in the colonies dramatically. The construction of queen cells may also reduce the population of the mites as well (Aggarwal and Kapil 1988). Thus far, reports on the occurrences of *E. wongsirii* outside of its native host are scarce, coming only from debris of *A. dorsata* colonies in Sabah, Malaysia, which were found to have *E. wongsirii* along with *Tropilaelaps* and *Varroa* (Koeniger et al. 2002).

Recent molecular and biogeographical studies of *A. florea* and *A. andreniformis* suggest that populations of the dwarf honeybees in Asia may not be as homogenous as previously thought (cf. Chaps. 2 and 4). This finding raises questions about the genetic homogeneity of the *Euvarroa* associated with these bees. Morin and Otis (1993) found some morphological differences in *E. sinhai* (e.g. number of setae on metatarsus I–IV) collected from India/Sri Lanka compared to specimens collected in Thailand. There is a great possibility of finding genetic and morphological variations among populations of both *E. sinhai* and *E. wongsirii* from different locations in the distribution areas of their hosts in Asia. This line of study will further our knowledge of *Euvarroa* diversity in Asia.

16.4 *Tropilaelaps* Mites

Twenty-five years ago, Burgett and Akratanakul (1985) predicted that in the near future *Tropilaelaps clareae* would play a major role in the destruction of commercial honeybee colonies far greater than that caused by *A. woodi* and *Varroa* mites. Since then, *T. clareae* has been reported in countries beyond its primary host distribution (Kumar et al. 1993; Matheson 1996; Sammataro et al. 2000; Otis and Kralj 2001), but still no account of its presence in Europe and the Americas has been reported. Because of the absence of *T. clareae* in western countries and the great interest in *Varroa* mites, the importance of *Tropilaelaps* mites has not been sufficiently recognised by bee researchers. Nevertheless, after the introduction of *A. mellifera* into Asia, cross-infection of *T. clareae* from its original Asian honeybee hosts occurred resulting in a significant loss to commercial honey production (De Jong et al. 1982; Burgett et al. 1983; Bailey and Ball 1991). Burgett and Akratanakul were not far off their prediction, particularly when one considers Asian apicultural industries.

Recently, many western bee researchers have expressed concerns about the cross-infestation of *Tropilaelaps* from its original host to *A. mellifera* in Europe and the Americas (Matheson 1996, 1997; Waite 2003; Sammataro 2004; Baker et al. 2005). The trepidation of *Tropilaelaps* was soon recognised by the Office International des Epizooties (OIE) in Paris who declared *Tropilaelaps* to be emerging parasite for honeybees worldwide in 2004. The relatively rapid developmental time of *Tropilaelaps* when it is established and starts reproducing in the brood cells of the host concern many beekeepers and researchers. When a colony is infected with *Tropilaelaps*, the mite populations can build up quickly and lead to the sudden collapse of colonies (Sammataro et al. 2000). Moreover, albeit the mites cannot feed and damage adult bees (because the mouthparts of *Tropilaelaps* are only suitable for feeding on soft tissue – Griffiths 1988), they are phoretic and can hitch-hike with adult bees foraging outside the hives, which can distribute the mites to other bees and colonies.

In Asia, reports on colony losses of *A. mellifera* due to *Tropilaelaps* infestations, specifically *T. clareae*, are not uncommon (Laigo and Morse 1968; Burgett et al. 1983, 1990; Tangkanasing et al. 1988; Otis and Kralj 2001). Beekeepers describe deformed adult bees with shortened abdomens, missing legs, and wrinkled wings as common pathological symptoms for colonies parasitised by *T. clareae*. Other symptoms include brood malformation and bees which crawl rather than fly (Sammataro 2004). Two recent separate reports considered the possibility of *Tropilaelaps* as a potential vector for the Deformed Wing Virus (Dainat et al. 2009; Forsgren et al. 2009), which is suggested as a culprit for the maladies observed in adult bees from infected colonies.

There are many review articles and books available (cf. De Jong et al. 1982; Morse and Nowogrodzki 1990; Sammataro et al. 2000; Webster and Delaplane 2001) that deal with much of the life history and host–parasite relationships between *Tropilaelaps* and its hosts. Because there have been some new and exciting discoveries regarding the diversity of *Tropilaelaps* species and their host specificities in recent years, we focus this part of the chapter on the morphological and genetic variations of *Tropilaelaps* and their associations with their honeybee hosts to clarify and give the reader a better understanding of the identity of this emerging threat to honeybees. A brief description of the *Tropilaelaps* life cycle is given in Sect. 16.4.2, but readers should consult Sammataro et al. (2000) and Oldroyd and Wongsiri (2006) for more comprehensive details.

16.4.1 Identity of **Tropilaelaps clareae** *and its Host Ranges*

Tropilaelaps mites (Family Laelapidae) are obligate ectoparasites that feed on the haemolymph of larval honeybees. Both sexes of the mites are elongated and their bodies are covered with numerous, short, spine-like setae. *Tropilaelaps* are reddish-brown in colour, though the males are less sclerotised. Compared to *Varroa* mites, *Tropilaelaps* are smaller, flatter and more oval-shaped than round. The type species of the genus *Tropilaelaps* is *T. clareae* Delfinado and Baker 1961. *T. clareae* was first described from *A. mellifera* colonies in the Philippines and from rats living near bee colonies (Delfinado and Baker 1961). Its primary host species was later revealed to be *A. dorsata* from the Philippines (Bharadwaj 1968; Laigo and Morse 1968). Although current molecular evidence suggests that "*A. dorsata*", host of *T. clareae* in the Philippines, is a distinct but closely related species, *A. breviligula* (Lo et al. 2010; cf. Chap. 1). *T. clareae* has been reported throughout most of the *A. dorsata* distribution in Asia (Matheson 1996; Anderson and Morgan 2007).

In 1982, Delfinado-Baker and Baker described a second *Tropilaelaps* species, *T. koenigerum*, from *A. dorsata* colonies in Sri Lanka. Later on, *T. koenigerum* was reported in colonies of *A. dorsata* in India, Thailand and Borneo (Delfinado-Baker and Baker 1982; Koeniger et al. 2002; Tangjingjai et al. 2003) and in colonies of *A. laboriosa* in Nepal (Delfinado-Baker et al. 1985), which lives sympatrically with *T. clareae* in the same colonies. There are also reports on the occurrence of *T. clareae* on *A. cerana* and *A. florea* in Asia, though the impact on these honeybees is not severe (Delfinado-Baker 1982; Aggarwal 1988; Sihag 1988; Delfinado-Baker et al. 1989; Abrol and Putatunda 1995; Woyke 2005; Anderson and Morgan 2007).

Before 2007, it was assumed that the mite that had become established in *A. mellifera* colonies in South Asia and the Oriental region and spread into Iran, Afghanistan, Kenya, South Korea and the Western Pacific Island of New Guinea was *T. clareae* (Burgett et al. 1983; Woyke 1984; Matheson 1996; Delfinado-Baker and Aggarwal 1987b; Kumar et al. 1993; Anderson 1994; Sammataro et al. 2000; Otis and Kralj 2001). However, Anderson and Morgan (2007), by examining the genetic variations, morphological variations and host associations of *Tropilaelaps* mites and their endemic hosts in Asia, showed that the once well-recognised "*T. clareae*" that infected *A. dorsata* and *A. mellifera* colonies throughout Asia and beyond was in fact comprised of two distinct species – *T. mercedesae*, which colonised *A. dorsata* and the introduced *A. mellifera* inhabiting mainland Asia and Indonesia (excluding Sulawesi Island), and *T. clareae* (s. str.), which parasitised *A. breviligula*, *A. dorsata binghami* (Sulawesi Island) and *A. mellifera* on the Philippine Islands. Also stemming from this work is the recognition of another *Tropilaelaps* species, *T. thaii*, collected from *A. laboriosa* in the mountainous eastern region of the Himalayas in northern Vietnam.

Adult females of *T. mercedesae* can be distinguished morphologically from *T. clareae* (s. str.) by their sheer size (a longer and wider dorsal plate) and variations in the shape of the apex of the epigynial plate, which varies from bluntly to sharply pointed, whereas in *T. clareae* s. str. it is always bluntly pointed. Adult males of *T. mercedesae* are also significantly larger than males of *T. clareae* (s. str.); however, comparative morphological studies of the nymphal stages of the two mites have not been carried out. In addition to the morphological characters used in distinguishing the four *Tropilaelaps* species, the amplified fragments of the mitochondrial DNA (mtDNA) of cytochrome *c* oxidase subunit I (COI) and a region between nuclear internal transcribed spacers and rDNA (ITS1-5.8S-ITS2) gene sequences of *T. mercedesae*, *T. clareae* s. str., *T. koenigerum* and *T. thaii* are proved to be useful as molecular markers for differentiating the species utilising the RFLP technique (Anderson and Morgan 2007).

Adult females of *T. koenigerum* are relatively more easily distinguishable morphologically from *T. mercedesae* and *T. clareae* (s. str.) by the presence of a pear-shaped anal plate, whereas in the latter two mites the anal plates are more rectangular (Delfinado-Baker and Baker 1982). In adult males of *T. mercedesae* and *T. clareae* (s. str.), the moveable digit of the chelicerae, which acts as a spermatodactyl organ appears a long "corkscrew-like" coiled structure, whereas in *T. koenigerum* the spermatodactyl is shorter, not coiled and has a "pig-tail-like" loop at the apex (Delfinado-Baker and Baker 1982). Specific molecular diagnostic markers have also been developed using ITS sequence and RAPD primers to assist in taxonomic identification of *T. mercedesae* (described under *T. clareae*) and *T. koenigerum* in Thailand (Tangjingjai et al. 2003). Adult females of *T. thaii* possess somewhat pear-shaped or bell-shaped anal plates, more or less similar to *T. koenigerum*, but can be distinguished from other *Tropilaelaps* species by the absence of a subapical tooth on the moveable chela, and its unique RFLP profile compared to other *Tropilaelaps* (Anderson and Morgan 2007).

The taxonomy of *Tropilaelaps*' primary host, *A. dorsata*, is in question as to whether it is comprised of only a single species. Historically, Ruttner (1988) recognised four subspecies of *A. dorsata* based on a morphometric study – *A. d. dorsata*, *A. d. laboriosa*, *A. d. breviligula* and *A. d. binghami*, which are geographically isolated, except for the sympatry of *A. d. dorsata* and *A. d. laboriosa* in the Himalayan areas. It is now suggested by a number of authors that the once widely recognised single species is a complex species based on current DNA evidence and differential mating times (Arias and Sheppard 2005; Raffiudin and Crozier 2007; Lo et al. 2010; cf. Chap. 1). As in the case of *Varroa* mites and populations of *A. cerana* in Asia, the biogeography of *Tropilaelaps* spp. and their hosts reflect a pattern of coevolution between the two groups – distributions of the giant honeybee species complex are broadly congruent with the distributions of the *Tropilaelaps* species. *A. dorsata* s. str., which is distributed in mainland Asia and Indonesia, is parasitised by *T. clareae* and sometimes with *T. koenigerum* living sympatrically in the same colonies, whereas *A. breviligula* and *A. d. binghami* harbour *T. clareae* in the Philippines and Sulawesi. One colony of *A. laboriosa* in North Vietnam was found to be infected by its unique mite, *T. thaii*. However, *T. mercedesae* was also found in the same colony where *T. thaii* was present. This might be a result of a cross-infection of *T. mercedesae* from *A. dorsata* s. str. cohabitating with *A. laboriosa* in the Himalayan trail.

The notion of the congruence of the patterns of distributions of the *A. dorsata* species complex and their *Tropilaelaps* mites provides evidence for co-evolution and may prove useful in future experimental studies on the host specificities and level of pathogenicity of different *Tropilaelaps* species on other *Apis*. This would be particularly so in the cavity-nesting species, as has been demonstrated with *Varroa* in *A. cerana* and *A. mellifera* (Anderson 1994; Anderson and Sukarsih 1996), which may lead to the understanding of *Tropilaelaps* biology that relates to their host immune systems and/or defence mechanisms.

16.4.2 Tropilaelaps *Life Cycle*

The life cycle of *Tropilaelaps* is similar to that of *Varroa*, except that *Tropilaelaps* can invade both worker and drone brood cells (Kapil and Aggarwal 1987, 1989), which can increase the mite population in a bee colony drastically compared to *Varroa* (Sammataro et al. 2000). The following description of the *Tropilaelaps* life cycle is summarised from previous literature (e.g. Sammataro et al. 2000; Oldroyd and Wongsiri 2006) under circumstances where *T. clareae* (according to D. Anderson in Oldroyd and Wongsiri 2006) infected an *A. mellifera* colony. Studies of the life cycles of different *Tropilaelaps* species on their native hosts are yet to be investigated.

In an established bee colony, a gravid female mite enters a brood cell before capping, feeds on larval haemolymph for about 2 days or less and then lays her first egg. Three or four eggs can be found in a brood cell per reproductive cycle. Normally, the first egg will develop into an adult male. The developmental time from egg to adult takes about 6–7 days. The mother mite and her offspring will emerge from the brood with the adult bee and can enter other brood cells directly (for the mother mite) or mate with the opposite sex in the bee colony (for the virgin offspring). In a period when the host queen bee does not lay eggs, adult *Tropilaelaps* will adopt a phoretic stage and stay on the bees' sclerites until there are brood cells to parasitise, though no more than 3 days (Koeniger and Muzaffar 1988; Rinderer et al. 1994; Wilde 2000). This might explain the mechanism by which *A. dorsata* and its related giant honeybee species have a reduced load of *Tropilaelaps* through seasonal migration (Kavinseksan et al. 2003).

16.5 Tracheal Mites of the Genus *Acarapis*

Acarapis woodi (Prostigmata; Tarsonemidae) invades the tracheal system of adult honeybees and feeds on the haemolymph by piercing the tracheal wall (Hirschfelder and Sachs 1952). In North America, colonies that are infested with *A. woodi* suffer symptoms such as brood decline, decreases in worker bees and low honey production (Royce and Rossignal 1989; Morse and Nowogrodzki 1990; Bailey and Ball 1991; Mussen 2001). In serious cases, an *A. woodi* population can overwhelm and kill the entire colony. In *A. mellifera*, the entire life cycle of *A. woodi* from egg to adult can be completed in about 2 weeks with males maturing a couple of days earlier than females, which contributes to the explosion of *A. woodi* population in a colony in a short period of time (Pettis and Wilson 1996; Wilson et al. 1997). All life stages of *A. woodi* are spent living inside the tracheal system of the bees, except for the adult females that sometimes venture out to find a new host (Sammataro and Needham 1996). Most dispersing female mites are attracted to the prothoracic spiracle of the adult bees and will eventually reside and lay her eggs in the trachea (Hirschfelder and Sachs 1952; Phelan et al. 1991). Adult female mites can survive outside of the host for only a few hours (Hirschfelder and Sachs 1952).

Since the discovery of *A. woodi* in the Isle of Wight, England, during the early twentieth century (Clark 1985; Sammataro 1995), there have been few reports on the occurrence of *A. woodi* in Asian honeybees. Until now, there are reports from the subcontinent of India–Pakistan where *A. woodi* was found infesting *A. cerana* and *A. dorsata* (Dhaliwal and Sharma 1974; Adlakha 1976; Delfinado-Baker et al. 1989; Abrol 2000). Accounts on the demise of managed *A. mellifera* colonies in Asia as a result of *A. woodi* have never been reported. We hypothesise two possible explanations, based on previous studies of the tracheal mite's natural history and biology, for the absence or the lack of reports on *A. woodi* in most parts of Asia. First, *A. woodi* has already resided in *A. mellifera* colonies in apiaries throughout Asia and might possibly have cross-infected other Asian honeybee species already, particularly the congeneric cavity-nesting *A. mellifera* and *A. cerana*.

However, because of the small size of the mites which cannot be observed with the naked eye, beekeepers cannot detect the mites at the initial stage of a mite infestation. When the colony progressively deteriorates, beekeepers notice the stress signs of the colonies such as declining populations and weak worker bees with deformed wings and metasoma (Sammataro 1995). These might be attributed to other causes that are more visibly conspicuous such as endemic *Varroa* or *Tropilaelaps* mites. Methods for the diagnosis of tracheal mites are also tedious, time-consuming and require experience to perform (Ragsdale and Furgala 1987; Fichter 1988; Ragsdale and Kjer 1989; Shimanuki and Knox 1991; Grant et al. 1993). Hence, reports of *A. woodi* in Asia might have been overlooked because the maladies can superficially resemble symptoms that are caused by other agents. This hypothesis is more or less comparable to the explanations for the lack of finding *A. woodi* in Europe before the early twentieth century (Eickwort 1988; Sammataro et al. 2000).

The second hypothesis is that, because *A. woodi* can increase its population size dramatically during the winter in the temperate areas of America and Europe, where most of the worker bees stay close together for a period of time, it causes the most damage (Eischen 1987; Otis and Scott-Dupree 1992; Sammataro et al. 1994). This situation presents a potential problem for the tracheal mite when it parasitises bees that are kept in a tropical climate, since the temperature during the "winter" in the tropics is not cold enough to inhibit the bees from foraging or performing other tasks. The "lack of winter" may be the reason why *A. woodi* is not established and causing difficulties for Asian beekeepers. Though, there is no direct evidence to suggest that seasonal change and ambient temperature have direct or indirect effects on the reproduction ability of the mites in the tropics.

There are other mites congeneric with *A. woodi* that live exclusively externally on the surface of the thorax and wing base of *A. mellifera* and *A. cerana* in Asia (Delfinado-Baker et al. 1989; De Guzman et al. 2001). *A. dorsalis* and *A. externus* are thought to be closely related to *A. woodi*. Thus far, there are no reports on the damage caused by these two mites on honeybee colonies (Bailey and Ball 1991; Sammataro et al. 2000).

16.6 Non-Parasitic Mites Associated with Asian Honeybees

Parasitic mites, described in the earlier sections, of this chapter represent only a minor fraction of the diversity of Acari associations with honeybees. Most Acari found in the nests of honeybees usually have a saprophagous lifestyle (Eickwort 1990) and feed on fungus-infected debris in the hives, dead bees and sometimes pollen (kleptophages). Three main orders of non-parasitic mites are commonly found in honeybee colonies: "Astigmata", Prostigmata and Mesostigmata. Mites of the order "Astigmata", currently placed in the order Oribatida (Krantz and Walter 2009), are the most abundant in the colonies of honeybees (Eickwort 1990). Two subfamilies of the family Acaridae, particularly Forcelliniinae and Horstiinae, are frequently found (O'Connor 1982, 1988; Delfinado-Baker and Baker 1987). Although primarily a myrmecophilous species, *Forcellinia faini* (Forcelliniinae), can be found on the hive floors of *A. cerana* colonies in Thailand (Fain and Gerson 1990). Among the prostigmatid mites, except for the parasitic tarsonemid mite on honeybee, *A. woodi*, members of this family are fungivorous and occur facultatively in honeybee colonies (Lindquist 1986). Generally, prostigmatid mites associated with social insects are known to have relatively short life-cycles and are commonly phoretic (Lindquist 1986; Eickwort 1990). The fore-tarsal claws of these mites are usually enlarged to hold onto the setae of their hosts. Sumangala and Haq (2002) reported that *Pseudoacarapis indoapis* (Tarsonemidae) is found in *A. cerana* colonies in India feeding on fungal debris and stored pollen.

Mites in the order Mesostigmata are primarily free-living predators, but many lineages have evolved parasitic lifestyles with other arthropods, e.g. *Varroa* mites (Hunter and Rosario 1988). Many mesostigmatid mites found in honeybees are free-living and feed on other saprophagous mites, insects, fungi and pollen. *Neocypholaelaps indica* (Family Ameroseiidae) is a common facultative kleptophage of *A. cerana*, *A. florea* and *A. dorsata* in Asia (Delfinado-Baker et al. 1989; Needham et al. 2001). The primary habitat of *N. indica* is on subtropical and tropical tree flowers where it feeds on pollen; however, during a foraging trip of a worker bee, this mite can "hitch-hike" on the mesosoma or metasoma to be brought back to the colonies, where the mite then feeds on stored pollen (Delfinado-Baker et al. 1989; Haq et al. 2001). Another Ameroseiidae species is also reported to be found associated with *A. cerana*, i.e. genus *Afrocypholaelaps* (Delfinado-Baker et al. 1989).

Works on the diversity and life history of the non-parasitic Asian honeybee mites are progressing more slowly than the corresponding parasitic groups. This may result from the non-commercial importance of the non-parasitic mites so that they receive less attention from bee researchers. Further investigations into population diversity of Asian honeybees will undoubtedly result in the discovery of additional unknown non-parasitic mites associated with honeybees in Asia.

16.7 Conclusion

As much as we know about the life history and biology of honeybee mites, we still have little information regarding variations in the genetics, behaviour and pathology of mites from different locations. The past decade has provided beekeepers and researchers with a fruitful "glimpse" into the complexity and diversity of the Asian honeybee mites. We are starting to learn about the genetic and reproductive variations of the Asian mites observed in different geographical regions that have impact on beekeeping industries in both western and eastern regions of the world. Future genetic and experimental work focusing on the population level of the mites can be of immense importance to the discovery of new genetic-types of mites and their corresponding levels of pathogenicity on their native host populations and on *A. mellifera*. Besides, studies of the Asian honeybee populations can reveal significant information regarding where to probe for genetically distinct types of mite populations. Microbes, particularly viruses associated with the mites, are also another subject of interest that will expand our knowledge for understanding the interactions among the hosts, parasites and pathogens. Biogeographical studies of the patterns of co-evolution between the honeybee and mite populations could well benefit from including the pathogen perspective to elucidate a clearer understanding of these complex honeybee–mite associations in Asia.

Appendix

A key to the female parasitic mesostigmatid mites of Asian honeybees (modified from Oldroyd and Wongsiri 2006)

	Morphological characters	Mite species
(1)	(a) Body elongated (considerably longer than wide)	Genus *Tropilaelaps* (2)
	(b) Body broadly elliptical, as wide or wider than long; small reversed triangular-shaped anal shield	Genus *Varroa* (5)
	(c) Body broadly pear-shaped or triangular, approximately same width and length; anal shield more or less rectangulate, about 1/3–1/4 as long as body length	Genus *Euvarroa* (8)
(2)	(a) Anal plate rectangular	(3)
	(b) Anal plate pear-shaped	(4)
(3)	(a) Apex of epigynial plate varies, from bluntly pointed to sharply pointed; female body size: length 978.8 ± 31.5 μm and width 542.5 ± 23.6 μm; male body size: length 920.9 ± 19.5 μm and width 523.2 ± 19.2 μm	*Tropilaelaps mercedesae*[a] (parasite of *A. dorsata* and *A. laboriosa*, and *A. mellifera* on Mainland Asia, Indonesia (except for Sulawesi) and Papua New Guinea)
	(b) Apex of the epigynial plate always bluntly pointed; female body size: length 881.9 ± 24.1 μm and width 484.4 ± 14.5 μm; male body size: length 856.6 ± 19.2 μm and width 500.9 ± 9.8 μm	*Tropilaelaps clareae*[a] (parasite of *A. brevilligua* and *A. dorsata binghami*, and *A. mellifera* on the Philippines and Sulawesi)
(4)	(a) Subapical tooth on moveable chela present	*Tropilaelaps koenigerum* (found primarily parasitising on *A. dorsata* and *A. laboriosa*)
	(b) Subapical tooth on moveable chela absent	*Tropilaelaps thaii* (parasite of *A. laboriosa*)
(5)	(a) Peritremes long looping up from ventral side, extending beyond the lateral margin and thus sometimes visible from dorsal surface	*Varroa rindereri* (primarily found parasitising *A. koschevnikovi*)
	(b) Peritremes not extending beyond lateral margin and not visible from dorsal surface	(6)
(6)	(a) Setae of the lateral margin long and slender	*Varroa underwoodi* (primarily found parasitising *A. dorsata*, *A. laboriosa* and *A. breviligula*)
	(b) Setae shorter and stout	(7)
(7)	(a) Body size ratio (width to length) 1.2–1.3:1	*Varroa jacobsoni*[b] (parasitise on Mainland *A. cerana* and *A. mellifera* worldwide)
	(b) Body size ratio ≥ 1.4:1	*Varroa destructor*[b] (parasitise on Sundaland *A. cerana*, including *A. nigrocincta* on Sulawesi)

(continued)

	Morphological characters	Mite species
(8)	(a) Body pear-shaped with 39–40 μm long lanceolate setae on rounded posterior margin	*Euvarroa sinhai* (primarily found parasitising on *A. florea*)
	(b) Body triangular with 47–54 μm long lanceolate setae on wide posterior margin	*Euvarroa wongsirii* (primarily found parasitising *A. andreniformis*)

[a]Measurements of *T. mercedesae* and *T. clareae* body size (length and width of the dorsal plate) are taken from Anderson and Morgan (2007). Readers should not depend on the body size of these mites exclusively to use as diagnostic characters of species. DNA sequencing and/or RFLPs profiles of the mtDNA COI and nuclear ITS1-5.8S-ITS2 gene sequences (Anderson and Morgan 2007) are appropriate markers for differentiating the species

[b]Body size ratio is not a reliable characteristic to differentiate between *V. jacobsoni* and *V. destructor*. Molecular diagnostics – DNA sequencing and/or RFLPs – of the mtDNA COI gene should be used to confirm the identification of the species (Anderson and Trueman 2000; Warrit et al. 2004, 2006)

References

Abrol DP (2000) Beekeeping with *Apis cerana* in Jammu and Kashmir: present status and future prospects. Bee World 81:149–152

Abrol DP, Putatunda BN (1995) Discovery of the ectoparasitic mite *Tropilaelaps koenigerum* Delfinado-Baker and Baker (Acari: Laelapidae) on *Apis dorsata* F, *A. mellifera* and *A. cerana* F in Jammu and Kashmir, India. Curr Sci 68:10

Adlakha RL (1976) Acarine disease of adult honey bees in India. Am Bee J 116(324):344

Aggarwal K (1988) Incidence of *Tropilaelaps clareae* on three *Apis* species in Hisar (India). In: Needham GR, Page RE Jr, Delfinado-Baker M, Bowman CE (eds) Africanized honey bees and bee mites. Ellis Horwood, Chichester, pp 396–403

Aggarwal K, Kapil RP (1988) Observations on the effect of queen cell construction on *Euvarroa sinhai* infestation in drone brood of *Apis florea*. In: Needham GR, Page RE Jr, Delfinado-Baker M, Bowman CE (eds) Africanized honey bees and bee mites. Ellis Horwood, Chichester, pp 404–408

Akratanakul P, Burgett M (1976) *Euvarroa sinhai* Delfinado and Baker (Acarina: Mesostigmata): a parasitic mite of *Apis florea*. J Apic Res 15:11–13

Anderson DL (1994) Non-reproduction of *Varroa jacobsoni* in *Apis mellifera* colonies in Papua New Guinea and Indonesia. Apidologie 25:412–421

Anderson DL (2004) Implication of the recent findings on parasitic mites for *Apis mellifera* beekeeping in Asia. In: Camaya EN, Cervancia CR (eds) Bees for new Asia. Proceedings of 7th Asian Apiculture Association Conference, Los Banos, pp 19–21

Anderson DL, Fuchs S (1998) Two genetically distinct populations of *Varroa jacobsoni* with contrasting reproductive abilities on *Apis mellifera*. J Apic Res 37:69–78

Anderson DL, Morgan MJ (2007) Genetic and morphological variation of bee-parasitic *Tropilaelaps* mites (Acari: Laelapidae): new and re-defined species. Exp Appl Acarol 43:1–24

Anderson DL, Sukarsih D (1996) Changed *Varroa jacobsoni* reproduction in *Apis mellifera* colonies in Java. Apidologie 27:461–466

Anderson DL, Trueman JWH (2000) *Varroa jacobsoni* (Acari: Varroidae) is more than one species. Exp Appl Acarol 24:165–189

Arias MC, Sheppard WS (2005) Phylogenetic relationships of honey bees (Hymenoptera: Apinae: Apini) inferred from nuclear and mitochondrial DNA sequence data. Mol Phylogenet Evol 37:25–35

Bailey L, Ball BV (1991) Honey bee pathology, 2nd edn. Academic, San Diego

Baker RA, Hick A, Chmielewski W (2005) Aspects of the history and biogeography of the bee mites *Tropilaelaps clareae* and *T. koenigerum*. J Apic Sci 49:13–19

Bharadwaj RK (1968) A new record of the mite *Tropilaelaps clareae* from *Apis dorsata* colonies. Bee World 49:115

Boecking O, Spivak M (1999) Behavioral defenses of honey bees against *Varroa jacobsoni* Oud. Apidologie 30:141–158

Boot WJN, Tan NQ, Dien PC, Vanhuan L, Vandung N, Long LT, Beetsma J (1997) Reproductive success of *Varroa jacobsoni* in brood of its original host, *Apis cerana*, in comparison to that of its new host, *A. mellifera* (Hymenoptera: Apidae). Bull Entomol Res 87:199–226

Büchler R, Drescher W, Tornier I (1992) Grooming behaviour of *Apis cerana, A. mellifera* and *A. dorsata* and its effect on the parasitic mite *Varroa jacobsoni* and *Tropilaelaps clareae*. Exp Appl Acarol 16:313–319

Burgett M, Akratanakul P (1985) *Tropilaelaps clareae*, a little known honey bee brood mite. Am Bee J 125:112–114

Burgett M, Akratanakul P, Morse RA (1983) *Tropilaelaps clareae*: a parasite of honey bees in south-east Asia. Bee World 64:25–28

Burgett M, Rossingol PA, Kitprasert C (1990) A model of dispersion and regulation of brood mite *Tropilaelaps clareae* parasitism in the giant honeybee *Apis dorsata*. Can J Zool 68:1423–1427

Camazine S (1986) Differential reproduction of the mite, *Varroa jacobsoni* (Mesostigmata: Varroidae), on Africanized and European honey bees (Hymenoptera: Apidae). Ann Entomol Soc Am 79:801–803

Clark KJ (1985) Mites (Acari) associated with the honey bee, *Apis mellifera* L. (Hymenoptera: Apidae), with emphasis on British Columbia. Thesis, Simon Fraser University, Burnaby

Colin ME, Vandame R, Jourdan P, Di Pasquale S (1997) Fluvalinate resistance of *Varroa jacobsoni* (Acari: Varroidae) in Mediterranean apiaries of France. Apidologie 28:375–384

Cook VA (1987) The spread of *Varroa jacobsoni* and *Tropilaelaps clareae*. Bee World 68: 163–164

Crane E (1988) Africanized bee, and mites parasitic on bees, in relation to world beekeeping. In: Needham GR, Page RE Jr, Delfinado-Baker M, Bowman CE (eds) Africanized honey bees and bee mites. Ellis Horwood, Chichester, pp 1–9

Dainat B, Ken T, Berthoud H, Neumann P (2009) The ectoparasitic mite *Tropilaelaps mercedesae* (Acari: Laelapidae) as a vector of honeybee viruses. Insectes Soc 56:40–43

de Guzman LI, Delfinado-Baker M (1996) A new species of *Varroa* (Acari: Varroidae) associated with *Apis koschevnikovi* (Apidae: Hymenoptera) in Borneo. Int J Acarol 22:23–27

de Guzman LI, Rinderer TE (1999) Identification and comparison of *Varroa* species infesting honey bees. Apidologie 30:85–95

de Guzman LI, Rinderer TE, Stelzer JA (1997) DNA evidence of the origin of *Varroa jacobsoni* Oudemans in the Americas. Biochem Genet 34:327–335

de Guzman LI, Rinderer TE, Stelzer JA, Anderson D (1998) Congruence of RAPD and mitochondrial DNA markers in assessing *Varroa jacobsoni* genotypes. J Apic Res 37:49–51

de Guzman LI, Burgett M, Rinderer TE (2001) Biology and life history of *Acarapis dorsalis* and *Acarapis externus*. In: Webster TC, Delaplane KS (eds) Mites of the honey bee. Dadant, Hamilton, IL, pp 17–27

De Jong D (1997) Mites: *Varroa* and other parasites of brood. In: Morse RM, Flottum PK (eds) Honey bee pests, predators, and diseases, 3rd edn. Root, Medina, Ohio, pp 281–327

De Jong D, Soares AEE (1997) An isolated population of Italian bees that has survived *Varroa jacobsoni* Oud infestation without treatment for over 12 years. Am Bee J 137:742–747

De Jong D, Morse RA, Eickwort GC (1982) Mite pests of honey bees. Ann Rev Entomol 27:229–252

Delfinado M, Baker EW (1961) *Tropilaelaps*, a new genus of mite from the Philippines (Laelapidae s. lat.). Fieldiana Zool 44:53–56

Delfinado M, Baker EW (1974) Varroidae, a new family of mites on honeybees (Mesostigmata: Acarini). J Wash Acad Sci 64:4–10

Delfinado-Baker M (1982) New records for *Tropilaelaps clareae* from colonies of *Apis cerana indica*. Am Bee J 122:382
Delfinado-Baker M (1988) Variability and biotypes of *Varroa jacobsoni* Oudemans. Am Bee J 128:567–568
Delfinado-Baker M, Aggarwal K (1987a) A new *Varroa* (Acari: Varroidae) from the nest of *Apis cerana* (Apidae). Int J Acarol 13:233–237
Delfinado-Baker M, Aggarwal K (1987b) Infestation of *Tropilaelaps clareae* and *Varroa jacobsoni* in *Apis mellifera ligustica* colonies in Papua New Guinea. Am Bee J 127:443
Delfinado-Baker M, Baker EW (1982) A new species of *Tropilaelaps* parasitic on honey bees. Am Bee J 122:416–417
Delfinado-Baker M, Baker EW (1987) Notes on mites new to beehives in Puerto Rico and North America. Am Bee J 127:365–366
Delfinado-Baker M, Houck MA (1989) Geographical variation in *Varroa jacobsoni* (Acari, Varroidae): application of multivariate morphometric techniques. Apidologie 20:345–358
Delfinado-Baker M, Underwood BA, Baker EW (1985) The occurrence of *Tropilaelaps* mites in brood nests of *Apis dorsata* and *A. laboriosa* in Nepal, with descriptions of the nymphal stages. Am Bee J 122:416–417
Delfinado-Baker M, Baker EW, Phoon ACG (1989) Mites (Acari) associated with bees (Apidae) in Asia, with description of a new species. Am Bee J 129:609–613
Dhaliwal HS, Sharma PL (1974) Foraging range of the Indian honeybee. J Apic Res 13:137–141
Donzé G, Guerin PM (1994) Behavioral attributes and parental care of *Varroa* mites parasitizing honeybee brood. Behav Ecol Sociobiol 34:305–319
Donzé G, Guerin PM (1997) Time-activity budgets and space structuring by the different life stages of *Varroa jacobsoni* in capped brood of the honey bee, *Apis mellifera*. J Insect Behav 10:371–393
Eguaras E, Marcangeli K, Oppendisano N, Fernandez N (1995) Mortality and reproduction of *Varroa jacobsoni* in resistant colonies of honey bees (*Apis mellifera*) in Argentina. Bee Sci 3:174–178
Eickwort GC (1988) The origin of mites associated with honey bees. In: Needham GR, Page RE Jr, Delfinado-Baker M, Bowman CE (eds) Africanized honey bees and bee mites. Ellis Horwood, Chichester, pp 327–384
Eickwort GC (1990) Associations of mites with social insects. Annu Rev Entomol 35:469–488
Eickwort GC (1994) Evolution and life-history patterns of mites associated with bees. In: Houck MA (ed) Mites: ecological and evolutionary analysis of life history patterns. Chapman & Hall, New York, pp 218–251
Eischen FA (1987) Overwintering performance of honey bee colonies heavily infested with *Acarapis woodi* (Rennie). Apidologie 18:293–304
Eischen FA (1998a) *Varroa* control problems: some answers. Am Bee J 138:107–108
Eischen FA (1998b) *Varroa's* response to fluvalinate in the Western US. Am Bee J 138:293
Elzen PJ, Eischen FA, Baxter JR, Pettis J, Elzen GW, Wilson WT (1998) Fluvalinate resistance in *Varroa jacobsoni* from several geographic locations. Am Bee J 138:674–676
Fain A, Gerson U (1990) Notes on two astigmatid mites (Acari) living in beehives in Thailand. Acarologia 31:381–384
Fichter BL (1988) ELISA detection of *Acarapis woodi*. In: Needham GR, Page RE Jr, Delfinado-Baker M, Bowman CE (eds) Africanized honey bees and bee mites. Ellis Horwood, Chichester, pp 526–529
Forsgren E, de Miranda JR, Isaksson M, Wei S, Fries I (2009) Deformed wing virus associated with *Tropilaelaps mercedesae* infesting European honey bee (*Apis mellifera*). Exp Appl Acarol 47:87–97
Fries I, Camazine S (2001) Implications of horizontal and vertical pathogen transmission for honey bee epidemiology. Apidologie 32:199–214
Fries I, Hauzhen W, Wei S, Jin CS (1996) Grooming behavior and damaged mites (*Varroa jacobsoni*) in *Apis cerana* and *Apis mellifera ligustica*. Apidologie 27:3–11
Garrido C, Rosenkranz P, Paxton RJ, Gonçalves L (2003) Temporal changes in *Varroa destructor* fertility and haplotype in Brazil. Apidologie 34:535–541

Gerson U, Mozes-Koch R, Cohen E (1991) Enzyme levels used to monitor pesticide in *Varroa jacobsoni*. J Apic Res 30:17–20
Grant GM, Dl N, Olsen PE, Rice WA (1993) The ELISA detection of tracheal mites in whole honey bee samples. Am Bee J 133:652–655
Griffith DA, Bowman CE (1981) World distribution of the mites *Varroa jacobsoni*, a parasite of honeybee. Bee World 62:154–163
Griffiths DA (1988) Functional morphology of the mouthparts of *Varroa jacobsoni* and *Tropilaelaps clareae* as a basis for the interpretation of their life-styles. In: Needham GR, Page RE Jr, Delfinado-Baker M, Bowman CE (eds) Africanized honey bees and bee mites. Ellis Horwood, Chichester, pp 479–486
HDOA: Hawaii Department of Agriculture (2010) Hawaii. http://hawaii.gov/hdoa/pi/ppc/varroa-bee-mite-page/?searchterm=varroa. Accessed 23 May 2010
Haq MA, Halliday RB, Walter DE, Proctor HC, Norton RA (2001) Life cycle and behavior of the coconut mite *Neocypholaelaps stridulans*. Acarology. In: Proceedings of 10th international congress, CSIRO, Melbourne, pp 361–365
Hirschfelder H, Sachs H (1952) Recent research on the acarine mite. Bee World 33:201–209
Hunter PE, Rosario RMT (1988) Associations of Mesostigmata with other arthropods. Annu Rev Entomol 33:393–417
Kapil RP, Aggarwal K (1987) Some observations on the concurrent parasitization of *Apis florea* by *Tropilaelaps clareae* and *Euvarroa sinhai*. Exp Appl Acarol 3:267–269
Kapil RP, Aggarwal K (1989) Observations on reproduction and seasonal population trends of *Euvarroa sinhai* (Mesostigmata: Varroidae) in India. In: Channabasavanna GP, Viraktamath CA (eds) Progress in acarology. Oxford and IBH, New Delhi, pp 277–281
Kavinseksan B, Wongsiri S, de Guzman LI, Rinderer TE (2003) Absence of *Tropilaelaps* infestation from recent swarms of *Apis dorsata* in Thailand. J Apic Res 42:49–50
Koeniger G, Koeniger N, Anderson A, Lekprayoon C, Tingek S (2002) Mites from debris and sealed brood cells of *Apis dorsata* colonies in Sabah (Borneo) Malaysia, including a new haplotypes of *Varroa jacobsoni*. Apidologie 33:15–24
Koeniger N, Muzaffar N (1988) Lifespan of the parasitic honeybee mite *Tropilaelaps clareae* on *Apis cerana*, *Apis dorsata*, and *Apis mellifera*. J Apic Res 27:207–212
Koeniger N, Koeniger G, de Guzman LI, Lekprayoon C (1993) Survival of *Euvarroa sinhai* Delfinado and Baker (Acari, Varroidae) on workers of *Apis cerana* Fabr., and *Apis mellifera* L. in cages. Apidologie 24:403–410
Krantz GW, Walter DE (2009) A manual of acarology. Texas Tech University Press, Lubbock, TX
Kraus B, Hunt G (1995) Differentiation of *Varroa jacobsoni* Oud populations by random amplification of polymorphic DNA (RAPD). Apidologie 26:283–290
Kumar NR, Kumar R, Mbaya I, Mwangi RW (1993) *Tropilaelaps clareae* found on *Apis mellifera* in Africa. Bee World 74:101–102
Laigo FM, Morse RA (1968) The mite *Tropilaelaps clareae* in *Apis dorsata* colonies in the Philippines. Bee World 49:116–118
Le Conte Y, Arnold G, Trouiller J, Masson C, Chappe B (1989) Attraction of the parasitic mite *Varroa* on the drone larvae of honey bees by simple aliphatic esters. Science 245:638–639
Lekprayoon C, Tangkanasing P (1991) *Varroa wongsirii*, a new species of bee mite from Thailand. Int J Acarol 17:255–258
Lekprayoon C, Tangkanasing P (1993) Comparative morphology of *Euvarroa sinhai* and *Euvarroa wongsirii*: parasites of *Apis florea* and *Apis andreniformis*. In: Conner LJ, Rinderer TE, Sylvester AH, Wongsiri S (eds) Asian apiculture. Wicwas, Cheshire, pp 427–433
Lindquist EE (1986) The world genera of Tarsonemidae (Acari: Heterostigmata): A morphological, phylogenetic, and systematic revision, with a reclassification of family-group taxa in the Heterostigmata. Mem Entomol Soc Can 136:1–517
Lipsitch M, Herre EA, Nowak MA (1995a) Host population structure and the evolution of virulence: a "law of diminishing returns". Evolution 49:743–748

Lipsitch M, Nowak MA, Ebert D, May RM (1995b) The population dynamics of vertically and horizontally transmitted parasites. Proc R Soc Lond Ser B 260:321–327

Lo N, Gloag RS, Anderson D, Oldroyd BP (2010) A molecular phylogeny of the genus *Apis* suggests that the giant honey bee of the Philippines, *A. breviligula* Maa, and the Plains honey bee of southern India, *A. indica* Fabricius, are valid species. Syst Entomol 35:226–233

Lodesani M, Colombo M, Spreafico M (1995) Ineffectiveness of Apistan treatment against the mite *Varroa jacobsoni* Oud. in several districts of Lombardy (Italy). Apidologie 26:67–72

Matheson A (1996) World bee health update 1996. Bee World 77:45–51

Matheson A (1997) Country records for honey bee diseases, parasites and pests. In: Morse RM, Flottum PK (eds) Honey bee pests, predators, and diseases, 3rd edn. Root, Medina, OH, pp 587–602

Moretto G, Gonçalves LS, De Jong DD, Bichuette MZ (1991) The effects of climate and bee race on *Varroa jacobsoni* Oud in several districts of Lombardy (Italy). Apidologie 26:67–72

Morin CE, Otis GW (1993) Observations on the morphology and biology of *Euvarroa wongsirii* (Mesostigmata: Varroidae), a parasite of *Apis andreniformis* (Hymenoptera: Apidae). Int J Acarol 19:167–172

Moritz RFA, Haenel H (1984) Restricted development of the parasitic mite *Varroa jacobsoni* Oud in the Cape honey bee *Apis mellifera capensis* Esch. Z Angew Entomol 97:91–95

Morse RA, Nowogrodzki R (eds) (1990) Honey bee pests, predators, and diseases, 2nd edn. Cornell University Press, Ithaca

Mossadegh MS (1990) Development of *Euvarroa sinhai* (Acari: Mesostigmata), a parasitic mite of *Apis florea*, on *A. mellifera* worker brood. Exp Appl Acarol 9:73–78

Mossadegh MS, Birjandi AK (1986) *Euvarroa sinhai* Delfinado and Baker (Acarina: Mesostigmata): a parasitic mite on *Apis florea* in Iran. Am Bee J 126:684–685

Mussen EC (2001) Introduction, spread and economic impact of tracheal mites in North America. In: Webster TC, Delaplane KS (eds) Mites of the honey bee. Dadant, Hamilton, IL, pp 43–56

Navajas M, Anderson DL, de Guzman LI, Huang ZY, Clement J, Zhou T, Le Conte YL (2010) New Asian types of *Varroa destructor*: a potential new threat for world apiculture. Apidologie 41:181–193

Needham GR, Gerson U, Sammataro D (2001) Mite biology. In: Webster TC, Delaplane KS (eds) Mites of the honey bee. Dadant, Hamilton, IL, pp 1–16

O'Connor BM (1982) Evolutionary ecology of astigmatid mites. Annu Rev Entomol 27:385–409

O'Connor BM (1988) Coevolution in astigmatid mite-bee associations. In: Needham GR, Page RE Jr, Delfinado-Baker M, Bowman CE (eds) Africanized honey bees and bee mites. Ellis Horwood, Chichester, pp 339–346

Oldroyd BP (1999) Coevolution while you wait: *Varroa jacobsoni*, a new parasite of western honeybees. Trends Ecol Evol 14:312–315

Oldroyd BP, Wongsiri S (2006) Asian honey bees. Harvard University Press, Cambridge

Otis GW (1996) Distributions of recently recognized species of honey bees (Hymenoptera: Apidae; *Apis*) in Asia. J Kans Entomol Soc 69:311–333

Otis GW, Kralj J (2001) Parasitic mites not present in North America. In: Webster TC, Delaplane KS (eds) Mites of the honey bee. Dadant, Hamilton, IL, pp 251–272

Otis GW, Scott-Dupree CD (1992) Effects of *Acarapis woodi* on overwintering colonies of honeybees (Hymenoptera: Apidae) in New York. J Econ Entomol 85:40–46

Oudemans AC (1904) Note VIII. On a new genus and species of parasitic Acari. Notes Leyden Mus 24:216–222

Peng CYS, Fang YZ, Xu SY, Ge LS (1987) The resistance mechanism of the Asian honey bee *Apis cerana* Fabr. to an ectoparasitic mite, *Varroa jacobsoni* Oudemans. J Invertebr Pathol 49:54–60

Pettis JS, Wilson WT (1996) Life history of the honey bee tracheal mite (Acari: Tarsonemidae). Ann Entomol Soc Am 89:368–374

Phelan LP, Smith AW, Needham GR (1991) Mediation of host selection by cuticular hydrocarbons in the honey bee tracheal mite *Acarapis woodi* (Rennie). J Chem Ecol 17:463–473

Raffiudin R, Crozier RH (2007) Phylogenetic analysis of honey bee behavioral evolution. Mol Phylogenet Evol 43:543–552
Ragsdale D, Furgala B (1987) A serological approach to the detection of *Acarapis woodi* parasitism in honey bees using an enzyme-linked immunosorbent assay. Apidologie 18:1–10
Ragsdale D, Kjer KM (1989) Diagnosis of tracheal mite (*Acarapis woodi* Rennie) parasitism of honey bees using a monoclonal based enzyme-linked immunosorbent assay. Am Bee J 129:550–553
Rath W (1999) Co-adaptation of *Apis cerana* Fabr. and *Varroa jacobsoni* Oud. Apidologie 30:97–110
Rath W, Delfinado-Baker M, Drescher W (1991) Observations on the mating behavior and sex ratio phoresy and dispersal of *Tropilaelaps clareae* Acari Laelapidae. Int J Acarol 17:201–208
Rinderer TE, Oldroyd BP, Lekprayoon C, Wongsiri S, Boonthai C, Thapa R (1994) Extended survival of the parasitic mite *Tropilaelaps clareae* on adult workers of *Apis mellifera* and *Apis dorsata*. J Apic Res 33:171–174
Ritter W, Michel P, Bartholdi M, Schwendemann A (1990) Development of tolerance to *Varroa jacobsoni* in bee colonies in Tunisia. In: Ritter W (ed) Proceedings of international symposium on recent research in bee pathalogy, Apimondia, pp 54–59
Royce LA, Rossignal PA (1989) Honey bee mortality due to tracheal mite parasitism. Parasitology 100:147–151
Ruttner F (1988) Biogeography and taxonomy of honeybees. Springer, Berlin
Sammataro D (1995) Studies on the control, behavior, and molecular markers of the tracheal mites (*Acarapis woodi* (Rennie)) of honey bees (Hymenoptera: Apidae). Thesis, Ohio State University, Ohio
Sammataro D (2004) *Tropilaelaps* infestation of honey bees (*Tropilaelaps clareae*, *T. koenigerum*). In: Office International des Epizooties (OIC) Manual of diagnostic tests and vaccines for terrestrial animals, 5th edn, vol 2. OIC, Paris, pp 992–995
Sammataro D, Needham GR (1996) Host-seeking behavior of tracheal mites (Acari: Tarsonemidae) on honey bees (Hymenoptera: Apidae). Exp Appl Acarol 20:121–136
Sammataro D, Cobey S, Smith BH, Needham GR (1994) Controlling tracheal mites (Acari: Tarsonemidae) in honey bees (Hymenoptera: Apidae) with vegetable oil. J Econ Entomol 87:910–916
Sammataro D, Gerson U, Needham G (2000) Parasitic mites of honey bees: life history, implications, and impact. Annu Rev Entomol 45:519–548
Sasagawa H, Matsuyama S, Peng CYS (1999) Recognition of parasite, hygienic allo-grooming behavior induced by parasitic *Varroa* mites in the Japanese honey bee, *Apis mellifera japonica*. In: Proceedings of 13th international congress, IUSSI, Adelaide, pp 415
Shimanuki H, Knox D (1991) Diagnosis of honey bee disease. USDA Agricultural Handbook Ah-690
Sihag RC (1988) Incidence of *Varroa*, *Euvarroa* and *Tropilaelaps* mites in the colonies of honey bees *Apis mellifera* L. in Haryana (India). Am Bee J 128:212–213
Smith DR, Hagen RH (1996) The biogeography of *Apis cerana* as revealed by mitochondrial DNA sequence data. J Kans Entomol Soc 69:294–310
Smith DR, Villafuerte L, Otis G, Palmer MR (2000) Biogeography of *Apis cerana* F. and *A. nigrocincta* Smith: Insights from mtDNA studies. Apidologie 31:265–279
Sumangala K, Haq MA (2002) Ecobiology of *Pseudacarapis indoapis* Lindquist (Acari: Tarsonemidae) 2: ontogeny and breeding behavior. J Entomol Res 26:83–88
Tangkanasing P, Wongsiri S, Vongsamanode S (1988) Integrated control of *Varroa jacobsoni* and *Tropilaelaps clareae* in bee hives. In: Needham GR, Page RE Jr, Delfinado-Baker M, Bowman CE (eds) Africanized honey bees and bee mites. Ellis Horwood, Chichester, pp 409–412
Tangjingjai W, Verakasala P, Sittipraneed S, Klinbunga S, Lekprayoon C (2003) Genetic differences between *Tropilaelaps clareae* and *Tropilaelaps koenigerum* in Thailand based on ITS and RAPD analyses. Apidologie 34:514–524
Tewarson NC, Singh A, Engles W (1992) Reproduction of *Varroa jacobsoni* in colonies of *Apis cerana indica* under natural and experimental conditions. Apidologie 23:161–171

Waite R (2003) Exotic pest legislation. Br Beekeep Assoc 143:7
Warrit N, Hagen TAR, Smith DR, Cakmak I (2004) A survey of *Varroa destructor* strains on *Apis mellifera* in Turkey. J Apic Res 43:190–191
Warrit N, Smith DR, Lekprayoon C (2006) Genetic subpopulations of *Varroa* mites and their *Apis cerana* hosts in Thailand. Apidologie 37:19–30
Webster TC, Delaplane KS (2001) Mites of the honey bee. Dadant, Hamilton, IL
Wilde J (2000) Is it possible to introduce *Tropilaelaps clareae* together with imported honey bee queens to Europe? Pszczeln Zesz Nauk 44:155–162
Wilson WT, Pettis JS, Henderson CE, Morse RA (1997) Tracheal mites. In: Morse RM, Flottum PK (eds) Honey bee pests, predators, and diseases, 3rd edn. Root, Medina, OH, pp 255–277
Woyke J (1984) Survival and prophylactic control of *Tropilaelaps clareae* infesting *Apis mellifera* colonies in Afghanistan. Apidologie 15:421–434
Woyke J (2005) Aktualne wiadomości o biologii I zwalczaniu roztocza *Tropilaelaps clareae*. Pasieka 3:43–47
Yoder JA, Sammataro D, Peterson JA, Needham GR, Bruce WA (1999) Water requirements of adult females of the honey bee parasitic mite, *Varroa jacobsoni* (Acari: Varroidae) and implications for control. Int J Acarol 25:329–335
Zhang ZQ (2000) Notes on *Varroa destructor* (Acari: Varroidae) parasitic on honeybees in New Zealand. Syst Appl Acarol Spec Publ 5:9–14

Chapter 17
Colony Defence and Natural Enemies

Stefan Fuchs and Jürgen Tautz

17.1 Robbers and Thieves

Although showing considerable differences in many respects, all the Asian honeybee species share the basic feature with the western honeybee to nest in colonies containing several thousand individuals. The amassing of thousands of adult insects alone in a more or less stationary place constitutes an attractive target to predators. In the open-nesting dwarf honeybee, *Apis florea*, a bee mass of 6,300 individuals amounts to 0.144 kg of organic nutrition. This is by far exceeded by nests of the giant honeybee, *A. dorsata*, with 36,600 individuals, which are about fivefold heavier than *A. florea*, and adding up to a 4.3 kg biomass; while a colony of the cavity-nesting eastern honeybee, *A. cerana*, with 9,200 individuals still constitutes a considerable amount of 0.420 kg biomass.

This, however, is only a part of the attraction of honeybee colonies, as the cells of the wax combs, with areas of 396 cm^2 in *A. florea*, 5,650 cm^2 in *A. cerana* and 6,356 cm^2 in *A. dorsata*, may contain 0.6 to 1.8 larvae and pupae per worker bee (in *A. florea* and *A. cerana*, respectively), which provide highly valuable proteins (Seeley et al. 1982; Dyer and Seeley 1991). Furthermore, the collection of considerable amounts of organic resources such as pollen and honey, with a high content of proteins or sugar, considerably enhances the nutritional value of a colony.

In view of the amount and variety of nutrition offered by a honeybee colony, it is not surprising that numerous potential predators, parasites, commensals and other pests set out to take a share of this concentrated high-value organic material. These include a wide array of taxa, notably mammals, birds and various arthropods that may target different parts of these commodities. Some only occasionally try to snatch

S. Fuchs
Institut für Bienenkunde (Polytechnische Gesellschaft), FB Biowissenschaften, Goethe-Universität Frankfurt am Main, Karl von Frisch Weg 2 61440, Germany
e-mail: s.fuchs@bio.uni-frankfurt.de

J. Tautz
BEE Group, Biozentrum, Universität Am Hubland, Würzburg 97074, Germany
e-mail: Tautz@biozentrum.uni-wuerzburg.de

H.R. Hepburn and S.E. Radloff (eds.), *Honeybees of Asia*,
DOI 10.1007/978-3-642-16422-4_17, © Springer-Verlag Berlin Heidelberg 2011

small amounts, while others extract significant parts of their diet from this rich resource, or even show specific adaptations which facilitate overcoming the defensive measures of the honeybees.

The most notorious honey-lovers are bears (*Helarctus malayanus*, *Melursus ursinus* and *Ursus thibetanus*) and martens (*Martes flavigula* and *M. foina*) (Seeley et al. 1982; Thapa et al. 2000; Oldroyd and Wongsiri 2006). Bears may use their strong claws to break into nest cavities, or even attack the strongly defensive colonies of *A. dorsata*. They aim to feast predominantly on the combs that are filled with brood and honey, and thus their attacks are most destructive to the colonies. Martens also attack entire colonies, but have also been observed to prey on flying bees.

Occasionally, monkeys (*Macaca* sp.) have been observed to rob the nests of *A. florea* or *A. cerana* (Seeley et al. 1982; Thapa et al. 2000). There are few reliable data on the impact of predators on honeybee colonies, but tree shrews appear to be a notable threat to dwarf honeybee colonies in Thailand (Seeley et al. 1982). Other vertebrates of various taxa are occasionally observed to prey on honeybees, among these rats, honey badgers, foxes, lizards and toads (Caron 1978; Novogrodzki 1990). However, there is little doubt that exploitation by humans, which use all nest components, including the wax for nutrition and various other purposes, is a prominent threat to colonies of all honeybee species.

Various insect-preying birds, such as drongos and swifts, have been observed to opportunistically prey on flying honeybees (Ambrose 1978; Seeley 1994; Thapa and Wongsiri 2003). However, a range of birds has become regular predators on bees. Some display very specific behaviours. Bee-eaters, after snatching a forager bee, swish their beaks back and forth against a tree branch on which they are perching – obviously to remove the sting: a behaviour termed "bee-rubbing" (Fry 1984). In the blue-throated bee-eater (*Merops virididis*), it has been observed that the sting pheromone that is rubbed on the branch attracts other foragers, which then fall easy prey to the bird (Koeniger et al. 2010).

Honeybees may form a considerable part of the diet in bee-eaters, for example, the crop content of *Merops orientalis* and *M. superciliosus* could contain considerable proportions of up to 30% of *A. florea* (Fry 1984). Birds are mostly interested in capturing individual bees and rarely attack colonies. However, the blue-bearded bee-eater may provoke counter attacks in *A. dorsata* by passing close to the colonies and feed on individuals that pursue them to a perch (Kastberger and Sharma 2000). Similarly, flocks of *M. orientalis* were reported to launch coordinated attacks at *A. dorsata* nests at low ambient temperatures when the bees' flying performance is impaired and they can be picked easily from the ground (Thapa et al. 2000). The highly specialised oriental honey buzzard (*Pernis ptylorhynchos*) also preys on nests of open-nesting honeybees, *A. dorsata* and *A. laboriosa* (Thapa and Wongsiri 2003; *A. andreniformis* – Koeniger et al. 2010), but differs as attacks are aimed at the combs to feed on brood and honey. During their specialised attacks, sometimes performed in pairs, they first arouse and mislead the defending worker force before attacking the combs. Honeyguides (*Indicatoridae*) are specialised wax feeders, but mainly feed on abandoned nests; the two species occurring in Asia do not constitute a major threat to honeybees (Oldroyd and Wongsiri 2006).

Besides vertebrates, other arthropods impose considerable predation pressure. A variety of different arthropods, such as mantids, crab spiders, orb-weavers and assassin bugs, have been observed to be opportunistic feeders (Seeley et al. 1982; Fuchs personal observation), causing no serious harm to colonies. In contrast, from the related Hymenoptera, two equally social groups, wasps and ants, are of major impact, either by posing a constant drain on the bees' work force or by attacking and robbing entire colonies. In particular, the larger vespine wasps of the genus *Vespa* play a major role as predators of Asian honeybees (Matsuura 1988); an array of hornets which attack Asian bees are listed in Oldroyd and Wongsiri (2006). Some of the giant wasps specialise on social bees and are outstanding in their effectiveness to deplete entire colonies. Wasps commonly prey on individual flying foragers during foraging. Quite commonly, wasps catch departing or homing foragers close to the colonies ("bee-hawking"), and open-nesting dwarf and giant honeybees as well as cavity-nesting bees are targeted (Seeley et al. 1982; Ono et al. 1987; Tan et al. 2007; Kastberger et al. 2008).

Some wasps hawk honeybees at cavity-nesting colonies (*Vespa multimaculata* – Koeniger et al. 1996; *V. velutina* – Shah and Shah 1991) and employ a specialised strategy in which they hover in front of a colony entrance facing outward to catch returning foragers ("goal-keeping behaviour", Koeniger et al. 1996). One *V. multimaculata* caught 14 *A. nuluensis* within 12 h, and *V. velutina* caught 17.8 in 151 (11.2%) hawking attempts (Tan et al. 2007), which could deplete a colony by 20–30% (Sakagami and Akahira 1960).

V. affinis has been observed to prey on drones in drone congregation areas, with the drones confusing them for a queen (Koeniger et al. 1994; cf. Chap. 8). Hornet attacks are not restricted to single flying individuals, but may even be fatal for a colony. A few *V. tropica* wasps can overpower an *A. florea* colony within 3 h by successively seizing and killing nest workers and gradually depleting the workforce (Seeley et al. 1982). By recruiting nestmates, *V. mandarinia* hornets are able to mass-attack entire *A. cerana* colonies and, having killed the bees, raid pupae and larvae. In order to focus their attacks, they use pheromones to mark the besieged colonies (Ono et al. 1995).

Ants are an almost omnipresent danger to honeybee colonies and may take anything for nutrition: live and dead bees, larvae, pupae and honey (Akratanakul 1987). The predominant ant species to endanger bees are the widespread tree-living weaver ants. In particular, *Oecophylla smaragdina* is a major threat to colonies of dwarf honeybees, who share the same habitat, but may also occasionally drive cavity-nesting *A. cerana* from their nests (Dyer 1991). These ants try to seize and pull down flying bees from a perch and several ants may be needed to overcome a bee. Other species, such as *Monomorium indicum*, *Diacamma rugosum* and *Pheidologeton* sp., were observed to molest *A. cerana* colonies, but without causing apparent harm. Few ant species, such as *Dolichoderus bituberculatus*, attack *A. dorsata* colonies (Seeley et al. 1982).

Only some of the most obvious and well documented examples have been mentioned, and without any doubt there are numerous other species which occasionally attack single honeybees or colonies. However, even this short overview may substantiate the threat from thieves and robbers who are trying to get their share of the spoils. However, bees are not helplessly exposed to these attacks, but have developed various means to meet such challenges. Due to strong selection pressure, these defence measures have taken various forms and led to sophisticated counter-strategies that are matched for the kind of predators. These will be the focus of the following sections.

Besides predators, honeybee colonies are also exploited by parasites, commensals, macroparasites and microparasites. Although these also include interesting and highly sophisticated interactions, they are not dealt with here. Bee mites and bee diseases are however reviewed in separate chapters of this book (cf. Chaps. 15 and 16).

17.2 Nesting and Defence Styles

It has been argued that predation has been a major selective force that is shaping the lifestyles of honeybees and, indeed, the striking differences among the Asian honeybees, dwarf, giant and cavity-nesting species, provide excellent insight into the different basic strategies to counter predation (Seeley et al. 1982; Dyer 1991; Dyer and Seeley 1991).

When facing an enemy, there are two opposing primary strategies: to retreat or to (counter-) attack. In honeybees, both are found – with differing emphasis between colonies (Kastberger et al. 2009a) and between the different honeybee species, taking very diverse forms in relation to varieties in predation patterns. Retreat and attack strategies may also occur in combination or mixed forms and may be switched due to different intensities of the predators' attacks. The propensity for attack or retreat strategies is highly dependent on nest structures and placements; the basic relationships between the ease of detection, approach and consumption have been outlined by Dyer (1991).

In the cavity-nesting species, the primary defence measure is to live in cavities. These offer strong protection by restricting accessibility to the nest and at the same time constitute a limited and valuable resource in themselves, to be actively defended if in danger. Cavity-nesting is assumed to be a derived "primitive" trait (Lindauer 1956), but because this is shared with many social Hymenoptera, this view has been questioned (Koeniger 1976; cf. Chap. 2).

Open-nesting bees are exposed to environmental stresses, such as rain, wind and temperature changes, and need to defend the entire nest surface against predators. It needs to be emphasised, however, that cavity- and open-nesting are not entirely separate modes, as during a cavity-nesting bees' lifecycle, reproductive or migrating swarms interrupt the cavity-sheltered mode, and will require some defence adaptations in order to reduce vulnerability during this open-nesting stage. Moreover, *A. cerana* occasionally creates open nests if no cavities are available. Matsuura (2003) – (cited from Sugahara and Sakamoto 2009) – reported that about 10% of *A. cerana* colonies in Japan are open-nesting.

In the open-nesting bees, the dwarf honeybees rely predominantly on retreat strategies, such as hiding by placing their small nests in bushes and preferring cover by leaves, with little emphasis on attack options. If challenged, they tend to abandon their nests. In contrast, the open-nesting giant honeybees prefer inaccessible, but often conspicuous, nesting sites in high trees or cliffs. In case of any attacks, they resort to exceptionally strong counter-attacks. This dichotomy in the predominant defence strategy may be related to body size (Dyer 1991), but may also be a selective force driving worker body sizes into the disparate "dwarf" and "giant" directions (Koeniger et al. 2010; cf. Chap. 2).

17.3 Bee Weapons: Stings, Mandibles, Legs and Wings

On an individual level, honeybees possess limited means and actions for defence. As for other insects, mandibles can be used for biting; however, honeybee mandibles are rather a multi-purpose tool that is also used for a range of other tasks and not specifically shaped for fighting. Nevertheless, they are useful for grabbing, and small predators such as ants may be crushed. Furthermore, legs may be used for grabbing, and thus aiding their defence actions. Wings can also be used to produce air blasts, which can be useful in defence actions to blow away intruders (Seeley et al. 1982; Dyer 1991; Yang et al. 2010).

These are, however, rather general and unimpressive faculties which are by far outclassed by the one highly sophisticated and specialised defence weapon that bees are famous for, which is the sting. As in the western honeybee, stinging is the main defence against vertebrate predators in all Asian honeybee species. All possess barbed stingers which they stab into vertebrate skin where they become anchored, while the bee is detached from the sting apparatus by autotomy. Without the bee, the sting is less likely to be removed by the victim and will continue to inject poison that is driven by movements of the detached sting apparatus, containing a nerve ganglion and a sophisticated structure of muscles and sclerotised plates.

There are no complete investigations on the sting apparatus for all Asian species, but some studies cover *A. cerana*, *A. florea* and *A. dorsata* as examples for cavity-nesting, dwarf and giant honeybees (Weiss 1978; Jayasvati 1989). From this limited evidence, it appears that there are only minor differences in the general morphology of the stinging apparatus. However, the size of the sting apparatus varies approximately in proportion to body size differences. Lancet length is 2.10 mm, 1.55 mm and 1.06 mm in *A. dorsata*, *A. cerana* and *A. florea*, respectively (Ramya and Rajagopal 2008). There is some variation in the numbers and expression of barbs on the stinger sheath and lancet, which appear to give better fixing of the stings in the open-nesting bees, *A. florea* and *A. dorsata*, in comparison to the cavity-nesting bee, *A. cerana*.

Although there is a surprising paucity of studies, the composition of sting poison is apparently fairly similar in all of the species investigated, with the most prominent active ingredients being the peptide mellittin and phospholipase. However, some minor components may differ (Natzir et al. 1999) from *A. mellifera*. The amino acid sequences of melittin show differences between *A. cerana*, *A. dorsata* and *A. florea* in five positions, which are of relevance in phylogenetic, but not in functional considerations (Kreil 1975).

Honeybee venom toxicity is predominantly aimed at vertebrates (Schmidt 1990) and is apparently similar among honeybee species. It does not differ if the bee venoms of *A. dorsata*, *A. florea* and *A. cerana* are applied to mice (~3 mg/kg mice). However, the amount of venom per worker bee is much higher in *A. dorsata* as compared to *A. cerana* or *A. florea* (218 μg, 43 μg, 27 μg, respectively) – even exceeding that of *A. mellifera* (138 μg). The percentage of poison to total body weight is 0.65%, 0.28% and 0.28%, respectively, and the higher percentage in *A. dorsata* is in line with the high defensiveness of this giant honeybee (Schmidt 1995). It has been estimated that the LD_{50} of *A. mellifera* bee stings for non-hypersensitive humans are between 500 and 1,500 stings (Vetter and Visscher 1998), which would be equal to about 600 stings from *A. dorsata*, 3,200 from *A. cerana* and 5,100 from *A. florea*.

17.4 Socially Organised Defence

The stunning and outstanding effectiveness of nest defence in honeybees is not so much based on the faculties of individual bees, as on the social coordination between thousands of individuals. Biting, grabbing and wing-buzzing of a single bee are hardly effectual, but the combined efforts of many individuals add up to astounding performances. In particular, social coordination opens up new possibilities on a higher level of colonial, behavioural and structural patterns that are not available to an individual. The great fascination of honeybee defence lies within this social dimension, and the Asian *Apis* species offer a unique opportunity to study the variations of defence patterns and special adaptations in response to the specific predator environments to which they are exposed.

17.4.1 Stinging

Even stinging, for which individual foraging bees might already be respected, achieves entirely new dimensions of awe if embedded within the patterns of a socially coordinated attack. In *A. dorsata*, such attacks are characterised by signs of unrest in the colonies if disturbed, which build up until entire squads of bees drop from the free-hanging colonies in several defence waves. The waves of bees will follow retreating intruders up to several kilometres or continue to search the vicinity of the nest for hours until they return to the nest (Lindauer 1956; Morse and Laigo 1969).

Such counter-attacks are likely to follow the general scheme for bee attacks as proposed by Breed et al. (2004), which describes the basic behavioural sequence for socially organised defence (Fig. 17.1). The typical chain of increasing defence responses would follow a sequence of primary perception and orientation of the target by a single bee, leading to either discrimination or identification. If this confirms an enemy situation, the bee is further alerted, approaches the target and, after secondary discrimination or identification based on close contact cues, may attack. Threat behaviour may provide an alternative; indeed an astounding portion of honeybee behaviour is shaped to frighten an enemy away, instead of engaging in outright attacks. At any of these steps, bees may disengage if stimulus intensity does not exceed the continuation thresholds.

This sequence alone would be apt to ensure a basic pattern of effective defence organisation. Assuming different behavioural thresholds, it could even explain some modulation of total defence efforts to a threat in a social context, based on the independent actions of the individuals. However, the main boost in the system's responsiveness comes from its built-in feedback loops. By alerting and recruiting, further nestmates join the sequence – possibly even skipping the initial steps of the sequence – and an entire colony might catapult into a completely different behavioural state. In this, the workforce now acts as a coherent unit that is governed by mechanisms of social regulation, opening up the domain of integrated, higher-level actions way beyond the scope of the individual. These actions in turn are subject to selection, to be shaped into the organised group defence for which bees are respected and feared.

Among others, Morse and Laigo (1969) give some impressive accounts for stinging defence actions in *A. dorsata*, which they describe as the "most ferocious stinging insect on earth". Within a certain threshold, colonies may be approached and cautiously investigated, but when disturbance by animals is mounting, social

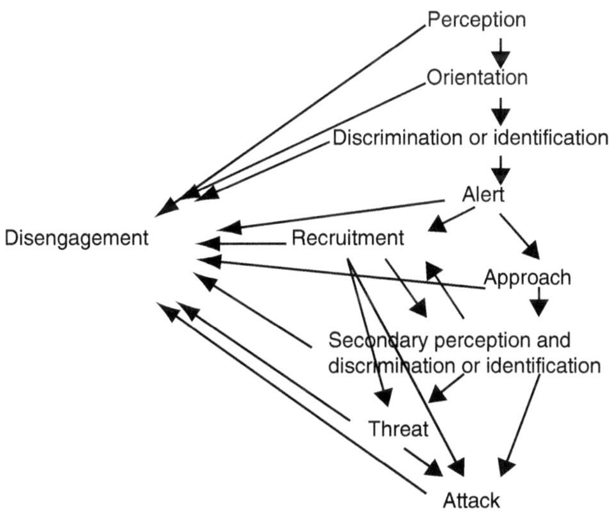

Fig. 17.1 Organisation of social defence: Pathways of perception and interactions leading to social attack in honeybees (from Breed et al. 2004)

chain reactions lead fairly suddenly to integrated attack patterns. Bees from the surface start to move down to the lower edge of the comb, from where they drop into the air in successive defence waves and seek the disturber in bands of flying defenders – actively seeking open areas between the plants in clouds of bees to detect intruders.

The most prominent feedback loop is activation by alarm pheromones. Although mandibular pheromones play some role, the main alarm pheromones are associated with the sting sheath of the sting apparatus (Cassier et al. 1994). An exception is *A. laboriosa*, which apparently relies on a mandibular pheromone for alarming (Blum et al. 2000). In almost all of the Asian bees investigated thus far, the major active constituent is isopentyl-acetate (Morse et al. 1967; Koeniger et al. 1979; cf. Chap. 9). This volatile substance is released when the sting shaft is exposed and alerts other bees to an enemy. Thus, attacking bees, if returning to the nest, will trigger off further guards to join in defence, until up to 10% of the total workforce might become airborne. Koeniger (1975) reported that between 50 and 5,000 workers of *A. dorsata* were activated by single alarmed workers who were returning to the colony.

The volatile isopentyl-acetate, evaporating within a few minutes, is particularly appropriate as a fast-acting colony defence activator. By autotomy, the barbed sting of successful defenders becomes fixed in a vertebrate enemy's skin, and the exuded alarm pheromone then guides others to these vulnerable spots. Interestingly, in two open-nesting species, *A. dorsata* and *A. florea*, a second active component, 2-decen-1-yl-acetate, is present (Veith et al. 1978; Koeniger et al. 1979), which is far less volatile and seemingly acts as a more persistent enemy-marker pheromone in these bees. Following this scent, fleeing predators and humans are reported to be pursued by *A. dorsata* for several hundred metres or even kilometres (Lindauer 1956; Morse and Laigo 1969). At 50 m distance from a nest, a man stung by a single bee was attacked by 500 more, and the numbers of attackers increased even further at 1.6 km distance from the nest, decreasing slowly only at a distance of 3.2 km away (Koeniger 1975). The pheromonal marking is so persistent that bees recognise the smell even the next day (Lindauer 1956).

A. dorsata has 10–20 times more isopentyl-acetate in its sting sheath than other bees (Morse et al. 1967; Koeniger et al. 1979). This implies a very strong feedback loop in the dynamic of the social defence organisation of this bee. It also underlines how colony-level characteristics of defence reactions may be shaped by simple variables, such as the amount of alarm substances in single bees; in this case towards a very strong dynamic that is tuned for high responsiveness, due to a high amplification factor. The resulting integrated social attacks pose a serious threat to vertebrate intruders, including humans (Morse and Laigo 1969). Lindauer (1956) reported the deaths of three water buffalos in Sri Lanka due to an attack by *A. dorsata*. Kastberger and Sharma (2000) recorded about 700–800 worker bees taking part in attacks against a blue-bearded bee-eater (*Nyctyornis athertoni*) that was approaching a nest of *A. dorsata*. Even highly specialised bee predators, such as the honey buzzard (*Pernis ptylorhynchos*) might be stung to death by *A. dorsata* in spite of protective feather shielding on its head (Thapa and Wongsiri 2003). However, a main effect likely comes from scaring away enemies through the coordinated attacks of hundreds of individuals, even if only a few stings are actually administered (Koeniger 1975). In fact, threatening is an alternative because lost bees are costly to the colony, and *A. cerana* is known for its reluctant use of stinging (Sakagami and Akahira 1960). Threatening is fortified by the continuous mock-attacking and biting of moribund *A. mellifera* bees, even after stinging, until their death due to sting loss injury (Cunard and Breed 1998), as has similarly been observed in *A. dorsata* by Koeniger (1975).

In contrast to the highly defensive open-nesting *A. dorsata* (Fig. 17.2), the stinging defence in cavity-nesting species, such as *A. cerana*, is far less pronounced. Living in cavities offers the option of retreating as a typical first response to slight disturbances (Seeley et al. 1982), and bees cease flights due to a predator's approach (Tan et al. 2007). Only if a colony is massively disturbed will it respond by stinging attacks. However, due to comparatively undeveloped sting barbs and characteristic "revolving" stinging behaviour, *A. cerana* avoids autotomy and the cost of the stinging bee's death (Sakagami and Akahira 1960), thus simultaneously providing a comparatively low feedback loop of sting pheromones in the colony's attack dynamic. Further, absconding is frequent if colonies are often disturbed (Sakagami 1960; Punchihewa 1994). Apparently, in Borneo, *A. koschevnikovi* is even less defensive as compared to *A. cerana* (Mathew and Mathew 1988), but

Fig. 17.2 Workers blocking the nest attachment line of the colony to the tree trunk in the open-nesting bees, *Apis dorsata*, so as to prevent access by ground-walking predators (photo: N. Koeniger)

there is very little known about defensiveness in the other cavity-nesting bees of Asia: *A. nuluensis* and *A. nigrocincta*.

Although the open-nesting dwarf honeybees predominantly rely on retreat strategies such as hiding and absconding (Seeley et al. 1982), they are nevertheless capable of issuing decisive stinging counter-attacks. *A. florea* is fairly docile if not provoked, but will first attack vertebrates in small groups of workers, which are later re-enforced by hundreds of workers if more persistently disturbed (Seeley et al. 1982). Stings fix to the skin with their pronounced barbs and provoke further stinging through the stimulus 2-decen-1-yl-acetate: the "enemy-marking pheromone" (Koeniger et al. 1979). Their stings are relatively painless to humans and may not repel decisive attacks by tree shrews, monkeys or humans (Seeley et al. 1982). *A. andreniformis* is far more prone to attacking and may possess a different alarm pheromone (Wongsiri et al. 1996).

17.4.2 Grasping, Pulling, Biting and Other Means

As with other insects, bees have mandibles, legs and wings as the main mechanical tools for dealing with the world, and which may also be used for defence. Again, the comparative ineffectiveness of these actions if employed as individuals is multiplied within the social context and new defensive means emerge in their combined use. Stinging is the predominant weapon in vertebrate defence, though *A. cerana* may use their mandibles also for biting (Sakagami 1960). Stinging, however, is of limited value in defence against a wide range of arthropod predators (Schmidt 1995). Here, bees rely on the more conventional weapons, gaining advantage from their socially coordinated exertion.

There is a wide range of solitary insects and other arthropods that are occasionally trying to prey on bees or their stores, and bees have little difficulty in deterring them due to their higher numbers. The more serious threats come from other social insects, namely ants and wasps, which, like bees, can make use of the advantages of sociability, join efforts and in particular may recruit nestmates for reinforcement to attack in great numbers. While vertebrate attacks are often deleterious to bee colonies, ants and wasps are a less conspicuous, but no less serious, threat to honeybees – only rarely destroying the colonies, but causing a constant drain. In dealing with these predators, Asian bees have developed highly refined social actions so as to keep at par with the dangers.

The prominent species of ants that are endangering Asian bees is the weaver ant, *Oecophylla smaragdina* (Seeley et al. 1982; Duangphakdee et al. 2005, 2009), which shares the tree environment, particularly with dwarf honeybees. Usually, these ants perch along the flight path of the bees, trying to grab single forager bees with their mandibles when leaving or returning. The small *A. florea* forager bees have little chance of breaking free if grabbed by a single *O. smaragdina* worker. However, the bigger *A. cerana* falls victim to the ants only if there are several of them grabbing a worker in a joint effort. In addition, several *A. cerana* workers can overcome an ant, grabbing the legs and tearing them apart; two *A. cerana* may suffice to kill a weaver ant. *A. florea* can also overcome *O. smaragdina* should they happen to fall on the nest, where they are mobbed by 10–15 workers and crushed (Seeley et al. 1982).

Ants and bees are effectively matched in fighting abilities, using similar means. Encounters are mainly decided by size differences and the relative numbers of opponents engaged. However, bees have one advantage over ants: that they can fly. This allows them to use the air route for trafficking from and to the nest, independent of the walking routes that ants are restricted to. Thus they are able to rely predominantly on isolating their nests from the ant environment, thereby minimising actual conflicts in spite of a situation of constant siege. In cavity-nesting bees, if any openings are sealed, the only remaining point of possible attack is at the flight entrance. This remaining conflict area can be structurally improved by carefully smoothing the surface at the nest entrance so as to weaken the grip of the ants and diminish their fighting chances. For defence, the entrance is staffed by vigilant guard bees, facing outward to ward off intruding ants with their mandibles. In addition, specific behaviours were observed making use of the wings. Guards may turn around and use air blasts to deter the ants or hover in buzzing flights above attackers to distract them (Seeley et al. 1982).

Most of the nest surface of the free-hanging nests of open-nesting bees is completely safe from ants, and bees can enter and leave without coming close to any ants. However, nests are accessible where they are attached to the substrate. In the case of the giant honeybees, this may be in the form of an extended line fixed to overhanging cliffs or the lower side of big branches. At the boundary, worker bees line up side by side in several rows, forming a permanent barrier effective against any walking insects (Seeley et al. 1982). Due to their preferential nesting in high and remote places, ant attacks are not very intense, but have been observed to occur, for example, by the arboreal ant species, *Dolichoderus tuberculata* and *Polyrhachis armata*. However, they seem to be easily kept at bay by giant honeybees (Seeley et al. 1982).

In contrast, dwarf honeybees nesting in shrubs formed by bushes or small trees live in the midst of "ant land". In Thailand, 37% of 76 *A. florea* nests were under attack from weaver ants (Seeley et al. 1982). As nests completely enclose the thin branches to which they are attached at the upper side of the comb, the accessible nest boundary is minimised and restricted to the outlet of the branches. Here, dwarf bee workers form similar permanent defence lines, as do giant honeybees, guarding this potential way of access. In order to strengthen these possible access points, dwarf bees create a further barrier by covering the branches with rings of sticky plant resins and deliberately use the material to spread it over opponents so as to immobilise them (Seeley et al. 1982; cf. Chap. 6). This resin substance of *A. florea* and *A. andreniformis* has a strong repellent effect on *O. smaragdina* ants (Duangphakdee et al. 2005, 2009). *A. florea* eagerly maintains the integrity of this defence line and takes great care to maintain these barriers, even fortifying them if challenged by ants (Duangphakdee et al. 2005, 2009). They react immediately if the barrier is bridged by leaves, either falling on the nest or otherwise touching the nest. In this case, they persistently push with their heads and mandibles, either to remove the object or at least to keep it at some distance, and specialised workers take turns over many hours (SenSarma et al. 2000; Fig. 17.3).

Dwarf honeybee nests are challenged from all sides by ants and other insects trying to snatch single bees from nearby perching places on leaves or twigs. Dwarf bees are therefore extremely watchful in their immediate environment and attack such perching predators by trying to grab them with their legs and pull them onto the nest, where they are then mobbed and overcome by the crowd. In order to bridge the distance, ants may pile up one on top of the other to form bee ladders, which eventually collapse due to their weight if piling up too high (Fig. 17.4). Thus, Asian honeybees try to minimise conflicts with ants by isolating terrestrial access to the nest as far as possible, by creating structural barriers and forming ordered defence lines at accessible points – decisively attacking any ant that is coming close, which could otherwise recruit other ants if successful.

The second arthropod group that seriously threaten Asian bees are large vespine wasps. In contrast to ants, wasps also attack on the wing and cannot be as easily excluded by the blocking of walk-ways. Furthermore, some of these wasps outclass the bees that they prey on, in size, and are virtually sting-proof due to the strong armouring of their body surfaces. Relying on animal protein to raise their brood, together with communication systems for recruiting nestmates, these groups are true "wolves" for bees.

Commonly, wasps prey on flying foragers and there is little possibility of a single bee defending itself once caught. However, wasps frequently try to catch leaving or returning foragers at the colony. Bees try to decrease the risk of being caught. When one or two *V. velutina* wasps are bee-hawking in front of a colony in the

Fig. 17.3 Defending nest integrity in *Apis florea* by pushing off bridges. Head-pushing is performed by a specialised group of workers, which take turns lasting a minute to almost an hour (photo: S. Fuchs)

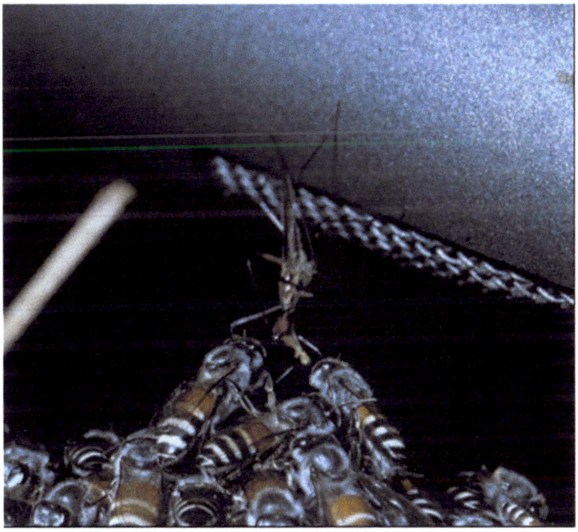

Fig. 17.4 *Apis florea* workers piling up, one on top of the other, to reach an assassin bug (*Reduviidae*), perched from the microphone that is fixed above the colony, to snatch single workers. Bee heaps become unstable if too high and collapse (photo: S. Fuchs)

"goal-keeper" manner (Koeniger et al. 1996), flight activity decreases and almost ceases completely after 12 min of wasp exposure. In addition, returning foragers make a rapid straight-line flight path into the flight entrance, nearly halving their time of exposure to the wasp.

Overcoming an entire colony is more difficult. Bees may use grabbing and mobbing in defence, overcoming even significantly bigger wasps due to their numbers. In *A. cerana*, the numbers of guards increased approximately eightfold in response to *V. velutina* predation and guards formed tongue-like groups covering the area below the entrance in order to increase their chances to grab and ball a wasp (Tan et al. 2007). Wasps thus avoid approaching too close to a hive entrance or an open-nesting bees' protective curtain. Nevertheless, wasps might still continuously kill individual bees until a colony is weakened and no longer able to defend itself. *V. tropica* was observed in bands of two or three, successively killing single workers perching above an *A. florea* colony. *A. florea* was unsuccessful in mobbing wasps, which, due to their size, could retreat in spite of 12 workers grabbing them. On the second day, the rest of the bees absconded, leaving the brood to the wasps.

Not all bees are as defenceless to wasp attacks. In Japan, *A. cerana* colonies are besieged by the enormous *Vespa mandarinia japonica*, reaching body sizes of 50 mm in length, about five to six times that of the worker bees. On top of this physical superiority of the wasps, they mark colonies by a scent from their abdominal Van der Vecht gland so as to attract further wasps to strengthen the attack force. *A. cerana* recognises this scent and increases the numbers of guard bees at the colony entrance. Attacked by a *V. mandarinia* wasp, the guard bees first retreat into the nest. If the wasp tries to follow, it meets a superior force of defenders, intensely mobbing the wasp and entirely enclosing it in a ball of bees containing about 500 workers – killing the wasp. Interestingly, this killing is not by stinging, because the stings cannot effectively penetrate the sclerotised armouring of the wasps. Instead, the bees apparently create such high temperatures within the ball that the wasps are killed by heat. Temperatures within the balls were 47°C, still below the upper thermal limit of 48–50°C in *A. cerana*, but higher than the thermal limit of 44–46°C in *V. mandarinia* (Ono et al. 1995).

Similar heat-balling in *A. cerana japonica* was described against *V. simillima xanthoptera*, where bees generated temperatures of 46°C within the balls, while the upper thermal limit of the wasps was 45–47°C (Ono et al. 1987). In China, when *A. cerana* is defending against *V. velutina*, about 30 workers ball the wasps within less than 1 s, and temperatures within the balls reach 45°C within 3 min, reaching the upper lethal thermal of 45°C for the wasps, while *A. cerana* survives at 50°C (Tan et al. 2005). Abrol (2006) reported that *A. cerana* killed about 12% of attacking *V. magnifica* by balling, with temperatures ranging between 45 and 46.8°C within balls.

Heat-balling is not restricted to *A. cerana* and was also documented in *A. nuluensis* (Koeniger et al. 1996). Tethered *V. multimaculata* brought into contact with guard bees at the hive entrance were rapidly covered by approximately 40 bees. Inner-ball temperature was not measured, but the balls felt warm on the hand (>37°C), and the bees showed no stinging behaviour. Heat-balling may also occur in open-nesting bees. Kastberger and Stachl (2003) documented that *A. dorsata* will grasp a tethered wasp as soon as it is brought into contact with the bees on the nest surface. The wasp was balled and killed within 10 min, during which time the recorded inner-ball temperatures were between 45 and 48°C. Heat-balling is likely to occur in most of the other Asian *Apis* species, although documentation is incomplete.

However, it is not entirely clear if heat is the only factor contributing to the death of the wasps within the balls. Abrol (2006) reported that during balling, *A. cerana* guards would seize, bite, pull, hold and sting *V. magnifica* and reported some degree of mutilation, particularly of the wings, in the killed wasps. Stinging may not be successful against the big *V. mandarinia*, but successful stinging was found in *V. simillima* that had been balled by *A. mellifera* (Ono et al. 1987).

Sugahara and Sakamoto (2009) determined a further factor contributing to the death of balled wasps. In their study, they recorded that *V. mandarinia* could survive temperatures of 47°C for 10 min, while inner-ball temperatures remained at temperatures below 45.9°C, insufficient to kill the wasps. However, they found that the CO_2 level within balls would increase to 3.6%, at the same time lowering the wasp's lethal thermal limit by 2°C, without affecting the upper thermal limit of the bees. They thus concluded that, in combination with heat, CO_2 is a major factor in killing wasps.

A recent reinvestigation of balling behaviour in the Cyprian subspecies of *A. mellifera* defending against *V. orientalis* supported the idea that temperature within the balls would be inadequate to explain the killing success. Inner-ball temperatures in *A. mellifera* are 1.5–3.3°C lower than in *A. cerana* (Ono et al. 1987; Tan et al.

2005). In *A. m. cypriaca*, the temperature within balls was 44°C, clearly below the upper lethal temperature of the wasps (approximately 50°C) which is about that of the worker bees. In a detailed examination, it was found that the killing of the wasps was accomplished by still another method: the balling bees would asphyxiate the hornets by imposing pressure on the abdomen, interfering with the wasp's breathing movements and blocking their spiracles (Papachristoforou et al. 2007).

Thus heat might be a less prominent factor in *A. mellifera* wasp-balling as compared to *A. cerana*. It appears that thermal insulation and heating up, intoxication by increased CO_2 levels and possibly asphyxiation by pressure on the abdomen all contribute to the killing of wasps. Additional factors involved throughout the genus still need to be determined.

Heat generation during fights is an obvious by-product of agitation, accompanying any event in which an intruder is mobbed by a group of nest defenders. Under most circumstances, this would add to the stress of the opponent during attack. From simple mobbing of small predators to dense balling, turning on heat and increasing CO_2 levels provide outstanding examples of how the social dimension opens up completely new means for defending, which are then perfected in their own right.

17.5 Colonial Displays

Coordinated stinging defence and balling are socially coordinated behaviours in which hundreds or thousands of individuals may participate. However, Asian bees have particularly impressed observers by conspicuous defence-related behavioural coordination in a much grander style, involving high proportions of the workers in activities, simultaneously giving the appearance that the entire colony is a coherent acting unit.

Since Roepke (1930) first mentioned coordinated waves of excitation moving over the combs of *A. dorsata*, there have been various accounts of conspicuous coordinated displays over the years. On closer examination, these proved to consist of several fairly distinct behavioural patterns, all sharing coordination between neighboured workers, but separated by different motor patterns, conditions of release and some degree of species-specific variation. The main dividing lines between these behaviours are the releasing stimuli, either by substrate vibration and air movement or by the perception of visual object movements, the vertical or transverse direction of their body axis movement, and the different relation to attack–retreat strategies. At the same time, some of these behaviours are apparently interrelated or show transitions according to the functional context, which still need to be clarified in more detail (cf. Chap. 18).

In early observations, variations in intensity of disturbance and unaccounted mixtures of different stimuli created variable outcomes and overlaps between these behaviour patterns. Fairly commonly, they were jointly addressed as "shimmering" (Butler 1954), and in past and recent literature, the different behaviour patterns were frequently misnamed or confused. By now, with more and more precise investigations available, the relations and functions of these behaviours can be outlined more clearly. Although knowledge is still incomplete, we will try here to summarise their characteristics, tentative functions and possible relationships. To date, at least four different behaviours can be distinguished: "hissing", "rearing-up", "abdomen-shaking" and "body-twisting". "Hissing" and "rearing-up" are both elicited by mechanical stimulation and constitute elements of defence against vertebrates. In contrast, "abdomen-shaking" and "body-twisting" are elicited exclusively by moving visual stimuli and are closely related subforms of "shimmering" behaviour, originally described by Butler (1954), serving to deter various flying predators, predominantly wasps.

17.5.1 Hissing Behaviour

Hissing behaviour was first described in *A. cerana* by Sakagami (1960) but erroneously likened to "shimmering" earlier described by Butler (1954) in *A. dorsata*. Later, the behaviour was investigated in *A. cerana* in more detail by Koeniger and Fuchs (1973) and was termed "hissing behaviour". In this behaviour, individual *A. cerana* workers, in a slight and inconspicuous movement, draw their entire body closer to the substrate, either to the comb or the bees they are sitting on. During this movement they brace their wings slightly, by about 45°, and produce a short 0.5–1 s burst of wing vibration sound, consisting of a 0.5–6 kHz energy spectrum lacking harmonic structure. This behaviour is contagious, such that neighbouring workers will be induced to perform the same behaviour within a short time delay. In effect, a hissing wave is produced which spreads at approximately 25 cm/s among the bees on a comb and may wander over the edge to the bees on the opposite side of the comb or even spread to other combs if these are connected by bee bridges. The excitation can be propagated by short-range air movements, without direct bee-to-bee contact or without a connecting comb. The joint activity of the bees produces the characteristic hiss, lasting until the hissing wave is exhausted. Series of more than 10 successive hisses may follow each other in 3–5 s intervals (Sakagami 1960), which points to an individual latency of hissing as a prerequisite for wave-like propagation.

Hissing behaviour is most readily released by substrate vibrations, such as by knocking or poking a colony or tapping the supporting branch of a open-nesting colony, and also by locally applied gusts of air. Substrate vibrations which are effective are in a low-frequency range, and sine waves exceeding 600 Hz, soon become ineffective (Fuchs and Koeniger 1974). In contrast, moving visual stimuli are not effective at all. Thus, the array of stimuli points to reactions that are tuned to approaching vertebrates, rather than flying predators. Hissing was mentioned to occur in *A. florea* and *A. cerana* by Seeley et al. (1982), was investigated in more detail in *A. florea* by SenSarma et al. (2002) and is widespread in the *Apis*.

Colony displays may serve to create the impression of the entire nest being a coherent organism rather than an aggregation of thousands individuals, and integrated actions such as "shimmering" waves can underline this by mimicking total movements of this entire structure. Hissing behaviour has thus been interpreted in terms of a threat, with the hissing sound serving as a warning signal (Ghatge 1949; Fuchs and Koeniger 1974; Akratanakul 1977; Seeley et al. 1982). Hissing sounds are common aposematic signals that are understood as a warning across many species (Kirchner and Röschard 1999). There are no conclusive experiments proving an effect of hissing behaviour on potential predators, but Koeniger and Fuchs (1973) demonstrated that sun bears, kept in an zoo (*Heliarctos malayanus*), could be temporarily repelled if hissing sounds were played back inside a box in which they had first found a honeycomb, thus providing at least anecdotal evidence that predators could be deterred.

Hissing alters the colonial state. After hissing, bees are rather less defensive than aroused (Fuchs and Koeniger 1974). In particular, nest activities of the bees cease, and foraging flights commence (SenSarma et al. 2002). In the open-nesting *A. florea*, during hissing, the nest bees on the surface draw closer to the comb surface and nests appear smaller, and, due to the absence of surface movements, are clearly less conspicuous. Thus, rather than a counter-attack strategy, hissing is related to a retreat strategy, by which bee nests become less noticeable and foragers are protected from individual predation.

17.5.2 "Rearing-up"

A second less widely recognised behaviour, which is superficially similar to "hissing", has rarely been mentioned and was observed only in colonies of the open-nesting species. On gross disturbances, workers in the colony turn their heads towards the colony and stretch their abdomens outwards. This behaviour is accompanied by wing vibrations and sound production, although of a lower frequency. However, it is set apart from hissing by the body movement where, after stimulation, the individual bees draw their thorax closer to the comb and simultaneously stretch their abdomen outward. It still needs to be clarified whether this is accompanied by exposure of the sting apparatus and/or the release of alarm pheromones. The movement pattern is contagious between neighbouring bees as it spreads over the colony and results in an increase of the colony's outline and a rough surface structure.

A detailed description of the behaviour in *A. florea* was given by Pirk et al. (2002), who, however, did not propose a particular name for the pattern, but which might be termed "testudo behaviour". Kastberger and Sharma (2000) described spreading waves of abdomen stretching in *A. dorsata*, which differs from "shimmering" behaviour by the slow speed of the propagation, and introduced the term "rearing-up", which we will maintain here.

"Rearing-up" as a colony display increases rather than decreases the colony's outline, and a small *A. florea* colony reacting to a disturbance resembles an angry vertebrate ruffling its fur, which, together with the sound production, is likely to have some impact on potential predators (Fig. 17.5). In relation to "hissing", "rearing-up" and subsequent attacking appears to be more easily elicited in *A. andreniformis* than in *A. florea*, whose colonies are in general more bent on attacking intruders (Oldroyd and Wongsiri 2006). In *A. dorsata*, "rearing-up" apparently corresponds to the "defensive roar" described by Morse and Laigo (1969), which precedes attacks from the bees and certainly does not fail to warn off humans to retreat before getting stung.

"Hissing" and "rearing-up" are both elicited by substrate concussions or air gusts; the movement patterns are related, and both are accompanied by sound production. However, a main difference between the behaviours is their different relationships to defensive or counter-attack strategies. While hissing has a distinctly defensive connotation, quieting the nest and rendering it inconspicuous, "rearing-up", with its outward extension of the sting apparatus, is closely connected to the organisation of counter-attack behaviour. "Rearing-up" typically occurs at higher stimulus intensities or after more persistent stimulation. Thus, in spite of their positioning at opposite ends of the defence/attack axis, the two behaviours connect in a reaction cascade, where more intense stimulation after the initial defensive hissing responses leads to attack-related "rearing-up" behaviour. An apparent relationship between the response types of "releaser" or "retreater" colonies to mechanical stimulation in *A. mellifera* (Kastberger et al. 2009a, b) might be a further reflection of the two different strategies as possible reactions to variable levels of the same threat.

If disturbance is further continued, "rearing-up" is followed by a general commotion among the bees, a loosening of the bee curtain and the organising into attack flights if an enemy is recognised. This results in bees moving down and hanging at the

Fig. 17.5 Display of "rearing-up" behaviour in a colony of *Apis florea* after disturbance by air gusts. *Left*: before stimulation, *right*: after stimulation

lower rim of the comb, where they then start, in attacking waves, to drive away any intruder (Morse and Laigo 1969). As a result of the extension of the bee curtain along the bottom of the comb, the optical outline of the colony is augmented. This may take quite impressive dimensions, as demonstrated in the apparent area of an *A. dorsata* colony more than doubling (Koeniger et al. 2010). This enhancement of the appearance of colony size presumably adds to the threatening effects of "rearing-up", and the defensive roar might contribute to repel intruders. A particularly interesting effect of these bee beards was observed in *A. dorsata* by Koeniger and Koeniger et al. (2010), who reported that the oriental honey buzzard, *ptylorhynchos*, could be misled to erroneously attack the extended bee curtain, and thus fail to successfully prey on the comb and its contents. This observation provides remarkable evidence as to the functionality of colony displays, where direct proof is generally sparse.

Both "hissing" and "rearing-up" have a common response to vibration stimuli, nest concussions and air gusts, which are likely connected to the approach of vertebrate predators. Both involve the emission of perceptible auditory signals. They can be understood as integrated elements of a graded defence cascade, aimed against vertebrates; reaching from retreat-related "hissing" behaviour to counter-attack-related "rearing-up" behaviour; ultimately leading to the organisation of a bee curtain extension and the release of attack waves.

17.5.3 "Shimmering" and "Abdomen-Shaking"

"Abdomen-shaking" was described by Sakagami (1960) in *A. cerana* as the violent, pendulum-like, lateral swaying of the abdomen, with the behaviour frequently performed by several groups of worker bees. Later, Schneider and Kloft (1971) gave a more detailed description of the movement pattern and recognised its function in group defensive behaviour ("Gruppenverteidigungsverhalten") against hornets. The behaviour corresponds to that described by Butler (1954) in *A. cerana*, which he termed "shimmering".

Koeniger and Fuchs (1973) analysed film sequences and confirmed the basic movement pattern consisting of approximately two to five successive lateral movements in parallel to the ground surface, which first increase and later decrease in amplitude, each lasting approximately 0.09 s. The behaviour is silent and not accompanied by any conspicuous emission of sound, other than a slight rustling of the wings. In several experiments, they confirmed that the behaviour is released exclusively by the movement of dark objects against a light background. In contrast, presentation of motionless objects (including insects), light flashes and chemical stimuli did not release any reaction. In particular, "abdomen-shaking" could not be released by ground vibrations or concussions, which constitute a fundamental difference of "body-shaking" from both "hissing" and "rearing-up" behaviour. The visual stimuli which release "abdomen-shaking" in *A. cerana* were tested by an optical dummy and showed that black circles of a vision at an angle of approximately 8°, moving at 140° per second over the vision field, were most effective. In particular, the reaction was surprisingly persistent and after a continuous stimulation of over 1 h, 20% of the initially participating bees were still found reacting.

In nature, the behaviour can be observed whenever flying predators of appropriate size and speed approach the bees. The synchronised movements of neighbouring bees suggest that body-shaking is contagious and, as a result of slight delays, waves of body-shaking can be observed travelling over groups of animals. The origin of the body-shaking waves is usually where the moving stimulus is at its closest; but, as the waves do not travel far, it is likely that the behaviour is mainly restricted to workers which perceive some of the stimulus by themselves.

Although body-shaking was observed to effectively discourage various insects, such as butterflies or flies, from landing, the behaviour is mainly aimed to deter predatory wasps (Butler 1954; Schneider and Kloft 1971; Koeniger and Fuchs 1973; Seeley et al. 1982; Abrol 2006; Tan et al. 2007). In the open-nesting species, efficient defence strategies to protect the freely exposed colony are a vital necessity, and this visual display may serve as an efficient initial barrier. "Abdomen-shaking" was observed in *A. florea* and *A. andreniformis* (Butler 1954; Koeniger and Fuchs 1973; SenSarma et al. 2002; Oldroyd and Wongsiri 2006), while the giant honeybees, *A. dorsata* and *A. laboriosa*, showed the related "shimmering" display of "body-twisting" described below.

Body-shaking also appears widespread in cavity-nesting species. It was long known and first described in the cavity-nesting *A. cerana* (Butler 1954; Sakagami 1960) and is expressed even more in *A. koschevnikovi* (Mathew and Mathew 1990) and *A. nuluensis* (Koeniger et al. 1996). There are no reports related to *A. nigrocincta*, but it would be an unlikely exception. In the cavity-nesting bees, body-shaking apparently defends the nest entrance against wasp intruders, as was reported in *A. nuluensis* defending against *V. multimaculata* (Koeniger et al. 1996). Occasionally, body-shaking at the nest entrance can be observed in *A. mellifera* when besieged by hornets (personal observation). However, body-shaking in cavity-nesting bees might also be of high relevance during unsheltered periods in reproductive swarming or migration, although no reports are available referring to these situations. Thus an effective colony surface defence is not restricted to permanently open-nesting species, but may be a prerequisite for this status.

17.5.4 "Shimmering" and "Body-Twisting"

"Shimmering" waves in the giant honeybees, *A. dorsata* and *A. laboriosa*, are an extraordinarily conspicuous and eye-catching colonial display and were described early by Roepke (1930). Even from a long distance, waves of excitation can be observed travelling across the extended comb surface of these open-nesting giant honeybees. Accordingly, these waves have attracted considerable attention, and impressive video examples are available in Kastberger et al. (1998), Kastberger and Sharma (2000) and Woyke et al. (2008). The characteristic spiralling patterns of the spreading waves closely correspond to other self-organising excitation patterns (Camazine et al. 2003) found in both animate and inanimate systems that are governed by contagious actions followed by latency periods. "Shimmering" patterns thus represent a stunning visualisation of self-organised action in social insects (cf. Chap. 18).

As has been consistently reported, the typical stimulus eliciting "shimmering" waves in giant honeybees is the visual disturbance by flying insects, such as wasps, butterflies or birds, and the behaviour can easily be elicited by moving a dummy object close to the bees. Particularly, responsive individuals play a crucial role in triggering the social waves (Schmelzer and Kastberger 2010). In a detailed study, Kastberger et al. (2008) provided convincing analytical evidence that these waves are indeed capable of driving away preying hornets, and thus for the first time giving direct proof for the efficiency of a honeybee colony-level display.

Although the stimulus releasing the behaviour clearly indicates its close relationship to body-shaking, "shimmering" waves in the giant honeybees differ from "shimmering" by "abdomen-shaking" as found in the other honeybee species, by the distinctly different body movement during which, instead of a lateral swaying of the abdomen, the abdomen is flipped upward. Closer video analysis by Woyke et al. (2008) revealed that the movement is not limited to the abdomen, but includes a rotation of the entire body around a horizontal cross-axis through the thorax, by which the head is simultaneously brought closer to the ground. They thus proposed the term, "defence body-twisting".

Although "abdomen-shaking" and "body-twisting" are fairly diverse motor patterns, the difference might not be very fundamental. In "abdomen-shaking", the abdomen is also somewhat raised upward, and this component of the movement may have become the predominant movement in the giant honeybees. Restricting movement to the dorso-ventral component could have been for economical reasons or advantageous for coordination over the fairly large comb areas. Whatever the reasons, a common origin of both motor patterns would be in accordance with a close relationship of the "shimmering" patterns across the species.

On the other hand, the dorso-ventral rotation, with a flipping of the abdomen, closely resembles the motor pattern of "rearing-up". As discussed above, "rearing-up" is unrelated to "shimmering", as it is elicited by vibrations and the response mainly addresses ground-moving vertebrate predators. One may, however, consider the possibility that bird predation might have contributed to connecting the action patterns. Birds constitute a major threat to *A. dorsata* nests. As moving objects, they constitute visual stimuli, and, being visually oriented predators, are unlikely to be confused or scared off by the communal "shimmering" display as are wasp predators. Instead, as highly sophisticated predators of a comparatively large body size, birds would require more substantial defence actions, such as anti-vertebrate counter-attack activities of the rearing-up response and the successive initiation of stinging raids; this could have worked to merge the honeybee defence responses.

It has been noted that in the giant honeybee's "body-twisting", the bees expose their Nasanov glands (Kastberger et al. 1998). This interesting observation underlines the role of the displays in social coordination between colony workers. Exposure of the Nasanov gland secretion ensures cohesion between the workers, rather than alerting them for counter-attack (Kastberger et al. 2008). This has also been observed in the "abdomen-shaking" of *A. nuluensis* (Koeniger et al. 1996) elicited by *V. multimaculata* attacks that functions to attract additional guards.

17.6 Communication About Predators

Colonial displays serve to deter or confuse attacking predators, with the main effect of the entire colony appearing as a contiguously acting unit. At the same time, these displays may function in an internal context to coordinate defence actions of the individuals. Spreading waves of contagious behaviour confer information to partaking individuals, even if they may not have perceived the releasing stimulus themselves. The well-described build-up of defensive actions in *A. dorsata*, from first "shimmering" responses until the coordination of coherent attack waves (Lindauer 1956; Morse and Laigo 1969; Koeniger 1975), clearly depicts a sequence in which colonial displays retroact on the individuals' behavioural states, modifying the colony-level arousal state until decisive action is taken. This information, spread between individuals, may involve not only tactile and vibrational stimuli, but also chemical communication, as in *A. dorsata*'s "body-twisting" or in *A. nuluensis*' "abdomen-shaking", in which the exposure of the Nasanov gland apparently releases pheromones (Koeniger et al. 1996; Kastberger et al. 1998).

A specifically sophisticated case of inner-colony transfer of information in relation to potential predators has recently been described in the piping–hissing sequence in *A. florea* (SenSarma et al. 2002), which also demonstrates how pre-existing colony-level behavioural components can be combined for building higher-level information-handling mechanisms. In this behaviour, single foragers who encounter a predator-like stimulus during their return from collection flights engage in impetuous vibrational signalling. They press their thoraces to the comb surface and, over several seconds, emit a series of faintly audible piping sounds, which prove to be a fairly energetic signal in substrate vibration recordings (Fig. 17.6). These vibrations are perceived by adjacent workers on the nest surface, which then react by hissing behaviour. From these initial workers, a hissing wave spreads over the comb and may encompass substantial parts of, or even the entire, colony. The colonial reaction is pattern-sensitive, as hissing is released more readily by subsequent piping within the piping series (Fig. 17.7).

In this two-step mechanism, single or few foragers are able to transfer information they have picked up in the field about a potential threat to the nest, and in a second step, this information is rapidly communicated to high proportions of the nestmates. The two steps of this reaction chain are completely separated by the stimulus modalities involved. Foragers detect enemies either by vision or possibly odour. However, the hissing reaction of the workers on the nest is released by sensitivity to nest vibration and not by vision. They do not react by hissing, even if a predator is very close and moving, if the movement or proximity does not cause nest vibrations. However, the nest vibrations applied by the returning workers through piping behaviour cause an

Fig. 17.6 An alerted *A. dorsata* colony, in which the guard bees have built a curtain, formed by chaining worker bees that are hanging from the lower rim of the comb, and thereby more than doubling the colony's original outline (solid part of the nest, ruler size 60 cm, photo: N. Koeniger, from Koeniger et al. 2010)

identical hissing reaction, as if the nest were directly disturbed by a predator, followed by a cessation of nest activity. There are some indications that this sequence of behaviour is not restricted to *A. florea*, as piping–hissing sequences have been recorded in *A. cerana* (S. Biswas personal communication).

The hissing reaction to the piping sound is the same reaction that the bees show to nest vibrations from outside sources, being the percussions from approaching enemies. It is thus likely that, in the evolution of the system, the bees tapped into the pre-existing connection between danger and nest concussions by using vibration signals to communicate danger. Besides the hissing sound, the most obvious reaction of the worker bees is the cessation of all activity. Stopping activity is an unspecific reaction that bees generally show to artificial vibrations applied to combs (Frings and Little 1957). Carpenter ants apply drumming series to the ground, which similarly mimic nest percussions due to their disturbance and immobilise recipient workers, while alerting them to react more strongly to any additional stimuli that may indicate an intruder (Fuchs 1976). Stop signals were emitted by bee foragers that had encountered dangers at the feeding sites and decreased foraging (Nieh 2010). Thus it seems plausible that the connection to ground vibrations indicating danger may be the common root for the various worker piping signals, which tend to induce a cessation of current activities in the recipients ("stop signal": Nieh 1993) and may also function as modulatory signals (Thom et al. 2003).

In this particular case, the energetic worker piping of returning foragers obviously carries a strong message of imminent danger. *A. florea* workers detect and

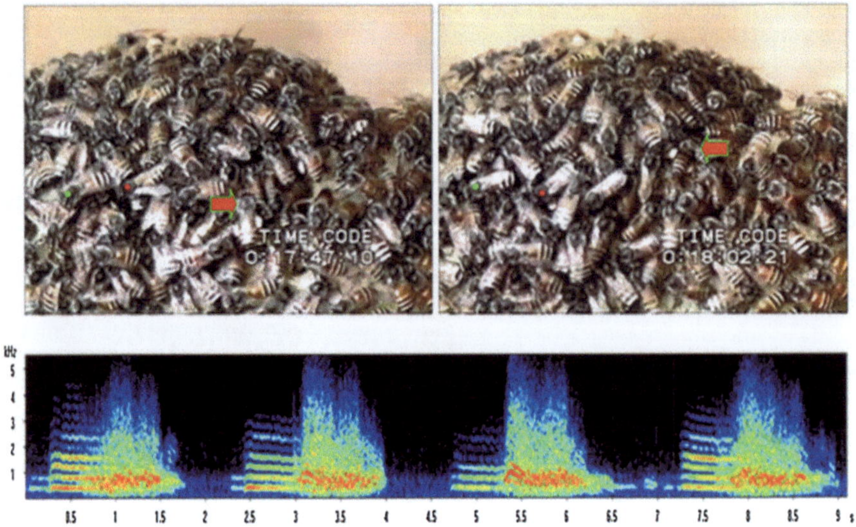

Fig. 17.7 A piping–hissing sequence in *Apis florea*. A pollen forager has just landed (*upper left, arrow*) and crawls forward, emitting a series of 10 piping sounds. In order to transmit the vibration, it presses the thorax to the comb (*upper right, arrow*). The workers on the comb react by stopping activities and draw closer to the comb, emitting a hissing sound while pulling their wings together. This can be clearly seen in the two bees that are marked by a red or green spot, respectively. The sonogram (*below*) shows a series of piping–hissing sequences. The harmonic structure of the foragers' piping overlaps in time with the noisy hissing response of the hive workers (photo: S. Fuchs)

report enemies over distances of more than 10 m from the nest, thus truly functioning as far-reaching sensory organs for the colony (Seeley 1994). The fact that honeybees collect and communicate information about food sources inside the nests has been known since von Frisch's (1967) studies and is regarded as an outstanding example of invertebrate communication skills. Collecting and communicating information on the equally essential environmental conditions and threats to the colony has, however, been widely overlooked and may reveal fascinating details in future research.

References

Abrol D (2006) Defensive behaviour of *Apis cerana* F. against predatory wasps. J Apic Sci 50:39–46

Akratanakul P (1977). The natural history of the dwarf honey bee, *Apis florea* F., in Thailand. Thesis, Cornell University, Ithaca

Akratanakul P (1987) Honey bee diseases and enemies in Asia: a practical guide. FAO, Rome

Ambrose JT (1978) Birds. In: Morse RA (ed) Honey bee pests, predators, and diseases. Cornell University Press, Ithaca, pp 215–226

Blum MS, Fales HM, Morse RA, Underwood BA (2000) Chemical characters of two related species of giant honeybees (*Apis dorsata* and *A. laboriosa*): possible ecological significance. J Chem Ecol 26:801–807

Breed M, Guzman-Novoa E, Hunt G (2004) Defensive behavior of honey bees: organization, genetics and comparison with other bees. Annu Rev Entomol 49:271–98

Butler CG (1954) The world of the honeybee. Collins, London

Camazine S, Deneubourg JL, Franks NR, Sneyd J, Theralauz G, Bonabeau E (2003) Self-organization in biological systems. Princeton University Press, Princeton

Caron DW (1978) Marsupials and mammals. In: Morse RA (ed) Honey bee pests, predators, and diseases. Cornell University Press, Ithaca, pp 227–256

Cassier P, Tel-Zur D, Lensky Y (1994) The sting sheath of honey bee workers (*Apis mellifera* L): structure and alarm pheromone secretion. J Insect Physiol 40:23–23

Cunard SJ, Breed MD (1998) Post-stinging behavior of worker honey bees (Hymenoptera: Apidae). Ann Entomol Soc Am 91:754–757

Duangphakdee O, Koeniger N, Koeniger G, Wongsiri S, Deowanish S (2005) Reinforcing a barrier – a specific social defense of the dwarf honeybee (*Apis florea*) released by the weaver ant (*Oecophylla smaragdina*). Apidologie 36:505–511

Duangphakdee O, Koeniger N, Deowanish S, Hepburn HR, Wongsiri S (2009) Ant repellent resins of honeybees and stingless bees. Insectes Soc 56:333–339

Dyer FC (1991) Coadaptation of colony design and worker performance in honey bees. In: Smith DR (ed) Diversity in the genus *Apis*. Westview, Boulder, pp 213–245

Dyer FC, Seeley TD (1991) Nesting behavior and the evolution of worker tempo in four honey bee species. Ecology 72:156–170

Frings H, Little F (1957) Reactions of honeybees in the hive to simple sounds. Science 125:122

Fry CH (1984) The bee-eaters. T&AD Poyser, Town Head House, Calton

Fuchs S (1976) The response to vibrations of the substrate and reactions to the specific drumming in colonies of carpenter ants (*Camponotus*, Formicidae, Hymenoptera). Behav Ecol Sociobiol 1:155–184

Fuchs S, Koeniger N (1974) Schallerzeugung im Dienst der Kolonieverteidigung bei *Apis cerana*. Apidologie 5:271–287 [in German]

Ghatge A (1949) Some research work on *Apis florea*. Indian Bee J 11:5–6

Jayasvati S (1989) Scanning electron microscopy analysis of honey bee *Apis florea* F., *A. dorsata* F., *A. cerana* F. *A. mellifera* L. and *A. andreniformis* stings in Thailand. Rec Adv Toxicol Res 2:193–204

Kastberger G, Sharma DK (2000) The predator-prey interaction between blue-bearded bee eaters (*Nyctyornis athertoni* Jardine and Selby 1839) and giant honey bees (*Apis dorsata* Fabricius 1798). Apidologie 31:727–736

Kastberger G, Stachl R (2003) Infrared imaging technology and biological applications. Behav Res Methods Instrum Comput 35:429–439

Kastberger G, Raspotnik G, Biswas S, Winder O (1998) Evidence of Nasanov scenting in colony defence of the giant honeybee *Apis dorsata*. Ethology 104:23–37

Kastberger G, Schmelzer E, Kranner I (2008) Social waves in giant honeybees repel hornets. PLoS ONE 3(9):e3141. doi:10.1371/journal.pone.0003141

Kastberger G, Thenius R, Stabentheiner A, Hepburn R (2009a) Aggressive and docile colony defence patterns in *Apis mellifera*. A retreater–releaser concept. J Insect Behav 22:65–85

Kastberger G, Weihmann F, Hoetzl T (2009b) Complex social waves of giant honeybees provoked by a dummy wasp support the special-agent hypothesis. J Integr Commun Biol 2:1–2

Kirchner WH, Röschard J (1999) Hissing in bumblebees: an interspecific defence signal. Insectes Soc 46:239–243

Koeniger N (1975) Observations on alarm behaviour and colony defence of *Apis dorsata*. In: Noirot C, Howse PE, LeMasne G (eds) Pheromones and defensive secretions in social insects. IUSSI, Dijon, pp 153–154

Koeniger N (1976) Neue Aspekte der Phylogenie innerhalb der Gattung *Apis*. Apidologie 7:357–366 [in German]

Koeniger N, Fuchs S (1973) Sound production as colony defence in *Apis cerana*. In: Proceedings of 7th international congress, IUSSI, London, pp 199–204

Koeniger N, Weiss J, Maschwitz U (1979) Alarm pheromones of the sting in the genus *Apis*. J Insect Physiol 25:467–476

Koeniger N, Koeniger G, Mardan M (1994) Mimicking a honey bee queen? *Vespa affinis indosiniensis* Perez 1910 hunts drones of *Apis cerana* F. Apidologie 27:353–359

Koeniger N, Koeniger G, Gries M, Tingek S, Kelitu A (1996) Observations on colony defense of *Apis nuluensis* Tingek, Koeniger and Koeniger 1996, and predatory behaviour of the hornet *Vespa multimaculata* Perez, 1910. Apidologie 27:314–352

Koeniger N, Koeniger G, Tingek S (2010) Honey bees of Borneo. Natural History, Borneo

Kreil G (1975) The structure of *Apis dorsata* melittin: phylogenetic relationships between honey bees as deduced from sequence data. FEBS Lett 54:100–102

Lindauer M (1956) Über die Verständigung bei indischen Bienen. Zeit Vergl Physiol 38:521–557 [in German]

Mathew S, Mathew K (1988) The "red" bees of Sabah. Int Bee Res Assoc Newsl Beekeep Trop Subtrop Countries 12:1

Mathew K, Mathew S (1990) Further observations on characters and behaviour of *Apis koschevnikovi*, the "red bee" of Sabah. Bee World 71:61–66

Matsuura M (1988) Ecological studies on vespine wasps (Hymenoptera: Vespidae) attacking honeybee colonies. Appl Entomol Zool 23:428–440

Matsuura M (2003) Biology and control of social wasps and bees in urban environments 3. Honeybee Sci 24:193–205

Morse RA, Laigo FM (1969) *Apis dorsata* in the Philippines. Philipp Assoc Entomol 1:1–96

Morse RA, Shearer DA, Bosh SR, Benton AW (1967) Observations on alarm substances in the genus *Apis*. J Apic Res 6:113–118

Natzir R, Teranishi H, Kitagawa M, Kasuya M (1999) A novel venom protein of the Asian bee (*Apis cerana indica*) with affinity to human α1-microglobulin. Allergol Int 48:121–129

Nieh JC (1993) The stop signal of honey bees: reconsidering its message. Behav Ecol Sociobiol 33:51–56

Nieh JC (2010) A negative feedback signal that is triggered by peril curbs honey bee recruitment. Curr Biol 20:310–315

Novogrodzki R (1990) Amphibians and reptiles. In: Morse RA (ed) Honey bee pests, predators, and diseases. Cornell University Press, Ithaca, pp 236–242

Oldroyd BP, Wongsiri S (2006) Asian honeybees. Harvard University Press, Cambridge

Ono M, Okada I, Sasaki M (1987) Heat production by balling in the Japanese honeybee *Apis cerana* japonica as a defensive behaviour against the hornet, *Vespa simillima xanthoptera* (Hymenoptera: Vespidae). Experientia 43:1031–1032

Ono M, Igarashi T, Ohno E, Sasaki M (1995) Unusual thermal defence by a honeybee against mass attack by hornets. Nature 377:334–336

Papachristoforou A, Rortais A, Zafeiridou G, Theophilidis G, Garnery L, Thrasyvoulou A, Arnold G (2007) Smothered to death: hornets asphyxiated by honeybees. Curr Biol 17:R795–R796

Pirk CWW, Hepburn HR, Radloff SE, Erlandson J (2002) Defense posture in the dwarf honeybee, *Apis florea*. Apidologie 33(3):289–294

Punchihewa RWK (1994) Beekeeping for honey production in Sri Lanka. Sri Lanka Department of Agriculture, Peradeniya

Ramya J, Rajagopal D (2008) Morphology of the sting and its associated glands in four different honey bee species. J Apic Res 47:46–52

Roepke DW (1930) Beobachtungen an indischen Bienen. Med Landbow Wageningen 34:1–22 [in German]

Sakagami SF (1960) Preliminary report on the specific difference of behaviour and other ecological characters between European and Japanese honeybees. Acta Hymenopt 1:171–198

Sakagami SF, Akahira Y (1960) Studies on the Japanese honeybee, *Apis cerana cerana* Fabricius. VIII. Two opposing adaptations in the post-stinging behaviour of honeybees. Evolution 14:29–40
Schmelzer E, Kastberger G (2010) 'Special agents' trigger social waves in giant honeybees (*Apis dorsata*). Naturwissenschaften 96:1431–1441
Schmidt JO (1990) Hymenopteran venoms: striving toward the ultimate defense against vertebrates. In: Evans DL, Schmidt JO (eds) Insect defenses: adaptive mechanisms and strategies of prey and predators. State University New York Press, New York, pp 387–463
Schmidt JO (1995) Toxicology of venoms from the honeybee genus *Apis*. Toxicon 33:917–927
Schneider P, Kloft W (1971) Beobachtungen zum Gruppenverteidigungsverhalten der Östlichen Honigbiene *Apis cerana* Fabr. Z Tierpsychol 29:337–342 [in German]
Seeley TD (1994) Honey bee foragers as sensory units of their colonies. Behav Ecol Sociobiol 34:51–62
Seeley TD, Seeley RH, Akratanakul P (1982) Colony defense strategies of the honeybees in Thailand. Ecol Monogr 52:43–63
SenSarma M, Fuchs S, Tautz J (2000) Debris removal by head pushing in *A. florea* Fabr. Honeybees. Naturwissenschaften 87:241–243
SenSarma M, Fuchs S, Tautz J (2002) Worker piping triggers hissing for coordinated colony defence in the dwarf honeybee *Apis florea*. Zoology 105:215–223
Shah FA, Shah TA (1991) *Vespa velutina*, a serious pest of honey bees in Kashmir. Bee World 72:161–164
Sugahara M, Sakamoto F (2009) Heat and carbon dioxide generated by honeybees jointly act to kill hornets. Naturwissenschaften 96:1133–1136
Tan K, Hepburn HR, Radloff SE, Yu Y, Liu Y, Zhou D, Neumann P (2005) Heat-balling wasps by honeybees. Naturwissenschaften 92:492–495
Tan K, Radloff SE, Li JJ, Hepburn HR, Yang MX, Zhang LJ, Neumann P (2007) Bee-hawking by the wasp, *Vespa velutina*, on the honeybees *Apis cerana* and *A. mellifera*. Naturwissenschaften 94:469–472
Thapa R, Wongsiri S (2003) Flying predators of the giant honey bees, *Apis dorsata* and *Apis laboriosa* in Nepal. Am Bee J 143:540–542
Thapa R, Wongsiri S, Manandhar DN (2000) Current status of predators and diseases of honey bees in Nepal. In: Proceedings of 7th international conference on tropical bees, Chiang Mai, pp 221–226
Thom C, Gilley DC, Tautz J (2003) Worker piping in honey bees (*Apis mellifera*): the behaviour of piping nectar foragers. Behav Ecol Sociobiol 53:199–205
Veith HJ, Weiss J, Koeniger N (1978) A new alarm pheromone (2-decen 1-yl-acetate) isolated from the stings of *Apis dorsata* and *Apis florea* (Hymenoptera: Apidae). Experientia 34: 423–424
Vetter RS, Visscher PK (1998) Bites and stings of venomous arthropods. Int J Dermatol 37: 481–496
von Frisch K (1967) The dance language and orientation of bees. Harvard University Press, Cambridge
Weiss J (1978) Vergleichende Morphologie des Stachelapparates bei den vier *Apis*-Arten (Hymenoptera: Apidae). Apidologie 9:19–32 [in German]
Wongsiri S, Lekprayoon C, Thapa R, Rinderer TE, Sylvester HA, Oldroyd BP, Booncham U (1996) Comparative biology of *Apis andreniformis* and *Apis florea* in Thailand. Bee World 77:23–35
Woyke J, Wilde J, Wilde M, Sivaram V, Cervancia C, Nagaraja N, Reddy M (2008) Comparison of defense body movements of *Apis laboriosa*, *Apis dorsata dorsata* and *Apis dorsata breviligula* honey bees. J Insect Behav 21:481–494
Yang M, Radloff S, Tan K, Hepburn R (2010) Anti-predator fan-blowing in guard bees, *Apis mellifera capensis* Esch. J Insect Behav 23:12–18

Chapter 18
Self-Assembly Processes in Honeybees: The Phenomenon of Shimmering

Gerald Kastberger, Frank Weihmann, and Thomas Hoetzl

18.1 Introduction

18.1.1 General Aspects of Giant Honeybee Biology

Honeybees (genus *Apis*) are thought to have evolved in Southeast Asia in the last 10–20 million years. With the exception of the Western honeybee, *Apis mellifera*, the most recent form of honeybee which occurs naturally in Eurasia and Africa, all other honeybee species are distributed in Southeast Asia (Kastberger 1999; cf. Chap. 3). They comprise three main groups: the evolutionarily older forms, the dwarf honeybees and the giant honeybees, both open-nesting forms and the evolutionarily younger, cavity-dwelling form of *Apis cerana* from which the Western honeybees probably evolved (cf. Chap. 2). Cavity-dwelling species possess defence traits that can be considered as residual homologues of their open-nesting ancestors.

Two species of giant honeybees exist: the lowland form, *Apis dorsata*, and the Himalayan honeybee *Apis laboriosa* (Roepke 1930; Lindauer 1956; Underwood 1990; Oldroyd and Wongsiri 2006; Hepburn and Hepburn 2007; Tan 2007). Giant honeybees live in the open in single-comb nests, which are attached to tree branches, rocks or buildings. Depending on the season, colonies occur either diffusely dispersed or in aggregations of up to 300 at a single site and migrate up to twice a year (Koeniger and Koeniger 1980). One and the same queen may return to traditional nesting sites over several years. This extraordinary re-orientation capacity of giant honeybee colonies has recently been described in detail (Neumann et al. 2000; Paar et al. 2000, 2004; Sulzer et al. 2010).

Honeybee colonies have vital needs to defend the energy-rich resources of their nests. Predatory pressure can be extreme and has seemingly induced the evolution of defence strategies in both groups of open-nesting and cave-dwelling species

G. Kastberger, F. Weihmann, and T. Hoetzl
Institute of Zoology, University of Graz, Graz, Austria
e-mail: gerald.kastberger@uni-graz.at; frank.weihmann@uni-graz.at; thohoe@aon.at

H.R. Hepburn and S.E. Radloff (eds.), *Honeybees of Asia*,
DOI 10.1007/978-3-642-16422-4_18, © Springer-Verlag Berlin Heidelberg 2011

(Butler 1962; Morse and Laigo 1969; Seeley et al. 1982; Oldroyd et al. 1995; Breed et al. 2004). There are important differences in the predation pressures between the various groups of honeybees. Their open-nesting habit exposes giant honeybees to a variety of predators, particularly birds (Kastberger and Sharma 2000) and wasps (Seeley et al. 1982; Kastberger et al. 2001) and, therefore, seem to have a broader array of defence strategies than cavity-dwelling species.

18.1.2 Defence Lines in **Apis dorsata**

In support of their open-nesting lifestyle, giant honeybees have evolved a set of defence strategies that keep predatory animals, birds and wasps, in particular, off the nest. The defence traits in giant honeybees can be divided into two categories. The first line of defence includes behaviours with repelling impact, organised collectively by individuals still in the nest (a) the aposematic colouration of the abdomen into yellow and black stripes (termed "warntracht") as characteristics of hymenopterans and (b) the "rearing-up" behaviours released by strong mechanical stimuli, characteristic of eusocial wasps and honeybees. Here, two further traits are especially notable among giant honeybees: "colony aggregation" (Seeley et al. 1982; Kastberger and Sharma 2000; Kastberger et al. 2001; for *A. mellifera*, see Oldroyd et al. 1995) and "shimmering behaviour" (Roepke 1930; Sakagami 1960; Butler 1962; Kloft and Schneider 1969; Morse and Laigo 1969; Schneider and Kloft 1971; Koeniger and Fuchs 1975; Seeley et al. 1982; Kastberger et al. 2008a).

The second line of defence brings the defenders into regular physical contact with intruders. Some of these behaviours are displayed by single individuals rather than collectively. In cavity-nesting honeybees, entrance guards focus primarily on checking incoming traffic. The guard cohorts of giant honeybees may be grossly categorised as equivalents to the guards of the Western honeybee (Breed et al. 1990, 1992, 2004; Stabentheiner et al. 2002; Arechavaleta-Velasco et al. 2003).

Guards, however, can also be recruited for collective defence against intruders. Regarding wasps, honeybees have evolved a killing strategy termed heat balling, documented in *A. cerana* (Ono et al. 1995; Ken et al. 2005), *A. mellifera* (Tan et al. 2007), *A. florea* (Seeley et al. 1982) and *A. dorsata* (Kastberger et al. 2001; Kastberger and Stachl 2003). Honeybee colonies pose a deadly peril to vespine, ropalidiine and polybiine wasps (Das and Gupta 1989) that touch the nest surface. A group of immediately recruited bees seize such intruders, draw them into the bee curtain and heat-ball them to death.

The obviously most spectacular defence action refers to the recruitment and release of flying defenders (Seeley et al. 1982; Kastberger and Sharma 2000; Kastberger et al. 2001; Oldroyd and Wongsiri 2006), which chase vertebrate disturbers or predators away from the nest by counter-attacking them through their stinging behaviour. This defence strategy is enhanced by the fact that honeybees have evolved a barbed stinger that is lost in the attack against birds and mammals, as altruistic self-sacrificing weaponry. After recurring threats by bee-eaters, honey buzzards (Kastberger and Sharma 2000; Kastberger et al. 2001), or by human honey hunters, non-guard defenders are mobilised. In giant honeybees, such additional guards (synonymous to the working terms of "soldiers", "defenders", "flying guards"; see Breed et al. 2004; Kastberger et al. 2008b) are recruited in sub-surface layers of the curtain and mobilised to emerge to the surface. There they wait for a successive stimulus by which they are released as flying guards within a fraction of a second. This defence capacity has brought the giant honeybees the reputation of being the most ferocious stinging insects on earth (Ruttner 1988).

18.1.3 Aims of this Chapter

We first document the physical properties of the "bee curtain" of *A. dorsata*, the substrate on which shimmering takes place. Based on individual bees and their interactions, the basic dynamic properties of the bee curtain are assessed and some functional aspects of the bee curtain are discussed.

Second, the mechanistic, "proximate" details of the nature of shimmering is considered with respect to the following questions: Which mechanisms are involved in generating, synchronising and spreading shimmering? In which way does the bee curtain undulate during shimmering? Which kind of cues are adequate stimuli that synchronise and spread the waves? Which bees participate in shimmering? Can we distinguish age cohorts? Is shimmering generated by special trigger bees, and are they formed as social subgroups?

Third, the evolutionary adaptiveness of shimmering is discussed. We have formulated "ultimate" hypotheses to explain why shimmering may benefit the colony. The "information" hypothesis assumes that shimmering provides a mechanoceptive cue for intrinsic colony use. The "confusion", "misguidance" and "repellence" hypotheses refer to the question why shimmering may safeguard the colony against potential predatory wasps, birds and mammals.

The research summarised in this chapter refers to two methodological approaches of observation and measurement: one assessing the interactions of bees and predators under naturally occurring conditions undisturbed by experimenters and the other assessing behaviours under experimentally controlled conditions.

The main results of this chapter address the anti-predatory goals of shimmering. Shimmering in giant honeybees is supposed to be an analogue of a "La-Ola" or the "Mexican-wave" process and can be added to the prominent wave-like examples of self-organisation in biological systems (Camazine et al. 2003). These "defence waves" (Kastberger et al. 2008a) are triggered visually, but the spreading process is supposed to be mainly under mechanoceptive control. It is supposed that shimmering also follows simple rules, which scale up individual traits to swarm intelligence (Kennedy and Eberhart 2001; Camazine et al. 2003). The conventional view, which would meet the principles of self-organisation, is that a shimmering wave should be released by bees nearest to the source of disturbance. This surmise has been addressed in this chapter by the "everywhere"-hypothesis. Shimmering would then be formed as a sophisticated collective behaviour of all colony members by uniform control of their internal parameters, such as sensory thresholds, the capacities for decision making, the memorisation of past disturbances, and sensory and refractory fittings.

We will further show that predatory wasps are deterred through visual cues (large-scale shimmering) when they are nearer than 50 cm to the giant honeybee nest. Small-scale shimmering has the capacity to confuse wasps in close contact with the honeybee nest. Shimmering forces the wasps to alter their hunting strategy; instead of foraging on bees directly from the nest, the hornets have to move away from the nest into a zone where they are only able to catch free-flying bees. We conclude that shimmering is a pivotal trait that allows giant honeybees to maintain their open-nesting life style, which evolved millions of years ago. The fine-tuned interplay between the bee colony and wasps during shimmering is discussed in terms of a co-evolutionary arms race between predator and prey.

18.2 Proximate Mechanisms of Shimmering

18.2.1 The Functional Architecture of the Bee Curtain in Giant Honeybees

A giant honeybee nest measures up to 2 m horizontally and may comprise over 100,000 individuals, with honey stores of more than 50 kg. A nest consists of the central comb with the honey and pollen stores and the reproductive cells, and the colony members that cover the comb in multiple layers. This cover of adult colony members is loosely fixed to the comb and is termed the "bee curtain" (Morse and Laigo 1969).

18.2.1.1 Functional Regions of the Bee Curtain

The bee curtain is the entity of adult bees of the colony and is a multi-layered structure on both sides of the comb. There are different regions in the nest, which differ in the way bees occur in the curtain (Fig. 18.1). Most of the day, the bees are quiescent and remain in the nest without any movement, except for the region where the foragers depart, arrive and dance. This is the interface between inside and outside of the nest and is called the "mouth" zone. The size of this zone is highly variable and has its maximal extension under normal temperature conditions in the early afternoon when the flight traffic is maximal.

Curtain bees share a variety of work loads involving locomotor activity such as foraging, feeding, dancing or guarding but are also involved in non-locomotor

Fig. 18.1 Functional regions of a giant honeybee nest (*Apis dorsata*). *Left image*, a nest in Chitwan (Nepal) as observed at 10:00 in the morning on January 2009 under foggy and relatively cold conditions (22°C). The whole colony was in a quiescent state. *Right image*, the same colony in the early afternoon under sunny conditions with high foraging activity. The mouth zone, marked by the *yellow circle*, was clearly discernable from bees in the rest of the nest which were quiescent

activities such as fanning, abdominal flipping or providing additional honey stores (Ruttner 1988). In these "quiescent" regions, the curtain bees normally hang seemingly motionless in a fixed habit in vertical position, with their heads up and linked with their neighbours by their legs (Butani 1950; Butler 1962; Morse and Laigo 1969). The quiescent roofing bees form a protective integument by which the internal nest climate, in particular the temperature and humidity, can be controlled within narrow limits. They also protect the nest from mechanical disturbances such as wind or rain and, by their light-reflecting wings, from too strong exposition to the sun. Lastly, they form a reserve force, which can be recruited and mobilised for defence purposes.

There are only two conditions under which this quiescent state is changed: Firstly, during the periodic mass flight activities that happen regularly up to five times on a daily basis (Kastberger et al. 1996b), this state leads to peeling off the outer layer of bees on the nest. Secondly, when circumstances demand defence and the colony has to fend off external attackers that come into close distance to the nest. Under this condition, groups of bees, mobilised through a collective alarm system, may start a great variety of defence actions such as shimmering, heat-balling or releasing flying guards (Kastberger et al. 2010).

The bee curtain is not generally fixed to the comb; loose regions provide traffic at the comb surface, necessary to organise food and brood stores. Here, the queen lays eggs, the worker bees feed the larvae, evaporate water from nectar and store honey and pollen. On the comb, the curtain bees probably leave cavities to provide free access for worker bees and the queen, to enable reproduction and food storing activities and the production of comb cells. The bee curtain stays connected with the comb not only at the upper part of the attachment zone, but is fixed to the comb at several points.

18.2.1.2 Curtain Properties

The bee curtain shows attributes of a "superorganism" when it contracts and expands (Kastberger and Kaefer 2002) in response to external stimulation or internal arousal. It is a changeable and reactive substrate and acts as a communication platform for its members in regards of visual, pheromonal and mechanoceptive pathways.

Dynamic-mesh hypothesis. A consequence of this network architecture is how the curtain bees are connected. Preliminary observations have shown that arousal of the colony provokes an expansion of the mesh. It is assumed that this expansion provides excess space facilitating the emerging of recruited defender bees to the surface. For the detection of the mobilisation of defender bees, we used infrared thermography (Kastberger et al. 2001; Kastberger and Kaefer 2002). This technique allowed us to formulate the *dynamic-mesh hypothesis*, which was tested by assessing the dynamics of the bee curtain, whether it expanded or contracted, measuring the mesh widths, and the occurrence of cavities and go-through-channels. Expansion of the nest occurs before defender bees emerge from inner layers (Fig. 18.2).

Vertical-thread hypothesis. The curtain bees, in particular those in the surface layers, are likely to be assembled in vertical columns and attached to their nest mates preferentially in a towards-the-comb direction rather than in left-to-right rows. The bee curtain can be compared with a set of threads hanging down from a solid structure rather than with a homogenously coupled three-dimensional matrix. The data gained from propagation experiments (Weihmann et al. 2009) support this *vertical-thread hypothesis* and assumes that the coupling in left–right directions is weaker than in up–down directions. Preliminary experiments (Kastberger and Kaefer 2002) showed that the mechanical coupling between the curtain bees is not even but directional, so that mechanical impulses have different directional characteristics regarding the three dimensions of space in the nest (left–right, up–down and to-and-off-the-comb directions).

18.2.2 The Nature of Shimmering

Giant honeybees produce shimmering waves at the surface of their nests during which groups of surface bees display wave-like patterns that spread from the centre to the peripheral parts of the rim and attachment zones of the nest, even across its "mouth" zone (Morse and Laigo 1969; Kastberger et al. 2008a). Shimmering is not only a single-wave event, but mostly produced repetitively. The spatial and temporal patterns of waving display varying strength levels and repetition rates and are adaptive in shaping visual cues for external addressees. It has been shown that shimmering reduces the chance of predation by wasps (Kastberger et al. 2008a), and possibly deters big game such as rhinos or elephants, in order to protect the nest from molestation. The giant honeybees have evolved defence lines with flying guards against birds (Kastberger and Sharma 2000).

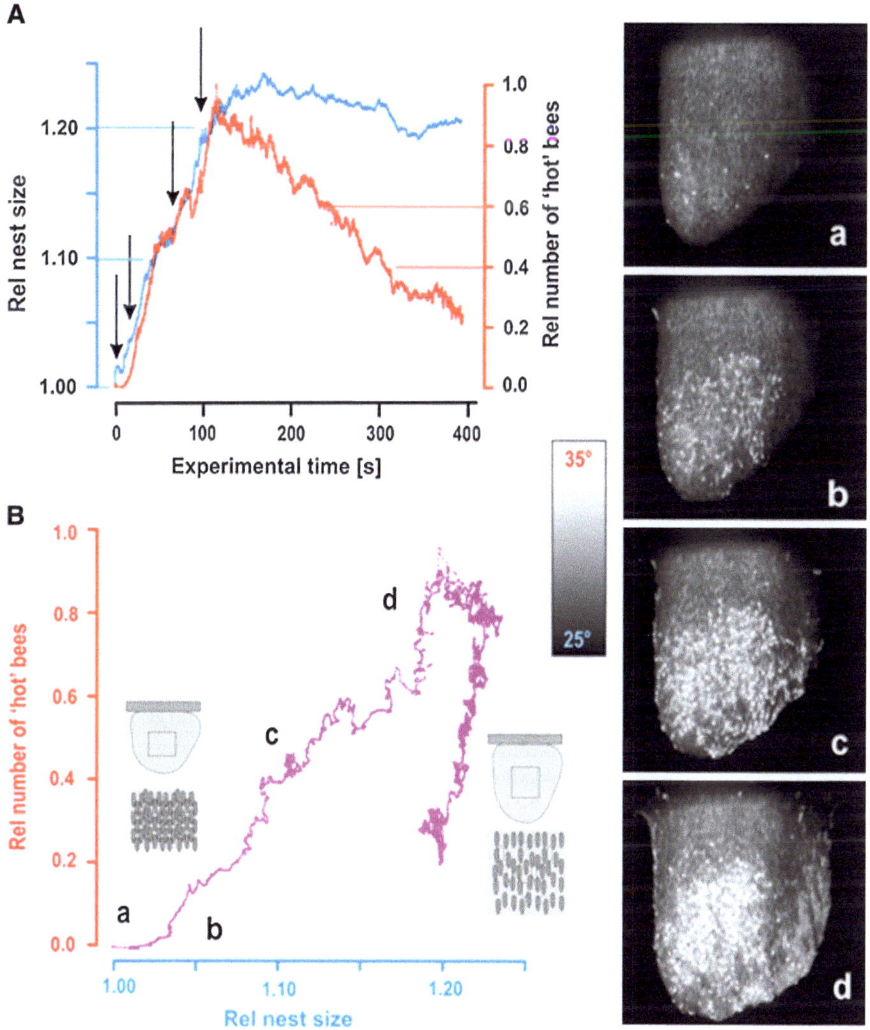

Fig. 18.2 Dynamics of the bee curtain in *Apis dorsata* in response to mechanical disturbance as assessed from infrared recordings (Kastberger and Kaefer 2002). (**A**) The *blue line* shows the relative nest size over time. At time zero, the real nest size was 1.0. It increased in size by nearly 25% due to four mechanical disturbances (marked with *arrows*) when small pieces of wood were thrown into the nest. The *red line* gives the relative number of "hot" bees, which emerged at the surface from deeper nest layers, as documented in the infrared images (a–d) on the *right side*; the value 1.0 refers to the maximum summarised area in the IR image marked as "hot" bees. The *grey scale* defines the temperature between 25°C (ambient temperature) and 35°C. Single individuals transport nest heat of 35°C to the surface. Note that such bees recruited from the inner nest layers are white-coded (i.e. "hot") from head to abdomen. Guard bees at the surface ready to take-off have only their thoracic muscles heated up. (**B**) Correlation of the relative number of "hot" bees (*ordinate*) with the relative nest size (*abscissa*). The *sketches* show how the mesh of the bee curtain changes. The increased mesh width facilitates the emergence of bees from deeper nest layers to the nest surface

Shimmering behaviour is an example of synchronised rhythmic activities that can be observed in the animal kingdom, such as the synchronised swaying in male fiddler crabs with their specific pincers to attract females (Koga et al. 1998) or the synchronised alarm drumming in *Camponotus* ants (Hölldobler and Wilson 1990; for a further list, see Camazine et al. 2003). Wave-like behaviours that also belong to the group of synchronised rhythmic activities are much rarer and are hardly to be associated with schooling behaviours such as in birds or fishes (Reynolds 1987). Only three waving processes have been reported in the literature so far that can be viewed as an epiphenomenon of self-organisation (Camazine et al. 2003): first, the body spiral waves of the slime mould *Dictyostelium discoideum* as the result of cAMP initiation during feeding activities (Bonner 1982, for summary see Camazine et al. 2003); second, the synchronised flashing of fireflies (Buck and Buck 1980; Buck et al. 1981), which have reproductive goals and lastly, the Mexican waves in human aggregations (Farkas et al. 2002). The slime moulds co-ordinate the wave-like processes by chemical principles, the fire-flies and humans, for example in football stadiums, do this solely visually after predefined logistic rules (Farkas et al. 2002).

18.2.2.1 Occurrence of Shimmering

Typically, shimmering waves are produced by giant honeybee colonies when a wasp hovers in front of the nest. These activities of abdominal movement were categorised (Kastberger et al. 2008a) into small-scale and large-scale waves (Fig. 18.3). In small-scale activities, only some tens of bees raised their abdomens. The respective signals were found to be local and non-repetitive and peaked after 200–250 ms. Large-scale waves spread over the nest, reaching their maximum activity typically after 400 ms and were repetitive.

Wave-like processes also occur without the presence of a threatening wasp. Foraging nestmates departing from or arriving at the nest are mostly the source for the generation of such shimmering in the absence of predators. However, large-scale waves preferentially occur in the presence of predatory threats (Fig. 18.3).

In particular in the morning, under the sun's first rays, or under humid and relatively cold conditions, the nests hardly show strong shimmering behaviour. However, they exhibit small cells of shimmering activity with a handful of bees, and regularly show spontaneous raising of abdomens as singular events. It is still unclear whether this stochastic abdominal thrusting behaviour is a preliminary phase for shimmering. One argument for this surmise is that this "flickering" behaviour occurs not diffusely over the whole nest, but in zones where an increased number of bees show this stochastic activity at a higher rate (Fig. 18.4).

18.2.2.2 Evoking Shimmering by Dummies and Tethered Wasps

A sample wave. A typical shimmering wave lasts less than 1 s, irrespective of whether the colony was provoked (Kastberger et al. 2008a; Schmelzer and

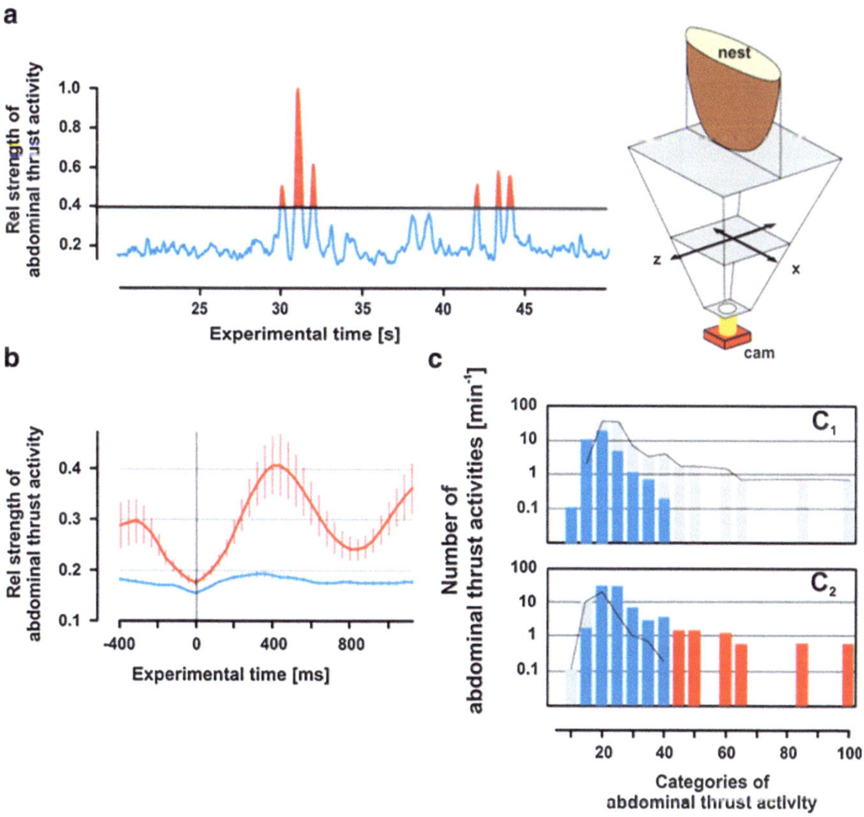

Fig. 18.3 Occurrence of shimmering under the presence of wasps. (**a**) Continuous assessment of shimmering activity of the experimental giant honeybee nest whilst a free flying predatory hornet (*Vespa* sp.) was present in front of it; the ordinate gives the relative strength of abdominal thrust activity, the value 1.0 refers to the maximum strength as observed during 300 s after the onset of waves; the threshold value of 0.4 discerns *small-scale* (*blue-line* area) from *large-scale* (*red* area) waves; (**b**) Mean time course of large-scale (*red lines*) and small-scale (*blue lines*) waves; *curves* show arithmetical means, *thin vertical lines* denote SEs; abscissa, real experimental time, time zero defines the onset of the abdominal thrust activity; (**c**) the rate of abdominal thrust activity per min of the experimental nest in two behavioural contexts (**c_1**) "undisturbed by a hornet" and (**c_2**) "disturbed by a hornet"; abscissa gives the categories of abdominal thrust activity as percentage of maximal waving strength; the *blue columns* refer to small-scale waves, the *grey columns* and the *thin lines* in the background give the respective rates of the opposite behavioural context for comparison. The distributions of the rates of abdominal thrust activities differed between c_1 and c_2 insofar as the proportions of the occurrences of waves regarding both states varied from one category to the other. Furthermore, in the presence of wasps the abdominal thrust activities show generally higher rates. The *sketch* on the *upper right row* shows the recording scenario. For more details, cf. Kastberger et al. (2008a, b)

Kastberger 2009) by a wasp predator (Fig. 18.3b) or by other visual cues (Fig. 18.5b). So, we can document a wave by a sequence of at least 20 frames in steps of 40 ms. Figure 18.6 shows one characteristic shimmering wave of a

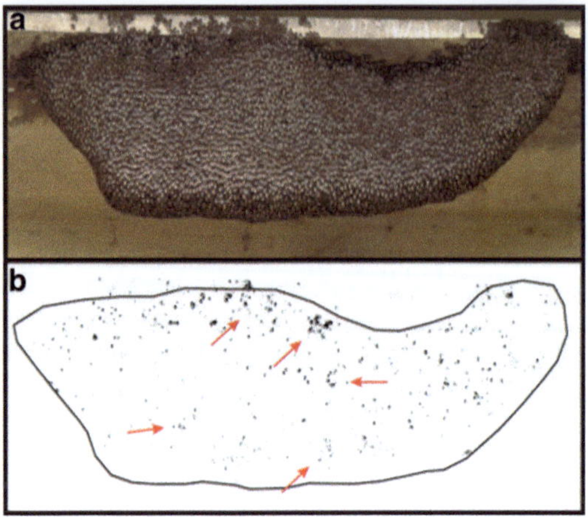

Fig. 18.4 "Flickering behaviour" in giant honeybee nests. (**a**) Experimental nest in Chitwan (Nepal) in January 2009. In the morning, under foggy humid and relatively cold conditions, the nests showed hardly any shimmering behaviour. Only exhibited small areas with shimmering activities with a handful of bees were observed. Periodically, individual bees spontaneously raised their abdomens as singular events. Such "flickering" behaviour occurs not diffusely across the whole nest, but there are zones where more bees show this stochastic activity at a higher rate. (**b**) The quiescent colony was traced over 2.63 s and showed some prominent zones with increased flickering activities (some of which are marked by *red arrows*). The *black spots* denote single bees which "flickered" with their abdomens during the observation period

tethered-wasp experiment as documented by a series of "differential" images superimposed by a sequence of charts (a1–a8). The blotches displayed differences in pixel luminance between two subsequent frames, which predominantly referred to groups of surface bees that lifted their abdomens during the shimmering activity (Kastberger et al. 2008a).

Complex Waves. In an experiment in Chitwan (Nepal) Kastberger et al. (2010) investigated the arousal of an experimental nest with a dummy wasp. It was moved by a miniature cable car with constant velocity along a horizontal line 10 cm in front of the experimental nest near its upper attachment zone (Fig. 18.7). Compared to observations of free-flying wasps, such experiments have the advantage that shimmering can be initiated under more controlled conditions with respect to position, direction and velocity of the wasp in front of the nest. The sample nest had a specific and individual history; it originated from merging of two neighbouring nests after massive disturbance by a honey hunter. The concave region on the right derived from a formerly separate nest. The thicker nest part on the left had honey cells, and after the honey hunting it was converted to the mouth zone.

At first sight, the shimmering wave in Fig. 18.7 seems to contradict the principle that a typical shimmering wave lasts a fraction of a second, irrespective of whether the colony is provoked by a wasp or by other visual cues. Figure 18.7 displays only the reaction of the colony to a single movement of the dummy wasp from right to left. This truncated analysis roughly refers to 53 frames or 1 s. It starts at frame 694

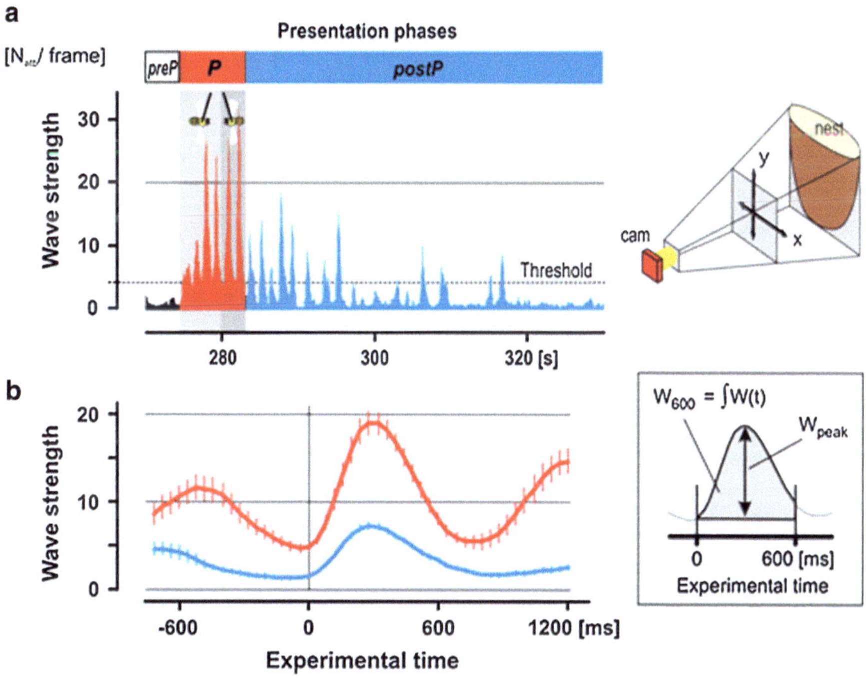

Fig. 18.5 Occurrence of shimmering in the presence of a tethered wasp. (**a**) Sequence of shimmering waves recorded from 270 to 330 s of the experiment; *red curves* signify the *P*-phase of experiments (the phase of the experiment when the tethered wasp was presented); it was moved from left to right (*bright grey background*) and from right to left (*darker grey background*), as well as after its disappearance (*blue curves* for the *postP*-phase of experiment after the disappearance of the tethered wasp). The ordinate parameter $N_{atb}/frame$ gives the number of abdomen-thrusting bees. (**b**) Means and SEs (*vertical bars*) of shimmering responses of *P*- (*red lines*) and *postP*- (*blue lines*) state. Time zero defines the start of shimmering. The ordinate gives the waving strength as the number of bees per frame that thrust their abdomens upwards. Shimmering waves in the presence of a tethered wasp exhibit higher waving peaks (W_{peak}, see *inset*), with more bees participating in 600 ms (W_{600}, see *inset*) and higher repetition rates than waves after the disappearance of the wasp. For more information, cf. Schmelzer and Kastberger (2009)

in the middle of the concave zone of the nest and ends at the left side of the nest at frame 747. It demonstrates the complexity of a shimmering process as it usually builds up a compound wave with several smaller waves. Partial wave processes may interfere and add up to form an overall moving, shimmering wave, which in the documented experiments has strikingly accompanied the dummy like a shadow.

Temporal structure, strength and repetition rate of shimmering. A characteristic sequence of shimmering activity in a single tethered wasp experiment has been exemplified in Fig. 18.5. It concerns three phases of stimulation: In the “pre-presentation” phase (preP-phase), the colony showed normal flight traffic without any shimmering activity. In the “presentation” phase (P-phase), the dummy wasp was presented in front of the nest and provoked repetitive shimmering waves of

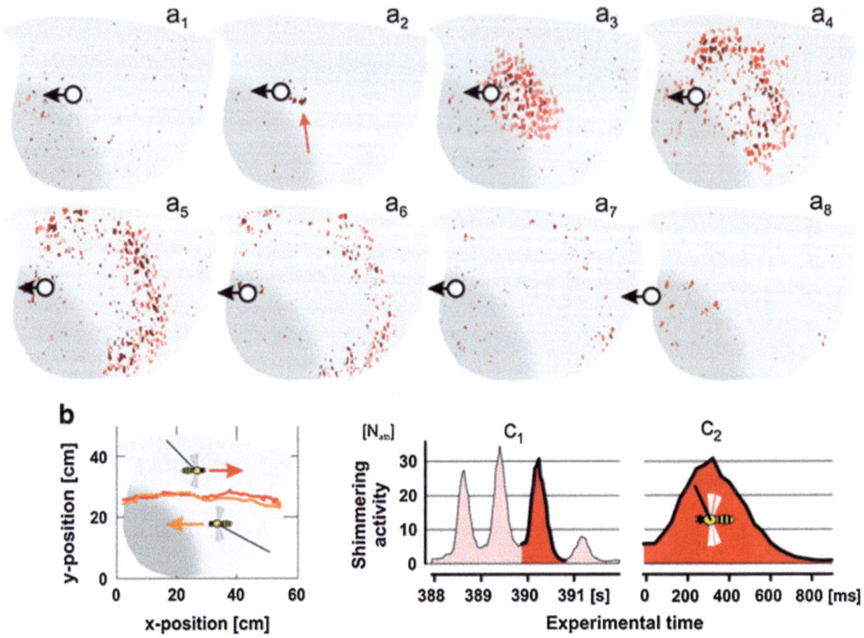

Fig. 18.6 A sample shimmering wave of the giant honeybee test colony provoked by a tethered wasp as described in Fig. 18.5. ($\mathbf{a_1}$–$\mathbf{a_8}$). Twenty-two images calculated using the pixel differences between two subsequent frames, disclose abdominal movements of nest bees, assembled and superimposed in eight charts ($\mathbf{a_1}$) three images recorded prior to the onset of the selected shimmering wave; ($\mathbf{a_2}$) one image tracing the group of bees that had started the wave (*red arrow*); ($\mathbf{a_2}$–$\mathbf{a_7}$) six charts with three images each display the subsequent spreading of the wave over the nest surface; ($\mathbf{a_8}$) three images after the sample wave had decayed. Successive frames were superimposed and coloured from *dark to bright red*. *Darker grey* areas on the *left bottom* side of the charts signify the "mouth zone"; complimentary brighter *grey* areas signify the "quiescent zone" of the nest. The positions of the tethered wasp are given by full *white circles*; the direction of it movement is marked by *arrows*. (**b**) Single session of stimulation in which the tethered wasp was moved from the left to the right side (*red symbols*) of the experimental nest, and then back to the left (*orange symbols*); during this experiment, the vertical position of the wasp and its distance from the nest (15 cm) were kept within narrow limits. (**c**) Time plots of waving strength of the sample waves; *ordinate*, calibrated in terms of the number of abdomen-thrusting bees per frame (N_{atb}); red colouration ($\mathbf{c_1}$, $\mathbf{c_2}$) refers to the sample wave displayed in (**a**); abscissa, the experimental time in s ($\mathbf{c_1}$) and the time course of the sample wave in ms ($\mathbf{c_2}$). For more information, cf. Schmelzer and Kastberger (2009)

progressive strength (Fig. 18.8c). The strength of the response escalated in the course of five successive waves, which proves that the giant honeybee colony became sensitised over time in their readiness for shimmering. Lastly, in the "post-presentation" phase (postP-phase), after the disappearance of the dummy, the shimmering activity decayed in strength and repetitiveness (Fig. 18.8b). Shimmering activity had a stronger waving strength and repetition rate when the colony was exposed to the tethered wasp, compared with the state of the colony after the disappearance of the tethered wasp. After the disappearance of the wasp, the waves

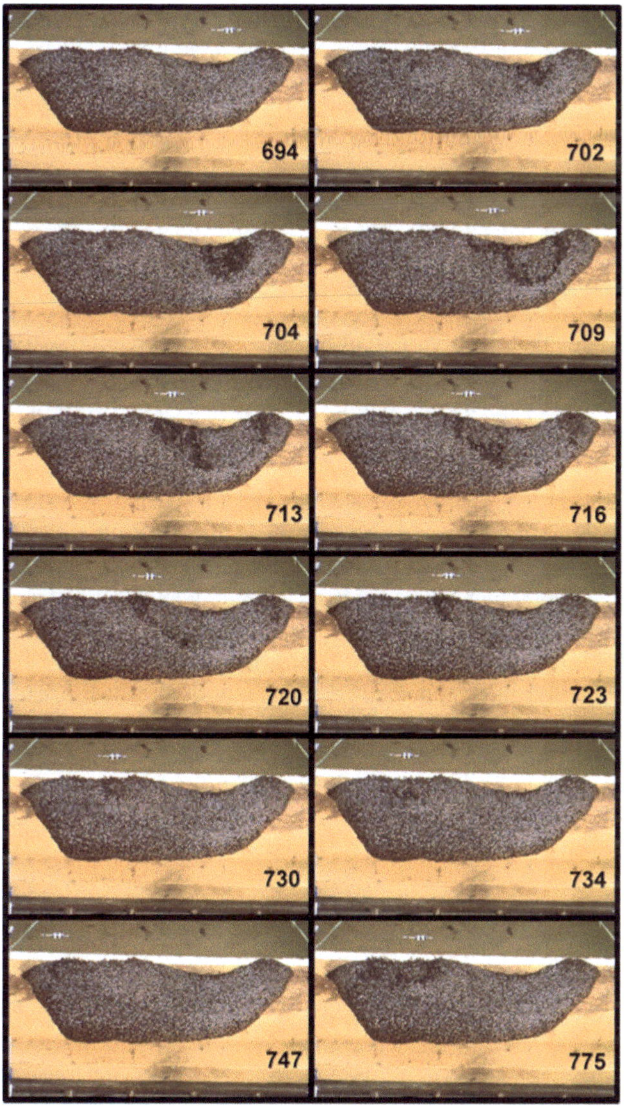

Fig. 18.7 Complex waves in response of the giant honeybee colony to a single movement of a dummy wasp from the right to the left. This truncated analysis shows 53 frames or 1 s, starting at frame 694 in the middle of the concave zone of the nest and ending at the left side of the nest at frame 747. The sequence demonstrates how the waving process comprises a compound wave with several smaller waves. Partial wave processes may interfere and add up to an overall moving shimmering wave, which here strikingly accompanied the dummy wasp to the left side of the nest like a shadow. While the dummy was moved to the left, the waves originated perpendicular, but slightly behind the dummy in the middle of the concave zone. The waving process was not homogenous (frame 702) but built up as a compound wave with several smaller waves. While the right wings of these smaller waves merged, forming a wider shimmering area (frame 704), the left side of the initial wave accompanied the dummy to the left with a lag of some milliseconds. At

died away over 30–60 s and had a much lower wave strength and lower repetition rate. The time course of the mean wave strength of shimmering waves in repetitive series is displayed in Figs. 18.5b and 18.8b. Here, shimmering activities can be separated into two categories, large-scale waves of the presentation phase and small-scale waves of the postP-phase, after the presentation of the tethered wasp (Schmelzer and Kastberger 2009).

18.2.2.3 Generation of Shimmering

Shimmering is evoked by a variety of disturbances, but seemingly predominantly through visual cues (cf. Chap. 17). We investigated test colonies under controlled disturbance using dummies that mimicked wasps hovering, with different velocities, at different distances from the nest, and analysed the temporal and spatial facets of the colony responses.

The conventional surmise (for summary, see Oldroyd and Wongsiri 2006) is that shimmering waves should be triggered by bees nearest to the disturbing source. However, more recent investigations (Kastberger et al. 1997, 2001, 2010; Schmelzer and Kastberger 2009) let us to propose another view, that defence waves are triggered by experienced bees, who react faster than others do. This "trigger-centre" hypothesis (Fig. 18.8a) was proved by tracing distinct groups of surface bees, which preferentially raised their abdomens, even if the dummy wasp had changed its relative position to the nest.

Spatial distribution of trigger zones. The nest co-ordinates of the initial group of bees, which had started the wave, were distributed closely around the "mouth" zone (Fig. 18.8f) but not the mouth zone itself or far distally at the periphery of the nest. Furthermore, the positions of the trigger centres of the waves in the P-phase were not congruent with that in the postP-phase of the tethered wasp stimulation. This means that some groups of bees, which were not identical with those that had triggered the waves in the presence of the wasp, had a residual tendency to start waves on their own without visual cues after the preceding arousal.

A more detailed analysis of the trigger centre topology is given in Fig. 18.8e and is based on the sectoral and segmental partitioning of the nest. It was found that the waves were mainly initialised at the centre of the nest and largely in a region adjacent to the mouth zone. In other words, the distributions differ slightly but significantly: in the P-phase, the waves were triggered by bee groups located in a

---

Fig. 18.7 (continued) this moment (frame 709), a new wave started slightly in front of the dummy, its left side merging with the residual left part of the initial wave (frames 713–716). Consequently, the new core area of the resulting waving activity fell behind the dummy wasp. From this time onwards, all successive waves were reinforced regarding their left parts, while their right parts, which were more distant to the dummy, tended to vanish immediately after their emergence (frames 720–723). Thus, the remaining left portions of the successive waves (730–747) added up to form an overall moving shimmering wave, which strikingly accompanied the dummy to the left side of the nest like a shadow. For more information, cf. Kastberger et al. (2010)

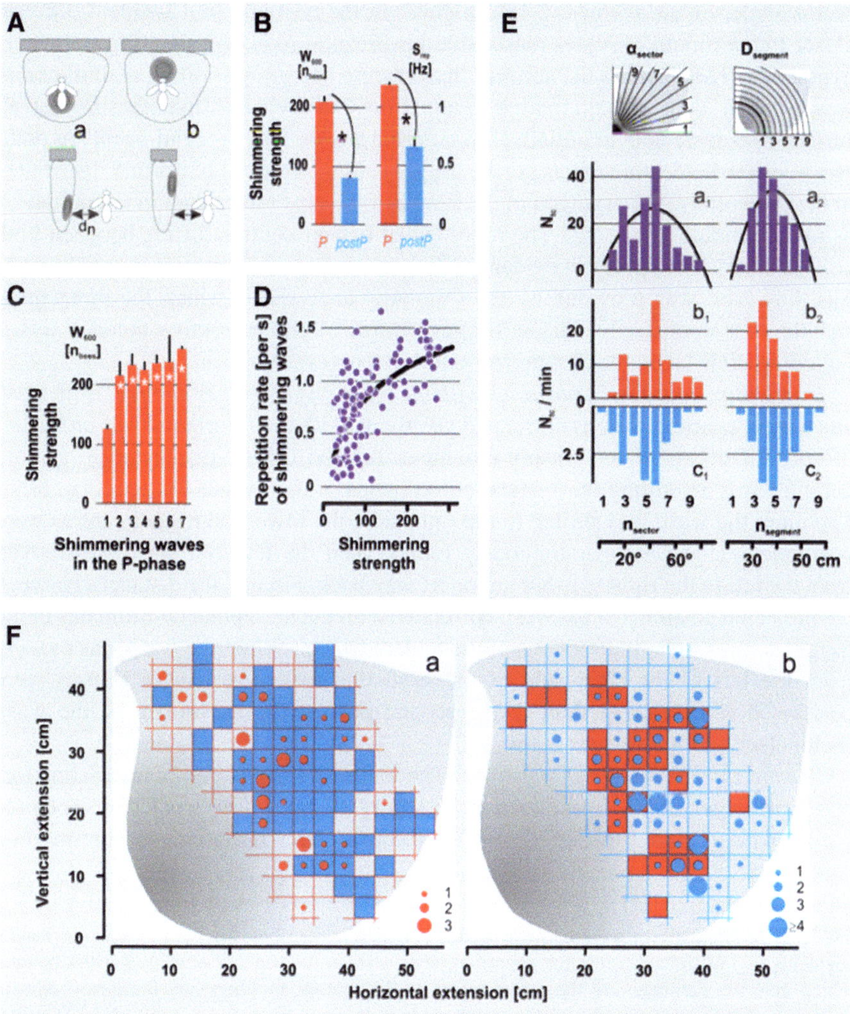

Fig. 18.8 Principles of the generation of shimmering in a giant honeybee nest. (**A**) Sketch demonstrating two hypotheses how shimmering waves are triggered (a) the "everywhere" hypothesis, (b) the special-agent hypothesis (Schmelzer and Kastberger 2009). (**B**) Waving strength and repetition rate of shimmering of the test colony of *Apis dorsata* in the *P*-phase and *postP*-phase of the experiment, shimmering strength was assessed by W_{600} (for definition, cf. Fig. 18.5). Shimmering waves in the *P*-phase have greater waving strengths and higher repetition rates than those in the *postP*-phase. (**C**) waving strength of the *initial* waves assessed in the *P*-phase (abscissa number 1), of the *second* (number 2) and *third* waves etc., irrespective of the direction in which the tethered wasp was moved; *stars* indicate that the initial waves of an arousal experiment differ from the sequential ones. Note the progressive escalation of shimmering power; columns and vertical bars signify means ± SE. (**D**) Correlation between the repetition rate of shimmering waves and the shimmering strength (W_{600}); data refer to both arousal states (*P*-phase and *postP*-phase). (**E**) Distribution of the "trigger centres" on the nest of the test colony of *Apis dorsata*. (a) *ordinate*, number of trigger centres (N_{tc}) in the whole observation session during and after the presence of

sector angle, on an average, 2.13° "higher" in the nest and in a segment 3.27 cm closer to the mouth provides reasonable justification to accept the "special-agent" hypothesis. This favours the surmise that specific bee groups initialise shimmering waves rather than the alternative assumption ("everybody" hypothesis) that all surface bees were able to initialise shimmering waves. The "special-agent" hypothesis is supported by three facts (1) the trigger centres were not uniformly distributed in the test colony, but arranged in the centre of the nest rather than in the periphery or in the "mouth" zone. (2) The topology of trigger centres differs between both arousal states. The strongest evidence (3) that validates the "special-agent" hypothesis, however, was provided by the following analysis by which the threatening signals were compared with the trigger centres of the respective defence waves regarding their locations (Schmelzer and Kastberger 2009).

Correlating the wasp's positions with the locations of trigger centres. Both wasp and trigger centre positions correlated positively with their horizontal co-ordinates, in both directions of the wasp's movment, but the colony responded in a more complex way regarding the vertical co-ordinates of the trigger centres (Fig. 18.9). The more the wasp was shifted to the right side, the lower the trigger centres were positioned. This was unambiguously found when the tethered wasp was moved from the left to the right side but not on its way back. Finally, the distances between the projected positions of the wasp correlated with the horizontal co-ordinates of the trigger centres in both directions of the tethered wasp. While the wasp was moving from the left to the right side of the nest, the respective trigger centres were positioned further away than in the second part of the movement of the wasp (Schmelzer and Kastberger 2009).

Special agents trigger shimmering. The tethered wasp experiments (Schmelzer and Kastberger 2009) questioned the validity of both the "trigger" hypotheses ("everywhere" and "special-agents") concerning the initiation of shimmering.

Fig. 18.8 (continued) the tethered wasp (coded in *violet*) as distributed in the 11 sectors (a_1) and 9 segments (a_2) of the nest area; for definition, see *charts* at the top and *abscissas* at the bottom, which give the numbers and the mean angles of the sectors, numbers and the mean segment distances; curves give polynomial regressions (Schmelzer and Kastberger 2009). (b–c) Distributions of trigger centres in the presence of the tethered wasp (arousal state in the *P*-phase of experiment, coded in *red*) and after the disappearance of the tethered wasp (arousal state in the *postP*-phase coded in *blue*) regarding the divisions into "sectors" (b_1–c_1) and "segments" (b_2–c_2); ordinates give the number of trigger centres per minute (N_{tc}/min). To test the lag between the occurrence of shimmering activity on the nest regarding angles (sector analysis) and distance (segment analysis, in cm), the best fit of both distributions of the *P*- and *postP*-phase was assessed by shifting the data of the *P*-phase compared to that of the *postP*-phase (cf. Schmelzer and Kastberger 2009). (**F**) Initial traces of individual shimmering waves in the presence of a tethered wasp in the *P*- and *postP*-phase of the experiments. *Charts* (a, b) show the spatial representation of the "trigger centres" plotted onto a grid with 3.41 cm unit square length. (a) *Full red circles* denote the scores of "trigger centres" per grid unit area in the arousal state in the *P*-phase in three classes (see *inset*); *full blue* areas signify that trigger centres were also observed in the consecutive arousal state in the *postP*-phase. (b) *Blue full circles* give the scores of trigger centres grid unit area in the *postP*-phase in four classes (see *inset*); the *full red squares* signify that trigger centres were also observed in the previous *P*-phase

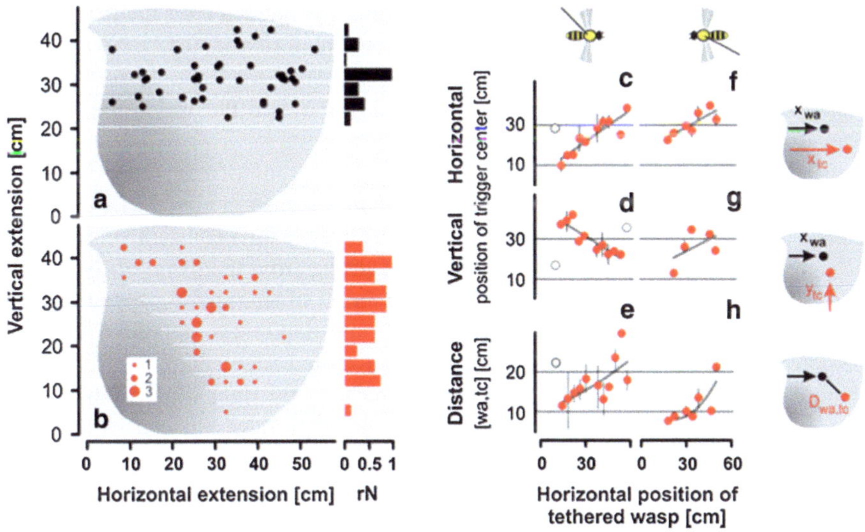

Fig. 18.9 Relationship between the position of a tethered wasp and its movement direction and the topology of the trigger centres of shimmering. (**a**) Positions of the tethered wasp at the moment of the onset of shimmering waves 15 cm in front of the experimental nest. (**b**) Number of waves initialised at the respective grid square during the presence of the wasp and summarised in three classes (see *inset* in **b**). *Black horizontal bars* on the right give the weighted distribution of trigger centres regarding the vertical grid steps (cf. Fig. 18.8); the data (rN) were weighted (*from top to down*) in relation to the progressively smaller region of "quiescence" (indicated by the *white grid lines*) and a larger mouth zone (*dark-grey coded area*). (**c–h**) give the correlations between horizontal position of the tethered wasp (x_{wa}) at the time of the onset of wave (**c**, **f**) with the co-ordinates of the trigger centres x_{tc} and (**d**, **g**) y_{tc}, and (**e**, **h**) with the distances between the tethered wasp and the trigger centre $D_{[wa,\ tc]}$; correlations refer to the shifting of the tethered wasp from left to right (**c–e**) and vice versa (**f–h**). For further information, cf. Schmelzer and Kastberger (2009)

The data suggest a dynamic topology of trigger centre arrangement as a species-specific feature of giant honeybee nest organisation in general. The cohorts that had initialised the shimmering waves were found to be arranged with three characteristic aspects (1) Their locations varied throughout the tethered wasp experiment and they depended on stimulation conditions, such as the direction of visual cue movement, and on the arousal state of the colony. (2) Shimmering was predominantly initiated in the centre of the nest, preferably along a circle adjacent and peripheral to the "mouth" zone. (3) There was some tendency that shimmering waves followed the tethered wasp on its way across the nest. However, the locations of trigger centres did not reveal the experimental path of the tethered wasp at all, because there was a discrepancy in their vertical alignment to the projected trajectory of the experimental wasp.

In other words, the bees that triggered shimmering in the experiment (Schmelzer and Kastberger 2009) were not necessarily arranged nearest to the threatening cue. In conclusion, these observations strongly suggest that in *A. dorsata* nests, cohorts with specialised abilities are tasked with the initiating shimmering waves. It is reasonable to suggest that those specialised workers located nearest to the threatening source start the wave. This also means that the majority of surface bees remain followers in the cascading chain of shimmering. They may simply forward the signal received from their neighbours that have initiated the wave due to their capacity of responding faster to threatening signals.

The sample waves of Fig. 18.10 illustrate the proof of the "special-agent" hypothesis. Two waves were provoked by a dummy wasp mounted to a fine wire connected to a long bamboo stick and moved by hand with a constant velocity and 10 cm in front of the experimental nest near starting from the distal of the mouth zone. It took around 40 frames (800 ms) until nuclei of shimmering appeared near the dummy on the left side of the nest. The tiny darker cycloids in frame 044 demonstrate clearly that the emerging wave originated simultaneously from more than eight centres. A heart-shaped wave resulted; only the right wings of the partial waves formed a singular wave front, which spread to the right end of the nest, towards the edge of the mouth. The second wave demonstrated here, started at frame 089 at a singular activity cente. It built up a strong and homogenous wave that spread in all directions. Although seemingly homogeneous, this wave was nevertheless composed of a series of smaller waves.

18.2.2.4 Propagation Mechanisms of Shimmering

In shimmering, individual bees in the nest respond to disturbing visual cues such as living or dummy wasps by raising their abdomens at more than 90°. This kind of abdominal shaking (Sakagami 1960) is observed predominantly in bees of the roofing layer of the curtain in all regions of the nest, and less by those in subsurface layers. The shaking behaviour of individual bees is transferred to neighbouring bees within less than 40 ms, resulting in a temporal and spatial pattern that emanates from one or more centres, radiating over the nest within half a second in linear or circular waves. When threatening visual cues are continuous, many waves with gradually increased intensity are generated in succession.

A frame-by-frame-analysis of individual participants of shimmering. In Fig. 18.11 the characteristic features of abdominal shaking of individual bees can be observed in steps of 40 ms in the original recordings (left-sided images) and by movement detection (Weihmann et al. 2009) using image analysis software (red areas in right-sided images). At the onset of an abdominal shaking period, only those curtain bees that were definitely involved in the ongoing defence response, showed a single wing stroke of 40–80 ms duration. After this wing stroke, abdominal thrusting started, but thrusting behaviour in neighbouring bees only occurred sequentially with time lags of 10–40 ms. The abdomens of the active bees were lifted by more than 90°: some bees bend their abdomens in a posture similar to that of *A. mellifera* workers during "Sterzeln". There are bees that do not regularly and actively participate in the shimmering process. The non-active candidate in Fig. 18.11 was possibly a member of a sub-surface position in the bee curtain.

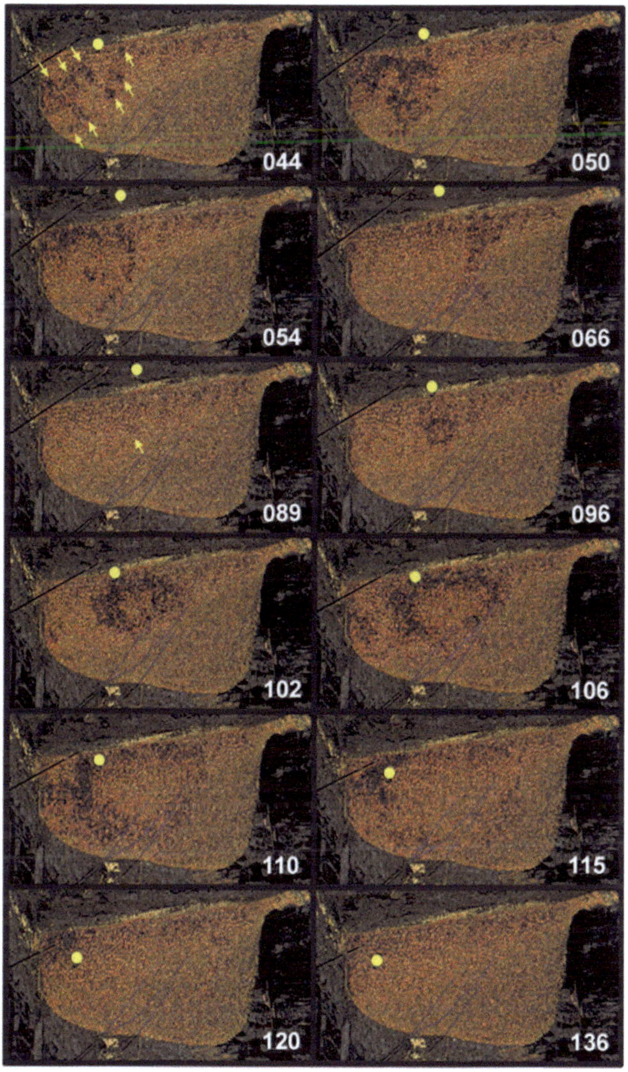

Fig. 18.10 Proof for the *special-agent* hypothesis of shimmering. A selected sequence of 12 images recorded by an HDTV camera (frame rate: 50 Hz) illustrates two defence waves provoked by a dummy wasp stimulation (from an experimental nest of *Apis dorsata*, in Chitwan Nepal, January 2009). *Yellow full circles* indicate the position of the dummy. This figure displays 12 images (from frame 044 to 136) out of 92 frames (1.84 s). The images were converted into false colours to enhance the wave fronts of shimmering. In frame 44 the initial phase of a wave is displayed showing at least 8–10 nuclei of waves (*yellow arrows*), which merged into a single wave front that ceased after frame 066 on the right side of the nest. In frame 089 a single wave centre emerged (*yellow arrow*) that gave rise to a single concentric wave in the middle of the nest; this wave lasted over 600 ms (until frame 120)

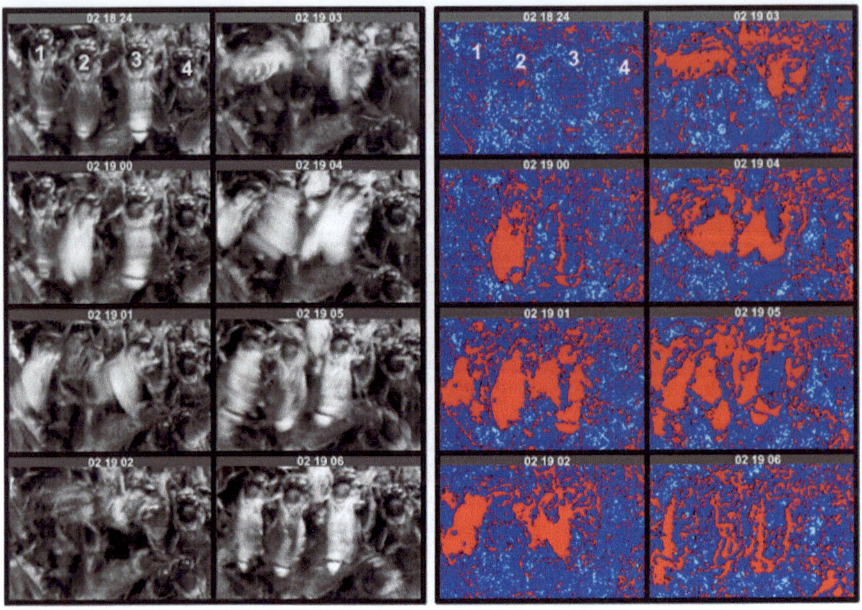

Fig. 18.11 Sequence of abdominal thrusting of curtain bees. Shimmering was recorded at an experimental nest in Assam (India) in 1995.The wave came from the left and spread to the right side of the images, affecting the bees coded with the numbers 1–4. *Left side*, original video frames, recorded every 40 ms from frame 02-18-24 to frame 02-19-01; *right side*, *blue* and *red* coded "differential" images displaying the difference in pixel luminance between two sequential original images (*blue*, subthreshold level; *red*, suprathreshold difference in pixel luminance). In frame 02-18-24 all four focus bees were in quiescent state. At frame 2 (02-19-00), the wave arrived from the left, and bee 2 responded promptly with thrusting her abdomen while bee 1 was still quiescent and bee 3 showed the early state of abdominal shaking, bending the abdomen and slightly unfolding the wings. In frame 3 (02-19-01) the abdominal shaking continued with bee 2 and 3, while bee 1 followed as the last in the group; it took further five frames (200 ms) until frame 02-19-06 until the abdominal thrusting of these three bees came to an end. Note that in frame 5 (02-19-03, the angle of the abdomen of bee 1 during its thrusting is clearly visible; the gap between the last tergites is open in all three bees. Bee 4 neither actively participated the shimmering process, nor was she pushed by their neighbours (see "differential" images); her darker expression in the image is due to her sub-surface position in the bee curtain

"Bucket-bridging" hypothesis. Shimmering is carried out predominantly, but not exclusively, by surface bees. In particular, in powerful waves, sub-surface bees also participate to some degree. The question still remains how the wave-like patterns in shimmering are produced? Is it a simple domino-like process by which the bees are stimulated to shake their abdomens just shortly after their adjacent neighbours had started? This basic mechanism is termed "bucket-bridging" because it reminds us of fire-fighting in mediaeval towns where queues of people co-operated to quickly pass over jugs of water to douse the fire.

Bucket-bridging agents use two self-organising rules (Camazine and Sneyd 1991; Camazine et al. 2003): they have to switch on a specific behavioural programme, in particular abdominal shaking behaviour, which provokes the raising of the abdomen by more than 90°, while all legs remain on the substrate; this substrate is provided by the bees of the sub-surface layer of the bee curtain. Rule one may be modified by the strength of the trigger stimulus in two ways the agent bee could control: first, the angle by which the abdomen is raised and, second, the number of immediate shakings produced as the time course of a damped oscillation.

Rule two of bucket-bridging defines the starting time of the behavioural programme. Here, the individual bee has to orient herself, obviously visually or mechanoceptically, with respect to her neighbours. That means, when one of her neighbour bees starts lifting her abdomen, she also starts her programme. The modifying aspect here is that several neighbours regularly encircle the focus bee, which may be active in shimmering. The focus bee would have the options to select, prefer or neglect specific neighbours for various reasons.

It seems that this "bucket-bridging" hypothesis can be proved for some fractions of a shimmering wave. Figure 18.12 illustrates such a condition, and shows how a wavefront of at least three bees width is propagated from the right side of the image to the left within only 66 ms. Obviously, it takes more than twice the time for a wave to pass a surface bee. In this example, most bees at the surface participated but a small number had not.

If we prove the bucket-bridging concept of shimmering, we have to define a reaction time by which the follower agent bee lags behind her leading neighbour. If the curtain bees at the nest surface rely solely on visual perception of her neighbours to start the abdominal activity, it may well take 100 ms to begin the behavioural programme. In *A. dorsata* nests, there are typically 18 bees arranged in a horizontal line of 25 cm on the outermost surface layer of the nest. At this individual reaction time, the honeybee colonies would achieve velocities of wave spreading of 1 m in 7.2 s. Comparing the classical Mexican waves in football stadiums, which utilise this bucket-bridging mechanism, the respective waves circle around a football stadium in at least 20 s. Humans have reaction times of several hundreds of ms and are, therefore, much slower than giant honeybees.

Here lies the basic discrepancy in the hypothecial bucket-bridging mechanism in shimmering. Shimmering waves have a typical active time course over only a fraction of a second, during which they increase and decrease in strength. In this time interval, they may spread over 1 m. At the same density of bees per metre as given above, this would mean that a single bee would have only 13.9 ms reaction time to join the behavioural programme of her neighbour. This is more than seven times faster than the bucket-bridging process if calculated on the basis of a 100 ms individual reaction time. This seems quite impossible even for honeybees, which do have astonishingly fast response capacities in flight manoeuvres. Therefore, other strategies must exist besides bucket-bridging, which make the shimmering wave spread so fast.

"Saltatoric-jumping" hypothesis. The second, possibly non-alternative strategy, is that the same specialised agents which were proven to exist in the bee curtain for the generation of wave processes, also participate in the spreading process. Their attribute is that they are stimulated to trigger the waves solely by visual cues.

Fig. 18.12 Evidence for the bucket-bridging hypothesis of shimmering. The sequence of images illustrate a shimmering wave in which abdominal thrusting was propagated from the right to the left side of the nest. The *upper four rows* display four successive black-and-white images in steps of 16.7 ms (from frame 674 to 689), which took 250 ms. The images were recorded at an experimental nest in Chitwan Nepal, January 2009 with a high-resolution camera at 60 Hz and inverted for better demonstration of the single bees at the surface of the bee curtain. Two rows of surface bees are coded in *yellow* from 1 to 19, first defined in frame 674. This enables the re-identification of these bees in the course of the shimmering wave, which spread from the right to the left side. Note, the wave front is here shaped as a *brighter zone*. It took about 120–160 ms for a wave to pass a surface bee. Some bees (such as numbers 1, 11 and 14) did not participate in the wave and kept their wings folded, but were passively moved by the wave. Most bees participated the wave and showed abdomen thrusting activities, during which they kept their wings unfolded. The *bottom graph* shows frame 680 enlarged. Note that most bees opened their gaps between the tergite 6 and 7, which is the *white stripe* between both tergites in the *inverted image*, indicating Nasonov scenting. Because the stimulation was started at frame 0 the bees had already achieved this state of arousal where their Nasonov glands were exposed

In addition, abdominal shaking produces a local mechanical oscillation in the bee curtain, which may actually synchronise the surrounding cluster of bees to raise their abdomens as well. This may well happen in a bucket-bridging process, by which the surface bees participate and produce wave patterns, which spread concentrically from a trigger centre in all directions (Figs. 18.7 and 18.10). In this way, the command for joining may jump from one synchronised cluster of surface bees to another, depending on the existence of sub-centres with special agents. This should be distributed in a matrix over the whole nest (Fig. 18.8f) but should remain at the same places, of the specialised agents, throughout repetitive shimmering. This type of spreading mechanism is termed as "saltatoric-jumping" (Kastberger et al. 2008a). The examples delivered in this chapter illustrate that shimmering utilises both principles, saltatoric-jumping and bucket-bridging.

18.2.2.5 Motivation of Participants in Shimmering

Nasonov gland exposure during shimmering. Nearly all patterns of social behaviour of Hymenoptera are regulated, or at least influenced, by pheromones (Bossert and Wilson 1963; Pain 1973; Avitabile et al. 1975). Accordingly, Kastberger et al. (1996a, 1998) proved that defence waving in *A. dorsata* is also subjected to pheromonal control, in which Nasonov scenting is obviously involved. Nasonov glands occur in all *Apis* species as well as in female *Bombus* and meliponine queens (Duffield et al. 1984). The Nasonov glands are multicellular units at the anterior margin of the seventh abdominal tergite comprising a groove as a scent reservoir, which lacks a definite closing mechanism (Jacobs 1925; Renner 1960). Consequently, the extension of the region between abdominal tergites 6 and 7 exposes the black groove in the gap (Figs. 18.11–18.13). This can be taken as a clearcut indication mark for Nasonov scent release, because this kind of posture automatically results in the reservoir opening and the emission of scent.

Wing-fanning in shimmering. The typical way of Nasonov scenting, as known from *A. mellifera* and *A. cerana*, is exposure of the gland accompanied by wing-fanning (Renner 1960; Sakagami 1960). Roepke (1930) had doubts about the existence of Nasonov glands in *A. dorsata*, because he was not able to observe wing-fanning in this species. However, we had observed this behaviour several times in *A. dorsata* when they were disturbed. During shimmering, the bees stroke the wings once before they shake their abdomens (Fig. 18.11). This kind of wing movement would be ineffective in spreading the scent, but it is possible to homologise this single wing stroke in *A. dorsata* with the characteristic fanning behaviour in *A. mellifera* or *A. cerana* during scenting (which is called "Sterzeln" – von Frisch 1967). On the other hand, for *A. dorsata* the forceful thrust of the abdomen itself could be a proper mechanism for distributing the scent by hurling it around the individual.

During shimmering, most bees stroke the wings (Fig. 18.11), thrust their abdomens and open their intertergital gaps between abdominal segments 6 and 7. This makes black grooves (white grooves in inverse images such as Fig. 18.11) visible

Fig. 18.13 Nasonov activity during shimmering. (**A**) scheme of Nasonov gland anatomy; (**B**) *left image*, a surface bee before the an abdominal thrust with the gap between tergite 6 and 7 opened; *right image*, the same bee during abdominal shaking, bending her abdomen upwards; (**C**) widths of gaps between tergite 6 and 7 assessed in 3 sets of selected surface bees that were active in subsequent shimmering waves; (a) before stimulation with the dummy wasp (6 bees/30 frames in steps of 120 ms) (b) 1 min after the onset of stimulation (6/13); (c) 2 min after the onset of stimulation (8/16). The ordinate shows the widths of the inter-tergital gaps in mm; widths of more than 0.5 mm indicate opened gaps (see the schemata of the abdomen for opened and closed status). (**D**) Two characteristic time plots (in seconds) of gap widths, a_3 for bee 3 in (a) and c_5 for bee 5 in (c). (**E**) The relative number of bees with opened and closed gaps between tergite 6 and 7, as observed before presenting the dummy wasp (**C**, a), during the onset of stimulation (**C**, b) and during continuous stimulation (**C**, c). The observation started at 13.20 h and continued for more than 40 min. For further details, cf. Kastberger et al. (1996a, b)

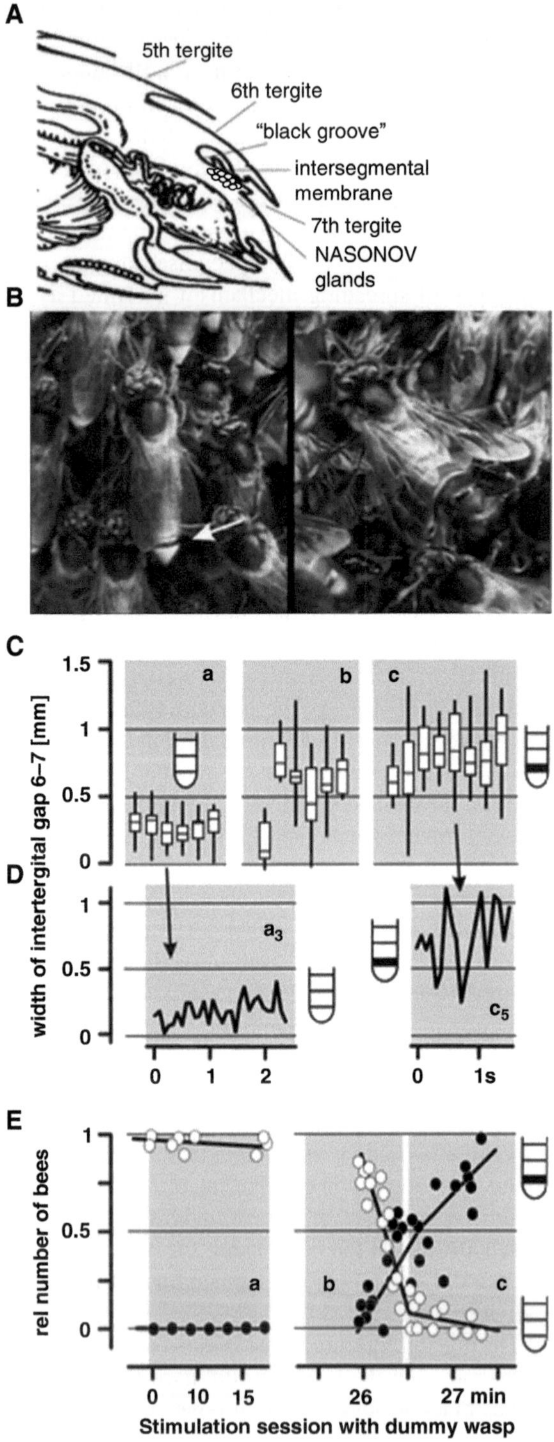

behind the white intersegmental membranes, indicating the exposure of the Nasonov glands. After the abdominal thrust, the intertergital gaps are frequently closed again. The Nasonov glands are recurrently exposed shortly after the start of stimulation by the dummy wasp. Finally, the bees kept their abdomens lifted upwards, exposing the Nasonov glands, with the seventh tergite directed downwards. The longer the stimulation session continued, the more bees kept the Nasonov glands exposed. The bees, which participated in shimmering during the continuous display of the dummy wasp, had significantly larger intertergital gap widths than quiescent bees (Fig. 18.13), which were not involved in shimmering. Before the dummy wasp was presented, the temporal course of the opening and closing of the intertergital gaps of the surface bees was affected by breathing. In general, the bees kept their Nasonov reservoirs closed. However, during continuous stimulation, the fluctuations in gap width increased to more than fourfold. In the course of 25 min of observation, before the dummy wasp was exposed in front of the nest, the number of bees with open intertergital gaps was very low: not a single individual was found with a gap clearly open in the sample investigated. The percentage of bees with open gaps significantly increased after the onset of the stimulus. During continuous stimulation, almost all the bees which participated in the defence wave kept their Nasonov glands exposed.

18.3 Shimmering is a Prominent Example for Predator–Prey Co-evolution

The spatial and temporal patterns of shimmering display varying strength levels and repetition rates and are adaptive in shaping visual cues for external addressees. It has been shown that shimmering reduces the chance of predation by wasps (Kastberger et al. 2008a), and it may also deter mammals. The giant honeybees have evolved defence strategies that pit flying guards against birds (Kastberger and Sharma 2000).

18.3.1 Shimmering as a Colony Response to Predatory Hornets

Kastberger et al. (2008a) investigated several flight parameters of predatory hornets for eliciting shimmering in the vicinity of giant honeybee colonies. Shimmering behaviour was measured by its wave strength and categorised by two parameters of the hornets' flights in the pre-wave phase (Fig. 18.14). It was found that when the hornets were close to the nest, immediately prior to the onset of the wave, the strength of the wave was maximal. If the hornets were farther away, the waves were much smaller and aroused fewer bees. The data demonstrate that wasps elicit strong waves if they fly at moderate speed, but hovering wasps were able to "creep" nearer to the bees without evoking large-scale waves.

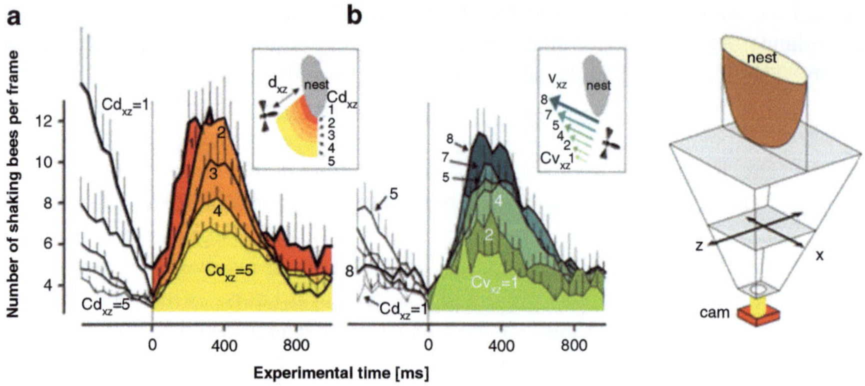

Fig. 18.14 Shimmering strength in giant honeybees in response to the behaviour of a free-flying hornet. The diagrams display the time courses of shimmering in response to approaching hornets. The measurement conditions are explained by the *sketch* on the far *right*; the camera (*cam*) recorded the *x–z* projection; therefore, the distances of the hovering wasp from the bee nest (d_{xz}) were addressed as projection on the *x–z* plane. The waving strength (number of abdomen-shaking bees per frame) depended on the hornet's distances from the nest d_{xz} (**a**) and on the hornet's flight velocities v_{xz} (**b**); time zero defines the onset of the waves; (**a**) five d_{xz} classes ($Cd_{xz} = 1–5$; coded in *yellow* to *red*) and (**b**) eight v_{xz} classes ($Cv_{xz} = 1–8$; coded in *green* to *blue*) of hornet flight episodes were considered; d_{xz} and v_{xz} class values were assessed from hornets in the 400 ms interval prior to the start of shimmering. Curves show arithmetical means, *thin vertical lines* denote SEs. For further details, cf. Kastberger et al. (2010)

The arousal state of a giant honeybee colony in the presence of predatory wasps is also expressed by the repetitiveness of shimmering (Figs. 18.14 and 18.15), which increased exponentially the nearer the hornet came to the nest. Regarding flight velocity, the repetitiveness of shimmering was seemingly complex and had a bell-shaped distribution (Fig. 18.15g). This means that hornets at moderate flight speed provoked a linear increase of repetitiveness with their flight speed, but faster hornets provoked decreasing repetitiveness of shimmering. This complex relationship results from the fact that the flight speed itself is a function of the hornet's proximity to the nest. On average, slower hornets flew closer to the nest than faster ones. Furthermore, most of the hornets flew within a mean hovering distance of 52 cm and slower than 50 cm s^{-1} when they provoked shimmering (Fig. 18.15a, c). If the hornets were further away from the nest, shimmering waves were less repetitive, although the hornets were fast, and their effect on the honeybees to elicit shimmering was consequently less. In other words, repetitiveness of shimmering is primarily a function of proximity (Kastberger et al. 2008a).

18.3.2 Hornet Behaviour in Response to Shimmering

An anti-predatory impact of shimmering on wasps can only be evidenced on the basis of a specific response to shimmering. In other words, if shimmering is

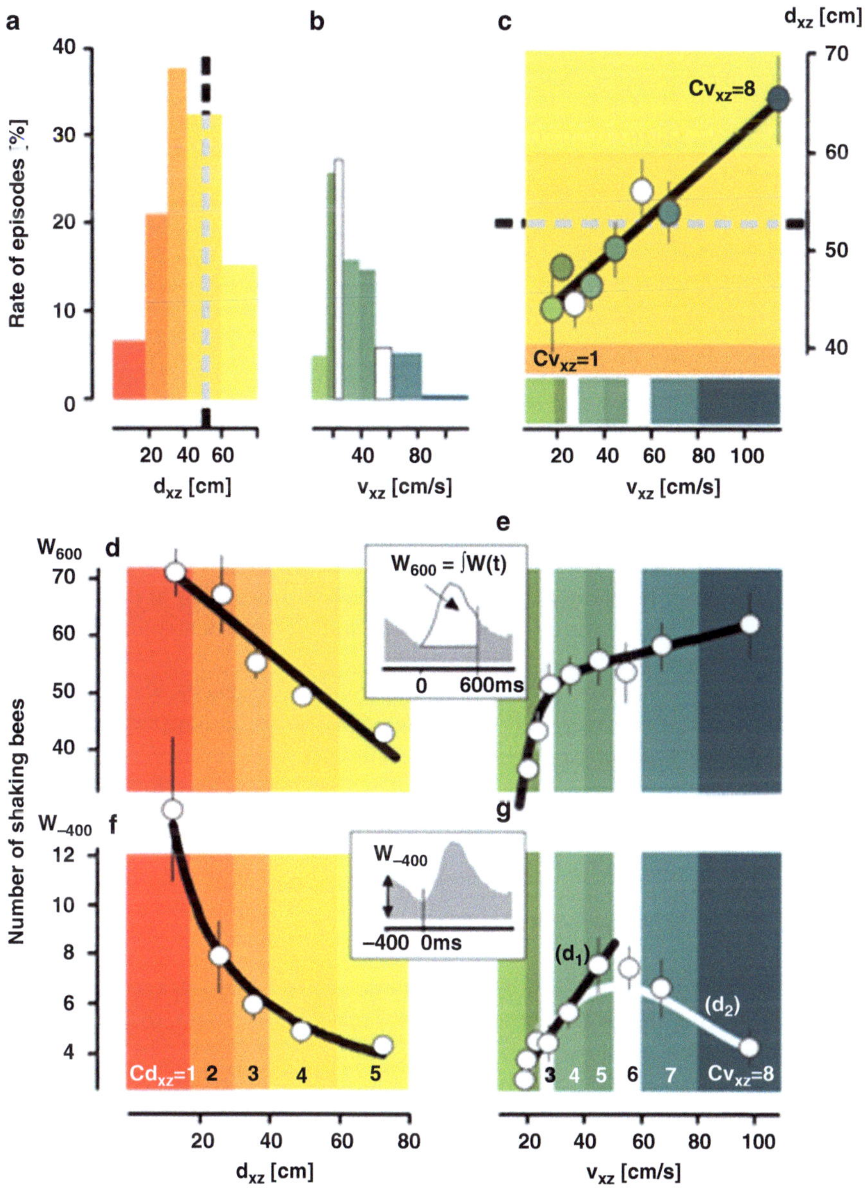

Fig. 18.15 The effects of hornet behaviour on waving strength (W) and repetitiveness of shimmering in giant honeybees. (**a**) Percentage of hornet episodes ($n = 317$) classified into five d_{xz} classes (**b**) and eight v_{xz} classes of hornets flights behaviour. (**c**) Relationship between the flight velocities v_{xz} (*abscissa*) of individual hornets and their distances d_{xz} to the nest (*ordinate*); *dashed lines* (**a** and **c**) define the average hovering distance of the hornets (cf. Fig. 18.15). *Red to yellow areas* denote five d_{xz} classes, and *green to blue areas* eight v_{xz} classes of hornet flights as used in Fig. 18.13. The waving strength W_{600} gives the numbers of bees that shook their abdomens over 15

supposed to improve colony defence against predators, it should be verified for each member of the nest that the probability of getting grabbed by the wasp is lowered. If this is true, the hunting success of the wasp would also be decreased. Convincing evidence for this kind of impairment is the predator's response of being confused, misguided or repelled.

18.3.2.1 Does Shimmering Confuse Wasps?

Kastberger et al. (2008a) sorted the flight trajectories of the predatory wasps according to two classes of arousal conditions for the hovering wasps, small-scale and large-scale shimmering (Fig. 18.16). The shimmering waves strongly influenced the flight trajectories of the wasps in both cases. Even small-scale shimmering lowered the direction fidelity of the predatory hornet (Fig. 18.16c). During shimmering, the turning tendencies deviated much more strongly, so that its direction fidelity dropped with the onset of shimmering. Without any directional preference, the hornet turned away from those bees that it was previously chasing. Large-scale shimmering caused the hornets to accelerate (Fig. 18.17a, b), but the acceleration pulses terminated after 400 ms, which happened just at the peak time of shimmering. In other words, large-scale shimmering waves stimulate wasps to accelerate for some hundreds of milliseconds.

Summarising, shimmering was found to confuse predatory wasps in close vicinity of the nest (Kastberger et al. 2008a). Small-scale shimmering lowers the directional fidelity of the wasps; large-scale shimmering urges them to accelerate. Consequently, both effects drive the wasps away from the target region to which the wasps had directed their predation flights.

18.3.2.2 Does Shimmering Repel Wasps?

Kastberger et al. (2008a) observed that the hornets hovered on average at a distance of 52 cm in front of the experimental nest (Figs. 18.15a, c and 18.18a). At this distance, hornets elicited only weak waves; they were not affected by shimmering and stayed neutral: on average, they neither approached nor left the nest site. When they were outside the mean hovering distance, hornets tended to approach the nest

---

Fig. 18.15 (continued) frames (definition see *inset*); W_{600} depends on the hornet's distance from the nest d_{xz} (**d**) and on the hornet's flight velocity v_{xz} (**e**). Time zero in the *insets* defines the start of the shimmering waves (cf. Fig. 18.3). (**d** and **f**) show five d_{xz} classes and (**e** and **g**) eight v_{xz} classes of hornet flights (see **a–c** and Fig. 18.13); *open circles* are arithmetical means, *thin vertical and horizontal lines* denote SE; *thick lines* are regressions of the mean values calculated from 201 waves with 317 flight episodes of two wasps (W_{-400}, the amplitude of the waving strength 400 ms before the onset of the consecutive wave (see *inset*), giving the residual shimmering strength under repetitive conditions; W_{-400} declines with d_{xz} and shows an overall nonlinear relation for Cv_{xz} = 1–8. For further details cf. Kastberger et al. (2010)

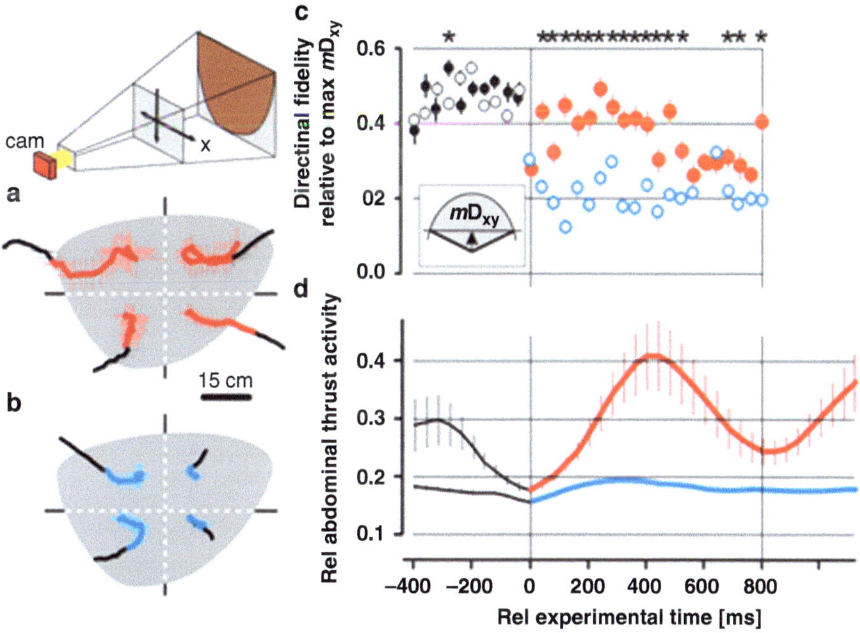

Fig. 18.16 "Confusion" of hornets hovering in front of the experimental honeybee nest during large-scale (*red colour code*) and small-scale (*blue colour code*) episodes (for measurement conditions see *schematic in the upper row left*). (**a** and **b**) Mean trajectories of the approaching hornet; the sectors between the *dashed lines* define the four divisions of the mean pre-wave flight directions of the hornets for pooling the *x*- and *y*-values of the positions of the approaching hornets in 40 ms intervals. *Thick lines* are arithmetical means of *x*- and *y*-values of the hornet's position. *Horizontal* and *vertical bars* show SEs. Note that the trajectories before the onset of the waves (shown by *black thick lines*) are straighter than after the onset of the waves (shown by *red and blue thick lines*). (**c**) Hornet flight behaviour under the influence of large-scale (*full red circles*) and small-scale (*open blue circles*) shimmering waves; *ordinate*, mD_{xy} (see *inset* for definition of the mean vector length mD_{xy}; *grey segment*, the defined turning range), calibrated between 0 and 1, the resulting scalar rel mD_{xy} is a measure of direction fidelity of the hovering wasp (*circles* and *vertical bars*, arithmetical mean ± SE). The data show that the direction fidelity of predatory hornets is lowered by shimmering waves, stronger by small-scale waves than by large-scale waves; stars indicate significant differences between the responses of the hornet to large-scale ($n = 20$) and small-scale ($n = 33$) waves per time interval. (**d**) The mean time courses of large-scale (*red line*) and small-scale (*blue line*) waves; *ordinate*, relative abdominal thrust activity (mean ± SE); *abscissa*, the relative experimental time (**c**, **d**) at time zero shows the onset of shimmering; pre-wave sessions are shown in *grey* or *black*, shimmering sessions are coded *blue* or *red* (**a–d**). For further details, cf. Kastberger et al. (2010)

under the influence of shimmering. However, when the hornets were closer than the mean hovering distance, they elicited bigger waves and turned away from the honeybee nest as soon as the shimmering wave became strong. Considering the change in distance of the hornet within 1 s after the onset of waving, the responses correlated linearly with the proximity of the hornets to the nest prior to the shimmering wave (Fig. 18.18b). Summarising, shimmering has anti-predatory

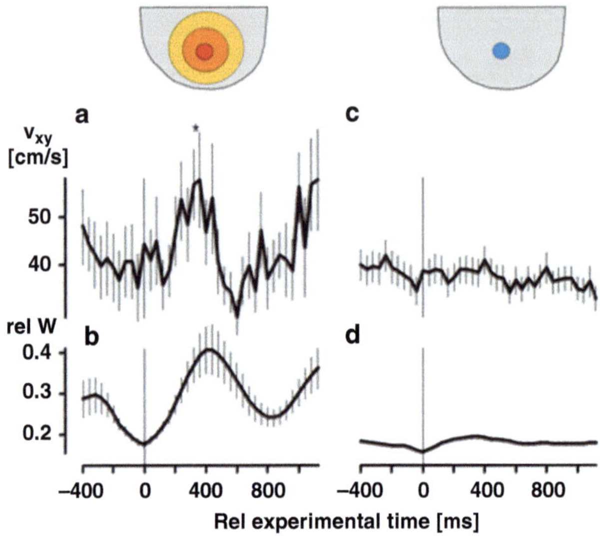

Fig. 18.17 Large-scale wave stimulate hornets to accelerate. Hornets' responses are shown regarding flight velocity v_{xy} in response to large-scale (**a**, **b**) and small-scale (**c**, **d**) shimmering waves (for measurement conditions cf. Fig. 18.15). Large-scale and small-scale waves were categorised according to a threshold of 40% of the maximal waving strength. The peak in flight velocity v_{xy} of large-scale wave episodes (**a**) from 240 to 360 ms after the onset of a wave differs significantly from pre-wave v_{xy} values (*$P < 0.05$, One-way Repeated Measures ANOVA; 15 episodes). Small-scale waves (**c** and **d**) do not affect the flight speed v_{xy} of the hornet. Note that the acceleration pulse of the hornet (**a**) coincides with the time courses of large-scale waves (**b**). *Abscissa*, experimental time in ms; time zero defines the start of shimmering; lines connect arithmetical means, *vertical bars* are SEs. The *schematics* above the graphs symbolise large-scale waves (*red-orange areas*) spreading across the nest while small-scale waves (*blue area*) remain local. For further details, cf. Kastberger et al. (2010)

features because it lowers the chances of the wasps of preying on the curtain bees on the surface of the giant honeybee nests and because it distracts predatory wasps from seizing bees. Hornet behaviour is indicative of confusion (Fig. 18.16) and avoidance (Fig. 18.18) behaviour. The results documented in Fig. 18.18 suggest that shimmering plays a role in repelling predatory hornets but only when close to the nest.

18.4 Reviewing the Ultimate Goals of Shimmering in Giant Honeybees

If wasps were repelled from feeding directly on the curtain bees, it would be sufficient to organise local groups of surface bees to confuse or misguide them (Eibl-Eibesfeldt 1978; Immelmann 1982). Bee-hawking wasps (Tan et al. 2007) should then learn that it is impossible to catch bees directly from the nest, and they

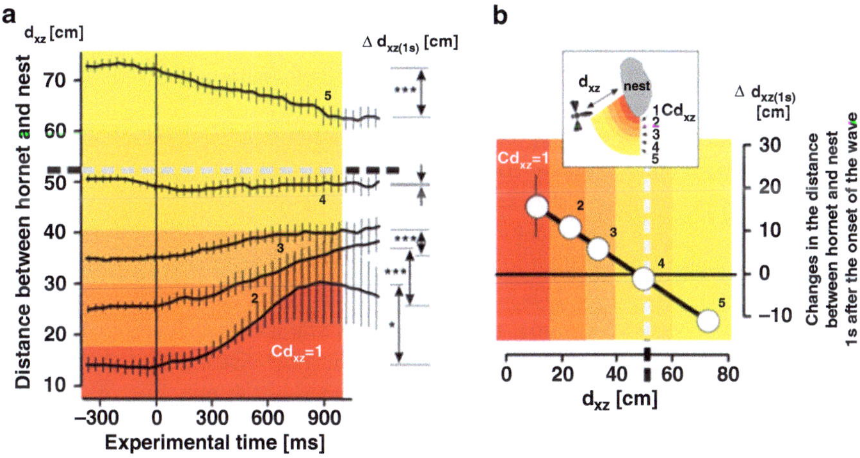

Fig. 18.18 The repellence effect of shimmering on approaching hornets. The hornets' responses to shimmering depend on the distance from the bee nest d_{xz} (for measurement conditions, cf. Fig. 18.13). Hornet behaviours were categorised in five distance classes $Cd_{xz} = 1$–5 (see *inset*). *Thick lines* connect the arithmetical means, *thin vertical lines* denote SE. (**a**) Wasp behaviour was monitored for 1,600 ms, starting 400 ms prior to time 0; $\Delta d_{xz\ (1\ s)}$ values give the changes in the position of the hornet regarding its distance to the honeybee nest within 1 s after the onset of the wave (significance levels: $*P < 0.05$, $**P < 0.01$, $***P < 0.001$; *t*-test); the *horizontal dashed line* gives the average hovering distance ($d_{hov} = 52.10 \pm 0.53$ cm; $n = 9757$ images). (**b**) Correlation between $\Delta d_{xz\ (1\ s)}$ and d_{xz}, the *thick line* gives the regression of means ($P < 0.001$; 326 episodes); positive values of $\Delta d_{xz\ (1\ s)}$ at $Cd_{xz} = 1$–3 represent movements of the hornets away from the bee nest and indicate avoidance responses; the response shown for $Cd_{xz} = 4$ is neutral, and the negative values of $\Delta d_{xz\ (1\ s)}$ at $Cd_{xz} = 5$ outside d_{hov} illustrate that the hornets usually approached the nest when shimmering started. For further details, cf. Kastberger et al. (2010)

would try to find easier prey near the honeybee nest, such as homing or departing worker bees. Any Mexican wave-like synchronisations of hundreds or even thousands of honeybees would then be seemingly too much ado for this defence purpose. Why then have giant honeybees evolved shimmering, an extraordinarily complex trait of group defence, unique in the whole animal kingdom? The findings of Kastberger et al. (2008a) allow us to assume that the goal of waving must be associated with the open-nesting lifestyle of the giant honeybees.

18.4.1 The Bee's View of Shimmering

For the open-nesting giant honeybees, it was essential to evolve defence traits that effectively support their lifestyle during the last five million years (Ruttner 1988) of co-existence with their wasp predators. Kastberger et al. (2008a) proved that the first line of defence of the giant honeybees against wasps includes shimmering behaviour. The reciprocal interactions between giant honeybees and hornets, during their attempts to catch bees from the nest and during the subsequent shimmering of the honeybees, are far more complex than mere stimulus–response behaviours. It seems extremely unlikely that the finely shaped, mutually adjusted behaviours (Fig. 18.20) have developed by chance as a kind of general flexibility in response. In view of the observations and experiments of Tan et al. (2007) with *A. cerana* and *A. mellifera* colonies in the same apiary, these mutual responses between giant honeybees and wasps are highly suggestive of co-evolutionary adaptation in a predator–prey relationship. The giant honeybees as potential prey, have acquired the ability to continuously signal to their wasp predators through shimmering "to keep a distance and do not expect a free meal". Thus, the visual cue of shimmering may have a combined impact of signalling vigilance of the prey and of unprofitability (Caro 2005). However, the capacity of shimmering goes beyond this goal, as shimmering has been proven to actively repel hornets to prevent predation (Kastberger et al. 2008a).

18.4.1.1 Sheltering the Nest

The capacity of shimmering to repel wasps is limited to a distance of half a metre from the nest (Fig. 18.20), which equals the mean hovering distance. This restriction in the defensive coverage of colonies of giant honeybees is likely to be associated with the obvious ultimate goal of colony defence to generate a safety zone around the nest that should keep predatory wasps away from the nest, preventing them from catching bees directly from the nest surface. If, nevertheless, the wasps succeed in breaching this shelter zone they should not stay long.

This is exactly what was observed during shimmering: when wasps approached the nest, the honeybee colony continuously generated shimmering waves (Fig. 18.14) that repeatedly confused (Figs. 18.16 and 18.17) and repelled (Fig. 18.18) the wasps. Shimmering evidently benefits the honeybee colony, because it lowers, actually to zero, the hunting success of predatory wasps. The chance to observe wasps trying to seize giant honeybees from the surface of the nests is quite rare. We observed, over years, hundreds of trials by hovering wasps to catch surface bees, but we have never observed any successful catch.

On the other hand, the hornets hardly elicited shimmering when they were more than 50 cm away from the nest. The question here is whether and why giant honeybees do not recognise hornets outside the distance of 50 cm as a threat. It is known that *A. dorsata* colonies, which mobilise their guards for a potential counter attack, recognise predatory birds at much greater distances from the nest than the mean hovering distance of wasps. One of the experimental *A. dorsata* nests we had observed in Chitwan instantaneously released hundreds of flying defenders when a kite approached, although this bird was still far more than 20 m away. Thus, it seems that giant honeybee colonies have developed specific distance measures for predatory wasps and birds.

18.4.1.2 Shimmering as Part of the General Defence

It is the current view (Koeniger and Fuchs 1975; Seeley et al. 1982; Oldroyd and Wongsiri 2006) that shimmering has evolved as a colony defence behaviour in giant honeybees. It belongs to those categories of strategies (Eibl-Eibesfeldt 1978) that pose practically no risk to the defenders. Shimmering also consumes far less energy than strategies that involved physical contact with the enemy. Kastberger et al. (2008a) presented the first quantitative proof that shimmering is an anti-predatory response of giant honeybee colonies to the presence of hornets and demonstrated fine-tuned prey–predator interactions. Shimmering waves become stronger and more frequent, the nearer a predatory hornet comes to the nest and the faster the hornet flew. In turn, hornets were more affected by shimmering, the nearer they came to the honeybee colony. In this intermezzo, substantial evidence was gained that shimmering has an anti-predatory impact on wasps. While local small-scale (Fig. 18.16) and large-scale (Fig. 18.17) shimmering may confuse wasps in different ways, which had come within reach of the nest, large-scale shimmering that might spread over the whole honeybee nest does have the capacity to repel predatory wasps, but only within a restricted limit away from the bee nest (Fig. 18.18). These data provide new insights into the complex spatial and temporal patterns of interaction between bee-hawking hornets and giant honeybees.

18.4.1.3 Confusing and Repelling Predatory Wasps

As demonstrated in Fig. 18.16, local small-scale waves increased the deviation of turning angles of predatory wasps by lowering their direction fidelity, when they hovered close to a giant honeybee nest (<20 cm). Here, they may well discern single bees, but such local waves make it difficult for the wasps to concentrate upon them. This aspect is termed "confusion", an anti-predatory strategy often cited as an important mechanism in predatory interactions (Eibl-Eibesfeldt 1978; Landeau and Terborgh 1986; Krause and Ruxton 2002; Jeschke and Tollrian 2007) as the reduced attack-to-kill ratio experienced by a predator resulting from an inability to single out and attack individual prey in a group. However, "confusion" has been proven only for very few predatory interactions (Jeschke and Tollrian 2007), such as the observation that small-scale waves, as a local response of giant honeybees on the nest surface, would suffice to confuse wasps and to prevent predation (Fig. 18.16). Conversely, large-scale waves were typically provoked by wasps that were further away from the nest and flew faster. For the hornets, large-scale shimmering causes much less "confusion" by lowering direction fidelity, but do confuse them by forcing them to accelerate (Fig. 18.17).

18.4.1.4 Risk-Minimising and Cost-Efficiency

Shimmering minimises costs and risks for the individual honeybees and for the whole colony. Infrared recordings show that during shimmering, surface bees actively raise their abdomens, do not heat up (Kastberger et al. 2001) and would not require much energy. This finding suits shimmering as a non-habituating trait. On the other hand, risk minimisation is obviously being governed by pheromonal control. It is likely that the release of Nasonov scent (Fig. 18.13) conveys to the colony the message to stay at the nest site and to perform social action in the community.

18.4.1.5 Enforcing Innate Avoidance in Wasps

We propose that the wave-like character of shimmering has evolved not primarily to confuse wasps, but, as shown above, to effectively repel wasps. This striking capacity probably has two adaptive aspects. Firstly, shimmering may reinforce innate and not habituating fixed action patterns of avoidance in wasps; and secondly, waving processes may also enforce the innate avoidance of the addressee. We have gained evidence (cf. video S4 in Kastberger et al. 2008a) that, when the wave of abdomen-thrusting bees spreads over the nest, the wave front stays "behind" the wasp. It seems that it actually press-gangs the wasps away from the place they originally wanted to prey on. Subsequently, the wasp is strongly inclined to retreat and fly away from the "threatening" wave front.

18.4.1.6 Does Shimmering Provide Pheromonal Cues for Potential Predators?

Associated with the concept of a shelter zone of giant honey bee nests, another question arises: Does shimmering deliver only visual cues to the wasps or does it also utilise pheromonal channels? It is known that shimmering is linked to chemical scenting (Kastberger et al. 1996a, 1998), but there are arguments that make chemical scenting extremely unlikely to trigger the avoidance response of wasps. First, the release of alarm pheromones in honeybees is accompanied by protrusion of the sting (Crewe and Hastings 1976; Breed et al. 2004). Stinging activities do not occur during shimmering. Moreover, alarm pheromones of honeybees do not prevent hornets from hunting bees (Ono et al. 1995; Oldroyd and Wongsiri 2006). Second, shimmering is accompanied by the release of Nasonov rather than sting/alarm pheromone. Nasonov scent is a social pheromone and signals bees to "find the nest entrance" (Butler et al. 1970) or "stay together" (Kastberger et al. 1998), thus preventing single bees from changing their roles into those of guard bees (flying defenders). However, there are reasons that make it extremely unlikely that Nasonov pheromone triggers the avoidance response of an approaching hornet. Firstly, the exposure of Nasonov glands has only been observed after a series of shimmering episodes (Kastberger et al. 1998), but hornets were disturbed by shimmering from the first wave onwards (Fig. 18.12). Additionally, and more importantly, the delay of the avoidance reaction of the wasp after the onset of shimmering is less than 100 ms, and is therefore, by several orders of magnitude, faster than the exposure of Nasonov glands and also faster than the obvious spreading of the pheromone would take. Summarising, the hornet's avoidance behaviour appears to be triggered solely by visual cues of shimmering.

Finally, a possible function of Nasonov scent as a direct defence strategy, similar to the allomones from various exocrine glands of arthropods (Blum 1981), is unlikely as the Nasonov gland's construction does not resemble a typical defensive gland, lacking any closing mechanism and any muscles for a forceful emission of the secretion (Renner 1960). Nonetheless, some components of the Nasonov lure (e.g. citral) detected in *A. mellifera* (Pickett et al. 1980; Cassier and Lensky 1994) and *A. cerana* are found in many defensive secretions of arthropods (Chada et al. 1962; Blum 1981).

18.4.2 The Wasps' View of Shimmering

Theoretically, there is a fundamental problem for a predator–prey relationship if a defence action of a potential prey, such as shimmering, does not lead to any physical contact with the enemy. Of course, such traits are less risky for the defenders, but they are obviously less dangerous for the predators, which may learn to ignore "non-perilous" signals of the potential prey. However, observations of Kastberger et al. (2008a) clearly demonstrate that repetitious shimmering efficaciously repels the same hornet again and again. Any habituation effect in hornets can be excluded. Obviously, they cannot ignore shimmering, although they repeatedly try, without showing any sign of habituation, to hunt their prey. Although the wasps decelerate as they approach the nest, shimmering interrupts their landing operations, and elicits confusion (Figs. 18.16 and 18.17) and avoidance (Fig. 18.18) reactions. Both reactions would take the wasps away from the site of the prey. Mostly, they are repelled from the nest, at least half a metre or more, from where they start the next hunting episode.

Because of their persistent and non-habituating bee-hawking quirks, it is assumed that wasps envisage honeybee nests as a prey of extraordinary attractiveness. To avoid widespread wasp predation, honeybees have acquired cavity-nesting abilities (in Southeast Asia: *A. cerana*, *A. nuluensis*; in Eurasia and Africa: *A. mellifera*). In particular, *A. cerana* and *A. nuluensis* have strong defence strategies against bee-hawking wasps and, at their nest entrances, they also exhibit shimmering against wasps, although at a far lower level than giant honeybees (Koeniger et al. 1996; Tan et al. 2007). European honeybees (*A. mellifera*) have acquired far less effective abilities to thwart the predation of wasps. This can be demonstrated in direct comparison with *A. cerana* (Tan et al. 2007), because the European honeybee has been introduced from Europe to South East Asia, where it fails to cope with widespread wasp predation. This highlights a lack of adaptation in the predator–prey relationship. Therefore, the co-evolution between *A. cerana* and their autochthone bee-hawking wasp predators must have been very intense (Ono et al. 1995; Ken et al. 2005; Tan et al. 2007).

18.4.2.1 Response to Shimmering Is Tightly Dependent upon Shimmering Strength

The data of Fig. 18.18 verify the *shimmering-repels-wasp* hypothesis (Kastberger et al. 2008a) that assumes that bee-hawking hornets show an avoidance response to shimmering when they come too close to the honeybee nest. Summarising, two aspects have been proved in support of the *shimmering-repels-wasp* hypothesis. Firstly, shimmering was proved to repel free-flying wasps, and the wasps respond with avoidance reactions (Kastberger et al. 2008a). The wasps alter their flight routes and are mostly driven off the nest (Figs. 18.15–18.17). Secondly, the special-agent attributes entail that shimmering is not triggered directly at the precise projected position of the moving wasp but slightly behind, and not ahead of the wasp. This kind of interaction between wasps and honeybees has the very real potential of shimmering bees to "wipe away" free-flying wasps from the nest area (cf. video S4 in Kastberger et al. 2008a).

However, the validity of the *shimmering-repels-wasp* hypothesis is restricted to the condition that the predatory wasps have to stay inside the zone, which is defined as the "mean hovering distance" (Fig. 18.18). When the hornets were outside this zone the repelling goal of shimmering was reversed. Then shimmering even attracted the wasps, which were inclined to approach the nest as a region of prey.

In addition, the responsiveness of hornets to shimmering is even more complex, which is illustrated by the fact that the reactivity of the wasps correlated with both the increasing and decreasing parts of the wave (Fig. 18.19). This finding suggests that the hornets are not stimulated by the shimmering wave to move away from the nest (Kastberger et al. 2008a). The data specify the responsiveness of the hornets in the following way: When the number of abdomen-thrusting bees increases, the hornets display more pronounced avoidance reactions by turning strongly and flying away faster from the nest; when shimmering declines, the hornets reduce and eventually terminate their shimmering-specific avoidance reactions.

An hypothesis opposing the *shimmering-repels-wasp* hypothesis is the *proximity-avoidance* hypothesis. It proposes that hornets are deterred by honeybee nests and would avoid its vicinity. However, there are two types of evidences that disprove the *proximity-avoidance* hypothesis. Firstly, shimmering actually attracts predatory wasps to the nest outside the mean hovering distance (Fig. 18.18), and secondly, bee-hawking hornets incessantly approach honeybee nests, again and again, to prey on them, without showing the slightest sign of habituation. Nevertheless, the factor "proximity to honeybee nest" apparently modulates the responsiveness of wasps to shimmering, essentially because the honeybee colony also alters its defence response depending upon the distance of the intruder.

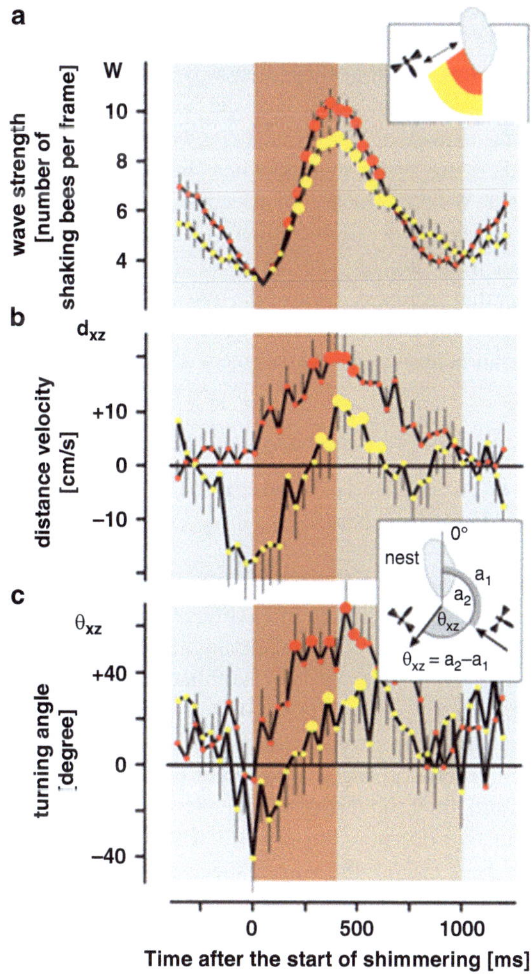

Fig. 18.19 The hornets' behaviours during shimmering. Measurement conditions are described in Fig. 18.13. (**a**) The time courses of shimmering and (**b** and **c**) of the hornets' flights. Honeybee and hornet behaviours were synchronised to the start of the shimmering waves (abscissas give the time in ms after the start of the shimmering waves). The hornets' behaviours are shown in terms of distance velocity v_{dxz} (**b**) and turning angle θ_{xz} (**c**); see *inset*. Two classes of hornets were defined according to their distance to the nest in the 400 ms interval prior to shimmering: $d_{xz} < 45$ cm (*red circles*, $n = 84$ "reactive" episodes), and $d_{xz} > 45$ cm (*yellow circles*, $n = 65$ "reactive" episodes). Different *brown-shaded areas* define two test intervals in relation to the time course of shimmering (*brown-shaded*: 0–400 ms, *grey-brown shaded*: 400–1,000 ms). *Circles* and *bars* give arithmetical mean ± SE. *Big full* (*red* or *yellow*) *circles* show significant differences between values in relation to the starting time of the wave at $t = 0$ ms. For further details, cf. Kastberger et al. (2010)

Some predators obviously ignore the shimmering cues while approaching the nest. In particular, specialised birds such as the bee-eater, *Nyctornis athertoni*, or the honey buzzard *Pernis ptilorhynchus* find their own ways to obtain prey (Kastberger and Sharma 2000; Kastberger et al. 2001). For the bees, this is a disturbing, although frequently occurring, situation. The giant honeybees have evolved defence behaviours against birds using flying guards, where hundreds of bees are successively released from the nest within a fraction of a second. In this case, it is possible that the absolute size and the characteristic movement patterns of threatening predators are the cues that trigger mass attacks (Kastberger et al. 2010). There is evidence that a wasp-sized predator that "refuses" to leave the nest could also trigger the mass release of flying guards. The question of whether such "ignoring" of shimmering by the predator would be an adaptive cue that triggers a mass attack is subject to further investigations.

18.4.2.2 Choice of an Alternative Strategy: Hunting Outside the Shelter Zone of Giant Honeybee Nests

Hornets are continuously attracted to honeybee nests by their rich resources of protein and sugar and even by shimmering (Fig. 18.18). Nevertheless, shimmering also effectively prevents these potential predators from collecting bees from the nest surface, particularly due to its non-habituative behaviour. In thousands of episodes recorded from several honeybee colonies, not a single case of a successful hunt by a hornet catching a bee from the nest curtain was observed (Kastberger et al. 2008a). However, hornets can successfully hunt bees if they focus on ingoing and outgoing bees within their hovering range. Here, bees are unprotected by the colony-bound collective defence, but are still able to escape by dodging and fast flight. Most flying bees escape the wasps successfully (Kastberger et al. 2008a), either flying off the nest at maximal speed or landing as fast as possible on the nest. The hunting success of wasps is about 3%, not a big loss for the bee colony, but still probably a major benefit to the wasps.

18.4.3 Does Shimmering Provide Colony-Intrinsic Information?

18.4.3.1 Nasonov Scenting

The association of abdominal shaking in shimmering along with exposing the Nasonov glands in *A. dorsata* implies a function for Nasonov scenting in defence behaviour. The theory of Nasonov scenting classes the Nasonov secretions generally with behaviours of attraction rather than of defence (in *A. mellifera*: Renner 1960; Free et al. 1984). This is true in clustering, during swarming (Morse and Boch 1971; Avitabile et al. 1975), in orientation to food sources (Free and Williams 1972), and to marking the hive entrance (Butler et al. 1970); also, the findings of

Free et al. (1984) are in agreement with this interpretation because they demonstrated that synthetic Nasonov lure is effective in attracting swarming bees. A role for social attraction by the Nasonov scent has also been described for *A. cerana* (Naik et al. 1988; Abdullah et al. 1990; Matsuyama et al. 1996; Sasagawa et al. 1996).

A function of the Nasonov scent in colony defence in *A. mellifera* has not been mentioned in reviews (cf. Free 1987) nor has been definitely excluded (Crewe 1976). However, Nasonov scenting has been described from different honeybee species, including *A. mellifera,* as not being associated with fanning (Sakagami 1960; Ferguson and Free 1981). *Apis mellifera* is said to scent "violently" (Sakagami 1960) when irritated and ready to sting (Jacobs 1925). Renner (1960) reported that disturbed *A. mellifera* workers flew off from the hive and exposed the Nasonov glands when re-entering the colony. *Apis cerana* individuals are known to form small groups and expose their scent glands when being attacked by the Asian giant hornet, *Vespa mandarina* (Matsuura and Sakagami 1973; Ruttner 1988).

In *A. dorsata*, the role of the Nasonov glands, with respect to shimmering behaviour, can be discussed under three aspects. Firstly, a possible role for direct defence can be excluded (see Sect. 18.2.2.5). Secondly, a possible function of information transfer to neighbouring bees in the colony in order to induce abdominal shaking is also implausible, because these social waves have to be passed on by each participant within only tens of milliseconds. The spreading of the scent must be regarded as much slower, in particular if diffusion is the primary transfer mechanism (Bossert and Wilson 1963), which meets the experimental conditions of shimmering in still air. Therefore, the preferential direction for spreading the defence wave, as displayed at the time of observation, cannot be linked to any wind or air convection. Although abdominal thrusting itself, combined with the wing strokes, might produce some wind stream on the nest surface, Nasonov scenting itself is an unlikely trigger for shimmering. In particular, it would not be suitable for switching on or off phasic reactions to threatening cues if, in the course of repetitive waving, the scent gradient at the nest surface is steadily enhanced. On the other hand, chemicals for recruiting bees for aggressive alarm are associated with other glands such as mandibular and sting glands (Free 1987).

A third assumption, scenting in order to maintain group stability in case of danger, is most likely consistent with the theory of Nasonov scenting (Free 1987). The Nasonov scent may bring the curtain bees into readiness for performing abdominal shaking and averting individual reactions. The main biological purpose of the Nasonov scent in defence behaviour may be to maintain cohesiveness rather than leave the nest and eventually include more and more bees performing waving, maximising the defensive response.

Considering all known aspects of Nasonov scenting, Kastberger et al. (1998) proposed a wider view of the biological role of the Nasonov pheromone. The secretion may first act as a social group pheromone, similar to that recently postulated by Schmidt et al. (1993), for attracting and recruiting single bees to form an alliance as a prerequisite to promote very different patterns of social behaviour. This is true for orientational aspects of foraging bees, for clustering (cf. Free 1987), but also for aspects of colony defence.

18.4.3.2 Shimmering Affects the Nest in the Three-Dimensions of Space

The question is whether abdominal shaking only matters to nestmates at the surface of the nest or also those in sub-surface layers. If bees in deeper nest layers were influenced, considerable consequences would arise. One of these consequences is that the colony members in the regions of shimmering must be kept informed about external threatening cues. The possibilitiy that bees are capable of receiving information via mechanical cues in the nest is well known (Sandeman et al. 1996).

A possible answer lies in the analysis of the coupling between the curtain members. Knowledge about this key attribute of shimmering-making is fundamental for the understanding of how shimmering is distributed over the nest and of why shimmering has evolved. The state of coupling between the curtain members can be assessed by the time course of the movement of curtain bees in the three dimensions of space, which are specifically defined for giant honeybee nests as follows: x- (horizontal), y- (vertical) and z- (towards and off the nest surface) directions (Fig. 18.20). We analysed (Weihmann et al. 2009) high-definition video sequences to provide the database for the x- and y-components of the movement of surface bees to obtain a survey about the wave pattern from the whole nest. Additionally, we extracted the z-component of the movement of single surface bees in two ways (Fig. 18.20): first by laser vibrometry of single bees, and second by stereo-tracking all individuals displayed in the frame. Dummy wasps, moved 10 cm in front of the nest surface in different directions at standardised velocities, were presented to stimulate the colonies to produce forceful and complex shimmering responses.

We showed that shimmering causes massive z-movements of several millimetres at the nest surface (Weihmann et al. 2009). At the wave front, the bees, also those which did not participate in shimmering, move outwards first and then inwards in a damped oscillation at 5–10 Hz at the same moment the nest portions of the bee curtain affected by the wave front shifted aside en masse. The movement of the surface bees is proven to occur in the three dimensions of space. This is convincing evidence that shimmering also affects the sub-surface layers of the bee curtain. Furthermore, we found that the z-component is differently linked to wave strength, dependent on the direction of the wave propagation. This is an obvious consequence of the specificity of the functional architecture of the bee curtain. The bees hang with their heads up and the abdomens down, and cling with their legs on the subjacent layers of bees and form a movable matrix. The bee curtain does not obviously match the view of a homogenously coupled three-dimensional bee matrix, but rather supports the *vertical-string hypothesis* (see Sect. 18.2.1.1), which proposes that the bee curtain could be compared with a set of strings hanging down from a solid structure. The propagation properties would then resemble that of the classical curtain of our homes. In particular, the findings of the occurrence of

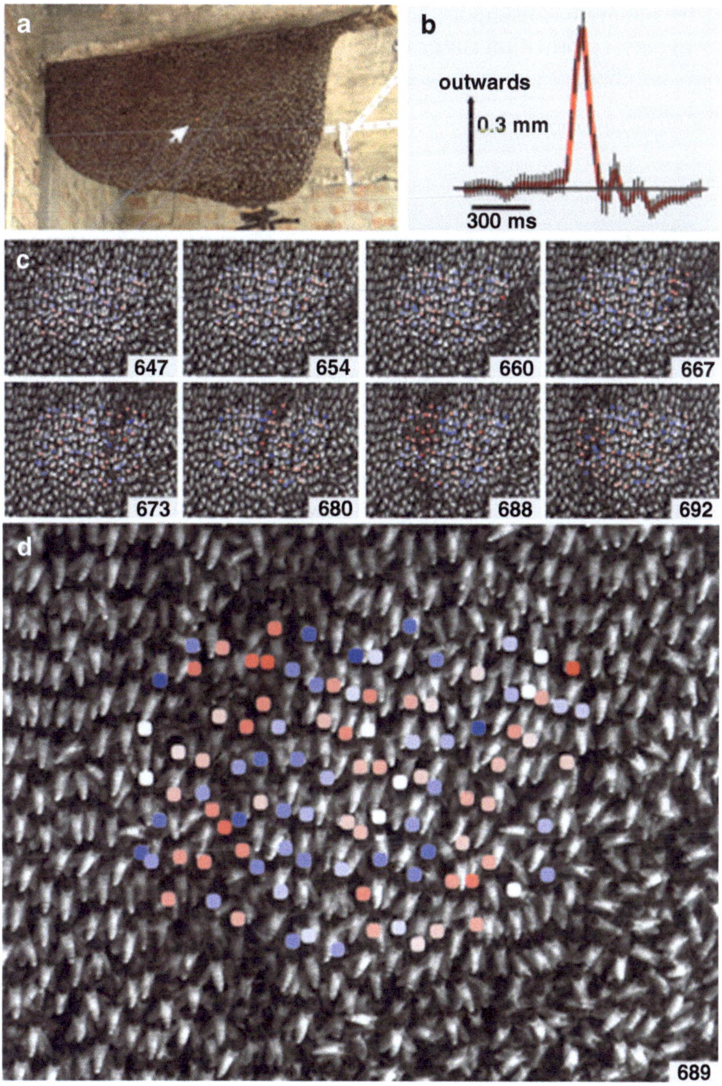

Fig. 18.20 Measurement of the *z*-component of shimmering in giant honey bee nests. (**a**) Experimental nest in Chitwan (Nepal) in January 2009. The *z*-component of a single bee was tracked by Laser vibrometry. The *red spot* (see *white arrow*) marks the laser point. (**b**) The data recorded from 51 waves which spread from the top to the bottom to the laser point at the thorax of the target bee. The abscissa is scaled in ms, the ordinate in mm. Data beneath the *black horizontal line* show movements in outward direction regarding the nest surface. Before the wave itself had arrived at the reference point it elevated the target bee slightly outwards the nest. The wave front strongly affected the bee, which came 1 mm or more towards the nest within 100 ms. After that, the bee came back again towards the comb, and reached positions even closer the comb than before. This process terminated in a damped oscillation. (**c**) Video sequences (frame 647–692) recorded by two high-resolution black and white cameras at 60 Hz allowing stereo tracking of the positional co-ordinates (*x*–*y*–*z*) of hundreds of bees. Here, 100 bees around the vibrometer spot were

the massive z-component support the view that shimmering provides mechanoceptive cues to code information for nest members and that it has the capacity to alert the colony members of sub-surface layers of the bee curtain at least at these spots where it occurs.

18.4.4 Is Shimmering Associated with Recruiting Flying Guards?

Kastberger et al. (2001) observed that wave processes also occur after massive disturbances of the nest site by mechanical stimuli. The response of the colony to such mechanical disturbance is striking and may lead to an overall arousal of the curtain bees of all layers and on both sides.of the nest. This behaviour is termed "rearing-up" (Kastberger et al. 2001) and is associated with abdomen thrusting behaviour similar to shimmering. However, the "rearing-up" response is associated with a series of collective behaviours in the bee curtain. There is a collective contraction of parts of the bee curtain (Kastberger and Kaefer 2002), with a successive "recruiting" wave originating from the centre of disturbance and spreading to all other parts of the nest, and finally an increase in the size of the nest, in particular with the forming of a protuberance of the lower parts of the nest. This particular change of the nest shape indicates that the recruitment of flying guards is under way (Kastberger et al. 2001). This series of processes are dramatically seen in film documents (Kastberger et al. 2001), but can only partly be quantified by imaging methods.

The example in Fig. 18.21 illustrates a tranquil nest in Chitwan (Nepal) in January 2009, where some small "classical" shimmering waves occurred periodically. A wind blast abruptly disturbed the nest and the colony reacted by massively contracting the mouth zone. Consecutively, a weak wave slowly spread from the mouth zone to the top of the nest in a broad front. This wave process is still unexplored and probably represents a second type of shimmering and pertains to the curtain members of all layers. After a further concentric shimmering wave that occurred distally, left of the mouth zone, the whole nest increased in size and formed a protuberance in the bottom of the nest over a period of several tens of seconds. Sometimes, such protuberances have a smaller cone-like shape expressing a state of extraordinary arousal with flying guards mobilised for counter-attack (Kastberger et al. 2001). If further disturbances occur, flying guards start preferentially from this protuberance site (Kastberger et al. 2001).

Fig. 18.20 (continued) screened. The outward z-positions of the bees were scaled in *red colours*, the inward z-positions in *blue* colours; *red* and *blue* colours are graded in 14 steps; the *bright red* colour encodes the maximum elevation +3 mm) off the nest surface, the *bright blue* colour the minimum elevation (−3 mm) towards the comb. Note that the wave spread from the right to the left side. The wave front itself displays *bright red* colours, followed by the *bright blue sinks*. For further details, cf. Weihmann et al. (2009). (**d**) gives the frame 689 in greater details

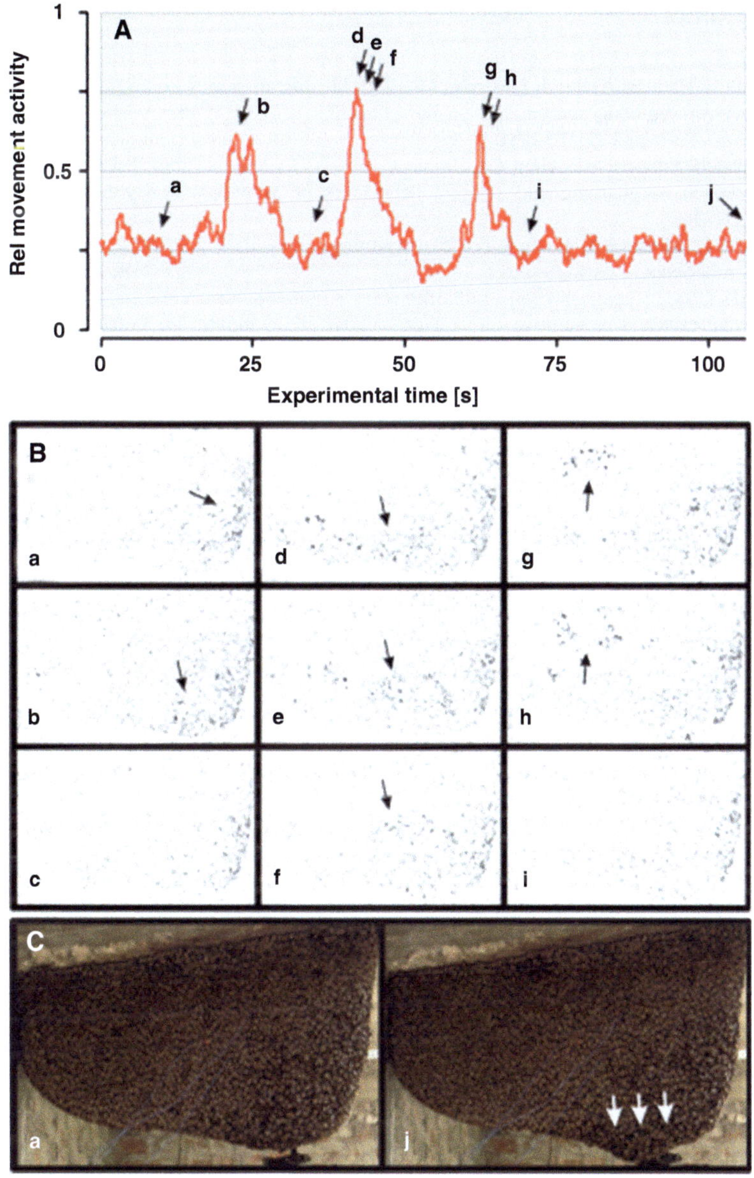

Fig. 18.21 Shimmering in displayed in combination with other defence behaviours. Measurement of a shimmering process in giant honeybee nests using movement analysis. (**A**) differential activities image analysis shown for an experimental nest in Chitwan (Nepal) in January 2009. (**A**) Tracking of movement activities using the changes in pixel luminance in sequential images. (**a–j**) refer to the respective differential images in (**B**). In (**B**, a) the quiescent state of the bee nest is documented and the *arrow* marks the increased locomotor activities in the mouth zone; the first peak in (**A**) and traced in (**B**, b) was caused by a wind blast as an external mechanical stimulus. Consequently, the mouth region of the nest contracted as a collective response that has been

References

Abdullah N, Hamzah A, Ramli J, Mardan M (1990) Identification of Nasonov pheromones and the effects of synthetic pheromones on the clustering activity of the Asiatic honeybee (*Apis cerana*). Pertanika 13:189–194

Arechavaleta-Velasco ME, Hunt GJ, Emore C (2003) Quantitative trait loci that influence the expression of guarding and stinging behaviours of individual honey bee. Behav Genet 33:357–364

Avitabile A, Morse RA, Boch R (1975) Swarming honey bees guided by pheromones. Ann Entomol Soc Am 68:1079–1082

Blum MS (1981) Chemical defenses of arthropods. Academic, New York

Bonner JT (1982) Comparative biology of cellular slime molds. In: Loomis WF (ed) The development of *Dictyostelium discoideum*. Academic, New York, pp 1–33

Bossert WH, Wilson EO (1963) An analysis of olfactory communication among animals. J Theor Biol 5:443–496

Breed MD, Robinson GE, Page RE Jr (1990) Division of labor during honey bee colony defense. Behav Ecol Sociobiol 27:395–402

Breed MD, Smith TA, Torres A (1992) Guard honey bees: role in nestmate recognition and replacement. Ann Entomol Soc Am 85:633–637

Breed MD, Guzmán-Novoa E, Hunt GJ (2004) Defensive behavior of honey bees: organisation, genetics, and comparison with other bees. Annu Rev Entomol 49:271–298

Buck J, Buck E (1980) Flash synchronization as tool and as enabler in firefly courtship competition. Am Nat 116:591–593

Buck J, Buck E, Case JF, Hanson FE (1981) Control of flashing in fireflies. V. Pacemaker synchronization in Pteroptyx cribellata. Flash synchronization as tool and as enabler in firefly courtship competition. J Comp Physiol A 144:287–298

Butani DK (1950) An *Apis dorsata* colony in New Delhi. Indian Bee J 12:115

Butler CG (1962) The world of the honeybee. Collins, London

Butler CG, Fletcher DJC, Watler D (1970) Hive entrance finding by honeybee (*Apis mellifera*) foragers. Anim Behav 18:78–91

Camazine S, Sneyd J (1991) A model of collective nectar source selection by honey bees: self-organization through simple rules. J Theor Biol 149:547–571

Camazine S, Deneubourg JL, Franks NR, Sneyd J, Theraula G, Bonabeau E (2003) Self-organization in biological systems. Princeton University Press, Princeton

Caro T (2005) Antipredator defenses in birds and mammals. University of Chicago Press, Chicago

Cassier P, Lensky Y (1994) The Nassanov gland of the workers of the honeybee (*Apis mellifera* L.): ultrastructure and behavioural function of the terpenoid and protein components. J Insect Physiol 40:577–584

Chada MS, Eisner T, Monro A, Meinwald J (1962) Defense mechanisms of arthropods. VII. Citronellal and citral in the mandibular gland secretions of the ant, *Acanthomyops claviger* (Roger). J Insect Physiol 8:175–179

Fig. 18.21 (continued) termed "rearing up". While this response was clearly detected as a contraction response, it was caused by a slightly increased movement activity below the arrow in (**B**, b). Several seconds afterwards a slow wave arising from the lower rim of the nest spread upwards. This wave was traced in the images (**B**, d–f); several further seconds afterwards a "normal" shimmering wave occurred in the region in the upper left corner of the nest (**B**, g–h). Afterwards, the whole nest increased in size and formed a protuberance at the bottom of the nest (compare the nest size between state a and j in **C**). Such excrescences (see *arrows* in **C**, j) may have a cone-like shape and express a state of extraordinary arousal with flying guards mobilised for counter-attack

Crewe RM (1976) Aggressiveness of honeybees and their pheromone production. S Afr J Sci 72:209–212

Crewe RM, Hastings H (1976) Production of pheromones by workers of *Apis mellifera adansonii*. J Apic Res 15:149–154

Das BP, Gupta VK (1989) The social wasps of India and the adjacant countries (Hymenoptera: Vespidae), Oriental Insects Monograph 11. Association for the study of oriental insects, Gainesville, FL

Duffield RM, Wheeler JW, Eickwort GC (1984) Sociochemicals of bees. In: Bell WJ, Carde RT (eds) Chemical ecology of insects. Chapman and Hall, London, pp 387–428

Eibl-Eibesfeldt J (1978) Grundriss der vergleichenden Verhaltensforschung. Piper, München [in German]

Farkas I, Helbing D, Vicsek T (2002) Social behaviour: Mexican waves in an excitable medium. Nature 419:131–132

Ferguson AW, Free JB (1981) Factors determining the release of Nasanov pheromone by honeybees at the hive entrance. Physiol Entomol 6:15–19

Free JB (1987) Pheromones of social bees. Chapman and Hall, London

Free JB, Williams IH (1972) The role of the Nasonov gland pheromone in crop communication by honeybees (*Apis mellifera* L.). Behaviour 41:314–318

Free JB, Pickett JA, Ferguson AW, Simpkins JR, Williams C (1984) Honeybee Nasonov lure. Bee World 65:175–181

Hepburn HR, Hepburn C (2007) Bibliography of the giant honeybees, *Apis dorsata* Fabricius (1793) and *Apis laboriosa* F Smith (1871). Apidologie 38:219–220

Hölldobler B, Wilson EO (1990) The ants. Harvard University Press, Cambridge

Immelmann K (1982) Wörterbuch der Verhaltensforschung. Paul Parey, Berlin Hamburg [in German]

Jacobs W (1925) Das Duftorgan von *Apis mellifera* und ähnliche Hautdrüsen. Z Morphol Ökol Tiere 3:1–80 [in German]

Jeschke JM, Tollrian R (2007) Prey swarming: which predators become confused and why? Anim Behav 74:387–393

Kastberger G (1999) The magic trees of Assam – Documentary film about the biology of the giant honeybee *Apis dorsata*. National Geographic, ZDF, ORF & epo-film, Wien

Kastberger G, Kaefer H (2002) Cascadic coordination in defense behaviour of the giant honeybee *Apis dorsata*. In: Proceedings of the 27th Göttingen neurobiology conference, p 260

Kastberger G, Sharma DK (2000) The predator-prey interaction between blue-bearded bee-eaters (*Nyctyonis athertoni*) and giant honeybees (*Apis dorsata*). Apidologie 31:727–736

Kastberger G, Stachl R (2003) Infrared imaging technology and biological applications. Behav Res Methods Instrum C 35:429–439

Kastberger G, Raspotnig G, Winder O (1996a) The Nasonov secretion of the Asian giant honeybee *Apis dorsata*: evidence for a role in colony defense. Verh Dtsch Zool Ges 89:57

Kastberger G, Winder O, Hoetzl T, Raspotnig G (1996b) Behavioural features of a periodic form of massed flight activity in the giant honeybee *Apis dorsata*. Apidologie 27:381–395

Kastberger G, Biswas S, Habibulla M (1997) Temporal and spatial patterns of colony alarming in *Apis dorsata*. Verh Dtsch Zool Ges 90:293

Kastberger G, Raspotnig G, Biswas S, Winder O (1998) Evidence of Nasonov scenting in colony defense of the giant honeybee *Apis dorsata*. Ethology 104:27–37

Kastberger G, Winder O, Steindl K (2001) Defense strategies in the giant honeybee *Apis dorsata*. Proc Dtsch Zool Gesells 94:1–7

Kastberger G, Schmelzer E, Kranner I (2008a) Social waves in giant honeybees repel hornets. PLoS ONE 3(9):e3141. doi:10.1371/journal.pone.0003141

Kastberger G, Thenius R, Stabentheiner A, Hepburn R (2008b) Aggressive and docile colony defense patterns in *Apis mellifera*. A retreater-releaser concept. J Insect Behav 22:65–85

Kastberger G, Weihmann F, Hoetzl T (2010) Complex social waves of giant honeybees provoked by a dummy wasp support the special-agent hypothesis. Commun Integr Biol 3(2):179–180

Ken T, Hepburn HR, Radloff SE, Yusheng Y, Yiqiu L, Danyin Z, Neumann P (2005) Heat-balling wasps by honeybees. Naturwissenschaften 92:492–495

Kennedy J, Eberhart RC (2001) Swarm intelligence. Morgan Kaufmann/Academic, California

Kloft W, Schneider P (1969) Gruppenverteidigungsverhalten bei wildlebenden Bienen (*Apis cerana* Fabr.) in Afghanistan. Naturwissenschaften 56:219 [in German]

Koeniger N, Fuchs S (1975) Zur Kolonieverteidigung der asiatischen Honigbienen. Tierpsychologie 37:99–106 [in German]

Koeniger N, Koeniger G (1980) Observations and experiments on migration and dance communication of *Apis dorsata* in Sri Lanka. J Apic Res 19:21–34

Koeniger N, Koeniger G, Gries M, Tingek S, Kelitu A (1996) Observations on colony defense of *Apis nuluensis* Tingek, Koeniger and Koeniger, 1996 and predatory behavior of the hornet, *Vespa multimaculata* Pérez, 1910. Apidologie 27:341–352

Koga T, Backwell PRY, Jennions MD, Christy JH (1998) Elevated predation risk changes mating behaviour and courtship in a fiddler crab. Proc Biol Sci 265:1385–1390

Krause J, Ruxton GD (2002) Living in groups. Oxford University Press, Oxford

Landeau L, Terborgh J (1986) Oddity and the 'confusion effect in predation. Anim Behav 34:1372–1380

Lindauer M (1956) Über die Verständigung bei indischen Bienen. Zeit Vergl Physiol 38:521–557

Matsuura M, Sakagami SF (1973) A binomic sketch of the giant hornet, *Vespa mandarina*, a serious pest of Japanese apiculture. J Fac Sci Hokkaido Univ Ser VI Zool 19:125–162

Matsuyama S, Suzuki T, Sasaki M, Sasagawa H (1996) The Nasonov pheromone of the Japanese honeybee, *Apis cerana japonica* Rad. In: Proceedings of the 20th international congress of entomology, Florenz, p 403

Morse RA, Boch R (1971) Pheromone concert in swarming honeybees (Hymenoptera: Apidae). Ann Entomol Soc Am 64:1414–1417

Morse RA, Laigo FM (1969) *Apis dorsata* in the Philippines. Philippine Association of Entomologists, Laguna

Naik DG, Gadre RV, Kapadi AH, Singh MK, Suryanarayana MC, Kshirsagar KK (1988) Nasonov gland pheromone of the Indian honeybee, *Apis cerana indica* F. J Apic Res 27:205–206

Neumann P, Koeniger N, Koeniger G, Tingek S, Kryger P, Moritz RFA (2000) Home-site fidelity in migratory honeybees. Nature 406:474–475

Oldroyd BP, Wongsiri S (2006) Asian honey bees. Harvard University Press, Cambridge

Oldroyd B, Smolenski A, Lawler S, Estoup A, Crozier R (1995) Colony aggregations in *Apis mellifera* L. Apidologie 26:119–130

Ono M, Igarashi T, Ohno E, Sasaki M (1995) Unusual thermal defense by a honeybee against mass attack by hornets. Nature 377:334–336

Paar J, Oldroyd B, Kastberger G (2000) Giant honey bees return to their nest sites. Nature 406:475

Paar J, Oldroyd BP, Huettinger E, Kastberger G, Schaeffer S (2004) Genetic structure of an *Apis dorsata* population: the significance of migration and colony aggregation. J Hered 95:119–126

Pain J (1973) Pheromones and hymenoptera. Bee World 54:1–14

Pickett JA, Williams IH, Smith MC (1980) Nasonov pheromone of the honeybee, *Apis mellifera* L. (Hymenoptera: Apidae). Part I. Chemical characterization. J Chem Ecol 6:425–434

Renner M (1960) Das Duftorgan der Honigbiene und die physiologische Bedeutung ihres Lockstoffes. Zeit Vergl Physiol 43:411–468 [in German]

Reynolds CW (1987) Flocks, herds, and schools: a distributed behavioural model. Comput Graph 21:25–34

Roepke W (1930) Beobachtungen an indischen Honigbienen, insbesondere an *Apis dorsata* F. Meded Landbouw Wageningen 34:1–28

Ruttner F (1988) Biogeography and taxonomy of honeybees. Springer, Berlin

Sakagami SF (1960) Preliminary report on the specific difference of behaviour and other ecological characters between European and Japanese honeybees. Acta Hymenopterol 1:171–198

Sandeman DC, Tautz J, Lindauer M (1996) Transmission of vibration across honeycombs and its detection by bee leg receptors. J Exp Biol 199:2585–2594

Sasagawa H, Matsuyama S, Yamaoka R, Sasaki M (1996) The oriental orchid lures the Japanese honeybee with Nasonov pheromone mimics. In: Proceedings of the international congress of entomology, Florenz, p 439

Schmelzer E, Kastberger G (2009) 'Special agents' trigger social waves in giant honeybees (*Apis dorsata*). Naturwissenschaften 96:1431–1441

Schmidt JO, Slessor KN, Winston ML (1993) Roles of Nasonov and queen pheromones in attraction of honeybee swarms. Naturwissenschaften 80:573–575

Schneider P, Kloft W (1971) Beobachtungen zum Gruppenverteidigungsverhalten der östlichen Honigbiene *Apis cerana* Fabr. Z Tierpsychol 29:337–342, in German

Seeley TD, Seeley RH, Aratanakul P (1982) Colony defense strategies of the honeybees in Thailand. Ecol Monogr 52:43–63

Stabentheiner A, Kovac H, Schmaranzer S (2002) Honeybee nestmate recognition: the thermal behaviour of guards and their examinees. J Exp Biol 205:2637–2642

Sulzer W, Kastberger G, Muick M, Hirschmugl M, Huettinger E (2010) Multitemporal crop and landcover analysis in Chitwan (Nepal) by means of remote sensing. Correlation with the distribution of giant honey bee (*Apis dorsata*) colonies. In: Kaufmann V, Sulzer W (eds) Proceedings of the 10th international symposium on high mountain remote sensing cartography. Grazer Schriften der Geographie und Ramuforschung 45, pp 63–75

Tan NQ (2007) Biology of *Apis dorsata* in Vietnam. Apidologie 38:221–229

Tan K, Radloff SE, Li JJ, Hepburn HR, Yang MX, Zhang LJ, Neumann P (2007) Bee-hawking by the wasp *Vespa velutina*, on the honeybees *Apis cerana* and *A. mellifera*. Naturwissenschaften 94:469–472

Underwood BA (1990) Seasonal nesting cycle and migration patterns of the Himalayan honey bee *Apis laboriosa*. Natl Geogr Res 6:276–290

von Frisch K (1967) The dance language and orientation of bees. Harvard University Press, Cambridge

Weihmann F, Hoetzl T, Kastberger G (2009) 3-D patterning of social waves in the giant honeybee *Apis dorsata*. Proc Dtsch Zool Gesells, Regensburg

Chapter 19
Interspecific Interactions Among Asian Honeybees

Ming-Xian Yang, Ken Tan, Sarah E. Radloff, and H.R. Hepburn

19.1 Introduction

The Asian continent is the richest in the world in terms of honeybee diversity and includes a number of indigenous species: *Apis cerana*, *Apis florea*, *Apis andreniformis*, *Apis dorsata*, *Apis laboriosa*, *Apis nigrocincta*, *Apis nuluensis* and *Apis koschevnikovi*, as well as the introduced *Apis mellifera*, which is widely used for honey production. When these *Apis* species occur sympatrically, they can interact in various ways (Koeniger 1982). Worker bees of different species may rob each other's nests and compete for food or nesting sites, while drones may interfere with each other during mating flights. Furthermore, parasites or diseases of one species may transfer to another. Interspecific interactions among the *Apis* species have no doubt played a role in their evolution. Even though current interspecific interactions may differ from those of the past – both before or during the process of speciation – it is still an interesting and potentially important topic that deserves investigation.

M.-X. Yang
Department of Zoology and Entomology, Rhodes University, Grahamstown 6140, South Africa
and
Currently, Department of Zoology, Sichuan Agricultural University, Yaan, People's Republic of China
e-mail: damoguyan2006@yahoo.com.cn

K. Tan
Eastern Bee Research Institute of Yunnan, Agricultural University, Heilongtan, Kunming, Yunnan Province, People's Republic of China
e-mail: eastbee@public.km.yn.cn

S.E. Radloff
Department of Statistics, Rhodes University, Grahamstown 6140, South Africa
e-mail: s.radloff@ru.ac.za

H.R. Hepburn
Department of Zoology and Entomology, Rhodes University, Grahamstown 6140, South Africa
e-mail: r.hepburn@ru.ac.za

H.R. Hepburn and S.E. Radloff (eds.), *Honeybees of Asia*,
DOI 10.1007/978-3-642-16422-4_19,

19.2 Nest Site Competition

In Asia, the different honeybee species have adopted different evolutionary strategies in adapting to their environments and, according to body size and nesting habits, they can be divided into three groups: dwarf honeybees, giant honeybees and cavity-nesting honeybees (Arias and Sheppard 2005; Oldroyd and Wongsiri 2006; cf. Chap. 1). Given that each group has a distinct nesting behaviour (cf. Chap. 6), nest site competition between them can rarely be observed. In this section, therefore, only competition within each group is discussed.

19.2.1 Nest Site Competition in the Dwarf Honeybees

The dwarf honeybees include two species, *A. florea* and *A. andreniformis*, and both naturally occur in tropical and sub-tropical regions of Asia (Wongsiri et al. 1996; Hepburn and Radloff 2011; cf. Chap. 3). *A. florea* extends from the Middle East, eastwards to peninsular Malaysia, whereas *A. andreniformis* is distributed from the Philippines to China and Myanmar, with a partial overlap in Southeast Asia (Otis 1996; Wongsiri et al. 1996; Hepburn and Radloff 2011). Our interest lies in whether they compete for nest sites in the limited areas where they overlap.

These two honeybee species are superficially similar in many respects and it took many years for honeybee biologists to define them as unequivocally separate biological species (Smith 1858; Maa 1953; Wu and Kuang 1987; Ruttner 1988; Wongsiri et al. 1990; Hepburn et al. 2005). Both species build single, exposed combs on the thin branches of bushes, shrubs or small trees (Wongsiri et al. 1996) and, in western Asia, often nest in small caves or in sheltered areas of buildings (Dutton and Free 1979; Whitcomb 1984). Although it has been reported that the two species may also nest at different altitudes: *A. andreniformis* in high mountainous and forest areas at about 1,600 m altitude, while *A. florea* is common in lowlands below 1,000 m (Wongsiri et al. 1996), an analysis of the complete distribution of the species shows that there is no significant difference in their altitudinal distributions (Hepburn and Radloff 2011; cf. Chap. 3).

However, the nests of *A. andreniformis* appear higher on average (about 6 m above the ground) than those of *A. florea* (about 4 m) (Rinderer et al. 2002), so that nesting competition between them can be inferred to happen only occasionally. Rinderer et al. (2002) found that when these two species occur together in the same area, they tend to avoid each other. The alleged avoidance between these two species, although still controversial (Koeniger personal communication), may make sense if they evolved the ability to recognise each other during the course of speciation and mutual adaptation. Interestingly, Rinderer et al. (2002) reported that both species of dwarf honeybees have a tendency to form aggregations of colonies in spatial distribution, but not as intensely as colonies of *A. dorsata*.

19.2.2 Nest Competition in the Giant Honeybees

A. dorsata and *A. laboriosa* form the group *Megapis* or commonly giant honeybees. *A. dorsata* is distributed mainly in tropical areas, while *A. laboriosa* naturally occurs in mountainous regions at altitudes between 1,500 and 4,000 m, particularly in the Himalayas (Sakagami et al. 1980; Ruttner 1988; Underwood 1990a, b). The former species has a tendency to aggregate, that is 100 or even more colonies may be crammed in a single tree (Deodikar et al. 1977; Seeley et al. 1982; Dyer and Seeley 1991), and has a habit of seasonal migration. The latter also forms colony aggregations but only on cliffs (Roubik et al. 1985; Kuang and Kuang 2002; Joshi et al. 2004). According to Underwood (1990a, b), *A. laboriosa* never nests on the limbs of trees. And, although they also have seasonal migration behaviour, which results in a temporary sympatry with *A. dorsata*, it occurs during the non-nesting phase of *A. laboriosa*. From this, we can conclude that nest competition between these two giant honeybees does not occur today; we can only speculate as to the past.

19.2.3 Nest Competition in the Cavity-Nesting Honeybees

Cavity-nesting honeybees include *A. cerana*, *A. nigrocincta*, *A. nuluensis*, *A. koschevnikovi* and *A. mellifera*, all of which are native to Southeast Asia with the exception of *A. mellifera*. The four Asian cavity-nesting honeybees began their divergence from a presumed *A. cerana* prototype some 2 million years ago (Smith 1991; Arias and Sheppard 2005). Even so, the habitats of each species are very different. For example, *A. nuluensis* is confined to the highlands on the island of Borneo, and it is only known from the Crocker Range in Sabah (Tingek et al. 1996). The Sulawesian honey bee, *A. nigrocincta*, is confined to the islands of Sulawesi, Sangihe and Mindanao (Otis 1996). *A. koschevnikovi* has a comparatively wider distribution area: from Java, Sumatra, peninsular Malaysia to southern Thailand; however, because this bee requires rainforest habitat, it is now rare outside of Borneo owing to deforestation (Hadisoesilo et al. 2008). *A. cerana* occurs on the mainland of Asia as well as the islands of the South China Sea (Radloff et al. 2010). For these combined reasons there are few reports about nesting competition among these cavity-nesting honeybees. Interestingly, all the cavity-nesting honeybees, with the exception of *A. cerana*, occur mainly on islands in the South China Sea. The considerable geographic isolation of the islands has undoubtedly played a very important role in their speciation. It seems that the only practical places to investigate possible nest competition among these species are Borneo, with three cavity-nesting bees: *A. koschevnikovi*, *A. nuluensis* and *A. cerana*, and Sulawesi, where *A. cerana* and *A. nigrocincta* are sympatric.

19.3 Social Parasitism

Social parasitism in honeybees is generally understood to mean the phenomenon of worker bees joining neighbouring colonies by drifting or direct invasion (Neumann et al. 2001a). Social parasitism is widespread in social insects but has been studied only in *A. cerana* and *A. florea* among Asian honeybees. Nanork et al. (2006a) found that in queenright *A. cerana* colonies, 2–6% of workers are non-natal. These drifted workers, do not, however, have active ovaries, suggesting that in queenright colonies social parasitism is not pervasive. In queenless colonies, however, there were significantly more non-natal workers with activated ovaries (72.7%) than natal workers (36.3%). Non-natal workers also had a significantly higher reproductive success than natal workers. The same phenomenon has been observed in the dwarf honeybees, *A. florea* (Nanork et al. 2006b; Chapman et al. 2009). In *A. florea* colonies, when a colony becomes queenless, worker bees have a higher tendency for parasitising other colonies, preferring queenless to queenright colonies as their hosts for reproduction; as a result, queenless colonies are heavily parasitised with the eggs of non-natal workers (Nanork et al. 2006b). It has been suggested that social parasitism is present more or less in all honeybees species: 2–4% of the workers are non-natal, although these unrelated workers are thought to arise via orientation errors while retuning from foraging trips (Chapman et al. 2009).

Although social parasitism has only been observed intraspecifically in honeybees, interspecific parasitism has yet to be investigated. *A. cerana*, however, was observed in a colony of *A. mellifera* for a short period before subsequently flying away (Denis Anderson personal communication) and *A. cerana* workers have been seen several times on nests of *A. florea* (Duangphakdee, Hepburn and Phiancharoen personal communication). The same phenomenon has been reported in *A. mellifera capensis* invading colonies of *A. mellifera scutellata* by Neumann et al. (2001b). Furthermore, during the long history of evolution, parasitism might have played a role in nest competition and/or nest avoidance in speciation.

19.4 Food Competition

Besides possible competition for habitat and reproduction, the species also compete for food resources when they occur in the same area. The performance of different bees in competition is of significance in speciation and/or coevolution. When different honeybees compete for food, body size is an important factor and the smaller bees are usually more aggressive in defending floral resources than larger bees, probably because smaller bees have more restricted foraging ranges than the larger ones (Ruttner 1988).

Koeniger and Vorwohl (1979) investigated the interactions of *A. florea*, *A. cerana*, *A. dorsata* and stingless bees, *Trigona*, by using an artificial feeding dish. They found that smaller bees generally attacked larger ones; *A. dorsata*, however, was attacked only by *A. cerana*, and never by the two smaller species. At times, only one species remained while the others stayed away, although a final "winner" was unpredictable. Ruttner (1988) concluded that honeybees with larger bodies enjoy more choices, usually avoid disastrous fighting and can shift to other, more distant food resources.

In Nepal, Partap (1998) investigated the impact of the introduction of *A. mellifera* colonies on the foraging behaviour of *A. cerana*. Foraging competition was studied by counting the number of *A. cerana* foragers on several flowers during the presence of, and after removal of, *A. mellifera* colonies (Table 19.1). The results indicated that *A. cerana* foragers spend more time visiting flowers in the absence of *A. mellifera*. They also spend more handling time on flowers, visit more flowers per trip and collect more pollen. Furthermore, more *A. cerana* foragers were seen on the flowers when the competition from *A. mellifera* was removed (Partap 1998; cf. Chap. 11).

Similarly, Dhaliwal and Atwal (1970) studied food competition between *A. cerana* and *A. mellifera*. Firstly, the two species were fed at separate feeders, not mixing, and showing no hostile behaviour; but, as the feeders were brought nearer to each other, the bees became more and more aggressive. When *A. mellifera* workers were freely alighting on both feeders, *A. cerana* workers were hesitant to do so and the latter were often stung by the former, with some dying, whereas no *A. mellifera* died. Finally, *A. mellifera* workers formed a ring around the feeder so that *A. cerana* workers could not alight to feed. The results indicate that *A. mellifera* was more successful in eliminating *A. cerana* in a foraging context.

Interestingly, as suggested by this experiment, honeybees can distinguish their nestmates outside of the nest, and so can jointly compete for food. Kalmus (1941) found that even different strains of the same species can distinguish one another. Two colonies of differently coloured *A. mellifera* bees, Caucasians and Italians, were trained to feeders and behaved aggressively to each other. We can, therefore, infer that during speciation, the newly forming species could probably recognise their own nestmates and fight others, which in turn could be expected to facilitate speciation. Stout and Goulson (2001) found that bumblebees and honeybees both recognise odour cues deposited on flowers by previous visitors. Both bumblebees and honeybees avoid flowers previously visited by each other when foraging on *Melilotus officinalis*, that is bumblebees avoided flowers recently visited by honeybees and vice versa. How do honeybees avoid serious competition among

Table 19.1 Number of *A. cerana* foragers during the presence of and after removal of *A. mellifera* (Partap 1998)

Crop	Number of *A. cerana* foragers		Difference significance (*P*)
	During the presence of *A. mellifera*	After the removal of *A. mellifera*	
Mustard	12.6 ± 1.2	20.8 ± 1.3	0.01
Broadleaf mustard	12.3 ± 1.3	18.3 ± 2.1	0.01
Cauliflower	18.4 ± 1.1	28.3 ± 0.8	0.01
Radish	11.7 ± 0.9	16.2 ± 1.2	0.01

different species? Different species have different strategies. When we discuss this topic, two important decisive factors must be mentioned: energy consumption and body temperature of foragers.

Firstly, body size and the length of the proboscis surely play an important role in competition, and the size of a forager may determine which floral resources are available to it (cf. Chap. 10). The giant *A. dorsata*, for example, can fly further than smaller honeybees and thereby enjoy larger foraging areas. Because they have longer glossae, they can collect nectar from some deep flower corolla tubes but are not able to gain access to some very small flowers. In the history of coevolution, therefore, flowers of highly specialised morphology have developed nectaries for specific pollinators, and foragers of the different species specialise to some extent on particular floral resources (Oldroyd et al. 1992).

Secondly, different honeybee species have different flight designs. Some researches have intensively investigated the flight designs of honeybees (Hepburn et al. 1998a, b, 1999; Radloff et al. 2003), and have suggested that several factors can integrate into an excess power index (EPI) that determines the flying ability of honeybees (Hepburn et al. 1998b; cf. Chap. 13). According to this index, the drones of Asian honeybees can be statistically divided into two groups: dwarf honeybee drones and other drones. As for workers, the EPI can divide the Asian honeybees into three groups. It was suggested that prowess of flight in drones is driven by the need to compete and mate with queens flying high in the air, while worker bees forage nectar and pollen on flowers (Radloff et al. 2003).

Dyer and Seeley (1991) reported that among Asian species, *A. cerana* shows a disproportionately high mass-specific metabolic rate and their foragers make many more trips per day in the same habitat than foragers of the other species. Last but not least, different species differ in the time and the temperatures at which they initiate their collecting trips (cf. Chap. 12). *A. cerana* colonies start work earlier in the day than *A. mellifera* workers, can endure lower ambient temperatures, and are more industrious in collecting nectar from scattered flowers, while *A. mellifera* workers tend to prefer big flower patches (Kuang and Kuang 2002). Oldroyd et al. (1992) investigated foraging competition among four species: *A. dorsata*, *A. cerana*, *A. andreniformis* and *A. florea* in Thailand on inflorescences of the king palm *Archontophoenix alexandrea*, which produces copious quantities of pollen overnight. Only the earliest visitors can collect the large amount of nectar available just before dawn. In order of appearance, *A. cerana* comes first, followed by *A. dorsata* shortly after dawn, and less than an hour later they were replaced by *A. florea*, *A. andreniformis* and some stingless bees.

It may be difficult to understand the above-mentioned phenomena without some analysis of two important factors: body temperature and energy consumption. Honeybee biologists have noticed that a thoracic temperature threshold is absolutely crucial for a forager to initiate flight (Dyer and Seeley 1987). A forager can increase its thoracic temperature by producing metabolic heat if the colony temperature is lower than the ambient temperature. The cavity-nesting species have the advantage of maintaining a high nest temperature, which explains why *A. cerana* foragers can begin collecting before dawn, earlier than the open-nesting species. Fighting and searching for food are high energy consumption tasks, and the bigger the body size, the more energy required. This may be the reason why the giant honeybees can fly further and exploit other flowers rather than fighting the smaller bees.

19.5 Robbing

Robbing is an act, or series of acts, by which bees from one colony pilfer or steal honey from other colonies (Ribbands 1953; Atwal and Dhaliwal 1969). This differs fundamentally from food competition, which happens extramurally, because robber bees enter the nests of other colonies, kill bees and take their stores. The smaller the colony, the more susceptible it is to the loss of stores and death of workers (Hepburn and Radloff 1998). Usually, every colony has some guard bees at the entrance to fight against intruders, and they are able to distinguish their nestmates by their colony-specific odours (Ribbands 1954, 1955).

Robbing usually occurs in times of dearth, when there is not enough available nectar (Hepburn and Radloff 1998). However, robbing may occur at any time when the nectar flow is interrupted or the colonies become weak or diseased (Atwal and Dhaliwal 1969). Atwal and Dhaliwal (1969) investigated robbing behaviour between *A. cerana* and *A. mellifera* and found that *A. cerana* are more prone to robbing than *A. mellifera*. But, Breed et al. (2007) suggested that robbing may be more characteristic of *A. mellifera* than other species. They compared nestmate recognition in several Asian honeybee species, *A. florea*, *A. andreniformis*, *A. dorsata* and *A. cerana*, and found that none of these species displayed strong aggressive responses to conspecific non-nestmates. This result indicates that *A. mellifera* has a more strongly developed response to conspecific non-nestmates than other *Apis* species. This conclusion explains in part what happens in China, when *A. cerana* and *A. mellifera* colonies are kept at the same apiaries. They rob each other during times of dearth, and it has been reported that *A. cerana* is more likely to initiate robbing, but usually lose when *A. mellifera* retaliates (Yang 2001). Numerous *A. cerana* colonies were killed in this way and lost territory in some areas (Yang 2001). Other interspecific instances of robbing were reported by Koeniger (1976) between *A. florea* and *A. mellifera* and by Atwal and Dhaliwal (1969) between *A. dorsata* and *A. cerana*.

Robbing can also be a means of transmitting bee diseases and parasites as shown by Atwal and Dhaliwal (1969) who reported that at a mixed apiary in India, under natural conditions, *A. mellifera* were free from acarine disease, but after robbing some weak *A. cerana* colonies, 70–80% of the bees in two *A. mellifera* colonies were infested. They also found that mites could be transmitted under experimental conditions from diseased *A. cerana* colonies to healthy *A. mellifera* colonies.

Although mites occur on several *Apis* species (Koeniger et al. 1983; Delfinado-Baker et al. 1985; Kuang and Kuang 2002), interspecific transmission has seldom been reported, with the exception of *Varroa* mites from *A. cerana* to *A. mellifera*

(Crane 1990), and *Neocypholaelaps indica* from *A. cerana* to *A. florea* and *A. dorsata* via foraging on the flowers (Koeniger et al. 1983; cf. Chap. 16).

19.6 Intervention of Mating

Among the most interesting of the interspecific interactions between the Asian species of *Apis* are those arising from the intervention of mating. Honeybee queens mate with drones in the air to reduce the probability of inbreeding. Because queens of all honeybee species have similar pheromones by which drones locate the virgin queens, particularly 9-keto-2-(*E*)-decenoic acid (9-ODA) (Butler et al. 1967; Shearer et al. 1970; Koeniger 1976), drones from one species might fly after queens of another species and try to mate with them. How queens avoid interspecific mating has long been a puzzle (cf. Chap. 8).

Intervention of mating among the Asian honeybee species has been widely investigated, and it has been suggested that three factors can lead to mating isolation: differences in male genitalia (Koeniger and Koeniger 1991), different drone congregation areas (DCA) (Koeniger and Koeniger 2000) and different mating times (Koeniger and Koeniger 2000). All of these factors are prezygotic barriers against interspecific mating and, if interspecific mating really occurs or was achieved by artificial insemination, there are also postzygotic barriers that prevent the development of hybrid offspring.

19.6.1 Male Genitalia

There are obvious differences in structure of the endophalli among the drones of *Apis*, which can undoubtedly lead to reproductive isolation (Fig. 8.14; cf. Chap. 8). Differences in body size between queens and drones and differences in drone genitalia among species also occur. Because mating occurs in the air, drones have to fly fast enough to catch a virgin queen on the wing. Only strong drones can seize the opportunity to mate with a queen, and queens and drones from different species cannot mate because of morphological differences. The weights of drones and queens of the nine species of honeybee are listed in Table 19.2. There are some crucial species-specific factors that determine the failure of interspecific mating (Table 19.2).

According to available data, *A. mellifera* drones produce the greatest number of spermatozoa (Table 19.2). It is somewhat strange that the drone of *A. mellifera* is heavier than that of the giant honeybee. The spermatozoa of *A. mellifera* are also longer than those of *A. dorsata*. The queens of *A. dorsata* and *A. nigrocinta* have higher mating frequencies than the queens of other *Apis* species.

Furthermore, there are some other species-specific organs that can prevent interspecific mating between species. For example, the drones of dwarf honeybees,

Table 19.2 Mean ± SD weights (mg), numbers of sperm, sperm length (μm) and mating frequency in *Apis* species

Species	Drone weight	Number of sperm per drone ($\times 10^6$)	Queen weight	Number of sperm in spermatheca of queen ($\times 10^6$)	Sperm length	Mating frequency
A. andreniformis	70.8 ± 3.0	0.13	112	1.3	–	10.5
A. florea	77.6 ± 2.6	–	86	1.1	205.81	7.9
A. cerana	83.4 ± 8.9	1.0 ± 0.1	122	1.4	267.07	14.1
A. koschevnikovi	105.5 ± 5.6	1.7 ± 0.16	170	2.1	–	13.3
A. nuluensis	107.0 ± 6.7	1.3 ± 0.1	–	–	–	–
A. nigrocincta	–	–	–	–	–	40.3
A. mellifera	211.1 ± 11.8	12.7	202	4.7	262.69	11.6
A. dorsata	155.7 ± 8.5	–	290	3.9	218.69	44.2
A. laboriosa	–	–	–	–	–	28.4

Data from Woyke (1975), Ruttner (1988), Koeniger et al. (1996a, d), Koeniger and Koeniger (2000), Baer (2005)

A. florea and *A. andreniformis*, have a basitarsus on the hind legs, which serve to clasp the hind legs of the queen during mating. *A. koschevnikovi* drones, on the other hand, have a hairy fringe on the margin of the tibia of the hind leg, which also strengthens their connexion with the queen during copulation (Rinderer et al. 1989).

19.6.2 Drone Congregation Area Differences

Without exception, all honeybee species mate on the wing. Drones from many colonies gather in a drone congregation area to form a drone cloud waiting for virgin queens. Different species and even subspecies have different DCAs. *A. mellifera* drones form their clouds at heights between 5 and 40 m depending on the weather. *A. mellifera carnica* drones form DCAs higher than those of *A. mellifera ligustica* (Koeniger and Koeniger 2001). DCAs of *A. cerana* are usually near the top of big trees (Punchihewa et al. 1990). The locations of some species are shown in Fig. 8.4.

19.6.3 Mating Times

Although different species of honeybees occurring in the same area tend to rear their new queens around the same season, given suitable weather and food resources, the species have differing mating times. The mating times of several sympatric species in some areas are listed in Table 19.3. In Sri Lanka, where several honeybee species occur, *A. florea* mates earlier than *A. cerana* and *A. dorsata*, while

Table 19.3 Mating time separation of sympatric honeybee species

Locality	Sri Lanka	Thailand	Sabah, Borneo
Species	Koeniger and Wijayagunesekera (1976)	Rinderer et al. (1993)	Koeniger et al. (1996d)
A. andreniformis	–	12:15–13:45	12:00–13:45
A. florea	12:00–14:30	14:00–16:45	–
A. cerana	16:15–17:15	15:15–17:30	14:00–16:15
A. koschevnikovi	–	–	16:45–18:30
A. dorsata	18:00–18:45	18:15–18:45	18:15–19:05

Koeniger and Koeniger (2000)

in Thailand, *A. andreniformis* is the earliest, and in Sabah Province of Malaysia, several species have mating times similar to those that they have in Thailand.

As we can see from Table 19.3, the same species in different locations may differ in mating times, but they do have a clear mating sequence when they occur with other species: the dwarf species, *A. andreniformis* and *A. florea,* mate early, followed by cavity-nesting and middle-sized honeybees, *A. cerana* and *A. koschevnikovi*. The drones of *A. dorsata* perform mating flights at dusk at all locations (Koeniger et al. 1994b).

Different male genitalia, different DCAs, combined with different mating times strongly indicate that the Asian honeybees have solved the mating intervention problem in the process of speciation. However, the balance can easily be broken when *A. mellifera* is present, having the same flight time, reaction to the sex attractant and congregation areas as *A. cerana*.

Although *A. mellifera* drones can actually mate with *A. cerana* queens, there is a noxious effect on the queen. A young *A. cerana* queen was found with its damaged sting chamber firmly blocked by the mating sign of an *A. mellifera* drone (Ruttner and Maul 1983). Thus, it can be concluded that no pre-mating barrier exists between these two species as is the case between other species (Ruttner 1988). Some researchers found that *A. mellifera* drones fly into the DCA of *A. cerana* and actually copulate with *A. cerana* queens (Yoshida 1994). In China, it has been reported that when commercial *A. mellifera* apiaries were first established, there was a significantly higher loss rate of *A. cerana* virgin queens during their mating flights. It has even been suggested that because of these phenomena, the two species can be regarded in an incomplete stage of speciation (Ruttner 1988). Moreover, Koeniger (1976) also inferred that mating intervention by *A. mellifera* may exist with regard to *A. florea* as well.

19.6.4 Artificial Insemination

That interspecific mating can actually occur under natural conditions invites the question: what happens after such matings and is any hybrid offspring produced? None of the eggs hatch because of postzygotic barriers between the species.

Artificial insemination between *A. cerana* and *A. mellifera* has been performed by several researchers (Ruttner 1969; Ruttner and Maul 1983; Woyke 1973; Koeniger et al. 1996b; Koeniger and Koeniger 2000; Phiancharoen et al. 2004), but no hybrids have been obtained thus far. Ruttner (1988) described the detailed developmental process in eggs laid by the queen after heterospecific instrumental insemination. The heterospecific spermatozoa can enter the spermatheca, survive there and fertilise eggs. Twenty-four hours after fertilisation, cleavage is observed to the blastula stage in the zygote. After that, however, the cell walls start to disintegrate and nuclei migrate into the secondary periplasm to accumulate in the antero-ventral part of the zygote and ultimately degenerate. Thus, no hybrid larva or imago ever develops.

Yoshida (1994) used mixed semen of *A. cerana* and *A. mellifera* drones to inseminate *A. mellifera* virgin queens. By using different mixed ratios of the two specific spermatozoa, he found that on hatching, the eggs produced only *A. mellifera* workers, and that interspecific fertilisation resulted in non-viable larvae. Koeniger (1996b) reported that interspecific hybrids between *A. cerana* and *A. koschevnikovi* produced by artificial insemination have low fertility and the hybrid colonies are probably non-viable.

Phiancharoen et al. (2004) used spermatozoa from drones of four species (*A. mellifera*, *A. cerana*, *A. florea* and *A. dorsata*) to respectively inseminate *A. mellifera* queens. They studied the survival rate of each specific sperm type and the frequency of eggs fertilised by each specific spermatozoon. The results showed that nearly 100% of *A. cerana* and *A. mellifera* spermatozoa were still viable 4 weeks after insemination, but the motility of *A. florea* and *A. dorsata* spermatozoa decreased to a large extent: 83.4 and 61.2%, respectively, after 3 days and that only a small proportion were still viable in the queens' spermathecae. As for the fertilisation rate, 57% of *A. mellifera* eggs were fertilised by *A. mellifera* spermatozoa, 40% of eggs fertilised by *A. cerana* and *A. florea* spermatozoa, while less than 20% by *A. dorsata* spermatozoa. The fluid in the queen's spermatheca played an important role in the survival rate and fertilisation success rate of the heterospecific spermatozoa, but no interspecific hybrid offspring emerged.

19.7 The Impact of the Introduction of *A. mellifera* to Asia

With the development of a beekeeping industry, honeybees, particularly *A. mellifera*, were introduced into many areas of Asia for such bee products as honey, pollen, royal jelly, propolis, etc. However, as the business benefits from the introduction of *A. mellifera* colonies grew, many problems emerged. As mentioned above, these included foraging competition, mating interference, robbing, and the transmission of disease. The introduction of *A. mellifera* colonies has also had an enormous impact on the native honeybee species in some areas of Asia [Japan: Sakagami (1959); India: Atwal and Sharma (1971); China: Ji et al. (2003), Yu and Han (2003) and Yang (2005); Nepal: Partap (1998)].

A. mellifera was first introduced into China in the 1920s (Kuang and Kuang 2002). On introduction, this western honeybee proved adaptable to a new environment and produced higher yields of bee products, especially royal jelly and propolis, of which the latter cannot be collected from *A. cerana* colonies because of their extremely low productivity. Since then, *A. mellifera* began to be widely adopted in Chinese beekeeping.

While enjoying the high profits of these bees, the negative aspects have been widely neglected and few if any had realised the strong impact of *A. mellifera* both on the environment and the local honeybees, especially *A. cerana*, until the 1980s. An investigation was launched and conducted by the *A. cerana* Association of China. The results showed that *A. cerana* has become extinct in the Daxinganling forest areas in the northeast and in XinJiang province in the northwest. In the Northeast Plain and North-China Plain areas, all *A. cerana* bees in man-made hives have absconded (Yang et al. 1982). In the entire northeast zone, *A. cerana* bees can only be found in wild and man-made hives in the Changbai mountain areas.

In the past, millions of *A. cerana* colonies were kept on the plains of the drainage area of the Yangtze River, but are now hard to find. In the southern provinces such as Jiangxi, Hunan, Fujian, Guangdong, Guangxi and Hainan, there are still many *A. cerana* colonies, however, their distribution area has shrunk greatly. Compared with those areas above, *A. cerana* colonies in the southwest are in a better condition, particularly in mountainous areas where many *A. cerana* bees can be found living in tree holes, caves and man-made hives in Yunnan province and Tibet (Yang et al. 1982).

In conclusion, the introduction of *A. mellifera* into China caused great losses of *A. cerana* colonies. The population of *A. cerana* colonies is now estimated at not more than one million, a decrease of some 60% following the introduction of *A. mellifera*, and their distribution has shrunk by 75% (Yang 2005).

As early as 1959, Sakagami noticed the impact of *A. mellifera* on *A. cerana* in Japan. In Nepal, Partap (1998) reported that plants and fruits were short of pollinators because of the population decrease of *A. cerana* caused by the introduction of *A. mellifera*. Even in Europe, with the rapid development of beekeeping at the beginning of twentieth century, many beekeepers preferred to raise some subspecies such as *A. mellifera ligustica* and introduced them from other areas, which caused the local extinction of native subspecies (Ruttner 1969, 1988). Moritz et al. (2005) recognised the severe disaster caused by the introduction of *A. mellifera* to tropical ecological systems and pointed out that the local honeybees or other pollinators suffered from the introduced species through food competition or diseases. This results in a reduction of biodiversity and an imbalance of the whole ecological system.

During the mating season, the virgin queens of both *A. cerana* and *A. mellifera* can attract heterospecific drones (Yang 2001; Ji et al. 2003; Wang et al. 2003). However, *A. mellifera* drones, which are much stronger fliers than *A. cerana* drones, can trap the *A. cerana* queens, although they cannot always mate with them successfully because of the differences in copulatory organs (Fig. 8.14). Their encirclement behaviour can inhibit successful mating between *A. cerana* queens and drones. In some areas with very many *A. mellifera* colonies, most of the virgin

A. cerana queens were trapped by *A. mellifera* drones, and only 16% of *A. cerana* queens were able to successfully mate.

More than 80% of *A. mellifera* queens could successfully mate with conspecific drones, although there was interference by *A. cerana* drones (Wang et al. 2003). This resulted in the population decline of *A. cerana* in some areas in recent years. In some areas they are threatened because the declining population is insufficient to support the community. The decrease or extinction of the native honeybees is a definite threat to the balance of ecology and some plant species will also become extinct because of insufficient pollination (Yang 2005).

19.8 Mixed-Species Colonies

The cavity-nesting honeybee species share several common morphological and behavioural characters and can be kept in the same colonies with heterospecific queens. Thus far, three types of mixed-species colonies: *A. cerana* with *A. koschevnikovi*, *A. cerana* with *A. nuluensis* and *A. cerana* with *A. mellifera* have been successfully organised experimentally. Recently, in Thailand, a super-mixed colony of *A. florea*, *A. mellifera*, *A. cerana* and *A. dorsata* was set up, but only lasted several weeks before all absconded together. No biological research has been done with this kind of super-mixed colony (Phiancharoen personal communication).

19.8.1 Mixed Colonies of A. cerana *and* A. koschevnikovi

Mixed colonies of *A. cerana* workers with an *A. koschevnikovi* queen were organised by Koeniger et al. (1996c). They grafted young larvae of *A. cerana* and *A. koschevnikovi* into artificial queen cells and inserted them into queenless colonies of both *A. cerana* and *A. koschevnikovi*. Not unexpectedly, all colonies preferred to rear conspecific larvae, but *A. cerana* colonies seemed more selective than *A. koschevnikovi* colonies against heterospecific larvae. Only 4% (4 out of 102) of *A. koschevnikovi* queens successfully emerged from *A. cerana* colonies, while 21% (30 out of 140) *A. cerana* larvae developed into adult queens in an *A. koschevnikovi* colony.

To set up mixed colonies, nearly emerging virgin queens in queen cells of either of the two species were introduced into heterospecific queenless colonies. In *A. koschevnikovi* colonies, all the *A. cerana* queen cells were destroyed and the queens killed; a few (4 out of 18) *A. koschevnikovi* queens were, however, accepted by the *A. cerana* colonies and three of them succeeded in mating and laying eggs. Although these queens were in heterospecific colonies, they mated with their own specific drones. Interestingly, the drones of *A. koschevnikovi* can also maintain their own species-specific mating times even when they were reared in *A. cerana* colonies (Koeniger et al. 1994a). The mated *A. koschevnikovi* queens laid eggs and the emerged bees were successfully reared by *A. cerana* worker bees, and thus the *A. cerana* host colonies were gradually transformed into *A. koschevnikovi* colonies.

19.8.2 Mixed Colonies of **A. cerana** *and* **A. nuluensis**

De Guzman et al. (1996) set up a mixed colony of *A. cerana* and *A. nuluensis*, with brood combs and adult bees from a single colony of *A. nuluensis*, placed into a queenless *A. cerana* colony, 200 km away. It was unusual that the adult workers did not attack each other. The authors investigated only the *Varroa* mites in this mixed colony and *V. jacobsoni* Oudemans and *V. underwoodi* were found. There have been no further reports regarding this kind of mixed colony.

19.8.3 Mixed Colonies of **A. cerana** *and* **A. mellifera**

A. cerana and *A. mellifera* are very closely related and very similar both in morphology and behaviour, to the extent that there was once doubt if they were distinct species (Ruttner and Maul 1983). Researchers and beekeepers have long wanted to hybridise them. For example, Atwal and Sharma (1968) introduced *A. mellifera* queens into *A. cerana* colonies and found that the introduction was successful if the *A. cerana* workers were no more than a week old. The introduced queen laid eggs in the host colonies, the eggs hatched into larvae, *A. cerana* workers attended them and they pupated and emerged as adults. Once the *A. mellifera* workers assumed field duties, they worked in harmony with the host *A. cerana* workers.

Studies show that young worker bees may lack pheromones and can be accepted by other colonies (Pham-Delegue et al. 1993; Laloi et al. 2001). It is therefore possible to exchange brood combs between colonies: firstly, previously prepared empty combs were added to strong *A. mellifera* colonies which were then checked every day until the combs were filled with eggs so that the emergence date for adults could be calculated. These brood combs were then kept in nurse colonies until the adult bees were just about to emerge and then removed and introduced into queen-right *A. cerana* colonies. When they emerged, the numbers of adult workers of the two species were about even, and no fighting was seen on opening the mixed-species colonies, nor were any dead *A. mellifera* workers found at the entrances.

19.8.4 Queen Rearing

Tan et al. (2006) studied queen rearing in mixed colonies to assess the effects of food on the development of offspring. *A. cerana* larvae were grafted for queen rearing into two of these mixed-species colonies. Similarly, *A. cerana* larvae and *A. mellifera* larvae were also grafted conspecifically as controls. The success rate of *A. cerana* queen rearing in the test colonies was 64.5%, surpassing all previous attempts at interspecific queen rearing, in which single-species host colonies were used (Oschmann 1965; Dhaliwal and Atwal 1970; Oku and Ono 1990; Potichot et al. 1993). After emergence, all virgin queens obtained from the three groups were measured morphometrically. The *A. cerana* queens from the mixed-species colonies differed significantly in size and pigmentation from the *A. cerana* control queens and closely approximated the *A. mellifera* queens. It is inferred that these changes in the *A. cerana* queens reared in the mixed-species colonies can be attributed to feeding by heterospecific nurse bees and/or chemical differences in royal jelly, with the data showing a strong impact of environment on the development of queens. The results further suggest that in honeybees, the cues for brood recognition can be learned by heterospecific workers after eclosion.

19.8.5 Retinue Behaviour

The retinue behaviour of worker bees of *A. cerana cerana* and *A. mellifera ligustica* in two types of mixed-species colonies was studied by Yang et al. (2010a). In mixed-species colonies that were headed by an *A. cerana* queen, almost equal numbers of *A. cerana* and *A. mellifera* workers attended the *A. cerana* queen; while in mixed colonies headed by an *A. mellifera* queen, significantly fewer *A. cerana* workers were attracted than *A. mellifera* workers. The pheromones 9-keto-2-(*E*)-decenoic acid (9-ODA), 9-hydroxy-2-(*E*)-decenoic acid (9-HDA) and 10-hydroxy-2-(*E*)-decenoic acid (10-HDA) of the queens were significantly different (Table 19.4) and the workers did not show avoidance behaviour to either heterospecific queen. Both species of workers were attracted by the queens and engaged in retinue behaviour (Figs. 19.1 and 19.2), suggesting that the retinue response was not related to a specific queen pheromone or colony environment. This non-specific queen retinue behaviour in the mixed colonies indicates that the queen pheromones can be transmitted amongst the workers from the two species without any obstacles. We conclude that retinue behaviour itself,

Table 19.4 Mean ± SD weight (μg) of mandibular gland pheromone components of *A. cerana* and *A. mellifera* queens

Species	HOB[a]	9-ODA[b]	9-HDA[c]	10-HDAA[d]	10-HDA[e]
A. mellifera	38.71 ± 7.05	244.13 ± 30.27	147.47 ± 21.54	13.71 ± 9.50	25.91 ± 10.29
A. cerana	49.98 ± 17.49	232.07 ± 27.55	31.92 ± 10.80	6.17 ± 3.50	4.26 ± 0.99

[a]Methyl hydroxybenzoate
[b]9-Keto-2-(*E*)-decenoic acid
[c]9-Hydroxy-2-(*E*)-decenoic acid
[d]10-Hydroxydecanoic acid
[e]10-Hydroxy-2-(*E*)-decenoic acid

Fig. 19.1 *A. cerana* queen attended by *A. mellifera* and *A. cerana* worker bees in an *A. cerana* queen-led mixed colony

Fig. 19.2 *A. mellifera* queen attended by *A. cerana* and *A. mellifera* worker bees in an *A. mellifera* queen-led mixed colony

as well as the pheromones of the queens that induce this behaviour, are primitive, conserved traits that preceded speciation in apine bees.

19.8.6 Ovary Activation

The workers in mixed-species colonies show different degrees of ovarian activation. *A. cerana* workers showed significantly greater ovarian activation in queenright mixed-species colonies than in conspecific queenright colonies. There was significantly greater ovary activation in *A. cerana* workers in mixed-species colonies that were headed by *A. mellifera* queens than *A. mellifera* workers in mixed-species colonies that were headed by *A. cerana* queens. *A. mellifera* workers in conspecific queenless colonies showed significantly greater ovarian activation than those in mixed-species queenless colonies (Fig. 19.3).

Quantification of the chemical components of mandibular gland pheromones of queens of the two species showed that they are similar. The combined results show that, although queen signals have been preserved in both species, the threshold of queen pheromone necessary to suppress ovary activation in *A. cerana* is higher than that for *A. mellifera* (Tan et al. 2009).

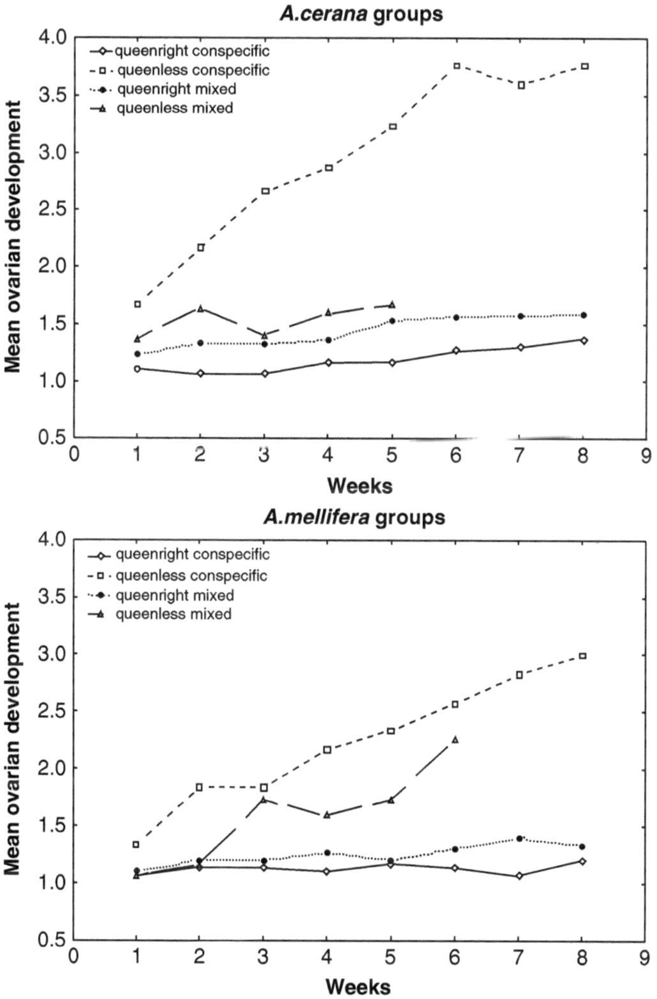

Fig. 19.3 Mean ovarian activation scores for ten workers in conspecific and mixed-species colonies of the honeybees, *A. mellifera* and *A. cerana*, over 8 weeks

19.8.7 Interspecific Communication

Among the most interesting of the interactions between *A. cerana* and *A. mellifera* workers in the same colony is interspecific communication. It is well known that honeybees have a dance language by which information about food resources can be transferred from successful foragers to nestmates (von Frisch 1967; Dyer 2002). The question arises: can the dance followers of one species understand the dances performed by the foragers of the other species? Although the structure of the dance language is very similar among species of honeybees (Lindauer 1956), studies have shown that the dance language not only differs among species in the genus of *Apis*, but different races of the same species may also have dialects (Steche 1957; Sarma et al. 2004).

Lindauer (1956) observed the Asian species *A. cerana*, *A. florea*, and *A. dorsata* and reported that there were differences in the distance at which dancers changed from round dances to waggle dances. The transition distance was much shorter for the Asian species: *A. florea* started waggling when the feeder was only 5 m from the hive. Lindauer (1956) and Boch (1957) also reported interspecific/interracial differences in the dance tempo (number of dance circuits per 15 s) at a given distance. For the same distance, different races or species would execute a different number of circuits per unit time.

Thus, the concept of dialects in the honeybee dance language was established, which basically pointed to two differences in the dances by different species and races: firstly in the flight distance at which the dancers start performing waggle dances instead of round dances, and secondly in the circuit duration of the waggle dance performed by dancers for a given flight distance. We, therefore, understand that, although the structure of the dance language is very similar among species of honeybees, communication of the distance component of the message varies both intraspecifically and interspecifically.

It is, however, not known whether different honeybee species would attend interspecific waggle dances and, if so, whether they can decipher such dances. Indeed, in the days when *A. cerana* had not yet been accepted as a species, Lindauer (1956) thought that it was indeed a species and suggested that the waggle dances of *A. cerana* and *A. mellifera* should be heterospecifically intelligible. So far, two reports have tried to answer this question, and both have found that *A. cerana* foragers can decode the dances of *A. mellifera* to successfully locate an indicated food source, by using mixed-species colonies of *A. cerana* and *A. mellifera* (Su et al. 2008; Tan et al. 2008). Tan et al. (2008) found that *A. mellifera* foragers can also be recruited to the experimental feeder by *A. cerana* dancers (Table 19.5).

19.8.8 Comb Building Cooperation

Cooperation in comb building in mixed colonies has also been investigated (Yang et al. 2010b; Figs. 19.4 and 19.5) using two types of cell size (*A. cerana* and

Table 19.5 Conspecific and heterospecific recruitment of new foragers of *A. cerana* and *A. mellifera*

	Feeder distance (m)	No. of new recruited bees to the feeder in 120 min			
		A. mellifera dancer		*A. cerana* dancer	
		Mean ± SD	Precision %	Mean ± SD	Precision %
A. mellifera recruited ($n = 3$ colonies)	110	2.0 ± 1.0	22.2	3.7 ± 3.5	33.3
	130	3.3 ± 0.6	37.0	4.7 ± 0.6	42.4
	170	2.3 ± 2.3	25.9	1.7 ± 0.6	15.2
	210	1.3 ± 2.3	14.8	1.0 ± 1.7	9.1
A. cerana recruited ($n = 3$ colonies)	110	25.0 ± 8.7	31.9	9.0 ± 2.6	20.6
	130	34.3 ± 22.7	43.8	16.0 ± 7.6	36.6
	170	10.0 ± 7.6	12.8	16.0 ± 13.4	36.6
	210	9.0 ± 9.6	11.5	2.7 ± 2.3	6.1

Fig. 19.4 Comb building by a mixed chain of *A. cerana* and *A. mellifera* workers

A. mellifera) foundation, made from the waxes of these two species, inserted into mixed colonies. The mixed colonies did not discriminate between the wax types, but the *A. cerana* cell size foundation was modified during comb building by both species of workers. In pure *A. cerana* colonies, workers did not accept any foundation, but were apparently stimulated by *A. mellifera* workers to secrete wax and build on the foundation in mixed colonies. The task of comb building was actually

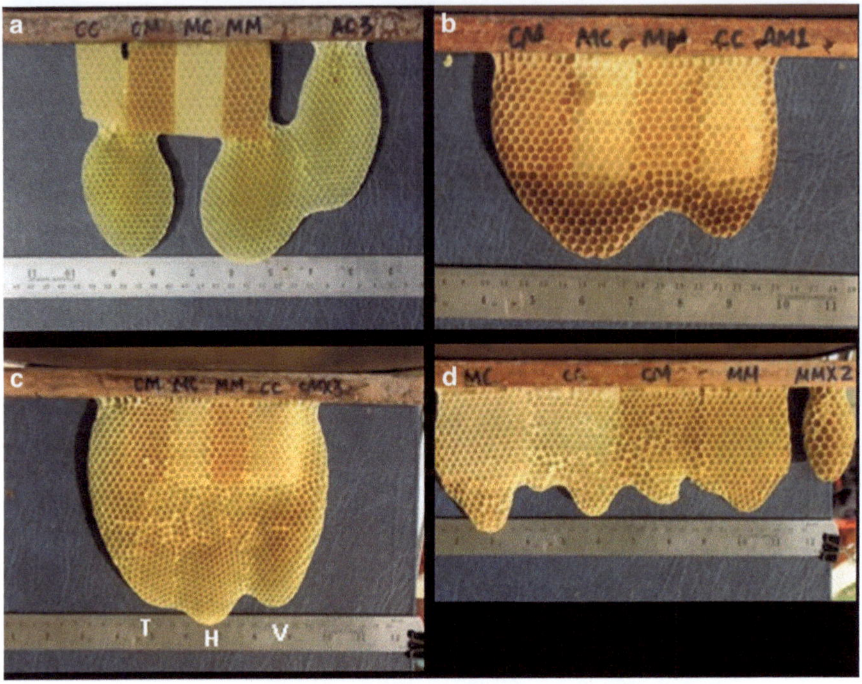

Fig. 19.5 Combs built in the four types of colonies. (**a**) pure *A. cerana,* (**b**) pure *A. mellifera,* (**c**) *A. cerana* queen-headed and (**d**) *A. mellifera* queen-headed colonies. Abbreviations on the top bars are CC: *A. cerana* cell-size foundation made from *A. cerana* wax; CM: *A. cerana* cell-size foundation made from *A. mellifera* wax; MM: *A. mellifera* cell-size foundation made from *A. mellifera* wax; MC: *A. mellifera* cell-size foundation made from *A. cerana* wax; cell direction patterns of newly built combs, H: horizontal, T: tilted, V: vertical

performed by small groups of workers of both species. In this way, the two species cooperated in comb building and could construct nearly normal combs, even though they contained many irregular cells.

19.8.9 Thermoregulation

A. cerana and *A. mellifera* normally display different strategies in cooling their nests, raising the question whether they would coordinate their efforts to achieve stable thermoregulation in mixed-species colonies. The results of experiments to assess this showed that the normal temperatures in the brood area in mixed colonies are more similar to those of pure *A. cerana* colonies than pure *A. mellifera* colonies. Under heat stress, *A. cerana* workers are more sensitive and initiate fanning earlier than *A. mellifera* workers. In mixed colonies, the former become the main force for

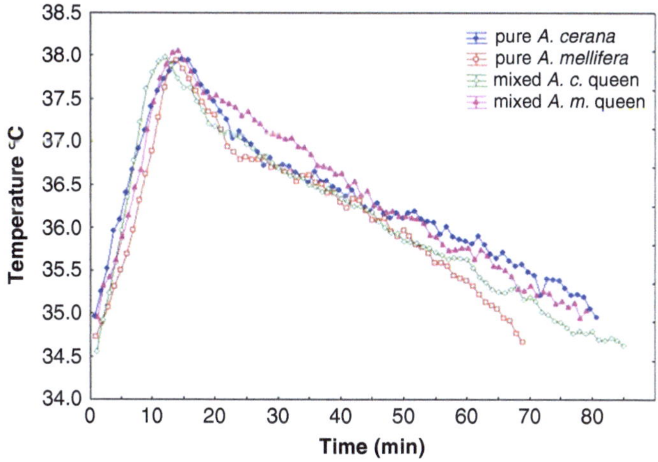

Fig. 19.6 Thermoregulation efficiency of the colonies

thermoregulation. When worker bees of both species were fanning together at the entrance, they retained their own species-specific postures, but due to a significantly smaller number of *A. mellifera* workers engaged in fanning, the cooling efficiency of mixed colonies was closest to that of pure *A. cerana* colonies (Fig. 19.6; Yang et al. 2010c).

19.8.10 Defence Behaviour

When vespine wasps hawk honeybees at their nest entrances, alerted and poised guard bees of *A. cerana* and *A. mellifera* in the mixed colonies have average thoracic temperatures slightly above 24°C. *A. cerana* workers assume their species-specific wing shimmering behaviour and raise their body temperatures up to about 29°C, while *A. mellifera* guard bees neither show significant body temperature increases nor wing shimmering. However, when faced with persistent hawking wasps, guard bees of both species raise their thoracic temperatures and form a ball around the wasp (Fig. 19.7). The core temperature of the mixed-species balls was about 45°C, which is not significantly different from the heat ball made up by only pure species. *A. cerana* bees engulf the wasp tightly in the inner core of the ball while *A. mellifera* can be seen more likely roaming at the outer edge. This result shows that the defence behaviours of the two species are based on their own species-specific adaptations (Tan et al. 2011).

In conclusion, mixed-species colonies offer a unique probe to study interspecific relations between species of honeybees. Behaviours in the mixed colonies confirm that *A. cerana* and *A. mellifera* are indeed very closely related species and that several behavioural patterns were conserved after speciation. Furthermore, it

Fig. 19.7 Heat-balling a wasp by the mixed-species colonies

provides us more information about these two societies and it may prove useful in finding solutions to the problems that have followed the introduction of western bees to Asia.

References

Arias MC, Sheppard WS (2005) Phylogenetic relationships of honey bees (Hymenoptera: Apinae; Apini) inferred from nuclear and mitochondrial DNA sequence data. Mol Phylogenet Evol 37:25–35

Atwal AS, Dhaliwal GS (1969) Robbing between *Apis indica* F. and *Apis mellifera* L. Am Bee J 109:462–463

Atwal AS, Sharma OP (1968) The introduction of *Apis mellifera* queens into *Apis indica* colonies and the associated behavior of the two species. Indian Bee J 30:41–56

Atwal AS, Sharma OP (1971) The dominance of *Apis mellifera* over *Apis indica*. Am Bee J 111:343

Baer B (2005) Sexual selection in *Apis* bees. Apidologie 36:187–200

Boch R (1957) Rassennmässige Unterschiede bei den Tänzen der Honigbiene (*Apis mellifica* L.). Z Vgl Physiol 40:289–320

Breed MD, Deng XB, Buchwald R (2007) Comparative nestmate recognition in Asian honeybees, *Apis florea*, *Apis andreniformis*, *Apis dorsata*, and *Apis cerana*. Apidologie 38:411–418

Butler CG, Calam DH, Callow RK (1967) Attraction of *Apis mellifera* drones by the odours of the queens of two other species of honeybees. Nature 213:423–424

Chapman NC, Nanork P, Gloag RS, Wattanachaiyingcharoen W, Beekman M, Oldroyd BP (2009) Queenless colonies of the Asian red dwarf honeybee (*Apis florea*) are infiltrated by workers form other queenless colonies. Behav Ecol 20:817–820

Crane E (1990) Bees and beekeeping: science, practice and world resources. Heinemann, Oxford

de Guzman LI, Rinderer TE, Whiteside RR (1996) A scientific note on the occurrence of *Varroa* mites on adult worker bees of *Apis nuluenis* in Borneo. Apidologie 27:429–430

Delfinado-Baker M, Underwood BA, Baker EW (1985) The occurrence of *Tropilaelaps* mites in brood nests of *Apis dorsata* and *Apis laboriosa* in Nepal, with descriptions of the nymphal stages. Am Bee J 125:703–706

Deodikar GB, Ghatge AL, Phadke RP, Mahindre DB, Kshirsagar KK, Muvel S, Thakar SS (1977) Nesting behaviour of Indian honeybees. III. Nesting behaviour of *Apis dorsata* Fabricius. Indian Bee J 39:1–12

Dhaliwal GS, Atwal AS (1970) Interspecific relations between *Apis cerana indica* and *Apis mellifera*. J Apic Res 9:53–59

Dutton RW, Free JB (1979) The present status of beekeeping in Oman. Bee World 4:176–185

Dyer FC (2002) The biology of the dance language. Annu Rev Entomol 47:917–949

Dyer FC, Seeley TD (1987) Interspecific comparisons of endothermy in honeybees (*Apis*): deviations from the expected size-related patterns. J Exp Biol 127:1–26

Dyer FC, Seeley TD (1991) Nesting behavior and the evolution of worker tempo in four honey bee species. Ecology 72:156–170

Hadisoesilo S, Raffiudin R, Susanti W, Atmowidi T, Hepburn C, Radloff SE, Fuchs S, Hepburn HR (2008) Morphometric analysis and biogeography of *Apis koschevnikovi* Enderlein (1906). Apidologie 39:495–503

Hepburn HR, Radloff SE (1998) Honeybees of Africa. Springer, Berlin

Hepburn HR, Radloff SE (2011) Biogeography of the dwarf honeybees, *Apis andreniformis* and *Apis florea*. Apidologie (in press)

Hepburn HR, Radloff SE, Steele GR, Brown R (1998a) Dimensional aspects of flight efficiency in honeybee populations of Africa. J Apic Res 37:147–154

Hepburn HR, Youthed C, Illgner P, Radloff SE, Brown RE (1998b) Production of aerodynamic power in mountain honeybees (*Apis mellifera*). Naturwissenschaften 85:389–390

Hepburn HR, Radloff SE, Fuchs S (1999) Flight machinery dimensions of honeybees, *Apis mellifera*. J Comp Physiol B 169:107–112

Hepburn HR, Radloff SE, Otis GW, Fuchs S, Verma LR, Tan K, Chaiyawong T, Tahmasebi G, Wongsiri S (2005) *Apis florea*: morphometrics, classification and biogeography. Apidologie 36:359–376

Ji R, Xie BY, Yang GH, Li DM (2003) From introduced species to invasive species: a case study on the Italian bee *Apis mellifera* L. Chin J Ecol 22:70–73

Joshi SR, Ahmad F, Gurung MB (2004) Status of *Apis laboriosa* populations in Kaski district, western Nepal. J Apic Res 43:176–180

Kalmus H (1941) The defence of a source of food by honeybees. Nature 148:228

Koeniger N (1976) Interspecific competition between *Apis florea* and *Apis mellifera* in the tropics. Bee World 57:110–112

Koeniger N (1982) Interactions among the four species of *Apis*. In: Breed MD, Michener CD, Evans HE (eds) The biology of social insects. Westview, Boulder, pp 59–64

Koeniger N (1996) Interspecific instrumental insemination of *Apis cerana* Fabricius, 1793, *Apis koschevnikovi* Buttel-Reepen, 1906 and *Apis dorsata* Fabricius, 1793. Apidologie 27:302–304

Koeniger N, Koeniger G (1991) An evolutionary approach to mating behaviour and drone copulatory organs in *Apis*. Apidologie 22:581–590

Koeniger N, Koeniger G (2000) Reproductive isolation among species of the genus *Apis*. Apidologie 31:313–339

Koeniger N, Koeniger G (2001) Behavioural mating barriers among sympatric species of the genus, *Apis*. In: Proceedings of the 7th international conference on tropical bees, Chiang Mai, Thailand, pp 80–85

Koeniger N, Vorwohl G (1979) Competition for food among four sympatric species of Apini in Sri Lanka: *Apis dorsata, Apis cerana, Apis florea* and *Trigona iridipenis*. J Apic Res 18:95–109

Koeniger N, Wijayagunesekera HNP (1976) Time of drone flight in the three Asian honeybee species (*Apis cerana, Apis florea, Apis dorsata*). J Apic Res 15:67–71

Koeniger N, Koeniger G, Delfinado-Baker M (1983) Observations on mites of the Asian honeybee species (*Apis cerana, Apis dorsata, Apis florea*). Apidologie 14:197–204

Koeniger N, Koeniger G, Tingek S (1994a) Crossfostered drones of *Apis cerana* (Fabricius, 1793) and *Apis koschevnikovi* (Buttel-Reepen, 1906) fly at their species specific mating times. Insectes Soc 41:73–78

Koeniger N, Koeniger G, Tingek S, Kalitu A, Mardan M (1994b) Drones of *Apis dorsata* (Fabricius 1793) congregate under the canopy of tall emergent trees in Borneo. Apidologie 25:249–264

Koeniger G, Koeniger N, Tingek S (1996a) A scientific note on weight and sperm numbers of drones of *Apis nuluensis* Tingek, Koeniger and Koeniger, 1996. Apidologie 27:427–428

Koeniger N, Koeniger G, Tingek S (1996b) Interspecific instrumental insemination of *Apis cerana* Fabricius, 1793, *Apis koschevnikovi* v. Buttel-Reepen, 1906, and *Apis dorsata* Fabricius 1793. Apidologie 27:302–304

Koeniger N, Koeniger G, Tingek S, Kelitu A (1996c) Interspecific rearing and acceptance of queens between *Apis cerana* Fabricius, 1793 and *Apis koschevnikovi* Buttel-Reepen, 1906. Apidologie 27:371–380

Koeniger N, Koeniger G, Gries M, Tingek S, Kelitu A (1996d) Reproductive isolation of *Apis nuluensis* Tingek, Koeniger and Koeniger, 1996 by species-specific mating time. Apidologie 27:353–359

Kuang BY, Kuang HO (2002) The biology of the honeybee. Kunming Science and Technology, Kunming [in Chinese]

Laloi D, Gallois M, Roger B, Pham-Delegue MH (2001) Changes with age in olfactory conditioning performance of worker honey bees (*Apis mellifera*). Apidologie 32:231–242

Lindauer M (1956) Über die Verständigung bei indischen Bienen. Z Vgl Physiol 38:521–557

Maa T (1953) An enquiry into the systematics of the tribus *Apidini* or honey bees (Hymenoptera). Treubia 21:525–640

Moritz RFA, Härtel S, Neumann P (2005) Global invasions of the western honeybee (*Apis mellifera*) and the consequences for biodiversity. Ecoscience 12:289–301

Nanork P, Chapman NC, Wongsiri S, Lim J, Gloag RS, Oldroyd BP (2006a) Social parasitism by workers in queenless and queenright *Apis cerana* colonies. Mol Ecol 16:1107–1114

Nanork P, Wongsiri S, Lim J, Gloag RS, Oldroyd BP (2006b) The reproductive dilemmas of queenless red dwarf honeybee (*Apis florea*) workers. Behav Ecol Sociobiol 61:91–97

Neumann P, Pirk CWW, Hepburn HR, Radloff SE (2001a) A scientific note on the natural merger of two honeybee colonies (*Apis mellifera capensis*). Apidologie 32:113–114

Neumann P, Radloff SE, Moritz RFA, Hepburn HR, Reece SL (2001b) A social parasitism by honeybee workers (*Apis mellifera capensis* Escholtz): host finding and resistance of hybrid host colonies. Behav Ecol 12:419–428

Oku N, Ono M (1990) Preliminary attempts to rear larvae of the Japanese honeybee, *Apis cerana japonica*, in an *Apis mellifera* colony and in the laboratory using *A. mellifera* royal jelly. Honeybee Sci 3:121–124 [in Japanese]

Oldroyd BP, Wongsiri S (2006) Asian honey bees. Harvard University Press, Cambridge

Oldroyd BP, Rinderer TE, Wongsiri S (1992) Pollen resource partitioning by *Apis dorsata, A. cerana, A. andreniformis* and *A. florea* in Thailand. J Apic Res 31:3–7

Oschmann H (1965) Can *Apis mellifica* bees rear brood of *A. cerana*? Die Biene Giessen 101:150–151 [in German]

Otis GW (1996) Distributions of recently recognized species of honey bees (Hymenoptera: Apidae; *Apis*) in Asia. J Kans Entomol Soc 69:311–333

Partap U (1998) Foraging competition between *Apis cerana* and *Apis mellifera* and its impact on crop pollination. In: Proceedings of the 4th Asian apiculture association international conference, Kathmandu, pp 206–208
Pham-Delegue MH, Trouiller J, Caillaud CM, Roger B, Masson C (1993) Effect of queen pheromone on worker bees of different ages: behavioural and electrophysiological responses. Apidologie 24:267–281
Phiancharoen M, Wongsiri S, Koeniger N, Koeniger G (2004) Instrumental insemination of *Apis mellifera* queens with hetero- and conspecific spermatozoa results in different sperm survival. Apidologie 35:503–511
Potichot S, Wongsiri S, Dietz A (1993) Attempts in queen rearing of *Apis cerana* larvae in *Apis mellifera* colonies and *Apis mellifera* larvae in *Apis cerana* colonies. In: Connor LJ, Rinderer TE, Sylvester HA, Wongsiri S (eds) Asian apiculture. Wicwas, Cheshire, pp 128–133
Punchihewa RWK, Koeniger N, Koeniger G (1990) Congregation of *Apis cerana indica* Fabricius 1798 drones in the canopy of trees in Sri Lanka. Apidologie 21:201–208
Radloff SE, Hepburn HR, Koeniger G (2003) Comparison of flight design of Asian honeybee drones. Apidologie 34:353–358
Radloff SE, Hepburn C, Hepburn HR, Fuchs S, Hadisoesilo S, Tan K, Engel MS, Kuznetsov V (2010) Population structure and classification of *Apis cerana*. Apidologie 41:589–601
Ribbands CR (1953) The behavior and social life of honeybees. Bee Research Association, London
Ribbands CR (1954) Communication between honeybees. I. The response of crop-attached bees to the scent of their crop. Proc R Entomol Soc Lond A 29:141–144
Ribbands CR (1955) The scent perception of the honeybee. Proc R Entomol Soc Lond B 143:367–379
Rinderer TE, Koeniger N, Tingek S, Mardan M, Koeniger G (1989) A morphological comparison of the cavity dwelling honeybees of Borneo *Apis koschevnikovi* (Buttel-Reepen, 1906) and *Apis cerana* (Fabricius, 1793). Apidologie 20:405–411
Rinderer TE, Oldroyd BP, Wongsiri S, Sylvester HA, de Guzman LI, Potichot S, Sheppard WS, Buchman SL (1993) Time of drone flight in four honeybee species in southeastern Thailand. J Apic Res 32:27–33
Rinderer TE, Oldroyd BP, de Guzman LI, Wattanachaiyingcharoen W, Wongsiri S (2002) Spatial distribution of the dwarf honey bees in an agroecosystem in southeastern Thailand. Apidologie 33:539–543
Roubik DW, Sakagami SF, Kudo I (1985) A note on distribution and nesting of the Himalayan honey bee *Apis laboriosa* Smith (Hymenoptera: Apidae). J Kans Entomol Soc 58:746–774
Ruttner F (1969) The cause of hybridization barrier between *Apis mellifera* L. and *Apis cerana* Fabricius. In: Proceedings of the 12th international beekeeping congress, Munchen, p 56
Ruttner F (1988) Biogeography and taxonomy of honeybees. Springer, Berlin
Ruttner F, Maul V (1983) Experimental analysis of reproductive interspecies *Apis mellifera* L. and *Apis cerana* Fabricius. Apidologie 14:309–327
Sakagami SF (1959) Displacement of native *Apis cerana* by *Apis mellifera* in Japan. J Anim Ecol 28:51
Sakagami SF, Matsumura T, Ito K (1980) *Apis laboriosa* in Himalaya, the little known world's largest honeybee (Hymenoptera: Apidae). Insecta Matsumurana 19:47–77
Sarma MS, Esch H, Tautz J (2004) A comparison of the dance language in *Apis mellifera carnica* and *Apis florea* reveals striking similarities. J Comp Physiol A 190:49–53
Seeley TD, Seeley RH, Akratanakul P (1982) Colony defense strategies of the honeybees in Thailand. Ecol Monogr 52:43–63
Shearer DA, Boch R, Morse RA, Laigo FM (1970) Occurrence of 9-oxodec-trans-2-enoic acid in queens of *Apis dorsata*, *Apis cerana* and *Apis mellifera*. J Insect Physiol 16:1437–1441
Smith DR (ed) (1991) Diversity of the genus *Apis*. Westview Press, Boulder
Smith F (1858) Catalogue of the hymenopterous insects collected at Sarawak, Borneo; Mount Ophir, Malacca; and at Singapore, by A.R. Wallace. Proc Linn Soc Lond 2:42–130

Steche W (1957) Beitraege zur Analyse der Bienentaenze. Insectes Soc 4:305–318
Stout JC, Goulson D (2001) The use of conspecific and interspecific scent marks by foraging bumblebees and honeybees. Anim Behav 62:183–189
Su S, Cai F, Si A, Zhang S, Tautz J, Chen S (2008) East learns from west: Asiatic honeybees can understand dance language of European honeybees. PLoS ONE 3:e2365
Tan K, Hepburn HR, Shaoyu H, Radloff SE, Neumann P, Fang X (2006) Gigantism in honeybees: *Apis cerana* queens reared in mixed-species colonies. Naturwissenschaften 93:315–320
Tan K, Yang MX, Radloff SE, Hepburn HR, Zhang ZY, Luo LJ, Li H (2008) Dancing to different tunes: heterospecific deciphering of the honeybee waggle dance. Naturwissenschaften 95:1165–1168
Tan K, Yang MX, Radloff SE, Pirk CWW, Crewe RM, Phiancharoen M, Hepburn HR, Oldroyd BP (2009) Worker reproduction in mixed-species colonies of honey bees. Behav Ecol 20:1106–1110
Tan K, Yang MX, Wang ZW, Li H, Zhang ZY, Radloff SE, Hepburn HR (2011) Cooperative wasp-killing by mixed-species colonies of honeybees, *Apis cerana* and *Apis mellifera*. J Insect Behav (Submitted)
Tingek S, Koeniger G, Koeniger N (1996) Description of a new cavity-nesting species of *Apis* (*Apis nuluensis* n. sp.) from Sabah, Borneo, with notes on its occurrence and reproductive biology (Insecta: Hymenoptera: Apoidea; Apini). Senckenb Biol 76:115–119
Underwood BA (1990a) Seasonal nesting cycle and migration patterns of the Himalayan honey bee *Apis laboriosa*. Natl Geogr Res 6:276–290
Underwood BA (1990b) The behavior and energetics of high-altitude survival by the Himalayan honey bee, *Apis laboriosa*. Thesis, Cornell University, Ithaca
von Frisch K (1967) The dance language and orientation of bees. Harvard University Press, Cambridge
Wang QF, Li WS, Zhang QM, Wu SS (2003) The natural mating interference between *A. cerana* and *A. mellifera*. Entomol Know 40:164–167 [in Chinese]
Whitcomb RP (1984) The biology of *Apis* spp in Oman with special reference to *Apis florea* Fabricius. Thesis, University of Durham, Durham
Wongsiri S, Limbipichai K, Tangkanasing P, Mardan M, Rinderer T, Sylvester HA, Koeniger G, Otis GW (1990) Evidence of reproductive isolation confirms that *Apis andreniformis* (Smith, 1858) is a separate species from sympatric *Apis florea* (Fabricius, 1787). Apidologie 21:47–52
Wongsiri S, Lekprayoon C, Thapa R, Thirakupt K, Rinderer TE, Sylvester HA, Oldroyd BP, Booncham U (1996) Comparative biology of *Apis andreniformis* and *Apis florea* in Thailand. Bee World 77:23–35
Woyke J (1973) Instrumental insemination of *Apis cerana indica* queens. J Apic Res 12:151–158
Woyke J (1975) Natural and instrumental insemination of *Apis cerana indica* in India. J Apic Res 19:153–159
Wu YR, Kuang B (1987) Two species of small honeybee: a study of the genus *Micrapis*. Bee World 68:153–155
Yang GH (2001) Chinese honeybee. China Agriculture Science and Technology, Beijing [in Chinese]
Yang GH (2005) Harm of introducing the western honeybee *Apis mellifera* L. to the Chinese honeybee *Apis cerana* F. and its ecological impact. Acta Entomol Sin 48:401–406 [in Chinese]
Yang GH, Xu SY, Sun QH (1982) The distribution and variation in Chinese honeybee. Apic China 3:17–19, 4:19, 6:13–14 [in Chinese]
Yang MX, Tan K, Radloff SE, Pirk CWW, Hepburn HR (2010a) Hetero-specific queen retinue behavior of worker bees in mixed-species colonies of *Apis cerana* and *Apis mellifera*. Apidologie 41:54–61
Yang MX, Tan K, Radloff SE, Phiancharoen M, Hepburn HR (2010b) Comb construction in mixed-species colonies of honeybees *Apis cerana* and *Apis mellifera*. J Exp Biol 213:1659–1664

Yang MX, Wang Z-W, Li H, Zhang Z-Y, Tan K, Radloff SE, Hepburn HR (2010c) Thermoregulation in mixed-species colonies of honeybees (*Apis cerana* and *Apis mellifera*). J Insect Physiol 56:706–709
Yoshida T (1994) Interspecific instrumental insemination using mixed semen of Japanese and European honeybees (Hymenoptera: Apidae). Appl Entomol Zool 29:464–467
Yu LS, Han SM (2003) Effect of habitat and interspecific competition on *Apis cerana cerana* colony distribution. Chin J Appl Ecol 14:553–556 [in Chinese]

Chapter 20
Bibliography of the Asian Species of Honeybees

H.R. Hepburn and Colleen Hepburn

20.1 Commentary

This bibliography of the literature on species of honeybees native to Asia is an extension of the recently published bibliographies on *Apis andreniformis* (Hepburn and Hepburn 2009), *Apis cerana* (Hepburn and Hepburn 2006), *Apis dorsata* and *Apis laboriosa* (Hepburn and Hepburn 2007a), *Apis florea* (Hepburn and Hepburn 2005) and *Apis koschevnikovi* (Hepburn and Hepburn 2007b). It was compiled from nearly 3,550 references, written by about 3,000 authors and published in over 700 different periodicals, conference proceedings, theses, reports, books and patents covering the period 1787–2009. The bibliography was prepared by obtaining all references to Asian honeybees published in Zoological Record (1864–2003) and Apicultural Abstracts (1950–2004) as well as the species-specific bibliographies of other workers cited in the Hepburn bibliographies.

Subsequently, copies of most of the original publications were obtained, and the references cited therein were traced in continuous iterations into the past until no new references were found. For the period 2005–2009, particular use was made of the website Google Scholar™, which is supported by various other citation websites, the most reliable of which are the British Library (BL Direct), US National Library of Medicine and the National Institutes of Health (http://www.pumed.gov), Institut de l'Information Scientifique et Technique of the Centre National de la Recherche Scientifique (INIST-CNRS) and Ingenta, a division of Publishing Technology plc (http://www.ingentaconnect.com). Recent references in Chinese are available at the website Wanfang Data, an affiliate of the Chinese Ministry of Science and Technology, but unfortunately, the citations are incomplete.

Oddly enough, Google™ often gives more download choices for citations of the same article than does Google Scholar™. In some cases, incomplete citations (particularly in Chinese) are given because complete information could not always be obtained, and they are marked with an asterisk. Similarly, several citations were

H.R. Hepburn and C. Hepburn
Department of Zoology and Entomology, Rhodes University, Grahamstown 6140, South Africa
e-mail: r.hepburn@ru.ac.za; c.hepburn@ru.ac.za

H.R. Hepburn and S.E. Radloff (eds.), *Honeybees of Asia*,
DOI 10.1007/978-3-642-16422-4_20, © Springer-Verlag Berlin Heidelberg 2011

only available as online versions, and in these cases, only their "doi" (digital object identifier) tags are given. As to the completeness of reference capture by Google Scholar™, this is entirely dependent on the references that have been and are being cited by contemporary authors in the contemporary literature. References that were published in the period 1990–2009 included in this bibliography are available on Google Scholar™ with few exceptions. However, if one delves further into the past (1980–1990), about 30% of the references in the current bibliography are not yet to be found on Google Scholar™, and nearly 50% for the previous decade. These results do not reflect on the efficacy of Google Scholar™ to capture references but indicates what time span of the past is regarded as relevant to contemporary authors (assuming that they actually know the literature in detail). However, Google Scholar™ does not list the journals that they survey, so it is not possible to know how thorough their searches are. As a corollary to this, it is regrettable that the application of the Primacy of Discovery is being increasingly ignored in many contemporary publications.

Each reference was entered in a computer database, and keywords for subject matter and countries for all publications were noted. Finally, references obtained through the iterative searches to older sources that lack full bibliographic details are indicated with an asterisk at the end of the entry. Of an estimated 100,000 publications on *Apis*, only about 4% of this literature concerns the Asian honeybee species: for the dwarf species, *A. andreniformis*, this is 2.3% and for *A. florea* 16.5%; for the cavity-nesting species, *A. cerana*, 56.3%, *A. koschevnikovi* 1.8%, *A. nigrocincta* <1% and *A. nuluensis* <1%; for the giant honeybees, *A. dorsata*, 20% while *A. laboriosa* 1.8%. That over 90% of all publications on the Asian species of honeybees have appeared in the last 50 years attests to a growing interest in them.

Accessing this small but rich heritage is difficult because most publications have not been captured by any of the enormous websites and computer search engines. This situation is exacerbated by an apparently widespread belief in the superiority of contemporary research over that of the pre-computerized past. Hopefully, this bibliography will serve to (1) provide access to historically important and too often forgotten publications; (2) indicate the relative need for further investigations in specific areas; (3) provide an information base for honeybee biologists and developmental agencies and (4) encourage honeybee scientists conversant in Asian languages to translate at least the titles of the relevant literature into English as the scientific lingua franca of the day. Crane (1993) estimated that the total literature on Asian honeybee was between 1,500 and 2,000 publications. However, after obtaining translations of works in Asian languages (Chinese, Japanese, Thai, Korean, Vietnamese and Russian) over the last decade and after visits to and discussions with colleagues in Asia, the total literature is now close to 4,000 publications. This corpus provides a firm footing for more intensive studies of the biology of Asian honeybees in future.

Turning to species sister groups, the directions in research on the dwarf honeybee *A. florea* reflect a predominant interest in general bee-keeping matters (bee botany, foraging and honey), presumably because this bee has some commercial importance. The literature on basic biology mainly refers to behaviour, distribution, classification, physiology and reproduction, while the more recent disciplines like population biology, genetics and pheromones are rapidly increasing. The references also demonstrate that there are many unanswered questions concerning population biology, molecular biology and climatology. Given the widespread occurrence of absconding and migration in both dwarf honeybee species, it is perhaps surprising that these phenomena have not yet been investigated more intensively. The extremely small literature on *A. andreniformis* is entirely concentrated on basic biology. It is difficult to draw many hard conclusions about human demography and publications on the dwarf honeybees, but numerically, the literature is dominated by works emanating from China, India and Thailand. The language distribution for works cited in this bibliography is approximately as follows: English 81%, Chinese 11%, Japanese 5%, German 2% and all other languages less than 1%.

Among the cavity-nesting species, the literature on the very commercially important wax bee, *A. cerana*, is the largest by far for any Asian species. Collectively, the *cerana* literature is dominated by bee-keeping matters such as foraging, honey, disease, pests, bee plants and general bee-keeping. In earlier decades, emphasis was placed on description of races, behaviour, reproduction and basic biology. More recently, great strides have been made in studies of pheromones, biochemistry, DNA and population biology. The lion's share of such works emanate from China, Japan and India. *A. koschevnikovi* was first described just over a century ago (Enderlein 1906; von Buttel-Reepen 1906) and, except for two papers, remained unstudied until its rediscovery towards the end of the last century (Mathew and Mathew 1988; Tingek et al. 1988). These two reports stimulated much recent research, bringing references on *A. koschevnikovi* above a hundred or so. This literature is all directed towards basic biology. References on *A. nigrocincta* and *A. nuluensis* are directed to basic biology and together comprise fewer than a hundred publications.

The literature on the giant honeybees, *A. dorsata* and *A. laboriosa*, is similar in nature to the dwarf species in that the former is far better known than the latter. Directions of research on *A. dorsata* reflect an even distribution on bee-keeping material, such as foraging, bee botany, pollination and honey on the one hand, and basic biological studies such as classification, genetics, biogeography, reproduction and queens on the other. Research has primarily derived from India and, to a lesser extent, Indonesia, Vietnam, Sri Lanka and Pakistan. The much smaller literature on *A. laboriosa* derives mainly from Nepal.

Acknowledgements We wish to acknowledge the kind advice and assistance of Dr Pamela Munn, formerly of the International Bee Research Association (Cardiff, United Kingdom); Prof Niko Koeniger at the Institut für Bienenkunde (Oberursel, Germany); Dr M Sasaki, University of Tamagawa (Tokyo, Japan) and DQTam, Bee Research and Development Centre (Hanoi, Vietnam) for providing access to literature. We owe special debts to Prof Hu Fuliang, Zhejiang University (Hangzhou, China), and Dr MX Yang (Kunming, China) for translations from Chinese; Dr V Kuznetsov[†], Institute of Biology and Pedology, Russian Academy of Sciences (Vladivostock, Russia), from Russian; PH Thai, Bee Research and Development Centre (Hanoi, Vietnam), from Vietnamese; Prof Jun Nakamura, University of Tamagawa (Tokyo, Japan), from Japanese; and Dr Mananya Phiancharoen, King Mongkut's University of Technology Thonburi, (Bangkok,

Thailand), from Thai. The earlier bibliographies in Apidologie were published at the suggestion of, and greatly improved by, the exacting bibliographic skills of Christiane Courant, the then Managing Editor of Apidologie.

References (* Indicates the reference is not complete)

Abdullah N, Hamzah A, Ramli J, Mardan MB (1990) Identification of Nasonov pheromones and the effects of synthetic pheromones on the clustering activity of the Asiatic honeybee (*Apis cerana*). Pertanika 13:189–194, Keywords: *cerana*/glands/pheromones/Malaysia

Abdullah A, Pengembangan Y, Mandiri U (2001) Forest-dependent community development through *Apis cerana* beekeeping progam. In: Proceedings of the 37th international apiculture congress, Apimondia, Durban (CD Rom), Keywords: beekeeping/*cerana*/Indonesia

Abrol DP (1986a) Differential dilution of nectar by subtropical bees. Environ Ecol 4:188–189, Keywords: honey/India

Abrol DP (1986b) Ecophysiological adaptations between pollinating bees and their flowers. Environ Ecol 4:161–162, Keywords: *cerana*/*dorsata*/*florea*/pollination/India

Abrol DP (1987a) Activity time budgets and pollination potential of *Apis cerana indica* workers. Environ Ecol 5:707–709, Keywords: *cerana*/pollination

Abrol DP (1987b) Interrelationship and path coefficient analysis of environmental factors influencing pollination activity of *Apis cerana* F on *Prunus persica* L. Trop Ecol 28:147–154, Keywords: *cerana*/pollination

Abrol DP (1989) Studies on ecology and behaviour of insect pollinators frequenting strawberry blossoms and their impact on yield and fruit quality. Trop Ecol 30:96–100, Keywords: behaviour/*cerana*/ecology/pollination/India

Abrol DP (1990) Energetics of nectar production in some apple cultivars as a predictor of floral choice by honeybees. Trop Ecol 31:116–122, Keywords: *cerana*/foraging/nectar/physiology/India

Abrol DP (1991a) Foraging strategies of honeybees in pollinating apple flowers. J Anim Morphol Physiol 38:109–113, Keywords: *cerana*/foraging/pollination/India

Abrol DP (1991b) Path analysis of environmental factors influencing daily flight activity of *Apis dorsata* F. Environ Ecol 2:819–824, Keywords: *dorsata*/foraging/general biology/India

Abrol DP (1992a) Bioenergetics in beeflower interrelationship and analysis of foraging behaviour. Korean J Apic 7:39–66, Keywords: behaviour/foraging/physiology/India

Abrol DP (1992b) Diel pattern of *Apis cerana indica* in *Asparagus officinalis* L. Sci & Cult 58:25, Keywords: *cerana*/foraging/India

Abrol DP (1992c) Energetics of nectar production in some strawberry cultivars as a predictor of floral choice by honeybees. J Biosci 17:41–44, Keywords: *cerana*/*mellifera*/nectar/physiology/pollination/India

Abrol DP (1992d) Foraging in honeybees *Apis cerana indica* F and *A dorsata* F (Hymenoptera: Apidae) activity and weather conditions. J Indian Inst Sci 72:395–401, Keywords: *cerana*/*dorsata*/foraging/India

Abrol DP (1992e) Oxygen consumption of foraging bees. J Anim Morph Physiol 39:27–42, Keywords: foraging/India

Abrol DP (1992f) Possible immune mechanism in indigenous honeybee *Apis cerana indica* F against 'Thai sac brood virus'. J Anim Morphol Physiol 39:165–166, Keywords: brood/*cerana*/disease/India

Abrol DP (1993a) Honeybee-virus infection and immunity. Korean J Apic 8:116–118, Keywords: *cerana*/disease/India

Abrol DP (1993) Intervarietal differences in nectar production affecting pollinator attractiveness and fruit set in almond. In: Proceedings of international symposium on pollination in the tropics, Bangalore, pp 120–122, Keywords: *cerana*/*mellifera*/nectar/pollination/India

Abrol DP (1994a) Comparative ecology of European and Asian honeybee species in subtropical and temperate climates. 1. Beekeeping and honey resources in subtropical temperate transition climate of Jammu. Korean J Apic 9:11–20, Keywords: bee botany/beekeeping/*cerana*/*dorsata*/*florea*/honey/physical factors/India
Abrol DP (1994b) Ecology, behaviour and management of social wasp *Vespa velutina* Smith (Hymenoptera. Vespidae), attacking honeybee colonies. Korean J Apic 9:5–10, Keywords: behaviour/*cerana*/ecology/*mellifera*/natural enemies/India
Abrol DP (1995a) Haemocytes of honey bees, *Apis mellifera* L and *Apis cerana* F (Hymenoptera: Apidae) and their alteration by mite parasitosis. Indian Bee J 57:6–7, Keywords: *cerana*/honey/*mellifera*/pests/physiology
Abrol DP (1995b) Occurrence of *Varroa jacobsoni* (Acari: Mesostigmata) Oudemans associated with honey bees in Jammu, India. Sci & Cult 61:205, Keywords: *cerana*/honey/pests/India
Abrol DP (1996a) Effect of mite parasitosis on blood counts of honey bees *Apis mellifera* L and *Apis cerana indica* F (Hymenoptera: Apidae). Insect Environ 2:18–19, Keywords: *cerana*/honey/*mellifera*/pests
Abrol DP (1996b) Incidence of *Varroa jacobsoni* Oudemans in *Apis cerana* F colonies in Jammu, India. Indian Bee J 58:15–17, Keywords: *cerana*/natural enemies/pests/India
Abrol DP (1996c) Sunflower pollination: abundance and diversity of pollinating insects and their effect on seed yield. Indian Bee J 58:60–63, Keywords: *cerana*/*dorsata*/*florea*/genetics/pollination/India
Abrol DP (1996d) *Varroa jacobsoni* Oudemans as a carrier of microorganisms. Insect Environ 2:20–21, Keywords: *cerana*/disease/pests/India
Abrol DP (1997) Diseases of honeybees in India *vis-a-viz* world overview. In: Mishra RC (ed) Perspectives in Indian Apiculture. Agro Bot, Bikaner, pp 304–333, Keywords: *cerana*/disease/*mellifera*/pests/predators/India
Abrol DP (1998a) Environmental factors influencing flight activity in honey bees, *Apis cerana* Fab. and *Apis mellifera* L (Hymenoptera: Apidae). Indian Bee J 60:71–75, Keywords: *cerana*/flight/foraging/honey/*mellifera*/India
Abrol DP (1998b) Food acquisition efficiency as a determinant of honeybee foraging decisions. Indian Bee J 60:133–136, Keywords: bee botany/*dorsata*/foraging/honey/India
Abrol DP (2000) Beekeeping with *Apis cerana* in Jammu and Kashmir: present status and future prospects. Bee World 81:149–152, Keywords: beekeeping/*cerana*/India
Abrol DP (2001a) *Acacia modesta*. Bee World 82:90–91, Keywords: bee botany/*cerana*/*dorsata*/*florea*/*mellifera*/India/Pakistan
Abrol DP (2001b) Proc Indian Natl Sci Acad B54:161–163, Keywords: bee botany/behaviour/*cerana*/*dorsata*/ecology/*florea*/India
Abrol DP (2004a) Foraging in honeybees *Apis cerana indica* F and *A dorsata* F (Hymenoptera: Apidae) activity and weather conditions. J Indian Inst Sci, Biol Sci 72:39–40, Keywords: *cerana*/climate/*dorsata*/foraging/physical factors/India
Abrol DP (2004b) Introduction of *Apis mellifera* in Jammu and Kashmir: present status and future prospects. Bee World 85:37–40, Keywords: bee botany/beekeeping/*cerana*/disease/*mellifera*/India
Abrol DP (2006a) Defensive behaviour of *Apis cerana* F against predatory wasps. Korean J Apic 9:5–10, Keywords: *cerana*/defensiveness/predators
Abrol DP (2006b) Diversity of pollinating insects visiting litchi flowers (*Litchi chinensis* Sonn) and path analysis of environmental factors influencing foraging behaviour of four honeybee species. J Apic Res 45:180–187, Keywords: *cerana*/*florea*/*mellifera*/pollination/India
Abrol DP (2006c) Factors influencing flight activity of *Apis florea* F, an important pollinator of *Dacus carota* L. J Apic Res 45:2–6, Keywords: *florea*/foraging
Abrol DP (2006d) Plant–insect interactions. In: Gupta SK (ed) Biology and breeding of crucifers. CRC Press, Boca Raton, pp 129–144, Keywords: *cerana*/*dorsata*/*florea*/*mellifera*/pollination
Abrol DP (2006e) Two decades of development in apiculture in Jammu and Kashmir - a brief review. Korean J Apic 21:145–156, Keywords: beekeeping/*cerana*/*dorsata*/*florea*/*mellifera*

Abrol DP (2007) Differences in sugar metabolism between the honey bees *Apis mellifera* and *Apis cerana* impact floral attractiveness. J Apic Res 46:181–190, Keywords: *cerana*/*mellifera*/physiology

Abrol DP, Abrol BL (1995) Important plant species for apiculture in Jammu Region and their impact on beekeeping. J Anim Morphol Physiol 42:57–64, Keywords: bee botany/beekeeping/*cerana*/*dorsata*/*florea*/*mellifera*/India

Abrol DP, Bhat AA (1987) Influence of atmospheric conditions and floral rewards on diurnal activity pattern, flower visitation rates and pollinating effectiveness of *Apis cerana indica* Fabr. foragers. Res Dev Rpt 4:13–15, Keywords: *cerana*/foraging/pollination

Abrol DP, Bhat AA (1989) Influence of behaviour of *Apis cerana indica* Fabr. on cross pollination in apple. Indian Bee J 51:103–104, Keywords: behaviour/*cerana*/pollination/India

Abrol DP, Bhat AA (1990) Studies on 'Thai sac brood virus' affecting indigenous honeybee *Apis cerana indica* Fab. colonies prospects and future strategies. J Anim Morphol Physiol 37:101–108, Keywords: brood/*cerana*/disease/India

Abrol DP, Kakroo SK (1996) Studies on seasonal activity and control of wax moths (*Galleria mellonella* L and *Achroia grisella* F) attacking combs of four honeybee species. Mysore J Agric Sci 30:365–373, Keywords: beeswax/*cerana*/combs/*dorsata*/*florea*/*mellifera*/natural enemies/India

Abrol DP, Kakroo SK (1997) Observations on concurrent parasitism by mites on four honeybee species in India. Trop Agric 74:150–156, Keywords: *cerana*/*dorsata*/*florea*/*mellifera*/pests/India

Abrol DP, Kakroo SK (1998a) Pests, predators and pathogens of honey bee, *Apis cerana* in Jammu, India. Indian Bee J 60:147–150, Keywords: *cerana*/disease/honey/natural enemies/India

Abrol DP, Kakroo SK (1998b) Studies on seasonal activity and control of predatory wasps attacking honey bee colonies. Indian Bee J 60:15–19, Keywords: *cerana*/honey/*mellifera*/natural enemies/India

Abrol DP, Kapil RP (1991) Foraging strategies of honeybees and solitary bees as determined by nectar sugar components. Proc Indian Natl Sci Acad, Biol Sci 57:127–132, Keywords: foraging/honey/India

Abrol DP, Katwal DR (1996) Occurrence of *Varroa jacobsoni* Oudemans in Jammu, India on honeybee. Insect Environ 2:48, Keywords: *cerana*/pests/India

Abrol DP, Khan AA (2002) Characterization of ecotypes of Indian honey bee (*Apis cerana* F) in the Kashmir region of North West Himalaya. Korean J Apic 17:7–20, Keywords: *cerana*/classification/India

Abrol DP, Putatunda BN (1995) Discovery of the ectoparasitic mite *Tropilaelaps koenigerum* Delfinado-Baker and Baker (Acari: Laelapidae) on *Apis dorsata* F, *A mellifera* L and *A cerana* F in Jammu and Kashmir, India. Curr Sci 68:90, Keywords: *cerana*/*dorsata*/*mellifera*/natural enemies/pests/India

Abrol DP, Sihag RC (1997) Pollination energetics. In: Sihag RC (ed) Pollination biology: basic and applied principles. Rajendra Scientific, Hisar, pp 75–97, Keywords: *cerana*/*dorsata*/*florea*/*mellifera*/physiology/pollination/India

Abrol DP, Sihag RC, Kapil RP (1988) Physiological adaptations of subtropical bees in relation to energy intakes. Res Dev Rpt 5:68–70, Keywords: *dorsata*/physiology

Abrol DP, Kitroo A, Bhat AA, Sharma B, Khanna YP (1998) Interspecific variations in haemolymph cation contents in Apis mellifera L and Apis cerana Fab. of different age groups. Indian Bee J 60:157–158, Keywords: cerana/mellifera/physiology/India

Adams HA, Southey BR, Robinson GE, Rodriguez-Zas SL (2008) Meta-analysis of genome-wide expression patterns associated with behavioral maturation in honey bees. BMC Genomics 9:503. doi:101186/1471-2164-9-503, Keywords: cerana/genetics/mellifera

Adlakha RL (1971) Preliminary studies on the mating behaviour of *Apis mellifera* and *Apis cerana* queens and their age of starting oviposition. In: Proceedings of the 23rd international apiculture congress, Apimondia, Moscow, pp 420–425, Keywords: beekeeping/behaviour/*cerana*/*mellifera*/oviposition/physical factors/queens/reproduction/India/Russia

Adlakha RL (1976) Acarine disease of adult honey bees in India. Am Bee J 116(324):344, Keywords: *cerana*/disease/*dorsata*/*florea*/honey/*mellifera*/pests/India

Adlakha RL, Dhaliwal HS (1979) Insect pollination of seed cauliflower (*Brassica oleracea* var *botrytis*) with particular reference to the role of honey bees. Indian Bee J 41:13–16, Keywords: *cerana*/*dorsata*/honey/*mellifera*/natural enemies/pests/pollination

Adlakha RL, Sharma OP (1971) Interspecific introduction of queens (*Apis mellifera* queens into *A indica* nuclei). In: Proceedings of the 23rd international apiculture congress, Apimondia, Moscow, p 402, Keywords: *cerana*/*indica*/interspecific interactions/*mellifera*/queens/India

Adlakha RL, Sharma OP (1974) Intraspecific robbing. Am Bee J 114:248–249, Keywords: behaviour/*cerana*/interspecific interactions/*mellifera*/India

Adlakha RL, Sharma OP (1975) *Apis mellifera* vs *Apis indica*. Glean Bee Cult 103:160, Keywords: *cerana*/interspecific interactions/*mellifera*/India

Adlakha RL, Sharma OP (1976) *Stylops* (Strepsiptera) parasites of honey bees (*Apis cerana*) in India. Am Bee J 116:66, Keywords: *cerana*/honey/natural enemies/parasites/India

Adorada JL, Cervancia CR (1993) Development of a pollen trap for the Asian honey bee (*Apis cerana*). In: Connor LJ, Rinderer TE, Sylvester HA, Wongsiri S (eds) Asian apiculture. Wicwas, Cheshire, CT, pp 272–280, Keywords: beekeeping/*cerana*/honey/pollen/Philippines

Aemprapa S, Wongsiri S (2000) Thai sac brood virus situation in Thailand. In: Proceedings of the 4th asian apiculture association international conference, Kathmandu, pp 55–56, Keywords: brood/*cerana*/disease/*mellifera*/Thailand

Agashe SN, Rangaswamy BE (1997) Melittopalynological studies of honey samples from Bangardka, Dakshina Kannada District, Karnataka. Indian Bee J 59:8–11, Keywords: cerana/honey/pollen

Aggarwal K (1988) Incidence of Tropilaelaps clareae on three *Apis* species in Hisar (India). In: Needham GR, Page RE, Delfinado-Baker M, Bowman CE (eds) Africanized honey bees and bee mites. Ellis Horwood, Chichester, pp 396–403, Keywords: absconding/*cerana*/combs/disease/*dorsata*/*florea*/honey/migration/pests/swarming/India

Aggarwal K, Kapil RP (1988a) Observations on the effect of queen cell construction on *Euvarroa sinhai* infestation in drone brood of *Apis florea*. In: Needham GR, Page RE, Delfinado Baker M, Bowman CE (eds) Africanized honey bees and bee mites. Ellis Horwood, Chichester, pp 404–458, Keywords: absconding/beekeeping/brood/classification/combs/disease/drones/*florea*/general biology/honey/migration/pests/queens/swarming/India

Aggarwal K, Kapil RP (1988b) Seasonal population dynamics of *Tropilaelaps clareae* (Acari: Laelapidae) in *Apis dorsata* colonies. Prog Acarol 2:283–286, Keywords: disease/*dorsata*/pests/predators/India

Aggarwal K, Mathur RB (1990) *Euvarroa sinhai* and some aspects of its association with its host *Apis florea*. In: Proceedings of the 11th international congress, IUSSI, New Delhi, pp 457–458, Keywords: disease/*florea*/pests/India

Ahlawat DS, Kasana VK, Malik MS, Mishra RC (1997) Effects of chemical additives in increasing degree of repellency and effective duration of carbonyl compounds against *Apis florea* F under field conditions. Indian Bee J 59:226–229, Keywords: biochemistry/*florea*/organic chemistry/pollen/India

Ahmad R (1981) A guide to honeybee management in Pakistan. Pamphlet, Miscellaneous Publications of the Agricultural Research Council of Pakistan, p 38. Keywords: beekeeping/*cerana*/*dorsata*/*mellifera*/Pakistan

Ahmad R (1987a) Honeybee pollination of important entomophilous crops. Pakistan Agricultural Research Council (PARC), Islamabad, Keywords: *cerana*/pollination

Ahmad R (1987) Queen rearing of the rock bee *Apis dorsata* F. In: Proceedings of the 31st international apiculture congress, Apimondia, Warsaw, p 182, Keywords: *dorsata*/queens

Ahmad R (1988a) Competition between *Apis mellifera* and other *Apis* species in Pakistan. Honeybee Sci 9:150–151, [in Japanese]. Keywords: *cerana*/interspecific interactions/*mellifera*/Pakistan

Ahmad R (1988) Honey bee parasitic mites and their control in Pakistan. In: Proceedings of the 2nd Australian and international bee congress, Brisbane, pp 88–90, Keywords: *cerana*/honey/*mellifera*/pests/Pakistan

Ahmad R (1988) Low cost comb foundation embossing press for *Apis cerana* F and *Apis mellifera* L. In: Proceedings of the 2nd Australian and international bee congress, Brisbane, p 232, Keywords: beekeeping/beeswax/*cerana*/*mellifera*/Pakistan

Ahmad R (1988) Studies on management of the rock bee, *Apis dorsata* F. in Pakistan. In: Proceedings of the 2nd Australian and international beekeeping congress, Brisbane, pp 187–188, Keywords: beekeeping/*dorsata*/Pakistan

Ahmad R (1989a) A note on the migration of *Apis dorsata* in the Andaman and Nicobar Islands. Bee World 70:62–65, Keywords: absconding/*dorsata*/migration/swarming/India

Ahmad R (1989b) Methods to control migration by *Apis dorsata* colonies in Pakistan. Bee World 70:160–162, Keywords: absconding/*dorsata*/environmental effects/migration/physiology/queens/swarming/Pakistan

Ahmad R (1992) Present status of beekeeping in Pakistan. In: Verma LR (ed) Honeybees in mountain agriculture. Oxford & IBH, New Delhi, pp 211–219, Keywords: beekeeping/*cerana*/natural enemies/Pakistan

Ahmad R (1993) Beekeeping with the Indian hive bee *Apis cerana indica* F in the Andaman and Nicobar Islands. Indian Bee J 55:1–8, Keywords: beekeeping/*cerana*/India

Ahmad R (2000) Biological control of *Varroa jacobsoni* in Pakistan. In: Proceedings of the 4th Asian apiculture association international conference, Kathmandu, p 67, Keywords: *cerana*/*mellifera*/pests/Pakistan

Ahmad R, Abbas SR (1985) Some observations on *Apis dorsata* Fab. in Andaman and Nicobar Islands, India. Indian Bee J 47:46–47, Keywords: combs/*dorsata*/environmental effects/India

Ahmad M, Aslam M (2002) Pollinators visiting carrot (*Dacus carota* L) seed crop. J Res (Sci), Bahauddin Zakariya Univ, Multan. Pakistan 13:31–35, Keywords: *cerana*/*mellifera*/pollination/Pakistan

Ahmad R, Muzaffar N (1984) Modern beekeeping. Pakistan Agricultural Research Council, Islamabad, [Urdu version].* Keywords: beekeeping/*cerana*/*dorsata*/*mellifera*/India/Pakistan

Ahmad R, Muzaffar N (1987) Some biological observations on the wild colonies of the oriental bee *Apis cerana* F in Pakistan. In: Proceedings of the 31st international apiculture congress, Apimondia, Warsaw, pp 182–183, Keywords: amalgamation/*cerana*/environmental effects/migration/queens/Pakistan

Ahmad R, Waghchoure ES (1984) Studies of migration of honeybee colonies and its effect on honey yield. Pak J Zool 16:113–117, Keywords: beekeeping/*cerana*/*dorsata*/*florea*/honey/migration/Pakistan

Ahmad R, Camphor E, Ahmad M (1983a) Factors affecting honey yield of the oriental bee, *Apis cerana* in Pakistan. Pak J Agric Res 4:190–197, Keywords: *cerana*/honey/Pakistan

Ahmad R, Muzaffar N, Ali Q (1983b) Biological control of the wax moths *Achroia grisella* (F) and *Galleria mellonella* (L) (Lep, Galleridae) by augmentation of the parasite *Apanteles galleriae* Wilk (Hym, Braconidae) in Pakistan. Apiacta 18:15–20, Keywords: beeswax/*cerana*/natural enemies/Pakistan

Ahmad R, Camphor E, Ahmad M (1984a) Migration of the eastern bee (*Apis cerana* F.) colonies and its effect on honey yield. Pak J Agric Res 5:65–68, Keywords: beekeeping/*cerana*/honey/pests/Pakistan

Ahmad R, Jilani G, Muzaffar N (1984b) Domestication of *Apis dorsata* F. (Hymenoptera: Apidae). Apiacta 19:69–70, Keywords: beekeeping/*dorsata*/Pakistan

Ahmad I, Siddiqui MKJ, Ray PK (1987) Pesticide burden on some insects of economic importance in Lucknow. Environ Monit Assess 9:25–28, Keywords: *cerana*/insecticides/India

Ahmad F, Partap U, Joshi SR, Gurung MB (2002) Why the HKH region needs the Himalayan honeybee *Apis cerana*. ICIMOD, Kathmandu, Nepal, Keywords: beekeeping/*cerana*/Nepal

Ahmad F, Joshi SR, Gurung MB (2003) The Himalayan cliff bee *Apis laboriosa* Smith and the honey hunters of Kaski. International Centre for Integrated Mountain Development, Kathmandu, p 52, Keywords: behaviour/bee-keeping/combs/*laboriosa*/Nepal

Ahmadi AA (1988) Parasites of *Apis florea* in Iran. Bee World 69:159–161, Keywords: classification/disease/pests/Iran

Aichholz R, Lorbeer E (1999) Investigation of combwax of honeybees with high-temperature gas chromatography and high-temperature gas chromatography - chemical ionization mass spectrometry I. High-temperature gas chromatography. J Chromatogr A 855:601–615, Keywords: *andreniformis*/beeswax/*cerana*/classification/*dorsata*/*florea*/*laboriosa*/*mellifera*

Aichholz R, Lorbeer E, Pechhacker H, Hüttinger E (2000) The chemical compounds of beeswax from *Apis* species. In: Proceedings of the 4th Asian apiculture association international conference, Kathmandu, pp 152–154, Keywords: beeswax/*cerana*/*dorsata*/*florea*/*laboriosa*/*mellifera*/India/Nepal/Thailand

Akahira Y, Sakagami SF (1958) Contemporary beekeeping with the Japanese honeybee in Kyushu, southern Japan. Z Bienenforsch 4:87–96, [in German]. Keywords: beekeeping/*cerana*/Japan

Akahira Y, Sakagami SF (1959a) A biometrical study on the Japanese honey bee, observations upon some populations of Kyûshû (studies on the Japanese honey bee, *Apis cerana cerana* Fabr. IX). J Hokkaido Gakugei Univ 2:353–362, Keywords: *cerana*/classification/honey/population biology/Japan

Akahira Y, Sakagami SF (1959b) Notes on the difference in some external characteristics between Japanese and European honey-bees (studies on the Japanese honey-bee, *Apis cerana cerana* Fabr. III). Annot Zool Jpn 32:35–42, Keywords: *cerana*/classification/*mellifera*/Japan

Akahira Y, Sakagami SF (1959c) Observations on the variability of wing venation in the honeybees. J Fac Sci Hokkaido Univ VI 14:175–184, Keywords: *cerana*/classification/morphology/Japan

Akahira Y, Sakagami SF (1959d) On the present status of beekeeping with the Japanese honeybee on Kyûshû. Studies on the Japanese honeybee Apis cerana indica Fabr. Z Bienenforsch 4:87, [in German]. Keywords: beekeeping/*cerana*/Japan

Akimoto T (2000) Winter temperature in the exposed nest of Japanese honeybee, *Apis cerana japonica*. Honeybee Sci 21:31–34, [in Japanese]. Keywords: *cerana*/nests/physiology/Japan

Akratanakul P (1976) Honey bees in Thailand. Am Bee J 116(120–121):124–126, Keywords: *cerana*/*dorsata*/*florea*/honey/Thailand

Akratanakul P (1977) The natural history of the dwarf honey bee, *Apis florea* F in Thailand. Thesis, Cornell University, Keywords: absconding/bee botany/behaviour/brood/classification/combs/defense/disease/environmental effects/foraging/general biology/honey/migration/pests/queens/swarming/workers/Thailand

Akratanakul P (1983) On bees and beekeeping. P Aktatanakul, Bangkok, [in Thai]. Keywords: beekeeping/Thailand

Akratanakul P (1985) Beekeeping for rural development. FAO-UNDP, Rome, Keywords: beekeeping/*cerana*/*mellifera*/physiology

Akratanakul P (1987) Beekeeping in Asia. FAO Agric Serv Bull, Rome, 68/4:1–112, Keywords: beekeeping/cerana/dorsata/florea/mellifera

Akratanakul P (1987) Honeybee diseases and enemies in Asia: a practical guide. FAO Agric Serv Bull, Rome, 68/5:1–51, Keywords: cerana/disease/dorsat/florea/mellifera/natural enemies

Akratanakul P (2001) Apiculture development in Thailand. In: Proceedings of the 7th international conference on tropical bees, Chiang Mai, pp 395–398, Keywords: beekeeping/*cerana*/development/*dorsata*/*florea*/*mellifera*/Thailand

Akratanakul P, Burgett MD (1976) *Euvarroa sinhai* Delfinado and Baker (Acarina: Mesostigmata): a parasitic mite of *Apis florea*. J Apic Res 15:11–13, Keywords: disease/pests

Akre RD, Mayer D (2004) Bees and vespine wasps. Bee World 75:29–37, Keywords: beekeeping/*cerana*/*dorsata*/*florea*/*mellifera*/physiology/Thailand

Al Ghamdi AA (2006) Scanning electron microscopic studies on antennal sensilla organs of adult honey bee workers in genus *Apis* (Hymenoptera: Apidae). Bull Entomol Soc Egypt 83:1–11, Keywords: biogeography/*florea*/Arabia/Saudi Arabia

Alam MK (1997) Reproductive biology of *Thyrsostachys regia* (Munro) Bennet (Bambusoideae: Poaceae). Bangladesh For Sci 26:25–30, Keywords: bee botany/*dorsata*/general biology/ Bangladesh

Alam MZ, Quadir MA (1986) Role of honeybee in fruit and seed setting of bottlegourd, *Lagenaria siceraria* (Mol) Standl. Punjab Veg Grower 21:32–34, Keywords: *cerana*/pollination/Bangladesh

Alam MZ, Quadir MA, Ali MS (1987) Pollinating behaviour of honeybee. Apis indica F. and its influence on seed production of cauliflower. Bangladesh Hort 15:25–30, Keywords: *cerana*/ pollination

Alam MZ, Alam MB, Quadir MA (1991) Comparative evaluation of some pollinating methods on onion seed and subsequent bulb production. Bangladesh J Bot 20:125–130, Keywords: *cerana*/ pollination/Bangladesh

Albert S, Schmitz H (2002) Characterization of major royal jelly-like DNA sequence in *Apis dorsata*. J Apic Res 41:75–82, Keywords: biochemistry/*dorsata*/organic chemistry/physiology

Albertová V, Su S, Brockmann A, Gadau J, Albert S (2005) Organization and potential function of the *mrjp3* locus in four honeybee species. J Agric Food Chem 53:8075–8081, Keywords: *cerana*/*dorsata*/*florea*/*mellifera*

Alexander BA (1990) A phylogenetic analysis of honey bees (Hymenoptera: Apidae: *Apis*). In: Proceedings of the 11th international congress, IUSSI, Bangalore, pp 120–121, Keywords: *andreniformis*/*cerana*/dance language/*dorsata*/*florea*/honey/*koschevnikovi*/*laboriosa*/*mellifera*/ phylogeny/India

Alexander BA (1991) A cladistic analysis of the genus *Apis*. In: Smith DR (ed) Diversity in the genus *Apis*. Westview Press, Boulder, Colorado, pp 1–28. Keywords: *cerana*/*dorsata*/*florea*/ genetics/*mellifera*/phylogeny

Alexander BA (1991b) Phylogenetic analysis of the genus *Apis* (Hymenoptera: Apidae). Ann Entomol Soc Am 84:137–149, Keywords: *cerana*/*dorsata*/*florea*/*mellifera*/phylogeny

Al'ghamdi A, Al'mugharabi J (2000) *Apis florea* in Saudi Arabia. In: Proceedings of the 4th Asian apiculture association international conference, Kathmandu, pp 44–45, Keywords: classification/*florea*/Saudi Arabia

Ali M (1991) Chemical evaluation of the wax of giant honeybee (*Apis dorsata* Fabr.). Indian Bee J 53:91–95, Keywords: combs/*dorsata*/honey/morphology/India

Ali T, Ali SI (1989) Pollination biology of *Calotropis procera* subsp *hamiltonii* (Asclepiadaceae). Phyton 29:175–188, Keywords: bee botany/*florea*/general biology

Ali MS, Sattar A, Kumar M (2006) Foraging behavior of honey bees and associated insects on Jatropha (*Jatropha curcas*) flowers under the climatic condition of north Bihar. Environ Ecol 24:581–583, Keywords: *cerana*/*dorsata*/*florea*/foraging/India

Alizadeh A, Mossadegh MS (1994) Stonebrood and some other fungi associated with *Apis florea* in Iran. J Apic Res 33:213–218, Keywords: classification/disease/pests/Iran

Allen MF, Ball BV (1996) The incidence and world distribution of honey bee viruses. Bee World 77:141–162, Keywords: biogeography/*cerana*/disease/*dorsata*/*florea*/honey/*mellifera*

Aluri RJS (1990) Some pollen and nectar resource plants of *Apis florea* and *A cerana indica*. In: Proceedings of the 11th international congress, IUSSI, Bangalore, p 429, Keywords: *cerana*/ *florea*/nectar/pollen/India

Aluri RJS, Subba Reddi C, Rama Das K (1998) Temporal dioecism and pollination by wasps and bees in *Allophylus serratus* (Roxb) Radlk (Sapindaceae). Plant Species Biol 13:1–5, Keywords: bee botany/*cerana*/*dorsata*/*florea*/pollination

Amano K (2000) Non-stinging honeybees induced by gamma radiation. In: Proceedings of the 4th Asian apiculture association international conference, Kathmandu, pp 109–111, Keywords: *cerana*/*mellifera*/stings/Japan

Ambrose DP, Livingstone D (1987) Biology of *Acanthaspis siva* Distant, a polymorphic assassin bug (Insecta, Heteroptera, Reduviidae). Mitt Zool Mus Berl 63:321–330, Keywords: biology/*cerana*/natural enemies/India

Ambrose DP, Livingstone D (1990) Polymorphism in *Acanthaspis siva* Distant (Reduviidae: Heteroptera), a predator of the Indian honey bee. J Bombay Nat Hist Soc 87:218–222, Keywords: *cerana*/honey/natural enemies/India

Amornsak W, Sonong A, Phoosuwan K (2001) An inventory of antennal sensilla of the dwarf honey bee worker, *Apis florea* F (Hymenoptera: Apidae). In: Proceedings of the 7th international conference on tropical bees, Chiang Mai, pp 143–150, Keywords: honey/morphology/Thailand

Amparo J (1993) A comparative karyomorphology of *Apis cerana* F from colonies in Bicol and Quezon. Thesis, University of the Philippines, Keywords: *cerana*/genetics/Philippines

Anderson DL (1989) Mites and pathogens of the European honey bee *Apis mellifera* and the eastern hive bee *Apis cerana* in Papua New Guinea. DSIR, Entomology Division, Auckland. Keywords: *cerana*/honey/*mellifera*/natural enemies/pests/New Guinea/Papua New Guinea

Anderson DL (1990) The disease and pest situation in the Pacific Region; the threat of exotic diseases and pests from Papua New Guinea; diseases and disorders of queen bees. Austral Beekeep 92:227–231, Keywords: *cerana*/disease/*mellifera*/natural enemies/pests/queen/New Guinea/Papua New Guinea

Anderson DL (1991) *Apis cerana* and its associated mites and pathogens in Papua New Guinea. In: Proceedings of international symposium on recent research on bee pathology, Ghent, pp 196–201, Keywords: *cerana*/natural enemies/pests/Papua New Guinea

Anderson DL (1995) Viruses of *Apis cerana* and *Apis mellifera*. In: Kevan PG (ed) The Asiatic hive bee: apiculture, biology, and role in sustainable development in tropical and subtropical Asia. Enviroquest Ltd, Cambridge, Ontario, pp 161–170. Keywords: biology/*cerana*/disease/*mellifera*/physiology/tropics/India/Thailand

Anderson DL (1996) Bee parasites in New Guinea - a threat to Australia. In: Proceedings of the 3rd Asian apiculture association international conference, Hanoi, p 198, Keywords: *cerana*/parasites/Australia/New Guinea

Anderson DL (1996) Control of bee mites in Irian Jaya. In: Proceedings of the 3rd Asian apiculture association international conference, Hanoi, pp 183–186, Keywords: beekeeping/*cerana*/*mellifera*/pests

Anderson DL (2001) *Varroa*-bee relationships. What they tell us about controlling *Varroa* mites on the European honey bee. In: Proceedings of the 37th international apiculture congress, Apimondia, Durban (CD Rom), Keywords: *cerana*/honey/*mellifera*/pests

Anderson C (2002) Self-organization in relation to several similar concepts: are the boundaries to self-organization indistinct? Biol Bull 202:247–255, Keywords: *dorsata*/*florea*/general biology

Anderson DL, Morgan MJ (2007) Genetic and morphological variation of bee-parasitic *Tropilaelaps* mites (Acari: Laelapidae): new and re-defined species. Exp Appl Acarol 43:1–24, Keywords: *dorsata*/*laboriosa*/*mellifera*

Anderson DL, Trueman JWH (2000) *Varroa jacobsoni* (Acari: Varroidae) is more than one species. Exp Appl Acarol 24:165–189, Keywords: *cerana*/pests/Asia

Anderson DL, Trueman JWH (2001) *Varroa jacobsoni* is more than one species. In: Proceedings of the 37th international apiculture congress, Apimondia, Durban (CD Rom), Keywords: *cerana*/DNA/pests

Anderson DL, Halliday RB, Otis GW (1997) The occurrence of *Varroa underwoodi* (Acarina: Varroidae) in Papua New Guinea and Indonesia. Apidologie 28:143–147, Keywords: *cerana*/*mellifera*/*nigrocincta*/pests/Indonesia/New Guinea/Papua New Guinea

Anderson C, Theraulaz G, Deneubourg JL (2002) Self-assemblages in insect societies. Insectes Soc 49:99–110, Keywords: behaviour/defense/*dorsata*/*florea*

Anh M, Dang V (1984) Honeybee foulbrood diseases in Vietnam. Agricultural Publisher, Hanoi. Keywords: *cerana*/disease/Vietnam

Anita M (1999) Comparative residual contact toxicity of insecticides to the Indian honeybee *Apis cerana indica*. Asian Bee J 1:20–35, Keywords: behaviour/*cerana*/foraging/genetics/insecticides/India

Anjum F, Siddiqui MKJ, Sultana H, Oadri SSH (1988) Some metabolic changes induced by DDT and piperonyl butoxide in honey bee *Apis cerana indica*. Comp Physiol Ecol 13:85–88, Keywords: *cerana*/honey/insecticides/India

Ankinovich GB (1957) An experiment on acclimatizing Indian bees and their use for pollinating agricultural crops. Dokl Timiryazevsk Sel' skokhoziaistvennaia Akad 30:307–310, [in Russian]. Keywords: *cerana*/foraging/interspecific interactions/*mellifera*/physical factors/seasonal variation/China

Anonymous (1942) The *A. dorsata* colony. Indian Bee J 4:64, Keywords: dorsata/general biology/India

Anonymous (1959) Bee flora of Karnataka and Kerala. Indian Bee J 21:90–92, Keywords: bee botany/cerana/India

Anonymous (1976) Honeybees in Kashmir. Indian Bee J 38:39–41, Keywords: cerana/dorsata/honey/India

Anonymous (1979) National symposium on pollination ecology and applied palynology. Haryana Agriculture University Hisar, pp 1–50, Keywords: bee botany/*cerana*/*dorsata*/ecology

Anonymous (1982) Pollination trial in Chinese cabbage. Nepal Lumle Agricultural Centre, pp16–18, Keywords: bee botany/*cerana*/*dorsata*/Nepal

Anonymous (1989) Bibliography on Asiatic honeybees and Indian apiculture (publications up to 1989). Indian Bee J 51:150–153, Keywords: bibliography/cerana/dorsata/honey

Anonymous (1992) Abstracts of papers and bibliography on Asian honeybees 1979–1991. International conference Asian honey bees and bee mites, Chulalongkorn University, Bangkok, Keywords: bibliography/*cerana*/*dorsata*/environmental effects/honey/disease/pests/predators/Thailand

Anonymous (1996) *Apis cerana* standard hive design with Langstroth type. Report, Directorate Reforestation, Jakarta, Keywords: *cerana*/nests/Indonesia

Anonymous (1998) Honeybee diseases – on the partial amendment of domestic infectious diseases control law. Honeybee Sci 19:97, Keywords: beekeeping/cerana/disease/mellifera/Japan

Anonymous (1999) An encounter with Apis dorsata in Gujurat. Beekeep Dev 52:5, Keywords: dorsata/defense/India

Anonymous (2002) Observations on the effects of artificial insemination on Apis cerana cerana. Apic Sci Tech 1:8–10 [in Chinese]. Keywords: breeding/cerana/queens/China

Ansary AB, Hossain M, Bari MN (2001) Analysis of honey and incidence of diseases, pests and predators in the colony of honeybee, *Apis cerana* (F). Bangladesh J Entomol 11:21–30, Keywords: *cerana*/disease/honey/pests/predators/Bangladesh

Aoki H (1970) Effect of royal jelly on the breeding of a Daphnia (*Daphnia magna* Straus). Collect Breed 32:282, [in Japanese]. Keywords: breeding/*cerana*/royal jelly/Japan

Apicultural Abstracts (1950-2004) Vol 1–55: International Bee Research Association, Cardiff.

Apicultural Society of Korea, The (1999) Proceedings of international symposium on apicultural science, Apicultural Society of Korea, Seoul. Keywords: beekeeping/*cerana*/Korea

Arai J, Akiyama K, Sakai S, Doguchi M, Suzuki M, Ogawa S (1960) Some notes on Japanese honey. Jpn Bee J 13:101–107, [in Japanese]. Keywords: biology/*cerana*/honey/Japan

Areekul S, Rojanavongse V (1984) Wild flora of the Asiatic honeybee in Doi Ang Khang, Kasetsart. J Nat Sci 18:79–91, [in Thai]. Keywords: bee botany/*cerana*/Thailand

Areekul S, Rojanavongse V, Oulamoon A, Intorn C (1979) Research on apiculture and insect pollinators in the highlands of northern Thailand. Kasetsart University, pp 1-237. Keywords: *cerana*/*mellifera*/pollination/Thailand

Arias MC, Sheppard WS (2005a) Phylogenetic relationships of honeybees (Hymenoptera: Apinae: Apini) inferred from nuclear and mitochondrial DNA sequence data. Mol Phylogen Evol 37:25–35, Keywords: biochemistry/*cerana*/*dorsata*/honey/organic chemistry

Arias MC, Sheppard WS (2005b) Phylogenetic relationships of honeybees (Hymenoptera: Apinae: Apini) inferred from nuclear and mitochondrial DNA sequence data. Mol Phylogenet Evol 37:25–35, Keywords: *cerana/mellifera*

Arias MC, Tingek S, Kelitu A, Sheppard WS (1996) *Apis nuluensis* Tingek, Koeniger and Koeniger, 1996 and its genetic relationship with sympatric species inferred from DNA sequences. Apidologie 27:415–422, Keywords: biogeography/*cerana*/DNA/*koschevnikovi*/*mellifera*/*nuluensis*/phylogeny

Armbruster L (1958) Combs of Indian bees. Arch Bienk 35:23–31, [in German]. Keywords: combs/*florea*/India

Arun K, Srivastava S, Chandel YS (1999) Symptomatology and behaviour of Thai sac brood virus (TSBV) infected honey bee colonies. Insect Environ 5:103–104, Keywords: *cerana*/*mellifera*/virus/India

Arun K, Srivastava S, Kashyap NP (2001) Infestation phenology of *Tropilaelaps clareae* Delfinado and Baker and *Acarapis woodi* (Rennie) in *Apis mellifera* L. and *Apis cerana* Fab. J Entomol Res 25:41–46, Keywords: *cerana*/*mellifera*/mites/India

Arya PS (1982) Effect of mode of pollination on seed setting in chicory (*Cichorium intybus* L). Udyanika 5:33–36, Keywords: *cerana*/pollination/India

Arya DR, Sihag RC, Yadav PR (1994) Diversity, abundance and foraging activity of insect pollinators of sunflower (*Helianthus annuus* L) at Hisar (India). Indian Bee J 56:172–178, Keywords: bee botany/biogeography/*dorsata*/foraging/India

Asada S (1993) Nesting habit of *Apis cerana japonica* in Shonan area of Kanagawa Prefecture. Honeybee Sci 14:165–169, [in Japanese]. Keywords: *cerana*/nests/Japan

Aseo SC, Laude RP (1992) Alkaline phosphatase polymorphism in *Apis mellifera* L and *Apis cerana* F in Laguna, Philippines. J Apic Res 31:46, Keywords: biochemistry/*cerana*/*mellifera*/Philippines

Aseo SC, Laude RP (1993) Alkaline phosphatase polymorphism in *Apis mellifera* and *Apis cerana* in Los Baños, Laguna. In: Connor LJ, Rinderer TE, Sylvester HA, Wongsiri S (eds) Asian apiculture. Wicwas Press, Cheshire, Connecticut, pp 57-63. Keywords: biochemistry/*cerana*/*mellifera*/Philippines

Ashmead WH (1904) Remarks on honey bees. Proc Entomol Soc Wash 6:120–123, Keywords: classification/*florea*/honey

Askendari H (1970) The biology of the little honeybee (*Micrapis florea* F) in Khuzestan, Iran. Thesis, Ahwaz Univ, Iran [in Persian]. Keywords: absconding/*florea*/general biology/migration/swarming/Iran

Aso K, Watanabe T, Yamao K (1960) Studies on honey I. On the sugar composition of honey. J Agric Res 11:101–108, Keywords: *cerana*/honey/Japan

Assareh MH, Siapoosh A, Mozafari M (1985) Improving pollination in alfalfa flowers by *Apis florea*. Ahwaz Univ, Iran, p73. Keywords: bee botany/*florea*/Iran

Atim AB, Yaakob AM (1989) Nectar dynamics of *Hevea brasiliensis* and its application to beekeeping management under rubber in Malaysia. In: Proceedings of the 4th international conference on apiculture in tropical climates, Cairo, pp 249–258, Keywords: beekeeping/*cerana*/nectar/physical factors/Malaysia

Atim AB, Daud MN, Yaakob AM (1986) Relationship of nectar flow on colony development and honey yield of *Apis cerana* under *Hevea brasiliensis* in Malaysia. J Nat Rubber Res 1:176–186, Keywords: beekeeping/*cerana*/honey/nectar/physiology/Malaysia

Attathom T (1984) A new sacbrood virus strain of the eastern honeybee, *Apis cerana*. J Nat Res Coun Thai 16:1–8, Keywords: *cerana*/disease/Thailand

Attri BS, Sharma PL (1969) Toxicity of some new insecticides to the Indian honey bee *Apis indica*. Pesticides, India 3:27–29, Keywords: *cerana*/honey/insecticides/India

Atwal AS (1964) Introduction of black Italian queens in *indica* hives. Apic Ital 31:100–101, [in Italian]. Keywords: *cerana*/*mellifera*/queens

Atwal AS (1967) Acarine disease problem of honey bees in the Punjab. Tech Report, College of Agriculture, Punjab Agricultural University, pp 1–39. Keywords: *cerana*/disease/*dorsata*/*florea*/honey/*mellifera*/pests/India

Atwal AS (1967) Studies on the acarine disease problems in *Apis indica*. In: Proceedings of the 21st international apiculture congress, Maryland, pp 70–71, Keywords: beekeeping/behaviour/*cerana*/disease/*dorsata*/*florea*/interspecific interactions/pests/India

Atwal AS (1970) Insect pollinators of crops. Punjab Agricultural University Press, Ludhiana. Keywords: *cerana*/pollination/India

Atwal AS (1971) Acarine disease problem of the Indian honey bee, *Apis indica* F. Am Bee J 111:134–135, Keywords: behaviour/*cerana*/disease/honey/*mellifera*/pests/India

Atwal AS (2000) Essentials of beekeeping and pollination. Kalyani Publishers, Ludhiana. Keywords: beekeeping/*cerana*/*mellifera*/pollination/India

Atwal AS (2002) The probability of transmission and spread of diseases, parasites, and hybridization between *Apis mellifera* L and *Apis cerana*. Asian Bee J 4:49–55, Keywords: *cerana*/disease/*mellifera*/parasites/India

Atwal AS, Dhaliwal GS (1969a) Robbing between *Apis indica* F and *Apis mellifera* L. Am Bee J 109:462–463, Keywords: behaviour/*cerana*/interspecific interactions/*mellifera*/India

Atwal AS, Dhaliwal GS (1969b) Some behavioural characteristics of *Apis indica* F. and *Apis mellifera* L. Indian Bee J 31:1–8, Keywords: aggressiveness/behaviour/*cerana*/homing instinct/*indica*/*mellifera*/natural enemies/nests/oviposition/India

Atwal AS, Goyal NP (1969) Communication in the Indian honeybee *Apis indica* F. In: Proceedings of the 22nd international apiculture congress, Apimondia, Munich, p 107, Keywords: beekeeping/*cerana*/communication/India

Atwal AS, Goyal NP (1974) The over summering of foreign honey bee *Apis mellifera* queens in *Apis indica* colonies and the associated behaviour of the two species. Indian Bee J 30:41–56, Keywords: beekeeping/behaviour/*cerana*/honey/*mellifera*/queens/seasonal variation/India

Atwal AS, Sharma OP (1967) The introduction of *Apis mellifera* queens into *Apis indica* colonies and the associated behavior of the two species. In: Proceedings of the 21st international apiculture congress, Maryland, p 77, Keywords: beekeeping/behaviour/*cerana*/combs/interspecific interactions/*mellifera*/oviposition/queens/India

Atwal AS, Sharma OP (1968) The introduction of *Apis mellifera* queens into *Apis indica* colonies and the associated behavior of the two species. Indian Bee J 30:41–56, Keywords: beekeeping/behaviour/*cerana*/combs/interspecific interactions/*mellifera*/oviposition/queens/India

Atwal AS, Sharma OP (1970) Acarine disease of adult honeybees: prevention and control. Indian Fmg 20:39-40. Keywords: disease/pests/India

Atwal AS, Sharma OP (1970b) Studies on the performance of five strains of *Apis mellifera* L as compared with *Apis indica* F under Nagrota conditions. J Res (Ludhiana) 7:477–486, Keywords: *cerana*/*mellifera*/India

Atwal AS, Sharma OP (1971) The dominance of *Apis mellifera* over *Apis indica*. Am Bee J 111:343, Keywords: behaviour/*cerana*/interspecific interactions/*mellifera*/India

Atwal AS, Bains SS, Singh B (1970) Bee flora for four species of *Apis* at Ludhiana. Indian J Entomol 32:330–334, Keywords: bee botany/*cerana*/*dorsata*/*florea*/*mellifera*/India

Australia FCAAA Inc (1993) Minutes of meeting held on 22/4/1993 Canberra. Austral Bee J 74:17–27, Keywords: biogeography/*cerana*/islands/Australia

Avetisyan GA (1960) On certain problems of evolution, geographical distribution, protection and utilisation of species and races of bees. In: Proceedings of the 17th international apiculture congress, Apimondia, Rome, Italy, pp 223–229, Keywords: biogeography/*cerana*/phylogeny/Russia

Avetisyan GA (1978) Apiculture. Apimondia Publishing House, Bucharest, Keywords: beekeeping/*cerana*

Avetisyan GA, Cherevko YA (1971) Effect of inbreeding and interline hybridization of inner, outer and economic characteristics of far-eastern bees. In: Proceedings of the 23rd international apiculture congress, Apimondia, Moscow, pp 368–374, Keywords: *cerana*/genetics/Russia

Ayyar TV (1940) Handbook of economic entomology for south India. Superintendent Government Press, Madras, Keywords: beekeeping/*cerana*/*dorsata*/*florea*/India

Azmi HK, Ali SZ (1998) Use of invertebrates as traditional drugs among the tribals of Chhattisgarh (MP). J Living World 5:22–29, Keywords: *cerana*/interspecific interactions/India

Baer B (2005) Sexual selection in *Apis* bees. Apidologie 36:187–200, Keywords: *cerana*/*dorsata*/genetics/*laboriosa*/reproduction

Bagnères AG, Huang Z, Navajas M, Salvy M, Christides JP, Le Conte Y (2006) Chemical signature and camouflage of *Varroa* in *Apis cerana* and *A. mellifera*: an adaptive coevolutionary phenomenon. IUSSI, French section, Avignon.* Keywords: *cerana*/*mellifera*/pests/pheromones

Bährmann R (1961a) Comparative morphological investigations on *Apis cerana* and *Apis mellifera*. Arch Geflügelz Kleintierkd 10:244–257, [in German]. Keywords: *cerana*/*mellifera*/morphology

Bährmann R (1961b) The structure of the copulatory organ of the four species of *Apis*. Liepiz Bienenztg 75:18–20, [in German]. Keywords: *cerana*/*dorsata*/drones/*florea*/*mellifera*/morphology

Bährmann R (1967) Comparative morphology of workers and drones of *A. cerana*. Z Bienenforsch 9:32–46, [in German]. Keywords: *cerana*/drones/morphology

Bai ARK, Reddy CC (1975) Ovary development and egg laying in *Apis cerana indica* workers. J Apic Res 14:149–152, Keywords: *cerana*/laying workers/oviposition/physiology/reproduction/India

Bai ARK, Reddy CC (1977a) Inhibition of acetylcholinesterase as a criterion to determine the degree of insecticide poisoning in *Apis cerana indica*. J Apic Res 16:112–114, Keywords: *cerana*/insecticides/India

Bai ARK, Reddy CC (1977b) Laboratory toxicity of some insecticides to *Apis cerana indica*. J Apic Res 16:161–162, Keywords: *cerana*/insecticides/India

Bailey L (1974) An unusual type of *Streptococcus pluton* from the Eastern hive bee, *Apis cerana*. J Invertebr Pathol 23:246–247, Keywords: *cerana*/disease/*mellifera*/nests

Bailey L (1981) Honeybee pathology. Academic, London, Keywords: *cerana*/disease/*dorsata*/*florea*/*mellifera*

Bailey L, Ball BV (1978) *Apis* iridescent virus and "clustering disease" of *Apis cerana*. J Invertebr Pathol 31:368–371, Keywords: *cerana*/disease

Bailey L, Ball BV, Woods RD (1976) An iridovirus from bees. J Gen Virol 31:459–461, Keywords: *cerana*/disease/*mellifera*

Bailey L, Carpenter JM, Woods RD (1982) A strain of sacbrood virus from *Apis cerana*. J Invertebr Pathol 39:264–265, Keywords: *cerana*/virus

Bajpai M (1987) Digestive enzymes of worker larvae of *Apis cerana indica* (Hymenoptera: Apoidea). In: Proceedings of the 31st international apiculture congress, Apimondia, Warsaw, p 183, Keywords: biochemistry/*cerana*/larvae/physiology

Bakar AA, Napi DM, Malik YA (1986) Relationship of nectar flow on colony development and honey yield of *Apis cerana* under *Hevea brasiliensis* in Malaysia. J Nat Rubber Res 1:176–186, Keywords: beekeeping/*cerana*/honey/nectar/physiology/Malaysia

Bakker DR (2002) Foraging and habitat selection by two species of honey bee near Lore Lindu National Park in Sulawesi, Indonesia. Thesis, University of Guelph, Keywords: breeding/*cerana*/*dorsata*/foraging/honey/Indonesia

Baldensperger PJ (1928) Bees in their natural state and with beekeepers. Bee World 9:173–174, Keywords: beekeeping/*cerana*/disease

Baltazar CR (1966) Genus *Apis* Linnaeus. Pacif Insects Monogr 8:408–410, Keywords: biogeography/*cerana*/classification/*dorsata*/*florea*/*mellifera*

Banerjee SK, Bharadwaj RK (1974) The nervous system and muscle innervations of *Apis cerana indica* (Hymenoptera: Apidae). J Nat Hist 8:19–28, Keywords: *cerana*/physiology

Banerjee A, Saxena P (1983) Digestive enzymes in the alimentary canal and associated glands of foragers of *Apis cerana indica*. In: Proceedings of the 2nd international conference on

apiculture in tropical climates, New Delhi, pp 337–343, Keywords: biochemistry/*cerana*/foraging/glands/physical factors/India
Bangyu K, Tan K (1996) Honeybee genetic diversity and its value to mountain farming. Exp Mtg agrobiodiversity HKH Region, Chengdu, China.* Keywords: *cerana*/genetics/China
Bangyu K, Hsiao K, Tan K (2000) Effect of synthetic queen pheromone on the behaviour of *Apis cerana*. In: Proceedings of the 4th Asian apiculture association international conference, Kathmandu, pp 24–25, Keywords: behaviour/*cerana*/queens
Baptist BA, Punchihewa RWK (1983) A preliminary analysis of the principal factors which will affect apiary honey production in Sri Lanka. In: Proceedings of the 2nd international conference on apiculture in tropical climates, New Delhi, pp 87–99,Keywords: beekeeping/*cerana*/*dorsata*/*florea*/honey/physical factors/Sri Lanka
Basavanna GPC (1979) Role of insect pollinators in sunflower production. Curr Res 8:1–3, Keywords: *cerana*/pollination/India
Basavanna GPC (1994) Thai sacbrood disease of the Indian honey bee, *Apis cerana indica*. In: Proceedings of a Workshop, Bangalore, pp 3–7, Keywords: *cerana*/disease/honey/*mellifera*/India
Basavarajappa S (1998) Status of natural colonies of *Apis dorsata* in Maidan region of Karnataka, India. Indian Bee J 60:143–146, Keywords: *dorsata*/general biology/India
Baskaran S, Bai MK (1998) Insect visitors of wood apple, *Limonia acidissima* L (Rutaceae). Insect Environ 4:88, Keywords: bee botany/*cerana*/India
Baskaran S, Bai MK (2001) Relation of temperature and humidity with the activity of the Asian hive bee, *Apis cerana indica*. Bionotes 3:8, Keywords: *cerana*/foraging/humidity/temperature
Baskaran S, Vigneswari A, Kasthuribai M (2001) Composition and relative abundance of the flower visitors of bhendi, *Abelmoschus esculentus* Linn. Insect Environ 7:17, Keywords: *cerana*/*dorsata*/*florea*/pollination
Basu PK (1957) Bees 'stung'. Indian Bee J 19:7, Keywords: beekeeping/*dorsata*/India
Batra SWT (1967) Crop pollination and the flower relationships of the wild bees of Ludhiana, India. J Kans Entomol Soc 40:164–177, Keywords: bee botany/*florea*/India
Batra SWT (1995) Biology of *Apis laboriosa* Smith, a pollinator of apples at the high altitude in the Greater Himalaya of Garhwal, India. United States Department of Agriculture (USDA), Agricultural Research Service, Tektran, Keywords: *laboriosa*/pollination
Batra SWT (1996) Biology of *Apis laboriosa* Smith, a pollinator of apples at high altitude in the Greater Himalaya Range of Garhwal, India (Hymenoptera: Apidae). J Kans Entomol Soc 69:177–181, Keywords: bee botany/environmental effects/general biology/*laboriosa*/India
Batra SWT (1997) Fruit-pollinating bees of the Garhwal Himalaya, UP, India. In: Proceedings of the 7th international symposium on pollination, Lethbridge, pp 325–328, Keywords: *cerana*/*dorsata*/*florea*/*mellifera*/pollination/India
Bawa SR, Marwaha RK (1975a) Apical cells in testis of the honeybee, *Apis indica*. Indian J Entomol 37:211–212, Keywords: cells/*cerana*/cytology/histology/India
Bawa SR, Marwaha RK (1975b) Cytochemical studies of the secretion of accessory glands and seminal vesicles of the honeybee *Apis cerana indica*. J Apic Res 14:15–19, Keywords: *cerana*/drones/glands/India
Bawa SR, Marwaha RK (1975c) The sperm bundles of honeybee *Apis cerana indica* Fabr. Experientia 31:684–686, Keywords: *cerana*/drones
Bediker E (1930) On the Japanese honeybee. Opyt Paseka 3(4):130–133, [in Russian]. Keywords: *cerana*/Japan
Bee Development Section (2002) Beekeeping: an introduction. His Majesty's Government, Kathmandu, [in Nepali]. Keywords: beekeeping/*cerana*/*dorsata*/*laboriosa*/*mellifera*/Nepal
Bee World (1967) Bibliography on honeybees other than *Apis mellifera*. Bee World 48:8–15, Keywords: bibliography/*cerana*/*dorsata*/*florea*/*mellifera*
Bee World (1971) *Apis cerana* at risk (Editorial). Bee World 52:141–142, Keywords: *cerana*
Beekeeping Institute of Jiangxi Province (1975) Studies on the prevention and cure of sacbrood in *Apis cerana cerana*. Apic China 1:31–33, [in Chinese]. Keywords: *cerana*/disease/China

Beekman M, Oldroyd BP (2008) When workers disunite: intraspecific parasitism by eusocial bees. Ann Rev Entomol 53:19–37, Keywords: *cerana/florea/mellifera*/reproduction

Beekman M, Gloag RS, Wattanachaiyingchareon W, Oldroyd B (2008) Dance precision of *Apis florea* – clues to the evolution of the honeybee dance language. Behav Ecol Sociobiol 62:1259–1265, Keywords: dance language/*florea*

Beenet Asia (1993) Proceedings of Beenet Asia, Universiti Pertanian, Selangor, Malaysia, pp 1–149, Keywords: beekeeping/*cerana*/Asia/China/Indonesia/Malaysia/Philippines/Thailand/Vietnam

Beeson CFC (1941) The ecology and control of the forest insects of India and the neighbouring countries. The Government of India, Dehra Dun, Keywords: absconding/*dorsata*/ecology/*florea*/honey/migration/swarming/India

Behura SK (2007) Analysis of nuclear copies of mitochondrial sequences in honeybee (*Apis mellifera*) genome. Mol Biol Evol 24:1492–1505, Keywords: *cerana/dorsata*/genetics/*koschevnikovi/mellifera*

Belavadi VV, Parvathi C (1993) Landing and departure rules used by honeybees on cardamom (*Elettaria cardamomum* Maton). In: Proceedings of international symposium on pollination in tropics, Bangalore, pp 73–76, Keywords: *cerana/dorsata*/foraging/pollination/tropics/India

Belavadi VV, Venkateshalu VHR (1997) Significance of style in cardamom corolla tubes for honeybee pollinators. Curr Sci 73:287–290, Keywords: *cerana/dorsata*/pollination/India

Bellier J (1906) Beeswax from Annam. Ann Chim Anal 11:366–368, Keywords: beeswax/*cerana*/Vietnam

Benton F (1881a) Letters from Frank Benton, from the Isle of Ceylon. Glean Bee Cult 9:168–169, Keywords: beekeeping/*cerana/dorsata/florea*/Sri Lanka

Benton F (1881b) Letter to Jones: *Apis dorsata* not to be found in Java. Am Bee J 17:145–146, Keywords: *dorsata*/Indonesia

Benton F (1881c) Letter to the editor: *Apis dorsata* and *Apis zonata*. Am Bee J 17:45, Keywords: *dorsata*

Benton F (1881) On Indian honeybees particularly *Apis dorsata*. Bienenztg 95* [in German]. Keywords: *dorsata*/India

Benton F (1885) *Apis dorsata* and the bees of the Orient. Am Bee J 21:151, Keywords: *dorsata*

Benton F (1895a) *Apis dorsata*, giant bees in India. Benton, Jamestown, Keywords: *dorsata*/India

Benton F (1895b) The honey bee. Benton, Washington, Keywords: *dorsata*/honey

Benton F (1899) *Apis dorsata*. A correction by Mr Benton. Glean Bee Cult 27:47–48, Keywords: *dorsata*/India

Benton AW, Morse RA (1968) Venom toxicity and proteins of the genus *Apis*. J Apic Res 7:113–118, Keywords: biochemistry/*cerana/dorsata/florea/laboriosa/mellifera*

Berkeley AGM (1978) Reports on honeybee project to Directorate of Agriculture and Durham University (cited from Whitcombe, 1984).* Keywords: beekeeping/*florea*/Oman

Bérubé C (1999) Himalayan ceranaid: development assistance to preserve and promote *Apis cerana* beekeeping in Nepal. Am Bee J 139:707–710, 784–787. Keywords: beekeeping/*cerana/laboriosa*

Betts AD (1949) Bee laboratory: research notes. Bee World 30:67, Keywords: drones/*florea*/queens

Bhaaradwaj RK (1968) A new record of the mite *Tropilaelaps clareae* from *Apis dorsata* colonies. Bee World 49:115, Keywords: disease/*dorsata*/pests/predators/India

Bhalla OP, Verma AK, Dhaliwal HS (1983a) Foraging activity of insect pollinators visiting stonefruits. J Entomol Res 7:91–94, Keywords: *cerana*/foraging/pollination/India

Bhalla OP, Verma AK, Dhaliwal HS (1983b) Insect visitors of mustard bloom (*Brassica campestris* var sarson), their number and foraging behaviour under midhill conditions. J Entomol Res 7:15–17, Keywords: behaviour/*cerana*/foraging/pollination/India

Bhambure CS (1962) A pollen trap design for Indian hive bees (*Apis indica* F). Indian Bee J 24:50–52, Keywords: *cerana*/nests/pollen/India

Bhambure CS (1980) Importance of pollen from *Apis dorsata* bee colonies. In: Proceedings of the 2nd international conference on apiculture in tropical climates, New Delhi, p 672, Keywords: bee botany/*dorsata*/India

Bhambure CS, Kshirsagar KK (1978) Occurrence of bee viral disease in *Apis cerana indica* F in Maharashtra area (India). Indian Bee J 40:66, Keywords: *cerana*/disease/India

Bhamburkar BL, Peshkar LN (1990) A simple hiving method for the little bee *Apis florea.* In: Proceedings of the 11th international congress, IUSSI, New Delhi, pp 475–476, Keywords: beekeeping/*florea*/India

Bhandari VC (1983) Biometrical studies on *Apis florea* F and *Apis dorsata* F of northwestern India. Thesis, Himachal Pradesh University, Simla, Keywords: classification/*dorsata*/*florea*/India

Bhandari R (2004) International Centre for Himalayan biodiversity. In: Proceedings of the 1st symposium issues concerning developing countries' international trade in honey, Hanoi, p 202, Keywords: *cerana*/conservation/*dorsata*/*florea*/honey/*mellifera*

Bharadwaj RK, Banerjee SK (1969) Anatomy and histology of the circulatory system of *Apis cerana indica.* J Apic Res 8:159–167, Keywords: *cerana*/physiology/India

Bhaskar V (1992) Pollination biology and fertilization in *Santalum album* L (Santalaceae). Flora (Jena) 187:73–78, Keywords: biology/*cerana*/*dorsata*/*florea*/pollination/India

Bhaskar V, Mahadevaiah YN (2002) Rock bee pollination in tamarind (*Tamarindus indica* L). In: Proceedings of the 11th international congress, IUSSI, Bangalore, p 437, Keywords: bee botany/*dorsata*/India

Bhat AA (1987) Relative toxicity of some insecticides to adults of the beetle, *Adoretus* spp and to the honeybee, *Apis cerana indica* Fabr. Indian J Plant Prot 15:97–99, Keywords: *cerana*/insecticides/India

Bhat AA (1993) Sexual maturity and sperm recovery in *Apis cerana indica* F drones. J Insect Sci 6:253–255, Keywords: *cerana*/drones/India

Bhat AA (1994) Efficiency of crossing Kashmir and Himachal strains of Indian honey bee, *Apis cerana indica* Fab. J Insect Sci 7:106–107, Keywords: breeding/*cerana*/classification/drones/genetics/geographical variation/honey/morphology/queens/India

Bhat AA, Mishra RC (1989) On flight behaviour of *Apis cerana indica* Fabr. drones. J Anim Morphol Physiol 36:93–98, Keywords: behaviour/*cerana*/drones/flight/India

Bhat NS, Kandoria JL, Goyal NP (1987) Quantity and quality of nectar in some south Indian bee plants. Indian Bee J 49:28–29, Keywords: bee botany/*cerana*/India

Bhat NS, Arpita R, Prakash NA, Jaganath G (1996) The role of insects in the pollination and seed set of *Peltophorum pterocarpum* (DC) (Fam: Leguminosae). Insect Environ 2:100–101, Keywords: *cerana*/*dorsata*/pollination/India

Bhatia R (1995) Relative abundance of insect visitors on flowers of major subtropical fruits in Himachal Pradesh and their effect on fruit set. Indian J Agric Sci 65:907–912, Keywords: bee botany/*florea*/India

Bhatnagar S (1975) Floral polymorphism in sympatric populations of *Calotropis procera* (Ait) R Br. Acta Bot Indica 3:43–46, Keywords: bee botany/*dorsata*/florea/general biology

Bhikuningputro W, Woelaningsih S, Santianawaty I (1975) Pollen collected by honeybees (*Apis cerana*). Research Report, Universitas Gadjah Mada, Indonesia, pp1–18 [in Indonesian]. Keywords: *cerana*/pollen/Indonesia

Bhupen (1943) Feeding *Apis indica.* Indian Bee J 5:149, Keywords: beekeeping/cerana/India

Bhusaro NV, Mate DM, Makde KH, Haribhau K (2005) Pollen of *Apis* honey from Maharashtra. Grana 44:216–224, Keywords: *florea*/pollen

Bhuyan M, Bhattacharayya PR (1998) Honey bee foraging upon sunflower (*Helianthus annuus*) under the agro-climatic conditions of Jorhat, Assam, India. Indian Bee J 60:87–90, Keywords: *cerana*/foraging/honey/India

Bhuyaneswari K, Uthamasamy S (1994) Toxicity of alphamethrin to beneficial insects in cotton ecosystem. J Insect Sci 7:228–229, Keywords: *cerana*/insecticides/India

Bi J, Wang Z, Hongazhen G (1996) Study on the flower mass of black locust and management of nectariferous forest in Tai Hang Mountain. Sci Salvae Sinica 32:509–515, [in Chinese]. Keywords: bee botany/beekeeping/*cerana*/China

Bienefeld K, Reinsch N, Thakur RK (2001) Selection for uncapping of *Varroa* infested brood cells in the honeybee (*Apis mellifera*). In: Proceedings of the 37th international apiculture congress, Apimondia, Durban (CD Rom), Keywords: breeding/brood/cells/*cerana*/*mellifera*/pests/India

Bijjur S, Kulkarni KA, Lingappa S (1991) Safety of *Heliothis armigera* (Hubner) NPV to some nontarget beneficial organisms. Indian J Entomol 53:475–478, Keywords: *cerana*/disease/India

Bilash GD (1981) Bee selection activity and selection criteria in Soviet Union. In: Proceedings of the 28th international apiculture congress, Apimondia, Acapulco, pp 228–234, Keywords: breeding/*cerana*/geographical variation/*mellifera*/Russia

Bingham CT (1897) The fauna of British India, including Ceylon and Burma, Hymenoptera Vol I (wasps and bees). Taylor and Francis, London, Keywords: *cerana*/classification/*dorsata*/*florea*/India/Myanmar/Sri Lanka

Binh NX (2004) Inspection and control honey residues in Vietnam. In: Proceedings of the 1st symposium issues concerning developing countries' international trade in honey, Hanoi, pp 8–13, Keywords: beekeeping/*cerana*/honey/*mellifera*/India

Binh PT, Tan NQ (1994) Beekeeping with *Apis cerana* in southern Vietnam and its economic efficiency. National Agricultural Publishing House, Hanoi, [in Vietnamese]. Keywords: beekeeping/*cerana*/Vietnam

Birdar AP, Shaila HM, Patil DR (2001) Honey bees menace on grape. Insect Environ 7:121–122, Keywords: *dorsata*/*florea*/foraging

Bisht DS (1966) Studies on the behaviour of *Apis indica* F under Delhi conditions with special reference to brood rearing and pollen gathering activities. Thesis, Indian Agricultural Research Institute, New Delhi, Keywords: behaviour/brood/*cerana*/foraging/pollen/India

Bisht DS (1975) Pollen gathering activity of honey bees and *Halictus* sp in roses. Indian Bee J 37:1–3, Keywords: *cerana*/*florea*/honey/pollen/India

Bisht DS (1983) Flight energetics of Indian honey bee, *Apis cerana indica*. J Entomol 45:1–5, Keywords: *cerana*/flight/honey/physiology

Bisht DS, Naim M (1979) Pollen gathering capacity of the Indian honeybee, *Apis cerana* F. Indian J Entomol 41:295–296, Keywords: *cerana*/foraging/pollen/India

Bisht DS, Pant NC (1968a) Studies on brood rearing activity of Indian honey bee *Apis indica* F under Delhi conditions. Indian J Entomol 30:223–228, Keywords: brood/*cerana*/honey/India

Bisht DS, Pant NC (1968b) Studies on the pollen gathering activity of the Indian honeybee, *Apis cerana indica* F under Delhi conditions. Indian J Entomol 30:163–168, Keywords: *cerana*/foraging/pollen/India

Bisht DS, Naim M, Sinha RBP (1973) Mating behaviour of queen bee of *Apis cerana indica*. Entomol Newsl 3:57, Keywords: *cerana*/reproduction/India

Bisht DS, Naim M, Mehrotra KN (1980) Studies on the role of honeybees in rapeseed production. In: Proceedings of the 2nd international conference on apiculture in tropical climates, New Delhi, pp 491–496, Keywords: *cerana*/*florea*/physical factors/pollination/India

Bisht DS, Naim M, Sinha RBP (1980b) A simple and hygienic method of honey extraction from rock bee. Apis dorsata Fabr. colony. Indian J Entomol 42:292–294, Keywords: beekeeping/*dorsata*/honey/India

Bisht DS, Naim M, Mehrotra KN (1982) Comparative efficiency of Jeolikote villager and Newton beehives for maintaining *Apis cerana indica*. Indian J Entomol 44:368–372, Keywords: beekeeping/*cerana*/India

Blum MS, Fales HM, Morse RA, Underwood BA (2000) Chemical characters of two related species of giant honeybees (*Apis dorsata* and *A. laboriosa*): possible ecological significance. J Chem Ecol 26:801–807, Keywords: *dorsata*/*laboriosa*/pheromones/Nepal

Boecking O (1993) Comparison of the defense mechanisms of *Apis cerana* and *Apis mellifera* against the *Varroa* mite. Allg Dtsc Imkerztg 27:23–27, [in German]. Keywords: aggressiveness/behaviour/*cerana*/*mellifera*/pests

Boecking O (1999) Sealing up and non-removal of diseased and *Varroa jacobsoni* infested drone brood cells is part of the hygienic behaviour in *Apis cerana*. J Apic Res 38:159–168, Keywords: behaviour/brood/cells/*cerana*/drones/pests

Boecking O, Spivak M (1999) Behavioral defenses of honey bees against *Varroa jacobsoni* Oud. Apidologie 30:141–158, Keywords: aggressiveness/behaviour/*cerana*/defense/honey/*mellifera*/pests

Boecking O, Rath W, Drescher W (1993a) Behavioral strategies of *Apis mellifera* and *Apis cerana* against *Varroa jacobsoni*. Int J Acarol 19:173–177, Keywords: behaviour/*cerana*/*mellifera*/pests

Boecking O, Rath W, Drescher W (1993b) Grooming and removal behaviour strategies of *Apis mellifera* and *Apis cerana* against *Varroa jacobsoni*. Am Bee J 133:117–119, Keywords: behaviour/*cerana*/*mellifera*/pests

Boecking O, Rosenkranz P, Sasaki M (1999) The pore in the hard conical *Apis cerana* drone capping results from a spinning process. Apidologie 30:513–519, Keywords: cells/*cerana*/drones/nests

Boecking O, Woo KS, Ritter W (2000) Introduction to *Varroa* disease. In: Proceedings of the 4th Asian apiculture association international conference, Kathmandu, pp 64–66, Keywords: *cerana*/disease/*dorsata*/*mellifera*/morphology/pests/Asia

Booncham U (1995) Niche differences among four sympatric species of honey bees in dry-evergreen forest. Thesis, Chulalongkorn University [in Thai]. Keywords: *cerana*/*dorsata*/*florea*/honey/*mellifera*/Thailand

Booncham U, Wongsiri S, Thirakupt K (1995) Niche differences between *A florea* and *A andreniformis* in dry-evergreen forest Thailand. In: Proceedings of international conference on tropical bees environment, Pedu Lake, Kedah, p 33, Keywords: *andreniformis*/ecology/*florea*/Thailand

Boonithee A, Juntawong N, Pechhacker H, Hüttinger E (1991) Floral visits to select crops by four *Apis* species and *Trigona* sp in Thailand. In: Proceedings of the 6th international symposium on pollination, Tilburg, pp 74–80, Keywords: *cerana*/*dorsata*/*florea*/foraging/*mellifera*/pollination/Thailand

Boot WJ, Calis JNM (1996) Reproductive strategies of mites parasitizing honeybee brood. In: Proceedings of the 3rd Asian apiculture association international conference, Hanoi, pp 164–167, Keywords: *andreniformis*/brood/*cerana*/*dorsata*/*florea*/*mellifera*/pests

Boot WJ, Calis JNM, Beetsma J, Hai DM, Lan NK, Toan TV, Trung LQ, Minh NH (1996) The phenomenon of non-reproduction in worker cells as a *Varroa* tolerance factor involves natural selection of the mites. Apidologie 27:282–284, Keywords: breeding/cells/*cerana*/*mellifera*/pests

Boot WJ, Nguyen QT, Pham CD, Van Huan L, Van Dung N, Le TL, Beetsma J (1997a) Reproductive success of *Varroa jacobsoni* in brood of its original host, *Apis cerana*, in comparison to that of its new host, *A mellifera* (Hymenoptera: Apidae). Bull Entomol Res 87:119–126, Keywords: brood/*cerana*/*mellifera*/pests

Boot WJ, Van Der Blom J, Beetsma J (1997b) Workers of *Apis cerana* reproduce during short periods of queenlessness. Entomol Ber 57:149–153, Keywords: *cerana*/laying workers/queenlessness/reproduction

Boot WJ, Calis JNM, Beetsma J, Dong MH, Nguyen KL, Tran VT, Le QT, Nguyen HM (1999) Natural selection of *Varroa jacobsoni* explains the different reproductive strategies in colonies of *Apis cerana* and *Apis mellifera*. Exp Appl Acarol 23:133–144, Keywords: breeding/*cerana*/*mellifera*/pests

Bradbear NJ (1986) Honeybees in Bhutan: a report. International Bee Research Association, IBRA, London, p 14, Keywords: *cerana*/*dorsata*/*laboriosa*/migration/Bhutan

Bradbear NJ (1988a) The Asian hive bee *Apis cerana*. Beekeep 55:250–252, Keywords: beekeeping/*cerana*/nests

Bradbear NJ (1988b) World distribution of major honeybee diseases and pests. Bee World 69:15–39, Keywords: biogeography/*cerana*/disease/*dorsata*/*florea*/*mellifera*/natural enemies/pests

Bradbear NJ (2001) Beekeeping for people living in countries under stress: Afghanistan and Iraq. In: Proceedings of the 7th international conference on tropical bees, Chiang Mai, pp 421–425, Keywords: beekeeping/*florea*/*mellifera*/Afghanistan/Iraq

Bradbear NJ, Clements FA (1989) Honeybees in Bhutan. In: Proceedings of the 4th international conference on apiculture in tropical climates, Cairo, pp 504–505, Keywords: *cerana*/*dorsata*/*laboriosa*/migration/physical factors/Bhutan

Bradbear NJ, Reddy MS (2000) Sustainable beekeeping development in Karnataka. In: Proceedings of the 4th Asian apiculture association international conference, Kathmandu, pp 266–270, Keywords: beekeeping/*cerana*/*mellifera*/physiology/India

Brahmachary RL, Saha L, Mondal AK, Dasgupta P, Paul P (1980) *Apis florea* and dammar bees (*Trigonidae*): as the pollinating agents of certain fruit trees in Bengal. In: Proceedings of the state-level seminar on beekeeping, West Bengal, India, pp 63–66, Keywords: bee botany/beekeeping/*florea*/India

Brand-Garnys EE, Sprenger J (1988) Beeswax new aspects of a classic raw material for cosmetics. Seifen, Öle, Fette 114:547–552, [in German]. Keywords: beeswax/*cerana*/*florea*

Breed MD, Guzman-Novoa E, Hunt GJ (2004) Defensive behavior of honey bees: organization, genetics and comparisons with other bees. Ann Rev Entomol 49:271–298, Keywords: aggressiveness/*andreniformis*/*cerana*/*dorsata*/*florea*/genetics/honey/*mellifera*/*nuluensis*

Breed MD, Xiao-Bao D, Buchwald B (2007) Comparative nestmate recognition in Asian honey bees, *Apis florea*, *Apis andreniformis*, *Apis dorsata*, and *Apis cerana*. Apidologie 38:411–418, Keywords: behaviour/*cerana*/*dorsata*/*florea*

Bridgnell E (1946) *Apis dorsata* nests in a tree. Indian Bee J 8:164, Keywords: combs/*dorsata*/general biology/India

Briscoe AD (2002) Homology modeling suggests a functional role for parallel amino acid substitutions between bee and butterfly red- and green-sensitive opsins. Mol Biol Evol 19:983–986, Keywords: biochemistry/*nigrocincta*

British Bee Journal (1986) Brief history of Japanese beekeeping. Br Bee J 114:33, Keywords: beekeeping/*cerana*/history/Japan

Brockmann A, Brückner D (2001) On the evolution of sex pheromone communication in the genus *Apis*. In: Proceedings of international conference, IUSSI, Berlin, p 69. Keywords: *cerana*/communication/*dorsata*/drones/*florea*/*mellifera*/pheromones/phylogeny

Brodmann J, Twele R, Francke W, Yi-bo L, Xi-qiang S, Ayasse M (2009) Orchid mimics honey bee alarm pheromone in order to attract hornets for pollination. Curr Biol 19:1368–1372, Keywords: *cerana*/*mellifera*/pollination/predators

Brown B, Hamer I (1983) An unusual recording of honeybees (*Apis florea*) in Abu Dhabi. Bull Emir Nat Hist (Abu Dhabi) 19:28, Keywords: classification/*florea*/Abu Dhab

Büchler R, Drescher W, Tournier I (1992) Grooming behaviour of *Apis cerana*, *Apis mellifera* and *Apis dorsata*, reacting to *Varroa jacobsoni* and *Tropilaelaps clareae*. Exp Appl Acarol 16:313–319, Keywords: behaviour/*cerana*/*dorsata*/*mellifera*/pests

Büchner G (1906) On Indian Ghedda-wax. Chem-Ztg 43:528–529, [in German]. Keywords: beeswax/*cerana*/*dorsata*/*florea*/*mellifera*

Budiansky S (1983) Is yellow rain simply bees' natural excreta? Nature 303:3, Keywords: *cerana*/*dorsata*/physiology

Bugharuang L (1996) Analysis of genetic variation of *Apis cerana* in Thailand by use of polymerase chain reaction amplified mitochondrial DNA sequences. Thesis, Chulalongkorn University, Keywords: *cerana*/DNA/population biology/Thailand

Bulletin AQIS (2000) A late-night liaison with bees. AQIS Bull 13:6, Keywords: *dorsata*/Australia

Bunker A (1887) A letter from our old friend A Bunker. Glean Bee Cult 15:499–450, Keywords: biogeography/*dorsata*/Myanmar

Burenin N, Lebedev V (1981) Among beekeepers of Vietnam. Pchelovodstvo 9:28–29, [in Russian]. Keywords: beekeeping/*cerana*/*dorsata*/*florea*/foraging/*mellifera*/Vietnam

Burgett MD (1973) The occurrence of glucose oxidase in the honeys of social Hymenoptera. Thesis, Cornell University, Keywords: biochemistry/*cerana*/*dorsata*/*florea*/honey/*mellifera*

Burgett MD (1974) Glucose oxidase: a food protective mechanism in social Hymenoptera. Ann Entomol Soc Am 67:545–546, Keywords: biochemistry/*cerana*/*dorsata*/*florea*/*mellifera*

Burgett MD (1990) Bakuti – a Nepalese culinary preparation of giant honey bee brood. Fd Insects Newsl 3(3):1, Keywords: brood/honey/*laboriosa*/Nepal

Burgett MD (2000) Honey hunters of the Sundarbans. Beekeep Dev 56:6–7, Keywords: brood/*cerana*/*dorsata*/*florea*/honey/*mellifera*/physiology/Bangladesh/India/Thailand

Burgett MD, Titayavan M (1993) Brood thermoregulation by the giant honey bee (*Apis dorsata* F). Nat Hist Bull Siam Soc 41:93–98, Keywords: brood/*dorsata*/thermoregulation

Burgett MD, Rossignol PA, Kitprasert C (1990) A model of dispersion and regulation of brood mite (*Tropilaelaps clareae*) parasitism on the giant honeybee (*Apis dorsata*). Can J Zool 68:1423–1427, Keywords: brood/disease/*dorsata*/pests/predators

Burgett MD, Titayavan M, Sukumalanand P (1998) Brood thermoregulation by the Eastern honey bee (*Apis cerana* F.). Nat Hist Bull Siam Soc 46:141–147, Keywords: *cerana*/physiology/Thailand

Burgett MD, Titayavan M, Sukumalanand P (2007) The drone mating flight of the eastern honeybee (*Apis cerana* F.): duration, temporal period and inter-flight period. J Apic Res 44:36–37, Keywords: *cerana*/drones/flight/Thailand

Burgos JRT, Manila AC, Cervancia CR (2001) Honeys from three bee species in Guinobatan Albay, the Philippines. In: Proceedings of the 7th international conference on tropical bees, Chiang Mai, pp 153–155, Keywords: *cerana*/*dorsata*/honey/*mellifera*/Philippines

Butani DK (1950) An *Apis dorsata* colony in New Delhi. Indian Bee J 12:115, Keywords: absconding/*dorsata*/general biology/migration/swarming/India

Butler CG (1954) The world of the honeybee. Collins, London, Keywords: absconding/behaviour/*cerana*/combs/*dorsata*/*florea*/Sri Lanka

Butler CG (1957) The honeybee. Eugen Diederich, Dusseldorf, [in German]. Keywords: biology/*cerana*/*dorsata*/*florea*/*mellifera*

Butler CG (1966) The effect of queen extracts from various social insects on queen rearing and ovary development of worker honey bees. Z Bienenforsch 8:143–147, [in German]. Keywords: beekeeping/brood/*florea*/honey/*mellifera*/pheromones/queens/reproduction

Butler CG (1974) The world of the honeybee. Collins, London, Keywords: absconding/*cerana*/*dorsata*/*florea*/general biology/*mellifera*/migration/swarming

Butler CG, Calam DH, Callow RK (1967) Attraction of *Apis mellifera* drones by the odours of the queens of two other species of honeybees. Nature 213:423–424, Keywords: *cerana*/drones/*florea*/*mellifera*/pheromones/queens

Byragi-Reddy T, Reddi CS (1994) Pollination ecology of *Vitex negundo* (Verbenaceae). Proc Indian Natl Sci Acad, Biol Sci 60:57–65, Keywords: bee botany/*cerana*/ecology/pollination/India

Cai MQ (1993) Methods for obtaining a high yield of *Astragalus sinicus* honey from *Apis cerana cerana*. Apic Sci Tech 1:9, [in Chinese]. Keywords: beekeeping/*cerana*/honey/China

Callow RK (1963) Chemical and biochemical problems of beeswax. Bee World 44:95–101, Keywords: beeswax/*cerana*/*dorsata*/*florea*/*mellifera*/wax

Camargo JMF (1972) Preliminary notes on the reproductive system of *Apis dorsata* and *A. florea* queens. Homeuagem à Warwick E Kerr, Faculdade de Filosofia, Ciências e Letras, Rio Claro, Brazil, pp 47–56 [in Portuguese]. Keywords: *cerana*/*dorsata*/*florea*/*mellifera*/queens/reproduction

Camazine S (1989) Beekeeping in South Korea. Bee World 70:66–74, Keywords: beekeeping/*cerana*/*mellifera*/Korea

Cameron SA (1991) A new tribal phylogeny of the Apidae inferred from mitochondrial DNA sequences. In: Smith DR (ed) Diversity in the genus *Apis*. Westview, Boulder, CO, pp 71–88, Keywords: *cerana*/DNA/*dorsata*/*florea*/genetics/*mellifera*/mtDNA/phylogeny

Cameron SA, Mardulyn P (2001) Multiple molecular data sets suggest independent origins of highly social behaviour in bees (Hymenoptera: Apidae). Syst Biol 50:194–214, Keywords: *dorsata*/evolution

Can TT, Tam DQ, Mulder V (1989) Beekeeping in Vietnam. In: Proceedings of the 4th international conference on apiculture in tropical climates, Cairo, pp 506–507, Keywords: absconding/bee botany/beekeeping/*cerana*/*mellifera*/physical factors/Vietnam

Cao FH (1992) Discussion on small migratory beekeeping with *Apis cerana cerana*. Apic Sci Tech 4:9–10, [in Chinese]. Keywords: beekeeping/*cerana*/China

Cao LY (1993) Discussion on the prevention and cure of sacbrood in *Apis cerana cerana*. Apic Sci Tech 1:1, [in Chinese]. Keywords: beekeeping/*cerana*/disease/China

Cao MH (1995) The form of strong colonies of *Apis cerana cerana* in the wild. Apic Sci Tech 3:13, [in Chinese]. Keywords: *cerana*/colony phenology/populations/China

Cao FH (1996) The use of movable frame hives without beeswax foundation for *Apis cerana cerana*. J Bee 11:17, [in Chinese]. Keywords: beekeeping/beeswax/*cerana*/China

Carlisle E (1955) Biometrical investigations of some European and other races of honeybees. Bee World 36:41–45, Keywords: *cerana*/classification/*dorsata*/*florea*/morphology/subspecies

Carlson DA, Roubik DW, Milstrey K (1991) Distinctive hydrocarbons among giant honeybees, the *Apis dorsata* group (Hymenoptera: Apidae). Apidologie 22:169–181, Keywords: biochemistry/*dorsata*/honey/organic chemistry

Castets J (1893) The honeybees of southern India. Rev Quest Sci 34:465–488, [in French]. Keywords: *florea*/general biology/India

Center Honeybee Science Research, University T (2001) Publication list of Prof Dr I Okada. Honeybee Sci 18:117–128, Keywords: bibliography/*cerana*/Japan

Cervancia CR (1996) Beekeeping in the Philippines. In: Proceedings of the 3rd Asian apiculture association international conference, Hanoi, pp 220–224, Keywords: *andreniformis*/beekeeping/*cerana*/disease/*dorsata*/*mellifera*/pests/Philippines

Cervancia CR (1997) Trends in apiculture and pollination biology in the Philippines. Honeybee Sci 18:29–35, (in Japanese). Keywords: biology/*cerana*/*mellifera*/pollination/Philippines

Cervancia CR (1998) Trends in apiculture and pollination biology in the Philippines. Asian Apic Assoc Newsl 7:7–10, Keywords: beekeeping/biology/*cerana*/pollination/Philippines

Cervancia CR (2003) Philippines: haven for bees. Honeybee Sci 24:129–134, [in Japanese]. Keywords: beekeeping/*cerana*

Cervancia CR, Manila AC (2001) Some aspects on the floral biology and pollination of calamondin, *Citrus madurensis* (Lour). In: Proceedings of the 7th international conference on tropical bees, Chiang Mai, pp 293–294, Keywords: biology/*cerana*/*mellifera*/pollination/Philippines

Chahal BS (1993) Promotion of beekeeping in Punjab. In: Proceedings of the 1st national conference of beekeeping, Chandigarh, pp 100–110, Keywords: *cerana*/disease/*mellifera*/India

Chahal BS, Gatoria GS (1983) Import of exotic bees and spread of bee diseases. Indian Bee J 45:11–13, Keywords: beekeeping/*cerana*/India

Chahal BS, Brar HS, Gatoria GS, Jhajj HS (1986) Aggressive behaviour of *Apis florea* towards *Apis mellifera* in hive robbing and in foraging. J Apic Res 25:134–138, Keywords: behaviour/classification/defense/ecology/*florea*/foraging/*mellifera*/India

Chaiyawong T, Jayavasti S, Suwanapong G, Wongsiri S (2001) Scanning electron microscopic analysis of honey bees, *Apis nigrocincta*, *A. cerana*, *A. koschevnikovi* and *A. mellifera* stings. In: Proceedings of the 7th international conference on tropical bees, Chiang Mai, pp 121–128, Keywords: *cerana*/honey/*koschevnikovi*/*mellifera*/*nigrocincta*/stings

Chakrabarti WB, Chaudhari AB (1972) Honey production and behaviour pattern of the honey bee. Sci Cult 38:257–290, Keywords: behaviour/*cerana*/*dorsata*/*florea*/honey/*mellifera*/India

Chakrabarti AK, Sinha SN (1980) Bee pollination and its impact on cauliflower seed production. In: Proceedings of the 2nd international conference on apiculture in tropical climates, New Delhi, pp 649–655, Keywords: *cerana*/physical factors/pollination/India

Chakrabati AK (1987) Sundarbans honey and the mangrove swamps. J Bombay Nat Hist Soc 84:133–137, Keywords: bee botany/*dorsata*/honey/India

Chakrabati AK, Chaudhari AB (1972) Wild life biology of the Sundarbans forests: honey production and behaviour pattern of the honeybee. Sci Cult 57:173–174, Keywords: absconding/bee botany/behaviour/combs/*dorsata*/general biology/honey/migration/swarming/India

Chanchao C (2009) Properties and antimicrobial activity of *Apis dorsata* honey from Thailand. Pak J Med Sci 25:313–318, Keywords: *dorsata*/honey/Thailand

Chanchao C, Deowanish S, Wongsiri S (2000) *Apis cerana* queen breeding, the unique Thai technique for the unique traditional culture. J Multidisc Res 13:28–33, Keywords: breeding/*cerana*/queens

Chand Hari E, Singh R, Hameed SF (1994) Population dynamics of honeybees and insect pollinators on Indian mustard, *Brassica juncea* L. J Entomol Res 18:233–239, Keywords: *cerana*/*dorsata*/*florea*/*mellifera*/pollination/India

Chand Hari E, Yaquin E, Singh R (2002) Biological development of *Apis mellifera* L and *Apis cerana indica* Fab. during different seasons under north Bihar conditions. Shashpa 9:133–137, Keywords: *cerana*/development/*mellifera*/India

Chand H, Singh R (1995) Effect of pollination by *Apis cerana* Fabr. on yield of mustard, *Brassica juncea* Linn. Indian Bee J 57:173–174, Keywords: *cerana*/pollination/India

Chandel YS (2000) Comparative performance of *A mellifera* L *vis a vis Apis cerana* Fab. on toria (*Brassica campestris* var *toria*) in mid-hill zone of Himachal Pradesh, India. Indian J Agric Res 34:264–267, Keywords: *cerana*/*mellifera*/pollination/India

Chandel RS, Gupta PR (1992) Toxicity of diflubenzuron and penfluron to immature stages of *Apis cerana indica* F and *Apis mellifera* L. Apidologie 23:465–473, Keywords: *cerana*/insecticides/*mellifera*/India

Chandel RS, Thakur RK, Bharadwaj NR, Pathania N (2004) Onion seed crop pollination: a missing dimension in mountain horticulture. Acta Hortic 631:79–86, Keywords: *cerana*/*dorsata*/*florea*/pollination

Chandra BCS, Geetha L, Abraham VA, Karanth P, Thomas K, Srinivasan MV, Gadagkar R (1998) Uniform discrimination of pattern orientation by honeybees. Anim Behav 56:1391–1398, Keywords: *cerana*/flight/physiology

Chandran K, Suryanarayana MC (1970) A note on entomophily in *Dendrophthoe falcata* and its usefulness to the Indian honeybee. J Apic Res 9:45–47, Keywords: bee botany/*cerana*/*dorsata*/*florea*/insecticides/India

Chandran K, Raian P, Joseph D, Suryanarayana MC (1983) Studies on the role of honeybees in the pollination of cardamom. In: Proceedings of the 2nd international conference on apiculture in tropical climates, New Delhi, pp 497–504, Keywords: *cerana*/*dorsata*/*florea*/physical factors/pollination/India

Chang YD, Woo KS, Lee HR (1993) Taxonomy, ecology and control of honey bee mites in Korea. Korean J Apic 8:35–47, Keywords: *cerana*/classification/ecology/honey/*mellifera*/pests/Korea

Chang SY, Hsieh FK, Chen CT, Ho KK (2001a) Pollination ecology of *Apis cerana* Fab. and *Apis mellifera* L on plum. Formosan Entomol 21:197–208, Keywords: *cerana*/ecology/*mellifera*/pollination/China

Chang YD, Lee MY, Mah Y (2001b) Pollination on strawberry in the vinyl house by *Apis mellifera* L and *A cerana* Fab. Acta Hortic 561:257–262, Keywords: *cerana*/*mellifera*/pollination

Chaoqun H, Ying J, Weihua S, Mingyang L, Piezhen Z, Dalong Z (1999) Genomic DNA map of Chinese bee and Italian bee. Apic China 50:8–10, Keywords: *cerana*/DNA/*mellifera*/China

Chapman NC, Nanork P, Reddy MS, Baht NS, Beekman M, Oldroyd BP (2008) Nestmate recognition by guards of the Asian hive bee (*Apis cerana*). Insectes Soc 55:382–386, Keywords: *cerana*/communication

Chapman NC, Nanork P, Gloag RS, Wattanachaiyingcharoen W, Beekman M, Oldroyd BP (2009) Queenless colonies of the Asian red dwarf honey bee (*Apis florea*) are infiltrated by workers from other queenless colonies. Behav Ecol 20:817–820, Keywords: *florea*/queenlessness/Thailand

Chaturvedi PL (1969) A tip for migratory beekeepers of Kumaon region. Indian Bee J 31:48–51, Keywords: beekeeping/*cerana*/India

Chaudhari RK (1977a) Bee forage in Punjab Plains (India) 1 Pathankot and adjacent villages. Indian Bee J 39:5–20, Keywords: bee botany/*cerana*/foraging/*mellifera*/India

Chaudhari RK (1977b) Peculiar foraging behaviour of honeybees in Kangra (India). Indian Bee J 39:24–26, Keywords: behaviour/*cerana*/foraging/*mellifera*/India

Chaudhari RK (1978) Floral fidelity in the Indian honey bee (*Apis cerana indica* F). Indian Bee J 40:33–35, Keywords: *cerana*/foraging/honey/*mellifera*/India

Chaudhari AB, Chakrabati AK (1989) Sundarbans mangrove – ecology and wildlife. Jugal Kishore, Dehra Dun, Keywords: *dorsata*/ecology/*mellifera*/India

Chaudhary OP (1999) Melliferous floral sources, its impact and future prospects on beekeeping in the "honey basket of India". In: Proceedings of the 37th international apiculture congress, Apimondia, Durban (CD Rom), Keywords: bee botany/beekeeping/*cerana*/pollen

Chaudhary DK, Mattu VK (1997) Incidence of *Varroa jacobsoni* Oud on *Apis mellifera* and *Apis cerana* in Himachal Pradesh, India. Am Bee J 137:597–598, Keywords: *cerana*/*mellifera*/pests/India

Chaudhary OP, Rakesh K (2000) Studies on honeybee foraging and pollination in cardamom (*Elettaria cardamomum* Maton). J Spices Arom Crops 9:37–42, Keywords: *cerana*/foraging/pollination/India

Chaudhary JL, Singh BP (1989) Bee pasturage and a pattern for seasonal management for Amethi, Uttar Pradesh. Indian Bee J 51:109–111, Keywords: bee botany/*cerana*/foraging/management/India

Chaudhary B, Choomsai MLA, Menon MGR (1973) Insect pollination in some vegetable crops. Haryan J Hortic Sci 2:56–62, Keywords: *cerana*/*dorsata*/*florea*/pollination/India

Chaudhary JL, Singh BP, Suryanarayana MC (1989) Beekeeping with *Apis cerana* in India. Khadi Gram 35:428–435, Keywords: beekeeping/*cerana*/India

Chaudhary OP, Chawda SS, Kumar R (1994) Large scale honey production from *Apis florea* in the Kutch region of Gujarat, India. Bee World 75:94–95, Keywords: classification/*florea*/honey/India

Chaudhary DK, Singh B, Singh PP (2002a) Diurnal variations in foragers activity on litchi flowers. J Appl Zool Res 13:97–100, Keywords: *cerana*/*dorsata*/foraging/India

Chaudhary DK, Singh B, Singh PP (2002b) Foraging pattern of insect pollinators of litchi (*Litchi chinensis*). Rau J Res 12:60–62, Keywords: *cerana*/*dorsata*/*florea*/foraging/India

Chaudhary DK, Singh B, Singh PP (2002c) Population dynamics of honeybees foraging on litchi flowers. J Entomol Res 26:71–75, Keywords: *cerana*/*dorsata*/*florea*/foraging/pollination/populations/India

Chauhan PS (1985) Studies on pollination ecology of apple orchards by hymenopterous insects in Narkanda and Matiana areas of Himachal Pradesh. Thesis, Himachal Pradesh University, Keywords: aggressiveness/*cerana*/ecology/*mellifera*/pollination/India

Chen DH (1981a) First report of investigations on the virus of 'large larvae disease' of Chinese honeybees. Zhongguo Yangfeng 6:21, [in Chinese]. Keywords: *cerana*/disease/larvae/China

Chen JX (1981b) The hive and honeybee. Agricultural Publishing House, Peping, [in Chinese]. Keywords: beekeeping/*cerana*/*mellifera*/nests/China

Chen Y (1993) Apiculture in China. Agricultural Scitech Publishing House of China, Beijing, [in Chinese]. Keywords: *andreniformis*/bee botany/biogeography/*cerana*/classification/*dorsata*/*florea*/*laboriosa*/*mellifera*/China

Chen SN (1995) Absconding in *Apis cerana cerana*. Apic Sci Tech 2:16, [in Chinese]. Keywords: absconding/*cerana*/China

Chen XG (1996a) The best method to prevent swarming and absconding of *Apis cerana cerana*. Apic Sci Tech 5:17, [in Chinese]. Keywords: absconding/beekeeping/*cerana*/swarming/China
Chen YH (1996b) Reasons for and the prevention of swarming in *Apis cerana cerana*. J Bee 7:18, [in Chinese]. Keywords: beekeeping/*cerana*/swarming/China
Chen CT, Hsieh FK (2001) Effect of honeybee pollination on the yield and fruit quality of strawberry variety 'Toyonoka' (*Fragaria* x *ananassa* Duch). Plant Prot Bull (Taipei) 43:117–127, [in Chinese]. Keywords: *cerana*/*mellifera*/pollination/China
Chen PP, Wongsiri S, Jamyanya T, Rinderer TE, Vongsamanode S, Matsuka M, Sylvester HA, Oldroyd BP (1998) Honey bees and other edible insects used as human food in Thailand. Am Entomol 44:25–29, Keywords: honey/Thailand
Chen XX, Hu Y, He JH, Chen W (1999) Induction of the antibacterial substances in the Chinese honeybee *Apis cerana* Fabricius. Entomol Knowl 36:215–218, Keywords: *cerana*/chemistry/disease/*mellifera*/China
Chen YW, Wang CH, An J, Ho KK (2000) Susceptibility of the Asian honey bee, *Apis cerana*, to American foulbrood, *Paenibacillus* larvae. J Apic Res 39:169–175, Keywords: *cerana*/disease/honey/larvae/*mellifera*/China
Chen S, Wang Q, Yu J (2001) Analysis of the effect of climate and temperature on rearing bees in the Mishan area (China). Apic China 52:6–9, Keywords: *cerana*/foraging/physical factors/physiology/China
Cheng J, Shi J, Shangguan FZ, Dafni A, Deng ZH, Luo YB (2009) The pollination of a self-incompatible, food-mimic orchid, *Coelogyne fimbriata* (Orchidaceae), by female *Vespula* wasps. Ann Bot 104:565–571, Keywords: *cerana*/pollination
Cherian MC (1942) Longevity and flight range of the Indian honeybee. *Apis indica*. Indian J Entomol 4:239, Keywords: *cerana*/foraging/India
Cherian MC, Mahadevan V (1945) The longevity of Indian bee (*Apis indica*). Indian Bee J 8:41–43, Keywords: biology/*cerana*/India
Cherian MC, Mahadevan V, Ramachandran S (1966) Preliminary studies on the pollen carrying capacity of the Indian Bee (*Apis cerana* F). Indian Bee J 8:109–111, Keywords: *cerana*/foraging/pollen/India
Cherian MC, Ramachandran S, Mahadevan V (1974) Studies in bee behaviour. Indian Bee J 9:116–124, Keywords: behaviour/*cerana*/India
Chinh PH (1989) Control of sacbrood disease in *Apis cerana* Fabr.: biological methods. In: Proceedings of the 1st Asia-Pacific conference of entomology, Chiang Mai, Keywords: *cerana*/disease/Thailand/Vietnam
Chinh PH (1991) *Apis dorsata* and rafter beekeeping in Minhai and Kiengiang provinces. Vietnam Bee J 2:1–5, [in Vietnamese]. Keywords: beekeeping/*dorsata*/Vietnam
Chinh PH (1994) Production of honey and beeswax in Sekong Province, southern Laos. In: J de Beer (ed) Our life depends on nature. Novib, The Hague, pp 19–21, 41, 59–61, Keywords: beeswax/*cerana*/honey/Laos
Chinh PH (2000) Thai sac brood virus disease control in Vietnam. In: Proceedings of the 4th Asian apiculture association international conference, Kathmandu, pp 57–59, Keywords: brood/*cerana*/disease/Vietnam
Chinh PH, Lap PV (1994) Closed population breeding ProgProgam for honeybee *Apis cerana* in Vietnam. In: Proceedings of the 1st National conference of beekeeping, Hanoi, pp 31–35, Keywords: beekeeping/breeding/*cerana*/population biology/Vietnam
Chinh PH, Tan NQ (2004) Honey quality for export. In: Proceedings of the 1st symposium issues concerning developing countries' international trade in honey, Hanoi, pp 27–33, Keywords: *cerana*/honey/*mellifera*/Vietnam
Chinh PH, Thai PH, Minh NH, Tan NQ (1995) Raftering: a traditional technique for honey and wax production from *Apis dorsata* in Vietnam. Beekeep Dev 36:8–9, Keywords: beekeeping/combs/*dorsata*/honey/Vietnam
Chinh PH, Lap PV, Thang DV, Hanh PD, Ngoc PV, Chinh TX, Toan TV (1996) Evaluation of a closed population breeding progamme with *Apis cerana cerana* after six years of selection. In:

Proceedings of the 3rd Asian apiculture association international conference, Hanoi, p 47, Keywords: absconding/breeding/*cerana*/disease/populations/swarming

Chinh PH, Tan NQ, Thai PH, Mulder V (1996) Some studies on biological characteristics of *Apis dorsata* in *Melaleuca* swamp forests in southern Vietnam. In: Proceedings of the 3rd Asian apiculture association international conference, Hanoi, pp 57–61, Keywords: beekeeping/combs/*dorsata*/general biology/honey/physiology/queens/reproduction/Vietnam

Chinh PH, Tan NQ, Thai PH (2004) The biological characteristics of *Apis dorsata* Fabr. in Vietnam. In: Proceedings of the 7th Asian apiculture association international conference, Los Banos, pp 75–85, Keywords: *dorsata*/Vietnam

Chinh TX, Boot WJ, Sommeijer MJ (2004b) Production of sexuals in the honeybee species *Apis cerana* Fabricius 1793 (Apidae: Apini) in North Vietnam. J Apic Res 37:191–198, Keywords: *cerana*/reproduction/Vietnam

Chinh TX, Boot WJ, Sommeijer MJ (2004) Reproductive strategies in sympatric species of eusocial bees. In: Proceedings of the 1st symposium issues concerning developing countries' international trade in honey, Hanoi, pp 74–80, Keywords: biogeography/*cerana*/*mellifera*/queens/reproduction/swarming

Chinniah C, Mohanasundaram M (1995) Record of new ascid mites (Ascidae: Acari) infesting insects in Tamil Nadu, India. Entomon 21:157–163, Keywords: *cerana*/pests/India

Cho SY, Cho YT (1955) Beekeeping. National Federation of Korean Beekeeping Association, Seoul, [in Korean]. Keywords: beekeeping/*cerana*/Korea

Cho SC, Huang ZY, Green DR, Smith DR, Zhang J (2006) Evolution of the complementary sex-determination gene of honey bees: balancing selection and trans-species polymorphisms. Genome Res 16:1366–1375, Keywords: *cerana*/*dorsata*/genetics/*mellifera*

Choi SY (1984) Brief report on the status of Korean beekeeping. In: Proceedings of the expert consultation on beekeeping *Apis mellifera* in tropical and sub-tropical Asia, FAO, Rome, pp 170–190, Keywords: beekeeping/*cerana*/*mellifera*/Asia/Korea

Choi SY, Lee ML (1988) Studies on the diurnal flight activities of honeybees on peach flowers. Korean J Apic 3:81–89, Keywords: *cerana*/flight/foraging/pollination/Korea

Choudhury B, Choomsai MLA, Menon MGR (1973) Insect pollination in some vegetable crops. Haryan J Hortic Sci 2:56–62, Keywords: *cerana*/*dorsata*/*florea*/pollination/India

Chouhan U, Pardasani KR (2008) A maximum parsimony model to reconstruct phylogenetic network in honey bee evolution. Int J Biol Med Sci 3:220–224, Keywords: *cerana*/*dorsata*/*florea*/*mellifera*/phylogeny

Chung AYC, Khen CV, Unchi S, Binti M (2002) Edible insects and entomophagy in Sabah, Malaysia. Malay Nat J 56:131–144, Keywords: *dorsata*/Malaysia

Cipriani L (1966) The Andaman Islanders. Weidenfeld & Nicolson, London. Keywords: beekeeping/*dorsata*/honey/India

Ciu J (1993) Distribution of *Apis cerana cerana* as a resource in the Changbai mountains. Apic Sci Tech 1:28, [in Chinese]. Keywords: biogeography/*cerana*/mountains/China

Claridge FM (1921) Differentiating races of bees. Bee World 2:119–120, Keywords: classification/*dorsata*/general biology

Clarke WF (1882) Specimens of *Apis dorsata*. Am Bee J 18:405–406, Keywords: *dorsata*

Cockerell TDA (1920) Supplementary notes on the social bees of the Philippine Islands. Philipp J Sci 16:631–632, Keywords: *cerana*/islands/Philippines

Collison CH (2001) The pathological effects of the tracheal mite on its host. In: Webster TC, Delaplane KS (eds) Mites of the honey bee. Dadant, Hamilton, pp 57–71, Keywords: *cerana*/disease/*dorsata*/honey/pathology/pests

Connor LJ, Rinderer TE, Sylvester HA, Wongsiri S (eds) (2001) Asian apiculture. Wicwas, Cheshire, CT, Keywords: *andreniformis*/biology/*cerana*/*dorsata*/*florea*/honey/*laboriosa*/pests

Cori E (1877) The large bee, *Apis dorsata*, of Java Part I. Am Bee J 13:406–407, Keywords: *dorsata*/Indonesia

Cori E (1878) The large bee, *Apis dorsata*, of Java Part II. Am Bee J 14:39–41, Keywords: *dorsata*/Indonesia

Corlett RT (2001) Pollination in a degraded tropical landscape: a Hong Kong case study. J Trop Ecol 17:155–161, Keywords: *cerana*/pollination/China

Corlett RT (2004) Flower visitors and pollination in the Oriental (Indomalayan) Region. Biol Rev 79:497–532, Keywords: *cerana*/foraging/pollination

Cornuet JM, Garnery L (1991) Mitochondrial DNA variability in honeybees and its phylogeographic implications. Apidologie 22:627–642, Keywords: *cerana*/DNA/*dorsata*/*florea*/*mellifera*/phylogeny

Cornuet JM, Garnery L, Solignac M (1991) Putative origin and function of the intergenic region between COI and COII of *Apis mellifera* L mitochondrial DNA. Genetics 128:393–403, Keywords: *cerana*/DNA/*dorsata*/*florea*/*mellifera*

Corpus Raros LA (2004) Two news species and a new record of Bak from the Philippines (Acari: Cheyletidae). Int J Acarol 26:321–328, Keywords: *cerana*/pests/Philippines

Craighead L (2000) Bears of the world. Voyager, Stillwater, Keywords: *cerana*/*dorsata*/disease/pests/predators

Crane E (1967) Bibliography on honeybees other than *Apis mellifera*. Bee World 48:8–15,18, Keywords: bibliography/*cerana*/*dorsata*/*florea*/*mellifera*

Crane E (1968) Mites infecting honeybees in Asia. Bee World 49:113–114, Keywords: *cerana*/*dorsata*/*florea*/pests/Asia

Crane E (1975) Honey: a comprehensive survey. Heinemann & International Bee Research Association, London, Keywords: bee botany/*cerana*/*dorsata*/*florea*/honey/*mellifera*

Crane E (1978a) Bibliography of tropical apiculture 3. Beekeeping in the Indian sub-continent (with Afghanistan and Iran). International Bee Research Association, London, Keywords: beekeeping/bibliography/*cerana*/tropics/Afghanistan/India/Iran

Crane E (1978b) Bibliography of tropical apiculture Part 11. The Asiatic hive bee Apis cerana, International Bee Research Association, London, Keywords: bibliography/*cerana*/nests

Crane E (1978c) Bibliography of tropical apiculture Part 13. The little honeybee. International Bee Research Association, London, p 10, Keywords: bibliography

Crane E (1984) Places to visit: Japan. Bee World 65:139–141, Keywords: *cerana*/Japan

Crane E (1987a) Bibliography of tropical apiculture. Satellite bibliographies S28 Beekeeping and bee research in India. International Bee Research Association, London, Keywords: bee botany/beekeeping/bibliography/*cerana*/*dorsata*/*florea*/*mellifera*/India

Crane E (1987b) Bibliography of tropical apiculture Satellite bibliographies S32 *Apis cerana*: laboratory studies. International Bee Research Association, London, Keywords: bibliography/*cerana*

Crane E (1989) The origination of *Apis cerana* beekeeping and its determining factors. In: Proceedings of the 4th international conference on apiculture in tropical climates, Cairo, pp 425–437, Keywords: beekeeping/*cerana*/history/physical factors

Crane E (1990) Bees and beekeeping: science, practice and world resources. Heinemann Newnes, Oxford, Keywords: beekeeping/biology/*cerana*/*dorsata*/*florea*/honey/*mellifera*/pollen

Crane E (1991) *Apis* species of tropical Asia as pollinators, and some rearing methods for them. In: Proceedings of the 6th international symposium on pollination, Tilburg, pp 29–48, Keywords: bibliography/*cerana*/*dorsata*/*florea*/pollination/Asia

Crane E (1992) Beekeeping in mountain life support systems. In: Verma LR (ed) Honeybees in mountain agriculture. Oxford & IBH, New Delhi, pp 17–27, Keywords: beekeeping/*cerana*/*dorsata*/*florea*/Asia

Crane E (1993) Current status of research on Asian honey bees. In: Connor LJ, Rinderer TE, Sylvester HA, Wongsiri S (eds) Asian apiculture. Wicwas, Cheshire, CT, pp 19–41, Keywords: absconding/behaviour/cells/*cerana*/combs/*florea*/geographicalvariation/honey/interspecific interactions/nectar/nests/pollen/swarming/India/Pakistan

Crane E (1995a) History of beekeeping with *Apis cerana* in Asia. In: Kevan PG (ed) The Asiatic hive bee: apiculture, biology, and role in sustainable development in tropical and subtropical Asia. Enviroquest, Cambridge, ON, pp 3–18, Keywords: beekeeping/biology/*cerana*/history/nests/physiology/Asia

Crane E (1995b) Removing water from honey. In: Kevan PG (ed) The Asiatic hive bee: apiculture, biology, and role in sustainable development in tropical and subtropical Asia. Enviroquest, Cambridge, ON, pp 233–243, Keywords: beekeeping/biology/*cerana*/environmental effects/honey/nests/physiology/Asia

Crane E (1999) The world history of beekeeping and honey hunting. Duckworth, London, Keywords: beekeeping/*cerana*/*dorsata*/*florea*/history/honey/Asia

Crane E (2001) Traditional beekeeping in Asia. In: Proceedings of the 7th international conference on tropical bees, Chiang Mai, pp 3–8, Keywords: beekeeping/*cerana*/*dorsata*/*florea*/*mellifera*/Asia

Crane E (2001b) The rock art of honey hunters. International Bee Research Association, Cardiff, Keywords: *dorsata*/honey

Crane E, Walker P (1993) Bibliography on Asian honey bees 1979–1991. In: Connor LJ, Rinderer TE, Sylvester HA, Wongsiri S (eds) Asian apiculture. Wicwas, Cheshire, CT, pp 651–680, Keywords: bibliography/*cerana*/*dorsata*/*florea*/honey/Asia

Crane E, Van Luyen V, Mulder V (1993a) Traditional management of *Apis cerana* using moveable-frame hives in Vietnam. Honeybee Sci 14:157–164, [in Japanese]. Keywords: beekeeping/*cerana*/Vietnam

Crane E, Van Luyen V, Mulder V (1993b) Traditional management of *Apis cerana* using movable comb hives in Vietnam. Bee World 74:75–85, Keywords: beekeeping/*cerana*/Vietnam

Cronin EW, Sherman PW (1976) A resource based mating system: the orange-rumped honeyguide. Living Bird 15:5–32, Keywords: disease/*dorsata*/pests/predators/reproduction/Nepal

Culliney TW (1983) Origin and evolutionary history of the honeybees *Apis*. Bee World 64:29–38, Keywords: *cerana*/*dorsata*/*florea*/history/*mellifera*/phylogeny

Culture B (1999) Peoples Republic of China. Bee Culture 127:50, Keywords: *cerana*/honey/*mellifera*/China

Cunanan HC (1982) Beekeeping in the Philippines: an industry. Thesis, Asian Institute of Management, Makati, Keywords: beekeeping/*cerana*/Philippines

Dai YG (1995) The prevention and cure of sacbrood in *Apis cerana cerana*. Apic China 5:19, [in Chinese]. Keywords: beekeeping/*cerana*/disease/China

Dai YG (1996a) Combining colonies of *Apis cerana cerana*. Apic China 4:4, [in Chinese]. Keywords: beekeeping/*cerana*/China

Dai YG (1996b) Summering and autumn and winter management of *Apis cerana cerana* in Guangdong Province. Apic China 3:15, [in Chinese]. Keywords: beekeeping/*cerana*/seasonal variation/China

Dale D (1975) Studies on the inhibition of cholinesterase from Indian honey bee. Apis indica. Pesticides 9:25, Keywords: *cerana*/honey/insecticides/India

Dale D, Kulkami AP (1970) Estimation of nucleic acids in the Indian honeybee, *Apis indica* F. Agric Res J Kerala 8:121–122, Keywords: *cerana*/DNA/physiology/India

Dale D, Mehrotra KN (1971) Molecular weights of cholinesterases from the Indian honeybee *Apis indica*, housefly *Musca domestica* and the desert locust *Schistocerca gregaria*. Curr Sci 40:369–371, Keywords: biochemistry/*cerana*/India

Dale D, Mehrotra KN (1976) Studies on the inhibition of honey bee cholinesterase by carbamates. Entomon 1:39–42, Keywords: biochemistry/*cerana*/honey/insecticides/India

Dale D, Mehrotra KN (1983) Physiological studies on the cholinesterase enzyme from the Indian honeybee *Apis cerana indica*. In: Proceedings of the 2nd international conference on apiculture in tropical climates, New Delhi, pp 350–355, Keywords: biochemistry/*cerana*/insecticides/physical factors/India

Damus MS (1995) A morphometric and genetic analysis of honey bee (*Apis cerana* F) samples from Malaysia: population discrimination and relationships. Thesis, University of Guelph, Keywords: *cerana*/classification/honey/Malaysia

Damus MS, Otis GW (1996) Morphometric analysis of cavity-nesting honey bees (*Apis cerana* and *A. nigrocincta*) of Malesia. In: Proceedings of the 3rd Asian apiculture association international conference, Hanoi, pp 42–46, Keywords: *cerana*/honey/*nigrocincta*/China/Japan/Malaysia

Damus MS, Otis GW (1997) A morphometric analysis of *Apis cerana* F and *Apis nigrocincta* Smith populations from southeast Asia. Apidologie 28:309–323, Keywords: *cerana*/classification/*nigrocincta*/populations/Asia/Indonesia/Malaysia/Philippines

Dandekar AB (1991) Studies on the external morphology of *Apis dorsata* F. worker bee. Thesis, University of Poona, Keywords: *dorsata*/morphology

Dang V (1991) Some methods of prevention and control of sacbrood disease in *Apis cerana*. Ngành Ong 3:10–13, [in Vietnamese]. Keywords: brood/*cerana*/disease/Vietnam

Danilova LV (1960) Some morphological and peculiarities of the Far Eastern wild bees (*Apis indica*). Pchelovodstvo 11:26–27, [in Russian]. Keywords: *cerana*/morphology/Russia

Darade MN, Singh BP, Chakrawarti P (1989) Flight range of *Apis cerana*. Indian Bee J 51:59–61, Keywords: *cerana*/flight

Das B (1946) The little bee as it appears to a layman. Indian Bee J 8:62–65, Keywords: beekeeping/India

Das Gupta KC (1945) The romance of scientific beekeeping. Khadi Pratisthan, Calcutta, Keywords: beekeeping/India

Das PK, Rahman A (2000) Brood rearing activity of *Apis cerana indica* F in Assam. Crop Res (Hisar) 19:469–473, Keywords: brood/*cerana*/India

Dashad SS, Sharma JK (1993) Relative abundance of insect visitors on blooms of certain cultivars of apple (*Malus domestica* Borkh). Crop Res (Hisar) 6:491–498, Keywords: *cerana*/*dorsata*/pollination/India

Dasiah J (1938) Hand book on bee culture. London Mission, Nagercoil, Keywords: beekeeping/*cerana*/India

Dass PK, Rahman A (1999) Biometrical changes during post-embryonic development of *Apis cerana* Fab. workers in Assam. Indian Bee J 61:31–36, Keywords: *cerana*/physiology

Dathe R (1883) Apis dorsata. Bienenztg Nörd 19:234, Keywords: *dorsata*

Davies KL, Stpiczynska M, Turner MP (2006) A rudimentary labellar speculum in *Cymbidium lowianum* (Rchb.f.) and *Cymbidium devonianum* Paxton (Orchidaceae). Ann Bot 97:975–984, Keywords: *cerana*/pollination

Davis AR, Prabhuswamy HP (2000) Influence of cell type and mandibular dentition on the ability of captive honey bee workers and drones (*Apis cerana indica* Fabr. and *Apis mellifera* L) to uncap brood cells: biological implications for parasitic mites (*Varroa* spp). Can J Zool 78:666–679, Keywords: brood/cells/*cerana*/drones/honey/*mellifera*/pests

Dayanandan S, Attygalla DNC, Abeygunasekera AWW, Gunatilleke IAUN, Gunatilleke CVS (1990) Phenology and floral morphology in relation to pollination of some Sri Lankan dipterocarps. In: Bawa KS, Hadley M (eds) Reproductive ecology of tropical forest plants. Parthenon, Paris, pp 103–134, Keywords: *cerana*/*dorsata*/ecology/morphology/pollination/Sri Lanka

De Fluiter HJ (1939) Observations on Javanese Hymenoptera I, *Apis dorsata* F. Entomol meded Nederl-Ind 5:45–49, Keywords: combs/*dorsata*

de Guzman LI, Forbes M, Cervancia CR, Rinderer TE, Somera S (1992) *Apis andreniformis* Smith in Palawan, Philippines. J Apic Res 31:111, Keywords: *andreniformis*/morphology

de Guzman LI, Burgett MD, Rinderer TE (2002) Biology and life history of *Acarapsis dorsalis* and *Acarapis externus*. In: Webster TC, Delaplane KS (eds) Mites of the honey bee. Dadant, Hamilton, pp 17–27, Keywords: biology/*cerana*/history/honey/pests/Japan

De Jong D (1988) *Varroa jacobsoni* does reproduce in worker cells of *Apis cerana* in South Korea. Apidologie 19:241–243, Keywords: cells/*cerana*/pests/Korea

de Jong W (2000) Micro-differences in local resource management: the case of honey in west Kalimantan, Indonesia. Human Ecol 28:631–639, Keywords: beekeeping/*cerana*/*dorsata*/*florea*/honey/Borneo/Indonesia

De Jong D, Morse RA, Eickwort GC (1982a) Mite pests of honey bees. Ann Rev Entomol 27:229–252, Keywords: *cerana*/disease/*dorsata*/honey/pests/predators

De Jong D, Roma D, Goncalves LS (1982b) A comparative analysis of shaking solutions for the detection of *Varroa jacobsoni* on adult honeybees (*Apis cerana*). Apidologie 13:297–303, Keywords: behaviour/*cerana*/pests

de la Rua P, Simon UE, Moritz RFA, Tilde AC (1999) MtDNA variations in *Apis cerana* from the Philippines. Apidologie 30:459–460, Keywords: *cerana*/DNA/mtDNA/Philippines

de la Rua P, Simon UE, Tilde AC, Moritz RFA, Fuchs S (2000) MtDNA variations in *Apis cerana* populations from the Philippines. Heredity 84:124–130, Keywords: *cerana*/DNA/mtDNA/populations/Philippines

de Mol GA (1933) Harvest of wax and honey in Borneo, vol 9. Landbouw, Buitenzorg, pp 80–86, [in Dutch]. Keywords: beeswax/cerana/honey

De Vries CA (1988) Beekeeping in the Konto River project area: experiences with the Javanese honey bee (*Apis cerana javanica*). Project Working Papers Series, DGRRL, Indonesia 17:1–81, Keywords: beekeeping/cerana/honey/Indonesia

De Vries R (1990) The history of honey and wax production of the giant honeybee (*A. dorsata*) in Sumatra and Kalimantan, Indonesia. In: Proceedings of the 1st NECTAR Seminar, Amsterdam, pp75–85, Keywords: bee botany/combs/*dorsata*/honey/Indonesia

Deh-Feng M, Wen-Cheng H (1981) Apiculture in the new China. Bee World 62:163–166, Keywords: beekeeping/*cerana*/China

Delfinado-Baker M (1982) New records for *Tropilaelaps clareae* from colonies of *Apis cerana indica*. Am Bee J 122:382, Keywords: *cerana*

Delfinado-Baker M (1990) Associations of mites (Acari) with bees (Apidae). In: Proceedings of the 11th international congress, IUSSI, Bangalore, pp 455–456, Keywords: *andreniformis*/*cerana*/*dorsata*/*florea*/*laboriosa*/*mellifera*/pests/India

Delfinado-Baker M, Aggarwal K (1987) A new *Varroa* (Acari: Varroidae) from the nest of *Apis cerana* (Apidae). Int J Acarol 13:233–237, Keywords: *cerana*/pests

Delfinado-Baker M, Baker EW (1974) Varroidae, a new family of mites on honey bees (Mesostigmata: Acarina). J Wash Acad Sci 64:4–10, Keywords: *andreniformis*/disease/*florea*/honey/pests

Delfinado-Baker M, Baker EW (1983) A new species of Neocypholaelaps (Acari: Meropidae) from brood combs of the Indian honey bee. Apidologie 14:1–7, Keywords: brood/*cerana*/combs/honey/pests/Pakistan

Delfinado-Baker M, Knox D (1982) Infestations of *Apis cerana indica* colonies by *Acarapis woodi* and *Varroa jacobsoni*. Am Bee J 122:592, Keywords: *cerana*/pests/Myanmar (Burma)/Pakistan

Delfinado-Baker M, Underwood BA, Baker EW (1985) The occurrence of *Tropilaelaps* mites in brood nests of *Apis dorsata* and *A. laboriosa* in Nepal, with descriptions of the nymphal stages. Am Bee J 125:703–706, Keywords: brood/*dorsata*/laboriosa/mites/nests/Nepal

Delfinado-Baker M, Baker EW, Phoon ACG (1989) Mites (Acari) associated with bees (Apidae) in Asia, with description of a new species. Am Bee J 129:609–613, Keywords: *andreniformis*/*cerana*/classification/*florea*/*koschevnikovi*/pests/Asia

Delfinado-Baker M, Rath W, Boecking O (1992) Phoretic bee mites and honeybee grooming behaviour. Int J Acarol 18:315–322, Keywords: behaviour/*cerana*/*mellifera*/pests

Delima AS, Robberts EJ (1987) Beekeeping in the Philippines: a feasibility study for small farmers. Netherlands Development Organization, Manila, Philippines, Keywords: beekeeping/*cerana*/physiology/Philippines

Delos Reyes F (1994) Total protein banding pattern of absconding and non-absconding colonies of *Apis cerana* F. Thesis, University of the Philippines, Keywords: absconding/biochemistry/*cerana*/Philippines

Demetria L (1994) Allozyme variation in natural populations of *Apis cerana* F (Apidae: Hymenoptera) from Albay and Quezon. Thesis, University of the Philippines, Keywords: biochemistry/*cerana*/populations/Philippines

Deng CY, Zhou GH (1991) Selective breeding of *Apis cerana cerana* in southern Guizhou Province. Guizhou Anim Husb Vet Med 4:7–9, [in Chinese]. Keywords: breeding/*cerana*/China

Deodikar GB (1959a) Geographical distribution based on morphological characters (structure of genitalia and crossability among different bee species). Indian Bee J 23:60–61, Keywords: biogeography/*cerana*/classification/*dorsata*/drones/*florea*/genetics/India

Deodikar GB (1959b) Some taxonomic problems in honeybees: The concept of supra-generic grouping. Bee World 40:121–124, Keywords: *cerana*/classification/*dorsata*/*florea*/*mellifera*

Deodikar GB (1960) Some indications on the phylogenetic interrelations among bees. Bee Genet Inf Bull 1:15–16, Keywords: *cerana*/*dorsata*/*florea*/foraging/*mellifera*/phylogeny

Deodikar GB (1962) Some aspects of bee botany. Indian Bee J 24:3–22, Keywords: bee botany/*cerana*/India

Deodikar GB (1964) Bees, beekeeping and apicultural research in India. Indian Bee J 26:12–22, Keywords: beekeeping/*cerana*/*dorsata*/*florea*/India

Deodikar GB (1970) A preliminary survey of beekeeping in Kashmir. Indian Bee J 32:12–22, Keywords: beekeeping/*cerana*/India

Deodikar GB (1974) Improvement of Indian honey bees. Indian J Genet Plant Breed 34A:263–266, Keywords: breeding/*cerana*/honey/India

Deodikar GB, Suryanarayana MC (1973) Bees enhance crop yields. Indian Fmg 23:31–36, Keywords: beekeeping/*cerana*/*dorsata*/*florea*/pollination/India

Deodikar GB, Suryanarayana MC (1977) Pollination in the service of increasing farm production in India. Adv Pollen-Spore Res 2:60–82, Keywords: *cerana*/pollination/India

Deodikar GB, Thakar SS (1953) A pollen study of major honey yielding plants of Mahabaleshwar hills. Village Industries Committee, Poona, p 8, Keywords: bee botany/*cerana*/honey/pollen/India

Deodikar GB, Thakar SS (1966a) Cytogenetics of Indian honeybees and bearing on taxonomic and breeding problems. Indian J Genet Plant Breed 26A:386–393, Keywords: absconding/breeding/*cerana*/classification/genetics/swarming/India

Deodikar GB, Thakar SS (1966b) Utilization and improvement of the local floras as bee pasturage. Indian Bee J 28:11–16, Keywords: bee botany/*cerana*/foraging/India

Deodikar GB, Thakar SS, Shah PN (1959) Cyto-genetic studies in Indian honeybees I. Somatic chromosome complement in Apis indica and its bearing on evolution and phylogeny. Proc Indian Acad Sci 49B:196–207, Keywords: *cerana*/*florea*/genetics/phylogeny

Deodikar GB, Thakar SS, Tonapi KV (1961) Evolution in the genus *Apis*. Indian Bee J 23:86–91, Keywords: *cerana*/classification/*dorsata*/*florea*/genetics/*mellifera*/phylogeny

Deodikar GB, Seethalakshmi TS, Suryanarayana MC (1976) Floral biology of sunflower with special reference to honey bees. J Palynol 12:115–125, Keywords: biology/*cerana*/foraging/honey/pollination/India

Deodikar GB, Ghatge AL, Phadke RP, Mahindre DB, Kshirsagar KK, Muvel S, Thakar SS (1977) Nesting behaviour of Indian honeybees III. Nesting behaviour of Apis dorsata Fab. Indian Bee J 39:1–12, Keywords: absconding/behaviour/classification/combs/defense/migration/queens/swarming/India

Deowanish S (2002) Diversity in *Apis cerana* analyzed by RFLP and PCR-RFLP of mitochondrial DNA. Thesis, University of Tamagawa, Keywords: *cerana*/genetics/mtDNA/Japan

Deowanish S, Matsuka M, Nakamura J (1996) Mitochondrial variation in *Apis cerana* by PCR-RFLP. In: Proceedings of the 3rd Asian apiculture association international conference, Hanoi, p 47, Keywords: beekeeping/*cerana*/mtDNA/physiology/variation/China/Japan/Thailand/Vietnam

Deowanish S, Nakamura J, Matsuka M, Kimura K (1996b) MtDNA variation among subspecies of *Apis cerana* using restriction fragment length polymorphism. Apidologie 27:407–413, Keywords: *cerana*/DNA/mtDNA/China/Japan/Korea/Nepal/Philippines/Thailand/Vietnam

Deowanish S, Wattanachaiyingcharon W, Wongsiri S, Oldroyd BP, Leepitakrat S, Rinderer TE, Sylvester HA (2001) Biodiversity of dwarf honey bees in Thailand. In: Proceedings of the 7th

international conference on tropical bees, Chiang Mai, pp 97–103, Keywords: *andreniformis*/bee botany/combs/*florea*/general biology/honey/Thailand
Deowanish S, Matsuka M, Nakamura J (2002) A study of diversity of *Apis cerana* in Thailand using PCRRFLP of mtDNA. In: Proceedings of the 13th international congress, IUSSI, Adelaide, p 138, Keywords: *cerana*/diversity/mtDNA/Thailand
Deshmukh SD (1977) Effects of bee visits and climate on sunflower pollination. Sci Cult 43:543–545, Keywords: *cerana*/*dorsata*/*florea*/physical factors/pollination/India
Deshmukh AK, Rao GM, Karve AD (1985) Studies on the effect of honey bee pollination on the yield of safflower. Indian Bee J 47:1–2, Keywords: *cerana*/*dorsata*/*florea*/honey/pollination/India
Deshmukh PB, Mathai AT, Pillai KS (1989) Preliminary laboratory test on the efficacy of a new pyrethroid insecticide cyhalothrin. Pestology 13:15–17, Keywords: *cerana*/insecticides/India
Devakumar C, Baskaran S, Mukerjee SK (1986) Isolation of n-triacontanol from Indian beeswax and its effect on dry matter of rice. Indian J Agric Sci 56:744–747, Keywords: beeswax/*cerana*/India
Devanesan S (1983) Sac brood bee disease in Kumaon Region, Uttar Pradesh. Indian Honey 6:2–5, Keywords: brood/*cerana*/disease/India
Devanesan S, Jacob A (2001) Thai sacbrood virus disease of Asian honeybee *Apis cerana indica* Fab. in Kerala, India. In: Proceedings of the 37th international apiculture congress, Apimondia, Durban (CD Rom), Keywords: *cerana*/disease/India
Devi CS (1996) Studies on the foraging behaviour, pollination requirements and pesticide toxicity of honey bees in relation to rapeseed. Thesis, Manipur University, Keywords: behaviour/*cerana*/*dorsata*/foraging/honey/insecticides/pollination/India
Devkota FR, Upreti G, Thapa RB, Shakya SM, Partap U (2003) Impact of honeybee pollination on productivity and quality of broccoli seed under Chitwan conditions. J Inst Agric Anim Sci 24:85–89, Keywords: *cerana*/foraging/*mellifera*
Devy S, Davidar P (2003) Pollination systems of trees in Kakachi, a mid-elevation wet evergreen forest in western Ghats, India. Am J Bot 90:650–657, Keywords: *cerana*/*dorsata*/*florea*/pollination/India
Dewan SMAL (1987) Beekeeping with *Apis cerana* in Bangladesh. In: Proceedings of the 31st international apiculture congress, Apimondia, Warsaw, p 185, Keywords: beekeeping/*cerana*/Bangladesh
Dewan SMAL (2001) Foraging and resource utilization of *Apis* spp (Hymenoptera: Apidae) of Bangladesh. Thesis, University of Wales, Cardiff, Keywords: absconding/beekeeping/biogeography/*cerana*/*dorsata*/*florea*/foraging/migration/Bangladesh/Iran
Dewan SMAL, Paxton RJ (1995) Patterns of foraging and honey production for *Apis cerana* in Bangladesh. In: Kevan PG (ed) The Asiatic hive bee: apiculture, biology, and role in sustainable development in tropical and subtropical Asia. Enviroquest, Cambridge, ON, pp 153–158, Keywords: beekeeping/biology/*cerana*/foraging/honey/physiology/nests/Asia/Bangladesh
Dey S, Raghuvarman A (1983) Occurrence of resilin in the lens cuticle of the honey bee, *Apis cerana indica*. Indian Zool 7:31–34, Keywords: biochemistry/*cerana*/honey/morphology
Dey S, Raghuvarman A (1986) Effect of light on brain neurosecretory cells of the honeybee *Apis cerana indica* and its significance in screening pigment movement of the compound eye. Curr Sci 55:808–809, Keywords: cells/*cerana*/hormones/physiology
Dey S, Varman AR (1984) Ascorbic acid in the compound eye of the honey bee, *Apis cerana indica*. Indian Biol 16:46–48, Keywords: *cerana*/honey/physiology/India
Deyto RC, Cervancia CR (2009) Floral biology and pollination of ampalaya (*Momordica charantia* L.). Phillip Agric Sci 92:8–18, Keywords: *cerana*/*mellifera*/pollination
Dhaduti SG, Mathad SB (1980) Effect of nuclear polyhedrosis virus of the armyworm *Mythimna (Pseudaletia) separata* on colonies of *Apis cerana indica*. J Apic Res 19:77–78, Keywords: *cerana*/disease/India
Dhaliwal GS (1970a) Impact of chemical control on honeybees. Pesticides 4:17–19, Keywords: *cerana*/insecticides/India

Dhaliwal GS (1970b) Observations on pollen carrying capacity of *Apis indica* and *Apis mellifera* L. Indian J Entomol 32:173–174, Keywords: *cerana*/foraging/*mellifera*/India

Dhaliwal JS (1975) Insect pollination in ber (*Zizyphus mauritiana* Lamk). Curr Sci 44:527, Keywords: bee botany/*cerana*/*dorsata*/India

Dhaliwal HS (1980) Ecology and relative efficiency of principal insect pollinators of *Brassica* seed crop. Thesis, Himachal Pradesh University, Keywords: *cerana*/*dorsata*/*florea*/mellifera/pollination

Dhaliwal GS, Atwal AS (1969) Comparative aggressiveness of the four species of *Apis* under controlled conditions. Indian Bee J 31:17–21, Keywords: aggressiveness/biogeography/*cerana*/*dorsata*/*florea*/interspecific interactions/*mellifera*/queens/India

Dhaliwal GS, Atwal AS (1970) Interspecific relations between *Apis cerana indica* and *Apis mellifera*. J Apic Res 9:53–59, Keywords: aggressiveness/*cerana*/interspecific interactions/*mellifera*/queens

Dhaliwal JS, Atwal AS (1976) Note on the effect of air temperature, relative humidity and wind velocity on bees visiting berseem at Ludhiana. Indian J Agric Sci 46:50–51, Keywords: *dorsata*/environmental effects/foraging/morphology/India

Dhaliwal GS, Bhalla OP (1983) On the foraging ecology of *Apis cerana indica*. In: Proceedings of the 2nd international conference on apiculture in tropical climates, New Delhi, pp 513–527, Keywords: *cerana*/ecology/foraging/physical factors/pollination/India

Dhaliwal GS, Sharma PL (1972) Homing ability and the foraging range of the Indian honeybee, *Apis indica* Fabricius. Indian J Entomol 35:225–228, Keywords: *cerana*/foraging/India

Dhaliwal GS, Sharma PL (1973) The foraging range of the Indian honeybee on two crops. J Apic Res 12:131–134, Keywords: *cerana*/foraging/India

Dhaliwal GS, Sharma PL (1974) Foraging range of the Indian honeybee. J Apic Res 13:137–141, Keywords: *cerana*/foraging/India

Dhuley JN, Mujumdar SM, Mujumdar AM, Naik SR (1996) Laboratory studies on contact toxicity of aureofungin in *Apis cerana indica* Fabr. Indian Bee J 58:207–208, Keywords: *cerana*/insecticides/India

Dhuyo AR, Munshi GH, Naqvi SMSH, Rizvi SNH, Rustamani MA (1988) Insect pollinator complex of cotton crop *Gossypium hirsutum* L at Tandojam. Pak Cottons 30:45–47, Keywords: bee botany/*florea*/Pakistan

Dien C, Huan LV, Dung NV, Long LT, Beetsma J, Boot WJ, Tan NQ (1997) Reproductive success of *Varroa jacobsoni* in brood of its original host, *Apis cerana* in comparison to that of its new host, *A. mellifera* (Hymenoptera: Apidae). Bull Entomol Res 87:119–126, Keywords: *cerana*/*mellifera*/pests/Vietnam

Dietermann V, Pirk CWW, Zheng HQ, Hepburn C, Hepburn HR, Crewe RM, Radloff SE, Hu FL (2008) Self-assessment in insects: Honeybee queens know their own strength. PLoS ONE 3(1): e1412

Dietz A (1986) Evolution. In: Rinderer TE (ed) Bee genetics and breeding. Academic, London, pp 3–21, Keywords: breeding/*cerana*/*dorsata*/*florea*/genetics/*mellifera*/phylogeny

Dietz A (1992) Honeybees of the world. In: Graham JM (ed) The hive and the honey bee. Dadant and Sons, Hamilton, IL, pp 23–71, Keywords: *cerana*/*dorsata*/*florea*/honey/*mellifera*/nests

Dinabandhoo CL, Dogra GS (1979) Dichlorvos for the control of acarine disease of Indian honey bee, *Apis cerana* Fab. Indian Bee J 41:1–4, Keywords: *cerana*/disease/honey/insecticides/India

Dinabandhoo CL, Dogra GS (1980) Incidence and control of ectoparasitic mite, *Pyemotes herfsi* Oudemans of the Indian honey bee, *Apis cerana* Fab. Am Bee J 120:44, 46–47. Keywords: *cerana*/honey/pests/India

Dinabandhoo CL, Dogra GS (1982) The pyemotids (Acarina: Pyemotidae) and their significance in apiculture. Indian Bee J 44:29–32, Keywords: *cerana*/pests/India

Dinabandhoo CL, Dogra GS (1983) The acarine induced decline in Indian honeybee *Apis cerana indica* F and its control in Kulu Valley. In: Proceedings of the 2nd international congress on apiculture in tropical climates, New Delhi, pp 435–441, Keywords: *cerana*/pests/India

Dinh QT, Nguyen VT, Le QT, Tran TD, Phung HC, Le TL (2001) Publications of Vietnam bee research 1990-1999. Bee Research and Development Center, Hanoi, Keywords: beekeeping/*cerana*/Vietnam

Diniz-Filho JAF, Malapsina O, Pignata MIB (1993) Geographic variation in *Apis cerana indica* F: a spatial autocorrelation analysis of morphometric patterns. J Apic Res 32:65–72, Keywords: biogeography/*cerana*/classification/geographical variation/India

Diwan VV (1971) Occurrence of a new bacterial disease of Indian honeybees *Apis indica* F. Curr Sci 40:196–197, Keywords: *cerana*/disease

Diwan VV (1980) Honey bee behaviour during total solar eclipse. Indian Bee J 42:97–99, Keywords: behaviour/*cerana*/foraging/honey/physical factors/India

Diwan VV, Rao SK (1969) Foraging behaviour of *Apis indica* F on the flowers of *Synadenium grantii* Hook. Indian Bee J 31:22, Keywords: bee botany/behaviour/*cerana*/foraging/India

Diwan VV, Salvi SR (1965) Some interesting behavioural features of *Apis dorsata* Fab. Indian Bee J 27:52, Keywords: bee botany/*dorsata*/India

Dixit KK (1956) Differential response to form and pattern in two species of Indian honeybees. J Bombay Nat Hist Soc 54:107–117, Keywords: *cerana*/dance language/*florea*/foraging/India

Dobrovsky TM (1950) Number of wing hooks varies by species. Bees 4:90, Keywords: *cerana*/*dorsata*/*florea*/*mellifera*/morphology

Dong BY (1984a) Studies on sacbrood virus of *Apis cerana cerana*. Apic China 2:6–8, [in Chinese]. Keywords: *cerana*/disease/China

Dong BY (1984b) Studies on the serological relationships between sacbrood virus of *Apis cerana cerana* and *Apis mellifera ligustica* and experiments on cross-infectivity. Apic China 3:8–9, [in Chinese]. Keywords: *cerana*/disease/*mellifera*/China

Dong X (1996) Observations on the habits of *Apis cerana cerana* collecting pollen. Apic China 3:16, [in Chinese]. Keywords: beekeeping/*cerana*/foraging/pollen/China

Dong X (1999) DNA extraction from *Apis cerana*. Apic China 50:3–4, [in Chinese]. Keywords: *cerana*/DNA/China

Dong X, Zhou D, Liu Y, Zhao L (1996) Study on the habit of *Apis cerana* collecting pollens. In: Proceedings of the 3rd Asian apiculture association international conference, Hanoi, p 88, Keywords: *cerana*/foraging/pollen/China

Dong X, Wang JM, Tan K (1998) Cultivation of two *Elshlotzia* nectar plant species. J Yunnan Agric Univ 13:321–323, [in Chinese]. Keywords: beekeeping/*cerana*/nectar/China

Dong X, Shi X, Wang J, Chen Y, Zhang Y (2002) Genetic divergences of *Apis cerana* from Yunnan, China and Malaysia were established by RAPD analysis. Apic China 53:9–10, 18. Keywords: *cerana*/genetics/China/Malaysia

Douglas JC (1884) A handbook of beekeeping. Indian Government Telegraph Department, Calcutta, Keywords: beekeeping/*cerana*/*dorsata*/*florea*

Douglas JC (1886) The hive-bees indigenous to India and the introduction of the Italian bee. J Asiat Soc Bengal Sci 55:83–96, Keywords: combs/*florea*/*mellifera*/India

Dreller C, Kirchner WH (1994) Hearing in the Asian honeybees *Apis dorsata* and *Apis florea*. Insect Soc 41:291–299, Keywords: *dorsata*/*florea*/physiology

Dreller C, Kirchner WH (1995) The sense of hearing in honey bees. Bee World 76:6–17, Keywords: behaviour/honey/physiology

Drieberg C (1920) Beekeeping in Ceylon. Bee World 2:38, Keywords: beekeeping/*cerana*/*dorsata*/*florea*/Sri Lanka

Drieberg C (1922) Indian bees. Bee World 3:292, Keywords: combs/defense/*florea*/honey/India

Drieberg C (1923) Bee-keeping for beginners. Bull Dept Agric (Ceyl), p 20. Keywords: beekeeping/combs/*florea*/Sri Lanka

Droege G (1962) Bibliography of German scientific apicultural literature.* Humboldt University, Berlin, Keywords: bibliography

Drory E (1888) From my notebook: notes on bees from a trip around the world. Bienenvater 20:20–30, [in German]. Keywords: combs/*florea*

Du ZL (1989) Scanning electron microscopic studies of the antennal sensillae of worker honeybees, *Apis cerana cerana*. Acta Entomol Sinica 32:166–169, [in Chinese]. Keywords: *cerana*/morphology/physiology/China

Du ZL (1993) Comparisons of the ultrastructure of the subpharyngeal gland with and without active secretion in *Apis cerana cerana* in autumn. Apic China 2:6, [in Chinese]. Keywords: *cerana*/glands/ultrastructure/China

Du ZL, Fu HL (1991) Ultrastructural changes in the midgut of worker honey bees infected with sacbrood virus. In: Proceedings of international symposium on recent research on bee pathology, Ghent, pp 176–177, Keywords: *cerana*/cytology/disease/glands/histology/honey/physiology/China

Du ZL, Li N (1991) The ultrastructure of the wax gland cell of the worker honey bee, *Apis cerana*. Acta Entomol Sinica 34:289–291, [in Chinese]. Keywords: beeswax/*cerana*/glands/honey/ultrastructure/China

Du ZL, Wu JX (1990) Electron microscope observations on the labial glands of the worker honey bee (*Apis cerana*). Acta Sci Nat Univ Pekin 26:467–473, [in Chinese]. Keywords: *cerana*/glands/ultrastructure/China

Du ZL, Zhang ZB (1984) An electron microscopic observation on the hypopharyngeal glands of the worker honey bee *Apis cerana*. Acta Entomol Sinica 27:392–395, [in Chinese]. Keywords: *cerana*/glands/honey/China

Du ZL, Zhang ZB (1985) Ultrastructural change in the hypopharyngeal glands of worker honey bees (*Apis cerana*) infected with sacbrood virus. Zool Res 6:155–162, [in Chinese]. Keywords: *cerana*/disease/glands/honey/China

Du ZL, Zhang XP (1990) The ultrastructure of the mandibular glands of the worker honey bee (*Apis cerana*). Acta Sci Nat Univ Pekin 26:232–236, [in Chinese]. Keywords: *cerana*/glands/honey/ultrastructure/China

Duan JY (1994) How to scout and catch wild colonies of *Apis cerana cerana*. Apic China 1:26, [in Chinese]. Keywords: beekeeping/*cerana*/China

Duangphakdee O, Koeniger N, Deowanish S, Hepburn HR, Wongsiri S (2009a) Ant repellent resins of honeybees and stingless bees. Insectes Soc 56:333–339, Keywords: defensiveness/*florea*/Thailand

Duangphakdee O, Radloff SE, Pirk CWW, Hepburn HR (2009b) Sun angle time windows for absconding by the dwarf honeybee, *Apis florea*. J Insect Physiol 55:1009–1012, Keywords: absconding/behaviour/*florea*/Thailand

Duddukuri GR, Kumar PS, Kumar VB, Athota RR (1997) Immunosuppressive effect of honey on the induction of allergen-specific humoral antibody response in mice. Int Arch Allerg Immunol 114:385–388, Keywords: *dorsata*/immunology

Dulta PC (1986) Comparative morphometric and biochemical studies on flight muscles of genus *Apis* and its role in apple pollination. Thesis, Himachal Pradesh University, Keywords: *cerana*/*dorsata*/flight/*florea*/physiology/pollination/India

Dulta PC, Verma LR (1987a) Comparative biometric studies on flight muscles of honeybees in the genus *Apis*. J Apic Res 26:205–209, Keywords: *cerana*/*dorsata*/flight/*florea*/*mellifera*/physiology/India

Dulta PC, Verma LR (1987b) Role of insect pollinators on yield and quality of apple fruit. Indian J Hortic 44:274–279, Keywords: *cerana*/*dorsata*/*florea*/pollination/India

Dulta PC, Verma LR (1989) Biochemical studies on flight muscles of the genus *Apis*. J Apic Res 28:136–141, Keywords: *cerana*/*dorsata*/flight/*florea*/*mellifera*/physiology/India

Dulta PC, Rana BS, Verma LR, Mattu VK (1988) Absconding behaviour of the Indian honey bee. Indian Bee J 50:67. Keywords: absconding/behaviour/brood/*cerana*/honey/migration/pests/pollen/predators/India

Dung PX, Rang NT (1996) Economic effectiveness of family beekeeping in the outskirts of Namdinh City. In: Proceedings of the 3rd Asian apiculture association international conference, Hanoi, p 150, Keywords: beekeeping/*cerana*/Vietnam

Dung NV, Long LT, Lan NK (1993) *Apis cerana* development time in northern Vietnam. In: Connor LJ, Rinderer TE, Sylvester HA, Wongsiri S (eds) Asian apiculture. Wicwas, Cheshire, CT, pp 222–225, Keywords: brood/*cerana*/physiology/Vietnam

Dung PX, Phan NH, Doi DT, Phuong BT (1996) Heating equipment experiments to reduce water content in honey. In: Proceedings of the 3rd Asian apiculture association international conference, Hanoi, p 81, Keywords: beekeeping/*cerana*/environmental effects/honey/*mellifera*/Vietnam

Dung PX, Thai PH, Trung LQ (1996) Some comparative characteristics between *Apis florea* and *Apis andreniformis* in Minh Hai Province. In: Proceedings of the 3rd Asian apiculture association international conference, Hanoi, p 62, Keywords: *andreniformis*/classification/combs/*florea*/reproduction/Vietnam

Dung PX, Trung LQ, Hap D, Matsuka M (1996) Initial results of programme on conservation for sustainable beekeeping using honeybee resources in Vietnam. In: Proceedings of the 3rd Asian apiculture association international conference, Hanoi, p 147, Keywords: *andreniformis*/*cerana*/*dorsata*/ecology/*florea*/*laboriosa*/Vietnam

Dunn KJ (1992) Exotic Asian bee detected in Torres Strait. Bee Briefs 9:18–19, Keywords: biogeography/*cerana*/Australia/Papua New Guinea

DuPraw EJ (1964) Non-Linnean taxonomy. Nature 202:849–852, Keywords: *cerana*/classification/*dorsata*/*florea*/*mellifera*

DuPraw EJ (1965a) Non-Linnean taxonomy and the systematics of honeybees. Syst Zool 14:1–24, Keywords: *cerana*/classification/*dorsata*/*florea*/*mellifera*

DuPraw EJ (1965b) The recognition and handling of honeybee specimens in non-Linnean taxonomy. J Apic Res 4:71–84, Keywords: *cerana*/classification/*dorsata*/*florea*/*mellifera*

Dutta TR, Ahmed R, Abbas SR (1983) The discovery of a plant in the Andaman Islands that tranquillizes *Apis dorsata*. Bee World 64:158–163, Keywords: bee botany/*cerana*/*dorsata*/environmental effects/India

Dutta TR, Ahmed R, Abbas SR (1984) The discovery of a plant in the Andaman Islands that tranquillizes *Apis dorsata*. Indian Bee J 46:43–45, Keywords: bee botany/*cerana*/*dorsata*/environmental effects/India

Dutta TR, Ahmed R, Abbas SR, Vasudeva-Rao MK (1985) Plants used by Andaman aborigines in gathering rock-bee honey. Econ Bot 39:130–138, Keywords: bee botany/*dorsata*/honey/India

Dutton RW, Free JB (1979) The present status of beekeeping in Oman. Bee World 60:176–185, Keywords: absconding/beekeeping/brood/classification/combs/*florea*/general biology/honey/migration/swarming/Oman

Dutton RW, Simpson J (1977) Producing honey with *Apis florea* in Oman. Bee World 58:71–76, Keywords: beekeeping/classification/combs/disease/*florea*/honey/pests/Oman

Dutton RW, Mjeni AM, Whitcombe RP (1982) Honeybees in Oman. Office of the adviser for conservation of the environment, Muscat and Durham University, p 38, Keywords: *florea*/Oman

Dyer FC (1985a) Mechanisms of dance orientation in the Asian honey bee *Apis florea*. J Comp Physiol A157:183–198, Keywords: behaviour/classification/*florea*/foraging/honey/Thailand

Dyer FC (1985b) Nocturnal orientation by the Asian honey bee, *Apis dorsata*. Anim Behav 33:769–774, Keywords: *dorsata*/foraging/honey

Dyer FC (1987a) Memory and sun compensation by honey bees. J Comp Physiol A160:621–633, Keywords: *cerana*/flight/honey/*mellifera*/India

Dyer FC (1987b) New perspectives on the dance orientation of the Asian honey bees. In: Menzel R, Mercer A (eds) Neurobiology and behavior of honey bees. Springer, Berlin, pp 54–65, Keywords: *cerana*/dance language/*dorsata*/flight/*florea*

Dyer FC (1990) Comparative analysis of the honey bee dance language: phylogeny, function, mechanism. In: Veeresh GK, Mallik B, Viraktamath CA (eds) Social insects and the environment. Oxford & IBH, New Dehli, pp 99–100, Keywords: *cerana*/dance language/*dorsata*/*florea*/honey/*mellifera*/phylogeny

Dyer FC (2002) The biology of the dance language. Ann Rev Entomol 47:917–949, Keywords: biology/*cerana*/dance language/*dorsata*/*florea*/*mellifera*/migration/swarming

Dyer FC, Seeley TD (1987) Interspecific comparisons of endothermy in honeybees (*Apis*): deviations from the expected size-related patterns. J Exp Biol 127:1–26, Keywords: *cerana*/*dorsata*/flight/*florea*/*mellifera*/physiology/Thailand

Dyer FC, Seeley TD (1991a) Dance dialects and foraging range in three Asian honey bee species. Behav Ecol Sociobiol 28:227–233, Keywords: biogeography/*cerana*/dance language/*dorsata*/*florea*/foraging/honey

Dyer FC, Seeley TD (1991b) Nesting behavior and the evolution of worker tempo in four honey bee species. Ecol 72:156–170, Keywords: behaviour/biogeography/*cerana*/colony phenology/*dorsata*/*florea*/honey/*mellifera*/nests/physiology/workers/Thailand

Dyer FC, Seeley TD (1994) Colony migration in the tropical honeybee *Apis dorsata* F. (Hymenoptera: Apidae). Insectes Soc 41:129–140, Keywords: dance language/*dorsata*/migration/nests/Thailand

Easwaramoorthy S, Jayaraj S (1987) Lack of susceptibility of Indian honey bee, *Apis cerana indica* Fabr. to two granulosis viruses infecting sugarcane borers. Indian Bee J 49:1–2, Keywords: *cerana*/disease/honey/India

Echigo T, Takanaka T, Yatsunami K (1986) Comparative studies on chemical composition of honey, royal jelly and pollen loads. Bull Fac Agric Tamagawa Univ 26:1–8, Keywords: *cerana*/honey/pollen/royal jelly/Japan

Eickwort GC (1988) The origins of mites associated with honey bees. In: Needham GR, Page RE, Delfinado-Baker M, Bowman CE (eds) Africanized honey bees and bee mites. Ellis Horwood, Chichester, pp 327–338, Keywords: *cerana*/*dorsata*/*florea*/honey/*mellifera*/pests

El Shafie HAF, Mogga JB, Basedow T (2002) Studies on the possible competition for pollen between the honey bee, *Apis mellifera sudanensis*, and the imported dwarf honey bee *Apis florea* (Hym, Apidae) in north-Khartoum (Sudan). J Appl Entomol 126:557–562, Keywords: bee botany/ecology/*florea*/foraging/honey/*mellifera*

Eltz T (2004) Spatio-temporal variation of apine bee attraction to honey baits in Bornean forests. J Trop Ecol 20:317–324, Keywords: absconding/bee botany/beekeeping/biogeography/*cerana*/classification/combs/*dorsata*/ecology/environmental effects/evolution/foraging/general biology/honey/migration/morphology/swarming/Borneo/Brunei/Indonesia/Malaysia/Singapore/Sri Lanka

Enderlein G (1906) New honeybees and contribution to the distribution of the genus *Apis*. Stettin Entomol Ztg 67:331–334, [in German]. Keywords: biogeography/*cerana*/*dorsata*/*florea*/*mellifera*

Engel MS (1999) The taxonomy of recent and fossil honey bees (Hymenoptera: Apidae: *Apis*). J Hymenoptera Res 8:165–196, Keywords: *andreniformis*/*cerana*/classification/*dorsata*/*florea*/honey/*koschevnikovi*/*mellifera*/*nigrocincta*

Engel MS (2001) The honey bees of Thailand (Hymenoptera: Apidae). Nat Hist Bull Siam Soc 49:113–116, Keywords: *andreniformis*/*cerana*/*dorsata*/*florea*/history/honey/*mellifera*/Thailand

Engel MS (2003) A replacement name in *Apis* (Hymenoptera: Apidae). J Kans Entomol Soc 76:71, Keywords: classification/*dorsata*

Engel MS, Schultz TR (1997) Phylogeny and behavior in honeybees (Hymenoptera: Apidae). Ann Entomol Soc Am 90:43–53, Keywords: *cerana*/*mellifera*/phylogeny/behaviour

Entomology Institute of Guangdong Province (1979) A review of experiments on scotch-cousin mating in *Apis cerana cerana*. Apic China 1:20–21, [in Chinese]. Keywords: *cerana*/reproduction/China

Fabricius JC (1787) Supplemental list of insects setting out recently discovered species together with their generic characteristics, specific distinguishing features, emendations and remarks. Hafniae, Proft, Copenhagen [in Latin]. Keywords: classification/*florea*

Fabricius JC (1793) Systematic description of nature in accordance with the three natural kingdoms, according to classes, orders, genera and species with their distinguishing characteristics, alternative names and habitats. Hafniae, Proft Copenhagen [in Latin]. Keywords: classification/*dorsata*

Fahrenhorst H (1975) Identification of chromosome numbers in all four species of *Apis*. In: Proceedings of the 25th international apiculture congress, Apimondia, Grenoble, p 111, Keywords: *cerana*/*dorsata*/*florea*/genetics/*mellifera*

Fahrenhorst H (1977a) Chromosome number in the tropical honeybee species *Apis dorsata* and *Apis florea*. J Apic Res 16:56–58, Keywords: *cerana*/*dorsata*/*florea*/genetics/*mellifera*

Fahrenhorst H (1977b) Uniform chromosome numbers (n=16) in the four species of *Apis*. Apidologie 8:89–100, [in German]. Keywords: *cerana*/*dorsata*/drones/*florea*/genetics/*mellifera*

Fain A, Gerson U (1993) Notes on two astigmatic mites (Acari) living in beehives in Thailand. Acarology 31:381–384, Keywords: *cerana*/pests/Thailand

Fajardo AC, Cervancia CR (2002) Pollen foraging behaviour of *Apis cerana* Fabricius. In: Proceedings of the 7th Asia apiculture association conference, Los Banos.* Keywords: bee botany/behaviour/*cerana*/foraging/pollen/Philippines

Fang TD (1956) Migration, defense and ventilation of hives in Chinese bees. In: Proceedings of the 16th international apiculture congress, Apimondia, Vienna, pp 38–39, Keywords: aggressiveness/beekeeping/*cerana*/migration/China

Fang SZ (1993) Effective cost management for *Apis cerana cerana*. Apic Sci Tech 2:11, [in Chinese]. Keywords: beekeeping/*cerana*/China

Fang WA (1981) Spring management for controlling swarming by Chinese honeybees (*Apis cerana*). Zhongguo Yangfeng 2:5, [in Chinese]. Keywords: beekeeping/*cerana*/swarming/China

Fang WF (1994) Studies on the relationship between royal jelly production and the arrangements of frames in *Apis cerana cerana*. Apic Sci Tech 2:2, [in Chinese]. Keywords: beekeeping/*cerana*/royal jelly/China

Fang WF (1995a) Bee space and frame design for hives of *Apis cerana cerana*. Apic China 1:22, [in Chinese]. Keywords: beekeeping/*cerana*/China

Fang WF (1995b) Effects of stimulative feeding on royal jelly yield in queen cells and the acceptance rate of queen cells in *Apis cerana cerana*. Apic Sci Tech 5:2, [in Chinese]. Keywords: beekeeping/cells/*cerana*/queens/royal jelly/China

Fang WF (1995c) Studies on royal jelly production in queen cups of *Apis cerana cerana*. Apic Sci Tech 1:3, [in Chinese]. Keywords: beekeeping/*cerana*/queens/royal jelly/China

Fang WF (1996) Manufacture of the FWF type hive for *Apis cerana cerana*. Apic Sci Tech 6:52–54, [in Chinese]. Keywords: beekeeping/*cerana*/nests/China

Fang WF (2002) Analysis of the chemical constituents of beeswax from *Apis cerana cerana* and *Apis mellifera ligustica*. Spectrum 7:175–176, [in Chinese]. Keywords: beeswax/*cerana*/*mellifera*/China

Fang WF, He AH (1995) Bee space in hives of *Apis cerana cerana*. J Fujian Agric Univ 24:227–230, [in Chinese]. Keywords: beekeeping/*cerana*/nests/China

Fang WF, Yao WM (1994) Relative head weight and royal jelly gland weight in workers of *Apis cerana cerana*. J Fujian Agric Coll 23:105–107, [in Chinese]. Keywords: beekeeping/*cerana*/glands/morphology/royal jelly/workers/China

Fang WF, Chen AP, Ma GY (1994) Study on the age distribution of royal jelly secretion in *Apis cerana cerana*. J Bee 5:3–5, [in Chinese]. Keywords: behaviour/biogeography/*cerana*/royal jelly/China

Fang YZ (1995d) The present situation of bee disease in beekeeping with *Apis cerana* in China. In: Kevan PG (ed) The Asiatic hive bee: apiculture, biology, and role in sustainable development in tropical and subtropical Asia. Enviroquest, Cambridge, ON, pp 191–192, Keywords: beekeeping/biology/*cerana*/disease/nests/pests/physiology/Asia/China

FAO (1986) Tropical and sub-tropical apiculture. FAO, Rome, Keywords: beekeeping/*cerana*/*dorsata*/*florea*/tropics

Fatima K, Ramanujam CGK (1989) Pollen analysis of two multifloral honeys from Hyderabad, AP. J Phytol Res 2:167–172, Keywords: *cerana*/honey/pollen/India

Fatima K, Ramanujam CGK (1993) Pollen and chemical characterization of *Apis cerana* and *Apis florea* honeys from Hyderabad urban complex. J Palynol 29:59–67, Keywords: *cerana*/*florea*/honey/pollen/India

Fernando EFW (1978) Studies on apiculture in Sri Lanka. 1. Characteristics of some honeys. J Apic Res 17:44–46, Keywords: *cerana*/*dorsata*/*florea*/honey/Sri Lanka

Fernando EFW (1979a) Some biometrical features of *Apis cerana* F from Sri Lanka. Indian Bee J 41:5–8, Keywords: *cerana*/classification/geographical variation/Sri Lanka

Fernando EFW (1979) The ecology of honey production in Sri Lanka. In: Beekeeping in rural development: unexploited beekeeping potential in the tropics with particular reference to the Commonwealth. Commonwealth Secretariat, London, pp 115–125, Keywords: beekeeping/*cerana*/*dorsata*/ecology/*florea*/honey/physiology/tropics/Sri Lanka

Filatkin GN (1992) Reservation and rational utilisation of the native honey bee (*Apis cerana cerana* F) in the Far East of the USSR. In: Readings in memory of AI Kurentsov, Institute for Biology and Soil Sciences, Far Eastern Branch of the Russian Academy of Sciences, Vladivostok, pp 117–124 [in Russian]. Keywords: *cerana*/honey/physiology/Russia

Fleming RL, Traylor MA (1964) Further notes on Nepal birds. Fieldiana: Zool 35:426–432, Keywords: combs/*laboriosa*/Nepal

Fletcher TB (1914) Some south Indian insects. Superintendent Government Press, Madras, Keywords: beekeeping/*florea*/India

Focke E (1968) The pollen spectrum of Chinese honeys. Z Bienenforsch 9:195–206, Keywords: *cerana*/honey/pollen/China

Forbes MF, Cervancia CR (1993) Foraging behaviour of *Apis cerana* F and *Apis mellifera* L (Apidae: Hymenoptera) in Majayjay, Laguna. In: Proceedings of international symposium on pollination in the tropics, Bangalore, pp 110–112, Keywords: behaviour/*cerana*/foraging/*mellifera*/pollination/tropics/India/Philippines

Forbes MF, Cervancia CR (1994) Foraging behaviour of *Apis cerana* F and *Apis mellifera* L (Apidae: Hymenoptera) in Majayjay. Philipp J Sci 123:21–27, Keywords: behaviour/*cerana*/foraging/*mellifera*/Philippines

Foster KR, Ratnieks FL (2001) The effect of sex-allocation bias on the evolution of worker policing in hymenopteran societies. Am Nat 158:615–623, Keywords: behaviour/biology/breeding/*cerana*/communication/DNA/ecology/*florea*/genetics/honey/interspecific interactions/*mellifera*/phylogeny/populations/pupae/queenlessness/queens/reproduction

Francis BR, Blanton WE, Nunamaker RA (1985) Extractable surface hydrocarbons of workers and drones of the genus *Apis*. J Apic Res 24:13–26, Keywords: *cerana*/chemistry/*dorsata*/drones/*florea*/*mellifera*/workers

Franssen CJH (1931) Beekeeping in Java and biology of *Apis indica*. Natuurhist Maandbl 20:44–48, 56–64, 71–74. Keywords: beekeeping/biology/*cerana*/Indonesia

Franssen CJH (1932) Notes on *Micrapis florea* F. Natuurk Tijdschr Ned-Indie 92:55–63, Keywords: *andreniformis*/biogeography/biology/cells/*cerana*/combs/*dorsata*/environmental effects/*florea*/nests/queens/swarming/workers/Indonesia

Free JB (1970) Insect pollination of crops. Academic, London, Keywords: *cerana*/*dorsata*/flight/*florea*/*mellifera*/pollination

Free JB (1981) Biology and behaviour of the honeybee *Apis florea*, and possibilities for beekeeping. Bee World 62:46–59, Keywords: absconding/beekeeping/behaviour/brood/combs/classification/defense/disease/*florea*/foraging/general biology/honey/migration/pests/pheromones/queens/reproduction/swarming/India/Oman

Free JB (1982a) Bees and mankind. JB Free, London, Keywords: beekeeping/biology/*cerana*/*dorsata*/*florea*/*mellifera*

Free JB (1982b) The biology and behaviour of the honeybee *Apis florea*. In: Jaisson P (ed) Social insects in the tropics, vol 1. Presses de Université Paris-Nord, Paris, pp 181–187, Keywords: absconding/behaviour/combs/ecology/*florea*/foraging/general biology/migration/swarming/Oman

Free JB (1987) Pheromones of social bees. Chapman & Hall, London, Keywords: behaviour/*cerana*/*dorsata*/*florea*/*mellifera*/pheromones

Free JB, Spencer-Booth Y (1961) Effect of temperature on *Apis indica* workers. Nature 190:933, Keywords: *cerana*/*mellifera*/physiology/workers

Free JB, Williams IH (1979) Communication by pheromones and other means in *Apis florea* colonies. J Apic Res 18:16–25, Keywords: behaviour/*florea*/pheromones

Friedmann H (1974) The Asian honeyguides. J Bombay Nat Hist Soc 71:426–432, Keywords: *cerana*/disease/*dorsata*/pests/predators

Fries I (1996) *Nosema ceranae* – a new microsporidian parasite of *Apis cerana*. Apidologie 27:279–280, Keywords: *cerana*/disease

Fries I, Feng F (1995) Crossinfectivity of *Nosema apis* in *Apis mellifera* and *Apis cerana*. In: proceedings of the 34th international apiculture congress, Apimondia, Bucharest, pp 151–155, Keywords: *cerana*/disease/*mellifera*

Fries I, Huazhen W, Wei S, Jin CS (1995) Grooming behavior and damaged mites in *Apis cerana* and *Apis mellifera*. Apidologie 26:324–325, Keywords: behaviour/*cerana*/*mellifera*/pests

Fries I, Huazhen W, Wei S, Jin CS (1996a) Grooming behavior and damaged mites (*Varroa jacobsoni*) in *Apis cerana cerana* and *Apis mellifera ligustica*. Apidologie 27:3–11, Keywords: behaviour/*cerana*/*mellifera*/pests

Fries I, Feng F, da Silva A, Slemenda SB, Pieniazek NJ (1996b) *Nosema ceranae* n. sp. (Microspora, Nosematidae), morphological and molecular characterization of a microsporidian parasite of the Asian honey bee *Apis cerana* (Hymenoptera, Apidae). Eur J Protist 32:356–365, Keywords: *cerana*/disease/parasite

Friese H (1902) On comb building in Asian species of *Apis*. Allg Z Entomol 7:198–200, [in German]. Keywords: *cerana*/combs/*dorsata*/*florea*

Friese H (1918) Wissenschaftliche Ergebnisse einer Forschungsreise nach Ostindien. VII. Bienen aus Sumatra, Java, Malakka und Ceylon. Zool Jb Abt System. Ökol Geogr 41:489–510, [in German]. Keywords: *cerana*/biology

Fuchikawa T, Shimizu I (2007) Effects of temperature on circadian rhythm in the Japanese honeybee, *Apis cerana japonica*. J Insect Physiol 53:1179–1187, Keywords: *cerana*/physiology/Japan

Fuchs S, Koeniger N (1974) Sound production as a means of defense of the honeybee colony (*Apis cerana*). Apidologie 5:271–287, [in German]. Keywords: aggressiveness/*cerana*/physiology

Fuchs S, Moritz RFA (1998) Evolution of extreme polyandry in the honeybee. Behav Ecol Sociobiol 9:269–275, Keywords: evolution/reproduction

Fuchs S, Koeniger N, Tingek S (1996) The morphometric position of *Apis nuluensis* Tingek, Koeniger and Koeniger, 1996 within cavity-nesting honey bees. Apidologie 27:397–405, Keywords: biogeography/*cerana*/classification/honey/*koschevnikovi*/*nigrocincta*/*nuluensis*/China/Sri Lanka

Fuchs S, Tu Long L, Anderson DL (2000) A scientific note on the genetic distinctness of *Varroa* mites on *Apis mellifera* L and on *Apis cerana* Fabr. in north Vietnam. Apidologie 31:459–460, Keywords: *cerana*/*mellifera*/pests/Vietnam

Fuchs S, Sen Sarma M, Werber C, Tautz J (2001) The cry of the honeybee *Apis florea*. Proceedings of the 2001 meeting of the European sections, IUSSI, Berlin, Germany, p 164, Keywords: behaviour/defense/*florea*/India

Fujiwara S (1990) Notes on the biology of *Apis cerana japonica* and efforts on behalf of its conservation in northern Japan. Honeybee Sci 11:21–26, [in Japanese]. Keywords: biology/*cerana*/ecology/Japan

Fujiwara S, Miura H, Kumagai T, Sawaguchi T, Naya S, Goto T, Asanuma H, Suzuki K (1994) Drone congregation of *Apis cerana japonica* (Radoszkowski, 1877) above large trees. Apidologie 25:331–337, Keywords: *cerana*/drones/Japan

Fujiwara S, Imai K, Fujiwara M, Kawashima T, Kobayashi S (2000) A potent antibacterial protein in royal jelly. J Biol Chem 265:11333–11337, Keywords: biochemistry/*cerana*/royal jelly/Japan

Fukuda H (1983) The relationship between work efficiency and population size in a honeybee colony. Res Pop Ecol, Kyoto Univ 25:249–263, Keywords: brood/*cerana*/colony phenology/foraging/morphology/oviposition/Japan

Fukuda H, Moriya K (1969) The weight of crop content in foraging honeybee workers. Annot Zool Jpn 42:80–90, Keywords: *cerana*/foraging/nectar/physiology/pollen/workers/Japan

Fukuda H, Sakagami SF (1968) Worker brood survival in honeybees. Res Pop Ecol, Kyoto Univ 10:31–39, Keywords: brood/*cerana*/Japan

Fukuda H, Sekiguchi K (1966) Seasonal change of the honeybee worker longevity in Sapporo, north Japan, with notes on some factors affecting the life-span. Jpn J Ecol 16:206–212, Keywords: *cerana*/colony phenology/workers/Japan

Fukuda M (1988) The Japanese honeybee and an Eastern orchid. Honeybee Sci 9:127–130, [in Japanese]. Keywords: bee botany/*cerana*/Japan

Gadau J, Wilson EO, Fewell J (2009) Organization of insect societies: from genome to socio-complexity. Harvard University Press, Cambridge, Keywords: *cerana*/*mellifera*/physiology

Gan YY, Otis GW, Mardan MB, Tan SG (1991) Allozyme diversity in Asian *Apis*. In: Smith DR (ed) Diversity in the genus *Apis*. Westview, Boulder, CO, pp 117–130, Keywords: biochemistry/*cerana*/*dorsata*/*florea*/genetics

Ganayev AI (1965) Indian bees in the Far East. Pchelovodstvo 11:11–12, [in Russian]. Keywords: biogeography/*cerana*/Russia

Gandheker SS (1959) Some common bee enemies in India. Glean Bee Cult 87:360–361, 363. Keywords: *cerana*/*dorsata*/*florea*/natural enemies/India

Gandheker SS (1962) Honeybees and their castes. Glean Bee Cult 90:142–144, Keywords: combs/*dorsata*/India

Ganeshaiah KN, Shaanker RU, Bawa KS (2001) Theme II: Tropical forests: structure, diversity and function - Part A. Plant-pollinator interaction. In: Proceedings of international conference on tropical ecosystem: structure, diversity and human welfare, Banglore, pp 15–18, Keywords: *dorsata*/foraging/pollination

Gani MO (2001) The giant honeybee (*Apis dorsata*) and honey hunting in Sundarbans reserved forests of Bangladesh. In: Proceedings of the 37th international apiculture congress, Apimondia, Durban, CD-ROM, Keywords: *dorsata*/honey/Bangladesh

Ganin GN (2000) Hymenoptera. In: Ishayev T (ed) The red data book of the Khabarovsk Territory rare and endangered species of plants and animals, 2nd edn. Institute of Water and Ecological Problems, Far East Branch of the Russian, Academy of Sciences, Khabarovsk, pp 221–231 [in Russian]. Keywords: biogeography/*cerana*/Russia

Garg A (1997) Apiculture, and its role in economy: a case study from Bhimtal. J Econ Taxon Bot 21:53–65, Keywords: beekeeping/*cerana*/India

Garg A, Nair PKK (1993) Honey bee pollen load as a bioindicator of bee pasturage in Bhimtal area of Western Himalayas, India. J Palynol 29:89–107, Keywords: *cerana*/honey/mountains/pollen/India

Garg A, Nair PKK (1994) Seasonal analysis of honey bee pollen loads from a temperate zone in Kumaon Himalaya in Uttar Pradesh, India. J Palynol 30:1–34, Keywords: bee botany/*cerana*/honey/mountains/pollen/India

Garg AK, Sethi GR, Prasad H (1977) Pollen feeding by adults of *Apis dorsata* Fabricius in captivity. Entomol Newslt 7:41–42, Keywords: bee botany/*cerana*/*dorsata*/physiology

Garg R, Kashyap NP (1998) Mites and other enemies of honeybees in India. In: Mishra RC (ed) Perspectives in Indian apiculture. Agro Botanica, New Delhi, pp 264–303, Keywords: *cerana*/*dorsata*/*florea*/*laboriosa*/*mellifera*/pests/India

Garnery L, Vautrin D, Cornuet JM, Solignac M (1991) Phylogenetic relationships in the genus *Apis* inferred from mitochondrial DNA sequence data. Apidologie 22:87–92, Keywords: *cerana*/*dorsata*/*florea*/*mellifera*/mtDNA/phylogeny

Gassparian S (1977) Studies on *Apis indica cerana* in eastern part of Iran. In: Proceedings of the 26th international apiculture congress, Apimondia, Adelaide, pp 293–296, Keywords: biogeography/*cerana*/Australia/Iran/Pakistan

Gassparian S (1978) Studies of *Apis cerana indica* in the eastern part of Iran. Quart Bull Fac Sci Tehran Univ 9:71–78, [in Persian]. Keywords: *cerana*/Iran

Gassparian S, Heydari S (1971) Genetic studies on local bee colonies of Nadjafabad when given queens from abroad. In: Proceedings of the 23rd international apiculture congress, Apimondia, Moscow, pp 361–364, Keywords: *cerana*/classification/genetics/*mellifera*/queens/Iran

Gatoria GS, Singh Y, Jhajj HS (1998) Beekeeping in the changing agricultural scenario for rural uplift. In: Mishra RC (ed) Perspectives in Indian apiculture. Agro Botanica, New Delhi, pp 14–39, Keywords: beekeeping/India

Ge FC (1992) Survey on the resources of *Apis cerana cerana* in the Changbai mountain area. Apic Hubei 2:21–22, [in Chinese]. Keywords: beekeeping/*cerana*/mountains/China

Ge FC (1995a) The long history of culturing *Apis cerana*, the Changbai mountain bee (2). Apic China 1:28* [in Chinese]. Keywords: beekeeping/*cerana*/history/China

Ge FC (1995b) The long history of culturing *Apis cerana*, the Changbai mountain bee (3). Apic China 2:22*, [in Chinese]. Keywords: beekeeping/*cerana*/history/China

Ge FC (1995c) The long history of culturing *Apis cerana*, the Changbai mountain bee (4). Apic China 3:27*, [in Chinese]. Keywords: beekeeping/*cerana*/history/China

Ge FC (1995d) The long history of culturing *Apis cerana*, the Changbai mountain bee (5). Apic China 4:27*, [in Chinese]. Keywords: beekeeping/*cerana*/history/China

Ge FC (1995e) The long history of culturing *Apis cerana*, the Changbai mountain bee (6). Apic China 5:22*, [in Chinese]. Keywords: beekeeping/*cerana*/history/China

Ge FC (1996a) The long history of culturing *Apis cerana*, the Changbai mountain bee (8). Apic China 1:21*, [in Chinese]. Keywords: beekeeping/*cerana*/history/China

Ge FC (1996b) The long history of culturing *Apis cerana*, the Changbai mountain bee (9). Apic China 2:32*, [in Chinese]. Keywords: beekeeping/*cerana*/history/China

Ge FC (1996c) The long history of culturing *Apis cerana*, the Changbai mountain bee (10). Apic China 3:29*, [in Chinese]. Keywords: beekeeping/*cerana*/history/China

Ge FC (1996d) The long history of culturing *Apis cerana*, the Changbai mountain bee (11). Apic China 4:31*, [in Chinese]. Keywords: beekeeping/*cerana*/history/China

Ge FC (1996e) The long history of culturing *Apis cerana*, the Changbai mountain bee (12). Apic China 5:28*, [in Chinese]. Keywords: beekeeping/*cerana*/history/China

Ge FC (1996f) The long history of culturing *Apis cerana*, the Changbai mountain bee (13). Apic China 6:28*, [in Chinese]. Keywords: beekeeping/*cerana*/history/China

Ge FC (1997) The long history of culturing *Apis cerana*, the Changbai mountain bee (14). Apic China 2:25*, [in Chinese]. Keywords: beekeeping/*cerana*/history/China

Ge FC, Chen DH (1993a) The history of *Apis cerana cerana* in the Changbai mountain area (1). Apic China 1:22*, [in Chinese]. Keywords: beekeeping/*cerana*/history/mountains/China

Ge FC, Chen DH (1993b) The history of *Apis cerana cerana* in the Changbai mountain area (2). Apic China 2:18*, [in Chinese]. Keywords: beekeeping/*cerana*/history/mountains/China

Ge FC, Xue YB, Niu QS (2000) Natural recovery of Chinese bee populations of Changbai mountains. In: Proceedings of the 4th Asian apiculture association international conference, Kathmandu, p 26, Keywords: *cerana*/mountains/populations

Ge FC, Li Y, Bai J, Ji L, Yan D (2002) Distribution and production efficiency of *Apis cerana* in the Changbai mountains. Apic China 53:4–7, [in Chinese]. Keywords: beekeeping/biogeography/*cerana*/geographical variation/*mellifera*/mountains/China

Gerson U, Fain A (1991) A new species of Bak (Acari: Cheyletidae) from Thailand, with a key to species. Acarolology 32:17–21, Keywords: *cerana*/pests/Thailand

Gerstäcker CEA (1862) On the geographical distribution and variation of honeybees with observations on foreign honeybees of the old world. Kraemer, Potsdam, [in German]. Keywords: biogeography/*cerana*/*dorsata*/*florea*/*mellifera*

Gerstäcker CEA (1863) On the geographical distribution and varieties of the honeybee, with remarks upon the exotic honeybees of the old world. Ann Mag Nat Hist 11:270–283, 333-347. Keywords: biogeography/classification/*dorsata*

Ghalay DR (1996) Beekeeping in Bhutan. In: Proceedings of the 3rd Asian apiculture association international conference, Hanoi, pp 208–210, Keywords: beekeeping/*cerana*/*dorsata*/*florea*/pests/Bhutan

Ghatge A (1949) Some research work on Apis florea. Indian Bee J 11:5–6, 9, Keywords: absconding/combs/drones/florea/general biology/migration/swarming/India

Ghatge A (1951) The bees of India. Indian Bee J 13:88, Keywords: beekeeping/*dorsata*/*florea*/India

Ghatge AL (1956) A scientific approach to (wild) rockbees. Ujjain (Madya Bharat) All India Khadi & Village Industries Board, p 17, Keywords: beekeeping/*dorsata*/India

Ghazali HM, Sinica MK (1986) Coconut honey: the effect of storage temperature on some of its physicochemical properties. J Apic Res 25:109–112, Keywords: *cerana*/honey/physiology/Malaysia

Ghosh CC (1915) Bee-keeping. Superintendent Government Printing, Calcutta, p 87, Keywords: beekeeping/*florea*/India

Ghosh CC (1919) Bee-keeping in India. In: Proceedings of the 3rd entomological meeting, Calcutta, pp 770–782, Keywords: beekeeping/*cerana*/*florea*/India

Ghosh CC (1936) Bee-keeping. Misc Bull No 6: Imperial Council of Agricultural Research, Delhi, p 91, Keywords: beekeeping/India

Giebel W (1969) The endopeptidase spectrum of the honeybee: comparative examinations of larvae and adult workers and drones and of the adult queen. In: Proceedings of the 22nd international apiculture congress, Apimondia, München, p 123, Keywords: beekeeping/biochemistry/*cerana*/drones/larvae/queens/workers

Giebel W, Zwilling R, Pfleiderer G (1971) The evolution of endopeptidases XII The proteolytic enzymes of the honeybee (*Apis mellifica* L). Comp Biochem Physiol 38B:197–210, Keywords: biochemistry/*cerana*/*mellifera*/phylogeny

Gill SS, Kaushik HD, Sharma SK (2001) Effect of modes of pollination on fruit set and insect pollinators of phalsa (*Grewia subinaequalis* D C). Res Crops 2:193–196, Keywords: *florea*/*dorsata*/*mellifera*/pollination

Gilliam M, Lorenz BJ, Prest DB, Camazine S (1993) *Ascosphaera apis* from *Apis cerana* from South Korea. J Invertebr Pathol 61:111–112, Keywords: *cerana*/disease/Korea

Giordani G, Scardovi V (1970) The bifid bacteria as characteristic inhabitants of different species in the genus *Apis*. J Apic Res 9:49–51, Keywords: *cerana*/disease/*dorsata*/*mellifera*/Philippines

Girish PP (1981) Role of bees in the pollination of summer squash (*Cucurbita pepo* Linne) with special reference to *Apis cerana* (Fabricius) (Hymenoptera: Apidae). Thesis, University of Agricultural Sciences, Bangalore, Keywords: *cerana*/pollination/India

Glaiim MK (1992) First definite record of *Apis florea* in Iraq. Beekeep Dev 24:3, Keywords: classification/*florea*/Iraq

Goh RA (1996) The potential of *Acacia mangium* forest plantation for a beekeeping industry in Sarawak, east Malaysia. In: Proceedings of the 3rd Asian apiculture association international conference, Hanoi, p 157, Keywords: beekeeping/*cerana*/pests/Malaysia

Goldsworthy D (1978) Honey collecting ceremonies on the east coast of north Sumatra. In: Kartomi MJ (ed) Studies in Indonesian music. Centre of Southeast Asian Studies, Monash University, pp 1–44, Keywords: *dorsata*/honey/honey hunters/Indonesia/Sumatra

Gondal SMA, Hashmi AA (1976a) Brood rearing in relation to the age of the queen in *Apis cerana indica*. J Apic Res 15:128–129, Keywords: breeding/brood/*cerana*/queens/Pakistan

Gondal SMA, Hashmi AA (1976b) Effect on egg production in *Apis cerana indica* queens on feeding them with royal jelly or crushed larvae. J Apic Res 15:130–132, Keywords: *cerana*/larvae/oviposition/queens/royal jelly/Pakistan

Gong FQ (1992a) Discussion on a perfect hive for *Apis cerana cerana* (1). J Bee 2:5–6, [in Chinese]. Keywords: beekeeping/*cerana*/nests/China

Gong FQ (1992b) Discussion on a perfect hive for *Apis cerana cerana* (2). J Bee 3:15, [in Chinese]. Keywords: beekeeping/*cerana*/nests/China

Gong FQ, Ning SR (1993a) Principles and methods of rearing *Apis cerana cerana* by supering (1). J Bee 7:17, [in Chinese]. Keywords: beekeeping/breeding/*cerana*/China

Gong FQ, Ning SR (1993b) Principles and methods of rearing *Apis cerana cerana* by supering (2). J Bee 8:16, [in Chinese]. Keywords: beekeeping/breeding/*cerana*/China

Gong FQ, Ning SR (1993c) Principles and methods of rearing *Apis cerana cerana* by supering (3). J Bee 9:18, [in Chinese]. Keywords: beekeeping/breeding/*cerana*/China

Gong FQ, Ning SR (1993d) Principles and methods of rearing *Apis cerana cerana* by supering (4). J Bee 10:15, [in Chinese]. Keywords: beekeeping/breeding/*cerana*/China

Gong FQ, Ning SR (1993e) Principles and methods of rearing *Apis cerana cerana* by supering (5). J Bee 11:19, [in Chinese]. Keywords: beekeeping/breeding/*cerana*/China

Gong FQ, Ning SR (1993f) Principles and methods of rearing *Apis cerana cerana* by supering (6). J Bee 12:14, [in Chinese]. Keywords: beekeeping/breeding/*cerana*/China

Gong FQ, Ning SR (1994) Studies on the secretion of royal jelly in *Apis cerana cerana*. J Bee 11:3, [in Chinese]. Keywords: *cerana*/glands/royal jelly/China

Gong FQ, Ning SR (1995) Design and making GN-multifunctional hives for *Apis cerana cerana*. J Bee 1:3, [in Chinese]. Keywords: beekeeping/*cerana*/China

Gong Y (1978) The Chinese honeybee. Apic China 1:10–13, [in Chinese]. Keywords: *cerana*/classification/China

Gong Y (1981) A review of Chinese apicultural history. Apic China 6:1–4, [in Chinese]. Keywords: beekeeping/*cerana*/history/China

Gong Y, Zhang Q, Su R (1987) Observation and discussion on Chinese bee mating biology. In: Proceedings of the 31st international apiculture congress, Apimondia, Warsaw, pp 184–186, Keywords: biology/*cerana*/reproduction/China

Gong YF (1983) The natural beekeeping conditions and honeybee races in China. J Fujian Agric Coll 12:241–249, Keywords: absconding/bee botany/beekeeping/behaviour/classification/disease/migration/pests/swarming/China

Gong YF (1984) Selective breeding against sacbrood in *Apis cerana cerana*. Apic China 3:1–2, [in Chinese]. Keywords: breeding/*cerana*/disease/China

Gong YF, Zhang QK (2000) The taxonomy and evolution of honeybees. Fujian Science Publishing House, Fuzhou, [in Chinese]. Keywords: classification/evolution/*florea*/China

Gong YF, Lin SG, Zhang QK (1982) Report on experimental artificial insemination of *Apis cerana cerana*. Apic China 5:1–3, [in Chinese]. Keywords: breeding/*cerana*/China

Gong YF, Zhang WJ, Su R (1987b) Observations and discussion of copulation in *Apis cerana cerana*. Apic China 1:2–3, [in Chinese]. Keywords: *cerana*/drones/reproduction/China

Gornostayev GN (1989) On insects included in the red data book of the USSR and Russian Federation. In: Amirkhanov AM (ed) Rare and conservation-needing animals. Contributions to the Red Data Book: a collection of scientific papers. Laboratory of Hunting Management and Nature Reserves, Central Research, Moscow, pp 111–118, [in Russian]. Keywords: *cerana*/conservation/Russia

Gornostayev GN (2001) The wax bee *Apis cerana* Fabricius, 1793. In: Pavlov DS (ed) The red data book of the Russian Federation: Animals. Astrel, Moscow, pp 167–168, [in Russian]. Keywords: beeswax/*cerana*/Russia

Gosh CC (2002) Bee keeping. Bulletin 46: Agricultural Research Institute, Pusa, India.* Keywords: beekeeping/*cerana*/*florea*/India

Govindappa DA, Shivamurthy GP (1975) The pollination mechanism in *Balanophora abbreviata* Blume. Ann Bot 39:977–978, Keywords: bee botany/classification/*florea*/India

Goyal NP (1974) *Apis cerana indica* and *Apis mellifera* as complementary to each other for the development of agriculture. Bee World 55:98–101, Keywords: *cerana*/*mellifera*/physiology/India

Goyal NP (1978) Performance of *Apis mellifera* and *A. cerana indica* as observed in the Punjab plains. Indian Bee J 40:3–5, Keywords: absconding/*cerana*/*mellifera*/nests/populations/swarming

Goyal NP (1993) Role of honeybees in improving agricultural productivity. In: Proceedings of the 1st National conference of beekeeping, Chandigarh, pp 61–65, Keywords: beekeeping/*cerana*/India

Goyal NP, Atwal AS (1969) Communication in Indian honey bee, *Apis indica* F. In: Proceedings of the 22nd international apiculture congress, Apimondia, München, p 419, Keywords: *cerana*/communication/India

Goyal NP, Atwal AS (1971) Comparative communication behaviour among *Apis indica* F and *Apis mellifera* L. In: Proceedings of the 23rd international apiculture congress, Apimondia, Moscow, Keywords: beekeeping/behaviour/*cerana*/communication/dance language/*mellifera*/India

Goyal NP, Atwal AS (1973) Role of honey bees in sunflower pollination. Indian Bee J 35:40–45, Keywords: *cerana*/*dorsata*/*florea*/honey/*mellifera*/pollen/pollination/India

Goyal NP, Atwal AS (1975) Studies on the relation of the population of insect pollinators with seed production in lucerne (*Medicago sativa*). Indian J Ecol 2:58–61, Keywords: bee botany/*florea*/India

Goyal NP, Atwal AS (1976) Comparative communication behaviour among *Apis cerana indica* and *Apis mellifera*: dancing behaviour. Indian Bee J 38:1–12, Keywords: behaviour/*cerana*/communication/dance language/*mellifera*/India

Goyal NP, Atwal AS (1977) Wing beat frequencies of *Apis cerana indica* and *Apis mellifera*. J Apic Res 16:47–48, Keywords: *cerana*/flight/*mellifera*

Goyal NP, Rana BS (1994) Abundance and foraging behaviour of honeybees (*Apis cerana indica* F and *Apis mellifera* L) on selected midhill flora. Indian Bee J 56:72–75, Keywords: behaviour/*cerana*/foraging/*mellifera*/India

Goyal NP, Singh M, Kandoria JL (1989) Role of insect pollination in seed production of carrot, *Dacus carota* Linn. Indian Bee J 51:89–93, Keywords: *cerana*/*dorsata*/*florea*/pollination/India

Greenspan RJ (2003) Methods for identifying compounds for motion sickness, vertigo and other disorders related to balance and the perception of gravity. US Patent, Issued: April 22:2003: Inventors: Greenspan; Ralph J (Coronado, CA), Assignee: Appl No:669751: Neurosciences Research Foundation, Inc San Diego, CA, USA, Keywords: behaviour/*cerana*/*florea*/*mellifera*/physical factors

Gregory P (2001) Report on visit to Philippines 23rd Nov - 15th Dec 2000. An audit of the potential for a Bees Abroad Development Project, unpublished report, IBRA: AA674/02. Keywords: *cerana*/*dorsata*/*florea*/honey hunting/*mellifera*

Grewal GS, Sidhu AS (1978) Insect pollinators of some curcurbits in Punjab. Indian J Agric Sci 48:79–83, Keywords: bee botany/*dorsata*/*florea*/*mellifera*/India

Grewal GS, Sidhu AS (1983) Studies on insect pollination in *Cucumis melo* Linn. In: Proceedings of the 2nd international conference on apiculture in tropical climates, New Delhi, pp 537–544, Keywords: bee botany/ecology/*florea*/India

Gries M, Koeniger N, Tingek S, Kalitu A (1996) The patient and the impatient grooms: drone mating flight behavior in two cavity nesting species in Borneo. In: Proceedings of the 3rd Asian apiculture association international conference, Hanoi, p 57, Keywords: *cerana*/drones/flight/*koschevnikovi*/nests/reproduction

Gries M, Koeniger N, Koeniger G, Tingek S, Kelitu A (1999) Different mating strategies among Asian honeybee drones. In: Proceedings of the 13th international congress, IUSSI, Adelaide, p 189, Keywords: *cerana*/*dorsata*/drones/*florea*/reproduction

Griffiths DA, Bowman CE (1981) World distribution of the *Varroa jacobsoni*, a parasite of honeybees. Bee World 62:154–163, Keywords: biogeography/*cerana*/*dorsata*/*florea*/*mellifera*/pests

Grimaldi D, Engel MS (2004) Evolution of the insects. Cambridge University Press, Cambridge, Keywords: *cerana*/*dorsata*/evolution

Grimaldi D, Underwood BA (1986) *Megabraula*, a new genus for two new species of Braulidae (Diptera), and a discussion of braulid evolution. Syst Entomol 11:427–438, Keywords: disease/evolution/*laboriosa*/pests/predators/Nepal

Grobov OF (1974) Gamasoid mites as parasites of honeybees. Veterinariya 8:78–82, [in Russian]. Keywords: *cerana*/*dorsata*/parasites/pests

Grobov OF (1981) The role of specific and interspecific differences of honey bees in relation to varroatosis. In: Petrov SG (ed) Varroatosis of bees. Moscow Society of Naturalists, Moscow (Nauka), pp 23–27, [in Russian]. Keywords: *cerana*/honey/pests/Russia

Gu PL (1992a) How to use the biological characteristics of *Apis cerana cerana*. J Bee 2:10, [in Chinese]. Keywords: beekeeping/biology/*cerana*/China

Gu WL (1992b) Techniques for keeping *Apis cerana cerana* and *Apis mellifera ligustica* in the same apiary. J Bee 10:15, [in Chinese]. Keywords: beekeeping/*cerana*/*mellifera*/China

Guanhang Y, Qinghal S (1993) Temperature regulation by *Apis cerana cerana*. In: Connor LJ, Rinderer TE, Sylvester HA, Wongsiri S (eds) Asian apiculture. Wicwas, Cheshire, CT, pp 215–221, Keywords: *cerana*/physiology/seasonal variation/China

Guo DS, Zeng ZJ, Sun LX (2009) Introduction of foreign DNA into *Apis cerana cerana* eggs by electroporation. J Anhui Agric Univ 36:2*, [in Chinese]. Keywords: *cerana*/eggs/genetics/China

Guo SC (1996) Measures for the prevention and cure of sacbrood in *Apis cerana cerana*. Apic China 3:20, [in Chinese]. Keywords: beekeeping/*cerana*/disease/China

Gupta D, Chandra K (1945) The romance of scientific beekeeping. Khadi Pratisthan, Calcutta, Keywords: beekeeping/*florea*/India

Gupta GA (1967) *Varroa jacobsoni*: a mite of *Apis indica*. Bee World 48:17–18, Keywords: *cerana*/pests/India

Gupta JK (1999) Spider, *Neoscona hadensis* (Tikader) (Argiopidae): an enemy of honey bees. Pest Manag Econ Zool 7:175–176, Keywords: *cerana*/honey/predators/India

Gupta JK, Dogra GS (1998) Constraints and thrust areas for the development of apiculture in India. In: Dogra RC (ed) Perspectives in Indian Apiculture. Agro Botanica, New Delhi, pp 1–13, Keywords: beekeeping/India

Gupta JK, Reddy MCM (1992) Foraging intensity of insects and nectar of wild cherry, *Prunus puddum* Roxb. Entomon 17.91–94, Keywords: *cerana*/foraging/*mellifera*/nectar/India

Gupta JK, Sharma PL (1997) Effect of endosulfan and oxydemeton methyl sprays on the activity of insect pollinators on cauliflower. Pest Manag Econ Zool 5:67–69, Keywords: *cerana*/insecticides/*mellifera*/India

Gupta JK, Sharma PL (2000) Impact of number of bee visits on cauliflower pollination. In: Proceedings of the 4th Asian apiculture association international conference, Kathmandu, pp 183–185, Keywords: beekeeping/*cerana*/*mellifera*/pollination/India

Gupta JK, Kumar J, Mishra RC (1984a) Nectar sugar production and honeybee foraging activity in different cultivars of cauliflower, *Brassica oleracea* var *botrytis*. Indian Bee J 46:21–22, Keywords: *cerana*/foraging/nectar/India

Gupta JK, Mishra RC, Kumar J (1984b) *Plectranthus* as forage for *Apis cerana indica* F and *Apis mellifera* L. Apidologie 15:75–82, Keywords: bee botany/*cerana*/foraging/*mellifera*/India

Gupta JK, Kumar J, Kumar R (1990a) Observations on the foraging activity of honey bees and nectar production in 'toon', *Toona ciliata* Roem. Indian Bee J 51:34–35, Keywords: *cerana*/foraging/honey/nectar/India

Gupta JK, Reddy MCM, Kumar J (1990b) Pattern of nectar secretion in wild cherry, *Prunus puddum* Roxb, and the associated foraging behaviour of *Apis cerana indica* F and *Apis mellifera* L. Apidologie 21:11–16, Keywords: behaviour/*cerana*/foraging/*mellifera*/nectar/India

Gupta JK, Rana BS, Sharma HK (2000) Pollination of kiwifruit in Himachal Pradesh. In: Proceedings of the 4th Asian apiculture association international conference, Kathmandu, pp 169–170, Keywords: beekeeping/*cerana*/*mellifera*/pollination/India

Gupta M (1985) Olfacto-gustatory response of *Apis florea* Fab. to aliphatic straight chain ketones. Indian Bee J 47:30–33, Keywords: biochemistry/*florea*/organic chemistry/physiology/India

Gupta M (1986a) A quantitative study and ultrastructure of flagellar sensillae of *Apis florea* F (Hymenoptera: Apidae). Proc Indian Acad Sci, Anim Sci 95:595–603, Keywords: *florea*/physiology/morphology/India

Gupta M (1986b) Conditioning of *Apis florea* F as a function of duration of exposure to chemical vapours at Rc50 and Rc60 dilutions. Beitr Entomol 36:225–229, Keywords: behaviour/*florea*/physiology/India

Gupta M (1986c) Olfactory response of *Apis florea* Fabr. to some cyclic ketones. Indian Bee J 48:44–46, Keywords: biochemistry/*florea*/organic chemistry/physiology/India

Gupta M (1987a) Behaviour of *Apis florea* Fabr. to some ketones and essential oils under semi-field conditions. Indian Bee J 49:29–32, Keywords: behaviour/biochemistry/*florea*/organic chemistry/pheromones/India

Gupta M (1987b) Efficacy of ketones on the foraging behaviour of *Apis florea* F in field conditions. Apidologie 18:121–128, Keywords: behaviour/biochemistry/*florea*/foraging/organic chemistry/India

Gupta M (1987) Essential oils: a new source of bee repellents. Chem Ind, March 162–163, Keywords: biochemistry/*florea*/organic chemistry/pheromones/India

Gupta M (1987d) Olfacto-gustatory response of *Apis florea* Fabr. to some essential oils. Indian Bee J 49:5–10, Keywords: biochemistry/*florea*/organic chemistry/pheromones/physiology/India

Gupta M (1987e) Olfactory response of *Apis florea* F to aliphatic straight chain ketones. J Anim Morph Physiol 34:117–130, Keywords: biochemistry/*florea*/organic chemistry/physiology/India

Gupta M (1987f) Olfactory response of *Apis florea* F to some essential oils. J Apic Res 26:310, Keywords: biochemistry/*florea*/organic chemistry/pheromones/physiology/India

Gupta M (1988a) Olfacto-gustatory response of *Apis florea* F to some halogen substituted ketones. Indian J Anim Res 22:55–59, Keywords: biochemistry/*florea*/organic chemistry/physiology/India

Gupta M (1988b) Olfactory response of *Apis florea* F to some phenolic ketones. Comp Physiol Ecol 13:64–68, Keywords: biochemistry/*florea*/organic chemistry/physiology/India

Gupta M (1989a) Conditioning of *Apis florea* F worker bees as a function of exposure to chemical vapours at 0.05 N concentration. Zool Jahrb Abt Allg Zool Physiol Tiere 93:215–218, Keywords: behaviour/*florea*/India

Gupta M (1989b) Effect of repellents on the foraging behaviour of *Apis florea* F under field conditions. J Res Haryana Agric Univ 19:6–15, Keywords: behaviour/biochemistry/*florea*/foraging/organic chemistry/physiology/India

Gupta M (1989c) Olfacto-gustatory response of *Apis florea* to some repellent ketones. Zool Jahrb Abt Allg Zool Physiol Tiere 93:105–111, Keywords: biochemistry/*florea*/organic chemistry/physiology/India

Gupta M (1989d) Olfactometric evaluation of the response of *Apis florea* to some repellent ketones. Zool Jahrb Abt Allg Zool Physiol Tiere 93:97–104, Keywords: biochemistry/*florea*/organic chemistry/physiology/India

Gupta M (1992) Scanning electron microscopic studies of antennal sensilla of adult worker *Apis florea* F (Hymenoptera: Apidae). Apidologie 23:47–56, Keywords: *florea*/morphology/physiology/India

Gupta M (1993) Adaptation of workers of *Apis florea* F as a function of duration of effectivity of compounds. Korean J Apic 8:157–164, Keywords: behaviour/biochemistry/*florea*/organic chemistry/pheromones/India

Gupta M, Kapil RP (1984) Olfactory response of workers of *Apis florea* F to some repellents in semi-field conditions. In: Proceedings of the 5th international symposium on pollination, Versailles, pp 65–78, Keywords: bee botany/biochemistry/*florea*/organic chemistry/physiology/workers/India

Gupta M, Mohla RS (1986) Aliphatic straight chain ketones as potential bee repellents. Chem Ind, May:327–328, Keywords: biochemistry/*florea*/organic chemistry/India

Gupta M, Pal HM, Pal B (1986) Olfacto-gustatory screening of n-propionaldehyde and n-butyraldehyde to *Apis florea* Fabr. in laboratory conditions. Indian Bee J 48:51–52, Keywords: biochemistry/*florea*/organic chemistry/physiology/India

Gupta PR, Chandel RS (1995) Effects of diflubenzuron and penfluron on workers of *Apis cerana indica* F and *Apis mellifera* L. Apidologie 26:3–10, Keywords: *cerana*/insecticides/*mellifera*/workers/India

Ha TD (1991) Scientific and technical strategy for the beekeeping development in Vietnam. In: Proceedings of the 1st National conference of entomology, Hanoi.* Keywords: beekeeping/*cerana*/*mellifera*/physiology/Vietnam

Haas JI, Probasco G (2009) Compositions and methods for inhibiting a honey bee pathogen infection or controlling a hive infestation. US Patent Appl 20090104288, Keywords: disease

Hachinohe Y, Onishi N (1954) Studies on the length of proboscis of the worker honeybee. Bull Natl Inst Agric Sci Ser G (Anim Husb) 8:9–16, Keywords: *cerana*/classification/morphology/Japan

Haddad N, Fuchs S, Hepburn HR, Radloff SE (2009) *Apis florea* in Jordan: source of the founder population. Apidologie 40:508–512, Keywords: biogeography/*florea*/Jordan

Hadisoesilo S (1989) Daily flight activity of *Apis cerana* in Aek Nauli, north Sumatra and Kuok, Riau. Bull Penelitian Kehutanan 5:231–242, [in Indonesian]. Keywords: *cerana*/flight/foraging/Indonesia

Hadisoesilo S (1990a) A preliminary study on the swarm of *Apis cerana* at Kuok, Riau Province, Indonesia. Bull Penelitian Kehutanan 5:357–363, Keywords: brood/cells/*cerana*/colony phenology/combs/*mellifera*/morphometrics/queens/swarming/workers/Indonesia/Pakistan

Hadisoesilo S (1990b) Honey quality from different apiaries in Java, Sumatra, and Irian Jaya. Bull Penelitian Kehutanan 6:69–80, [in Indonesian]. Keywords: *cerana*/honey/*mellifera*/Indonesia/Irian Jaya

Hadisoesilo S (1990c) Preferences of *Apis cerana* for different concentrations of sugar syrup. Bull Penelitian Kehutanan 6:49–57, [in Indonesian]. Keywords: *cerana*/foraging/Indonesia

Hadisoesilo S (1997) A comparative study of two species of cavity-nesting honey bees of Sulawesi, Indonesia. Thesis, University of Guelph, Keywords: *cerana*/honey/*nigrocincta*/Indonesia

Hadisoesilo S (2001) Diversity in traditional techniques for enticing *Apis dorsata* colonies in Indonesia. In: Proceedings of the 37th international apiculture congress, Apimondia, Durban, CD-ROM, Keywords: biogeography/*dorsata*/Indonesia

Hadisoesilo S (2001) Tingku, a traditional "management" technique for *Apis dorsata binghami* in central Sulawesi, Indonesia. In: Proceedings of the 7th international conference on tropical climates, Chiang Mai, pp 309–312, Keywords: beekeeping/*dorsata*/Indonesia

Hadisoesilo S, Kuntadi TH (1989) Moisture, sucrose, and hydroxymethylfurfural contents of *Apis dorsata* honey in Riau Province. Bull Penelitian Kehutanan 5:143–152, [in Indonesian]. Keywords: *dorsata*/honey/Indonesia

Hadisoesilo S, Otis GW (1995) Morphometric analysis reveals two distinct species of cavity-nesting honey bees (Hymenoptera: Apidae) in Sulawesi, Indonesia. J Kans Entomol Soc 68:399–407, Keywords: *cerana*/honey/morphology/*nigrocincta*/Indonesia

Hadisoesilo S, Otis GW (1996) Drone flight times confirm the species status of *Apis nigrocincta* Smith, 1861 to be a species distinct from *Apis cerana* F, 1793 in Sulawesi, Indonesia. Apidologie 27:361–369, Keywords: *cerana*/drones/flight/*nigrocincta*/reproduction/Indonesia

Hadisoesilo S, Otis GW (1998) Differences in drone cappings of *Apis cerana* and *Apis nigrocincta*. J Apic Res 37:11–15, Keywords: cells/*cerana*/drones/*nigrocincta*

Hadisoesilo S, Purwanto DB (1987) Construction of emergency queen cells and establishment of new laying queens in *Apis cerana* colonies at Dramaga Experimental Garden, Bogor. Bull Penelitian Kehutanan 3:111–118, Keywords: cells/*cerana*/queens/Indonesia

Hadisoesilo S, Shanti EN, Kuntadi TH (2001) Drone flight times of *Apis cerana* and *Apis mellifera* in the island of Java. In: Proceedings of the 37th international apiculture congress, Apimondia, Durban (CD Rom), Keywords: *cerana*/drones/flight/islands/*mellifera*/Indonesia

Hadisoesilo S, Raffiudin R, Susanti W, Atmowidi T, Hepburn C, Radloff SE, Fuchs S, Hepburn HR (2008) Morphometric analysis and biogeography of *Apis koschevnikovi* Enderlein (1906). Apidologie 39:495–503, Keywords: biogeography/*koschevnikovi*/morphology

Hadorn H (1946) Observations on wild bees in Sumatra. Schweiz Bienenztg 69:309–314, [in German]. Keywords: behaviour/*cerana*/*dorsata*/*florea*/nests/Indonesia

Hadorn H (1977) The Indian giant honeybee *Apis dorsata*. Schweiz Bienenztg 10:398–404, Keywords: absconding/*dorsata*/migration/swarming/Indonesia

Hagmeister A (1973) Immunological comparison of the haemolymph proteins of the four existing *Apis* species: *mellifera*, *cerana*, *florea* and *dorsata*. In: Proceedings of the 7th international congress, IUSSI, London, pp 151–156, Keywords: biochemistry/*cerana*/disease/*dorsata*/*florea*/*mellifera*/physiology

Hagmeister A (1973) Immunological investigations on the occurrence (except in oocytes) of vitellogenic protein and its interspecific differences in the genus *Apis*. Thesis, Universität Münster, [in German]. Keywords: biochemistry/*cerana*/*dorsata*/*florea*/*mellifera*

Halling LA, Oldroyd BP, Patimus B, Wattanaachaiyingchaeron W, Barron AB, Nanork P, Wongsiri S (2002) Worker policing in the bee *Apis florea*. Behav Ecol Sociobiol 49:509–513, Keywords: behaviour/*florea*/Thailand

Hamakawa H (1968) Some observations on the foraging behaviour of honeybees, *Apis mellifera* and *A. cerana*, in the flowers of *Parthenocissus tricuspidata*. Jpn Bee J 240:106–110, [in Japanese]. Keywords: bee botany/behaviour/*cerana*/foraging/*mellifera*/Japan

Hamakawa H (1986) Foraging behavior of honeybees (*Apis cerana* and *Apis mellifera*) and blackish hornet visiting the flowers of buckwheat (*Fagopyrum esculentum*). In: Proceedings of the 30th international apiculture congress, Apimondia, Nagoya, pp 371–373, Keywords: *cerana*/foraging/*mellifera*/Japan

Hamakawa H (1986b) Foraging behaviour of honeybees (*Apis cerana* and *A mellifera*) and *Vespula flaviceps* visiting flowers of buckwheat (*Fagopyrum esculentum*). Honeybee Sci 7:53–56, [in Japanese]. Keywords: bee botany/behaviour/*cerana*/foraging/*mellifera*/Japan

Hamakawa H, Morimoto H (1967) Studies on honeybee foragers. Apis mellifera and A. cerana foraging behaviour April to November. Jap J Zootech Sci 38(Suppl):111, [in Japanese]. Keywords: behaviour/*cerana*/foraging/*mellifera*/Japan

Hameed SF, Adlakha RL (1973a) Preliminary comparative studies on the brood rearing activities of *Apis mellifera* L and *Apis cerana indica* F in Kulu Valley. Indian Bee J 35:27–35, Keywords: breeding/brood/*cerana*/*mellifera*/India

Hameed SF, Adlakha RL (1973b) Scarabaeid beetles a new menace to beekeepers in Kulu Valley. Indian Bee J 35:48, Keywords: *cerana*/*mellifera*/natural enemies/pests/pollen/India

Hameed SF, Singh B (1989) Effect of *Varroa jacobsoni* Oud infestation on foraging and brood rearing in *Apis cerana indica* Fabr. Indian Bee J 51:135–136, Keywords: breeding/brood/*cerana*/foraging/pests/rearing/India

Hameed SF, Singh B (1990) Effectiveness of various fumigants against *Varroa jacobsoni* mites in *Apis cerana indica* F colonies. J Entomol Res 14:75–77, Keywords: *cerana*/insecticides/pests/India

Hameed SF, Singh SP (1998) Pesticide bee poisoning in changing scenario of Indian agriculture. In: Mishra RC (ed) Perspectives in Indian apiculture. Agro Botanica, New Delhi, pp 334–362, Keywords: *cerana*/insecticides/India

Hamlyn-Harris R (1902) *Apis dorsata* F considered in the light of domestication. Entomol Rec J Var 14:321, Keywords: *dorsata*

Han SM, Yang GH (1991) Scanning electron microscope studies of sensilla on the antenna, strigil and corbicula of the worker honeybee *Apis cerana cerana*. Apic China 2:22, [in Chinese]. Keywords: *cerana*/morphology/physiology/ultrastructure/China

Hänel H, Koeniger N (1986) Possible regulation of the reproduction of the honey bee mite *Varroa jacobsoni* (Mesostigmata: Acari) by a host's hormone, juvenile hormone III. J Insect Physiol 32:791–798, Keywords: *cerana*/honey/hormones/*mellifera*/reproduction
Hänel H, Ruttner F (1985) The origin of the pore in the drone cell capping of *Apis cerana* Fabr. Apidologie 16:157–164, Keywords: cells/*cerana*/drones/nests
Hanh PD, Kustermann T (1996) Family beekeeping: a contribution to the protection of National Parks. In: Proceedings of the 3rd Asian apiculture association international conference, Hanoi, pp 148–150, Keywords: beekeeping/*cerana*/Vietnam
Hannan MA (2000) Beekeeping and apicultural products in Bangladesh. Honeybee Sci 21:154–158, [in Japanese]. Keywords: beekeeping/*cerana*/*dorsata*
Haq MA (2001) Life cycle and behaviour of the coconut mite *Neocypholaelaps stridulans* (Evans) (Acari: Ameroseiidae) in India. In: Proceedings of the 10th international congress of Acarology, Canberra, pp 361–365, Keywords: behaviour/*cerana*/*mellifera*/pests/India
Hara A (1980) *Varroa jacobsoni* - an ectoparasitic mite of the honeybee. Honeybee Sci 1:17–20, [in Japanese]. Keywords: *cerana*/pests/Japan
Hara M (1981) Japanese honeybees in 'Kumanoji' (Wakayama Prefecture). Honeybee Sci 2:157–160, [in Japanese]. Keywords: beekeeping/*cerana*/Japan
Hara M (1987) *Apis cerana japonica* in Tostukawa Village, Nara Prefecture. Honeybee Sci 8:11–16, [in Japanese]. Keywords: beekeeping/*cerana*/Japan
Hara M (1991) Beekeeping by the Lord Naiaki Tokugawa. Honeybee Sci 12:163–169, [in Japanese]. Keywords: beekeeping/*cerana*/history/Japan
Hara M (2002) Honeybees in Sri Lanka. Honeybee Sci 17:77–80, [in Japanese]. Keywords: beekeeping/*cerana*/*mellifera*/Sri Lanka
Hari C, Yaquin E, Singh R (2002) Biological development of *Apis mellifera* L and *Apis cerana indica* Fab. during different seasons under north Bihar conditions. Shashpa 9:133–137, Keywords: biology/*cerana*/*mellifera*/physiology/India
Hasan SB, Deo PG, Majumder SK (1986) Relative toxicity of insecticides to household and beneficial insects. Indian Bee J 48:42–44, Keywords: *cerana*/insecticides/India
He CX, Liu YS (1996) Analysis of and countermeasures for white-head pupa disease in *Apis cerana cerana*. J Bee 8:18*, [in Chinese]. Keywords: beekeeping/*cerana*/disease/China
Hedge KK (1942) Collection of rock bee honey in Bombay Presidency. Indian Bee J 4:21–23, Keywords: beekeeping/*dorsata*/honey/India
Hedicke H (1926) Contribution to the apine fauna of the Philippines. Dtsch Entomol Z 5:413–423, [in German]. Keywords: biogeography/cerana/Philippines
Hepburn HR (2009) Wax discrimination in the lesser honeyguide, *Indicator minor*. Ostrich 80:119–120, Keywords: *cerana*/*dorsata*/*florea*/*mellifera*/waxes
Hepburn HR, Hepburn C (2005) A bibliography on *Apis florea*. Apidologie 36:377–378, Keywords: bibliography/*florea*
Hepburn HR, Hepburn C (2006) Bibliography of *Apis cerana*. Apidologie 37:651, Keywords: bibliography/*cerana*
Hepburn HR, Hepburn C (2007a) Bibliography of the giant honeybees *Apis dorsata* Fabricius (1793) and *Apis laboriosa* F. Smith (1871). Apidologie 38:219–220, Keywords: bibliography/*dorsata*/*laboriosa*
Hepburn HR, Hepburn C (2007b) Bibliography of *Apis koschevnikovi* Enderlein (1906). Apidologie 38:507, Keywords: bibliography/*koschevnikovi*
Hepburn HR, Hepburn C (2009) Bibliography of *Apis andreniformis*. Apidologie 40:3, Keywords: *andreniformis*/bibliography
Hepburn HR, Radloff SE (2004) The wing coupling apparatus and the morphometric analysis of honeybee populations. S Afr J Sci 100:565–570, Keywords: *cerana*/*dorsata*/general biology
Hepburn HR, Radloff SE, Verma S, Verma LR (2001a) Morphometric analysis of *Apis cerana* populations in the southern Himalayan Region. Apidologie 32:435–448, Keywords: *cerana*/classification/physical factors/populations

Hepburn HR, Smith DR, Radloff SE, Otis GW (2001b) Infraspecific categories of *Apis cerana*: morphometric, allozymal and mtDNA diversity. Apidologie 32:3–23, Keywords: biochemistry/biogeography/*cerana*/classification/DNA/genetics/morphology/mtDNA
Hepburn HR, Radloff SE, Otis GW, Fuchs S, Verma LR, Tan K, Chaiyawong T, Tahmasebi GH, Wongsiri S (2005) *Apis florea*: multivariate morphometric analysis and biogeography. Apidologie 36:359–376, Keywords: classification/*florea*/morphology/Afghanistan/Bangladesh /Cambodia/China/India/Indonesia/Iran/Laos/Malaysia/Myanmar/Nepal/Oman/Pakistan/Saudi Arabia/Sri Lanka/Thailand/UAE/Vietnam
Hepburn HR, Duangphakdee O, Radloff SE (2009a) Comb wax salvage by the red dwarf honeybee, *Apis florea*. J Insect Behav 23:159–164, Keywords: behaviour/*florea*/wax
Hepburn HR, Radloff SE, Phiancharoen M, Duangphakdee O (2009b) Interspecific utilisation of wax in comb building by honeybees. Naturwissenschaften 96:719–723, Keywords: *cerana*/*dorsata*/*florea*/*mellifera*/waxes
Hepburn R, Duangphakdee O, Phiancharoen M, Radloff S (2010) Comb wax salvage by the red dwarf honeybee, *Apis florea* F. J Insect Behav 23:159–164, Keywords: *florea*/wax
Hermann HR (1971) Sting autotomy, a defensive mechanism in certain social Hymenoptera. Insectes Soc 18:111–120, Keywords: *cerana*/*dorsata*/stings
Higgs JS, Wattanachaiyingcharoen W, Oldroyd BP (2009) Scientific note on a genetically-determined color morph of the dwarf honey bee, *Apis andreniformis*. Apidologie 40:513–514, Keywords: *andreniformis*/genetics/Thailand
Higgs JS, McHale M, Oldroyd BP (2010) A scientific note on a rapid method for the molecular discrimination of *Apis andreniformis* and *A. florea*. Apidologie 41:96–98, Keywords: *andreniformis*/classification/*florea*
Hiran S, Pande TK, Pani S, Gupta R, Vishwanathan KA (1994) Rhabdomyolysis due to multiple honeybee stings. Postgrad Med J 70:937, Keywords: defense/dorsata/honey/India
Hirashima Y (1991) Natural history of wild bees and honeybees. Honeybee Sci 12:49–54, [in Japanese]. Keywords: combs/*dorsata*/honey/Japan
Hiratsuka Y (1920) Bees and beekeeping in Formosa. Bee World 2:103, Keywords: beekeeping/*cerana*/China/Taiwan
Ho KK, Chen YC (2001) Susceptibility of the Asian honey bee (*Apis cerana*) to American foulbrood (*Paenibacillus larvae*). In: Proceedings of the 37th international apiculture congress, Apimondia, Durban (CD Rom), Keywords: *cerana*/disease/larvae/Korea
Hoekman J (1929) The honeybee in the Dutch East Indies. Ind Boerd 20:407–411, Keywords: *cerana*/Indonesia
Hon CC (1962) Investigation of Taiwan beekeeping. Plant Prot Bull (Taipei) 4:28–29, Keywords: beekeeping/*cerana*/China
Hongsongkeith S (1988) Beekeeping. Thesis, Faculty of Agricultural Technology, King Mongkut's Institute of Technology, Bangkok [in Thai]. Keywords: beekeeping/*cerana*/Thailand
Hooker JD (1854) Himalayan journals. John Murray, London. Keywords: combs/*laboriosa*/Nepal
Hooper D (1904) Indian bees'-wax. Agric Ledg 7:201–238, Keywords: bee botany/beeswax/*cerana*/*dorsata*/*florea*/India
Hooper D, Buchner G (1906) On Indian beeswaxes. Anal Bioanal Chem 45:657–658, [in German]. Keywords: *cerana*/*dorsata*/*florea*/nests
Hoopingarner R, Connor LJ (1999) Biology of parasitic Asian bee mites and their honey bee hosts. In: Hoopingarner R, Connor LJ (eds) Apiculture for the 21st Century. Wicwas, Cheshire, CT*, Keywords: biology/*cerana*/*dorsata*/*florea*/honey/pests/Asia
Horibe K (1991) My experience with honeybees. Honeybee Sci 12:175–178, [in Japanese]. Keywords: beekeeping/*cerana*/Japan
Horn E (1975) Mechanisms of gravity processing by leg and abdominal gravity receptors in bees. J Insect Physiol 21:673–679, Keywords: *cerana*/*florea*/*mellifera*/physical factors/physiology
Horne C, Smith F (1872) Notes on the habits of some hymenopterous insects of the north-west provinces of India. Trans Zool Soc Lond 7:161–185, Keywords: disease/pests/India

Hoshiba H (1981) Sex determination and cytogenetical studies of honeybees. Honeybee Sci 2:5–10, [in Japanese]. Keywords: *cerana*/cytology/genetics/histology
Hoshiba H, Akio K (1978) Karyological study of honeybees. J Apic Res 17:106–109, Keywords: *cerana*/genetics
Hoshiba H, Okada I (1986) G-banding analyses of male chromosomes in *Apis cerana* and *A. mellifera ligustica*. Apidologie 17:101–105, Keywords: *cerana*/genetics/*mellifera*/Japan
Hoshiba H, Okada I, Kusanagi A (1981) The diploid drone of *Apis cerana japonica* and its chromosomes. J Apic Res 20:143–147, Keywords: *cerana*/drones/genetics/Japan
Hossain ABME, Sharif M (1988) Beekeeping potentials of Bangladesh I Studies on the important honey and pollen sources of northern Chittagong Region. Bangladesh J Zool 16:23–30, Keywords: beekeeping/*cerana*/*dorsata*/*florea*/honey/*mellifera*/pollen/Bangladesh
Hossain ABME, Sharif M (1991) Control of mite infestations of hives of *Apis cerana* Fabr. (Hymenoptera: Apidae). Bangladesh J Zool 19:101–106, Keywords: *cerana*/pests/Bangladesh
Hossain ABME, Sharif M, Laskar MR (1989) Beekeeping potentials of Bangladesh: II Annual honey production by *Apis cerana* L at Mirsarai area of northern Chittagong Region. Bangladesh J Zool 17:75–82, Keywords: beekeeping/*cerana*/honey/nests/Bangladesh
Hossain ABME, Sharif M, Baqui AH (1992) Nest site selection of the Asian hive bee, *Apis cerana* F, in Bangladesh. J Apic Res 31:45–46, Keywords: breeding/*cerana*/nests/Bangladesh
Hossain ABME, Sharif M, Baqui MA (1993) Appropriate time for colony nursing and honey extraction of *Apis cerana* Fab. during honey flow season. Bangladesh J Zool 21:35–39, Keywords: *cerana*/honey/Bangladesh
Hsieh FK (2001) Beekeeping in Taiwan. Honeybee Sci 12:159–162, [in Japanese]. Keywords: beekeeping/*cerana*/*mellifera*/China
Hsieh FK, Chen CT, Chang CP, Chang SY (2002) Foraging activities and numerical changes of honeybees on buckwheat, rape and pear. Plant Prot Bull (Taipei) 44:1–13, [in Chinese]. Keywords: *cerana*/foraging/*mellifera*/pollination/China
Hu S, Shuai Q, Shao X, Sui W (2000) Study of the extraction and analysis of bee larva extraction and determination of flavones. Apic China 51:7–9, [in Chinese]. Keywords: *cerana*/larvae/China
Hu Y-L, Chen XX, Chen WL, He J-H (1999) The antibacterial substances in the Chinese honeybee, *Apis cerana* Fabricius: purification and partial characterization. J Zhejiang Agric Univ 25:307–310, Keywords: *cerana*/China
Huang BY (1986) Varieties of *Apis* in China. Apic China 5:7–9, [in Chinese]. Keywords: *cerana*/classification/China
Huang Q, Zeng ZJ (2009) Nepotism in swarming honeybees (*Apis cerana cerana*). Chinese Bull Entomol 46:1, * Keywords: *cerana*/swarming/China
Huang Q, Zeng ZJ, Yan WY, Huang K (2009) Decision making plasticity of swarming honeybees (*Apis cerana cerana*). Res J Biotech 4.* Keywords: *cerana*/genetics/swarming
Huang W (1990) The effects of cell size on morphological characteristics of worker honey bees. Natl Enemies Insects 12:129–131, [in Chinese]. Keywords: cells/*cerana*/classification/honey/*mellifera*/morphology/China
Huang WC, Yang GH, Chen SB (1963) Preliminary study of the biological characters of *Apis cerana cerana*. Sci Agric Sinica 1:31–32, [in Chinese]. Keywords: biology/*cerana*/China
Huang WC, Yang GH, Chen SB (2002) Pilot study on the biological characteristics of *Apis cerana cerana*. China Agric Sci 1:43–44, [in Chinese]. Keywords: biology/*cerana*/China
Huang ZY, Kuang HO, Kuang BY, Qin Y (2001) Juvenile hormone and division of labour in *Apis cerana*. J Apic Res 40:3–6, Keywords: behaviour/*cerana*/hormones
Hukusima S, Inoue A (1964) Respiration rates of Japanese and European honeybees during various stages of development. Kansai-Byochyngai-Kenkyakai-Ho 6:36–44, [in Japanese]. Keywords: *cerana*/interspecific interactions/*mellifera*/physiology/Japan
Hunter-Fujita FR, Mossadegh MS, Taylor SC, Lawson PA, Ball BV (2001) Microbiological investigations of the dwarf honey bee *Apis florea*. In: Proceedings of the 37th international apiculture congress, Apimondia, Durban (CD Rom), Keywords: disease/*florea*/honey/pests

Husain SW (1938) How honey is extracted from the combs of the giant bee of India. Bee World 19:139–140, Keywords: absconding/combs/*dorsata*/environmental effects/honey/migration/swarming/India

Husain SW (1939a) Bee sting. Indian Bee J 1:96–97, Keywords: defense/*dorsata*/India

Husain SW (1939b) Honey from the giant bee. Indian Bee J 1:79–82, Keywords: *dorsata*/honey/India

Huson DH (2005) Splits Tree: a progam for analysing and visualizing evolutionary data. Bioinformatics 141:68–73, Keywords: *cerana*/*dorsata*/evolution

Hussain SA, Ali S (1980) Beehive predation by wasps (genus *Vespa*) and its possible benefit to honeyguides (Indicatoridae) in Bhutan. J Bombay Nat Hist Soc 76:157–159, Keywords: disease/*dorsata*/pests/predators/Bhutan

Hussain SW (1939) Present position of beekeeping in India. Indian Bee J 1:53–55, Keywords: beekeeping/*cerana*/natural enemies/India/Malaysia

Ibrahim R, Yusoh MR (1989) Control of wasps and weaver ants in *Apis cerana* apiaries in west Malaysia. In: Proceedings of the 4th international conference on apiculture in tropical climates, Cairo, pp 377–378, Keywords: *cerana*/natural enemies/physical factors/Malaysia

Ichino H, Okada I (1994) Japanese honey bees living in hornet nest. Honeybee Sci 15:123–124, [in Japanese]. Keywords: *cerana*/honey/natural enemies/nests/Japan

ICIMOD (1989) Honeybees in mountain agriculture. In: Proceedings of international expert meeting on beekeeping development in the Hindu Kush-Himalayan region, ICIMOD, Kathmandu,* Keywords: beekeeping/*cerana*/physiology/Nepal

ICIMOD (1994a) Exploration of genetic diversity, *Apis cerana*. ICIMOD, Kathmandu*, Keywords: *cerana*/genetics

ICIMOD (1994b) Promotion and development of beekeeping through preservation of indigenous *Apis cerana*. ICIMOD, Kathmandu*, Keywords: beekeeping/*cerana*/physiology

ICIMOD (1995) Promotion and development of beekeeping through preservation of indigenous *Apis cerana*. ICIMOD, Kathmandu*, Keywords: beekeeping/*cerana*/physiology/Nepal

ICIMOD (1996) Promotion and development of beekeeping through preservation of indigenous *Apis cerana*. ICIMOD, Kathmandu*, Keywords: beekeeping/*cerana*/physiology

Ihtishamulhaq (1999) Effect of honeybees' species on the physico-chemical parameters of honey. Sarhad J Agric 15:61–63, Keywords: cerana/dorsata/florea/honey/mellifera/Pakistan

Ikuse M, Sahashi N, Maeda H (1981) Observations on pollen grains in Japanese honeys. Honeybee Sci 2:97–104, [in Japanese]. Keywords: *cerana*/honey/pollen/Japan

Ikuta A, Sasaki M (1996) Ventilative fanning of Japanese and European honey bees. *Apis cerana japonica* Rad and *A. mellifera* L. Honeybee Sci 17:14–18, [in Japanese]. Keywords: behaviour/*cerana*/honey/*mellifera*/Japan

Ikuta H (1931) The investigation of Japanese beeswax. Anal Chem 3:430–437, Keywords: beeswax/*cerana*/Japan

Ikuta H (1933) The investigation of Japanese beeswax (III): composition of hydroxy fatty acid. Anal Chem 5.* Keywords: beeswax/*cerana*/Japan

Imai K (1991) Comparison of honey sac contents and nectar carrying capacity in *Apis cerana japonica* and *A. mellifera*. Honeybee Sci 12:107–110, [in Japanese]. Keywords: *cerana*/honey/*mellifera*/nectar/physiology/Japan

Imjongjirak C, Klinbunga S, Sittipraneed S (2005) Cloning, expression and genomic organization of major royal jelly proteins 1 and 2 of the honey bee (*Apis cerana*). J Biochem Mol Biol 38:49–57, Keywords: *cerana*/genetics/protein/Thailand

Imongjirak C, Ngewsra L, Pramual C, Insuan S, Pala-or K, Klinbunga S, Sittipraneed S (2004) Development of a simple and rapid method for typing population origins of the honey bee (*Apis cerana*) in Thailand. J Apic Res 43:17–20, Keywords: *cerana*/DNA/honey/physiology/populations/Thailand

India (1979) National symposium on pollination ecology and applied palynology, abstracts. Haryana Agricultural University, pp 1–50, Keywords: *cerana*/*dorsata*/ecology/*florea*/*mellifera*/pollination

Journal IB (1942) Editorial. Indian Bee J 4:64, Keywords: beekeeping/*florea*/India

Indian Bee Journal (1976) Honey bees in Kashmir. Indian Bee J 38:39–41, Keywords: *cerana*/*dorsata*/*florea*/honey/India

Indian Bee Journal (1981) Work on bee pollination in India. Indian Bee J 43:140–144, Keywords: *cerana*/pollination/India

Indian Bee Journal (1989) Bibliography on Asiatic honey bees and Indian apiculture (publications up to 1989). Indian Bee J 51:150–153, Keywords: bibliography/*cerana*/*dorsata*/*florea*/honey

Inoka WA, Karunaratne P, Edirisinghe JP (2002) Bee diversity and floral hosts in selected habitats of the Peradeniya University Park. Ceylon J Sci, Biol Sci 30:21–36, Keywords: biogeography/*cerana*/*dorsata*/*florea*/genetics/Sri Lanka

Inoue A (1962) Preliminary report on the rearing of Japanese honeybee queens in colonies of the European honeybee. Indian Bee J 24:73–74, Keywords: breeding/*cerana*/interspecific interactions/*mellifera*/queens/Japan

Inoue A (1964) Studies on the respiration metabolism of honeybee: I. Change of respiration rate during larval development of the Japanese and European worker honeybees. Bee Sci 5:3–6, Keywords: *cerana*/*mellifera*/physiology/workers/Japan

Inoue A, Nakajima T (1986) A comparison of the venom compositions in *Apis cerana japonica* and *Apis mellifera*. In: Proceedings of the 30th international apiculture congress, Apimondia, Nagoya, pp 441–443, Keywords: biochemistry/*cerana*/*mellifera*/Japan

Inoue H (1985) On two species of *Achroia* Hüber (Pyralidae, Galleriinae) from Japan. Jap Heterocerists' J 149:377–378, Keywords: *cerana*/natural enemies/Japan

Inoue H, Nakajima T, Okada I (1987) The venomous components in the worker and queen bees during their maturation and seasonal generation. Jpn J Sanit Zool 38:211–217, Keywords: *cerana*/queens/workers/Japan

Inoue N, Inoue A (1995) Traditional beekeeping in the southern part of Kii Peninsula in Japan. Honeybee Sci 16:153–162, [in Japanese]. Keywords: beekeeping/*cerana*/Japan

Inoue T (1957) The Japanese honeybees: important honey plants in Japan. Glean Bee Cult 85:304–305, Keywords: bee botany/*cerana*/honey/Japan

Inoue T (1959) Ancient beekeeping of Japan. Glean Bee Cult 87:14–17, Keywords: beekeeping/*cerana*/history/Japan

Inoue T (1981a) Japanese native honeybees on Tsushima Island. Honeybee Sci 2:19–22, [in Japanese]. Keywords: *cerana*/islands/Japan

Inoue T (1981b) Native honeybees past and present in a village in Kyûshû. Honeybee Sci 2:17–18, [in Japanese]. Keywords: *cerana*/population biology/Japan

Inoue T (1998) Pollination biology in a lowland dipterocarp forest in Sarawak. Malaysia: I. Characteristics of the plant-pollinator community in a lowland dipterocarp forest. Am J Bot 85:1477–1501, Keywords: bee botany/*dorsata*/general biology/Borneo/Malaysia

Inoue T (2002) An analysis of apid bee richness in Sumatra. In: Proceedings of the 11th international congress, IUSSI, Bangalore, pp 581–582, Keywords: *cerana*/*dorsata*/foraging/nests/India/Indonesia

Inoue T, Adri SS (1990) Nest site selection and reproductive ecology of the Asian honey bee, *Apis cerana indica*, in central Sumatra. In: Sakagami SF, Ohgushi R, Roubik DW (eds) Natural history of social wasps and bees in equatorial Sumatra. Hokkaido University Press, Sapporo, pp 219–232, Keywords: *andreniformis*/breeding/cells/*cerana*/*dorsata*/ecology/genetics/history/honey/*mellifera*/nests/population biology/queens/swarming/workers/Indonesia

Inoue T, Kato M, Kakutani T, Suka T, Itino T (1990a) Insect-flower relationships in the temperate deciduous forest of Kibune, Kyoto: an overview of the flowering phenology and the seasonal pattern of insect visits. Contrib Biol Lab Kyoto Univ 27:377–463, Keywords: bee botany/*cerana*/ecology/Japan

Inoue TS, Salmah S, Sakagami SF, Yamane SK, Kato M (1990b) An analysis of anthophilous insects in central Sumatra. In: Sakagami SF, Ohgushi R, Roubik DW (eds) Natural history of social wasps and bees in equatorial Sumatra. Hokkaido University Press, Sapporo, pp 201–217, Keywords: *cerana*/history/Indonesia

Institute of Bee Science Research (1996) Beekeeping handbook. China Academy of Agricultural Science, Beijing, [in Chinese]. Keywords: beekeeping/*cerana*/*mellifera*/China

Institute of Honeybee Science, Tamagawa University (1991) A comprehensive list of references on Japanese honeybees, Apis cerana japonica. Honeybee Sci 12:77–86, [in Japanese]. Keywords: bibliography/cerana/Japan

Insuan S (2001) Genetic variation of giant honeybee *Apis dorsata,* Fabricius, 1793 in Thailand analysed by DNA markers. Thesis, Chulalongkorn University, Keywords: biochemistry/classification/*dorsata*/honey/organic chemistry/Thailand

Insuan S, Deowanish S, Klinbunga S, Sylvester HA, Wongsiri S (2007) Genetic differentiation of the giant honey bee (*Apis dorsata*) in Thailand analysed by mitochondrial genes and microsatellites. Biochem Genetics 45:345–361, Keywords: biochemistry/*dorsata*/genetics

International Bee Research Association, IBRA (1989) Research to provide guidelines for future development of beekeeping in Bangladesh. International Bee Research Association, Cardiff, Keywords: bee botany/beekeeping/*cerana*/foraging/physiology/Bangladesh

International Conference on the Asian honey bees and bee mites (1992) Abstracts of papers and bibliography on Asian honey bees 1979–1991, Keywords: bibliography/*cerana*/*dorsata*/environmental effects/*florea*/honey/pests/Asia/Thailand

International Union for the Study of Social Insects (IUSSI) (1992) 13th Meeting of the German-speaking Section of IUSSI. Blaubeuren, Heinrich Fabri Institute of Tübingen University, p 166, Keywords: biology/*cerana*/*dorsata*/*florea*/*mellifera*

Islam MN (2000) Beekeeping in Bangladesh. In: Proceedings of the 4th Asian apiculture association international conference, Kathmandu, pp 271–272, Keywords: beekeeping/*cerana*/*dorsata*/*florea*/Bangladesh

Ismail MM, Rashid ZBA (1993) The production and marketing of wild honey in Malaysia. In: Connor LJ, Rinderer TE, Sylvester HA, Wongsiri S (eds) Asian apiculture. Wicwas, Cheshire, pp 297–307, Keywords: *dorsata*/honey/Malaysia

Itioka T, Inoue T, Kato M, Nagamitsu T, Momose K, Sakai S, Yumoto T, Mohamad SU, Hamid AA, Yamane SK (2002) Six- year population fluctuation of the giant honeybee *Apis dorsata* F (Hymenoptera: Apidae) in a tropical lowland dipterocarp forest in Sarawak. Ann Entomol Soc Am 94:545–549, Keywords: *dorsata*/general biology/honey/Malaysia

Iwasaki Y, Ihara M (1994) Japanese honey bees in Ina Valley (Nagano Prefecture). Honeybee Sci 15:7–18, [in Japanese]. Keywords: beekeeping/*cerana*/honey/physical factors/swarming/Japan

Jacobs W (1924) The scent organ of *Apis mellifera* and related dermal secretory organs of social and solitary Apidae. Ztschr Morphol u Oekol Tiere 3:1–80, [in German]. Keywords: *florea*/*mellifera*/pheromones

Jacobs W (1927) Tree-working by bears. Trop Natuur 16:121, [in Dutch]. Keywords: disease/*dorsata*/pests/predators

Jadhav LD (1981) Role of insects in the pollination of onion (*Allium cepa* L) in the Maharashtra State, India. Indian Bee J 43:61–63, Keywords: *cerana*/*dorsata*/*florea*/pollination/India

Jadhav LD (1982) Effect of solar eclipse on the foraging activity of honey bees. BIOVD 8:157, Keywords: *dorsata*/*florea*

Jadhav LD, Ajiri DS (1981) Insect pollinators of onion (*Allium cepa*) in Ahmednagar District of Maharashtra. India Indian Bee J 43:109, Keywords: *cerana*/*dorsata*/*florea*/pollination/India

Jaffar ASBU (1987) Experiments on selection of *Apis cerana* colonies for honey yield. Thesis, University Pertanian, Keywords: breeding/*cerana*/honey/Malaysia

Jagannadham B, Goyal NP (1983) Morphological and behavioural characteristics of honeybee workers reared in combs with larger cells. In: Proceedings of the 2nd international conference on apiculture in tropical climates, New Delhi, pp 238–253, Keywords: cells/*cerana*/classification/combs/*mellifera*/morphology/physical factors/workers/India

Jain KL (1993) Electrophysiological study of the stimulatory effects of sugars on gustatory receptors in honeybee and solitary bee species. Indian J Comp Anim Physiol 11:65–69, Keywords: physiology/India

Jain KL (1994) Stimulating effects of sugars in flabellar receptors of *Apis dorsata* F foragers. Indian J Exp Biol 32:267–270, Keywords: *dorsata*/foraging/physiology/India

Jain KL, Guman S (1992) Sublethal effects of some insecticides on the feeding responses of *Apis* species. J Res Haryana Agric Univ 22:68–70, Keywords: *cerana*/*dorsata*/*florea*/insecticides/*mellifera*/India

Jain KL, Puri G (1991) Effect of soil moisture on foraging activity of bees and seed yield of alfalfa. Forage Res 17:132–137, Keywords: *dorsata*/*florea*/foraging/*mellifera*/India

Jain KL, Rani U (1996) Age related morphometric changes of hypopharyngeal glands in honeybees. Entomon 21:143–146, Keywords: *dorsata*/*florea*/morphology/India

Jain KL, Singh S, Saini K (1998) Comparison of nutritional value as determinant of honey bee preference for pollen sources. Indian Bee J 60:137–140, Keywords: bee botany/drones/*florea*/honey/*mellifera*/physiology/India

Jain PC (1967) Morphometrics of bees *Apis dorsata* F, *Apis indica* F, *Apis florea* F, *Megachile lanata* F, *Trigona smithii* B (Hymenoptera, Apidae) and their foraging habits at Udaipur. Thesis, University of Udaipur, Keywords: *cerana*/classification/*dorsata*/*florea*/foraging/India

Jamieson BR (2000) Detections of *Apis cerana* in the Austral port of Brisbane. Can Beekeep 21:130, Keywords: *cerana*/Australia

Jander R, Jander U (1970) The phylogeny of geotaxis in bees. Z Vergl Physiol 66:355–368, Keywords: behaviour/*cerana*/*dorsata*/*florea*/*mellifera*/phylogeny

Japan HSRC (1999) Bees and beekeeping in Asia and the tropics. Farm Jpn 33:32–33, Keywords: beekeeping/*cerana*/tropics

Japanese Kokai (1985) Stable royal jelly powder. Unexamined Patent Application No 85-37942, p 5 [in Japanese]. Keywords: *cerana*/royal jelly/Japan

Jayarathnam K (1972) *Hevea brasiliensis* as a source of honey. J Palynol 6:101–103, Keywords: bee botany/*cerana*/honey/India

Jayavasti S (1989) Scanning electron microscopy analysis of honey bee (*Apis florea* F, *Apis dorsata* F, *Apis cerana* F, *Apis mellifera* L and *Apis andreniformis* S) stings in Thailand. In: Proceedings of the 1st Asia-Pacific conference of entomology, Chiang Mai, pp 89–98, Keywords: *andreniformis*/defense/*dorsata*/*florea*/honey/*mellifera*/Thailand

Jayavasti S, Wongsiri S (1992) Scanning electron microscopy analysis of honey bees (*Apis florea*, *Apis dorsata*, *Apis cerana*, *Apis mellifera*, *Apis andreniformis* and *Apis koschevnikovi*) stings. Rec Adv Toxic Res 2:193–204, Keywords: *andreniformis*/*cerana*/*dorsata*/*florea*/honey/*koschevnikovi*/*mellifera*/stings

Jayavasti S, Wongsiri S (1993) Stings of honeybees in Asia. Honeybee Sci 14:105–109, [in Japanese]. Keywords: *andreniformis*/*cerana*/*dorsata*/*florea*/honey/*koschevnikovi*/*mellifera*/morphology/stings

Jhajj HS, Goyal NP (1979a) Comparative behaviour of pollen foragers of *Apis cerana indica* and *Apis mellifera*. J Apic Res 18:279–284, Keywords: behaviour/*cerana*/foraging/*mellifera*/pollen/India

Jhajj HS, Goyal NP (1979) Comparative studies on the flower constancy of *Apis cerana indica* F and *Apis mellifera* L. In: Proceedings of the 4th international symposium on pollination, Maryland, pp 333–339, Keywords: *cerana*/foraging/*mellifera*/pollination/India

Jhajj HS, Goyal NP (1979c) Reversal in the normal behavioural sequence of honeybee foragers with carbon dioxide anaesthesia. Indian Bee J 41:17–18, Keywords: behaviour/*cerana*/foraging/*mellifera*/India

Jhajj HS, Goyal NP (1986) Comparative flower constancy of *Apis cerana indica* F and *Apis mellifera* L on removing their colonies to a new site. In: Kapil RP (ed) Pollination biology: an analysis. Inter-India, New Delhi, pp 73–77, Keywords: *cerana*/foraging/*mellifera*/India

Jhajj HS, Gatoria GS, Bakhetia DRC (1996) Three decades of beekeeping research in Punjab. National Agricultural Technology Information Centre, Ludhiana, pp 23–27, Keywords: beekeeping/India

Jhansi P, Ramanujam CGK (1987) Pollen analysis of extracted and squeezed honey of Hyderabad. Geophytology 17:237–240, Keywords: *cerana*/honey/pollen/India

Jhansi P, Ramanujam CGK (1990) Pollen of some honey samples from Andhra Pradesh. Asian J Plant Sci 2:19–26, Keywords: *cerana*/honey/pollen/India

Jhansi P, Kalpana TP, Ramanujam CGK (1991) Pollen analysis of rock bee summer honeys from the Prakasam district of the Andhra Pradesh, India. J Apic Res 30:33–40, Keywords: bee botany/*dorsata*/honey/India

Jhansi P, Kalpana TP, Ramanujam CGK (1994) Pollen analysis of some *Apis cerana* Fabr. honeys from Andhra Pradesh, India. Apidologie 25:289–296, Keywords: bee botany/*cerana*/foraging/honey/pollen/India

Ji T (1998) Beekeeping and the beehive. Apic China 2:28–30, [in Chinese]. Keywords: beekeeping/*cerana*/China

Jia SC, Liu YS (1995) Methods to prevent the rearing of brood by *Apis cerana cerana* in autumn. J Bee 9:21, [in Chinese]. Keywords: breeding/brood/*cerana*/China

Jie WF, Yu LS, Hong JI, Ding J, Liu J, Hui Li, Huang SS (2009) Comparison of pollination behaviour of *Apis mellifera ligustica* and *Apis cerana cerana* in greenhouses. J Bee [in Chinese].* Keywords: *cerana*/*mellifera*/pollination

Jing TD (1991) The drones management of migratory beekeeping. Apic Sci Tech 13.* Keywords: beekeeping/*cerana*/drones/China

Jitendra-Mohan KV (1973) Some observations on the bee pollination in sunhemp (*Crotalaria juncea*) and their importance in breeding work. Jute Bull 35:214–216, Keywords: bee botany/breeding/*cerana*/pollination/India

Jonathan KH, Raju AJS (2009) Insect pollination and self-planting seed dispersal strategy in the true viviparous mangrove tree species, *Ceriops tagal* (Perr.) C.B. Robinson (Rhizophoraceae). J Threat Taxa 1:133–140, Keywords: *cerana*/*florea*/pollination

Jonathan KH, Raju AJS, Branham KS, Devi DS (2009) Interactions between insect pollinators and the ornamental tree, *Tecoma stans* (L.). J Threat Taxa 1:126–127, Keywords: *cerana*/*dorsata*/pollination

Jones HL (1899) Apis dorsata Glean Bee Cult 27:572, Keywords: biogeography/*dorsata*/Malaysia

Jong FS (1995) Research for the development of sago palm (*Metroxylon sagu* Rottb) cultivation in Sarawak. Thesis, Wageningen Agricultural University, Keywords: *dorsata*/pollination/Malaysia

Jordan RA, Brosemer RW (1974) Characterization of DNA from three bee species. J Insect Physiol 20:2513–2520, Keywords: *cerana*/DNA/*florea*/*mellifera*

Jose VT, Thankamony S, Toms J, Kothandaraman R (1999) Economics of beekeeping with *Apis cerana indica* and *A. mellifera* in rubber smallholdings of Kerala. Indian J Nat Rubber Res 12:49–54, Keywords: beekeeping/*cerana*/*mellifera*/India

Joseph D, Mohandas N (1985) Role of honey bees in the pollination of cardamom. Indian Bee J 47:12–13, Keywords: *cerana*/honey/pollination/India

Joshi MA (1982) Honey and beeswax trade in India. Indian Bee J 44:1–6, Keywords: combs/*dorsata*/*florea*/honey/*mellifera*/India

Joshi M, Agarwal HC (1977) Site of cholesterol absorption in some insects. J Insect Physiol 23:403–404, Keywords: biochemistry/*cerana*

Joshi MA, Dange SG, Wakode MT (1995a) Bibliography on Asiatic honey bees and Indian apiculture (publications up to 1990). Indian Bee J 57:52–54, Keywords: bibliography/*cerana*/*dorsata*/*florea*/honey/*mellifera*/India

Joshi MA, Dange SG, Wakode MT (1995b) Bibliography on Asiatic honey bees and Indian apiculture (publications up to 1993). Indian Bee J 57:200–204, Keywords: bibliography/*cerana*/*dorsata*/*florea*/honey/*mellifera*/India

Joshi MA, Dange SG, Wakode MT (1996a) Bibliography on Asian bees and apiculture (publications up to 1994). Indian Bee J 58:31–32, Keywords: bibliography/*cerana*/*dorsata*/*florea*/*mellifera*

Joshi MA, Dange SG, Wakode MT (1996b) Bibliography on Asian bees and apiculture (publications up to 1995). Indian Bee J 58:92–94, Keywords: bibliography/*cerana*/*dorsata*/*florea*/*mellifera*

Joshi MA, Dange SG, Wakhle DM (1996c) Bibliography on Asian bees and apiculture (publications up to 1996). Indian Bee J 58:235–240, Keywords: bibliography/*cerana*/*dorsata*/*florea*/*mellifera*

Joshi MA, Lakshmi K, Suryanarayana MC (1998a) Melittopalynological investigations on *Apis* and *Trigona* honeys collected in and around Pune, Maharashtra. Indian Bee J 60:90 98, Keywords: bee botany/*dorsata*/*florea*/honey/India

Joshi MA, Rao GM, Suryanarayana MC (1998b) *Nyctanthes arbortristis* - a bee forage source for *Apis dorsata*. Indian Bee J 60:20–21, Keywords: bee botany/*dorsata*/India

Joshi MA, Singh TSMS, Rao GM, Suryanarayana MC (1998c) Common arboreal bee forage plants in India - *Samanea saman* (Jacq) Merr. Indian Bee J 60:26–27, Keywords: bee botany/*cerana*/*dorsata*/India

Joshi NK, Verma SK (1985) Studies on sacbrood disease in *Apis cerana indica* Fab. in Kumaon Hills of UP, India. Indian Bee J 47:39–41, Keywords: *cerana*/disease/India

Joshi SR (2000) Indigenous beekeeping techniques in Dadeldhura, Nepal. In: Proceedings of the 4th Asian apiculture association international conference, Kathmandu, pp 252–258, Keywords: bee botany/beekeeping/*cerana*/Nepal

Joshi SR, Pechhacker H, Scheurer S, Shreshta KK (2000) Honeydew honey in Nepal. In: Proceedings of the 4th Asian apiculture association international conference, Kathmandu, pp 140–144, Keywords: bee botany/*cerana*/*dorsata*/honey/*mellifera*/Nepal

Joshi SR, Pechhacker H, William A, von der Ohe W (2000b) Physico-chemical characteristics of *Apis dorsata*, *A. cerana* and *A. mellifera* honey from Chitwan District, central Nepal. Apidologie 31:367–375, Keywords: *cerana*/*dorsata*/honey/*mellifera*/Nepal

Joshi SR, Pechhacker H, von der Ohe W (2001) Physico-chemical properties of *Apis dorsata* honeys from Chitwan District, Nepal. In: Proceedings of the 7th international conference on apiculture in tropical climates, Chiang Mai, pp 71–76, Keywords: bee botany/*dorsata*/honey/Nepal

Joshi SR, Ahmad F, Gurung MB (2004) Status of *Apis laboriosa* populations in Kaski district, western Nepal. J Apic Res 43:176–180, Keywords: biogeography/general biology/*laboriosa*/Nepal

Jothi BD, Tandon PL (1993) Insect pollinator activity on ber (*Zizyphus mauritiana* Lamk). In: Proceedings of international symposium on pollination in the tropics, Bangalore, pp 210–211, Keywords: bee botany/ecology/*florea*/India

Jyothi JVA (1989) Ecological and physiological studies on greater waxmoth, *Galleria mellonella* L. Thesis, Bangalore University, Keywords: combs/disease/*dorsata*/pests/predators/India

Jyothi JVA (1994a) Incidence of the mite *Neocypholaelaps indica* (Evans) in the colonies of Indian hive bee (*Apis cerana indica* F) at Bangalore (India). Indian Bee J 56:32–34, Keywords: *cerana*/nests/pests/India

Jyothi JVA (1994b) Visitation frequency and abundance of *Apis cerana indica* F on mango (*Mangifera indica* L) at Bangalore, India. Indian Bee J 56:35–36, Keywords: *cerana*/pollination/India

Jyothi JVA (2003) Pollen foraging activity of *Apis cerana* F and *Apis mellifera* L under tropical conditions of Karnataka, India. Environ Ecol 21:417–419, Keywords: *cerana*/ecology/foraging/*mellifera*/pollen/India

Jyothi JVA, Reddy CC (1992) Rate of larval mobility and orientation of greater wax moth, *Galleria mellonella* L. GBOSBU 19:168–170, Keywords: beeswax/*cerana*/flight/natural enemies/India

Jyothi JVA, Reddy CC (1993a) Influence of different comb parts as diets on the life cycle of greater wax moth (*Galleria mellonella* L). Indian Bee J 55:29–35, Keywords: combs/disease/*dorsata*/pests/physiology/predators/India

Jyothi JVA, Reddy CC (1993b) Physical and chemical properties of the comb of giant honeybee, *Apis dorsata* F. Geobios (Jodhpur) 20:269–271, Keywords: combs/*dorsata*/India

Jyothi JVA, Reddy CC (1993) The effect of phoretic mite *Neocypholaelaps indica* on pollen gathering activity of honeybee *Apis cerana indica*. In: Proceedings of international symposium

on pollination in the Tropics, Bangalore, pp 96–98, Keywords: behaviour/*cerana*/pests/pollen/pollination/tropics/India
Jyothi JVA, Reddy CC (1996) The effect of phoretic mite, *Neocypholaelaps indica* on pollen gathering activity of honeybee, *Apis cerana indica*. GBOSBU 23:74–76, Keywords: behaviour/*cerana*/pests/pollen/India
Jyothi JVA, Reddy MS, Reddy CC (1990) Incidence of infestation of greater wax moth in *Apis dorsata* colonies. In: Proceedings of the 11th international conference, IUSSI, Bangalore, pp 219–223, Keywords: combs/disease/*dorsata*/pests/predators/India
Kaal J, Velthuis HHW, Jongeleen F, Beetsma J (eds) (1992) Traditional bee management as a basis for beekeeping development in the tropics. In: Proceedings of the 1st NECTAR seminar, Royal Tropical Institute, Amsterdam pp 1–87, Keywords: beekeeping/*cerana*/*dorsata*/nectar/physiology/tropics/Indonesia/Vietnam
Kafle GP (1979) Elementary beekeeping. Rupayan, Kathmandu, [in Nepali]. Keywords: beekeeping/*cerana*/Nepal
Kafle GP (1984) Bee forage in Kathmandu Valley. Nepalese J Agric 15:89–99, Keywords: bee botany/*cerana*/foraging/Nepal
Kafle GP (1992a) Modern beekeeping (Nepal). Rupayan, Kathmandu, [in Nepali]. Keywords: beekeeping/*cerana*/Nepal
Kafle GP (1992b) Salient features of beekeeping in Nepal. In: Verma LR (ed) Honeybees in mountain agriculture. Oxford & IBH, New Delhi, pp 155–162, Keywords: beekeeping/*cerana*/natural enemies/Nepal
Kahono S, Nakamura K, Amir M (1999) Seasonal migration and colony behaviour of the tropical honeybee *Apis dorsata* F (Hymenoptera: Apidae). Treubia 31:283–297, Keywords: absconding/*dorsata*/migration/swarming
Kakkar KL (1981) Foraging behaviour of insect pollinators of cauliflower bloom. Indian J Ecol 8:126–130, Keywords: behaviour/*cerana*/foraging/pollination/India
Kakkar KL (1983) Prospects of bees as pollinating agents of cauliflowers. In: Proceedings of the 2nd international conference on apiculture in tropical climates, New Delhi, pp 545–556, Keywords: *cerana*/physical factors/pollination/India
Kakkar KL (1993) Comparative pollination efficiency of introduced colonies of Indian honey bee, *Apis cerana* Fabr. and European honey bee, *Apis mellifera* Linn. on apple. In: Proceedings of international symposium on pollination in the tropics, Bangalorem,* Keywords: *cerana*/honey/*mellifera*/pollen/pollination/tropics
Kakkar KL (1994) Repellent effect of synthetic pyrethroids on Indian honey bee, *Apis cerana indica* Fab. J Insect Sci 7:191–193, Keywords: *cerana*/honey/insecticides/India
Kakkar KL (2000) Effect of pollination strategies on fruit set of apple in Himachal Pradesh. In: Proceedings of the 4th Asian apiculture association international conference, Kathmandu, pp 166–168, Keywords: *cerana*/foraging/*mellifera*/pollen/pollination/India
Kakkar KL, Sharma PL (1991) Studies on the role of honey bee, *Apis cerana indica* F in the pollination of cauliflower, *Brassica oleracea* var *botrytis*. Indian J Entomol 53:66–69, Keywords: *cerana*/honey/pollination/India
Kakutani T (1992) Nectar secretion of *Cayratia japonica* and visiting pattern of *Apis cerana japonica*. Honeybee Sci 13:27–34, [in Japanese]. Keywords: *cerana*/nectar/Japan
Kakutani T, Inoue T, Kato M (1989) Nectar secretion pattern of the dish-shaped flower, *Cayratia japonica* (Vitaceae), and nectar utilization patterns by insect visitors. Res Pop Ecol Kyoto Univ 31:381–400, [in Japanese]. Keywords: bee botany/*cerana*/nectar/Japan
Kakutani T, Inoue T, Kato M, Ichihashi H (1990) Insect-flower relationships on the campus of Kyoto University, Kyoto: an overview of the flowering phenology and the seasonal pattern of insect visits. Contrib Biol Lab Kyoto Univ 27:465–521, [in Japanese]. Keywords: bee botany/*cerana*/colony phenology/ecology/Japan
Kalita H, Rahman A (1995) Evaluation of toxicity of certain insecticides to *Apis cerana indica* F. J Agric Sci Soc NE India 8:97–99, Keywords: *cerana*/insecticides/India
Kallapur SK (1950) An experiment in the collection of wild honey. Indian Bee J 12:122–124, Keywords: bee botany/combs/*dorsata*/honey/India

Kalpana TP, Ramanujam CGK (1989) A melittopalynological investigation of Nawabpet mandal of Ranga Reddy district, AP. J Swamy Bot C 6:57–64, Keywords: bee botany/*florea*/honey/India

Kalpana TP, Ramanujam CGK (1990) Melittopalynology of *Prosopis juliflora* honeys from Ranga Reddy district (AP) and its significance. Indian Bee J 52:44–46, Keywords: *cerana*/*florea*/honey/palynology/India

Kalpana TP, Ramanujam CGK (1994) *Carum copticum* a major source of winter honeys in Ranga Reddy district, Andhra Pradesh. Proc Indian Nat Sci Acad, Biol Sci 60:583–593, Keywords: bee botany/*dorsata*/*florea*/honey/India

Kalpana TP, Ramanujam CGK (1996) Sugarcane honey and its significance. Curr Sci 70:261–262, Keywords: bee botany/*cerana*/honey/India

Kalpana TP, Ramanujam CGK (1997) Melittopalynology, bee plants and beekeeping potential in some coastal districts of Andhra Pradesh, India. Indian Bee J 59:1–8, Keywords: bee botany/beekeeping/*cerana*/honey/India

Kalpana TP, Fatima K, Ramanujam CGK (1990) Pollen analysis of *Apis cerana* and *Apis florea* honeys from Adikmet area, Hyderabad. Proc Indian Acad Sci, Pl Sci 100:183–193, Keywords: bee botany/*cerana*/*florea*/honey/pollen/India

Kameda T (2004) Molecular structure of crude beeswax studied by solid-state 13C NMR. J Insect Sci 4:29–34, Keywords: beeswax/*cerana*/Japan

Kameda T, Tamada Y (2009) Variable-temperature 13C solid-state NMR study of the molecular structure of honeybee wax and silk. Int J Biol Macromol 44:64–69, Keywords: beeswax/*cerana*/silk/Japan

Kamei A, Kosuge K, Kosone K, Kanayama A (2002) The habitat of the honey bee in Yokohama. Honeybee Sci 23:12–16, [in Japanese]. Keywords: *cerana*/*mellifera*/nests

Kandemir I, Moradi MG, Özden B, Özkan A (2009) Wing geometry as a tool for studying the population structure of dwarf honey bees (*Apis florea* Fabricius 1876) in Iran. J Apic Res 48:238–246, Keywords: *florea*/morphology/Iran

Kang LC (1980) Honeybees in Malaysia. Nat Malay 5:26–33, Keywords: *cerana*/*dorsata*/honey/Malaysia

Kang-Chen L, Rou-Su C (1975) Preliminary investigations on bee mites in Taiwan. J Agric Res China 24:50–56, [in Chinese]. Keywords: behaviour/*cerana*/pests/China

Kang-Chen L, Rou-Su C (1977) The preliminary investigations on bee mites in Taiwan. Chin Bee J 1:32, [in Chinese]. Keywords: *cerana*/pests/China

Kannan R (1996) Prospects for beekeeping at the Palni Hills and associated foothill plains with *Apis cerana indica*. In: Proceedings of international workshop revive *Apis cerana indica* Kodaikanal, pp 15–19, Keywords: beekeeping/*cerana*/India

Kanwar U, Brar N (1982a) Cytochemical studies on the venom gland of the honeybee, *Apis dorsata*. Res Bull Punjab Univ Sci 33:51–54, Keywords: biochemistry/*dorsata*/morphology/organic chemistry

Kanwar U, Brar N (1982b) Morphological and cytological studies in the venom gland of the honeybee, *Apis dorsata*. Res Bull Punjab Univ Sci 33:37–41, Keywords: biochemistry/*dorsata*/morphology/organic chemistry

Kapil RP (1956) Variations in the biometrical characters of the Indian honeybee (*Apis indica* F). Indian J Entomol 18:440–457, Keywords: *cerana*/classification/India

Kapil RP (1957) The length of life and the brood-rearing cycle of the Indian bee. Bee World 38:258–263, Keywords: bee botany/brood/*cerana*/colony phenology/workers/India

Kapil RP (1958) Morphology and histology of the salivary gland system of honeybee, *Apis indica* Fab. (Apidae, Hymenoptera). Arch Bienkd 35:17–22, Keywords: *cerana*/glands/*mellifera*/morphology/India

Kapil RP (1959a) Anatomy and histology of the alimentary canal of honeybee, *Apis indica* Fab. (Apidae, Hymenoptera). Zool Anz 163:306–323, Keywords: *cerana*/morphology/physiology/India

Kapil RP (1959b) Variation in developmental period of the Indian bee. Indian Bee J 21:3–6, Keywords: brood/*cerana*/India

Kapil RP (1960) Observations on the temperature of the cluster of *Apis indica*. Apicolt Ital 27:79–83, [in Italian]. Keywords: brood/*cerana*/physiology/India

Kapil RP (1962a) Anatomy and histology of the female reproductive system of *Apis indica* F (Hymenoptera, Apidae). Insectes Soc 9:145–163, Keywords: *cerana*/drones/reproduction/India

Kapil RP (1962b) Anatomy and histology of the male reproductive system of *Apis indica* Fab. (Apidae, Hymenoptera). Insectes Soc 9:73–90, Keywords: *cerana*/reproduction/India

Kapil RP (1971) A hive for the Indian honeybees. Apiacta 6:107–109, Keywords: beekeeping/*cerana*/nests/India

Kapil RP (1974) Ecotypes and breeding of honeybees. Indian J Genet Pl Breed 34 A:267–269, Keywords: breeding/*cerana*/geographical variation/India

Kapil RP, Aggarwal K (1987) Some observations on the concurrent parasitization of *Apis florea* by *Tropilaelaps clareae* and *Euvarroa sinhai*. Exp Appl Acarol 3:267–269, Keywords: disease/*florea*/pests/India

Kapil RP, Aggarwal K (1988) Observations on reproduction and seasonal population trends of *Euvarroa sinhai* (Mesostigmata: Varroidae) in India. Prog Acarol 2:277–281, Keywords: disease/*florea*/pests/reproduction/India

Kapil RP, Brar HS (1971) Foraging behavior of *Apis florea* F in relation to *Brassica campestris* var *toria* L. In: Proceedings of the 23rd international apiculture congress, Apimondia, Moscow, pp 335–339, Keywords: bee botany/*cerana*/*dorsata*/*florea*/foraging/India

Kapil RP, Kumar S (1975) Foraging activity of *Apis dorsata* Fab. on *Brassica juncea* Hook and Thomas. J Bombay Nat Hist Soc 71:327–331, Keywords: bee botany/dorsata/foraging/India

Kapil RP, Lamba DPS (1974) Toxicity of some important insecticides to *Apis cerana* Fab. Indian J Entomol 36:6–10, Keywords: *cerana*/insecticides/India

Kapil RP, Lamba DPS, Brar HS (1971a) Integration of bee behaviour with aphid control for seed production of *Brassica campestris* var *toria*. Indian J Entomol 33:221–223, Keywords: behaviour/*cerana*/pollination/India

Kapil RP, Grewal GS, Kumar S, Atwal AS (1971b) Insect pollinators of rapeseed and mustard. Indian J Entomol 33:61–66, Keywords: *cerana*/pollination/India

Kapil RP, Grewal GS, Kumar S, Atwal AS (1974) Insect pollinators of alfalfa, *Medicago sativa* L. Indian J Entomol 36:214–220, Keywords: bee botany/*florea*/pollination/India

Kapoor KS, Dhaliwal HS (1989a) Comparative energetics of *Apis cerana indica* Fabr. and *A. mellifera* L on *Brassica oleracea* L var *botrytis*. Indian Bee J 51:101–103, Keywords: *cerana*/*mellifera*/physiology/pollination/India

Kapoor KS, Dhaliwal HS (1989b) Comparative foraging strategies of *Apis cerana indica* Fabr. and *Apis mellifera* L on cauliflower. Indian Bee J 51:99–101, Keywords: *cerana*/foraging/*mellifera*/India

Karlsson T (1990) The natural nest of the Asian hive bee (*Apis cerana*) in Bangladesh: a minor field study. Swed Univ Agric Sci 134:32, Keywords: *cerana*/nests/Bangladesh

Karlsson T, Svensson B (1991) Appropriate technology for profitable beekeeping in the Philippines. HB Bikonsult, Sala, Sweden. Keywords: beekeeping/*cerana*/Philippines

Karpowicz J (1980) Special Report -*Apis florea* in Dhofar. Oman Ministry of Agriculture and Fisheries and Durham University, unpublished report, Keywords: classification/*florea*/Oman

Karunaratne WAIP, Edirisinghe JP (2008) Diversity of bees at different altitudes in the Knuckles Forest Reserve. Ceylon J Sci (Biol Sci) 37:61–72, Keywords: biogeography/*cerana*/Sri Lanka

Kasahara R, Nakamura J, Matsuka M (2002) Significance of caulking behaviour of honeybees, *Apis mellifera* and *Apis cerana*. In: Proceedings of the 14th international congress, IUSSI, Sapporo. Keywords: behaviour/*cerana*/environmental effects/*mellifera*/nests/physiology/Japan

Kasana VK, Malik MS, Malik OP (1998a) Status of honeybee repellents: research and application. In: Mishra RC (ed) Perspectives in Indian apiculture. Agro Botanica, New Delhi, pp 363–386, Keywords: *florea*/insecticides

Kasana VK, Malik DS, Suryanarayana MC (1998b) Evaluation of carbonyl compounds and essential oils as repellents against *Apis florea* Fab. under field conditions. Indian Bee J 60:83–85, Keywords: *florea*/pheromones/India

Kasana VK, Malik DS, Malik MS, Mishra RC (2000) Effect of chemical additives in increasing degree of repellency and effective duration of carbonyl compounds against *Apis dorsata* visiting *Foeniculum vulgare*. Indian Bee J 62:41–45, Keywords: biochemistry/*dorsata*/organic chemistry/India

Kashyap NP, Kumar A (2001) Distribution and infestation phenology of *Tropilaelaps clareae* and *Acarapis woodi* in *Apis mellifera* L and *Apis cerana* F in northern hills of India. In: Proceedings of the 37th international apiculture congress, Apimondia, Durban (CD Rom), Keywords: biogeography/*cerana*/*mellifera*/pests/India

Kastberger G Attenborough R, Reddish P (1999) The magic trees of Assam. National Geographic, ZDF, ORF & epo-film Wien, Keywords: *dorsata*/migration

Kastberger G, Biswas S (1998) Mechanisms in the cascadic alarming response of a giant honeybee colony. Zool 101:1–48, Keywords: defense/*dorsata*/India

Kastberger G, Sharma DK (2000) The predator-prey interaction between blue-bearded bee-eaters (*Nyctyornis athertoni*) and giant honeybees (*Apis dorsata*). Apidologie 31:727–736, Keywords: disease/*dorsata*/pests/predators/India

Kastberger G, Winder O, Hoetzl T, Raspotnig G (1996) Behavioural features of aperiodic form of massed flight activity in the giant honeybee *Apis dorsata*. Apidologie 27:381–395, Keywords: behaviour/defense/*dorsata*/foraging/India

Kastberger G, Kaefer H (2002) Cascadic coordination in defense behaviour of the giant honeybee *Apis dorsata*. In: Elsner N, Eysel U (eds) Proceedings of the 27th Göttingen Neurobiology Conference, p 260, Keywords: *dorsata*/defense

Kastberger G, Winder O, Biswas S, Raspotnig G (1998) Evidence of Nasonov scenting in colony defence of the giant honeybee *Apis dorsata*. Ethology 104:27–37, Keywords: *dorsata*/pheromones/India

Kastberger G, Attenborough R, Reddish P (1999) The magic trees of Assam. National Geographic: video, Keywords: absconding/*dorsata*/migration/swarming

Kastberger G, Sharma DK, Hüttinger E, Kranner G (2001) Morphometric studies of *Apis dorsata* in Assam, India. In: Proceedings of the 7th international conference on apiculture in tropical climates, Chiang Mai, p 129, Keywords: classification/*dorsata*/India

Kastberger G, Radloff SE, Kranner G (2003) Individuality of wing patterning in giant honeybees (Apis laboriosa). Apidologie 34:311, Keywords: dorsata/wings/morphometrics

Kastberger G, Schmelzer E, Kranner I (2008) Social waves in giant honeybees repel hornets. PLoS ONE 3(9):e3141. doi:10.1371/journal.pone.0003141, Keywords: dorsata/wings/morphometrics

Kastberger G, Weihmann F, Hoetzl T (2009) Complex social waves of giant honeybees provoked by a dummy wasp support the special-agent hypothesis. J Integr Comm Biol 3:2 March/April 2010, Keywords: dorsata/defense

Kasturi Bai AR, Reddy CC (1977) Laboratory toxicity of some insecticides to *Apis cerana indica*. J Apic Res 16:161–162, Keywords: *cerana*/foraging/insecticides/*mellifera*/pollen/India

Kato M (1994) Caste-related and age-related toxic activities of honeybee venom on the same species of honeybees. Honeybee Sci 15:119–122, [in Japanese]. Keywords: biochemistry/*cerana*/*mellifera*/Japan

Kato M, Kakutani T, Inoue T, Itino T (1990) Insect flower relationship in the primary beech forest of Ashu, Kyoto: an overview of the flowering phenology and the seasonal pattern of insect visits. Contrib Biol Lab Kyoto Univ 27:309–375, Keywords: bee botany/*cerana*/colony phenology/ecology/Japan

Kavinseksan B (1994) Production and quality of royal jelly from *Apis cerana*. Thesis, Chulalongkorn University, Keywords: *cerana*/queens/royal jelly/Thailand

Kavinseksan B, Wongsiri S, de Guzman LI, Rinderer TE (2003) Absence of *Tropilaelaps* infestation from recent swarms of *Apis dorsata* in Thailand. J Apic Res 55:101–109, Keywords: disease/*dorsata*/ecology/pests/predators/Thailand

Kawagoe T, Suzuki N (2002) Floral sexual dimorphism and flower choice by pollinators in a nectarless monoecious vine *Akebia quinata* (Lardizabalaceae). Ecol Res 17:295–303, Keywords: *cerana*/pollination/Japan

Keeling CI, Slessor KN, Koeniger N, Koeniger G, Punchihewa RWK (2000) Quantitative analysis of the mandibular gland components of the dwarf honey bee (*Apis florea*). Apidologie 31:293–299, Keywords: classification/*florea*/honey/pheromones/Sri Lanka

Keeling CI, Otis GW, Hadisoesilo S, Slessor KN (2001) Mandibular gland component analysis in the head extracts of *Apis cerana* and *Apis nigrocincta*. Apidologie 32:243–252, Keywords: *cerana*/morphology/*nigrocincta*/pheromones/Indonesia

Kellogg CR (1929) Notes on the tongue length of the Chinese bee (*Apis indica*). Proc Nat Hist Soc, Fukien Christian Univ 2:57–59, Keywords: *cerana*/classification/morphology/China

Kellogg CR (1930) The use of biometry to distinguish between *Apis indica* and *Apis mellifica*. Bee World 11:127–130, Keywords: *cerana*/classification/*mellifera*/China

Kellogg CR (1936) Biometric studies of the Chinese honey bee (*A. indica*). Arch Bienkd 17:36–38, [in German]. Keywords: *cerana*/classification/honey/China

Kellogg CR (1938) Biometrical studies on the wings of the Chinese honey bee. Fukien Christian Univ Sci J 1:1–8, Keywords: *cerana*/classification/honey/China

Kellogg CR (1941a) Some characteristics of oriental honeybee (*Apis indica* F), in China. J Econ Entomol 34:717–719, Keywords: absconding/aggressiveness/beekeeping/behaviour/brood/*cerana*/classification/colony phenology/combs/*mellifera*/migration/morphology/swarming/China/India

Kellogg CR (1941b) The Chinese Bee. Bee World 11:82, Keywords: *cerana*/China

Kellogg CR (1967) Honey in Chinese history entomological excerpts from south-eastern China (Fukien Province). In: Kellogg CR (ed) Aborigines: silkworm, honeybees and other insects. Claremont Manor, Claremont, CA, pp 1–89, Keywords: *cerana*/history/honey/China

Kellogg CR (1968) Entomological excerpts from south eastern China. Claremont Manor, Claremont, Keywords: *cerana*/China

Kellogg CR, Tang CC (1963) Honey and its uses in China. Am Bee J 13:176–178, Keywords: *cerana*/honey/China

Kellogg CR, Yung LL (1929) Studies in the variation in number of hooks on the hind wings of the Chinese honey bee (*Apis indica*). Proc Nat Hist Soc, Fukien Christian Univ 2:57–59, Keywords: *cerana*/honey/morphology/China

Ken T, Warrit N, Smith DR (2007) Mitochondrial DNA diversity of Chinese *Apis cerana*. Apidologie 38:238–246, Keywords: cerana/mtDNA/China

Ken T, Neumann P, Yu Y, Zhang L, Yang M, Hepburn HR, Radloff SE (2009a) Intra- and interspecific brood recognition in the honeybees *Apis cerana* and *A. mellifera*. Apidologie 40:184–191, Keywords: behaviour/brood/*cerana*/*mellifera*/China

Ken T, Yang M, Radloff SE, Pirk CWW, Crewe RM, Phiancharoen M, Hepburn HR, Oldroyd BP (2009b) Worker reproduction in mixed-species colonies of honey bees. Behav Ecol 20:1106–1110, Keywords: brood/*cerana*/*mellifera*/China

Kenta T, Isagi Y, Nakagawa M, Yamashita M, Nakashizuka T (2004) Variation in pollen dispersal between years with different pollination conditions in a tropical emergent tree. Mol Ecol 13:3575–3584, Keywords: bee botany/classification/*dorsata*

Kerr WE (1960) Evolution and communication in bees and its role in speciation. Evolution 14:386–387, Keywords: evolution/general biology

Kevan PG (1989) Beekeeping with *Apis cerana* in tropical and sub-tropical Asia. International Bee Research Association, London, Keywords: beekeeping/*cerana*/Asia

Kevan PG (1993a) Pollination in modern agriculture: changing practices, new and hybrid crops. In: Connor LJ, Rinderer TE, Sylvester HA, Wongsiri S (eds) Asian apiculture. Wicwas, Cheshire, CT, pp 398–409, Keywords: *cerana*/*laboriosa*/*mellifera*/pollination

Kevan PG (1993b) Wind or insects: pollination of coconut (*Cocos nucifera*) in the Maldive Islands. In: Connor LJ, Rinderer TE, Sylvester HA, Wongsiri S (eds) Asian apiculture. Wicwas, Cheshire, CT, pp 372–377, Keywords: *cerana*/islands/physical factors/pollination/Maldive Islands

Kevan PG (1995a) Integrating science and development to successful apiculture. In: Kevan PG (ed) The Asiatic hive bee: apiculture, biology, and role in sustainable development in tropical and subtropical Asia. Enviroquest, Cambridge, ON, pp 251–256, Keywords: beekeeping/biology/*cerana*/physiology/nests

Kevan PG (ed) (1995b) The Asiatic hive bee: apiculture, biology, and role in sustainable development in tropical and subtropical Asia. Enviroquest, Cambridge, ON, Keywords: beekeeping/biology/*cerana*/physiology/nests

Kevan PG, Lack AJ (1985) Pollination in a cryptically dioecious plant *Decaspermum parviflorum* (Lam) AJ Scott (Myrtaceae) by pollen collecting bees in Sulawesi, Indonesia. Biol J Linn Soc 25:319–330, Keywords: bee botany/*dorsata*/Indonesia

Kevan PG, Punchihewa RWK, Greco CF (1995) Foraging range for *Apis cerana* and its implications for honey production and apiary management. In: Kevan PG (ed) The Asiatic hive bee: apiculture, biology, and role in sustainable development in tropical and subtropical Asia. Enviroquest, Cambridge, ON, pp 223–228, Keywords: beekeeping/biology/*cerana*/foraging/honey/nests/physiology/Sri Lanka

Press K (1999) Revival of *Apis cerana* in south India. Beekeep Dev 52:12–13, Keywords: beekeeping/*cerana*/India

Khajuria DR, Dogra GS (1990) Studies on the reproductive system of foraging workers of the Indian honey bee, *Apis cerana indica* Fab. Indian Bee J 52:21–23, Keywords: *cerana*/foraging/honey/reproduction/workers/India

Khajuria DR, Dogra GS (1991) Endocrine control of oocyte maturation and vitellogenesis in the laying workers of the Indian hive bee (*Apis cerana indica* Fab.). Indian Bee J 53:9–16, Keywords: *cerana*/physiology/India

Khalique A, Kader NN (1967) Studies on waxes Part I. The constituents of beeswax. Sci Res (Dacca). Bangladesh 4:64–72, Keywords: beeswax/*cerana*/Pakistan

Khan AH, Mohammad A (1950) Natural crossing in cotton in Western Punjab IV. Agents of natural crossing Agron J 42:236–238, Keywords: bee botany/*florea*/India

Khan BM (1992) Honeybee resources in North West Frontier Province of Pakistan. In: Verma LR (ed) Honeybees in mountain agriculture. Oxford & IBH, New Delhi, pp 59–65, Keywords: biogeography/*cerana*/*dorsata*/*florea*/Pakistan

Khan BM (1995) Comparative study on pollination effect of honeybee species *Apis cerana* and *Apis mellifera* on the fruit yield of toria (*Brassica napus*) in Peshawar, Pakistan. In: Kevan PG (ed) The Asiatic hive bee: apiculture, biology, and role in sustainable development in tropical and subtropical Asia. Enviroquest, Cambridge, ON, pp 151–152, Keywords: biology/*cerana*/*mellifera*/nests/physiology/pollination/Pakistan

Khan BM, Chaudhary MI (1988) Comparative assessment of honey bees and other insects with self-pollination of sarson in Peshawar. Pakistan J For 38:231–238, Keywords: *cerana*/honey/pollination/Pakistan

Khan BM, Chaudhary MI (1995a) Comparative assessment of honeybees and other insects with self-pollination of sarson (*Brassica campestris*) in Peshawar. In: Kevan PG (ed) The Asiatic hive bee: apiculture, biology, and role in sustainable development in tropical and subtropical Asia. Enviroquest, Cambridge, ON, pp 147–150, Keywords: biology/*cerana*/nests/physiology/pollination/Pakistan

Khan BM, Chaudhary MI (1995b) Testing of acaricidal fumigants for the control of tracheal mite disease of honeybees in Pakistan. In: Kevan PG (ed) The Asiatic hive bee: apiculture, biology, and role in sustainable development in tropical and subtropical Asia. Enviroquest, Cambridge, ON, pp 181–183, Keywords: biology/*cerana*/disease/insecticides/nests/pests/physiology/Pakistan

Khan BM, Shahid M, Chaudhary MI (1986) Effect of honey bee pollination on the fruit setting and yield of loquat. Pakistan J For 36:73–77, Keywords: *cerana*/honey/pollination/Pakistan

Khan BM, Chaudhary MI, Khawaja P (1987) Acarine disease of honey bees and its control in Pakistan. Pakistan J For 37:109–114, Keywords: *cerana*/disease/honey/pests/Pakistan

Khan MR, Razzaque MA (1981) A microbiological approach to the causes of migration of honey bees from their hives. Bangladesh J Bot 10:53–57, Keywords: absconding/*cerana*/honey/migration/nectar/Bangladesh

Khan MS, Kumari B, Rohilla HR, Kaushik HD, Arora RK (2004) Analysis of insecticide residues in honeys from apiary (*Apis mellifera*) and wild honeybee (*Apis dorsata* and *Apis florea*) colonies in India. J Apic Res 43:79–82, Keywords: biochemistry/*dorsata*/honey/organic chemistry/India

Khatri B (1990) Tree beekeeping in Laos. Newsl Beekeep Trop Sub-trop Countries 17:4–5, Keywords: beekeeping/*cerana*/Laos

Khosla PK, Singh G, Sehgal RN (1982) Pollen plant pollinator interactions and pollination energetics of *Bombax ceiba*. In: Khosla PK (ed) Improvement of forest biomass. Indian Society of Tree Scientists, Solan, Himachal Pradesh, pp 395–401, Keywords: bee botany/*cerana*/*dorsata*/*florea*/physiology/pollen/pollination/India

Kiang YT (1979) Beekeeping in Taiwan. Am Bee J 119:363–367, Keywords: beekeeping/*cerana*/China

Kiew R (1993) Characteristics of Malaysian flowers visited by *Apis cerana*. In: Connor LJ, Rinderer TE, Sylvester HA, Wongsiri S (eds) Asian apiculture. Wicwas, Cheshire, CT, pp 361–368, Keywords: bee botany/*cerana*/pollen/Malaysia

Kiew R (1995) Bee botany in tropical Asia, with special reference to peninsular Malaysia. In: Kevan PG (ed) The Asiatic hive bee: apiculture, biology, and role in sustainable development in tropical and sub-tropical Asia. Enviroquest, Cambridge, ON, pp 117–124, Keywords: bee botany/biology/*cerana*/environmental effects/nests/physiology/Asia/Bangladesh/China/India/Malaysia/Myanmar (Burma)/Sri Lanka/Thailand

Kiew R, Muid M (1989) Bees and palms in peninsular Malaysia. Principes 33:74–77, Keywords: bee botany/*cerana*/*dorsata*/Malaysia

Kim BN, Kim TJ, Cheigh HS (1991) Organic acids and fatty acids of honey harvested in Kangwon area. J Korean Soc Food Nutr 20:52–58, [in Korean]. Keywords: *cerana*/honey/*mellifera*/Korea

Kimura K, Okumura T, Ninaki O, Kidwell MG, Suzuki K (1993) Transposable elements in commercially useful insects: I. Southern hybridization study of silkworms and honeybees using Drosophila probes Jpn J Genet 68:63–71, Keywords: *cerana*/DNA/*mellifera*/Japan

Kimura K, Amano K, Yoshiyama M, Matsuka M, Nakamura J, Deowanish S (2001) A possible origin of the Japanese honeybee. In: Proceedings of the 7th international conference on tropical bees, Chiang Mai, pp 131–136, Keywords: biogeography/*cerana*/mtDNA/phylogeny/Japan/Korea

Kirchner WH, Dreller C (1993) Acoustical signals in the dance language of the giant honeybee *Apis dorsata*. Behav Ecol Sociobiol 33:67–72, Keywords: *dorsata*/foraging/physiology

Kirchner WH, Dreller C, Grasser A, Baidya D (1996) The silent dances of the Himalayan honeybee, *Apis laboriosa*. Apidologie 27:331–339, Keywords: foraging/*laboriosa*/Nepal

Kiselev V (2002) The bees from Primorski Krai. Dtsch Bienen J 10:18–19, [in German]. Keywords: *cerana*/*mellifera*/Russia

Kitroo A, Abrol DP (1996a) Effect of flowering phenology, bloom density and floral rewards on the attractiveness of insects visiting litchi flowers. Indian Bee J 58:1–5, Keywords: bee botany/*cerana*/*dorsata*/*florea*/*mellifera*/India

Kitroo A, Abrol DP (1996b) Studies on pollen carrying capacity and pollination efficiency of honey bees visiting litchi flowers. Indian Bee J 58:55–57, Keywords: *cerana*/*dorsata*/*florea*/foraging/honey/*mellifera*/pollen/pollination/India

Kittaka Y, Yoshida T (2001) Reproduction of *Varroa jacobsoni* parasitizing newly-emerged *Apis mellifera* drones introduced into an *Apis cerana japonica* colony. Honeybee Sci 2:31–36, [in Japanese]. Keywords: *cerana*/drones/*mellifera*/pests/reproduction/Japan

Kloft W, Schneider P (1969) Group defensive behaviour in *Apis cerana* living wild in Afghanistan. Naturwissenschaften 56:219, [in German]. Keywords: aggressiveness/behaviour/*cerana*/natural enemies/Afghanistan

Kloft WJ, Djalal AS, Drescher W (1976) Comparative study of the food distribution in workers of various races of *Apis mellifica*, using 32P as a tracer. Apidologie 7:49–60, [in German]. Keywords: biogeography/*cerana*/communication/*mellifera*/subspecies/workers

Kobayashi M, Ishikawa S, Yoshida T (2000) Studies on Japanese honeybees as pollinators for rabitteye blueberries, *Vaccinium ashei*. Honeybee Sci 21:159–163, [in Japanese]. Keywords: *cerana*/pollination/Japan

Koeniger G (1986) Reproduction and mating behaviour. In: Rinderer TE (ed) Genetics of the honey bee. Academic, New York, pp 255–275, Keywords: *cerana*/*dorsata*/*florea*/genetics/honey/*mellifera*/reproduction

Koeniger G (1988) Mating behavior of honey bees. In: Needham GR, Page RE, Delfinado-Baker M, Bowman CE (eds) Africanized honey bees and bee mites. Ellis Horwood, Chichester, pp 167–172, Keywords: biogeography/*cerana*/*dorsata*/*florea*/honey/*mellifera*/pests/reproduction/ Sri Lanka

Koeniger G (1990a) Multiple mating, mating sign and mate selectivity in honey bees. Adv Invertebr Reprod 5:483–487, Keywords: *cerana*/*dorsata*/*florea*/honey/reproduction

Koeniger G (1991) Diversity in *Apis* mating systems. In: Smith DR (ed) Diversity in the genus *Apis*. Westview, Boulder, CO, pp 199–212, Keywords: *cerana*/*dorsata*/*florea*/genetics/*mellifera*/reproduction

Koeniger G, Koeniger N (1990a) Evolution of reproductive behaviour in honey bees. In: Veeresh GK, Mallik B, Viraktamath CA (eds) Social insects and the environment. Oxford & IBH, Bombay, pp 101–102, Keywords: *cerana*/*dorsata*/*florea*/honey/*mellifera*/phylogeny/reproduction

Koeniger G, Koeniger N (1990b) Investigations on the origin of polyandry in *Apis*. In: Pfannenstiel HD (ed) Verhandlungen Deutsch Zoologische Gesellschaft. Gustav Fischer, Stuttgart, p 618, [in German]. Keywords: *cerana*/*dorsata*/*florea*/*mellifera*/reproduction

Koeniger G, Koeniger N (2002) Morphology and multiple function of the mating sign in *Apis* species. In: Proceedings of the 14th international congress, IUSSI, Sapporo, p 212, Keywords: *cerana*/drones/*koschevnikovi*/*mellifera*/morphology/reproduction/Japan

Koeniger G, Mardan MB, Ruttner F (1990a) Male reproductive organs of *Apis dorsata*. Apidologie 21:161–164, Keywords: *dorsata*/morphology

Koeniger G, Koeniger N, Mardan MB, Punchihewa RWK, Otis GW (1990b) Numbers of spermatozoa in queens and drones indicate multiple mating of queens in *Apis andreniformis* and *Apis dorsata*. Apidologie 21:281–286, Keywords: *dorsata*/drones/queens/reproduction/Malaysia/Sri Lanka

Koeniger G, Koeniger N, Mardan MB, Otis GW, Wongsiri S (1991) Comparative anatomy of male genital organs in the genus *Apis*. Apidologie 22:539–552, Keywords: *andreniformis*/*cerana*/drones/*florea*/*koschevnikovi*/*mellifera*/Malaysia/Thailand

Koeniger G, Koeniger N, Mardan MB, Wongsiri S (1993a) Variance in weight of sexuals and workers within and between four *Apis* species (*Apis florea*, *Apis dorsata*, *Apis cerana* and *Apis mellifera*). In: Connor LJ, Rinderer TE, Sylvester HA, Wongsiri S (eds) Asian apiculture. Wicwas, Cheshire, CT, pp 104–109, Keywords: *cerana*/*dorsata*/*florea*/*mellifera*/physiology/workers/Thailand

Koeniger G, Koeniger N, Tingek S (1994a) Crossfostered drones of *Apis cerana* (Fabricius, 1793) and *Apis koschevnikovi* (Von Buttel-Reepen, 1906) fly at their species specific mating times. Insectes Soc 41:73–78, Keywords: brood/*cerana*/drones/interspecific interactions/*koschevnikovi*/populations/reproduction/Malaysia

Koeniger G, Koeniger N, Tingek S (1996a) Interspecific instrumental insemination of *Apis cerana* Fabricius, 1793. Apidologie 27:302–304, Keywords: breeding/*cerana*/*dorsata*/interspecific interactions/*koschevnikovi*

Koeniger G, Koeniger N, Tingek S (1996b) A scientific note on weight and sperm numbers of drones of *Apis nuluensis* Tingek, Koeniger and Koeniger, 1996. Apidologie 27:427–428, Keywords: *andreniformis*/*cerana*/drones/*koschevnikovi*/*mellifera*/*nuluensis*/Malaysia

Koeniger G, Koeniger N, Tingek S, Kalitu A (1996) Heterospecific queen rearing, queen acceptance and cross insemination of *Apis cerana* and *Apis koschevnikovi*. In: Proceedings of the 3rd Asian apiculture association international conference, Hanoi, p 40, Keywords: breeding/*cerana*/genetics/*koschevnikovi*/queens

Koeniger G, Koeniger N, Tingek S, Kelitu A (1999) Interspecific insemination and hybrids between *Apis cerana* and *Apis koschevnikovi*. In: Proceedings of the 13th international congress, IUSSI, Adelaide, p 259, Keywords: *cerana*/genetics/*koschevnikovi**

Koeniger G, Koeniger N, Tingek S (2001) Diversity of the mating sign and its function in the genus *Apis*. In: Proceedings of the 7th international conference on tropical bees, Chiang Mai, pp 87–90, Keywords: *andreniformis*/*cerana*/*dorsata*/drones/*florea*/genetics/*koschevnikovi*/*mellifera*/reproduction/Malaysia

Koeniger G, Koeniger N, Anderson DL, Lekprayoon C, Tingek S (2001b) Mites from debris and sealed brood cells of *Apis dorsata* colonies in North Vietnam. Apidologie 32:41–50, Keywords: brood/combs/disease/*dorsata*/pests/predators/Vietnam

Koeniger G, Koeniger N, Anderson DL, Lekprayoon C, Tingek S (2002) Mites from debris and sealed brood cells of *Apis dorsata* colonies in Sabah (Borneo) Malaysia, including a new haplotype of *Varroa jacobsoni*. Apidologie 33:15–24, Keywords: brood/combs/disease/*dorsata*/pests/predators/Borneo/Malaysia

Koeniger G, Koeniger N, Tingek S, Phiancharoen M (2005a) Variance in spermatozoa number among *Apis dorsata* drones and among *Apis mellifera* drones. Apidologie 36:279–284, Keywords: *dorsata*/drones/reproduction

Koeniger N (1975) Observations on alarm behaviour and colony defense of *Apis dorsata*. In: Proceedings of the symposium on pheromones defensive secretions of social insects, IUSSI, Dijon, pp 153–154, Keywords: behaviour/defense/*dorsata*/pheromones

Koeniger N (1976) Interspecific competition between Apis florea and Apis mellifera in the tropics. Bee World 57:100,110–112, Keywords: classification/ecology/florea/mellifera

Koeniger N (1976b) New phylogenetic aspects of the genus *Apis*. Apidologie 7:357–366, [in German]. Keywords: *cerana*/*dorsata*/*florea*/*mellifera*/phylogeny

Koeniger N (1976) The Asiatic honeybee *Apis cerana*. In: Proceedings of the 1st international conference on apiculture in tropical climates, London, pp 47–49, Keywords: biology/*cerana*

Koeniger N (1977) Keeping the three *Apis* species from Asia, *Apis cerana*, *Apis dorsata*, *Apis florea*, in bee flight rooms. Insectes Soc 24:286–289, [in German]. Keywords: *cerana*/*dorsata*/flight/*florea*

Koeniger N (1982) Interactions among the four species of the genus *Apis*. In: Breed MD, Michener CD, Evans HE (eds) The biology of social insects. Westview, Boulder, CO, pp 59–64, Keywords: biology/*cerana*/*dorsata*/*florea*/interspecific interactions/*mellifera*

Koeniger N (1987) The Asian honeybee (*Apis cerana*) and its mite *Varroa jacobsoni*. Imkerfreund 42:303–306, [in German]. Keywords: *cerana*/natural enemies/pests/Indonesia/Sri Lanka

Koeniger N (1990) Coevolution of honey bees and their parasitic mites. In: Proceedings of the 11th international congress, IUSSI, New Delhi, pp 130–131, Keywords: disease/honey/pests/*mellifera*/India

Koeniger N (1990) The parasite host relationships of *Varroa jacobsoni* and *Apis* species. In: Proceedings of the 11th international congress, IUSSI, Bangalore, pp 453–454, Keywords: *cerana*/*koschevnikovi*

Koeniger N (1995) Biology of the Eastern honeybee *Apis cerana* (Fabricius 1773). In: Kevan PG (ed) The Asiatic hive bee: apiculture, biology, and role in sustainable development in tropical and subtropical Asia. Enviroquest, Cambridge, ON, pp 29–39, Keywords: biology/*cerana*/physiology/nests

Koeniger N (1999) Biology of parasitic Asian bee mites and their honeybee hosts. In: Hoopingarner R, Connor LJ (eds) Apiculture for the 21st century. Wicwas, Cheshire, Keywords: *cerana*/disease/*dorsata*/general biology/honey/pests/predators

Koeniger N, Fang YZ (1991) Longevity of *Tropilaelaps clareae* in *Apis cerana*, *Apis dorsata* and *Apis mellifera*. Apic China 1:36–38, [in Chinese]. Keywords: *cerana*/*dorsata*/*mellifera*/pests/workers/China

Koeniger N, Fuchs S (1972) Communication by sound production in the *Apis cerana* colony. Naturwissenschaften 59:169, [in German]. Keywords: *cerana*/communication/*dorsata*/*florea*/*mellifera*/physiology

Koeniger N, Fuchs S (1973) Sound production as colony defence in *Apis cerana* Fabr. In: Proceedings of the 7th international congress, IUSSI, London, UK, pp 199–204, Keywords: aggressiveness/*cerana*/natural enemies/physiology

Koeniger N, Fuchs S (1975) Colony defence in Asian honeybees. Z Tierpsychol 37:99–106, [in German]. Keywords: *cerana*/defense/*dorsata*/*florea*/interspecific interactions/*mellifera*/physical factors/swarming

Koeniger N, Koeniger G (1980) Observations and experiments on migration and dance communication of *Apis dorsata* in Sri Lanka. J Apic Res 19:21–34, Keywords: absconding/*dorsata*/environmental effects/foraging/general biology/migration/swarming/Sri Lanka

Koeniger N, Koeniger G (1991) An evolutionary approach to mating behaviour and drone copulatory organs in *Apis*. Apidologie 22:581–590, Keywords: *andreniformis*/behaviour/*cerana*/drones/*florea*/*mellifera*/reproduction

Koeniger N, Koeniger G (2000) Reproductive isolation among species of the genus *Apis*. Apidologie 31:313–339, Keywords: evolution/queens/reproduction/Borneo/Malaysia/Sri Lanka/Thailand

Koeniger N, Koeniger G (2001) Behavioural mating barriers among sympatric species of the genus *Apis*. In: Proceedings of the 7th international conference on tropical bees, Chiang Mai, pp 80–85, Keywords: *andreniformis*/*cerana*/*dorsata*/drones/*florea*/*koschevnikovi*/*laboriosa*/*mellifera*/reproduction/Pakistan/Sri Lanka/Thailand

Koeniger N, Muzaffar N (1988) Lifespan of the parasitic honeybee mite, *Tropilaelaps clareae*, on *Apis cerana*, *dorsata* and *mellifera*. J Apic Res 27:207–212, Keywords: *cerana*/*dorsata*/*mellifera*/natural enemies/pests/Pakistan

Koeniger N, Vorwohl G (1979) Competition for food among four sympatric species of Apini in Sri Lanka (*Apis dorsata*, *Apis cerana*, *Apis florea* and *Trigona iridipennis*). J Apic Res 18:95–109, Keywords: bee botany/biogeography/*cerana*/*dorsata*/*florea*/foraging/interspecific interactions/Sri Lanka

Koeniger N, Wijayagunesekera HNP (1976) Time of drone flight in the three Asiatic honeybee species (*Apis cerana*, *Apis florea*, *Apis dorsata*). J Apic Res 15:67–71, Keywords: biogeography/*cerana*/*dorsata*/drones/flight/*florea*/reproduction/Sri Lanka

Koeniger N, Weiss J, Ritter W (1975) Capture and management in cages of the colonies of giant honeybees *Apis dorsata*. In: Proceedings of the 25th international apiculture congress, Grenoble, pp 300–303, Keywords: beekeeping/*dorsata*/honey

Koeniger N, Weiss J, Maschwitz U (1979) Alarm pheromones of the sting in the genus *Apis*. J Insect Physiol 25:467–476, Keywords: aggressiveness/*cerana*/*dorsata*/*florea*/*mellifera*/pheromones/stings

Koeniger N, Koeniger G, Wijayagunesekera HNP (1981) Observations on the adaptation of *Varroa jacobsoni* to its natural host *Apis cerana* in Sri Lanka. Apidologie 12:37–40, [in German]. Keywords: *cerana*/drones/pests/Sri Lanka

Koeniger N, Koeniger G, Delfinado-Baker M (1983) Observations on mites of the Asian honeybee species (*Apis cerana*, *Apis dorsata*, *Apis florea*). Apidologie 14:197–204, Keywords: *cerana*/*dorsata*/drones/*florea*/pests/Indonesia/Sri Lanka

Koeniger N, Koeniger G, Tingek S, Mardan MB, Rinderer TE (1988) Reproductive isolation by different time of drone flight between *Apis cerana* Fabricius, 1793 and *Apis vechti* (Maa, 1953). Apidologie 19:103–106, Keywords: *cerana*/drones/flight/reproduction/Sri Lanka

Koeniger N, Koeniger G, Wongsiri S (1989) Mating and sperm transfer in *Apis florea*. Apidologie 20:413–418, Keywords: disease/drones/*florea*/pests/reproduction

Koeniger N, Koeniger G, de Guzman LI, Lekprayoon C (1993b) Survival of *Euvarroa Sinhai* Delfinado and Baker (Acari, Varroidae) on workers of *Apis cerana* Fabr., *Apis florea* Fabr. and *Apis mellifera* L in cages. Apidologie 24:403–410, Keywords: biogeography/*cerana*/*florea*/*mellifera*/natural enemies/pests/workers/Thailand

Koeniger N, Koeniger G, Mardan MB (1994b) Mimicking a honeybee queen? *Vespa affinis indosinensis* Pérez 1910 hunts drones of *Apis cerana* F 1793. Ethology 98:149–153, Keywords: *cerana*/drones/natural enemies/queens/Malaysia

Koeniger N, Koeniger G, Tingek S, Kalitu A, Mardan M (1994c) Drones of *Apis dorsata* (Fabricius 1793) congregate under the canopy of tall emergent trees in Borneo. Apidologie 25:249–264, Keywords: *dorsata*/drones/Borneo/Malaysia

Koeniger N, Koeniger G, Tingek S, Kelitu A (1996d) Interspecific rearing and acceptance of queens between *Apis cerana* Fabricius, 1793 and *Apis koschevnikovi* Buttel-Reepen, 1906. Apidologie 27:371–380, Keywords: breeding/brood/cells/*cerana*/interspecific interactions/*koschevnikovi*

Koeniger N, Koeniger G, Gries M, Tingek S, Kelitu A (1996e) Observations on colony defense of *Apis nuluensis* Tingek, Koeniger and Koeniger, 1996 and predatory behaviour of the hornet, *Vespa multimaculata* Pérez, 1910. Apidologie 27:341–352, Keywords: aggressiveness/behaviour/*cerana*/glands/natural enemies/*nuluensis*

Koeniger N, Koeniger G, Gries M, Tingek S, Kelitu A (1996f) Reproductive isolation of *Apis nuluensis* Tingek, Koeniger and Koeniger, 1996 by species-specific mating time. Apidologie 27:353–360, Keywords: *cerana*/drones/flight/*koschevnikovi*/*nuluensis*/reproduction/Malaysia

Koeniger N, Koeniger G, Steinmentz M, Punchihewa RWK (1999) Behaviour, ovary development and oviposition of *Apis florea* worker bees after queen removal. In: Proceedings of the 13th international congress, IUSSI, Adelaide, p 260, Keywords: behaviour/brood/*florea*/queens/reproduction

Koeniger N, Koeniger G, Vorwohl G, Tingek S (2001c) Comparing the weight of pollen loads collected by four Asian honeybee species from the same plant. Apidologie 32:406, Keywords: *cerana*/*dorsata*/mellifera/foraging/*koschevnikovi*/pollen

Koeniger N, Koeniger G, Mardan MB, Punchihewa RWK, Otis GW (2003) Numbers of spermatozoa in queens and drones indicate multiple mating of queens in *Apis andreniformis* and *Apis dorsata*. Apidologie 21:282–286, Keywords: *dorsata*/drones/queens/reproduction

Koeniger N, Koeniger G, Gries M, Tingek S (2005b) Drone competition at drone congregation areas in four *Apis* species. Apidologie 36:211–221, Keywords: *cerana*/*dorsata*/mellifera/koschevnikovi/drones/ecology/foraging/reproduction

Kohli N (1958) Bee flora of northern India. Indian Bee J 20:113–118, Keywords: bee botany/*cerana*/*dorsata*/*florea*/India

Koltermann R (1973) Genetic determination of learning behaviour in the honeybee, and its modification by training. In: Proceedings of the 11th international congress, IUSSI, Bangalore, pp 206–211 [in German]. Keywords: behaviour/*cerana*/*mellifera*

Koltermann R (1973b) Retroactive memory inhibition in *Apis mellifica* and *Apis cerana* (Apidae) after learning two kinds of information in succession. J Comp Physiol 84:299–310, [in German]. Keywords: behaviour/*cerana*/*mellifera*/pheromones/physiology

Koltermann R (1973c) Race specific and species specific scent evaluation in honeybees and their ecological adaptation. J Comp Physiol 85:327–360, [in German]. Keywords: *cerana*/foraging/*mellifera*/pheromones

Komeili AB (1990) Beekeeping in Iran. Bee World 71:12–24, Keywords: beekeeping/*cerana*/*florea*/Iran

Kong IF (1982) A summary of Chinese beekeeping history. In: Selected references of the Chinese honey bee. Gansu Provincial Apicultural Research Institute, Tianshui, pp 33–41, [in Chinese]. Keywords: beekeeping/*cerana*/history/honey/China*

Konishi M (1992) Origin of beekeeping and development of hives. Honeybee Sci 13:55–59, Keywords: beekeeping/*cerana*/history/physiology/Japan

Koschevnikov GA (1926) Indian honeybee in the Far East. Opyt Paseka 9:4–5, [in Russian]. Keywords: biogeography/*cerana*/Russia
Koschevnikov GA (1931) The life history of the honeybee. Leningrad State Agricultural Publishing House, Moscow, [in Russian]. Keywords: biology/*cerana*/history/Russia
Kozin RB (1976) Pollination of entomophilous agricultural crops by bees. Amerind, New Delhi, Keywords: *cerana*/*dorsata*/*florea*/*mellifera*/pollination
Kranner G (2000) Visualization of multiple influences on ocellar flight control in giant honeybees with the data-mining tool, Viscovery SOMine. Behav Res Meth Instrum Comp 32:157–168, Keywords: *dorsata*/foraging/morphology
Kraus B, Velthuis HHW, Tingek S (1998) Temperature profiles of the brood nests of *Apis cerana* and *Apis mellifera* colonies and their relation to varroatosis. J Apic Res 37:175–181, Keywords: brood/*cerana*/*mellifera*/nests/physiology/Malaysia
Kraus B, Neumann P, van Praagh J, Moritz RFA (2004) Sperm limitation and the evolution of extreme polyandry in honeybees (*Apis mellifera* L). Behav Ecol Sociobiol 55:494–501, Keywords: *andreniformis*/evolution/*florea*/*mellifera*/reproduction
Kraus B, Koeniger N, Tingek S, Moritz RFA (2005a) Temporal genetic structure of a drone congregation area of the giant Asian honeybee (*Apis dorsata*). Naturwissenschaften 92:1–4, Keywords: *dorsata*/drones/genetics
Kraus B, Koeniger N, Tingek S, Moritz RFA (2005b) Using drones for estimating colony number by microsatellite DNA analysis of haploid males in *Apis*. Apidologie 36:223–229, Keywords: biochemistry/*dorsata*/drones/general biology/organic chemistry
Kreil G (1973) Structure of melittin isolated from two species of honey bees. FEBS Lett 33:241–244, Keywords: biochemistry/*cerana*/*florea*/honey/*mellifera*
Kreil G (1975) The structure of *Apis dorsata* melittin: phylogenetic relationships between honeybees as deducted from sequence data. FEBS Lett 54:100–102, Keywords: biochemistry/*cerana*/*dorsata*/*florea*/*mellifera*
Kronvall G (1980) Traditional beekeeping in India. Bitidningen 79:53–54, Keywords: beekeeping/*cerana*/India
Kshirsagar KK (1967) Mites on the Indian honeybee. Bee World 48:84–85, Keywords: *cerana*/pests/India
Kshirsagar KK (1968) Further observations on phoretic bee mites *Neocypholaelaps indica* (Evans). Indian Bee J 30:68–69, Keywords: *cerana*/pests/India
Kshirsagar KK (1969) Preliminary observations on the biometry of *Apis dorsata* Fab. queen. Indian Bee J 31:12–16, Keywords: classification/*dorsata*/queens/India
Kshirsagar KK (1971) Comparative biometric studies on Indian bees: II Preliminary observations on the biometry of *Apis florea* F queen. Indian Bee J 33:7–13, Keywords: classification/*florea*/queens/India
Kshirsagar KK (1973) Comparative biometric studies on Indian honeybees: III Preliminary observations on biometry of *Apis cerana* F queen. Indian Bee J 35:21–26, Keywords: *cerana*/classification/queens/India
Kshirsagar KK (1976) Studies on Indian Apidae with special reference to Indian hive bee, *Apis cerana* F. Thesis, University of Poona, Keywords: *cerana*/nests/India
Kshirsagar KK (1981) Morphometric studies on *Apis cerana indica* F worker. III Effect of geographical position on morphometric characters Indian Bee J 43:1–5, Keywords: *cerana*/classification/geographical variation/physical factors/India
Kshirsagar KK (1982) Current incidence of honeybee diseases and parasites in India. Bee World 63:162–164, Keywords: *cerana*/disease/parasites/pests/India
Kshirsagar KK (1983a) Diseases of bees in India. Indian Bee J 45:39–42, Keywords: *cerana*/disease/India
Kshirsagar KK (1983) Morphometric studies on the Indian hive bee *Apis cerana indica* F I: Morphometric characters useful in identification of intraspecific taxa. In: Proceedings of the 2nd international conference on apiculture in tropical climates, New Delhi, pp 254–261, Keywords: *cerana*/classification/morphology/nests

Kshirsagar KK (1983c) Spread of sacbrood disease in Uttar Pradesh. Indian Bee J 45:41–42, Keywords: *cerana*/disease/India

Kshirsagar KK, Godbole SH (1974) Indian Streptococcus pluton Indian Bee J 36:39, Keywords: *cerana*/disease/*mellifera*/India

Kshirsagar KK, Percy AP (1966) Further observations on the mites on Indian honeybee *Apis indica* in Kodaikanal. Indian Bee J 28:25–26, Keywords: behaviour/*cerana*/pests/India

Kshirsagar KK, Phadke RP (1985) Occurrence and spread of Thai sacbrood disease in *Apis cerana*. In: Proceedings of the 3rd international conference on apiculture in tropical climates, Nairobi, pp 149–151, Keywords: *cerana*/disease/physical factors/India/Nepal/Thailand

Kshirsagar KK, Ranade DR (1981) Morphometric characterization of Indian hive bee *Apis cerana indica* F. (Apidae, Hymenoptera), worker. J Univ Poona Sci Technol 54:101–120, Keywords: *cerana*/classification/morphometrics/nests/India

Kshirsagar KK, Chauhan RM, Singh YK (1974) Detection of Nosema *Apis* Zander in *Apis cerana indica* in Uttar Pradesh (India). Indian Bee J 36:23–24, Keywords: *cerana*/disease/India

Kshirsagar KK, Mahindre DB, Salvi SR, Mittal MC (1975) Occurrence of a viral disease in Indian hive bee *Apis cerana indica* F: a preliminary report. Indian Bee J 37:19–20, Keywords: *cerana*/disease/India

Kshirsagar KK, Mittal MC, Chauhan RM (1981a) Occurrence of sacbrood disease in *Apis cerana indica*. Indian Bee J 43:44, Keywords: *cerana*/disease/India

Kshirsagar KK, Singh BP, Chauhan RM (1981b) Queen cells and queen bees in *Apis dorsata* F. Indian Bee J 43:129, Keywords: combs/*dorsata*/queens/India

Kshirsagar KK, Saxena UC, Chauhan RM (1982) Occurrence of sacbrood disease in *Apis cerana indica* F in Bihar, India. Indian Bee J 44:8–9, Keywords: brood/*cerana*/disease/India

Kshirsagar KK, Muvel KS, Mittal MC, Phadke RP (1983) Some observations on the behaviour of *Apis florea* F. In: Proceedings of the 2nd international conference on apiculture in tropical climates, New Delhi, pp 356–366, Keywords: behaviour/classification/combs/environmental effects/*florea*/reproduction/India

Kshirsagar KK, Mittal MC, Mahindre DB, Salvi SR, Chauhan RM (1983) Seasonal fluctuations in the incidence of acarine disease in bees and colony losses in India. In: Proceedings of the 2nd international conference on apiculture in tropical climates, New Delhi, pp 442–453, Keywords: *cerana*/disease/pests/physical factors/India

Kshirsagar KK, Salvi SR, Mahindre DB, Chauhan RM (1983) Menthol as an effective acaricidal fumigant. In: Proceedings of the 2nd international conference on apiculture in tropical climates, New Delhi, pp 454–456, Keywords: *cerana*/insecticides/physical factors/India

Kuang BY (1981a) Studies of *Apis florea*. J Bee 4:11–12, [in Chinese]. Keywords: *florea*/general biology/China

Kuang BY (1981b) Study on *Apis dorsata* and *A. florea* in Yunnan. Yunnan Agric Sci 6:43–47, [in Chinese]. Keywords: *dorsata*/ecology/*florea*/China

Kuang BY (1983) Morphological identification of 2 species of large honeybee. Chinese Apic 4:1, Keywords: classification/*dorsata*/*laboriosa*/morphology/China

Kuang BY (1986) The species of *Apis* in China. Apic China 5:7–9, [in Chinese]. Keywords: *cerana*/*dorsata*/*florea*/mellifera/China

Kuang BY (1988a) Comparative studies on the biology of six honeybee species. Apic Sci Tech 1:12–14, [in Chinese]. Keywords: biology/*cerana*/*mellifera*/*florea*/*dorsata*/*laboriosa*/andreniformis/China

Kuang BY (1988b) Taxonomy and evolution of Apini. Beekeep Hubei 1:19–20, [in Chinese]. Keywords: beekeeping/*cerana*/classification/*dorsata*/*florea*/phylogeny/China

Kuang BY (2002) Evolution and classification of Apini. J Bee 2:1–2, [in Chinese]. Keywords: *cerana*/classification/*dorsata*/*florea*/phylogeny/China

Kuang BY, Li YQ (1985) The genus *Apis* in China. Chinese Beekeep 76:7–9, [in Chinese]. Keywords: biogeography/*cerana*/*dorsata*/laboriosa/florea/China

Kuang BY, Li YQ (1988) Studies on Apini. Sci Agric Sinica 21:85–90 [in Chinese]. Keywords: biochemistry/biogeography/*cerana*/*dorsata*/ecology/*laboriosa*/*florea*/morphology/organic chemistry/China

Kuang BY, Tan K (1996) Honeybee genetic diversity and its value to mountain farming. In: Proceedings of the 3rd Asian apiculture association international conference, Hanoi, pp 23–33, Keywords: *andreniformis*/beekeeping/biogeography/*cerana*/*dorsata*/*florea*/genetics/interspecific interactions/*laboriosa*/*mellifera*/pollination/China

Kuang BY, Tan K (1997) Technique for managing the strong potency of *A. cerana* colony for high honey yield. In: Proceedings of the 35th international apiculture congress, Apimondia, Antwerp, p 82, Keywords: beekeeping/*cerana*/honey/China

Kuang BY, Kuang HO, Tan K (2000) Effect of synthetic queen pheromone on the behaviour of *Apis cerana*. In: Proceedings of the 4th Asian apiculture association international conference, Kathmandu, pp 24–25, Keywords: behaviour/*cerana*/queens/China/Philippines

Kuang BY, Tan K, Liu YQ (2002) Honeybee biology. Yunnan Science Publishing House, Kunming, [in Chinese]. Keywords: beekeeping/biology/*cerana*/*dorsata*/*florea*/*mellifera*/China

Kuang HO, Liu YQ, Tan K (1998) Synthetic queen pheromone affects honey production of *A. cerana* and *A. mellifera* colonies. J Yunnan Agric Univ 13:315–318. Keywords: *cerana*/honey/*mellifera*/pheromones/queens/China

Kuang HO, Liu Y, Tan K (2000) Effect of synthetic queen pheromone on honey production of *Apis mellifera* and *Apis cerana*. In: Proceedings of the 4th Asian apiculture association international conference, Kathmandu, pp 26–28, Keywords: *cerana*/honey/*mellifera*/pheromones/queens/China

Kubota A (1984) Comparison of pollen plants foraged by Japanese and European honeybees. Honeybee Sci 5:157–158 [in Japanese]. Keywords: bee botany/*cerana*/foraging/*mellifera*/pollen/Japan

Kullol SK, Sreekala AK, Pandurangan AG (2009) Floral biology of *Impatiens trichocarpa* Hook. f. (Balsaminaceae) an endemic balsam of Western Ghats. Indian J Sci Tech 2:30–34. Keywords: *cerana*/*dorsata*/*florea*/pollination/India

Kuman NR, Bhalla OP, Kumar R (1993) Studies on phenological influences on the semen of *Apis* species. In: Connor LJ, Rinderer TE, Sylvester HA, Wongsiri S (eds) Asian apiculture, Wicwas Press, Cheshire, Connecticut, pp 211–214. Keywords: *cerana*/drones/India

Kumar A, Kumar J (1994) Variability in tolerance to some insecticides in *Apis mellifera* L and *Apis cerana indica* F colonies. Entomon 19:67–72. Keywords: *cerana*/insecticides/*mellifera*/India

Kumar J, Gupta JK (1993) Nectar sugar production and honeybee foraging activity in 3 species of onion (*Allium* species). Apidologie 24:391–396. Keywords: *cerana*/foraging/nectar/pollination/India

Kumar J, Kumar A (1991) Intercolony and interspecific variation for fat and nucleic acid contents in *Apis mellifera* L and *Apis cerana indica* F. J Apic Res 30:21–23. Keywords: biochemistry/*cerana*/*mellifera*/physiology/India

Kumar J, Mishra RC, Gupta JK (1985) The effect of mode of pollination on *Allium* spp. with observations on insects as pollinators. J Apic Res 24:62–66. Keywords: *cerana*/pollination/India

Kumar J, Mishra RC, Gupta JK, Dogra GS (1985) Pollination requirements of some peach cultivars. Indian Bee J 47:3–6. Keywords: *cerana*/pollination/India

Kumar J, Patyal SK, Mishra RC (1986) Evaluation of some essential oils as repellents to the Indian honeybee *Apis cerana indica* 1. Gustatory repellency and toxicity J Apic Res 25:256–261, Keywords: *cerana*/chemistry/India

Kumar J, Mishra RC, Gupta JK (1989a) Effect of honey bee pollination on onion (*Allium cepa* L) seed production. Indian Bee J 51:3–5, Keywords: *cerana*/honey/pollination/India

Kumar J, Rao KVK, Gupta PR, Dogra GS (1989b) Temporal distribution of bees (Hymenoptera: Apoidea) on spring flowering crops in midhills of Himachal Pradesh, India. Indian Bee J 51:55–58, Keywords: biogeography/*cerana*/*dorsata*/*florea*/pollination/India

Kumar M, Singh R (2001a) Effect of honeybee foraging distance on the pollination of sunflower. J Entomol Res 25:37–39, Keywords: *cerana*/foraging/*mellifera*/pollination/India

Kumar M, Singh R (2001b) Initiation-cessation and period of foraging activity of honeybees on sunflower. J Entomol Res 25:141–143, Keywords: *cerana*/foraging/*mellifera*/India

Kumar M, Singh R, Chand H (2001) Effect of different modes of pollination on quantitative and qualitative characters in sunflower seeds (*Helianthus annuus* L). Shashpa 8:157–160, Keywords: *cerana*/*mellifera*/pollination/India

Kumar M, Singh R, Chand H (2002) Foraging activity of *Apis cerana indica* and *Apis mellifera* visiting sunflower (*Helianthus annuus* L). Shashpa 9:31–34, Keywords: *cerana*/foraging/*mellifera*/India

Kumar M, Singh R, Chand H (2003) Foraging rate and foraging speed of *Apis cerana indica* Fab. and *Apis mellifera* L in sunflower (*Helianthus annuus* L). Shashpa 10:27–32, Keywords: *cerana*/foraging/*mellifera*/India

Kumar M, Hari C, Singh R, Ali MS (2004) Effect of different modes of honeybee pollination on oil content in seeds of sunflower (*Helianthus annuus* L). J Entomol Res 26:219–221, Keywords: *cerana*/*mellifera*/pollination/India

Kumar NR (2000) Beekeeping: self-employment opportunity for mountain women. In: Proceedings of the 4th Asian apiculture association international conference, Kathmandu, pp 245–247, Keywords: beekeeping/*cerana*/India

Kumar NR, Kumar R (1997a) Insect pollination of apple in mid-hills of Himachal Pradesh. Indian Bee J 59:179–180, Keywords: *cerana*/*mellifera*/pollination/India

Kumar NR, Kumar R (1997b) Successful management of a laying worker colony. Am Bee J 137:647, Keywords: beekeeping/*cerana*/laying workers/India

Kumar NR, Kumar R (1998) Evaluation of honey hoarding quality in the Asian honey bee populations. Pest Manag Econ Zool 6:101–103, Keywords: behaviour/*cerana*/geographical variation/honey/populations

Kumar NR, Shivali K (1999) Cytochemical characterisation of the royal jelly producing hypopharyngeal glands in honey bee. Pest Manag Econ Zool 7:111–114, Keywords: *cerana*/glands/honey/morphology/royal jelly

Kumar NR, Kumar R, Bhalla OP (1994a) Pine marten attacks the colonies of *Apis cerana indica* F at Nauni (India). Indian Bee J 56:64–67, Keywords: *cerana*/natural enemies/India

Kumar R (1993) Bee pasturage of Khetri. Indian Bee J 55:40–42, Keywords: beekeeping/bee botany/*cerana*/foraging/India

Kumar R (1997a) A check-list of sources of bee forage of Pialitown, South 24-Paraganas District, West Bengal, India. Indian Bee J 59:241–242, Keywords: bee botany/*cerana*/foraging/India

Kumar R (1997) A novel strategy for conservation of the native honeybee *A. cerana* Fabr. In: Proceedings of the national seminar on institutional environment, Punjab University, Chandigarh, Keywords: *cerana*/ecology/India

Kumar R (1997c) Nosema *Apis* Zander (Protozoa, Sporozoa) infection in honey bee, *Apis cerana* Fabr., populations. Pest Manag Econ Zool 5:71–73, Keywords: *cerana*/disease/honey/populations/India

Kumar R, Chaudhary OP (1993) Bee plants in India. Khadi Gramodyog 39:844–854, Keywords: bee botany/India

Kumar R, Kumar NR (1995) Mites in close environment of honey bees. Insect Environ 1:5–7, Keywords: *cerana*/honey/*mellifera*/pests/India

Kumar R, Kumar NR (1996) Traditional hives of *A. cerana* Fabr. Insect Environ 3:9, Keywords: beekeeping/*cerana*/India

Kumar R, Kumar NR (1997c) Observations on *Varroa jacobsoni* Oudemans (Varroidae: Mesostigmata) parasitizing *Apis cerana* Fabr. and *Apis mellifera* Linn. Indian Bee J 59:224–225, Keywords: *cerana*/*mellifera*/pests/India

Kumar R, Kumar NR (2000) Effects of modernised wall hives on performance of *Apis cerana*. In: Proceedings of the 4th Asian apiculture association international conference, Kathmandu, pp 129–131, Keywords: beekeeping/*cerana*/nests/India

Kumar R, Kumar NR (2000) Queen-rearing and royal-jelly production in Asian honeybee *Apis cerana*. In: Proceedings of the 4th Asian apiculture association international conference, Kathmandu, pp 145–147, Keywords: beekeeping/*cerana*/queens/royal jelly/India

Kumar R, Lenin JK (2001) Studies on foraging behaviour of honey bees and pollination in orange, *Citrus reticulata* Blanco. Indian Bee J 60:106–107, Keywords: behaviour/*cerana*/*florea*/foraging/honey/*mellifera*/pollination/India

Kumar R, Wakode MT (1997) A note on the brood rearing activity of *Apis cerana indica* F at Pune, India. Indian Bee J 59:163–164, Keywords: bee botany/breeding/brood/*cerana*/colony phenology/India

Kumar R, Kumar NR, Bhalla OP (1993) A comparative study on the response of *Apis* species to predation by *Vespa* species. In: Connor LJ, Rinderer TE, Sylvester HA, Wongsiri S (eds) Asian apiculture. Wicwas, Cheshire, CT, pp 238–242, Keywords: *cerana*/natural enemies

Kumar R, Chaudhary OP, Lenin JK (1994b) Studies on the foraging behaviour of honeybees and their role as pollinators of sunflower (*Helianthus annuus* L). Indian Bee J 56:207–210, Keywords: bee botany/behaviour/foraging/India/natural enemies/pollination/India

Kumar R, Kumar NR, Verma LR (1996a) A novel eco-friendly hive for the native honey bee *Apis cerana*. Insect Environ 2:89–90, Keywords: beekeeping/*cerana*/honey/nests/India

Kumar R, Lenin JK, Chandran K (1996b) Honey bee pollination of *Psidium guajava* L. Indian Bee J 58:206–207, Keywords: *cerana*/honey/pollination/India

Kumar R, Lenin JK, Chandran K (1997) Studies on floral biology and foraging behaviour of honey bees on coconut palm, *Cocos nucifera* L. Indian Bee J 59:74–75, Keywords: behaviour/biology/*cerana*/*florea*/foraging/honey/*mellifera*/India

Kumar S, Mishra PK, Yazdani SS (1996c) Foraging activity of honeybees on litchi (*Litchi chinensis* (Gaertn) Sonn) flowers. J Res Birsa Agric Univ 8:105–109, Keywords: *cerana*/foraging/pollination/India

Kumari R (1987) Morphometry of Indian hive bee, *Apis cerana indica* F of Andaman Islands. Thesis, Himachal Pradesh University, Keywords: *cerana*/classification/islands/morphology/nests/India

Kumari S (1988) Reproductive biology and in vitro preservation of honeybee *Apis mellifera* L and *Apis cerana* F semen. Thesis, Himachal Pradesh University, Keywords: biology/*cerana*/drones/*mellifera*/reproduction/India

Kumbhakarni CG (1964) Cytological studies in Hymenoptera: Part I. Cytology of parthenogenesis in the honeybee, *Apis dorsata*. Indian J Exp Biol 2:65–68, Keywords: *dorsata*/morphology

Kundu BC, Basak KC, Sarcar PB (1959) Jute in India. Leonard Hill, London, Keywords: beekeeping/*cerana*/India

Kung YF Summary of China beekeeping history* [in Chinese]. Keywords: beekeeping/*cerana*/history/China

Kuntadi TH (1989) Nesting habit of *Apis cerana* F in nature. Bull Penelitian Kehutanan 5:21–26, [in Indonesian]. Keywords: behaviour/*cerana*/nests/Indonesia

Kuntadi TH (1990) Nest characters of *Apis cerana* in three different altitudes. Bull Penelitian Kehutanan 5:345–355, [in Indonesian]. Keywords: *cerana*/nests/physical factors/Indonesia

Kurentsov AI (1964) On the protection of some beneficial and relic insect species of the Ussuri fauna. In: Khomentovskiy AS (ed) Nature reservation in the Far East (of the USSR) No 2. Far East Branch of the Sibirian Division of the Academy of Sciences of the USSR, Vladivostok, pp 103–111, [in Russian]. Keywords: *cerana*/conservation/Russia

Kurentsov AI (1973) On the protection of rare and relic insect species of the Far East of the USSR. In: Mirzoyan SA (ed) On the insect protection. Academy of Sciences of the Armenian SSR, Erevan, pp 51–60, [in Russian]. Keywords: *cerana*/ecology/Russia

Kuriakose G, Sinicau PA, Shivanna KR (2009) Domestication of cardamom (*Elettaria cardamomum*) in Western Ghats, India: divergence in productive traits and a shift in major pollinators. Ann Bot 103:727–733, Keywords: *cerana*/*dorsata*/pollination

Kustantiny A, Pane R, Sitadewi H (1988) Beekeeping with *Apis cerana* in Indonesia. In: Proceedings of the 2nd Australian and international bee congress, Brisbane, pp 184–186, Keywords: beekeeping/*cerana*/Indonesia

Kustermann T, Hanh PD (1996) Family beekeeping - an important component of natural resource conservation progams in Vietnam. In: Proceedings of the 3rd Asian apiculture association international conference, Hanoi, pp 110–114, Keywords: bee botany/beekeeping/*cerana*/ecology/Vietnam

Kuznetsov VN (1997) On the necessity of conservation of the Chinese wax bee, *Apis cerana cerana* F (Hymenoptera, Apidae), in the Primorski Territory. In: Plant and animal world of the Far East (of Russia) No 3. Ussuriysk State Pedagogical Institute, Ussuryisk, pp 57–65* [in Russian]. Keywords: beeswax/*cerana*/Russia

Kuznetsov VN (1999) On the state of populations of the Chinese wax bee *Apis cerana cerana* F (Hymenoptera, Apidae) in the Primorski Territory. In: Urusov VM (ed) The study and formation of landscapes of the Russian Far East and Siberia: a collection of scientific papers. Pacific Institute of Geography, Far Eastern Branch of the Russian Academy of Sciences, Vladivostok, pp 202–212, Keywords: beeswax/biogeography/*cerana*/population biology/populations/Russia

Kuznetsov VN (2002) The Chinese wax bee *Apis cerana cerana* F (Hymenoptera, Apidae) in the Primorski Territory. Institute for Biology and Soil Sciences, Far Eastern Branch of the Russian Academy of Sciences, Vladivostok, [in Russian]. Keywords: beeswax/biogeography/biology/*cerana*/Russia

Kwon YJ, Huh EY (1992a) Electron morphometric classification of the native honey bees from Korea. Part I. Discriminant analysis on the wing characters. Korean J Apic 7:1–31, [in Korean]. Keywords: *cerana*/classification/honey/morphology/Korea

Kwon YJ, Huh EY (1992b) Electron morphometric classification of native honey bees from Korea. Part II. Discriminate analysis for different populations based on total characters. Korean J Appl Entomol 32:30–41, [in Korean]. Keywords: *cerana*/classification/honey/population biology/populations/morphology/Korea

Kwon YJ, Huh EY (1992c) Electron morphometric classification of native honey bees from Korea. Part III. Discriminant analysis for different localities based on total characters. Korean J Appl Entomol 32:42–50, [in Korean]. Keywords: *cerana*/classification/honey/locality/morphology/Korea

Kwon YJ, Huh EY (1992d) Electron morphometric classification of native honey bees from Korea. Part IV. Discriminate analysis for 5 sublocal populations in Chirisan based on the total characters. Korean J Apic 7:107–117, [in Korean]. Keywords: *cerana*/classification/honey/population biology/populations/morphology/Korea

Kwon YJ, Huh EY (1992e) Electron morphometric classification of native honey bees from Korea. Part V. Cluster analysis by canonical function score. Korean J Syst Zool 8:189–200, [in Korean]. Keywords: *cerana*/classification/honey/morphology/Korea

Kwon YJ, Huh EY (1992f) Electron morphometric classification of native honey bees from Korea. Part VI. Cluster analysis by arithmetic mean. Korean J Apic 7:118–129, [in Korean]. Keywords: *cerana*/classification/honey/morphology/Korea

Lai DZ (1982) Studies on sacbrood virus tissue cultures of *Apis cerana cerana*. Apic China 2:23–24, [in Chinese]. Keywords: *cerana*/disease/China

Lai DZ, Shi BL, Li GX (1982) Studies of tissue culture of sac brood virus from *Apis cerana cerana*. Zhongguo Yangfeng 2:23–24, [in Chinese]. Keywords: brood/*cerana*/disease/China

Lai YS (1960) Chinese beekeeping handbook. Division of Apiculture, Gungzhu, [in Chinese]. Keywords: beekeeping/*cerana*/*mellifera*/China

Laigo FM, Morse RA (1968) The mite *Tropilaelaps clareae* in *Apis dorsata* colonies in the Philippines. Bee World 49:116–118, Keywords: disease/*dorsata*/pests/predators/Philippines

Lakshmi K, Mohana-Rao G (1998) Guava. Bee World 79:135–137, Keywords: *cerana*/*dorsata*/*florea*/pollination

Lakshmi K, Suryanarayana MC (1996) Common arboreal sources of bee forage in India *Pterospermum xylocarpum*. Indian Bee J 58:209–210, Keywords: bee botany/*dorsata*/foraging/India

Lakshmi K, Suryanarayana MC (1997a) A contribution to the study of the kapok tree, *Ceiba pentandra* (L) Gaert, as a source of forage to honey bees. Indian Bee J 59:165–169, Keywords: *cerana*/foraging/honey/pollen/India

Lakshmi K, Suryanarayana MC (1997b) Melittoplaynology of forest honeys from Chintapalli Hills, Andhra Pradesh, India. Proc Indian Natl Sci Acad 63:581–596, Keywords: bee botany/*cerana*/*dorsata*/honey/pollen/India

Lakshmi K, Suryanarayana MC (1997c) Microscopical analysis of rock bee honeys from Nallamalai forest of Andhra Pradesh, India. J Palynol 33:263–272, Keywords: *dorsata*/honey/India

Lakshmi K, Suryanarayana MC (1997d) Sources of bee forage of *Apis dorsata* Fabr. in forest areas of Andhra Pradesh, India. Indian Bee J 59:11–23, Keywords: bee botany/*dorsata*/drones/India

Lakshmi K, Suryanarayana MC (1999) Palynological analysis of *Apis dorsata* Fab. honeys from deciduous forests of Andhra Pradesh, India. Asian Bee J 1:64–72, Keywords: *cerana*/*dorsata*/honey/India

Lakshmi K, Suryanarayana MC (2001) Melittopalynology of red sanders (*Pterocarpus santalinus* L) honeys. Indian Bee J 59:218–224, Keywords: *dorsata*/honey/India

Lakshmi K, Rao GM, Suryanarayana MC (2000) Cashew. Bee World 81:30–33, Keywords: bee botany/*cerana*/*dorsata*/*florea*/India

Lal KB (1944) Summering of hill bees in the plains of India. Indian Bee J 6:155, Keywords: *cerana*/seasonal variation/India

Lal KB (1945) The distribution of *Apis indica* F in the plains of India. Indian Bee J 7 (133–134):144, Keywords: biogeography/*cerana*/India

Lan SQ (1994) Summer and autumn management of *Apis cerana cerana* in southern China. J Bee 7:22, [in Chinese]. Keywords: beekeeping/*cerana*/China

Lan SQ, Liu YS (1995) The breeding potential of *Apis cerana cerana* colonies during autumn in southern China. J Bee 8:41, [in Chinese]. Keywords: beekeeping/breeding/*cerana*/China

Langstroth LL (1878) *Apis dorsata*. Am Bee J 14:417–418, Keywords: *dorsata*

Laoaroon S (1998) Genetic differentiation among Thai honeybees Apis cerana using microsatellite variation and nuclear ribosomal RNA genes. Thesis, Chulalongkorn University, Keywords: *cerana*/DNA/genetics/RNA/Thailand

Lap PV, Chinh TX (1996) Artificial insemination of *Apis cerana cerana* (Fab.). In: Proceedings of the 3rd Asian apiculture association international conference, Hanoi, pp 34–39, Keywords: breeding/*cerana*/drones/physiology/Vietnam

Lap PV, Chinh PH, Chinh TX (1992) Some biological characteristics of *A. cerana* queen bees in Vietnam. In: Proceedings of international conference on asian honeybees and bee mites, Bangkok, Keywords: *cerana*/pests/physiology/queens/reproduction/Thailand/Vietnam

Lap PV, Chinh PH, Ha TD, Chinh TX, Hanh PD, Ngoc PV (1993) Some biological characteristics of *Apis cerana* queen bees in Vietnam. In: Connor LJ, Rinderer TE, Sylvester HA, Wongsiri S (eds) Asian apiculture. Wicwas Press, Cheshire, Connecticut, pp 117–123, Keywords: biology/*cerana*/queens/Vietnam

Latif A, Hussain M (1955) Double-queen colonies of *Apis indica* F at Sialkot. In: Proceedings of the 7th Pakistan science conference, Pakistan, Keywords: *cerana*/queens/India/Pakistan*

Latif A, Qayyum A, Manzoor ul Haq (1956) Role of Apis indica F in the pollination of Egyptian clover (Trifolium alexandrinum Linn). Pak J Sci Res 8:48–50, Keywords: *cerana*/pollination/Pakistan

Latif A, Qayyum A, Manzoor ul Haq (1956) Researches on the composition of Pakistan honey. Pak J Sci Res 8:163–166, Keywords: *cerana*/*dorsata*/*florea*/honey/Pakistan

Latif A, Qayyum A, Manzoor ul Haq (1956) Two-queen system in bee-hives of *Apis indica* F at Lyallpur. In: Proceedings of the 7th Pakistan science conference, Pakistan, Keywords: beekeeping/*cerana*/queens/India/Pakistan*

Latif A, Qayyum A, Manzoor ul Haq (1956) Two-queen system in relation to colony development. In: Proceedings of the 7th Pakistan science conference, Pakistan, Keywords: beekeeping/*cerana*/physiology/queens/Pakistan*
Latif A, Qayyum A, Abbas M (1960a) The role of *Apis indica* in the pollination of 'toria' and 'sarson' (*Brassica campestris* var *campestristoria* and *dichotoma*). Bee World 41:283–286, Keywords: *cerana*/pollination/Pakistan
Latif A, Qayyum A, Manzoor ul Haq (1960) Multiple and two-queen systems in Apis indica F. Bee World 41:201–209, Keywords: beekeeping/*cerana*/queens/swarming/Pakistan
Latif A, Qayyum A, Tufail M (1960c) Studies on Pakistan beeswax. Bee World 41:153–156, Keywords: beeswax/*cerana*/*dorsata*/*florea*/Pakistan
Laude VT, Naegel L, Horn H (1991) The physicochemical properties of some Philippine honeys. Apidologie 22:371–380, Keywords: *cerana*/honey/Philippines
Lavrekhin FA (1935) Contribution to the study of variation of the honey bee 2. A comparative biometric characteristic of the sexual appendages of the drones belonging to different forms of the genus Apis. Zool Zh 14:655–663, [in Russian]. Keywords: *cerana*/classification/*dorsata*/drones/*florea*/honey/*mellifera*/morphology/reproduction
Lavrekhin FA (1947a) Expedition to maritime territory to study the Indian bee. Pchelovodstvo 6:23–28, [in Russian]. Keywords: biogeography/*cerana*/Russia
Lavrekhin FA (1947b) Peculiarities in the structure and behaviour of the Indian bee. Pchelovodstvo 7:31–36, [in Russian]. Keywords: behaviour/*cerana*/morphology/Russia
Lavrekhin FA (1960a) Comparative observations on the flight activity of drones. Pchelovodstvo 37:43–45, [in Russian]. Keywords: *cerana*/drones/flight/*mellifera*/Russia
Lavrekhin FA (1960) First experiments of introducing the wild bees of Ussuriland (*Apis indica* F) into the European part of the Soviet Union. In: Proceedings of the 17th International Apiculture Congress, Rome, pp 237–238, [in German]. Keywords: biogeography/*cerana*/queens/Russia
Lavrekhin FA, Pankova SV (1969) Biology of the honeybee family. Izdatel'stvo 'Kolos', Moskva, [in Russian]. Keywords: biology/*cerana*/*dorsata*/*florea*/*mellifera*
Lazar M (1995) *Apis cerana* nesting in the open. Indian Bee J 57:146, Keywords: *cerana*/nests
Le Conte Y (2004) *Apis florea*, the little honeybee of Southeast Asia. Abeilles Fleurs 651:1516, [in French]. Keywords: beekeeping/*florea*/India
Lee CJ, Shen J-Y, Part S-C, Shim JH (2003) Chemical analysis of cuticular hydrocarbons in *Apis mellifera* L and *Apis cerana* F. Korean J Appl Entomol 42:9–13, Keywords: *cerana*/chemistry/*mellifera*
Lee CY, Smith NL, Underwood BA, Morse RA (1990) Honey protein from different bee species in relation to apple juice clarification activity. Am Bee J 130:478–479, Keywords: biochemistry/*cerana*/honey/*mellifera*/Korea
Lee HS, Woo KS (1983) A comparative study of apid bee (Hym: Apidae) fauna between Suwon and Simwon. Korean J Apic 8:111–115, Keywords: *cerana*/ecology/Korea
Lee ML (1993) Morphological and biochemical characteristics of *Apis cerana* in South Korea. In: Connor LJ, Rinderer TE, Sylvester HA, Wongsiri S (eds) Asian apiculture. Wicwas Press, Cheshire, Connecticut, pp 161–172, Keywords: biochemistry/*cerana*/classification/DNA/morphology/Korea
Lee ML, Choi SY (1986) Biometrical studies on the variation of some morphological characters in Korean honeybees. Apis cerana F and A. mellifera L. Korean. J Apic 1:5–23, [in Korean]. Keywords: *cerana*/classification/*mellifera*/morphology/Korea
Lee ML, Woo KS (1991) Enzyme polymorphism of *Apis cerana* Fabr. in Korea. Honeybee Sci 12:58–60, [in Japanese]. Keywords: biochemistry/*cerana*/Korea
Lee ML, Yim YH, Kim SS, Woo KS, Suh DS (1989) Malate dehydrogenase and nonspecific esterase polymorphism in *Apis mellifera* and *Apis cerana* in South Korea. Korean J Apic 4:68–74, [in Korean]. Keywords: biochemistry/*cerana*/*mellifera*/Korea
Lee SK (1981) Beekeeping in Korea. In: Proceedings of the 28th international apiculture congress, Apimondia, Acapulco, pp 193–196, Keywords: beekeeping/*cerana*/Korea

Leelamanit W, Neelasaewee S, Boonyom R, Panyim S, Hayashi T, Yasue H, Amano K (2004) The NADH dehydrogenase genes of *Apis mellifera*, *A cerana*, *A dorsata*, *A laboriosa* and *A florea*: sequence comparison and genetic diversity. J Anim Genet 31:3–12, Keywords: cerana/dorsata/florea/genetics/laboriosa/mellifera

Leelamma PJ (1988) Biometry of the Indian hive bee, Apis cerana F of South India. Thesis, Himachal Pradesh University, Keywords: *cerana*/classification/nests/India

Legewie H (1925) On the theory of honeybee colonies. Zoomorphology 4:246–300, Keywords: *dorsata*/colonies/evolution

Lekprayoon C, Tangkanasing P (1991) *Euvarroa wongsirii* a new species of bee mite from Thailand. Int J Acarol 17:255–258, Keywords: *andreniformis*/classification/disease/*florea*/pests/Thailand

Lekprayoon C, Tangkanasing P (1993) Comparative morphology of *Euvarroa sinhai* and *E. wongsirii*: parasites of *Apis florea* and *A. andreniformis*. In: Connor LJ, Rinderer TE, Sylvester HA, Wongsiri S (eds) Asian apiculture. Connecticut, Wicwas Press, Cheshire, pp 427–433, Keywords: *andreniformis*/disease/*florea*/morphology/pests/Thailand

Lekprayoon C, Wongsiri S, Jayavasti S, Leepitakrat S (1994) Survival of the parasitic honeybee mites *Tropilaelaps clareae* on adult *Apis florea* and *Apis andreniformis* workers. In: Proceedings of the 9th Acarological Conference, Ohio, Keywords: *andreniformis*/disease/*florea*/pests/workers/Thailand

Lelej AS (1995) *Apis* L. In: Lehr PH (ed) A key to insects of the Far East of Russia Vol 4 Megaloptera, Raphidioptera, Neuroptera, Mecoptera, Hymenoptera, Pt 1. Nauka, St Petersburg, p 580, [in Russian]. Keywords: *cerana*/classification/Russia

Lelyakov AD (1929) Mass apiculture in the Far East (of Russia). Khabarovsk, Vladivostok (Knizhnoye Delo), [in Russian]. Keywords: beekeeping/*cerana*/Russia

Lensky Y, Slabezki Y, Gal H, Gerson U, Dechmani U, Jirasavetakul U (eds) (1996) Integrated control of Varroa mite (Varroa jacobsoni Oud) in Apis mellifera colonies in Thailand. Final Report, German–Israel Agricultural Research Agreement for Developing Countries, Rehovot, Israel, p 96, Keywords: *cerana*/drones/*mellifera*/Thailand

Leong CT (ed) (1987) Apiculture. In: Annual Report of the Research Branch, Department of Agriculture for the year 1986: Ministry of Agriculture and Community Development, Sarawak, Malaysia, p 160, Keywords: absconding/beekeeping/*cerana*/natural enemies/Indonesia/Malaysia

Lew YS (1993) A historical review of beekeeping in Korea in the middle of the Chosun Dynasty II. Korean J Apic 8:56–86, [in Korean]. Keywords: beekeeping/*cerana*/history/Korea

Lewis D (1982) Incompatibility, stamen movement and pollen economy in a heterostyled tropical forest tree, *Cratoxylum formosum* (Guttiferae). Proc R Soc Lond B 214:273–283, Keywords: bee botany/*cerana*/*florea*/pollen

Li BY (1983) Research on the Chinese honeybee (*Apis cerana* Fab.) in Yunnan Province. Report, Institute of Honeybee Science Research, Kunming, [in Chinese]. Keywords: *cerana*/China*

Li HL, Wang HY, Gao QK, Cheng JA (2009a) Spatio-temporal expressed analysis of two kinds of chemical communication-related proteins in worker bees of *Apis cerana cerana*. J Agric Biotech 17:1, [in Chinese]. Keywords: *cerana*/protein/communication/China*

Li HZ, Zhang ZL, Wang Z, Tan K (2009b) Morphology of Yunnan *Apis dorsata* and *Apis laboriosa*. Apic China 60:4, [in Chinese]. Keywords: *dorsata*/*laboriosa*/morphology*

Li JH, Li X, Wang A (2004a) Honeybee species in China. Am Bee J 144:49–54, Keywords: *andreniformis*/*cerana*/*dorsata*/*florea*/*mellifera*/*laboriosa*/China

Li JK, Huang SQ (2009) Effective pollinators of Asian sacred lotus (*Nelumbo nucifera*): contemporary pollinators may not reflect the historical pollination syndrome. Ann Bot 104:845–851, Keywords: *cerana*/*mellifera*/pollination

Li JH, Li SW (1989) Differences of isozymes in two species of *Apis*. In: Proceedings of the 32nd international apiculture congress, Apimondia, Rio de Janeiro, pp 65–66, Keywords: biochemistry/*cerana*/*mellifera*/China

Li JX (1992a) How to deal with swarms of *Apis cerana cerana* caught in the wild. J Bee 3:12, [in Chinese]. Keywords: beekeeping/*cerana*/China

Li JZ (1990) Keeping *Apis cerana cerana* scientifically. Agric Sci Tech Inf 8:28, [in Chinese]. Keywords: beekeeping/*cerana*/China

Li JZ (1992b) Dealing with a colony of *Apis cerana cerana* with laying workers. J Bee 1:15, [in Chinese]. Keywords: *cerana*/laying workers/workers/China

Li SF (1993a) Techniques for measuring *Apis cerana cerana* colonies raised in small migratory beekeeping in the southeast of Sichuan Province. J Bee 4:15, [in Chinese]. Keywords: beekeeping/*cerana*/China

Li SJ, Zhang DX, Li L, Chen ZY (2004b) Pollination ecology of *Caesalpinia crista* (Leguminosae: Caesalpinioideae). Acta Bot Sinica 46:271–278, Keywords: *cerana*/ecology/pollination/China

Li SW, Li JH (1986) A comparative study of esterase isozymes in 6 species of *Apis*. J Peking Univ 4:53–57, [in Chinese]. Keywords: biochemistry/*cerana*/*dorsata*/*florea*/*mellifera*/*laboriosa*/*andreniformis*/China

Li SW, Yang QM, Meng YP, Chang JT (1982) Studies on the haemolymph protein pattern of two species of honeybees. Apis mellifera and A. cerana. Acta Entomol Sinica 25:185–190, [in Chinese]. Keywords: *cerana*/*mellifera*/physiology/proteins

Li SW, Meng YP, Chang JT (1985) Studies on isozymes of honeybees. Acta Entomol Sinica 28, Keywords: biochemistry/*cerana*/China*

Li SW, Meng YP, Chang JT, Li JH, He SY, Kuang BY (1986) A comparative study of esterase isozymes in 6 species of *Apis* and 9 genera of Apoidea. J Apic Res 25:129–133, Keywords: *andreniformis*/biochemistry/*cerana*/*dorsata*/*florea*/*laboriosa*/*mellifera*/China

Li SW, Meng YP, Chang JT, Li JH (1987) A comparative study of esterase isozymes in Hymenoptera. Acta Entomol Sinica 30:266–270, [in Chinese]. Keywords: *andreniformis*/biochemistry/*cerana*/*dorsata*/*florea*/*laboriosa*/*mellifera*/China

Li SW, Meng YP, Chang JT, Li JH (1988) Studies on four isozymes of two species of honey bee, *Apis mellifera* and *Apis cerana*. Acta Entomol Sinica 31:15–19, [in Chinese]. Keywords: biochemistry/*cerana*/honey/*mellifera*/China

Li SY (1995) Why *Apis cerana cerana* collect less pollen than *Apis mellifera ligustica*. J Bee 5:21, [in Chinese]. Keywords: *cerana*/foraging/*mellifera*/pollen/China

Li WS (1991a) One important reason why breeding *Apis cerana cerana* is difficult in plain and upland areas. Apic Hubei 3:3–4, [in Chinese]. Keywords: breeding/*cerana*/China

Li WS (1993b) Mating interference and its after effects between *Apis cerana cerana* and *Apis mellifera ligustica*. J Bee 7:3, [in Chinese]. Keywords: beekeeping/*cerana*/interspecific interactions/*mellifera*/reproduction/China

Li WS, Yu LS (1990) Production and biological characteristics of diploid drones of *Apis cerana cerana*. J Bee 12:3–6, [in Chinese]. Keywords: biology/*cerana*/drones/China

Li XY (2002) Arranging the nest of *Apis cerana cerana* according to biological characteristics. J Bee 2:34, [in Chinese]. Keywords: beekeeping/biology/*cerana*/China

Li YQ (1984) Discussions on the taxonomy of six species of *Apis*. J Bee 4:19–21, [in Chinese]. Keywords: *andreniformis*/*cerana*/classification/*dorsata*/*florea*/*laboriosa*/*mellifera*/China

Li ZK (1993c) Methods for improving royal jelly yields in *Apis cerana cerana*. Apic Sci Tech 4:32, [in Chinese]. Keywords: beekeeping/*cerana*/royal jelly/China

Li ZS (1991b) How to rear strong colonies of *Apis cerana cerana*. Bull Biol 12:36, [in Chinese]. Keywords: beekeeping/*cerana*/colony phenology/China

Li ZX, Liu XM (1989) Rearing techniques for twin queens in the same hive without isolation. Agric Sci Tech Gansu 1:30–31, [in Chinese]. Keywords: *cerana*/queens/China

Li ZY, Xi GS, Yan HL (2009c) Microstructure of drone brain of *Apis cerana cerana*. Chinese Bull Entomol 46:121, [in Chinese]. Keywords: *cerana*/morphology/China*

Li ZY, Xue TB Zhao HY (2009) Amplification and sequence analysis of mtDNA fragment from non-coding region to COI gene region of Chinese bee (*Apis cerana cerana*) in Changbai mountain areas. Hubei Agric Sci 48(5), [in Chinese]. Keywords: *cerana*/genetics/mtDNA*

Li ZY, Xue YB, Zhao HY, Chang S (2009) Comparison of *Varroa* mites (Acari: Varroidae) infesting *Apis cerana cerana* and *Apis mellifera* in Changbai mountain. J Bee 5, [in Chinese]. Keywords: *cerana*/*mellifera*/mites*

Liang BJ (1981) Beekeeping in beehouses - a modern approach for *Apis cerana cerana*. J Bee 2:6–7, [in Chinese]. Keywords: beekeeping/*cerana*/China

Liang JY, Lai YS (1980) Description of external morphological characters in *Apis cerana cerana*. Apic China 4:10–13, [in Chinese]. Keywords: *cerana*/morphology/China

Liebig G (2001) The situation with *Varroa* can the problem be solved by breeding *Varroa*-resistant honey bees? Neue Bienenzucht 28:104–106, [in German]. Keywords: breeding/*cerana*/honey/pests

Lien M (2003) Morphology and ultrastructure of the endophallus and musculature in the bees Apis cerana, Apis mellifera and Apis florea. Thesis, Katholieke Universiteit Leuven, Keywords: *cerana*/drones/*florea*/*mellifera*/morphology/ultrastructure

Lim CK (1954) Beekeeping in Singapore. Glean Bee Cult 82(649–650):657, Keywords: beekeeping/*cerana*/Singapore

Limbipichai K (1990) Morphometric studies on the eastern honey bee (Apis cerana Fabricius) in Thailand and the Malaysian peninsula. Thesis, Chulalongkorn University, Keywords: *cerana*/morphometrics/Thailand

Limbipichai K, Tangkanasing P, Mardan MB, Rinderer TE (1990) Evidence of reproductive isolation confirms *Apis andreniformis* (Smith, 1858) is a separate species from sympatric *Apis florea* (Fabricius, 1787). Apidologie 21:47–52, Keywords: *andreniformis*/classification/evolution/*florea*/honey/Malaysia/Thailand

Lin GL (1984) Studies on the esterase isoenzyme IEF of the honeybee, *Apis cerana cerana*. Apic China 2:32–33, [in Chinese]. Keywords: biochemistry/*cerana*/China

Lin HT (1991) Danger of *Galleria mellonella* and *Achroia grisella* to *Apis cerana cerana*. J Bee 6:23–24, [in Chinese]. Keywords: beekeeping/*cerana*/pests/China

Lin JY (1989) Characteristics of and new rearing methods for *Apis cerana cerana*. Today Sci Tech 1:3–4, [in Chinese]. Keywords: beekeeping/breeding/*cerana*/China

Lin ML (1995) The prevention of swarming and migration of *Apis cerana cerana*. Apic China 1:16, [in Chinese]. Keywords: beekeeping/*cerana*/migration/swarming/China

Lin YC, Shew SY, Kong IIII (1977) Physicochemical characteristics of certain Taiwan honeys. Formosan Sci 31:34–39, Keywords: *cerana*/honey/China

Lin ZC (1994) Analysis of the management benefits of rearing *Apis cerana cerana* scientifically. J Bee 7:3, [in Chinese]. Keywords: beekeeping/breeding/*cerana*/China

Lindauer M (1952) A contribution to the question of division of labour in honeybee colonies. Z Vergl Physiol 37:263–324, [in German]. Keywords: behaviour/*cerana*/*dorsata*/*florea*/*mellifera*

Lindauer M (1956a) Communication among the honeybees and stingless bees of India. Bee World 38(3–14):34–39, Keywords: absconding/brood/*cerana*/communication/dance language/*dorsata*/drones/*florea*/migration/swarming/India/Sri Lanka

Lindauer M (1956b) Communication among the honeybees and stingless bees of India. Z Vergl Physiol 38:521–557, [in German]. Keywords: absconding/*cerana*/communication/dance language/*dorsata*/*florea*/India

Lindauer M (1961) Communication among social bees. Harvard University Press, Cambridge, Massachusetts, Keywords: *cerana*/communication/dance language/*dorsata*/*florea*

Lindquist EE (1968) An unusual new species of *Tarsonemus* (Acarina: Tarsonemidae) associated with the Indian honey bee. Can Entomol 100:1002–1006, Keywords: *cerana*/honey/pests/India

Lingappa S, Jagadish A, Shivaramu K, Prabhuswamy HP (1985) Relative intrinsic toxicity of seven insecticides to foragers of the Indian hive bee, *Apis cerana indica* F. Insect Sci Appl 6:567–568, Keywords: *cerana*/foraging/insecticides/nests/India

Lipinski Z (2001) Essence and mechanism of nest abandonment by honeybee swarms. Lipinski, Olsztyn, Poland, Keywords: absconding/biology/*cerana*/*dorsata*/*florea*/hormones/*mellifera*/migration/swarming

Liu AZ, Kress WJ, Wang H, Li DZ (2002) Insect pollination of *Musella* (Musaceae), a monotypic genus endemic to Yunnan, China. Plant Syst Evol 235:135–146, Keywords: *cerana*/*florea*/pollination/China

Liu CM, Shen WY, Chen KY (1998) Comparison of growth and development of the greater wax moth *Galleria mellonella* reared on natural diets and an artificial diet. Wuyi Sci J 14:136–138, Keywords: beeswax/*cerana*/natural enemies/physiology/China

Liu CS (1993a) Observations on the biological characteristics of *Apis cerana cerana* in different areas. J Bee 3:14, [in Chinese]. Keywords: biology/*cerana*/China

Liu CS, Lai YS (1990) Swarming in *Apis cerana cerana*. Apic Sci Tech 3:2–4, [in Chinese]. Keywords: *cerana*/swarming/China

Liu CS, Lai YS (1998) Nest temperature and humidity during development in *Apis cerana cerana*. Apic China 2:8–9, [in Chinese]. Keywords: *cerana*/physical factors/physiology/China

Liu F (1998) Foraging behaviour of *Apis cerana*. J Yunnan Agric Univ 13:407–410, [in Chinese]. Keywords: behaviour/*cerana*/foraging/China

Liu FL, Zhang XW (2000) Spatial distribution of foragers, *Apis cerana cerana*. Zool Res 21:173–175, [in Chinese]. Keywords: biogeography/*cerana*/foraging/China

Liu FL, Kuang B, He D (2000) Relationship between character diversity and environment factors of *Apis cerana* in Yunnan. In: Proceedings of the 4th Asian apiculture association international conference, Kathmandu, p 27, Keywords: *cerana*/classification/genetics/China

Liu FL, Zhang XW, He D, Kuang B (2000) Special distribution future of *Apis cerana* forage in mountain regions. In: Proceedings of the 4th Asian apiculture association international conference, Kathmandu, p 27, Keywords: biogeography/*cerana*/foraging/mountains/China

Liu GX (1996a) Methods for accelerating breeding in *Apis cerana cerana* 1. Apic Sci Tech 1:14, [in Chinese]. Keywords: beekeeping/breeding/*cerana*/China

Liu J (1964a) The determination of the weight and morphology of the Chinese honeybee in Guizhou Province. Apic China 3:80–81, [in Chinese]. Keywords: *cerana*/morphology/China

Liu LH (1993b) Autumn management of *Apis cerana cerana*. Apic Sci Tech 4:31, [in Chinese]. Keywords: beekeeping/*cerana*/China

Liu LH (1993c) Discussion on *Apis cerana cerana*. Apic Sci Tech 3:23, [in Chinese]. Keywords: beekeeping/*cerana*/China

Liu R, Yan Z (2009) The organization and communication in social insects. Bull Biol 44:6, Keywords: *cerana*/China*

Liu X (1985) Advancing Chinese apiculture. In: Proceedings of the 3rd international conference apiculture on tropical climates, Nairobi, pp 93–95, Keywords: beekeeping/*cerana*/China

Liu Y (1997) Breeding Apis cerana cerana. J Bee 3:18, [in Chinese]. Keywords: beekeeping/breeding/*cerana*/China

Liu Y, Liu F (2010) Post-ingestive effect of plant phenolics on the feeding behaviour of the honeybee *Apis cerana*. Physiol Entomol. doi:10.1111/j.1365-3032.2009.00716.x, Keywords: *cerana*/physiology

Liu Y, Liang R, Lai YS (1982) A report on the Chinese honeybee in Guangdong Province. Report, Institute of Honeybee Science Research, Kunming. [in Chinese]. Keywords: *cerana*/classification/China*

Liu YS (1995) Management of *Apis cerana cerana* during the middle and downstream areas of the Changjiang River. J Bee 6:21, [in Chinese]. Keywords: beekeeping/*cerana*/China

Liu YS (1996b) Spring management of *Apis cerana cerana* in the middle and down-stream areas of the Changjiang River. J Bee 3:16, [in Chinese]. Keywords: beekeeping/*cerana*/China

Liu YW (1962) Characteristics of high yields of royal jelly and their correlations in *Apis cerana cerana*. China Agric Sci 2:43–47, [in Chinese]. Keywords: beekeeping/*cerana*/glands/royal jelly/China

Liu YW (1964b) The relationship between morphological characteristics and productivity of *Apis cerana cerana*. Apic China 5:133–136, [in Chinese]. Keywords: *cerana*/honey/morphology/China

Liu YW (1981) Techniques for producing royal jelly from Chinese honeybees (*Apis cerana*). Zhongguo Yangfeng 3:20–21, [in Chinese]. Keywords: beekeeping/*cerana*/royal jelly/China

Liu Z (1986) Beekeeping in the Guangdong Province of China and some observations on the Chinese honey bee *Apis cerana* and the European honey bee *Apis mellifera ligustica*. Am Bee J 126:748–752, Keywords: beekeeping/*cerana*/honey/*mellifera*/China

Liu ZS (1982) Relationships between some biological characteristics and environmental factors for *Apis cerana cerana*. J Bee 3:8–11, [in Chinese]. Keywords: biology/*cerana*/China

Liu ZS, Lai YS (1980) Studies on the development of *Apis cerana cerana* in relation to nest temperatures. Apic China 4:10–13, [in Chinese]. Keywords: *cerana*/nests/physiology

Lo NH, Gloag RS, Snderson DL, Oldroyd BP (2010) A molecular phylogeny of the genus *Apis* suggests that the giant honey bee of the Philippines, *A. breviligula* Maa, and the plains honey bee of southern India, *A. indica* Fabricius, are valid species. Syst Entomol 35:226–233, Keywords: classification/*dorsata*/*cerana*

Lockhart PJ, Cameron SA (2001) Trees for bees. Trends Ecol Evol 16:84–88, Keywords: *dorsata*/evolution/*mellifera*/*nigrocincta*

Lord WG, Nagi SK (1987) *Apis florea* discovered in Africa. Bee World 68:39–40, Keywords: classification/*florea*/Sudan

Louveaux J (1984) The honeybee in its relationships with cultivated plants. In: Pesson P, Louveaux J (eds) Pollination and plant production. Inst Nat Recher Agron, Paris, pp 527–555, [in French]. Keywords: bee botany/beekeeping/*cerana*/*dorsata*/*florea*/*mellifera*

Lu DR (1993) Winter management of *Apis cerana cerana*. J Bee 10:16, [in Chinese]. Keywords: beekeeping/*cerana*/seasonal variation/China

Lu DR (1994) Prevention and cure of sacbrood in *Apis cerana cerana* in the Ninxia area. Apic China 10:25, [in Chinese]. Keywords: *cerana*/disease/China

Ludlow F (1957) The gorges of the Tsangpo 3. Gard Chron 142:309, Keywords: defense/*laboriosa*/China/Nepal

Luo QH, Zhou T, Wang Q, Dai PL, Wu YY (2010) Advances in research on classification and major species of bee mites. Sci Agric Sinica 43:585–593, [in Chinese]. Keywords: *andreniformis*/*cerana*/*dorsata*/*florea*/*koschevnikovi*/*laboriosa*/*mellifera*/mites/*nigrocintca*/*nuluensis*

Luo S, Zhang D, Renner SS (2006) *Oxalis debilis* in China: distribution of flower morphs, sterile pollen and polyploidy. Ann Bot 98:459–464, Keywords: *cerana*/pollination

Luo YX, Chen HS (1990) Studies on wax moths and their danger to *Apis cerana cerana* in Guangdong Province. Apic Hubei 4:11–14, [in Chinese]. Keywords: beekeeping/beeswax/*cerana*/pests/China

Luo YX, Lai YS, Luo MS (1995) Studies on the harmful effects of wax moths in comb honey of *Apis cerana cerana* and its prevention and cure. J Bee 11:3, [in Chinese]. Keywords: beekeeping/beeswax/*cerana*/honey/pests/China

Lusby DA, Lusby EW (1992) Suggested biological manipulative treatment for control of honeybee mites. Apiacta 27:109–117, Keywords: cells/*cerana*/morphology/pests

Lwin NN (1994) The morphology of Apis florea (Linnaeus, 1758) and quality of its honey. Thesis, University of Yangon, Myanmar, Keywords: classification/*florea*/honey/morphology/Myanmar

Ma DF, Huang WC (1981) Apiculture in the new China. Bee World 62:163–166, Keywords: beekeeping/*cerana*/China

Maa TC (1944) On the classification and phylogeny of the Chinese honeybees. Entomol Shaow 1:4–5, [in Chinese]. Keywords: *cerana*/classification/*dorsata*/*florea*/phylogeny/China

Maa TC (1953) An enquiry into the systematics of the tribus Apidini or honey bees (Hymenoptera). Treubia 21:525–640, Keywords: *andreniformis*/*cerana*/classification/*dorsata*/*florea*/honey/*koschevnikovi*/*mellifera*

Maa TC, Shao GC (1947) Variability of structure and color of Apis cerana Fabr., with reference to its related species. Bulletin of Taiwan Agricultural Research Institute, Taipei 6:23–49, Keywords: *cerana*/morphology

Madegowda C (2009) Traditional knowledge and conservation. Econ Polit Wkly 21:65–69, Keywords: *cerana*/*dorsata*/*florea*/honey

Madhusoodnan KJ, Dandin SB (1981) Flower biology of cardamom (*Elettaria cardamomun* (L) Maton) in relation to the foraging behaviour of honey bees (*Apis* spp). Indian Bee J 43:104–108, Keywords: behaviour/biology/*cerana*/foraging/honey/pollination/India

Mah Y (2004) Quality control of honey and its relative products in Korea with reference to honey trade. In: Proceedings of the 1st symposium issues concerning developing countries international trade in honey, Hanoi, pp 60–68, Keywords: *cerana*/honey/*mellifera*/Korea

MAHA M (1936) Bee-keeping in Malaya. Malayan Agric Hort Assoc Mag 6:63–65, Keywords: beekeeping/*cerana*/Malaysia

Mahajan DM, Salunkhe IB, Gunvale VR (2000) Pollen analysis of *Apis florea* Fabr. honeys from the south-western Satpura (Maharashtra). Indian For 126:159–164, Keywords: bee botany/*florea*/honey/India

Mahajan U (1984) Physicochemical analysis of honeys of Shimla Hills (Himachal Pradesh). Thesis, Himachal Pradesh University, Keywords: *cerana*/honey/India

Mahanta M, Rahman A (1997) Studies on pollination of litchi (*Litchi chinensis* Sonn) by honey bees (*Apis cerana* F). Indian Bee J 59:74–75, Keywords: *cerana*/honey/pollination

Mahavir G (1999) Relative abundance and diversity of different pollinating insects on sunflower hybrids (*Helianthus annuus* L.). Ann Biol Ludhiana 15:113–115, Keywords: *dorsata*/*florea*/*mellifera*/pollination

Mahavir S (1982) Olfactory physiology of Apis florea F with reference to repellents. Thesis, Haryana Agricultural University, Hisar, Keywords: biochemistry/*florea*/organic chemistry/physiology/India

Mahindre DB (1968) *Apis dorsata* – the giant bee of India. Glean Bee Cult 96:102–108, Keywords: *dorsata*/India

Mahindre DB (1983a) A suitable hive for *Apis cerana*. Indian Bee J 45:80–81, Keywords: beekeeping/*cerana*/nests/India

Mahindre DB (1983b) Handling rock bee colonies. Indian Bee J 45:72–73, Keywords: beekeeping/*dorsata*/India

Mahindre DB (1997a) Improved methods of honey harvest from rockbee. Indian Bee J 59:95–98, Keywords: beekeeping/*dorsata*/honey/India

Mahindre DB (1997b) *Apis dorsata* – are they migratory? Indian Bee J 59:184–188, Keywords: absconding/combs/disease/*dorsata*/honey/migration/pests/predators/swarming/India

Mahindre DB (1997c) Feeding pollen stores of *Apis dorsata* to other honeybees. Indian Bee J 59:243–245, Keywords: bee botany/*cerana*/*dorsata*/honey

Mahindre DB (1998) Income generated from another form of beekeeping – *Apis dorsata* honey harvesting. Indian Bee J 60:109–110, Keywords: beekeeping/*dorsata*/honey/India

Mahindre DB (2000) Developments in the management of *Apis dorsata* colonies. Bee World 81:155–163, Keywords: beekeeping/*dorsata*/general biology/India

Mahindre DB (2004) *Apis dorsata* F – the manageable bee. Agrotech, Pune, Keywords: beekeeping/*dorsata*/India

Mahindre DB, Muvel KS, Kshirsagar KK, Thakar SS, Phadke RP, Deodikar GB (1977) Nesting behaviour of *Apis dorsata*. In: Proceedings of the 26th international apiculture congress, Adelaide, pp 299–300, Keywords: behaviour/combs/*dorsata*

Mahindru N, Singh G, Grewal GS (1998) Comparative abundance and foraging behaviour of insect pollinators of rape, *Brassica juncea* L and role of *Apis mellifera* L in crop pollination. Indian Bee J 11:34–37

MaiShihah A (1987) Initial research on the behaviour of Apis cerana in selecting pollen and nectar sources. Thesis, Universiti Pertanian, Selangor, [in Malay]. Keywords: behaviour/*cerana*/foraging/nectar/pollen/Malaysia/Nepal

MaiShihah A (1991) Foraging activities of Apis cerana in a coconut plantation. Thesis, Universiti Pertanian, Selangor, Keywords: *cerana*/foraging/Malaysia

Makarenkov V, Legendre P, Desdevises Y (2003) Modelling phylogenetic relationships using reticulated networks. Zool Scr 33:89–96, Keywords: *andreniformis*/*cerana*/*dorsata*/*florea*/*koschevnikovi*/*mellifera*/phylogeny

Makhdzir M, Osman MS (1980) Beekeeping in coconut smallholdings in Pontian, Johore, West Malaysia. In: Proceedings of the 2nd international conference on apiculture in tropical climates, New Delhi, Keywords: beekeeping/*cerana*/physical factors/India/Malaysia*

Makhmoor HD, Ahmad H (2001) Biometric studies on four species of honey bees in Jammu Region, India. Indian Bee J 60:141–142, Keywords: *cerana*/classification/*dorsata*/*florea*/honey/*mellifera*/India

Makhmoor HD, Kotwal DR, Singh R, Rahman S (1994) Problem of robbing and its management in *Apis mellifera* L (Apidae: Hymenoptera). Res Dev Rep 11:48–51, Keywords: beekeeping/behaviour/*dorsata*/*mellifera*/India

Malabanan JM (1994) Mites associated with Apis mellifera (Linnaeus) and Apis cerana (Fabricius) in the southern Tagalog Region. Thesis, University of the Philippines, Keywords: *cerana*/*mellifera*/pests/Philippines

Malaipan S (1972) A revision of the genus of Apis of Thailand (Hymenoptera: Apidae). Thesis, Kasetsart University, Keywords: *cerana*/*dorsata*/*florea*/Thailand

Malaviya DR, Pandey KC, Roy AK, Kaushal P (1999) Role of honey bees in seed setting of Egyptian clover. Crop Improv 26:204–207, Keywords: *florea*/*mellifera*/pollination

Malik B, Raman IS (1990) Does floral heliotropism of *Brassica juncea* influence visitation pattern of *Apis dorsata*. In: Proceedings of the 11th international congress, IUSSI, Bangalore, pp 432–433, Keywords: bee botany/*dorsata*/foraging/India

Mall P, Rathore RRS (1998) Investigations on the foraging pattern of Indian honey bee, *Apis cerana* F. Pest Manag Econ Zool 6:151–154, Keywords: *cerana*/foraging/honey/India

Man EH (1883) The aboriginal inhabitants of the Andamanese Islands. Mittal, New Delhi, Reprinted 1975, Keywords: *dorsata*/India

Manandhar H (2000) Study on incidence of parasites and diseases of *Apis cerana* and *Apis mellifera* in Nepal. In: Proceedings of the 4th Asian apiculture association international conference, Kathmandu, pp 80–81, Keywords: *cerana*/disease/*mellifera*/parasites/pests/Nepal

Manley MJD (1977) Rustaq bee-keeping development centre reports 1–6. Directorate of Agriculture at Durham University, pp 1–6, Keywords: beekeeping/brood/*florea*/Oman

Manley MJD (1978) Beekeeping in northern Oman. Al Mawared al Tabei'eiah 4:11–14, Keywords: beekeeping/*florea*/Oman

Manley MJD (1980) Technical aspects of extension work in northern Oman with reference to importation and distribution of movable frame hives. In: Proceedings of the 2nd international conference on apiculture in tropical climates, New Delhi, pp 172–178, Keywords: beekeeping/classification/environmental effects/*florea*/Oman

Mann GS, Sagar P (1987) Activity and abundance of flower visiting insects of loquat, *Eriobotrya japonica* (Thunb). Indian J Hort 44:123–125, Keywords: bee botany/*cerana*/*dorsata*/foraging/pollination/India

Mann GS, Singh G (1980) Period of activity and comparative abundance of flower visiting insects on pear at Ludhiana (Punjab). Entomon 5:65–66, Keywords: bee botany/*cerana*/foraging/pollination/India

Mann GS, Singh G (1981a) A note on activity and abundance of flower visiting insects of peach (*Prunus persica* L) at Ludhiana (Punjab). Prog Hort 13:25–27, Keywords: bee botany/*cerana*/foraging/pollination/India

Mann GS, Singh G (1981b) Activity and abundance of flower visiting insects of almond (*Prunus amygdalus* Batsch) at Ludhiana (Punjab). J Bombay Nat Hist Soc 78:617–618, Keywords: bee botany/*cerana*/foraging/pollination/India

Mann GS, Singh G (1983) Activity and abundance of pollinators of plum at Ludhiana (Punjab). Am Bee J 123:595, Keywords: bee botany/*cerana*/*dorsata*/*florea*/foraging/pollination/India

Mann GS, Tanda AS (1985) Activity and abundance of insects visiting grape flowers in Punjab, India. In: Proceedings of the 3rd international conference on apiculture in tropical climates,

Nairobi, p 139, Keywords: bee botany/*dorsata*/environmental effects/foraging/pollination/India

Mann SK, Narang AS, Atwal AS (1970) Sample collection of *A. dorsata* honey and its analysis. J Food Sci Tech 7:142–143, Keywords: *cerana*/*dorsata*/honey/India

Manoj K, Ramashrit S (2001a) Effect of honeybee foraging distance on the pollination of sunflower. J Entomol Res 25:37–39, Keywords: *cerana*/foraging/*mellifera*/pollination/India

Manoj K, Ramashrit S (2001b) Initiation-cessation and period of foraging activity of honeybees on sunflower. J Entomol Res 25:141–143, Keywords: *cerana*/foraging/*mellifera*/India

Manoj K, Singh R (2001a) Preference for competing flora with sunflower for foraging of honeybees. J Appl Biol 11:70–71, Keywords: *cerana*/foraging/*mellifera*/India

Manoj K, Singh R (2001b) Effect of honeybee pollination on seed yield and germinability in sunflower. J Appl Biol 11:72–74, Keywords: *cerana*/foraging/*mellifera*/pollination/India

Manoj K, Hari C, Ramashrit S, Ali MS (2002a) Effect of different modes of honeybee pollination on oil content in seeds of sunflower (*Helianthus annuus* L.). J Entomol Res 26:219–221, Keywords: *cerana*/*mellifera*/pollination/India

Manoj K, Singh R, Hari C (2002b) Foraging activity of *Apis cerana indica* and *Apis mellifera* visiting sunflower (*Helianthus annuus* L.). Shashpa 9:31–34, Keywords: *cerana*/foraging/*mellifera*/India

Manzoor ul Haq, Muhammad F (1980) Insect pollination of sunflower (Helianthus annuus) in Pakistan. J Apic Res 19:83–87, Keywords: bee botany/*dorsata*/*florea*/pollination/Pakistan

Manzoor ul Haq, Rafie UD, Ghaffar A (1978) Effect of insect pollination on fruit bearing in Kinnow mandarin (Citrus reticulata), and physical and chemical properties of the fruit. J Apic Res 17:47–49, Keywords: bee botany/*dorsata*/*florea*/pollination/Pakistan

Mao HS (1989) Experiments on the prevention of sacbrood disease using traditional Chinese herbal drugs and Western medicines. J Bee 2:26–28, Keywords: beekeeping/*cerana*/disease/China

Mar TT (1996) Morphological study of Apis dorsata and the analysis of its honey quality. Thesis, University of Yangon, Keywords: classification/*dorsata*/honey/morphology/Myanmar

Mardan MB (1987) Colony size, nesting sites and cavity-size preference of A. cerana feral colonies in coconut growing areas of peninsular Malaysia. In: Proceedings of the 31st International Apiculture Congress, Apimondia, Warsaw, p 185, Keywords: *cerana*/colony phenology/morphology/nests/Malaysia

Mardan MB (1987b) Extruding liquid gobbets by Asian giant honeybees as thermoregulation mechanism. In: Eder J, Rembold H (eds) Chemistry and biology of social insects. J Peperny, Munich, p 243, Keywords: behaviour/*dorsata*/general biology/Malaysia

Mardan MB (1989) Thermoregulation in the Asiatic giant honeybee Apis dorsata (Hymenoptera: Apidae). Thesis, University of Guelph, Keywords: behaviour/*dorsata*/Malaysia

Mardan MB (1993) Rafter beekeeping with Asiatic giant honeybee (*Apis dorsata*) in Vietnam. Beenet Online 1:12–13, Keywords: beekeeping/*dorsata*/Vietnam

Mardan MB (1995) Water requirement of *Apis dorsata* to maintain the nest temperature during summer. Indian Bee J 57:56, Keywords: *dorsata*/environmental effects

Mardan MB, Ashaari AH (1990) Some ecological and physiological factors affecting the thermoregulation of the Asiatic giant honeybee (*A. dorsata* F). In: Proceedings of the 11th international congress, IUSSI, Bangalore, pp 725–726, Keywords: behaviour/*dorsata*/honey/India/Malaysia

Mardan MB, Kevan PG (1989) Mass defecation, yellow rain, and thermoregulation by the giant honeybee *Apis dorsata*. Nature 341:191, Keywords: behaviour/*dorsata*/physiology/Malaysia

Mardan MB, Kevan PG (2002) Critical temperatures for survival of brood and adult workers of the giant honeybee, *Apis dorsata* (Hymenoptera: Apidae). Apidologie 33:295–301, Keywords: behaviour/brood/*dorsata*/environmental effects/nests/workers/Malaysia

Mardan MB, Kiew R (1985) Flowering periods of plants visited by honeybees in two areas of Malaysia. In: Proceedings of the 3rd international conference on apiculture in tropical climates, Nairobi, pp 209–216, Keywords: bee botany/*cerana*/physical factors/Malaysia

Mardan MB, Osman MS (1983) Beekeeping in coconut small holdings in Pontian, Johor, west Malaysia. In: Proceedings of the 2nd international conference on apiculture in tropical climates, New Delhi, pp 179–186, Keywords: beekeeping/*cerana*/physical factors/Malaysia

Mardan MB, Hamid AH, Emby Z, Marasidi AR, Ismail MM (1989) Some aspects of honey collecting from colonies of *Apis dorsata* in peninsular Malaysia. In: Proceedings of the 3rd international conference on apiculture in tropical climates, Cairo, pp 442–448, Keywords: beekeeping/*dorsata*/environmental effects/honey/Malaysia

Martin C, Le Conte Y, Jeanne F (2001) Resistance of the honeybee (*Apis mellifera*) to *Varroa* (*Varroa destructor*). Bull Tech Apic 28:61–72, [in French]. Keywords: *cerana*/disease/*mellifera*/pests

Martin SJ (2001) Biology and life history of *Varroa* mites. In: Webster TC, Delaplane KS (eds) Mites of the honey bee. Dadant, Hamilton, pp 131–148, Keywords: *andreniformis*/biology/*cerana*/*dorsata*/*florea*/history/honey/*koschevnikovi*/*laboriosa*/pests

Martin SJ, Jones GR (2004) Conservation of biosynthetic pheromone pathways in honeybees *Apis*. Naturwissenschaften 91:232–236, Keywords: *andreniformis*/*cerana*/ecology/*florea*/pheromones

Masato O (1996) Prey–predator interaction between Japanese honey bee, *Apis cerana japonica* and giant hornet, *Vespa mandarinia japonica*. Honeybee Sci 17:27–30, [in Japanese]. Keywords: *cerana*/honey/natural enemies/Japan

Maskey M (1989) Inter-relationship between bees and flowers in Kathmandu Valley. In: Proceedings of the 4th international conference on apiculture in tropical climates, Cairo, pp 116–126, Keywords: bee botany/*cerana*/physical factors/Nepal

Maskey M (1992) Mountain women and beekeeping in Nepal. In: Verma LR (ed) Honeybees in mountain agriculture. Oxford & IBH, New Delhi, pp 119–130, Keywords: bee botany/beekeeping/*cerana*/Nepal

Massé DM (1950) Honey bees in French Indo-China. Am Bee J 90:17, Keywords: combs/honey/Vietnam

Mataac R (1984) Queen rearing of the native bee *Apis cerana indica* Fab. during rainy season. Cent Luzon State Univ Tech J, Keywords: breeding/*cerana*/queens/Philippines*

Mathew KP (1990) Tropical beekeeping manual. Department of Agriculture, Sabah, Malaysia, p 52, Keywords: beekeeping/*cerana*/*koschevnikovi*/Malaysia

Mathew S, Mathew K (1988) The 'red' bees of Sabah. Newsl Beekeep Trop Subtrop Countries 12:10, Keywords: *koschevnikovi*

Mathpal Y (1984) Newly discovered rock paintings in central India showing honey collection. Bee World 65:121–126, Keywords: bee botany/beekeeping/*dorsata*/history/honey/India

Mathur S, Kumar S (2001) Reproductive biology of *Bacopa monnieri*. J Genet Breed 55:101–109, Keywords: bee botany/beekeeping/general biology/genetics/India

Matsuka M (1991) Nature and use of pollen loads. Honeybee Sci 12:34–38, Keywords: *cerana*/pollen/Japan

Matsuka M (1996) Country report: Japan. In: Proceedings of the 3rd Asian apiculture association international conference, Hanoi, p 214, Keywords: beekeeping/*cerana*/*mellifera*/Japan

Matsuka M (1999) Bee products in Japan. Farm Japan 33:23–27, Keywords: beeswax/biochemistry/*cerana*/honey/*mellifera*/nests/pheromones/pollen/royal jelly/Japan

Matsuka M, Iwauchi Y (2001) Germination of pollen in honeybee's pollen load. Collect Breed 32:265–268, [in Japanese]. Keywords: *cerana*/pollen/Japan

Matsuka M, Nakamura J (1990) Oxytetracycline residues in honey and royal jelly. J Apic Res 29:112–117, Keywords: *cerana*/honey/royal jelly/Japan

Matsuka M, Sakai T (1989) Bee pollination in Japan with special reference to strawberry production in greenhouses. Bee World 70:55–61, Keywords: *cerana*/pollination/Japan

Matsuka M, Tsurata T, Sasaki M (1987) Temperature as a causative factor of the seasonal colour dimorphism in Apis cerana japonica. In: Proceedings of the 31st International Apiculture Congress, Apimondia, Warsaw, p 186, Keywords: *cerana*/morphology/physiology/Japan

Matsuka M, Tsurata T, Sasaki M (1989) Temperature as a causative factor of the seasonal colour dimorphism of *Apis cerana japonica*. Apidologie 20:149–155, Keywords: *cerana*/physiology/Japan

Matsuka M, Tsurata T, Sasaki M (1995) Temperature as a causative factor of the seasonal colour dimorphism in *Apis cerana japonica*. In: Kevan PG (ed) The Asiatic hive bee: apiculture, biology, and role in sustainable development in tropical and subtropical Asia. Enviroquest Ltd, Cambridge, ON, pp 55–57, Keywords: biology/brood/*cerana*/islands/morphology/nests/physiology/Japan

Matsuka M, Verma LR, Wongsiri S, Shreshta KK, Partap U (eds) (2000) Asian bees and beekeeping: progress of research and development. In: Proceedings of the 4th Asian apiculture association international conference, Kathmandu, pp 1–274, Keywords: beekeeping/*cerana*/*dorsata*/*florea*/physiology

Matsuura M (1983) Notes on the nesting habits of the three honeybee species in Sumatra. Honeybee Sci 4:117–122, [in Japanese]. Keywords: aggressiveness/*cerana*/*dorsata*/*florea*/nests/Sumatra

Matsuura M (1984) Comparative biology of the five Japanese species of the genus *Vespa* (Hymenoptera, Vespidae). Bull Fac Agric Mie Univ 69:1–131, Keywords: biology/*cerana*/natural enemies/Japan

Matsuura M (1988) Ecological study on vespine wasps (Hymenoptera: Vespidae) attacking honeybee colonies. I Seasonal changes in the frequency of visits to apiaries by vespine wasps and damage inflicted, especially in the absence of artificial protection. Appl Entomol Zool 23:428–440, Keywords: *cerana*/natural enemies/Japan

Matsuura M (1990) In: Sakagami SF, Ohgushi RI, Roubik DW (eds) Natural history of social wasps and bees in equatorial Sumatra. Hokkaido University Press, Sapporo, p 312, Keywords: biology/*cerana*/history/natural enemies/Indonesia

Matsuura M (1993) Life history patterns of vespine wasps. Honeybee Sci 14:73–78, [in Japanese]. Keywords: biology/*cerana*/history/pests/Japan

Matsuura M (1998) Cooking of honey bee brood in east Java. Honeybee Sci 19:149–154, [in Japanese]. Keywords: brood/*cerana*/honey/Indonesia

Matsuura M (2003) Biology and control of wasps and bees in urban environments. II Changes in abundance of *Apis mellifera* and *A. cerana* in urban areas, with a note on the temporary occurrence of the giant honeybee, *A. dorsata* in Kawasaki City, Kanagawa Prefecture, Japan. Honeybee Sci 24:97–109, [in Japanese]. Keywords: beekeeping/*cerana*/*dorsata*/general biology/*mellifera*/Japan

Matsuura M, Sakagami SF (1973) A bionomic sketch of the giant hornet, *Vespa mandarinia*, a serious pest for Japanese apiculture. J Fac Sci Hokkaido Univ VI 19:125–162, Keywords: *cerana*/natural enemies/Japan

Matsuyama S, Suzuki T, Sasagawa H (1996) Studies on semiochemicals in the Japanese (*Apis cerana japonica* Rad) and the European honeybees (*Apis mellifera* L). In: Proceedings of the 3rd Asian apiculture association international conference, Hanoi, p 41, Keywords: *cerana*/glands/*mellifera*/pheromones/Japan

Matsuyama S, Suzuki H, Sasagawa H (2000) Chemical analysis of worker pheromone component of Asian honeybees. In: Proceedings of the 4th Asian apiculture association international conference, Kathmandu, p 24, Keywords: *cerana*/*dorsata*/*florea*/*mellifera*/pheromones

Matsuyama S, Suzuki T, Sasagawa H (2001) Semio-chemicals in the Japanese honeybee, Apis cerana japonica Rad. Zool Sci 14:49, Keywords: *cerana*/pheromones/Japan

Mattu VK (1982) Morphometric and behavioural studies on Indian honeybee (Apis cerana indica F). Thesis, Himachal Pradesh University, Keywords: behaviour/*cerana*/classification/morphology/India

Mattu VK (1992) Scope and strategies for apicultural development in Himachal Pradesh, India. In: Verma LR (ed) Honeybees in mountain agriculture. Oxford & IBH, New Delhi, pp 181–192, Keywords: beekeeping/*cerana*/physiology/India

Mattu VK, Chaudhary DK (1993) Relative abundance and foraging activity of insect pollinators on almond flowers. In: Proceedings of the International Symposium Pollination in Tropics, Bangalore, pp 218–221, Keywords: *cerana*/foraging/*mellifera*/pollination/tropics/India

Mattu VK, Sharma N (2000) Predatory behaviour of *Vespa velutina* and *Vespa mandarina* attacking honeybee colonies. In: Proceedings of the 4th Asian apiculture association international conference, Kathmandu, p 71, Keywords: behaviour/*cerana*/*mellifera*/predators/India

Mattu VK, Verma LR (1983) Comparative morphometric studies on introduced European bee *Apis mellifera* L and Indian honeybee *Apis cerana indica* F in Himachal Pradesh. In: Proceedings of the 2nd international conference on apiculture in tropical climates, New Delhi, pp 262–277, Keywords: *cerana*/classification/*mellifera*/physical factors/India

Mattu VK, Verma LR (1983b) Comparative morphometric studies on the Indian honeybee of the northwest Himalayas. 1Tongue and antenna. J Apic Res 22:79–85, Keywords: *cerana*/classification/morphology/mountains/India

Mattu VK, Verma LR (1984a) Comparative morphometric studies on the Indian honeybee of the northwest Himalayas. 2 Wings. J Apic Res 23:3–10, Keywords: *cerana*/classification/morphology/mountains/India

Mattu VK, Verma LR (1984b) Comparative morphometric studies on the Indian honeybee of the northwest Himalayas. 3 Hind leg, tergites and sternites. J Apic Res 23:117–122, Keywords: *cerana*/mountains

Mattu VK, Verma LR (1984c) Morphometric studies on the Indian honeybee *Apis cerana indica* F effect of seasonal variations. Apidologie 15:63–73, Keywords: *cerana*/classification/seasonal variation/India

Mattu VK, Verma LR (1985) Studies on the annual foraging cycle of *Apis cerana indica* F in Simla Hills of northwest Himalayas. Apidologie 16:1–18, Keywords: *cerana*/foraging/mountains/India

Mattu VK, Mattu N, Verma LR, Lakhanpal TN (1989) Pollen spectrum of honeys from *Apis cerana* colonies in Himachal Pradesh, India. In: Proceedings of the 4th international conference on apiculture in tropical climates, Cairo, pp 146–153, Keywords: *cerana*/honey/physical factors/pollen/India

Mattu VK, Mattu N, Verma LR (1990) Foraging activity of Apis cerana F and Apis mellifera L on Plectranthus bloom in relation to the sugar concentration in its nectar. In: Proceedings of the 11th International Congress, IUSSI, Bangalore, p 746, Keywords: *cerana*/foraging/*mellifera*/nectar/India

Mattu VK, Chaudhary DK, Kumar L (1994) Foraging ecology of *Apis cerana* F and *A. mellifera* L in pollinating stone fruit crops. Pest Manage Econ Zool 2:35–39, Keywords: *cerana*/ecology/foraging/*mellifera*/pollination/India

Maul V (1969) The cause of the hybridization barrier between Apis mellifera L and Apis cerana Fabr. (= syn A. indica Fabr.). 2 Egg fertilization and embryonic development. In: Proceedings of the 22nd International Apiculture Congress, Apimondia, Munich, p 151, Keywords: beekeeping/*cerana*/genetics/interspecific interactions/*mellifera*/physiology

Maung MN (1984a) Notes on *Apis dorsata* and *Tropilaelaps clareae* in Burma. Am Bee J 127:504–505, Keywords: *cerana*/*dorsata*/pests/Myanmar (Burma)

Maung MN (1984b) Study on traditional method of keeping Indian honey bees in Burma and keeping with modern method. Pamphlet, Bureau of Life Science. Burma Research Association, Rangoon, Burma, p 10, Keywords: beekeeping/*cerana*/honey/Myanmar (Burma)

Maung MN, Zmarlicki C (1982) Control of mites in European bees in Burma. Am Bee J 122:638–639, Keywords: *cerana*/*dorsata*/pests/Myanmar (Burma)

Maxwell-Lefroy H, Howlett FM (1909) Indian insect life, a manual of the insects of the plains. Thacker, Spink and Company, London, Keywords: beekeeping/*florea*/India

McEvoy MV, Underwood BA (1988) The drone and species status of the Himalayan honeybee, *Apis laboriosa* (Hymenoptera: Apidae). J Kans Entomol Soc 61:246–249, Keywords: classification/*dorsata*/drones/honey/*laboriosa*/Nepal

McNaughton A, Sotha M (2009) Cambodia honey value chain assessment. USAID, Cambodia, Keywords: *dorsata*/honey/Cambodia

Meeprome W (1987) Effect of temperature on toxicity of some insecticides on the European honey bee (Apis mellifera L) and the Asian hive bee (Apis cerana F). Thesis, Chulalongkorn University, [in Thai]. Keywords: *cerana*/honey/insecticides/*mellifera*/nests/physiology/Thailand

Mehrotra KN, Bisht DS (1981) Twenty five years of apicultural research at IARI. I apiculture in relation to agriculture. Indian Bee J 43:118–123, Keywords: beekeeping/*cerana*/*mellifera*/India

Mehrotra KN, Bisht DS (1984) Current status of the biology, physiology and behaviour of the Indian hive bee, *Apis cerana indica* Fab. Indian Bee J 46:36–42, Keywords: biology/*cerana*/physiology/India

Mehrotra KN, Phokela A (1980) Properties of acetylcholinesterase from the Indian honey bee, *Apis cerana indica* Fabricius. J Entomol Res 4:113–118, Keywords: *cerana*/honey/insecticides/India

Mehta VK (1947) An uncommon sight. Indian Bee J 9:60, Keywords: *florea*/foraging/reproduction/India

Mehta VK (1948) *Apis florea*. Indian Bee J 10:79–81, Keywords: biogeography/*cerana*/combs/*dorsata*/*florea*/migration/India

Menzel R, Freudel H, Rühl U (1973) Intraspecific differences in the learning behaviour of the honeybee. Apidologie 4:1–24, [in German]. Keywords: behaviour/*cerana*/interspecific interactions/*mellifera*

Meznar ER, Gadau J, Koeniger N, Rueppell O (2010) Comparative linkage mapping suggests a high recombination rate in all honeybees. J Heredity. doi:10.1093/jhered/esq002, Keywords: *florea*/genetics

Mian HR (1991) Substitute queens can frequently keep colonies of *Apis cerana* strong. Apic Sci Tech 4:11, [in Chinese]. Keywords: beekeeping/*cerana*/queens/China*

Miao XQ (2000) Study of amylase activity in the alimentary canal of *Apis cerana cerana*. Apic China 51:3–5, [in Chinese]. Keywords: biochemistry/*cerana*/physiology/China

Miao XQ, Li JH, Wu ZH, Zhang QK, Sun LX, Chen HS (2000) Band distribution of digestive enzymes in alimentary canal of bee and its alpha-amylase activities. J Fujian Agric Univ 29:95–98, Keywords: biochemistry/biogeography/*cerana*/*mellifera*/physiology

Miao XQ, Zhang QK, Wu ZH, Fu ZM, Wang JD (2001) Comparison of respiratory metabolism between Italian and Chinese worker bees during the pupal stage. J Fujian Agric Univ 30:524–527, [in Chinese]. Keywords: *cerana*/*mellifera*/physiology/China

Michener CD (1974) The social behaviour of the bees: a comparative study. The Belknap Press of Harvard University Press, Cambridge, MA, Keywords: behaviour/biology/*cerana*/*dorsata*/*florea*/*mellifera*

Michener CD (1979) Biogeography of the bees. Ann Mo Bot Gard 66:277–347, Keywords: biogeography/*cerana*/*dorsata*/*florea*/*mellifera*

Michener CD (1990) Classification of the Apidae. Univ Kans Sci Bull 54:75–164, Keywords: *cerana*/classification/*dorsata*/*florea*/*mellifera*

Michener CD (2000) The bees of the world. Johns Hopkins University Press, Baltimore, Keywords: *cerana*/*dorsata*/*florea*/*mellifera*

Michener CD (2008) The bees of the world, 2nd edn. Johns Hopkins University Press, Baltimore, Keywords: *cerana*/*dorsata*/*florea*/*mellifera*

Millen TW (1942a) *Apis dorsata* queens. Indian Bee J 4:6–9, Keywords: *dorsata*/queens

Millen TW (1942b) Bee breeding, laying workers and their progeny. Indian Bee J 4:94–95, Keywords: breeding/*cerana*/*dorsata*/laying workers/workers/India

Millen TW (1943) Bee breeding. What bees can we keep on the plains of India? Indian Bee J 5:36–39, Keywords: breeding/*cerana*/India

Milne PS (1957) Acarine disease in *Apis indica*. Bee World 38:156, Keywords: *cerana*/disease

Minh HV (1969) The chemical composition of honey produced by bees of the species *Apis dorsata*. Thesis, University of the Philippines, Keywords: *dorsata*/honey/Philippines

Minh HV, Mendoza BV, Laigo FM (1971) The chemical composition of honey produced by *Apis dorsata*. J Apic Res 10:91–97, Keywords: *dorsata*/honey/Philippines

Ministry of Agriculture of People's Republic of China (1993) Keeping of the Chinese honeybee, *Apis cerana cerana* in a movable-frame hive. Honeybee Sci 14:123–128, [in Japanese]. Keywords: beekeeping/*cerana*/nests/China

Mishra PK, Yazdani SS (1997) Population dynamics of forager honey bees on litchi, *Litchi chinensis* (Gaertn) Sonn, blooms in north Bihar. Indian Bee J 59:154–157, Keywords: *cerana*/*dorsata*/*florea*/foraging/honey/*mellifera*/population biology/India

Mishra RC (1993) Beekeeping research set-up, achievements and future strategies. In: Proceedings of the 1st national conference on beekeeping, Chandigarh, pp 41–49, Keywords: beekeeping/*cerana*/India

Mishra RC (1993) Handling, processing and storage of honey. In: Proceedings of the 1st national conference beekeeping, Chandigarh, pp 88–97, Keywords: *cerana*/honey/India

Mishra RC (1997) Perspectives in Indian apiculture. Agro Botanica, Bikaner, Keywords: bee botany/beekeeping/*cerana*/disease/*dorsata*/pests/predators/India

Mishra RC (ed) (1998) Perspectives in Indian apiculture. Agro Botanica, New Delhi, Keywords: beekeeping/*cerana*/India

Mishra RC, Dogra GS (1983) Post-embryonic development of *Apis cerana indica* F. worker bee. In: Proceedings of the 2nd international conference on apiculture in tropical climates, New Delhi, pp 278–288, Keywords: biochemistry/brood/*cerana*/physical factors/physiology/India

Mishra RC, Dogra GS (1983) Postembryonic changes in the endocrine and nervous systems of *Apis cerana indica* F. In: Proceedings of the 2nd international conference on apiculture in tropical climates, New Delhi, pp 289–300, Keywords: brood/*cerana*/hormones/physical factors/physiology/India

Mishra RC, Kaushik HD (1992) Bee flora and beekeeping maps of India. Changing Village 11:88–107, Keywords: bee botany/*cerana*/India

Mishra RC, Kaushik HD (1998) Bee flora and beekeeping maps of India. In: Mishra RC (ed) Perspectives in Indian apiculture. Agro Botanica, New Delhi, pp 40–65, Keywords: bee botany/*cerana*/*mellifera*/India

Mishra RC, Kumar J (1987) Importance of beekeeping in social forestry. In: Khosla PK, Kohli RK (eds) Social forestry for rural development. Indian Society of Tree Scientists, pp 180–206, Keywords: beekeeping/*cerana*/India*

Mishra RC, Sihag RC (1987) Apicultural Research in India. Report, all India co-ordinated project on honeybee research and training at Haryana Agricultural University, p 120, Keywords: bibliography/*cerana*/*dorsata*/*florea*/India

Mishra RC, Singh BP (1987) Comparative food imbibing, carrying capacity and consumption by *Apis* spp. In: Proceedings of the 31st international apiculture congress, Apimondia, Warsaw, pp 186–187, Keywords: *cerana*/*dorsata*/*florea*/physiology/India

Mishra RC, Singh BP (1987) Residual repellent effect of *Azidirachata indica* Juss seed oil to *Apis cerana indica* F. In: Proceedings of the 31st international apiculture congress, Apimondia, Warsaw, p 187, Keywords: *cerana*/reproduction/India

Mishra RC, Verma AK (1982) Relative toxicity of some insecticides to *Apis cerana indica* F workers. Indian Bee J 44:69–71, Keywords: *cerana*/insecticides/workers/India

Mishra RC, Dogra GS, Gupta PR (1976) Some observations on insect pollinators of apple. Indian Bee J 38:20–22, Keywords: *cerana*/pollination/India

Mishra RC, Dogra GS, Gupta PR (1977) An attempt to domesticate rock bee. Apis dorsata F. Indian Bee J 39:21, Keywords: beekeeping/*dorsata*/India

Mishra RC, Dogra GS, Gupta PR (1980) Some observations on iridovirus of bees. Indian Bee J 42:9–10, Keywords: *cerana*/disease/India

Mishra RC, Rather AJ, Kumar J (1984) Laboratory experiments on acceptance and sustenance value of sugar, apple juice and brown sugar to *A. c. indica*. Indian Bee J 46:13–14, Keywords: biology/*cerana*/India

Mishra RC, Gupta JK, Kumar J (1987a) Nectar secretion, amount and type of nectar sugars and insect foraging in *Woodfordia floribunda* Salisb. Apidologie 18:173–178, Keywords: bee botany/*cerana*/foraging/nectar/India

Mishra RC, Kumar J, Gupta JK (1987b) Effect of mode of pollination on fruit characteristics of okra, *Abelmoschus esculentus* (L) Moench. Proc Indian Natl Sci Acad B 53:157–160, Keywords: *cerana*/pollination/India

Mishra RC, Kumar J, Gupta JK (1988) The effect of mode of pollination on yield and oil potential of *Brassica campestris* L var *sarson* with observations on insect pollinators. J Apic Res 27:186–189, Keywords: *cerana*/pollination/India

Mishra RK (1983) Digestive physiology of honeybee *Apis cerana indica*. In: Proceedings of the 2nd international conference on apiculture in tropical climates, New Delhi, pp 367–371, Keywords: *cerana*/physical factors/physiology/India

Miyamoto S (1958) Biological studies on Japanese bees. X Differences in flower relationships between Japanese and European honeybees. Science reports of the Hyogo University of Agriculture, Series Agriculture 3:99–107, Keywords: *cerana*/foraging/*mellifera*/Japan

Miyano AT, Abe H (2001) Observation on the nest of *Apis cerana japonica* attacked by the ant, *Tetramorium spitum*. Honeybee Sci 2:17–18, [in Japanese]. Keywords: *cerana*/predators/Japan

Miyoshi N (1993) Pollen morphology of honey plants by scanning electron microscope. 1 Woody plants. Honeybee Sci 14:81–89, [in Japanese]. Keywords: *cerana*/honey/morphology/pollen/Japan

Mo YG (1995) Production of royal jelly by *Apis cerana cerana*. Apic Sci Tech 4:11, [in Chinese]. Keywords: beekeeping/*cerana*/royal jelly/China

Mochi S, Asakawa S (1993) For multiple uses of forests – a case study on apiculture in Mamurogawa Town, Yamagata prefecture. Honeybee Sci 14:115–122, [in Japanese]. Keywords: beekeeping/*cerana*/Japan

Mogga JB (1994) Comparison between the daily flight activity of *Apis mellifera yemenitica* R and *Apis florea* F in Khartoum. Apidologie 25:490–491, Keywords: *florea*/foraging/*mellifera*/Sudan

Mogga JB, Ruttner F (1988) *Apis florea* in Africa: source of the founder population. Bee World 69:100–103, Keywords: classification/*florea*/Sudan

Mohamad BM, Mardan MB (1985) Effect of the presence of *Apis cerana* colonies on cashew fruit set. In: Proceedings of the 3rd international conference on apiculture in tropical climates, Nairobi, pp 140–144. Keywords: *cerana*/physical factors/pollination/Malaysia

Mohana-Rao G, Singh TSMS (1997) Current trends in beekeeping in South India. Indian Bee J 59:33–36, Keywords: beekeeping/*cerana*/disease/India

Mohana-Rao G, Suryanarayana MC (1983) Potentialities for bee pollination of crops in UP. Indian Bee J 45:58–61, Keywords: *cerana*/pollination/India

Mohana-Rao G, Lazar M, Suryanarayana MC (1981) Foraging behaviour of honeybees in sesame (*Sesamum indicum* L). Indian Bee J 43:97–100, Keywords: behaviour/*cerana*/foraging/India

Mohana-Rao G, Subba Rao K, Jasperkinse CY (1996) Studies on *Apis cerana* Fabr. for Thai sacbrood disease tolerance strain. In: Proceedings of international workshop revive *Apis cerana* Kodaikanal, pp 33–37. Keywords: *cerana*/disease/India

Mol GA (1933) Collecting wax and honey in the lake region of western Borneo. Landbw Buitenzorg 2:80–86, [in Dutch]. Keywords: beekeeping/combs/*dorsata*/honey/Borneo

Moltas-Colting L, Cervancia CR (2004) Nesting behaviour of giant honeybee (*Apis dorsata*) in northern Luzon, Philippines. In: Proceedings of the 7th Asian Apiculture Association Internation Conference, Los Banos, pp 155–158. Keywords: behaviour/combs/*dorsata*/Philippines

Momose K, Yumoto T, Nagamitsu T, Kato M, Nagamasu H, Sakai S, Harrison R, Itioka T, Hamid AA (1998) Pollination biology in a lowland dipterocarp forest in Sarawak, Malaysia. I

Characteristics of the plant-pollinator community in a lowland dipterocarp forest. Am J Bot 85:1477–1501, Keywords: *dorsata*/pollination/Borneo

Mondal M, Mitra K (1980) Pollen analysis of honey from Sundarbans (W Bengal). Geophytology 10:137–139, Keywords: bee botany/*dorsata*/honey/India

Montalbo MD, Laude RP (2001) Isozyme and morphometric analysis of natural populations of the Asian honey bee (*Apis cerana* Fabricius) from different geographical areas in the Philippines. In: Proceedings of the 7th international conference on tropical bees, Chiang Mai, pp 333–336. Keywords: biochemistry/biogeography/*cerana*/classification/honey/population biology/populations/Philippines

Morimoto H (1965) Investigations on a method of measuring abdominal size in worker honeybees, *Apis mellifera ligustica* and *Apis cerana cerana*. J Apic Res 4:17–21, Keywords: *cerana*/classification/*mellifera*/morphology

Morimoto H (1967) A method of measuring physical characteristics in the honeybees. In: Proceedings of the 21st International Apiculture Congress, Maryland, pp 86–87, Keywords: beekeeping/*cerana*/classification/morphology/Japan

Morimoto H (1968) The use of the labial palpus as a measure of proboscis length in worker honeybees *Apis mellifera ligustica* and *Apis cerana cerana*. J Apic Res 7:147–150, Keywords: *cerana*/classification/*mellifera*/morphology/Japan

Morin CE, Otis GW (1993) Observations on the morphology and biology of *Euvarroa wongsirii* (Mesostigmata: Varroidae), a parasite of *Apis andreniformis* (Hymenoptera: Apidae). Int J Acarol 19:167–172, Keywords: *andreniformis*/disease/general biology/morphology/pests

Moritz RFA (1986) Genetics of bees other than *Apis mellifera*. In: Rinderer TE (ed) Bee genetics and breeding. Academic Press, London, pp 121–154, Keywords: breeding/*cerana*/*dorsata*/*florea*/genetics/*mellifera*

Moritz RFA, Kryger P, Koeniger G, Koeniger N, Estoup A, Tingek S (1995) High degree of polyandry in *Apis dorsata* queens detected by DNA microsatellite variability. Behav Ecol Sociobiol 37:357–363, Keywords: biochemistry/classification/*dorsata*/drones/organic chemistry/queens/reproduction/Indonesia/Malaysia

Moritz RFA, Haddad N, Bataieneh A, Shalmon B, Hefetz A (2009) Invasion of the dwarf honeybee *Apis florea* into the near East. Biol Invas. doi:101007/s10530-009-9527-z, Keywords: biogeography/florea

Morris AJ (1943) My experience with *dorsata* bees. Indian Bee J 5:46–47, Keywords: beekeeping/*dorsata*

Morse RA (1970a) Annotated bibliography on *Apis florea*. Bibl Bee Res Ass 10:16, Keywords: bibliography/*florea*

Morse RA (1970b) Species of Asian *Apis* and their potential as pollinators. In: Proceedings of the 9th Pollination Conference, Arkansas, pp 1–91, Keywords: *cerana*/*dorsata*/*florea*/pollination

Morse RA (1978) Honey bee pests, predators and diseases. Cornell University Press, Ithaca, Keywords: *cerana*/disease/*dorsata*/*florea*/honey/*mellifera*/natural enemies/pests/predators

Morse RA (1982) Bee disease consultancy in Burma. Report, Food and agriculture organization of the United Nations, Rome, Keywords: *cerana*/disease/Myanmar (Burma)

Morse RA (1982b) The beekeeping program in Burma. Glean Bee Cult 110(454):455, 467, Keywords: beekeeping/*cerana*/Myanmar (Burma)

Morse RA (1982c) Beekeeping in Nepal. Glean Bee Cult 110(507):510, 527, Keywords: beekeeping/*cerana*/*dorsata*/Nepal

Morse RA (1983) Cell orientation and comb strength in honey bee colonies. Glean Bee Cult 111:10–19, Keywords: combs/*dorsata*/foraging

Morse RA, Benton AW (1967) Venom collection from species of honeybees in Southeast Asia. Bee World 48:19–29, Keywords: bee botany/biochemistry/biogeography/*cerana*/*dorsata*/*florea*/*mellifera*/India/Philippines/Thailand

Morse RA, Laigo FM (1968a) Beekeeping in the Philippines. Farm Bull Uttar Pradesh Coll 27:1–56, Keywords: beekeeping/*cerana*/*dorsata*/Philippines
Morse RA, Laigo FM (1968b) Honeybees in the Philippines. Philipp Biota 3:3–4, Keywords: biology/*cerana*/*dorsata*/Philippines
Morse RA, Laigo FM (1968c) The mite *Tropilaelaps clareae* in *Apis dorsata* colonies in the Philippines. Bee World 49:116–118, Keywords: *dorsata*/pests/Philippines
Morse RA, Laigo FM (1969a) The potential and problems of beekeeping in Philippines. Bee World 50:9–14, Keywords: beekeeping/*cerana*/*dorsata*/Philippines
Morse RA, Laigo FM (1969b) The Philippine spine-tailed swift, *Chaetura dubia* McGregor as a honey bee predator. Philipp Entomol 1:138–143, Keywords: *cerana*/honey/natural enemies/Philippines
Morse RA, Laigo FM (1969c) Control of the bee mites, *Varroa jacobsoni* Oudemans and *Tropilaelaps clareae* Delfinado and Baker with chlorobenzylate. Philipp Entomol 1:144–148, Keywords: *cerana*/insecticides/pests/Philippines
Morse RA, Laigo FM (1969d) *Apis dorsata* in the Philippines. Philippine Association of Entomologists Inc, Laguna, Keywords: absconding/behaviour/bibliography/classification/combs/defense/disease/*dorsata*/drones/foraging/general biology/migration/morphology/pests/predators/queens/reproduction/swarming/Philippines
Morse RA, Shearer DA, Boch R, Benton AW (1967) Observations on alarm substances in the genus *Apis*. J Apic Res 6:113–118, Keywords: aggressiveness/biogeography/*cerana*/*dorsata*/*florea*/*mellifera*/pheromones/Philippines/Thailand
Morton J (1987a) Lychee. In: Morton JF (ed) Fruits of warm climates. Julia F Morton, Miami, pp 249–259, Keywords: *cerana*/*dorsata*/*mellifera*/physical factors/pollination
Morton J (1987b) Jambolan. In: Morton JF (ed) Fruits of warm climates. Julia F Morton, Miami, pp 375–378, Keywords: bee botany/*dorsata*/environmental effects/pollination
Mossadegh MS (1990) Nesting behaviour of *Apis florea* F (Hymenoptera: Apidae) in Khuzestan, Iran. In: Proceedings of the 11th International Congress, IUSSI, New Delhi, pp 669–670, Keywords: absconding/beekeeping/behaviour/brood/combs/drones/*florea*/honey/migration/queens/swarming/Iran
Mossadegh MS (1991a) Development of *Euvarroa sinhai* (Acarina: Mesostigmata), a parasitic mite of *Apis florea*, on *A. mellifera* worker brood. Exp Appl Acarol 9:73–78, Keywords: brood/disease/pests/Iran
Mossadegh MS (1991b) Geographical distribution, levels of infestation and population density of the mite *Euvarroa sinhai* Delfinado and Baker (Acarina: Mesostigmata) in *Apis florea* F colonies in Iran. Apidologie 22:127–134, Keywords: classification/disease/ecology/*florea*/general biology/pests/Iran
Mossadegh MS (1993a) Fresh news of *Euvarroa*. Bee World 74:44–45, Keywords: disease/*florea*/pests/Iran
Mossadegh MS (1993b) New geographical distribution line of *Apis florea* in Iran. In: Connor LJ, Rinderer TE, Sylvester HA, Wongsiri S (eds) Asian apiculture. Wicwas Press, Cheshire, CT, pp 64–66, Keywords: classification/environmental effects/*florea*/Iran
Mossadegh MS, Bandpay V (1993) Plants visited by *Apis florea* in Iran. In: Connor LJ, Rinderer TE, Sylvester HA, Wongsiri S (eds) Asian apiculture. Wicwas, Cheshire, CT, pp 369–371, Keywords: bee botany/*florea*/pollination
Mossadegh MS, Birjandi AK (1986) *Euvarroa Sinhai* Delfinado & Baker (Acarina: Mesostigmata): a parasite mite on *Apis florea* F in Iran. Am Bee J 126:684–685, Keywords: absconding/beekeeping/brood/classification/disease/*florea*/migration/pests/swarming/Iran
Mossadegh MS, Alizadeh A, Eslamizadeh R (1997) A pathogenicity study of Aspergillus spp on *Apis florea* F and fungal agents causing brood diseases and some other fungi associated with *Apis mellifera meda* Skorikov in Iran. J Agric Sci Isl Azad Univ 3:6–7, Keywords: brood/disease/*florea*/pests/Iran

Muhammad S, Gondal A, Manzoor ul Haq (1973) Studies on the role of *Apis indica* F in the pollination of cauliflower (*Brassica oleracea* var *botrytis* Linn) and radish (*Raphanus sativus* Linn). Sinicad Univ Res J Sci Ser 7:87–93, Keywords: *cerana*/pollination/India
Mujumdar S, Kshirsagar KK (1986) Morphometric characterization of *Apis dorsata* Fabr. workers. Indian Bee J 48:25–29, Keywords: classification/*dorsata*/morphometrics/workers/India
Mujumdar SM, Mohana Rao G (1996) Spread of Thai sac brood disease in India and revival of *Apis cerana indica*. In: Proceedings of international workshop revive *Apis cerana indica* Kodaikanal, pp 20–26. Keywords: brood/*cerana*/disease/India
Mulder V (1990) Beekeeping in Vietnam. Newsl Beekeep Trop Subtrop Countries, Pamphlet 16, pp 1–16, Keywords: beekeeping/*cerana*/Vietnam
Mulder V (1991) Traditional beekeeping using *Apis cerana* in Vietnam. In: Proceedings of the NECTAR seminar, Amsterdam, pp 61–74, Keywords: beekeeping/*cerana*/nectar/Vietnam
Mulder V (1992a) Honey and wax production from *Apis dorsata* and *Apis florea* in submerged *Melaleuca* forests in Vietnam. In: Beetsma J (ed) Bees and forests in the tropics. NECTAR, Wageningen, pp 27–41, Keywords: beekeeping/combs/ecology/honey/Vietnam
Mulder V (1992b) Honey and wax production in submerged *Melaleuca* forests in Vietnam. BOS Newlett 11:15–29, Keywords: bee botany/combs/honey/Vietnam
Mulder V, Heri V, Wickham T (2000) Traditional honey and wax collection with *Apis dorsata* in the Upper Kapuas Lake Region, West Kalimantan. Borneo Res Bull 31:1–15, Keywords: bee botany/beekeeping/combs/*dorsata*/honey/wax/Borneo
Munaan A (1997) Research station for coconut cropping systems. Indonesian Agric Res Dev J 19:43–49, Keywords: *cerana*/pollination/Indonesia
Murrell DC, Nash WT (1981) Nectar secretion by toria (*Brassica campestris* L v *toria*) and foraging behaviour of three *Apis* species on toria in Bangladesh. J Apic Res 20:34–38, Keywords: behaviour/*cerana*/*dorsata*/*florea*/foraging/nectar/pollination/Bangladesh
Murthy VA, Venkataramanan R (1985) Contribution to the biology of the pseudoscorpion, *Ellingsenius indicus* Chamberlain. Indian Bee J 47:34–35, Keywords: biology/*cerana*/natural enemies/India
Murthy VA, Venkataramanan R (1986) *Ellingsenius indicus* (Arachnida; Chelonethi) as a tool to the assessment of the settling nature of honey bee (*Apis cerana indica*) colony in a new habitat. Indian Bee J 48:54–55, Keywords: *cerana*/honey/natural enemies/India
Murtillo MT (1994) Allozyme variation among geographically isolated populations of *Apis cerana* F from Laguna and Palawan. Thesis, University of the Philippines, Keywords: *cerana*/population biology/Philippines
Muthappa BM (1979) Rock bee culture in forests. Myforest 15:189–199, Keywords: absconding/beekeeping/*dorsata*/general biology/honey/migration/swarming/India
Muttoo RN (1939a) Editorial: [*Apis dorsata*]. Indian Bee J 1:1, Keywords: *dorsata*/India
Muttoo RN (1939b) Tree apiaries. Indian Bee J 1(3):43, Keywords: *dorsata*/India
Muttoo RN (1944) Beekeeping in India, its past, present and future. Indian Bee J 6:54–77, Keywords: beekeeping/*cerana*/*dorsata*/*florea*/history/India
Muttoo RN (1946) Hygienic method of honey collection from wild bee colonies in the forests. Indian Bee J 8:79–80, Keywords: *dorsata*/honey/India
Muttoo RN (1950) Honeybees and fruit crops. Indian J Hort 7:17–20, Keywords: *cerana*/pollination/India
Muttoo RN (1951) The correct scientific nomenclature for our Indian hive bees. Indian Bee J 13:150–153, Keywords: *cerana*/interspecific interactions/*mellifera*/nests/queens
Muttoo RN (1952) UP Government beekeeping station, Jeolikot, District Nainital, UP – A summary of 14 years work (1938–1952). Indian Bee J 14:40–51, Keywords: beekeeping/*cerana*/*dorsata*/*florea*/India
Muttoo RN (1956) Beekeeping for beginners. Farming Bulletin Indian Council of Agricultural Research, New Delhi, pp 1–48, Keywords: beekeeping/cerana/florea/India
Muttoo RN (1956) Facts about beekeeping in India. Bee World 37:125–133,154–157, Keywords: bee botany/beekeeping/behaviour/brood/*cerana*/combs/*dorsata*/*florea*/honey/nests/India

Muttoo RN (1956c) Starting with bees twenty years ago. Indian Bee J 17:50–52, Keywords: beekeeping/*cerana*/India
Muttoo RN (1956d) Problem of rock bee honey. Indian Bee J 18:172, Keywords: *dorsata*/honey/India
Muttoo RN (1957a) Some so called 'peculiarities' of behaviour of Indian honeybees as compared to the European bees. Indian Bee J 19:62–64, Keywords: absconding/behaviour/*cerana*/*dorsata*/environmental effects/general biology/migration/swarming/India
Muttoo RN (1957b) Bee eaters. Indian Bee J 19:7, Keywords: *cerana*/natural enemies/India
Muzaffar N (1982) Honeybee an efficient pollinator. Prog Fmg 2:5–7, Keywords: *cerana*/pollination/Pakistan
Muzaffar N (1988) Studies on the oriental bee (*Apis cerana* F) and its management in Pakistan. Thesis, Quaid-i-Azam University, Keywords: beekeeping/*cerana*/Pakistan
Muzaffar N (1992) Beekeeping an income-generating cottage industry for rural women in Pakistan. In: Verma LR (ed) Honeybees in mountain agriculture. Oxford & IBH, New Delhi, India, pp 113–117, Keywords: beekeeping/*cerana*/Pakistan
Muzaffar N (2000) Research work on Oriental honeybee, *Apis cerana*, in Pakistan. In: Proceedings of the 4th Asian Apiculture Association International Conference, Kathmandu, p 18, Keywords: beeswax/*cerana*/*dorsata*/honey/*mellifera*/Pakistan
Muzaffar N (2000) Chemical control of *Vespa* spp attacking *Apis* spp in Pakistan. In: Proceedings of the 4th Asian Apiculture Association International Conference, Kathmandu, p 73, Keywords: *cerana*/*mellifera*/predators/Pakistan
Muzaffar N (2000) Beekeeping in Pakistan: present status and economics. In: Proceedings of the 4th Asian Apiculture Association International Conference, Kathmandu, p 225, Keywords: beekeeping/*cerana*/*dorsata*/*florea*/*mellifera*/Pakistan
Muzaffar N, Ahmad R (1989) Insect and plant sources of honeydew honey in Pakistan. In: Proceedings of the 4th international conference on apiculture in tropical climates, Cairo, pp 127–133, Keywords: biogeography/*cerana*/*dorsata*/*florea*/honey/interspecific interactions/physical factors/Pakistan
Muzaffar N, Ahmad R (1989) Distribution and competition of *Apis* spp in Pakistan. In: Proceedings of the 4th International Conference on Apiculture in Tropical Climates, Cairo, pp 449–452, Keywords: biogeography/*cerana*/*dorsata*/*florea*/interspecific interactions/*mellifera*/physical factors/population biology/Pakistan
Muzaffar N, Ahmad R (1990) Comparative performance of the Oriental bee *Apis cerana* F and the European bee *A. mellifera* L. Pak J Zool 22:203–206, Keywords: *cerana*/*mellifera*/Pakistan
Muzaffar N, Ahmad R (1991a) Investigations on the seasonal incidence and control of the acarine mite *Acarapis woodi* (Rennie) in *Apis cerana* F colonies in Pakistan. Pak J Zool 23:139–143, Keywords: *cerana*/pests/Pakistan
Muzaffar N, Ahmad R (1991b) Integrated management of the Oriental honeybee *Apis cerana* F in Pakistan. Pak J Zool 23:167–168, Keywords: beekeeping/*cerana*/Pakistan
Muzaffar N, Ahmad R (1991c) Some honeybee flora for *Apis* spp in Pakistan. Pak J Zool 23:201–218, Keywords: bee botany/*cerana*/*dorsata*/Pakistan
Nadre KR, Suryanarayana MC, Rao GM (1996) Effect of honey bee pollination on foundation seed production of sunflower cv LDMRSH3. Indian Bee J 58:17–18, Keywords: *cerana*/honey/pollination/India
Nagarajan B, Krishnamoorthy M, Padmini S, Daniel A (2010) Nectar robbery in bird pollinated mangroves. Curr Sci 98:603, Keywords: *dorsata*/nectar
Nagai T, Sakai M, Inoue R, Inoue H, Suzuki N (2001) Antioxidative activities of some commercial honeys, royal jelly, and propolis. Fd Chem 75:237–240, Keywords: *cerana*/honey/propolis/royal jelly
Nagamitsu T, Inoue T (1999) Differences in pollen sources of *Apis cerana* and *Apis mellifera* at a primary beech forest in central Japan. J Apic Res 38:71–78, Keywords: *cerana*/*mellifera*/pollen/Japan

Nagaraja N, Brockmann A (2009) Drones of the dwarf honey bee *Apis florea* are attracted to (2E)-9-oxodecenoic acid and (2E)-10-hydroxydecenoic acid. J Chem Ecol 35:653–655, Keywords: drones/*florea*/pheromones

Nagaraja N, Rajagopal D (1999) Colony establishment, nesting and foraging activity of little honey bee, *Apis florea* F (Hymenoptera: Apidae). J Entomol Res 23:331–338, Keywords: combs/foraging/honey/India

Nagaraja N, Rajagopal D (2000) Foraging and brood rearing activity of rock bee, *Apis dorsata*. J Entomol Res 24:243–248, Keywords: beekeeping/brood/*dorsata*/foraging/India

Naik DG, Gadre RV, Kapadi AH, Singh MK, Suryanarayana MC, Kshirsagar KK (1988) Nasonov gland pheromone of the Indian honeybee, *Apis cerana indica*. J Apic Res 27:205–206, Keywords: *cerana*/pheromones/India

Naik DG, Kapadi AH, Singh MK, Suryanarayana MC (1989) Lure development for Indian honey bees *Apis cerana indica* Fabr. Indian Bee J 51:47–50, Keywords: *cerana*/honey/pheromones/physiology/India

Naik DG, Bhongle AS, Kapadi AH, Suryanarayana MC, Chawda SS, Chaudhary OP (1995) Antennal sensilla of adult worker *Apis cerana indica*. J Apic Res 34:205–208, Keywords: *cerana*/morphology/physiology/India

Naik DG, Katke S, Chawda SS, Thomas D (1997a) 2-Heptanone as a repellent for *Apis cerana*. J Apic Res 36:151–154, Keywords: *cerana*/chemistry/pheromones/India

Naik DG, Katke S, Chawda SS, Thomas D (1997b) Repellency of 2-heptanone to *Apis cerana*. Indian Bee J 59:80–81, Keywords: *cerana*/chemistry/foraging/pheromones/India

Naik DG, Banhatti P, Chawda SS, Thomas D (1999) Ethylthiocyclopentane as a repellent to *Apis cerana*. Indian Bee J 61:43–48, Keywords: *cerana*/India

Naik DG, Banhatti P, Chawda SS, Thomas D (2003) *Fagara budrunga* fruit extract as an attractant for *Apis cerana*. J Apic Res 42:48–49, Keywords: bee botany/*cerana*/India

Naik DG, Dandge C, Deshpande P (2008) Chemical examination of *Swertia densifolia* leaf essential oil and its repellent activity towards *Apis cerana indica* F. J Apic Res 47:63–67, Keywords: *cerana*/physiology

Naim M (1983) The queen bee of Indian *Apis cerana* and its effect on colony performance. Indian Bee J 45:69–70, Keywords: *cerana*/colony phenology/oviposition/queens/India

Naim M (1984) Field behaviour of honeybee foragers. Indian J Entomol 46:335–339, Keywords: behaviour/*cerana*/flight/foraging/India

Naim M, Bisht DS (1972) A simple method of disinfection of honeybee combs against wax moth. Indian Bee J 34:70–71, Keywords: beeswax/*cerana*/combs/natural enemies/India

Naim M, Bisht DS (1979) Pollen gathering capacity of the Indian honeybee, *Apis cerana indica* Fabr. Indian J Entomol 41:295–296, Keywords: *cerana*/foraging/pollen/India

Naim M, Bisht DS (1984) The life span and the egg-laying potential of a queen bee *Apis cerana indica*. Indian J Entomol 46:380–382, Keywords: *cerana*/oviposition/queens/workers/India

Naim M, Bisht DS (1989) Fidelity and time spent on flowers by *Apis cerana* F foragers. Indian J Entomol 51:222–223, Keywords: *cerana*/foraging/India

Naim M, Phadke KG (1972) Flight range of *Apis cerana indica* when enticed to a feeding dish. J Apic Res 11:167–169, Keywords: *cerana*/flight/foraging/India

Naim M, Phadke KG (1976) Bee flora and seasonal activity of *Apis cerana indica* at Pusa (Bihar). Indian Bee J 38:13–19, Keywords: bee botany/*cerana*/*florea*/foraging/interspecific interactions/India

Naim M, Singh PB (1996) Recent advances in apiculture. In: Lal OP (ed) Recent advances in Indian entomology. APC, New Delhi, pp 346–352, Keywords: beekeeping/*cerana*/India

Naim M, Bisht DS, Sinha RBP (1976) Percentage of honeybee visitation to mustard fields. Entomol Newsl 6:5–6, Keywords: *cerana*/*dorsata*/*florea*/foraging/pollination/India

Naim M, Rao PJ, Phadke KG (1985a) Effect of queenless colony condition on some of the haemolymph constituents of workers and drones of *Apis cerana indica*. Indian J Entomol 47:27–31, Keywords: biochemistry/*cerana*/drones/physiology/queenlessness/workers/India

Naim M, Phadke KG, Rao PJ (1985b) Morphometrics and some behavioural aspects of *Apis cerana indica* workers, under queenless colony condition. Indian J Entomol 47:179–182, Keywords: behaviour/*cerana*/classification/queenlessness/morphometrics/workers/India

Naim M, Bisht DS, Rao PJ (1988) Haemolymph protein pattern of drones and queens of the Indian honeybee. Apis cerana F. J Entomol Res 12:9, Keywords: drones/queens/India

Nair PV, Pillai MS, Nair CSB (1950) Studies in the chemistry and utilisation of Travancore minor forest products. Part II. Clarification, storage, and composition of forest honey. Bulletin of the Central Research Institute, University of Travancore Series A 1:69–78, Keywords: *cerana*/*dorsata*/honey/India

Nakajima J, Matsuka M (1993) Terramycin and honeybee products. Honeybee Sci 14:97–104, [in Japanese]. Keywords: *cerana*/*mellifera*/Japan

Nakajima T (1983) The pattern analysis of biologically active amines in some hymenopteran venoms by high performance liquid chromatography. Jpn J Sanit Zool 34:61–71, Keywords: biochemistry/*cerana*/*mellifera*/Japan

Nakamura J (1987) Beekeeping in Nepal report on the Chepang tribe development programme. Honeybee Sci 8:124–133, [in Japanese]. Keywords: beekeeping/*cerana*/physiology/Nepal

Nakamura J (1992) Role of Himalayan bee concerning modernization of beekeeping in Nepal. Honeybee Sci 13:65–70, [in Japanese]. Keywords: beekeeping/*cerana*/Nepal

Nakamura J (1993) Regulatory system in a honeybee colony toward resource deterioration. Thesis, Tamagawa University, Keywords: absconding/*cerana*/environmental effects/foraging/migration/oviposition/queens/Japan/Nepal/Thailand

Nakamura J (1994) Absconding as an adaptive behaviour of the Asian honey bee, *Apis cerana*. Bulletin of the Faculty of Agriculture, Tamagawa University 34:81–102, [in Japanese]. Keywords: absconding/amalgamation/behaviour/*cerana*/environmental effects/honey/migration/Japan/Thailand

Nakamura J (1995) Brood cannibalism in colonies of Asian honeybee, *Apis cerana*, as response to resource fluctuation. Tamagawa Univ Res Rev 1:33–48, Keywords: absconding/behaviour/biochemistry/brood/*cerana*/environmental effects/larvae/Japan

Nakamura J (1996a) Why do *Apis cerana* abscond? Honeybee Sci 17:71–76, [in Japanese]. Keywords: absconding/*cerana*/migration/Japan

Nakamura J (1996b) Combs of honey bees are vertically long. Honeybee Sci 17:159–163, [in Japanese]. Keywords: *cerana*/combs/honey/Japan

Nakamura J (1999) Problems of beekeeping in Asia. Fmg Jpn 33(Special Issue):15–17, Keywords: beekeeping/*cerana*/*dorsata*/*florea*/tropics/Japan

Nakamura J, Matsuka M (1992) Honeybees trim their brood on the comb. In: Proceedings of the International Conference on Asian Honey Bees and Bee Mites, Chulalongkorn University, Bangkok, p 84, Keywords: brood/*cerana*/combs/honey/pests/Japan

Nakamura J, Ohsawa H (1999) Cumulative titles lists and indexes for "Honeybee Science" Volumes 1 to 20. Honeybee Sci 20:157–188, Keywords: beekeeping/bibliography/*cerana*/history/*mellifera*/Japan

Nakamura J, Yoshida T (1996) Effect of frame shape on the comb utilization by honeybees, *Apis cerana japonica*. In: Proceedings of the 3rd Asian Apiculture Association International Conference, Hanoi, p 62, Keywords: beekeeping/*cerana*/combs/physiology/Japan

Nakamura J, Wongsiri S, Sasaki M (1991) *Apis cerana* in Samui Island and its beekeeping. Honeybee Sci 12:27–30, [in Japanese]. Keywords: absconding/beekeeping/*cerana*/islands/migration/Thailand

Nakamura J, Wongsiri S, Sasaki M (1992) Beekeeping and tourism on Samui. Beekeep Dev No 24:4–5, Keywords: beekeeping/*cerana*/Thailand

Nakamura J, Matsuka M, Wongsiri S (1993) Allured absconding of *Apis cerana* colonies. In: Proceedings of the 33rd international apiculture congress, Apimondia, Beijing, pp 62–63, Keywords: absconding/*cerana*

Nakamura J, Matsuka M, Wongsiri S (1993b) Honeybees trim their brood on the comb. In: Connor LJ, Rinderer TE, Sylvester HA, Wongsiri S (eds) Asian apiculture. Wicwas, Cheshire, CT, pp 226–230, Keywords: absconding/behaviour/brood/*cerana*/larvae/oviposition/Thailand

Nakamura T (1985) Quality standards of royal jelly for medical use. In: Proceedings of the 30th International Apiculture Congress, Apimondia, Nagoya, pp 462–464, Keywords: *cerana*/royal jelly/Japan

Nandi NC, Mahabal A (1974) Nesting behaviour of *Apis dorsata* F (Hymenoptera: Apidae) in relation to local wind directions around Poona, India. Indian Bee J 36:19–20, Keywords: behaviour/combs/*dorsata*/environmental effects/India

Nanork P, Deowanish S, Wongsiri S (2001) Mitochondrial DNA variability of dwarf honey bee (*Apis florea* Fabricius (1787)) in Thailand by using PCR-RFLP technique. In: Proceedings of the 7th International Conference on Tropical Bees, Chiang Mai, pp 341–345, Keywords: classification, genetics, honey, Thailand

NAQS News (1999) The birds and bees. NAQS News 7:3, Keywords: *cerana*/queens/Australia

Narayana N (1970) Studies in Indian honeys and bees waxes. Central Bee Research Institute, Poona, Keywords: bee botany/beeswax/*cerana*/*dorsata*/*florea*/honey/*mellifera*/India

Narayanan ES, Sharma PL, Phadke KG (1960a) Studies on requirements of various crops for insect pollinators I. Insect pollinators of saunf (*Foeniculum vulgare*) with particular reference to the honeybees at Pusa, (Bihar). Indian Bee J 22:23–30, Keywords: *cerana*/*florea*/pollination/India

Narayanan ES, Sharma PL, Phadke KG (1960b) Studies on biometry of the Indian bees I. Tongue length and number of hooks on the hind wings of Apis indica F Indian Bee J 22:58–63, Keywords: *cerana*/classification/morphology/India

Narayanan ES, Sharma PL, Phadke KG (1961a) Studies on biometry of the Indian bees III. Tongue length and number of hooks on the hind wings of Apis indica F collected from Madras State Indian Bee J 23:3–9, Keywords: *cerana*/classification/morphology/India

Narayanan ES, Sharma PL, Phadke KG (1961b) Studies on biometry of the Indian bees IV. Tongue length and number of hooks on the hind wings of Apis indica F collected from Uttar Pradesh Indian Bee J 23:69–74, Keywords: *cerana*/classification/morphology/physical factors/workers/India

Narayanan ES, Sharma PL, Phadke KG (1961c) Studies on requirements of various crops for insect pollinators of berseem Egyptian clover (*Trifolium alexandrinum* L) with particular reference to honeybees and their role in seed setting. Indian Bee J 23:23–30, Keywords: bee botany/*cerana*/*dorsata*/*florea*/pollination/India

Naresh M, Gurdip S, Grewal GS (1998) Comparative abundance and foraging behaviour of insect pollinators of raya. Brassica juncea L and role of Apis mellifera L in crop pollination J Insect Sci 11:34–37, Keywords: *florea*/*mellifera*/pollination

Nargis S, Srimathi P, Krishnasamy V (2000) Influence of supplementary hand pollination on seed yield in *Oldenlandia umbellata*, nectar sources for *Apis florea* F. Indian Bee J 62:29–33, Keywords: *florea*/pollination

Nash WT, Murrell DC (1981) Beekeeping in Bangladesh. Am Bee J 121:352–356, Keywords: beekeeping/*cerana*/*dorsata*/*florea*/Bangladesh

Nath A, Sharma DC (1985) Pharmacodynamics of endosulfan in the Indian honey bee *Apis cerana indica* F. Indian J Exp Biol 23:363–365, Keywords: *cerana*/honey/insecticides/India

Nath A, Sharma ID, Sharma DC, Kumar A (1984) Toxicity of endosulfan to Indian honey bee: *Apis cerana indica* F. In: Proceedings of the 5th International Symposium on Pollination, Versailles, pp 195–199, Keywords: *cerana*/honey/insecticides/pollination/India

Nath A, Sharma DC, Sharma ID (1985) Metabolism of endosulfan in the Indian honey bee. Experientia 41:1411–1412, Keywords: *cerana*/honey/insecticides/physiology/India

Nathan SS, Murugan K, Kumar NS, Jeyabalan D, Muthuraman S (1999) Pollination potential of honeybee *Apis florea* Linn in relation to biochemical profiles of host plants. Zoos' Print J 14:3–4, Keywords: bee botany/India

Natzir R, Teranishi H, Kitagawa M, Kasuya M (1999) A novel venom protein of the Asian bee (*Apis cerana indica*) with an affinity to human 1-microglobulin. Allergol Int 48:121–128, Keywords: biochemistry/*cerana*/*dorsata*/*nigrocincta*

Nayak KG, Davidar P (2009) Pollinator limitation and the effect of breeding systems on plant reproduction in forest fragments. Acta Oecol 36:191–196, Keywords: *cerana*/*dorsata*/*florea*/pollination

Nazir R, Reshi Z, Wafa BA (2008) Reproductive ecology of medicinally important Kashmir Himalayan species of *Digitalis* L. Pl Species Biol 23:59–70, Keywords: *cerana*/ecology/India

Needham GR, Gerson U, Sammataro D (2001) Mite biology. In: Webster TC, Delaplane KS (eds) Mites of the honey bee. Dadant, Hamilton, pp 1–16, Keywords: *andreniformis*/biology/*cerana*/*dorsata*/*florea*/honey/*koschevnikovi*/*laboriosa*/pests/*nigrocincta*/*nuluensis*

Neena SJP (1984) RNA-content during the post-embryonic development of *Apis cerana indica* F (Hymenoptera: Apidae). Zool Orient 1:45–46, Keywords: brood/*cerana*/physiology/RNA/India

Neena SJP (1985a) Changes in the free amino acids content during the postembryonic development of *Apis cerana indica* (Hymenoptera: Apidae). Insect Sci Appl 6:687–689, Keywords: biochemistry/brood/*cerana*/physiology/India

Neena SJP (1985b) Quantitative changes in the total proteins during the postembryonic development of *Apis cerana indica* (Hymenoptera: Apidae). Indian Zool 9:23–26, Keywords: biochemistry/brood/*cerana*/physiology/India

Neena SJP (1986) Cholesterol content in the postembryonic stages of *Apis cerana indica* (Hymenoptera: Apidae). Ann Entomol 4:11–13, Keywords: biochemistry/brood/*cerana*/India

Neena SJP (1987) Free fatty acids during the post-embryonic development of *Apis cerana indica* Fabr. (Hymenoptera: Apidae). Indian Bee J 49:11–12, Keywords: biochemistry/brood/*cerana*/physiology/India

Neena SJP, Sharma R (1997) Changes in the ovaries during postembryonic stages in the development of workers of *Apis cerana indica* F. J Entomol Res 21:119–122, Keywords: brood/*cerana*/physiology/reproduction/workers/India

Neena SJP, Sharma R, Kaur B (1995) Changes in the circulatory system during the postembryonic developmental stages of workers and drones of *Apis cerana indica* F. J Entomol Res 19:191–195, Keywords: brood/*cerana*/drones/physiology/workers/India

Neeraj K, Hari C, Ramashrit S (2001) Effect of method of grafting on queen production in *Apis mellifera* and *Apis cerana indica*. Shashpa 8:47–49, Keywords: *cerana*/*mellifera*/queens/reproduction/India

Nehru CR, Thankamony S, Jayarathnam K, Levi JPM (1984) Studies on off seasonal bee forage in rubber plantations. Indian Bee J 46:23–25, Keywords: *cerana*/foraging/India

Nehru CR, Joseph PML, Jayarathnam K, Sethuraj MR, Panikkar AON, Kothandaraman R (1995) Studies on the management of the Indian honey bee, *Apis cerana indica* F and the European honey bee, *Apis mellifera* L in rubber plantations. Rubb Board Bull 27:5–6, Keywords: beekeeping/*cerana*/disease/foraging/honey/*mellifera*

Nekrutenko YP, Pesenko YA, Tanasiychuk VN (1987) Insects in the Red Data Book of the USSR. Zool Zh 66:198–210, [in Russian]. Keywords: *cerana*/ecology/Russia

Nepal Lumle Agricultural Centre, The (1982) Pollination trial in Chinese cabbage. In: Annual Report 1980–1981: Lumle Agricultural Centre, Nepal, pp 16–18, Keywords: *cerana*/*dorsata*/*florea*/pollination/Nepal

Nessa Z, Muttalib MA, Begum A (1980) Studies on pollen collection pattern of honey bee. Bangladesh J Agric 5:75–82, Keywords: bee botany/*cerana*/honey/pollen/Bangladesh

Nessa Z, Muttalib MA, Begum A (1983) A preliminary study on nosema disease of honeybee and its control. In: Proceedings of the 2nd International Conference on Apiculture in Tropical Climates, New Delhi, pp 457–460, Keywords: *cerana*/disease/physical factors/Bangladesh/India

Nessa Z, Muttalib MA, Begum A (1983) Waxmoth and its control in Bangladesh. In: Proceedings of the 2nd International Conference on Apiculture in Tropical Climates, New Delhi, pp 461–465, Keywords: *cerana*/natural enemies/physical factors/Bangladesh/India

Neumann P, Koeniger N, Koeniger G, Tingek S, Kryger P, Moritz RFA (2000) Home-site fidelity in migratory bees. Nature 406:474–475, Keywords: absconding/biochemistry/combs/*dorsata*/genetics/migration/organic chemistry/swarming/workers/Malaysia

Neupane KR, Dhakal DD, Thapa RB, Gautam DM (2007) Foraging preference of indigenous hive bee. Apis cerana F. to selected horticultural crops. IAAS Res Adv 1:85–88, Keywords: *cerana*/foraging/Nepal

Newton LV (1917) The domestication of the Indian honey bee. Agric J India 12:44–57, Keywords: beekeeping/*cerana*/honey/India

Ng SC, Corlett RT (2000) Comparative reproductive biology of six species of *Rhododendron* (Ericaceae) in Hong Kong, South China. Can J Bot 78:221–229, Keywords: bee botany/biology/*cerana*/China

Nguyen QT, Dang TH (2002) Socio-economic factors in traditional rafter beekeeping with *Apis dorsata* in Vietnam. Bee World 83:165–170, Keywords: beekeeping/*dorsata*/Vietnam

Nguyen QT, Pham TB (1994) Harmony or conflict? *Apis mellifera* and *Apis cerana* in southern Vietnam. Beekeep Dev 32:4–5, Keywords: beekeeping/*cerana*/*mellifera*/Vietnam

Nguyen QT, Phung HC, Thai PH, Mulder V (1900) Rafter beekeeping with *Apis dorsata*: some factors affecting the occupation of rafters by bees. J Apic Res 36:49–54, Keywords: beekeeping/*dorsata*/Vietnam

Nguyen-Van T (1991) Some experiences of control of European foul brood and sac brood diseases in *Apis cerana* in Tien Giang Province. Ngành Ong 4:5–8, [in Vietnamese]. Keywords: brood/*cerana*/disease/Vietnam

Nie WM, Gao QK, Cheng JA (2009) Cloning of cDNA encoding odorant binding protein ASP2 in working bee's antenna of *Apis cerana cerana* and its prokaryotic expression. Sci Agric Sinica 42:891–899, [in Chinese]. Keywords: *cerana*/physiology

Niem NV, Trung LQ (2000) Morphological comparison of three Asian native honey bees (*Apis cerana, A. dorsata, A. florea*). Biotropia 14:10–16, Keywords: *cerana*/*dorsata*/*florea*/honey/morphology/Vietnam

Niem NV, Spooner-Hart R, White B (1996) Relationship between several measured characteristics of drone honey bees (*Apis mellifera*). In: Proceedings of the 3rd Asian apiculture association international conference, Hanoi, pp 50–55, Keywords: beekeeping/biochemistry/*cerana*/drones/honey/*mellifera*/physiology/morphology/Vietnam

Nixon M (1982) Preliminary world maps of honey bee diseases and parasites. Bee World 63:23–42, Keywords: *cerana*/disease/*dorsata*/*florea*/honey/*mellifera*/natural enemies/parasites

Nixon M (1983a) Diseases and parasites of honeybees, their distribution and introduction into new areas. In: Singh KG (ed) Exotic plant quarantine pests and procedures for introduction of plant materials. ASEAN Plant Quarantine Centre and Training Institute, Serdang, pp 207–221, Keywords: biogeography/*cerana*/disease/*dorsata*/*florea*/natural enemies/parasites/pests/Malaysia

Nixon M (1983b) World maps of Varroa jacobsoni and Tropilaelaps clareae with additional records for honeybee diseases and parasites previously mapped. Bee World 64:124–131, Keywords: *cerana*/disease/*dorsata*/*florea*/*mellifera*/parasites/pests

Nogge G (1974) The geographical boundary between western and eastern honeybees. Allg Dtsch Imkerztg 8:165, [in German]. Keywords: biogeography/*cerana*/*mellifera*/Afghanistan/Iran

Noirot C, Howse PE, le Masne G (1975) Pheromones and defensive secretions in social insects. In: Proceedings of a Symposium, IUSSI, Dijon, p 248, Keywords: *cerana*/*dorsata*/*florea*/*mellifera*/pheromones

Noor MJ, Khan M, Camphor ES (2009) Palynological analysis of pollen loads from pollen sources of honeybees in Islamabad, Pakistan. Pak J Bot 41:495–501, Keywords: *cerana*/*dorsata*/*florea*/foraging/Pakistan

Nowicke JW, Meselson M (1984) Yellow rain a palynological analysis. Nature 309:205–206, Keywords: *cerana/dorsata*/physiology/pollen/Thailand
Nunamaker RA, Wilson WT, Ahmad R (1984) Malate dehydrogenase and non-specific esterase isoenzymes of *Apis florea*, *A. dorsata*, and *A. cerana* as detected by isoelectric focusing. J Kans Entomol Soc 7:591–595, Keywords: biochemistry/biogeography/*cerana/dorsata/florea*/Pakistan
Oadri SSH, Sultana H, Anjum F (1982) Selective toxicity of organophosphorus and carbamate pesticides to honey bee and freshwater fish. Int Pest Control 24:124–126, Keywords: *cerana*/honey/insecticides/India
Ochi T (1985) Japanese honeybees in Ehime Prefecture. Honeybee Sci 6:31–38, [in Japanese]. Keywords: beekeeping/*cerana*/Japan
O'Connor R, Peck ML (1978) Venoms of Apidae. In: Bettini S (ed) Arthropod venoms. Springer, Berlin, pp 613–659, Keywords: biochemistry/*cerana/dorsata/florea/mellifera*
Ohri D, Khoshoo TN (1981) Cytogenetics of garden *Gladiolus* pollination mechanism and breeding system. Proc Indian Natl Sci Acad B 47:510–515, Keywords: breeding/*cerana*/pollination/India
Ohtani T (1989) The working style of honeybee workers. Honeybee Sci 10:49–58, [in Japanese]. Keywords: *cerana*/foraging/workers/Japan
Ohtomo H (1998) On the partial amendment of domestic animal infectious diseases control law. Honeybee Sci 19:98, [in Japanese]. Keywords: beekeeping/*cerana*/disease/*mellifera*/Japan
Oian Z (1982) The causes of gnawing combs by Chinese honeybees, and prevention measures. Zhongguo Yangfeng 1:21, [in Chinese]. Keywords: behaviour/*cerana*/combs/China*
Okabe KT, Nakamura T, Gotoh T (2000) Preliminary study of mites associated with the Japanese honey bee, *Apis cerana japonica* Radoszkowski (Hymenoptera: Apidae). J Acarol Soc Jpn 9:23–30, Keywords: *cerana*/honey/pests
Okada I (1958) Introducing modern hives to Japanese beekeeping. Anim Husb 12:1477–1482, [in Japanese]. Keywords: beekeeping/*cerana*/Japan
Okada I (1959) Natural enemies of the Japanese honeybee. Heredity 13:22–26, [in Japanese]. Keywords: *cerana*/natural enemies/Japan
Okada I (1961) Notes on the habits of a giant hornet (*Vespa mandarinia* Smith) as a noxious insect against honey bee. Bull Fac Agric Tamagawa Univ 2:73–89, Keywords: *cerana*/honey/natural enemies/Japan
Okada I (1970a) About the Japanese honeybee: studies on honeybees. Bull Fac Agric Tamagawa Univ 32:259–282, [in Japanese]. Keywords: biology/*cerana*
Okada I (1970b) Studies on the honeybee. Collect Breed 32:259–282, Keywords: breeding/*cerana*/Japan
Okada I (1975) Science of the honeybee. Tamagawa University Press, Tokyo, [in Japanese]. Keywords: biology/*cerana*/Japan
Okada I (1979) Honeybees. Zootech Entomology Tamagawa University Tokyo pp 205–286, Keywords: *cerana*/Japan
Okada I (1983) Observations on the robbing behaviour of honeybees, with special reference to European and Japanese species. Honeybee Sci 4:29–36, Keywords: behaviour/*cerana*/inter-specific interactions/*mellifera*/Japan
Okada I (1984) Aggressive behaviour of honeybees against hornets in Japan. Honeybee Sci 5:105–112, [in Japanese]. Keywords: aggressiveness/behaviour/*cerana*/natural enemies/Japan
Okada I (1985a) Honeybees and their various clusters. Insectarium 22:4–10, [in Japanese]. Keywords: behaviour/*cerana*/Japan
Okada I (1985b) The Japanese honeybee in nature. Heredity 39:58–68, [in Japanese]. Keywords: biology/*cerana*/Japan
Okada I (1986a) Biological characteristics of the Japanese honeybee, *Apis cerana japonica*. In: Proceedings of the 30th international apiculture congress, Apimondia, Nagoya, pp 119–122, Keywords: biology/*cerana*/natural enemies/Japan
Okada I (1986b) Honeybees from my album. I Okada, Tokyo, Keywords: biology/*cerana*/Japan

Okada I (1986c) Japanese and European honeybees in Japan. Honeybee Sci 7:49–52, [in Japanese]. Keywords: *cerana*/*mellifera*/morphology/Japan

Okada I (1988) Three species of wax moths in Japan. Honeybee Sci 9:145–149, [in Japanese]. Keywords: beeswax/*cerana*/natural enemies/Japan

Okada I (1989) Various methods of drone production, and abnormal drones in honeybee colonies. Heredity 43:47–53, [in Japanese]. Keywords: *cerana*/drones/Japan

Okada I (ed) (1990a) A book on the Japanese honeybee (*Apis cerana japonica* Rad). I Okada, Tokyo, [in Japanese]. Keywords: biology/*cerana*/Japan

Okada I (1990b) Notes on various excretions by the Japanese honeybee, *Apis cerana japonica* Rad. Honeybee Sci 11:151–154, [in Japanese]. Keywords: *cerana*/physiology/Japan

Okada I (1991a) Memorandum on Japanese honey bees I. Honeybee Sci 12:13–26, [in Japanese]. Keywords: bee botany/*cerana*/classification/honey/natural enemies/Japan

Okada I (1991b) Memorandum of Japanese honeybees II. Honeybee Sci 12:61–76, [in Japanese]. Keywords: beekeeping/*cerana*/colony phenology/genetics/honey/queens/Japan

Okada I (1993) Swarm of the Japanese honeybee, *Apis cerana japonica* Radoszkowski. Honeybee Sci 14:61–72, [in Japanese]. Keywords: *cerana*/swarming/Japan

Okada I (1995) Recent status of native Japanese and introduced European honey bees in Sagamihara. Honeybee Sci 16:67–68, Keywords: *cerana*/honey/interspecific interactions/*mellifera*/queens/Japan

Okada I (1996) Dialogs with bees and honey plants. Honeybee Sci 17:14–18, [in Japanese]. Keywords: bee botany/*cerana*/honey/Japan

Okada I (1997a) Honeybee Science: Publication list of Professor Dr I Okada. Honeybee Sci 18:117–128, [in Japanese]. Keywords: bibliography/*cerana*/Japan

Okada I (2000) Some observations on Japanese honeybees kept in an apiary. Honeybee Sci 21:27–36, [in Japanese]. Keywords: beekeeping/*cerana*/Japan

Okada I, Obata H (1963) Time course of royal jelly accumulation. Chikusan-no-kenkyu 17:475–478, [in Japanese]. Keywords: *cerana*/royal jelly/Japan

Okada I, Ono M (1994) The Japanese honeybee, *Apis cerana japonica* Rad and spiders. Honeybee Sci 15:69–74, [in Japanese]. Keywords: beekeeping/*cerana*/pests/Japan

Okada I, Sakai T (1960) A comparative study on natural comb of the Japanese and European honeybee, with special reference to cell number. Bull Fac Agric Tamagawa Univ 1:1–11, [in Japanese]. Keywords: *cerana*/combs/*mellifera*/Japan

Okada I, Sakai T, Hasegawa M (1956) Notes on some morphological characters of Japanese honeybee. Kontyû 24:145–154, Keywords: *cerana*/morphometrics/Japan

Okada I, Sakai T, Obata H (1958) On the habits of the Japanese honeybee. In: Proceedings of the 17th international beekeeping congress, Rome, p 22, Keywords: beekeeping/biology/*cerana*/Japan

Okada I, Ono M, Kurihara T, Nakamura C (1984) Observations on winter kill of Japanese and European honeybees. Honeybee Sci 5:159–166, [in Japanese]. Keywords: *cerana*/*mellifera*/seasonal variation/Japan

Okada S (1997b) Effects of pollination by Japanese and European honeybee to peach trees in plastic film greenhouses. Honeybee Sci 18:49–54, Keywords: *cerana*/*mellifera*/pollination/Japan

Okimoto T, Sasaki M (1990) Dance communication and foraging range of the Japanese honeybee, *Apis cerana japonica*. Honeybee Sci 11:117–120, [in Japanese]. Keywords: *cerana*/communication/dance language/foraging/Japan

Oku N, Ono M (1990) Preliminary attempts to rear larvae of the Japanese honeybee, *Apis cerana japonica*, in an *Apis mellifera* colony and in the laboratory using *A. mellifera* royal jelly. Honeybee Sci 11:121–124, [in Japanese]. Keywords: *cerana*/larvae/*mellifera*/royal jelly/Japan

Oldroyd BP, Wongsiri S (2006) Asian honey bees. Harvard University Press, Cambridge, Keywords: *andreniformis*/biology/*cerana*/*dorsata*/*florea*/*koschevnikovi*/*nigrocincta*/*nuluensis*/*mellifera*/*laboriosa*

Oldroyd BP, Rinderer TE, Wongsiri S (1992) Pollen resource partitioning by *Apis dorsata*, *A. cerana*, *A. andreniformis* and *A. florea* in Thailand. J Apic Res 31:3–7, Keywords: *andreniformis*/bee botany/biogeography/*cerana*/*dorsata*/*florea*/foraging/pollen/Thailand
Oldroyd BP, Sylvester HA, Wongsiri S, Rinderer TE (1994) Task specialization in a wild bee, *Apis florea* (Hymenoptera: Apidae) revealed by RFLP banding. Behav Ecol Sociobiol 34:25–30, Keywords: behaviour/*florea*/Thailand
Oldroyd BP, Smolenski AJ, Cornuet JM, Wongsiri S, Estoup A, Rinderer TE, Crozier RH (1995) Levels of polyandry and intracolonial genetic relationships in *Apis florea*. Behav Ecol Sociobiol 37:329–335, Keywords: *florea*/genetics/queens/reproduction/Thailand
Oldroyd BP, Smolenski AJ, Cornuet J-M, Wongsiri S, Estoup A, Rinderer TE, Crozier RH (1996) Levels of polyandry and intracolonial genetic relationships in *Apis dorsata* (Hymenoptera: Apidae). Ann Entomol Soc Am 89:276–283, Keywords: biochemistry/*dorsata*/drones/organic chemistry/queens/reproduction/Thailand
Oldroyd BP, Clifton MJ, Wongsiri S, Rinderer TE, Sylvester HA, Crozier RH (1997) Polyandry in the genus *Apis*, particularly *Apis andreniformis*. Behav Ecol Sociobiol 40:17–26, Keywords: *andreniformis*/behaviour/*cerana*/*dorsata*/*florea*/*mellifera*/reproduction
Oldroyd BP, Clifton MJ, Parker K, Wongsiri S, Rinderer TE, Crozier RH (1998) Evolution of mating behavior in the genus *Apis* and an estimate of mating frequency in *Apis cerana* (Hymenoptera: Apidae). Ann Entomol Soc Am 91:700–709, Keywords: *cerana*/*koschevnikovi*/*nuluensis*/phylogeny/reproduction
Oldroyd BP, Osborne KE, Mardan MB (2000) Colony relatedness in aggregations of *Apis dorsata* Fabricius (Hymenoptera: Apidae). Insectes Soc 47:94–95, Keywords: *dorsata*/general biology/genetics
Oldroyd BP, Halling LA, Good G, Wattanachaiyingcharoen W, Barron AB, Nanork P, Wongsiri S, Ratnieks FL (2002) Worker policing and worker reproduction in *Apis cerana*. Behav Ecol Sociobiol 50:371–377, Keywords: behaviour/*cerana*/laying workers/reproduction
Oldroyd BP, Reddy MS, Chapman NC, Thompson GL, Beekman M (2006) Evidence for reproductive isolation between two colour morphs of cavity-nesting honeybees (*Apis*) in south India. Insectes Soc 53:428–434, Keywords: *cerana*/reproduction
Oldroyd BP, Gloag RS, Even N, Wattanachaiyingcharoen W, Beekman M (2008) Nest-site selection in the open-nesting honey bee *Apis florea*. Behav Ecol Sociobiol 62:1643–1653, Keywords: *florea*/nesting/Thailand
Olek A, Steinkraus KH, Mattick LR, Muego K, Underwood BA, Morse RA (1987) Carbohydrate composition of two Nepalese honeys produced by *Apis laboriosa* and *Apis cerana*. J Apic Res 26:203–204, Keywords: *cerana*/honey/*laboriosa*/Nepal
Olsson J (1989) *Apis cerana*: management methods for prevention of absconding. In: Proceedings of the 4th international conference on apiculture in tropical climates, Cairo, pp 453–457, Keywords: absconding/beekeeping/*cerana*/pests/physical factors/India
Olsson J (1992) Honey bee disease in south India. Beekeep Dev 23:6, Keywords: *cerana*/disease/honey/India
Olsson J (1995) The problem of absconding in *Apis cerana* beekeeping in south India and management for its prevention. Indian Bee J 57:22–30, Keywords: absconding/beekeeping/*cerana*/natural enemies/India
Onlamai C, Thanee N (2004) Some ecological aspects of little honeybee (*Apis florea* F) and type of sugar contents in honey in Northeast Thailand. Pak J Biol Sci 7:658–661, Keywords: *florea*/honey/Thailand
Ono M (1992) The Asian honey bees (*Apis* spp). Honeybee Sci 13:19–22, Keywords: *cerana*/*dorsata*/*florea*/honey/Asia
Ono M (1996) Prey-predator interaction between Japanese honeybee, *Apis cerana japonica* and giant hornet, *Vespa mandarinia japonica*. Honeybee Sci 17:27–30, Keywords: *cerana*/natural enemies/Japan
Ono M, Sasaki M (1993) Defensive strategy of *Apis cerana japonica* against attack by *Vespa* spp. In: Connor LJ, Rinderer TE, Sylvester HA, Wongsiri S (eds) Asian apiculture. Wicwas, Cheshire, CT, pp 231–237, Keywords: *cerana*/natural enemies/Japan

Ono M, Tsurata T (1987) Variation of body colour in the Japanese honeybee *Apis cerana japonica*. Honeybee Sci 8:113–114, Keywords: *cerana*/classification/morphology/Japan
Ono M, Okada I, Sasaki M (1987) Heat production by balling in the Japanese honeybee, *Apis cerana japonica* as a defensive behavior against the hornet, *Vespa simillima xanthoptera* (Hymenoptera:Vespidae). Experientia 43:1031–1032, Keywords: aggressiveness/behaviour/*cerana*/natural enemies/Japan
Ono M, Okada I, Sasaki M (1988) Heat production by balling in the Japanese honeybee, *Apis cerana japonica*, as a defensive behavior against the hornet, *Vespa simillima xanthoptera* (Hymenoptera: Vespidae). Honeybee Sci 9:19–22, [in Japanese]. Keywords: aggressiveness/behaviour/*cerana*/natural enemies/Japan
Ono M, Ohno E, Sasaki M (1991) Foraging and mating strategies of the giant hornet, *Vespa mandarinia japonica* Radoszkowski. In: Proceedings of the 22nd international ethology conference, Kyoto, Keywords: *cerana*/foraging/natural enemies/reproduction/Japan
Ono M, Igarashi T, Ohno E, Sasaki M (1995) Unusual thermal defence by a honeybee against mass attack by hornets. Nature 377:334–336, Keywords: aggressiveness/behaviour/*cerana*/natural enemies/physiology/Japan
Ono Y (1991) Toward preservation of arboreal honey plants. Honeybee Sci 12:170–174, Keywords: bee botany/*cerana*/honey/Japan
Oottamadilok K (1983) Cost of production of honeybee products in the northern region of Thailand. Thesis, University of Chulalongkorn, [in Thai]. Keywords: beekeeping/*cerana*/Thailand
Oppitz M (1991) Beekeeping and the hunt for wild honey among the Magars of Nepal. Bee World 72:165–169, Keywords: beekeeping/honey/*laboriosa*/Nepal
Oschmann H (1961) A beekeeping journey in the People's Republic of China. Arch Geflügelz Kleintierkd 10:235–255, Keywords: beekeeping/*cerana*/China
Oschmann H (1965) Can *Apis mellifica* bees rear brood of *A. cerana*? Biene Giessen 101:150–151, [in German]. Keywords: brood/*cerana*/combs/interspecific interactions/larvae/*mellifera*/oviposition/queens/China
Otanes FQ (1950) Beekeeping in the Philippines. Plant Indus Dig (Manila) 13:2–8, Keywords: beekeeping/*cerana*/Philippines
Otis GW (1989a) Malaysian bees and beekeeping Part 1: The bees of Asia. Can Beekeep 14:204–205, Keywords: beekeeping/*cerana*/*florea*/Malaysia
Otis GW (1989b) Malaysian bees and beekeeping Part 2: Beekeeping with the Eastern hive bee, *Apis cerana*. Can Beekeep 14:224–226, Keywords: aggressiveness/beekeeping/*cerana*/nests/Malaysia
Otis GW (1990) Diversity of *Apis* in Southeast Asia. In: Proceedings of the 11th international congress, IUSSI, Bangalore, pp 104–105, Keywords: *andreniformis*/*cerana*/*dorsata*/*florea*/genetics/*koschevnikovi*/*laboriosa*/*mellifera*/India
Otis GW (1991) A review of the diversity of species within *Apis*. In: Smith DR (ed) Diversity in the genus *Apis*. Westview, Boulder, CO, pp 29–49, Keywords: *andreniformis*/biogeography/*cerana*/*dorsata*/*florea*/genetics/*laboriosa*/*mellifera*
Otis GW (1994) Revised distribution of three recently recognized species of honey bees in Asia. Honeybee Sci 15:167–170, [in Japanese]. Keywordkoschevnikovi/nigrocincta/nuluensis/classification/honey/Bangladesh/Cambodia/China/India/Indonesia/Iraq/Laos/Myanmar/Oman/Pakistan/Thailand/Vietnam
Otis GW (1996) Distributions of recently recognized species of honey bees (Hymenoptera: Apidae; *Apis*) in Asia. J Kans Entomol Soc 69:311–333, Keywords: *andreniformis*/biogeography/*cerana*/*dorsata*/*florea*/honey/*koschevnikovi*/*laboriosa*/*nigrocincta*
Otis GW, Hadisoesilo S (1990) Honey bee survey of south Sulawesi. J Penelitian Kehutanan 4:1–3, Keywords: biogeography/*cerana*/classification/*dorsata*/honey/pests/Indonesia
Otis GW, Hadisoesilo S (1996) *Apis nigrocincta*, a previously unrecognized honey bee species from Indonesia: Lessons in speciation. In: Proceedings of the 3rd Asian apiculture association

international conference, Hanoi, pp 2–6, Keywords: *cerana*/classification/drones/honey/*nigrocincta*/Indonesia/Japan/Philippines/Sri Lanka

Otis GS, Hadisoesilo S (1999) Insights into honeybee biology from *Apis nigrocincta* of Indonesia. In: Hoopingarner R, Connor LJ (eds) Apiculture for the 21st century. Wicwas, Cheshire, Keywords: biology/*nigrocincta*/Indonesia

Otis GW, Kralj J (2001) Parasitic brood mites not present in North America. In: Webster TC, Delaplane KS (eds) Mites of the honey bee. Dadant, Hamilton, pp 251–272, Keywords: *andreniformis*/brood/*cerana*/*dorsata*/*florea*/honey/*koschevnikovi*/*laboriosa*/*nigrocincta*/*nuluensis*/pests

Otis GW, Mardan MB, McGee K (1990) Age polyethism in *Apis dorsata*. In: Proceedings of the 11th international congress, IUSSI, Bangalore, p 378, Keywords: behaviour/*dorsata*/India

Otis GW, Patton K, Tingek S (1995) Piping by queens of *Apis cerana* Fabricius 1793 and *Apis koschevnikovi* von Buttel-Reepen 1906. Apidologie 26:61–65, Keywords: *cerana*/*koschevnikovi*/*mellifera*/queens

Otis GW, Hadisoesilo S, Sugiyani, Cooper AM (2001) Using social bees as indicators of habitat quality in Sulawesi, Indonesia. In: Proceedings of the 7th international conference on tropical bees, Chiang Mai, pp 313–317, Keywords: *cerana*/*dorsata*/*nigrocincta*/Indonesia

Otis GW, Koeniger N, Rinderer TE, Hadisoesilo S, Yoshida T, Tingek S, Wongsiri S, Mardan M (2001) Comparative mating flight times of Asian honey bees. In: Proceedings of the 7th international conference on tropical bees, Chiang Mai, pp 137–141, Keywords: *andreniformis*/*cerana*/drones/flight/*florea*/honey/*koschevnikovi*/*nigrocincta*/*nuluensis*/reproduction/Indonesia/Japan/Malaysia/Sri Lanka/Thailand

Otsubo F (1990) Apiculture in Tsushima Islands. Honeybee Sci 11:59–62, [in Japanese]. Keywords: beekeeping/*cerana*/islands/Japan

Ovsenko G (1945) Bees of the Ussuri Taiga. Pchelovodstvo 2:40–43, [in Russian]. Keywords: *cerana*/Russia

Özkan A, Gharleko MM, Özden B, Kandemir I (2009) Multivariate morphometric study on Apis florea distributed in Iran. Turk J Zool 33:93–102, Keywords: florea/morphology/Iran

Paar J (2001) Migratory behaviour and colony aggregation in giant honeybees. Thesis, Universität Graz, Keywords: absconding/*dorsata*/migration/swarming

Paar J, Oldroyd BP, Kastberger G (2000) Giant honeybees return to their nest sites. Nature 406: 475, Keywords: absconding/biochemistry/combs/*dorsata*/genetics/honey/migration/organic chemistry/swarming/India

Paar J, Oldroyd BP, Hüttinger E, Kastberger G (2002) Drifting of workers in nest aggregations of the giant honeybee *Apis dorsata*. Apidologie 33:553–561, Keywords: *dorsata*/general biology/genetics/workers

Paar J, Oldroyd BP, Hüttinger E, Kastberger G (2004a) Genetic structure of an Apis dorsata population: the significance of migration and colony aggregation. J Hered 95:119–126, Keywords: absconding/dorsata/general biology/genetics/migration/swarming

Paar J, Oldroyd BP, Hüttinger E, Kastberger G (2004b) Levels of polyandry in *Apis laboriosa* from Nepal. Insectes Soc 51:212–214, Keywords: genetics/*laboriosa*/reproduction/Nepal

Pal N, Thomas D (2001) Role of KVIC in beekeeping and rural development in India. In: Proceedings of the 37th international apiculture congress, Apimondia, Durban (CD Rom), Keywords: beekeeping/*cerana*/physiology/India

Palanichamy PS, Baskaran S, Mohandoss A (1995) Insect pollination of the moringa plant *Moringa concanensis* Nimmo Linn. Environ Ecol 13:47–51, Keywords: bee botany/pollination/India

Paliwal GN (1993) Studies on the neuroendrocrine and reproductive systems in the rock honeybee *Apis dorsata* (F) (Hymenoptera: Apidae).Thesis, Nagpur University, Keywords: *dorsata*/physiology/reproduction

Paliwal GN (1996) Eco-friendly management of giant honeybee *Apis dorsata* in India. In: Proceedings of the 3rd Asian apiculture association conference, Hanoi, p148, Keywords: beekeeping/*dorsata*/India

Paliwal GN, Tembhare DB (1998) Surface ultrastructure of the sting in the rock honeybee *Apis dorsata* F (Hymenoptera: Apidae). Entomon 23:203–209, Keywords: defense/*dorsata*/honey/morphology

Palmer K, Oldroyd BP (2001) Mating frequency in *Apis florea* revisited (Hymenoptera, Apidae). Insectes Soc 48:40–43, Keywords: *florea*/reproduction

Panchabhavi KS, Devaiah MA (1977) A note on the activities of insect pollinators on sunflower during winter and summer seasons at Bangalore. Curr Res 6:88–89, Keywords: *cerana*/*dorsata*/*florea*/foraging/pollination/India

Panchabhavi KS, Rao KJ (1978) Note on the effect of mixed cropping of niger of the activities of insect pollinators and seedfilling of sunflower in Karnataka. Indian J Agric Sci 48:254–255, Keywords: bee botany/pollination/India

Panchabhavi KS, Devaiah MA, Basavanna GPC (1976) The effect of keeping *Apis cerana indica* F colonies on the seed set of sunflower, *Helianthus annuus* Linn. Mysore J Agric Sci 10:631–636, Keywords: *cerana*/pollination/India

Panda P, Sontakke BK, Sarangi PK (1988) Preliminary studies on the effect of bee (*Apis cerana indica* Fabr.) pollination on yield of sesamum and niger. Indian Bee J 50:63–64, Keywords: *cerana*/pollination/India

Panda P, Sontakke BK, Swain LK, Mohapatra RN (1989a) Laboratory studies on the contact toxicity of some insecticides against *Apis cerana indica* Fabr. Indian Bee J 51:50–52, Keywords: *cerana*/insecticides/India

Panda P, Padhi J, Nanda UK, Mohanty NC (1989b) Record of *Varroa jacobsoni* and *Euvarroa Sinhai* on *Apis cerana indica* Fabr. at Bhubaneshwar, Orissa. India Indian Bee J 51:59, Keywords: *cerana*/natural enemies/pests/India

Panda P, Nanda UK, Mohapatra H, Padhi J (1989c) Insect pollination in some oilseed crops in Orissa, India. Indian Bee J 51:97–98, Keywords: *cerana*/pollination/India

Panda P, Padhi J, Nanda UK, Senapati B (1990) Pesticides and honeybees. Orissa J Agric Res 3:291–294, Keywords: *cerana*/insecticides/India

Panda P, Panda B, Sontakke BK (1991) Foraging behaviour of honey bees on sunflower (*Helianthus annuus* L) sown on different dates. Indian Bee J 53:44–49, Keywords: behaviour/*cerana*/foraging/honey/India

Panda P, Sontakke BK, Panda B (1993a) Effect of different modes of pollination on yield of sunflower and niger. J Insect Sci 6:75–77, Keywords: *cerana*/pollination/India

Panda P, Sontakke BK, Panda B (1993b) Foraging behaviour of honey bee species on different varieties of niger, *Guizotia abyssinica* Cass in Orissa. J Insect Sci 6:104–106, Keywords: behaviour/*cerana*/*dorsata*/*florea*/foraging/honey/interspecific interactions/India

Panda P, Rath LK, Padhi J, Panigrahi D (1995) Relative abundance and foraging behaviour of common bee species on niger in Phulbani District, Orissa, India. Indian Bee J 57:10–14, Keywords: behaviour/*cerana*/*dorsata*/*florea*/foraging/India

Panda P, Sarangi PK, Padhi J, Panda SK (1997) Thai sacbrood prevention is the only cure. Indian Bee J 59:32, Keywords: *cerana*/disease/India

Pande YD, Bandopadhyay S (1985a) The foraging behaviour of honey bees on flowers of pigeon pea (*Cajanas cajan*) in Agartala, Tripura. Indian Bee J 47:13–15, Keywords: behaviour/*cerana*/foraging/honey/India

Pande YD, Bandopadhyay S (1985b) Effect of fenitrothion on the foraging activity of honey bees on *Cajanus cajan* in Tripura. Indian Bee J 47:42–43, Keywords: *cerana*/foraging/honey/insecticides/India

Pandey RS (1967) *Varroa jacobsoni*: a new mite infesting honeybee (*Apis indica*) colonies in India. Bee World 48:16–17, Keywords: *cerana*/natural enemies/pests/India

Pandey RS (1974a) Honey collection in relation to migration schedules of wild honey bees in Uttar Pradesh. India Am Bee J 114:379, Keywords: beekeeping/biogeography/*cerana*/*dorsata*/*florea*/honey/migration/India

Pandey RS (1974b) *Varroa jacobsoni*: a new mite infesting honeybee (*Apis indica*) colonies in India. Am Bee J 114:379, Keywords: *cerana*/pests

Pandey RS (1977) Behavior of the Indian honey bee in double brood chamber hives. Am Bee J 117:627, Keywords: behaviour/brood/*cerana*/honey/queens/India

Pandey RS, Yadav RPS (1970) Pollination of litchi (*Litchi chinensis*) by insects with special reference to honeybees. J Apic Res 9:103–105, Keywords: *cerana*/pollination/India

Pane R, Syamsi E, Sofyan S (1996) Beekeeping development progams. In: Proceedings of the 3rd Asian apiculture association international conference, Hanoi, pp 122–127, Keywords: beekeeping/*cerana*/*mellifera*/physiology/Indonesia

Pant DD, Chaturvedi SK (1982) Anthecological study of some insect pollinators under SEM. Beitr Biol Pflanz 60:89–92, Keywords: bee botany/morphology/India

Pant DD, Nautiyal DD, Chaturvedi SK (1982a) Insect pollination in some Indian Glumiflorea. Beitr Biol Pflanz 57:229–235, Keywords: bee botany/*cerana*/*dorsata*/*florea*/pollen/pollination/India

Pant DD, Nautiyal DD, Chaturvedi SK (1982b) Pollination ecology of some Indian asclepiads. Phytomorphology 32:302–313, Keywords: bee botany/*cerana*/*dorsata*/ecology/*florea*/pollination/India

Panzer MF (1915) On the beekeeping and the honey bees in the Ussuri Territory. Russ Pchelovod Listok (Moscow) 8:35–40, [in Russian]. Keywords: beekeeping/*cerana*/honey/Russia

Pardede D, Randriani E, Marfuati A (1986) The pollination of coconut at the Pakuwon sub-research Institute for coconut. Pemberitaan, Penelitian Tanaman Industri, Indonesia 12:26–35 [in Indonesian]. Keywords: cerana/pollination/Indonesia

Parmar C (1976) Pollination and fruit set in phalsa (*Grewia asiatica* L). Agr Agroind J 9:12–14, Keywords: bee botany/*cerana*/*dorsata*/*florea*/foraging/pollination/India

Partap T, Verma LR (2000) Asian bees and beekeeping: issues and initiatives. In: Proceedings of the 4th Asian apiculture association international conference, Kathmandu, pp 3–14, Keywords: bee botany/beekeeping/biogeography/*cerana*/*dorsata*/ecology/*florea*/genetics/interspecific interactions/*laboriosa*/*mellifera*/mountains/Bangladesh/China/India/Japan/Korea/Thailand

Partap U (1999a) Conservation of endangered Himalayan honeybee, *Apis cerana* for crop pollination. Asian Bee J 1:44–49, Keywords: *cerana*/ecology/pollination

Partap U (1999b) Pollination management of mountain crops through beekeeping: trainers' resource book. ICIMOD, Kathmandu, Keywords: beekeeping/*cerana*/*mellifera*/pollination/China/India/Nepal/Pakistan

Partap U (2000) Foraging behaviour of *Apis cerana* on sweet orange (*Citrus sinensis* var Red Junar) and its impact on fruit production. In: Proceedings of the 4th Asian apiculture association international conference, Kathmandu, pp 174–177, Keywords: beekeeping/behaviour/*cerana*/foraging/pollination/Nepal

Partap U (2000) Pollination of strawberry by the Asian hive bee, *Apis cerana*. In: Proceedings of the 4th Asian apiculture association international conference, Kathmandu, pp 178–182, Keywords: *cerana*/nests/pollination/Nepal

Partap U (2000) Foraging competition between *Apis cerana* and *Apis mellifera* and its impact on crop pollination. In: Proceedings of the 4th Asian apiculture association international conference, Kathmandu, pp 206–208, Keywords: *cerana*/foraging/interspecific interactions/*mellifera*/pollination/Nepal

Partap U, Partap T (1997) Managed crop pollination – the missing dimension of mountain agricultural productivity. Discussion Paper Series – mountain farming systems, ICIMOD, No 97/1, Keywords: *cerana*/*mellifera*/pollination/Nepal

Partap U, Partap T (2002) Warning signals from the apple valleys of the Hindhu Kush-Himalayas productivity concerns and pollination problems. ICIMOD, Kathmandu, Keywords: beekeeping/*cerana*/pollination/Bhutan/China/India/Nepal/Pakistan

Partap U, Verma LR (1993) Pollination of vegetable crops using the Asian hive bee, *Apis cerana*, in the Kathmandu Valley, Nepal. In: Proceedings of the international symposium on pollination tropics, Bangalore, pp 184–186, Keywords: *cerana*/nests/pollination/India/Nepal

Partap U, Verma LR (1994) Pollination of radish by *Apis cerana*. J Apic Res 33:237–241, Keywords: *cerana*/foraging/pollination/Nepal

Partap U, Verma LR (1996) Impact of food resource availability on growth and development of *Apis cerana* colonies in the Kathmandu Valley of Nepal. Ecoprint 3:21–29, Keywords: bee botany/*cerana*/physiology/Nepal

Partap U, Verma LR (1997) Comparative performance, behaviour, and productivity of *Apis cerana* ecotypes from Kathmandu and Jumla Regions. Ecoprint 4:43–48, Keywords: behaviour/*cerana*/geographical variation/Nepal

Partap U, Shukla AN, Verma LR (2000) Pollination of peach and plum by *Apis cerana*. In: Proceedings of the 4th Asian apiculture association international conference, Kathmandu, pp 171–173, Keywords: beekeeping/*cerana*/pollination/India/Nepal

Partap U, Shukla AN, Verma LR (2000) Comparative foraging behaviour of *Apis cerana* and *Apis mellifera* in pollinating peach and plum flowers in Kathmandu Valley, Nepal. In: Proceedings of the 4th Asian apiculture association international conference, Kathmandu, pp 193–198, Keywords: behaviour/*cerana*/foraging/*mellifera*/Nepal

Partap U, Shukla AN, Verma LR (2000) Comparative attractiveness of broad-leaf mustard, cauliflower and radish to *Apis cerana* in Kathmandu Valley of Nepal. In: Proceedings of the 4th Asian apiculture association international conference, Kathmandu, pp 203–205, Keywords: *cerana*/pollination/Nepal

Parthiban M, Baskaran S, Mohandoss A (1994) Insect pollination of the chrysanthemum, *Chrysanthemum indicum* Linn. Environ Ecol 12:9–13, Keywords: biogeography/*cerana*/*florea*/pollination/India

Parvathi C, Shadakshari YG, Belavadi VV, Chandrappa HM (1993) Foraging behaviour of honeybees on cardamom (*Elettaria cardamomum* Maton). In: Proceedings of the international symposium on pollination in tropics, Bangalore, pp 99–103, Keywords: behaviour/*cerana*/*dorsata*/foraging/pollination/India

Patel BK, Upadhyaya VR, Muralidharan CM, Judal GS (1988) Effects of various insecticides on honey bee, *Apis florea* Fabricius in ber (*Zizyphus mauritiana* Lamk). Curr Sci 57: 1199–1200, Keywords: biochemistry/classification/honey/insecticides/organic chemistry/India

Patel BK, Marvania TG, David BV (2000) A method to evaluate the toxicity (LC) of phosphine gas from magnesium phosphide to beneficial insects. J Appl Zool Res 11:63–65, Keywords: *cerana*/insecticides/India

Patil BS, Shashidhar V (2001) Foraging behaviour of two species of honey bees on sesamum. Karnataka J Agric Sci 14:796–798, Keywords: behaviour/*cerana*/foraging/honey/*mellifera*/pollination/India

Patil BS, Viraktamath S (2000) Pollinator fauna of sesamum in Dharwad. Insect Environ 5:171–172, Keywords: bee botany/*dorsata*/India

Patil BS, Viraktamath S, Hat ARS (2000) Effect of Bee-Q and Bee-here on pollinators and yield of sesamum. Insect Environ 5:151–152, Keywords: *cerana*/chemistry/*dorsata*/*florea*/foraging/*mellifera*/pollination/India

Patinawin S (1989) Ultrastructure of the honey bee spermatozoa [*A. cerana*]. In: Proceedings of the 1st Asia-Pacific conference of entomology, Chiang Mai, pp 83–88, Keywords: *cerana*/drones/honey/ultrastructure

Patinawin S (1993) High resolution electron microscopy of *Apis cerana indica* metaphase chromosomes. In: Connor LJ, Rinderer TE, Sylvester HA, Wongsiri S (eds) Asian apiculture. Wicwas, Cheshire, CT, pp 124–127, Keywords: *cerana*/drones/genetics/Thailand

Patinawin S, Wongsiri S (1993) Male genitalia of honey bees. In: Connor LJ, Rinderer TE, Sylvester HA, Wongsiri S (eds) Asian apiculture. Wicwas Press, Cheshire, Connecticut, pp 110–116, Keywords: *andreniformis*/*cerana*/*dorsata*/drones/*florea*/honey/*mellifera*

Patinawin S, Wongsiri S (1994) Scanning electron microscopy of cervical lobes of *Apis* drones. J Apic Res 33:65–68, Keywords: *andreniformis*/*cerana*/*dorsata*/drones/*florea*/*mellifera*/phylogeny

Patra K, Suwanda O (1988) Beekeeping with *Apis cerana* in Indonesia. Scout Movement Apiary Centre & Deputy of Natural Resources Development, Jakarta, Keywords: beekeeping/*cerana*/physiology/Indonesia

Patyal SK, Kumar J (1989) Olfactogustatory repellency of some essential oils to the Indian honey bee, *Apis cerana indica* Fabr. Indian Bee J 51:5–8, Keywords: *cerana*/chemicals/honey/physiology/India

Pechhacker H, Juntawong N (1994) *Apis mellifera* versus *Apis cerana* in the north of Thailand. Beekeep Dev 30:6–7, Keywords: beekeeping/*cerana*/interspecific interactions/*mellifera*/Thailand

Pechhacker H, Hüttinger E, Dippelreiter K, Shreshta KK (2000) Temperature measurements of different types of beehives in Nepal. In: Proceedings of the 4th Asian apiculture association international conference, Kathmandu, pp 125–128, Keywords: beekeeping/*cerana*/physiology/nests/Nepal

Pechhacker H, Hüttinger E, Dippelreiter K, Shreshta KK (2000) Size of traditional beehives. In: Proceedings of the 4th Asian apiculture association international conference, Kathmandu, pp 131–132, Keywords: beekeeping/*cerana*/morphology/nests/Nepal

Pechhacker H, Joshi SR, Chatt A (2001) Traditional beekeeping in rural areas: a comparison between *Apis cerana* beekeeping in Nepal and *Apis mellifera* beekeeping in Morocco. Bee World 82:99–104, Keywords: beekeeping/*cerana*/*mellifera*/Nepal

Peddy PVR, Jacob TK (2001) Comparative foraging behaviour of the Indian honeybee, *Apis cerana* F and European bee. Apis mellifera L. Insect Environ 6:165, Keywords: behaviour/*cerana*/foraging/*mellifera*/India

Peng YS (1988) The resistance mechanism of the Asian honey bee (*Apis cerana*) to the mite *Varroa jacobsoni*. In: Needham GR, Page RE, Delfinado-Baker M, Bowman CE (eds) Africanized honey bees and bee mites. Ellis Horwood, Chichester, pp 426–429, Keywords: *cerana*/disease/honey/pests

Peng YS, Fang Y (1988) Removal of the mite *Varroa jacobsoni* from the brood of the European honey bee *Apis mellifera* mites. In: Needham GR, Page RE, Delfinado-Baker M, Bowman CE (eds) Africanized honey bees and bee mites. Ellis Horwood, Chichester, pp 430–433, Keywords: brood/*cerana*/honey/pests

Peng YS, Fang Y, Xu S, Ge L (1987a) The resistance mechanism of the Asian honeybee, *Apis cerana* Fabr., to an ectoparasitic mite, *Varroa jacobsoni* Oudemans. In: Proceedings of the 31st international apiculture congress, Apimondia, Warsaw, pp 193–194, Keywords: *cerana*/disease/pests/China

Peng YS, Fang Y, Xu S, Ge L (1987b) The resistance mechanism of the Asian honey bee, *Apis cerana* Fabr., to an ectoparasitic mite, *Varroa jacobsoni* Oudemans. J Invertebr Pathol 49:54–60, Keywords: *cerana*/disease/honey/pests

Peng YS, Fang Y, Xu S, Ge L, Nasr ME (1987c) Response of foster Asian honeybee (*Apis cerana* Fabr.) colonies to the brood of European honeybee (*Apis mellifera* L) infested with parasitic mite, *Varroa jacobsoni* Oudemans. J Invertebr Pathol 49:259–264, Keywords: brood/*cerana*/*mellifera*/pests

Peng YS, Nasr ME, Locke SJ (1989) Geographical races of *Apis cerana* Fabricius in China and their distribution. Review of recent Chinese publications and a preliminary statistical analysis. Apidologie 20:9–20, Keywords: biogeography/*cerana*/classification/geographical variation/subspecies/China

Percy AP, Kshirsagar KK, Chandran K (1968) Further observations on the phoretic bee mites *Neocypholaelaps indica* (Evans). Indian Bee J 30:68–69, Keywords: *cerana*/pests/India

Perk J (1973) Division of labour in *Apis cerana*. Thesis, Institut Bienenkunde, Oberursel [in German]. Keywords: behaviour/*cerana*

Perti ON, Pandey P (1967) Study of some physicochemical constants of *Apis dorsata* honey. Part II. Specific gravity. Indian Bee J 29:1–3, Keywords: bee botany/*dorsata*/honey/India

Pesenko YA, Lelei AS, Radchenko VG, Filatkin GN (1989) Chinese wax-bee, *Apis cerana cerana* F (Hymenoptera, Apidae) in the Far East of the USSR. Entomol Obozr 68:527–548, [in Russian]. Keywords: biology/*cerana*/*mellifera*/Russia

Pesenko YA, Lelei AS, Radchenko VG, Filatkin GN (1990) The Chinese wax bee (*Apis cerana cerana*) in the Far East of the USSR. In: Proceedings of the 10th Congress All-Union Entomological Society, Leningrad, pp 99–101 [in Russian]. Keywords: beeswax/biogeography/biology/*cerana*/Russia

Pesenko YA, Lelei AS, Radchenko VG, Filatkin GN (1990b) The Chinese wax bee *Apis cerana cerana* F (Hymenoptera, Apidae) in the Soviet Far East. Entomol Rev 69:21–46, Keywords: beeswax/*cerana*/Russia

Petersen S (2001a) Adventures in beekeeping 'the Akha way'. Am Bee J 141:635–639, Keywords: beekeeping/*cerana*/*dorsata*/Laos

Petersen S (2001) Sustainable beekeeping development in Laos. In: Proceedings of the 7th international conference on tropical bees, Chiang Mai, pp 427–431, Keywords: beekeeping/*cerana*/*dorsata*/*florea*/*mellifera*/physiology/Laos

Petrov P (1992) Distribution and phylogeny of the honey bee genus *Apis* (Hymenoptera, Apidae). Usp Sovrem Biol 112:359–372, [in Russian]. Keywords: biogeography/*cerana*/*dorsata*/*florea*/honey/*mellifera*/phylogeny

Pettis JS (2001) Biology and life history of tracheal mites. In: Webster TC, Delaplane KS (eds) Mites of the honey bee. Dadant, Hamilton, pp 29–41, Keywords: biology/*cerana*/*dorsata*/history/honey/pests

Pflugfelder J, Koeniger N, Svatos A, Crewe R (2004) Fatal combat of queens in the genus *Apis*: verification of a fighting pheromone and its interspecific efficacy. In: Proceedings of the 1st European conference of apidolgy, Udine, pp 3–4, Keywords: defense/pheromones/queens

Phadke KG, Naim M (1974) Observations on the honeybee visitation to the litchi (*Nephelium litchi*) blossoms at Pusa (Bihar, India). Indian Bee J 36:9–12, Keywords: bee botany/*cerana*/*dorsata*/*florea*/India

Phadke KG, Pant NC, Sinha RBP (1967) The foraging behaviour of *Apis indica* F (Hymenoptera, Apidae) visiting flowers of sesame (*Sesamum orientale* L). Indian J Entomol 29:157–163, Keywords. *cerana*/foraging/India

Phadke RP (1961) Some physico-chemical constants of Indian beeswaxes. Bee World 42:149–153, Keywords: beeswax/*cerana*/*dorsata*/*florea*/India

Phadke RP (1962) Physicochemical composition of major unifloral honeys from Mahabaleshwar (Western Ghats). Indian Bee J 24:59–65, Keywords: *cerana*/honey/India

Phadke RP (1964) Nectar concentration in *Carvia callosa*. Indian Bee J 26:22–25, Keywords: *cerana*/honey/nectar/India

Phadke RP (1967a) Studies on Indian honeys. 1. Proximate composition and physicochemical characteristics of Indian multifloral apiary honeys from *Apis indica* bee. Indian Bee J 29:14–26, Keywords: *cerana*/honey/India

Phadke RP (1967b) Studies on Indian honeys. 2. Proximate composition and physicochemical characteristics of unifloral honeys of Mahabaleshwar. Indian Bee J 29:33–46, Keywords: *cerana*/honey/India

Phadke RP (1968) Studies on Indian honeys. 3. Proximate composition and physico-chemical characterizations of honeys from the wild honey bees *Apis dorsata*, *Apis florea* and *Trigona*. Indian Bee J 30:3–8, Keywords: biogeography/*cerana*/*dorsata*/*florea*/honey/India

Phadke RP (1983) Sac brood disease in India. Indian Bee J 45:43–45, Keywords: *cerana*/disease/India

Phadke RP (1986) Beekeeping in India- present status, constraints and plan of action for further development. Indian Bee J 48:29–33, Keywords: beekeeping/*cerana*/India

Phadke RP, Nair KS (1973) Studies on Indian honeys. 5. Distinguishing characteristics of the apiary honey from the wild variety. Indian Bee J 35:36–39, Keywords: *cerana*/honey/India

Phadke RP, Phadke RS (1975) Physicochemical characteristics of waxes from Indian honey bees. II. Viscosity and refractive index. Indian Bee J 37:15–18, Keywords: beeswax/*cerana*/honey/India

Phadke RP, Nair KS, Nandekar KU (1969) Indian beeswaxes. I. Their physico-chemical constants. Indian Bee J 31:52–55, Keywords: beeswax/*cerana*/*dorsata*/*florea*/*mellifera*/India

Phadke RP, Nair KS, Nandekar KU (1971) Indian beeswaxes. II. The nature of their chemical constituents. Indian Bee J 33:3–5, Keywords: beeswax/*cerana*/*dorsata*/*florea*/India

Pham VL, Pham CH, Vu TC (2001) Situation of sac-brood disease infection on *Apis cerana* beehives in some northern provinces of Vietnam. In: Proceedings of the 37th international apiculture congress, Apimondia, Durban (CD Rom), Keywords: *cerana*/disease/Vietnam

Phiancharoen M, Wongsiri S, Koeniger N, Koeniger G (2004) Instrumental insemination of *Apis mellifera* queens with hetero- and conspecific spermatozoa results in different sperm survival. Apidologie 35:503–511, Keywords: *cerana*/*dorsata*/*florea*/*mellifera*/reproduction

Phiancharoen M, Wongsiri S, Hepburn HR (2011) Queen production and instrumental insemination of *Apis florea* queens. Apidologie (in press), Keywords: *florea*/reproduction/Thailand

Phin J (1884) A dictionary of practical apiculture: giving the correct meaning of nearly five hundred terms intended as a guide to uniformity of expression amongst bee-keepers. Industrial Publication, New York, Keywords: *cerana*/*dorsata*/*florea*

Phoon ACG (1983) Beekeeping in Malaysia. Pertanika 6:3–17, Keywords: bee botany/beekeeping/*cerana*/Malaysia

Phoon ACG (1985) Pollination and fruit production of carambola, *Averrhoa carambola*, in Malaysia. In: Proceedings of the 3rd international conference on apiculture in tropical climates, Nairobi, pp 129–133, Keywords: *cerana*/physical factors/pollination/Malaysia

Pickett JA, Williams IH, Martin AP (1982) Pheromonal component from the sting of the honeybee, Apis mellifera L (Hymenoptera: Apidae). J Chem Ecol 18:163–175, Keywords: *cerana*/*dorsata*/honey/pheromones

Pirk CWW, Hepburn HR, Radloff SE, Erlandson J (2002) Defense posture in the dwarf honeybee, *Apis florea*. Apidologie 33:289–294, Keywords: behaviour/defense/*florea*/Thailand

Poilane F (1926) Notes on the honeybees of Indochina. Bull Econ Indochine (Hanoi) 2:461–464, [in French]. Keywords: cerana/dorsata/florea/Cambodia/Laos/Vietnam

Plettner E, Otis GW, Wimalaratne PDC, Winston ML, Slessor KN, Pankiw T, Punchihewa RWK (1997) Species and caste determined mandibular gland signals in honeybees (*Apis*). J Chem Ecol 23:363–377, Keywords: behaviour/*cerana*/*dorsata*/morphology/pheromones/queens/workers

Pokhrel S (2008) The ecological problems and possible solutions of beekeeping in hills and Terai of Chitwan, Nepal. J Agric Environ 9:23–33, Keywords: absconding/beekeeping/*cerana*/*mellifera*/Nepal

Pokhrel S (2009) Comparative benefits of beekeeping enterprise in Chitwan, Nepal. J Agric Environ 10:46–59, Keywords: beekeeping/Nepal*

Polous H (1969) The introduction and establishment of little bee (*Micrapis florea*) at Ahwaz Agricultural College and first account on their migration. Thesis, Ahwaz Agricultural College, Keywords: absconding/*florea*/migration/swarming/Iran

Poltev VI (2004) Epizootology of varroatosis in the honey bees. In: Khazbiyevich LM (ed) Varroatosis of honey bees. Nauka, Moscow, pp 5–9, [in Russian]. Keywords: *cerana*/honey/parasites/Russia

Pootong S (1999) Investigation of genetic variation among Thai honeybee *A. cerana* using mitochondrial DNA control region. Thesis, Chulalongkorn University, Keywords: *cerana*/genetics/mtDNA/Thailand

Potichot S (1988) Production and artificial insemination of queen bees (*Apis cerana* Fabricius). Thesis, Chulalongkorn University [in Thai]. Keywords: breeding/*cerana*/queens/Thailand

Potichot S, Wongsiri S (1987) Research on Chinese bee, *Apis cerana cerana*, queen rearing. Report, Chulalongkorn University, Keywords: breeding/*cerana*/queens

Potichot S, Vongsamanode S, Lekprayoon C, Wongsiri S (1989) Comparison of queen production of *Apis mellifera* and *Apis cerana*. In: Proceedings of the 1st Asia-Pacific conference of entomology, Chiang Mai, pp 59–62, Keywords: *cerana*/*mellifera*/queens/Thailand

Potichot S, Wongsiri S, Dietz A (1993) Attempts in queen rearing of *Apis cerana* larvae in *Apis mellifera* colonies and *Apis mellifera* larvae in *Apis cerana* colonies. In: Connor LJ, Rinderer TE, Sylvester HA, Wongsiri S (eds) Asian apiculture. Wicwas, Cheshire, CT, pp 128–133, Keywords: breeding/*cerana*/interspecific interactions/larvae/*mellifera*/queens/Thailand

Pourashgar D (1979) Investigations on morphological characteristics of Iranian native honeybees. In: Proceedings of the Apiculture em Clima Quente: Simpósio Internacional, Florianópolis, Santa Clara, pp 95–96, Keywords: biogeography/*cerana*/*florea*/*mellifera*/morphology/Iran

Prakash R, Kumaraswami T (1984) Toxicity of some insecticides to Indian bee, *Apis cerana* Fab. Indian Bee J 46:15–17, Keywords: *cerana*/insecticides

Prakash S, Bhat NS, Naik MI, Swamy BCH (2008) Influence of pollen supplement and food substitute on brood rearing and foraging activity of Indian honey bee *Apis cerana* Fabricius. Mysore J Agric Sci 42:405–409, Keywords: *cerana*/foraging/nutrition/India

Pramual C (1994) DNA analysis of genetic diversity of *Apis cerana* Fabricius in Thailand. Thesis, Chulalongkorn University, Keywords: *cerana*/DNA/genetics/Thailand

Prasad D, Hameed SF, Singh R, Yazdani SS, Singh B (1989a) Effect of bee pollination on the quantity and quality of rape crop (*Brassica juncea* Cross). Indian Bee J 51:45–47, Keywords: *cerana*/pollination/India

Prasad D, Hameed SF, Singh R, Singh B (1989b) Foraging behaviour of insect pollinators on brown mustard, *Brassica juncea*, in Bihar, India. Indian Bee J 51:131–133, Keywords: behaviour/foraging/pollination/India

Prasad VG, Verghese A (1990) Integration of apiculture in horticulture. In: Proceedings of the 11th international congress, IUSSI, Bangalore, pp 477–478, Keywords: *cerana*/*dorsata*/pollination/India

Priti (1998) Abundance and pollination efficiency of insect visitors of onion blossom. Indian Bee J 60:75–78, Keywords: bee botany/dorsata/florea/mellifera/pollination/India

Priti SRC (1997) Diversity, visitation frequency, foraging behaviour and pollinating efficiency of insect pollinators visiting cauliflower (*Brassica olearacea* L var *botrytis* cv 1606 Hazipur Local) blossoms. Indian Bee J 59:230–237, Keywords: behaviour/*dorsata*/*florea*/foraging/*mellifera*/pollination

Priti SRC (1998b) Diversity, visitation frequency, foraging behaviour and pollinating efficiency of different insect pollinators visiting carrot, *Dacus carota* L var HC-1 blossoms. Indian Bee J 60:1–8, Keywords: behaviour/*dorsata*/*florea*/foraging/*mellifera*/pollination/India

Priti MRC, Sihag RC (2001) Role of insect pollination in seed production of radish (*Raphanus sativus* L). Seed Res 29:231–234, Keywords: bee botany/*dorsata*/*florea*/foraging/*mellifera*/pollination/India

Punchihewa RWK (1992a) Mating behaviour of *Apis cerana* in Sri Lanka is advantageous for controlled natural breeding. Apidologie 23:348–349, Keywords: behaviour/breeding/*cerana*/reproduction/Sri Lanka

Punchihewa RWK (1992) Observations and experiments on mating of *Apis cerana indica* in Sri Lanka. Thesis, Institut für Bienenkunde, FB Biologie, Universität Frankfurt am Main, Frankfurt [in German]. Keywords: *cerana*/reproduction/Sri Lanka

Punchihewa RWK (1994) Beekeeping for honey production in Sri Lanka: management of Asiatic hive honeybee *Apis cerana* in its natural tropical monsoonal environment. Sri Lanka Department of Agriculture and Canadian International Development Agency, Peradeniya, Keywords: beekeeping/*cerana*/honey/nests/physiology/Sri Lanka

Punchihewa RWK, Koeniger N, Kevan PG, Gadawski RM (1985) Observations on the dance communication and natural foraging ranges of *Apis cerana*, *Apis dorsata* and *Apis florea* in

Sri Lanka. J Apic Res 24:168–175, Keywords: behaviour/*cerana*/*dorsata*/*florea*/foraging/Sri Lanka
Punchihewa RWK, Koeniger N, Howpage D (1990a) Absconding behaviour of *Apis cerana* in Sri Lanka. In: Proceedings of the 11th international congress, IUSSI, Bangalore, pp 106–107, Keywords: absconding/beeswax/behaviour/brood/*cerana*/nests/oviposition/pests/physical factors/queens/Sri Lanka
Punchihewa RWK, Koeniger N, Koeniger G (1990b) Mating behaviour of *Apis cerana* in Sri Lanka. In: Proceedings of the 11th international congress, IUSSI, Bangalore, pp 108, Keywords: behaviour/*cerana*/drones/reproduction/Sri Lanka
Punchihewa RWK, Koeniger N, Koeniger G (1990c) Congregation of *Apis cerana indica* Fabricius 1798 drones in the canopy of trees in Sri Lanka. Apidologie 21:201–208, Keywords: *cerana*/drones/Sri Lanka
Punchihewa RWK, Koeniger N, Kevan PG (2000) Management of swarming in *Apis cerana*. In: Proceedings of the 4th Asian apiculture association international conference, Kathmandu, pp 114–115, Keywords: beekeeping/*cerana*/swarming/Sri Lanka
Pundir RS (1971) Hiving Apis florea. Indian Bee J 33:63, Keywords: biogeography/*cerana*/combs/*florea*/nests/queens/India
Punjabi AA, Saraf SK, Singh G (1969) Studies on the pollen carrying capacity of the hill strain of the Indian honey bee *Apis indica* F in Kashmir, in relation to its body weight. Indian Bee J 31:9–11, Keywords: bee botany/brood/*cerana*/foraging/honey/pollen/India
Purna CRS, Atluri JB, Subba Reddy C (1995) *Sterculia foetida* as a pollen source for the Indian honeybee, *Apis cerana indica*. Indian Bee J 57:18–19, Keywords: *cerana*/pollen
Purwanto DB (1999) Some observations on beekeeping management with *Apis dorsata* in Belitung, south Sumatra, Indonesia. Thesis, Georg-August University, Keywords: beekeeping/*dorsata*/Indonesia
Purwanto DB, Hadisoesilo S, Kasno, Koeniger N, Lunderstadt J (2002) Sunggau system: a sustainable method of honey production from Indonesia with the Asian giant honeybee *Apis dorsata*. In: Proceedings of the 7th international conference on apiculture in tropical climates, Chiang Mai, pp 201–206, Keywords: beekeeping/*dorsata*/honey/Indonesia
Putra RE, Nakamura K (2009) Foraging ecology of a local wild bee community in an abandoned Satoyama system in Kanazawa, central Japan. Entomol Res 39:99–106, Keywords: *cerana*/foraging/Japan
Pyramam K, Wongsiri S (1986) Bee flora for four species of Apis in Thailand. J Sci Res (Chulalongkorn) 11:95–103, Keywords: bee botany/biology/cerana/dorsata/florea/mellifera/Thailand
Qayyum HA, Ahmad N (1967) Biology of *Apis dorsata* F. Pak J Sci 19:109–113, Keywords: *dorsata*/general biology
Qian ZB (1994) Summering management for *Apis cerana cerana* in the south of Jiangsu Province. Apic China 4:16, [in Chinese]. Keywords: beekeeping/*cerana*/seasonal variation/China*
Qiao Q (1995) Annotation of the ancient apicultural literature. Sci Tech China, [in Chinese]. Keywords: beekeeping/*cerana*/history/China*
Qinghal S (1993) A brief history of Chinese apiculture, [in Chinese]. Keywords: beekeeping/*cerana*/history/China*
Qiu CH (1993a) Artificial queen rearing in *Apis cerana cerana*. Apic Sci Tech 3:14, [in Chinese]. Keywords: beekeeping/breeding/*cerana*/queens/China*
Qiu CH (1993b) Methods for rearing *Apis cerana cerana*. Apic Sci Tech 4:14, [in Chinese]. Keywords: beekeeping/breeding/*cerana*/China*
Qiu CH (1995a) An approach for improving the rearing of *Apis cerana cerana*. Apic Sci Tech 3:17, [in Chinese]. Keywords: beekeeping/breeding/*cerana*/China*
Qiu CH (1995b) Methods to obtain the high yield product of *Dendranthema indicum* honey by *Apis cerana cerana* in southern Jiangxi Province. Apic Sci Tech 5:15, [in Chinese]. Keywords: beekeeping/*cerana*/honey/China*

Quinones AC (1981) Artificial production of honeybee queen by grafting. Cent Luzon State Univ Tech J 1:25–27, Keywords: beekeeping/*cerana*/queens/Philippines

Quinones AC (1995) Beekeeping with *Apis cerana* – research needs and problems: The example of the Philippines. In: Kevan PG (ed) The Asiatic hive bee: apiculture, biology, and role in sustainable development in tropical and subtropical Asia. Enviroquest, Cambridge, ON, pp 263–270, Keywords: beekeeping/biology/*cerana*/*dorsata*/*florea*/natural enemies/nests/physiology/Philippines

Radloff SE, Hepburn HR (2002) Multivariate morphometric analysis of the *Apis cerana* populations of southern Asia. In: Proceedings of the 14th international congress, IUSSI, Sapporo, p 209, Keywords: *cerana*/classification/morphometrics/population biology/Japan

Radloff SE, Hepburn HR, Otis GW (2001) Flight machinery of honey bees (Hymenoptera: Apidae, *Apis*). In: Proceedings of the 7th international conference on tropical bees, Chiang Mai, pp 163–168, Keywords: *andreniformis*/*cerana*/*dorsata*/flight/*florea*/honey/*koschevnikovi*/*mellifera*/*nigrocincta*

Radloff SE, Hepburn HR, Bangay L (2003a) Quantitative analysis of intercolonial and intracolonial morphometric variance in honeybees, *Apis mellifera* and *Apis cerana*. Apidologie 34:339–352, Keywords: *cerana*/classification/*mellifera*/population biology

Radloff SE, Hepburn HR, Koeniger G (2003b) Comparison of flight design of Asian honeybee drones. Apidologie 34:353–358, Keywords: *andreniformis*/*cerana*/*dorsata*/drones/flight/*florea*/*koschevnikovi*/*laboriosa*/*nigrocincta*/*nuluensis*

Radloff SE, Hepburn HR, Fuchs S (2005a) The morphometric affinities of *Apis cerana* of the Hindu Kush and Himalayan regions of western Asia. Apidologie 36:25–30, Keywords: *cerana*/morphometrics/Afghanistan/India

Radloff SE, Hepburn HR, Hepburn C, Fuchs S, Otis GW, Sein MM, Aung HL, Pham HT, Dam DQ, Nuru AM, Ken T (2005b) Multivariate morphometric analysis of *Apis cerana* of southern mainland Asia. Apidologie 36:127–139, Keywords: *cerana*/classification/morphometrics

Radloff SE, Hepburn HR, Fuchs S, Otis GW, Hadisoesilo S, Hepburn C, Ken T (2005c) Multivariate morphometric analysis of the *Apis cerana* populations of oceanic Asia. Apidologie 36:475–492, Keywords: biogeography/*cerana*/classification/morphometrics/populations

Radloff SE, Hepburn C, Hepburn HR, Fuchs S, Hadisoesilo S, Tan K, Engel MS, Kuznetsov V (2010) Population structure and classification of *Apis cerana*. Apidologie 41: doi:10.1051/apido/2010008, Keywords: biogeography/*cerana*/classification/morphology/populations

Radoszkowski OI (1877) Hymenoptera of Korea. Hor Soc Entomol Ross 21:428–436, [in French]. Keywords: *cerana*/Korea

Raffiudin R, Crozier RH (2001) Phylogenetic analyses of *Apis* based on inosotol 1:4:5-trisphosphate receptor gene sequences. In: Proceedings of the 7th international conference on tropical bees, Chiang Mai, pp 353–357, Keywords: *andreniformis*/*cerana*/*dorsata*/*florea*/*nigrocincta*/Indonesia/Malaysia/Thailand

Raghavarao D (1970) Use of a transformation of chi square distribution with one degree of freedom in analysis of variance: application to an entomological problem. J Res (Ludhiana) 7:244–247, Keywords: biogeography/*cerana*/environmental effects/interspecific interactions/*mellifera*

Rahman A (1990) Bee forage plants of Jorhat (Assam, India). Indian Bee J 52:47–50, Keywords: bee botany/*cerana*/India

Rahman A, Rahman S (2000) Effect of honeybee (*Apis cerana indica*) pollination on seedset and yield of buckwheat (*Fagopyrum esculentum*). Indian J Agric Sci 70:168–169, Keywords: *cerana*/pollination/India

Rahman KA (1940) Insect pollination of toria (*B. napus*) and sarson (*B. campestris*) at Faisalabad. Indian J Agric Sci 5:135–145, Keywords: *cerana*/pollination/India

Rahman KA (1942) Honey-bee in the service of the fruit grower. Punjab Fruit J 6:1328–1330, Keywords: beekeeping/pollination/India

Rahman KA (1944) Distribution of *Apis indica* F. Indian Bee J 6:155, Keywords: biogeography/*cerana*/India

Rahman KA (1945) The Indian honeybee (*Apis indica* F) (Hymenoptera) at Lyallpur. Proc R Entomol Soc Lond (A) 20:33–42, Keywords: biogeography/*cerana*/India

Rahman KA, Sharma PL (1945) Wintering of *Apis indica* F at Lyallpur. Indian Fmg 6:17–18, Keywords: *cerana*/seasonal variation/India

Rahman KA, Singh S (1940a) Beekeeping in India. Indian Fmg 1:10–17, Keywords: beekeeping/*cerana*/*dorsata*/*florea*/India

Rahman KA, Singh S (1940b) Italian versus Indian bees in the Punjab. Indian Bee J 2:20–22, Keywords: *cerana*/*mellifera*/India

Rahman KA, Singh S (1940c) The editor's clip on Indian honeybee. Apis indica F. Indian Bee J 2:93, Keywords: *cerana*/India

Rahman KA, Singh S (1946) Size of the cell of the brood comb of the Indian honey bees. Indian Bee J 8:154–159, Keywords: biogeography/brood/cells/*cerana*/combs/*dorsata*/*florea*/honey/*mellifera*/morphology/India

Rahman KA, Singh S (1947) Preliminary studies on the bionomics of the Indian honeybee *Apis indica* F. Indian Bee J 9:6–8, Keywords: biology/*cerana*

Rahman KA, Singh S (1950) Variation in the tongue length of the honeybee. Indian J Entomol 10:63–73, Keywords: *cerana*/classification/*florea*/morphology/India

Rahman MF, Siddiqui MKJ, Anjum F, Oadri SSH, Sidky MM, Osman FH (1989) Acute toxicity and anti-acetylcholinesterase potential of some biphenyl derivatives to non-target species. Indian J Exp Biol 27:138–140, Keywords: *cerana*/insecticides/India

Rahman S, Rahman A (1993) Comparative pollen gathering activity of *Apis cerana indica* F and *Apis mellifera* L, under Johrat conditions of Assam (India). Indian Bee J 55:42–46, Keywords: *cerana*/foraging/*mellifera*/pollen/India

Rahoo GM, Lohar MK, Munshi GH, Nizamani SM (1985) The insect pollinators of sunflower, *Helianthus annuus* L, in Pakistan. Pakistan J Agric, Agric Eng, Vet Sci 1:79–80, Keywords: *cerana*/pollination/Pakistan

Rai KM, Gupta BP (1983) Role of honey bees and other insects as pollinators in apple and pear. Indian Bee J 45:56–57, Keywords: *cerana*/honey/pollination/India

Rai KM, Hamid MT, Joshi R, Tewari HC (1988) Honey bee and pollination in apple. Prog Hortic 20:353–354, Keywords: *cerana*/honey/pollination/India

Raj D, Rana VK (1993) Foraging behaviour of *Apis mellifera* L and *Apis cerana indica* F at the hive entrance during rapeseed bloom. J Entomol Res 17:105–110, Keywords: behaviour/*cerana*/foraging/*mellifera*/nests/India

Raj D, Rana VK (1994) Time spent by *Apis mellifera* L and *Apis cerana indica* F foragers on rapeseed bloom. J Entomol Res 18:335–339, Keywords: *cerana*/foraging/*mellifera*/India

Raj D, Rana VK, Devi N (1993a) Comparative studies on pollen loads and pollen constancy of *Apis mellifera* L and *Apis cerana indica* F on rapeseed. J Entomol Res 17:43–48, Keywords: *cerana*/foraging/*mellifera*/pollen/India

Raj J (1990) Instant introduction of queen honey bees. In: Proceedings of the 11th international congress, IUSSI, Bangalore, p 419, Keywords: *cerana*/honey/*mellifera*/queens/India

Raj J (1990) Keep dangerous rock bees (*Apis dorsata*) away from human habitation. In: Proceedings of the 11th international congress, IUSSI, Bangalore, p 618, Keywords: defense/*dorsata*/India

Raj J, Basavanna GPC (1980a) Influence of brood on the attraction and mobility of worker honeybees (*Apis cerana* F). Mysore J Agric Sci 14:224–226, Keywords: absconding/brood/*cerana*/queens/workers/India

Raj J, Basavanna GPC (1980b) Orientation of workers in colonies of honeybee (*Apis cerana* F). Mysore J Agric Sci 14:227–229, Keywords: *cerana*/flight/workers/India

Raj J, Channabasavanna GP (1983) Queen recognition, rearing and colony building by the Indian honeybee (*Apis cerana indica*). In: Proceedings of the 2nd international conference on apiculture in tropical climates, New Delhi, pp 378–384, Keywords: breeding/*cerana*/physical factors/queens/India

Raj J, Nanda KTG, Gopala GHS (1993) Influence of fermented cane juice syrup on activities of the honeybee (*Apis cerana indica*). In: Proceedings of an international symposium on pollination in tropics, Bangalore, pp 246–247, Keywords: *cerana*/foraging/pollination/India

Raj JS (2004) Physico-chemical and melissopalynological characteristics of Nepalese honey. Thesis, Universitäts bibliothek der Universität für Bodenkultur Wien, Vienna [in German]. Keywords: *cerana*/honey/*mellifera*/Nepal

Rajagopal D, Kenchcharaddi RN (1996) Current status and management of Thai sac brood virus disease of Indian honey bee *Apis cerana* Fab. in Karnataka. In: Proceedings of an international workshop revive *Apis cerana indica* Kodiakanal, pp 49–51, Keywords: beekeeping/brood/*cerana*/disease/honey/India

Rajagopal D, Veeresh GK, Chikkadevaiah T, Nagaraja N, Kencharaddi RN (1999) Potentiality of honeybees in hybrid seed production of sunflower (*Helianthus annuus*). Indian J Agric Sci 69:40–43, Keywords: *cerana*/*dorsata*/*florea*/*mellifera*/pollination

Rajagopal D, Kenchcharaddi RN, Nagaraja N (2000) Utilisation of honeybees in sunflower hybrid seed production. In: Proceedings of the 4th Asian apiculture association international conference, Kathmandu, p 186, Keywords: *cerana*/*dorsata*/*florea*/*laboriosa*/*mellifera*/pollination/India

Rajashekharappa BJ (1972) Elevating technique for swarm prevention in honey bees. Am Bee J 112:45, 51, Keywords: beekeeping/brood/*cerana*/combs/drones/honey/queens/swarming/India

Rajashekharappa BJ (1976) Probable mechanism of recognition by a colony of honey bees (*Apis cerana indica*) of the presence of its queen. Curr Sci 5:133, Keywords: *cerana*/honey/pheromones/queens/India

Rajashekharappa BJ (1979) Queen recognition, and the rearing of new queens, by honeybee colonies (*Apis cerana*). J Apic Res 18:173–178, Keywords: behaviour/breeding/*cerana*/pheromones/queens/India

Rajashekharappa BJ, Basavanna GPC (1978) Queen rearing and occurrence of broodless cells in larger colonies of honeybees (*Apis cerana* Fab.). Curr Res, *Univ Agric* Sci, *Bangalore* 7:93–95, Keywords: *cerana*/queens/India

Rajashekharappa BJ, Channabasavanna GP (1979) Importance of mated queen in comb expansion by the honeybee *Apis cerana*. Mysore J Agric Sci 13:51–54, Keywords: *cerana*/combs/pheromones/queens/India

Raju AJS (1998) Bee pollination in *Azidiractha indica* (Meliaceae). Indian Bee J 60:86, Keywords: bee botany/*cerana*/pollination/India*

Raju AJS (ed) (2009) Bioresources conservation and management. Today and Tomorrow, Keywords: *cerana*/*dorsata*/*florea*/foraging/pollination/India*

Raju AJS, Ezradanam V (2002) Pollination ecology and fruiting behaviour in a monoecious species, *Jatropha curcas* L (Euphorbiaceae). Curr Sci 83:1398, Keywords: behaviour/ecology/pollination/India

Raju AJS, Naidu SA (1999) Foraging activity of *Apis cerana indica* on *Mimosa pudica* L. Insect Environ 5:13, Keywords: *cerana*/foraging/India

Raju AJS, Naidu SA (2000) Floral biology and *Apis cerana* revisited (Hymenoptera, Apidae). Insect Environ 6:141, Keywords: bee botany/biology/*cerana*/India

Raju AJS, Raju VK, Victor P, Naidu SA (2001) Floral ecology, breeding system and pollination in *Antigonon leptopus* L (Polygonaceae). Pl Sp Biol 16:159–164, Keywords: bee botany/biology/breeding/*cerana*/*dorsata*/ecology/*florea*/pollination/India

Raju AJS, Rao SP, Jonathan KH (2006) Andromonoecy, insect pollination and fruiting behaviour in *Acacia caesia* (L.) Willd. (Mimosaceae) in the Eastern Ghats. Curr Sci 91:939–942, Keywords: *cerana*/*florea*/pollination/India

Rakshpal R, Singh A (1971) Free amino acids in the acid glands and poison sacs of some social Hymenoptera. Indian J Entomol 33:42–44, Keywords: biochemistry/*cerana*/*dorsata*/*florea*/glands/India

Rakshpal R, Singh A (1972) Free amino acids in the haemolymph of some insects. Indian J Entomol 34:216–218, Keywords: biochemistry/*cerana*/physiology/India

Ram K, Kumar NR (2001) Management of sac brood inflicted honeybee colonies. J Appl Zool Res 12:164–167, Keywords: beekeeping/brood/*cerana*/disease/India

Ramachandran S (1939) Beekeeping in south India. Report, Department of Agriculture, Superintendent Government Press, Madras, Bull 37:1–12, Keywords: beekeeping/*cerana*/India

Ramachandran S, Cherian MC (1952) Beekeeping in south India. Bulletin, Department of Agriculture, Superintendent Government Press, Madras, Keywords: beekeeping/*cerana*/*florea*/India

Ramachandran S, Mahadevan V (1950) Preliminary studies on brood-rearing and hive-population of *Apis indica* F. Indian Bee J 12:154–159, Keywords: bee botany/behaviour/brood/*cerana*/colony phenology/India

Ramanan RV, Ghai S (1984) Observations on the mite *Neocypholaelaps indica* Evans and its relationship with the honey bee *Apis cerana indica* Fabricius and the flowering of *Eucalyptus* trees. Entomon 9:291–292, Keywords: *cerana*/honey/pests/India

Ramanujam CGK, Kalpana TP (1992) *Tamarindus indica* L: an important forage plant for *Apis florea* F in south central India. Apidologie 23:403–413, Keywords: bee botany/*cerana*/foraging/India

Ramanujam CGK, Kalpana TP (1993) Pollen analysis of honeys from Kondevaram apiaries of East Godavari District, Andhra Pradesh. BIOVD 19:11–19, Keywords: bee botany/*cerana*/honey/pollen/India

Ramanujam CGK, Kalpana TP (1994) Summer pollen sources for honey bees in the coastal belt of East Godavari District, Andhra Pradesh. Geophytology 23:249–252, Keywords: bee botany/*cerana*/*florea*/honey/pollen/India

Ramanujam CGK, Khatjia F (1991) Melittopalynology of the agricultural tracts in Guntur District, Andhra Pradesh. J Indian Inst Sci 71:25–34, Keywords: bee botany/*cerana*/honey/India

Ramanujam CGK, Reddy PR, Kalpana TP (1992) Pollen analysis of apiary honeys from East Godavari District, AP. J Indian Inst Sci 72:289–299, Keywords: bee botany/*cerana*/honey/pollen/India

Rambo WE (1899) Apis dorsata. Glean Bee Cult 27:466, Keywords: *dorsata*

Ramireddy PV, Jacob TK (1999) Queen caging – a technique to manage TSBV disease in Indian honeybee. Insect Environ 5:90, Keywords: *cerana*/disease/queens

Rana BS (1985) Biological and economic characteristics of *Apis cerana indica* F in relation to selection and breeding. Thesis, Himachal Pradesh University, Keywords: breeding/*cerana*/genetics/oviposition/reproduction/India

Rana BS (1987) Selective breeding, mating behaviour and sacbrood virus disease in Indian hive bee, *Apis cerana indica* F. Thesis, Himachal Pradesh University, Keywords: behaviour/breeding/*cerana*/disease/nests/reproduction/India

Rana BS, Goyal NP (1991) Field toxicity of methyl demeton and dimethoate to the foragers of honeybee (*Apis cerana indica* Fab.) visitors to *Brassica chinensis* L. Indian Bee J 53:73–77, Keywords: *cerana*/foraging/insecticides/India

Rana BS, Goyal NP (1994) Comparative brood rearing activity of *Apis mellifera* and *Apis cerana indica* at Nauni (Solan), midhills of Himachal Pradesh. Indian Bee J 56:24–28, Keywords: breeding/brood/*cerana*/*mellifera*/India

Rana BS, Gupta JK (1997) Pollination studies on plum, *Prunus salicina* Lindl, cv Santa Rosa at Nauni, Himachal Pradesh, India. Indian Bee J 59:214–217, Keywords: *cerana*/pollination/India

Rana BS, Verma LR (1994) Hoarding behaviour and lifespan of workers of *Apis mellifera* and *Apis cerana*. J Apic Res 33:205–208, Keywords: behaviour/*cerana*/*mellifera*/physiology/workers/India

Rana BS, Verma LR, Mattu VK (1983) Selection and breeding of *Apis cerana indica* F. In: Proceedings of the 15th international congress of genetics, New Delhi, Keywords: breeding/*cerana*/genetics/India*

Rana BS, Garg ID, Khurana SMP, Verma LR, Agrawal HO (1986) Thai sacbrood virus of honeybees (*Apis cerana indica* F) in northwest Himalayas. Indian J Virol 2:127–131, Keywords: *cerana*/disease/mountains/India

Rana BS, Garg ID, Khurana SMP, Ball BV, Verma LR, Agrawal HO (1987) Sacbrood virus disease in *Apis cerana indica* F in South-East Asia. In: Eder J, Rembold H (eds) Chemistry and biology of social insects. Peperny, Munich, pp 640–641, Keywords: biology/*cerana*/disease/India

Rana BS, Kumar S, Khurana SMP, Verma LR (1988) Antiserum for the Thai sacbrood virus of honey bees. Indian Bee J 50:40–41, Keywords: *cerana*/disease/honey/India

Rana BS, Verma LR, Mattu VK (1993a) Morphological variations in the inbred lines of Indian honey bee *Apis cerana indica* Fabr. Pest Manag Econ Zool 1:112–114, Keywords: breeding/*cerana*/classification/honey/morphology/India

Rana BS, Gupta JK, Goyal NP (1997a) Survey of bee flora and beekeeping in Himachal Pradesh. Indian Bee J 59:144–149, Keywords: bee botany/beekeeping/*cerana*/*mellifera*/India

Rana BS, Gupta JK, Sharma HK (2000) Status of beekeeping in Himachal Pradesh. In: Proceedings of 4th Asian apiculture association international conference, Kathmandu, pp 33–36, Keywords: beekeeping/*cerana*/*mellifera*/migration/India

Rana RS (1989) Aggressive and hoarding behaviour of *Apis mellifera* and *Apis cerana* F and their role in the pollination of plum and apple blooms. Thesis, Himachal Pradesh University. Keywords: aggressiveness/behaviour/*cerana*/*mellifera*/pollination/India

Rana RS, Verma LR (2001) Hoarding behaviour and lifespan of workers of *Apis mellifera* and *Apis cerana*. J Apic Res 33:205–208, Keywords: behaviour/*cerana*/*florea*/*mellifera*/workers

Rana RS, Verma LR, Mattu VK (1998) Behavioural responses in *Apis mellifera* L and *Apis cerana* F to alarm pheromones. Pest Manage Econ Zool 6:143–146, Keywords: aggressiveness/*cerana*/*mellifera*/pheromones/India

Rana VK, Raj D, Devi N (1993b) Comparative hive performance of *Apis mellifera* L and *A. cerana indica* F on rapeseed. J Entomol Res 17:61–63, Keywords: *cerana*/*mellifera*/nests/India

Rana VK, Raj D, Kaushik R (1997b) Comparative foraging activity of *Apis mellifera* L and *Apis cerana indica* F on rapeseed bloom. J Entomol Res 21:59–64, Keywords: *cerana*/foraging/*mellifera*/India

Rana VK, Kapoor KS, Raj D (1999) Comparative pollinating activities of *Apis cerana indica* F and *Apis mellifera* L in cauliflower (*Brassica oleracea* var *botrytis*). J Entomol Res 23:141–148, Keywords: *cerana*/*mellifera*/pollination/India

Ranabhat NB, Tamrakar AS (2008) Study on seasonal activity of predatory wasps attacking honeybee *Apis cerana* Fab. colonies in southern belt of Kaski District, Nepal. J Nat Hist Mus 23:125–128, Keywords: *cerana*/predators/Nepal

Ranabhat NB, Tamrakar AS (2009) Natural enemies of honey bees *Apis cerana* F and associated problems in beekeeping in Kaski, W Nepal. J Nat Hist Mus 24:89–95, Keywords: beekeeping/*cerana*/natural enemies/Nepal

Rangarajan AV (1971) Occurrence of the mite. Varroa jacobsoni Oudemans on the Indian honey bee in South India. Sci Cult 37:292, Keywords: *cerana*/honey/pests/India

Rangarajan AV, Mahadevan NR, Iyemperumal S (1974) Note on the time of visit of pollinating honeybees to sunflower. Indian J Agric Sci 44:66–67, Keywords: *cerana*/*florea*/foraging/pollination/India

Rani U, Jain KL (1997) Age related differences in taste responses to sugars in *Apis* species. Indian J Exp Biol 35:470–472, Keywords: behaviour/*cerana*/*dorsata*/*florea*/physiology/India

Rao GM (1991) Studies on the floral biology and pollination requirement of scented methi (*Trigonella corniculata* Linn). Indian Bee J 53:39–43, Keywords: biology/*cerana*/pollination/India

Rao GM (1997) Bee pollination of crops – the Indian scenario. Indian Bee J 59:91–94, Keywords: *cerana*/*dorsata*/*florea*/*mellifera*/pollination/India

Rao GM, Lazar M (1983) Studies on bee behaviour and pollination in onion (*Allium cepa* L). In: Proceedings of 2nd international conference on apiculture in tropical climates, New Delhi, pp 580–589, Keywords: behaviour/*cerana*/*florea*/physical factors/pollination/India

Rao GM, Singh TSMS (1997) Current trends in beekeeping in South India. Indian Bee J 59:33–36, Keywords: beekeeping/*cerana*/*mellifera*/India

Rao GM, Suryanarayana MC (1983) Potentialities for bee pollination of crops in UP. Indian Bee J 45:58–62, Keywords: *cerana*/pollination/India

Rao GM, Suryanarayana MC (1989) Effect of honey bee pollination on seed yield in onion (*Allium cepa* L). Indian Bee J 51:9–11, Keywords: *cerana*/honey/pollination/India

Rao GM, Suryanarayana MC (1990) Studies on the foraging behaviour of honey bees and its effect on the seed yield in niger. Indian Bee J 52:31–33, Keywords: behaviour/*cerana*/foraging/honey/India

Rao GM, Lazar M, Suryanarayana MC (1981) Foraging behaviour of honeybees in sesame (*Sesamum indicum* L). Indian Bee J 43:97–100, Keywords: behaviour/*cerana*/foraging/India

Rao GM, Lazar M, Suryanarayana MC, Thakar SS (1984) Studies on floral biology and pollination of litchi (*Litchi chinensis* Sonn). Indian Bee J 46:7–11, Keywords: biology/*cerana*/pollination/India

Rao GM, Nadre KR, Suryanarayana MC (1995) Studies on pollination of male sterile line for foundation seed production in hybrid sunflower. Indian Bee J 57:170–173, Keywords: *cerana*/pollination/India

Rao GM, Nadre KR, Suryanarayana MC (1996a) Studies on the utility of honey bees on production of foundation seed of cotton cv NCMHH20. Indian Bee J 58:13–15, Keywords: *cerana*/honey/pollination/India

Rao GM, Mujumdar SM, Singh TSMS, Rao KS (1996b) Studies on *Apis cerana* Fabr. for Thai sacbrood disease resistance stock at Neyyattinkara, Kerala, India. Indian Bee J 58:64–66, Keywords: breeding/*cerana*/disease/India

Rao GM, Mujumdar SM, Rao KS, Dhas JRK (1996c) Studies on management of *Apis cerana* Fabr. colonies in the coastal area of Orissa, India. Indian Bee J 58:117–121, Keywords: beekeeping/*cerana*/India

Rao GM, Atluri JB, Reddi CS (2001) Intermittent mass blooming, midnight anthesis and rock bee pollination in *Pterocarpus santalinus* (Fabaceae). Nordic J Bot 21:271–276, Keywords: bee botany/*dorsata*/India

Rao KS, Choudhary OP, Rao KV (1994) How to make and use a hive frame holder. Am Bee J 134:29–30, Keywords: beekeeping/*cerana*/*mellifera*/nests/India

Rao MRN (1987) Prospects of beekeeping in Andaman and Nicobar Islands. Indian Fmg 36:8–9, Keywords: beekeeping/*cerana*/islands/India

Rao PJ, Tiwari LD (1983) Haemolymph protein pattern of drones and workers of the Indian honeybee *Apis cerana indica*. In: Proceedings of the 2nd international conference on apiculture in tropical climates, New Delhi, pp 372–377, Keywords: biochemistry/*cerana*/drones/physical factors/physiology/workers/India

Rao PVS, Mohanasundaram M, Subramaniam TR (1972) *Odontomantis micans* Stal (Mantidae: Dictyoptera) as an enemy of *Apis indica* F, the Indian honey bee in Coimbatore. Indian Bee J 34:72, Keywords: *cerana*/honey/natural enemies/India

Rao SP, Raju AJ (2002) Pollination ecology of the red sanders *Pterocarpus santalinus* (Fabaceae), an endemic and endangered tree species. Curr Sci 83:1144–1148, Keywords: bee botany/*dorsata*/ecology/India

Rao SPC, Atluri JB, Rao CS (1985) *Sterculia foetida* as a pollen source for the Indian honey bee, *Apis cerana indica*. Indian Bee J 57:18–19, Keywords: bee botany/*cerana*/honey/pollen/pollination/India

Rao SPC, Atluri JB, Subba Reddi C (1997) *Crataeva magna* (Lour) DC as a pollen source for the honey bees. Indian Bee J 59:85–87, Keywords: *cerana*/honey/pollen/pollination/India

Rappard FW (1936) An attack of *Apis dorsata* F. Trop Natuur 25:160, [in Dutch]. Keywords: defense/*dorsata*

Rasmidatta A, Suwanapong G, Wongsiri S (1999) Ultrastructure of the compound eye of the giant honeybee, *Apis dorsata*. Asian Bee J 1:60–63, Keywords: *dorsata*/morphology

Rath HW (1991) Investigations on the parasitic mites *Varroa jacobsoni* Oud and *Tropilaelaps clareae* Delfinado Baker and their hosts *Apis cerana* Fabr., *Apis dorsata* Fabr., and *Apis mellifera* L. Thesis, Rheinische Friedrich-Wilhelms-Universität, Bonn [in German]. Keywords: *cerana*/*dorsata*/*mellifera*/pests/Thailand

Rath W (1989) Survival and reproduction of *Varroa jacobsoni* on *Apis cerana*. In: Proceedings of the 1st Asia-pacific conference entomology, Chiang Mai, p 277, Keywords: *cerana*/pests/reproduction/Thailand

Rath W (1991b) Experiences with bees and parasitic mites in Thailand. Allg Dtsch Imkerztg 25:12–15, [in German]. Keywords: *cerana*/*dorsata*/*florea*/pests/Thailand

Rath W (1992a) The key for *Varroa*: *Apis cerana* drones and their cell cappings. Allg Dtsch Imkerztg 26:12–14, [in German]. Keywords: cells/*cerana*/drones/pests

Rath W (1992b) The key to *Varroa*: the drones of *Apis cerana* and their cell cap. Am Bee J 132:329–331, Keywords: *cerana*/drones/pests/Thailand

Rath W (1993) Aspects of preadaption in *Varroa jacobsoni* while shifting from its original host *Apis cerana* to *Apis mellifera*. In: Connor LJ, Rinderer TE, Sylvester HA, Wongsiri S (eds) Asian apiculture. Wicwas, Cheshire, CT, pp 417–426, Keywords: *cerana*/*mellifera*/pests/Thailand

Rath W (1999) Coadaptation of *Apis cerana* Fabr. and *Varroa jacobsoni* Oud. Apidologie 30:97–110, Keywords: *cerana*/pests/Thailand

Rath W, Delfinado-Baker M (1991) Analysis of *Tropilaelaps clareae* populations in the debris of *Apis dorsata* and *Apis mellifera* in Thailand. In: Proceedings of an international symposium recent research bee pathology, Ghent, pp 86–89, Keywords: *dorsata*/*mellifera*/mites/Thailand

Rath W, Drescher W (1990a) Response of *Apis cerana* Fabr. towards brood infested with *Varroa jacobsoni* Oud and infestation rate of colonies in Thailand. Apidologie 21:311–321, Keywords: brood/*cerana*/pests/Thailand

Rath W, Drescher W (1990b) *Varroa jacobsoni* and rate of *V. jacobsoni* infestation in *A. cerana* colonies in Thailand. Honeybee Sci 11:99–104, [in Japanese]. Keywords: brood/*cerana*/pests/Thailand

Rath W, Delfinado-Baker M, Drescher W (1991) Observations on the mating behavior, sex ratio, phoresy and dispersal of *Tropilaelaps clareae* (Acari: Laelapidae). Int J Acarol 17:201–208, Keywords: behaviour/disease/*dorsata*/genetics/pests/predators/reproduction/Thailand

Rathi A, Sihag RC (1993) Differential attractiveness of *Brassica campestris* L and *Cajanus cajan* L (Millsp) to two honey bee species. In: Proceedings of an international symposium *on pollination in tropics*, Bangalore, pp 113–119, Keywords: *cerana*/*florea*/foraging/honey/pollination/India

Rathore RRS, Wells H (1995) Training Asian rock bees (*Apis dorsata*) to forage at specific locations. Indian Bee J 57:8–9, Keywords: behaviour/*dorsata*/pheromones/workers/India

Rathore RRS, Reddy MLK, Sachan GC (1987) Effect of insecticidal sprays on the visits of Indian honey bee. Apis cerana indica Fab. on radish crop. Cruciferae Newsl 12:93, Keywords: *cerana*/honey/insecticides/India

Ratnam R (1939) A preliminary study in the biometric variations in the Indian honeybees. Madras Agric J 27:432–439, Keywords: *cerana*/classification/*dorsata*/*florea*/India

Ratnieks FL (1990) Worker policing in social insects. In: Proceedings of the 11th international congress, IUSSI, Bangalore, pp 365–366, Keywords: behaviour/*cerana*/*mellifera*/India

Rattanawannee A, Chanchao C, Wongsiri S (2007) Morphometric and genetic variation of small dwarf honeybees *Apis andreniformis* Smith, 1858 in Thailand. Insect Sci 14:451–460, Keywords: *andreniformis*/genetics/morphology/Thailand

Rauala TS (1972) Pollination studies in cauliflower (*Brassica oleracea* var *botrytis* L). J Res Punjab Agric Univ 9:580–585, Keywords: *cerana*/*dorsata*/*florea*/pollination/India

Rawat BS (1959) Successful beekeeping. Rawat Apiaries, Almora, [in Hindi]. Keywords: beekeeping/*cerana*/*dorsata*/*florea*/*mellifera*/India
Rawat BS (1982) Bee farming in India. Rawat Apiaries, Almora, [in Hindi]. Keywords: beekeeping/*cerana*/*dorsata*/*florea*/*mellifera*/India
Ray S, Pant NC, Mehrotra KN (1968) Phosphohexose isomerase (D-glucose-6-phosphate ketol-isomerase) from the Indian honeybee, *Apis indica*. Indian J Entomol 30:37–44, Keywords: biochemistry/*cerana*/flight/physiology/India
Rayment T (1935) A cluster of bees. The Endeavour, Sydney, Keywords: beekeeping/*florea*
Razmadze IL, Revenok VA (1980) Beekeeping in Vietnam. Pchelovodstvo 9:28–29, [in Russian]. Keywords: beekeeping/*cerana*/Vietnam
Reddi CS, Reddi PS (1982) The pollination biology of *Ipomoea kentrokaulos* (Convolvulaceae). Indian Bot Rep 1:1–4, Keywords: bee botany/biology/*cerana*/*florea*/pollination/India
Reddi CS, Raddi CEUB, Reddi NS (1981) Breeding structure and pollination ecology of *Tribulus terrestris*. Proc Indian Natl Sci Acad B 47:185–193, Keywords: bee botany/breeding/*cerana*/ecology/*florea*/pollination/India
Reddi CS, Raddi CEUB, Reddi NS (1983) Floral reward and honeybee visitation rates in the soapnut tree. In: Proceedings of the 2nd international conference on apiculture in tropical climates, New Delhi, pp 600–608, Keywords: bee botany/*cerana*/*florea*/foraging/India
Reddi CS, Raddi CEUB, Reddi NS, Reddi PS (1983b) Reproductive ecology of *Sapindus emarginatus* Vahl (Sapindaceae). Proc Indian Natl Sci Acad B 49:57–72, Keywords: bee botany/biogeography/*cerana*/ecology/*florea*/India
Reddi CS, Raju AJS, Atluri JB (1997) Bee pollination in tamarind trees. Indian Bee J 59:178–179, Keywords: *cerana*/pollination/India
Reddy C (1979a) Effect of insecticides on the activities of digestive amylase and protease of honey bee *Apis cerana indica*. Indian Bee J 41:9–11, Keywords: biochemistry/*cerana*/honey/insecticides/India
Reddy C (1979b) Inhibition of magnesium activated adenose triphosphatase as the criterion to determine the degree of organochlorine insecticide poisoning of honey bees, *Apis cerana indica*. Indian Bee J 41:19–23, Keywords: *cerana*/honey/insecticides/India
Reddy C, Reddy MS (1987) The influence of pollen and nectar on the rate of foraging activity of *Apis cerana indica* F. In: Proceedings of the 31st international apiculture congress, Apimondia, Warsaw, p 184, Keywords: *cerana*/foraging/nectar/pollen/India
Reddy CC (1977) Biochemical and physiological studies on the effects of pesticides on honeybee *Apis cerana indica*. Thesis, Bangalore University. Keywords: *cerana*/insecticides/India
Reddy CC (1979) The influence of various meteorological factors on the foraging behavior of *Apis cerana indica*. In: Proceedings of the 4th international symposium on pollination, Maryland, pp 341–345, Keywords: *cerana*/foraging/physical factors/pollination/India
Reddy CC (1980) Observations on the annual cycle of foraging and brood rearing by *Apis cerana indica* colonies. J Apic Res 19:17–20, Keywords: breeding/brood/*cerana*/colony phenology/foraging/India
Reddy CC (1983a) Effects of pesticides on the level of ions and amino acids in the haemolymph of worker *Apis cerana indica* Fab. Indian Bee J 45:105–107, Keywords: *cerana*/insecticides/India
Reddy CC (1983) Studies on the nesting behaviour of *Apis dorsata* F. In: Proceedings of the 2nd international conference on apiculture in tropical climates, New Delhi, pp 391–397, Keywords: absconding/behaviour/combs/*dorsata*/environmental effects/general biology/migration/nests/swarming/India
Reddy CC (1983b) Foraging index of Indian honeybee *Apis cerana indica* F. In: Proceedings of the 2nd international conference on apiculture in tropical climates, New Delhi, pp 595–599, Keywords: bee botany/*cerana*/foraging/physical factors/India
Reddy CC (1995c) Management practices and migratory beekeeping. In: Kevan PG (ed) The Asiatic hive bee: apiculture, biology, and role in sustainable development in tropical and subtropical Asia. Enviroquest, Cambridge, ON, pp 209–215, Keywords: beekeeping/biology/*cerana*/physiology/nests/India

Reddy CC, Reddy MS (1986) Effects of insecticides on the rate of oxygen consumption in *Apis cerana indica* Fabr. workers. Indian Bee J 48:48–50, Keywords: *cerana*/insecticides/physiology/workers/India

Reddy CC, Reddy S, Ramesh SP (1986) Studies on the colony population of *Apis dorsata* Fabr. Indian Bee J 48:57–58, Keywords: biogeography/combs/*dorsata*/general biology/honey

Reddy EV, Reddy C (2006) Oral and dermal toxicity of some insecticides to Indian honey bee, *Apis cerana* F. J Entomol Res 30:47–49, Keywords: *cerana*/honey/insecticides/physiology

Reddy MLK (1997) Relative toxicity of some insecticides to honey bee (*Apis cerana indica* Fab.). J Res ANGRAU 25:26–28, Keywords: *cerana*/honey/insecticides/India

Reddy MS (1988) Studies on the factors of selection of nesting site by *Apis dorsata* F. Thesis, Bangalore University, Keywords: *dorsata*/combs/genetics/nests/India

Reddy MS (1999) Revival of beekeeping in Karnataka. Beekeep Dev 52:14–15, Keywords: beekeeping/*cerana*/*dorsata*/*florea*/honey/India

Reddy MS, Reddy CC (1987) Effects of insecticides on the rate of oxygen consumption in *Apis cerana indica* F workers. In: Proceedings of 31st international apiculture congress, Apimondia, Warsaw, p 189, Keywords: *cerana*/insecticides/physiology/workers/India

Reddy MS, Reddy CC (1989) Height-dependent nest site selection in *Apis dorsata* Fabr. Indian Bee J 51:105–106, Keywords: combs/*dorsata*/genetics/nests/India

Reddy PVR (2001) Floral calendar for European bee, *Apis mellifera* L in Kodagu, Karnataka. Insect Environ 6:166–167, Keywords: bee botany/*cerana*/*mellifera*/India

Reddy PVR, Jacob TK (1998) Foraging behaviour of honey bees in guava. Insect Environ 4:96, Keywords: behaviour/*cerana*/*dorsata*/*florea*/foraging/honey/*mellifera*/pollination/India

Reddy PVR, Jacob TK (2001) Comparative foraging behaviour of Indian honey bee, *Apis cerana* F and European bee. Apis mellifera L. Insect Environ 6:165, Keywords: behaviour/*cerana*/foraging/honey/*mellifera*

Rehman KA, Singh S (1940) Italian versus Indian bee. Indian Bee J 2:1–2, Keywords: *cerana*/*mellifera*/India

Ribbands CR (1953) The behaviour and social life of honeybees. Bee Research Association, London, Keywords: bee botany/behaviour/*cerana*/*dorsata*/*mellifera*

Richards AJ (2001) Does low biodiversity resulting from modern agricultural practice affect crop pollination and yield? Ann Bot 88:165–172, Keywords: *cerana*/pollination

Rieko U (2000) Traditional beekeeping of the Japanese honey bee in Tono city. Iwate prefecture. A case study of traditional beekeeping in the Tohuku region. Honeybee Sci 21:97–106, [in Japanese]. Keywords: beekeeping/*cerana*/Japan

Rietschel P (1937) Morphology and function of the sting in colony building bees and wasps. Zoomorphology 33:313–357, [in German]. Keywords: defense/*dorsata*/*laboriosa*/morphology

Rinderer TE (1986) Bee genetics and breeding. Academic, New York, Keywords: breeding/*cerana*/*dorsata*/*florea*/genetics/*mellifera*

Rinderer TE, Koeniger N, Tingek S, Mardan M, Koeniger G (1989) A morphological comparison of the cavity dwelling honeybees of Borneo *Apis koschevnikovi* (Buttel-Reepen, 1906) and *Apis cerana* (Fabricius, 1793). Apidologie 20:405–411, Keywords: *cerana*/classification/*koschevnikovi*/morphometrics/Borneo

Rinderer TE, Oldroyd BP, Sylvester HA, Wongsiri S, de Guzman LI (1992) Evolution of bee dances. Nature 360:305, Keywords: *andreniformis*/behaviour/*cerana*/dance language/*dorsata*/drones/*florea*/*mellifera*/phylogeny/reproduction

Rinderer TE, Oldroyd BP, Wongsiri S, Sylvester HA, de Guzman LI, Potichot S, Sheppard WS, Buchmann SL (1993) Time of drone flight in four honey bee species in south-eastern Thailand. J Apic Res 32:27–33, Keywords: *andreniformis*/biogeography/*cerana*/*dorsata*/drones/flight/*florea*/honey/reproduction/Thailand

Rinderer TE, Oldroyd BP, Lekprayoon C, Wongsiri S, Boonthai C, Thapa R (1994) Extended survival of the parasitic honey bee mite *Tropilaelaps clareae* on adult workers of *Apis mellifera* and *Apis dorsata*. J Apic Res 33:171–174, Keywords: disease/*dorsata*/honey/pests/workers/Thailand

Rinderer TE, Oldroyd BP, Wongsiri S, Sylvester HA, de Guzman LI, Stelzer JA, Riggio RM (1995) A morphological comparison of the dwarf honey bees of southeastern Thailand and Palawan, Philippines. Apidologie 26:387–394, Keywords: *andreniformis*/*florea*/morphometrics/Philippines/Thailand

Rinderer TE, Marx BD, Gries M, Tingek S (1996a) A scientific note on stratified foraging by Sabahan bees on the yellow flame tree (*Peltophorum pterocarpum*). Apidologie 27:423–425, Keywords: *andreniformis*/*cerana*/*dorsata*/foraging/interspecific interactions/*koschevnikovi*/Malaysia

Rinderer TE, Wongsiri S, Kuang BY, Liu JS, Oldroyd BP, Sylvester HA, de Guzman LI (1996b) Comparative nest architecture of the dwarf honey bees. J Apic Res 35:19–26, Keywords: *andreniformis*/combs/*florea*/honey/China/Thailand

Rinderer TE, Oldroyd BP, de Guzman LI, Wattanachaiyingcharon W, Wongsiri S (2002) Spatial distribution of the dwarf honey bees in an agroecosystem in southeastern Thailand. Apidologie 33:539–543, Keywords: *andreniformis*/ecology/*florea*/Thailand

Riply SD (1950) Birds from Nepal, 1947–1949. J Bombay Nat Hist Soc 49:417, Keywords: combs/*laboriosa*/natural enemies/predators/Nepal

Rishni ND (1983) On the occurrence of some diseases and predators of *Apis cerana indica* F in Kashmir. In: Proceedings of the 2nd international conference on apiculture in tropical climates, New Delhi, pp 466–472, Keywords: *cerana*/disease/natural enemies/physical factors/predators/India

Rising D, Babcock D, Blasdale B, Abraham H, Ostwald W, Sherman HC (1906) The viscosity of soap solutions as characteristic of the purity of fats. Anal Bioanal Chem 45:654–657, [in German]. Keywords: biochemistry/organic chemistry

Ritter WT, Sakai T, Takeuchi K (1980) Development and control of varroatosis in Japan. Diagnosis and therapy for varroatosis. In: Proceedings of the 29th international apiculture congress, Apimondia, Budapest, pp 69–71 [in German]. Keywords: *cerana*/pests/physiology/Japan

Roberts OD, Islip HT (1922) The constants of Indian beeswax. Analysis 47:246–251, Keywords: beeswax/*cerana*/*dorsata*/*florea*

Robinson WS (1988) Beekeeping with *Apis cerana* in Sri Lanka. Bee World 69:125–130, Keywords: beekeeping/*cerana*/Sri Lanka

Roepke W (1923) A queen of *Apis dorsata* F. Entomol Berich 6:156–159, [in German]. Keywords: *dorsata*/queens

Roepke W (1930) Observations on Indian honeybees. Meded. Landbouwhogesch Wageningen 34:1–28, Keywords: absconding/biology/*cerana*/*dorsata*/*florea*/Indonesia

Roonwal ML (1956) Behaviour of the rock bees, *Apis dorsata* Fabr., during a partial solar eclipse in India. Proc Nat Inst Sci India 22:281–286, Keywords: behaviour/*dorsata*/foraging/India

Rosenkranz P, Engels W (1994) Genetic and environmental influences on the duration of preimaginal worker development in eastern (*Apis cerana*) and western (*Apis mellifera*) honey bees in relation to varroatosis. Rev Bras Genet 17:383–391, Keywords: brood/*cerana*/genetics/honey/*mellifera*/pests/physiology/India

Rosenkranz P, Tewarson NC (1992) Experimental infection of *Apis cerana indica* worker brood with *Varroa* females. Apidologie 23:365–367, Keywords: brood/*cerana*/hormones/*mellifera*/pests/India

Rosenkranz P, Tewarson NC, Singh A, Engels W (1993a) Differential hygienic behaviour towards *Varroa jacobsoni* in capped worker brood of *Apis cerana* depends on alien scent adhering to the mites. J Apic Res 32:89–93, Keywords: behaviour/brood/*cerana*/pests/pheromones/India

Rosenkranz P, Tewarson NC, Rachinsky A, Strambi A, Strambi C (1993b) Juvenile hormone titer and reproduction of *Varroa jacobsoni* in capped brood stages of *Apis cerana indica* in comparison to *Apis mellifera ligustica*. Apidologie 24:375–382, Keywords: brood/*cerana*/hormones/*mellifera*/pests/reproduction/India

Rothenbuhler WC, Kulincevic JM, Kerr WE (1958) Bee genetics. Ann Rev Entomol 3:413–438, Keywords: *dorsata*/*florea*/genetics/*mellifera*/queens

Roubik DW (1989) Ecology and natural history of tropical bees. Cambridge University Press, Cambridge, Keywords: *cerana*/*dorsata*/ecology/*florea*/history/*mellifera*

Roubik DW (1990) Nesting and defense in tropical honeybees and stingless bees. In: Proceedings of the 9th international congress social insects and the environment, IUSSI, Bangalore, pp 662–663, Keywords: combs/defense/nests

Roubik DW (1995) Pollination of cultivated plants in the tropics. FAO, Rome, Keywords: bee botany/*cerana*/*dorsata*/*florea*/pollination

Roubik DW (1996) Wild bees of Brunei Darassalam. In: Edwards DS, Berth WE, Choy SC (eds) Tropical rain forest research – current issues. Kluwer, Dordrecht, pp 59–66, Keywords: *cerana*/*dorsata*/Brunei/Malaysia

Roubik DW, Sakagami SF, Kudo I (1985) A note on distribution and nesting of the Himalayan honeybee *Apis laboriosa* Smith (Hymenoptera: Apidae). J Kans Entomol Soc 58:746–749, Keywords: biogeography/combs/*dorsata*/environmental effects/honey/*laboriosa*/nests/Nepal

Roubik DW, Inoue T, Hamid AA (1995) Canopy foraging by two tropical honeybees: bee height fidelity and tree genetic neighborhoods. Tropics (Kuching) 5:81–93, Keywords: *dorsata*/foraging/genetics/Malaysia

Rowley FA (1976) The sugars of some common Philippines nectars. J Apic Res 15:19–22, Keywords: *cerana*/nectar/Philippines

Roy P (1999) Bee diversity across a tropical tract: honey bee and people in India. Beekeep Dev 52:3–4, Keywords: beekeeping/*cerana*/*dorsata*/*florea*/honey/*laboriosa*/*mellifera*/India

Rozalski RJ, Sakurai H, Tsuchida K (1996) Esterase and malate dehydrogenase isozymes analysis in the population of honeybee, *Apis cerana japonica* and *Apis mellifera*. Jpn J Entomol 64:910–917, Keywords: biochemistry/*cerana*/*mellifera*/population biology/Japan

Rozalski RJ, Tsuchida K, Sakurai H (1997) A preliminary report of PCR amplifications of *Apis cerana japonica* genomic DNA. Appl Entomol Zool 32:259–261, Keywords: *cerana*/DNA/Japan

Rubin Reyes M, Cervancia CR (1999) Floral visitors and pollination of Chinese mustard, *Brassica campestris* L. Philipp J Sci 128:31–37, Keywords: *cerana*/pollination/Philippines

Ruttner F (1969) The cause of the hybridization barrier between *Apis mellifera* L and *Apis cerana* Fabr. (= syn *A. indica* Fabr.) 1. Experiments with natural and artificial insemination. In: Proceedings of the 22nd international apiculture congress, Apimondia, Munich, p 170, Keywords: beekeeping/breeding/*cerana*/drones/genetics/interspecific interactions/*mellifera*/queens/China

Ruttner F (1971) Morphological and biological characterization of economically interesting bee species and races. In: Proceedings of the 23rd international apiculture congress, Apimondia, Moscow, pp 383–386, Keywords: biology/*cerana*/*mellifera*/morphology/subspecies

Ruttner F (1972) Races and strains of honeybees of special scientific interest *Apis cerana* and *Apis mellifera*. Allg Dtsch Imkerztg 6:61,63, 65, [in German]. Keywords: *cerana*/*mellifera*/subspecies

Ruttner F (1973) Drones of *Apis cerana* at a congregation area. Apidologie 4:41–44, [in German]. Keywords: *cerana*/drones/*mellifera*

Ruttner F (1975) A metatarsal clasping structure in the *Apis* drone. Entomol Germ 2:22–29, Keywords: *cerana*/*dorsata*/drones/*florea*/*mellifera*/morphology/reproduction

Ruttner F (1985) Characteristics and geographic variability of *Apis cerana* (Fabr.). In: Proceedings of the 30th international apiculture congress, Apimondia, Nagoya, pp 130–133, Keywords: *cerana*/classification/geographical variation/mountains/Japan

Ruttner F (1986) Geographical variability and classification. In: Rinderer TE (ed) Bee genetics and breeding. Academic, London, pp 23–56, Keywords: breeding/*cerana*/classification/*dorsata*/*florea*/genetics/geographical variation/*mellifera*

Ruttner F (1987a) Taxonomy of honeybees. In: Eder J, Rembold H (eds) Chemistry and biology of social insects. Peperny, München, pp 59–62, Keywords: biology/*cerana*/classification/*dorsata*/*florea*/*mellifera*

Ruttner F (1987b) The evolution of honeybees. In: Menzel R, Mercer A (eds) Neurobiology and behavior of honeybees. Springer, Berlin, pp 8–20, Keywords: *cerana*/drones/*florea*/*mellifera*/ phylogeny

Ruttner F (1988) Biogeography and taxonomy of honeybees. Springer, Berlin, Keywords: biogeography/*cerana*/classification/*dorsata*/*florea*/*mellifera*/subspecies

Ruttner F (1992) Natural history of honey bees. Ehrenwirth, Munich, [in German]. Keywords: biology/*cerana*/classification/*dorsata*/*florea*/history/honey/*mellifera*/subspecies

Ruttner F, Kaissling KE (1968) Interspecific effect of the sex attractant of *Apis mellifica* and *A. cerana*. Z Vergl Physiol 59:362–370, [in German]. Keywords: *cerana*/drones/interspecific interactions/*mellifera*/pheromones

Ruttner F, Maul V (1969) Investigations on mating barriers between *Apis mellifera* and *Apis cerana*. In: Proceedings of the 22nd international apiculture congress, Apimondia, Munich, pp 515–516 [in German]. Keywords: *cerana*/*mellifera*/reproduction

Ruttner F, Maul V (1983) Experimental analysis of reproductive interspecies isolation of *Apis mellifera* L and *Apis cerana* Fabr. Apidologie 14:309–327, Keywords: breeding/*cerana*/development/drones/interspecific interactions/*mellifera*/physiology/queens/reproduction

Ruttner F, Woyke J, Koeniger N (1972) Reproduction in *Apis cerana* 1. Mating behaviour. J Apic Res 11:141–146, Keywords: behaviour/*cerana*/reproduction

Ruttner F, Pourashgar D, Kauhausen D (1985) Honeybees of Iran 1. Apis florea Fabricius. Apidologie 16:119–137, [in German]. Keywords: classification/Iran

Ruttner F, Wilson EC, Snelling R, Vorwohl G, Kauhausen D (1986) The evolution of honeybee wing venation. Apidologie 17:348–350, Keywords: *cerana*/*dorsata*/*florea*/*mellifera*/morphology/phylogeny

Ruttner F, Mossadegh MS, Kauhausen-Keller D (1995) Distribution and variation of size of *Apis florea* F in Iran. Apidologie 26:477–486, Keywords: classification/ecology/*florea*/morphology/ Iran

Sabato BC (1997) Utilization of rock bee resource in Orissa for honey production. Indian Bee J 59:47–48, Keywords: *dorsata*/honey/India

Sadakathulla S (1991) Evaluation of honey bee visits on coconut genotypes. Cocos 9:47–50, Keywords: *cerana*/*florea*/genetics/honey/pollination/India

Sagar P (1981) Role of insects in cross pollination of fennel crop at Ludhiana. J Res Punjab Agric Univ 18:388–392, Keywords: bee botany/*dorsata*/*florea*/*mellifera*/India

Sahay S (1980) Pollen and nectar feeding by foragers on *Duranta plumieri* Jacq. In: Proceedings of the 5th international palynological conference, Cambridge, p 351, Keywords: *cerana*/*dorsata*/ *florea*/foraging/nectar/pollen

Sahu RC, Thapa RB, Mandal SM, Shrestha KK, Sapkota M (2007) Effect of endosulfan on the mortality of forager honeybees (*Apis mellifera* L. and *Apis cerana* F.). IAAS Res Adv 1:147–149, Keywords: *cerana*/insecticide/*mellifera*/Nepal

Saini MS, Aggarwal R, Dhillon SS (1983) Functional morphology of the wing coupling apparatus and its bearing on the taxonomy of genus *Apis* Linnaeus. Uttar Pradesh J Zool 3:10–14, Keywords: biogeography/*cerana*/classification/*dorsata*/flight/*florea*/*mellifera*/morphology

Saitanu K, Tsai C, Kondo F (1993) Natural bacterial inhibitor in honey: detection of sulfadiazine-like substance. In: Connor LJ, Rinderer TE, Sylvester HA, Wongsiri S (eds) Asian apiculture. Wicwas, Cheshire, CT, pp 348–357, Keywords: *cerana*/honey/*mellifera*/Thailand

Saji O (1995) Ethnoecology of bee and man – traditional beekeeping in the southern hilly region of the Aizu Basin. Honeybee Sci 16:69–76, [in Japanese]. Keywords: beekeeping/*cerana*/Japan

Saji O (1996) Ethnoecology of bee and man II. Traditional beekeeping in Ishizuchi rim region on Shikoku Island. Honeybee Sci 17:49–60, [in Japanese]. Keywords: beekeeping/*cerana*/islands/ swarming/Japan

Sajwani A, Farooq SA, Patzelt A, Eltayeb EA, Bryant VM (2007) Melissopalynological studies from Oman. Palynology 31:63–79, Keywords: florea/mellifera/pollen

Sakagami SF (1953) Division of labour in a small colony made up of workers all the same age. J Fac Sci Hokkaido Univ VI 11:343–400, [in German]. Keywords: behaviour/*cerana*/workers/Japan

Sakagami SF (1954) Occurrence of an aggressive behaviour in queenless hives, with considerations on the social organisation of honeybee. Insectes Soc 1:331–343, Keywords: aggressiveness/behaviour/*cerana*/*mellifera*/queenlessness/reproduction

Sakagami SF (1958) The false-queen: fourth adjustive response in dequeened honeybee colonies. Behaviour 13:280–296, Keywords: behaviour/biology/*cerana*/flight/laying workers/oviposition/queenlessness/Japan

Sakagami SF (1959a) Regularities in the pattern of comb utilization in the honeybees, illustrated by the life history of a Japanese honeybee colony forced to live in a Langstroth hive (Studies on the Japanese honeybee, *Apis cerana cerana* Fabr. VII). Insectes Ecol 7:117–128, [in Japanese]. Keywords: biology/*cerana*/colony phenology/combs/history/nests/Japan

Sakagami SF (1959b) Some interspecific relations between Japanese and European honeybees. J Anim Ecol 28:51–68, Keywords: biology/*cerana*/interspecific interactions/*mellifera*/Japan

Sakagami SF (1960) Preliminary report on the specific difference of behaviour and other ecological characters between European and Japanese honeybees. Acta Hymenopt (Tokyo) 1:171–198, Keywords: absconding/aggressiveness/beeswax/behaviour/biology/cells/*cerana*/communication/dance language/*dorsata*/*florea*/laying workers/*mellifera*/migration/nectar/pheromones/physical factors/physiology/queens/subspecies/China/India/Indonesia/Japan/Sri Lanka

Sakagami SF (1992) Evolution of brood rearing habits in bees. Honeybee Sci 13:9–18, [in Japanese]. Keywords: breeding/brood/*cerana*/evolution/phylogeny/Japan

Sakagami SF, Akahira Y (1958) Comparison of ovarian size and number of ovarioles between the workers of Japanese and European honeybees (Studies on the Japanese honeybee, *Apis indica cerana* Fabricius I). Kontyû 26:103–109, Keywords: *cerana*/laying workers/morphology/queenlessness/reproduction/workers/Japan

Sakagami SF, Akahira Y (1960) Studies on the Japanese honeybee. Apis cerana cerana Fabricius VIII. Two opposing adaptations in the post stinging behavior of honeybees. Evolution 14:29–40, Keywords: behaviour/*cerana*/*mellifera*/morphology/physiology/stings

Sakagami SF, Fukuda H (1968) Life tables for worker honeybees. Res Pop Ecol Kyoto Univ 10:127–139, Keywords: *cerana*/colony phenology/population biology/workers

Sakagami SF, Kouta S (1958) An attempt to rear the Japanese bee in a framed hive (Studies on the Japanese honeybee, *Apis cerana cerana* Fabr. IV). J Fac Sci Hokkaido Univ VI 14:1–8, Keywords: beekeeping/*cerana*/nests/Japan

Sakagami SF, Yoshikawa K (1973) Additional observations on the nest of the dwarf honeybee, *Apis florea* (Hymenoptera, Apidae). Kontyû 41:217–219, Keywords: combs/*florea*/Thailand

Sakagami SF, Matsumura T, Ito K (1980) *Apis laboriosa* in the Himalaya, the little known world largest honeybee (Hymenoptera, Apidae). Insecta Matsumurana 19:47–77, Keywords: combs/*laboriosa*/Himalayas

Sakai S, Hoshiba H, Hoshiba E (1986) Integrated taxonomic information on honeybees. In: Proceedings of the 30th international apiculture congress, Apimondia, Nagoya, pp 134–139, Keywords: *cerana*/classification/Japan

Sakai S, Momose K, Yumoto T, Nagamitsu T, Nagamasu H, Hamid AA, Nakashizuka T (1999) Plant reproductive phenology over four years including an episode of general flowering in lowland mixed dipterocarp forests of South-East Asia. Biol J Linn Soc 75:233–248, Keywords: bee botany/*dorsata*

Sakai T (1956) Morphological studies on the difference among some strains of the honeybee. Jpn Bee J 9:115–125, [in Japanese]. Keywords: *cerana*/classification/*mellifera*/morphology/Japan

Sakai T (1958) Morphological studies on the drone honeybee. Jpn Bee J 11:40–46, [in Japanese]. Keywords: *cerana*/classification/drones/*mellifera*/morphology/Japan

Sakai T (1989) *Apis cerana* apiculture in Japan. In: Proceedings of the center for integrated mountain development international conference, ICIMOD, Kathmandu, p 24, Keywords: beekeeping/*cerana*/physiology/Japan

Sakai T (1989b) Defensive behaviour of honeybees against *Vespa simillima xanthoptera* attacking *Apis cerana* and *A. mellifera* colonies. Honeybee Sci 10:73–78, [in Japanese]. Keywords: aggressiveness/behaviour/*cerana*/*mellifera*/natural enemies/Japan

Sakai T (1992) *Apis cerana* beekeeping in Japan. In: Verma LR (ed) Honeybees in mountain agriculture. Oxford & IBH, New Delhi, pp 235–238, Keywords: absconding/beekeeping/behaviour/*cerana*/interspecific interactions/*mellifera*/seasonal variation/Japan

Sakai T (1993) Outlining the research project on the bionomic studies of the Japanese honeybee, *Apis cerana* Radoszkowski. Honeybee Sci 14:1–2, [in Japanese]. Keywords: biology/*cerana*/Japan

Sakai T, Hoshiba H (2001) How to keep honeybees. Collect Breed 39:262–264, [in Japanese]. Keywords: beekeeping/*cerana*/Japan

Sakai T, Matsuka M (1982) Beekeeping and honey resources in Japan. Bee World 63:63–71, Keywords: beekeeping/*cerana*/honey/Japan

Sakai T, Ono M (1990) Comparative studies on the bionomics of *Apis mellifera* L and *A. cerana japonica* Radoszkowski I. Differences in the seasonal foraging activity. Bull Fac Agric Tamagawa Univ 30:73–86, [in Japanese]. Keywords: biology/*cerana*/foraging/interspecific interactions/*mellifera*/Japan

Sakai T, Ono M (1991) Comparative studies on bionomics of *Apis mellifera* L and *A. cerana japonica* Radoszkowski (Hymenoptera: Apidae) II. Pollen gathering and foraging preferences. Bull Fac Agric Tamagawa Univ 31:169–178, [in Japanese]. Keywords: biology/*cerana*/foraging/*mellifera*/pollen/Japan

Sakai T, Sasaki M (1989) *Tarsonemus indoapis* (Acarina: Tarsonemidae) from the Japanese honeybee, *Apis cerana japonica*. Honeybee Sci 10:37–38, [in Japanese]. Keywords: *cerana*/pests/Japan

Sakai T, Sasaki M, Tanaka K (1983) Utilization of pollen loads from honeybees for artificial pollination of fruit trees, with special reference to the washing method used and long-term preservation. Honeybee Sci 4:81–82, [in Japanese]. Keywords: *cerana*/pollen/pollination/Japan

Sakai T, Ono M, Kobayashi S, Sasaki M (1993) Comparative studies on the bionomics of *Apis mellifera* L and *A. cerana japonica* Radoszkowski (Hymenoptera: Apidae) at the same apiary. Honeybee Sci 14:13–22, [in Japanese]. Keywords: biology/*cerana*/foraging/*mellifera*/Japan

Salmah S, Inoue T, Sakagami SF (1990) An analysis of apid bee richness (Apidae) in central Sumatra. In: Sakagami SF, Ohgushi R, Roubik DW (eds) Natural history of social wasps and bees in equatorial Sumatra. Hokkaido University Press, Sapporo, pp 139–174, Keywords: *andreniformis*/*cerana*/*dorsata*/history/nests/Sumatra

Salvi SR (1975) Seasonal management of the Indian hive bees at Pathankot, Punjab: a preliminary report. Indian Bee J 37:10–14, Keywords: beekeeping/*cerana*/nests/India

Saman-Wanakij L (1953) Beekeeping. L Saman-Wanakij, Bangkok, [in Thai]. Keywords: beekeeping/*cerana*/*dorsata*/*florea*/*mellifera*/Thailand

Sammataro D, Gerson U, Needham GR (2000) Parasitic mites of honey bees: life history, implications, and impact. Ann Rev Entomol 45:519–548, Keywords: *andreniformis*/*cerana*/*dorsata*/*florea*/history/honey/*koschevnikovi*/*laboriosa*/*mellifera*/pests

Sandhu AS, Singh S (1960) The biology and brood rearing activities of the little honeybee (*Apis florea* Fabricius). Indian Bee J 22:27–35, Keywords: beekeeping/brood/classification/*florea*/general biology/India

Sandhu DK, Waraich MK (1985) Yeasts associated with pollinating bees and flower nectar. Microb Ecol 11:51–58, Keywords: *cerana*/*dorsata*/*florea*/*mellifera*/nectar/pests/India

Sanford MT (2001) Introduction, spread and economic impact of *Varroa* mites in North America. In: Webster TC, Delaplane KS (eds) Mites of the honey bee. Dadant, Hamilton, pp 149–162, Keywords: *cerana*/honey/pests

Sannasi A, Rajulu G, Sundara G (1971) 9-Oxo-trans-2-decenoic acid in the Indian honeybees. Life Sci 10:195–201, Keywords: *cerana*/*dorsata*/*florea*/*mellifera*/pheromones

Santharam G, Basha AA, Balasubramanian M (1982) Lack of susceptibility of honey bees to three insect viruses. Indian Bee J 44:35, Keywords: *cerana*/disease/honey/India

Saraf SK (1985a) Beekeeping in Gurez, Jammu and Kashmir, India. Indian Bee J 47:36–37, Keywords: beekeeping/*cerana*/swarming/India

Saraf SK (1985b) Buckwheat (*Fagopyrum sagittatum*), a major nectar source in Gurez (India). Indian Bee J 47:48, Keywords: beekeeping/*cerana*/nectar/India

Saraf SK, Wali JL (1972) Preliminary studies on the egg laying capacity of the queen bee of *Apis indica* F (hill strain) in Kashmir. Indian Bee J 34:27–31, Keywords: *cerana*/oviposition/queens/India

Sarunggalo ZL, Soekarto ST, Budijanto S (2002) Study on decreasing the melting point of beeswax (*Apis cerana*) in the production of low-calorie margarine. J Food Tech Ind 13, Keywords: beeswax/*cerana**

Sarwan BS, Sohi BS (1985) Phytosociological studies on *Apis mellifera* L and *Apis cerana* F in Punjab, India. Indian Bee J 47:15–18, Keywords: *cerana*/interspecific interactions/*mellifera*/India

Sasagawa H, Kuwahara Y (1990) Comparative study of hydrocarbon profiles on the Japanese and European honeybee. In: Proceedings of the 11th international congress, IUSSI, Bangalore, pp 421–422, Keywords: *cerana*/chemistry/*mellifera*/India/Japan

Sasagawa H, Matsuyama S, Yamaoka R, Sasaki M (1996) An Oriental orchid, Kinnryou-hen (*Cymbidium floribundum* Rolfe: Cf) mimics Nasonov pheromone of the Japanese honeybee (*Apis cerana japonica* Rad). In: Proceedings of the 3rd Asian apiculture association international conference, Hanoi, pp 40–41, Keywords: bee botany/*cerana*/glands/*mellifera*/pheromones/Japan

Sasagawa H, Matsuyama S, Leal WS, Peng YS (2000) Hygienic grooming behaviour induced by parasitic *Varroa* mites in *Apis cerana japonica*. In: Proceedings of the 4th Asian apiculture association international conference, Kathmandu, p 70, Keywords: behaviour/*cerana*/pests/Japan

Sasagawa H, Kadowaki T, Matsuyama S, Hirai Y, Peng YS (2002) Semiochemically mediated recognition in honey bee societies. In: Proceedings of the 14th international congress, IUSSI, Sapporo, p 15, Keywords: *cerana*/honey/*mellifera*/pests/pheromones/Japan

Sasaki M (1989) The reason why *Apis cerana japonica* is resistant to the *Varroa* mite. Honeybee Sci 10:28–36, [in Japanese]. Keywords: *cerana*/pests/Japan

Sasaki M (1990) Photoperiodic regulation of honeybee mating-flight time: exploitation of innately phase-fixed circadian oscillation. Adv Invertebr Reprod 5:503–508, Keywords: *cerana*/drones/foraging/flight/queens/reproduction/seasonal variation/Japan

Sasaki M (1991) Absconding dance in *Apis cerana japonica*: a slow and long tail-waggling motivates the whole colony to translocate. In: Veeresh GK, Mallik B, Viraktamath CA (eds) Social insects and the environment. Oxford & IBH, New Delhi, pp 125–126, Keywords: absconding/behaviour/biogeography/brood/*cerana*/combs/dance language/foraging/honey/migration/nests/physiology/queens/Japan

Sasaki M (1992) Deception of the Japanese honey bee, *Apis cerana japonica*, by the Oriental orchid *Cymbidium pumilum*. Honeybee Sci 13:167–172, [in Japanese]. Keywords: bee botany/*cerana*/honey/pheromones/Japan

Sasaki M (1994) Comparative evaluation of southern and northern *Apis cerana* ecotypes and points which should be considered for future breeding. Honeybee Sci 15:99–106, [in Japanese]. Keywords: breeding/*cerana*/geographical variation/China/Japan

Sasaki M (1999a) Diversity of information delivery systems in honey bees. Honeybee Sci 20: 119–126, [in Japanese]. Keywords: behaviour/*cerana*/communication/dance language/genetics/honey/physiology/queens/Japan

Sasaki M (1999b) Ecological characteristics of Asiatic hive bee and its possibility as apicultural species. Fmg Jpn 33:18–22, Keywords: biology/*cerana*/nests/Japan

Sasaki M (2000) Wonders of the Japanese honeybee, biology of northernmost *Apis cerana*. Kaiyusha Publishers Co Ltd, Tokyo, [in Japanese]. Keywords: biology/*cerana*/Japan

Sasaki M (2002) Wonders of the Japanese honeybee, *Apis cerana japonica*. In: Proceedings of the 14th international congress, IUSSI, Sapporo, p 28, Keywords: *cerana*/genetics/natural enemies/Japan

Sasaki M, Okada I (1988) Feral nest of *Apis cerana japonica* constructed in an open space. Honeybee Sci 9:77–78, [in Japanese]. Keywords: *cerana*/nests/Japan

Sasaki M, Irokawa T, Sato M (1989a) Humoral control of the queen pheromone (9-ODA) biosynthesis in honeybees: induction in queenless worker and in worker mandibular gland implanted into queen. Bull Fac Agric Tamagawa Univ 29:11–21, Keywords: *cerana*/glands/hormones/pheromones/queens/queenlessness/Japan

Sasaki M, Ono M, Yoshida T (1989) Recent topics on the biology of the north-adapted eastern honeybee, *Apis cerana japonica*. In: Proceedings of the advanced course in beekeeping *A. cerana* in tropical and sub-tropical Asia, Keywords: beekeeping/biology/*cerana*/Japan*

Sasaki M, Ono M, Asada S, Yoshida T (1991) Oriental orchid (*Cymbidium pumilum*) attracts drones of the Japanese honeybee (*Apis cerana japonica*) as pollinators. Experientia 47:1229–1231, Keywords: *cerana*/drones/pheromones/pollination/Japan

Sasaki M, Ono M, Fukuda M (1993a) Pollination ecology of the Japanese honey bee, *Apis cerana japonica*, and the oriental orchid, *Cymbidium pumilum*. In: Connor LJ, Rinderer TE, Sylvester HA, Wongsiri S (eds) Asian apiculture. Wicwas Press, Cheshire, Connecticut, pp 391–397, Keywords: bee botany/*cerana*/drones/ecology/honey/pheromones/pollination/China/Japan

Sasaki M, Takahashi H, Sato Y (1993b) Comparison of the dance dialect and foraging range between *Apis mellifera* and northernmost subspecies of *A. cerana* in Japan. Honeybee Sci 14:49–54, [in Japanese]. Keywords: *cerana*/dance language/foraging/*mellifera*/Japan/Sri Lanka/Thailand

Sasaki M, Ono M, Yoshida T (1995) Some biological aspects of the north-adapted eastern honeybee, *Apis cerana japonica*. In: Kevan PG (ed) The Asiatic hive bee: apiculture, biology and role in sustainable development in tropical and subtropical Asia. Enviroquest, Cambridge, ON, pp 59–78, Keywords: aggressiveness/behaviour/biology/cells/*cerana*/classification/drones/flight/interspecific interactions/*mellifera*/morphology/nests/pests/pheromones/physiology/queens/reproduction/swarming/workers/Japan

Sasaki M, Ikuta A, Nakamura J (1996) Ventilative fanning and its regulation in Japanese and European honeybees. In: Proceedings of the 3rd Asian apiculture association international conference, Hanoi, p 56, Keywords: behaviour/*cerana*/*mellifera*/physiology/Japan

Sasaki M, Fujiwara M, Numaguchi Y (2000) Proboscis extension reflex (PER) for bee odour and non-volatile surface wax components: analysis of nestmate discrimination. In: Proceedings of the 4th Asian apiculture association international conference, Kathmandu, p 37, Keywords: beeswax/behaviour/*cerana*/morphology/pheromones/Japan

Sato M, Ono M, Shimomaki S (1984) Notes on laying workers in the Japanese honeybee, *Apis cerana japonica*, with special reference to the number of ovarioles. Honeybee Sci 5:1–6, [in Japanese]. Keywords: *cerana*/laying workers/pheromones/queenlessness/reproduction/workers/Japan

Sattigi HN (2001) Nesting behaviour of rock bee, *Apis dorsata* F: preference of nesting structure and nest supports. Karnataka J Agric Sci 14:650–655, Keywords: behaviour/combs/*dorsata*/nests/India

Sattigi HN, Kulkani KA (2001a) Colonization by rock bee in Dharwad: a transition belt of Karnataka. Karnataka J Agric Sci 14:646–649, Keywords: combs/*dorsata*/ecology/India

Sattigi HN, Kulkani KA (2001b) Abundance of rock bee colonies during different months at Dharwad. Karnataka J Agric Sci 14:656, Keywords: combs/*dorsata*/seasonal variation/India

Sattigi HN, Kulkani KA (2001c) Studies on decolonization by rock bee, *Apis dorsata* F under Dharwad conditions. Karnataka J Agric Sci 14:660–663, Keywords: combs/*dorsata*/India

Sattigi HN, Kulkani KA (2001d) Nesting behaviour of rock bee selection of nesting height. Karnataka J Agric Sci 14:664–667, Keywords: behaviour/combs/*dorsata*/genetics/nests/India

Sattigi HN, Lingappa S (1993) Foraging activities of Indian honeybee, *Apis cerana* Fabricius under Dharwad conditions. Karnataka J Agric Sci 6:352–354, Keywords: *cerana*/foraging/India

Sattigi HN, Lingappa S (1994) Performance of Indian hive bee, *Apis cerana* F in Dharwad. Karnataka J Agric Sci 7:360–362, Keywords: beekeeping/*cerana*/nests/India

Sattigi HN, Kulkani KA, Lingappa S (1994) Studies on wax recovery from natural combs of rock bee *Apis dorsata* Fabricius. Karnataka J Agric Sci 7:481–484, Keywords: combs/*dorsata*/India

Sattigi HN, Kulkani KA, Lingappa S (1995a) A survey on the colony distribution of rock bees, *Apis dorsata* F on different colour backgrounds. Karnataka J Agric Sci 8:89–90, Keywords: biogeography/combs/*dorsata*/morphology/India

Sattigi HN, Kulkani KA, Lingappa S (1995b) Wax recovery from deserted natural combs of rock bee. Apis dorsata Fabr. Indian Bee J 57:14, Keywords: combs/*dorsata*/India

Satyanarayana AR, Seetharam A (1982) Studies on the method of hybrid seed production in oilseed sunflower (*Helianthus annuus*) 3. Role and activity of insect visitors in pollination and seed set. Seed Sci Tech 10:13–17, Keywords: *cerana*/*dorsata*/*florea*/pollination/India

Saubolle BR, Bachmann A (1979) Beekeeping: an introduction to modern beekeeping in Nepal. Sahayogi Prakashan, Kathmandu, Keywords: beekeeping/*cerana*/Nepal

Saville NM (2000) Small-scale beeswax processing in remote western Nepal. In: Proceedings of the 4th Asian apiculture association international conference, Kathmandu, pp 155–158, Keywords: beekeeping/beeswax/*cerana*/Nepal

Saville NM (2001) Battling with bee brood disease in *Apis cerana* in W Nepal - some findings. In: Proceedings of the 7th international conference on tropical bees, Chiang Mai, pp 243–248, Keywords: brood/*cerana*/disease/Nepal

Saville NM, Upadhyaya SN (2000) Indigenous knowledge of beekeeping in Jumla, western Nepal. In: Proceedings of the 4th Asian apiculture association international conference, Kathmandu, pp 248–251, Keywords: beekeeping/*cerana*/nests/Nepal

Saville NM, Upadhyaya VR, Pradhan S (2000) Effect of hive design on internal hive temperature: a new application of temperature data loggers. In: Proceedings of the 4th Asian apiculture association international conference, Kathmandu, pp 120–124, Keywords: *cerana*/nests/physiology/Nepal

Sawada M (1986) Patterns of man–honeybee relationships with special reference to the traditional beekeeping of *Apis cerana cerana* in Japan. Kikan-Jinrui-Gaku 17:61–125, [in Japanese]. Keywords: beekeeping/*cerana*/swarming/Japan

Sawaguchi T (1992) Observations on drone congregation areas of the Japanese honey bee. Honeybee Sci 13:38–40, [in Japanese]. Keywords: *cerana*/drones/honey/Japan

Saxena KB, Singh L, Gupta MD (1990) Variation for natural out-crossing in pigeon pea. Euphytica (Hist Arch) 46:143–148, Keywords: bee botany/classification/*dorsata*

Saxena P, Banerjee A (1987) Effect of incubation time on the activity of gut sucrase of *Apis cerana indica* foragers (Hymenoptera: Apoidea). In: Proceedings of the 31st international apiculture congress, Apimondia, Warsaw, pp 187–189, Keywords: biochemistry/*cerana*/physiology/India

Schabacker H (1972) Haemolymph proteins and oogenetic activity in workers of the four *Apis* species. Thesis, Universität Münster [in German]. Keywords: biochemistry/*cerana*/*dorsata*/*florea*/*mellifera*/physiology/workers

Scheuermeier U (1988) Beekeeping in Nepal. Schweiz Bienenztg 111:123–127, [in German]. Keywords: beekeeping/*cerana*/Nepal

Schlüns H (2004) Evolutionary aspects of mating and reproduction in honeybees. Thesis, Martin-Luther-Univ, Halle-Wittenberg, Keywords: *cerana/dorsata/florea/mellifera*/reproduction

Schlüns H, Moritz RFA, Lattorf HMG, Koeniger G (2005) Paternity skew in seven species of honeybees (Hymenoptera: Apidae: *Apis*). Apidologie 36:201–209, Keywords: *cerana/dorsata/mellifera/andreniformis/florea/laboriosa/koschevnikovi*/reproduction

Schmelzer E, Kastberger G (2009) Special agents' trigger social waves in giant honeybees (*Apis dorsata*). Naturwissenschaften 96:1431–1441, Keywords: defense/*dorsata*

Schmidt JO (1982) Biochemistry of insect venoms. Ann Rev Entomol 27:339–368, Keywords: biochemistry/*cerana/dorsata/florea/mellifera*

Schmidt JO (1995) Toxicology of venoms from the honeybee genus *Apis*. Toxicon (Oxford) 33:917–927, Keywords: biochemistry/*cerana/dorsata/florea/mellifera*

Schmidt JO, Schmidt PJ, Starr CK (1985) Investigating the giant honeybee, *Apis dorsata*, in Sabah. Am Bee J 125:749–751, Keywords: disease/*dorsata*/drones/honey/pests/predators/Borneo/Malaysia

Schmidt JO, Morgan ED, Oldham NJ, do Nascimento RR, Dani FR (1997) (Z)-11-Eicosen-1-ol, a major component of Apis cerana venom. J Chem Ecol 23:1929–1939, Keywords: *andreniformis*/biochemistry/*cerana/dorsata/florea/koschevnikovi/mellifera*/pheromones/stings

Schmidt-Hempel P (1998) Parasites in social insects. Princeton University Press, New Jersey, Keywords: *cerana/dorsata/florea/mellifera*/natural enemies/parasites

Schmitz J, Moritz RFA (1998) Sociality and the rate of rDNA sequence evolution in wasps (Vespidae) and honeybees (*Apis*). J Mol Evol 47:606–612, Keywords: biochemistry/*dorsata*/evolution/organic chemistry

Schneider G (1908) On a primitive forest bee (*Apis dorsata*) F. Z Wiss Insektenbiol 12:447–453, [in German]. Keywords: *dorsata*/general biology

Schneider P, Djalal AS (1970) Distribution and rearing of the Eastern honeybee (*Apis cerana*) in Afghanistan. Apidologie 1:329–341, [in German]. Keywords: beekeeping/biogeography/breeding/*cerana*/Afghanistan

Schneider P, Kloft W (1971) Observations on group defensive behaviour in the Eastern honeybee. Z Tierpsychol 29:337–342, [in German]. Keywords: aggressiveness/behaviour/*cerana*/natural enemies/Afghanistan

Sechrist EL (1950) Notes on the queen cells of the giant bee. Bees 4:76–77, Keywords: combs/*dorsata*/queens

Seeley TD (1982) Colony defense strategies of honeybees in Thailand. In: Breed MD, Michener CD, Evans HE (eds) The biology of social insects. Westview Press, Boulder, Colorado, pp 285–287, Keywords: aggressiveness/biology/*cerana/dorsata/florea*/Thailand

Seeley TD (1983) The ecology of temperate and tropical honeybee societies. Am Sci 71:264–272, Keywords: *cerana/dorsata*/ecology/*florea*/Thailand

Seeley TD (1985) Honeybee ecology: a study of adaptation in social life. Princeton University Press, New Jersey, Keywords: biology/*cerana/dorsata*/ecology/*florea/mellifera*

Seeley TD, Heinrich B (1981) Regulation of temperature in the nests of social insects. In: Heinrich B (ed) Insect thermoregulation. Wiley, New York, pp 159–234, Keywords: *cerana/dorsata/florea/mellifera*/nests/physiology

Seeley TD, Seeley RH, Akratanakul P (1982) Colony defense strategies of honeybees in Thailand. Ecol Monogr 52:43–63, Keywords: aggressiveness/*cerana/dorsata/florea*/Thailand

Sekariah PC, Seth RN (1959) Studies on the bacterial flora of the intestinal tract of the honey bee (*Apis indica*) II. On the occurrence of *Bacillus thuringiensis,* Berliner, in the foulbrood of honeybees in India. Indian Vet J 36:19–23, Keywords: *cerana*/disease/honey/India

Sekhar MR, Singh SP, Mehra RB, Govil JN (2001) Effect of barrier crop on yield, out-crossing and pollinator activity in pigeon pea (*Cajanus cajan* (L) Millsp). Indian J Genet Pl Breed 61:125–128, Keywords: bee botany/beekeeping/*dorsata*/genetics/India

Seligmann CG, Seligmann BZ (1911) The Veddas. Cambridge University Press, Cambridge, Keywords: *cerana*/India

SenSarma M, Fuchs S, Tautz J (2000) Debris removal by head-pushing in *A. florea* Fabr. honeybees. Naturwissenschaften 87:241–243, Keywords: behaviour/*florea*/India

SenSarma M, Fuchs S, Werber C, Tautz J (2002) Worker piping triggers hissing for coordinated colony defense in the dwarf honeybee *Apis florea*. Zoology 105:215–223, Keywords: behaviour/defense/*florea*/India

SenSarma M, Esch H, Tautz J (2004) A comparison of the dance language in *Apis mellifera carnica* and *Apis florea* reveals striking similarities. J Comp Physiol A 190:49–53, Keywords: dance language/*florea*/*mellifera*

Senthilnathan S, Murugan N, Kumar NS, Jeyabalan D, Muthuraman S (1999) Pollination potential of honeybee *Apis florea* Linn in relation to biochemical profiles of host plants. Zoos' Print J 14:3–4, Keywords: bee botany/*florea*/India

Sevast'yanov VD (1957) Correlation of the size of certain features of *Apis indica* and *Apis mellifera*. Nauchn Ezheg Odesskogo Univ Biol Fak 1956:247–248, [in Russian]. Keywords: *cerana*/classification/*mellifera*/morphology

Sgorbati B, Scardovi V, Leblanc DJ (1986) Related structures in the plasmid profiles of *Bifidobacterium asteroids,* B. indicum and B. globosum. Microbiologica 9:443–454, Keywords: *cerana*/disease/*dorsata*/pests/predators/Philippines

Shaanker RU, Ganeshaiah KN, Krishnan S, Ramya R, Meera C, Aravind NA (2005) Livelihood gains and ecological costs of non-timber forest product dependence: assessing the roles of dependence, ecological knowledge and market structure in three contrasting human and ecological settings in south India. Environ Conserv 31:242–253, Keywords: bee botany/beekeeping/behaviour/biogeography/combs/disease/*dorsata*/ecology/environmental effects/evolution/general biology/genetics/honey/pests/predators/morphology/India

Shafer ABA, Williams GR, Shutler D, Rogers REL, Stewart DT (2009) Cophylogeny of *Nosema* (Microsporidia: Nosematidae) and bees (Hymenoptera: Apidae) suggests both cospeciation and a host-switch. J Parasitol 95:198–203, Keywords: disease/*cerana*/*mellifera*

Shafi M (1986) Pest and disease records Pakistan, a new mite pest of hive bees. Q Newsl Asia Pacif Pl Prot Com 29:31, Keywords: *cerana*/disease/nests/pests/Pakistan

Shah AM (1980) Beekeeping in Kashmir. In: Proceedings of the 2nd international conference on apiculture in tropical climates, New Delhi, pp 197–204, Keywords: beekeeping/*cerana*/physical factors/India

Shah AM (1988) All Kashmir beekeepers federation (Thai sacbrood). Indian Bee J 50:24, Keywords: *cerana*/disease/India

Shah FA (1974) Wintering of bees in Kashmir. Indian Bee J 36:13–15, Keywords: *cerana*/seasonal variation/India

Shah FA (1975a) Some facts about beekeeping in Kashmir. Bee World 56:103–108, Keywords: beekeeping/*cerana*/India

Shah FA (1975b) Some notes on beekeeping in Kodai Hills, Tamil Nadu. Indian Bee J 37:25–30, Keywords: beekeeping/*cerana*/India

Shah FA (1977) Beekeeping, Shah Beekeepers. Kursu, Rajbagh, Srinagar, Kashmir [in Urdu]. Keywords: beekeeping/cerana/India

Shah FA (1980b) Storing and mailing queen bees in single cage. Indian Bee J 42:18–19, Keywords: beekeeping/*cerana*/queens/India

Shah FA (1983) Fundamentals of beekeeping. Shah Beekeepers, Srinagar, Kashmir, Keywords: beekeeping/*cerana*/India

Shah FA (1984) The origin of beekeeping in Kashmir. Bee World 65:12–18, Keywords: beekeeping/*cerana*/history/India

Shah FA, Shah TA (1976) A note on the bee activity and bee flora of Kashmir. Indian Bee J 38:29–33, Keywords: bee botany/*cerana*/India

Shah FA, Shah TA (1978a) Effect of infestation with *Acarapis woodi* (Rennie) on the development of bee colonies in *Apis cerana indica*. Indian Bee J 40:63–64, Keywords: *cerana*/pests/physiology/India

Shah FA, Shah TA (1978b) Control of acarine in the Kashmir Valley. Indian Bee J 40:70–71, Keywords: *cerana*/pests/India
Shah FA, Shah TA (1979) Use of pollen supplement in *Apis cerana* colonies in Kashmir. Indian Bee J 41:24–25, Keywords: beekeeping/*cerana*/India
Shah FA, Shah TA (1980a) Early life, mating and egg laying of *Apis cerana* queens in Kashmir. Bee World 61:137–140, Keywords: *cerana*/oviposition/queens/reproduction/India
Shah FA, Shah TA (1980b) Flight range of *Apis cerana* from Kashmir. Indian Bee J 42:48, Keywords: bee botany/*cerana*/flight/foraging/India
Shah FA, Shah TA (1981a) *Apis* iridescent virus. Indian Bee J 43:44–45, Keywords: *cerana*/disease/India
Shah FA, Shah TA (1981b) Egg-laying capacity of Kashmir *Apis cerana*. Bee World 62:114–115, Keywords: *cerana*/oviposition/queens/India
Shah FA, Shah TA (1982a) Multiple queen system in *Apis cerana* in Kashmir. Indian Bee J 44:11–12, Keywords: beekeeping/*cerana*/queens/India
Shah FA, Shah TA (1982b) The role of Kashmir bee in exploiting beekeeping potential in India. Indian Bee J 44:37–42, Keywords: beekeeping/*cerana*/India
Shah FA, Shah TA (1986) Search for *Nosema apis* in Kashmir *Apis cerana*. Indian Bee J 48:55–56, Keywords: *cerana*/disease/India
Shah FA, Shah TA (1987) Thai sacbrood disease of *Apis cerana*. Br Bee J 115:237–240, Keywords: *cerana*/disease/India
Shah FA, Shah TA (1988a) Worker bee development period in Kashmir variety of *Apis cerana*. Indian Bee J 50:41–42, Keywords: brood/*cerana*/physiology/India
Shah FA, Shah TA (1988b) Behaviour of Kashmir *Apis cerana* after queen loss in the presence and absence of brood. Indian Bee J 50:71, Keywords: behaviour/*cerana*/India
Shah FA, Shah TA (1989) Comparative foraging behaviour of *Apis cerana* Fabr. and *Apis mellifera* L during *Isodon rugosus* (Wall) Codd bloom in Kashmir Valley. Indian Bee J 51:138–139, Keywords: behaviour/*cerana*/foraging/*mellifera*/India
Shah FA, Shah TA (1991) *Vespa velutina*, a serious pest of honey bees in Kashmir. Bee World 72:161–164, Keywords: *cerana*/honey/natural enemies/India
Shah NK (1986a) Hisalu (*Rubus ellipticus* Linn) a potential source of nectar to hive bees (*Apis cerana indica* Fabr) in northwestern Himalayas. Indian Bee J 48:70 Keywords: bee botany/*cerana*/mountains/nectar/nests/India
Shah NK (1986b) Hisalu (Rubus ellipticus Linn) a potential source of nectar to hive bees (Apis cerana indica Fabr.) in northwestern Himalayas. Indian Bee J 48:70, Keywords: bee botany/*cerana*/pollen/India
Shah TA (1985) Seasonal incidence of *Apis* iridescent virus and internal parasites in *Apis cerana* in Kashmir, India. Indian Bee J 47:43–44, Keywords: *cerana*/disease/parasites/India
Shah TA, Shah FA (1990) Mating interference between *Apis cerana* F and *Apis mellifera* L. Indian Bee J 52:19–20, Keywords: *cerana*/interspecific interactions/*mellifera*/reproduction/India
Shahid M (1992) Beekeeping in the North West Frontier Province of Pakistan. In: Verma LR (ed) Honeybees in mountain agriculture. Oxford & IBH, New Delhi, pp 193–209, Keywords: bee botany/beekeeping/*cerana*/*dorsata*/*florea*/Pakistan
Shankar-Reddy M, Reddy CC, Devaiah PB (1993) Impact of Thai sacbrood disease of honeybees on pollination. In: Proceedings of international symposium on pollination in tropics, Bangalore, pp 244–245, Keywords: *cerana*/disease/*dorsata*/*florea*/pollination/India
Shao RY, Wang LH, Lu Q, Hou GS, Yu WJ, Hu CN, He TF (1998) Preliminary report on the experiment of microinjection of Chinese honeybee egg. J Fujian Agric Univ 27:463–466, Keywords: *cerana*/oviposition/China
Sharma A (1988) Biometry of Asian hive bees *Apis cerana* F of central and western Himalayas. Thesis, Himachal Pradesh University, Keywords: *cerana*/classification/mountains
Sharma DK, Chetri G, Gogoi H, Kastberger G (2000) Amylase and esterase activity as markers in larvae of *Apis dorsata*. In: Proceedings of the 4th Asian apiculture association international conference, Kathmandu, p 50, Keywords: biochemistry/brood/*dorsata*/organic chemistry/India

Sharma GP, Gupta BL, Kumkarni CG (1961) Cytology of spermatogenesis in the honeybee (*Apis indica* F). J R Microscop Soc 79:337–351, Keywords: *cerana*/cytology/drones/histology/India
Sharma HK, Gupta JK (1993) Diversity and density of bee flora of Solan Region of Himachal Pradesh (India). Indian Bee J 55:9–20, Keywords: bee botany/*cerana*/genetics/India
Sharma HK, Gupta JK (1996) Seasonal variation in colour, weight and crude protein content of pollen loads of hive bees. Indian Bee J 58:125–128, Keywords: *cerana*/morphology/nests/physiology/pollen/seasonal variation/India
Sharma HK, Gupta JK (2001) Impact of bee pollination on sustainable production in apple orchards. Pest Manag Econ Zool 9:45–48, Keywords: beekeeping/*cerana*/pollination/India
Sharma HK, Rana BS (2001) Foraging behaviour of insect pollinators on pollination of cherry. In: Proceedings of the 7th international conference on tropical bees, Chiang Mai, pp 281–283, Keywords: behaviour/*cerana*/foraging/*mellifera*/pollination/India
Sharma HK, Gupta JK, Rana BS (2000b) Diurnal activity of *Apis cerana* F and *A. mellifera* L on different flora during spring and honey flow period. Pest Manage Econ Zool 8:151–154, Keywords: beekeeping/*cerana*/foraging/honey/*mellifera*/India
Sharma HK, Gupta JK, Rana BS (2000) Resource partitioning among *Apis mellifera* and *Apis cerana* under mid-hill conditions of Himachal Pradesh. In: Proceedings of the 4th Asian apiculture association international conference, Kathmandu, pp 213–215, Keywords: bee botany/*cerana*/*mellifera*/India
Sharma HK, Verma LR, Gupta JK (2000) Traditional beekeeping with *Apis cerana* in Kullu Valley of Himachal Pradesh. In: Proceedings of the 4th Asian apiculture association international conference, Kathmandu, pp 259–261, Keywords: beekeeping/*cerana*/India
Sharma HK, Gupta JK, Thakur JR (2001) Pollination studies on apple and pear. In: Proceedings of the 7th international conference on tropical bees, Chiang Mai, pp 265–270, Keywords: *cerana*/*mellifera*/pollination/India
Sharma ID, Nath A (1996) Metabolism of 1-naphthyl-N-methyl carbamate (carbaryl) by bacterial isolates from honey bees and the effect of bacterial inoculations on carbaryl tolerance in bees. J Appl Bacteriol 81:235–241, Keywords: *cerana*/disease/honey/insecticides/*mellifera*/physiology/India
Sharma ID, Nath A (2000) Accelerated degradation of carbaryl by acclimated bacterial isolates from *Apis cerana* bees. In: Proceedings of the 4th Asian apiculture association international conference, Kathmandu, pp 92–95, Keywords: *cerana*/insecticides/India
Sharma ID, Nath A (2001) Metabolic transformation of carbaryl by bacteria inhabiting honey bees and their role in inducing tolerance in bees. In: Proceedings of the 7th international conference on tropical bees, Chiang Mai, pp 383–392, Keywords: *cerana*/honey/insecticides/India
Sharma KK, Bhattacharya A, Sushil SN (1999) Indian lac insect, *Kerria lacca*, as an important source of honeydew. Bee World 80:115–118, Keywords: *cerana*/*florea*/honey/India
Sharma N (1983) Melissopalynology and survey of honey plants in Simla Hills (Himachal Pradesh). Thesis, Himachal Pradesh University, Keywords: bee botany/*cerana*/honey/India
Sharma OP, Thakur AK (1982) Beekeeping in Kangra Valley. Indian Fmg 31:33–35, Keywords: beekeeping/*cerana*/India
Sharma OP, Kumar A, Srivastava S (1992) Behavioural reaction of *Apis cerana* (Hymenoptera, Apidae) to the mite, *Tropilaelaps clareae* (Acari, Laelapidae). Acta Entomol Bohemoslov 89:317–321, Keywords: *cerana*/pests/India
Sharma PC (1983) Morphometric studies on *Apis florea* F and *Apis dorsata* F of Himachal Pradesh and Punjab. Thesis, Himachal Pradesh University, Shimla, Keywords: *dorsata*/*florea*/morphometrics/India
Sharma PL (1944) Queen introduction. Indian Bee J 6:152–154, Keywords: *cerana*/queens/India
Sharma PL (1945) Distribution of *Apis indica* F in the plains of northern India. Indian Bee J 1:53–54, Keywords: biogeography/*cerana*/India
Sharma PL (1948) Studies on seasonal activities of *Apis indica* F at Lyallpur. Indian Bee J 10:20–23, Keywords: *cerana*/colony phenology/India

Sharma PL (1951) Bee work done in the Punjab (I) up to 1950. Indian Bee J 13:27–35, Keywords: bibliography/*cerana*/*dorsata*/*florea*/India

Sharma PL (1958) Brood rearing activity of *Apis indica* F and egg laying capacity of its queen. Indian Bee J 20:166–173, Keywords: breeding/brood/*cerana*/geographical variation/oviposition/physical factors/queens/India

Sharma PL (1960a) Observations on the swarming and mating habits of the Indian honeybee. Bee World 41:121–125, Keywords: cells/*cerana*/*mellifera*/physical factors/queens/reproduction/supersedure/swarming/India

Sharma PL (1960b) Experiments with *Apis mellifera* in India. Bee World 41:230–232, Keywords: *cerana*/*mellifera*/queens/India

Sharma PL (1961) The honeybee population among insects visiting temperate zone fruit flowers and their role in setting fruit. Bee World 42:6–8, Keywords: *cerana*/pollination/population biology/India

Sharma PL (1972) Homing ability and the foraging range of the Indian honey bee *Apis indica* Fabricius. Indian J Entomol 34:225–227, Keywords: *cerana*/foraging/honey/India

Sharma PL, Sharma AC (1950a) Preliminary studies on the weight of the Indian honeybee, *Apis indica* F, hill variety. Indian Bee J 12:4–5, Keywords: *cerana*/physiology/India

Sharma PL, Sharma AC (1950b) Influence of numbers in a colony on the honey gathering capacity of bees (*Apis indica* hill variety). Indian Bee J 12:106–107, Keywords: *cerana*/foraging/honey/India

Sharma PL, Dhaliwal HS, Kakkar KL (1974) Insect visitors and pollinators of cauliflower (*Brassica oleracea* var *botrytis*) seed crop bloom. Himachal J Agric Res 2:74–78, Keywords: *cerana*/*dorsata*/*florea*/pollination/India

Sharma PL, Dogra GS, Mishra RC (1979) Evaluation of methods of control of predatory wasps, *Vespa* spp in apiaries. Indian Bee J 41:10–16, Keywords: *cerana*/natural enemies/India

Sharma SK, Singh JR (1999) Pollinating efficiency of *Apis dorsata* F and *A. florea* F on carrot (*Dacus carota* L). Indian Bee J 61:75–78, Keywords: bee botany/*dorsata*/*florea*/India

Sharma SK, Singh JR, Ombir (2000) Comparative pollination efficiency of three Apis spp on Brassica campestris var "sarson" BSH1. Crop Res (Hisar) 19:125–127, Keywords: bee botany/*dorsata*/*florea*/*mellifera*/India

Sharma SK, Singh JR, Kaushik H (2001b) Efficiency of two wild honey bees (*Apis* spp) as pollinators of onion (*Allium cepa* L). Haryana J Hort Sci 30:134–135, Keywords: bee botany/*dorsata*/*florea*/pollination/India

Sharma SK, Singh JR, Mahl JC (2001c) Foraging behaviour of Apis spp in semi-arid subtropical climate 3. On flowers of mustard, onion, carrot, berseem and sunflower. Crop Res (Hisar) 21:332–334, Keywords: behaviour/*dorsata*/environmental effects/*florea*/foraging/*mellifera*/India

Sharma V, Thakur ML (1999) Morphometric characterization of giant honeybee *Apis dorsata* Fab (Hymenoptera: Apidae) from Doon Valley. Ann For 7:125–135, Keywords: *dorsata*/morphometrics/India

Sharma V, Thakur ML (1998) Some aspects of foraging behaviour of *Apis florea* Fabr. on *Ammi majus* Linn in Doon Valley (Uttar Pradesh). Indian J For 21:115–118, Keywords: behaviour/*florea*/foraging/India

Shearer DA, Boch R, Morse RA, Laigo FM (1970) Occurrence of 9-oxodec-trans-2-enoic acid in queens of *Apis dorsata*, *Apis cerana*, and *Apis mellifera*. J Insect Physiol 16:1437–1441, Keywords: *cerana*/*dorsata*/drones/*mellifera*/pheromones/queens

Sheikh MS, Chetri G (2000) Distribution of honey bees in Assam. Environ Ecol 18:236–240, Keywords: *dorsata*/*florea*/India

Shelar DG, Suryanarayana MC (1981) Preliminary studies on pollination of coriander (*Coriandrum sativum* L). Indian Bee J 43:110–111, Keywords: *cerana*/*florea*/pollination/India

Shelar DG, Suryanarayana MC (1983) Effect of pollination by honeybee (*Apis cerana indica* Fabricius) on the yield of lucerne. Indian J Agric Sci 53:190–191, Keywords: *cerana*/*florea*/pollination/India

Shen LR, Zhang CX, Cheng JA (2001) Cloning and sequencing of gene encoding hyaluronidase from the venom of *Apis cerana cerana*. Entomol Sinica 8:353–360, Keywords: biochemistry/*cerana*/genetics/*mellifera*/China
Shende SG, Phadke RP (1995) Beekeeping in India – history, present status and future. Indian Bee J 57:31–37, Keywords: beekeeping/*cerana*/history/India
Shenglu C, Songkun S, Xuezhen L (2002) An introduction to high-yielding royal jelly production methods in China. Bee World 83:69–77, Keywords: *cerana*/*mellifera*/royal jelly
Sheppard WS (1985) Electrophoretic variation in the honey bees. Apis Isozyme Bull 18:69, Keywords: biochemistry/*cerana*/*dorsata*/*florea*/honey/*mellifera*
Sheppard WS (1986) Genetic variation and differentiation in honey bees (*Apis*). Thesis, University of Illinois, Keywords: *cerana*/*dorsata*/*florea*/genetics/Sri Lanka
Sheppard WS (1989a) Ribosomal RNA diversity in *Apis*. Entomol Soc Am Ann Meet, San Antonio, Keywords: biochemistry/biogeography/*cerana*/*dorsata*/organic chemistry/Arabia*
Sheppard WS (1989) Ribosomal RNA diversity in *Apis*. In: Proceedings of the Entomological Society of America annual meeting, San Antonio, Texas, Keywords: biochemistry/biogeography/*cerana*/*dorsata*/*florea*/genetics/*mellifera*/phylogeny/RNA/Arabia*
Sheppard WS, Berlocher SH (1989) Allozyme variation and differentiation among four *Apis* species. Apidologie 20:419–431, Keywords: biochemistry/biogeography/*cerana*/classification/*dorsata*/*mellifera*/*floea*/evolution/organic chemistry
Sheppard WS, McPheron BA (1991) Ribosomal DNA diversity in Apidae. In: Smith DR (ed) Diversity in the genus *Apis*. Westview Press, Boulder, Colorado, pp 89–103, Keywords: *cerana*/DNA/*dorsata*/*florea*/genetics/*laboriosa*/*mellifera*
Sheppard WS, McPheron BA (1993) Honey bee molecular systematics and DNA sequence data: a review of recent studies. In: Connor LJ, Rinderer TE, Sylvester HA, Wongsiri S (eds) Asian apiculture. Wicwas, Cheshire, CT, pp 185–193, Keywords: *cerana*/DNA/*dorsata*/*florea*/honey/*mellifera*
Shi D (1992a) Management of *Apis cerana cerana* during honey flows. J Bee 6:14, [in Chinese]. Keywords: beekeeping/*cerana*/honey/China*
Shi D (1992b) Swarm prevention in *Apis cerana*. J Bee 4:12, [in Chinese]. Keywords: beekeeping/*cerana*/swarming/China*
Shi YH, Qiao XK (1989) Techniques for rearing *Apis cerana cerana* scientifically. J Bee 1:12–13, [in Chinese]. Keywords: beekeeping/breeding/brood/*cerana*/China
Shield J, Lee BT (2000) Potential for migration of Asian honeybee, *Apis cerana*, from Lae, Papua New Guinea, to Australia – options for minimising the risk. Austral Beekeep 102:51–52, Keywords: *cerana*/migration/queens/Australia/New Guinea/Papua New Guinea
Shimamori K (1987) On the biology of *Apanteles galleriae*, a parasite of the two species of wax moths. Honeybee Sci 8:107–112, Keywords: beeswax/biology/*cerana*/natural enemies/Japan
Shimanuki H (1993) Current status and future prospects for the control of bee mites. In: Connor LJ, Rinderer TE, Sylvester HA, Wongsiri S (eds) Asian apiculture. Wicwas, Cheshire, CT, pp 43–48, Keywords: *andreniformis*/*cerana*/*dorsata*/*florea*/*koschevnikovi*/*laboriosa*/pests
Shimizu I, Kawai Y, Taniguchi M, Aoki S (2001) Circadian rhythm and cDNA cloning of the clock gene period in the honeybee *Apis cerana cerana*. Zool Sci (Tokyo) 18:779–789, Keywords: *cerana*/DNA/seasonal variation/Japan
Shimizu I, Taniguchi S, Aoki S (2002) Circadian rhythm in the honeybee *Apis cerana japonica* - behavioural and molecular approach. In: Proceedings of the 14th international congress, IUSSI, Sapporo, p 31, Keywords: *cerana*/DNA/seasonal variation/Japan
Shimizu M, Arakawa T (1993) Biogenic amines in brains of the honeybee and venom glands of the parasitoid wasps. Honeybee Sci 14:170–172, [in Japanese]. Keywords: biochemistry/*cerana*/glands/physiology/Japan
Shimoji M (1981) Keeping Japanese honeybees in south Wakayama. Honeybee Sci 2:37–38, [in Japanese]. Keywords: beekeeping/*cerana*/natural enemies/Japan

Shimomaki S (1991) Activities of Japan honey cooperative association during 20 years. Honeybee Sci 12:179–182, [in Japanese]. Keywords: beekeeping/*cerana*/history/honey/Japan
Shindo M (1970) Black carpenter ants fed on royal jelly of honeybee. Collect Breed 32:276–279, [in Japanese]. Keywords: *cerana*/royal jelly/Japan
Shipolini RA (1984) Biochemistry of bee venom. In: Tu AT (ed) Handbook of natural toxins, vol 2. Marcel Dekker, New York, pp 49–85, Keywords: biochemistry/*cerana*/*dorsata*/*florea*/*mellifera*
Shiroiwa B (1991) My memoirs of beekeeping. Honeybee Sci 12:183–186, [in Japanese]. Keywords: beekeeping/*cerana*/history/Japan
Shivakoti GP, Bista S (2000) Major constraints in the performance of *Apis mellifera* and *Apis cerana* under Khumaltar conditions. In: Proceedings of the 4th Asian apiculture association international conference, Kathmandu, pp 29–32, Keywords: behaviour/*cerana*/colony phenology/disease/environmental effects/*mellifera*/pests/queens/Nepal
Shivapuri CN (1955) Bees of India. Glean Bee Cult 83:26–27, Keywords: beekeeping/*cerana*/*dorsata*/*florea*/India
Shivrana S, Jain KL (1994) Effects of low doses of pyrethroid feeding on honeybee mortality and their body sugar levels. Indian Bee J 56:145–151, Keywords: biochemistry/*dorsata*/organic chemistry/insecticides/India
Shiying Y, Saxena KB, Wen P, Guangtian W, Ziping H (2002) First report of natural outcrossing in pigeonpea from China. Int Chickpea Pigeonpea Newsl 9:37–38, Keywords: *cerana*/*dorsata*/*florea*/pollination/China
Shrestha JB, Shreshta KK (2000) Beekeeping in Nepal: problems and potentials. In: Proceedings of the 4th Asian apiculture association international conference, Kathmandu, pp 262–265, Keywords: bee botany/beekeeping/biochemistry/disease/organic chemistry/pests/physiology/Nepal
Shrestha JB, Shukla AN, Joshi SR, Pechhacker H, Hüttinger E (2000) Improvement of *Apis cerana* beekeeping using the ICIMOD bee project as an example. In: Proceedings of the 4th Asian apiculture association international conference, Kathmandu, pp 273–274, Keywords: beekeeping/*cerana*/Nepal
Shreshta KK (1996) A decade of experience with *Apis cerana* in Nepal. In: Proceedings of the 3rd Asian apiculture association international conference, Hanoi, pp 115–121, Keywords: beekeeping/*cerana*/*florea*/*mellifera*/Nepal
Shreshta KK (1996) Country Report: Nepal. In: Proceedings of the 3rd Asian apiculture association international conference, Hanoi, p 216, Keywords: *cerana*/disease/*dorsata*/*florea*/honey/*laboriosa*/*mellifera*/Nepal
Shreshta KK (2000) Role of ICIMOD in the promotion and development of *Apis cerana* beekeeping. In: Proceedings of the 4th Asian apiculture association international conference, Kathmandu, pp 241–244, Keywords: beekeeping/*cerana*/physiology/Nepal
Shreshta KK, Shrestha NCTD (2000) Study on Thai sac brood virus disease of *Apis cerana* in Nepal. In: Proceedings of the 4th Asian apiculture association international conference, Kathmandu, pp 60–63, Keywords: brood/*cerana*/disease/Nepal
Shreshta KK, Verma LR (1992) Beekeeping research and training in Hindu Kush-Himalayan region: future perspectives. In: Verma LR (ed) Honeybees in mountain agriculture. Oxford and IBH, New Delhi, pp 251–257, Keywords: beekeeping/*cerana*/India
Shrestha M (1997) Laboratory experience in Nepal: honey analysis and honey composition. In: Sommerijer MJ, Beetsma J, Boot WJ, Robberts EJ, De Vries R (eds) Perspectives for honey production in the tropics. Netherlands Expertise Centre for Tropical Apicultural Resources, Bennekom, pp 127–132, Keywords: *cerana*/honey/Nepal
Shrestha M (2000) Physical and chemical properties of Nepalese honey. In: Proceedings of the 4th Asian apiculture association international conference, Kathmandu, pp 137–139, Keywords: *cerana*/*dorsata*/honey/*laboriosa*/*mellifera*/Nepal

Shylesha AN, Rajagapal BK (1996a) Bioecology of greater wax moth *Galleria mellonella* L a major pest of *Apis cerana* colonies. J Hill Res 9:251–254, Keywords: beeswax/*cerana*/natural enemies/India

Shylesha AN, Rajagapal BK (1996b) Biology of *Apanteles galleriae* Wilkinson (Braconidae: Hymenoptera) a larval parasite of wax moth and its hyperparasite, *Nesolynx flavipes* Ashmead (Eulophidae: Hymenoptera). J Hill Res 9:419–421, Keywords: beeswax/biology/*cerana*/natural enemies/parasites/India

Siddappaji C, Channabasavanna GP (1983) Role of honeybees in the pollination of cardamom *Elettaria cardamomum* (L) Maton. In: Proceedings of the 2nd international conference on apiculture in tropical climates, New Delhi, pp 640–648, Keywords: *cerana*/physical factors/pollination/India

Siddappaji C, Parameswar NS (1973) Slug (*Limas* spp), a new pest of Indian bee *Apis cerana* (F) and its control. Curr Res 2:40–41, Keywords: *cerana*/combs/natural enemies/India

Sidhu AS, Singh S (1961) Studies on agents of cross pollination in cotton. Indian Cott Grow Rev 15:341–353, Keywords: *cerana*/*dorsata*/*florea*/pollination/India

Sidhu DS, Kumar N, Kaur SP (1982) Neutral and phospholipids from the poison reservoir of the sting apparatus of *Apis dorsata* Fabr. (Hymenoptera). J Sci Res 4:41–43, Keywords: biochemistry/defense/*dorsata*/organic chemistry/India

Sidik D (1996) The comfortability of the hive wood for *Apis cerana* Fabr. In: Proceedings of the 3rd Asian apiculture association international conference, Hanoi, pp 74–76, Keywords: beekeeping/*cerana*/nests/Indonesia

Sihag RC (1982) Problems of waxmoth (*Galleria mellonella* L) infestation on giant honeybee (*Apis dorsata* Fab.) colonies in Haryana. Indian Bee J 44:107–109, Keywords: combs/disease/*dorsata*/honey/pests/India

Sihag RC (1983a) Prospects of beekeeping in Haryana. Indian Bee J 45:8–11, Keywords: beekeeping/*cerana*/India

Sihag RC (1983b) Honey bee forage for grape (*Vitis vinifera* L) juices during dearth period. Indian Bee J 45:111–112, Keywords: *cerana*/environmental effects/foraging/honey/India

Sihag RC (1984a) Influence of environmental factors on the pollination activity of bees. Environ Ecol 2:149–152, Keywords: *cerana*/environmental effects/pollination/India

Sihag RC (1984b) The role of temperature in the evolution of bee plant mutualism. Indian Bee J 46:28–32, Keywords: bee botany/environmental effects/evolution/foraging/India

Sihag RC (1985) Floral biology, melittophily and pollination ecology of cultivated cruciferous crops. In: Varghese TM (ed) Recent advances in pollen research. Allied, New Delhi, pp 241–268, Keywords: biology/*cerana*/ecology/*mellifera*/pollen/pollination/India

Sihag RC (1986) Insect pollination increases seed production in cruciferous and umbelliferous crops. J Apic Res 25:121–126, Keywords: bee botany/*dorsata*/*florea*/India

Sihag RC (1987a) A simple apparatus for sampling honeybees from wild colonies. Bee World 68:179, Keywords: beekeeping/India

Sihag RC (1987b) Effect of pesticides and bee-pollination on seed yield of some crops in India. J Apic Res 27:49–54, Keywords: *cerana*/insecticides/India

Sihag RC (1988) Characterization of the pollinators of cultivated cruciferous and leguminous crops of subtropical Hissar, India. Bee World 69:153–158, Keywords: *cerana*/pollination/India

Sihag RC (1990) Differential attractiveness of two cruciferous species to honeybees. In: Proceedings of the 11th international congress, IUSSI, Bangalore, pp 434-436, Keywords: *cerana*/foraging/India

Sihag RC (ed) (1990b) Pollination biology environmental factors and pollination. Rajendra Scientific, Hisar, Keywords: *cerana*/*dorsata*/environmental effects/*florea*/*mellifera*/physical factors/pollination

Sihag RC (1993) Behaviour and ecology of the subtropical carpenter bee, *Xylocopa fenestrata* F 6. Foraging dynamics, crop hosts and pollination potential. J Apic Res 32:94–101, Keywords: bee botany/behaviour/*dorsata*/ecology/*florea*/foraging/India

Sihag RC (1998) Wild *Apis* and non-*Apis* bees; their behaviour, management and utilization for crop pollination. In: Mishra RC (ed) Perspectives in Indian apiculture. Agro Bot, New Delhi, pp 248–260, Keywords: *andreniformis*/beekeeping/*cerana*/*dorsata*/*florea*/*koschevnikovi*/*laboriosa*/pollination
Sihag RC (2000) Ecobiology of the little honeybee (*Apis florea*) in semi and subtropical climates of India. In: Proceedings of the 4th Asian apiculture association international conference, Kathmandu, pp 46–49, Keywords: absconding/bee botany/environmental effects/*florea*/foraging/general biology/migration/swarming/India
Sihag RC (2000) Foraging ecology of *Apis cerana* and *Bombus tunicatus* in pollinating balsam flowers. In: Proceedings of the 4th Asian apiculture association international conference, Kathmandu, p 163, Keywords: beekeeping/*cerana*/ecology/foraging/*mellifera*/pollination/India
Sihag RC (2000) Management of bees for pollination. In: Proceedings of the 4th Asian apiculture association international conference, Kathmandu, pp 163–165, Keywords: beekeeping/*cerana*/*mellifera*/pollination
Sihag RC, Abrol DP (1986) Correlation and path-coefficient analysis of environmental factors influencing flight activity of *Apis florea* F. J Apic Res 25:202–208, Keywords: environmental effects/*florea*/foraging/India
Sihag RC, Kapil RP (1984) Foraging strategies of honeybees as determined by quality and quantity of nectar. In: Proceedings of the 5th international symposium on pollination, Versailles, pp 51–59, Keywords: bee botany/*dorsata*/*florea*/foraging/honey/India
Sihag RC, Khatkar S (1999) Foraging pattern of three honeybee species on eight cultivars of oilseed crops 1. Diurnal foraging during the entire blooming period of the crops. Trop Agric 7:245–252, Keywords: *dorsata*/*florea*/foraging/*mellifera*/India
Sihag RC, Mishra RC (1995) Crop pollination and *Apis cerana*. In: Kevan PG (ed) The Asiatic hive bee: apiculture, biology, and role in sustainable development in tropical and subtropical Asia. Enviroquest, Cambridge, ON, pp 135–141, Keywords: biology/*cerana*/nests/physiology/pollination/India
Sihag RC, Rathi A (1994) Diversity, abundance, foraging behaviour and pollinating efficiency of different bees visiting pigeon pea (*Cajanus cajan* L Millsp) blossoms. Indian Bee J 56:187–201, Keywords: bee botany/behaviour/*dorsata*/*florea*/foraging/pollination/India
Sihag RC, Verma S (2002) Acute contact and acute stomach toxicity of five organophosphate insecticides to the honey bee, *Apis florea*. Asian Bee J 4:13–19, Keywords: biochemistry/*florea*/insecticides/organic chemistry/India
Sihag RC, Kumar S, Gahlawat S (1998) Acute contact toxicity of endosulfan and carbaryl insecticides to two honeybee species, *Apis mellifera* L and *Apis dorsata* F. Korean J Apic 13:55–60, Keywords: biochemistry/*dorsata*/insecticides/*mellifera*/organic chemistry/India
Sihanuntavong D (1997) Genetic variation in mitochondrial genes of honey bee *Apis cerana* in Thailand. Thesis, Chulalongkorn University, Keywords: *cerana*/DNA/genetics/honey/Thailand
Sihanuntavong D, Sittipraneed S, Klinbunga S (1999) Mitochondrial DNA diversity and population structure of the honey bee, *Apis cerana*, in Thailand. J Apic Res 38:211–219, Keywords: *cerana*/DNA/genetics/honey/population biology/Thailand
Sim YG, Han YY, Koag DY, Choi YE (1996) Studies on diurnal activity of honey bees in the peony flower. Korean J Apic 11:1–7, [in Korean]. Keywords: *cerana*/foraging/Korea
Simmons LW (2001) Sperm competition and its evolutionary consequences in the insects. Princeton University Press, New Jersey, Keywords: *dorsata*/ecology/evolution/*florea*/reproduction
Simpson J (1960) Male genitalia of Apis species. Nature 185:56, Keywords: *cerana*/*dorsata*/drones/*florea*/*mellifera*
Singer RB (2002) The pollination mechanism in Trigonidium obtusum Lindl (Orchidaceae: Maxillariinae): sexual mimicry and trap-flowers. Ann Bot 89:157–163, Keywords: bee botany/*cerana*/pollination

Singh A (1981) Morphometric studies on *Apis dorsata* F and *Bombus heamorrhoidalis* S of Simla Hills (Himachal Pradesh).Thesis, Himachal Pradesh University, Keywords: classification/*dorsata*/morphology/India

Singh B, Chauhary DK (2002) Insect visitors of ber (*Ziziphus mauritiana* L). Insect Environ 8:14, Keywords: *cerana*/*dorsata*/*florea*/pollination/India

Singh BP, Kshirsagar KK (1987) Communication behaviour in *Apis cerana indica* F. In: Proceedings of the 31st international apiculture congress, Apimondia, Warsaw, pp 189–190, Keywords: behaviour/*cerana*/communication/dance language/India

Singh BP, Phadke RP, Mittal MC (1983) Beekeeping potential and pattern of management in Uttar Pradesh (India). Indian Bee J 45:35–37, Keywords: beekeeping/*cerana*/India

Singh BP, Singh MK, Chakrawarti P (1989a) Influence of sugar concentration on load carrying capacity of Indian honey bees. Indian Bee J 51:52–53, Keywords: *cerana*/flight/foraging/honey/physiology/India

Singh G (1971a) *Plectranthus rugosus* Wall the major honey plant in Kashmir Valley. Indian Bee J 33:58–59, Keywords: bee botany/*cerana*/foraging/honey/India

Singh G (1971b) Sucrose solution imbibing capacity of *Apis indica* F (hill strain) in relation to various sucrose solution concentrations and body weight. Indian Bee J 33:51–54, Keywords: *cerana*/foraging/India

Singh G (1972) Defensive behaviour of *Apis indica* F (Hill Strain) against predatory hornets in Kashmir. Indian Bee J 34:65–69, Keywords: aggressiveness/behaviour/*cerana*/natural enemies/India

Singh G (2000) Clustering behaviour of *Apis cerana indica* in Kashmir Valley during dearth and colony depletion. Indian Bee J 36:22–23, Keywords: behaviour/*cerana*/environmental effects/India

Singh G, Kashyap RK, Khan MS, Sharma SK (2001) Effect of pollination mode on yield and quality of hybrid seeds of sunflower, *Helianthus annuus* L. Seed Sci Technol 29:567–574, Keywords: bee botany/*dorsata*/pollination/India

Singh GD (1957a) Departmental collection of honey and wax, Bhinga Range, Baharaich Forest Division, Uttar Pradesh. Indian For 83:113–117, Keywords: bee botany/combs/*dorsata*/honey/India

Singh JP (1969) Effect of thiodon on honeybees (*Apis indica* Fabr.). Indian Bee J 30:9–10, Keywords: *cerana*/*dorsata*/*florea*/insecticides/India

Singh JP, Dharamwal SS (1970) The role of honeybees in seed setting on onion at Pantnagar, District Nanital, UP, India. Indian Bee J 32:23–27, Keywords: *cerana*/pollination/India

Singh JP, Talwar RC (1969) Acarine infestation of honey bees *Apis indica* Fabr. Trudy NauchnoIssledovatel'skogo Instituta Pchelovodstva. pp 234–251, [in Russian]. Keywords: *cerana*/honey/pests/India*

Singh MM (2008) Foraging behaviour of the Himalayan honeybee (*Apis cerana* F.) on flowers of *Fagopyrum esculentum* M. and its impact on grain quality and yield. Ecoprint 15:37–46, Keywords: *cerana*/foraging/pollination/Nepal

Singh MP (1984a) Morphometric studies on Indian honeybee, *Apis cerana indica* F of northeast Himalayas. Thesis, Himachal Pradesh University, Keywords: *cerana*/classification/geographical variation/morphology/mountains/India

Singh MP (1984b) Studies on the activity of some insect pollinators on jujube (*Zizyphus mauritiana* Lamk). Entomon 9:177–180, Keywords: bee botany/*cerana*/foraging/pollination/India

Singh MP (1989) Melissopalynology and identification of the Himalayan races of honeybees by computer assisted multivariate analysis. Thesis, Himachal Pradesh University, Keywords: *cerana*/classification/*dorsata*/*florea*/honey/pollen/subspecies/India

Singh MP, Devi CS (1998) Effect of environmental factors on the activities of *Apis cerana* F in rapeseed pollination. Insect Environ 4:50, Keywords: *cerana*/pollination/India*

Singh MP, Singh KI (1993) Effect of bee pollination on the yield of sunflower. In: Proceedings of the international symposium on pollination in tropics, Bangalore, pp 226–228, Keywords: *cerana*/pollination/India

Singh MP, Verma LR (1993) Morphometric comparison of three geographic populations of the northeast Himalayan *Apis cerana*. In: Connor LJ, Rinderer TE, Sylvester HA, Wongsiri S (eds) Asian apiculture. Wicwas, Cheshire, CT, pp 67–80, Keywords: *cerana*/classification/environmental effects/mountains/population biology/populations/morphometrics/India
Singh MP, Verma LR, Daly HV (1990) Morphometric analysis of the Indian honeybee in the northeast Himalayan Region. J Apic Res 29:3–14, Keywords: *cerana*/classification/morphometrics/mountains/India
Singh MP, Singh KM, Singh KI (1999a) Efficacy and economics of some insecticides against mustard aphid, *Liaphis erysimi* on rapeseed and their safety to predator and foraging bee. Indian J Entomol 61:11–18, Keywords: *cerana*/foraging/insecticides/India
Singh MP, Singh KI, Devi CS (2000) Role of *Apis cerana himalaya* pollination on yield and quality of rape seed and sunflower crops. In: Proceedings of the 4th Asian apiculture association international conference, Kathmandu, pp 186–189, Keywords: *cerana*/foraging/pollination/India
Singh MP, Singh KI, Devi CS (2000) Foraging behaviour of *Apis cerana himalaya* on sunflower and rape seed. In: Proceedings of the 4th Asian apiculture association international conference, Kathmandu, pp 199–202, Keywords: behaviour/*cerana*/foraging/India
Singh MT, Sharma PL, Dhaliwal HS (1974) Toxicity of insecticides to honeybee workers, *Apis cerana indica*. Pesticides 8:28–29, Keywords: *cerana*/insecticides/workers/India
Singh N, Singh JP (1985) Changes in the free amino acid content during the post-embryonic development of *Apis cerana indica* (Hymenoptera: Apidae). Indian Zool 9:23–26, Keywords: biochemistry/*cerana*/physiology/India
Singh N, Singh JP (1986) Cholesterol content in the post-embryonic stages of *Apis cerana indica* (Hymenoptera: Apidae). Ann Entomol 4:11–13, Keywords: biochemistry/*cerana*/India
Singh PS (1993) Major constraints in beekeeping. In: Proceedings of the 1st national conference on beekeeping, Chandigarh, pp 50–60, Keywords: beekeeping/*cerana*/India
Singh R, Garg R (1994) Relative abundance of insect visitors and pollination requirements of citrus. Pest Manag Econ Zool 2:143–146, Keywords: bee botany/*dorsata*/pollination/India
Singh R, Pal HM, Gupta M, Malik OP (1993) Screening of some chemical compounds as repellents to honey bees I. Effect of carbonyl compounds on Apis florea F visitors to Brassica campestris var toria. Indian Bee J 55:47–50, Keywords: biochemistry/*florea*/organic chemistry/India
Singh R, Pal HM, Gupta M, Malik OP (1995) Screening of some carbonyl compounds as repellents to honey bees II. Effect of carbonyl compounds on Apis florea F visitors to Ziziphus mauritiana var dichotoma. Korean J Apic 10:131–134, Keywords: biochemistry/*florea*/organic chemistry/India
Singh RP, Singh PN (1988) Proline, an early index of sac brood disease of honey bee. Natl Acad Sci Lett (India) 11:129–131, Keywords: biochemistry/brood/*cerana*/disease/honey/India
Singh RP, Singh PN (1991) A study on the biochemical composition of fresh and stored mustard pollen and honey. Apiacta 26:38–44, Keywords: *cerana*/honey/pollen/India
Singh RP, Singh PN (1992) Impact of bee pollination on seed yield, carbohydrate composition and lipid composition of mustard seed. J Apic Res 31:128–133, Keywords: *cerana*/pollination/India
Singh RP, Singh PN (1996) Amino acid and lipid spectra of larvae of honey bee (*Apis cerana* Fabr.) feeding on mustard pollen. Apidologie 27:21–28, Keywords: biochemistry/brood/*cerana*/honey/physiology/pollen/India
Singh S (1940) 'Tearing down' of combs. Bee World 21:66, Keywords: *cerana*/combs/India
Singh S (1941) Two queens in a colony. Indian Bee J 3:56–57, Keywords: *cerana*/queens/India
Singh S (1947) The honeybees of India. In: Root AI (ed) The ABC and XYZ of bee culture. AI Root Company, Medina, OH, pp 560–562, Keywords: absconding/beekeeping/*cerana*/*dorsata*/*florea*/India
Singh S (1957b) Acarine disease in the Indian honey bee (*Apis indica* F). Indian Bee J 19:27–28, Keywords: *cerana*/disease/honey/pests/India

Singh S (1961) Appearance of American foulbrood disease in India honeybee (*Apis indica* Fabr.). Indian Bee J 23:46–50, Keywords: *cerana*/disease/India
Singh S (1962) Beekeeping in India. Indian Council of Agricultural Research, New Delhi, Keywords: absconding/beekeeping/*dorsata*/environmental effects/migration/swarming/India
Singh S (1964) Beekeeping and entomology in India. Indian J Entomol 25:497 511, Keywords: beekeeping/*cerana*/*florea*/physical factors/India
Singh S (1975a) Beekeeping in India. Indian Council of Agricultural Research, New Delhi, Keywords: beekeeping/*dorsata*/India
Singh S, Adlakha RL (1958) Acarine disease of adult honeybee. Indian Bee J 20:64–78, Keywords: *cerana*/disease/pests/India
Singh S, Adlakha RL (1960) Survey of acarine disease of adult honeybees in Simla Hills. Indian Bee J 22:3–6, Keywords: *cerana*/disease/pests/India
Singh S, Sharma PL (1943) Absconding among Indian bees, *Apis indica*. Indian Bee J 5:98–101, Keywords: absconding/behaviour/*cerana*/natural enemies/India
Singh S, Saini K, Jain KL (1999b) Quantitative comparison of lipids in some pollens and their phagostimulatory effects in honeybees. J Apic Res 38:87–92, Keywords: bee botany/biochemistry/*dorsata*/honey/organic chemistry/pollen/India
Singh T, Singh B (1981) A comparative study on the mandibular gland of workers of three species of *Apis* (Hymenoptera: Apidae). J Entomol Res 5:31–35, Keywords: biogeography/*cerana*/*dorsata*/*florea*/glands/morphology/workers/India
Singh T, Singh B (1982) Comparative morphological studies on the mandibular gland and mandibular groove in Hymenoptera Apocrita. J Anim Morphol Physiol 29:78–84, Keywords: *cerana*/*dorsata*/*florea*/glands/morphology/India
Singh TSMS, Suryanarayana MC (1997) Melittopalynological investigations on the honeys of Karnataka. Indian Bee J 59:76–79, Keywords: bee botany/*cerana*/honey/India
Singh WJK, Singh R, Hameed SF, Singh B (1989b) Field toxicity of some insecticides to *Apis cerana indica* Fabr. Indian Bee J 51:137, Keywords: *cerana*/insecticides/India
Singh Y (1974) *Nosema* in Indian honey bee. Indian Bee J 36:16, Keywords: *cerana*/disease/honey/India
Singh Y (1975b) *Nosema* in Indian honey bee (*Apis cerana indica*). Am Bee J 115:59, Keywords: *cerana*/disease/honey/India
Singh Y (1979a) Iridescent virus in the Indian honey bee *Apis cerana indica* F. Am Bee J 119:398, Keywords: *cerana*/honey/India
Singh Y (1979b) Pollination activity on strawberry at Jeolikote (District Nainital, India). Indian Bee J 41:17–19, Keywords: *cerana*/pollination/India
Singh Y (1981b) Studies on pollen gathering activity of Indian honey bees (*Apis cerana indica* Fabr.) under Saharanpur conditions. Prog Hortic 12:31–38, Keywords: *cerana*/foraging/honey/pollen/India
Singh Y (1983a) Beekeeping in Uttar Pradesh – a review. Indian Bee J 45:84–92, Keywords: beekeeping/*cerana*/India
Singh Y (1983b) Beekeeping in Uttar Pradesh – a review. In: Proceedings of the 2nd international conference on apiculture in tropical climates, New Delhi, pp 211–226, Keywords: beekeeping/*cerana*/physical factors/India
Singh Y, Verma SK (1981) Occurrence of acarine disease in *Apis cerana indica* in Uttar Pradesh (India). Indian Bee J 43:37–39, Keywords: *cerana*/disease/pests/India
Singh Y, Verma SK (1983) Preliminary observation on the effects of antibiotic feeding to *Apis cerana indica* F colonies. Indian Bee J 45:110, Keywords: beekeeping/*cerana*/India
Sinha SN, Chakrabarti AK (1983) Bee pollination and its impact on cauliflower seed production. In: Proceedings of the 2nd international conference on apiculture in tropical climates, New Delhi, pp 649–655, Keywords: biogeography/*cerana*/*dorsata*/*florea*/physical factors/pollination/India

Sinha SN (2000) Pollination requirement in sunflower hybrid seed production III. Effects of methods of pollination on seed setting and yield. Seed Res 28:113–118, Keywords: bee botany/*dorsata*/*florea*/*mellifera*/India

Sito AS (2004) Beekeeping and honey trading in the Philippines. In: Proceedings of the 1st symposium issues concerning developing countries international trade in honey, Hanoi, pp 69–74, Keywords: *cerana*/*dorsata*/*mellifera*/Philippines

Sittipraneed S, Laoaroon S, Klinbunga S (2001a) Genetic differentiation of the honey bee (*Apis cerana*) in Thailand: evidence from microsatellite polymorphism. J Apic Res 40:9–16, Keywords: *cerana*/DNA/honey/morphology/population genetics/Thailand

Sittipraneed S, Sihanuntavong D, Klinbunga S (2001b) Genetic differentiation of the honey bee (*Apis cerana*) in Thailand revealed by polymorphism of a large subunit of mitochondrial ribosomal DNA. Insectes Soc 48:266–272, Keywords: *cerana*/DNA/honey/mtDNA/Thailand

Sivaram V (2001) Honeybee flora and beekeeping in Karnataka State, India. In: Proceedings of the 37th international apiculture congress, Apimondia, Durban (CD Rom). Keywords: bee botany/ beekeeping/*cerana*/*dorsata*/*florea*/India

Sivaram V, Anita M, Reddy CC (1993) The preservation and management of wild honeybees for pollination. In: Proceedings of the international symposium on pollination in tropics, Bangalore, pp 251–253, Keywords: bee botany/beekeeping/combs/*dorsata*/ecology/India

Skorikov AS (1929) A new basis for the revision of the genus *Apis*. Rep Appl Entomol (Leningrad) 4:249–270, [in Russian]. Keywords: biogeography/*cerana*/classification/*florea*

Smith DR (1990) Mitochondrial DNA and biogeography of *Apis* species. In: Proceedings of the 11th international congress, IUSSI, Bangalore, pp 113–114, Keywords: *andreniformis*/biogeography/*cerana*/*dorsata*/*florea*/*mellifera*/mtDNA

Smith DR (ed) (1991a) Diversity of the genus *Apis*. Westview, Boulder, CO, Keywords: *andreniformis*/*cerana*/DNA/*dorsata*/*florea*/genetics/*mellifera*

Smith DR (1991b) Mitochondrial DNA and honey bee biogeography. In: Smith DR (ed) Diversity in the genus *Apis*. Westview, Boulder, CO, pp 131–176, Keywords: *andreniformis*/biogeography/*cerana*/*dorsata*/*florea*/genetics/honey/*mellifera*/mtDNA

Smith DR (1993) Mitochondrial DNA variation in *Apis* species and subspecies. In: Connor LJ, Rinderer TE, Sylvester HA, Wongsiri S (eds) Asian apiculture. Wicwas, Cheshire, CT, pp 173–183, Keywords: *andreniformis*/*cerana*/*dorsata*/*florea*/mtDNA/India/Japan/Malaysia/ Pakistan/Thailand

Smith DR (2001) Biogeography of *Apis cerana* and *A. nigrocincta*: insights from mtDNA sequence studies. In: Proceedings of the 7th international conference on trop bees, Chiang Mai, pp 20–26, Keywords: biogeography/*cerana*/mtDNA/*nigrocincta*/phylogeny/China/India/ Indonesia/Japan/Korea/Malaysia/Nepal/Philippines/Sri Lanka

Smith DR (2002) Biogeography of *Apis cerana*: southeast Asia and the Indo-Pakistan subcontinent. In: Proceedings of the 14th international congress, IUSSI, Sapporo, p 209, Keywords: biogeography/*cerana*/mtDNA/Cambodia/Japan/Laos/Myanmar (Burma)/Pakistan/Thailand/Vietnam

Smith DR, Hagen RH (1996) The biogeography of *Apis cerana* as revealed by mitochondrial DNA sequence data. J Kans Entomol Soc 69:294–310, Keywords: biogeography/*cerana*/ DNA/mtDNA/China/India/Indonesia/Japan/Korea/Malaysia/Philippines/Sri Lanka/Sundaland/Thailand

Smith DR, Hagen RH, Otis GW (1999) Phylogeny and biogeography of *Apis cerana* subspecies: testing alternative hypotheses. In: Hoopingarner R, Connor LJ (eds) Apiculture for the 21st century. Wicwas, Cheshire, CT, pp 60–68, Keywords: biogeography/*cerana*/DNA/phylogeny

Smith DR, Villafuerte L, Otis G, Palmer MR (2000) Biogeography of *Apis cerana* F and *A. nigrocincta* Smith: insights from mtDNA studies. Apidologie 31:265–280, Keywords: biogeography/*cerana*/DNA/mtDNA/*nigrocincta*

Smith DR, Palmer MR, Otis G, Damus M (2003) Mitochondrial DNA and AFLP markers support species status of *Apis nigrocincta*. Insectes Soc 50:185–190, Keywords: nigrocincta/mtDNA

Smith DR, Hepburn HR, Warrit N (2004) *Apis cerana* from Myanmar (Burma): unusual distribution of mitochondrial lineages. Apidologie 35:637–644, Keywords: biogeography/*cerana*/distribution/mtDNA/Myanmar (Burma)

Smith DR, Warrit N, Otis GW, Thai PH (2005) A scientific note on high variation in the non-coding mitochondrial sequences of *Apis cerana* from South East Asia. J Apic Res 44:197–198, Keywords: *cerana*/mtDNA

Smith FG (1953) Beekeeping in the tropics. Bee World 34:233–248, Keywords: beekeeping/biology/*cerana*/*dorsata*/*florea*

Smith FG (1958) Communication and foraging range of African bees compared with that of European and Asian bees. Bee World 39:249–252, Keywords: *cerana*/communication/*florea*/foraging/*mellifera*

Smith FG (1960) Beekeeping in the tropics. Longmans, London, Keywords: beekeeping/*cerana*/*dorsata*/*florea*/*mellifera*

Soloman Raju AJ, Nagalakshma S, Naidu SA (1999) Insect pollination in marking-nut tree. Insect Environ 5:108–109, Keywords: bee botany/*florea*/pollination/India

Soman AG (1990) A note on *Apis florea* storing honey in drone cells. Bee World 71:33–34, Keywords: combs/drones/*florea*/foraging/honey/India

Soman AG, Chawda SS (1996) A contribution to the biology and behaviour of the dwarf bee, *Apis florea* F and its economic importance in Kutch, Gujarat, India. Indian Bee J 58:81–88, Keywords: absconding/bee botany/behaviour/combs/classification/*florea*/foraging/general biology/ honey/migration/queens/swarming/India

Soman AG, Sawant SB (2001) Nesting of Indian hive bees in open. Indian Bee J 59:240, Keywords: *cerana*/nests/India

Son N (1996) The ancient southern part. Ho Chi Minh City Publishing House, Ho Chi Minh City, [in Vietnamese]. Keywords: beekeeping/*dorsata*/Vietnam

Sonezaki T (1990) Cuticular surface structure, a critical cue of the queen-queen recognition for the elimination of reproductive rivals in honeybees. In: Proceedings of the 11th international congress, IUSSI, Bangalore, p 420, Keywords: *cerana*/*florea*/*mellifera*/morphology/queens/Japan

Song JX (2009) Preliminary study on insect diversity of Tongbiguan nature reserve in Yunnan. Forest Invent Plan 34:3, [in Chinese]. Keywords: biogeography/*cerana*/China*

Song WF, Tan K, Liu XH (2009) Foraging comparison between *Tripterygium hypolaucum* Hutch and *Cuphea balsarnona* Cham by *Apis cerana*. J Bee 29:6, [in Chinese]. Keywords: *cerana*/foraging/China*

Songram O (1997) Genetic variation of *Apis cerana* in Thailand inferred by PCR-RFLP analysis of the mitochondrial ATPase6 - ATPase8 gene. Thesis, Chulalongkorn University. Keywords: *cerana*/DNA/Thailand

Songram O, Sittipraneed S, Klinbunga S (2006) Mitochondrial DNA diversity and genetic differentiation of the honeybee (*Apis cerana*) in Thailand. Biochem Genet 44:256–269, Keywords: *cerana*/mtDNA/Thailand

Sontakke BK, Dash AN (1996) Studies on the effects of insecticides and neem formulation against major pests of mustard and their safety to honey bees. Indian J Environ Toxicol 6:87–88, Keywords: biochemistry/disease/*florea*/insecticides/organic chemistry/pests/India

Sorthia BK, Chari MS (1985) Toxicity of some insecticides to honey bees *Apis florea* F and *Apis mellifera* L. J Entomol Res 9:195–197, Keywords: biochemistry/*florea*/insecticides/*mellifera*/organic chemistry/India

Sperlin A, Campbell R, Brosemer RW (1976) The hybridization of DNA from two species of honeybee. J Insect Physiol 22:373–376, Keywords: *cerana*/DNA/genetics/*mellifera*/Thailand

Speth K (1986) The top bar long hive. Newsl Beekeep Trop Sub-trop Countries 9:2–3, Keywords: beekeeping/*cerana*/nests/Nepal

Speth K (1987) Beekeeping in Nepal Part 2. Maandschr Bijent 89:325–327, [in Dutch]. Keywords: beekeeping/*cerana*/Nepal

Speth K (1988a) Beekeeping in Nepal Part 3. Maandschr Bijent 90:18–20, [in Dutch]. Keywords: beekeeping/*cerana*/Nepal

Speth K (1988b) Beekeeping in Nepal Part 4. Beekeeping and development programmes. Maandschr Bijent 90:142–145, [in Dutch]. Keywords: beekeeping/*cerana*/physiology/Nepal

Spieckermann A, Konig J, Dinslage E (1914) On the beeswax of three Asian honeybee species. Eur Food Res Technol 27:355–357, [in German]. Keywords: beeswax/*cerana*/*dorsata*/*florea*/nests

Spivak M, Boecking O (2001) Honey bee resistance to *Varroa* mites. In: Webster TC, Delaplane KS (eds) Mites of the honey bee. Dadant, Hamilton, pp 205–227, Keywords: *cerana*/disease/honey/pests

Srawan BS, Sohi BS (1985) Phytosociological studies on *Apis mellifera* L and *Apis cerana indica* F in Punjab, India. Indian Bee J 47:15–18, Keywords: bee botany/*cerana*/*mellifera*/India

Sreeramulu KR, Gavi G (1998) Thai sac brood virus (TSBV) and its management. Insect Environ 4:99, Keywords: beekeeping/*cerana*/diseases/virus

Srivastava AS, Awasthi GP (1962) The effect of temperature on the activity of honey bee head cholinesterase. In: Proceedings of the National Institute of Science (India) 32 B:23–24, Keywords: biochemistry/*cerana*/honey/morphology/physiology/India

Srivastava VS, Bhaaradwaj RK (1968) A practical method for collecting specimens of the rockbee, *Apis dorsata* Fabricius, its colony and queen. Entomologist 101:233–235, Keywords: beekeeping/*dorsata*/queens

Stace P (1999) Observations on the Asian bee, *Apis cerana*, in the Rego Area of Papua New Guinea. Austral Beekeep 101:149, Keywords: *cerana*/Papua New Guinea

Starr CK, Schmidt PJ, Schmidt JO (1987) Nest-site preferences of the giant honeybee, *Apis dorsata* (Hymenoptera: Apidae), in Borneo. Pan-Pac Entomol 63:37–42, Keywords: *cerana*/classification/combs/*dorsata*/honey/nests/Borneo/New Guinea/Papua New Guinea

Stitz J, Szebe P (1933) The evolution of the races of bees and their geographical distribution. Bee World 14:128–130, Keywords: biogeography/*cerana*/*dorsata*/evolution/*florea*/*mellifera*/phylogeny/subspecies

Strange JP (2001) A severe stinging and much fatigue – Frank Benton and his 1881 search for Apis dorsata. Am Entomol 147:112–116, Keywords: defense/*dorsata*

Strickland SS (1982) Honey hunting by the Gurungs of Nepal. Bee World 63:153–161, Keywords: *dorsata*/Nepal

Su R (1992) A preliminary study on flight duration of drone honey bees, *Apis cerana cerana*. J Fujian Agric Coll 21:316–318, [in Chinese]. Keywords: *cerana*/drones/flight/honey/reproduction/China

Su R (1993a) Effects of an *Apis cerana cerana* queen leaving its nest and laying eggs in a different mating colony. J Bee 8:3, [in Chinese]. Keywords: *cerana*/oviposition/queens/reproduction/China

Su R (1993b) Effect of mating nuclei on mating flights and egg laying of queens of Chinese bees (*A. cerana cerana*). J Bee 8:4–5, [in Chinese]. Keywords: *cerana*/flight/oviposition/queens/reproduction

Su R (1994) Reaction of *Apis cerana cerana* feeding at different distances and heights. J Fujian Agric Coll 23:234–236, [in Chinese]. Keywords: *cerana*/dance language/foraging/China

Su R, Lin X, Hu F (2000) The comparison of pollen morphs and nutrient components between corbicular tea (*Camellia sinensis*) pollen and bee bread. Apic China 51:3–6, Keywords: *cerana*/pollen/China

Su RM, Zhang QK (1990) Techniques for extracting and washing semen of *Apis cerana cerana*. Apic China 5:8–10, [in Chinese]. Keywords: *cerana*/drones/China

Su RM, Zhang QK (1991) Studies on the effects of sugar on semen energetics in *Apis cerana cerana*. J Bee 8:3–5, [in Chinese]. Keywords: *cerana*/drones/physiology/China

Su S, Chen S (2004) Advances in the genetic markers of honey bee. Apic China 55:8–10, [in Chinese]. Keywords: *cerana*/genetics/*mellifera*

Su S, Chen SL, Zhong BX, Albert S (2004) Cloning and sequence analysis of mrjp1 gene from *Apis cerana cerana*. Acta Genet Sinica 31:1248–1253, [in Chinese]. Keywords: *cerana*/DNA/genetics/*mellifera*/morphology/India

Su S, Albert S, Chen S, Zhong B (2005a) Molecular cloning and analysis of four cDNAs from the heads of *Apis cerana cerana* nurse honeybees coding for major royal jelly proteins. Apidologie 36:389–401, Keywords: *cerana*/DNA/proteins/China

Su S, Chen S, Zhong B, Zheng H, Albert S (2005b) Cloning and sequence analysis of cDNA encoding mrjp3 of *Apis cerana cerana*. Sci Agric Sinica 38:612–618, [in Chinese]. Keywords: *cerana*/DNA/genetics/China

Subbiah MS (1956) Studies on pollen substitutes for bees. Indian Bee J 18:136–142, Keywords: beekeeping/*cerana*/pollen/India

Subbiah MS, Mahadevan V (1957) *Vespa cincta* Fabr. – a predator of the hive bees and its control. J Vet Sci 27:153–154, Keywords: *cerana*/natural enemies/nests/India

Sudarsanam D, Murthy VA (1990) Phoretic association of the pseudoscorpion *Ellingsenius indicus* with *Apis cerana indica*. In: Proceedings of the 11th international congress, IUSSI, Bangalore, pp 721–722, Keywords: *cerana*/pests/India

Sudrajat A, Sulistianto A (1993) Beekeeping in Indonesia. Honeybee Sci 14:129–132, [in Japanese]. Keywords: beekeeping/*cerana*/Indonesia

Sugahara M (1997) Distribution of *Apis cerana japonica* in Moriguchi City, Osaka Prefecture. Honeybee Sci 18:17–20, [in Japanese]. Keywords: *cerana*/distribution/Japan

Sugahara M (1998) Feral colonies of Japanese honey bees in the northeastern part of Osaka Prefecture. Honeybee Sci 19:37–41, [in Japanese]. Keywords: *cerana*/honey/nests/population biology/Japan

Sugahara M (2000) Feral colonies of Japanese honey bees, *Apis cerana japonica* and their life history 2. Natural nests and swarming. Honeybee Sci 21:35–39, Keywords: *cerana*/history/honey/nests/population biology/seasonal variation/swarming/Japan

Sugahara M (2001a) Hybrids of *Cymbidium pimilum* can attract the Japanese honeybee, *Apis cerana japonica*. Honeybee Sci 22:83–86, [in Japanese]. Keywords: bee botany/*cerana*/genetics/pheromones/Japan

Sugahara M (2001b) *Cymbidium devonianum* also attracts the Japanese honeybee, *Apis cerana japonica*. Honeybee Sci 22:139–140, [in Japanese]. Keywords: bee botany/*cerana*/pheromones/Japan

Sugahara M (2002) Swarming of colonies of the Japanese honeybee *Apis cerana japonica* are attracted by *Cymbidium devonianum* and its hybrid, Miss Muffet. Honeybee Sci 23:121–122, [in Japanese]. Keywords: bee botany/*cerana*/pheromones/swarming/Japan

Sugahara M, Kondou K (2005) *Apis cerana japonica* Rad. killing a female *Promachus yesonicus* Bigot by making a spherical swarm around it. Hana abu (Kyota) 20:87–88, [in Japanese]. Keywords: aggressiveness/*cerana*/Japan

Sugahara M, Kusakabe A (2001) Feral colonies of Japanese honeybees, *Apis mellifera japonica* and their life history 2. Habitat and life cycle. Honeybee Sci 22:122–124, [in Japanese]. Keywords: biology/*cerana*/history/Japan

Sugahara M, Tsutsui K (1998) Foraging behaviour of *Apis cerana japonica* on Oriental orchids. Cymbidium kanran and C. virescens. Honeybee Sci 19:81–82, [in Japanese]. Keywords: bee botany/behaviour/*cerana*/foraging/Japan

Sugahara M, Shimizu Y, Minamoto T, Shimizu I (2006) The honeybee habitation in the Osaka prefecture (Japan). Honeybee Sci 27:19–22, [in Japanese]. Keywords: *cerana*/ecology/Japan

Sugimoto K (1989) Traditional hives for native Japanese honeybees (*Apis cerana japonica*) in Tsushima Island. Honeybee Sci 10:121–124, [in Japanese]. Keywords: *cerana*/islands/Japan

Suh YT, Shim JH (1988) Effects of insecticides on acetylcholinesterase activity in honeybees (*Apis cerana* and *A. mellifera*). Korean J Apic 3:48–60, [in Korean]. Keywords: *cerana*/insecticides/*mellifera*/Korea

Suh YT, Shim JH (1989) A study on the enzyme activities of a honeybee (*Apis cerana* F) associated with the degradation of some insecticides. Korean J Environ Agric 8:47–54, [in Korean]. Keywords: *cerana*/insecticides/Korea

Suka T, Tanaka H (2005) New mitochondrial CO 1 haplotypes and genetic diversity in the honeybee *Apis koschevnikovi* of the Crocker Range Park, Sabah, Malaysia. J Trop Biol Conser 1:1–7, Keywords: *koschevnikovi*/mtDNA/Borneo

Sukarsih DLA (1991) Infestation of honey bees (*Apis mellifera* and *A. cerana*) by *Varroa jacobsoni* and *Tropilaelaps clareae* after treatment with a mixture of sulfur and naphthalene. Penyakit Hewan 23:19–21, [in Indonesian]. Keywords: *cerana*/honey/insecticides/*mellifera*/pests

Sukartiko B (1981) Country report on beekeeping in Indonesia. In: Proceedings of the 28th international apiculture congress, Apimondia, Acapulco, pp 198–202, Keywords: beekeeping/*cerana*/Indonesia

Sulaksono S, Suryati T, Nismah T, Susilohadi RCH (1986) Biology of *Apis cerana* by means of stress on foraging activity. Proc Prosiding Lokakarya Pembudidayaan Lebah Madu Untuk Peningkatan Kesejahteraan Masyarakat, Jakarta, pp 49–64 [in Indonesian]. Keywords: biology/*cerana*/foraging/Indonesia

Sulistianto A (1990) Morphometric analysis of Indonesian honeybees. Thesis, University of Wales College of Cardiff, Keywords: *cerana*/classification/geographical variation/*koschevnikovi*/morphometrics/Indonesia

Sumangala K, Haq MA (2001) Survey of the mite fauna associated with *Apis* spp. in Kerala, southern India. In: Proceedings of the 10th international congress, Acarol, pp 565–568, Keywords: *cerana*/*dorsata*/*florea*/*mellifera*/mites/pests

Sumitani M, Lee JM, Hatakeyana M, Oishi K (2002) Cloning and characterization of Acmar I, a mariner-like element in the Asiatic honeybee, *Apis cerana japonica* (Hymenoptera, Apocrita). Arch Insect Biochem Physiol 50:183–190, Keywords: *cerana*/Japan

Sumoto S, Sasaki M (1991) On the morphology of blunt bent hair of the honey bee mandible. Honeybee Sci 12:111–114, [in Japanese]. Keywords: *cerana*/honey/morphology/Japan

Sun GF, Zhai SH, Guo LH, Xiao FH (2009) The floral syndrome and its pollination in *Cynanchum otophyllum* (Asclepiadaceae). Acta Bot Yunnanica 31:8–14, [in Chinese]. Keywords: bee botany/*cerana*/pollination

Sun QH (1982) The distribution of the Chinese honey bee and its variations II. A survey of the Chinese honey bee in the southern mountainous regions of Xizang. Zhongguo Yangfeng 4:19, [in Chinese]. Keywords: biogeography/*cerana*/honey/mountains/China*

Sun Y, Zhuo LH (2009) Study on pollinating insects of *Agapanthus africanus* and their foraging behaviour. Acta Agric Shanghai 25:1, [in Chinese]. Keywords: *cerana*/foraging/pollination/China*

Sun ZX (2001) The rare giant bee (*Apis dorsata*) and little bee (*A. florea*). Special Economic Propagation 4:9 [in Chinese]. Keywords: *dorsata*/*florea*/China*

Sundararaju D (2002) Foraging behaviour of pollinators on cashew. Cashew 14:17–20, Keywords: behaviour/*cerana*/foraging/pollination/India

Surdacki S (1973) The geographical range and distribution of the genus *Apis* L bees. Ann Univ Mariae Curie-Sklodowska Sect C Biol B 28:315–330, Keywords: behaviour/biogeography/classification/*dorsata*/*florea*/*mellifera*

Suryanarayana MC (1966) Studies on flora of Coorg with emphasis on the bee forage plants 1. General ecological and botanical features. Indian Bee J 28:59–75, Keywords: bee botany/*cerana*/*dorsata*/*florea*/foraging/India

Suryanarayana MC (1995) The story before and after formation of beekeeping development board. Indian Bee J 57:38–47, Keywords: beekeeping/*cerana*/India

Suryanarayana MC (1996a) Appropriate technologies for beekeeping with *Apis cerana*. In: Proceedings of international workshop revive *Apis cerana indica* Kodaikanal, pp 1–5, Keywords: beekeeping/*cerana*/India

Suryanarayana MC (1996b) Revival of *Apis cerana* beekeeping in Karnataka: a tour report. Indian Bee J 58:25–30, Keywords: beekeeping/*cerana*/India

Suryanarayana MC (1996c) Revival of *Apis cerana* beekeeping in Tamil Nadu: a tour report. Indian Bee J 58:157–168, Keywords: beekeeping/*cerana*/India

Suryanarayana MC, Mohana-Rao G (1993) Role of honey bees in the pollination biology of tropical crop plants. In: Proceedings of international symposium on pollination in tropics, Bangalore, pp 310–320, Keywords: biology/*cerana*/honey/pollination/India

Suryanarayana MC, Subba Rao K (1997) *A. cerana* F for Indian apiculture and its management technology. In: Mishra RC (ed) Perspectives in Indian apiculture. Agro Botanica, New Delhi, pp 66–118, Keywords: beekeeping/*cerana*/India

Suryanarayana MC, Rao GM, Singh TSMS (1990) Coconut palm – a pollen and nectar source to honey bees. Indian Bee J 52:41–43, Keywords: bee botany/*cerana*/*dorsata*/foraging/honey/nectar/pollen/India

Suryanarayana MC, Rao GM, Singh TSMS (1992) Studies on pollen sources for *Apis cerana* Fabr. and *Apis mellifera* L. bees at Muzaffarpur, Bihar, India. Apidologie 23:33–46, Keywords: bee botany/*cerana*/foraging/*mellifera*/pollen/India

Suryanarayana MC, Mohana-Rao G, Lakshmi K (1993) The sandal tree: a nectar source for *Apis cerana* and *Apis mellifera*. Bee World 74:108–110, Keywords: bee botany/*cerana*/foraging/*mellifera*/nectar

Sutakam K (1998) Relationship between honey bee (*Apis mellifera* L) population density and longan fruit setting. Thesis, Chiang Mai University, Keywords: *cerana*/*dorsata*/*florea*/honey/*mellifera*/pollination/population biology/Thailand

Suvakon C (1999) The efficiency of pollination by different bees: comparison between the Asian honey bee *Apis cerana* Fabr. and the European honey bee *Apis mellifera* L, some bumble bees and solitary bees. Thesis, Universität für Bodenkultur Wien, Keywords: bee botany/*cerana*/foraging/honey/*mellifera*/pollination

Suwanagul M, Wongsiri S, Wongsathuaythong S (1987) Comparative study of allergic reaction to honey bee venom by skin testing. J Sci Res (Chulalongkorn) 12:9–13, Keywords: biochemistry/*cerana*/*dorsata*/*florea*/honey/*mellifera*/Thailand

Suwanapong G (2000) Ultrastructure and pheromones of the mandibular glands of honeybee foragers in Thailand. Thesis, Chulalongkorn University, Keywords: *cerana*/*dorsata*/*florea*/glands/*mellifera*/morphology/pheromones/ultrastructure/Thailand

Suwanapong G, Wongsiri S (2004) Scanning electron microscopic study of the antennal sensillae of *Apis dorsata* Fabricius, 1793 workers. In: Proceedings of the 7th Asian apiculture association international conference, Manila, pp 149–154, Keywords: *dorsata*/morphology/workers/Thailand

Suzuki H, Iwanami Y, Okada I (1969) Studies on the relationship between pollen grains and honeybees. Jpn J Palynol 3:19–31, Keywords: *cerana*/pollen/Japan

Svensson B (1989) Bee management and company management. In: Kevan PG (ed) Beekeeping with *Apis cerana* in tropical and sub-tropical Asia. Enviroquest, Cambridge, ON, pp 217–221, Keywords: beekeeping/*cerana*

Swaminathan MS (1986) Tropical and subtropical apiculture. Report, vol 68, FAO Agricultural Services, Rome, pp 1–283, Keywords: beekeeping/*cerana*/India

Swaminathan R, Bharadwaj SC (1998) Role of insect pollinators in seed setting of sunflower (*Helianthus annuus* L) at Udaipur, Rajasthan, India. Indian Bee J 60:155–156, Keywords: bee botany/*dorsata*/pollination/India

Swamy H, Rajagopal D (2002) Association of *Varroa jacobsoni* and Thai sac brood virus disease of *Apis cerana* in Karnataka. Asian Bee J 4:19–26, Keywords: brood/*cerana*/disease/India

Syed IH (1972) Morphological studies of the larvae of *Apis dorsata*. Pak J Zool 4:69–71, Keywords: brood/*dorsata*/morphology/Pakistan

Sylvester HA, Wongsiri S (1986) Beekeeping and research needs in Thailand. Apiacta 21:119–125, Keywords: beekeeping/*cerana*/Thailand

Sylvester HA, Wongsiri S (1993) DNA analysis of genetic variation in Asian honey bees. In: Connor LJ, Rinderer TE, Sylvester HA, Wongsiri S (eds) Asian apiculture. Wicwas, Cheshire, CT, pp 156–160, Keywords: *andreniformis*/*cerana*/DNA/*dorsata*/*florea*/genetics/honey/*mellifera*/workers/Thailand

Sylvester HA, Limbipichai K, Wongsiri S, Rinderer TE, Mardan MB (1998) Morphometric studies of *Apis cerana* in Thailand and the Malaysian peninsula. J Apic Res 37:137–145, Keywords: *cerana*/classification/morphometrics/Malaysia/Thailand

Szabo TI (1988) Sri Lanka: the bee paradise. Am Bee J 128:405–409, Keywords: beekeeping/*cerana*/*dorsata*/*florea*/Sri Lanka

Szabo TI (1990) Morphometric characteristics of *Apis cerana* from Sri Lanka. Apidologie 21:505–509, Keywords: *cerana*/classification/morphometrics/Sri Lanka

Tadauchi O, Dawut A, Inoue H (2001) On image database file HANABACHI based on the Japanese bees. Esakia 41:149–154, Keywords: biology/*cerana*/Japan

Tahmasebi G, Ebadi R, Tajbadi N, Akhondi M, Faraji S (2002) The effects of geographical and climatic conditions on the morphological variation and separation of Iranian small honeybee (*Apis florea* F) populations. J Sci Tech Agric Nat Res 6:169–176, [in Persian]. Keywords: biogeography/classification/environmental effects/*florea*/general biology/morphology/Iran

Takahashi JI, Nakamura J (2003) A scientific note on levels of polyandry of 2 queens of the Himalayan giant honeybee, *Apis laboriosa*. Apidologie 34:191–192, Keywords: *laboriosa*/queens/reproduction/Nepal

Takahashi JI, Yoshida T (2002) Phylogenetic position of Tsushima Island honeybees: multivariate analysis using wing morphological characters of the worker. Honeybee Sci 23:115–120, [in Japanese]. Keywords: *cerana*/islands//morphometrics/phylogeny/Japan

Takahashi JI, Yoshida T (2003) Origin of the Japanese honeybee *Apis cerana japonica* inferred from mitochondrial DNA. Honeybee Sci 24:71–76, [in Japanese]. Keywords: *cerana*/DNA/mtDNA/Japan

Takahashi JI, Nakamura J, Sasaki M, Tingek A, Akimoto S (2002) New haplotypes for the non-coding region of mitochondrial DNA in cavity nesting honey bees *Apis koschevnikovi* and *Apis nuluensis*. Apidologie 33:25–31, Keywords: *cerana*/DNA/honey/*koschevnikovi*/mtDNA/nests/*nigrocincta*/*nuluensis*

Takahashi J, Yoshida T, Takagi T, Akimoto S, Woo KS, Deowanish S, Hepburn R, Nakamura J, Matsuka M (2007) Geographic variation in the Japanese islands of *Apis cerana* japonica and in *A. cerana* populations bordering its geographic range. Apidologie 38:335–340, Keywords: biogeography/*cerana*/Japan

Takahashi JI, Shimizu KS, Kimura K, Shimizu K (2009) Variable microsatellite loci isolated from the Asian honeybee, *Apis cerana* (Hymenoptera; Apidae). Mol Ecol Res 9:819–821, Keywords: *cerana*/genetics/Japan

Takanaka T (1982) Chemical composition of royal jelly. Honeybee Sci 3:69–74, [in Japanese]. Keywords: *cerana*/royal jelly/Japan

Takanaka T (1983) Proteins and peptides in royal jelly. Nippon Noge Kagakukai Shi. J Agric Chem Soc Jpn 57:1203–1209, Keywords: biochemistry/*cerana*/royal jelly/Japan

Takanaka T (1984) Studies on proteins and carboxylic acid in royal jelly. Bull Fac Agric Tamagawa Univ 24:101–149, [in Japanese]. Keywords: biochemistry/*cerana*/royal jelly/Japan

Takanaka T, Takanaka Y (2002) Royal jelly from *Apis cerana japonica* and *Apis mellifera*. Biosci Biotech Biochem 60:518–520, Keywords: *cerana*/*mellifera*/royal jelly/Japan

Takata T (1937) Old-style beekeeping in Korea. Bee World 18:132, Keywords: beekeeping/*cerana*/Korea

Takeba K, Kanzaki M, Murakami F, Matsumoto M (1984) Simplified analytical method for tetracycline residues in honey by height performance liquid chromatography. Rep Tokyo Metrop Res Lab Pub Health 35:187–191, [in Japanese]. Keywords: *cerana*/honey/insecticides/Japan

Takeuchi K (1993) Extinction of *Varroa* mites in the Japanese honeybee (*Apis cerana japonica* Rad) colony. Honeybee Sci 14:58–60, [in Japanese]. Keywords: *cerana*/pests/Japan

Takeuchi K, Hagiwara M, Wada K (2001) On the nutritive effects of royal jelly and pollen loads of honeybee to Drung darkling beetle. Collect Breed 32:280–281, [in Japanese]. Keywords: *cerana*/pollen/royal jelly/Japan

Taki T, Okabe K, Makino S, Yamaura Y, Sueyoshi M (2009) Contribution of small insects to pollination of common buckwheat, a distylous crop. Ann Appl Biol 155:121–129, Keywords: *cerana*/*mellifera*/pollination/Japan

Takuno Y (1992) Traditional beekeeping of Japanese honey bees in Chugoku mountainous area. Honeybee Sci 13:159–166, [in Japanese]. Keywords: beekeeping/*cerana*/honey/natural enemies/Japan

Takuno Y (1994) Traditional beekeeping on Tsushima Island. Honeybee Sci 15:59–68, [in Japanese]. Keywords: beekeeping/cells/*cerana*/islands/nests/swarming/ultrastructure/Japan

Takuno Y (1997) Traditional beekeeping in the mountainous region of Kii. Honeybee Sci 18:55–64, [in Japanese]. Keywords: beekeeping/*cerana*/Japan

Tam DQ (1997) Marketing honey from various honeybee species in Vietnam. In: Sommerijer MJ, Beetsma J, Boot WJ, Robberts EJ, De Vries R (eds) Perspectives for honey production in the tropics. Netherlands Expertise Centre for Tropical Apicultural Resources, Bennekom, Netherlands, pp 171–182, Keywords: beekeeping/*cerana*/*dorsata*/*florea*/honey/Vietnam

Tam DQ (2001) Honey marketing in Vietnam. In: Proceedings of the 37th international apiculture congress, Apimondia, Durban (CD Rom). Keywords: *andreniformis*/beekeeping/*cerana*/*dorsata*/*florea*/honey/Vietnam

Tam DQ, Dung PX (1996) Introduction of Vietnam apiculture. In: Proceedings of the 3rd Asian apiculture association international conference, Hanoi, pp 17–24, Keywords: *andreniformis*/beekeeping/beeswax/*cerana*/*dorsata*/*florea*/history/honey/*mellifera*/pests/Vietnam

Tam DQ, Niem NV, Trung LQ, Dau TT, Chinh PH, Long LT (eds) (2005) Publications of Vietnam bee research 1990–1999. Bee Research and Development Center, Hanoi, Keywords: beekeeping/behaviour/*cerana*/disease/*dorsata*/genetics/*laboriosa*/pests/predators/reproduction/Vietnam

Tamura T (1985) Royal jelly from the standpoint of clinical pharmacology. Honeybee Sci 6:117–124, [in Japanese]. Keywords: *cerana*/royal jelly/Japan

Tan HT, Rahman RA, Gan SH, Halim AS, Hssan AS, Sulaiman SA, Kirnpal-Kaur BS (2009a) The antibacterial properties of Malaysian tualang honey against wound and enteric microorganisms in comparison to manuka honey. BMC Complement Altern Med 9:34, Keywords: honey

Tan HW, Dong X, He SY (2009b) Amplification and sequence analysis of the mitochondrial 16S rRNA gene of *Apis cerana* from Yunnan Province. J Yunnan Agric Univ 24:3, [in Chinese]. Keywords: *cerana*/genetics/China*

Tan K (1991) China's beekeeping and the Journal of the Bee. Am Bee J 131:459–460, Keywords: beekeeping/*cerana*/China

Tan K (1996) Breeding techniques and selection for breeding of *Apis cerana* in Yunnan Province. In: Proceedings of the 3rd Asian apiculture association international conference, Hanoi, p 48, Keywords: breeding/*cerana*/China

Tan K, Kuang BY (1999) Comparison of the foraging behaviour between *A. cerana* and *A. mellifera*. In: Proceedings of the 36th international apiculture congress, Apimondia,Vancouver, p 69, Keywords: behaviour/*cerana*/foraging/*mellifera*/China

Tan K, Kuang BY (2000) Discussion on the eco-genetics of the honeybee. In: Proceedings of the 14th international congress, IUSSI, Sapporo, p 208, Keywords: *cerana*/ecology/genetics/China

Tan K, Liu YQ (2000) Nutrition techniques for different species of economically important insects. Yunnan Science, Kunming, [in Chinese]. Keywords: *cerana*/physiology/China

Tan K, Kuang BY, Zhang X (2000a) Ecology and genetics of the eastern honeybee *Apis cerana*. J Bee 26:4–5, [in Chinese]. Keywords: *cerana*/ecology/genetics/China

Tan K, Fuchs S, Koeniger N, Ruiguang Z (2000b) Morphometrical characterisation of *A. cerana* in Yunnan Province of China. Pszcz Zesz Nauk 44:339–340, Keywords: *cerana*/classification/morphometrics/China/Nepal/Vietnam

Tan K, Zhang X (2001) Morphological study of *Apis cerana* in Yunnan. J Bee 27:2–4, [in Chinese]. Keywords: *cerana*/morphology/China

Tan K, Fuchs S, Ruiguang Z (2002a) Morphometrical characterization of *Apis cerana* in Yunnan Province of China. In: Proceedings of the 14th international congress, IUSSI, Sapporo, p 208, Keywords: *cerana*/morphology/China/Japan/Korea/Myanmar (Burma)/Nepal/Thailand/Vietnam

Tan K, He SY, Zhang X (2002b) MtDNA variation of the Eastern honeybee *Apis cerana* in Yunnan. Apic China 53:4–6, [in Chinese]. Keywords: *cerana*/mtDNA/China

Tan K, He SY, Zhang X (2002c) Study on the correlation between morphometric characters and environmental factors of *Apis cerana* in Yunnan. J Bee 28:5–7, [in Chinese]. Keywords: *cerana*/classification/environmental effects/morphometrics/China

Tan K, He SY, Zhang X (2002) Using the Dra I enzyme to study mtDNA variation of *Apis cerana* in Yunnan. J Bee 28, [in Chinese]. Keywords: *cerana*/mtDNA/China*

Tan K, Yu YH, Zhang X (2002e) Resistance of *Apis cerana* colonies towards *Varroa jacobsoni* mites. Apic China 53:10–11, [in Chinese]. Keywords: *cerana*/disease/pests/China

Tan K, Fuchs S, Koeniger N, Ruiguang Z (2002f) Morphological characterization of *Apis cerana* (Fabr.) in the Province Yunnan of China. Apidologie 34:553–561, Keywords: biogeography/*cerana*/classification/morphometrics/China

Tan K, He SY, Zhang X, Zhou DY (2002) Study on mtDNA variation of *Apis cerana* in Yunnan. Apic China 53* [in Chinese]. Keywords: *cerana*/DNA/mtDNA/China/Laos/Thailand/Vietnam

Tan K, Zhang X, He SY, Zhou DY (2003) Morphometrics and biogeography of *Apis cerana* in Yunnan. J Yunnan Agric Univ 53:4–6, [in Chinese]. Keywords: biogeography/*cerana*/classification/morphometrics/China

Tan K, Hepburn HR, Radloff SE (2005) Heat-balling wasps by honeybees. Naturwissenschaften 92:492–495, Keywords: aggressiveness/*cerana*/*Vespa*

Tan K, Hepburn HR, Radloff SE (2006a) Gigantism in honeybees: *Apis cerana* queens reared in mixed-species colonies. Naturwissenschaften 93:315–320, Keywords: biology/*cerana*/queen

Tan K, Meixner MD, Fuchs S, Zhang X, He SY, Kandemir I, Sheppard WS, Koeniger N (2006b) Geographic distribution of the eastern honeybee, *Apis cerana* (Hymenoptera: Apidae) across ecological zones in China: morphological and molecular analyses. Syst Biodiv 4:473–482, Keywords: biogeography/*cerana*/DNA/ecology/China

Tan K, Warrit N, Smith DR (2007a) Mitochondrial DNA diversity of Chinese *Apis cerana.* Apidologie 38:238–246, Keywords: *cerana*/mtDNA

Tan K, Guo YH, Nicolson SW, Radloff SE, Song QS, Hepburn HR (2007b) Honeybee (*Apis cerana*) foraging responses to the toxic honey of *Tripterygium hypoglaucum* (Celastraceae): changing threshold of nectar acceptability. J Chem Ecol 33:2209–2217, Keywords: *cerana*/foraging/honey/*mellifera*

Tan K, Radloff SE, Li JJ, Hepburn HR, Yang MX, Lj Z, Neumann P (2007c) Bee-hawking by the wasp, *Vespa velutina*, on the honeybees *Apis cerana* and *A. mellifera*. Naturwissenschaften 94:469–472, Keywords: *cerana*/*mellifera*/natural enemies

Tan K, Fuchs S, Engel M (2008a) An adventitious distal abscissa in the forewing of honey bees. Apidologie 39:674–682, Keywords: *cerana*/morphology

Tan K, Yang MX, Radloff SE, Hepburn HR, Zhang ZY, Luo LJ, Li H (2008b) Dancing to different tunes: heterospecific deciphering of the honeybee waggle dance. Naturwissenschaften 95: 1165–1168, Keywords: *cerana*/*mellifera*/dance language

Tan K, Yang MX, Radloff S, Pirk CWW, Crewe RM, Phiancharoen M, Hepburn R, Oldroyd BP (2009) Worker reproduction in mixed-species colonies of **honeybees**. Behav Ecol 20:1106–1110, Keywords: *cerana*/laying workers/*mellifera*/reproduction

Tan NQ (2004) Social-economic factors in traditional rafter beekeeping with *Apis dorsata* in Vietnam. Bee World 83:165–170, Keywords: beekeeping/*dorsata*/Vietnam

Tan NQ (2008) Pollination ecology of *Melaleuca cajuputi*, *Nypa fruticans* and their flower visitors. J Apic Res 47:10–16, Keywords: *florea*/pollination

Tan NQ, Binh PT (1994) Harmony or conflict? *Apis mellifera* and *Apis cerana* in southern Vietnam. Beekeep Dev 32:4–7, Keywords: beekeeping/*cerana*/*mellifera*/physiology/Vietnam

Tan NQ, Huan LV, Dung NV (1993) Development time of eggs, open brood and capped brood of *Apis cerana* and *Apis mellifera* in southern Vietnam. In: Connor LJ, Rinderer TE, Sylvester HA, Wongsiri S (eds) Asian apiculture. Wicwas, Cheshire, CT, pp 243–248, Keywords: brood/*cerana*/*mellifera*/oviposition/physiology/Vietnam

Tan NQ, Dien PC, Huan LV (1996) Response of *Apis cerana* colonies towards *Varroa jacobsoni* introduced from *Apis mellifera* and that of their own. In: Proceedings of the 3rd Asian apiculture association international conference, Hanoi, pp 191–197, Keywords: *cerana*/*mellifera*/pests/Vietnam

Tan NQ, Mardan MB, Thai PH, Chinh PH (1996) Some reproductive biology of *Apis dorsata*. In: Proceedings of the 3rd Asian apiculture association conference, Hanoi, pp 15–22, Keywords: *dorsata*/drones/general biology/reproduction/Vietnam

Tan NQ, Chinh PH, Thai PH, Mulder V (1997) Rafter beekeeping with *Apis dorsata*: some factors affecting the occupation of rafters by bees. J Apic Res 36:49–54, Keywords: beekeeping/*dorsata*/Vietnam

Tan NQ, Mardan MB, Thai PH, Chinh PH (1999) Observations on multiple mating flights of *Apis dorsata* queens. Apidologie 30:339–346, Keywords: *dorsata*/foraging/queens/reproduction

Tanabe Y, Tamaki T (1985) Biochemical genetic studies on *Apis mellifera* and *Apis cerana*. In: Proceedings of the 30th international apiculture congress, Apimondia, Nagoya, pp 152–154, Keywords: beekeeping/*cerana*/genetics/*mellifera*/Japan

Tanabe Y, Tamaki Y, Nakano S (1970) Variations of esterase isozymes in seven species of bees and wasps. Jpn J Genet 45:425–428, Keywords: biochemistry/*cerana*/Japan

Tanaka H, Suka T, Roubik DW, Mohamed M (2001) Genetic differentiation among geographic groups of three honeybee species, *Apis cerana*, A. koschevnikovi and A. dorsata in Borneo. Nat Hum Activities 6:5–12, Keywords: biogeography/*cerana*/*dorsata*/genetics/honey/*koschevnikovi*/Borneo

Tanaka H, Roubik DW, Kato M, Liew F, Gunsalam G (2002) Phylogenetic position of *Apis nuluensis* of northern Borneo and phylogeography of *A. cerana* as inferred from mitochondrial DNA sequences. Insectes Soc 48:44–51, Keywords: *cerana*/mtDNA/*nuluensis*/phylogeny/Borneo

Tanaka H, Suka T, Kahono S, Samejima MM, Roubik DW (2003) Mitochondrial variation and genetic differentiation in honeybees (*Apis cerana, A. koschevnikovi* and *A. dorsata*) of Borneo. Tropics 13:107–117, Keywords: *cerana*/classification/*dorsata*/genetics/honey/*koschevnikovi*/mtDNA/Borneo

Tanda AS (1983) Assessing the role of honey bees in a field of Asiatic cotton (*Gossypium arboreum* L). Am Bee J 123:593–594, Keywords: *cerana*/*dorsata*/honey/*mellifera*/pollination/India

Tanda AS (1984a) Bee pollination increases yield of 2 interplanted varieties of Asiatic cotton (*Gossypium arboreum* L). Am Bee J 124:539–540, Keywords: *cerana*/*dorsata*/*mellifera*/pollination/India

Tanda AS (1984b) Foraging behaviour of three species of *Apis* on raya in relation to the sugar concentration in its nectar. Indian Bee J 46:5–6, Keywords: behaviour/*cerana*/*dorsata*/*florea*/foraging/nectar/India

Tanda AS (1985) Floral biology, pollen dispersal, and foraging behaviour of honeybees in okra (*Abelmoschus esculentum*). J Apic Res 24:225–227, Keywords: behaviour/biology/*cerana*/foraging/*mellifera*/pollen/pollination/India

Tanda AS, Goyal NP (1979a) Insect pollination in Asiatic cotton (*Gossypium arboreum*). J Apic Res 18:64–72, Keywords: *cerana*/*mellifera*/pollination/India

Tanda AS, Goyal NP (1979b) Preliminary observations on the effect of intervarietal bee pollination of desi cotton (*Gossypium arboreum* Linn). Indian J Entomol 41:281–282, Keywords: *cerana*/pollination/India

Tanda AS, Goyal NP (1979c) Some observations on the behavior of *Apis mellifera* Linn and *Apis cerana indica* Fabr. workers in a field of desi cotton (*Gossypium arboreum* Linn). Am Bee J 119:106, Keywords: *cerana*/foraging/*mellifera*

Tandon R, Shivanna KR, Mohan Ram HY (2001) Pollination biology and breeding system of *Acacia senegal*. Bot J Linn Soc 135:251–262, Keywords: bee botany/beekeeping/*dorsata*/general biology/pollination/India

Tandon R, Shivanna KR, Ram HYM (2010) Reproductive biology of some gum-producing Indian desert plants. In: Ramawat KG (ed) Desert plants: biology and biotechnology. Springer, Berlin, Keywords: bee botany/*cerana*/*dorsata*/pollination/India

Tang CC, Kellogg CR (1963) Honey and its history in China. Am Bee J 103:176–178, Keywords: *cerana*/history/honey/China

Tangjingjai W, Verakalasa P, Sittipraneed S, Klinbunga S, Lekprayoon C (2003) Genetic differences between *Tropilaelaps clareae* and *Tropilaelaps koenigerum* in Thailand. Apidologie 34:513–524, Keywords: disease/*dorsata*/pests/predators/Thailand

Tangkanasing P, Wongsiri S, Vongsamanode S (1988) Integrated control of *Varroa jacobsoni* and *Tropilaelaps clareae* in Thailand. In: Needham GR, Page RE, Delfinado-Baker M, Bowman CE (eds) Africanized honey bees and bee mites. Ellis Horwood, Chichester, pp 409–412, Keywords: *cerana*/honey/pests/Thailand

Taniguchi S (1954) Biological studies on the Japanese bees. 1. Comparative study of glossa. *science reports* of the *Hyogo University of Agriculture*, Series Agriculture 1:81–89, Keywords: *cerana*/classification/morphology/Japan

Tao T, Su SK, Miao YG, Yue WF, Du HH, Chen SL, Liu F, Zhanam Y (2008) Expression of apalbumin 1 of *Apis cerana cerana* in the larvae of silkworm, *Bombyx mori*. J Agric Fd Chem 56:9464–9468, Keywords: *cerana*/genetics/China

Tarpy DR, Nielsen DI (2002) Sampling error, effective paternity, and estimating the genetic structure of honey bee colonies (Hymenoptera: Apidae). Ann Entomol Soc Am 95:513–528, Keywords: *andreniformis*/*cerana*/*dorsata*/*florea*/genetics/*koschevnikovi*/*mellifera*/*nigrocincta*

Tarpy DR, Nielsen R, Nielsen DI (2004) A scientific note on the revised estimates of effective paternity frequency in *Apis*. Insectes Soc 51:203–204, Keywords: *cerana*/*dorsata*/*laboriosa*/reproduction

Tealdi C (1988) Honeybees in Indonesia. Ape Nostra Amica 10:16–19, [in Italian]. Keywords: *cerana*/*dorsata*/Indonesia

Tealdi C (1989) Bees in Malaysia. Ape Nostra Amica 11:15–19, [in Italian]. Keywords: biogeography/*dorsata*/Malaysia

Tembhare DB, Paliwal GN (1990) Cephalic neuroendocrine system of the rock honeybee, *Apis dorsata* (F) worker. In: Proceedings of the 11th international congress, IUSSI, Bangalore, pp 137–138, Keywords: *dorsata*/physiology/India

Tennent JE (1859) Ceylon: an account of the island: physical, historical, and topographical with notices of its natural history, antiquities and production. Longman, Green, Longman & Roberts, London, Keywords: biogeography/*cerana*/history/islands/Sri Lanka

Tew J (1980) Beekeeping technology Burma's basic reason for training beekeepers. Glean Bee Cult 108:461, Keywords: beekeeping/*cerana*/Myanmar (Burma)

Tewari GN, Singh K (1983a) Role of insect pollinators in vegetable seed production. Indian Bee J 45:51, Keywords: *cerana*/pollination/India

Tewari GN, Singh K (1983b) Studies on insect pollinators in relation to seed production in cauliflower (*Brassica oleracea* var *botrytis* L). Indian Bee J 45:54–55, Keywords: *cerana*/foraging/pollination/India

Tewarson NC (1987) Use of host hemolymph proteins, seasonal reproduction and a hypothesis on nutritional imprinting in the honey bee mite, *Varroa jacobsoni* on *Apis mellifera* and *Apis cerana*. In: Eder J, Rembold H (eds) Chemistry and biology of social insects. J Peperny, Munich, pp 688–689, Keywords: biochemistry/biology/*cerana*/honey/*mellifera*/natural enemies/pests/physiology/reproduction

Tewarson NC (1990) Few observations of *Varroa jacobsoni*, mite pest of honey bees in the colonies of *Apis cerana* and *mellifera*. In: Proceedings of the 11th international congress, IUSSI, Bangalore, p 460, Keywords: *cerana*/honey/*mellifera*/India

Tewarson NC, Singh A, Engels W (1992) Reproduction of *Varroa jacobsoni* in colonies of *Apis cerana indica* under natural and experimental conditions. Apidologie 23:161–171, Keywords: *cerana*/pests/reproduction/India

Thai PH (1994) The honey production of *A. dorsata* in Uminh forest. Sci Tech Beekeep 3:26–28, [in Vietnamese]. Keywords: beekeeping/*dorsata*/honey/Vietnam

Thai PH (2003) Investigation for indigenous honeybee in Vietnam. In: Proceedings of the 3rd Vietnamese-Hungarian conference on domestic animal production and aquaculture – quality and rural development, National Institute of Animal Husbandry, Hanoi, pp 16–19, Keywords: *andreniformis*/biogeography/*cerana*/*dorsata*/*florea*/*laboriosa*/Vietnam

Thai PH, Tan NQ, Chinh PH, Mulder V (1996) Rafter beekeeping with *Apis dorsata*: effect of rafter direction and material on the occupation of rafter by bees. In: Proceedings of the 3rd Asian apiculture association conference, Hanoi, pp 144–146, Keywords: beekeeping/combs/*dorsata*/Vietnam

Thailand, Department of Agricultural Extension (1986) The consultation of beekeeping development on *Apis cerana*, 16–17 Jan 1986. Phuket, Department of Agricultural Extension, Bangkok, Keywords: beekeeping/*cerana*/physiology/Thailand

Thakar SS (1967) Central Bee Research Institute, Khadi and Village Industries Commission Annual Report 1966–1967. Indian Bee J 29:43–62, Keywords: beekeeping/*cerana*/India

Thakar SS (1973) A preliminary note on hiving *Apis dorsata* colonies. Bee World 54:24–27, Keywords: beekeeping/combs/*dorsata*/India

Thakar SS (1974) Experiments on pollination of sunflower. Indian Bee J 36:36–38, Keywords: bee botany/*cerana*/*dorsata*/pollination/India

Thakar SS (1976a) Beekeeping technology: its pace and plans. Indian Bee J 38:23–28, Keywords: beekeeping/*cerana*/India

Thakar SS (1976b) Practical aspects of bee management in India with *Apis cerana indica*. In: Proceedings of the 1st international conference on apiculture in tropical climates, London, pp 51–59, Keywords: beekeeping/*cerana*/physical factors/India

Thakar SS (1976c) The beekeeping development and research programmes of the Khadi and Village Industries Commission, India. In: Crane E (ed) Apiculture in tropical climates. International Bee Research Association (IBRA), London, pp 125–134, Keywords: beekeeping/*cerana*/physical factors/physiology/India

Thakar SS, Deodikar GB (1966) Chromosome number in *Apis florea* Fab. Curr Sci 35:186, Keywords: *florea*/genetics

Thakur SS, Kashyap NP (1996) Wasp control by lure trap in apiary. Insect Environ 2:6–7, Keywords: *cerana*/natural enemies/India

Thakar SS, Kashyap NP (1997) Development and testing of a trap for *Vespa magnifica*. Indian Bee J 59:83–85, Keywords: beekeeping/*cerana*/natural enemies/physiology/India

Thakar SS, Tonapi KV (1961) Nesting behaviour of Indian honeybees I. Differentiation of worker, queen and drone cells on the combs of Apis dorsata Fab. Bee World 42:61–62, 72, Keywords: behaviour/cells/combs/*dorsata*/drones/nests/queens/India

Thakar SS, Tonapi KV (1962) Nesting behaviour of Indian honeybees II. Nesting habits and comb cell differentiation in Apis florea Fab. Indian Bee J 24:27–31, Keywords: absconding/behaviour/*cerana*/classification/combs/*florea*/migration/nests/swarming/India

Thakur AK, Garg R, Sharma OP (1981) Vertebrate enemies of bees and their control in Dhauladhar Mountains (Himachal Pradesh, India). Indian Bee J 43:112–113, Keywords: *cerana*/mountains/natural enemies/India

Thakur AK, Sharma OP, Garg R, Dogra GS (1982) Comparative studies on foraging behaviour of *Apis mellifera* and *A. cerana indica* on mustard. Indian Bee J 44:91–92, Keywords: bee botany/behaviour/*cerana*/foraging/*mellifera*/pollination/India

Thakur AK, Dhaliwal HS, Kumar J (1995a) Studies on pollen foraging behaviour of *Apis cerana* and *Apis mellifera* under midhill conditions of Himachal Pradesh. Indian Bee J 57:77–79, Keywords: behaviour/*cerana*/foraging/*mellifera*/pollen/India

Thakur AK, Dhaliwal HS, Kumar J (1995b) Pollen foraging ecology of *Apis cerana* F and *Apis mellifera* L under midhill conditions of Himachal Pradesh, India. Indian Bee J 57:127–128, Keywords: bee botany/*cerana*/ecology/foraging/*mellifera*/pollen/India

Thakur SS, Kumar J, Mishra RC (1995c) Floral morphology of some cultivars of almond (*Prunus amygdalus* Batsch) and foraging behaviour of honey bees on them. Indian Bee J 57:79–81, Keywords: behaviour/*cerana*/foraging/honey/*mellifera*/pollination/nectar/India

Thakur SS, Kumar J, Mishra RC (1995d) Nectar secretion, concentration and type of nectar sugars and foraging by insects on almond, *Prunus amygdalus* Batsch. Indian Bee J 57:1–5, Keywords: *cerana*/foraging/*mellifera*/nectar/pollination/India

Thakur SS, Mishra RC, Kumar J (1997) Bloom visiting insect and their foraging behaviour on almond, *Prunus amygdalus* Batsch. Indian Bee J 59:157–160, Keywords: behaviour/*cerana*/foraging/pollination/India

Thakur SS, Singh HB, Kashyap NP (2000) Comparative foraging efficiency of insect pollinators visiting turnip bloom. Insect Environ 5:168–169, Keywords: *cerana*/foraging/pollination/India

Thangaraja A, Senthilkumar N, Ganesan V (2001) Foraging dynamics of floral visitors of *Tectona grandis* L. f. (Verbenaceae). Insect Environ 7:133–134, Keywords: *cerana*/foraging/*mellifera*/pollination

Thapa R (1995) Comparative study on effects of some pesticides to Asian honey bee (*Apis cerana* Fabr.) and European honey bee (*Apis mellifera* L). Thesis, Chiang Mai University, Keywords: *cerana*/honey/insecticides/*mellifera*/Thailand

Thapa R (1998) Colony migration of the giant honeybee, *Apis dorsata* Fabr. Thesis, Chulalongkorn University, Keywords: absconding/*dorsata*/migration/swarming

Thapa R (2001) The Himalayan giant honeybee and its role in ecotourism development in Nepal. Bee World 82:141, Keywords: general biology/*laboriosa*/Nepal

Thapa R, Wongsiri S (1997) Toxicity of *Azadirachta* derivatives and synthetic pyrethroids on oilseed rape to *Apis cerana* (Hymenoptera; Apidae). In: Proceedings of the 1st international symposium on biopesticides, Bangkok, pp 82–86, Keywords: *cerana*/insecticides/Thailand

Thapa R, Wongsiri S (1999) Comparative toxicity effects of neem (*Azadirachta indica*) and some selective pesticides to forager honeybees; *Apis cerana* and *Apis mellifera* in caged field of oilseed rape. Asian Bee J 1:73–79, Keywords: *cerana*/foraging/insecticides/*mellifera*

Thapa R, Wongsiri S (2003) Flying predators of the giant honeybees; *Apis dorsata* and *Apis laboriosa* in Nepal. Am Bee J 143:540–542, Keywords: *dorsata*/honey/*laboriosa*/pests/predators/Nepal

Thapa R, Wongsiri S, Oldroyd BP, Prawan S (2000) Migration of *Apis dorsata* in northern Thailand. In: Proceedings of the 4th Asian apiculture association conference, Kathmandu, pp 39–43, Keywords: absconding/disease/*dorsata*/migration/natural enemies/pests/predators/swarming/Thailand

Thapa R, Wongsiri S, Manandhar H (2001) Current status of predators and diseases of honey bees in Nepal. In: Proceedings of the 7th international conference on tropical bees, Chiang Mai, pp 221–226, Keywords: *cerana*/disease/*dorsata*/honey/*laboriosa*/*mellifera*/natural enemies/predators/Nepal

Thapa R, Shrestha R, Manandhar DN, Bista S, Kafle B (2001) Beekeeping in Nepal. In: Proceedings of the 7th international conference on tropical bees, Chiang Mai, pp 409–413, Keywords: beekeeping/*cerana*/*dorsata*/*florea*/*laboriosa*/*mellifera*/Nepal

Thirakupt K (1988) Honey bee predation by and protection from bee-eating birds. In: Proceedings of the 2nd Australian and international bee congress, Brisbane, pp 90–91, Keywords: *cerana*/honey/*mellifera*/natural enemies/predators/Thailand

Thirakupt K (1989) Bee-eating birds and honey-bee predation. J Sci Res (Chulalongkorn) 14:85–90, Keywords: *cerana*/*dorsata*/*florea*/natural enemies/predators/Thailand

Thiyagesan K, Varadharajan M, Abirami S (2001a) Studies on central place foraging in the Indian honey bee, *Apis cerana indica*. J Apic Res 40:21–26, Keywords: *cerana*/foraging/honey/nectar/physiology/India

Thiyagesan K, Varadharajan M, Abirami S (2001b) Studies on risk-sensitive foraging in the Indian honey bee (*Apis cerana indica*). J Apic Res 40:16–20, Keywords: *cerana*/foraging/honey/India

Thomas D, Pal N (2001) New initiatives in bee management and productivity of honey bees. J Apic Res 40:27–32, Keywords: beekeeping/*cerana*/honey/India

Thomas D, Pal N, Subba Rao K (2001) Bee management and productivity of Indian honey bees. In: Proceedings of the 37th international apiculture congress, Apimondia, Durban (CD Rom). Keywords: beekeeping/*cerana*/*dorsata*/*florea*/honey/India

Thomas SG, Varghese A, Roy P, Bradbear N (2009) Characteristics of trees used as nest sites by *Apis dorsata* (Hymenoptera, Apidae) in the Nilgiri Biosphere Reserve India. Trop Ecol 25:559–561, Keywords: *dorsata*/nesting/India

Thompson CR (1940) Handling *Apis dorsata*. Indian Bee J 2:122–124, Keywords: beekeeping/*dorsata*/India

Thompson CR (1941) Collection of wild honey. Indian Bee J 3:28–31, Keywords: *dorsata*/honey/India

Tian XJ (1995) Apiculture in Yunnan Province, China. Bee World 76:106–108, Keywords: beekeeping/*cerana*/China

Tilde AC, Payawal PC (1992a) Commercial honey in the Philippines I. Pollen grain analysis. Philipp Agric 75:81–87, Keywords: *cerana*/*dorsata*/honey/*mellifera*/pollen/Philippines

Tilde AC, Payawal PC (1992b) Commercial honey in the Philippines II. Physical and chemical properties. Philipp Agric 75:89–91, Keywords: *cerana*/*dorsata*/honey/*mellifera*/Philippines

Tilde AC, Fuchs S, Koeniger N, Cervancia CR (2000) Morphometric diversity of *Apis cerana* Fabr. within the Philippines. Apidologie 31:249–264, Keywords: *cerana*/classification/genetics/morphometrics/Philippines

Tilde AC, Cervancia CR, Payawal PC (2001) Pollen spectra among natural populations of *Apis cerana* Fabricius in the Philippines. In: Proceedings of the 7th international conference on tropical bees, Chiang Mai, pp 177–181, Keywords: bee botany/*cerana*/foraging/pollen/population biology/Philippines

Ting JI, Chen J, Yin L, Bao WB, Chen GH (2007) Study on microsatellite genetic polymorphism of wild and domestic *Apis cerana cerana* populations from Jiangsu province. China Anim Husb Vet Med 12: [in Chinese]. Keywords: *cerana*/genetics/mtDNA/populations*

Ting JI, Yin L, Liu M, Chen GH (2009) Genetic diversity and genetic differentiation of six geographic populations of *Apis cerana cerana* (Hymenoptera: Apidae) in East China. Acta Entomol Sinica 52, [in Chinese]. Keywords: biogeography/*cerana*/genetics/populations/China*

Tingek S (1986) Life cycle in *Apis cerana*. Thesis, Universiti Pertanian, Selangor, Keywords: biology/*cerana*

Tingek S, Mardan M, Rinderer TE, Koeniger N, Koeniger G (1988) Rediscovery of *Apis vechti* (Maa, 1953): the Sabah honey bee. Apidologie 19:97–102, Keywords: *koschevnikovi*

Tingek S, Koeniger N (2002) Exploring the centre of honeybee diversity. Honeybee Sci 23: 145–152, Keywords: *cerana*/*dorsata*/genetics/*koschevnikovi*/*mellifera*/*nuluensis*

Tingek S, Koeniger N, Koeniger G (2002) Diversity of mating distance and mating behaviour of five sympatric honeybee species in Sabah (Borneo) Malaysia. In: Proceedings of the 14th international congress, IUSSI, Sapporo, p 211, Keywords: *andreniformis*/behaviour/*cerana*/*dorsata*/drones/genetics/*koschevnikovi*/*nuluensis*/reproduction/Japan/Malaysia

Tingek S, Koeniger G, Koeniger N, Gries M (2004) *Physocephala paralleliventris* Krober (Conopidae) a parasitic fly of three honeybee species in northern Borneo. Apidologie 35:549–550, Keywords: *cerana*/*dorsata*/pests/natural enemies/Borneo

Tirgari S (1971) On the biology and manipulation of *Apis* (*Micrapis*) *florea* F in Iran. In: Proceedings of the 23rd international beekeeping congress, Moscow, pp 330–332, Keywords: absconding/beekeeping/biology/classification/combs/environmental effects/*florea*/foraging/general biology/migration/queens/swarming/workers/Iran

Tirgari S (1971) Biology and behavioral characteristics of the Iranian dwarf honeybee (*Apis florea*). In: Proceedings of the 23rd international beekeeping congress, Moscow, 23:344–345 [in German]. Keywords: behaviour/*florea*/general biology/Iran

Tirgari S, Polous H, Foroughi H, Fakhari F, Nabavit MA (1969) On the occurrence and biology of little bee (*Micrapis florea* F) and the first account of its migration (Hymenoptera: Apidae). In: Proceedings of the national congress entomology and pest control (Iran), p 4 [in Persian]. Keywords: absconding/biogeography/biology/*florea*/migration/swarming/Iran

Tokuda Y (1922) A pore of the Japanese drone cocoon. Bee World 3:89, Keywords: cells/*cerana*/drones/nests/Japan

Tokuda Y (1924) Studies on the honeybee, with special reference to the Japanese honeybee. Trans Sapporo Nat Hist Soc 9:1–26, Keywords: *cerana*/drones/ecology/morphology/Japan

Tokuda Y (1935) Study on the honeybee, with special reference to the Japanese honeybee. Bull Imp Zootech Exp Sta 15:1–27, Keywords: biology/*cerana*/Japan

Tokuda Y (1955) Experimental studies on beeswax production by the honeybee colonies I. Beeswax yield by the honeybee colonies under sugar syrup feeding. Bull Natl Inst Agric Sci Ser G (Anim Husb) 11:111–133, [in Japanese]. Keywords: beeswax/biology/*cerana*/Japan

Tokuda Y (1971) On the biological characteristics of the indigenous Japanese bees. In: Proceedings of the 23rd international apiculture congress, Apimondia, Moscow, pp 334–335, Keywords: beekeeping/biology/*cerana*/Japan

Tomsik B (1960) Reproduction and development in the honeybee. Vcelarstvi 13:22–23, [in Czech]. Keywords: brood/*cerana*/physiology/reproduction/Korea

Tonapi GT (1958) Presence of propolis in the prothoracic spiracles of *Apis florea* Fab. Curr Sci 27:133–134, Keywords: combs/*florea*/biology/morphology

Tonapi GT (1959) The second thoracic spiracle in *Apis* (Hymenoptera Apidae). Curr Sci 28:291–292, Keywords: *cerana*/*dorsata*/*florea*/morphology

Tonapi GT (1983) Some morphophysiological aspects of bee biology. In: Proceedings of the 2nd international conference on apiculture in tropical climates, New Delhi, pp 398–406, Keywords: biology/*cerana*/*dorsata*/*florea*/physical factors/India

Torchio PF, Torchio DM (1975) Larvae of the Apidae (Hymenoptera, Apoidea) 1. Apini, *Apis*. Research Report, Agricultural *Experiment* Station, Utah State University 10:1–36, Keywords: biology/*cerana*/*dorsata*/*florea*/larvae/*mellifera*

Toumanoff C (1933) Documentation on beekeeping in Annam. Bull Econ Indochine 36:169–175, [in French]. Keywords: beekeeping/*cerana*/Vietnam

Toumanoff C (1939) The enemies of honeybees. Imprimerie d'Extreme-Orient, Hanoi, [in French]. Keywords: *cerana*/*dorsata*/*florea*/natural enemies

Toumanoff C (1940) Some notes on *Apis dorsata*. Microscope 4:101–110, Keywords: *dorsata*/Vietnam

Toumanoff C, Nanta J (1933) Enquiry on beekeeping in Tonkin. Bull Econ Indochine 36: 1015–1048, [in French]. Keywords: beekeeping/*cerana*/Vietnam

Towne WF (1985) Acoustic and visual cues in the dances of four honey bee species. Behav Ecol Sociobiol 16:185–187, Keywords: *cerana*/communication/dance language/*dorsata*/*florea*/honey/*mellifera*/physiology

Towne WF, Gould JL (1988) The spatial precision of the honey bees' dance communication. J Insect Behav 1:129–155, Keywords: *cerana*/communication/dance language/*dorsata*/*florea*/honey/*mellifera*

Trehan KN, Longia HS (1962) Some hints for the collection of *Apis dorsata* Fabr. Indian Bee J 24:53–54, Keywords: beekeeping/*dorsata*/India

Trung LQ, Dung PX, Ngan TX (1996) A discovery of *Apis laboriosa* F Smith (Hymenoptera, Apidae) in northwestern Vietnam. In: Proceedings of the 3rd Asian apiculture association international conference, Hanoi, pp 7–14, Keywords: aggressiveness/*andreniformis*/*cerana*/classification/combs/*dorsata*/*florea*/*laboriosa*/migration/nests/swarming/Vietnam

Tsang ACW (2006) The ecology of *Ilex* species in Hong Kong. Thesis, University of Hong Kong, Keywords: bee botany/*cerana*/ecology/China

Tsang ACW, Corlett RT (2005) Reproductive biology of the *Ilex* species (Aquifoliaceae) in Hong Kong, China. Can J Bot 83:1645–1654, Keywords: *cerana*/pollination/China

Tseng HN (1954) Beekeeping in China. Glean Bee Cult 82:216–217, Keywords: beekeeping/*cerana*/China

Tsurata T, Matsuka M, Sasaki M (1989) Temperature as a causative factor in the seasonal colour dimorphism of *Apis cerana japonica* workers. Apidologie 20:149–155, Keywords: *cerana*/environmental effects/morphology/physiology/workers/Japan

Tuan CN (1980) Methods of feeding Chinese honeybees. Hunan K'o Hsueh Chi Shu Ch'u Pan She, Ch'angsha, China, [in Chinese]. Keywords: beekeeping/*cerana*/China*

Tulloch AP (1980) Beeswax – composition and analysis. Bee World 61:47–62, Keywords: beeswax/biochemistry/*cerana*/*dorsata*/*florea*/*mellifera*

Turner GJ (1974) Possible transmission of *Puccinia polysora* by bees. Trans Br Mycol Soc 62:205–206, Keywords: *dorsata*/foraging/Malaysia

Turner PD (1969) Pests and diseases of oil palm in Thailand. FAO Plant Protect Bull 17:107–108, Keywords: bee botany/disease/*dorsata*/pests/predators/Thailand

Uchida T (2001) The seasonal population trends of the honey-bee in Sapporo, with some remarks on supersedure. Jap J Zool 11:387–394, Keywords: bee botany/*cerana*/colony phenology/population biology/seasonal variation/supersedure/Japan

Ueno S (1915) Corean beeswax. Analysis 40:343–344, Keywords: beeswax/*cerana*/Korea

Underwood BA (1986) The natural history of *Apis laboriosa* Smith in Nepal. Thesis, Cornell University, Keywords: absconding/bee botany/behaviour/combs/defense/foraging/honey/*laboriosa*/migration/nests/reproduction/swarming/Nepal

Underwood BA (1987) In search of the world's largest honeybee. Glean Bee Cult 115:234–236, 245, Keywords: combs/*laboriosa*/Nepal

Underwood BA (1990a) Seasonal nesting cycle and migration patterns of the Himalayan honeybee *Apis laboriosa*. Nat Geogr Res 6:276–290, Keywords: absconding/combs/*laboriosa*/migration/seasonal variation/swarming/Himalayas/Nepal

Underwood BA (1990b) The behaviour and energetics of high-altitude survival by the Himalayan honeybee, *Apis laboriosa*. Thesis, Cornell University, Keywords: absconding/biogeography/combs/ecology/environmental effects/foraging/honey/*laboriosa*/migration/physiology/swarming/Nepal

Underwood BA (1990c) The seasonal cycle of nesting and migration by the Himalayan honeybee, *Apis laboriosa*. In: Proceedings of the 11th international congress, IUSSI, Bangalore, p 117, Keywords: absconding/combs/ecology/honey/*laboriosa*/migration/seasonal variation/swarming/Nepal

Underwood BA (1990d) Time of drone flight of *Apis laboriosa* Smith in Nepal. Apidologie 21:501–504, Keywords: *dorsata*/drones/foraging/*laboriosa*/reproduction/Himalayas/Nepal

Underwood BA (1991a) Thermoregulation and energetic decision-making by the honeybees *Apis cerana*, *Apis dorsata* and *Apis laboriosa*. J Exp Biol 157:19–34, Keywords: behaviour/*cerana*/*dorsata*/environmental effects/foraging/*laboriosa*/physiology/Himalayas/Nepal

Underwood BA (1991b) The giant rock bees of the Himalayas. Imkerfreund 46:8–14, [in German]. Keywords: *laboriosa*/Himalayas/Nepal

Underwood BA (1992a) Impact of human activities on the Himalayan honeybee, *Apis laboriosa*. In: Verma LR (ed) Honeybees in mountain agriculture. Oxford & IBH, New Delhi, pp 51–57, Keywords: absconding/combs/environmental effects/*laboriosa*/migration/swarming/Himalayas/Nepal

Underwood BA (1992b) Notes on the orange-rumped honeyguide *Indicator xanthonotus* and its association with the Himalayan honeybee *Apis laboriosa*. J Bombay Nat Hist Soc 89:290–295, Keywords: honey/*laboriosa*/predators/natural enemies/Nepal

Underwood BA (1992c) Seasonal cycle of the Himalayan honeybee, *Apis laboriosa*. Honeybee Sci 13:1–8, [in Japanese]. Keywords: absconding/ecology/honey/*laboriosa*/migration/seasonal variation/swarming

Uno R (2000) Traditional beekeeping of Japanese honeybee in Tono City. Iwate Prefecture. A case study of traditional beekeeping in the Tohoku Region. Honeybee Sci 21:97–106, Keywords: beekeeping/*cerana*/Japan

Usha K, Kandasamy C (1986) Comparative toxicity of certain pesticides to the honeybee *Apis indica* F. Pest Manage 1:31–35, Keywords: *cerana*/insecticides/India

Ushirokita M (1997) Observation of a colony of *Apis cerana japonica* which survived an earthquake. Honeybee Sci 18:9–16, [in Japanese]. Keywords: *cerana*/environmental effects/ physical factors/Japan

Utami N (1986) Pollination in sago (*Metroxylon sagu*). Berita Biol 3:229–231, [in Indonesian]. Keywords: bee botany/*cerana*/pollination/Indonesia

Uthaisang W (1994) Preparation of DNA probe for the analysis of genetic variation in *Apis cerana*. Thesis, Chulalongkorn University, Keywords: *cerana*/DNA/genetics/Thailand

Vadgama MR, Kamat DN (1973a) Methylation studies on the salivary glands of some insects with different feeding habits. J Anat Soc India 22:5–9, Keywords: *cerana*/glands/physiology/India

Vadgama MR, Kamat DN (1973b) The secretion of mucopolysaccharides by the salivary gland cells of *Apis indica* and their possible physiological significance. Indian J Zool 1:32–38, Keywords: *cerana*/glands/physiology/India

Vaidya DN, Kumar S (1995) Repellent effect of some common insecticides to *Apis cerana indica* Fab. foraging on *Brassica campestris* L var brown sarson. Himachal J Agric Res 21:42–46, Keywords: *cerana*/foraging/insecticides/India

Vaidya DN, Kumar S (1997) Toxicity of some insecticides to *Apis cerana* Fabr. foragers on treated bloom of rapeseed, *Brassica campestris* L var brown sarson. Indian Bee J 59:141–143, Keywords: *cerana*/insecticides/India

Vaidya K, Bista M (2000) Honeybee fauna in Mahakali zone and their Ayurvedic relations. In: Proceedings of the 4th Asian apiculture association international conference, Kathmandu, pp 148–151, Keywords: bee botany/*cerana*/*dorsata*/*florea*/foraging/honey/*mellifera*/Nepal

Vaish OP, Aggarwal SC, Joshi MJ (1978) Frequency of insect visitors for pollen foraging on sunflower in relation to daily temperature and humidity. In: Proceedings of the 8th international sunflower conference, Minneapolis, pp 148–157, Keywords: bee botany/*dorsata*/environmental effects/foraging/pollination/India

Vallette P (1922) Apiculture in the colonies: Indo-China. In: Proceedings of the 6th international apiculture congress, Apimondia, Marseille, pp 57–58 [in French]. Keywords: beekeeping/ *cerana*/Cambodia/Laos/Vietnam

Valli E (1998) Golden harvest of the Raji. Nat Geogr 193:84–105, Keywords: *dorsata*/honey/ Nepal

Valli E, Summers D (1988) Honey hunters of Nepal. Thames and Hudson, London, Keywords: *dorsata*/honey/Nepal

van Cattenburch FMG (1864) Beekeeping in the Lampong District (of south Sumatra). Tijds Nijverheid Landbouw Nederland-Indie pp 253–262, [in Dutch]. Keywords: beekeeping/*cerana*/Indonesia*

Vandenberg JD (1990) Selected research needs in honey bee pathology. In: Proceedings of the 11th international congress, IUSSI, Bangalore, pp 630–631, Keywords: *cerana*/disease/honey/ *mellifera*/pests/India

Varma K, Joshi NK (1983) Studies on the role of honey bees in the pollination of cauliflower (*Brassica oleracea* var *botrytis*). Indian Bee J 45:52–53, Keywords: *cerana*/honey/pollination/ India

Vaskovsky II (1928) Indian bees in the Far East. Ussur Sadovod I Ogorodn 5:23–24, [in Russian]. Keywords: biogeography/*cerana*/Russia

Vasudeva R, Lakshmana R, Radhamani TR (1999) Differential bee visitation to male sterile and fertile lines in pearl millet (*Pennisetum glaucum* L). J Res ANGRAU 27:26–28, Keywords: *cerana*/pollination/India

Vats BR (1953) Breeding Italian bees in Kashmir. Kashmir Sci 3:23–24, Keywords: breeding/ *cerana*/interspecific interactions/*mellifera*/queens/India

Veeresh GK (1994) Importance of bees in pollination of crops. Proceedings of a Workshop, Bangalore, pp 14–16, Keywords: *cerana*/*mellifera*/pollination/India

Veeresh GK (1994) Thai sacbrood disease of the Indian honey bee and problems and prospects of establishing the European bee *Apis mellifera* in Karnataka. Proceedings of a Workshop, Bangalore, Keywords: *cerana*/disease/honey/*mellifera*/India

Veeresh GK, Sangappa HK (1993) Comparative foraging behaviour of four honeybee species on sunflower. In: Proceedings of international symposium on pollination in tropics, Bangalore, pp 190–191, Keywords: behaviour/*cerana*/*dorsata*/*florea*/foraging/*mellifera*/pollination/tropics/India

Veeresh GK, Malik B, Vishwanathan H (1990a) Diversity of *Apis* in Southeast Asia. In: Veeresh GK, Malik B, Viraktamath CA (eds) Social insects and the environment. Oxford & IBH, New Delhi, pp 725–726, Keywords: *cerana*/*dorsata*/*florea*/genetics

Veeresh GK, Malik B, Viraktamath CA (eds) (1990b) Social insects and the environment. In: Proceedings of the 11th international congress, IUSSI, New Delhi, p 765, Keywords: beekeeping

Veith HJ, Weiss J, Koeniger N (1978) A new alarm pheromone (2-decen-1-yl-acetate) isolated from the stings of *Apis dorsata* and *Apis florea* (Hymenoptera: Apidae). Experientia 34:423, Keywords: aggressiveness/*cerana*/*dorsata*/*florea*/interspecific interactions/pheromones/stings

Velthuis HHW (1976) Egg laying, aggression and dominance in bees. In: Proceedings of the 15th international congress of entomology, Washington, pp 436–449, Keywords: aggressiveness/*cerana*/defense/*dorsata*/*mellifera*/oviposition

Velthuis HHW, Clement JL, Morse RA, Laigo FM (1971) The ovaries of *Apis dorsata* workers and queens from the Philippines. J Apic Res 10:63–66, Keywords: *dorsata*/queens/reproduction/workers/Philippines

Venkata Rau RN (1946) My final report of the observations made on the *Apis dorsata* hive built in the Ramakrishna Mission students home in Mylapore, Madras. Indian Bee J 8:160–163, Keywords: beekeeping/defense/*dorsata*/genetics/reproduction/India

Venkatasubbaya N (1938) Variations in the tongue length of the honeybees. J Mysore Agric Exp Union 17:65–72, Keywords: *cerana*/classification/morphometrics/India

Venkatesh G, Kumar BSS (1996) Adaptive strategies of giant honeybee *Apis dorsata* F. (Hymenoptera: Apidae) to some environmental conditions. Environ Ecol 14:920–925, Keywords: *dorsata*/environmental effects/India

Venkatesh G, Raj D (1994) Distribution of bee flora, their abundance and blooming period with special reference to *Apis dorsata* F. In: Sarma TA, Saini SS, Trivedi ML, Sharma M (eds) Current researchers in plant sciences. Department of Zoology, Punjabi University, Patiala, pp 191–196, Keywords: bee botany/biogeography/*dorsata*/foraging/India

Venkatesh G, Reddy CC (1989) Rates of swarming and absconding in the giant honeybee, *Apis dorsata* F. Proc Indian Acad Sci 98:425–430, Keywords: absconding/*dorsata*/honey/migration/swarming/India

Venkatesh G, Reddy CC (1990) Influence of comb age on migratory behaviour of *Apis dorsata* F. In: Proceedings of the 11th international congress, IUSSI, Bangalore, pp 64–66, Keywords: absconding/behaviour/combs/*dorsata*/migration/swarming/India

Venkatesh G, Sharma R, Neena KB (1995) Changes in fat bodies during the postembryonic developmental stages of workers and drones of *Apis cerana indica* F. J Entomol Res 19:175–179, Keywords: biochemistry/brood/*cerana*/drones/physiology/workers/India

Verghese A, Prasad VG (1983) A note on the foraging activity of the honeybee *Apis cerana indica* F. In: Proceedings of the 2nd international congress on apiculture in tropical climates, New Delhi, pp 659–660, Keywords: *cerana*/foraging/India

Verghese A, Veeraraju PV (1990) Association of *Apis cerana* with *Apis florea* and *Eristalinus arvorum* during mango bloom. In: Proceedings of the 11th International Congress, IUSSI, Bangalore, p 479, Keywords: *cerana*/*florea*/pollination/India

Verhagen R (1971a) Apicultural observations in West Pakistan. Abeilles Fleurs 207:5–7, [in French]. Keywords: beekeeping/*cerana*/Pakistan

Verhagen R (1971b) Beekeeping observations in Afghanistan. Abeilles Fleurs 208:4–5, [in French]. Keywords: beekeeping/*cerana*/Afghanistan

Verma LR (1970) A comparative study of temperature regulation in *Apis mellifera* L and *Apis indica* F. Am Bee J 110:390–391, Keywords: *cerana*/*mellifera*/physiology/India

Verma LR (1984) Beekeeping in northern India – major constraints and potentials. Proceedings of the FAO Expert Consultation on Beekeeping with Apis mellifera in Tropical and Sub-tropical Asia, Bangkok, pp 148–155, Keywords: beekeeping/*cerana*/*mellifera*/India/Thailand

Verma LR (1985) Sac brood in *Apis cerana*. IBRA Newsl Beekeep Trop Sub-trop Count 6:4, Keywords: brood/*cerana*/disease/India

Verma LR (1986a) Biological and economic characters of *Apis cerana indica* F. Honeybee Sci 7:151–157, [in Japanese]. Keywords: biology/*cerana*/India

Verma LR (1986b) Comparative comb structure, temperature regulation and foraging behaviour of *Apis cerana indica* F and *Apis mellifera* L. In: Eder J, Rembold H (eds) Chemistry and biology of social insects. Peperny, Munich, p 540, Keywords: behaviour/biology/*cerana*/combs/foraging/*mellifera*/physiology/Thailand

Verma LR (1986) Studies on genetic improvement of Indian honeybee *Apis cerana indica* F by selection and breeding. Final Report, Government of India, New Delhi, pp 1–153, Keywords: breeding/*cerana*/genetics/India

Verma LR (1986) Studies on genetic improvement of Indian honeybee *Apis cerana indica* F by selection and breeding. Proceedings of the group monitoring workshop on DST founded projects in environmental biology, entomology, reproductive biology and sociobiology, Ministry of Science and Technology, New Delhi, pp 11–14, Keywords: biology/breeding/*cerana*/genetics/India

Verma LR (1987) Biology of *Apis cerana* F in relation to beekeeping development Progamme in Asia. In: Proceedings of the 31st international apiculture congress, Apimondia, Warsaw, pp 190–192, Keywords: beekeeping/biology/*cerana*/physiology

Verma LR (1987) Current status of parasitic mites in relation to beekeeping with *Apis cerana* F and *Apis mellifera* in India. In: Proceedings of the FAO Workshop on parasitic bee mites and their control, Pulawy, pp 195–198, Keywords: beekeeping/*cerana*/*mellifera*/pests/India

Verma LR (1987) Pollination ecology of apple orchards by hymenopterous insects in Matiana-Narkanda temperate zone. Final Report, Ministry of Environment and Forests, Government of India, pp 1–118, Keywords: *cerana*/*dorsata*/ecology/*florea*/pollination/India

Verma LR (1988) *Apis cerana*: biometric, genetic and behavioural aspects. In: Proceedings of international workshop on advanced course in beekeeping *Apis cerana* tropical and sub-tropical Asia, Serdang, Keywords: beekeeping/behaviour/*cerana*/classification/genetics/India/Malaysia

Verma LR (1989) Apiculture as a component of hill farming systems: case of Himachal Pradesh. ICIMOD *Internal Report, Himachal Experiences in Mountain Agriculture. Policies* and Strategies. Keywords: beekeeping/*cerana*/India

Verma LR (1989) Some practical aspects of beekeeping with *Apis cerana* in Asia. In: Proceedings of the 4th international conference on apiculture in tropical climates, Cairo, pp 438–441, Keywords: beekeeping/*cerana*/physical factors

Verma LR (1990a) Beekeeping in Bhutan problems and prospects. In: Proceedings of *international* expert meeting on apiculture *development*, ICIMOD, Kathmandu, Keywords: beekeeping/*cerana*/physiology/Bhutan

Verma LR (ed) (1990b) Beekeeping in integrated mountain development. Oxford & IBH, New Delhi, Keywords: beekeeping/*cerana*/*dorsata*/*florea*/mountains/physiology

Verma LR (1990c) Genetic diversity and conservation of *Apis cerana* in Hindu Kush-Himalayas. In:Proceedings of the 11th international congress, IUSSI, Bangalore, pp 118–119, Keywords: *cerana*/ecology/genetics/*mellifera*/India/Nepal

Verma LR (1992a) Apiculture in Bhutan: problems and prospects. In: Verma LR (ed) Honeybees in mountain apiculture. Oxford & IBH, New Delhi, pp 163–179, Keywords: absconding/*cerana*/*dorsata*/*florea*/*laboriosa*/*mellifera*/natural enemies/swarming/Bhutan

Verma LR (1992b) Beekeeping and pollination ecology of mountain crops. In: Verma LR (ed) Honeybees in mountain agriculture. Oxford & IBH, New Delhi, pp 69–89, Keywords: beekeeping/*cerana*/ecology/pollination/Bhutan/India/Nepal/Pakistan
Verma LR (ed) (1992c) Honeybees in mountain agriculture. Oxford & IBH, New Delhi, Keywords: beekeeping/*cerana*/*dorsata*/*florea*/mountains
Verma LR (1992d) Species and genetic diversity in Himalayan honeybees. In: Verma LR (ed) Honeybees in mountain agriculture. Oxford & IBH, New Delhi, pp 39–49, Keywords: *cerana*/classification/genetics/geographical variation/islands/India/Japan
Verma LR (1993a) Declining genetic diversity of *Apis cerana* in Hindu Kush-Himalayan Region. In: Connor LJ, Rinderer TE, Sylvester HA, Wongsiri S (eds) Asian apiculture. Wicwas Press, Cheshire, Connecticut, pp 81–88, Keywords: *cerana*/genetics/*mellifera*/pests/queens/Bhutan/China/India/Japan/Korea/Myanmar (Burma)/Nepal/Pakistan
Verma LR (1993b) Potentials of *Apis cerana* as pollinator for increasing global food production. In: Proceedings of *international symposium on pollination in tropics*, Bangalore, pp 296–301, Keywords: beekeeping/*cerana*/pollination/India
Verma LR (1993c) Strategies for conservation of Asian hive bee, *Apis cerana*. Asian Apic Assoc Newsl, pp 2–3, Keywords: *cerana*/ecology/nests*
Verma LR (1994a) A framework for research and development on beekeeping with Asian hive bee *Apis cerana*. Honeybee Sci 15:19–24, [in Japanese]. Keywords: beekeeping/*cerana*/physiology/nests/Japan
Verma LR (1994b) Potentials of *Apis cerana* as pollinator for increasing global food production. Honeybee Sci 15:163–166, [in Japanese]. Keywords: beekeeping/*cerana*/pollination/India
Verma LR (1995a) *Apis cerana*: biometric, genetic and behavioural aspects. In: Kevan PG (ed) The Asiatic hive bee: apiculture, biology, and role in sustainable development in tropical and subtropical Asia. Enviroquest, Cambridge, ON, pp 41–53, Keywords: behaviour/biology/breeding/brood/*cerana*/disease/genetics/geographical variation/nests/physiology/India
Verma LR (1995b) Pollination ecology of apple orchards by hymenopterous insects in Matiana-Narkanda temperate zone. In: Kevan PG (ed) The Asiatic hive bee: apiculture, biology, and role in sustainable development in tropical and subtropical Asia. Enviroquest, Cambridge, ON, pp 143–146, Keywords: beekeeping/biology/*cerana*/ecology/nests/physiology/pollination/India
Verma LR (1995c) *Apis cerana* in global beekeeping. Apiacta 30:51–55, Keywords: beekeeping/*cerana*
Verma LR, Chauhan P (1985) Distribution, abundance and diversity of insect pollinators in apple orchards of Shimla Hills. Indian J Ecol 12:286–292, Keywords: biogeography/*cerana*/*dorsata*/*florea*/genetics/pollination/India
Verma LR, Dulta PC (1986) Foraging behaviour of *Apis cerana indica* and *Apis mellifera* in pollinating apple flowers. J Apic Res 25:197–201, Keywords: behaviour/*cerana*/foraging/*mellifera*/physical factors/pollination/India
Verma LR, Edwards DK (1971) Metabolic acclimatization to temperature, and temperature tolerance, in *Apis mellifera* and *Apis cerana indica* kept in India. J Apic Res 10:105–108, Keywords: *cerana*/*mellifera*/physiology/India
Verma LR, Jindal KK (1997) Fruit crops pollination. Kalyani, Ludhiana, Keywords: *cerana*/*dorsata*/*florea*/pollination/India
Verma LR, Mattu VK (1982) Ecotypes studies on *Apis cerana indica* F. In: Breed MD, Michener CD, Evans HE (eds) Biology of social insects. West View, Boulder, CO, p 413, Keywords: biology/*cerana*/geographical variation/India
Verma LR, Partap U (1993) The Asian hive bee, *Apis cerana*, as a pollinator in vegetable seed production. Report, International centre for integrated mountain development (ICIMOD), Kathmandu, Keywords: *cerana*/nests/pollination/India
Verma LR, Partap U (1994) Foraging behaviour of *Apis cerana* on cauliflower and cabbage and its impact on seed production. J Apic Res 33:231–236, Keywords: behaviour/*cerana*/foraging/pollination/India

Verma LR, Rana RS (1994) Further studies on the behaviour of *Apis cerana* and *Apis mellifera* foraging on apple flowers. J Apic Res 33:175–179, Keywords: behaviour/breeding/*cerana*/foraging/*mellifera*/pollination/India

Verma LR, Mattu VK, Singh MP (1984) Races of Indian honeybee in Himalayas. In: Proceedings of the 15th international congress of entomology, Hamburg, p 508, Keywords: *cerana*/classification/mountains/subspecies

Verma LR, Rana BS, Mattu VK (1988a) Correlations among economic and biological characters of *Apis cerana* Fabr. Indian Bee J 50:9–11, Keywords: beekeeping/biology/*cerana*/India

Verma LR, Rana BS, Mattu VK (1988b) Economic and biological characters of the Indian honey bee, *Apis cerana* Fabr.: effect of seasonal variations. Indian Bee J 50:33–35, Keywords: biology/*cerana*/honey/seasonal variation/India

Verma LR, Kafle GP, Sharma A, Mattu VK (1989) Biometry of *Apis cerana* of Nepal Himalayas. In: Proceedings of the 4th international conference on apiculture in tropical climates, Cairo, pp 458–465, Keywords: *cerana*/classification/mountains/physical factors/Nepal

Verma LR, Rana BS, Mattu VK (1990a) Mating behaviour of *Apis cerana* Fab. and *Apis mellifera* Linn in the Shimla Hills of the northwest Himalayas. J Insect Sci 3:130–132, Keywords: behaviour/*cerana*/*mellifera*/mountains/reproduction/India

Verma LR, Rana BS, Verma S (1990b) Observations on *Apis cerana* colonies surviving from Thai sacbrood virus infestation. Apidologie 21:169–174, Keywords: *cerana*/disease/India

Verma LR, Sakai T, Matsuka M (1991) Foundation of Asian Apicultural Association. Honeybee Sci 12:1–4, [in Japanese]. Keywords: beekeeping/*cerana*/*dorsata*/*florea*/*mellifera*/Asia

Verma LR, Mattu VK, Daly HV (1994) Multivariate morphometrics of the Indian honeybee in the northwest Himalayan Region. Apidologie 25:203–223, Keywords: *cerana*/classification/morphometrics/mountains

Verma LR, Kumar NR, Kumar R (1997a) Brood rearing activities of *Apis cerana* Fabr. populations: effect of seasonal variations. Pest Manage Econ Zool 5:21–24, Keywords: breeding/brood/*cerana*/colony phenology/geographical variation/population biology/seasonal variation/India

Verma LR, Kumar NR, Kumar R (2000) Evaluation and selection of *Apis cerana* populations for brood-rearing efficiency. In: Proceedings of the 4th Asian apiculture association international conference, Kathmandu, pp 17–19, Keywords: breeding/brood/*cerana*/colony phenology/oviposition/population biology/India

Verma LR, Kumar R, Kumar NR (2000) Beekeeping needs of farming communities in the Hindu-Kush-Himalayan Region. In: Proceedings of the 4th Asian apiculture association international conference, Kathmandu, pp 225–229, Keywords: beekeeping/*cerana*/India

Verma OP, Pathak VN (1984) Role of insects in secondary spread of pearl millet ergot. Phytophylactica 16:257–258, Keywords: *cerana*/pollination/India

Verma S (1988) Chromosomal analysis of mitotic and meiotic divisions of four species of genus *Apis* (honeybee). Proceedings of the international symposium recent advances in cytogenetic research, Kurukshetra, p 1, Keywords: *cerana*/*dorsata*/*florea*/genetics/India

Verma S (1993d) Biodiversity in Asian honeybees: cytological perspectives. In: Connor LJ, Rinderer TE, Sylvester HA, Wongsiri S (eds) Asian apiculture. Wicwas, Cheshire, CT, pp 89–93, Keywords: genetics/*dorsata*/honey/reproduction/India

Verma S, Dayal K (1994) Pseudomeiosis in the testes of drones produced by laying workers of *Apis cerana* F. Indian Bee J 56:29–31, Keywords: *cerana*/drones/genetics/laying workers/physiology/workers

Verma S, Karol A (1992) Chromosomal analysis of the eggs and ovaries of the laying workers of *Apis cerana* F. Apidologie 23:285–289, Keywords: *cerana*/drones/genetics/laying workers/physiology/workers

Verma S, Phogat KPS (1982a) Seasonal incidence of *Apis* iridescent virus in *Apis cerana indica* Fab. in Uttar Pradesh, India. Indian Bee J 44:36–37, Keywords: *cerana*/disease/environmental effects/seasonal variation/India

Verma S, Phogat KPS (1982b) Studies on the effect of water soluble vitamin ‘C’ on brood rearing and comb building activities of worker honey bees (*Apis cerana indica* F). Indian Bee J 44:73, Keywords: breeding/brood/*cerana*/combs/environmental effects/honey/India

Verma S, Phogat KPS (1994) Impact of pollination by honeybee (*Apis cerana indica*) on the yield gain of radish under valley conditions of Himalayan hills of UP (India). Indian Bee J 56: 183–186, Keywords: bee botany/*cerana*/India

Verma S, Singh K (1999) Morphometrics of developmental stages of honey bee mite, *Neocypholaelaps indica* Evans. J Insect Sci 12:75–76, Keywords: *cerana*/classification/honey/*mellifera*/pests/India

Verma SC, Desh R (2000) Incidence of wax moth (*Galleria mellonella* and *Achroia grisella* F) and their parasitization by *Apanteles galleriae* Wilk in *Apis cerana* colonies in Kangra Valley. Himachal J Agric Res 26:44–49, Keywords: beeswax/*cerana*/pests/natural enemies/India

Verma SC, Raj D, Devi N, Srivastava S (1997b) Efficacy of *Apanteles galleriae* against wax moths in *Apis cerana* and *Apis mellifera* colonies. Indian Bee J 59:211–213, Keywords: beeswax/*cerana*/*mellifera*/natural enemies/India

Verma SC, Desh R, Nirmala D (1999) Extent of parasitisation of wax moths by *Apanteles galleriae* Wilk, in *Apis cerana* colonies at Palampur. J Insect Sci 12:75–76, Keywords: beeswax/*cerana*/pests/natural enemies/India

Verma SK (1983) Studies on the foraging behaviour of *Apis cerana indica* Fab. in Jeolikote (Nainital, India). Indian Bee J 45:5–7, Keywords: behaviour/*cerana*/foraging/India

Verma SK (1987d) Preliminary studies on the effect of honey bees on the yield of greater cardamom. Indian Bee J 49:25–26, Keywords: *cerana*/honey/pollination/India

Verma SK (1988c) Pollen collection and brood rearing activities of *Apis cerana indica* Fabr. in Jeolikote, Uttar Pradesh. Indian Bee J 50:12–13, Keywords: bee botany/breeding/brood/*cerana*/foraging/pollen/India

Verma SK (1995d) Studies on the control of greater wax moth, *Galleria mellonella* L in *Apis cerana* F colonies with the biological insecticide, Dipel. Indian Bee J 57:121–123, Keywords: beeswax/*cerana*/insecticides/pests/natural enemies/India

Verma SK, Joshi NK (1983) Studies on insect pollinators in relation to seed production in cauliflower (*Brassica oleracea* var *botrytis* L). Indian Bee J 45:52–53, Keywords: *cerana*/pollination/India

Verma SK, Joshi NK (1985) Intensity of bee diseases in Indian hive bee (*Apis cerana indica* Fab.) in Uttar Pradesh. Indian Bee J 47:24–26, Keywords: *cerana*/disease/nests/India

Verma SK, Joshi NK (1988a) Immunity to Thai sacbrood virus in *Apis cerana indica* Fabr. Indian Bee J 50:12, Keywords: *cerana*/disease/India

Verma SK, Joshi NK (1988b) *Syntretomorpha szyboi* Papp – a new hymenopterous parasite of Indian hive bee, *Apis cerana indica* Fabr. Indian Bee J 50:39–40, Keywords: *cerana*/natural enemies/nests/pests/India

Veth PJ (1876) The honeybees of Java. Apiculture 20:269–273, [in French]. Keywords: *cerana*/Indonesia

Veth PJ (ed) (1881) Central Sumatra, travels and investigations of the Sumatra expedition 1877–1879. Brill, Leiden, Keywords: biogeography/*cerana*/Indonesia

Vidano C (1984) Remarks and impressions of beekeeping and other branches of agriculture in China. Part 2. Shanghai and Guangzhou Zones. Apic Mod 75:55–73, [in Italian]. Keywords: beekeeping/*cerana*/interspecific interactions/*mellifera*/China

Vien PN (1984) Preliminary studies on some brood diseases of *Apis cerana* Fabr. in northern Vietnam and their control methods. Thesis, Agricultural University No 1, Hanoi, Keywords: brood/*cerana*/disease/Vietnam

Viljakainen L, Evans JD, Hasselmann M, Rueppell O, Tingek S, Pamilo P (2009) Rapid evolution of immune proteins in social insects. Mol Biol Evol 26:1791–1801, Keywords: *andreniformis*/biochemistry/*cerana*/*dorsata*/evolution/*koschevnikovi*/Borneo

Villafuerte L (1993) Esterase polymorphism in a population of *Apis cerana* F (Hymenoptera: Apoidea) from different localities. Thesis, University of the Philippines, Keywords: biochemistry/biogeography/*cerana*/morphology/population biology/Philippines

Villafuerte L (1999) Mitochondrial DNA (mtDNA) polymorphism in the Asian honeybee (*Apis cerana* F) in the Philippines. Thesis, University of the Philippines, Los Banos, Keywords: *cerana*/DNA/morphology/mtDNA/Philippines

Villafuerte L, Laude RP, Smith DR (2001) Mitochondrial DNA (mtDNA) polymorphism in the Asian honeybee, *Apis cerana* F in the Philippines. In: Proceedings of the 7th international conference on tropical bees, Chiang Mai, pp 349–352, Keywords: *cerana*/DNA/genetics/morphology/mtDNA/Philippines

Viraktamath S (1989a) Incidence of greater wax moth *Galleria mellonella* (L) in three species of honey bees. Indian Bee J 51:139–140, Keywords: beeswax/*cerana*/*dorsata*/*florea*/honey/natural enemies/pests/India

Viraktamath S (1989b) Nesting behaviour of rock bees, *Apis dorsata* Fabr. in Raichur, Karnataka. Indian Bee J 51:133–134, Keywords: absconding/behaviour/combs/*dorsata*/migration/queens/swarming/India

Viraktamath S, Lingappa S, Somasekhar (1990) Performance of *Apis cerana* in Raichur, Karnataka State, India. In: Proceedings of the 11th international congress, IUSSI, Bangalore, pp 221–222, Keywords: brood/*cerana*/pests/India

Visetson S, Milne M, Milne J (2001) Toxicity of 4, 11-selinnadien-3-one from nutsedge (Cyperus rotundus L.) tuber extracts to diamondback moth larvae (Plutella xylostella L.), detoxification mechanisms and toxicity to non target species. Kasetsart J Nat Sci 35:284–292, Keywords: *florea*/insecticides

Vishwanathan H (1950a) Notes on *Apis dorsata* queen cells. Indian Bee J 12:55, Keywords: cells/combs/*dorsata*/queens/nests/India

Vishwanathan H (1950b) Temperature reading on *Apis dorsata* combs. Indian Bee J 12:72, Keywords: combs/*dorsata*/environmental effects/India

von Buttel-Reepen H (1906) Contributions to the systematics biology as well as the historical and geographical distribution of honeybees (*Apis mellifica* L) their variability and other *Apis* species. Mitt Zool Mus Berl 3:117–201, [in German]. Keywords: biogeography/biology/*cerana*/classification/*dorsata*/*florea*/history/*mellifera*

von Buttel-Reepen H (1921) Contribution on the systematics, biology as well as the historical and geographical distribution of honeybees. Arch Bienkd 3:19–67, [in German]. Keywords: biogeography/biology/*cerana*/classification/*dorsata*/*florea*/history/*mellifera*

von Frisch K (1955) Lindauer's observations and experiments on Indian bees. Sitz Bayer Akad Wiss 10:209–216, [in German]. Keywords: *cerana*/communication/dance language/*dorsata*/*florea*/foraging

von Frisch K (1967) The dance language and orientation of bees. Belknap Press of Harvard University, Cambridge, Keywords: behaviour/communication/dance language/foraging

von Hasselt AL (1882) Description of the people of central Sumatra. In: Veth PJ (ed) Central Sumatra, travels and investigations of the Sumatra expedition 1877–1897. J Brill, Leiden, Keywords: Sumatra

Vorwohl G (1968) Natural diastase deficiency in honeys from *Apis cerana*. Z Bienenforsch 9:232–236, [in German]. Keywords: biochemistry/*cerana*/honey

Vorwohl G (1996) Microscopical examination of Vietnamese honeys. In: Proceedings of the 3rd Asian apiculture association international conference, Hanoi, pp 81–86, Keywords: bee botany/*cerana*/honey/Vietnam

Wadhi SR (1961) Beekeeping in central Nepal. Indian Bee J 23:54–55, Keywords: beekeeping/*cerana*/Nepal

Wakhle DM (1981) The versatile pollen. Indian Bee J 43:114–117, Keywords: *cerana*/pollen/India

Wakhle DM (1987) Rendering of beeswax. Bee Sci News 3:2–5, Keywords: beekeeping/beeswax/*cerana*/India

Wakhle DM (1998) Beekeeping technology: production, characteristics and uses of honey and other products. In: Mishra RC (ed) Perspectives in Indian apiculture. Agro Botanica, Bikaner, New Delhi, pp 150–186, Keywords: beekeeping/*cerana*/honey/*mellifera*/India

Wakhle DM, Choudhary OP (1994) Potential of honey in food industry in India. In: Proceedings of the food and agro products and packaging symposium, World Trade Fair, New Delhi, p 735, Keywords: beekeeping/*cerana*/honey/India

Wakhle DM, Desai DB (1983) Enzymes in Indian honeys. In: Proceedings of the 2nd international conference on apiculture in tropical climates, New Delhi, pp 703–710, Keywords: biochemistry/*cerana*/*dorsata*/*florea*/honey/physical factors/India

Wakhle DM, Desai DB (1991) Estimation of antibacterial activity of some Indian honeys. Indian Bee J 53:80–90, Keywords: *cerana*/*dorsata*/honey/India

Wakhle DM, Kamble KD (2001) Technology and value addition of honey. Central Bee Research Training Institute, Pune, Keywords: *dorsata*/honey/India

Wakhle DM, Pal N (2001) Honey and hive products in India: present status. In: Proceedings of the 7th international conference on tropical bees, Chiang Mai, pp 41–47, Keywords: biochemistry/*cerana*/disease/*dorsata*/honey/*mellifera*/nests/pollen/royal jelly/India

Wakhle DM, Phadke RP (1995) Design of honey processing unit. Part 1. Indian Bee J 57:144–146, Keywords: *cerana*/honey/*mellifera*/India

Wakhle DM, Nair KS, Phadke RP (1978) Effect of bee pollination on the oil and protein content in the seeds of sunflower (*Helianthus annuus* L). Indian Bee J 40:1–2, Keywords: *cerana*/pollination/India

Wakhle DM, Phadke RP, Nair KS (1983) Study on enzyme activities of pollen and honey bees (*Apis cerana* Fab.). Indian Bee J 45:3–5, Keywords: biochemistry/*cerana*/honey/pollen/India

Wakhle DM, Phadke RP, Nair KS (1985) Bee venom extraction from *Apis cerana*. Bee Sci News 3:4–5, Keywords: biochemistry/*cerana*/India

Wakhle DM, Phadke RP, Pais DVE, Nair S (1996) Design of honey processing unit. Part 2. Indian Bee J 58:5–9, Keywords: *cerana*/honey/*mellifera*/India

Walker AK, Joshi NK, Verma SK (1990) The biosystematics of *Syntretomorpha szaboi* Papp (Hymenoptera: Braconidae: Euphorinae) attacking the Oriental honey bee, *Apis cerana* Fabricius (Hymenoptera: Apidae), with a review of braconid parasitoids attacking bees. Bull Entomol Res 80:79–83, Keywords: *cerana*/honey/natural enemies/pests/India

Wallace AR (1869) The Malay Archipelago: the land of the orang-utan, and the bird of paradise a narrative of travel, with studies of man and nature. Macmillan, London, Keywords: beekeeping/*dorsata*/Indonesia

Walrecht BJJR (1963) Remarks on comb building by *Apis dorsata*. Entomol Ber 23:32–36, [in Dutch]. Keywords: combs/*dorsata*/nests

Wang DK, Sun GF, Zhai SH, Guo LH, Xiao FH (2009a) The floral syndrome and its pollination in *Cynanchum otophyllum* (Asclepiadaceae). Acta Bot Yunnanica 31:1, [in Chinese]. Keywords: bee botany/*cerana*/foraging/pollination/China*

Wang FM (1996a) A basic method for strengthening colonies of *Apis cerana cerana*. Apic China 4:17, [in Chinese]. Keywords: beekeeping/*cerana*/China*

Wang GX, Miura R, Kusanagi T (1995) The enantiostyly and the pollination biology of *Monochoria korsakowii* (Pontederiaceae). Acta Phytotaxon Geobot 46:55–65, Keywords: bee botany/biology/*cerana*/pollination/Japan

Wang GX, Yamasue Y, Itoh K, Kusanagi T (1998) Outcrossing rates as affected by pollinators and the heterozygote advantage of *Monochoria korsakowii*. Aquat Bot 62:135–143, Keywords: bee botany/*cerana*/foraging/pollination

Wang JD (1985) A report on the discovery of *Malpighamoeba mellificae* in the Chinese bee, *Apis cerana cerana*, and its prevention and cure. Zhongguo Yangfeng 4:1–2, [in Chinese]. Keywords: *cerana*/disease/China

Wang JD (1989a) Experiments to improve the ripening of honey and protection of colonies of *Apis cerana cerana* against disease. J Bee 3:4–7, [in Chinese]. Keywords: *cerana*/disease/honey/China

Wang JD, Liang Q (1993) Keeping *Apis cerana cerana* in earth pits and holes. J Fujian Agric Coll 22:476–479, [in Chinese]. Keywords: beekeeping/*cerana*/China

Wang JD, Wang MG (1989) A preliminary report on increasing the ripening of honey and disease control in *Apis cerana cerana*. Condens Oil Catal 3:4–7, [in Chinese]. Keywords: *cerana*/disease/honey/China

Wang JD, Liang Q, Zhang WH (1985) First discovery of Malpighian tubule amoebiasis in *Apis cerana cerana* and experiments on its prevention and cure. Apic China 4:1–2, [in Chinese]. Keywords: *cerana*/disease/China

Wang LH, Zhang QK, Hou GS, Lou JN, Lou KP, Yu WJ (1999) Freezing preservation technique of honeybee semen. J Fujian Agric Univ 28:204–209, Keywords: *cerana*/drones/China

Wang R, Dong J (1999) Cloning and sequencing of ATPase8 and ATPase6 genes of the mitochondrial DNA of *Apis cerana cerana*. Apic China 3:15–17, Keywords: *cerana*/DNA/genetics/mtDNA/China

Wang RW (1998) Research development on the mtDNA of honeybees. Apic China 49:30–31, [in Chinese]. Keywords: *cerana*/genetics/mtDNA/physiology/China

Wang RW, Yang GH, Kuang YH (1992) Factors influencing hypopharyngeal gland activity in *Apis cerana* workers. Acta Entomol Sinica 35:500–502, [in Chinese]. Keywords: *cerana*/glands/workers/China

Wang S, Kuang HO, Liu FL (2000) Biological diversity of *Apis cerana* hive, Keywords: biology/*cerana*/genetics/nests/China*

Wang WH (1991a) Discussion of a multiple-queen colony with mother and daughter in the same hive of *Apis cerana cerana*. Apic Sci Tech 3:12, [in Chinese]. Keywords: *cerana*/nests/queens/China*

Wang WH (1991b) The safe transport of *Apis cerana cerana*. Apic Sci Tech 4:10, [in Chinese]. Keywords: beekeeping/*cerana*/China*

Wang WH (1992a) Coiled-wing and white-head disease in *Apis cerana cerana* and *Apis mellifera ligustica*. Apic Sci Tech 2:17, [in Chinese]. Keywords: *cerana*/disease/*mellifera*/China*

Wang WS (1993) Methods for catching wild swarms of *Apis cerana cerana*. Apic China 4:20, [in Chinese]. Keywords: beekeeping/*cerana*/China*

Wang X, Wang ZP (1988) Studies on the chromosome nucleus body of *Apis cerana* Fabricius. Apic China 2:2–4, [in Chinese]. Keywords: *cerana*/genetics/China

Wang YY, Zhao ZM, Zhang YS, Zhao ZF (2009b) Experimental report of Honeywell Taisui active bacteria liquid on honeybee colonies. J Bee 29:6, [in Chinese]. Keywords: *cerana*/disease/China*

Wang Z, Li ZY, Zhu HW, Li JL (2009c) Nutritional crossbreed technology of honeybees and its application in genetics and breeding. J Bee 29:7, [in Chinese]. Keywords: breeding/*cerana*/genetics/China*

Wang ZP (1989b) Report on the chromosomes of *Apis dorsata* Fabricius and *Apis florea* Fabricius. Apic Sci Tech 1:36, [in Chinese]. Keywords: *dorsata*/*florea*/genetics

Wang ZW (1996b) Discussion on a hive for *Apis cerana cerana*. Apic Sci Tech 2:24, [in Chinese]. Keywords: beekeeping/*cerana*/nests/China

Wang ZZ (1981) Discussions on the mechanisms of colony disturbance in *Apis cerana cerana*. J Bee 3:6–8, [in Chinese]. Keywords: aggressiveness/*cerana*/China

Wang ZZ (1992b) Rearing techniques for two colonies of *Apis cerana cerana* in one hive. Apic China 6:25, [in Chinese]. Keywords: beekeeping/breeding/*cerana*/nests/queens/China*

Wanitwattana S (1940) Beekeeping in Thailand. J Chulalongkorn Univ, Bangkok, pp 79–90, [in Thai]. Keywords: beekeeping/cerana/dorsata/florea/Thailand*

Wanitwattana S (1944) Honeybees in Thailand. The Royal Institute, Bangkok, pp 154–165, [in Thai]. Keywords: *cerana*/*dorsata*/*florea*/Thailand*

Ward DJ (1953) Madhu Kosh [sweetness + receptacle for honeycomb]. Am Bee J 93:240–241, Keywords: beekeeping/*cerana*/natural enemies/India

Waring C, Jump DR (2004) Beekeeping around the world: Rafter beekeeping in Cambodia with *Apis dorsata*. Bee World 85:14–18, Keywords: beekeeping/*dorsata*/*laboriosa*/Cambodia

Warncke K (1979) Contributions to the bee fauna of Iran 4. The genus Apis. Boll Mus Civ Storia Nat Venezia 30:157–158, [in German]. Keywords: bee botany/*cerana*/*florea*/foraging/*mellifera*/Iran

Warrit N (2002) Biogeography of *Varroa* mites on their native host, *Apis cerana* in the Malay Peninsula. In: Proceedings of the 14th international congress, IUSSI, Sapporo, p 210, Keywords: biogeography/*cerana*/*mellifera*/mtDNA/pests/Japan

Warrit N, Smith DR, Lekprayoon C (2006) Genetic subpopulations of *Varroa* mites and their *Apis cerana* hosts in Thailand. Apidologie 37:19–30, Keywords: *cerana*/pests/populations

Warth AH (1956) The chemistry and technology of waxes. Reinhold, New York, Keywords: beeswax/biochemistry/*cerana*/*dorsata*/*mellifera*

Watanabe C (1987) Occurrence of *Apanteles galleriae* (Hymenoptera, Braconidae), parasite of wax moth in Japan. Kontyû 55:165, Keywords: beeswax/*cerana*/natural enemies/pests/Japan

Watanabe K (1965) On the 'origin' of honeybees and beekeeping in Japan. In: Proceedings of the 20th international apiculture congress, Bucharest, pp 714–715, Keywords: beekeeping/*cerana*/history/Japan

Watanabe K (1974) Modern beekeeping. Nihon Yoho Shinkokai, Gifu, Keywords: beekeeping/*cerana*/Japan

Watanabe K (1981) Beekeeping and bee research in Japan before 1900. Honeybee Sci 1:277–280, [in Japanese]. Keywords: beekeeping/*cerana*/history/Japan

Watanabe K (1984) Beekeeping and bee-researchers in Japan before the 19th century. Honeybee Sci 2:75–86, [in Japanese]. Keywords: beekeeping/*cerana*/history/Japan

Watanabe K (1986) On 'Hachimitsu-ichiran' – an illustrated booklet in 1872. In: Proceedings of the 30th international apiculture congress, Apimondia, Nagoya, pp 359–362, Keywords: beekeeping/*cerana*/history/Japan

Watt G (1966) The commercial products of India being an abridgement of 'the dictionary of the economic products of India'. Today and Tomorrow's Printers and Publishers, New Delhi, Keywords: *cerana*/*florea*/India

Wattanachaiyingcharoen W, Oldroyd BP, Good G, Halling LA, Ratnieks FL, Wongsiri S (2002) Lack of worker reproduction in the giant honeybee *Apis dorsata* Fabricius. Insectes Soc 49:80–85, Keywords: *dorsata*/honey/reproduction/Thailand

Wattanachaiyingcharoen W, Oldroyd BP, Wongsiri S, Palmer K, Paar J (2003) A scientific note on the mating frequency of *Apis dorsata*. Apidologie 34:85–86, Keywords: *dorsata*/reproduction/Thailand

Wcislo WT, Tierney SM (2009) Behavioural environments and niche construction: the evolution of dim-light foraging in bees. Biol Rev 84:19–37, Keywords: behaviour/*dorsata*/flight/foraging

Webster TC, Peng YS (1988) The evolution of food-producing glands in eusocial bees (Apoidea, Hymenoptera). J Evol Biol 2:165–176, Keywords: *cerana*/*dorsata*/*florea*/glands/*mellifera*/phylogeny

Wehner R (2009) Martin Lindauer: obituary. J Comp Physiol A 195:217–220, Keywords: *cerana*/communication/dance language/*dorsata*/*florea*

Wei CZ (1994) Reasons for and prevention of absconding in *Apis cerana cerana*. Apic China 2:23, [in Chinese]. Keywords: absconding/beekeeping/*cerana*/China*

Weiss J (1978) Comparative morphology of the sting apparatus in the genus *Apis*. Apidologie 9:19–31, [in German]. Keywords: *cerana*/*dorsata*/*florea*/*mellifera*/morphology

Wells H, Ohtani T (1996) Foraging behavior of the Japanese honeybee in an artificial condition. Nat Human Act 1:45–50, Keywords: behaviour/*cerana*/foraging/Japan

Wells H, Rathore RRS (1994) Foraging ecology of the Asian hive bee, *Apis cerana indica*, within artificial flower patches. J Apic Res 33:219–230, Keywords: behaviour/*cerana*/ecology/foraging/nests

Wells H, Rathore RRS (1995) Discriminant conditioning of foragers in the Asian honey bees *Apis cerana* and *A. dorsata*. Ecol Entomol 20:374–379, Keywords: behaviour/*cerana*/*dorsata*/foraging/honey/pheromones

Weng AC (1991) Controlling sacbrood in *Apis cerana cerana* by sterilization and without removing larvae and using drugs. J Bee 6:24, [in Chinese]. Keywords: beekeeping/*cerana*/disease/larvae/China*

Wenger AMR (1949) A remarkable observation on the Indian honeybee versus the yellowthroated marten from Java. Treubia 20:31–33, Keywords: *cerana*/natural enemies/predators/Indonesia

Werber C, Sen Sarma M, Fuchs S, Tautz J (2001) Piping and hissing in the dwarf honeybee *Apis florea*: an alerting dialog? Apidologie 32:499–501, Keywords: behaviour/defense/*florea*/India

Weymeersch M (1999) Synthesis of systematic knowledge of the genus *Apis* L, 1758 (Hymenoptera: Apidae). Thesis, University of Gembloux, [in French]. Keywords: *andreniformis*/*cerana*/classification/*dorsata*/*florea*/*koschevnikovi*/*laboriosa*/*nigrocincta*/*nuluensis*/phylogeny

Wheeler WM (1908) Vestigial instincts in insects and other animals. Am J Psychol 19:1–13, Keywords: *dorsata*/nests

Whitcombe RP (1978) Reports on *Apis florea* to Directorate of Agriculture, Durham University, Durham, Keywords: beekeeping/*florea*/Oman

Whitcombe RP (1980) Honeybees in the Sultanate of Oman. Directorate of Agriculture, Oman, [in Arabic]. Keywords: classification/*florea*/foraging/Oman

Whitcombe RP (1981) Temperature regulation and foraging activity of two species of honey bee (*Apis mellifera* and *A. florea*) in Oman. Joint Honey Production Project Apicultural Research Programme Report, Durham University/Khabura Development Project, Keywords: environmental effects/*florea*/foraging/honey/*mellifera*/physiology/Oman

Whitcombe RP (1982) Ants (Formicidae), especially those associated with honeybees (*Apis* spp), from the Sultanate of Oman. Durham University/Khabura Development Project Report 2C, Keywords: brood/disease/*florea*/pests/Oman

Whitcombe RP (1982b) Experiments with a hive for little bees: some observations on manipulating colonies of *Apis florea* in Oman (Part 1). Indian Bee J 44:57–63, Keywords: beekeeping/*florea*/Oman

Whitcombe RP (1982c) Experiments with a hive for little bees: some observations on manipulating colonies of *Apis florea* in Oman (Part 2). Indian Bee J 44:93–105, Keywords: beekeeping/*florea*/Oman

Whitcombe RP (1984a) The Bedouin bee. ARAMCO World M 35:34–41, Keywords: *florea*/general biology/Oman

Whitcombe RP (1984b) The biology of *Apis* spp in Oman with special reference to *Apis florea* Fab. Thesis, University of Durham, Keywords: classification/combs/ecology/environmental effects/*florea*/general biology/Oman

Whitcombe RP (1985) A preliminary bibliography on bees and honey in Arabia, Sultanate of Oman. Khabura Development Project Report, Keywords: bibliography/*florea*/honey/Oman

Whitcombe RP (1985) Aspects of the biology and management of *Apis florea* in Oman. In: Proceedings of the 3rd international conference on apiculture in tropical climates, Nairobi, pp 96–103, Keywords: beekeeping/environmental effects/*florea*/general biology/Oman

Whitfield JB, Cameron SA (1998) Hierarchical analysis of variation in the mitochondrial 16S rRNA gene among Hymenoptera. Mol Biol Evol 15:1728–1743, Keywords: biochemistry/*cerana*/*dorsata*/*florea*/genetics/*koschevnikovi*/*mellifera*/morphology/organic chemistry

Wickham T (2000) Traditional honey and wax collection with *Apis dorsata* in the upper Kapuas Lake region, West Kalimantan. Borneo Res Bull, Keywords: beeswax/*dorsata*/honey/Borneo*

Widjaja MC (1992) The effect of tapping coconut trees on the activity and production of *Apis cerana* honey bees. Duta Rimba 18:37–39, [in Indonesian]. Keywords: beekeeping/*cerana*/honey/Indonesia

Widjaja MC (2001) Beekeeping in Perum Perhutani. In: Proceedings of the 37th international apiculture congress, Apimondia, Durban (CD Rom), Keywords: beekeeping/*cerana*/combs/queens/Indonesia

Widjaja MC, Purwanto DB (1989) The effect of size of comb foundation on comb area development of *Apis cerana* colonies in the forest district of Sukabumi, West Java. Bull Penelitian Hutan 4:35–43, [in Indonesian]. Keywords: *cerana*/combs/morphology/physiology/Indonesia

Widjaja MC, Purwanto DB, Desrianty (1992) Acacia mangium extrafloral nectar for honeybee consumption. Duta Rimba 18:2–5, Keywords: bee botany/cerana/foraging/nectar/Indonesia

Widjaja MC, Kustanti A, Oshida T (1996) Hive design for *Apis cerana*. In: Proceedings of the 3rd Asian apiculture association international conference, Hanoi, pp 71–73, Keywords: beekeeping/*cerana*/nests/Indonesia

Widjaja MC, Purwanto DB, Desrianty (1992) *Acacia mangium* extrafloral nectar for honeybee consumption. Duta Rimba 18:2–5, Keywords: bee botany/*cerana*/nectar/Indonesia

Wilde J, Woyke J, Wilde M (2001) Rearing honeybee queens on cups of *Apis mellifera* and *Apis dorsata* wax. In: Proceedings of the 7th international conference on apiculture in tropical climates, Chiang Mai, pp 207–209, Keywords: beekeeping/*dorsata*/honey/queens/*mellifera*/ Nepal

Wilde J, Woyke J, Wilde M (2003) Competition between drones and between spermatozoa of *Apis cerana* and *Apis mellifera*. In: Proceedings of the 6th Asian apiculture association international conference, Bangalore, pp 39–42, Keywords: *cerana*/drones/*mellifera*/reproduction/spermatozoa

Williams IH (1977) Behaviour of insects foraging on pigeon pea (*Cajanus cajan* (L) Millsp) in India. Trop Agric 54:353–363, Keywords: bee botany/behaviour/*cerana*/*dorsata*/foraging/ pollination/India

Williams IH (1983) The pollination of pigeon pea (*Cajanus cajan* (L) Millsp) in India. In: Proceedings of the 2nd international conference on apiculture in tropical climates, New Delhi, pp 661–666, Keywords: bee botany/*cerana*/*dorsata*/foraging/physical factors/pollination/India

Willis LG (1991) The phylogenetic relationships in the genus *Apis* as determined by the sequence of the cytochrome oxidase II region of mitochondrial DNA. Thesis, Simon Fraser University, Keywords: *cerana*/DNA/*dorsata*/*florea*/genetics/*mellifera*/mtDNA/phylogeny

Willis LG, Winston ML, Honda BM (1992) Phylogenetic relationships in the honeybee (genus *Apis*) as determined by the sequence of the cytochrome oxidase II region of mitochondrial DNA. Mol Phylogenet Evol 1:169–178, Keywords: *andreniformis*/*cerana*/DNA/*dorsata*/ *florea*/genetics/*koschevnikovi*/mtDNA/phylogeny/Malaysia/Thailand

Wilson EO (1971) The insect societies. Harvard University Press, Cambridge, Keywords: biology/ *cerana*/*dorsata*/*florea*/*mellifera*

Wilson GD (1995) Marketing pure bees honey in Sri Lanka. In: Kevan PG (ed) The Asiatic hive bee: apiculture, biology, and role in sustainable development in tropical and subtropical Asia. Enviroquest, Cambridge, Ontario, pp 245–248, Keywords: beekeeping/biology/*cerana*/honey/ nests/physiology/Sri Lanka

Wilson WT, Nelson DL, Clark KJ (2001) Chemical control of tracheal mites. In: Webster TC, Delaplane KS (eds) Mites of the honey bee. Dadant, Hamilton, pp 85–104, Keywords: *cerana*/ honey/pests/India

Won R, Li CY, Kim JW, Rhee HI (2009) Immunological characterization of honey major protein and its application. Food Chem 113:1334–1338, Keywords: biochemistry/*cerana*/honey/*mellifera*/Korea

Wongchawalit J, Yamamoto T, Nakai H, Kim YM, Sato N, Nishimoto M, Okuyama M, Mori H, Saji O, Chanchao C, Wongsiri S, Surarit R, Svasti J, Chiba S, Kimura A (2006) Purification and characterization of α-glucosidase I from Japanese honeybee (*Apis cerana japonica*) and molecular cloning of its cDNA. Biosci Biotech Biochem 70:2889–2898, Keywords: biochemistry/*cerana*/DNA/genetics

Wongiyasatid A, Hormchan P (2001) Comparison of the single and double flowered *Portulaca varieties* in bee attraction. Kasetsart J Nat Sci 35:8–13, Keywords: bee botany/*cerana*/*dorsata*/ foraging/pollination/India

Wongsiri S (1984) Beekeeping in China. Sci Soc Thailand J 39:418–420, [in Thai]. Keywords: beekeeping/*cerana*/*mellifera*/China

Wongsiri S (1985) Beekeeping in Thailand. Funny, Bangkok, [in Thai]. Keywords: beekeeping/ *cerana*/*dorsata*/*florea*/Thailand

Wongsiri S (1987) Regulation of *Apis cerana* absconding during a dearth period. In: Proceedings of the 31st international apiculture congress, Apimondia, Warsaw, p 192, Keywords: absconding/behaviour/*cerana*/environmental effects/Thailand

Wongsiri S (1988) Queen production. Pamphlet, Universiti Pertanian, Selangor, Malaysia, p 23, Keywords: beekeeping/*cerana*/queens/Thailand

Wongsiri S (1988) The resistance mechanism of the Asian hive bee (*Apis cerana*) to a bee mite *Tropilaelaps clareae*. In: Proceedings of the 18th international congress entomology, Vancouver, p 245, Keywords: *cerana*/disease/nests/pests/Thailand

Wongsiri S (1989) *Apis cerana* beekeeping problems in developing countries of Southeast Asia. Honeybee Sci 10:160–164, [in Japanese]. Keywords: beekeeping/*cerana*

Wongsiri S (1987) Regulation of *Apis cerana* absconding during a dearth period. In: Proceedings of the 31st International Apiculture Congress, Apimondia, Warsaw, p 192, Keywords: absconding/behaviour/*cerana*/environmental effects/honey/Thailand

Wongsiri S (1992) Beekeeping problems in developing countries of South East Asia. In: Verma LR (ed) Honeybees in mountain agriculture. Oxford and IBH, New Delhi, pp 239–248, Keywords: beekeeping/*cerana*

Wongsiri S (1995) Queen production of *Apis cerana*. In: Kevan PG (ed) The Asiatic hive bee: apiculture, biology, and role in sustainable development in tropical and subtropical Asia. Enviroquest, Cambridge, Ontario, pp 91–106, Keywords: biology/*cerana*/nests/oviposition/physiology/queens/reproduction/Thailand

Wongsiri S, Chen PP (1995) Effects of agricultural development on honey bees in Thailand. Bee World 76:3–5, Keywords: *cerana*/ecology/honey/interspecific interactions/*mellifera*/physiology/population biology/Thailand

Wongsiri S, Deowanish S (1999) Bees and beekeeping in Thailand. Honeybee Sci 20:135–137, [in Japanese]. Keywords: beekeeping/*cerana*/*dorsata*/*florea*/*mellifera*/Thailand

Wongsiri S, Matsuka M (1999) Tropical apiculture and organizations. Farming Jpn 19:28–33, Keywords: beekeeping/*cerana*/*mellifera*/China/Japan

Wongsiri S, Pyramam K (1987) Beekeeping techniques with *Apis cerana* F in poor areas of northeastern Thailand: experiments on the effect of standard Langstroth hives and modified Chinese hives with wax foundation sheets. J Sci Res 12:3–7, Keywords: beekeeping/beeswax/*cerana*/nests/Thailand

Wongsiri S, Tangkanasing P (1986a) *Apis cerana* F beekeeping in Thailand: problems and research needs. J Sci Res 11:1–6, Keywords: absconding/beekeeping/*cerana*/swarming/Thailand

Wongsiri S, Tangkanasing P (1986) Biology of honeybees. Chulalongkorn University, [in Thai]. Keywords: beekeeping/biology/*cerana*/Thailand

Wongsiri S, Tangkanasing P (1987) *Apis cerana* F beekeeping in Thailand: problems and research needs. In: Proceedings of the workshop on parasitic mites control, FAO, Pulawy, p 193, Keywords: beekeeping/*cerana*/Thailand

Wongsiri S, Tangkanasing P (1989) The resistance behavior of *Apis cerana* against *Tropilaelaps clareae*. In: Proceedings of the 1st Asia-Pacific conference of entomology, Chiang Mai, pp 25–34, Keywords: *cerana*/disease/pests/Thailand

Wongsiri S, Lai YS, Liu ZS (1986) Beekeeping in the Guangdong Province of China and some observations on the Chinese honey bee *Apis cerana* and the European honey bee *Apis mellifera ligustica*. Am Bee J 126:748–752, Keywords: beekeeping/*cerana*/disease/honey/*mellifera*/natural enemies/China

Wongsiri S, Tangkanasing P, Sylvester HA (1987a) Mites, pests and beekeeping with *Apis cerana* and *Apis mellifera* in Thailand. Am Bee J 127:500–503, Keywords: beekeeping/*cerana*/*mellifera*/natural enemies/pests/Thailand

Wongsiri S, Tangkanasing P, Vongsamanode S (1987b) Effectiveness of Asuntol (coumaphos), Perizin (coumaphos), Mitac (amitraz) and powder of sulphur with naphthalene for the control of bee mites (*Varroa jacobsoni* and *Tropilaelaps clareae*). In: Proceedings of the 31st international apiculture congress, Apimondia, Warsaw, pp 135–137, Keywords: *cerana*/insecticides/pests/Thailand

Wongsiri S, Wongsathuaythong S, Suwanagul M (1987c) Comparison of skin tests for honey bee allergy using venom extracts of *Apis mellifera*, *Apis cerana* and *Apis florea* in normal persons, atopic population and beekeepers. In: Proceedings of the 1st Asia-Pacific congress on animal, plant and microbials Toxins, Singapore, pp 603–615, Keywords: biochemistry/*cerana*/disease/*florea*/honey/*mellifera*/population biology/Singapore

Wongsiri S, Potichot S, Feng-Zhi CF (1988) Environmental factors necessary for *A. cerana* queen rearing in Thailand. In: Proceedings of the 4th international conference on apiculture in tropical climates, Cairo, pp 466–470, Keywords: beekeeping/breeding/*cerana*/environmental effects/physical factors/Thailand

Wongsiri S, Dietz A, Potichot S (1989) A comparison of instrumental insemination and naturally mated queen of *Apis cerana* Fabricius. In: Proceedings of the 32nd international apiculture congress, Apimondia, Rio de Janeiro, p 77, Keywords: breeding/*cerana*/queens/reproduction/Thailand

Wongsiri S, Potichot S, Chao FZ (1989) Queen rearing with *Apis cerana* in Thailand. In: Proceedings of the 4th international conference on apiculture in tropical climates, Cairo, pp 466–470, Keywords: breeding/*cerana*/physical factors/queens/reproduction/Thailand

Wongsiri S, Lai YS, Sylvester HA (1990a) Queen rearing with *Apis cerana*. Am Bee J 130:32–35, Keywords: breeding/*cerana*/queens/Thailand

Wongsiri S, Lekprayoon C, Potichot S (1990b) Biological control of the *Varroa* bee mite: mass rearing biological control agent by crossing the Chinese strain *Apis cerana cerana* with the Thai strain *Apis cerana indica* by artificial insemination. In: Proceedings of the workshop on agriculture biotech, Kasetsart University and USAID, Thailand, pp 1–7, Keywords: breeding/*cerana*/genetics/geographical variation/natural enemies/pests/oviposition/queens/China/Thailand

Wongsiri S, Limbipichai K, Tangkanasing P, Mardan MB, Rinderer TE, Sylvester HA, Koeniger G, Otis GW (1990c) Evidence of reproductive isolation confirms that *Apis andreniformis* (Smith, 1858) is a separate species from sympatric *Apis florea* (Fabricius, 1787). Apidologie 21:47–52, Keywords: *andreniformis*/classification/evolution/*florea*/morphology/Thailand

Wongsiri S, Rinderer TE, Sylvester HA (eds) (1991) Biodiversity of honey bees in Thailand. Chulalongkorn University, Keywords: *cerana*/*dorsata*/*florea*/honey/Thailand

Wongsiri S, Lekprayoon C, Potichot S (1993) Biological control of the *Varroa* mite: mass rearing a biological control agent by crossing the Chinese strain of *Apis cerana cerana* with the Thai strain *Apis cerana indica* by instrumental insemination. In: Connor LJ, Rinderer TE, Sylvester HA, Wongsiri S (eds) Asian apiculture. Wicwas, Cheshire, Connecticut, pp 148–155, Keywords: breeding/*cerana*/genetics/geographical variation/natural enemies/pests/oviposition/queens/China/Thailand

Wongsiri S, Chen PP, Thapa R (1995) Other uses for bee products in Thailand. Br Bee J 123:144–146, 148. Keywords: beekeeping/*cerana*/Thailand

Wongsiri S, Thapa R, Oldroyd BP, Burgett MD (1996) A magic bee tree: home of *Apis dorsata*. Am Bee J 136:196–199, Keywords: *dorsata*/combs/Thailand

Wongsiri S, Lekprayoon C, Thapa R, Thirakupt K, Rinderer TE, Sylvester HA, Oldroyd BP, Booncham U (1997) Comparative biology of *Apis andreniformis* and *Apis florea* in Thailand. Bee World 78:23–35, Keywords: *andreniformis*/*florea*/general biology/Thailand

Wongsiri S, Pyramam K, Leepitakrat S, Aemprapa S (1998a) Rubber: a potential major honey plant in Thailand. Bee World 80:187–190, Keywords: bee botany/*cerana*/*dorsata*/*florea*/foraging/honey/Thailand

Wongsiri S, Tangkanasing P, Lekprayoon C, Rinderer TE, Sylvester HA, Oldroyd BP, Deowanish S (eds) (1998b) Biodiversity of honey bees in Thailand. Department of Biology, Chulalongkorn University and National Science and Technology Development Agency, Bangkok, Keywords: biology/*cerana*/*dorsata*/*florea*/honey/Thailand

Wongsiri S, Thapa R, Kongpitak P (1998c) Longan: a major honey plant in Thailand. Bee World 79:23–28, Keywords: bee botany/*cerana*/*dorsata*/foraging/honey/*mellifera*/Thailand

Wongsiri S, Chanchao C, Deowanish S, Aemprapa S, Chaiyawong T, Petersen S, Leepitakrat S (2000) Honey bee diversity and beekeeping in Thailand. Bee World 81:20–29, Keywords: *andreniformis*/beekeeping/*cerana*/*dorsata*/*florea*/genetics/honey/*mellifera*/Thailand

Wongsiri S, Chanchao C, Lekprayoon C, Wattanasermkit K, Deowanish S, Leepitakrat S (2001) Honeybee diversity and management in the new millennium in Thailand. In: Proceedings of the 7th international conference on tropical bees, Chiang Mai, pp 9–14, Keywords: *andreniformis*/beekeeping/*cerana*/*dorsata*/*florea*/genetics/*mellifera*/Thailand

Wongsiri S, Deowanish S, Thapa R (2002) Some behaviour diversity of *Apis dorsata*, *Apis florea* and *Apis andreniformis*. In: Proceedings of the 14th international congress, IUSSI, Sapporo, p 211, Keywords: *andreniformis*/behaviour/classification/*dorsata*/*florea*/physiology/Thailand

Wongvilas S, Higgs JS, Beekman M, Wattanachaiyingcharoen W, Deowanish S, Oldroyd B (2010) Lack of inter-specific parasitism between the dwarf honeybees *Apis andreniformis* and *A. florea*. Behav Ecol Sociobiol. doi:10.1007/s00265-010-0932-1, Keywords: *andreniformis*/*florea*/pests/natural enemies

Woo KS (1991) Beekeeping in Korea, current status of Korean bee pests and their control. Asian Apic Assoc Newsl 1:8–9, Keywords: beekeeping/*cerana*/*mellifera*/pests/populations/Korea

Woo KS (1992a) New honeybee mite *Varroa underwoodi* on *Apis cerana* in South Korea. Honeybee Sci 13:173–174, [in Japanese]. Keywords: *cerana*/pests/Korea

Woo KS (1992b) The occurrence of a new honeybee mite *Varroa underwoodi* from Korea. Asian Apic Assoc Newsl 2:5–6, Keywords: *cerana*/pests/Korea

Woo KS (1993) The bionomics, distribution and control of bee mites (*Varroa jacobsoni*) in South Korea. In: Connor LJ, Rinderer TE, Sylvester HA, Wongsiri S (eds) Asian apiculture. Wicwas, Cheshire, Connecticut, pp 479–490, Keywords: biogeography/biology/*cerana*/pests/Korea

Woo KS (1996) Country report: Korea. In: Proceedings of the 3rd Asian apiculture association international conference, Hanoi, p 215, Keywords: *cerana*/*mellifera*/pests/Korea

Woo KS (2000) Integrated control methods of parasitic mites on honeybees. In: Proceedings of the 4th Asian apiculture association international conference, Kathmandu, pp 82–84, Keywords: *cerana*/*mellifera*/pests/Korea

Woo KS (2001) Bionomics of *Varroa underwoodi* on *Apis cerana cerana* in Korea. In: Proceedings of the 7th international conference on tropical bees, Chiang Mai, pp 227–229, Keywords: biology/*cerana*/pests/Korea

Woo KS, Lee JH (1993) A study on the mites inhabiting beehives in Korea I. Korean J Apic 8:140–156, [in Korean]. Keywords: *cerana*/pests/Korea

Woo KS, Lee JH (1996) Current status of honeybee mites in Korea. In: Proceedings of the 3rd Asian apiculture association international conference, Hanoi, pp 168–171, Keywords: *cerana*/*mellifera*/pests/Korea

Woo KS, Cho KS, Lew YS (1994) Analysis on the level of damage caused by honeybee mites. Korean J Apic 9:33–39, Keywords: *cerana*/pests/Korea

Woo KS, Lee JH, Cho KS, Cho YH (1995) Infestation status and control methods of honeybee pests (bee mites and wax moths) in Korea. Korean J Apic 10:35–48, Keywords: beeswax/*cerana*/*mellifera*/pests/Korea

Woo KS, Cha YH, Kim Y (2000) Abilities of *Apis mellifera* and *Apis cerana* to pollinate strawberry in plastic houses. Asian Bee J 2:31–35, Keywords: *cerana*/*mellifera*/pollination

Woyke J (1963a) What happens to diploid drone larvae in a honeybee colony? J Apic Res 2:73–76, Keywords: brood/*dorsata*/drones

Woyke J (1973b) Artificial insemination of *Apis cerana indica*. In: Proceedings of the 24th international apiculture congress, Apimondia, Buenos Aires, pp 335–338, Keywords: breeding/*cerana*/genetics/queens

Woyke J (1973b) Instrumental insemination of *Apis cerana indica* queens. J Apic Res 12:151–158, Keywords: breeding/*cerana*/genetics/*mellifera*/queens

Woyke J (1973c) Reproduction and artificial insemination of *Apis cerana indica* Fabr. Indian Bee J 35:10–20, Keywords: breeding/brood/*cerana*/drones/genetics/queens/reproduction
Woyke J (1975a) Eggs in comb cells of three species of Indian bees. In: Proceedings of the 25th internatinaol apiculture congress, Apimondia, Grenoble, pp 321–322, Keywords: beekeeping/cells/*cerana*/*dorsata*/*florea*/oviposition/queens/India
Woyke J (1975b) Natural and instrumental insemination of *Apis cerana indica* in India. J Apic Res 14:153–159, Keywords: breeding/*cerana*/genetics/reproduction/queens/India
Woyke J (1976) Brood-rearing efficiency and absconding in Indian honeybees. J Apic Res 15:133–143, Keywords: absconding/biogeography/breeding/brood/*cerana*/disease/*dorsata*/environmental effects/*florea*/honey/migration/nests/oviposition/pollen/queens/reproduction/swarming/workers/India
Woyke J (1977a) Brood rearing and absconding of tropical honeybees. In: Fletcher DJC (ed) African bees: taxonomy, biology and economic use. Apimondia, Pretoria, pp 96–102, Keywords: absconding/biology/breeding/brood/*cerana*/classification/*mellifera*/queens/reproduction
Woyke J (1977b) The density of bristles covering the wings as discrimination value between African and other races of honey bee. In: Fletcher DJC (ed) African bees: taxonomy, biology and economic use. Apimondia, Pretoria, pp 15–24, Keywords: biology/*cerana*/classification/*dorsata*/*florea*/honey/*mellifera*/morphometrics/subspecies
Woyke J (1978) Biology of reproduction and genetics of the honeybee (E21ENT28). Bee Culture Division, Warsaw Agricultural University, Poland, Keywords: absconding/biology/breeding/brood/cells/*cerana*/*dorsata*/drones/*florea*/genetics/oviposition/reproduction
Woyke J (1979) Sex determination in *Apis cerana indica*. J Apic Res 18:122–127, Keywords: *cerana*/genetics/physiology
Woyke J (1980) Evidence and action of cannibalism substance in *Apis cerana indica*. J Apic Res 19:6–16, Keywords: behaviour/*cerana*
Woyke J (1993a) Rearing and instrumental insemination of *Apis florea* queens. In: Connor LJ, Rinderer TE, Sylvester HA, Wongsiri S (eds) Asian apiculture. Wicwas, Cheshire, Connecticut, pp 206–210, Keywords: beekeeping/genetics/queens/reproduction/India
Woyke J (1993b) Size change of eggs during the incubation period in three Asian honey bee species. In: Connor LJ, Rinderer TE, Sylvester HA, Wongsiri S (eds) Asian apiculture. Wicwas, Cheshire, Connecticut, pp 197–205, Keywords: *cerana*/*dorsata*/*florea*/honey/morphology/oviposition
Woyke J (1994) Comparison of the size of eggs from *Apis mellifera* L queens and laying workers. Apidologie 25:179–187, Keywords: brood/morphology/oviposition/queens/reproduction/workers/India
Woyke J (1996) The background and history of apicultural scientific activities in Vietnam. In: Proceedings of the 3rd Asian apiculture association international conference, Hanoi, pp 26–32, Keywords: beekeeping/*cerana*/history/*mellifera*/pests/Afghanistan/Vietnam
Woyke J (1997) Expression of body colour patterns in three castes of four Asian honey bees. In: Proceedings of beenet Asia, Serdang, pp 85–95, Keywords: biochemistry/*cerana*/*dorsata*/*florea*/honey/morphology
Woyke J (2000) Differences in body colour expression between European and Asian honeybees. In: Proceedings of the 4th Asian apiculture association international conference, Kathmandu, pp 20–23, Keywords: *andreniformis*/*cerana*/*dorsata*/*florea*/genetics/*koschevnikovi*/*mellifera*/morphology/Afghanistan/China/India/Malaysia/Thailand/Vietnam
Woyke J (2001a) *Apis dorsata* drone flights, collection of semen from everted endophalli and instrumental insemination of queens. Apidologie 32:407–416, Keywords: *dorsata*/drones/flight/foraging/genetics/queens/reproduction
Woyke J (2001b) Eversion of drone endophallus and the probable mating process in *Apis dorsata*. In: Proceedings of the 7th international conference on apiculture in tropical climates, Chiang Mai, pp 189–194, Keywords: *dorsata*/drones/morphology/reproduction/India/Nepal/Thailand
Woyke J (2002) Different reactions of *Apis dorsata* and *Apis mellifera* to brood infestation by parasitic mites. In: Proceedings of the 3rd Asian Apiculture Association International

Conference, Hanoi, pp 172–175, Keywords: behaviour/brood/disease/*dorsata*/*mellifera*/natural enemies/pests/predators/China/India/Malaysia

Woyke J, Wongsiri S (1992) Occurrence and size of laying worker eggs in *Apis florea* colonies. J Apic Res 31:124–127, Keywords: brood/classification/*florea*/laying workers/morphology/oviposition/queens/reproduction/India

Woyke J, Ruttner F, Koeniger N (1972) Reproduction in *Apis cerana* 1. Mating behaviour. J Apic Res 11:141–146, Keywords: behaviour/*cerana*/reproduction

Woyke J, Ruttner F, Koeniger N (1973) Reproduction in *Apis cerana* 2. Reproductive organs and natural insemination. J Apic Res 12:21–34, Keywords: *cerana*/morphology/reproduction

Woyke J, Wilde J, Wilde M (2001a) A scientific note on *Apis laboriosa* winter nesting and brood rearing in the warm zone of the Himalayas. Apidologie 32:601–602, Keywords: beekeeping/brood/combs/*laboriosa*/nests/seasonal variation/Himalayas

Woyke J, Wilde J, Wilde M (2001) Coexistence of *Apis mellifera* and *Apis dorsata* workers in the same colonies. In: Proceedings of the 7th international conference on apiculture in tropical climates, Chiang Mai, pp 115–120, Keywords: behaviour/brood/*dorsata*/general biology/interspecific interactions/*mellifera*/physiology/reproduction/workers

Woyke J, Wilde J, Wilde M (2001) Swarming, migration and absconding of *Apis dorsata* colonies. In: Proceedings of the 7th international conference on apiculture in tropical climates, Chiang Mai, pp 183–188, Keywords: absconding/*dorsata*/migration/swarming/Nepal

Woyke J, Wilde J, Wilde M (2003a) Heritable versus environmental cues in recognition and acceptance of two species of honeybees: *Apis cerana* and *Apis dorsata* in one colony. In: Proceedings of the 6th Asian Apiculture Association International Conference, Bangalore, Keywords: behaviour/*cerana*/*dorsata*/environmental effects/genetics/interspecific interactions

Woyke J, Wilde J, Wilde M (2003b) Type of recognition cue as a basis to elaborate a method to keep two phylogenetically distant honey bee species *Apis dorsata* and *Apis cerana* in the same colony. Asian Bee J Special Issue, Proceedings of the 6th Asian apiculture association international conference, Bangalore, 5:27–33, Keywords: behaviour/*cerana*/*dorsata*/interspecific interactions

Woyke J, Wilde J, Wilde M (2003c) Flight activity reaction to temperature changes in *Apis dorsata*, *Apis laboriosa* and *Apis mellifera*. J Apic Res 47:73–80, Keywords: behaviour/*dorsata*/environmental effects/flight/foraging/*laboriosa*/*mellifera*

Woyke J, Wilde J, Wilde M (2003d) Periodic mass flights of *Apis laboriosa* in Nepal. Apidologie 34:121–128, Keywords: behaviour/flight/foraging/*laboriosa*/Nepal

Woyke J, Chanchao C, Wongsiri S, Wilde J, Wilde M (2003e) Size of eggs from queens of three Asian species and laying workers of *Apis cerana*. J Apic Sci 47:39–57, Keywords: *cerana*/*dorsata*/*florea*/laying workers/morphology/oviposition/queens/workers

Woyke J, Kruk C, Wilde J, Wilde M (2004a) Periodic mass flights of the giant honeybee, *Apis dorsata*. J Apic Res 43:181–186, Keywords: absconding/behaviour/*dorsata*/foraging/honey/migration/swarming/Nepal

Woyke J, Wilde J, Wilde M (2004b) Temperature correlated dorso-ventral abdomen flipping of *Apis laboriosa* and *Apis dorsata* worker bees. Apidologie 35:493–502, Keywords: behaviour/*dorsata*/environmental effects/*laboriosa*/morphology/Nepal

Woyke J, Wilde J, Reddy CC (2004c) Open-air nesting honey bees, *Apis dorsata* and *Apis laboriosa*, differ from the cavity-nesting *Apis mellifera* and *Apis cerana* in brood hygiene behaviour. J Invertebr Pathol 84:1–6, Keywords: behaviour/brood/*cerana*/disease/*dorsata*/honey/*laboriosa*/*mellifera*/nests

Woyke J, Wilde J, Wilde M, Sivaram V, Cervancia C, Nagaraja N, Reddy M (2008) Defense body movements of *Apis laboriosa*, *Apis dorsata dorsata* and *Apis dorsata brevigula* honey bees. J Insect Behav 21:481–494, Keywords: behaviour/defense/*dorsata*/*laboriosa*

Wu J (1989) Effects of yeast powder as a pollen substitute for rearing *Apis cerana cerana*. Apic China 6:11–13, [in Chinese]. Keywords: breeding/brood/*cerana*/pollen/China

Wu J (2003) The characteristics and application of several pollination honeybees. Apic China 54:24–25, [in Chinese]. Keywords: bee botany/biology/pollination/China

Wu PY (1986) Studies on the body fluid proteins of female *Apis cerana cerana*. Apic Sci Tech 4:1–4, [in Chinese]. Keywords: biochemistry/*cerana*/physiology/workers/China

Wu XB, Yan WY, Zeng ZJ (2009a) Emergency cells and breeding brood of the introduced *Apis* species in queenless colonies of *Apis mellifera ligustica* and *Apis cerana cerana*. Chin Bull Entomol 46:1, [in Chinese]. Keywords: breeding/brood/*cerana*/*mellifera*/queenlessness/reproduction*

Wu XB, Yan WY, Zeng ZJ (2009b) Study on the key technique of *Apis cerana cerana* to produce comb honey. J Anhui Agric Sci 37:10, [in Chinese]. Keywords: *cerana*/honey/China*

Wu YR, Kuang B (1986) A study of the genus *Micrapis* (Apidae). Zool Res 7:99–102, [in Chinese]. Keywords: classification/combs/China

Wu YR, Kuang B (1987) Two species of small honeybee: a study of the genus *Micrapis*. Bee World 68:153–155, Keywords: *andreniformis*/combs/defense/*florea*/China

Yakobson B, Pothichot S, Wongsiri S (1992) Possible transfer of *Nosema apis* from *Apis mellifera* to *Apis cerana*. In: Abstracts of papers of international conference on the Asian honeybees and bee mites, Bangkok, p 97, Keywords: *cerana*/disease/*mellifera*

Yamada K (1988) A note on Chinese beekeeping. Honeybee Sci 9:166–169, [in Japanese]. Keywords: beekeeping/*cerana*/China

Yamagami S (1981) Observation report on keeping Japanese honeybees. Honeybee Sci 2:23–26, [in Japanese]. Keywords: beekeeping/*cerana*/Japan

Yamaguchi H (1998) A consideration on the traditional beekeeping of Japanese honey-bee (*Apis cerana* Fabricius) in Tsushima Island. Honeybee Sci 19:129–136, Keywords: beekeeping/*cerana*/islands/Japan

Yan WY, Le Conte Y, Beslay D, Zeng ZJ (2009a) Identification of brood pheromone in the Chinese honeybee, *Apis cerana cerana* (Hymenoptera: Apidae). Sci Agric Sinica 42:2250–2254, [in Chinese]. Keywords: brood/*cerana*/pheromone/China

Yan WY, Zeng ZJ, Wu XB, Liu YB (2009b) A preliminary study on the ultrastructure and chemical components of the egg surface of the Chinese honeybee, *Apis cerana cerana*. Acta Entomol Sinica 52:1, [in Chinese]. Keywords: *cerana*/chemistry/morphology/China*

Yan Z (1984) The spread procedure of Chinese bee rearing for economic benefits. J Bee 4, [in Chinese]. Keywords: beekeeping/breeding/*cerana*/China*

Yan ZL, Chen GJ, Wang SJ (1996) The history and present situation on raising eastern bees in movable frames in China. In: Proceedings of the 3rd Asian apiculture association international conference, Hanoi, pp 106–109, Keywords: beekeeping/*cerana*/history/nests/China

Yang CH (1997a) How to find wild colonies of *Apis cerana cerana*. J Bee 3:19, [in Chinese]. Keywords: beekeeping/*cerana*/China*

Yang CH (1997b) A method for capturing wild colonies of *Apis cerana cerana*. J Bee 4:18, [in Chinese]. Keywords: beekeeping/*cerana*/China*

Yang CH (1997c) Subspecies and fine characteristics of *Apis cerana*. J Bee 6:17, [in Chinese]. Keywords: *cerana*/classification/morphology China*

Yang CH (1997d) Problems of making hives and their utilization by *Apis cerana cerana*. J Bee 7:22, [in Chinese]. Keywords: beekeeping/*cerana*/nests/China*

Yang G (1981a) An investigation of the Chinese honeybee in the Ganzi area of Szechwan Province. Apic China 3:26–28, [in Chinese]. Keywords: *cerana*/classification/China

Yang G (1983a) The honeybee resources of northwest China. Apic China 3:16–17, Keywords: *cerana*/classification/China

Yang G (1983b) A summary of the distribution and classification of Chinese honeybees. Apic China 3:17–19, [in Chinese]. Keywords: biogeography/*cerana*/classification/China

Yang G (1983c) The honeybee resources of northeast China. Apic China 5:19, [in Chinese]. Keywords: *cerana*/classification/China*

Yang G (1983d) Chinese bee breeding. Agricultural Publishing House, Peking, [in Chinese]. Keywords: beekeeping/breeding/*cerana*/China

Yang G (1986) The distribution of the eastern honeybee (*Apis cerana* Fab.) in China and an analysis of its subspecies. J Yunnan Agric Univ 12:202–206, [in Chinese]. Keywords: biogeography/*cerana*/classification/China

Yang G (1995) An investigation on and distribution of the characteristics of the eastern honeybee (*Apis cerana* Fab.) in the plateau of the northwest of Szechwan Province. Agric Sci Chin 28:202–206, [in Chinese]. Keywords: biogeography/*cerana*/classification/China
Yang GH (1973) Biological characters of *Apis cerana cerana*. Apic China 5:42–44, [in Chinese]. Keywords: biology/*cerana*/China
Yang GH (1979) Identification of sacbrood disease in *Apis cerana cerana*. Apic China 5:17–18, [in Chinese]. Keywords: *cerana*/classification/disease/China
Yang GH (1981b) Designing a standard hive for the Chinese honeybee. Zhongguo Yangfeng 6:13–20, [in Chinese]. Keywords: beekeeping/*cerana*/nests/China
Yang GH (1982) The resources of the Chinese honey bee and its utilization. In: Selected references on the Chinese honey bee, Gansu Provincial Apicultural Research Institute, Tianshui, pp 15–32, [in Chinese]. Keywords: bibliography/*cerana*/classification/honey/pollination/China*
Yang GH (1982) The survey of the Chinese honey bee in Ganzi area in Sichuan. In: Selected references on the Chinese honey bee, Gansu Provincial Apicultural Research Institute, Tianshui, pp 191–196, [in Chinese]. Keywords: bibliography/biogeography/*cerana*/classification/honey/China*
Yang GH (1983e) Movement and behaviour of *Apis cerana cerana*. J Bee 1:17–18, [in Chinese]. Keywords: behaviour/*cerana*/China
Yang GH (1983) 10-frame hives for *Apis cerana cerana*. Pamphlet (China National Standard), Standard Publication of China, [in Chinese]. Keywords: beekeeping/*cerana*/China*
Yang GH (1984a) The survey of the resource of the Chinese honey bee. Zhongguo Yangfeng 3:4–7, [in Chinese]. Keywords: *cerana*/classification/honey/China
Yang GH (1984b) The survey of the resources of the Chinese honey bee II. Zhongguo Yangfeng 6:16–19, [in Chinese]. Keywords: *cerana*/classification/honey/China
Yang GH (1990) Distribution of *Apis cerana* F and its subspecies in China. In: Proceedings of the international expert meeting, ICIMOD, Kathmandu, Keywords: biogeography/*cerana*/classification/physiology/China/Nepal*
Yang GH (1996) Selection and breeding of *Apis cerana cerana* in the north of China. In: Proceedings of the 3rd Asian apiculture association international conference, Hanoi, p 48, Keywords: beekeeping/breeding/*cerana*/China
Yang GH (2001) The biology of *Apis cerana*. GH Yang, Beijing, [in Chinese]. Keywords: beekeeping/behaviour/biogeography/biology/*cerana*/classification/China
Yang GH, Hu F (1964) Preliminary study on the quantity of queen laying eggs in *Apis cerana cerana*. Apic China 3:18, [in Chinese]. Keywords: *cerana*/oviposition/queens/China*
Yang GH, Li GH (1983) The distribution and variations of the Chinese honey bee V. The survey of the Chinese honey bee in the northeastern regions. Zhongguo Yangfeng 5:19, [in Chinese]. Keywords: biogeography/*cerana*/classification/honey/China*
Yang GH, Sun QH (1983) The distribution and variations of the Chinese honey bee IV The status of the Chinese honey bee in the northwestern regions. Zhongguo Yangfeng 3:16–17, [in Chinese]. Keywords: biogeography/*cerana*/classification/honey/China
Yang GH, Sun QH (1993) Studies on the ability of *Apis cerana cerana* to make ecological adjustments in the nest. Apic China 5:9, [in Chinese]. Keywords: behaviour/*cerana*/ecology/nests/physiology/China*
Yang GH, Xu SY (1982) The distribution and variations of the Chinese honey bee I. Zhongguo Yangfeng 3:17–19, [in Chinese]. Keywords: biogeography/*cerana*/classification/honey/morphology/physical factors/population biology/swarming/China
Yang GH, Zhang QH (1995) Distribution, variation and characteristics of *Apis cerana cerana* in the northwest altiplano of Sichuan Province. Sci Agric Sinica 28:202–206, [in Chinese]. Keywords: biogeography/classification/*cerana*/morphology/China
Yang GH, Wang RW, Du ZL (1991) Hypopharyngeal gland activity in workers of *Apis cerana cerana*. Acta Entomol Sinica 34:146–150, [in Chinese]. Keywords: *cerana*/glands/workers/physiology/China

Yang GH, Sun DJ, Xu JY (1999) Study on modulation ability of temperature, humidity and RC50 concentration in colonies of *Apis cerana cerana*. Sci Agric Sinica 3:96–101, [in Chinese]. Keywords: behaviour/*cerana*/physical factors/physiology/China
Yang H, Kang M, Guo X, Xu B (2010a) Cloning, structural features, and expression analysis of the gene encoding thioredoxin reductase 1 from *Apis cerana cerana*. Comp Biochem Physiol B. doi:10.1016/j.cbpb.2010.04.003, Keywords: cerana/DNA/genetics
Yang JM (1991a) Introduction of a simple method for queen rearing in *Apis cerana cerana*. J Bee 9:40, [in Chinese]. Keywords: beekeeping/breeding/*cerana*/queens/China*
Yang JM, Miao JQ (1992) Studies on a comparison of the properties of AchE in *Apis cerana cerana* and *Apis mellifera ligustica*. Acta Sci Natl Univ Pekin 28:717–721, [in Chinese]. Keywords: biochemistry/*cerana*/*mellifera*/China
Yang KF (1982) The characteristics of *Apis cerana*. Report, Kansu Apiculture Institute, Tien-Shui, pp 43–48, Keywords: bee botany/*cerana*/China
Yang L, Qin Y, Song D, Qi M (2007) Brain melatonin content and polyethism in adult workers of *Apis mellifera* and *Apis cerana* (Hym., Apidae). J Appl Entomol 131:734–739, Keywords: behaviour/biochemistry/*cerana*/*mellifera*/physiology
Yang MH, Gong WG (1991) Preliminary discussions on keeping *Apis cerana cerana* under earth in houses. Agric Sci Tech Gannan 2:25–27, [in Chinese]. Keywords: beekeeping/*cerana*/China
Yang MX, Wang ZW, Li H, Zhang ZY, Tan K, Radloff SE, Hepburn HR (2010b) Thermoregulation in mixed-species colonies of honeybees (*Apis cerana* and *Apis mellifera*). J Insect Physiol 56:706–709, Keywords: behaviour/*cerana*/*mellifera*/thermoregulation
Yang MX, Tan K, Radloff SE, Phiancharoen M, Hepburn HR (2010c) Comb construction in mixed-species colonies of honeybees *Apis cerana* and *Apis mellifera*. J Exp Biol 213:1659–1664, Keywords: *cerana*/comb/interspecific interactions/*mellifera*/nests
Yang MX, Tan K, Radloff SE, Pirk CWW, Hepburn HR (2010d) Hetero-specific queen retinue behavior of worker bees in mixed-species colonies of *Apis cerana* and *Apis mellifera*. Apidologie 41:54–61, Keywords: behaviour/*cerana*/*mellifera*/pheromones
Yang RJ (1988a) Investigation of nucleic acids and polypeptides of Chinese sac brood virus. Virol Sinica 3:54–58, [in Chinese]. Keywords: brood/*cerana*/disease/physiology/China
Yang S (1991) Chinese apicultural history. Encyc Chinese Agric 312–367, [in Chinese]. Keywords: beekeeping/*cerana*/history/China*
Yang SP (1988b) Knowledge of beekeeping and beekeeping techniques in ancient China. Ancient Agric 1:242–251, [in Chinese]. Keywords: beekeeping/*cerana*/history/China
Yang SX, Gao JY (2009) Flowering pattern and protandry in *Boesenbergia longiflora*. Chin J Pl Ecol 33:449–459, [in Chinese]. Keywords: *cerana*/*florea*/foraging/pollination
Yang YX, Shen LR, Gao QK, Hong HT, Cheng JA (2009) Preparation and quality detection on gene chip of head cDNA from *Apis cerana cerana*. J Zhejiang Univ Agric Life Sci 35:1, [in Chinese]. Keywords: *cerana*/DNA/China*
Yano M (1950) Illustrations of Japanese insects. Ishi T (ed) Hokuryukan, Tokyo, p 1494 [in Latin], Keywords: *cerana*/classification/subspecies/Japan
Yatap H (1998) The development of honey harvesting of the giant honeybee (*A. dorsata*) at Seginim subdistrict, South Bengkulu District, Bengkulu Province, Indonesia. Thesis, Bogor Agricultural University, [in Indonesian]. Keywords: beekeeping/*dorsata*/general biology/honey/Indonesia
Yatsunami K, Echigo T (1985) Antibacterial action of royal jelly. Bull Fac Agric Tamagawa Univ 25:13–29, Keywords: biochemistry/*cerana*/royal jelly/Japan*
Ye K (1992) Autumn and winter management of *Apis cerana cerana* in southern areas of China. Apic China 5:22, [in Chinese]. Keywords: beekeeping/*cerana*/seasonal variation/China*
Ye K, Luo MS (1982) A method of producing comb honey using Chinese honeybees. Zhongguo Yangfeng 1:2–3, [in Chinese]. Keywords: beekeeping/*cerana*/honey/China
Ye ZA (1993) Methods for rearing *Apis cerana cerana* step by step. J Bee 5:16, [in Chinese]. Keywords: beekeeping/breeding/*cerana*/China*

Yen DF, Ho KK (1977) A brief history of beekeeping in Taiwan. J Bee 1:31–32, Keywords: beekeeping/*cerana*/history/China

Yi Y, Liu ZA (1994) Clinical validation of Chinese herbal medicines for the prevention and cure of sacbrood in *Apis cerana cerana*. Human Anim Husb Vet Med 2:33–35, [in Chinese]. Keywords: biochemistry/*cerana*/disease/China

Yim YH, Jang YD (1990) Karyological comparison of *Apis mellifera* L and *Apis cerana* F. Korean J Apic 5:43–46, [in Korean]. Keywords: *cerana*/genetics/*mellifera*/Korea

Yim YH, Jang YD, Kim SS (1990) Comparative studies on isoenzymes of various parts of the body in *Apis mellifera* L and *Apis cerana* F. Korean J Apic 5:47–55, [in Korean]. Keywords: biochemistry/*cerana*/*mellifera*/Korea

Yim YH, Chang YD, Kim SS (1993) Karyotype comparison between *Apis mellifera* and *Apis cerana* in Korea. In: Connor LJ, Rinderer TE, Sylvester HA, Wongsiri S (eds) Asian apiculture. Wicwas, Cheshire, Connecticut, pp 94–98, Keywords: *cerana*/genetics/*mellifera*/Korea

Yin SG (1991) The discovery of a new species of *Varroa* in the nest of *Apis cerana cerana*. Apic Sci Tech 1:20–22, [in Chinese]. Keywords: *cerana*/pests/China

Yoneda M, Maita K (1995) Bears and their damage to beekeeping. Honeybee Sci 16:145–152, [in Japanese]. Keywords: beekeeping/*cerana*/natural enemies/Japan

Yoshida T (1990) Honey extraction from native Japanese honeybees (*Apis cerana japonica*) in the Tsushima Islands. Honeybee Sci 11:63–66, [in Japanese]. Keywords: *cerana*/honey/islands/Japan

Yoshida T (1994a) Difference in drone congregation areas of native *Apis cerana japonica* Rad and introduced *A. mellifera* L. Japanese J Entomol Zool 38:85–90, [in Japanese]. Keywords: *cerana*/drones/*mellifera*/reproduction/Japan

Yoshida T (1994b) Interspecific instrumental insemination using mixed semen of Japanese and European honeybees (Hymenoptera: Apidae). Appl Entomol Zool 29:464–467, Keywords: breeding/*cerana*/drones/genetics/interspecific interactions/*mellifera*/reproduction/Japan

Yoshida T (1995a) Differences in mating flight times and drone congregation areas of native *Apis cerana japonica* and introduced *Apis mellifera*. Honeybee Sci 16:57–66, [in Japanese]. Keywords: *cerana*/drones/flight/*mellifera*/reproduction/Japan

Yoshida T (1995b) *Apis cerana* in Taiwan. Honeybee Sci 16:132–134, [in Japanese]. Keywords: biogeography/*cerana*/China

Yoshida T (1995c) Comparative studies on the mating system of the Japanese honey bee, *Apis cerana japonica* Radoszkowski, and the European honey bee, *Apis mellifera* L (Hymenoptera: Apidae). Bull Fac Agric Tamagawa Univ 35:159–208, Keywords: *cerana*/*mellifera*/morphology/physiology/reproduction/Japan

Yoshida T (1996) Interspecific instrumental insemination using mixed semen of Japanese and European honeybees. Honeybee Sci 17:23–26, [in Japanese]. Keywords: breeding/*cerana*/drones/genetics/*mellifera*/reproduction/Japan

Yoshida T (1997a) Japanese honeybee, ecology and its rearing methods I. Honeybee Sci 18:1–8, [in Japanese]. Keywords: beekeeping/breeding/*cerana*/ecology/honey/Japan

Yoshida T (1997b) Japanese honeybee, ecology and its rearing methods III. Honeybee Sci 18:137–148, [in Japanese]. Keywords: beekeeping/breeding/*cerana*/ecology/Japan

Yoshida T (1997c) Japanese honey bee, ecology and rearing methods IV. Honeybee Sci 18:165–174, [in Japanese]. Keywords: beekeeping/behaviour/breeding/*cerana*/disease/ecology/flight/foraging/honey/*mellifera*/natural enemies/queens/reproduction/swarming/Japan

Yoshida T (1998) Japanese honey bee, ecology and rearing methods V. Honeybee Sci 19:27–36, [in Japanese]. Keywords: beekeeping/biology/breeding/*cerana*/ecology/honey/Japan

Yoshida T (1999a) Bees and beekeeping in Asia. Farm Japan 33:10–17, Keywords: *andreniformis*/beekeeping/biogeography/*cerana*/*dorsata*/*florea*/*koschevnikovi*/*laboriosa*/*nigrocincta*/*nuluensis*

Yoshida T (1999b) History of 20 volumes of "Honeybee Science". Honeybee Sci 20:145–156, [in Japanese]. Keywords: beekeeping/bibliography/*cerana*/history/*mellifera*/Japan

Yoshida T (2000a) Honey bees and beekeeping in Asia. Gestión Apícola 4:38–44, [in Spanish]. Keywords: *andreniformis*/biogeography/*cerana*/*dorsata*/*florea*/*koschevnikovi*/*laboriosa*/*nigrocincta*/*nuluensis*

Yoshida T (2000b) Parasitism and reproduction of *Varroa* in *Apis cerana japonica*. In: Proceedings of the 4th Asian apiculture association international conference, Kathmandu, p 64, Keywords: *cerana*/pests/reproduction/Japan

Yoshida T (2001) Coexistence of introduced *Apis mellifera* and indigenous *Apis cerana japonica* adaptation of European honeybee to new habitat. In: Proceedings of the 7th international conference on tropical bees, Chiang Mai, pp 113–114, Keywords: *cerana*/interspecific interactions/*mellifera*/reproduction/Japan

Yoshida T (2005) Investigations on the societies of Japanese honeybees. Tamagawa University Press, Tokyo, p 141, [in Japanese]. Keywords: beekeeping/*cerana*/Japan

Yoshida T, Kittaka Y (2000) Reproduction of *Varroa jacobsoni* introduced into *Apis cerana japonica* colony. In: Proceedings of the 4th Asian apiculture association international conference, Kathmandu, pp 70–74, Keywords: *cerana*/pests/reproduction/Japan

Yoshida T, Matsuka M (1983) Trade and consumption of royal jelly in Japan. Honeybee Sci 4:181–184, [in Japanese]. Keywords: *cerana*/royal jelly/Japan

Yoshida T, Saito J (1991) Mating success of both native *Apis cerana japonica* and introduced *A. mellifera* in sympatric condition. In: Veeresh GK, Mallik B, Viraktamath CA (eds) Social insects and the environment. Oxford and IBH, Bombay, pp 127–128, Keywords: *cerana*/environmental effects/*mellifera*/reproduction/Japan

Yoshida T, Sasaki M (1995) Identification and quality control of beeswax from Oriental and European honey bees. Honeybee Sci 16:167–174, [in Japanese]. Keywords: beeswax/biochemistry/*cerana*/classification/honey/*mellifera*/Japan

Yoshida T, Yamazaki M (1993) Difference in drone congregation areas of *Apis mellifera* and *A. cerana japonica* as a reproductive isolation mechanism. In: Connor LJ, Rinderer TE, Sylvester HA, Wongsiri S (eds) Asian apiculture. Connecticut, Wicwas, Cheshire, pp 99–103, Keywords: *cerana*/drones/flight/*mellifera*/reproduction/Japan

Yoshida T, Sasaki M, Yamazaki S (1987) Artificial control of the mating flight time of honeybees by photoperiodic treatment within a hive. In: Proceedings of the 31st international apiculture congress, Apimondia, Warsaw, p 194, Keywords: *cerana*/flight/nests/physiology/reproduction/Japan

Yoshida T, Ono M, Okada I (1993a) Keeping the Japanese honeybee, *Apis cerana japonica* Radoszkowski using a modified Langstroth style movable-frame hive. Honeybee Sci 14: 3–12, [in Japanese]. Keywords: beekeeping/*cerana*/nests/Japan

Yoshida T, Saito J, Iizuka A (1993b) Instrumental insemination of Japanese honeybee (*Apis cerana japonica*) queens. Honeybee Sci 14:55–57, [in Japanese]. Keywords: breeding/*cerana*/genetics/queens/reproduction/Japan

Yoshida T, Saito J, Kajigaya N (1994) The mating flight times of native *Apis cerana japonica* Radoszkowski and introduced *Apis mellifera* L in sympatric conditions. Apidologie 25: 353–360, Keywords: *cerana*/drones/flight/*mellifera*/reproduction/Japan

Yoshida T, Sasaki M, Mamazaki S (1995) Parasitism and reproduction of *Varroa* mite on the Japanese honeybee, *Apis cerana japonica*. In: Kevan PG (ed) The Asiatic hive bee: apiculture, biology, and role in sustainable development in tropical and subtropical Asia. Enviroquest, Cambridge, pp 171–175, Keywords: biology/*cerana*/natural enemies/nests/pests/physiology/reproduction/Japan

Yoshida T, Hanawa R, Deowanish S (1996) Differences of insemination technique for *Apis cerana japonica* and *Apis mellifera* queens. In: Proceedings of the 3rd Asian apiculture association international conference, Hanoi, p 56, Keywords: breeding/*cerana*/drones/genetics/*mellifera*/physiology/queens/reproduction/Japan

Yoshida T, Ono M, Okada I (1998) Keeping the Japanese honey bee *Apis cerana japonica* Radoszkowski using a modified Langstroth style movable frame hive. Bull Fac Agric Tamagawa Univ 29:41–55, [in Japanese]. Keywords: beekeeping/*cerana*/honey/nests/Japan

Yoshida Y, Yoshida T (2002) Response of honeybees to comb foundation and artificial queen cups made by various waxes. Honeybee Sci 12:31–33, Keywords: beeswax/*cerana*/combs/queens/nests/Japan

Yoshigaki S, Sasaki M (2000) Integrative defensive strategy against microbial infection in honeybee colonies. In: Proceedings of the 4th Asian apiculture association international conference, Kathmandu, pp 85–89, Keywords: *cerana*/disease/*mellifera*/Japan

Yoshimi K, Yoshida T (2001) Reproduction of *Varroa jacobsoni* parasitizing newly emerged *Apis mellifera* drones introduced into *Apis cerana* colonies. Honeybee Sci 22:31–36, [in Japanese]. Keywords: *cerana*/drones/*mellifera*/pests/reproduction/Japan

Yoshiyama M, Kimura K (2009) Bacteria in the gut of Japanese honeybee, *Apis cerana japonica*, and their antagonistic effect against *Paenibacillus* larvae, the causal agent of American foulbrood. J Invert Path 102:91–96, Keywords: biochemistry/*cerana*/disease/Japan

You LS, Chen SG (1993) Bionomics of *Bracteodes* sp attacking *Apis cerana* Fabricius. Entomol Knowl 30:107–111, [in Chinese]. Keywords: biology/*cerana*/natural enemies/China

Yu CX, Tan HZ (1991) Preliminary studies on the histology of the neurosecretory cells of the brain of *Apis cerana cerana* workers. Apic Sci Tech 3:2–4, [in Chinese]. Keywords: biochemistry/*cerana*/cytology/physiology/workers/China

Yu JS (1991) The effect of hive structure on colony development in *Apis cerana cerana*. Sci Tech 1:49–54, [in Chinese]. Keywords: beekeeping/*cerana*/colony phenology/nests/physiology/China

Yu LS (1996a) Studies on rearing *Apis cerana cerana* using an *Apis mellifera ligustica* hive. Apic China 6:8, [in Chinese]. Keywords: beekeeping/breeding/*cerana*/*mellifera*/nests/China*

Yu LS (1997) Studies and observations on comb construction and use in *Apis cerana cerana*. Apic China 1:5, [in Chinese]. Keywords: *cerana*/combs/China*

Yu LS, Cao YF, Zou YD, Bi SD (2006) Investigation on temperature change and adjustment in the colony of *Apis cerana cerana*. J Nanking Agric Univ 29:53–58, Keywords: behaviour/*cerana*/environmental effects/thermoregulation/China

Yu YS (1993) The production of royal jelly by *Apis cerana cerana*. Apic Sci Tech 2:14, [in Chinese]. Keywords: beekeeping/*cerana*/royal jelly/China*

Yu YS (1996b) Management for robbing between *Apis cerana cerana* and *Apis mellifera ligustica*. Apic Sci Tech 3:10, [in Chinese]. Keywords: beekeeping/behaviour/*cerana*/interspecific interactions/*mellifera*/China*

Yua XH (1979) Discussion on the determination of morphological characteristics in *Apis cerana cerana*. Apic China 4:20–25, [in Chinese]. Keywords: *cerana*/classification/morphology/China

Yuan G, Duan J, Wu Y, Xiao H (1983) The Chinese bee (*Apis cerana*) ten-frame hives. Pamphlet, Ministry of Agriculture, Animal Husbandry and Fisheries, Beijing, p 10, Keywords: beekeeping/*cerana*/nests/China

Yuan GB (1996) The present situation of apiculture in China. In: Proceedings of the 3rd Asian apiculture association international conference, Hanoi, pp 128–131, Keywords: beekeeping/*cerana*/China

Yumoto T (2000) Bird-pollination of three *Durio* species (Bombacaceae) in a tropical rain forest in Sarawak, Malaysia. Am J Bot 87:1181–1188, Keywords: bee botany/*dorsata*/Malaysia

Zayed A (2009) Bee genetics and conservation. Apidologie 40:237–262, Keywords: *cerana*/ecology/genetics

Zeng ZJ (1989) Ecology of honeybees. Acta Jiangxi Agric Univ 22–24, [in Chinese]. Keywords: *cerana*/ecology/China*

Zeng ZJ (1989b) Study on the ecology of honeybees. Apic Sci Tech 1:34–35, [in Chinese]. Keywords: *cerana*/ecology/*mellifera*/China

Zeng ZJ (1993a) Studies on sex ratio of honeybee. J Bee 3:3–5, [in Chinese]. Keywords: *cerana*/genetics/China

Zeng ZJ (1993) Study on the sex ratio in bees. In: Proceedings of the 33rd international apiculture congress, Apimondia, Beijing, Keywords: *cerana*/genetics/China*

Zeng ZJ (1996) The honeybee sex ratio theory and its application. In: Proceedings of the 3rd Asian apiculture association international conference, Hanoi, pp 49–50, Keywords: beekeeping/*cerana*/genetics/physiology/China

Zeng ZJ (1996) A new frontier subject: ecology of bees. In: Proceedings of the 3rd Asian apiculture association international conference, Hanoi, pp 131–133, Keywords: *cerana*/ecology/*mellifera*/China

Zeng ZJ, Xi FG (1990) Morphological studies of the subpharyngeal gland of *Apis mellifera ligustica* and *Apis cerana cerana*. Apic China 5:6–7, [in Chinese]. Keywords: *cerana*/glands/*mellifera*/morphology/China

Zha HG, Milne RI, Sun H (2010) Asymmetric hybridization in *Rhododendron agastum*: a hybrid taxon comprising mainly F_1s in Yunnan, China. Ann Bot 105:89–100, Keywords: *cerana*/pollination

Zhan YI, Liu F, Li WF, Song SUK (2009) Review of the intrinsic factors affecting the information transmission of honeybee foraging. Apic China 60:7, [in Chinese]. Keywords: *cerana*/communication/dance language/foraging/China*

Zhang CX, Tang XCD, Cheng JA (2008) The utilization and industrialization of insect resources in China. Entomol Res 38:38–47, Keywords: beekeeping/*cerana*/China

Zhang HR (1993a) How to rear *Apis cerana cerana* to harvest rape honey. Apic Sci Tech 1:3, [in Chinese]. Keywords: beekeeping/*cerana*/honey/China*

Zhang HR (1993b) Summer and winter management of *Apis cerana cerana*. Apic Sci Tech 2:12, [in Chinese]. Keywords: beekeeping/*cerana*/seasonal variation/China*

Zhang HR (1995a) Improved methods for queen rearing in *Apis cerana cerana*. Apic Sci Tech 4:8, Keywords: beekeeping/breeding/*cerana*/queens/China*

Zhang JG (1995b) Rearing species of anti-shock wintering in *Apis cerana cerana*. Apic Sci Tech 6:7–9, [in Chinese]. Keywords: beekeeping/breeding/*cerana*/seasonal variation/China

Zhang JG (1995c) The effect of wing-clipping on egg laying by queens of *Apis cerana cerana*. Apic China 3:19, [in Chinese]. Keywords: beekeeping/*cerana*/morphology/oviposition/queens*

Zhang JG (1996a) A method for making new combs from redundant combs for *Apis cerana cerana*. Apic Sci Tech 2:16, [in Chinese]. Keywords: beekeeping/beeswax/*cerana*/combs/China*

Zhang JG (1996b) Prevention and management of laying workers in *Apis cerana cerana*. Apic Sci Tech 5:17, [in Chinese]. Keywords: beekeeping/*cerana*/laying workers/workers/China*

Zhang JG (1996c) The effects of swarming leading to the old queen leaving the hive in *Apis cerana cerana*. Apic Sci Tech 4:17, [in Chinese]. Keywords: beekeeping/*cerana*/nests/queens/reproduction/swarming/China*

Zhang JG, Zhang GY (1995) The relationship between packages of *Apis cerana cerana* and their characteristics in winter and spring. Apic China 6:12, [in Chinese]. Keywords: beekeeping/*cerana*/seasonal variation/China*

Zhang JH (2002) Hybrid vigour of *Apis cerana cerana*. Apic China 1:18–19, [in Chinese]. Keywords: breeding/*cerana*/genetics/China

Zhang JW, Ren L (1995) Exploring the use of *Apis cerana cerana* in the Kuandian mountain area. J Bee 4:18, [in Chinese]. Keywords: beekeeping/*cerana*/mountains/China*

Zhang QK (1994) Discussion on the superior characteristics of two-queen colonies in *Apis cerana cerana*. Apic Sci Tech 5:2, [in Chinese]. Keywords: beekeeping/breeding/*cerana*/queens/China*

Zhang QK, Su RM (1991) Studies on the effects of pH and osmotic pressure on semen energetics in *Apis cerana cerana*. Apic China 5:2–4, [in Chinese]. Keywords: *cerana*/drones/physiology/China

Zhang QM, Wang ZX, Li WS, Huang SZ (1983) A preliminary report on the survey of the Chinese honey bee in the Dabie Shan area. Zhongguo Yangfeng 3:18–20, [in Chinese]. Keywords: biogeography/*cerana*/honey/China

Zhang RX (1997) Discussion on the exploration for pollen by *Apis cerana cerana*. Apic China 3:16, [in Chinese]. Keywords: beekeeping/*cerana*/foraging/pollen/China*

Zhang SF (1982) A preliminary report of the survey of the Chinese honey bee in Sichuan. In: Selected references on the Chinese honey bee, Gansu Provincial Apicultural Research Institute, Tianshui, pp 179–189, [in Chinese]. Keywords: bibliography/biogeography/*cerana*/honey/China*

Zhang SF, Shi WJ, Zhang CX, Cheng JA (2003) Cloning and expression of a cDNA encoding secapin from the venom of *Apis cerana cerana*. Chin J Biochem Mol Biol 19:149–154, Keywords: biochemistry/*cerana*/DNA/genetics/China

Zhang X, Tan K, Liu YQ (2000) Five honeybee nectar plants in the tropical region of Yunnan. J Bee 26:8–10, [in Chinese]. Keywords: bee botany/*cerana*/foraging/nectar/China

Zhang X, Huang W, Chen H, Luo YX (2001) Preliminary study on the use of hybrid superiority in Chinese honeybee races (*Apis cerana cerana* F). Apic China 52:7–9, [in Chinese]. Keywords: breeding/*cerana*/genetics/honey/subspecies/China

Zhang Y, Gao X, Huang J (2005) Comparative distribution and kinetic properties of acetylcholinesterase from *Apis mellifera* and *Apis cerana*. J Apic Res 44:146–150, Keywords: biochemistry/*cerana*/*mellifera*

Zhang YW, Qian Yu, Zhao JM, Guo YH (2009a) Differential effects of nectar robbing by the same bumble-bee species on three sympatric *Corydalis* species with varied mating systems. Ann Bot 104:33–39, Keywords: *cerana*/foraging/interspecific interactions/nectar/pollination

Zhang ZQ (2000) Notes on *Varroa* destructor (Acari: Varroidae) parasitic on honeybees in New Zealand. Syst Appl Acarol (Special Publ) 5:9–14, Keywords: *cerana*/*mellifera*/pests/Japan/Thailand

Zhang ZR (1993c) How to improve the honey yield of *Apis cerana cerana*. J Bee 6:16, [in Chinese]. Keywords: beekeeping/*cerana*/honey/China*

Zhang ZY, Li H, Ken T (2009b) Morphological and taxonomic study of *Apis cerana* in Wumeng mountain zones. J Yunnan Agric Univ 24:3, [in Chinese]. Keywords: *cerana*/classification/morphology/mountains/China*

Zhen-Ming J (1990) Apicultural research work in China. In: Proceedings of the international expert meeting, ICIMOD, Kathmandu, Keywords: beekeeping/*cerana*/China/Nepal*

Zhen-Ming J, Guanhang Y, Shuangxiu H, Shikui L, Zaijin R (1992) The advancement of apicultural science and technology in China. In: Verma LR (ed) Honeybees in mountain agriculture. Oxford and IBH, New Delhi, pp 133–147, Keywords: bee botany/beekeeping/*cerana*/classification/geographical variation/pests/subspecies/China

Zheng O, Lee A, He K (1983) Analysis of chemical constituents in Chinese Hainan beeswax by gas chromatography. Jinan Liyi Xuebao 8:49–52, Keywords: beeswax/biochemistry/*cerana*/China

Zheng SY (1963) Pilot study on filter paper chromatography of the amino acids of royal jelly of *Apis cerana cerana*. J Zhongshan Univ 3:107, [in Chinese]. Keywords: biochemistry/*cerana*/royal jelly/China*

Zheng YL (1994) Long term caging of queens leads to permanent sterilization in *Apis cerana cerana*. Apic Sci Tech 2:26, [in Chinese]. Keywords: *cerana*/queens/physiology/reproduction/China*

Zhou B, Hu B, Zue F, Zhang T, Jiang T (2003) The effect of temperature on the honeybee nest (*Apis cerana cerana*) on opening the hive cover. Apic China 54:8–10, [in Chinese]. Keywords: beekeeping/*cerana*/environmental effects/nests/physiology/China

Zhou BF, Bao XL (1995a) Effects of the juvenile hormone analogue ZR-512 on birth weight of queen bees of *Apis cerana cerana*. J Fujian Agric Univ 24:341–343, [in Chinese]. Keywords: *cerana*/hormones/physiology/queens/China

Zhou BF, Bao XL (1995b) Effects of the juvenile hormone analogue ZR-512 on queen development in *Apis cerana cerana*. Apic China 2:3–6, [in Chinese]. Keywords: *cerana*/hormones/physiology/queens/China

Zhou BF, Jiang WJ (1991) Studies on the nectar foraging ability of individual workers of *Apis cerana*. Apic China 4:3–6, [in Chinese]. Keywords: *cerana*/foraging/nectar/workers/China

Zhou BF, Zhang FR (1993) The effects of collecting venom by electric shocks on the longevity of *Apis cerana cerana*. Apic China 2:4–6, [in Chinese]. Keywords: biochemistry/*cerana*/workers/China

Zhou BF, Luo YX, Chen HS, Lai YS (1989) Study on the prevention and control of *Galleria mellonella* L (Lepidoptera: Pyralidae) in beehives. Nat Enemies Insects 11:39–43, [in Chinese]. Keywords: *cerana*/natural enemies/China

Zhou DY (1995) Management of *Apis cerana cerana* through the four seasons in the Yunnan altiplano. Apic Sci Tech 1:9, [in Chinese]. Keywords: beekeeping/*cerana*/colony phenology/seasonal variation/China*

Zhou T, Anderson DL, Huang ZY, Huang S, Yao J, Ken T, Zhang Q (2004) Identification of *Varroa* mites (Acari: Varroidae) infesting *Apis cerana* and *Apis mellifera* colonies in China. Apidologie 35:645–654, Keywords: *cerana*/classification/*mellifera*/pests/China

Zhu FG (1989a) Techniques for extracting hydroxytriacontane from the beeswax of *Apis cerana cerana*. J Shandong Agric Univ 20:73–74, [in Chinese]. Keywords: beeswax/biochemistry/*cerana*/China

Zhu T, Zhu C (1989) Studies on the pollination of *Amomum villosum* Lour by *Apis cerana cerana*. Chin Trad Herbal Mat 12:13–15, [in Chinese]. Keywords: *cerana*/foraging/pollination/China

Zhu XJ, Zhou B, Wu X, Xu X, Wang QZ (2009a) Morphometric genetic analysis: population differentiation of *Apis cerana cerana* in Damen Island. Apic China 60:1, Keywords: *cerana*/classification/genetics/islands/morphology/population biology/China*

Zhu XJ, Zhou BF, Wu XD, Xu XJ, Wang Q (2009b) Morphometric genetic analysis: population differentiation of *Apis cerana cerana* in Longxi mountain. J Fujian Agric Univ 38:1, [in Chinese]. Keywords: *cerana*/genetics/morphology/mountains/population biology/China*

Zhu YY (1989b) A method for producing drone pupae in *Apis cerana cerana*. J Bee 4:6–7, [in Chinese]. Keywords: beekeeping/*cerana*/drones/reproduction/China

Zhu YY (1996a) Techniques for producing pollen by *Apis cerana cerana*. Apic China 1:15, [in Chinese]. Keywords: beekeeping/*cerana*/pollen/China*

Zhu YY (1996b) The causes of swarming in *Apis cerana cerana*. J Bee 10:23, [in Chinese]. Keywords: behaviour/*cerana*/swarming*

Zhu YY, Wang CY (1991) Breeding techniques for *Apis cerana cerana* in winter. J Bee 11:40–41, [in Chinese]. Keywords: breeding/*cerana*/seasonal variation/China

Zhu YY, Wang CY (1994) Report on the prevention of white-head disease and wax moths in *Apis cerana cerana*. J Bee 11:6, [in Chinese]. Keywords: beekeeping/beeswax/*cerana*/disease/pests/China*

Zhuang DA (1978) Studies on the external morphological characteristics of workers of *Apis cerana cerana* in different areas of Guizhou Province. Apic China 4:19–25, [in Chinese]. Keywords: biogeography/*cerana*/classification/morphology/workers/China

Zhuang DA (1979) A report on selective breeding of *Apis cerana cerana*. Apic China 4:26–28, [in Chinese]. Keywords: breeding/*cerana*/genetics/China

Zhuang DA (1982) Studies on the morphological characteristics of the Chinese honey bee in various areas of Guizhou Province. In: Selected references on the Chinese honey bee, Gansu Provincial Apicultural Research Institute, Tianshui, pp 231–227, [in Chinese]. Keywords: bibliography/biogeography/*cerana*/classification/honey/morphology/China*

Zimsen E (1964) The type material of IC Fabricius. Munksgaard, Copenhagen, Keywords: *cerana*/classification

Zoological Record (1864–2003)

Zou XH (1995a) Management for weak colonies of *Apis cerana cerana*. J Bee 4:19, [in Chinese]. Keywords: beekeeping/*cerana*/China*

Zou XH (1995b) The use and management of weak colonies of *Apis cerana cerana*. Apic Sci Tech 3:16, [in Chinese]. Keywords: beekeeping/*cerana*/colony phenology/China*

Zou XH (1996) Swarming in *Apis cerana cerana*. J Bee 1:18, [in Chinese]. Keywords: behaviour/*cerana*/swarming*

Zuyin X (1991) An experiment to control swarming impulse in *Apis cerana* and to increase honey yield during the spring nectar flow. Apic China 50:3–5, [in Chinese]. Keywords: beekeeping/*cerana*/honey/nectar/swarming/China

Index

DOI 10.1007/978-3-642-16422-4,

C

D